ADVANCED TRANSPORT PHENOMENA

Advanced Transport Phenomena is ideal as a graduate textbook. It contains a detailed discussion of modern analytic methods for the solution of fluid mechanics and heat and mass transfer problems, focusing on approximations based on scaling and asymptotic methods, beginning with the derivation of basic equations and boundary conditions and concluding with linear stability theory. Also covered are unidirectional flows, lubrication and thin-film theory, creeping flows, boundary-layer theory, and convective heat and mass transport at high and low Reynolds numbers. The emphasis is on basic physics, scaling and nondimensionalization, and approximations that can be used to obtain solutions that are due either to geometric simplifications, or large or small values of dimensionless parameters. The author emphasizes setting up problems and extracting as much information as possible short of obtaining detailed solutions of differential equations. The book also focuses on the solutions of representative problems. This reflects the author's bias toward learning to think about the solution of transport problems.

L. Gary Leal is professor of chemical engineering at the University of California in Santa Barbara. He also holds positions in the Materials Department and in the Department of Mechanical Engineering. He has taught at UCSB since 1989. Before that, from 1970 to 1989 he taught in the chemical engineering department at Caltech. His current research interests are focused on fluid mechanics problems for complex fluids, as well as the dynamics of bubbles and drops in flow, coalescence, thin-film stability, and related problems in rheology. In 1987, he was elected to the National Academy of Engineering. His research and teaching have been recognized by a number of awards, including the Dreyfus Foundation Teacher-Scholar Award, a Guggenheim Fellowship, the Allan Colburn and Warren Walker Awards of the AIChE, the Bingham Medal of the Society of Rheology, and the Fluid Dynamics Prize of the American Physical Society. Since 1995, Professor Leal has been one of the two editors of the AIP journal *Physics of Fluids* and he has also served on the editorial boards of numerous journals and the Cambridge Series in Chemical Engineering.

CAMBRIDGE SERIES IN CHEMICAL ENGINEERING

Series Editor:

Arvind Varma, *Purdue University*

Editorial Board:

Alexis T. Bell, *University of California, Berkeley*
Edward Cussler, *University of Minnesota*
Mark E. Davis, *California Institute of Technology*
L. Gary Leal, *University of California, Santa Barbara*
Massimo Morbidelli, *ETH, Zurich*
Athanassios Z. Panagiotopoulos, *Princeton University*
Stanley I. Sandler, *University of Delaware*
Michael L. Schuler, *Cornell University*

Books in the Series:

E. L. Cussler, *Diffusion: Mass Transfer in Fluid Systems,* Second Edition

Liang-Shih Fan and Chao Zhu, *Principles of Gas-Solid Flows*

Hasan Orbey and Stanley I. Sandler, *Modeling Vapor-Liquid Equilibria: Cubic Equations of State and Their Mixing Rules*

T. Michael Duncan and Jeffrey A. Reimer, *Chemical Engineering Design and Analysis: An Introduction*

John C. Slattery, *Advanced Transport Phenomena*

A. Varma, M. Morbidelli, and H. Wu, *Parametric Sensitivity in Chemical Systems*

M. Morbidelli, A. Gavriilidis, and A. Varma, *Catalyst Design: Optimal Distribution of Catalyst in Pellets, Reactors, and Membranes*

E. L. Cussler and G. D. Moggridge, *Chemical Product Design*

Pao C. Chau, *Process Control: A First Course with MATLAB®*

Richard Noble and Patricia Terry, *Principles of Chemical Separations with Environmental Applications*

F. B. Petlyuk, *Distillation Theory and Its Application to Optimal Design of Separation Units*

Leal, L. Gary, *Advanced Transport Phenomena: Fluid Mechanics and Convective Transport*

Advanced Transport Phenomena

Fluid Mechanics and Convective Transport Processes

L. Gary Leal

CAMBRIDGE UNIVERSITY PRESS
Cambridge, New York, Melbourne, Madrid, Cape Town, Singapore, São Paulo, Delhi

Cambridge University Press
32 Avenue of the Americas, New York, NY 10013-2473, USA

www.cambridge.org
Information on this title: www.cambridge.org/9780521849104

© Cambridge University Press 2007

This publication is in copyright. Subject to statutory exception
and to the provisions of relevant collective licensing agreements,
no reproduction of any part may take place without
the written permission of Cambridge University Press.

First published 2007
Reprinted 2008

Printed in the United States of America

A catalog record for this publication is available from the British Library.

Library of Congress Cataloging in Publication Data

Leal, L. Gary.
Advanced transport phenomena : fluid mechanics and convective trasport processes / L. Gary Leal.
   p. cm. – (Cambridge series in chemical engineering)
Includes bibliographical references and index.
ISBN-13: 978-0-521-84910-4 (hardback)
ISBN-10: 0-521-84910-1 (hardback)
1. Fluid mechanics – Textbooks. 2. Transport theory – Textbooks. 3. Continuum mechanics – Textbooks. I. Title. II. Series.
QC145.2.L43 2007
660'.2842 – dc22   2006018348

ISBN 978-0-521-84910-4 hardback

Cambridge University Press has no responsibility for
the persistence or accuracy of URLs for external or
third-party Internet Web sites referred to in this publication
and does not guarantee that any content on such
Web sites is, or will remain, accurate or appropriate.

Contents

Preface		*page* xv
Acknowledgments		xix
1	**A Preview**	1
	A A Brief Historical Perspective of Transport Phenomena in Chemical Engineering	1
	B The Nature of the Subject	2
	C A Brief Description of the Contents of This Book	4
	Notes and References	11
2	**Basic Principles**	13
	A The Continuum Approximation	13
	1 Foundations	14
	2 Consequences	15
	B Conservation of Mass – The Continuity Equation	18
	C Newton's Laws of Mechanics	25
	D Conservation of Energy and the Entropy Inequality	31
	E Constitutive Equations	36
	F Fluid Statics – The Stress Tensor for a Stationary Fluid	37
	G The Constitutive Equation for the Heat Flux Vector – Fourier's Law	42
	H Constitutive Equations for a Flowing Fluid – The Newtonian Fluid	45
	I The Equations of Motion for a Newtonian Fluid – The Navier–Stokes Equation	49
	J Complex Fluids – Origins of Non-Newtonian Behavior	52
	K Constitutive Equations for Non-Newtonian Fluids	59
	L Boundary Conditions at Solid Walls and Fluid Interfaces	65
	1 The Kinematic Condition	67
	2 Thermal Boundary Conditions	68
	3 The Dynamic Boundary Condition	69
	M Further Considerations of the Boundary Conditions at the Interface Between Two Pure Fluids – The Stress Conditions	74
	1 Generalization of the Kinematic Boundary Condition for an Interface	75
	2 The Stress Conditions	76
	3 The Normal-Stress Balance and Capillary Flows	79
	4 The Tangential-Stress Balance and Thermocapillary Flows	84

Contents

    N   The Role of Surfactants in the Boundary Conditions at a Fluid Interface   89
    Notes and Reference   96
    Problems   99

3 Unidirectional and One-Dimensional Flow and Heat Transfer Problems   110
    A   Simplification of the Navier–Stokes Equations for Unidirectional Flows   113
    B   Steady Unidirectional Flows – Nondimensionalization and Characteristic Scales   115
    C   Circular Couette Flow – A One-Dimensional Analog to Unidirectional Flows   125
    D   Start-Up Flow in a Circular Tube – Solution by Separation of Variables   135
    E   The Rayleigh Problem – Solution by Similarity Transformation   142
    F   Start-Up of Simple Shear Flow   148
    G   Solidification at a Planar Interface   152
    H   Heat Transfer in Unidirectional Flows   157
        1   Steady-State Heat Transfer in Fully Developed Flow through a Heated (or Cooled) Section of a Circular Tube   158
        2   Taylor Diffusion in a Circular Tube   166
    I   Pulsatile Flow in a Circular Tube   175
    Notes   183
    Problems   185

4 An Introduction to Asymptotic Approximations   204
    A   Pulsatile Flow in a Circular Tube Revisited – Asymptotic Solutions for High and Low Frequencies   205
        1   Asymptotic Solution for $R_\omega \ll 1$ 206
        2   Asymptotic Solution for $R_\omega \gg 1$ 209
    B   Asymptotic Expansions – General Considerations   216
    C   The Effect of Viscous Dissipation on a Simple Shear Flow   219
    D   The Motion of a Fluid Through a Slightly Curved Tube – The Dean Problem   224
    E   Flow in a Wavy-Wall Channel – "Domain Perturbation Method"   232
        1   Flow Parallel to the Corrugation Grooves   233
        2   Flow Perpendicular to the Corrugation Grooves   237
    F   Diffusion in a Sphere with Fast Reaction – "Singular Perturbation Theory"   242
    G   Bubble Dynamics in a Quiescent Fluid   250
        1   The Rayleigh–Plesset Equation   251
        2   Equilibrium Solutions and Their Stability   255
        3   Bubble Oscillations Due to Periodic Pressure Oscillations – Resonance and "Multiple-Time-Scale Analysis"   260
        4   Stability to Nonspherical Disturbances   269
    Notes   282
    Problems   284

5 The Thin-Gap Approximation – Lubrication Problems   294
    A   The Eccentric Cylinder Problem   295
        1   The Narrow-Gap Limit – Governing Equations and Solutions   297

Contents

		2	Lubrication Forces	303
	B		Derivation of the Basic Equations of Lubrication Theory	306
	C		Applications of Lubrication Theory	315
		1	The Slider-Block Problem	315
		2	The Motion of a Sphere Toward a Solid, Plane Boundary	320
	D		The Air Hockey Table	325
		1	The Lubrication Limit, $\tilde{R}e \ll 1$	328
		2	The Uniform Blowing Limit, $p_R^* \gg 1$	332
			a $\tilde{R}e \ll 1$	334
			b $\tilde{R}e \gg 1$	336
			c Lift on the Disk	345
	Notes			346
	Problems			347

6 The Thin-Gap Approximation – Films with a Free Surface — 355

	A		Derivation of the Governing Equations	355
		1	The Basic Equations and Boundary Conditions	355
		2	Simplification of the Interface Boundary Conditions for a Thin Film	359
		3	Derivation of the Dynamical Equation for the Shape Function, $h(\mathbf{x}_s, t)$	360
	B		Self-Similar Solutions of Nonlinear Diffusion Equations	362
	C		Films with a Free Surface – Spreading Films on a Horizontal Surface	367
		1	Gravitational Spreading	367
		2	Capillary Spreading	371
	D		The Dynamics of a Thin Film in the Presence of van der Waals Forces	376
		1	Linear Stability	378
		2	Similarity Solutions for Film Rupture	381
	E		Shallow-Cavity Flows	385
		1	The Horizontal, Enclosed Shallow Cavity	386
		2	The Horizontal Shallow Cavity with a Free Surface	391
			a Solution by means of the classical thin-film analysis	392
			b Solution by means of the method of domain perturbations	396
			c The end regions	401
		3	Thermocapillary Flow in a Thin Cavity	404
			a Thin-film solution procedure	410
			b Solution by domain perturbation for $\delta = 1$	413
	Notes			418
	Problems			418

7 Creeping Flow – Two-Dimensional and Axisymmetric Problems — 429

	A		Nondimensionalization and the Creeping-Flow Equations	430
	B		Some General Consequences of Linearity and the Creeping-Flow Equations	434
		1	The Drag on Bodies That Are Mirror Images in the Direction of Motion	434
		2	The Lift on a Sphere That is Rotating in a Simple Shear Flow	436
		3	Lateral Migration of a Sphere in Poiseuille Flow	438
		4	Resistance Matrices for the Force and Torque on a Body in Creeping Flow	439

    C Representation of Two-Dimensional and Axisymmetric Flows in Terms of the Streamfunction   444
    D Two-Dimensional Creeping Flows: Solutions by Means of Eigenfunction Expansions (Separation of Variables)   449
        1 General Eigenfunction Expansions in Cartesian and Cylindrical Coordinates   449
        2 Application to Two-Dimensional Flow near Corners   451
    E Axisymmetric Creeping Flows: Solution by Means of Eigenfunction Expansions in Spherical Coordinates (Separation of Variables)   458
        1 General Eigenfunction Expansion   459
        2 Application to Uniform Streaming Flow past an Arbitrary Axisymmetric Body   464
    F Uniform Streaming Flow past a Solid Sphere – Stokes' Law   466
    G A Rigid Sphere in Axisymmetric, Extensional Flow   470
        1 The Flow Field   470
        2 Dilute Suspension Rheology – The Einstein Viscosity Formula   473
    H Translation of a Drop Through a Quiescent Fluid at Low Re 477
    I Marangoni Effects on the Motion of Bubbles and Drops   486
    J Surfactant Effects on the Buoyancy-Driven Motion of a Drop   490
        1 Governing Equations and Boundary Conditions for a Translating Drop with Surfactant Adsorbed at the Interface   493
        2 The Spherical-Cap Limit   497
        3 The Limit of Fast Adsorption Kinetics   503
    Notes   510
    Problems   512

8 Creeping Flow – Three-Dimensional Problems   524

    A Solutions by Means of Superposition of Vector Harmonic Functions   525
        1 Preliminary Concepts   525
            a Vector "equality" – pseudo-vectors   525
            b Representation theorem for solution of the creeping-flow equations   526
            c Vector harmonic functions   527
        2 The Rotating Sphere in a Quiescent Fluid   528
        3 Uniform Flow past a Sphere   529
    B A Sphere in a General Linear Flow   530
    C Deformation of a Drop in a General Linear Flow   537
    D Fundamental Solutions of the Creeping-Flow Equations   545
        1 The "Stokeslet": A Fundamental Solution for the Creeping-Flow Equations   545
        2 An Integral Representation for Solutions of the Creeping-Flow Equations due to Ladyzhenskaya   547
    E Solutions for Solid Bodies by Means of Internal Distributions of Singularities   550
        1 Fundamental Solutions for a Force Dipole and Other Higher-Order Singularities   551
        2 Translation of a Sphere in a Quiescent Fluid (Stokes' Solution)   554
        3 Sphere in Linear Flows: Axisymmetric Extensional Flow and Simple Shear   555

	4	Uniform Flow past a Prolate Spheroid	557
	5	Approximate Solutions of the Creeping-Flow Equations by Means of Slender-Body Theory	560
F		The Boundary Integral Method	564
	1	A Rigid Body in an Unbounded Domain	565
	2	Problems Involving a Fluid Interface	565
	3	Problems in a Bounded Domain	568
G		Further Topics in Creeping-Flow Theory	570
	1	The Reciprocal Theorem	571
	2	Faxen's Law for a Body in an Unbounded Fluid	571
	3	Inertial and Non-Newtonian Corrections to the Force on a Body	573
	4	Hydrodynamic Interactions Between Widely Separated Particles – The Method of Reflections	576
Notes			580
Problems			582

9 Convection Effects in Low-Reynolds-Number Flows — 593

A		Forced Convection Heat Transfer – Introduction	593
	1	General Considerations	594
	2	Scaling and the Dimensionless Parameters for Convective Heat Transfer	596
	3	The Analogy with Single-Solute Mass Transfer	598
B		Heat Transfer by Conduction ($Pe \to 0$)	600
C		Heat Transfer from a Solid Sphere in a Uniform Streaming Flow at Small, but Nonzero, Peclet Numbers	602
	1	Introduction – Whitehead's Paradox	602
	2	Expansion in the Inner Region	605
	3	Expansion in the Outer Region	606
	4	A Second Approximation in the Inner Region	611
	5	Higher-Order Approximations	613
	6	Specified Heat Flux	615
D		Uniform Flow past a Solid Sphere at Small, but Nonzero, Reynolds Number	616
E		Heat Transfer from a Body of Arbitrary Shape in a Uniform Streaming Flow at Small, but Nonzero, Peclet Numbers	627
F		Heat Transfer from a Sphere in Simple Shear Flow at Low Peclet Numbers	633
G		Strong Convection Effects in Heat and Mass Transfer at Low Reynolds Number – An Introduction	643
H		Heat Transfer from a Solid Sphere in Uniform Flow for $Re \ll 1$ and $Pe \gg 1$	645
	1	Governing Equations and Rescaling in the Thermal Boundary-Layer Region	648
	2	Solution of the Thermal Boundary-Layer Equation	652
I		Thermal Boundary-Layer Theory for Solid Bodies of Nonspherical Shape in Uniform Streaming Flow	656
	1	Two-Dimensional Bodies	659
	2	Axisymmetric Bodies	661
	3	Problems with Closed Streamlines (or Stream Surfaces)	662
J		Boundary-Layer Analysis of Heat Transfer from a Solid Sphere in Generalized Shear Flows at Low Reynolds Number	663

Contents

    K Heat (or Mass) Transfer Across a Fluid Interface for Large Peclet Numbers   666
        1 General Principles   666
        2 Mass Transfer from a Rising Bubble or Drop in a Quiescent Fluid   668
    L Heat Transfer at High Peclet Number Across Regions of Closed-Streamline Flow   671
        1 General Principles   671
        2 Heat Transfer from a Rotating Cylinder in Simple Shear Flow   672
    Notes   680
    Problems   681

10 Laminar Boundary-Layer Theory   **697**
    A Potential-Flow Theory   698
    B The Boundary-Layer Equations   704
    C Streaming Flow past a Horizontal Flat Plate – The Blasius Solution   713
    D Streaming Flow past a Semi-Infinite Wedge – The Falkner–Skan Solutions   719
    E Streaming Flow past Cylindrical Bodies – Boundary-Layer Separation   725
    F Streaming Flow past Axisymmetric Bodies – A Generalizaiton of the Blasius Series   733
    G The Boundary-Layer on a Spherical Bubble   739
    Notes   754
    Problems   756

11 Heat and Mass Transfer at Large Reynolds Number   **767**
    A Governing Equations ($Re \gg 1$, $Pe \gg 1$, with Arbitrary Pr or Sc numbers)   769
    B Exact (Similarity) Solutions for Pr (or Sc) $\sim O(1)$ 771
    C The Asymptotic Limit, Pr (or Sc) $\gg 1$ 773
    D The Asymptotic Limit, Pr (or Sc) $\ll 1$ 780
    E Use of the Asymptotic Results at Intermediate Pe (or Sc)   787
    F Approximate Results for Surface Temperature with Specified Heat Flux or Mixed Boundary Conditions   788
    G Laminar Boundary-Layer Mass Transfer for Finite Interfacial Velocities   793
    Notes   797
    Problems   797

12 Hydrodynamic Stability   **800**
    A Capillary Instability of a Liquid Thread   801
        1 The Inviscid Limit   804
        2 Viscous Effects on Capillary Instability   808
        3 Final Remarks   811
    B Rayleigh–Taylor Instability (The Stability of a Pair of Immiscible Fluids That Are Separated by a Horizontal Interface)   812
        1 The Inviscid Fluid Limit   816
        2 The Effects of Viscosity on the Stability of a Pair of Superposed Fluids   818
        3 Discussion   822

Contents

C	Saffman–Taylor Instability at a Liquid Interface		823
	1	Darcy's Law	823
	2	The Taylor–Saffman Instability Criteria	826
D	Taylor–Couette Instability		829
	1	A Sufficient Condition for Stability of an Inviscid Fluid	832
	2	Viscous Effects	835
E	Nonisothermal and Compositionally Nonuniform Systems		840
F	Natural Convection in a Horizontal Fluid Layer Heated from Below – The Rayleigh–Benard Problem		845
	1	The Disturbance Equations and Boundary Conditions	845
	2	Stability for Two Free Surfaces	851
	3	The Principle of Exchange of Stabilities	853
	4	Stability for Two No-Slip, Rigid Boundaries	855
G	Double-Diffusive Convection		858
H	Marangoni Instability		867
I	Instability of Two-Dimensional Unidirectional Shear Flows		872
	1	Inviscid Fluids	873
		a The Rayleigh stability equation	873
		b The Inflection-point theorem	875
	2	Viscous Fluids	876
		a The Orr–Sommerfeld equation	876
		b A sufficient condition for stability	877
Notes			878
Problems			880

Appendix A: Governing Equations and Vector Operations in Cartesian, Cylindrical, and Spherical Coordinate Systems   891

Appendix B: Cartesian Component Notation   897

Index   899

Preface

This book represents a major revision of my book *Laminar Flow and Convective Transport Processes* that was published in 1992 by Butterworth-Heinemann. As was the case with the previous book, it is about fluid mechanics and the convective transport of heat (or any passive scalar quantity) for simple Newtonian, incompressible fluids, treated from the point of view of classical continuum mechanics. It is intended for a graduate-level course that introduces students to fundamental aspects of fluid mechanics and convective transport processes (mainly heat transfer and some single solute mass transfer) in a context that is relevant to applications that are likely to arise in research or industrial applications. In view of the current emphasis on small-scale systems, biological problems, and materials, rather than large-scale classical industrial problems, the book is focused more on viscous phenomena, thin films, interfacial phenomena, and related topics than was true 14 years ago, though there is still significant coverage of high-Reynolds-number and high-Peclet-number boundary layers in the second half of the book. It also incorporates an entirely new chapter on linear stability theory for many of the problems of greatest interest to chemical engineers.

The material in this book is the basis of an introductory (two-term) graduate course on transport phenomena. It starts with a derivation of all of the necessary governing equations and boundary conditions in a context that is intended to focus on the underlying fundamental principles and the connections between this topic and other topics in continuum physics and thermodynamics. Some emphasis is also given to the limitations of both equations and boundary conditions (for example "non-Newtonian" behavior, the "no-slip" condition, surfactant and thermocapillary effects at interfaces, etc.). It should be noted, however, that though this course starts at the very beginning by deriving the basic equations from first principles, and thus can be taken successfully even without an undergraduate transport background, there are important topics from the undergraduate curriculum that are *not* included, especially macroscopic balances, friction factors, correlations for turbulent flow conditions, etc. The remainder of the book is concerned with how to solve transport and fluids problems analytically; but with a lot of emphasis on basic physics, scaling, nondimensionalization, and approximations that can be used to obtain solutions that are due either to geometric simplifications or large or small values of dimensionless parameters.

THE SCOPE OF THIS BOOK

No single book can encompass all topics, and the present book is no exception. We consider only laminar flows and transport processes involving laminar flows, for incompressible, Newtonian fluids. Specifically, we do not consider turbulent flows. We do not consider compressibility effects, nor do we consider numerical methods, except by means of a brief

Preface

introduction to boundary integral techniques for creeping flows. Further, we do not consider non-Newtonian flows, except for a few limited homework examples, nor even the basic constitutive equations for non-Newtonian fluids except briefly in the introductory chapter, Chapter 2, primarily in the context of thinking about why fluids may exhibit non-Newtonian behavior and hence what the limitations of the Newtonian fluid approximation may be. We do consider both flow and convective transport processes, but with the latter generally posed as a heat transfer problem. We shall see, however, that much of the same analysis and principles apply to mass transfer when there is a single solute. Finally, multicomponent mass transfer is not considered, and in the graduate transport sequence of classes would often be taught as a separate class.

The goal of this book is to provide a fundamental understanding of the governing principles for flow and convective transport processes in Newtonian fluids, and some of the modern tools and methods for "analysis" of this class of problems. By "analysis," I mean both what one can achieve from a qualitative point of view without actually solving differential equations and boundary conditions, as well as detailed analytic solutions obtained generally from an asymptotic point of view. There is a strong emphasis on the derivation of basic equations and boundary conditions, including those relevant to a fluid interface. I also focus on complete descriptions of the solutions of representative problems rather than an exhaustive summary of all possible problems. This is because of the importance that I place on learning how to think about transport problems, and how to actually solve them, rather than just being told that some problem exists with a certain solution, but without adequate details to really understand how to achieve that solution or to generalize from the current problem to a related but presently unanticipated extension.

An important tool that we develop in this book is the use of characteristic scales, nondimensionalization, and asymptotic techniques, in the analysis and understanding of transport processes. At the most straightforward level, asymptotic methods provide a systematic framework to generate approximate solutions of the nonlinear differential equations of fluid mechanics, as well as the corresponding thermal energy (or species transport) equations. Perhaps more important than the detailed solutions enabled by these methods, however, is that they demand an extremely close interplay between the mathematics and the physics, and in this way contribute a very powerful understanding of the physical phenomena that characterize a particular problem or process. The presence of large or small dimensionless parameters in appropriately nondimensionalized equations or boundary conditions is indicative of the relative magnitudes of the various physical mechanisms in each case, and is thus a basis for approximation via retention of the dominant terms.

There is, in fact, an element of truth in the suggestion that asymptotic approximation methods are nothing more than a sophisticated version of dimensional analysis. Certainly it is true, as we shall see, that successful application of scaling/nondimensionalization can provide much of the information and insight about the nature of a given fluid mechanics or transport process without the need either to solve the governing differential equations or even be concerned with a detailed geometric description of the problem. The latter determines the magnitude of numerical coefficients in the correlations between dependent and independent dimensionless groups, but usually does not determine the form of the correlations. In this sense, asymptotic theory can reduce a whole class of problems, which differ only in the geometry of the boundaries and in the nature of the undisturbed flow, to the evaluation of a single coefficient. When the body or boundary geometry is simple, this can be done by means of detailed solutions of the governing equations and boundary conditions. Even when the geometry is too complex to obtain analytic solutions, however, the general asymptotic framework is unchanged, and the correlation between dimensionless groups is still reduced to determination of a single constant, which can now be done (in principle) by means of a single experimental measurement.

Preface

It is important, however, not to overstate what can be accomplished by asymptotic (and related analytic) techniques applied either to fluid mechanics or heat (and mass) transfer processes. At most, these methods can treat limited regimes of the overall parameter domain for any particular problem. Furthermore, the approximate solutions obtained can be no more general than the framework allowed in the problem statement; that is, if we begin by seeking a steady axisymmetric solution, an asymptotic analysis will produce only an approximation for this class of solutions and, by itself, can guarantee neither that the solution is unique within this class nor that the limitation to steady and axisymmetric solutions is representative of the actual physical situation. For example, even if the geometry of the problem is completely axisymmetric, there is no guarantee that an axisymmetric solution exists for the velocity or temperature field, or if it does, that it corresponds to the motion or temperature field that would be realized in the laboratory. The latter may be either time dependent or fully three dimensional or both. In this case, the most that we may hope is that these more complex motions may exist as a consequence of instabilities in the basic, steady, axisymmetric solution, and thus that the conditions for departure from this basic state can be predicted within the framework of classical stability theories. The important message is that analytic techniques, including asymptotic methods, are not sufficient by themselves to understand fluid mechanics or heat transfer processes. Such techniques would almost always need to be supplemented by some combination of stability analysis or, more generally, by experimental or computational studies of the full problem.

I want to thank my many colleagues and students who have contributed to this work for many years. I would also like to thank the users of the first edition who made substantial suggestions for improvement. I look forward to the reader's reaction to this new version.

L. Gary Leal
Santa Barbara

Acknowledgments

I want to thank a number of people who contributed to this book. Most important among these were Professor G. M. Homsy, and several years of graduate students from my own classes at the University of California at Santa Barbara, who used this book in preprint form and provided much useful input on topics that required better explanation, typos, etc. In addition, these students had the first "opportunity" to work many of the problems at the end of each chapter, and this led to a number of important changes in the problem statements. I specifically appreciate their patience in this latter endeavor. I also owe a major debt of gratitude to number of faculty around the country, who had taught graduate transport classes from my previous book and provided detailed comments on the proposed contents and format of this new book. In addition, several of these individuals also contributed problems from their own classes, which they kindly allowed me to use in this new book. For this major contribution, I thank David Leighton from Notre Dame, John Brady from the California Institute of Technology, Roger Bonnecaze from the University of Texas at Austin, and James Oberhauser from the University of Virginia. In addition, Professor Howard Stone from Harvard University provided very useful notes on the dynamics of thin films from his own class, and also kindly read several of the new sections. Finally, I thank Cambridge University Press, and particularly Peter Gordon, for their patience in waiting for me to finish this book. The last 10 percent took at least 50 percent of the time! I take full responsibility for the contents of this book.

1

A Preview

A. A BRIEF HISTORICAL PERSPECTIVE OF TRANSPORT PHENOMENA IN CHEMICAL ENGINEERING

"Transport phenomena" is the name used by chemical engineers to describe the subjects of fluid mechanics and heat and mass transfer. The earliest step toward the inclusion of specialized courses in fluid mechanics and heat or mass transfer processes within the chemical engineering curriculum probably occurred with the publication in 1923 of the pioneering text *Principles of Chemical Engineering* by Walker, Lewis, and McAdams.[1] This was the first major departure from curricula that regarded the techniques involved in the production of specific products as largely unique, to a formal recognition of the fact that certain physical or chemical processes, and corresponding fundamental principles, are common to many widely differing industrial technologies.

A natural outgrowth of this radical new view was the gradual appearance of fluid mechanics and transport in both teaching and research as the underlying basis for many of the unit operations. Of course, many of the most important unit operations take place in equipment of complicated geometry, with strongly coupled combinations of heat and mass transfer, fluid mechanics, and chemical reaction, so that the exact equations could not be solved in a context of any direct relevance to the process of interest. Hence, insofar as the large-scale industrial processes of chemical technology were concerned, even at the unit operations level, the impact of fundamental studies of fluid mechanics or transport phenomena was certainly less important than a well-developed empirical approach (and this remains true today in many cases). Indeed, the great advances and discoveries of fluid mechanics during the first half of the twentieth century took place almost entirely without the participation (or even knowledge) of chemical engineers.

Gradually, however, chemical engineers began to accept the premise that the generally "blind" empiricism of the "lumped-parameter" approach to transport processes at the unit operations scale should at least be supplemented by an attempt to understand the basic physical principles. This finally led, in 1960, to the appearance of the landmark textbook of Bird, Stewart, and Lightfoot,[2] which not only introduced the idea of detailed analysis of transport processes at the continuum level, but also emphasized the mathematical similarity of the governing field equations, along with the simplest constitutive approximations for fluid mechanics and heat and mass transfer. The presentation of Bird *et al*. was primarily focused on results and solutions rather than on the methods of solution or analysis. However, the combination of the more fundamental approach that it pioneered within the chemical engineering community and the appearance of chemical engineers with very strong mathematics backgrounds produced the most recent transitions in our ways of thinking about and understanding transport processes.

Initially, this was focused largely on the use of asymptotic and numerical methods for detailed analysis and understanding of the important correlations between the dependent and independent dimensionless groups in flow and transport processes relevant to large-scale engineering applications. Although asymptotic approximation methods were initially the product of applied mathematicians, chemical engineers now played an extremely important role in their application to many transport processes and viscous flow phenomena. A critical component in this approach is nondimensionalization by means of characteristic scales to identify dominant physical balances and the use of this approach in approximating (simplifying) the governing equations and boundary conditions.

Another simultaneous development within chemical engineering was a major focus on flows at very low Reynolds number, at least partly motivated by the classic book *Low Reynolds Number Hydrodynamics* by Happel and Brenner,[3] originally published in 1965. Initially, the primary application of creeping-flow theory was to the analysis of suspensions, emulsions, and other particulate materials, in combination with the effects of Brownian motion that typically play an important role for particulates with length scales of 10 μm or less. More recently, the scale of many processes of interest has decreased, to the point that there is sometimes a need to incorporate "nonhydrodynamic" forces such as van der Waals forces that act over very short length scales. Furthermore, the recent development of microelectromechanical system technology, mainly focused on very small-scale flow systems, and many of the applications of fluid mechanics and biotransport, have provided a further emphasis on the importance of a fundamental understanding of viscous flow and transport phenomena. Finally, the relevance of interfacial phenomena and of non-Newtonian rheology associated with complex fluids such as polymeric liquids has additionally broadened the scope of what chemical engineers are likely to encounter as "transport phenomena."

Finally, we cannot overlook the development of computational tools for the solution of problems in fluid mechanics and transport processes. Methods of increasing sophistication have been developed that now enable quantitative solutions of some of the most complicated and vexing problems at least over limited parameter regimes, including direct numerical simulation of turbulent flows; so-called free-boundary problems that typically involve large interface or boundary deformations induced by flow; and methods to solve flow problems for complex fluids, which are typically characterized by viscoelastic constitutive equations and complicated flow behavior.

B. THE NATURE OF THE SUBJECT

The study of fluid mechanics and convective transport processes for heat or molecular species is an old subject. Provided that we limit ourselves to Newtonian fluids and to flow domains involving only solid boundaries, there is no question of understanding the underlying physical principles that govern a problem, at least from a continuum mechanical point of view. On a point-by-point basis, these are represented by the Navier–Stokes and thermal energy (or species transport) equations, with boundary conditions that are generally well established.[4] These equations and boundary conditions (at least for solid boundaries) have been known for more than a century. Yet these subjects are still extremely active topics of basic research, and, for the most part, this research is aimed at the discovery of new phenomena and principles of fluid motion, rather than the engineering application of previously discovered phenomena to new systems. Although the underlying physical principals are completely understood for this class of fluids, the macroscopic phenomena that are inherent in these physical principles can be extremely complex. From a mathematical point of view, this is largely because the governing equations (and the boundary conditions too at a fluid interface) are nonlinear. However, one need think only of the amazing variety

B. The Nature of the Subject

of fluid flow phenomena that are encountered in everyday experience to recognize the complexity allowed by the well-known fundamental principles.

Examples include the dynamics of waves and of breaking waves at a beach; the complex "mixing" flows created by a spoon moving through a cup of coffee; the dripping of water from a tap; the bathtub vortex as water drains from a tub; or the complicated coiling motion of honey dripping from a knife onto a slice of toast. We may also think of the changes in flow structure that can be caused by variations in the flow rate, even when the geometry of the boundaries and all the fluid properties are fixed. Likely familiar to most readers is the transition in pressure-driven flow through a circular tube, from a one-dimensional time-independent "laminar" flow at low flow rates to a fully three-dimensional time-dependent "turbulent" flow when the flow rate is increased beyond some critical value. Perhaps less familiar is the flow produced by the translational motion of a cylindrical body perpendicular to its axis through an otherwise stationary Newtonian fluid. The motion of the fluid in all circumstances is governed by the Navier–Stokes equations, yet the range of observable flow phenomena that occur as the velocity of the cylinder is increased is quite amazing, beginning at low speeds with steady motion that follows the body contour, and then followed by a transition to an asymmetric motion that includes a standing pair of recirculating vortices at the back side of the cylinder, which are at first steady and attached and then become unsteady and alternately detached as the velocity increases. Finally, at even higher speeds the whole motion in the vicinity of the cylinder and downstream becomes three dimensional and chaotic in the well-known regime of turbulent flow. All of these phenomena are inherent in the physical principles encompassed in the Navier–Stokes equations. It is "only" that the solutions of these equations (namely, the physical phenomena) become increasingly complex with increase in the velocity of the cylinder through the fluid. Generally, then, for Newtonian fluids, the basic objective is to understand the physics of the flow rather than the underlying physical principles.

In a sense, we can visualize the physics of a flow by carrying out laboratory experiments, or by observing natural flows directly. When this is not possible or is not feasible, the advent of increasingly powerful computers and numerical techniques sometimes allow a "computational experiment" to be carried out, based on the known governing equations and boundary conditions. It is unfortunate that a qualitative description, or even flow-visualization pictures, of complex phenomena do not translate immediately into "understanding." Obviously, if this were the case, it would be possible to provide students with a much more realistic picture of real phenomena than they can hope to achieve in the normal classroom (or textbook) environment. The difficulty with a qualitative description is that it can never go much beyond a case-by-case approach, and it would clearly be impossible to encompass all of the many flow or transport systems that will be encountered in technological applications. The present book does not provide anything like a catalog of physically interesting phenomena. Hopefully, the reader will have already encountered some of these in the context of undergraduate laboratory studies or personal experience. There is also at least one textbook[5] that attempts (with some success) to fill the gap between "analytic technique" and "physical phenomena" in fluid mechanics, and this can provide an important complement to the material presented here. In fluid mechanics, there is also a very useful video series available from Cambridge University Press for a very nominal cost, "Multi-Media Fluid Mechanics," which is an excellent source of visual exposure to real phenomena, coupled with useful physical explanations as well.[6] Finally, every student and teacher of fluid mechanics should examine the wonderful collection of photos in the book *An Album of Fluid Motions* by Van Dyke,[7] the series of articles "A Gallery of Fluid Motion," published annually in *Physics of Fluids*, and the recent compilation of highlights from these articles published as a book by Cambridge University Press, and also titled *A Gallery of Fluid Motion*.[8] The events depicted in these latter photographs provide a graphic reminder of the vast wealth of complex, important,

C. A BRIEF DESCRIPTION OF THE CONTENTS OF THIS BOOK

and interesting phenomena that are encompassed within fluid mechanics. Clearly the fluid mechanics and heat and mass transfer presented in the classroom or by any textbook only scratch the surface of this fascinating subject.

The material in this book is the basis of an introductory (two-term) graduate course on transport phenomena. It starts (in Chap. 2 of the book that is subsequently described in more detail) with a derivation of all of the necessary governing equations and boundary conditions in a context that is intended to focus on the underlying fundamental principles and the connections between this topic and other topics in continuum physics and thermodynamics. Some emphasis is also given to the limitations of both equations and boundary conditions (for example, "non-Newtonian" behavior, the "no-slip" condition, surfactant and thermocapillary effects at interfaces, etc.). It should be noted, however, that, though this course starts at the very beginning by deriving the basic equations from first principles and thus can be taken successfully even without an undergraduate transport background, there are important topics from the undergraduate curriculum that are *not* included, especially macroscopic balances, friction factors, correlations for turbulent flow conditions, etc.

The remainder of the book is more or less concerned with how to solve transport and fluids problems analytically, but with a lot of emphasis on basic physics, scaling and nondimensionalization, and approximations that can be used to obtain solutions that are due either to geometric simplifications or large or small values of dimensionless parameters. I am more specific in the following subsections, but it is important to note that *there is a strong emphasis on setting the problem up and extracting as much information as possible short of obtaining detailed solutions of differential equations*. The book reflects my bias that it is more important for students to see moderate numbers of problems with enough detail so that they can follow the analysis and the thinking behind the analysis from the beginning to the end. Although the problems chosen are obviously not going to be identical in most cases to a research problem that students may encounter later, they are chosen to expose students to many qualitative phenomena, and one may hope that with this background behind them they may be able to actually use the material in some future "application." At the minimum, they should be able to read and understand the research literature on any transport-related problem that arises in their later work, including an understanding of the approximations or limitations, etc. In what follows, I outline the content of the various chapters, including the most important ideas or concepts that I hope a student or reader will extract.

Chapter 2: The Basic Principles

This book begins with a detailed derivation of the governing equations and boundary conditions for fluid mechanics and convective transport processes. Some emphasis is placed on understanding the limitations of these equations and boundary conditions, including the origins of non-Newtonian behavior. We also consider, in some detail, the boundary conditions at a fluid interface and the role of surfactants when these are present. At several points in this chapter, we begin to think qualitatively about flows, and particularly what we may anticipate about flows that are driven by body forces (gravity) in the presence of density gradients and by capillary forces that are due either to gradients in the interface curvature or to surface-tension gradients.

Chapter 3: Unidirectional and One-Dimensional Flow and Heat Transfer Problems

This chapter is *primarily* concerned with the most general class of problems for which an "exact" analytic solution is possible. Thus it is used to review classical methods of solution

C. A Brief Description of the Contents of This Book

for linear partial differential equations, but that is not really the main point. The main point is to introduce the concepts of characteristic scales, nondimensionalization, dynamic similarity, diffusive time scales and their role in the transient evolution of flows and transport processes, and self-similarity for problems that do not exhibit characteristic scales. There is also a discussion of Taylor diffusion that does not exhibit an exact solution of the transport equations, but is an important and interesting problem of transport in a unidirectional flow with many applications. By the time we finish this chapter, there should be no doubt about how to nondimensionalize problems, how to solve problems that can be reduced to a linear form, and the reader will also have seen the first examples of using characteristic scales to think about transport problems.

Chapter 4: An Introduction to Asymptotic Approximations[9]

In this chapter, we discuss general concepts about asymptotic methods and illustrate a number of different types of asymptotic methods by considering relatively simple transport or flow problems. We do this by first considering pulsatile flow in a circular tube, for which we have already obtained a formal exact solution in Chap. 3, and show that we can obtain useful information about the high- and low-frequency limits more easily and with more physical insight by using asymptotic methods. Included in this is the concept of a boundary layer in the high-frequency limit. We then go on to consider problems for which no exact solution is available. The problems are chosen to illustrate important physical ideas and also to allow different types of asymptotic methods to be introduced:

(a) We consider viscous dissipation effects in shear flow and indicate what it may have to do with the use of a shear rheometer to measure viscosities.
(b) We consider flow in a tube that is slightly curved. This illustrates that the flows in straight and curved tubes are fundamentally different with potentially important implications for transport processes.
(c) We consider flow in a wavy-wall channel primarily to show how "domain perturbation methods" can be used to turn this problem into a simpler problem that we can solve as flow in a straight-wall channel with "slip" at the boundaries.
(d) We consider a simple model problem of transport inside a catalyst pellet with fast reaction to illustrate another example of a boundary-layer-type problem.
(e) Finally there is a longish section on the dynamics of a gas bubble in a time-dependent pressure field that introduces ideas about linear stability analysis and its connection to perturbation methods, resonance when the forcing and natural frequencies of oscillation match, and multiple-time-scale asymptotic methods to analyze resonant behavior.

Chapter 5: The Thin-Gap Aproximation – Lubrication Problems

One important class of problems for which we can obtain significant results at the first level of approximation is the motion of fluids in thin films. In this and the subsequent chapter, we consider how to analyze such problems by using the ideas of scaling and asymptotic approximation. In this chapter, we consider thin films between two solid surfaces, in which the primary physics is the large pressures that are set up by relative motions of the boundaries, and the resulting ideas about "lubrication" in a general sense.

(a) The basic ideas are introduced by use of the classic problem of the eccentric Couette problem, called the "journal-bearing problem" in the lubrication literature. This problem is advantageous because the thin-gap approximation is uniformly valid throughout the domain in the so-called narrow-gap limit.
(b) Following this, we derive the thin-film/lubrication equations from a more general point of view; one result of this general analysis is the famous Reynolds equation of lubrication

theory, but we consider how to analyze such problems from a fundamental point of view that can be adapted to many applications even when it may not be immediately obvious how to apply the Reynolds equation.
(c) In the last sections, we consider several examples:
   (1) the so-called "slider-block problem";
   (2) the motion of a sphere near a solid bounding wall, which leads to the conclusion that the sphere will not come into contact with the wall in finite time if it is moving under the action of a finite force and the surfaces are smooth;
   (3) an analysis of the dynamics of the disk on an *air hockey table*. This problem is amenable to "standard" lubrication theory when the blowing velocity is small enough (though still large enough to maintain a finite gap between the disk and the tabletop), but requires a boundary-layer-like analysis when the blowing velocity is large (even though the thin-film approximation is still valid).

Chapter 6: The Thin-Gap Approximation – Films with a Free Surface
The second basic class of thin-film problems involves the dynamics of films in which the upper surface is an interface (usually with air). In this case, the same basic scaling ideas are valid, but the objective is usually to determine the shape of the upper boundary (i.e., the geometry of the thin film), which is usually evolving in time.

A typical example is a spreading film on a solid substrate, and we begin with this class of problems. Analysis of this class of thin-film problems requires use of the interface boundary conditions derived in Chap. 2 and also revisits a number of examples of capillary and Marangoni flow problems that were discussed qualitatively in Chap 2. The governing equation for the thin-film shape function often takes the form of a "nonlinear diffusion equation," and this allows the scaling behavior of the thin-film dynamics to be deduced by means of a similarity transformation ("advanced" dimensional analysis), without necessarily solving the resulting nonlinear equation. For example, for the spreading of an axisymmetric film (or drop) on a solid substrate caused by capillary effects, we can deduce the famous Tanner's law that the radius of the contact circle should increase as $R(t) \sim t^{1/10}$ without solving equations. These are great examples for illustrating what we can get from seeking the form of self-similar solutions.

We then go on consider the role of van der Waals forces on the dynamics of a thin film. First we consider the *stability* of a horizontal fluid layer (which is bounded either above or below by a solid substrate) due to the coupled interactions of gravity, capillary forces, and van der Waals forces across the film. This allows us to introduce the ideas of a *linear stability analysis* and leads to interesting and important results. We then consider the actual rupture process of a thin film with van der Waals forces present. In particular, we show that the final stages of the rupture process, including the geometry of the film and the scaling of the process with time, can be analyzed again by means of a similarity transformation (without solving equations).

Finally, we consider a number of problems involving nonisothermal flows in a shallow cavity. The motion in this cavity may be due to buoyancy caused by differential heating of the end walls or to thermocapillary flow that is due to Marangoni stresses at the upper interface, again with differential heating at the end walls. These problems are idealized models for a number of important applications; for example, the latter case is a model for the "liquid bridge" in containerless processing of single crystals. The objective of analysis is the flow and temperature fields, but also the shape of the free surface. It is shown that the interface shape problems can be analyzed both by means of the classic thin-gap approach of preceding sections of this chapter and also by the method of "domain perturbations," first

C. A Brief Description of the Contents of This Book

introduced in Chap. 4. These latter problems focus again on the important issue of interfacial boundary conditions and the role of capillary and thermocapillary effects in flow.

Chapter 7: Creeping Flows (Two-Dimensional and Axisymmetric Problems)[10]

We begin, in this chapter and the next, with the class of flow problems for general geometries in which the dynamics is dominated by the balance between pressure gradients and the viscous terms in the Navier–Stokes equation. This class of problem is known collectively as "creeping" flows. In the first of these two chapters, we initially consider nondimensionalization, the role of the Reynolds number for this general class of problems, the concepts of quasi-steady flow, and some extremely important consequences of the fact that the governing equations in the creeping-flow limit are linear. The latter material is important beyond the several examples considered, because it forces the student to think about what can be said about the solution of linear problems without actually solving any equations. We then go on in this chapter to consider two-dimensional and axisymmetric problems that can be solved by introducing the concept of a streamfunction. This leads to a single fourth-order partial differential equation and the natural use of general eigenfunction expansions. The following specific problems are solved:

(a) 2D corner flows (scraping, mixing, etc.),
(b) uniform flow past a solid sphere (the classic Stokes problem),
(c) axisymmetric extensional flow in the vicinity of a solid sphere and the use of this result to derive the famous Einstein expression for the viscosity of a dilute suspension of spheres,
(d) the buoyancy-driven translation of a drop through a quiescent fluid including the fact that the shape is a sphere independent of the interfacial tension,
(e) motions of drops driven by Marangoni stress in a nonuniform temperature field,
(f) the effects of surfactants on the buoyancy-driven motion of a drop.

These problems are chosen because they illustrate important ideas and concepts in addition to simply solving problems. However, the analysis in this chapter is completely based on classical eigenfunction expansions.

Chapter 8: Creeping Flows (Three-Dimensional Problems)[11]

We begin this chapter in Sections A–C by discussing the construction of solutions to the creeping-flow equations by representing the solutions in terms of "vector harmonic functions." It is shown that one can literally write the solutions of a *whole class of problems* almost by inspection, thus eliminating the need for the laborious eigenfunction expansions of Chap. 7 even for the two-dimensional and axisymmetric problems for which they can be used, but also simultaneously obtaining the solutions for fully three-dimensional problems (e.g., a sphere in a linear shear flow). The main requirement is that the boundaries of the flow domain must correspond approximately to surfaces in a known analytic coordinate system. In this chapter, we consider problems that we can solve by using vector harmonic functions in a spherical coordinate system. The method is illustrated for a number of examples including both problems with axisymmetric symmetry that we could solve by using the methods of Chap. 7, and problems such as particle motion in a linear shear flow that we could not solve by using these methods. We conclude in Section C by considering the motion of drops in general linear (shearlike) flows, including an illustration of how to estimate the deformed shape of the drop in the flow.

In subsequent sections of this chapter we discuss the use of *fundamental solutions* of the creeping-flow equations to construct solutions for which the flow domain has a more general geometry. This includes "slender-body" theory for slender rodlike objects, and an introduction to a powerful method known as the "boundary-integral" technique that can be implemented numerically to solve virtually any creeping-flow problem, including those with complex or unknown (possibly evolving) geometries. Other sections illustrate general results that can be obtained because of the linearity of creeping-flow problems, with an emphasis on illustrating general physical phenomena for this class of problem.

Chapter 9: Convection Effects and Heat Transfer for Viscous Flows

Now that we have learned how to solve for the detailed velocity fields for at least one class of flow problems (creeping/viscous flows), we turn to a general introduction to convection effects for heat transfer (primarily) for this class of flows.

We begin by considering the nondimensional form of the thermal energy equation, leading to the recognition of the Peclet number (the product of the Reynolds number and the Prandtl number) as the critical independent parameter for "forced" convection heat transfer problems. At the end of this section, we briefly discuss the analogy with mass transfer in a two-component system, with the Schmidt number replacing the Prandtl number and the Sherwood number replacing the Nusselt number.

The limit $Pe \to 0$ yields the pure conduction heat transfer case. However, for a fluid in motion, we find that the pure conduction limit is not a uniformly valid first approximation to the heat transfer process for $Pe \ll 1$, but breaks down "far" from a heated or cooled body in a flow. We discuss this in the context of the "Whitehead" paradox for heat transfer from a sphere in a uniform flow and then show how the problem of forced convection heat transfer from a body in a flow can be understood in the context of a singular-perturbation analysis. This leads to an estimate for the first correction to the Nusselt number for small but finite Pe – this is the first "small" effect of convection on the correlation between Nu and Pe for a heated (or cooled) sphere in a uniform flow.

We then return briefly to consider the creeping-flow approximation of the previous two chapters. We do this at this point because we recognize that the creeping-flow solution is exactly analogous to the pure conduction heat transfer solution of the preceding section and thus should also not be a uniformly valid first approximation to flow at low Reynolds number. We thus explain the sense in which the creeping-flow solution can be accepted as a first approximation (i.e., why does it play the important role in the analysis of viscous flows that it does?), and in principle how it might be "corrected" to account for convection of momentum (or vorticity) for the realistic case of flows in which Re is small but nonzero.

We then go on to consider the generalization of the analysis of heat transfer problems for small Peclet numbers. *These generalizations clearly illustrate the power of the asymptotic method to provide insight into the form of correlations between dimensionless parameters, with a minimum of detailed analysis.* First, we show that the detailed analysis that we developed for a sphere actually can be applied with no extra work to obtain the first correction to Nu for bodies of *arbitrary shape* in a uniform flow (or where the body is sedimenting through an otherwise motionless fluid). Next, we consider heat transfer from a sphere in a shear flow. The purpose of this is to show that the same theoretical framework can be applied again, but that the form of the correlation between Nu and Pe changes if the nature of the flow is changed. Again, the analysis for a sphere in linear shear flow can be generalized with little additional work to obtain the correlation for any linear flow and for bodies of arbitrary shape.

The second half of this chapter considers the opposite limit in which $Pe \gg 1$. In this case, the superficial conclusion is that heat transfer must be dominated everywhere by convection. However, this cannot be true, as the only mechanism for heat transfer from the surface of a body to a surrounding fluid is by conduction. This leads to the concept of the

C. A Brief Description of the Contents of This Book

thermal boundary layer and a fundamentally different form for the correlation between Nu and Pe.

Chapter 10: Boundary-Layer Theory for Laminar Flows

The concept of a boundary layer is one of the most important ideas in understanding transport processes. It is based on the idea that transport systems often generate internal length scales so that dissipative effects (or diffusive effects in the case of heat transfer or mass transfer) continue to play an essential role even in the limit as the viscosity (or the diffusivity) becomes smaller and smaller. In this chapter, we continue the development of these ideas, first introduced at the end of the previous chapter, by considering their application to the approximate solution of fluid mechanics problems in the asymptotic limit of large Reynolds number. The chapter begins with a section on potential-flow theory, namely the solutions of the equations of motion when viscous effects are completely neglected. We find that the predictions that leave out viscous effects are fatally flawed for some problems such as flow past a circular cylinder, leading to the famous d'Alembert's paradox, which says that the drag on bodies at high Reynolds number is zero. This occurs mainly because potential-flow theory cannot predict the asymmetry that is responsible for boundary-layer separation and the dominance of "form" drag for nonstreamlined bodies. The next section of the chapter develops the key ideas of the asymptotic boundary-layer theory. This is first applied to the classic Blasius problem of flow past a horizontal flat plate and then considers the class of problems in which self-similar solutions of the boundary-layer equations are possible. This is followed by the Blasius series solution for flow past nonstreamlined bodies and the application of this theory to the problem of flow past a circular cylinder. This exposes a key result, which is the ability of boundary-layer theory to predict the onset of "separation" and thus to determine whether a two-dimensional body (such as an airfoil) is sufficiently streamlined to avoid "form" drag. We then consider the generalization of boundary-layer theory to axisymmetric geometries. Finally, we address the question of boundary layers on a free surface, such as an interface, by considering the application of boundary-layer concepts to the motion of a spherical bubble at high Reynolds number. This section is perhaps the most important one in the chapter from a pedagogical point of view, because it challenges most of the simplistic ideas that students may have from undergraduate transport courses, and forces them to see that boundary layers are applicable to a very broad class of problems. For example, the question of a boundary layer on a bubble forces students to reconsider the simplistic (and often incorrect) idea that a boundary layer exists because of the no-slip condition.

Chapter 11: Heat and Mass Transfer at Large Reynolds Number

In this chapter, we return to forced convection heat and mass transfer problems when the Reynolds number is large enough that the velocity field takes the boundary-layer form. For this class of problems, we find that there must be a correlation between the dimensionless transport rate (i.e., the Nusselt number for heat transfer) and the independent dimensionless parameters, Reynolds number Re and either Prandtl number Pr or Schmidt number Sc of the form

$$Nu = cRe^a Pr^b \quad \text{or} \quad Nu = cRe^a Sc^b.$$

The coefficient $a = 0.5$ for laminar flow conditions and $Re \gg 1$. On the other hand, the coefficient b depends on the maginitude of the Prandtl (or Schmidt) number and also changes depending on whether the boundary is a no-slip surface or a fluid interface. For example, for a no-slip surface, $b = 1/2$ in the limit Pr (or Sc) $\to 0$ but $b = 1/3$ for Pr (or Sc) $\to \infty$. By now, students can easily analyze and understand qualitatively the reasons for these changes, as well as the effect of changes in the fluid mechanics or thermal boundary conditions. The coefficient c is an order 1 number that depends on the geometry, but we show that

A Preview

very general solutions for "arbitrary" body shapes can be obtained by means of similarity transformations. Finally, we readdress the issue of the analogy between heat transfer and single-component mass transfer by considering the effects of finite interfacial velocities that must exist at a boundary that acts as a source or sink of material in the mass transfer problem but not in the thermal problem.

Chapter 12: Hydrodynamic Stability

All of the preceding chapters seek solutions for various transport and fluid flow problems, without addressing the stability of the solutions that are obtained. The ideas of linear stability theory are very important both within the transport area and also in a variety of other problem areas that students are likely to encounter. Too often, it is not addressed in transport courses, even at the graduate level. The purpose of this chapter is to introduce students to the ideas of linear stability theory and to the methods of analysis. The problems chosen are selected because it is possible to make analytic progress and because they are of particular relevance to chemical engineering applications. The one topic that is only lightly covered is the stability of parallel shear flows. This is primarily because it is such a subtle and complicated subject that one cannot do justice to it in this type of presentation (it is the subject of complete books all by itself).

We begin with *capillary instability* of a liquid thread. This is a problem that was discussed qualitatively already in Chap. 2. It is a problem with a physically clear mechanism for instability and thus provides a good framework for introducing the basic ideas of linear stability theory. This problem is one of several examples in which the viscosity of the fluid plays no role in determining stability, but only influences the rate of growth or decay of the infinitesimal disturbances that are analyzed in a linear theory.

Next, we turn to the classic problem of *Rayleigh–Taylor instability* for the gravitationally driven "overturning" of a pair of immiscible superposed fluids in which the upper fluid has a higher density than the lower fluid. This is another example of a problem in which the viscosity of the fluid is not an essential factor in its instability.

The third problem is known as the *Saffman–Taylor instability* of a fluid interface for motion of a pair of fluids with different viscosities in a porous medium. It is this instability that leads to the well-known and important phenomenon of viscous fingering. In this case, we first discuss Darcy's law for motion of a single-phase fluid in a porous medium, and then we discuss the instability that occurs because of the displacement of one fluid by another when there is a discontinuity in the viscosity and permeability across an interface. The analysis presented ignores surface-tension effects and is thus valid strictly for "miscible displacement."

Next we turn to the *stability of Couette flow* for parallel rotating cylinders. This is an important flow for various applications, and, though it is a shear flow, the stability is dominated by the centrifugal forces that arise because of centripetal acceleration. This problem is also an important contrast with the first two examples because it is a case in which the flow can actually be stabilized by viscous effects. We first consider the classic case of an inviscid fluid, which leads to the well-known criteria of Rayleigh for the stability of an inviscid fluid. We then analyze the role of viscosity for the case of a narrow gap in which analytic results can be obtained. We show that the flow is stabilized by viscous diffusion effects up to a critical value of the Reynolds number for the problem (here known as the Taylor number).

We then go on to consider three examples of instabilities that arise because of buoyancy and Marangoni effects in a nonisothermal system. This is preceded by a brief discussion of the Bousinesq approximation of the Navier–Stokes and thermal energy equations.

The first problem considered is the classic problem of *Rayleigh–Benard convection* – namely the instability that is due to buoyancy forces in a quiescent fluid layer that is heated from below. In this case, both viscous diffusion and thermal diffusion play a role in stabilizing the fluid, leading to the concept of a critical Rayleigh number for instability.

This is followed by an analysis of the related buoyancy-driven instabilities that can occur in a system in which the density is dependent on two "species" that diffuse at significantly different rates. This problem is known as the *"double-diffusive" stability problem*. It was originally analyzed in a geophysical context in which the two factors influencing the density are the temperature and the salinity of the fluid (hence in this context it is known as the thermohaline instability problem). However, it has many important applications in chemical engineering in which there are two "solutes" (or more, though a theory to describe this is not presented here) rather than salt and heat. Students often find this problem very interesting as an example of a situation in which instability may occur even though simple ideas suggest that it should not. An example is a fluid layer in which the density decreases with height, yet the system exhibits spontaneous buoyancy-driven convection that is due to the difference in transport rates of the two species.

Finally, we consider the problem of *Marangoni instability*; namely convection in a thin-fluid layer driven by gradients of interfacial tension at the upper free surface. This is another problem that was discussed qualitatively in Chap. 2, and is a good example of a flow driven by Marangoni stresses.

The last section in this chapter is a brief introduction to *stability of parallel shear flows*. We consider three basic issues: (i) Rayleigh's equation for inviscid flows, (ii) Rayleigh's necessary condition on an inflection point for inviscid instability, and (iii) a derivation of the Orr–Sommerfeld equation and Squire's theorem.

NOTES AND REFERENCES

1. W. H. Walker, W. K. Lewis, and W. H. McAdams, *Principles of Chemical Engineering* (McGraw-Hill, New York, 1923).
2. S. R. Bird, W. B. Stewart, and B. N. Lightfoot, *Transport Phenomena* (Wiley, New York, 1960).
3. J. Happel and H. Brenner, *Low Reynolds Number Hydrodynamics* (Noordhoff International, Leyden, The Netherlands, 1973).
4. This is not to say that there are no unresolved issues in formulating the basic principals for a continuum description of fluid motions. Effective descriptions of the constitutive behavior of almost all complex, viscoelastic fluids are still an important fundamental research problem. The same is true of the boundary conditions at a fluid interface in the presence of surfactants, and effective methods to make the transition from a pure continuum description to one which takes account of the molecular character of the fluid in regions of very small scale is still largely an open problem.
5. D. I. Tritton, *Physical Fluid Dynamics* (Van Nostrand Reinhold, London, 1977).
6. G. M. Homsy, H. Aref, K. S. Breuer, S. Hochgreb, J. R. Koseff, B. R. Munson, K. G. Powell, C. R. Robertson, and S. T. Thoroddsen, "Multi-Media Fluid Mechanics," CD-ROM, (Cambridge University Press, Cambridge, 2004).
7. M. Van Dyke, *An Album of Fluid Motion* (Parabolic Press, Stanford, CA, 1982).
8. M. Samimy, K. S. Breuer, L. G. Leal, and P. H. Steen, *A Gallery of Fluid Motion* (Cambridge University Press, Cambridge, 2004).
9. *Introductory note:* Most transport and/or fluids problems are not amenable to analysis by classical methods for linear differential equations, either because the equations are nonlinear (or simply too complicated in the case of the thermal energy equation, which is linear in temperature if natural convection effects can be neglected), or because the solution domain is complicated in shape (or in the case of problems involving a fluid interface having a shape that is *a priori* unknown). Analytic results can then be achieved only by means of approximations. One "approach" is to "simply" discretize the equations in some way and turn on the computer. Another is to use the family of approximations methods known as asymptotic approximations that lead to useful concepts such as boundary layers, etc. This course is about the latter approach. However, it is not just a

course about some mathematical methods, but rather the coupling between these methods and thinking physically about the problem at hand. Ultimately, our objective is to get as much useful insight about a problem as we can with as little detailed work as we can get away with. Asymptotic methods are seen in this sense as an extension of scaling–identifying the dominant physical effects in different parts of a domain, and we use this to deduce the most important results for many problems by setting them up, rather than by actually solving the detailed equations. An example is the well-known correlation between Nusselt number Nu and the Reynolds and Prandtl numbers for heat transfer at high Reynolds number. If you understand how to use scaling and asymptotic methods, you can show that the correlation must take the form

$$Nu = c\,Re^a\,Pr^b,$$

with coefficients a and b that can be obtained depending on whether Pr is large or small, and whether the surface is a solid surface or an interface without solving any differential equations. Only the $O(1)$ constant c cannot be determined without solving the equations because it depends on the geometry of the surface, but even there we are guaranteed that it must be an $O(1)$ number.

10. *Introductory note*: In the preceding two chapters, the basis of approximation is the special geometry of the flow (or transport) domain. Now we embark on the remaining chapters, all of which (except for the last chapter) are focused on approximations based on the dominance of specific physical mechanisms and the identification of these dominant mechanisms by means of scaling, nondimensionalization, and the magnitude of characteristic dimensionless parameters, such as Reynolds number, Peclet number, Prandtl number, etc. We typically assume that flows are laminar, and we generally seek steady (or quasi-steady) solutions, with only an occasional brief discussion about the stability of these solutions (i.e., under what circumstances may they actually be observed in the "laboratory"?). The last chapter of the book, which presents classical linear stability analysis of a number of problems of special interest in chemical engineering applications, therefore adds an important perspective to the material in this book.

11. From a superficial point of view, this chapter simply represents the generalization of the theory of viscous dominated flows to consider three-dimensional problems. However, it also introduces much more powerful and convenient mathematical methods, many of which can be used in other applications. Sections A–C are particularly important. Other sections represent more advanced (and thus elective) topics for coverage in class.

2

Basic Principles

We are concerned in this book primarily with a description of the motion of fluids under the action of some applied force and with convective heat transfer in moving fluids that are not isothermal. We also consider a few analogous mass transfer problems involving the convective transport of a single solute in a solvent.

It is assumed that the reader is familiar with the basic principles and equations that describe these processes from a continuum mechanics point of view. Nevertheless, we begin our discussion with a review of these principles and the derivation of the governing differential equations (DEs). The aim is to provide a reasonably concise and unified point of view. It has been my experience that the lack of an adequate understanding of the basic foundations of the subject frequently leads to a feeling on the part of students that the whole subject is impossibly complex. However, the physical principles are actually quite simple and generally familiar to any student with a physics background in classical mechanics. Indeed, the main problems of fluid mechanics and of convective heat transfer are not in the complexity of the underlying physical principles, but rather in the attempt to understand and describe the fascinating and complicated phenomena that they allow. From a mathematical point of view, the main problem is not the derivation of the governing equations that is presented in this second chapter, but in their solution. The latter topic will occupy the remaining chapters of this book.

A. THE CONTINUUM APPROXIMATION

It will be recognized that one possible approach to the description of a fluid in motion is to examine what occurs at the microscopic level where the stochastic motions of individual molecules can be distinguished. Indeed, to a student of physical chemistry or perhaps chemical engineering, who has been consistently exhorted to think in molecular terms, this may at first seem the obvious approach to the subject. However, the resulting many-body problem of molecular dynamics is impossibly complex under normal circumstances because the fluid domain contains an enormous number of molecules. Attempts to simulate such systems with even the largest of present-day computers cannot typically handle more than a few thousand molecules of simple shape and then only for a very short period of time.[1] Thus efforts to provide a mathematical description of fluids in motion could not have succeeded without the introduction of sweeping approximations. The most important among these is the so-called *continuum hypothesis*. According to this hypothesis, the fluid is modeled as infinitely divisible without change of character. This implies that all quantities, including the material properties such as density, viscosity, or thermal conductivity, as well as variables such as pressure, velocity, and temperature, can be defined at a mathematical point in an

Basic Principles

unambiguous way as the limit of the mean of the appropriate quantity over the (inevitable) molecular fluctuations.

The motivation for this approach, apart from an anticipated simplification of the problem, is that, in many applications of applied science or engineering, we are concerned with fluid motions or heat transfer in the vicinity of bodies, such as airfoils, or in confined geometries, such as a tube or pipeline, where the physical dimensions are very much larger than any molecular or intermolecular length scale of the fluid. The desired description of fluid motion is then at this larger, *macroscopic* level where, for example, an average of the forces of interaction between the fluid and the bounding surface may be needed, but not the instantaneous forces of interaction between this surface and individual molecules of the fluid.

Once the continuum hypothesis has been adopted, the usual macroscopic laws of classical continuum physics are invoked to provide a mathematical description of fluid motion and/or heat transfer in nonisothermal systems – namely, conservation of mass, conservation of linear and angular momentum (the basic principles of Newtonian mechanics), and conservation of energy (the first law of thermodynamics). Although the second law of thermodynamics does not contribute directly to the derivation of the governing equations, we shall see that it does provide constraints on the allowable forms for the so-called constitutive models that relate the velocity gradients in the fluid to the short-range forces that act across surfaces within the fluid.

The development of convenient and usable forms of the basic conservation principles and the role of the constitutive models and boundary conditions in a continuum mechanics framework occupy the remaining sections of this chapter. In the remainder of this section, we discuss the foundations and consequences of the continuum hypothesis in more detail.

1. Foundations

In adopting the continuum hypothesis, we assume that it is possible to develop a description of fluid motion (or heat transfer) on a much coarser scale of resolution than on the molecular scale that is still *physically equivalent* to the molecular description in the sense that the former could be derived, in principle, from the latter by an appropriate averaging process. Thus it must be possible to define any dependent macroscopic variable as an average of a corresponding molecular variable. A convenient average for this purpose is suggested by the utility of having macroscopic variables that are readily accessible to experimental observation. Now, from an experimentalist's point of view, any probe to measure velocity, say, whose dimensions were much larger than molecular, would automatically measure a *spatial* average of the molecular velocities. At the same time, if the probe were sufficiently small compared with the dimensions of the flow domain, we would say that the velocity was measured "at a point," in spite of the fact that the measured quantity was an average value from the molecular point of view. This simple example suggests a convenient definition of the macroscopic variables in terms of molecular variables, namely as *volume averages*, for example,

$$\mathbf{u} \equiv \langle \mathbf{w} \rangle \equiv \frac{1}{V} \int_V \mathbf{w} \, dV, \qquad (2\text{--}1)$$

where V is the averaging volume.[2]

It is important to remark that we shall never actually calculate macroscopic variables as averages of molecular variables. The purpose of introducing an explicit connection between the macroscopic and molecular (or microscopic) variables is that the conditions for $\langle \mathbf{w} \rangle$ to define a meaningful macroscopic (or continuum) point variable provide sufficient conditions

A. The Continuum Approximation

for validity of the continuum hypothesis. In particular, if $\langle \mathbf{w} \rangle$ is to represent a statistically significant average, the typical linear dimension of the averaging volume $V^{1/3}$ must be large compared with the scale δ that is typical of the microstructure of the fluid. Most frequently δ represents a molecular length scale. However, multiphase fluids such as suspensions may also be considered, and in this case δ is the largest microstructural dimension, say, the interparticle length scale or the particle radius. If at the same time $\langle \mathbf{w} \rangle$ is to provide a meaningful point variable in the macroscopic description, it must have a unique value at each point in space at any particular instant, and this implies that the linear dimension $V^{1/3}$ must be arbitrarily small compared with the macroscopic scale L that is characteristic of spatial gradients in the averaged variables (frequently this scale will be determined by the size of the flow domain). Thus, with macroscopic variables defined as volume averages of corresponding microscopic variables, the existence of an equivalent continuum description of fluid motions or heat transfer processes (that is, the validity of the continuum hypothesis) requires

$$\delta \ll V^{1/3} \ll L. \tag{2-2}$$

In other words, it must be possible to choose an averaging volume that is arbitrarily small compared with the macroscale L while still remaining very much larger than the microscale δ. Although the condition (2–2) will always be sufficient for validity of the continuum hypothesis, it is unnecessarily conservative because of the use of volume averaging in the definition (2–1) rather than the more fundamental ensemble average definition of macroscopic variables. Nevertheless, the preceding discussion is adequate for our present purposes.

2. Consequences

One consequence of the continuum approximation is the necessity to hypothesize two independent mechanisms for heat or momentum transfer: one associated with the transport of heat or momentum by means of the continuum or macroscopic velocity field $\mathbf{u}$, and the other described as a "molecular" mechanism for heat or momentum transfer that will appear as a surface contribution to the macroscopic momentum and energy conservation equations. This split into two independent transport mechanisms is a direct consequence of the coarse resolution that is inherent in the continuum description of the fluid system. If we revert to a microscopic or molecular point of view for a moment, it is clear that there is only a single class of mechanisms available for transport of any quantity, namely, those mechanisms associated with the motions and forces of interaction between the molecules (and particles in the case of suspensions). When we adopt the continuum or macroscopic point of view, however, we effectively split the molecular motion of the material into two parts: a molecular average velocity $\mathbf{u} \equiv \langle \mathbf{w} \rangle$ and local fluctuations relative to this average. Because we define $\mathbf{u}$ as an instantaneous spatial average, it is evident that the local net volume flux of fluid across any surface in the fluid will be $\mathbf{u} \cdot \mathbf{n}$, where $\mathbf{n}$ is the unit *normal* to the surface. In particular, the local fluctuations in molecular velocity relative to the average value $\langle \mathbf{w} \rangle$ yield no net flux of *mass* across any macroscopic surface in the fluid. However, these local random motions *will* generally lead to a net flux of heat or momentum across the same surface.

To illustrate this fact, we may adopt the simplest model fluid – the billiard-ball gas – and refer to the simple situation shown in Fig. 2–1. Here we consider a "fluid" made up of two species–namely, black billiard balls and white billiard balls, which are identical apart from their color. By "billiard-ball gas" we mean that the molecules are modeled as hard spheres that interact only when they collide. The motion of each billiard ball (or molecule) is stochastic and thus time dependent, but we assume that there is a nonzero, *steady* macroscopic velocity field $\mathbf{u}$. At an initial moment in time, we imagine a configuration in which the two species are separated by a surface in the fluid that is defined to be locally

Basic Principles

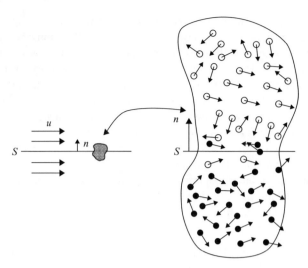

Figure 2–1. We consider a surface S drawn in a fluid that is modeled as a billiard-ball gas. Initially, when viewed at a macroscopic level, there is a discontinuity across the surface. The fluid above is white and the fluid below is black. The macroscopic (volume average) velocity is parallel to S so that $\mathbf{u} \cdot \mathbf{n} = 0$. Thus there is no transfer of black fluid to the white zone, or vice versa, because of the macroscopic motion $\mathbf{u}$. At the molecular (billiard-ball) level, however, all of the molecules undergo a random motion (it is the average of this motion that we denote as $\mathbf{u}$). This random motion produces no net transport of billiard balls across S when viewed at the macroscopic scale because $\mathbf{u} \cdot \mathbf{n} = 0$. However, it does produce a net flux of *color*. On average there is a net flux of black balls across S into the white region and vice versa. In a macroscopic theory designed to describe the transport of white and black fluid, this net flux would appear as a surface contribution and will be described in the theory as a *diffusive* flux. The presence of this flux would gradually smear the initial step change in color until eventually the average color on both sides of S would be the same mixture of white and black.

tangent to the macroscopic velocity $\mathbf{u}$ so that $\mathbf{u} \cdot \mathbf{n} = 0$ at each point on the surface S. The significance of the condition $\mathbf{u} \cdot \mathbf{n} = 0$ is simply that, for every gas molecule that passes across the surface in one direction, a second will, on average, pass across in the opposite direction. Thus the net flux of mass across the chosen surface will be zero, as well as the "convective" flux of any fluid property. For the system in Fig. 2–1, this means that there can be no net flux of black or white balls across the surface S due to the macroscopic motion described by $\mathbf{u}$. Clearly, however, the random translational motions of the molecules will lead to a net flux of black balls across this surface from the bottom to the top (Fig. 2–1), and an equal but oppositely directed flux of white balls from top to bottom until finally at some large time the average color on both sides of S will be the same, and the net flux of either black or white balls will be zero. If we wish to describe the time-dependent evolution of the concentration of black and white balls in a macroscopic (or continuum) theory, it is necessary not only to determine the net transport that is due to the mean (averaged) velocity $\mathbf{u}$, but also to include the transport of black and white balls that is due to the random translational motions of molecules across the surface, modeled as a "diffusion" of the quantity being transported. Evidently this would be true of any property associated with the fluid that has a gradient across the surface, though conventionally the word "diffusion" is used for the transport of molecular species such as the black and white balls in this case.

For example, the presence of a mean temperature gradient (a gradient of the mean molecular kinetic energy) normal to the surface means that each interchange of gas molecules would also transfer a net quantity of heat, even though the average *convective* flux of heat associated with $\mathbf{u}$ is zero because $\mathbf{u} \cdot \mathbf{n} = 0$ on the surface S we have chosen. It is this additional transport that is due to fluctuations in the molecular velocity about the continuum

A. The Continuum Approximation

velocity **u** that must be incorporated in a continuum description as a local "molecular" surface heat flux contribution. We emphasize that the split of heat transfer into a convective contribution associated with **u**, plus an additional molecular contribution, is due to the continuum description of the system. It is conventional to call the diffusive flux of heat "conduction" rather than diffusion, but this is only semantics and does not change the basic diffusive nature of the transport mechanism. *Obviously, the sum of the convective and molecular heat flux contributions in the continuum description must be identical to the total flux of heat due to molecular motions if the continuum description of the system is to have any value.*

Similarly, the continuum description of momentum transfer across a surface must also include both a convective part associated with **u** and a molecular part that is due to random fluctuations of the actual molecular velocity about the local mean value **u**. The billiard-ball gas again provides a convenient vehicle for descriptive purposes. In this case, if we consider a surface that is locally tangent to **u**, it is evident that there can be no transport of mean momentum ρ**u** across the surface that is due to the macroscopic motion itself; the momentum flux that is due to this mechanism is ρ**u**(**u** · **n**), and this is identically equal to zero for any surface on which **u** · **n** = 0. On the other hand, if $(\nabla u) \cdot$ **n** $\neq 0$, then the random interchange of gas molecules that is due to fluctuations in their velocity relative to **u** will lead to a net transport of momentum that must be included in the (macroscopic) continuum mechanics principle of linear momentum conservation as a local, molecular flux of momentum across any surface element in the fluid. In this case, the molecular transport of mean momentum has the effect of decreasing any existing gradient of **u** – crudely speaking, the slower-moving fluid on one side of the surface appears to be accelerated, whereas the faster-moving fluid on the other side is decelerated. Thus, from the continuum point of view, it appears that equal and opposite *forces* have acted across the surface, and the molecular flux of momentum is often described in the continuum description as a surface force per unit area and is called the *stress vector*. Whatever it is called, however, it is evident that the continuum or macroscopic description of the system that results from the continuum hypothesis can be successful only if the flux of momentum that is due to fluctuations in molecular velocity about **u** is modeled in such a way that the sum of its contributions plus the transport of momentum by means of **u** is equal to the actual molecular average momentum flux. As we shall see, the attempt to provide models to describe the "molecular" flux of momentum or heat in the continuum formulation, without the ability to actually calculate these quantities from a rigorous molecular theory, is the greatest source of uncertainty in the use of the continuum hypothesis to achieve a tractable mathematical description of fluid motions and heat transfer processes.

A second similar consequence of the continuum hypothesis is an uncertainty in the boundary conditions to be used in conjunction with the resulting equations for motion and heat transfer. With the continuum hypothesis adopted, the conservation principles of classical physics, listed earlier, will be shown to provide a set of so-called field equations for molecular average variables such as the continuum point velocity **u**. To solve these equations, however, the values of these variables or their derivatives must be specified at the boundaries of the fluid domain. These boundaries may be solid surfaces, the phase boundary between a liquid and a gas, or the phase boundary between two liquids. In any case, when viewed on the *molecular* scale, the "boundaries" are seen to be regions of rapid but continuous variation in fluid properties such as number density. Thus, in a molecular theory, boundary conditions would not be necessary. When viewed with the much coarser resolution of the macroscopic or continuum description, on the other hand, these local variations of density (and other molecular variables) can be distinguished only as discontinuities, and the continuum (or molecular average) variables such as **u** appear to vary smoothly on the scale L, right up to the boundary where some boundary condition is applied.

Basic Principles

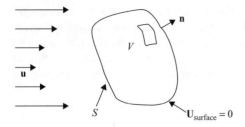

Figure 2–2. An arbitrarily chosen control volume of fixed position and shape immersed in a fluid with velocity **u**. The velocity of the surface of the control volume is zero, and thus there is a net flux of fluid through its surface. Equation (2–3) represents a mass balance on the volume V, with the left-hand side giving the rate of mass accumulation and the right-hand side the net flux of mass *into* V that is due to the motion **u**.

One consequence of the inability of the continuum description to resolve the region nearest the boundary is that the continuum variables extrapolated toward the boundary from the two sides may experience jumps or discontinuities. This is definitely the case at a fluid interface, as we shall see. Even at a stationary, solid boundary, the fluid velocity **u** may appear to "slip" when the fluid is a high-molecular-weight material or a particulate suspension.[3]

Here the situation is very similar to that encountered in connection with the need for continuum (constitutive) models for the molecular transport processes in that a derivation of appropriate boundary conditions from the more fundamental, molecular description has not been accomplished to date. In both cases, the knowledge that we have of constitutive models and boundary conditions that are appropriate for the continuum-level description is largely empirical in nature. In effect, we make an educated guess for both constitutive equations and boundary conditions and then normally judge the success of our choices by the resulting comparison between predicted and experimentally measured continuum velocity or temperature fields. Models derived from molecular theories, with the exception of kinetic theory for gases, are generally not available for comparison with the empirically proposed models. We discuss some of these matters in more detail later in this chapter, where specific choices will be proposed for both the constitutive equations and boundary conditions.

B. CONSERVATION OF MASS – THE CONTINUITY EQUATION

Once we adopt the continuum hypothesis and choose to describe fluid motions and heat transfer processes from a macroscopic point of view, we derive the governing equations by invoking the familiar conservation principles of classical continuum physics. These are conservation of mass and energy, plus Newton's second and third laws of classical mechanics.

The simplest of the various conservation principles to apply is conservation of mass. It is instructive to consider its application relative to two different, but equivalent, descriptions of our fluid system. In both cases, we begin by identifying a specific macroscopic body of fluid that lies within an arbitrarily chosen volume element at some initial instant of time. Because we have adopted the continuum mechanics point of view, this volume element will be large enough that any flux of mass across its surface that is due to random molecular motions can be neglected completely. Indeed, in this continuum description of our system, we can resolve only the molecular average (or continuum point) velocities, and it is convenient to drop any reference to the averaging symbol $\langle \rangle$. The continuum point velocity vector is denoted as **u**.[4]

In the first description of mass conservation for our system, we consider on arbitrarily chosen volume element (here called a *control volume*) of fixed position and shape as illustrated in Fig. 2–2. Thus, at each point on its surface, there is a mass flux of fluid $\rho \mathbf{u} \cdot \mathbf{n}$ through the surface. With **n** chosen as the outer unit normal to the surface, this mass flux will be negative at points where fluid enters the volume element and positive where it exits.

B. Conservation of Mass – The Continuity Equation

There is no reason, at this point, to assume that the fluid density ρ is necessarily constant. Indeed, conservation of mass requires the density inside the volume element to change with time in such a way that any imbalance in the mass flux in and out of the volume element is compensated for by an accumulation of mass inside. Expressing this statement in mathematical terms we obtain

$$\int_V \frac{\partial \rho}{\partial t} dV = -\int_A \rho \mathbf{u} \cdot \mathbf{n} dA, \qquad (2\text{–}3)$$

where V denotes the arbitrarily chosen volume element of fixed position and shape and A denotes its (closed) surface. Equation (2–3) is an integral constraint on the velocity and density fields within a given closed volume element of fluid. Because this volume element was chosen arbitrarily, however, an equivalent differential constraint at each point in the fluid can be derived easily. First, the well-known divergence theorem[5] is applied to the right-hand side of (2–3), which thus becomes

$$\int_V \left[\frac{\partial \rho}{\partial t} + \nabla \cdot (\rho \mathbf{u}) \right] dV = 0. \qquad (2\text{–}4)$$

Then we note that this integral condition on ρ and $\mathbf{u}$ can be satisfied for an arbitrary volume element only if the integrand is identically zero, that is,

$$\frac{\partial \rho}{\partial t} + \nabla \cdot (\rho \mathbf{u}) = 0. \qquad (2\text{–}5)$$

This is the *continuity equation*, which we now recognize as the pointwise constraint on ρ and $\mathbf{u}$ that is required by conservation of mass. To justify (2–5), we note that the only other way, in principle, to satisfy (2–4) would be if the integrand were positive within some portion of V and negative elsewhere in such a way that the nonzero contributions to the volume integral cancel. However, if this were the case, the freedom to choose an arbitrary volume element would lead to a contradiction. In particular, instead of the original choice of V, we could simply choose a new volume element that lies entirely within the region where the integrand is positive (or negative). Evidently, (2–4) is then violated, leading us to conclude that (2–5) must hold everywhere.

Although the derivation of the continuity equation by use of a *fixed* control volume is perfectly satisfactory, it is less obvious how to apply Newton's laws of mechanics in this framework. The familiar use of these principles from coursework in classical mechanics is that they are applied to describe the motion of a specific "body" subject to various forces or torques. To apply these same laws to a fluid (i.e., a liquid or a gas), we introduce the concepts of *material points* and a *material volume* (or material control volume) that we denote as $V_m(t)$. Now a material point is a continuum point that moves with the local continuum velocity of the fluid. A *material volume* $V_m(t)$, is a macroscopic control volume whose shape at some initial instant, $t = 0$, is arbitrary, that contains a fixed set of material points. Because the material volume contains a fixed set of such points, it must move with the local continuum velocity of the fluid at every point. Hence, as illustrated in Fig. 2–3, it must deform and change volume in such a way that the local flux of mass through all points on its surface is identically zero for all time (though, of course, there may still be *exchange* of molecules due to random molecular motion). Because mass is neither created nor destroyed according to the principle of mass conservation, the total mass contained

Basic Principles

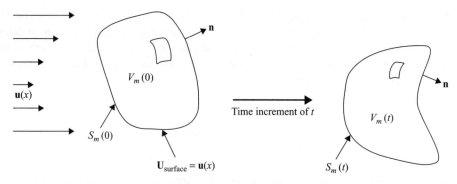

Figure 2–3. An arbitrarily chosen material control volume shown at some initial moment $t = 0$ and at a later time t, after which it has translated and distorted in shape because each point on its surface moves with the local fluid velocity **u**. Equation (2–6) represents a statement of mass conservation for this material volume.

within the material volume is constant, independent of time. This may be expressed in the mathematical statement

$$\frac{D}{Dt}\left[\int_{V_m(t)} \rho \, dV\right] = 0. \tag{2–6}$$

Here, the symbol D/Dt stands for the *convected or material* time derivative, which we shall subsequently discuss in some detail. In the context of (2–6) it is clear that D/Dt represents the time derivative of the total mass of material in the material volume $V_m(t)$. Alternatively, we could say that it is the time derivative of the total mass associated with the fixed set of material points that comprise $V_m(t)$. We shall see shortly that Eq. (2–6), which derives directly from the definition of a material volume for a fluid that conserves mass, is entirely equivalent to (2–3) or (2–4) and leads precisely to the pointwise continuity equation, (2–5). However, this cannot be seen easily without further discussion of the convected or material time derivative.

The first question that we may ask is the form of the relationship between D/Dt and the ordinary partial time derivative $\partial/\partial t$. The so-called "sky-diver" problem illustrated in Fig. 2–4 provides a simple physical example that may serve to clarify the nature of this relationship without the need for notational complexity. A sky diver leaps from an airplane at high altitude and begins to record the temperature T of the atmosphere at regular intervals of time as he falls toward the Earth. We denote his velocity as $-U_{diver}\mathbf{i}_z$, where $\mathbf{i}_z$ is a unit vector in the vertical direction, and the time derivative of the temperature he records as D^*T/Dt^*. Here, D^*/Dt^* represents the time rate of change (of T) measured in a reference frame that moves with the velocity of the diver. Evidently there is a close relationship between this derivative and the convected derivative that was introduced in the preceding paragraph. Let us now suppose, for simplicity, that the temperature of the atmosphere varies with the distance above the Earth's surface but is independent of time at any fixed point, say, $z = z^*$. In this case, the partial time derivative $\partial T/\partial t$ is identically equal to zero. Nevertheless, in the frame of reference of the sky diver, D^*T/Dt^* is not zero. Instead,

$$\frac{D^*T}{Dt^*} = -U_{diver}\frac{\partial T}{\partial z}. \tag{2–7}$$

The temperature is seen by the sky diver to vary with time because he is falling relative to the Earth's surface with a velocity U_{diver}, while T varies with respect to position in the direction of his motion.

B. Conservation of Mass – The Continuity Equation

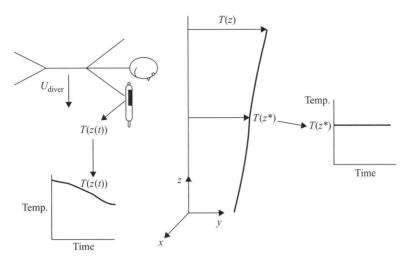

Figure 2–4. A sky diver falls with velocity U_{diver} from a high altitude carrying a thermometer and a recording device that plots the instantaneous temperature, as shown in the lower left-hand corner. During the period of descent, the temperature at any fixed point in the atmosphere is independent of time (i.e., the partial time derivative $\partial T/\partial t \equiv 0$). However, the sky diver is in an inversion layer and the temperature decreases with decreasing altitude. Thus the recording of temperature versus time obtained by the sky diver shows that the temperature decreases at a rate $DT/Dt^* = U_{diver}\partial T/\partial z$.. This time derivative is known as the Lagrangian derivative for an observer moving with velocity U_{diver}.

The relatively simple concept represented by the sky-diver example is easily generalized to provide a relationship between the convected derivative of any scalar quantity B associated with a fixed material point and the partial derivatives of B with respect to time and spatial position in a fixed (inertial) reference frame. Specifically, B changes for a moving material point both because B may vary with respect to time at each fixed point at a rate $\partial B/\partial t$ and because the material point moves through space and B may be a function of spatial position in the direction of motion. The rate of change of B with respect to spatial position is just ∇B. The rate at which B changes with time for a material point with velocity $\mathbf{u}$ is then just the projection of ∇B onto the direction of motion multiplied by the speed, which is $\mathbf{u} \cdot \nabla B$. It follows that the convected time derivative of any scalar B can be expressed in terms of the partial derivatives of B with respect to time and spatial position as

$$\boxed{\frac{DB}{Dt} = \frac{\partial B}{\partial t} + \mathbf{u} \cdot \nabla B.} \qquad (2\text{--}8)$$

It may now be evident why the *convected derivative* D/Dt is also known as the *material time derivative*. DB/Dt is, in fact, the time derivative of the quantity B for a fixed material point. Material points are often specified by the position vector $\mathbf{x}_0$ corresponding to their position at $t = 0$. The position vector of the material point $\mathbf{x}_0$ at an arbitrary time $t > 0$ is thus

$$\mathbf{x} = \mathbf{x}(\mathbf{x}_0, t) \equiv \mathbf{x}_0 + \int_0^t \mathbf{u}(\tau, \mathbf{x}_0) d\tau \qquad (2\text{--}9)$$

Any property of the fluid, say B, that is specified as a function of time for an arbitrary material point $\mathbf{x}_0$,

$$B(\mathbf{x}_0, t),$$

can be turned into the corresponding spatial description by use of (2–9):

$$B[\mathbf{x}(\mathbf{x}_0, t), t].$$

It is now clear that there are two distinct time derivatives:

$$\frac{\partial}{\partial t} \equiv \left(\frac{\partial}{\partial t}\right)_{\mathbf{x}} \quad \text{and} \quad \frac{D}{Dt} \equiv \left(\frac{\partial}{\partial t}\right)_{\mathbf{x}_0}.$$

Furthermore, the relationship between them can be obtained formally by application of the chain rule[6]:

$$\frac{DB}{Dt} = \left[\frac{\partial}{\partial t}(B(\mathbf{x}_0, t))\right]_{\mathbf{x}_0} = \left[\frac{\partial}{\partial t}(B(\mathbf{x}(\mathbf{x}_0, t), t))\right]_{\mathbf{x}_0}$$

$$= \frac{\partial B}{\partial x_i}\left(\frac{\partial x_i}{\partial t}\right)_{\mathbf{x}_0} + \left(\frac{\partial B}{\partial t}\right)_{\mathbf{x}},$$

$$\frac{DB}{Dt} = u_i \frac{\partial B}{\partial x_i} + \frac{\partial B}{\partial t}.$$

This is identical to (2–8), though now expressed in Cartesian component notation.[6] Time derivatives at fixed spatial position are often called the *Eulerian* time derivatives, whereas those taken at a fixed material point are known as *Lagrangian*. Although we have derived a simple relationship relating the convected or material derivative to the ordinary partial derivative at a fixed point, this cannot be applied directly to (2–6) without derivation of a general relationship, known as the Reynolds transport theorem.

Let us consider any scalar quantity $B(\mathbf{x}, t)$ that is associated with a moving fluid. Then the Reynolds transport theorem says

$$\boxed{\frac{D}{Dt}\left[\int_{V_m(t)} B(\mathbf{x}, t) dV\right] = \int_{V_m(t)} \left[\frac{\partial B}{\partial t} + \nabla \cdot (B\mathbf{u})\right] dV.} \qquad (2\text{--}10)$$

This is essentially a generalization of Leibnitz rule for differentiation of a one-dimensional integral with respect to some variable when both the integrand and the limits of integration depend on that variable. The proof of (2–10) is straightforward.[7] We first note that every point $\mathbf{x}(t)$ within a material control volume is a material point whose position is prescribed by (2–9). Hence, once the (arbitrary) initial shape of the material control volume is chosen (so that all initial values of $\mathbf{x}_0$ are specified), a scalar quantity B associated with any point within the material control volume can be completely specified as a function of time only, that is, $B[\mathbf{x}(t), t]$. Thus the usual definition of an ordinary time derivative can be applied to the left-hand side of (2–10), and we write

$$\frac{D}{Dt}\left\{\int_{V_m(t)} B[\mathbf{x}(t), t] dV\right\} \equiv \lim_{\delta t \to 0}\left\{\frac{1}{\delta t}\left[\int_{V_m(t+\delta t)} B(t + \delta t)\, dV - \int_{V_m(t)} B(t) dV\right]\right\}. \qquad (2\text{--}11)$$

To the quantity on the right-hand side of (2–11), we now add and subtract the term $\int_{V_m(t)} B(t + \delta t) dV$:

$$\frac{D}{Dt}\int_{V_m(t)} B dV = \lim_{\delta t \to 0}\left\{\frac{1}{\delta t}\left[\int_{V_m(t+\delta t)} B(t+\delta t) dV - \int_{V_m(t)} B(t+\delta t) dV\right]\right.$$

$$\left. + \frac{1}{\delta t}\left[\int_{V_m(t)} B(t+\delta t) dV - \int_{V_m(t)} B(t) dV\right]\right\}. \qquad (2\text{--}12)$$

B. Conservation of Mass – The Continuity Equation

The second term is just

$$\lim_{\delta t \to 0} \left[\int_{V_m(t)} \frac{B(t+\delta t) - B(t)}{\delta t} dV \right] \equiv \int_{V_m(t)} \frac{\partial B}{\partial t} dV. \tag{2-13}$$

The first term is simply rewritten as

$$\lim_{\delta t \to 0} \left[\frac{1}{\delta t} \int_{V_m(t+\delta t) - V_m(t)} B(t+\delta t) dV \right], \tag{2-14}$$

which shows that it is the integral of $B(t)$ over the differential volume element $V_m(t+\delta t) - V_m(t)$. To evaluate (2–14), we notice that any differential element of surface dA_m of $V_m(t)$ will move a distance $\mathbf{u} \cdot \mathbf{n} \delta t$ in a time interval δt. Thus, for sufficiently small δt, the differential volume element dV in (2–14) can be approximated as $(\mathbf{u} \cdot \mathbf{n}) \delta t \, dA_m$ and the volume integral over $V_m(t+\delta t) - V_m(t)$ then converted to an integral over the surface of $V_m(t)$. Thus,

$$\lim_{\delta t \to 0} \left[\frac{1}{\delta t} \int_{V_m(t+\delta t) - V_m(t)} B(t+\delta t) dV \right]$$
$$= \lim_{\delta t \to 0} \left[\frac{1}{\delta t} \int_{A_m(t)} B(t+\delta t) \mathbf{u} \cdot \mathbf{n} \delta t \, dA \right] = \int_{A_m(t)} B(t) \mathbf{u} \cdot \mathbf{n} \, dA, \tag{2-15}$$

where $A_m(t)$ is the surface of the material volume, and

$$\frac{D}{Dt} \int_{V_m(t)} B \, dV = \int_{V_m(t)} \frac{\partial B}{\partial t} dV + \int_{A_m(t)} B(\mathbf{u} \cdot \mathbf{n}) dA. \tag{2-16}$$

The proof of the Reynolds transport theorem in the form (2–10) is completed by application of the divergence theorem to the surface integral in (2–16). The physical interpretation of the Reynolds transport theorem, seen from (2–16), is that the total accumulation of a quantity B in a material control volume is the sum of the volume integral of the local accumulation of B at each fixed point in space, plus the total rate of entry through the surface of the control volume that is due to its motion through space. We may note that if we were to consider a volumetric region moving through space with some velocity $\mathbf{u}^*$ that differs from the fluid velocity $\mathbf{u}$, the accumulation of α within that region would take the same form as (2–16), but with $\mathbf{u}$ replaced with $\mathbf{u}^*$:

$$\frac{D^*}{Dt^*} \int_{V^*(t)} B \, dV = \int_{V^*(t)} \frac{\partial B}{\partial t} dV + \int_{A^*(t)} B(\mathbf{u}^* \cdot \mathbf{n}) dA, \quad \text{where} \quad \frac{D^*}{Dt^*} \equiv \frac{\partial}{\partial t} + \mathbf{u}^* \cdot \nabla.$$

We can now apply the Reynolds transport theorem in the form (2–10) to (2–6). In this case, the scalar property $B(\mathbf{x}, t)$ is just the fluid density, $\rho(\mathbf{x}, t)$. Thus, the mass conservation principle, (2–6), can be reexpressed in the form

$$\int_{V_m(t)} \left[\frac{\partial \rho}{\partial t} + \nabla \cdot (\rho \mathbf{u}) \right] dV = 0. \tag{2-17}$$

Because the initial choice of $V_m(t)$ is arbitrary, we obtain the same differential form for the continuity equation, (2–5), that we derived earlier by using a fixed control volume. Of course, the fact that we obtain the same form for the continuity equation is not surprising. The two derivations are entirely equivalent. In the first, conservation of mass is imposed by the requirement that the time rate of change of mass in a fixed control volume be exactly balanced by a net imbalance in the influx and efflux of mass through the surface. In particular, no mass is created or destroyed. In the second approach, we define the material volume element so that the mass flux through its surface is everywhere equal to zero. In this case, the condition that mass is conserved means that the total mass in the material volume element is constant. The differential form (2–5) of the statement of mass conservation, which we have called the continuity equation, is the main result of this section.

Basic Principles

Before leaving the continuity equation and mass conservation, there are a few additional remarks that we can put to good use later. The first is that (2–5) can be expressed in a precisely equivalent alternative form:

$$\boxed{\frac{1}{\rho}\frac{D\rho}{Dt} + \nabla \cdot \mathbf{u} = 0.} \tag{2–18}$$

Or, because the specific volume of the fluid is $V \equiv 1/\rho$, we can also write (2–18) as

$$\boxed{\frac{1}{V}\frac{DV}{Dt} = \nabla \cdot \mathbf{u}.} \tag{2–19}$$

The left-hand side of (2–19) is sometimes referred to as the rate of expansion or the rate of dilation of the fluid and provides a clear physical interpretation of the quantity $\nabla \cdot \mathbf{u}$ (or div $\mathbf{u}$).

The forms of the continuity equation (2–18) or (2–19) also lead directly to a simpler statement of the mass conservation principle that applies if it can be assumed that the density is constant, so that $D\rho/Dt = 0$. In this case, the fluid is said to be (i.e., is approximated as) *incompressible*. In general, the density is related to the temperature and pressure by means of an equation of state, $\rho = \rho(p, T)$. In an isothermal fluid, the incompressibility approximation is therefore a statement that the density is independent of the pressure. No fluid is truly incompressible in this sense. However, experience has shown that it is a good approximation if a dimensionless parameter, known as *Mach number*, M, is small:

$$M \equiv \frac{|\mathbf{u}|}{u_{sound}} \ll 1.$$

Here, $|\mathbf{u}|$ represents a characteristic velocity of the flow and u_{sound} is the speed of sound in the fluid at the same temperature and pressure. It may be noted that u_{sound} for air at room temperature and atmospheric pressure is approximately 300 m/s, whereas the same quantity for liquids such as water at 20°C is approximately 1500 m/s. Thus the motion of liquids will, in practice, rarely ever be influenced by compressibility effects. For nonisothermal systems, the density will vary with the temperature, and this can be quite important because it is the source of buoyancy-driven motions, which are known as *natural convection* flows. Even in this case, however, it is frequently possible to neglect the variations of density in the continuity equation. We will return to this issue of how to treat the density in nonisothermal flows later in the book.

In any case, if the fluid is isothermal and the density is approximated as a constant, the continuity equation takes the simpler form

$$\boxed{\nabla \cdot \mathbf{u} \equiv \text{div } \mathbf{u} = 0.} \tag{2–20}$$

Vector fields whose divergence vanishes are sometimes referred to as *solenoidal*. A more comprehensive discussion of the conditions for approximating the velocity field as solenoidal has been given by Batchelor.[8] These imply that, in cases in which the fluid is subjected to an oscillating pressure, the characteristic velocity in the Mach number condition should be interpreted as the product of the frequency times the linear dimension of the fluid domain, and that the difference in static pressures over the length scale of the domain must be small compared with the absolute pressure. Because our subject matter will frequently deal with incompressible, isothermal fluids, we shall often make use of (2–20) in lieu of the more general form, (2–5). In the presence of heat transfer, however, the density will not be constant because of temperature variations in the fluid, and the simple form (2–20) cannot be used even for an incompressible fluid unless additional approximations are made. We shall return to this point later in our presentation.

C. NEWTON'S LAWS OF MECHANICS

We have shown how a pointwise DE can be derived by application of the macroscopic principle of mass conservation to a material (control) volume of fluid. In this section, we consider the derivation of differential equations of motion by application of Newton's second law of motion, and its generalization from linear to angular momentum, to the same material control volume. It may be noted that introductory chemical engineering courses in transport phenomena often approach the derivation of these same equations of motion as an application of the "conservation of linear and angular momentum" applied to a fixed control volume. In my view, this obscures the fact that the equations of motion in fluid mechanics are nothing more than the familiar laws of Newtonian mechanics that are generally introduced in freshman physics.

We begin with *Newton's second law*, which may be stated in the form

$$\left\{\begin{array}{l}\text{the time rate of change}\\\text{of linear momentum}\\\text{of a given body,}\\\text{relative to an inertial}\\\text{reference frame}\end{array}\right\} = \left\{\begin{array}{l}\text{the sum of forces acting}\\\text{on the body}\end{array}\right\}. \qquad (2\text{--}21)$$

This can be applied directly to the material (control) volume of fluid, $V_m(t)$, which was introduced in the last section. As required for application of (2–21), this is a fixed body of material in the continuum sense. The resulting equation is

$$\frac{D}{Dt}\int_{V_m(t)} \rho \mathbf{u}\, dV = \begin{array}{l}\text{sum of forces}\\\text{acting on } V_m(t)\end{array}. \qquad (2\text{--}22)$$

The fact that the material control volume has a time-dependent shape does not lead to any complication of principle in applying Newton's second law. To proceed further, we must consider the types of forces that appear on the right-hand side of (2–22).

From the purely continuum mechanics viewpoint that we have now adopted, we recognize two kinds of forces acting on the material control volume. First are the *body forces*, associated with the presence of external fields, that are capable of penetrating to the interior of the fluid and acting equally on all elements (per unit mass). The most familiar body force is gravity, and we will be concerned exclusively with this single type of body force in this book. Another example is the action of an electromagnetic field when the fluid is an electrical conductor, which leads to an extension of fluid mechanics into the topic of magnetohydrodynamics (see Problem 2–11 at the end of the chapter). The second type of force is a *surface force*, which acts from the fluid outside the material control volume on the fluid inside and vice versa. In reality, there may exist short-range forces of molecular origin in the fluid. With the scale of resolution inherent in the use of the continuum approximation, these will appear as surface-force contributions in the basic balance (2–22). In addition, the surface-force terms will always include an *effective* surface-force contribution to simulate the transport of momentum across the boundaries of the material control volume that is due to random differences between the continuum velocity $\mathbf{u}$ and the actual molecular velocities. As explained earlier, the necessity for these latter effective surface-force contributions is entirely a consequence of the crude scale of resolution inherent in the continuum approximation, coupled with the discrete nature of any real fluid. Indeed, these latter contributions would be zero if the material were truly an indivisible continuum. We shall see that the main difficulties in obtaining pointwise DEs of motion from (2–22) all derive from the necessity for including surface forces (whether real or effective) whose molecular origin is outside the realm of the continuum description.

With the necessity for body and surface forces thus identified, we can complete the mathematical statement of Newton's second law for our material control volume:

$$\boxed{\frac{D}{Dt}\int_{V_m(t)} \rho \mathbf{u}\, dV = \int_{V_m(t)} \rho \mathbf{g}\, dV + \int_{A_m(t)} \mathbf{t}\, dA.} \qquad (2\text{--}23)$$

The left-hand side is just the time rate of change of linear momentum of all the fluid within the specified material control volume. The first term on the right-hand side is the net body force that is due to gravity (other types of body forces are not considered in this book). The second term is the net surface force, with the local surface force per unit area being symbolically represented by the vector $\mathbf{t}$. We call $\mathbf{t}$ the *stress vector*. It is the vector sum of all surface-force contributions per unit area acting at a point on the surface of $V_m(t)$.

Before proceeding further, let us return briefly to the derivation based upon a fixed control volume and "conservation of linear momentum." In this alternative approach, momentum is transported through the surface of the control volume by convection at a rate $\rho \mathbf{u}(\mathbf{u} \cdot \mathbf{n})$ at each point, and this is treated as an additional contribution to the rate at which linear momentum is accumulated or lost from the control volume. Of course, there is no term in (2–23) that corresponds to a flux of momentum across the surface of the material (control) volume. Because all points within $V_m(t)$ and on its surface $A_m(t)$ are material points, they move precisely with the local continuum velocity $\mathbf{u}$ and there is no flux of mass or momentum across the surface that is due to convection.

We may now attempt to simplify (2–23) to a differential form, as we did for the mass conservation equation, (2–6). The basic idea is to express all terms in (2–23) as integrals over $V_m(t)$, leading to the requirement that the sum of the integrands is zero because $V_m(t)$ is initially arbitrary. However, it is immediately apparent that this scheme will fail unless we can say more about the surface-stress vector $\mathbf{t}$. Otherwise, there is no way to express the surface integral of $\mathbf{t}$ in terms of an equivalent volume integral over $V_m(t)$.

We note first that $\mathbf{t}$ is not only a function of position and time, $\mathbf{t}(\mathbf{x}, t)$, as is the case with $\mathbf{u}$, but also of the orientation of the differential surface element through $\mathbf{x}$ on which it acts. The reader may well ask how this is known in the absence of a direct molecular derivation of a theoretical expression for $\mathbf{t}$ (the latter being outside the realm of continuum mechanics, even if it were possible in principle). The answer is that we can either deduce or derive certain general properties of $\mathbf{t}$, including its orientation dependence, from (2–23) by considering the limit as we decrease the material control volume progressively toward zero while holding the geometry (shape) of V_m constant. Let us denote a characteristic linear dimension of V_m as ℓ, with ℓ^3 defined to be equal to V_m. An estimate for each of the integrals in (2–23) can be obtained in terms of ℓ by use of the mean-value theorem. A useful preliminary step is to apply the Reynolds transport theorem to the left-hand side. Although this might, at first sight, seem to present new difficulties because $\rho \mathbf{u}$ is a vector, whereas the Reynolds transport theorem was originally derived for a scalar, the result given by (2–9) carries over directly, as we may see by applying it to each of the three scalar components of $\rho \mathbf{u}$ and then adding the results. Thus (2–23) can be rewritten in the form

$$\int_{V_m(t)} \left[\frac{\partial(\rho\mathbf{u})}{\partial t} + \nabla \cdot (\rho\mathbf{u}\mathbf{u}) - \rho\mathbf{g}\right] dV = \int_{A_m(t)} \mathbf{t}\, dA. \qquad (2\text{--}24)$$

C. Newton's Laws of Mechanics

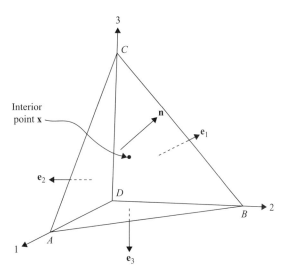

Figure 2–5. A tetrahedron, $ABCD$, is centered about the point $\mathbf{x}$. The surface ABC is arbitrarily oriented with respect to the Cartesian axes 123, and its area is denoted as ΔA_n. The unit outer normal to ABC is $\mathbf{n}$. The areas of surfaces BCD, ACD, and ABD are denoted, respectively, as ΔA_1, ΔA_2, and ΔA_3, with outer unit normals $\mathbf{e}_1$, $\mathbf{e}_2$, and $\mathbf{e}_3$ each parallel to the opposing coordinate axis but oriented in the negative direction.

Now, denoting the mean value over V_m or A_m by the symbol $\langle \ \rangle$, we can express (2–24) in the symbolic form

$$\langle \ \rangle \ell^3 = \langle \ \rangle \ell^2.$$

It is evident that, as $\ell \to 0$, the volume integral of the momentum and body-force terms vanishes more quickly than the surface integral of the stress vector. Hence, in the limit as $\ell \to 0$, (2–24) reduces to the form

$$\lim_{\ell \to 0} \left[\int_{A_m(t)} \mathbf{t}\, dA \right] \to 0. \quad (2\text{--}25)$$

This result is sometimes called *the principle of stress equilibrium*, because it shows that the surface forces must be in local equilibrium for any arbitrarily small volume element centered at any point $\mathbf{x}$ in the fluid. This is true independent of the source or detailed form of the surface forces.

Now it is clear that the stress vector at (or arbitrarily close to) a point $\mathbf{x}$ must depend not only on $\mathbf{x}$ but also on the orientation of the surface through $\mathbf{x}$ on which it acts, because otherwise the equilibrium condition (2–25) could not be satisfied. At first this dependence on orientation may seem to suggest that we would need a triply infinite set of components to specify $\mathbf{t}$ for all possible orientations of a surface through each point $\mathbf{x}$. Not only is this clearly impossible, but (2–25) shows that it is not necessary. Let us consider a surface with completely arbitrary orientation, specified by a unit normal $\mathbf{n}$, which passes near to point $\mathbf{x}$ but not precisely through it. Then, using this surface as one side, we construct a tetrahedron, illustrated in Fig. 2–5, centered around point $\mathbf{x}$, whose remaining three sides are mutually perpendicular. In the limit as the volume of this tetrahedral volume element goes to zero, the surface-stress equilibrium principle applies, and it is obvious that the surface-stress vector on the arbitrarily oriented surface (which now passes arbitrarily close to $\mathbf{x}$) must be expressible in terms of the surface-stress vectors acting on the three mutually perpendicular faces (which also pass arbitrarily close to $\mathbf{x}$). From this, we deduce that the specification of the surface-force vector on three mutually perpendicular surfaces through a point $\mathbf{x}$ (nine independent components in all) is sufficient to completely determine the surface-force vector acting on a surface of any arbitrary orientation at the same point $\mathbf{x}$.

Basic Principles

We can actually go one step further than this general observation and use the stress equilibrium principle applied to the tetrahedron to obtain a simple expression for the surface-stress vector **t** on an arbitrarily oriented surface at **x** in terms of the components of **t** on the three mutually perpendicular surfaces at the point **x**. We denote the stress vector on a surface with normal **n** as **t(n)**. The area of the surface with unit normal **n** is denoted as ΔA_n. Then, applying the surface-stress equilibrium principle to the tetrahedron, we have

$$\langle \mathbf{t(n)} \rangle \Delta A_n - \langle \mathbf{t(e_1)} \rangle \Delta A_1 - \langle \mathbf{t(e_2)} \rangle \Delta A_2 - \langle \mathbf{t(e_3)} \rangle \Delta A_3 = 0, \quad (2\text{--}26)$$

where the **e**'s are unit normal vectors for the three mutually perpendicular surfaces and $\langle \ \rangle$ represents the mean value of the indicated stress vector over the surface in question. The minus signs in (2–26) are a consequence of *Newton's third law*, according to which $\mathbf{t(e_1)} = -\mathbf{t(-e_1)}$. Now, ΔA_1 is the projected area of ΔA_n onto the plane perpendicular to the $\mathbf{e_1}$ axis. Thus,

$$\Delta A_i = \Delta A_n (\mathbf{n} \cdot \mathbf{e}_i) \text{ for } i = 1, 2, \text{ or } 3,$$

and (2–26) can be expressed in the form

$$\mathbf{0} = [\langle \mathbf{t(n)} \rangle - \langle \mathbf{t(e_1)} \rangle (\mathbf{n} \cdot \mathbf{e}_1) - \langle \mathbf{t(e_2)} \rangle (\mathbf{n} \cdot \mathbf{e}_2) - \langle \mathbf{t(e_3)} \rangle (\mathbf{n} \cdot \mathbf{e}_3)] \Delta A_n.$$

It follows, in the limit as $\ell \rightarrow 0$ (so that the tetrahedron collapses onto point **x**), that

$$\boxed{\mathbf{t(n)} = \mathbf{n} \cdot [\mathbf{e_1 t(e_1)} + \mathbf{e_2 t(e_2)} + \mathbf{e_3 t(e_3)}].} \quad (2\text{--}27)$$

The quantity in square brackets is a second-order tensor, formed as a sum of dyadic products of the vectors $\mathbf{t(e}_i)$ and $\mathbf{e}_i$ for $i = 1, 2,$ and 3. This second-order tensor is known as the *stress tensor*,

$$\boxed{\mathbf{T} = [\mathbf{e_1 t(e_1)} + \mathbf{e_2 t(e_2)} + \mathbf{e_3 t(e_3)}],} \quad (2\text{--}28)$$

and is denoted here by the symbol **T**. It follows from (2–27) and (2–28) that the surface-stress vector on any arbitrarily oriented surface through a point **x** is completely determined by specification of the nine independent components of the stress tensor **T** at that point.

We see from (2–27) and (2–28) that the second-order tensor **T** is just the "linear vector operator" that operates on the *unit normal* to a surface at a point P to produce the surface-stress vector acting at that point. Indeed, rewriting (2–27), we obtain

$$\boxed{\mathbf{t(x}_p; \mathbf{n}) = \mathbf{n} \cdot \mathbf{T(x}_p),} \quad (2\text{--}29)$$

where $\mathbf{x}_p$ is the position vector corresponding to an arbitrarily chosen point P in the flow domain. Furthermore, the components of **T** are just the components of the surface-force vectors acting on the three mutually perpendicular surfaces at $\mathbf{x}_p$, whose unit normal vectors are in the direction of the three coordinate axes. For example, the 21 component of **T** is the component in the 1 direction of the surface-force vector that acts on the surface with unit normal in the $\mathbf{e}_2$ direction. Although the derivation of (2–29) has been carried out with the notation of Cartesian vector and tensor analysis, it is evident that the components of **T** can be defined in terms of stress vector components for any orthogonal coordinates through point **x**, and the result, (2–29), is completely invariant to the choice of coordinate systems. The stress tensor **T** depends only on **x** and **t**, but not on **n**. Because knowledge of **T** at a

C. Newton's Laws of Mechanics

point enables us to determine the surface force acting on any surface through that point, **T** is said to represent *the state of stress* in the fluid at the given point.

With the relationship between the stress vector and the stress tensor in hand, DEs of motion can be derived from the macroscopic form (2–24) of Newton's second law of mechanics. Substituting (2–29) into (2–24) and applying the divergence theorem to the surface integral in the form

$$\int_{A_m(t)} \mathbf{n} \cdot \mathbf{T} \, dA = \int_{V_m(t)} (\nabla \cdot \mathbf{T}) \, dV,$$

we obtain

$$\int_{V_m(t)} \left[\frac{\partial (\rho \mathbf{u})}{\partial t} + \nabla \cdot (\rho \mathbf{u}\mathbf{u}) - \rho \mathbf{g} - \nabla \cdot \mathbf{T} \right] dV = 0. \tag{2–30}$$

Because the initial choice for $V_m(t)$ is arbitrary, it follows that the condition (2–30) can be satisfied only if the integrand is equal to zero at each point in the fluid, that is,

$$\boxed{\frac{\partial (\rho \mathbf{u})}{\partial t} + \nabla \cdot (\rho \mathbf{u}\mathbf{u}) = \rho \mathbf{g} + \nabla \cdot \mathbf{T}.} \tag{2–31}$$

Combining the first two terms in this equation with the continuity equation, we can also write the DEs of motion in the form

$$\boxed{\rho \left(\frac{\partial \mathbf{u}}{\partial \mathbf{t}} + \mathbf{u} \cdot \nabla \mathbf{u} \right) = \rho \mathbf{g} + \nabla \cdot \mathbf{T}.} \tag{2–32}$$

This is known as *Cauchy's* equation of motion. It is clear from our derivation that it is simply the differential form of Newton's second law of mechanics applied to a moving fluid.

It is, perhaps, well to pause for a moment to take stock of our developments to this point. We have successfully derived DEs that must be satisfied by any velocity field that is consistent with conservation of mass and Newton's second law of mechanics (or conservation of linear momentum). However, a closer look at the results, (2–5) or (2–20) and (2–32), reveals the fact that we have far more unknowns than we have relationships between them. Let us consider the simplest situation in which the fluid is isothermal and approximated as incompressible. In this case, the density is a constant property of the material, which we may assume to be known, and the continuity equation, (2–20), provides one relationship among the three unknown scalar components of the velocity **u**. When Newton's second law is added, we do generate three additional equations involving the components of **u**, but only at the cost of nine additional unknowns at each point: the nine independent components of **T**. It is clear that more equations are needed.

One possible source of additional relationships between **u** and **T** that we have not yet considered is the generalization of Newton's second law from linear to angular momentum. We may state the resulting principle, for a material control volume, in the form

$$\boxed{\frac{D}{Dt} \int_{V_m(t)} (\mathbf{x} \wedge \rho \mathbf{u}) \, dV = \text{sum of torques acting on the material control volume,}} \tag{2–33}$$

where **x** is the position vector associated with points within the material volume $V_m(t)$. The left-hand side of (2–33) is the rate of change of the angular momentum of the fluid inside the material (control) volume. There are, in principle, four sources of torque acting on the

Basic Principles

material control volume. The first two are simply the moments of the surface and body forces that act on the fluid:

$$\int_{A_m(t)} [\mathbf{x} \wedge \mathbf{t(n)}]dA, \tag{2-34a}$$

$$\int_{V_m(t)} (\mathbf{x} \wedge \rho \mathbf{g})dV. \tag{2-34b}$$

In addition, the possibility exists of body couples per unit mass $\mathbf{c}$ and surface couples per unit surface area $\mathbf{r}$ that are independent of the moments of surface and body forces. Thus the full statement of (2–33) takes the form

$$\frac{D}{Dt}\int_{V_m(t)} (\mathbf{x} \wedge \rho \mathbf{u})dV = \int_{A_m(t)} [\mathbf{x} \wedge (\mathbf{n} \cdot \mathbf{T}) + \mathbf{r}]dA + \int_{V_m(t)} [\mathbf{x} \wedge \rho \mathbf{g} + \rho \mathbf{c}]dV. \tag{2-35}$$

Clearly the presence of the surface-torque terms in (2–35) contributes additional unknowns, and the imbalance between unknowns and equations is not improved in the most general case by consideration of angular acceleration. However, in practice, there is no evidence of significant surface-torque contributions in real fluids, and we shall assume that $\mathbf{r} \equiv 0$. On the other hand, some fluids do exist on which the influence of body couples is significant. One commercially available set of examples is the so-called *Ferrofluids*, which are actually suspensions of fine iron particles that have either permanent or induced magnetic dipoles.[9] When such a fluid flows in the presence of a magnetic field, there is a body *torque* applied to each particle, but in the continuum approximation this is described as a continuously distributed body couple per unit mass. Thus a Ferrofluid is an example of a fluid in which $\mathbf{c} \neq 0$. In spite of the fact that fluids do exist in which body couples play a significant role, however, this is not true of the vast majority of fluids and none of the liquids or gases of common experience; water, air, oils, and so on. Let us suppose therefore that $\mathbf{c} = 0$. In this case, the angular acceleration principle reduces to the simpler form

$$\frac{D}{Dt}\int_{V_m(t)} (\mathbf{x} \wedge \rho \mathbf{u})dV = \int_{A_m(t)} \mathbf{x} \wedge (\mathbf{n} \cdot \mathbf{T})dA + \int_{V_m(t)} \mathbf{x} \wedge \rho \mathbf{g}dV. \tag{2-36}$$

We can explore the consequences of this equation by converting it to an equivalent differential form. To do this, we first apply Reynolds transport theorem to the left-hand side. This gives

$$\frac{D}{Dt}\int_{V_m(t)} (\mathbf{x} \wedge \rho \mathbf{u})dV = \int_{V_m(t)} \mathbf{x} \wedge \left[\frac{\partial(\rho \mathbf{u})}{\partial t} + \nabla \cdot (\rho \mathbf{u}\mathbf{u})\right]dV. \tag{2-37}$$

Also, on application of the divergence theorem, the surface integral in (2–36) becomes

$$\int_{A_m(t)} \mathbf{x} \wedge (\mathbf{n} \cdot \mathbf{T})dA = \int_{V_m(t)} [\mathbf{x} \wedge (\nabla \cdot \mathbf{T}) - \boldsymbol{\varepsilon} : \mathbf{T}]dV, \tag{2-38}$$

where $\boldsymbol{\varepsilon}$ is the third-order alternating tensor,[10]

$$\varepsilon_{ijk} \equiv \begin{cases} +1 & \text{if } (ijk) \text{ is an even permutation of } (123) \\ -1 & \text{if } (ijk) \text{ is an odd permutation of } (123), \\ 0 & \text{otherwise (some or all subscripts are equal)} \end{cases}$$

D. Conservation of Energy and the Entropy Inequality

and the symbol : indicates a double inner product of this tensor with $\mathbf{T}$.[11] Introducing (2–37) and (2–38) into (2–36), we thus obtain

$$\int_{V_m(t)} \left\{ \mathbf{x} \wedge \left[\frac{\partial(\rho \mathbf{u})}{\partial t} + \nabla \cdot (\rho \mathbf{u}\mathbf{u}) - \nabla \cdot \mathbf{T} - \rho \mathbf{g} \right] + \boldsymbol{\varepsilon} : \mathbf{T} \right\} dV = 0. \quad (2\text{–}39)$$

In view of the differential form of the equation of motion, Eq. (2–31), and the fact that $V_m(t)$ is arbitrary, we see that the angular acceleration principle, (2–33), requires that

$$\boldsymbol{\varepsilon} : \mathbf{T} = 0 \quad (2\text{–}40)$$

at all points in the fluid. Condition (2–40) requires that the stress tensor must be *symmetric*:

$$\boxed{\mathbf{T} = \mathbf{T}^T.} \quad (2\text{–}41)$$

Note that the condition of stress symmetry is not valid if there is a significant body couple per unit mass $\mathbf{c}$ in the field. In this case, we can easily show, following the same steps that we used in going from (2–35) to (2–40), that

$$\boxed{\boldsymbol{\varepsilon} : \mathbf{T} - \rho \mathbf{c} = 0.} \quad (2\text{–}42)$$

This gives a relationship between $\mathbf{c}$ and the off-diagonal components of $\mathbf{T}$, but the stress is clearly not symmetric. We hereafter restrict our attention to the case in which $\mathbf{c} = 0$.

We see that application of the angular acceleration principle does reduce, somewhat, the imbalance between the number of unknowns and equations that derive from the basic principles of mass and momentum conservation. In particular, we have shown that the stress tensor must be symmetric. Complete specification of a symmetric tensor requires only six independent components rather than the full nine that would be required in general for a second-order tensor. Nevertheless, for an incompressible fluid we still have nine apparently independent unknowns and only four independent relationships between them. It is clear that the equations derived up to now – namely, the equation of continuity and Cauchy's equation of motion – do not provide enough information to uniquely describe a flow system. Additional relations need to be derived or otherwise obtained. These are the so-called constitutive equations. We shall return to the problem of specifying constitutive equations shortly. First, however, we wish to consider the last available conservation principle, namely, conservation of energy.

D. CONSERVATION OF ENERGY AND THE ENTROPY INEQUALITY

We begin again by considering an arbitrary material control volume as it moves along with the fluid, and in this case we consider the change in its total energy with respect to time. In the simplest molecular description based on a hard-sphere gas model, this energy would be recognized as purely kinetic in nature, associated with the intensity of individual molecular motions. In the continuum approximation, however, we explicitly resolve only the molecular average velocity (denoted as $\mathbf{u}$ in the earlier developments of this chapter), and it is necessary to consider the total energy as consisting of two parts. First is the *kinetic energy* associated with the macroscopic or continuum velocity field $\mathbf{u}$, and second is a so-called internal energy term that encompasses all additional contributions including those that are due to the differences between the continuum velocity $\mathbf{u}$ at a point and the actual velocities of the molecules that occupy the continuum averaging volume that is centered at that point. In this description, the *internal energy* is a measure of the intensity of random molecular motion relative to the mean continuum velocity. Thus, from the continuum point of view, the total energy of an arbitrary material control volume is written as

$$\int_{V_m(t)} \left[\frac{\rho(\mathbf{u} \cdot \mathbf{u})}{2} + \rho e \right] dV,$$

where $\mathbf{u} \cdot \mathbf{u} = u^2$ is the local "speed" of the continuum motion and e is the internal energy (representing additional kinetic energy at the molecular level) per unit mass.[12]

The rate at which the total energy changes with time is determined by the *principle of energy conservation* for the material volume element, according to which

$$\frac{D}{Dt} \int_{V_m(t)} \left(\frac{\rho u^2}{2} + \rho e \right) dV = \begin{array}{c} \text{rate of work done} \\ \text{on the material} \\ \text{control volume} \\ \text{by external} \\ \text{forces} \end{array} + \begin{array}{c} \text{rate of internal energy} \\ \text{flux across the} \\ \text{boundaries of the} \\ \text{material control} \\ \text{volume.} \end{array} \quad (2\text{--}43)$$

We note that this conservation principle, for a closed system such as the material control volume, is precisely equivalent to the *first law of thermodynamics*, which we can obtain from it by integrating with respect to time over some finite time interval.

The terms on the right-hand side of (2–43) can be expressed in mathematical form, based on the following observations. First, work can be done on the material control volume only as a consequence of forces acting on it. In our continuum description, these are body forces and surface forces associated with the stress vector $\mathbf{t}(\mathbf{n})$. We recall that the surface forces appear, in part, as a consequence of our inability to fully resolve momentum transfer at the molecular level in a continuum description. It is not surprising, therefore, that work done in the macroscopic description may lead to changes in either the macroscopic kinetic energy or the internal energy representing changes in the intensity of motions at the molecular level. The motivation for a term in (2–43) that is associated with energy flux across the boundaries of the material control volume is very similar to that associated with the appearance of a surface force (or stress) in the linear momentum principle. In particular, there would be no local flux of kinetic or internal energy across the surface of a material control volume if the fluid were actually a continuous, infinitely divisible, and homogeneous medium, because the material control volume is defined as moving and deforming with the fluid in such a way that the local flux of mass across its surface is zero. However, random motions of molecules (which are not resolved explicitly in the continuum description) can contribute a net flux of internal energy across the surface, and this can only be included in the continuum energy balance (2–43) by the assumed existence of a *surface energy flux vector* $\mathbf{q}$. This surface energy flux is usually called the *heat flux vector*, in recognition of the fact that it is internal energy (or average intensity of molecular motion) that is being transferred across the surface by random molecular motion. Incorporating the rate of working terms that are due to surface and body forces, as well as a surface flux of energy term, we can write (2–43) in the mathematical form:

$$\frac{D}{Dt} \int_{V_m(t)} \left(\frac{\rho u^2}{2} + \rho e \right) dV = \int_{A_m(t)} [\mathbf{t}(\mathbf{n}) \cdot \mathbf{u}] dA + \int_{V_m(t)} (\rho \mathbf{g}) \cdot \mathbf{u} dV - \int_{A_m(t)} \mathbf{q} \cdot \mathbf{n} dS. \quad (2\text{--}44)$$

Here we have adopted the convention that a flux of heat into the material control volume is positive. The negative sign in the last term appears because $\mathbf{n}$ is the outer normal to the material control volume.

To obtain a pointwise DE from (2–44), we follow the usual procedure of applying the Reynolds transport theorem to the left-hand side and the divergence theorem to the

D. Conservation of Energy and the Entropy Inequality

right-hand side after first using (2–29) to express $\mathbf{t} \cdot \mathbf{u}$ as $\mathbf{n} \cdot (\mathbf{T} \cdot \mathbf{u})$. With all terms then expressed as volume integrals over the arbitrary material control volume $V_m(t)$, we obtain the differential form of the energy conservation principle:

$$\rho \frac{D}{Dt}\left(\frac{u^2}{2} + e\right) = \nabla \cdot (\mathbf{T} \cdot \mathbf{u}) + \rho \mathbf{g} \cdot \mathbf{u} - \nabla \cdot \mathbf{q}. \tag{2-45}$$

It appears from (2–45) that contributions from any of the terms on the right-hand side will lead to a change in the sum of kinetic and internal energy, but may not contribute separately to one or the other of these energy terms. However, this *is not* true as we may see by further examination. First, we may note that the Cauchy's equation of motion provides an independent relationship for the rate of change of kinetic energy. In particular, if we take the inner product of (2–32) with $\mathbf{u}$, we obtain

$$\frac{\rho}{2}\frac{Du^2}{Dt} = (\rho \mathbf{g}) \cdot \mathbf{u} + \mathbf{u} \cdot (\nabla \cdot \mathbf{T}). \tag{2-46}$$

This relationship is known as the *mechanical energy balance* and is a direct consequence of Newton's second law. Substituting for Du^2/Dt in (2–45) using (2–46) and recalling that $\mathbf{T}$ is symmetric, we obtain the so-called *thermal energy balance*:

$$\rho \frac{De}{Dt} = \mathbf{T} : \mathbf{E} - \nabla \cdot \mathbf{q}. \tag{2-47}$$

The second-order tensor $\mathbf{E}$ that appears in (2–47) is defined in terms of $\mathbf{u}$ as

$$\mathbf{E} \equiv \frac{1}{2}(\nabla \mathbf{u} + \nabla \mathbf{u}^T) \tag{2-48}$$

and is known as the *rate-of-strain tensor*. We shall later see the origins of this name. For now, we simply note that $\mathbf{E}$ is the symmetric part of the velocity gradient tensor, $\nabla \mathbf{u}$, that is,

$$\nabla \mathbf{u} \equiv \underbrace{\frac{1}{2}(\nabla \mathbf{u} + \nabla \mathbf{u}^T)}_{\text{symmetric}} + \underbrace{\frac{1}{2}(\nabla \mathbf{u} - \nabla \mathbf{u}^T)}_{\text{antisymmetric}} = \mathbf{E} + \boldsymbol{\Omega}. \tag{2-49}$$

The antisymmetric contribution to $\nabla \mathbf{u}$, which we have denoted in (2–49) as $\boldsymbol{\Omega}$, is known as the *vorticity tensor*. Again, more is said about the vorticity tensor later in this chapter.

Returning to (2–47), the term $\mathbf{T} : \mathbf{E}$ represents a contribution to the internal energy of the fluid because of the presence of mean motion (note that $\mathbf{E} \equiv 0$ if $\nabla \mathbf{u} \equiv 0$); that is, it represents a conversion from kinetic energy of the velocity field $\mathbf{u}$ to internal energy of the fluid – a process that is termed *dissipation* of kinetic energy to internal energy (or heat). The local rate of working that is due to body forces and surface forces may be seen from (2–46) to contribute directly to kinetic energy, but to lead to changes in internal energy only through dissipation. On the other hand, the surface energy (or heat) flux contribution to the total energy balance contributes directly to the change of internal energy, but only indirectly to the kinetic energy.

An alternative is to express the thermal energy balance, (2–47), in terms of the specific enthalpy h:

$$h \equiv e + (p/\rho). \tag{2-50}$$

Basic Principles

It can be seen from the definition (2–50) that

$$\rho \frac{De}{Dt} = \rho \frac{Dh}{Dt} - \frac{Dp}{Dt} + \frac{p}{\rho}\frac{D\rho}{Dt} = \rho \frac{Dh}{Dt} - \frac{Dp}{Dt} - p\nabla \cdot \mathbf{u}.$$

Hence, it follows that (2–47) can be expressed in terms of the specific enthalpy as

$$\boxed{\rho \frac{Dh}{Dt} = \mathbf{T}:\mathbf{E} - \nabla \cdot \mathbf{q} + \frac{Dp}{Dt} + p\nabla \cdot \mathbf{u}.} \qquad (2\text{–}51)$$

Although (2–47) and (2–51) are equivalent, it is generally more convenient for a flowing system to deal with the enthalpy rather than with the internal energy.

We may note that the energy conservation principle (or, equivalently, the first law of thermodynamics) has not improved the balance between the number of unknown, independent variables and differential relationships between them. Indeed, we have obtained a single independent scalar equation, either (2–47) or (2–51), but have introduced several new unknowns in the process, the three components of $\mathbf{q}$ and either the specific internal energy e or enthalpy h. A relationship between e or h and the thermodynamic state variables, say, pressure p and temperature θ, can be obtained provided that equilibrium thermodynamics is assumed to be applicable to a fluid element that moves with a velocity $\mathbf{u}$. In particular, a differential change in θ or p leads to a differential change in h for an equilibrium system:

$$dh = C_p d\theta + \left\{ \frac{1}{\rho} - \theta \left[\frac{\partial (1/\rho)}{\partial \theta} \right]_p \right\} dp.$$

Hence, for a fluid element moving with the fluid,

$$\frac{Dh}{Dt} = C_p \frac{D\theta}{Dt} + \left\{ \frac{1}{\rho} - \theta \left[\frac{\partial (1/\rho)}{\partial \theta} \right]_p \right\} \frac{Dp}{Dt},$$

and (2–51) can be expressed in terms of θ rather than h in the form

$$\boxed{\rho C_p \frac{D\theta}{Dt} = \mathbf{T}:\mathbf{E} + p\nabla \cdot \mathbf{u} - \nabla \cdot \mathbf{q} - \frac{\theta}{\rho}\left(\frac{\partial \rho}{\partial \theta}\right)_p \frac{Dp}{Dt}.} \qquad (2\text{–}52)$$

An alternative form for (2–52) can be written in terms of the heat capacity at constant volume by means of the general thermodynamic relationship

$$C_v = C_p + \frac{\theta}{\rho^2}\left(\frac{\partial p}{\partial \theta}\right)_\rho \left(\frac{\partial \rho}{\partial \theta}\right)_p.$$

However, this is less useful than (2–52) because it contains terms such as $(\partial p/\partial \theta)_\rho$, which are not small and are difficult to evaluate.

We shall see that the sum $p(\nabla \cdot \mathbf{u}) + \mathbf{T}:\mathbf{E}$ on the right-hand side of (2–52) represents the conversion of kinetic energy to heat, due to the internal friction within the fluid and is known as the *viscous dissipation* term. The last term on the left-hand side of (2–52) is related to the work required for compressing the fluid. Although this term is identically zero only for constant-pressure conditions (that is, the material is a solid or it is stationary so that $Dp/Dt = 0$), it is frequently small compared with other terms in (2–52) because the density at constant pressure is only weakly dependent on the temperature, and we shall generally adopt this approximation in the analyses of nonisothermal systems in later chapters.

We have seen that the energy conservation principle, applied to a material control volume of fluid, is equivalent to the first law of thermodynamics. A natural question, then, is whether

D. Conservation of Energy and the Entropy Inequality

any additional useful information can be obtained from the *second law of thermodynamics*. In its usual differential form the second law states

$$dS \geq \frac{dQ}{\theta},$$

where dS is the entropy change for the thermodynamic system of interest, dQ is the change in its total heat content caused by heat exchange with the surroundings, and θ is its temperature. When applied to a material control volume of fluid, this principle can be expressed in the form

$$\frac{D}{Dt}\int_{V_m(t)} (\rho s)dV + \int_{A_m(t)} \frac{\mathbf{n}\cdot\mathbf{q}}{\theta} dA \geq 0, \qquad (2\text{–}53)$$

where s is the entropy per unit mass of the fluid. The only mechanism for heat transfer from the surrounding fluid is molecular transport represented by the heat flux vector $\mathbf{q}$. The sign in front of the second term is a consequence of the fact that $\mathbf{n}$ is the outer unit normal. We easily obtain a differential form of the inequality (2–53) by applying the Reynolds transport theorem to the first term and the divergence theorem to the second term to show that

$$\int_{V_m(t)}\left[\rho\frac{Ds}{Dt} + \nabla\cdot\left(\frac{\mathbf{q}}{\theta}\right)\right] dV \geq 0.$$

This inequality can be satisfied for an arbitrary material control volume $V_m(t)$ only if

$$\boxed{\rho\frac{Ds}{Dt} + \nabla\cdot\left(\frac{\mathbf{q}}{\theta}\right) \geq 0.} \qquad (2\text{–}54)$$

We can obtain an inequality that is equivalent to (2–54) by using thermodynamics to express Ds/Dt in the form

$$\theta\rho\frac{Ds}{Dt} = \rho\frac{De}{Dt} - \frac{p}{\rho}\frac{D\rho}{Dt}$$

and then substituting for De/Dt from the thermal energy balance (2–47). The result for Ds/Dt is

$$\rho\frac{Ds}{Dt} = \frac{1}{\theta}[\mathbf{T}:\mathbf{E} + p(\nabla\cdot\mathbf{u}) - \nabla\cdot\mathbf{q}]. \qquad (2\text{–}55)$$

Then, because

$$\nabla\cdot\left(\frac{\mathbf{q}}{\theta}\right) = \frac{1}{\theta}\nabla\cdot\mathbf{q} - \frac{1}{\theta^2}\mathbf{q}\cdot\nabla\theta,$$

the inequality (2–54) can be combined with (2–55) to obtain

$$\boxed{\frac{1}{\theta}(\mathbf{T}:\mathbf{E} + p(\nabla\cdot\mathbf{u})) - \frac{\mathbf{q}\cdot\nabla\theta}{\theta^2} \geq 0.} \qquad (2\text{–}56)$$

Although there is no immediately useful information that we can glean from (2–56), we shall see that it provides a constraint on allowable constitutive relationships for $\mathbf{T}$ and $\mathbf{q}$. In this sense, it plays a similar role to Newton's second law for angular momentum, which led to the constraint (2–41) that $\mathbf{T}$ be symmetric in the absence of body couples. In solving fluid mechanics problems, assuming that the fluid is isothermal, we will use the equation of continuity, (2–5) or (2–20), and the Cauchy equation of motion, (2–32), to determine the velocity field, but the angular momentum principle and the second law of thermodynamics will appear only indirectly as constraints on allowable constitutive forms for $\mathbf{T}$. Similarly, for nonisothermal conditions, we will use (2–5) or (2–20), (2–32), and either (2–51) or

(2–52) to determine the velocity and temperature distributions, but neither the generalization of Newton's second law to angular momentum nor the second law of thermodynamics will appear directly. However, we are getting ahead of our story.

So far, we have seen that the basic macroscopic principles of continuum mechanics lead to a set of five scalar DEs – sometimes called the field equations of continuum mechanics – namely, (2–5) or (2–20), (2–32), and (2–51) or (2–52). On the other hand, we have identified many more unknown variables, $\mathbf{u}, \mathbf{T}, \theta, p$, and $\mathbf{q}$, plus various fluid or material properties such as ρ, C_p (or C_v), $(\partial \rho / \partial \theta)_p$, [or $(\partial \rho / \partial \theta)_p$], which generally require additional equations of state to be determined from p and θ if the latter are adopted as the thermodynamic state variables. Let us focus just on the independent variables $\mathbf{u}, \mathbf{T}, \theta, p$, and $\mathbf{q}$. Taking account of the symmetry of $\mathbf{T}$, these comprise 14 unknown scalar variables for which we have so far obtained only the five independent "field" equations that were just listed. It is evident that we require additional equations relating the various unknown variables if we are to achieve a well-posed problem from a mathematical point of view. Where are these equations to come from? Why is it that the fundamental macroscopic principles of continuum physics do not, in themselves, lead to a mathematical problem with a closed set of equations?

E. CONSTITUTIVE EQUATIONS

We have seen that the basic field equations of continuum mechanics are not sufficient in number to provide a mathematical problem from which to determine solutions for the independent field variables $\mathbf{u}, \mathbf{T}, \theta, p$, and $\mathbf{q}$. It is apparent that additional relationships must be found, hopefully without introducing more independent variables. In the next several sections, we discuss the origin and form of the so-called constitutive equations that provide the necessary additional relationships.

We begin with some general observations. In the first place, the idea that additional equations are necessary has so far been based on the purely mathematical statement that the field equations by themselves do not lead to a problem with a closed set of equations. Although this argument is powerful and certainly persuasive, it is also instructive to think about the problem from a more heuristic, physical point of view. In particular, if we first restrict ourselves to isothermal, incompressible conditions for which the relevant field equations are continuity, in the form of (2–20), and the Cauchy equations of motion, (2–32), we see that the only material property that appears explicitly is the density ρ. That is, according to (2–20) and (2–32) in the form that they stand, it appears that the only material property that distinguishes the motion of one fluid from another is the density. This is clearly at odds with experimental observation – we can find (or create by blending) a variety of fluids that have the same density within experimental error yet clearly demonstrate differences in flow properties. Consider, for example, the many grades of silicon oils that are sold commercially. These various grades differ in molecular weight, but their densities are all very nearly equal. Yet, if we were to simply pour a low- and a high-grade silicon oil from one container to another, we could not help but note a remarkable difference in the ease with which the fluids flow. The lowest grades would appear visually somewhat like water, whereas the highest grades would be more nearly akin to something like corn syrup. Quite apparently, there is something of the basic physics that is missing from the field equations alone. Similarly, if we consider a nonisothermal system in the absence of any mean motion, that is, $\mathbf{u} \equiv 0$, the thermal energy, (2–52), reduces to the form

$$\boxed{\rho C_p \frac{\partial \theta}{\partial t} = -\nabla \cdot \mathbf{q}.} \qquad (2\text{–}57)$$

F. Fluid Statics – The Stress Tensor for a Stationary Fluid

Not only are there more independent variables (θ and the three components of **q**) than equations (one!), but it would appear that the only material property relevant to energy transfer is ρC_p. Once again, simple observations would show that this is not enough to characterize the energy transfer processes in real materials.

Why is it that the basic conservation principles of continuum mechanics do not provide a complete problem statement, either from a mathematical or a physical point of view? The answer is that the fluids or materials that we wish to consider are not actually indivisible and homogeneous as presumed in continuum mechanics, but rather they have a definite molecular structure. Although this structure is not directly evident at the scale of resolution relevant to continuum mechanics, we have seen in the derivation of the basic field equations that it cannot be ignored altogether even in a purely continuum mechanical formulation. Instead, the differences between the continuum velocity (which we have seen is really an average of the molecular velocities "at" a point) and the instantaneous, local molecular velocities are manifest as apparent *surface force or stress, and surface energy or heat flux* contributions to the basic Newton's second law and principles of energy conservation. Indeed, in the absence of the stress tensor **T** and the heat flux vector **q**, as would be appropriate for a material with a completely continuous and homogeneous structure down to the finest possible scale of resolution, the basic field equations are completely adequate in number to determine all of the remaining independent field variables, **u**, θ, and p. It is the presence of **T** and **q**, reflecting the existence of transport processes at the molecular scale, that causes the field equations to contain more independent variables than there are equations. In view of this, we may anticipate that a full statement of the physics relevant to flowing fluids, whether isothermal or not, will require additional relationships between the surface stress and/or heat flux (representing molecular transport processes) and the macroscopic (or continuum) velocity and temperature fields. These relationships are known as the *constitutive equations* for the fluid.

But where do we get these additional equations? Because the underlying mechanisms responsible for the appearance of surface stress or surface heat flux in the continuum description are molecular, it is evident that continuum mechanics, by itself, can offer no basis to deduce what form these relationships should take. Thus, if we insist on a purely continuum mechanical approach, we must generally guess at the appropriate constitutive equations and then judge the correctness of our guess by comparisons between theoretically predicted velocity, temperature, or pressure fields and experimental measurements of the same quantities.[13] This is, in fact, the approach that was historically taken, and, in some ways, it is still the most successful approach. Fortunately, just about the simplest possible guess of equations relating **T** and **u**, or q and θ, turn out to provide an extremely good approximation for the large class of fluids (many liquids and all gases) that we know as *Newtonian*. We discuss the constitutive model for this class of fluids in more detail in Section G of this chapter. Regardless of the success of a particular constitutive equation, however, it is obvious that the status of constitutive equations in continuum mechanics is entirely different from the field equations that we derived in previous sections. The latter represent a deductive consequence of the basic laws of Newtonian mechanics and thermodynamics, whereas the former are never more than a guess, no matter how *educated*, in the absence of a fundamental molecular, statistical mechanical theory.[14]

F. FLUID STATICS – THE STRESS TENSOR FOR A STATIONARY FLUID

Let us begin our quest for specific constitutive equations by considering the special case of a stationary fluid ($\mathbf{u} \equiv \mathbf{0}$). In this case, the acceleration of a fluid element is zero, and the linear momentum equation, (2–32), reduces to a balance between body and surface forces,

$$\nabla \cdot \mathbf{T} + \rho \mathbf{g} = 0, \tag{2-58}$$

whereas the thermal energy equation reduces to the form (2–57). Although the equations are thus considerably simplified for a stationary fluid, the basic problem of requiring constitutive equations for $\mathbf{T}$ and $\mathbf{q}$ remains.

In this section, we consider an isothermal, stationary fluid. In this case, from thermodynamics, we know that the only surface force is the normal thermodynamic pressure, p. The pressure at a point P acts *normal* to any surface through P with a magnitude that is independent of the orientation of the surface. That is, for a surface with orientation denoted by the unit normal vector $\mathbf{n}$, the surface-force vector $\mathbf{t}(\mathbf{n})$ takes the form

$$\mathbf{t}(\mathbf{n}) = -\mathbf{n}p. \tag{2-59}$$

The minus sign in this equation is a matter of convention: $\mathbf{t}(\mathbf{n})$ is considered positive when it acts inward on a surface whereas $\mathbf{n}$ is the outwardly directed normal, and p is taken as always positive. The fact that the *magnitude* of the pressure (or surface force) is independent of $\mathbf{n}$ is "self-evident" from its molecular origin but also can be proven on purely continuum mechanical grounds, because otherwise the principle of stress equilibrium, (2–25), cannot be satisfied for an arbitrary material volume element in the fluid. The form for the stress tensor $\mathbf{T}$ in a stationary fluid follows immediately from (2–59) and the general relationship (2–29) between the stress vector and the stress tensor:

$$\mathbf{T} = -p\mathbf{I}. \tag{2-60}$$

In other words, in this case $\mathbf{T}$ is strictly diagonal:

$$\mathbf{T} = \begin{pmatrix} -p & 0 & 0 \\ 0 & -p & 0 \\ 0 & 0 & -p \end{pmatrix}.$$

Equation (2–60) is the constitutive equation for the stress in a stationary fluid.

Substituting (2–60) into the force balance (2–58), and noting that

$$\nabla \cdot \mathbf{T} = \nabla \cdot (-p\mathbf{I}) = -\nabla p,$$

we obtain the fundamental equation of fluid statics:

$$\rho \mathbf{g} - \nabla p = 0. \tag{2-61}$$

It follows that the presence of a body force leads to a nonzero gradient of pressure parallel to the body force even in a stationary fluid. Indeed, it is well know that the pressure increases with depth under the action of gravity. Provided the fluid density remains constant, the pressure increases linearly with depth

$$p(z) = p_0 + \rho g z \tag{2-62}$$

where p_0 is a reference pressure at the vertical position, $z = 0$, and z increases with depth. If we consider any arbitrary volume element from within a larger body of *stationary* fluid, it,

F. Fluid Statics – The Stress Tensor for a Stationary Fluid

is evident that there must be an exact balance between the total body force and the pressure forces acting on the volume element. This balance can be easily demonstrated for a volume element in the shape of a vertical cylinder of fluid, say between $z = z_1$ and $z = z_2$. The net pressure force acting on the ends acts upward and has a magnitude

$$[p(z_2) - p(z_1)]\pi R^2.$$

This is precisely equal to the total downward body force (ρg times the volume), namely,

$$\rho g(\pi R^2)(z_2 - z_1).$$

It is the increase of the *hydrostatic pressure* with depth that is responsible for the *buoyant force* on any body that is immersed within the fluid. If we consider a cylindrical body of radius R and length L that is oriented vertically, there is a net upward force that is due to the increase of the hydrostatic pressure with depth, which is the same as that acting on the column of fluid above, i.e.,

$$\rho g(\pi R^2)L.$$

The net downward force that is due to gravity is

$$\rho_s g(\pi R^2)L,$$

where ρ_s is the density of the body. Hence, the net force on the cylinder is

$$F = (\rho_s - \rho)g(\pi R^2)L,$$

which may be either up or down depending on whether the density ρ_s is smaller or larger, respectively, than the density of the fluid. Although this result was derived for a circular cylinder that is oriented vertically, it can be generalized to bodies of arbitrary shape.

For any arbitrarily shaped body of density ρ_s immersed in a fluid of density ρ, there is a net upward force that is due to the increase of hydrostatic pressure with depth, which is

$$\rho g \text{ (volume of the body)}.$$

When combined with the direct body force on the body that is due to gravity, this leads to a net "buoyancy" force,

$$(\rho - \rho_s)\,\mathbf{g}\text{ (volume of body)}, \qquad (2\text{–}63)$$

for a body of arbitrary shape. This is known as *Archimedes' principle*. It is important to note that the presence of a nonzero force on a body immersed in a fluid means that the body (and thus the fluid around it) will move unless the body is acted on by some additional force. The condition for zero motion is that the density of the body must equal that of the fluid,

$$\rho = \rho_s.$$

A body satisfying this condition is called "neutrally buoyant."

We can also interpret Eq. (2–61) as a necessary condition for the fluid to remain motionless. In particular, the hydrostatic balance between $\rho \mathbf{g}$ and ∇p can be satisfied *only* if the pressure gradient is colinear with (or in the same direction as) the body force, $\rho \mathbf{g}$. If there is a component of ∇p in any other direction, the fluid must flow. This observation allows us to examine certain fluid configurations to determine whether they are stable or will evolve by means of motion of the fluid to some other configuration. To see how this works, we can begin with an example for which the result will be obvious. Let us consider a horizontal layer of liquid, within an open cylindrical container, which we assume to initially have a surface elevation that is raised at the edges and lower in the middle, as sketched in Fig. 2–6.[15] Our intuition probably tells us that such a configuration cannot exist as a stable, stationary state with no fluid motion, but that fluid will flow from the outer edge toward the middle until

Basic Principles

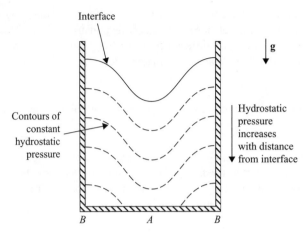

Figure 2–6. A sketch of the contours of hydrostatic pressure in a body of liquid within a cylindrical container when the upper interface is depressed in the center and raised at the walls as shown. As discussed in the text, the fact that the horizontal component of the pressure gradient is nonzero implies that this fluid configuration is unstable and will undergo spontaneous motion from the walls toward the center (i.e., in the direction of decreasing hydrostatic pressure in any horizontal plane) until the contours of hydrostatic pressure are all horizontal and the upper interface is also flat and horizontal.

the depth of the fluid is uniform everywhere in the container. What makes this happen? The hydrostatic pressures in the fluid layer create a *horizontal* pressure gradient, which cannot be balanced by $\rho \mathbf{g}$, and thus the fluid flows in such a way as to produce a fluid layer of uniform depth. Let us suppose that the pressure at all points on the upper free surface of the fluid layer is the ambient pressure of air at the conditions of the experiment[16] and that the fluid is motionless. Then, because the depth of the liquid is larger near the outer edge and smaller in the middle, it is clear that the hydrostatic pressure near the bottom of the container at point B (near the outer edge) would be larger than the hydrostatic pressure at point A. Hence, the fluid could not remain motionless in the configuration shown in Fig. 2–6, because there would then be a horizontal pressure gradient that would drive fluid motion from B toward A. This motion will continue until the depth is the same at every point, at which stage the horizontal hydrostatic pressure gradient will completely vanish. It may be noted that the hydrostatic pressure gradient in the horizontal direction will diminish as the surface flattens, and thus we may expect the flow rate to decrease with time. It should also be emphasized that we cannot calculate the actual flow by using the hydrostatic pressure gradients. The flow will actually cause changes in the pressure distribution, and thus the flow too will be different in detail from what we would calculate from the hydrostatic pressure gradient alone.

Another example of the qualitative usefulness of thinking about hydrostatic pressure distributions occurs for a drop that is initially placed on a flat surface, as illustrated in Fig. 2–7. We shall later see that interfacial tension at the interface between the drop and surrounding air, as well as line tension at the solid/liquid/gas contact line can play a role in the dynamics of such a drop. For now we assume that these factors play a negligible role[17] and concentrate on the role of the gravitational force on the drop, which we initially imagine to be motionless. As we have seen, this leads to a hydrostatic pressure distribution within the drop. With an initial configuration, such as that shown in Fig. 2–7, there would be a horizontal pressure gradient within the drop, with the largest hydrostatic pressure at the center where the drop is highest and the lowest pressures at the outer rim where the height goes to zero. The drop cannot therefore be motionless because the horizontal pressure gradient will produce motion of the fluid within the drop, causing the drop to spread on the

F. Fluid Statics – The Stress Tensor for a Stationary Fluid

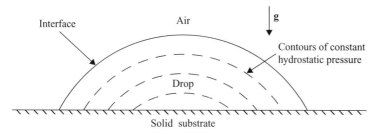

Figure 2–7. A sketch of the contours of constant hydrostatic pressure within a drop that is initially placed on a flat surface in air. As discussed in the text, the fact that the horizontal component of the pressure gradient is nonzero implies that this fluid configuration is unstable and will undergo a spontaneous "spreading" motion to decrease the hydrostatic pressure gradients in any horizontal plane. In this case, this process will continue indefinitely but at a decreasing rate as the contours of hydrostatic pressure become increasingly horizontal and flat. If the edge of the drop encounters container walls these will stop further spread and the drop will continue to move only until it has filled the container to a constant depth.

solid substrate. Assuming that interfacial tension and contact line forces can be neglected, as stated previously, the drop will continue to spread until it either encounters the bounding walls of a container (at which point it will again flow until it achieves a fluid layer of uniform depth) or until it thins to molecular dimensions when other nonhydrodynamic effects will come into play. Again, the rate of spread is expected to diminish with time as the horizontal component of the hydrostatic pressure gradient decreases.

A similar dynamics will occur at the "nose" or front of any fluid layer that has a density ρ_1 that exceeds the density ρ_2 of a surrounding fluid. An important example from geophysics is the so-called "gravity current" that occurs with a fluid configuration such as that sketched in Fig. 2–8. In the geophysical examples, fluid 1 may contain a heavier solid phase, but in any case, its density exceeds that of surrounding fluid 2. As a consequence there is a horizontal hydrostatic pressure gradient near the nose, from a larger pressure at A to a smaller pressure at B, and the heavy fluid, 1, must move from left to right, displacing the lighter fluid, 2, as it does. In this case, the fact that flow is accompanied by modifications of the pressure distribution in the fluid is very obvious, because the hydrostatic pressure distribution in fluid 2 has no horizontal component, even though it must clearly undergo motion in the horizontal direction as the gravity current propagates from left to right.

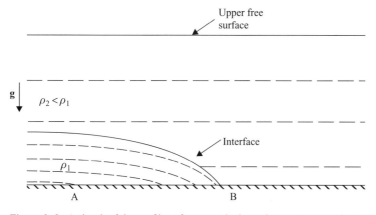

Figure 2–8. A sketch of the profiles of constant hydrostatic pressure near the "nose" of a gravity current. Because of the horizontal gradients of hydrostatic pressure within the nose region, it will propagate to the right, displacing the exterior fluid as it goes.

Basic Principles

To emphasize again, the preceding arguments are useful only in identifying systems in which flow will occur as a consequence of variations in the hydrostatic pressure that cannot be balanced by a body force. To analyze any details of the motion, we would have to solve the equations of motion to determine the velocity distribution, $\mathbf{u}(\mathbf{x}, t)$, and the pressure $\mathbf{p}(\mathbf{x}, t)$, which would be modified from the hydrostatic form because of the motion of the fluid.

G. THE CONSTITUTIVE EQUATION FOR THE HEAT FLUX VECTOR – FOURIER'S LAW

The simplicity of the constitutive equation for stress in a stationary fluid is due to the fact that the only surface force in this case is the thermodynamic pressure. The constitutive equation for the heat flux vector $\mathbf{q}$ is not so easy to obtain, even if we assume that the fluid is stationary, i.e., $\mathbf{u} \equiv 0$. In fact, we shall see later that, for many fluids, the form of the constitutive equation for $\mathbf{q}$ is expected to be the same whether the fluid is moving or not. However, here, for simplicity of presentation, we imagine $\mathbf{u} \equiv 0$. The governing thermal energy equation is then (2–57). To determine the constitutive equation for $\mathbf{q}$, we adopt a continuum approach in the sense that we essentially attempt to guess the relevant form. However, we first consider the process responsible for $\mathbf{q}$ from a qualitative, molecular point of view, with the goal of providing as much insight as possible for this guess.

Let us begin by recalling that $\mathbf{q}$ represents the flux of mean molecular kinetic energy (or heat) that is due to the purely random component of the motions of individual molecules. In a system with a nonzero continuum (average) velocity, heat may also be transported because of the mean motion – and this is called transport of heat by *convection*. However, here, we have assumed $\mathbf{u} = 0$. Let us consider the simplest molecular (or kinetic) fluid model, namely, that of a hard-sphere (or billiard-ball) gas in which the only molecular interactions are due to collisions and the random molecular motions are purely translational in character. If we focus our attention on an arbitrary surface within this fluid, it is clear that the only possible mechanism for a flux of heat is the random interchange of hard-sphere molecules across the surface and that this will lead to heat transfer only if there is a *nonzero gradient of temperature* (or average molecular kinetic energy) with a component normal to the surface. The fact that a molecule passing across the surface in one direction is accompanied on average by a second molecule going in the opposite direction is guaranteed by the continuum approximation and conservation of mass. Furthermore, in this case of a hard-sphere gas, it is evident that the net flux of heat across a surface will be proportional to the product of the magnitude of the temperature gradient normal to the surface, the mean free path between successive molecular collisions, and the frequency of the molecular exchange process. Of course, the molecular transport process for real fluids will be more complicated than for a billiard-ball gas. Nevertheless, we conclude from our considerations of that simple model fluid that

$$\mathbf{q} = \mathbf{q}(\nabla\theta, \text{ terms involving higher-order spatial derivatives of } \theta) \qquad (2\text{–}64)$$

for all real fluids; that is, the rate of heat transport by means of molecular motions is dependent on the magnitude of temperature gradients in the fluid (and quite possibly on higher-order spatial derivatives of θ as well). The right-hand side of (2–64) represents, in this case, either a function of $\nabla\theta$ or, possibly, a functional over the past "history" of $\nabla\theta$ for the fluid point of interest. Except for simple model materials like the hard-sphere gas, this is as much as we can deduce from our understanding of the molecular origins of $\mathbf{q}$. From this point, we must guess the constitutive form for $\mathbf{q}$ and ultimately judge

G. The Constitutive Equation for the Heat Flux Vector – Fourier's Law

the success of our guess by comparing measured and predicted temperature fields in real fluids.

A reasonable initial guess is the simplest assumption that is consistent with the relationship (2–64); namely, that the heat flux vector, $\mathbf{q}(\mathbf{x}, t)$, depends linearly on $\nabla\theta(\mathbf{x}, t)$:

$$\boxed{\mathbf{q} = -\mathbf{K} \cdot \nabla\theta.} \qquad (2\text{–}65)$$

Here, $\mathbf{K}$ is a second-order tensor that is known as the *thermal conductivity tensor*, and the constitutive equation is known as the *generalized Fourier heat conduction model* for the surface heat flux vector $\mathbf{q}$. The minus sign in (2–65) is a matter of convention; the components of $\mathbf{K}$ are assumed to be positive whereas a *positive* heat flux is defined as going from regions of high temperature toward regions of low temperature (that is, in the direction of $-\nabla\theta$).

The reader may well be curious why the particular linear form of (2–65) was chosen because at least one other vector function is linear in $\nabla\theta$, namely $\boldsymbol{\beta} \wedge (\nabla\theta)$, where $\boldsymbol{\beta}$ is a constant vector. To provide a complete answer, it is necessary to introduce two important principles that all constitutive relations are expected to obey. The first, which is frequently taken for granted, may be called *coordinate invariance*. This principle states simply that the *form* of a constitutive equation must be invariant under orthogonal coordinate transformations. Underlying this principle is the obvious fact that a change in orientation or sense of the coordinate system *cannot* influence the relevant physical processes and thus should not influence the form of the constitutive equation. The second invariance requirement of a constitutive equation is that it must also remain unchanged under a transformation in the frame of reference of the observer, even if the frame of the observer (or the fluid) is accelerating with respect to an inertial frame. This is usually thought of as being a consequence of the intuitive notion that the mechanical or thermal properties of a material element cannot depend on any motion of the person observing the material and is called the *principle of material objectivity*. Material objectivity is a stronger requirement than coordinate invariance, but is relevant only for constitutive equations that involve dynamical variables, such as $\mathbf{u}$.

Returning to the form of the constitutive equation for $\mathbf{q}$, we have seen that there are two distinct possibilities that are *linear* in $\nabla\theta$, namely (2–65) and $\boldsymbol{\beta} \wedge (\nabla\theta)$. In this case, the principle of coordinate invariance is sufficient to distinguish between these two possibilities. The reader who is experienced with vector and tensor analysis may immediately recognize that $\boldsymbol{\beta} \wedge (\nabla\theta)$ is not an acceptable form because it consists of the vector product (or cross product) of two vectors and is thus a *pseudo-vector*. A key property of a pseudo-vector is that it changes sign if we invert the coordinate axes from a right- to a left-handed coordinate system whereas a *true* vector is invariant to this transformation. In particular, if we define $\mathbf{L}$ as the coordinate transformation matrix ($\mathbf{L} \cdot \mathbf{L}^T = \mathbf{I}$ when the transformation is orthogonal), then a pseudo-vector transforms according to the rule $\mathbf{B} = (\det \mathbf{L})\mathbf{L} \cdot \overline{\mathbf{B}}$ whereas a true vector transforms according to $\mathbf{A} = \mathbf{L} \cdot \overline{\mathbf{A}}$. The vector formed as the cross product of the two vectors $\boldsymbol{\beta} \wedge (\nabla\theta)$ changes sign on inversion of coordinates and it is thus a pseudo-vector. The heat flux vector, on the other hand, is a *true* vector that is invariant to such changes of coordinate systems. One condition for satisfying coordinate invariance is that all terms in any equation involving vectors or tensors must have the same "parity" – that is, they must all be either true vectors or they must all be pseudo-vectors. Because $\mathbf{q}$ is a true vector, the only choice for the form of a constitutive equation that is linear in $\nabla\theta$ and involves $\nabla\theta$ only at the present moment in time and the same point in space as $\mathbf{q}$, is (2–65). Although the same result can be obtained formally by application of a coordinate transformation to the terms, $\mathbf{q}$, $\mathbf{K} \cdot \nabla\theta$, and $\boldsymbol{\beta} \wedge (\nabla\theta)$, we will be content here to accept the conclusion based on the qualitative arguments previously outlined.

Basic Principles

It is important to emphasize that the mathematical constraint imposed by coordinate invariance addresses only the selection of an allowable form of a constitutive equation, *given* the *physical assumption*, based on an educated guess, that there is a linear relationship between **q** and $\nabla\theta$. Whether the resulting constitutive equation captures the behavior of any real material is really a question of whether the *physical assumption* of linearity is an adequate approximation. In fact, in the generalized Fourier heat conduction model, Eq. (2–65), there are several additional physical assumptions that must be satisfied, besides linearity between **q** and $\nabla\theta$:

1. The surface heat transfer process is assumed to be *local*, in the sense that the flux associated with the fluid at some point depends on only the temperature gradient at the same point.
2. The surface heat transfer process is assumed to be *instantaneous*; the heat flux at a point depends on only the temperature gradient at that point at the same instant of time. In particular, there is no dependence on the thermal history of the fluid element that currently occupies the point in question.
3. The fluid is assumed to be *homogeneous*. The form of the relationship between the heat flux **q** and the temperature gradient $\nabla\theta$ is the same at all points. Furthermore, the only dependence of **q** on position **x** is due to the possible dependence of the so-called thermal conductivity tensor **K** on the thermodynamic state variables (say, p and θ) or the dependence of $\nabla\theta$ on spatial position.
4. When there is no temperature gradient, the surface heat flux is identically zero.

It should be emphasized that all the preceding points are simply assumptions underlying the assumed constitutive form (2–65). We can make no claim *on the basis of continuum mechanics alone* that these assumptions or the basic linearity of (2–65) will necessarily be satisfied by any real fluid.

Fortunately, in view of the simplicity of (2–65), comparison between predicted and measured data for the heat flux and temperature gradient shows that the general linear form does work extremely well for many common gases and liquids. However, the majority of these materials exhibit one additional characteristic that leads to further simplification of the constitutive form (2–65) – they are *isotropic*. This means that the magnitude of the heat flux at any point is dependent on only the *magnitude* of the temperature gradient, not on its *orientation* relative to axes fixed in the material. A common material that is *not* isotropic in this sense is wood, because a temperature gradient of given magnitude in wood generally produces a larger heat flux if it is oriented along the grain than it does if it is oriented across the grain. In the absence of motion, almost all common fluids will be isotropic (an exception is a liquid crystalline material).[18] If the fluid is made up of molecules and/or particles that are not spherical (or spherically symmetric), the orientations of these molecules or particles will generally be random at equilibrium as a consequence of random (Brownian) motions. Hence, when seen from the spatially averaged continuum viewpoint, such a fluid will be isotropic.

A mathematical statement of the property of *isotropy* is that the constitutive equation must be completely invariant to orthogonal rotations of the coordinate system. For the constitutive form (2–65), it can be shown that this condition will be satisfied if and only if

$$\boxed{\mathbf{K} = k\mathbf{I},} \qquad (2\text{–}66)$$

where k is a *scalar* property of the fluid that is known as the *thermal conductivity*.

H. Constitutive Equations for a Flowing Fluid – The Newtonian Fluid

Thus, for an isotropic fluid that exhibits a linear, instantaneous relationship between the heat flux and temperature gradient, the most general constitutive form for **q** is

$$\boxed{\mathbf{q} = -k\nabla\theta.} \qquad (2\text{--}67)$$

This is known as *Fourier's law of heat conduction*. We may note that the inequality (2–56) imposes a restriction on the sign of the thermal conductivity k. In particular, in the absence of fluid motion, (2–56) reduces to the simple form

$$-\frac{\mathbf{q}\cdot\nabla\theta}{\theta^2} \geq 0.$$

It follows from this inequality and the constitutive form (2–67) that

$$k\left(\frac{\nabla\theta}{\theta}\right)^2 \geq 0.$$

Hence, assuming $k \neq 0$, we see that the thermal conductivity must be positive, $k > 0$.

Although this simplified version of Fourier's heat conduction law is well known to be an accurate constitutive model for many real gases, liquids, and solids, it is important to keep in mind that, in the absence of empirical data, it is no more than an educated guess, based on a series of assumptions about material behavior that one cannot guarantee ahead of time to be satisfied by any real material. This status is typical of all constitutive equations in continuum mechanics, except for the relatively few that have been derived by means of a molecular theory.

H. CONSTITUTIVE EQUATIONS FOR A FLOWING FLUID – THE NEWTONIAN FLUID

In the previous sections, we discussed constitutive approximations for the stress and surface heat flux in a stationary fluid, where $\mathbf{u} \equiv 0$. In view of the molecular origins of **q**, there is no reason to expect that the basic linear form for its constitutive behavior should be modified by the presence of mean motion, at least for materials that are not too complicated in structure. Of course, this situation may be changed for materials such as polymeric liquids or suspensions, because in these cases the presence of motion may cause the structure to become anisotropic or changed in other ways that will affect the heat transfer process. We will return to this question in Section J.

The constitutive equation, (2–60), for the stress, on the other hand, will be modified for all fluids in the presence of a mean motion in which the velocity gradient $\nabla\mathbf{u}$ is nonzero. To see that this must be true, we can again consider the simplest possible model system of a hard-sphere or billiard-ball gas, which we may assume to be undergoing a simple shear flow,

$$\mathbf{u} = \dot{\gamma} y \mathbf{i}_x$$

with velocity in the x direction and a gradient of magnitude $\dot{\gamma}$ in the y direction. The parameter $\dot{\gamma}$ is called the *shear rate*. A flow of this type, which has the velocity in a single direction, is called *unidirectional*. In fact, in the next chapter, we will consider the general class of unidirectional flows. Now let us consider a surface in this fluid whose normal **n** is parallel to $\nabla\mathbf{u}$ (that is, $\mathbf{n} \equiv \mathbf{i}_y$). In this case, any interchange of molecules across the surface will result in a transfer of mean momentum; that is, the faster-moving fluid on one side of the surface will appear to be decelerated as one of its molecules is exchanged for a slower moving molecule from the other side of the surface, whereas the slower-moving fluid will appear to be accelerated by the same process. Hence, from the continuum point of view in

Basic Principles

which this transfer of momentum is modeled in terms of equivalent surface forces (stresses), we see that the surface-force vector in a moving fluid must generally have a component that acts *tangent* to the surface, and this is fundamentally different from the case of a stationary fluid in which the only surface forces are pressure forces that act *normal* to surfaces. We may also note, in the case of a hard-sphere gas, that the rate of momentum transfer by means of random molecular motions is proportional to $\nabla \mathbf{u}$. On the other hand, we know that $\mathbf{T}$ must reduce to the form (2–60) when $\nabla \mathbf{u} = 0$, both for a hard-sphere gas and for other, more complicated (real) fluids.

We conclude, based on the insight that we have drawn from the hard-sphere gas model and our general understanding of the molecular origins of $\mathbf{T}$, that

$$\mathbf{T} + p\mathbf{I} = \boldsymbol{\tau}\,(\nabla \mathbf{u}, \text{terms involving higher-order spatial derivatives}) \qquad (2\text{--}68)$$

for real fluids, where $\boldsymbol{\tau}\,(\nabla \mathbf{u}\ldots)$ could be either a function of current and local values of $\nabla \mathbf{u}$ or a functional that includes a dependence on both previous and current values. The second-order tensor $\boldsymbol{\tau}$ is usually known as the *deviatoric stress*.

To obtain a specific form for $\boldsymbol{\tau}(\nabla \mathbf{u}\ldots)$, we again require a guess. However, some general properties of $\boldsymbol{\tau}$ can be deduced that do not depend upon a specific constitutive form, and we begin by discussing these general properties. First, it is obvious from (2–58) and (2–68) that $\boldsymbol{\tau}(\nabla \mathbf{u}\ldots) = 0$ for $\nabla \mathbf{u} = 0$, provided that a large-enough time increment has passed after setting $\nabla \mathbf{u} = 0$. The requirement that $\boldsymbol{\tau} \to 0$ asymptotically for $t \gg 1$ is necessary because all fluids are not "instantaneous" in the sense that $\boldsymbol{\tau}$ depends on only the current values of $\nabla \mathbf{u}$. It is known, for example, that fluids exist where the deviatoric stress, $\boldsymbol{\tau}(\nabla \mathbf{u}\ldots)$, vanishes only if $\nabla \mathbf{u}$ has been zero for a finite period. Such fluids are said to possess a "memory" for past configurations and are typified by polymer solutions in which the molecular structure can return to an equilibrium state only by means of diffusion processes that require a finite period of time. A second general property of $\boldsymbol{\tau}$ is that it is symmetric in the absence of external body couples. This follows directly from (2–68) and the fact that $\mathbf{T}$ is symmetric in the same circumstances, as shown in Section C. A third general property is that $\boldsymbol{\tau}$ must depend explicitly on only the symmetric part of $\nabla \mathbf{u}$, rather than on $\nabla \mathbf{u}$ itself. We have already noted in (2–49) that the symmetric part of $\nabla \mathbf{u}$ is called the rate-of-strain tensor and is usually denoted as $\mathbf{E}$. It might seem, at first, that this third property would follow from the fact that $\boldsymbol{\tau}$ is symmetric, but this is not the case.

There are two proper explanations, one based on physical intuition and the other based on the principle of material objectivity. The latter is discussed in many books on continuum mechanics.[19] Here, we content ourselves with the intuitive physical explanation. The basis of this is that contributions to the deviatoric stress cannot arise from rigid-body motions – whether solid-body translation or rotation. Only if adjacent fluid elements are in relative (nonrigid-body) motion can random molecular motions lead to a net transport of momentum. We shall see in the next paragraph that the rate-of-strain tensor relates to the rate of change of the length of a line element connecting two material points of the fluid (that is, to relative displacements of the material points), whereas the antisymmetric part of $\nabla \mathbf{u}$, known as the vorticity tensor $\boldsymbol{\Omega}$, is related to its rate of (rigid-body) rotation. Thus it follows that $\boldsymbol{\tau}$ must depend explicitly on $\mathbf{E}$, but not on $\boldsymbol{\Omega}$:

$$\boxed{\boldsymbol{\tau} = \boldsymbol{\tau}(\mathbf{E}, \ldots,\,).} \qquad (2\text{--}69)$$

To prove our assertions about the physical significance of the rate-of-strain and vorticity tensors, we consider the relative motion of two nearby *material* points in the fluid P, initially at position $\mathbf{x}$ and Q, which is at $\mathbf{x} + \delta\mathbf{x}$. We denote the velocity of the material point P as

H. Constitutive Equations for a Flowing Fluid – The Newtonian Fluid

u and that of Q as $\mathbf{u} + \delta\mathbf{u}$. Now a Taylor series approximation can be used to relate **u** and $\mathbf{u} + \delta\mathbf{u}$, namely,

$$\mathbf{u} + \delta\mathbf{u} = \mathbf{u} + (\mathbf{E} + \boldsymbol{\Omega}) \cdot \delta\mathbf{x} + O(|\delta\mathbf{x}|^2). \tag{2–70}$$

The symbol $O(\ )$ is known as the order symbol. It indicates the magnitude of the error in truncating the Taylor series approximation after only two terms, which is negligible in this case since $|\delta\mathbf{x}|^2 \ll |\delta\mathbf{x}|$. It follows from (2–70) that the material point Q moves *relative* to the material point P with a velocity

$$\delta\mathbf{u} = \mathbf{E} \cdot \delta\mathbf{x} + \boldsymbol{\Omega} \cdot \delta\mathbf{x} + O(|\delta\mathbf{x}|^2). \tag{2–71}$$

Now, the length of the line element connecting P and Q is

$$|\delta\mathbf{x}| = (\delta\mathbf{x} \cdot \delta\mathbf{x})^{1/2},$$

and the rate of change in this length is thus proportional to

$$\delta\mathbf{x} \cdot \delta\mathbf{u} = \delta\mathbf{x} \cdot [\mathbf{E} \cdot \delta\mathbf{x} + \boldsymbol{\Omega} \cdot \delta\mathbf{x} + O(|\delta\mathbf{x}|^2)], \tag{2–72}$$

where $\delta\mathbf{u} = [D(\delta\mathbf{x})/Dt]$. However, because $\boldsymbol{\Omega}$ is antisymmetric,

$$\delta\mathbf{x} \cdot \boldsymbol{\Omega} \cdot \delta\mathbf{x} \equiv O,$$

so that

$$\frac{1}{2}\frac{D}{Dt}(|\delta\mathbf{x}|^2) = \delta\mathbf{x} \cdot \mathbf{E} \cdot \delta\mathbf{x} + O(|\delta\mathbf{x}|^3). \tag{2–73}$$

Thus the rate of change of the distance between two neighboring material points depends on only the rate-of-strain tensor **E**, i.e., on the symmetric part of $\nabla\mathbf{u}$. It can be shown in a similar manner that the contribution to the relative velocity vector $\delta\mathbf{u}$ that is due to the vorticity tensor $\boldsymbol{\Omega}$ is the same as the displacement that is due to a (local) rigid-body rotation with angular velocity $\boldsymbol{\omega}/2$, where

$$\boldsymbol{\omega} = \boldsymbol{\varepsilon} : \boldsymbol{\Omega}. \tag{2–74}$$

For future reference, we also note that

$$\boldsymbol{\omega} = \nabla \wedge \mathbf{u}. \tag{2–75}$$

The vector $\boldsymbol{\omega}$ is known as the vorticity vector. It can be calculated from **u** by means of either (2–74) or (2–75).

To proceed beyond the general relationship (2–69), it is necessary to make a guess of the constitutive behavior of the fluid. The simplest assumption consistent with (2–69) is that the deviatoric stress (at some point **x**) depends *linearly* on the rate of strain at the same point in space and time, that is,

$$\boldsymbol{\tau} = \mathbf{A} : \mathbf{E}. \tag{2–76}$$

Here, **A** is a fourth-order tensor that must be symmetric in its first two indices,

$$A_{ijkl} \equiv A_{jikl}, \tag{2–77}$$

because $\boldsymbol{\tau}$ is symmetric according to the constraint (2–41). The constitutive relation (2–76) is analogous to the generalized Fourier model (2–65) for the heat flux vector **q**. Like the generalized Fourier model, it assumes that the fluid is *local, instantaneous, homogeneous*, and invariant to rotations or inversions of the coordinate axes.

An additional physical assumption that is satisfied by many fluids is that the structure is *isotropic* even in the presence of motion. For an isotropic fluid, the constitutive equation is completely unchanged by rotations of the coordinate system. It can be shown, by use of the

Basic Principles

methods of tensor analysis, that the most general fourth-order tensor with this property[10] is

$$A_{ijpq} = \lambda \delta_{ij}\delta_{pq} + \mu(\delta_{ip}\delta_{jq} + \delta_{iq}\delta_{jp}) + \nu(\delta_{ip}\delta_{jq} - \delta_{iq}\delta_{jp}) \qquad (2\text{--}78)$$

where λ, μ, and ν are arbitrary scalar constants and δ_{ij} is the ij component of the identity tensor **I**, that is,

$$\delta_{ij} = \begin{cases} 1 & i = j \\ 0 & i \neq j \end{cases}, \quad \text{where } i, j = 1, 2, 3. \qquad (2\text{--}79)$$

Because the tensor **A** must also satisfy the symmetry condition, (2–77), it follows that

$$\nu \equiv 0.$$

Substituting (2–78) into (2–76), we see that the most general constitutive equation for the total stress **T** that is consistent with the linear and instantaneous dependence of the deviatoric stress $\boldsymbol{\tau}$ on **E**, plus the assumption of isotropy, is

$$\boxed{\mathbf{T} = (-p + \lambda \operatorname{tr} \mathbf{E})\mathbf{I} + 2\mu \mathbf{E}.} \qquad (2\text{--}80)$$

Expressed in component form using Cartesian tensor notation, this equation is

$$T_{ij} = (-p + \lambda E_{pp})\delta_{ij} + 2\mu E_{ij}.$$

Fluids for which this constitutive equation is an adequate model are known as *Newtonian fluids*. We have shown that the Newtonian fluid model is the most general form that is linear and instantaneous in **E** and isotropic. If the fluid is also incompressible,

$$\operatorname{tr} \mathbf{E} = \nabla \cdot \mathbf{u} = \mathbf{0},$$

and the constitutive equation further simplifies to the form

$$\boxed{\mathbf{T} = -p\mathbf{I} + 2\mu \mathbf{E}.} \qquad (2\text{--}81)$$

The coefficient μ that appears in this equation is known as the *shear viscosity* and is a property of the fluid.

We have, of course, said nothing about the physical reality of the assumptions of isotropy or of a linear, instantaneous dependence of **T** on **E**. It is possible, insofar as continuum mechanics is concerned, that no fluid would be found for which these are adequate assumptions. Fortunately, in view of the simplicity of the resulting constitutive model, (2–80) or (2–81), experimental observation shows that the Newtonian constitutive assumptions are satisfied for gases in almost all circumstances and for the majority of low- to moderate-molecular-weight liquids, providing that $\|\mathbf{E}\|$ is not extremely large and does not change too rapidly with respect to time. Polymeric liquids, suspensions, and emulsions do not generally satisfy the Newtonian assumptions and require much more complicated constitutive equations for **T**. We shall briefly discuss these latter fluids in the next two sections. An extremely important fact is that Newtonian fluids are also generally found to follow Fourier's law of heat conduction in the isotropic form, (2–67).

It has been emphasized repeatedly that continuum mechanics provides no guidance in the choice of a general constitutive hypothesis for either the heat flux vector **q** or the stress tensor **T**. On the other hand, it was noted earlier that (2–41) and (2–56), derived respectively from the law of conservation of angular momentum and the second law of thermodynamics, must be satisfied by the resulting constitutive equations. It thus behooves us to see whether these two constraints are satisfied for the Newtonian and the Fourier constitutive models that have been proposed in this and the preceding section. In the absence of external body couples, the constraint from angular momentum conservation requires only that the stress be symmetric, and this is obviously satisfied by the Newtonian model for any choice of λ and μ. The second constraint, from the second law of thermodynamics, requires

$$\mathbf{T}:\mathbf{E} + p(\nabla \cdot \mathbf{u}) - \frac{\mathbf{q} \cdot \nabla \theta}{\theta^2} \geq 0.$$

Substituting for $\mathbf{T}$ from (2–80) and for $\mathbf{q}$ from (2–67), we find

$$\left(\lambda + \frac{2}{3}\mu\right)(\operatorname{tr}\mathbf{E})^2 + 2\mu\left\{\left[\mathbf{E} - \left(\frac{1}{3}\operatorname{tr}\mathbf{E}\right)\mathbf{I}\right]:\left[\mathbf{E} - \left(\frac{1}{3}\operatorname{tr}\mathbf{E}\right)\mathbf{I}\right]\right\} + \frac{k}{\theta}(\nabla\theta)^2 \geq 0. \tag{2-82}$$

If we consider the special case of an isothermal, incompressible fluid, inequality (2–82) becomes

$$2\mu(\mathbf{E}:\mathbf{E}) \geq 0.$$

Obviously, if the Newtonian constitutive model for an incompressible fluid is to be consistent with the second law of thermodynamics, we require that the viscosity be nonnegative, that is,

$$\mu \geq 0. \tag{2-83}$$

This is, in fact, the most significant result that can be obtained for incompressible Newtonian fluids from the second-law inequality. If we do not restrict ourselves to incompressible or isothermal conditions, inequality (2–82) can be satisfied for arbitrary motions and arbitrary temperature fields only if

$$\left(\lambda + \frac{2}{3}\mu\right) \geq 0, \quad \mu \geq 0, \quad \text{and } k \geq 0. \tag{2-84}$$

The quantity $[\lambda + (2/3)\mu]$ is commonly called the bulk viscosity coefficient. Besides the inequalities (2–84), no further information appears to be attainable for a Newtonian fluid from the constraint (2–56).

I. THE EQUATIONS OF MOTION FOR A NEWTONIAN FLUID – THE NAVIER–STOKES EQUATION

Let us now return to the equations of motion for a Newtonian fluid. With the constitutive equation, (2–80) [or (2–81) if the fluid is incompressible], the continuity equation, (2–5) [or (2–20) if the fluid is incompressible], and the Cauchy equation of motion, (2–32), we have achieved a balance between the number of independent variables and the number of equations for an isothermal fluid. If the fluid is not isothermal, we can add the thermal energy equation, (2–52), and the thermal constitutive equation, (2–67), and the system is still fully specified insofar as the balance between independent variables and governing equations is concerned.

In this section, we combine the Cauchy equation and the Newtonian constitutive equation to obtain the *Navier–Stokes equation* of motion. First, however, we briefly reconsider the notion of pressure in a general, Newtonian fluid.

The physical significance of pressure, as it first appeared in the constitutive equation for stress in a stationary fluid, (2–60), is clear. This is the familiar pressure of thermodynamics. When a fluid is undergoing a motion, however, the simple notion of a normally directed surface force acting equally in all directions is lost. Indeed, it is evident on examining the

Basic Principles

Newtonian constitutive equation, (2–80), that the normal component of the surface force or stress acting on a fluid element at a point will generally have different values depending on the orientation of the surface. Nevertheless, it is often useful to have available a scalar quantity for a moving fluid that is analogous to static pressure in the sense that it is a measure of the local intensity of "squeezing" of a fluid element at the point of interest. Thus it is common practice to introduce a mechanical definition of pressure in a moving fluid as

$$\bar{p} \equiv -\frac{1}{3} \text{tr } \mathbf{T}. \tag{2-85}$$

This quantity has the following desirable properties. First, it is invariant under rotation of the coordinate axes (unlike the individual components of $\mathbf{T}$). Second, for a static fluid $-1/3 \cdot \text{tr } \mathbf{T} = p$, the thermodynamic pressure. And third, $\bar{p}$ has a physical significance analogous to pressure in a static fluid in the sense that it is precisely equal to the average value of the normal component of the stress on a surface element at position $\mathbf{x}$ over all possible orientations of the surface (alternatively, we may say that $1/3 \cdot \text{tr } \mathbf{T}$ is the average magnitude of the normal stress on the surface of an arbitrarily small sphere centered at point $\mathbf{x}$).

The definition (2–85) is a purely mechanical definition of pressure for a moving fluid, and nothing is implied directly of the connection for a moving fluid between $\bar{p}$ and the ordinary static or thermodynamic pressure p. Although the connection between $\bar{p}$ and p can always be stated once the constitutive equation for $\mathbf{T}$ is given, one would not necessarily expect the relationship to be simple for all fluids because thermodynamics refers to equilibrium conditions, whereas the elements of a fluid in motion are clearly not in thermodynamic equilibrium. Applying the definition (2–85) to the general Newtonian constitutive model, (2–80), we find

$$\bar{p} = p - \left(\lambda + \frac{2}{3}\mu\right) \nabla \cdot \mathbf{u},$$
$$\bar{\boldsymbol{\tau}} = \mathbf{T} + \left(\frac{1}{3}\text{tr } \mathbf{T}\right) \mathbf{I} = 2\mu \mathbf{E} - \frac{2}{3}\mu(\nabla \cdot \mathbf{u})\mathbf{I}. \tag{2-86}$$

Only if the fluid can be modeled as incompressible does the connection between $\bar{p}$ and p simplify greatly for a Newtonian fluid. In that case,

$$\bar{p} \equiv p; \quad \bar{\boldsymbol{\tau}} = \boldsymbol{\tau} = 2\mu \mathbf{E}. \tag{2-87}$$

So far, we have simply stated the Cauchy equation of motion and the Newtonian constitutive equations as a set of nine independent equations involving $\mathbf{u}$, $\mathbf{T}$, and p. It is evident in this case, however, that the constitutive equation, (2–80), for the stress [or equivalently (2–86)] can be substituted directly into the Cauchy equation to provide a set of equations that involve only $\mathbf{u}$ and $\bar{p}$ (or p). These combined equations take the form

$$\rho \left(\frac{\partial \mathbf{u}}{\partial t} + \mathbf{u} \cdot \nabla \mathbf{u}\right) = \rho \mathbf{g} - \nabla \bar{p} + \nabla \cdot (2\mu \mathbf{E}) - \frac{2}{3}\nabla \cdot [\mu(\text{div } \mathbf{u})\mathbf{I}]. \tag{2-88}$$

If the fluid can be approximated as incompressible and if the fluid is isothermal so that the viscosity μ can be approximated as a constant, independent of spatial position (note that

I. The Equations of Motion for a Newtonian Fluid – The Navier–Stokes Equation

any dependence of μ on pressure is generally weak enough to be neglected altogether), this equation takes the form

$$\rho\left(\frac{\partial \mathbf{u}}{\partial t} + \mathbf{u}\cdot\nabla\mathbf{u}\right) = \rho\mathbf{g} - \nabla p + \mu\nabla^2\mathbf{u}. \qquad (2\text{–}89)$$

This is the *Navier–Stokes* equation of motion for an incompressible, isothermal Newtonian fluid. Two comments are in order with regard to this equation. First, we shall always assume that ρ and μ are known–presumably by independent means – and attempt to solve (2–89) and the continuity equation, (2–20) for $\mathbf{u}$ and p. Second, the ratio μ/ρ, which is called the kinematic viscosity and denoted as ν, plays a fundamental role in determining the fluid's motion. In particular, it can be seen from (2–89) that the contribution to acceleration of a fluid element ($D\mathbf{u}/Dt$) that is due to viscous stresses is determined by ν rather than by μ.

Finally, we note that it is frequently convenient to introduce the concept of a *dynamic pressure* into (2–89). This is a consequence of the utility of having the pressure gradient that appears explicitly in the Navier–Stokes equation act as a driving force for motion. In the form (2–89), however, a significant contribution to ∇p is the static pressure variation $\nabla p = \rho\mathbf{g}$, which has nothing to do with the fluid's motion. In other words, nonzero pressure gradients in (2–89) do not necessarily imply fluid motion. Because of this it is convenient to introduce the so-called *dynamic pressure* P, such that $\nabla P = 0$ in a static fluid. This implies

$$-\nabla P \equiv \rho\mathbf{g} - \nabla p. \qquad (2\text{–}90)$$

Note that this equation requires that the sum $\nabla p - \rho\mathbf{g}$ be expressible in terms of the gradient of a scalar P. This is possible only if ρ is constant – that, is the fluid must be both incompressible and isothermal. In this case,

$$\rho\left(\frac{\partial \mathbf{u}}{\partial t} + \mathbf{u}\cdot\nabla\mathbf{u}\right) = -\nabla P + \mu\nabla^2\mathbf{u}. \qquad (2\text{–}91)$$

Equation (2–91) would seem to imply that the gravity force $\rho\mathbf{g}$ has no direct effect on the velocity distribution in a moving fluid, provided the fluid is incompressible and isothermal so that the density is constant. This is generally true. An exception occurs when one of the boundaries of the fluid is a gas–liquid or liquid–liquid interface. In this case, the actual pressure p appears in the boundary conditions (as we shall see), and the transformation of the body force out of the equations of motion by means of (2–90) simply transfers it into the boundary conditions. We shall frequently use the equations of motion in the form (2–91), but we should always keep in mind that it is the dynamic pressure that appears there.

Finally, it was stated previously that fluids that satisfy the Newtonian constitutive equation for the stress are often also well approximated by the Fourier constitutive equation, (2–67), for the heat flux vector. Combining (2–67) with the thermal energy, (2–52), we obtain.

$$\rho C_p \frac{D\theta}{Dt} = -\frac{\theta}{\rho}\left(\frac{\partial \rho}{\partial \theta}\right)_p \frac{Dp}{Dt} + p(\nabla\cdot\mathbf{u}) + (\mathbf{T}:\mathbf{E}) + \nabla\cdot(k\nabla\theta). \qquad (2\text{–}92)$$

Basic Principles

For an incompressible Newtonian fluid this take the simpler form

$$\rho C_p \frac{D\theta}{Dt} = -\frac{\theta}{\rho}\left(\frac{\partial \rho}{\partial \theta}\right)_p \frac{Dp}{Dt} + 2\mu(\mathbf{E}:\mathbf{E}) + \nabla \cdot (k\nabla\theta), \qquad (2\text{–}93)$$

which we shall use frequently in subsequent developments.

J. COMPLEX FLUIDS – ORIGINS OF NON-NEWTONIAN BEHAVIOR

Experience has shown that the Newtonian fluid model subject to the previously discussed assumptions, provides a very good approximation under most flow conditions for gases and small-molecule liquids such as water. The reason for this is that the random, thermally driven motions within such materials at moderate temperatures are sufficiently vigorous that they completely overcome any tendency of the forces associated with flow to produce a molecular configuration state that differs significantly from the isotropic, homogeneous state of statistical equilibrium. There is net displacement of mass associated with the mean (continuum-level) flow, but virtually no change in the "shapes," orientation distributions, or other statistical features of the configuration of the material at the molecular level.

There is, however, a vast body of materials whose behavior as liquids undergoing flow do *not* satisfy the assumptions for a Newtonian fluid. This includes many polymeric liquids, suspensions, multifluid blends, liquids containing surfactants that tend to form particle-like micelles when they are present at high concentrations, and many others. As we shall subsequently discuss from a qualitative point of view, these fluids exhibit more complicated macroscopic properties and have historically been lumped together under the general designation of *non-Newtonian fluids*. In the more recent literature, they have also been called *complex liquids*.[20]

This book is focused almost entirely on the dynamics of, and transport processes in, Newtonian fluids. Nevertheless, the class of complex or non-Newtonian fluids is extremely common, especially in chemical engineering applications, and it is probably useful to spend some time discussing what is responsible for the departures from Newtonian behavior. We shall see that the applicability of the Newtonian fluid model is based on a combination of the intrinsic characteristics of the fluid and the characteristics of the flow. Specifically, a fluid may exhibit behavior that can be described accurately by the Newtonian fluid model under one set of flow conditions but appear as non-Newtonian under a different set of conditions. One motivation for discussing the origins of non-Newtonian behavior is that this will help us to understand when we might expect the Newtonian fluid model to apply. A second is that it will give us a very crude and qualitative understanding of the behavior that is characteristic of non-Newtonian materials.

Let us begin by discussing a specific example, namely, a dilute solution of rigid rodlike macromolecules in a solvent. By dilute, we mean that the macromolecules are far enough apart that they do not interact directly in the flow. An analogous system is a suspension of rod-shaped colloidal particles. In both cases, we suppose that the macromolecules/particles are small enough to experience significant random rotational motion that is due to Brownian motion, so that their orientation and position in the absence of flow is a random function of time. To describe the microstructural state of the solution/suspension, we therefore require a probability density function $\Psi(\mathbf{p}, \mathbf{x}_p)$, which prescribes the probability of finding a particle with its principal axis oriented in a direction $\mathbf{p}$ at a position $\mathbf{x}_p$. In the following discussion, we assume that the macromolecules/particles are always uniformly dispersed with respect to spatial position. Hence the probability of a particle being at some position can be specified

J. Complex Fluids – Origins of Non-Newtonian Behavior

in terms of a concentration, say c, that is assumed to be independent of position. Then specification of the microstructural state requires only the orientation distribution function,

$$N(\mathbf{p}; \mathbf{x}, t).$$

The probability of finding a particle/macromolecule with an orientation within a solid angle $d\mathbf{p}$ of some orientation $\mathbf{p}$ is

$$N(\mathbf{p}; \mathbf{x}, t)d\mathbf{p}.$$

In general, this probability will be a function of both position $\mathbf{x}$ and time t. However, at equilibrium, all orientations are equally probable. Hence the equilibrium orientation distribution, normalized so that the total probability of orientation in some direction is unity, is just

$$N(\mathbf{p}; \mathbf{x}, t) = \frac{1}{4\pi}.$$

In general, we seek a description of the solution/suspension from a macroscopic-continuum perspective. Hence we generalize our notion of a material point to be a volumetric region that is small enough compared with the (macroscopic) length scales of a flow domain to be considered as a point, but large enough to contain a statistically meaningful number of particles/macromolecules. Any macroscopic property of the suspension/solution at a material point will then depend on the orientation distribution at that point and time. In the equilibrium state, the orientation probability is uniform, as already noted, and hence the material at equilibrium is statistically *isotropic*, in spite of the fact that individual particles/macromolecules are rodlike in shape.

Now, let us think qualitatively about what happens to the material when it is subjected to a flow. For simplicity, let us imagine that the flow is a simple shear flow

$$\mathbf{u} = \dot{\gamma} y\, \mathbf{i}_x, \qquad (2\text{--}94)$$

with a shear rate $\dot{\gamma}$. Simple shear flow is a good approximation to the flows generated in many rheometers – namely, the instruments designed to measure the flow properties of liquids – and thus it is an appropriate choice for the present discussion.

When an individual particle/macromolecule is subjected to a flow of the type (2–94), it will tend to rotate. In general, the rotational motion of a rod in a shear flow is complex, though a complete theory for rodlike ellipsoids had been developed already in the early 1900s by Jefferey (1922).[21] For present purposes, let us simply consider a rod that lies in the plane of the flow (i.e., the x–y plane). This rod will rotate. However, the rate of rotation is not uniform. If the rod is oriented in the y direction (across the flow), it is intuitively obvious that the hydrodynamic torque on it will be much larger than if the same rod were oriented with its axis in the x direction (parallel to the flow). Hence it will rotate more rapidly in the cross-flow orientation than when it is aligned with the flow. Even though the rod will continue to tumble over and over, it will spend relatively longer in orientations that are near alignment with the flow and less time in orientations that are across the flow. Thus the action of the flow is to produce a nonuniform orientation distribution, with larger N values in the flow-aligned direction and smaller values away from this direction. Of course, Brownian motion is still present – i.e., "rotational diffusion" – and this continues to try to maintain the orientation distribution in a uniform or completely random state. The orientation distribution that results can be viewed as a consequence of the competition between the flow-induced tendency toward a nonuniform orientation distribution and the tendency of Brownian diffusion to maintain a uniform state. The strength of the flow effect will clearly be proportional to $\dot{\gamma}$, whereas the diffusion process is characterized by a so-called rotational diffusivity D. Both of these quantities have units of inverse time. D, in particular, provides a measure of the

Basic Principles

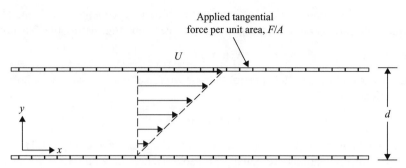

Figure 2–9. A number of simple flow geometries, such as concentric cylinder (Couette), cone-and-plate, and parallel disk, are commonly employed as rheometers to subject a liquid to shear flows for measurement of the fluid viscosity (see, e.g., Fig. 3–5). In the present discussion, we approximately represent the flow in these devices as the flow between two plane boundaries as described in the text and sketched in this figure.

time scale required for a system with a nonuniform initial orientation distribution to return to the uniform equilibrium state by means of Brownian motion in the absence of flow. The magnitude of flow-induced deviations from the uniform equilibrium state then depends on the dimensionless ratio

$$Wi \equiv \frac{\dot{\gamma}}{D}. \qquad (2\text{–}95)$$

This parameter is known in the rheological literature as the *Weissenberg number* (also sometimes – mistakenly – called the *Deborah number*).[22]

As noted previously, we expect the macroscopic properties to depend on the orientation distribution. Because the latter will, in general, change in the presence of flow, it follows that properties, such as the viscosity, as measured in a standard *shear rheometer*, will have different values at different shear rates. In such a rheometer, the fluid is between two plane solid boundaries (as sketched in Fig. 2–9). The shear flow is induced by the movement of one of the boundaries (in its own plane) relative to the other. The shear rate is determined by the differential velocity (U) divided by the gap width d. The force F required for maintaining the velocity U is then measured. For a *Newtonian* fluid, the ratio

$$\frac{(F/A)}{(U/d)}$$

is equal to the viscosity μ (here the ratio F/A is equal to τ_{xy} and the ratio U/d is equal to E_{xy}). The viscosity, being a material constant according to the model, is independent of the shear rate (U/d). Hence the ratio $(F/A)(U/d)(U/d)$ will be independent of U if the fluid is truly Newtonian. For the suspension/solution that we have been considering, experimental measurements show that the ratio $(F/A)/(U/d)$ actually decreases with increase of the shear rate, U/d, at least when U/d is large enough so that the magnitude of the Weissenberg number

$$Wi = \frac{(U/d)}{D}$$

is of the order of unity or larger. As the shear rate increases, the flow-induced tendency for alignment of the particles/macromolecules in the flow direction is enhanced. However, all else being equal, it seems intuitively obvious that the work (or force) required to produce a certain shear-rate will be decreased as the degree of flow alignment is increased, and thus the *apparent* viscosity, $(F/A)/(U/d)$ will decrease with increase of the shear rate U/d. This is illustrated pictorially in Fig. 2–10.

J. Complex Fluids – Origins of Non-Newtonian Behavior

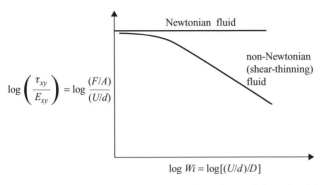

Figure 2–10. A qualitative sketch of the dependence of the ratio of the shear stress to the shear rate in a simple shear flow. The upper curve represents the result that would be obtained if the fluid were Newtonian. The lower curve illustrates the shear-thinning behavior that is described in the text for the solution of rodlike macromolecules (or the suspension of rigid-rod particles) when the shear rate is large enough to produce partial alignment of the macromolecules (the particles) in the flow direction.

A fluid that exhibits an apparent viscosity that decreases with increasing shear rate is often called a "shear thinning" fluid. Shear thinning is one easily measurable signal that a fluid cannot be described as a Newtonian fluid. It should be cautioned, however, that shear-rate-dependent apparent viscosity is but one way that non-Newtonian fluids may differ from a Newtonian fluid. Not all non-Newtonian fluids exhibit this particular property, though certainly every fluid that does shear thin is non-Newtonian. The reader may have noted that the word "apparent" has been used in describing the "viscosity" measured in our solution/suspension (or in any non-Newtonian fluid). This is because the measurement previously described, although clearly giving a direct measure of the material viscosity μ when the fluid is Newtonian, can only be said to measure the same ratio of macroscopic quantities, (F/A) and (U/d), when the fluid is not Newtonian. Although it is sometimes proposed, on a purely empirical basis, that we might describe the non-Newtonian behavior as reflecting a shear-rate-dependent viscosity, it is *not* true for any real non-Newtonian or complex fluid that the Newtonian constitutive equation, (2–81), can be replaced with

$$\mathbf{T} = -p\mathbf{I} + 2\mu(\dot{\gamma})\mathbf{E}, \tag{2–96}$$

as would be true if the ratio $(F/A)/(U/d)$ were measuring a true material function, $\mu(\dot{\gamma})$. In truth, constitutive equations required to describe the behavior of complex or non-Newtonian fluids are much more complex than the so-called "generalized Newtonian fluid" form (2–96), even for the relatively simple system made of rigid Brownian macromolecules in a Newtonian fluid that we have been discussing. Hence the measured quantity $(F/A)/(Ud)$ is not a true viscosity except when the fluid is Newtonian. It is nevertheless convenient to refer to it in terms of a shear-rate-dependent viscosity, and we add "apparent" to emphasize that it is not a true material constant. We shall briefly discuss constitutive models for non-Newtonian complex fluids in the next section. First, we should complete the discussion of our specific example.

The preceding description of the expected behavior of our solution/suspension of rodlike macromolecules/particles led to the realization that this is a fluid whose behavior is not that of a Newtonian fluid. However, there is more that we can learn about such fluids by taking the discussion somewhat further.

The first, and most obvious but critical point is that the solution/suspension is non-Newtonian in behavior because its statistical structure at the macromolecular/particulate level is modified when it is subjected to flow. This may be contrasted with typical small molecule liquids whose constitutive behavior is generally well approximated as Newtonian.

If we think about it, even the small molecule liquid, when subjected to flow, will experience forces at the molecular level that will tend to produce a nonisotropic statistical structure. In the simplest case, at least conceptually, this may be simply a tendency to produce a nonuniform orientation distribution for molecules that are not strictly spherical in shape. As in the case of our suspension/solution, the random motions corresponding to orientational diffusion will oppose this tendency. The difference between Newtonian fluids and complex fluids is the enormous change in the magnitude of the relevant diffusivities. In the case of our macromolecular solution or colloidal suspension, the characteristic time scales for relaxation from a nonequilibrium to an equilibrium configuration may be as large as seconds or even tens of seconds, depending on the solvent viscosity and the molecular weight or size of the rodlike macromolecules/particles. For a small-molecule liquid, on the other hand, the same time scale could be fractions of a microsecond or smaller. Hence, whereas a shear rate of order $1-10$ s^{-1} leads to Weissenberg numbers of order 1 or larger for the suspension/macromolecular solution, the magnitude of shear rates normally accessible in technological or naturally occurring flows are almost never large enough to produce anything other than extremely small values of Wi for a small-molecule liquid. Thus the microstructural state in the latter case tends to be completely denominated under all normal flow conditions by random, thermal motion, and there is no perceptible effect of the flow on the statistical structure of the fluid at the molecular level. Thus the fluid properties remain the same under flow conditions as they are at equilibrium (i.e., in the absence of flow) and the fluid can be approximated as Newtonian provided only that it is isotropic at equilibrium.[23]

An important conclusion from this discussion is that it is not appropriate to describe a particular fluid as Newtonian (or not) without coupling this statement to a characterization of the flow conditions (i.e., in the case of the simple shear flow previously discussed to the range of shear rates). Clearly it should be possible under especially severe flow conditions of high shear rate for a small-molecule liquid to exhibit non-Newtonian behavior. An example is some types of motor oil that appear as Newtonian under conventional conditions for shear rheometry (maximum shear rates in common commercial rheometers are generally less than 100 s^{-1}), but are subjected to extremely large shear rates in reciprocating piston engines and are then known to exhibit shear thinning unless they have additional (usually polymeric) additives included. Even a fluid like water is expected to exhibit non-Newtonian characteristics if subjected to extremely high shear or rapidly varying flow conditions. On the other hand, a much more common situation is that polymeric liquids or other complex fluids may be approximated as Newtonian fluids if they only experience flows with shear rates such that $Wi \ll 1$. Hence, to emphasize again, no fluid should be thought of as categorically Newtonian or non-Newtonian. The behavior depends on expected flow conditions.

Finally, let us return to review the various assumptions that were made in the derivation of the Newtonian constitutive model. These were, first, that the stress depends linearly on the strain rate $\mathbf{E}$. This is *not* true in our example of the solution/suspension of rods. Evidently, if we are given a fixed material configuration, the stress may depend linearly on $\mathbf{E}$, as in small-molecule liquids. In our case, however, the configuration also depends on $\mathbf{E}$ (i.e., on $\dot{\gamma}$ in shear flow), and thus the stress must depend on $\mathbf{E}$ in some nonlinear fashion. This is the source of shear thinning that was discussed earlier. The second condition required for the Newtonian fluid model to apply is that the stress is assumed to respond instantaneously to the *current* value of the strain rate at the particular material point of interest. This assumption is also clearly violated by the solution/suspension of rods. In general, following a change in shear rate, the orientation distribution must also change, and thus too the stress. However, the orientation distribution responds to a change in flow conditions in only a *finite* increment of time. This is most easily seen if the shear rate is suddenly put to zero. In a Newtonian fluid, the stress would instantaneously return to its equilibrium value. In the solution/suspension, however, this occurs over only the finite

J. Complex Fluids – Origins of Non-Newtonian Behavior

time scale D^{-1} that is required for the orientation distribution to return to its equilibrium state. As a consequence, the statistical structure (orientation distribution) and thus the stress at a material point at any instant of time must depend on the *past history* of flow of that material point. On the other hand, it is apparent that the configuration (and stress) at an instant of time will depend most strongly on the *recent* flow history and less strongly on the flow at earlier times. In the language of continuum mechanics, and "rheology", it is said that a material such as ours exhibits a "*memory*" for past shear-rate history, which "fades" on a time scale proportional to D^{-1}. The suspension always returns to the same configuration state of equilibrium and in that sense resembles an elastic solid. Fluids with this feature are called *viscoelastic* liquids. We note that the elasticity in this case comes from the entropic tendency of the orientation distribution to return to an isotropic equilibrium state. The individual particles/macromolecules are modeled as solid–rigid objects, hardly the direct source of any elasticity. Finally, in the Newtonian fluid model, it is assumed that the material is completely isotropic. In our example case, this is true at *equilibrium*. However, as soon as flow is added to the picture, the structure becomes *anisotropic* as the orientation distribution becomes nonuniform. The only one of the original assumptions that actually applies for complex/non-Newtonian fluids is the statement that the stress at a material point depends on only the material configuration at that material point. However, because material points move with finite velocity, the dependence of the microstructural state on past history *does* imply a dependence on the flow history along the fluid pathline that is followed by the material point. It may also be noted that the size of the region that comprises a material point will generally be larger for a complex fluid than for a small-molecule liquid. Hence the conditions for applicability of the continuum approximation will generally be somewhat more restrictive.

We have focused our discussion on the specific example of a dilute suspension of Brownian rodlike particles or macromolecules. However, as indicated at the beginning of this section, there are many examples of complex or non-Newtonian materials. Others that come immediately to mind include emulsions or blends of two immiscible fluids (generally with one dispersed as bubbles or drops within the other), polymeric liquids (either solutions or melts), and colloidal suspensions of solid spherical particles (which might be immersed in either a Newtonian or non-Newtonian liquid). It will not be difficult for the reader to think of other examples. The common characteristic is that the material contains or is made up of relatively large molecules, or particles, that, if perturbed from their equilibrium configurational state, require a relatively long time scale (of the order of milliseconds or longer) to return to the equilibrium state. This means that flows with moderate-velocity gradients can produce a significant departure from the equilibrium state. Because macroscopic properties depend on the configurational state of the material, all such fluids will exhibit non-Newtonian (i.e., flow-dependent) properties. In fact, though the various types of complex fluids are quite different at a detailed, microscopic level, there is a common conceptual framework that applies to all. This also suggests a certain commonality to the macroscopic material behavior, and quite likely, also some similarity in the form of constitutive models that could be used to describe this behavior (though, of course, one should expect that any material coefficients would be strongly dependent on the specific material.).

Let us consider the three additional examples just mentioned. First, we need to identify the features, in each case, that define the microstructural state. In the case of the emulsion or blend, the most important microscale feature that can be influenced by the flow is the orientation and shape of the disperse-phase bubbles or drops (the mean drop size and drop-size distribution will also generally be important and can be influenced by flow-induced drop breakup and coalescence events, but we will ignore this extra complication for purposes of our current discussion). At equilibrium, the drops will be spherical and the microstructure isotropic. For polymeric liquids, it is the statistical configuration of the polymer molecules (both individually and collectively) that can be influenced by flow. For example, in dilute solutions of flexible polymer molecules, the flow can induce a transformation from the random coil configuration, which exists at equilibrium, to a stretched and oriented threadlike conformation. More generally, flow will tend to produce an aligned and stretched statistical state, beginning from an isotropic state at equilibrium. Finally, for a suspension of solid spherical particles, it is obvious that individual particles cannot change shape. However, in the presence of a flow, hydrodynamic interactions between particles can change the nearest-neighbor probability density distribution from an isotropic distribution at equilibrium that is established by random thermal motion (diffusion), to a distribution with a higher probability in some directions than others.

In each case, the macroscopic properties will be changed by these flow-induced changes in the material microstructure. For all three materials, there is a "relaxation" mechanism that tends to drive the system back toward an *isotropic*, equilibrium state; the drops are spherical at equilibrium and are uniformly dispersed; the polymer molecules are characterized by a random configuration of chains and/or chain segments; and the colloidal particles are uniformly dispersed. The mechanism by which the system relaxes toward equilibrium obviously is different in each case. The drops achieve a spherical shape at equilibrium because of the action of interfacial or surface tension (as we shall discuss in more detail later in this chapter), and any tendency of flow to produce a nonisotropic distribution of neighboring drops (*n*-drop distribution function) is overcome by Brownian diffusion (on a time scale that increases with drop size). Polymer molecules undergo random fluctuations in configuration because of random thermal motions (Brownian) associated with the solvent and/or other polymer chains, and this leads to an isotropic equilibrium configuration of "undeformed" chains. Finally, the suspension of colloidal, spherical particles is subject to translational Brownian diffusion that drives the system back toward a uniformly dispersed state. The time scales for these "relaxation" processes are finite in each case: inversely proportional to the surface tension for emulsions, $\mu a/\sigma$, where μ is the suspending fluid viscosity, a is the undeformed drop radius, and σ is the interfacial tension; and proportional to an appropriate diffusivity D^{-1} in the other two cases. The degree of departure from equilibrium in each case is then determined by a Weissenberg number

$$Wi = \tau_{relxn} \cdot |\nabla \mathbf{u}|$$

where $|\nabla \mathbf{u}|$ is a measure of the magnitude of velocity gradients in the flow (i.e., it is the shear rate for a simple shear flow).

Hence we see that there is a great deal of the overall behavior that is common to all four of the specific examples that we have considered. Each is isotropic in the equilibrium state, maintained by a relaxation process (or more than one in some cases) that is characterized by a finite relaxation time, τ_{relxn}. In the presence of flow each becomes anisotropic in structure, with the degree of departure from equilibrium determined by a balance between the flow-induced deformation and the relaxation process(es) with the relative importance of each characterized by the ratio of time scales

$$Wi = \tau_{relxn}/(|\nabla \mathbf{u}|)^{-1}$$

that is known as the Weissenberg number. Each has a preferred equilibrium state – and so will be expected to exhibit *viscoelastic* behavior. Further, each will exhibit a so-called "fading memory" for past flow conditions and a nonlinear dependence of the macroscopic stress on $|\nabla \mathbf{u}|$. Although the various complex fluids are quite different in terms of their microscopic makeup, it should be clear that there is a great deal in common insofar as expected macroscopic flow behavior is concerned. It should also be clear that many fluids of common experience (indeed all fluids under sufficiently extreme flow conditions) will exhibit viscoelastic, non-Newtonian rheology, and this is true even if the microscale particles or

K. Constitutive Equations for Non-Newtonian Fluids

macromolecules are rigid. The "elasticity" in such cases is essentially "entropic" in origin – a measure of the "deterministic" tendency to revert to a completely random equilibrium structure.

In the next section, we briefly discuss some specific constitutive models and their derivation.

K. CONSTITUTIVE EQUATIONS FOR NON-NEWTONIAN FLUIDS

In the preceding section, we have seen that there are many fluids that cannot be approximated as Newtonian under normal flow conditions. An obvious question is whether successful generalizations of the Newtonian fluid model exist that can be used to solve flow and transport problems for this class of materials?

If we interpret this question as asking whether models exist for the general class of complex/non-Newtonian fluids that are known to provide accurate descriptions of material behavior under general flow conditions, the current answer is that such models do *not* exist. Currently successful theories are either restricted to very specific, simple flows, especially generalizations of simple shear flow, for which rheological data can be used to develop empirical models, or to very *dilute* solutions or suspensions for which the microscale dynamics is dominated by the motion deformation of single, isolated macromolecules or particles/drops.[24]

It should be emphasized that many constitutive models have been proposed especially for polymeric solutions and melts, and there is a great deal of current research that is aimed at both new models[25] and numerical analysis of fluid motions by use of the existing models.[26] The problem is that few have been carefully compared with the behavior of real fluids outside the highly simplistic flows of conventional rheometers, and then mainly under flow conditions in which the perturbations in material structure are weak. Thus there is currently no model that is known to provide quantitatively accurate or even qualitatively reliable descriptions of real complex fluids for a wide spectrum of flows.

Assuming this to be an accurate assessment of the current state of affairs, the reader may wonder at the existence of the present section of this book. Indeed, it is unique (with respect to the rest of the book) in the sense that we are unable to follow the general goal of introducing specific methods and ways of thinking that will allow us the opportunity to address new or unresolved problems of fluid motion. Nevertheless, the class of complex fluids is so important, and the lack of sections on "non-Newtonian" fluid mechanics appears as such a glaring omission from this textbook, intended as it is for scientists and engineers who will likely encounter complex fluids throughout their careers, that I believe that it is worthwhile to provide some additional discussion of past and current attempts to develop the necessary constitutive theories. The reader who is strictly concerned with Newtonian fluids may skip this section with no effect on understanding later sections of this book.

There are, in fact, two quite distinct approaches to the generation or derivation of new constitutive models. The first is the purely *continuum mechanics* approach that was so successful in obtaining the Newtonian fluid approximation. This approach has a long history, extending over at least the past 50–100 years in a form that resembles its current use and is responsible for the vast majority of constitutive models that have been proposed up to the present time. The steps are familiar. First, and most critical, is that a *guess* is made about the general nature of the proposed relationship: What independent dynamical variables are required (i.e., only the rate of strain in the case of the Newtonian fluid model) and what assumptions are to be imposed on the relationship between these variables and the macroscopic stress (e.g., linear, instantaneous, local, and isotropic in the case of a Newtonian fluid)? Given this input, the mathematical constraints of coordinate invariance and material objectivity are then used to determine admissible forms for the functional relationships that

Basic Principles

satisfy these assumptions.[27] The result will be a constitutive relationship that is mathematically acceptable, but the usefulness of this relationship depends on whether real fluids exist that satisfy the *physical* assumptions! In the continuum mechanics framework, the only way to tell is to compare predictions from the model with experimental observations, and this is a long, difficult, and, often, nearly impossible task.

In fact, the field of experimental rheology is concerned with the initial steps in this process. Measurements of stress components in shear-flow rheometers provide a basis for a first-level test of any model. There are a number of different geometric configurations that are used for such measurements: concentric cylinders that are rotated relative to one another; a pair of parallel disks that are rotated relative to one another; and a similar geometry in which one of the disks is replaced with a shallow-angle cone.[28] The advantage of these shear rheometers is that the flow can be approximated as the linear shear flow sketched in Fig. 2–9 (assuming that the fluid remains homogeneous) for both Newtonian and non-Newtonian fluids (we will discuss this in the next chapter). It is only the stresses resulting from this flow that are different. Because the form of the flow is known *a priori*, we do not need to solve a fluid mechanics problem to make predictions of the measurable stresses from a proposed constitutive model. If a proposed model cannot predict the measurable behavior in shear flow, it is certainly not a generally acceptable model (though, of course, it might still find use for other types of flow if the behavior in these flows was better). However, even if a continuum-based model performs perfectly in predicting shear-flow behavior, there is no guarantee that it will perform adequately in other flows. This must be tested. However, at this point the situation becomes extremely difficult because there is no other flow configuration in which the non-Newtonian behavior does not significantly alter the flow. Hence it is necessary to both measure and predict not only the stress components as before, but also the details of the flow. Not only are the experiments demanding, but it is difficult in many instances to even solve the equations of motion corresponding to a proposed constitutive model (either numerically or analytically) for comparison with the experiments. In a worst-case scenario, it is possible that a solution does not even exist because it is not known that the model represents the behavior of any realizable fluid.

Given the apparent arbitrariness of the assumptions in a purely continuum-mechanics-based theory and the desire to obtain results that apply to at least some real fluids, there has been a historical tendency to either relax the Newtonian fluid assumptions one at a time (for example, to seek a constitutive equation that allows quadratic as well as linear dependence on strain rate, but to retain the other assumptions) or to make assumptions of such generality that they must apply to some real materials (for example, we might suppose that stress is a functional over past times of the strain rate, but without specifying any particular form). The former approach tends to produce very specific and reasonable-appearing constitutive models that, unfortunately, do not appear to correspond to any real fluids. The best-known example is the so-called *Stokesian fluid*. If it is assumed that the stress is a nonlinear function of the strain rate $\mathbf{E}$, but otherwise satisfies the Newtonian fluid assumptions of isotropy and dependence on $\mathbf{E}$ only at the same point and at the same moment in time, it can be shown (see, e.g., Leigh[29]) that the most general form allowed for the constitutive model is

$$\mathbf{T} = (-p + \alpha)\mathbf{I} + \beta\mathbf{E} + \delta(\mathbf{E} \cdot \mathbf{E}). \quad (2\text{--}97)$$

Here, α, β, and δ are material coefficients that can depend on the thermodynamic state, as well as the invariants of $\mathbf{E}$, namely tr $\mathbf{E}$, det $\mathbf{E}$, and (tr$^2\mathbf{E}$ − tr $\mathbf{E}^2$). The Stokesian model appears to be a perfectly obvious generalization of the Newtonian fluid model. However, no real fluid has been found for which the model with $\delta \neq 0$ is an adequate approximation. We should, perhaps, not be surprised by this result as the examples in the all seem to suggest that the assumptions of "isotropy," plus instantaneous and linear dependence on $\mathbf{E}$, all seem to break down at the same time in real, complex fluids.

K. Constitutive Equations for Non-Newtonian Fluids

The other end of the spectrum, namely invoking extremely general assumptions, produces models of *excessive generality* that typically cannot provide detailed predictions of material behavior. The best-known example is the so-called *"simple" fluid* developed in the 1950s and 1960s by Collman, Truesdal, Noll, and others.[30] Although some useful results have been retained from this general theory, in the form of the "nth-order fluid approximation," for description of the first weak departures from the Newtonian fluid limit (i.e., the limit of small Weissenberg number, $Wi \ll 1$), it cannot generally provide any detailed predictions under more general flow conditions. We may conclude that the continuum mechanics approach, which worked fine at the level of a linear (Newtonian) material, is basically unsatisfactory for effective or productive treatment of constitutive nonlinearities. The necessary continuum-level assumptions are largely arbitrary, and their relationship with the material behavior of a complex fluid at a macromolecular or particulate level is unclear. In the rest of this section, we consider the second basic approach, which is "molecular" or "microscale" modeling.

The starting point for molecular/microscale modeling is a mathematical description of the material at the scale of the macromolecules or particles. This is followed, in principle, by rigorous statistical mechanical calculations to obtain (1) an equation relating the microstructural state of the fluid at a material point to the macroscale (or continuum) stress at that point; and (2) an equation or equations that describe the evolution of the microstructural state of the material from some initial equilibrium-state-due to flow. The limitations of this approach are quite different from those associated with the continuum mechanics approach. The most obvious is that the microscale description of the fluid may be too simple to provide a direct match with any specific example of fluids in this class. One example is the most common models of polymer molecules in a solution, which envision them as a freely jointed bead-spring chain, in which the hydrodynamic interactions with the solvent are approximated by (point) centers of friction (the beads) separated by springs (or rods) that do not interact with the solvent, with flexible joints at the position of the beads. An even simpler model is the dumbbell with just two beads and a spring, intended to represent the end-to-end vector of the polymer chain.[31] Although certainly different in detail from an actual polymer molecule, both of these models can potentially capture the tendency of the polymer chain (or chain segments) to orient and stretch in a flow. This may be sufficient for purposes of estimating the polymer contribution to the stress, without requiring a more detailed description of the polymer chain. However, it is clear that this simple model, with its simplistic descriptions of polymer chains, will not correspond precisely to any specific polymer-solvent system, and will not model all aspects of polymer dynamics.

The underlying assumption in the "molecular" modeling approach is that the primary, generic features of nonlinearity, finite relaxation time, and flow-induced anisotropy can be captured by means of simple mechanical models, and this is sufficient to provide a first-order approximation to the non-Newtonian rheology. One important point is that the model system, even a solution of bead-spring "molecules," is a "real" viscoelastic liquid in the sense that it is physically realizable (at least in principle). This is a fundamental improvement over the pure continuum mechanics approach, which may produce models that do not correspond to any physically realizable material. If the microscale description is overly simplified, the resulting constitutive models are at least a simplified version of the "exact" model. Furthermore, the discussion of the preceding section leads us to expect that the non-Newtonian behavior of complex fluids is likely to be rather generic and thus able to be modeled without a large number of details. At the very least, the molecular modeling approach should provide useful clues on the expected form of constitutive models for complex fluids (the exact models, whatever these may be, cannot be simpler or of a fundamentally different form). Furthermore, the molecularly derived models provide specific values of the material constants, and useful insight into what micromechanical

Basic Principles

features they depend on. Although it may be necessary to "fit" constants by comparisons with experimental data when the simplified models are applied to real systems, the theoretical formulae can then often be used successfully to extrapolate from measured values for one fluid to predicted values for a similar system (e.g., if the solvent viscosity is changed, or the molecular weight of the polymer, etc.).

Molecular models typically consist of a pair of equations: One relates the stress to the microscale structure, and the second describes the evolution of the microscale structure by means of the competing effects of flow and Brownian diffusion (or some other mechanism for relaxation of the system back toward an equilibrium state), i.e.,

$$\begin{aligned}(1) \quad & \boldsymbol{\tau} = \boldsymbol{\tau}\,(\text{structure}), \\ (2) \quad & \text{structure evolution in flow.}\end{aligned} \quad (2\text{--}98)$$

It is perhaps useful to briefly discuss the details of such a model for the specific and relatively simple example of the dilute solution or suspension of rigid, rodlike (i.e., axisymmetric) macromolecules or particles immersed in a Newtonian solvent-suspending fluid, which was introduced in the preceding section. We consider a material point that is subjected to a general linear flow, i.e., a flow that may be expressed in the form.

$$\mathbf{u} = \boldsymbol{\Gamma} \cdot \mathbf{x}, \quad (2\text{--}99)$$

where $\boldsymbol{\Gamma}$ is the velocity gradient tensor. It is convenient in what follows to express $\boldsymbol{\Gamma}$ as the sum of the strain-rate and vorticity tensors:

$$\mathbf{u} = (\mathbf{E} + \boldsymbol{\Omega}) \cdot \mathbf{x}. \quad (2\text{--}100)$$

As stated in the preceding section, we assume that the macromolecules/particles remain uniformly dispersed, so that it is only their rotational motion in the flow (2–100) that is relevant to the microstructural derivation of a constitutive model. It was shown, almost 80 years ago,[21] that this rotational motion can be expressed in terms of the time dependence of a unit vector $\mathbf{p}$ that is aligned along the principal axis of the rod as

$$\dot{\mathbf{p}} = \boldsymbol{\Omega} \cdot \mathbf{p} + G[\mathbf{E} \cdot \mathbf{p} - \mathbf{p}(\mathbf{p} \cdot \mathbf{E} \cdot \mathbf{p})], \quad (2\text{--}101)$$

where G is a shape factor that has the following values:

$G = 0$ for a sphere,
$G = 1$ for a long rod of infinite length-to-diameter ratio,
$G \equiv [(r^2 - 1)/(r^2 + 1)]$ for an axisymmetric ellipsoid with axis ratio r.

This is known as Jeffery's solution.[21] It may be noted that the last term in (2–101) is added to ensure that $|\mathbf{p}| = 1$. When Brownian motion is present, the orientation of a particle must be given statistically by means of an orientation probability distribution function

$$N(\mathbf{p}, t),$$

which satisfies a Fokker–Planck equation in the orientation space occupied by $\mathbf{p}$, i.e.,

$$\frac{DN}{Dt} + \nabla_P \cdot \left(N\dot{\mathbf{p}} - D\nabla_P N \right) = 0. \quad (2\text{--}102)$$

Here ∇_P is used to denote that the gradient operates with respect to the orientation space of $\mathbf{p}$,[32] and D is the diffusion coefficient for rotational Brownian motion.

We use the convected derivative D/Dt to remind us that this equation is to be applied to the ensemble of particles/macromolecules belonging to a fixed material point. The second term represents the effect of the flow and contains $\dot{\mathbf{p}}$ [Eq. (2–101)], whereas the last term represents the rotational diffusion process. Finally, we require an expression relating the bulk (i.e., continuum) stress to the orientation distribution. For a dilute suspension, we

K. Constitutive Equations for Non-Newtonian Fluids

can do this by calculating the contribution to the stress for a single isolated particle with an arbitrary orientation **p** and averaging over all possible orientations weighted with the orientation probability, $N(\mathbf{p}, t)$, for particles at the specified material point. The result[33] is

$$\sigma = -p\mathbf{I} + 2\mu\mathbf{E} + 2\mu\phi\left[2A\langle\mathbf{pppp}\rangle : \mathbf{E} + 2B\left(\langle\mathbf{pp}\rangle \cdot \mathbf{E} + \mathbf{E}\langle\mathbf{pp}\rangle\right) + C\mathbf{E} + F\langle\mathbf{pp}\rangle D\right], \quad (2\text{--}103)$$

where μ is the solvent viscosity, ϕ is the volume fraction of particles, A, B, C, and F are known material coefficients that depend on particle shape, and D is again the orientational diffusivity. The angle brackets denote an average over possible orientations, i.e.,

$$\langle\mathbf{pp}\rangle = \int_{\substack{all \\ orientations}} \mathbf{pp}N(\mathbf{p}, t)d\mathbf{p}. \quad (2\text{--}104)$$

Hence we see that the stress depends on only the second and the fourth moments of the orientation distribution, $\langle\mathbf{pp}\rangle$ and $\langle\mathbf{pppp}\rangle$. However, in general, it is necessary to know the complete orientation distribution function $N(\mathbf{p}, t)$ to calculate exact results for these moments by means of (2–104).

We can see that Eqs. (2–101) (2–104) are sufficient to calculate the continuum-level stress σ given the strain-rate and vorticity tensors **E** and Ω. As such, this is a *complete* constitutive model for the dilute solution/suspension. The rheological properties predicted for steady and time-dependent linear flows of the type (2–99), with $\Gamma = \Gamma(t)$, have been studied quite thoroughly (see, e.g., Larson[34]). Of course, we should note that the contribution of the particles/macromolecules to the stress is actually quite small. Because the solution/suspension is assumed to be dilute, the volume fraction ϕ is very small, $\phi \ll 1$. Nevertheless, the qualitative nature of the particle contribution to the stress is found to be quite similar to that measured (at larger concentrations) for many polymeric liquids and other complex fluids. For example, the apparent viscosity in a simple shear flow is found to shear thin (i.e., to decrease with increase of shear rate). These qualitative similarities are indicative of the generic nature of viscoelasticity in a variety of complex fluids. So far as we are aware, however, the full model has not been used for flow predictions in a fluid mechanics context. This is because the model is too complex, even for this simplest of viscoelastic fluids. The primary problem is that calculation of the stress requires solution of the full two-dimensional (2D) convection–diffusion equation, (2–102), at each point in the flow domain where we want to know the stress.

The fact that the stress depends only on moments of $N(\mathbf{p},t)$ and not on $N(\mathbf{p},t)$ itself would seem to provide an opportunity to simplify the model. Instead of calculating $N(\mathbf{p},t)$ and using that to calculate the moments, it would be much easier if we could calculate the moments directly. The procedure to derive differential equations for $\langle\mathbf{pp}\rangle$ or $\langle\mathbf{pppp}\rangle$ is actually quite simple. We simply multiply all terms in Eq. (2–102) by **pp** (or **pppp**) and integrate over all possible orientations. The result of this procedure for **pp** is a DE of the form

$$\frac{D\langle\mathbf{pp}\rangle}{Dt} = g(\langle\mathbf{pp}\rangle, \langle\mathbf{pppp}\rangle), \quad (2\text{--}105)$$

whereas multiplying by **pppp** and integrating gives

$$\frac{D\langle\mathbf{pppp}\rangle}{Dt} = h(\langle\mathbf{pppp}\rangle, \langle\mathbf{pppppp}\rangle). \quad (2\text{--}106)$$

Unfortunately, the equation for $\langle\mathbf{pp}\rangle$ involves the higher moment $\langle\mathbf{pppp}\rangle$, whereas that for $\langle\mathbf{pppp}\rangle$ involves $\langle\mathbf{pppppp}\rangle$. This is an example of a classical "closure" problem of statistical physics. The only way to utilize (2–105) and (2–106) in conjunction with (2–103) to form a simpler constitutive model is to introduce a so-called *closure approximation*. This is either an equation relating the fourth moment directly to the second, or the sixth to the fourth, etc., depending on the level at which we wish to truncate the hierarchy of moment equations. The simplest choice for truncation is the quadratic form

$$\langle \mathbf{pppp} \rangle = \langle \mathbf{pp} \rangle \langle \mathbf{pp} \rangle. \qquad (2\text{--}107)$$

If we substitute this into (2–105) and (2–103) we obtain a specific, closed constitutive model that involves only $\mathbf{E}$ and the so-called configuration (or moment) tensor, $\langle \mathbf{pp} \rangle$. Although this would seem to "solve the problem" of a relatively simple, constitutive model for the specific material, it must be emphasized that the closure assumption, (2–107), is only one arbitrary choice from a huge set of possible alternatives. There is, in fact, no guarantee that there is any complex fluid for which (2–107) is an adequate approximation. Indeed, the necessity to invoke an *ad hoc* mathematical approximation may produce a model that has distinctly different properties from those of the full unapproximated model. It may even exhibit pathologies that render it more or less useless. Again, the only way to verify the usefulness of the resulting model is by comparison with predictions from the exact model, but this is generally difficult and tedious, and there are many possible choices in lieu of (2–107). It is perhaps not surprising that there are extensive efforts in current research to develop rational or deductive paths for generating accurate closure approximations (see, e.g., Cintra and Tucker,[35] Chaubal and Leal[36]). Recent research has also attempted to develop alternative approaches to approximating the system (2–101)-(2–104). (Chaubal *et al.*,[35] Ottinger[36]).

In spite of the complexity of the complete constitutive model, (2–101)–(2–104), and the difficulties of deriving approximations to it, it does appear that the microscale/molecular modeling approach is capable of generating constitutive theories for complex fluids. Thus it is important to recognize some additional limitations of this approach so that the reader may appreciate why it has not yet been completely successful in providing constitutive models for the complete class of complex fluids. One intrinsic problem is that the resulting models are generally too complex to use with present-day computing power to provide the engineer a tool for solving realistic flow problems of the type that might be encountered in processing applications. Effective means of approximating the full model to a simpler form have not yet been developed.

A second issue is the derivation of the structural evolution equation. In the dilute suspension case, a complete and exact fluid dynamics description of the motion of individual particles is possible, and the structural evolution equation can then be derived by a rigorous statistical averaging procedure. However, in more general circumstances, this exact deductive procedure is not possible. In the case of nondilute suspensions of rigid, nonspherical, Brownian particles (perhaps the next simplest of complex fluids), we cannot solve the multiparticle fluid dynamics problem to obtain a formula for the rotation of particles in a flow, and, in addition, we also do not have an exact description for the multiparticle diffusion process that provides the relaxation mechanism for return to an equilibrium state. Hence we are forced to invoke models to approximate these processes, and this introduces major uncertainties into the structural evolution equation. For polymeric liquids, the situation is even more difficult. In this case, we do not even start with an exact description of a single-polymer chain, but instead must introduce a model even for a single chain. At higher concentrations in many polymer solutions, the "fluid" consists of an extremely complex system of interacting polymer molecules, which may be highly intertwined. These systems are known as "entangled." New models must be invoked to describe both the flow-induced orientation and stretching of chains and the complex diffusion process by which the system can return to an equilibrium state. Although major progress has been made (the contributions of the

L. Boundary Conditions at Solid Walls and Fluid Interfaces

French physicist, P. G. DeGennes, were responsible, in part for his achieving the Nobel prize in physics in 1991), there are still many open issues in modeling the microdynamics of entangled polymers, and we are only now approaching constitutive models that can be used for fluid mechanics predictions.[37]

It is likely, in the interim, while we await models from the molecular modeling perspective for the more "difficult" complex fluids, that the most success in predicting fluid mechanics results for non-Newtonian fluids will come from a hybrid approach – combining some elements of both continuum mechanics and molecular modeling to produce relatively simple *empirical* models. There is a great deal of current research focused on all aspects of constitutive model development; on numerical analysis of flow solutions based on these models; and on experimental studies of many flows. There are a number of books and references available, but this is a complicated field that really requires a textbook/class of its own. At this point, it is time to return from our little sojourn into the land of complex fluids and come back to the principle subject of Newtonian fluids.

L. BOUNDARY CONDITIONS AT SOLID WALLS AND FLUID INTERFACES

We are concerned in this book with the motion and transfer of heat in incompressible, Newtonian fluids. For this case, the equations of motion, continuity, and thermal energy,

$$\rho \left(\frac{\partial \mathbf{u}}{\partial t} + \mathbf{u} \cdot \nabla \mathbf{u} \right) = \rho \mathbf{g} - \nabla p + \mu \nabla^2 \mathbf{u} + \nabla \mu \cdot (\nabla \mathbf{u} + \nabla \mathbf{u}^T), \qquad (2\text{–}108)$$

$$\nabla \cdot \mathbf{u} = 0, \qquad (2\text{–}109)$$

$$\rho C_p \left(\frac{\partial \theta}{\partial t} + \mathbf{u} \cdot \nabla \theta \right) = -\frac{\theta}{\rho} \left(\frac{\partial \rho}{\partial \theta} \right)_p \frac{Dp}{Dt} + 2\mu (\mathbf{E} : \mathbf{E}) + \nabla \cdot (k \nabla \theta), \qquad (2\text{–}110)$$

provide a complete set from which to determine the velocity **u**, pressure p, and temperature θ within any homogeneous fluid regime. By homogeneous, we mean any region in which material properties such as density, viscosity, heat capacity, or thermal conductivity vary only "slowly" on length scales that are proportional to the size of the flow domain itself. There are, inevitably, regions within the fluid domain where fluid properties vary on much shorter length scales. These include the immediate vicinity of either a phase boundary (i.e., solid–liquid, liquid–gas, etc.) or a liquid–liquid interface between two homogeneous fluids that are immiscible. In a purely molecular theory, no special treatment would be required for encompassing these regions. However, at the continuum mechanics level of resolution, these transition zones appear as surfaces. Equations (2–108)–(2–110) apply right up to these bounding surfaces or interfaces. Material properties that are seen as varying rapidly, but continuously, across the transition zones in a molecular theory are approximated in the continuum mechanics description as suffering a discontinuous jump at the surface or interface from the value in one bulk phase to the other. To determine solutions of (2–108)–(2–110) for the velocity or temperature distributions in the "homogeneous" or bulk fluid domains, boundary conditions must be specified for these variables or their derivatives at these bounding surfaces or interfaces.

The obvious question is this: What conditions should be imposed? Without a molecular or microscopic theory for guidance, there is no deductive route to answer this question. The application of boundary conditions then occupies a position in continuum mechanics that is analogous to the "derivation" of constitutive equations in the sense that only a limited number of these conditions can be obtained from fundamental principles. The rest represent an educated "guess" based to a large extent on indirect comparisons with experimental data. In recent years, insights developed from molecular dynamics simulations of relatively simple molecular theories have also played a useful role. Our focus in this section is on boundary conditions at either a solid boundary or at a fluid–fluid interface.

However, let us begin from a slightly more general perspective. There are really two types of boundary conditions encountered in theoretical analyses of fluid flow or heat transfer problems when (2–108)–(2–110) are used. In particular, when we are interested in the temperature or velocity fields in the vicinity of an object of finite size in a much larger fluid domain, it is often a convenient and reasonable approximation to assume that the fluid domain is *unbounded*. This is particularly useful if the form of the temperature or velocity field far from the object of interest is known in advance. In this case, the form of the temperature or velocity field far from any boundaries is prescribed in lieu of boundary conditions at an actual wall or surface. An example is the translation of a heated sphere through a cooler, quiescent fluid that is held in a large container. Now, if the sphere is much smaller than the container, and if it is not close to any of the container boundaries, the velocity and temperature perturbations caused by the sphere will be relatively localized in the vicinity of the sphere and be independent of where the sphere is located in relation to the distant container walls. In this case, instead of solving for the temperature and velocity fields in the complete fluid domain, with boundary conditions applied at the sphere surface and container walls, an adequate approximation will be to treat the fluid domain around the sphere as though it were *unbounded* (i.e., infinite in extent) and then require that the temperature and velocity fields take the "ambient" form far from the sphere that would exist in the container in the absence of any disturbance from the sphere. Although the approximation of applying far-field boundary conditions "at infinity," in lieu of actual boundary conditions at some distant boundary, may at first seem questionable, the far-field conditions themselves are generally assumed to be known. Indeed, the unbounded fluid approximation is useful only if we know the undisturbed form of the temperature or velocity fields in advance.

The other class of boundary conditions is those applied at bounding surfaces, i.e., either at solid surfaces or at an interface if there are two (or more) distinct "homogeneous" fluids in the flow domain. We denote these surfaces with the generic symbol S. The transition between two bulk materials occurs over a finite but thin region. In the continuum description, we approximate this as a surface of discontinuity in material properties. An immediate question that may arise is what *surface* we should choose within the finite, but thin, surface or interface region for the purpose of applying the macroscale or continuum boundary conditions. For that matter, we may equally ask whether it makes any difference. From a purely geometrical point of view, there is *no difference* what choice we make for S as long as it remains within the interfacial/surface zone. This whole region is vanishingly thin, in any case, compared with the continuum scale of resolution L. Nevertheless, it has historically proven to be extremely convenient to adopt a specific convention. To explain this convention, let us initially limit ourselves to an interface separating two pure bulk fluids A and B. As we move across the interfacial region there is a rapid, but smooth variation of the density from ρ_A to ρ_B. However, from the continuum viewpoint, we model this as though there were simply the two homogeneous fluids of density ρ_A and ρ_B right up to the surface S, across which the density jumps discontinuously from ρ_A to ρ_B. The position of S is chosen so that the total mass is the same in either description of the system, i.e.,

$$\int_{-L}^{L} \rho(z)dz = L\rho_A + L\rho_B, \qquad (2\text{–}111)$$

where z is the coordinate direction normal to S, the interface is (locally) at $z = 0$, and the range $-L$ to L covers a large but finite region such that $\rho - \rho_A \to 0$ as $z \to L$ and $\rho - \rho_B \to 0$ as $z \to -L$. If a different choice were made for S, there would be either more or less mass in the idealized continuum description than in the real system, and it would be

L. Boundary Conditions at Solid Walls and Fluid Interfaces

necessary to "account for" this *excess* (or *deficit*) *mass* by assigning any difference to the surface itself, as a so-called *surface-excess* quantity. The convenience of the choice just described is that the surface-excess mass is zero for a pure two-fluid system. In fact, the exact same considerations can be applied at any phase or fluid–fluid boundary, including that between a solid and a fluid (i.e., either gas or liquid). By convention, we assume that the surface *S* where boundary conditions are applied is chosen such that there is no surface-excess mass in a pure fluid system.

1. The Kinematic Condition

Turning now to the question of boundary conditions, the solution of (2–108)–(2–110) requires both thermal boundary conditions relating the temperature or its derivatives and the velocity and its derivatives on the two sides of *S*. We begin with the so-called *kinematic* boundary condition, which derives from the principle of mass conservation at any boundary of the flow domain.

Let us denote the bulk-phase densities on the two sides of the interface as ρ and $\hat{\rho}$ and the fluid velocities as **u** and **û**. The orientation of surface *S* is specified in terms of a unit normal **n**. In general, the surface *S* is not a *material* surface. For example, if there is a phase transition occurring between the two bulk phases (e.g., a solid phase is melting or a liquid phase is evaporating), mass will be transferred across *S*. However, the surface *S* is not a source or sink for mass, and thus mass conservation requires that the net flux of mass to (or from) the surface must be zero.

It is probably useful to think of two specific situations. In one, there is no phase transformation occurring, and in this case *S* is in fact a *material* surface separating a viscous fluid and a second medium that may either be solid or fluid. Hence, in the absence of phase change at *S*, the normal component of velocity must be continuous across it and equal to the normal velocity of the surface:

$$\mathbf{u} \cdot \mathbf{n} = \hat{\mathbf{u}} \cdot \mathbf{n} \quad \text{at } S. \tag{2–112}$$

If the second phase is a solid wall, then $\hat{\mathbf{u}} = \mathbf{U}_{solid}$, which is assumed to be known. In a frame of reference fixed to a solid wall, $\hat{\mathbf{u}} \cdot \mathbf{n} = 0$, and in this frame of reference

$$\mathbf{u} \cdot \mathbf{n} = 0 \quad \text{at } S, \tag{2–113}$$

provided the wall is impermeable. This choice of reference configuration is often a convenient one when one of the materials is a solid. If the second medium is another fluid, then the kinematic condition, (2–112), provides a single relationship between the two normal velocity components, both of which are unknowns to be determined in the solution process.

A generalization of the condition (2–112) is required if there is an active phase transformation occurring at *S*, i.e., if the liquid is vaporizing or the solid is melting. In this case, we must distinguish between the bulk fluid velocities in the limit as we approach the interface, and the velocity of the interface itself, $\mathbf{u}^I \cdot \mathbf{n}$ (where the interface is specified still by the criteria of zero excess mass discussed earlier). The condition of conservation of mass then requires that

$$\rho(\mathbf{u} - \mathbf{u}^I) \cdot \mathbf{n} = \hat{\rho}(\hat{\mathbf{u}} - \mathbf{u}^I) \cdot \mathbf{n} \quad \text{at } S, \tag{2.114a}$$

where the velocity components are all still measured with respect to fixed, "laboratory" coordinates. The term on the left-handside is just the net mass flux of material from the first fluid to (or from) the interface, and because mass cannot accumulate on the surface *S*, this is balanced by an equal flux of mass away from (or to) *S* on the other side. It will be noted in this case that the normal velocity components $\mathbf{u} \cdot \mathbf{n}$ and $\hat{\mathbf{u}} \cdot \mathbf{n}$ are no longer equal. Suppose, for example, that one phase is a liquid with density ρ that is freezing to a solid phase with a different (generally larger) density $\hat{\rho}$. There is no motion within the solid phase, and thus the velocity $\hat{\mathbf{u}} \cdot \mathbf{n}$, measured in the "fixed" laboratory reference frame, is zero. Hence, to satisfy mass conservation, the liquid must flow toward the interface with a velocity

$$\mathbf{u} \cdot \mathbf{n} = \left(\frac{\rho - \hat{\rho}}{\rho}\right) \mathbf{u}^I \cdot \mathbf{n} \quad \text{at S.} \qquad (2.114\text{b})$$

In the absence of an active phase-transformation process, both sides of (2.114a) are zero, i.e. $\mathbf{u} \cdot \mathbf{n} = \mathbf{u}^I \cdot \mathbf{n}$ and $\hat{\mathbf{u}} \cdot \mathbf{n} = \mathbf{u}^I \cdot \mathbf{n}$, and the condition (2–114a) reduces to (2–112). It is important to emphasize that the *kinematic* condition is a direct consequence of mass conservation at S, and must always be satisfied, regardless of the specific fluid properties or any details of the flow.

2. Thermal Boundary Conditions

Next, we consider the thermal boundary conditions, as these are also relatively straightforward to obtain.

The critical assumption about the bounding surface or phase boundary is that it is maintained at thermal equilibrium, independent of whatever mechanisms may exist for transport to the surface or release of heat at the surface that is due to phase transformation. Hence it follows that the two bulk (or "homogeneous") materials will have the same temperature at S, i.e.,

$$\theta = \hat{\theta} \quad \text{at each point on } S. \qquad (2\text{–}115)$$

If one of the two materials is a solid on which the temperature is known, $\theta = \hat{\theta}_{solid}$, the condition (2–115) is sufficient to solve the thermal energy equation. If, on the other hand, surface S is an interface, or phase boundary, the condition (2–115) provides only one relationship between two unknown temperature fields, and an additional boundary condition is required.

This second condition is the local statement of conservation of thermal energy at the interface. In particular, assuming that there is no independent source of heat at the interface other than that associated with the possibility of a phase transition (which is included as we shall see), then conservation of thermal energy requires

$$\mathbf{j} \cdot \mathbf{n} = \hat{\mathbf{j}} \cdot \mathbf{n} \quad \text{at } S, \qquad (2\text{–}116)$$

where the vectors $\mathbf{j}$ and $\hat{\mathbf{j}}$ are the total heat flux vectors. In a fluid for which Fourier's law applies, these take the form

$$\begin{aligned}\mathbf{j} &= -k\nabla\theta + \rho(\mathbf{u} - \mathbf{u}^I)\, C_p(\theta - \theta_{ref}), \\ \hat{\mathbf{j}} &= -\hat{k}\nabla\hat{\theta} + \hat{\rho}(\hat{\mathbf{u}} - \mathbf{u}^I)\, \hat{C}_p(\hat{\theta} - \theta_{ref}).\end{aligned} \qquad (2\text{–}117)$$

With the convective term calculated with the heat capacity at constant pressure, the appropriate measure of heat content is the enthalpy $H = C_p \Delta\theta$ per unit mass.

In the absence of a phase transition at S, the boundary S is a *material* surface so that $\mathbf{u} \cdot \mathbf{n} = \hat{\mathbf{u}} \cdot \mathbf{n} = \mathbf{u}^I \cdot \mathbf{n}$, and there is no *convective* flux of heat across it. Hence the condition (2–116) can be written in the form

$$-k(\nabla\theta \cdot \mathbf{n}) = -\hat{k}(\nabla\hat{\theta} \cdot \mathbf{n}) \quad \text{at } S. \qquad (2\text{–}118)$$

Because $k \neq \hat{k}$ in most cases, the temperature gradient is discontinuous across the boundary S. In some cases, in which the second material is a solid, the heat flux at the solid surface may be known, rather than the temperature. In this situation, instead of specifying θ in the

L. Boundary Conditions at Solid Walls and Fluid Interfaces

fluid in terms of a known temperature at the solid surface, $\hat{\theta}_{solid}$, by means of (2–115) the temperature gradient is specified in the fluid in terms of the known heat flux in the solid:

$$-k\nabla\theta \cdot \mathbf{n} = \mathbf{Q}_{solid} \quad \text{at } S, \tag{2–119}$$

where the right-hand side is assumed to be known. Either of the conditions (2–115) or (2–118) with the right-hand side *specified* at a boundary is sufficient to determine the temperature in the fluid (at least up to an arbitrary reference temperature if all boundary conditions are of the gradient type), but both cannot be specified simultaneously unless the surface is an interface for which both the fields $\theta(\mathbf{x}, t)$ and $\hat{\theta}(\mathbf{x}, t)$ are unknown.

If there is a phase change at S, the conservation of energy condition, (2–116), takes a different form from (2–118). If we simply substitute (2–117) into (2–116), with $C_p \Delta \theta$ expressed in terms of the enthalpy H, we obtain

$$-k(\nabla\theta \cdot \mathbf{n}) + [\rho(\mathbf{u} - \mathbf{u}^I) \cdot \mathbf{n}]H = -\hat{k}(\nabla\hat{\theta} \cdot \mathbf{n}) + (\hat{\rho}(\hat{\mathbf{u}} - \mathbf{u}^I) \cdot \mathbf{n})\hat{H} \quad \text{at } S. \tag{2–120}$$

Using (2–114a), we can rearrange this to the form

$$-k(\nabla\theta \cdot \mathbf{n}) + \hat{k}(\nabla\hat{\theta} \cdot \mathbf{n}) = \hat{\rho}(\hat{H} - H)(\hat{\mathbf{u}} - \mathbf{u}^I) \cdot \mathbf{n} \quad \text{at } S. \tag{2.121a}$$

Alternatively, this could be written as

$$-k(\nabla\theta \cdot \mathbf{n}) + \hat{k}(\nabla\hat{\theta} \cdot \mathbf{n}) = \rho(\hat{H} - H)(\mathbf{u} - \mathbf{u}^I) \cdot \mathbf{n} \quad \text{at } S. \tag{2.121b}$$

The change of enthalpy in transforming from one phase to the other is known as the "latent heat" per unit mass, and may be denoted as either ΔH or L. The thermal energy balance at the interface thus requires that there be an imbalance in the heat flux to and from the interface by conduction that is just equal to the rate of release (or adsorption) of enthalpy as the material changes phase.

3. The Dynamic Boundary Condition

The third type of boundary condition at the surface S involving the bulk-phase velocities is known as the *dynamic* condition. It specifies a relationship between the *tangential* components of velocity, $[\mathbf{u} - (\mathbf{u} \cdot \mathbf{n})\mathbf{n}]$ and $[\hat{\mathbf{u}} - (\hat{\mathbf{u}} \cdot \mathbf{n})\mathbf{n}]$. However, unlike the kinematic and thermal boundary conditions, there is no fundamental macroscopic principle on which to base this relationship. The most common *assumption* is that the tangential velocities are *continuous* across S, i.e.,

$$\mathbf{u} - (\mathbf{u} \cdot \mathbf{n})\mathbf{n} = \hat{\mathbf{u}} - (\hat{\mathbf{u}} \cdot \mathbf{n})\mathbf{n} \quad \text{at } S, \tag{2–122}$$

and this is known as the *no-slip* condition. If the second medium is a solid, then again $\hat{\mathbf{u}} = \mathbf{U}_{solid}$ which is assumed to be known, and the condition (2–122) prescribes a specific value for the tangential velocity of the fluid. A convenient frame of reference in this latter case is one fixed to the solid wall, so that

$$\mathbf{u} - (\mathbf{u} \cdot \mathbf{n})\mathbf{n} = 0 \quad \text{at } S. \tag{2–123}$$

This is the most common form of the *no-slip* condition.

It is generally accepted, based on empirical evidence, that the no-slip condition applies under almost all circumstances for small-molecule (Newtonian) fluids at either solid surfaces or at a fluid–fluid interface and also applies under many circumstances for complex liquids, such as polymer solutions or melts. This assertion is based primarily on comparisons of predictions from solutions of the equations of motion, which incorporate the no-slip condition, and experimental data – we shall discuss one example of a problem for which this kind of comparison has been done in the next chapter. Here, we simply note that these comparisons with experiments are often between *macroscopic* quantities – such as overall flow rates – rather than detailed measurements of the velocities in the immediate vicinity of boundaries or interfaces. There has, however, also been work in recent years, based on molecular dynamics simulations of simple model fluids, in which detailed examination of averaged velocities in the immediate vicinity of solid walls supports the relevance of the no-slip at a solid wall.[38] We will frequently use the no-slip condition for the solution of Newtonian fluid flow problems in this book.

In spite of the frequent applicability of *no-slip* as the appropriate dynamical boundary condition, however, there is no compelling theoretical argument for it, nor can kinetic theory contribute much, in general, to understanding it because kinetic theory cannot cope with the problem of dense liquids adjacent to a rigid solid. Scientists, for nearly 300 years, have thus questioned whether "no-slip" is the fundamentally correct boundary condition at a solid surface, or whether it is simply an approximation from some more general condition, which just happens to work well for many fluids under many flow conditions.[39] A more general condition that is frequently suggested is the so-called *Navier-slip condition*. Before this condition is stated, it should be emphasized that it has no more fundamental foundation than the no-slip condition. It is only a somewhat more general condition, albeit still ad hoc in origin, which happens to include the no-slip condition as a limiting case. In its most general tensor form, the Navier-slip condition at a solid boundary is

$$\mathbf{u} - (\mathbf{u} \cdot \mathbf{n})\mathbf{n} - \hat{\beta}\{\mathbf{T} \cdot \mathbf{n} - [(\mathbf{T} \cdot \mathbf{n}) \cdot \mathbf{n}]\mathbf{n}\} = 0. \tag{2–124}$$

Here $\mathbf{n}$ is the unit normal to the boundary, $\mathbf{u}$ and $\mathbf{T}$ are the (continuum) velocity and stress, and $\hat{\beta}$ is an empirical parameter known as the "slip coefficient." The Navier-slip condition says, simply, that there is a degree of slip at a solid boundary that depends on the magnitude of the tangential stress. We note, however, that it is generally accepted that the slip coefficient is usually very small, and then the no-slip condition (2–123) appears as an excellent approximation to (2–124) for all except regions of very high tangential stress.

For a Newtonian fluid, an equivalent statement of the Navier-slip condition is

$$\mathbf{u} - (\mathbf{u} \cdot \mathbf{n})\mathbf{n} - \beta\{\mathbf{E} \cdot \mathbf{n} - [(\mathbf{E} \cdot \mathbf{n}) \cdot \mathbf{n}]\mathbf{n}\} = 0. \tag{2–125}$$

In this case, β clearly has units of length and is known as the "slip length." It is expected to be of molecular dimensions.

Although the Navier-slip condition has been largely ignored for many years in favor of the corresponding no-slip boundary condition, there has been a growing interest in the Navier-slip condition in recent years, even for Newtonian fluids, driven by both new experimental observations and by certain theoretical problems that arise from application of the "no-slip" condition. The best known of the theoretical problems arises when a contact line separating the two immiscible fluids moves with velocity $-U$ along a solid surface at which the no-slip condition is assumed to apply, as sketched in Fig. 2–11.

A more convenient, but entirely equivalent, problem for analysis is to consider the position and shape of the interface to be fixed, with the boundary translating at a velocity U. If we calculate the velocity and pressure fields for an incompressible, Newtonian fluid, assuming no-slip at the solid wall and the kinematic plus no-slip conditions at the interface, we find that the tangential stress component on the boundary exhibits a nonintegrable singularity as the distance to the contact point goes to zero, i.e.,

$$\mathbf{T} \cdot \mathbf{n} - [(\mathbf{T} \cdot \mathbf{n}) \cdot \mathbf{n}]\mathbf{n} \sim \frac{1}{r^m}$$

as $r \to 0$ with $m \geq 1$. We also find that the viscous dissipation rate is divergent. Both results lead to the conclusion that it would take an infinite force to maintain a finite relative velocity between the contact line and the boundary. One proposed resolution of this physically unacceptable result is to *assume* that the fluid slips at, and in the immediate vicinity of,

L. Boundary Conditions at Solid Walls and Fluid Interfaces

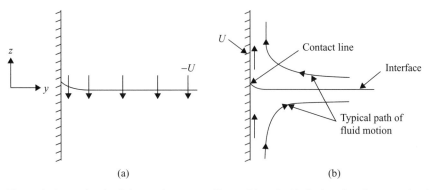

Figure 2–11. A sketch of the moving contact line problem. In (a) the interface is assumed to be moving with velocity $-U$ along the solid wall. In (b), the equivalent problem is shown in which the interface is viewed as fixed and the wall moving in its own plan with velocity U.

the contact line. If we impose the Navier-slip condition at the solid boundary, say in the form of (2–125) with β a very small but nonzero parameter, then the fluid will come very close to satisfying the no-slip condition everywhere except within a region of length proportional to β around the contact line, where the tangential stress (and hence also the velocity gradient) becomes extremely large so that the slip term in (2–125) is nonnegligible and the tangential velocity is significantly different from zero. It has been shown that this eliminates the singular behavior of the solution associated with a moving contact line.[40] The rationale is that even a Newtonian fluid may slip at a boundary if it is acted on by a sufficiently large tangential stress. Again, molecular dynamic simulations of a moving contact line also appear to be consistent with the presence of a local "slip" region.[38]

The example of slip in the vicinity of a moving contact line may appear a rather special circumstance, involving slip in a region of very high tangential stress. For the majority of Newtonian fluids and boundary materials, the idea that slip manifests itself under only extreme stress conditions is most likely correct. However, recent experimental results, some coming from the emerging literature on small-scale flows [microelectromechanical system (MEMS) devices], suggest that slip may be relevant under more general flow conditions for water at highly hydrophobic walls.[41] At the same time these observations may shed some light on the underlying physical factors that are responsible for the no-slip condition at a solid wall. For example, Watanabe et al.[42] showed that a Newtonian fluid (water) showed evidence of slip at the walls in flow through a pipe when the pipe was made from a class of materials that they termed "water repellent." By this they meant a material for which the contact angle θ_c with water was greater than $150°$ (see Fig. 2–12). To understand the significance of this, we must digress briefly to discuss the contact angle.

The contact angle exhibited at a stationery contact line, where a gas–liquid interface intersects a solid surface, is a *unique* characteristic of the three materials involved; namely, the gas, the liquid, and the solid. It reflects the nature of their interaction across the various surfaces that intersect at the contact line. It is known from thermodynamics that a fluid–fluid

Figure 2–12. A sketch of a three-phase contact line region at a solid wall illustrating the definition of the contact angle.

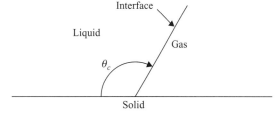

Basic Principles

interface (i.e., gas–liquid or liquid–liquid) can be characterized by its surface free energy per unit area.

This quantity is called the interfacial tension (sometimes also "surface tension" when the interface involves a liquid and a gas). It provides a measure of the work required for increasing the area of an interface (i.e., to form new interface by bringing molecules from the bulk fluids to the interface), and in this context it can also be viewed as the force per unit length acting at the boundaries of an element of surface that is required to keep it from shrinking in area (thus decreasing the surface free energy). We shall discuss the significance of this latter interpretation of interfacial tension in a more general fluid dynamical context in the next subsection in which we expand on our discussion of boundary conditions at a fluid–fluid interface. For present purposes, it is sufficient to note that we can apply the concept of interfacial tension, viewed as surface free energy per unit area, to the solid–liquid and/or solid–gas (or vapor) surfaces that meet the fluid–fluid interface at the contact line. If we then apply a force balance at the contact line *in the plane of the solid boundary*, we obtain the so-called Young equation, first proposed in 1805:

$$\gamma_{LG} \cos \theta_c = \gamma_{SG} - \gamma_{SL}, \tag{2-126}$$

where γ_{LG} is the interfacial tension at the liquid–gas interface, γ_{SG} and γ_{SL} are the corresponding interfacial energy densities at the solid–gas and solid–liquid surfaces, respectively, and θ_c is the contact angle. If we interpret interfacial tension loosely as specifying the work required to create a unit of surface area (as is the case at a fluid–fluid interface), then a *small* value of γ_{XY} indicates a strong *attractive* interaction between X and Y. Now, it can be seen from (2–126) that a small value of θ_c implies that $\gamma_{SG} > \gamma_{SL}$, whereas a large value of θ_c means that $\gamma_{SL} > \gamma_{SG}$. Hence, when θ_c is small, it implies that the liquid is strongly attracted to the solid, whereas the opposite is true if $\gamma_{SL} > \gamma_{SG}$. In fact, the issue of "adhesion" between a solid and liquid is more complex than suggested by this simplified discussion, and the Young equation is only qualitatively useful as it is nearly impossible to directly measure either γ_{SG} or γ_{SL}.

The conclusion from the experimental observations of Watanabe et al.,[42] when coupled with the preceding discussion about the contact angle, is that the strength of "adhesion" between a fluid and a wall is an important factor in determining the applicability of the slip versus no-slip boundary conditions. Indeed, more recent investigations[41] have used direct ("micro-particle image velocimetry") measurements of the velocity profile in small channels to confirm that highly hydrophobic (high-contact-angle) walls allow for slip and to provide an estimate of the slip length. For water with walls coated by hydrophobic octadecyltrichlorosilane, the slip length was found to be 1 μm (i.e., 10^{-4} cm). This implies that slip will tend to be important only for small-scale flows, for example, measurable only for channels smaller than about 1 mm even when the walls are highly hydrophobic ("repelling") for the liquid in question. Presumably $\beta \ll 10^{-4}$ for a wall that is not highly hydrophobic. In fact measurements in the same channel without the hydrophobic coating showed no evidence of slip even though the channel cross section was only 30×300 μm. For a solid boundary, we conclude that the no-slip approximation is generally quite good except in special regions of very high tangential stress, such as the vicinity of the moving contact line or with wall/fluid combinations in which the adherence between the wall and the fluid is very low (for water such walls are called hydrophobic).

There is also the possibility of slip at a fluid–fluid interface, especially for a pair of thermodynamically incompatible fluids. However, we are not aware of any evidence of slip at an interface between two "small-molecule" Newtonian fluids. One reason is that there are no examples of flow conditions analogous to the moving contact line at solid boundaries that can lead to very large tangential stress. Hence, because β is usually very small for Newtonian fluids, conditions that would produce any significant slip are absent, even assuming that

L. Boundary Conditions at Solid Walls and Fluid Interfaces

we should apply the Navier-slip condition in the first place. Although the potential exists for liquid pairs that exhibit relatively larger β in analogy with the liquid–solid surface, no examples have yet been found.

The body of evidence on Newtonian fluids is that significant departures from the no-slip conditions are uncommon. Thus, though we should always keep in mind the possibility of significant slip for some materials or in some local regions of the flow, we will follow the usual custom of applying no-slip conditions in the analysis of Newtonian fluid flows.

In spite of this, it is perhaps useful to briefly consider the conditions at solid boundaries and fluid interfaces for complex/non-Newtonian fluids. One reason for doing this is that it provides additional emphasis to the idea from the proceeding paragraphs that there will be conditions when the commonly applied no-slip condition breaks down. It should be stated, at the outset, that the question of slip or no-slip is still a matter of current research interest for complex fluids. Nevertheless, the occurrence of "slip" is generally accepted to be much more common for complex/non-Newtonian fluids than for Newtonian/small molecule liquids. In the latter case, we have seen that "slip" generally involves either extreme shear stresses or solid walls that exhibit extremely weak attractive interactions with the liquids, and the issue is primarily one of basic scientific interest. Polymer melts, on the other hand, commonly exhibit apparent manifestations of "slip" that play a critical role in the success or failure of certain types of commercial processing applications.[43]

It appears that slip can occur for polymer melts by means of two distinct mechanisms. One is believed to result from adhesive "failure" at the walls, in which "adsorbed" polymers are literally pulled off of the walls when the tangential stress exceeds some critical value. Empirical evidence from the polymer-processing literature indicates that this mechanism for slip can be "tuned" by the choice of wall materials. A second mechanism for "slip" is believed to be active for systems that exhibit strong adherence between the polymer and the walls. Because of the forces on the polymer caused by flow, the polymer, which exists in an "entangled" state under normal circumstances, can become (temporarily) disentangled, leaving an adsorbed layer of polymer on the wall, but creating a local region of effectively low viscosity between the polymer layer on the wall and the bulk (entangled) polymer fluid. In this case, there is empirical evidence that the fluid may *alternate* between apparent "slip" and "no-slip," suggesting that the entanglement/disentanglement process may be a dynamic one in which the fluid spontaneously alternates between being entangled and unentangled. In either case, when the polymer either loses adherence at the wall or becomes disentangled very near to the wall, there is a fundamental change in the "efficiency" with which the wall is able to inhibit the motion of the polymer, and the bulk polymer motion is consistent with a boundary condition of slip that is at least qualitatively similar to the Navier-slip condition discussed earlier.

Polymeric liquids at an *interface* between two immiscible polymers may also exhibit an "apparent" slip that is due to a similar disentanglement mechanism, driven again by the applied shear stress, but also facilitated by the incompatibility between the two bulk-phase polymers, which minimizes the amount of chain intermingling at the interface. Indeed, statistical mechanical theories of the interfacial region between two bulk polymers show that the condition of "slip" or at least "partial-slip" is likely to be relevant whenever the flow system is one involving relatively small length scales.[44] Two examples are (1) motions involving thin superposed fluid layers and (2) the motions of very small drops. The thermodynamic incompatibility of the two bulk polymers leads to an interfacial region with a modified polymer density. Slip comes from the fact that this interfacial fluid region has an apparent viscosity that is less than that of the two bulk fluids. Hence, when there is a velocity gradient across (i.e., normal to) this region, it tends to be higher than in the two bulk fluids and there is an apparent jump in the bulk-phase velocities across "the interface." The magnitude of this jump depends on the ratio of the viscosity of the interfacial region

Basic Principles

to that of the bulk fluids and on the width of the interfacial region. Hence if the width of the interfacial region becomes comparable with the length scales characteristic of the bulk fluids, the apparent slip associated with this jump is increased.

In summary, we have so far seen that there are two types of boundary conditions that apply at any solid surface or fluid interface: the *kinematic* condition, (2–117), deriving from mass conservation; and the *dynamic* boundary condition, normally in the form of (2–122), but sometimes also in the form of a Navier-slip condition, (2–124) or (2–125). When the boundary surface is a *solid wall*, then $\hat{u}$ is known and the conditions (2–117) and (2–122) provide a sufficient number of boundary conditions, along with conditions at other boundaries, to completely determine a solution to the equations of motion and continuity when the fluid can be treated as Newtonian.

M. FURTHER CONSIDERATIONS OF THE BOUNDARY CONDITIONS AT THE INTERFACE BETWEEN TWO PURE FLUIDS – THE STRESS CONDITIONS

When a bounding surface is a fluid–fluid interface instead of the surface of a solid, the kinematic and dynamic boundary conditions can be seen, from (2–112) and (2–122), to provide either two (or three) independent relationships between the unknown velocity vectors, **u** and **û**. However, there are a total of either four or six unknown components of **u** and **û** (the number depending on whether the flow is 2D or fully 3D), and thus additional conditions must be imposed at an interface to completely specify the solutions of the Navier–Stokes and continuity equations. In this section, we assume that there is no phase change at the interface.

One additional new feature, if the bounding surface is an interface, is that it will generally *deform* (i.e., change shape) when the fluids undergo motion. Hence we cannot *specify* the interface shape, but must determine the shape as part of the solution of a flow problem. In other words, the shape of any interface that appears in a flow problem must be considered as an additional unknown. If we go back and examine the kinematic and dynamic boundary conditions, we see that these also involve the interface shape implicitly, through the normal unit vector **n**. In fact, we shall see shortly that the kinematic condition in the form (2–112) is not complete, because it should involve the interface shape explicitly. This condition is to be applied at the interface, but if $\mathbf{u} \cdot \mathbf{n} = \hat{\mathbf{u}} \cdot \mathbf{n}$ are not zero, then the interface itself must be moving. Only two possibilities are consistent with this. The first is that the whole interface is translating with the *same velocity* $\mathbf{u} \cdot \mathbf{n}(= \hat{\mathbf{u}} \cdot \mathbf{n})$ at every point relative to the chosen reference frame. In this case, the shape of the interface does not change, and we could adopt an alternative reference frame such that $\mathbf{u} \cdot \mathbf{n}(= \hat{\mathbf{u}} \cdot \mathbf{n}) = 0$. The more likely case is that $\mathbf{u} \cdot \mathbf{n}(= \hat{\mathbf{u}} \cdot \mathbf{n}) \ne 0$ and varies from point to point on the interface. In this case, the shape of the interface must change in time, and we may anticipate that there should be a direct relationship between $\mathbf{u} \cdot \mathbf{n}(= \hat{\mathbf{u}} \cdot \mathbf{n})$ and the rate of change of the unknown function that describes the interface shape.

To incorporate the unknown interface shape into the boundary conditions, the most convenient approach is to introduce a scalar function $F(\mathbf{x}, t)$, which describes the interface shape in the sense that the interface is the set of points $\mathbf{x}_s$, such that

$$F(\mathbf{x}_s, t) \equiv 0. \tag{2-127}$$

It may be useful to introduce a couple of examples to illustrate the use of F to describe interface geometry. Let us start with the simplest (trivial) example of an interface that is a flat surface at equilibrium. This would correspond to the interface between two stationary fluids of different density, with the larger density fluid on the bottom. In this case, we can

M. Further Considerations of the Boundary Conditions

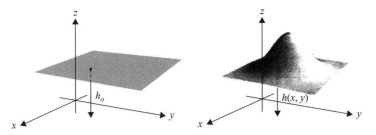

Figure 2–13. A sketch of a flat and deformed interface, showing the definition of the coordinates and the interface shape function as discussed in the text.

introduce a Cartesian coordinate system, with x and y in the plane of the interface and z normal to it, as illustrated in Fig. 2–13. The choice for F in this case is

$$F \equiv z - h_0 \quad \text{all } x, y.$$

Now, let us suppose that there is flow present in the two fluids. In this case, the interface shape is expected to change, again as illustrated in Fig. 2–13, and we would need to generalize to

$$F \equiv z - h(x, y, t) \quad \text{all } x, y,$$

if we have chosen to specify the problem (i.e., equations of motion plus boundary conditions) in the same Cartesian coordinates.

Now, in general, with the interface defined in terms of a function $F(\mathbf{x}, t)$, it is known from analytic geometry that the unit normal to S can be defined in terms of F as

$$\mathbf{n} = \pm \frac{\nabla F}{|\nabla F|}, \tag{2-128}$$

where the sign is chosen so that $\mathbf{n}$ is (by convention) the *outer* unit normal vector. For an interface that is a closed surface (such as the interface of a liquid drop) this means that the sign is chosen so that $\mathbf{n}$ points from the interface outward into the external fluid. For an open interface, such as the flat surface pictured in Fig. 2–13, we can choose $\mathbf{n}$ to be in either direction.

1. Generalization of the Kinematic Boundary Condition for an Interface

Now, because F is a scalar function that is always equal to zero at any point on the fluid interface, its time derivative following *any material point* on the interface [which means that there is no phase transformation occuring so that the velocity $\mathbf{u} = \hat{\mathbf{u}}$ on S according to (2–112) and (2–122)] is obviously equal to zero, that is

$$\frac{\partial F}{\partial t} + \mathbf{u} \cdot \nabla F = 0 \quad \text{for any material point on } S.$$

Thus, rearranging slightly, we obtain

$$\boxed{\frac{1}{|\nabla F|} \frac{\partial F}{\partial t} + \mathbf{u} \cdot \mathbf{n} = 0 \quad \text{on } S.} \tag{2-129}$$

This is the most general form of the kinematic condition. Obviously, in view of (2–112), it can be written in terms of either $\mathbf{u}$ or $\hat{\mathbf{u}}$. The reader should note that if the shape of the

Basic Principles

interface is independent of time and the interface is not translating relative to the origin of the coordinate system, $\mathbf{x} = 0$, then this condition again becomes (2–113), i.e.

$$\boxed{\mathbf{u} \cdot \mathbf{n} = 0 \quad \text{on } S.} \qquad (2\text{–}130)$$

2. The Stress Conditions

For motions of a single fluid involving solid boundaries, we have already noted that the no-slip and kinematic boundary conditions are sufficient to determine completely a solution of the equations of motion, provided the motion of the boundaries is specified. In problems involving two fluids separated by an interface, however, these conditions are not sufficient because they provide relationships only between the velocity components in the fluids and the interface shape, all of which are unknowns. The additional conditions necessary to completely determine the velocity fields and the interface shape come from a *force equilibrium* condition on the interface. In particular, because the interface is viewed as a surface of zero thickness, the volume associated with any arbitrary segment of the interface is zero, and the sum of all forces acting on this interface segment must be identically zero (to avoid infinite acceleration).

To formally express this force equilibrium condition, it is necessary to consider the nature of the forces that act on an interface. For this purpose, we must hypothesize that the interface is characterized by certain *constitutive* properties. In the absence of a molecular theory, the validity of this hypothesis can ultimately be tested only by comparison between predictions obtained by use of the boundary conditions that result from it and measurements of the macroscopic fields in real systems. It is fair to say that the appropriate interface formulation remains a topic of ongoing research, and the correct macroscopic conditions remain uncertain, particularly for fluid systems that involve surface-active solutes in addition to the two primary fluids. However, in this section and elsewhere throughout this book, we adopt the simplest possible description that is consistent with equilibrium thermodynamics – namely, we suppose that the interface can be entirely characterized by a *surface or interfacial tension* that is a function of the local thermodynamic state, such as temperature or pressure, and may also depend on the concentration and nature of any solutes, but is independent of whether the interface is undergoing any macroscopic motion or deformation. We should note that the simple model adopted here assumes implicitly that the interface has no macroscopically significant *dynamical* properties of its own, in spite of the fact that some research suggests that the interface may also have dynamical properties, such as a surface or interface viscosity, that are independent of the corresponding bulk-phase dynamical properties. Although the simple interface description adopted here may be incomplete, I do not believe that current knowledge provides a sufficiently strong motivation to adopt a more complex model. The reader who is interested in a more general prescription of interface properties may wish to refer to the classic paper by Scriven.[45]

According to the simple interface description, which involves only interfacial tension, the forces acting on any segment of an interface are of two kinds: First, there are the bulk pressure and stresses that act on the faces of the interface element and produce a net effect that is proportional to the surface area; and second, there is a tensile force that is due to surface or interfacial tension that acts in the plane of the interface at the edges of the surface element and is specified by means of the magnitude of the surface or interfacial tension as a force per unit length. If we assume that these are the only forces acting on the interface (this is akin to assuming that there are no dynamic forces in the interface itself, as, previously

M. Further Considerations of the Boundary Conditions

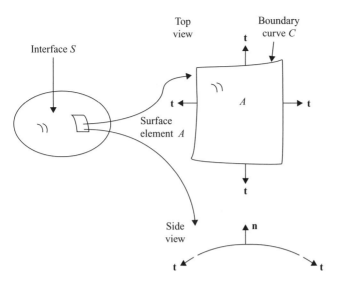

Figure 2–14. A sketch showing the top and side views of an arbitrary surface element, denoted as A, of a fluid interface. The unit normal vector is denoted as **n**, and the unit tangent that is perpendicular to the boundary curve C is denoted as **t**.

explained), it is not difficult to derive a differential, pointwise form for the force equilibrium condition.

Before actually doing this, it may be useful to remind ourselves briefly of the origin of surface tension, as seen from thermodynamics, and explain why it appears in a continuum mechanics theory as a tensile force per unit length in the interface. First, from a thermodynamic point of view, we may recall that interfacial tension is introduced as a measure of a free energy per unit surface area. Thus an increase in the area of the interface requires an increase in the free energy of the system (work). The necessity of doing work to create new interfacial area is a consequence of the fact that the molecules in the immediate vicinity of the interface experience a net force that tends to pull them back into the bulk liquid. This is particularly clear in the case of a gas–liquid interface, where the density of molecules in the gas is very much lower than in the bulk liquid, but it is generally true also at liquid–liquid interfaces. Thus, for example, if we consider the deformation of a drop in a flow, from spherical to some deformed shape, there is an increase in surface area, and this requires work in addition to that associated with the motions of the two fluids. However, in a macroscopic *mechanical* theory, the only way to include the rate of working required for changes in interface area is to assume the existence of a force per unit length that acts at the edges of an interfacial element. The magnitude of this force is γ, the so-called *interfacial tension*.

To express the force equilibrium condition in a mathematical form, we can now consider a force balance on an arbitrary surface element of a fluid interface, which we denote as A. A sketch of this surface element is shown in Fig. 2–14, as seen when viewed along an axis that is normal to the interface at some arbitrary point within A. We do *not* imply that the interface is flat (though it could be) – indeed, we shall see that curvature of an interface almost always plays a critical role in the dynamics of two-fluid systems. We denote the unit normal to the interface at any point in A as **n** (to be definite, we may suppose that **n** is positive when pointing upward from the page in Fig. 2–14) and let **t** be the unit vector that is normal to the boundary curve C and *tangent* to the interface at each point (see

Basic Principles

Fig. 2–14). With these conventions, the force equilibrium condition applied to A requires that

$$\iint_A (\mathbf{T} - \hat{\mathbf{T}}) \cdot \mathbf{n} \, dA + \int_C \gamma \mathbf{t} \, dl = 0. \tag{2-131}$$

Here, $\mathbf{T}$ is the total bulk stress in the fluid above the surface element A evaluated in the limit as we approach the interface, and $\hat{\mathbf{T}}$ is the stress in the second fluid, evaluated as we approach the interface from below. The negative sign associated with $\hat{\mathbf{T}} \cdot \mathbf{n}$ is a consequence of the fact that the unit normal from the interface into the second fluid is $-\mathbf{n}$.

Although the expression (2–131) is a perfectly general statement of the force balance at an interface, it is not particularly useful in this form because it is an overall balance on a macroscopic element of the interface. To be used in conjunction with the differential Navier-Stokes equations, which apply pointwise in the two bulk fluids, we require a condition equivalent to (2–131) that applies at each point on the interface. For this purpose, it is necessary to convert the line integral on C to a surface integral on A. To do this, we use an exact integral transformation (Problem 2–26) that can be derived as a generalization of Stokes' theorem:

$$\int_C \gamma \mathbf{t} \, dl = \iint_A (\mathrm{grad}_s \gamma) \, dA - \int_A \gamma \mathbf{n} (\nabla \cdot \mathbf{n}) \, dA, \tag{2-132}$$

where $\mathrm{grad}_s \equiv \nabla - \mathbf{n}(\mathbf{n} \cdot \nabla)$ is the component of the gradient operator in the local plane of the interface. The proof of this result is straightforward but is not pursued here. What should be noted, however, is the qualitative physical implication of (2–132). To do this, it is useful to think of A as corresponding to an arbitrarily small surface element centered on some point P, with $\mathbf{n}$ thus being the unit normal to the interface at (or arbitrarily near) P. Then we see that the tensile force associated with the action of interfacial tension [represented by the integral over C in (2–131)] contributes a net force at P in the *tangential* direction that is proportional to the gradient of γ at P, plus a net force *normal* to the interface that is proportional to γ times the curvature of the interface at P (that is, $\nabla \cdot \mathbf{n}$). Physically, if the surface element A is flat, the action of surface tension can produce a *net* contribution to the force balance only if $\mathrm{grad}_s \gamma \neq 0$. On the other hand, if the surface is curved, the simple sketch in Fig. 2–14 shows that surface tension acting at the edges of A can produce a net force contribution that is in the direction of the normal at P, and this contribution will be nonzero even if $\mathrm{grad}_s \gamma = 0$. These facts are reflected in the theorem (2–132) from vector calculus.

Now, combining (2–131) and (2–132), we obtain

$$\iint \left[(\mathbf{T} - \hat{\mathbf{T}}) \cdot \mathbf{n} + \mathrm{grad}_s \gamma - \gamma \mathbf{n}(\nabla \cdot \mathbf{n}) \right] dA = 0. \tag{2-133}$$

Thus, because A is an arbitrary surface element in the interface, it follows that

$$\boxed{(\mathbf{T} - \hat{\mathbf{T}}) \cdot \mathbf{n} + \mathrm{grad}_s \gamma - \gamma \mathbf{n}(\nabla \cdot \mathbf{n}) = 0} \tag{2-134}$$

at each point on the interface. This is the differential-stress balance that we seek. For application to the solution of flow problems, it is convenient to discuss separately the components in the normal and the tangential directions.

M. Further Considerations of the Boundary Conditions

3. The Normal-Stress Balance and Capillary Flows

To obtain the normal component, which is generally referred to as the *normal-stress balance*, we take the inner product of (2–134) with $\mathbf{n}$. Recalling that $\mathbf{T} = -p\mathbf{I} + \boldsymbol{\tau}$ and $\hat{\mathbf{T}} = -\hat{p}\mathbf{I} + \hat{\boldsymbol{\tau}}$, this gives

$$\boxed{\hat{p}_{tot} - p_{tot} + \{[(\boldsymbol{\tau} - \hat{\boldsymbol{\tau}}) \cdot \mathbf{n}] \cdot \mathbf{n}\} - \gamma(\nabla \cdot \mathbf{n}) = 0.} \qquad (2\text{–}135)$$

Here, p_{tot} and $\hat{p}_{tot}$ represent the actual *total* pressure in the exterior and interior fluids, including both dynamic and hydrostatic contributions. In crossing an interface, we see that the normal component of the total stress undergoes a jump equal to $\gamma(\nabla \cdot \mathbf{n})$. In the limiting case of *no motion* in the fluids, this implies that

$$\boxed{\hat{p}_{tot} - p_{tot} = \gamma(\nabla \cdot \mathbf{n}).} \qquad (2\text{–}136)$$

This equilibrium condition is known as the *Young–Laplace* equation. The physical significance of (2–136) is that the pressure *inside* a curved interface at equilibrium is larger than that *outside* by an amount that depends on the curvature $\nabla \cdot \mathbf{n}$ and γ. Now the curvature term $\nabla \cdot \mathbf{n}$ can be expressed as the sum of the two principle radii of curvature of S at any point $\mathbf{x}_s$ on the interface, that is,

$$\nabla \cdot \mathbf{n} = \frac{1}{R_1} + \frac{1}{R_2}. \qquad (2\text{–}137)$$

Hence, for a spherical bubble or drop,

$$\boxed{\hat{p}_{tot} - p_{tot} = \frac{2\gamma}{R},} \qquad (2\text{–}138)$$

and the internal pressure is seen to exceed the exterior pressure by $2\gamma/R$. Of course, this result is also well known from equilibrium thermodynamics.

We have obtained the Young–Laplace relationship from normal-stress balance (2–135) by invoking "the limiting case of no motion in the fluids." One might be tempted to suppose that the condition of "no motion" is independent of the interface, e.g., that it is determined by whether there is some source of fluid motion in the bulk-phase fluids away from the interface. However, this is not correct. In fact, if $(\nabla \cdot \mathbf{n}) \neq$ constant on the interface (i.e., independent of position on S), the Young–Laplace equation *cannot* be satisfied, and there *must* be fluid motion so that the balance of normal stresses includes viscous contributions. Utilizing (2–137), we see that the condition $(\nabla \cdot \mathbf{n}) =$ constant requires the sum of R_1^{-1} and R_2^{-1} to be constant on S. Examples of surfaces that satisfy this requirement are a sphere, where $R_1 = R_2 = R$ (the radius); a circular cylinder, where $R_1 = R$, $R_2 = \infty$; and a flat interface, where $R_1 = R_2 = \infty$.

We have said that the Young–Laplace equation cannot be satisfied unless $\nabla \cdot \mathbf{n} =$ constant. The "proof" is more or less trivial. Let us suppose that $\mathbf{u} = 0$. Then, according to the equation of fluid statics, (2–61),

$$\nabla p = \rho \mathbf{g}; \quad \nabla \hat{p} = \hat{\rho} \mathbf{g}.$$

For convenience, let us use a Cartesian coordinate representation of these equations. Then, if ρ, $\hat{\rho} =$ constant,

$$\begin{aligned} p &= -\rho g z + c, \\ \hat{p} &= -\hat{\rho} g z + \hat{c}, \end{aligned} \qquad (2\text{–}139)$$

Basic Principles

where we have assumed $\mathbf{g} = -g\mathbf{i}_z$. Hence,

$$\hat{p} - p = (\rho - \hat{\rho})gz + (\hat{c} - c). \tag{2-140}$$

Now if $\rho \neq \hat{\rho}$, there will exist a "buoyancy force" and the fluids will move until (or unless) all interfaces are normal to $\mathbf{g}$ (as discussed earlier in this chapter). In this case, we must apply the condition (2–135) at the interface simply because the fluid is moving. If, on the other hand, $\rho = \hat{\rho}$, then we see from (2–140) that $\hat{p} - p = $ constant in a motionless fluid, and this requires $\nabla \cdot \mathbf{n} = $ constant to satisfy the Young–Laplace limit of the normal-stress balance. Conversely, if $\nabla \cdot \mathbf{n} \neq$ constant on an interface, there will be (*must* be) motion even though $\rho = \hat{\rho}$ because there is no way to satisfy the normal-stress balance with a motionless fluid. In effect, we are saying that flows will be driven by surface tension whenever there are gradients in $\nabla \cdot \mathbf{n}$. Such flows are usually called *capillary flows*, and $\gamma(\nabla \cdot \mathbf{n})$ is sometimes referred to as the "capillary pressure." Generally, a system of two immiscible fluids (with $\gamma \neq 0$) of the same density can exist in an *equilibrium* state only if the configuration of the interface between them corresponds to a surface of constant curvature. Capillary flows are important in many applications, as we shall see shortly. Studies of capillary flows have received increased attention in recent years because they remain an important source of fluid motion even when $\mathbf{g} \approx 0$ as in the microgravity environment of space shuttle and/or space station.

Although the Young–Laplace equation can be applied in a rigorous sense only to systems in which there is no fluid motion, the concept that surface tension leads to a pressure jump across an interface with nonzero curvature can be used to qualitatively anticipate the nature of many capillary flows. This qualitative use of the Young–Laplace equation is similar in spirit to the use of the hydrostatic pressure distributions to anticipate the nature of gravity-driven flows, such as the gravity front, discussed earlier in this chapter.

Let us begin with a result that is familiar to all of us; namely, the fact that a bubble or drop that is stationary because of the absence of any external force (i.e., either $\rho = \hat{\rho}$ or $\mathbf{g} = 0$) will be *spherical*. This can be explained either as a consequence of the fact that the sphere minimizes surface area (and thus surface free energy for a given γ) or, equivalently, that the sphere is the shape that is consistent with a *constant* hydrostatic pressure difference between the two fluids. According to (2–136), any other shape would induce pressure gradients in addition to hydrostatic pressure gradients in one or both fluids, and thus the fluids could not remain in a static state – the ensuing motion would, in fact, drive the shape toward spherical. If, for example, a spherical drop is deformed slightly to an ellipsoid, the variations of interface curvature will produce capillary motions within the drop that drive it back toward a spherical shape. An illustrative example of the shape-relaxation process may be seen in Fig. 2–15, in which a drop is shown that is initially deformed into the shape of a prolate spheroid. As a consequence, the curvature at the ends is increased relative to that in the middle, and the internal pressure is correspondingly higher at the ends. However, this capillary-induced pressure gradient induces a motion of the fluid from the ends of the drop toward the middle, and this motion causes the drop to return to the equilibrium spherical shape.

The process just described, by which a nonspherical drop would be driven toward the spherical equilibrium shape by pressure gradients associated with variations in interface curvature, is but one example of a large number of situations in which fluid motions are actually caused by pressure gradients that are produced by variations in the curvature of a fluid interface.

The horizontal flat interface configuration between two fluids, the lower of which has a density ρ and the upper a density $\hat{\rho} < \rho$, is also a stable configuration for deformations that are not too large. We have seen earlier in this chapter that gravity forces will cause a humped configuration such as that shown in Fig. 2–13 to go back to a flat configuration. However, even if the two fluids have the same density, surface tension will also drive the

M. Further Considerations of the Boundary Conditions

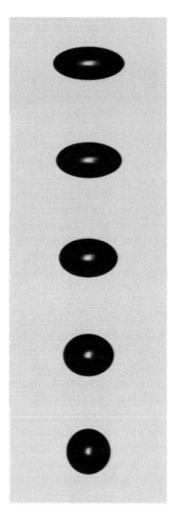
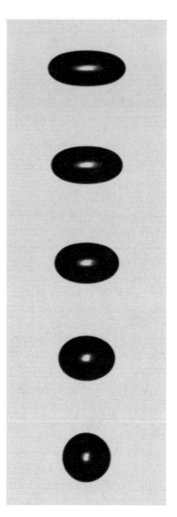

Figure 2–15. Photographs of the relaxation of a pair of initially deformed viscous drops back to a sphere under the action of surface tension. The characteristic time scale for this surface-tension-driven flow is $t_c \equiv \mu R(1+\lambda)/\gamma$. The properties of the drop on the left-hand side are $\lambda = 0.19, \mu_d = 5.5$ Pa s, $\mu = 29.3$ Pa s, $\gamma = 4.4$ mN/m, $R = 187$ μm, and this gives $t_c = 1.48$ s. For the drop on the right-hand side, $\lambda = 6.8$, $\mu_d = 199$ Pa s, $\mu = 29.3$ Pa s, $\gamma = 4.96$ mN/m, $R = 217$ μm, and $t_c = 9.99$ s. The photos were taken at the times shown in the figure. When compared with the characteristic time scales t_c, these correspond to exactly equal dimensionless times ($t^* = t/t_c$): (a) $t^* = 0.0$, (b) $t^* = 0.36$, (c) $t^* = 0.9$, (d) $t^* = 1.85$, (e) $t^* = 6.5$. It will be noted that the drop shapes are virtually identical when compared at the same characteristic times. This is a first illustration of the principle of dynamic similarity, which will be discussed at length in subsequent chapters.

system back to the flat configuration. In the lower fluid, the pressure is largest at the apex of the hump and this will tend to drive the fluid out of the hump, which will then flatten. The reverse is true in the upper fluid – the pressure minimum is at the apex and this will contribute also to flattening of the interface. One can also note that the flat configuration has a smaller surface area, and thus one can again argue for the flat configuration on the basis that the surface free energy is reduced when the hump is removed.

Another configuration that satisfies the Young–Laplace equation is the infinitely long cylindrical thread of constant radius, as indicated previously. However, this configuration *is unstable* to infinitesimal perturbations, which tend to cause it to break into a large number

Basic Principles

Figure 2–16. A sketch of the initial varicose shape of a fluid cylinder that may lead to a capillary-driven instability that causes the cylinder to break into a line of small droplets as described in the text.

of small spherical drops. We easily see the source of this instability qualitatively by considering an initial shape that has a small varicose wavelike departure from the cylinder of constant radius, as sketched in Fig. 2–16. In this case, depending on the wavelength of the perturbations of shape compared with the cylinder radius, there is an obvious mechanism for *growth* of the initial wavelike disturbance of shape. If the total surface curvature is such that the points of minimum radius produce a local maximum in $\nabla \cdot \mathbf{n}$, then there will be a local maximum in the internal pressure at the same point, which tends to drive fluid away from the region of minimum radius toward the regions of maximum radius. But this motion further decreases the radius at the waists and increases the radius between the waists and this produces a varicose shape of increasing amplitude with time that ultimately terminates when the waists approach zero radius and break, causing the original thread to disperse into a line of small drops. This process is known as *capillary instability* and is an important mechanism for breakup of liquid threads, elongated drops, and so forth.

Although one might initially guess that capillary instability should always occur if the thread has a varicose shape like that sketched in Fig. 2–16, this is not true. Instead, the instability occurs only for varicose shapes in which the wavelength λ exceeds a critical value, λ_{crit}. Specifically, short-wavelength perturbations of the shape are actually *stable* – i.e. a short-wave length perturbation will decay in amplitude with time, rather than grow. The reason is that the curvature along the axial direction has the opposite effect of the curvature in the radial direction on the capillary pressure distribution inside the thread. Again, the latter produces a higher pressure in the throats and a lower pressure in the regions of maximum thread radius. However, the z component of the curvature tends to produce a minimum in the internal pressure at the throats relative to that in the outer fluid, because the axial curvature there is convex, and a maximum in the internal pressure in the wide-gap region because the *axial* curvature there is concave. The importance of the axial curvature on the internal pressure is increased relative to the radial curvature as the wavelength is decreased. Hence, for $\lambda < \lambda_{crit}$, the disturbance in shape is driven back toward a constant-radius cylinder. We shall have more to say about the capillary instability later in Chap. 12 on hydrodynamic stability, where a quantitative analysis can be found that leads to detailed predictions of the growth or decay rates as functions of the wavelength of the initial perturbation.

One other capillary-flow problem that is related to the previous problems has to do with the breakup mechanism for viscous drops in flow of a viscous fluid. It is found experimentally that the breakup of large drops often occurs by means of a mechanism in which the drops are first deformed by the flow into an elongated threadlike shape and then break into smaller pieces by mechanisms that are caused by capillary flows. A common case scenario is that the drop is first elongated in some part of a blender or emulsifier where the flow is strong, but then breaks after it moves to a different region where the flow is weak. One might be tempted to assume that the capillary instability would be the source of breakup for an elongated drop and that the processes described previously would apply directly. However, the situation is more complicated because an elongated drop differs in a fundamental way from an infinite thread, and that is that it has closed ends. Let us imagine, for a moment, that the drop shape can be described as a circular cylinder of radius R, with hemispherical ends,

M. Further Considerations of the Boundary Conditions

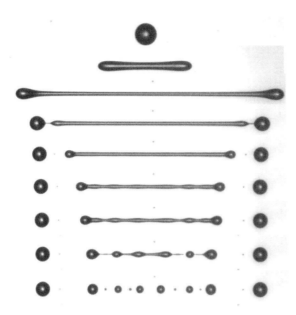

Figure 2–17. A series of photographs of the time-dependent breakup of a drop by the capillary-driven flow mechanism that is known as "end-pinching." The drop is initially stretched in a flow, and then the flow is stopped, allowing the drop shape to evolve toward breakup. The viscosity of the drop relative to the suspending fluid is 1.2.

also of radius R. If we estimate the internal pressure by using the Young–Laplace equation, it is evident that the pressure in the cylindrical region is γ/R larger than the external pressure, but the pressure in the hemispherical ends exceeds the external pressure by $2\gamma/R$. Hence the addition of the closed ends means that the elongated drop is not an equilibrium configuration, but that there will be a flow that pumps fluid from the ends toward the center of the drop. In this sense, the highly elongated drop appears similar to the spheroidal drop described earlier. However, the situation is actually somewhat more complicated. In the present case the capillary pressure gradient exists only near the ends of the drop. Hence the flow produced by that gradient is relatively localized at the ends. As fluid is pumped from the ends, the effect is to produce bulbous ends, as shown in Fig. 2–17, with little or no fluid initially going into the central threadlike region. In fact, it might seem at first that a potential equilibrium shape satisfying the Young–Laplace equation could be achieved by producing spherical ends that have a radius that is $2R$. In this case, the capillary pressure increment in both the central cylinder and in the spherical bulbous ends is γ/R. However, it turns out that there is no way to smoothly connect a sphere of radius $2R$ and a cylinder of radius R without going through a local transition zone where the total curvature exceeds $1/R$. Hence, in this local transition, the capillary pressure is a little larger than in either the central cylinder or the ends of the drop, and this tends (locally) to drive fluid out of this transition zone, both into the cylindrical region and back into the bulbous ends. This flow produces a local constriction or throat in which the capillary pressure is even larger, and this continues at an increasing rate until the throat actually pinches off. At this point the center fragment is a new now-shortened cylindrical thread, and the process repeats. The breakup process just described is called "end-pinching." While this is happening, however, the central part of the cylindrical thread remains essentially unperturbed – with little or no flow inside. This thread is subject to capillary instability triggered by infinitesimal perturbations of shape. Hence the fate of the central region depends on whether the instability leads to breakup before the drop is completely consumed by a sequence of end-pinching processes. As the drop becomes increasingly elongated, the time for the succession of end-pinching events to envelop the whole drop increases relative to the time scale for capillary instability, and for a very long drop, the majority of the drop may actually break up because of capillary instability.

Of course, all of the discussion of capillary flows in this section is qualitative. Although there are many instances in which this type of insight can be extremely useful, it is important to remember that quantitative details can only be obtained via solutions of the equations of motion.

4. The Tangential-Stress Balance and Thermocapillary Flows

To this point, we have considered only the component of the stress balance (2–134), in the direction normal to the interface. There will be, in general, two tangential components of (2–134), which we obtain by taking the inner product with the two orthogonal unit tangent vectors that are normal to **n**. If we denote these unit vectors as $\mathbf{t}_i$ (with $i = 1$ or 2), the so-called *shear-stress balances* can be written symbolically in the form,

$$0 = \left[(\boldsymbol{\tau} - \hat{\boldsymbol{\tau}}) \cdot \mathbf{n}\right] \cdot \mathbf{t}_i + (\text{grad}_s \gamma) \cdot \mathbf{t}_i \tag{2–141}$$

where $\text{grad}_s \equiv \nabla - \mathbf{n}(\mathbf{n} \cdot \nabla)$, as noted earlier.

We see that the shear- (i.e., tangential-) stress components are *discontinuous* across the interface whenever $\text{grad}_s \gamma$ is nonzero. Now, the interfacial tension for a two-fluid system, made up of two pure bulk fluids, is a function of the local thermodynamic state – namely, the temperature and pressure. However, it is much more sensitive to the temperature than to the pressure, and it is generally assumed to be a function of temperature only. If the two-fluid system is a multicomponent system, it is often the case that there may be a preferential concentration of one or more of the components at the interface (for example, we may consider a system of pure A and pure B, which are immiscible, with a third solute component C that is soluble in A and/or B but that is preferentially attracted to the interface), and then the interfacial tension will also be a function of the (surface-excess) concentration of these solute components. Both the temperature and the concentrations of adsorbed species can be functions of position on the interface, thus leading to spatial gradients of γ.

If we consider, first, the case with $\text{grad}_s \gamma \equiv 0$, we see that the tangential-stress balance requires *continuity* of the tangential stress. If the fluids are Newtonian, this condition can also be written in terms of the rate of strain, in the form

$$\left[(\mathbf{E} - \lambda \hat{\mathbf{E}}) \cdot \mathbf{n}\right] \cdot \mathbf{t}_{(i)} = 0, \tag{2–142}$$

where λ is the ratio of viscosities $\hat{\mu}/\mu$. Hence we see that, if $\lambda \neq 1$, the velocity gradient will be discontinuous even though the stress is continuous. In particular, if $\lambda \approx 0$, as if we had a bubble or a drop with a very low interior viscosity, then

$$\left[(\mathbf{E}) \cdot \mathbf{n}\right] \cdot \mathbf{t}_{(i)} \approx 0 \tag{2–143}$$

in the fluid with larger viscosity. When the interface is a gas–liquid interface, the condition (2–143) applies approximately to the liquid, and is often known as the "zero-shear-stress" condition (though it is strictly the velocity gradient that is approximately zero rather than the stress). If we consider that the fluids are not moving at all, then $\mathbf{E} = \hat{\mathbf{E}} = 0$, and the tangential-stress balance is satisfied *automatically*. Unlike the normal-stress balance, which requires that the Young–Lapace equation be satisfied in an equilibrium (or stationary) system, there is no equilibrium requirement stemming from (2–141) provided $\text{grad}_s \gamma = 0$ as we have assumed in writing (2–142).

The converse situation, however, is that the condition (2–141) absolutely requires that there be flow whenever $\text{grad}_s \gamma \neq 0$. A system with $\text{grad}_s \gamma \neq 0$ at some initial instant, but

M. Further Considerations of the Boundary Conditions

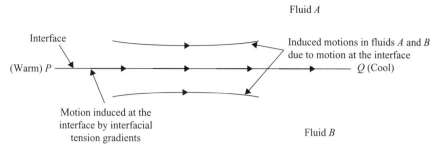

Figure 2–18. A sketch of the motion driven by gradients of temperature and thus of interfacial tension on a fluid interface. The net tensile-stress imbalance tends to pull the interface from the warm side toward the cool side (i.e., in the direction of increasing interfacial tension).

no sources of flow other than this, and no mechanism to maintain temperature gradients or solute concentration gradients at the interface, will undergo a flow that leads eventually to the equilibrium state, $\text{grad}_s \gamma = 0$. In the absence of other mechanisms for producing flow, the motion that occurs as a consequence of a nonzero gradient of γ can be dominant. The role of solute in establishing conditions with $\text{grad}_s \gamma \neq 0$ will be discussed shortly. Here, we focus on the most common and simplest situation, of surface-tension-driven flows that are caused by gradients in temperature. Such flows are usually called *thermocapillary flows* (or sometimes *Marangoni flows* for reasons that will become evident shortly). Under normal Earth's gravity, the dominant effect of nonisothermal conditions is often to produce *buoyancy*-driven flows, caused by the body-force terms in the equations of motion because of the dependence of the density on temperature. We shall study some examples of this type of flow later. In zero g, or microgravity conditions, buoyancy-driven flows are negligible, even for strongly nonisothermal conditions, but there can still be quite important flows driven by surface-tension gradients when there is an interface present.

Calculation of those flows, by means of the equations of motion, can be moderately complex. The simple part is generally to specify the information necessary to apply the boundary condition, (2–141). For this we require

$$\gamma = \gamma(\theta), \tag{2-144}$$

i.e., the interfacial tension as a function of temperature. In fact, the critical parameter is

$$\frac{d\gamma}{d\theta} = -\beta < 0. \tag{2-145}$$

This measures the sensitivity of γ to variations of θ and as indicated it is a negative quantity. This means that the interfacial tension *decreases* as the temperature increases, and vice versa as the temperature decreases. Then

$$\text{grad}_s \gamma = -\beta \, \text{grad}_s \theta. \tag{2-146}$$

The complexity of detailed calculations of thermocapillary flows is the coupling between the velocity and temperature fields, primarily by means of the boundary condition, (2–146). To determine $\text{grad}_s \gamma$ (and thus $\mathbf{u}$ by means of the equation of motion), we need to solve the thermal energy equation (2–110) to determine the temperature distribution. However, to do that, we generally need $\mathbf{u}$ because it appears as a coefficient in (2–110). We shall later consider some simple problems of this type.

It is useful to realize that we can understand a number of thermocapillary problems from a *qualitative* point of view by simply understanding how nonzero interfacial tension gradients lead to fluid motion. Suppose, for simplicity, we consider a flat surface element,

Basic Principles

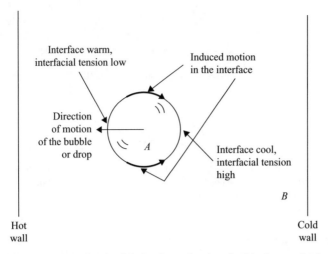

Figure 2–19. A sketch of the configuration described in the text for thermocapilllary-driven motion of a bubble or drop through a fluid at zero **g**.

which is depicted end-on in Fig. 2–18. Let us further suppose that the temperature varies along the surface element, being higher on the side denoted by P and lower at Q. The macroscopic effect of interfacial tension can be modeled as a tensile force per unit length acting at the edges of the surface element in the plane (locally) of the surface. In the system pictured, the interfacial tension will be lower on the side P where θ is higher, and higher at Q where the temperature is lower. Hence the surface element of the interface will feel an imbalance of tensile force, which will tend to make it move from point P toward Q, pulling the bulk fluids adjacent to the interface with it. If we keep this simple picture in mind we can at least qualitatively understand the behavior of most thermocapillary flows. Let us consider a couple of well-known problems that we will later analyze in complete detail.

The first is the thermocapillary-driven motion of either a gas bubble or a liquid drop through a second bulk liquid. For simplicity we may consider that we are at zero **g** so that there is no buoyancy-driven motion due to the density difference between the drop (or bubble) and the surrounding liquid. In two-phase systems consisting of one immiscible liquid A (or a gas) dispersed as drops (or bubbles) in a second liquid B, it is frequently important to be able to separate the system into bulk A and bulk B. In a gravitational environment, if there is a density difference between A and B, the simplest way to accomplish such a separation is by means of buoyancy-driven motion that causes the drops (or bubbles) to either rise up or sediment to the bottom of the vessel where they can coalesce and form a single bulk phase. In the absence of gravity, or if the two fluids happen to have precisely the same density so that there is no buoyancy, thermocapillary-driven motions of the bubbles or drops offer a useful alternative. To see how this works, let us consider a single drop or bubble at zero **g** suspended in a fluid that is heated on one side and cooled at the other. The situation is depicted in Fig. 2–19. Because of the conduction of heat, a temperature gradient will be established in the liquid, and this, in turn, will produce a temperature gradient at the interface of the drop or bubble. As a consequence, there will be a gradient in γ. The interfacial tension will be larger on the colder side of the drop or bubble and smaller on the warmer side. This means that, at all points on the interface, there will be a tensile-stress imbalance that tends to pull surface elements from the warm to the cool side of the drop or bubble. As a consequence the nearby exterior fluid will also be pulled along from the high- to the low-temperature direction, and the bubble or drop will "swim" in the opposite direction, i.e., toward the heated side of the container. In fact, this thermocapillary migration

M. Further Considerations of the Boundary Conditions

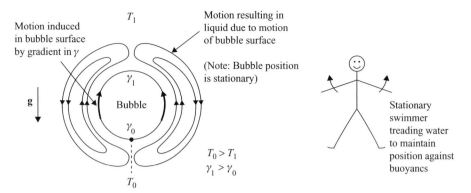

Figure 2–20. A sketch of the problem of Young *et al.* of a gas bubble that is held at a fixed position in a gravitational field because of the action of a surface-tension gradient at its surface (caused by a vertical gradient of temperature in the tank). The temperature decreases in the vertical direction. This means that the interfacial tension is higher at the top of the bubble and lower at the bottom. As a result, the interface is pulled from the bottom toward the top, dragging fluid with it, as shown by the sketch of fluid pathlines in the vicinity of the bubble. This causes the bubble to "swim" downward against the upward buoyant force produced by gravity, much as a swimmer pulling himself downward by the upward motion of his arms.

of bubbles or drops from the cooler to warmer fluid is one of the key mechanisms used in low-gravity space environments for multiphase separations. Of course, the same mechanism will also be present in a gravitational environment and will often compete with or assist buoyancy-driven motions of bubbles or drops in a suspending liquid.

A very striking example of thermocapillary effects, which we shall analyze in detail in Chap. 7, was discovered over 40 years ago by Young and co-workers (1959).[46] If we suspend a small gas bubble in a tank filled with liquid under normal gravitational conditions, the bubble will rise at some velocity. However, if we now heat the tank from below in a way that establishes a stable vertical temperature gradient, the thermocapillary effect will add a tendency for the bubble to swim down toward the bottom, as illustrated pictorially in Fig. 2–20. If the temperature gradient is large enough, these two effects (buoyancy-driven motion up and thermocapillary migration down) may cancel, causing the bubble to stay at a fixed position. The effect is analogous to a swimmer who stays (temporarily) at some underwater depth by treading downward to compensate for the effect of buoyancy.

A second example of thermocapillary flow is the famous instability problem originally studied by the Italian scientist, Marangoni, after whom this class of flows is sometimes named. This concerns the motion in a thin layer of liquid, with a gas–liquid interface at the top and a solid wall below that is heated above the ambient gas temperature. For a relatively small temperature difference across the liquid layer, $\Delta\theta$, there is no motion, only a linear decrease of the temperature as the heat is transferred by conduction. At some critical $\Delta\theta$, however, there is an abrupt transition to convection in the fluid layer. At first, one might assume that the onset of convection is due to buoyancy forces caused by the fact that the density of the heated fluid at the bottom of the fluid layer is lower than the density of the cooler fluid at the top. However, though buoyancy can indeed cause a stationary layer of fluid to undergo a transition to convective motion (and we will analyze the details of this process in Chap. 12), it is found that there is another more dominant mechanism for very *thin* fluid layers associated with thermocapillary effects at the upper interface. (We may also note that the buoyancy-driven mechanism would be completely absent at zero **g**, but the thermocapillary mechanism would be unchanged.) Of course, it is not possible to provide any quantitative results for the value of $\Delta\theta_{critical}$ or other features without carrying out a detailed analysis (as we shall do in Chap. 12). However, we can examine the problem

Basic Principles

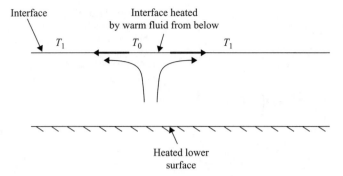

Figure 2–21. A sketch of an intial flow perturbation that can lead to instability and strong cellular motion due to Marangoni instability in a fluid layer that is heated from below.

qualitatively on the basis of our understanding of thermocapillary flows. The first question is whether a mechanism exists to explain the onset of convection by means of instability of the basic, stationary fluid layer that is heated from below. What is meant by instability here is whether an infinitesimal fluid motion, caused by some perturbation of the fluid layer, will increase in amplitude with time (and thus lead to significantly large fluid velocities) or decrease in amplitude because of the dissipative effect of viscosity. Let us consider a specific type of initial perturbation (or disturbance) that consists of a local cellular flow, as sketched in Fig. 2–21, that brings fluid up from the bottom in the center of the cell and returns fluid to the bottom at the periphery. For simplicity, we completely neglect any buoyancy effects. Then, if the upper surface were a solid wall, there would be no mechanism to sustain such a flow, and it would decay in amplitude with time until the system returned to the stationary base state. Because the upper surface in our case is an interface, however, we must pay careful attention to the role of thermocapillary effects. Indeed, if we examine these, we see that they add a mechanism that will tend to sustain the initial cellular flow, and, if strong enough, to lead to increased flow rates and instability in the sense described earlier. Now, the upwelling fluid in the center of our cell comes from the bottom of the fluid layer and hence arrives at the interface with a temperature that exceeds that of the fluid around it. Indeed, as the hot plume spreads across the layer, it is gradually cooled by interaction with the air above. Hence there is a temperature gradient established and a corresponding surface-tension gradient with the smallest values of γ at the center of the cell and the largest values at the periphery. However, this establishes a tensile-stress imbalance that tends to draw fluid across the top of the flow cell in the same direction as the initial perturbation. As $\Delta\theta$ across the layer is increased, the temperature and interfacial tension gradients at the interface are also increased, and the strength of the thermocapillary flow contribution is enhanced. Not surprisingly, above some critical $\Delta\theta$, the effect becomes strong enough to sustain and even enhance the initial perturbation, leading to growth of the velocities with time, and thus instability. One question that may occur to the reader is whether an initial perturbation of the type considered will ever occur? This question deserves a more detailed response than is possible here (see Chap. 12). The answer boils down to the fact that the amplitude of the disturbance is infinitesimal, by which we mean arbitrarily small. Any real system is subject at all times to *small* perturbations, and any arbitrarily chosen infinitesimal perturbation will always be present, including the one chosen previously for illustrative purposes. Hence, if there is *any* arbitrarily small disturbance that will grow at some value of $\Delta\theta$, the whole system is said to be "unconditionally unstable" because it is *impossible* to shield the system from an unstable perturbation that is arbitrarily small.

N. THE ROLE OF SURFACTANTS IN THE BOUNDARY CONDITIONS AT A FLUID INTERFACE

We have noted that the interfacial tension between two immiscible fluids can be modified because of the presence of solutes. Especially important in this regard are the solutes that are known as surfactants (or "surface-active agents"). These are typically molecules with two distinct chemical moieties, each of which (on its own) would be soluble in one of the two bulk fluids, and more or less insoluble in the other. When one of the two fluids is water, the portion of the surfactant that prefers the water is known as *hydrophilic*, whereas the part that prefers the other liquid is known as *hydrophobic*. Hence part of the surfactant molecule would like to be in fluid A and part in fluid B. The result is that there is a strong tendency for the surfactant to accumulate at the interface between A and B, where each part can be more satisfied with its chemical environment than if the surfactant were wholly in either fluid A or B. Not only do surfactants tend to accumulate at interfaces, but their presence generally results in a strong decrease in the interfacial tension relative to the value of a clean AB interface. The fact that the interfacial tension is decreased certainly makes qualitative sense if we recall the interpretation of γ as the surface free energy that measures the work required to achieve an increase of interfacial area.

There is a wide variety of materials that can act as surfactants at a given fluid–fluid interface. One class of particular relevance when one of the two fluids is liquid water is common soaps. In a system of two immiscible polymers, a diblock copolymer containing a segment of each of the two bulk-phase polymers will act as a surfactant, as each part of the copolymer will clearly prefer its own bulk phase. The vast majority of commercially available surfactants are intended for use in water. This is not only because water is such a common and important liquid, but also is a consequence of the fact that the interfacial tension of water-based systems is generally larger than for other liquid–liquid or liquid–gas interfaces. This means that a larger absolute change in interfacial tension is possible, and thus surfactants can have their most profound effects in water-based systems. In general, the addition of surfactant may introduce fundamental changes in the form of the stress balance, (2–134), at a fluid interface. When the surfactant coverage on the interface is high, or when two or more surfactants are present so that there are surfactant–surfactant interactions, the assumption inherent in (2–134) that the interface can be completely characterized by the interfacial tension is likely to break down. In particular, the interface may exhibit a finite *interfacial viscosity* (i.e., a rate-dependent resistance to shear or velocity gradients within the plane of the interface), or an even more complex *interfacial rheology*. When two or more surfactants are present, these may sometimes also interact in such a way as to undergo a phase transition with increasing concentration from a mobile "fluidlike" condition to a relatively immobile elastic film. The study of surfactant systems, and their phase and dynamical behavior at a fluid interface is a major area of research and study in colloid and surface physics, and is beyond the scope of the present book.[47]

Here, we consider only the simpler situation in which the surfactant is assumed to be relatively dilute so that it is mobile on the interface and contributes a change only in the interfacial tension, without any more complex dynamical or rheological effects. In this case, the boundary conditions derived for a fluid interface still apply. Specifically, the dynamic and kinematic boundary conditions, in the form (2–122) and (2–129), respectively, and the stress balance, in the form (2–134), can still be used. However, the interfacial tension, which appears in the stress balance, now depends on the local concentration of surfactant. We shall discuss how this concentration is defined shortly. First, however, we note that flows involving an interface with surfactant are qualitatively similar to thermocapillary flows. The primary difference is that the concentration distribution of surfactant on the interface is almost always dominated by convection and diffusion *within the interface*, whereas the interface *temperature* is assumed to be determined by heat transfer effects in the two bulk fluids [note from Eq. (2–112) that $\theta = \hat{\theta} \to \theta_s$]. Because a surfactant is often sparingly soluble in the bulk fluids, or even insoluble in some cases, transport processes in the bulk-phase fluids are often not significant factors in determining the concentration distribution at the interface.

We specify the concentration of a solute in a bulk fluid as mass per unit volume denoted as c. As before, we designate the position of the interface so that there is no "surface excess" of the two pure bulk solvents. The concentration of surfactant, which we may denote as c_A and c_B, respectively, in the two bulk fluids, will exhibit a peak in the vicinity of the "interface" between A and B. Hence, in a macroscopic continuum description, it is necessary to include a "surface excess" of surfactant "on" the interface. If we denote the local concentration as $c(\mathbf{x}, t)$, again defined as a mass per unit volume, then the total mass of surfactant per unit area of interface is

$$\int_{-L}^{L} c(x, y, z, t)dz,$$

where we denote the coordinate direction normal to the interface as z and introduce L as a distance normal to the interfaces where c can be approximated within some small error as either c_A or c_B. The total quantity of surfactant between $-L$ and L in the two bulk phases, estimated by use of the macroscopic continuum description, is

$$(c_A + c_B)L.$$

The difference,

$$\Gamma \equiv \int_{-L}^{L} c(x, y, z, t)dz - (c_A + c_B)L, \quad (2\text{--}147)$$

must be assumed, in a macroscopic (continuum) description, to reside "on" the interface, as a surface-excess concentration, Γ, defined as mass of surfactant per unit area.

An important principle from thermodynamics is the well-known *Gibbs adsorption isotherm*, which provides a relationship among the bulk-phase concentration c, the interfacial concentration Γ, and the interfacial tension (essentially the surface free energy per unit area):

$$\Gamma = -\frac{c}{RT}\frac{\partial \gamma}{\partial c}. \quad (2\text{--}148)$$

It should be noted that (2–148) holds at equilibrium, where c is the uniform bulk concentration. However, for systems with surfactant under flow, we shall assume that this expression is also valid but with Γ and γ both interpreted as local values and c representing the limit of the bulk concentration as we approach the same point on the fluid interface. For a surfactant, which tends to adsorb at the interface, we see that the interfacial tension is, in general, a function of Γ:

$$\gamma = \gamma(\Gamma). \quad (2\text{--}149)$$

However, if we are going to do anything with this, we need to discuss how Γ is related to c, as well as how Γ varies as a function of position along the interface. This is determined by the adsorptive–desorptive flux from the surrounding fluid, as well as the transport of surfactant both in the bulk fluids as well as along the interface.

A reasonable model for the net flux of material to and from the interface is to assume that the desorption rate is linear in the interface concentration Γ, whereas the adsorption rate

N. The Role of Surfactants in the Boundary Conditions at a Fluid Interface

is linear in the bulk concentration, as well as linear in the space remaining on the interface, in which the maximum possible interface concentration is Γ_∞. Hence the net flux to the interface is

$$-\mathbf{j} \cdot \mathbf{n} = \delta[c(\Gamma_\infty - \Gamma)] - \alpha\Gamma, \qquad (2\text{--}150)$$

where

$$\delta = \delta_0 \exp(-E_a/RT),$$
$$\alpha = \alpha_0 \exp(-E_d/RT),$$

R is the ideal gas constant, T is the absolute temperature, and E_a and E_d denote energies of activation for adsorption and desorption, respectively. At equilibrium, the net flux is equal to zero, and we can solve (2–150) for the interface concentration:

$$\Gamma_{eq} = \frac{\Gamma_\infty c}{b + c}, \quad \text{where } b \equiv \alpha/\delta. \qquad (2\text{--}151)$$

If E_a and E_d are constants equal to $E_{a,0}$ and $E_{d,0}$, then

$$b = b_0 = (\alpha_0/\delta_0) \exp\left[(E_{d,0} - E_{a,0})/RT\right],$$

and this is known as the *Langmuir adsorption isotherm*. On the other hand, nonideal interactions between the adsorbed surfactant can lead to energy barriers that depend on the local concentration. If these can be modeled as linear,

$$E_a = E_{a,0} + \nu_a \Gamma,$$
$$E_d = E_{d,0} + \nu_d \Gamma, \qquad (2\text{--}152)$$

with $\nu_i \equiv (\partial E_a/\partial \Gamma)_{\Gamma=\Gamma_{eq}}$, then the solution of (2–150) at equilibrium becomes

$$\Gamma_{eq} = \frac{\Gamma_\infty c}{b_0 \exp(-\chi \Gamma_{eq}/\Gamma_\infty) + c}, \quad \text{with } \chi \equiv \frac{(\nu_d - \nu_a)\Gamma_\infty}{RT}. \qquad (2\text{--}153)$$

This is known as the *Frumkin adsorption isotherm*. Obviously, for $\chi = 0$ it reduces to the Langmuir form. If χ is negative, $\nu_a > \nu_d$, and the cost of adsorption increases faster than desorption and relative to the case of constant E_i, the system is "repulsive." For $\chi > 0$, on the other hand, the system is "cohesive."

The isotherms (2–151) or (2–153) can now be combined with the Gibbs' expression, (2–148), to obtain the relationship between the equilibrium surface tension and the equilibrium interfacial concentration:

$$\gamma_{eq} = \gamma_s + RT\Gamma_\infty \left[\ln\left(1 - \frac{\Gamma_{eq}}{\Gamma_\infty}\right) + \frac{1}{2}\chi\left(\frac{\Gamma_{eq}}{\Gamma_\infty}\right)^2\right]. \qquad (2\text{--}154)$$

Here we denote the interfacial tension associated with the clean interface as γ_s. Again, this reduces to the Langmuir-based expression if $\chi = 0$. We note that this expression is used even if the system is undergoing flow and the interface concentration is not uniform. It is

Basic Principles

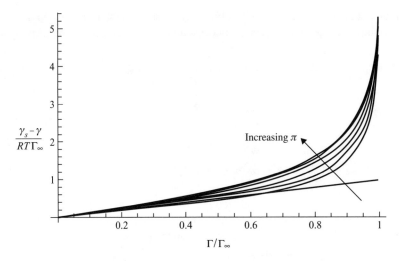

Figure 2–22. The equation of state for the dependence of the interfacial tension on the interface concentration of surfactant, plotted as $\{\ln[1 - (\Gamma/\Gamma_\infty)] + 0.5\chi(\Gamma/\Gamma_\infty)^2\}$ for $\chi = 0, 0.1, 0.5, 1.0, 1.5,$ and 2.0. Also shown is the linear approximation, (2–156), $(\gamma_s - \gamma)/RT\Gamma_\infty = \Gamma/\Gamma_\infty$.

assumed that the system is in "local" equilibrium so that the same exact expression holds with γ and Γ replacing γ_{eq} and Γ_{eq}, i.e.,

$$\gamma = \gamma_s + RT\Gamma_\infty \left[\ln\left(1 - \frac{\Gamma}{\Gamma_\infty}\right) + \frac{1}{2}\chi\left(\frac{\Gamma}{\Gamma_\infty}\right)^2 \right]. \qquad (2\text{–}155)$$

If the surfactant concentration Γ is sufficiently small, a linear relationship exists between γ and Γ, in which case (2–155) is typically written as

$$\gamma_s - \gamma = \Gamma RT. \qquad (2\text{–}156)$$

A comparison of the full nonlinear relationship (2–155) with the linear form (2–156) is shown in Fig. 2–22. Although (2–156) is often used in fluid dynamic calculations because of the simplicity that derives from the fact that it is linear, it can be seen from Fig. 2–22 that it requires $\Gamma/\Gamma_\infty \ll 1$. The strong reduction in the interfacial tension as $\Gamma \to \Gamma_\infty$ often has important consequences in flow systems, because the convection of surfactant on the interface often produces local regions of high concentration, but these will all be missed if the linear form (2–156) is used. Further discussion of surfactant adsorption at the fluid interface can be found either in classic textbooks[48] or the research literature (e.g., Pawar and Stebe).[49]

In general, the surfactant is distributed along the interface by a combination of convection and diffusion, as well as transport to and from the interface from the bulk solvents. However, in many cases, the solubility of a surfactant in the two solvents is very low, and a good approximation is that the transport from the solvents is negligible. In this case, it is said that the surfactant is an "insoluble surfactant," and the total quantity of surfactant on the interface is conserved. We have not previously derived a bulk-phase conservation equation to describe the transport of a solute in a solvent. Hence in this section we adopt the "insoluble" surfactant case, and follow Stone[50] in deriving a surfactant transport equation that relates only to convection and diffusion processes on the interface.

N. The Role of Surfactants in the Boundary Conditions at a Fluid Interface

The starting point is a conservation principle for the mass of surfactant within a material surface element that we can denote as $S_m(t)$. In the absence of any sources or sinks, either because of chemical reactions or a flux to or from the surrounding bulk-phase fluids, and, neglecting diffusion temporarily, we may write a "surface" mass balance in analogy with (2–6):

$$\frac{D}{Dt} \int_{S_m(t)} \Gamma \, dS = 0, \qquad (2\text{–}157)$$

where D/Dt is a material derivative for points on the interface. Now, it may be shown, by generalization of the Reynolds transport theorem to a surface rather than to a volume integral, that

$$\int_{S_m(t)} \left[\frac{\partial \Gamma}{\partial t} + \nabla_s \cdot (\Gamma \mathbf{u}) \right] dS = 0, \qquad (2\text{–}158)$$

where

$$\nabla_s \equiv \nabla - \mathbf{n}(\mathbf{n} \cdot \nabla)$$

is the gradient operator "in the plane" of the interface. Hence

$$\frac{\partial \Gamma}{\partial t} + \nabla_s \cdot (\Gamma \mathbf{u}) = 0 \qquad (2\text{–}159)$$

at any point on the interface. Finally, because $\mathbf{u}$ can be expressed as the sum of a component in the interface $\mathbf{u}_s$, and one normal to it,

$$\mathbf{u} = \mathbf{u}_s + (\mathbf{u} \cdot \mathbf{n})\mathbf{n}, \qquad (2\text{–}160)$$

it follows that

$$\frac{\partial \Gamma}{\partial t} + \nabla_s \cdot (\Gamma \mathbf{u}_s) + \Gamma(\nabla_s \cdot \mathbf{n})(\mathbf{u} \cdot \mathbf{n}) = 0. \qquad (2\text{–}161)$$

Hence, in the absence of diffusion, which has been neglected in deriving (2–161), we can see that there are two contributions to change in Γ at any point; one is simple convection with the interface velocity $\mathbf{u}_s$ and the second turns out to be "dilution" – in the sense that local changes in interfacial area must lead to changes in Γ. If we add a diffusive contribution, Eq. (2–161) becomes

$$\boxed{\frac{\partial \Gamma}{\partial t} + \nabla_s \cdot (\Gamma \mathbf{u}_s) + \Gamma(\nabla_s \cdot \mathbf{n})(\mathbf{u} \cdot \mathbf{n}) = D_s \nabla_s^2 \Gamma.} \qquad (2\text{–}162)$$

The diffusion term on the right-hand side assumes that the flux of surfactant that is due to random Brownian motions is linear in the gradient of Γ and that the process is isotropic. In writing a complete mass balance for any species, it is necessary to include the possibility of a mean flux that is due to random molecular motion in the presence of a gradient in concentration – just as it was necessary to hypothesize a heat flux vector in the conservation of energy balance. The relationship

$$\mathbf{j}_s = -D_s \nabla_s \Gamma \qquad (2\text{–}163)$$

is known as *Fick's law* and is the constitutive equivalent of the Fourier heat conduction law. The diffusivity that appears in (2–162) and (2–163) is called a "surface diffusivity" – and is generally believed to be distinct from the diffusivity for the same species in either of the bulk phase fluids – but it is extremely difficult to measure.

Basic Principles

Finally, if we assume that the surfactant is soluble in the bulk-phase fluids, we must add a flux to and from the interface to the surfactant mass balance, which will appear in (2–162) as a net source (or sink) *term*, i.e.,

$$\frac{\partial \Gamma}{\partial t} + \nabla_s \cdot (\Gamma \mathbf{u}_s) = D_s \nabla_s^2 \Gamma - \Gamma(\nabla_s \cdot \mathbf{n})(\mathbf{u} \cdot \mathbf{n}) + \mathbf{j} \cdot \mathbf{n}. \tag{2–164}$$

We have previously written an expression for $\mathbf{j} \cdot \mathbf{n}$ in Eq. (2–150), but this expression is in terms of the local bulk concentration evaluated at the interface, c, and thus to determine c we would need to solve bulk-phase transport equations. We will not pursue that subject here. However, when we use this material to solve flow problems, we will consider several cases for which it is not necessary to solve the full convection–diffusion equation for c. We will see that the concentration of surfactant tends to become nonuniform in the presence of flow – i.e., when $\mathbf{u} \cdot \mathbf{n}$ and $\mathbf{u}_s$ are nonzero at the interface. This tendency is counteracted by surface diffusion. When mass transfer of surfactant to and from the bulk fluids is added, this will often tend to act as an additional mechanism for maintenance of a uniform concentration Γ. This is because the rate of "desorption" from the interface will tend to be largest where Γ is largest, and the rate of "adsorption" largest where Γ is smallest.

Now, from a qualitative point of view, the role of surfactant is twofold: One is to simply reduce the interfacial tension everywhere by an amount that depends on the mean value of Γ, and the second is to produce Marangoni effects that are due to flow-induced nonuniformity in Γ. As a consequence, it is convenient to define Γ as being

$$\Gamma = \Gamma_{eq} + \Gamma', \tag{2–165}$$

where Γ_{eq} is the equilibrium concentration of surfactant (and is therefore a constant) and Γ' is the deviation in Γ relative to Γ_{eq} (so although Γ and Γ_{eq} are strictly positive if there is surfactant present, Γ' can be either positive or negative). We denote the interfacial tension γ at concentration Γ_{eq} as γ_{eq}. Hence, if we adopt the Langmuir form,

$$\gamma_{eq} = \gamma_s + RT\Gamma_\infty \ln\left(1 - \frac{\Gamma_{eq}}{\Gamma_\infty}\right) \quad \text{and} \quad \gamma = \gamma_s + RT\Gamma_\infty \ln\left(1 - \frac{\Gamma_{eq} + \Gamma'}{\Gamma_\infty}\right). \tag{2–166}$$

Alternatively, *in the linear case*, Eq. (2–156),

$$\gamma_{eq} = \gamma_s - \Gamma_{eq} RT \quad \text{and} \quad \gamma = \gamma_{eq} - \Gamma' RT. \tag{2–167}$$

Now the stress balance, (2–134), is

$$(\mathbf{T} - \hat{\mathbf{T}}) \cdot \mathbf{n} + \text{grad}_s \gamma - \gamma \mathbf{n}(\nabla \cdot \mathbf{n}) = 0.$$

To describe the effects of changes in the surfactant concentration for this condition, we need to calculate the derivative of γ with respect to Γ. Based on (2–166), the result is

$$\frac{1}{\gamma_s} \frac{d\gamma}{d\Gamma} = -\frac{RT}{\gamma_s}\left[\frac{1}{1 - (\Gamma_{eq} + \Gamma')/\Gamma_\infty}\right]. \tag{2–168}$$

In the limit $\Gamma/\Gamma_\infty \ll 1$, corresponding to (2–167), this reduces to

$$\frac{1}{\gamma_s} \frac{d\gamma}{d\Gamma} = -\frac{RT}{\gamma_s}.$$

The parameter

$$\beta = \frac{RT}{\gamma_s} \tag{2–169}$$

is sometimes called the "*Gibbs elasticity*." It provides a measure of the sensitivity of γ to Γ, but is generally useful primarily in the low concentration limit because the coefficient on the right-hand side of (2–168) can become very large as $\Gamma/\Gamma_\infty \to 1$.

Hence, if we now reexpress the stress balance, (2–134), so that it is explicit in Γ,

$$(\mathbf{T} - \hat{\mathbf{T}}) \cdot \mathbf{n} = \gamma_s \mathbf{n}(\nabla_s \cdot \mathbf{n}) + \beta \gamma_s \left[\Gamma_\infty \ln(1-\zeta)\mathbf{n}(\nabla_s \cdot \mathbf{n}) + \left(\frac{1}{1-\zeta}\right)\nabla_s \Gamma' \right] \quad (2\text{–}170)$$

where

$$\zeta \equiv (\Gamma_{eq} + \Gamma')/\Gamma_\infty.$$

If we consider only the dilute limit $\zeta \ll 1$, this reduces to the simpler form,

$$(\mathbf{T} - \hat{\mathbf{T}}) \cdot \mathbf{n} = \gamma_{eq} \mathbf{n}(\nabla_s \cdot \mathbf{n}) - \beta \gamma_s \{\Gamma'[\mathbf{n}(\nabla_s \cdot \mathbf{n})] - \nabla_s \Gamma'\}, \quad (2\text{–}171)$$

with an error on the right-hand side of $O(\Gamma'\zeta)$.

In this latter form, the first term on the right-hand side is simply the capillary pressure term from the Laplace equation, (2–136), now evaluated using the interfacial tension for the equilibrium system including surfactant, and the second term represents the influence of departures of the surfactant concentration from its equilibrium value. In the case of an insoluble surfactant, Γ' can be nonzero both because there are gradients in Γ either introduced by flow, or by nonuniform changes of surface area, or because the total surface area expands or contracts (for example, the surface of a bubble that expands or contracts, or the surface of a drop that undergoes a change of shape from spherical to nonspherical (say ellipsoidal) with a corresponding increase in surface area). It is convenient to use the form (2–171) for the stress balance because γ_{eq} is the interfacial tension that would be measured by standard methods for a given two-fluid system, whereas γ_s would not be easily accessible (one would have to either get clean samples of the two pure solvents or completely remove the surfactant from the system at hand, which is very difficult). In any case, unless the system departs significantly from equilibrium, it is the value γ_{eq} for a uniformly dispersed surfactant that would determine capillary effects.

In summary, then, we can incorporate surfactant effects into any problem of fluid motion with an interface by simply utilizing the usual interface boundary conditions, together with a constitutive relation between γ and Γ and a "transport" equation from which the distribution of Γ on the interface can be determined. We shall later analyze some problems in which surfactant effects play an important role. However, one qualitative effect that can be appreciated without the need for analysis is the effect of surfactant on the buoyancy-driven motion of a bubble or drop through a second fluid. It is observed experimentally that the rise velocity of a small drop or bubble is significantly decreased when surfactant is present compared with the velocity of the same volume bubble or drop in a system that is free of surfactant. We have seen that surfactant accumulates at the interface. In the absence of motion, it is distributed uniformly because of the presence of surface diffusion. However, if we consider the upward buoyancy-induced motion of the bubble or drop, fluid at the interface moves from the front of the bubble or drop toward the rear, and this motion carries the surfactant with it. The result, as a consequence of the balance between this convection from the front to the rear and surface diffusion, is a nonuniform concentration distribution of surfactant with higher concentrations at the rear of the bubble or drop and a lower concentration at the front. This balance of convection and diffusion is described by (2–162) or (2–164), depending on whether the surfactant is soluble or insoluble in the bulk fluids. The nonuniform concentration leads to a gradient of interfacial tension and hence to a "Marangoni" stress contribution to the tangential-stress balance that tends to immobilize the interface. The result is that the resistance to motion of the bubble or drop is

Basic Principles

increased, and, because the buoyancy force is fixed, it moves more slowly than if there were no surfactant present.

NOTES AND REFERENCES

1. J. P. Boon and S. Yip, *Molecular Hydrodynamics* (McGraw-Hill, New York, 1980); D. Frenkel and B. Smits, *Understanding Molecular Simulation: From Algorithms to Applications* (Elsevier Science, New York, 1996); M. P. Allen and D. J. Tildesley, *Computer Simulation of Liquids* (Oxford University Press, Oxford, 1989); J. M. Haile, *Molecular Dynamics Simulations: Elementary Methods* (Wiley, New York, 1997); D. C. Rapaport and D. C. Rapaport, *The Art of Molecular Dynamics Simulation* (Cambridge University Press, Cambridge, 1997).
2. The notation adopted in Eq. (2–1) is a convenient "shorthand" for the actual process of averaging, which might be more accurately represented as a sum over all of the molecules that exist in the averaging volume, i.e.,

$$\mathbf{u} = \frac{1}{V} \sum_i \mathbf{w}_i,$$

where the $\mathbf{w}_i$ represent the velocities of the individual molecular species. In an N-component fluid, this would be generalized to a mass average velocity,

$$\mathbf{v} = \sum_{j=1}^{N} \omega_j \mathbf{v}_j,$$

where ω_i is the mass fraction of species j in the averaging volume, and $\mathbf{v}_j$ is the volume-averaged species velocity,

$$\mathbf{v}_j = \frac{1}{V} \sum_i (\mathbf{w}_i)_j,$$

and the sum is over all molecules of species j in the averaging volume. The necessity for generalizing to a mass average velocity for a multicomponent fluid will be explained later in this chapter.
3. W. R. Schowalter, The behavior of complex fluids at solid boundaries, *J. Non-Newtonian Fluid Mech.* **29**, 25–36 (1988); S. Goldstein, *Modern Developments in Fluid Dynamics* (Dover, New York, 1965), Vol. 2, p. 676.
4. Strictly speaking, for a multicomponenet fluid, this should be the mass average velocity **v** as defined in Note 2. However, in this section we restrict our discussion to a single-component fluid for which the volume average velocity **u** and the mass average velocity **v** are equal.
5. Harold Jeffreys, Bertha S. Jeffreys, Bertha Swirles, and Bertha Swirles Jeffreys, *Methods of Mathematical Physics*, 3rd ed. (Cambridge University Press, Cambridge, 2000), Chap. 5; K. F. Riley, M. P. Hobson, and S. J. Bence, *Mathematical Methods for Physics and Engineering: A Comprehensive Guide* (Cambridge University Press, Cambridge, 1998).
6. Here, for simplicity, we utilize Cartesian component notation. For the reader who is not familiar with this notation, a brief summary is given in Appendix B at the end of the book.
7. The proof given in this text closely follows that given by S. Whitaker in his book *Introduction to Fluid Mechanics*, Chap. 3 (Prentice-Hall, Englewood Cliffs, NJ, 1968). Another extremely useful discussion of convected derivatives and the Reynolds transport theorem is given in Ref. 8.
8. G. K. Batchelor, *An Introduction to Fluid Dynamics* (Cambridge University Press, Cambridge, 1967), Section 3.6.
9. R. E. Rosensweig, Magnetic fluids, *Annual Rev. Fluid Mech.* **19**, 437–463 (1987); R. E. Rosensweig, *Ferrohydrodynamics* (Cambridge University Press, Cambridge, 1985).
10. R. Aris, *Vectors, Tensors and the Basic Equations of Fluid Mechanics* (Dover, New York, 1989).
11. If **A** and **B** are second-order tensors, the double inner product is a scalar, defined in terms of Cartesian index notation as $A_{ij}B_{ji}$. The double inner product $\boldsymbol{\varepsilon}:\mathbf{T}$ is thus a vector with components $\varepsilon_{ijk}T_{kj}$.

Notes and References

12. Although the internal energy will generally be dominated by the kinetic energy of the molecules (relative to **u**), there will also be contributions that are due to the intramolecular and intermolecular interaction energies.
13. One exception to this is dilute gases, for which kinetic theory can provide specific predictions for the form of the constitutive equations: see D. A. McQuarrie, *Statistical Mechanics* (Harper & Row, New York, 1976).
14. J. P. Hanson and I. R. McDonald, *Theory of Simple Liquids* (Elsevier Science, Amsterdam, 1987).
15. Such a configuration might be achieved by rotating the container about its central axis.
16. A fluid interface will be characterized by a property known as *interfacial tension* (which we will discuss in detail in the latter sections of this chapter). The assumption that the pressure in the liquid, evaluated at the interface, is the same as the ambient pressure in the air on the other side of the interface, is equivalent to assuming that the interfacial tension is zero.
17. Conditions for neglect of interfacial tension forces relative to gravity forces will be derived in Chap. 6.
18. A. M. Donald and A. H. Windle, *Liquid Crystalline Polymers* (Cambridge University Press, Cambridge, 1992); P. G. de Gennes and J. Prost, *The Physics of Liquid Crystals*, 2nd ed. (Oxford University Press, New York, 1993).
19. G. Astarita and G. Marrucci, *Principles of Non-Newtonian Fluid Mechanics* (McGraw-Hill, London, 1974); D. C. Leigh, *Nonlinear Continuum Mechanics* (McGraw-Hill, New York, 1968).
20. R. G. Larson, *The Structure and Rheology of Complex Fluids* (Oxford University Press, New York, 1999).
21. G. B. Jeffery, The motion of ellipsoidal particles immersed in a viscous fluid. *Proc. R. Soc.* London Ser. A **102**, 161–179 (1922).
22. There is considerable confusion in published works about the proper usage of the two dimensionless parameters, the Weissenberg number and the Deborah number. The former may be viewed as a measure of the strength of the flow, namely, the magnitude of the velocity gradient scaled with the principle relaxation time scale of the fluid, $Wi \equiv |\nabla \mathbf{u}| \tau_{relxn}$. In the case of a steady shear flow, this is just $\dot{\gamma} \tau_{relxn}$. The Deborah number, on the other hand is generally intended to provide a measure of the rate at which the velocity gradient changes, relative to the relaxation time of the fluid. In a homogeneous, but unsteady flow, such as an oscillatory shear flow, $\mathbf{u} = (\dot{\gamma} \sin \omega t) y \mathbf{i}_x$, there is a clear distinction between the Weissenberg number, which is still $Wi \equiv \dot{\gamma} \tau_{relxn}$, and the Deborah number, $De \equiv \omega \tau_{relxn}$. The confusion arises because most flows are nonhomogeneous, and thus steady flows may be unsteady from the point of view of a material element of the fluid that moves through the flow domain and encounters different velocity gradients at different points in space (and is thus unsteady from the Lagrangian point of view of a typical material element). In this case, the Weissenberg number might be thought of as varying from point to point, but still serve as a measure of the magnitude of the velocity gradient, $Wi \equiv |\nabla \mathbf{u}| \tau_{relxn}$, whereas the Deborah number should provide a measure of the Lagrangian rate at which $\nabla \mathbf{u}$ is changing, e.g., one might use a definition like $De \equiv [D|\nabla \mathbf{u}|/Dt] |\nabla \mathbf{u}|^{-1} \tau_{relxn}$. This subject is actually quite a subtle and complex issue. There are several places where one can find a more complete and sophisticated discussion; see, e.g., R. I. Tanner, *Engineering Rheology* (Oxford University Press, Oxford, 1985), pp. 190–1.
23. An exception is thus liquid-crystalline materials in a nematic or smetic state (Ref. 18).
24. Not surprisingly, the material properties in the latter case tend to be only weakly influenced by the polymer or particles, and the rheology is dominated by the suspending fluid.
25. M. Doi and S. F. Edwards, *The Theory of Polymer Dynamics* (Oxford University Press, Oxford, 1986); R. G. Larson, *The Structure and Rheology of Complex Fluids* (Oxford University Press, New York, 1999).
26. R. G. Owen and T. N. Phillips, *Computational Rheology* (Imperial College Press, London, 2002).
27. A. N. Beris and B. J. Edwards, *Thermodynamics of Flowing Systems with Internal Microstructure* (Oxford University Press, New York, Oxford, 1994).
28. A good summary of the subject of rheometry is given in a book of the same title: K. Walters, *Rheometry* (Chapman and Hall, London, 1974).
29. D. C. Leigh, *Nonlinear Continuum Mechanics* (McGraw-Hill, New York, 1968), pp. 70–6.
30. R. R. Huilgol, *Continuum Mechanics of Viscoelastic Liquids* (Wiley, Sons, New York, 1975).

31. M. Doi and S. F. Edwards, *The Theory of Polymer Dynamics* (Oxford University Press, Oxford, 1986); R. B. Bird, C. F. Curtiss, R. C. Armstrong, and O. Hassager, *Dynamics of Polymer Liquids*, Vol. 2: Kinetic Theory, 2nd ed. (Wiley, New York, 1987).
32. For example, if we specify **p** in terms of the polar and azimuthal angles of a spherical coordinate system(θ, ϕ), then

$$\nabla_P \equiv \mathbf{i}_\theta \frac{\partial}{\partial \theta} + \mathbf{i}_\phi \frac{1}{\sin\theta} \frac{\partial}{\partial \phi}.$$

33. E. J. Hinch and L. G. Leal, Constitutive equations in suspension mechanics. Part 2. Approximate forms for a suspension of rigid particles affected by Brownian rotations. *J. Fluid Mech.* **76**, 187–208 (1976).
34. See Chap. 6 of Ref. 20.
35. J. S. Cintra and C. L. Tucker, Orthotropic closuere approximations for flow-induced fiber orientation, *J. Rheol.* **39**, 1095–1122 (1995); C. V. Chaubal, A. Srinivasan, O. Egecioglu, and L. G. Leal, Smoothed particle hydrodynamics technique for the solution of kinetic theory problems. Part 1. Method, *J. Non-Newtonian Fluid Mech.* **70**, 125–54 (1997).
36. C. V. Chaubal and L. G. Leal, A closure approximation for liquid crystalline polymer models based on parametric density estimation, *J. Rheol.* **42**, 177–201 (1998); H. C. Ottinger, *Stochastic Processes in Polymeric Fluids* (Springer, Berlin, 1996).
37. R. S.Graham, T. C. B. McLeish and O. G. Harlen, Using the pom-pom equations to analyze polymer melts in exponential shear, *J. of Rheol.* **45**, 275–90 (2001); N. J. Inkson, T. C. B. McLeish, O. G. Harlen and D. J. Groves, Predicting low density polyethylene melt rheology in elongational and shear flows with "pom-pom" constitutive equations, *J. Rheol.* **43**, 873–96 (1999); G. W. M. Peters, J. F. M. Schoonen, F. P. T. Baaijens, and H. E. H. Meijer, On the performance of enhanced constitutive models for polymer melts in a cross-slot flow, *J. Non-Newtonian Fluid Mech.* **82**, 387–427 (1999); W. H. M.Verbeeten, G. W. M. Peters, and F. P. T. Baaijens, Differential constitutive equations for polymer melts: The extended pom-pom model, *J. Rheol.* **45**, 823–43 (2001).
38. P. A.Thompson and S. M. Troian, A general boundary condition for liquid flow at solid surfaces. *Nature (London)* **389**, 360–2 (1997) (and references therein).
39. A very interesting "history" of discussions about the no-slip condition is given as an appendix titled, "Note on the Conditions at the Surface of Contact of a Fluid with a Solid Body" in the following book: S. Goldstein, *Modern Developments in Fluid Dynamics* (Clarendon, Oxford, 1938) (reproduced in 1965 by Dover, New York).
40. E. B. Dussan V, The moving contact line: The slip boundary condition, *J. Fluid Mech.* **77**, 665–84 (1976); L. M. Hocking, A moving fluid interface. Part 2. The removal of the force singularity by a slip flow, *J. Fluid Mech.* **79**, 209–29 (1977); S. Somalinga and A. Bose, Numerical investigation of boundary conditions for moving contact line problems, *Phys. Fluids* **12**, 499–510 (2000).
41. D. C. Tretheway and C. D. Meinhart, Apparent fluid slip at hydrophobic microchannel walls, *Phys. Fluids* **14**, L9–L12 (2002); Chang-Hwan Choi, J. A. Westin, and K. S. Breuer, Apparent slip flows in hydrophilic and hydrophobic microchannels, *Phys. Fluids* **15**, 2897–902 (2003).
42. K. Watanabe, Yanuar, and H. Udagawa, Drag reduction of Newtonian fluid in a circular pipe with a highly water-repellent wall, *J. Fluid Mech.* **381**, 225–38 (1999).
43. M. M. Denn, Extrusion instabilities and wall slip, *Annual Rev. Fluid Mech.* **33**, 265–87 (2001).
44. P. G. de Gennes, *Physics of Polymer Surfaces and Interfaces*, edited by Isaac C. Sanchez (Butterworth-Heinemann, Burlington, MA (1992), Chap. 3, pp. 55–71; J. L. Goveas and G. H. Fredrickson (1992), Apparent slip at a polymer–polymer interface, *Eur. Phys. J.* **B2**, 79–92 (1998); R. Zhao and C. W. Macosko, Slip at polymer–polymer interfaces: Rheological measurements on coextruded multilayers, *J. Rheol.* **46**, 145–67 (2002).
45. L. E. Scriven, Dynamics of a fluid interface. *Chem. Eng. Sci.* **12**, 98–108 (1960).
46. N. O. Young, J. S. Goldstein, and M. J. Block, The motion of bubbles in a vertical temperature gradient, *J. Fluid Mech.* **6**, 350–6 (1959).
47. M. J. Rosen, *Surfactants and Interfacial Phenomena*, 2nd ed. (Wiley, New York, 1994).
48. A. W. Adamson and A. P. Gast, *Physical Chemistry of Surfaces* (Wiley, New York, 1997).
49. Y. Pawar and K. J. Stebe, Marangoni effects on drop deformation in an extensional flow: The role of surfactant physical chemistry. I. Insoluble surfactants, *Phys. Fluids* **8**, 1738–51 (1996).

Problems

50. H. A. Stone, A simple derivation of the time-dependent convective-diffusion equation for surfactant transport along a deforming interface, *Phys. Fluids A* **2**, 111–12 (1990).

PROBLEMS

Problem 2–1. Some Vector and Tensor Identities. Prove the following identities, where ϕ is a scalar, $\mathbf{u}$ and $\mathbf{v}$ are vectors, $\mathbf{T}$ is a second-order tensor, and $\boldsymbol{\varepsilon}$ is the third-order alternating tensor:

(a) $\nabla \cdot \phi \mathbf{v} = \phi \nabla \cdot \mathbf{v} + \mathbf{v} \cdot \nabla \phi$,
(b) $\nabla \wedge \phi \mathbf{v} = \phi \nabla \wedge \mathbf{v} + \nabla \phi \wedge \mathbf{v}$,
(c) $\nabla \cdot (\mathbf{u} \wedge \mathbf{v}) = \mathbf{v} \cdot \nabla \wedge \mathbf{u} - \mathbf{u} \cdot \nabla \wedge \mathbf{v}$,
(d) $\nabla \wedge (\mathbf{u} \wedge \mathbf{v}) = \mathbf{v} \cdot \nabla \mathbf{u} - \mathbf{u} \cdot \nabla \mathbf{v} + \mathbf{u}\nabla \cdot \mathbf{v} - \mathbf{v}\nabla \cdot \mathbf{u}$,
(e) $\nabla \wedge \nabla \phi = 0$,
(f) $\nabla \cdot \nabla \wedge \mathbf{v} = 0$,
(g) $\nabla \cdot \nabla = \nabla^2 = \partial^2/\partial x_i^2$,
(h) $\mathbf{v} \cdot \nabla \mathbf{v} = (\nabla \wedge \mathbf{v}) \wedge \mathbf{v} + \nabla(\tfrac{1}{2}\mathbf{v}^2)$,
(i) $\nabla^2 \mathbf{v} = \nabla(\nabla \cdot \mathbf{v}) - \nabla \wedge (\nabla \wedge \mathbf{v})$,
(j) $\nabla \cdot (\phi \mathbf{u}\mathbf{u}) = \phi \mathbf{u} \cdot \nabla \mathbf{u} + \mathbf{u}[\nabla \cdot (\phi \mathbf{u})]$,
(k) $\nabla \cdot (\mathbf{T} \wedge \mathbf{x}) = -\mathbf{x} \wedge (\nabla \cdot \mathbf{T}) + \boldsymbol{\varepsilon} : \mathbf{T}$,

Problem 2–2. Recall that $\nabla \phi$ at any point is perpendicular to the level surface of $\phi(\mathbf{x})$ that passes through that point:

(a) How would you determine the unit normal to the level surface of ϕ?
(b) Suppose $\mathbf{v}$ is a steady velocity field and $\mathbf{v} \cdot \nabla \phi = 0$. How does ϕ vary along a particle path that is determined by $\mathbf{v}$?

Problem 2–3. Consider the flow field

$$\mathbf{u}(\mathbf{r}, t) = (xy)^2 \mathbf{e}_x + z e^{-\alpha t} \mathbf{e}_y + \cos(2xz) \mathbf{e}_z.$$

Calculate

(a) the fluid acceleration from an Eulerian point of view $\mathbf{a}$,
(b) the fluid acceleration from a Lagrangian point of view, $\mathbf{A}$, for a fluid element that moves with the velocity $\mathbf{u}$,
(c) the velocity gradient tensor $\nabla \mathbf{u}$,
(d) the curl of the velocity $\nabla \wedge \mathbf{u}$.

Problem 2–4. Index Notation. Write the following expressions using index notation:

(a) $A = \nabla \cdot \mathbf{u}$ (dot product),
(b) $\mathbf{A} = \nabla \mathbf{u}$ (vector dyadic product),
(c) $\boldsymbol{\omega} = \nabla \wedge \mathbf{u}$ (curl),
(d) $\mathbf{C} = (\mathbf{x} \cdot \mathbf{y})\mathbf{z}$,
(e) $\mathbf{A}^T \cdot \mathbf{A} \cdot \mathbf{x} = \mathbf{A}^T \cdot \mathbf{b}$,

Problem 2–5. Pseudo-Vectors. Demonstrate that the pseudo-vector equation

$$\mathbf{c} = \mathbf{b} \wedge \mathbf{a}$$

is not invariant under an orthogonal transformation of coordinates. Assume that $\mathbf{a}$, $\mathbf{b}$, and $\mathbf{c}$ are all normal vectors that transform according to the rule $\hat{\mathbf{a}} = \mathbf{L} \cdot \mathbf{a}$, $\hat{\mathbf{b}} = \mathbf{L} \cdot \mathbf{b}$, and $\hat{\mathbf{c}} = \mathbf{L} \cdot \mathbf{c}$.

Problem 2–6.

(a) Let $\mathbf{E} = \tfrac{1}{2}[\nabla \mathbf{v} + (\nabla \mathbf{v})^T]$, where $(\cdot)^T$ means the transpose of a tensor (i.e., $G_{ij}^T = G_{ji}$), and $\nabla \cdot \mathbf{v} = 0$. Show that $\mathbf{E} : \mathbf{I} = 0$.

Basic Principles

(b) Suppose that u, v are scalar harmonic functions within a volume V with boundary S. Show that
$$\int_S \mathbf{n} \cdot u \nabla v \, dS = \int_S \mathbf{n} \cdot v \nabla u \, dS.$$

(c) Let $\mathbf{T}$ be a symmetric matrix. Show that $\boldsymbol{\varepsilon} : \mathbf{T} = 0$.

Problem 2–7. Vorticity Tensor. Consider two nearby material points, P and Q. In Section H, we demonstrated that the distance between these two points increases (or decreases) at a rate that depends on the rate-of-strain tensor $\mathbf{E}$. Show that the rate of angular rotation of the vector $\delta \mathbf{x}$ between these points depends on the vorticity tensor $\boldsymbol{\Omega}$.

Problem 2–8. The Divergence Theorem. Let $\mathbf{T}$ be a second-order tensor, and $\mathbf{f}$ be a vector. Given the divergence theorem
$$\int_S \mathbf{n} \cdot \mathbf{v} \, dS = \int_V \nabla \cdot \mathbf{v} \, dV,$$
where $\mathbf{v}$ is a vector, show that

(a) $\int_S \mathbf{n} \cdot \mathbf{T} \, dS = \int_V \nabla \cdot \mathbf{T} \, dV$,

(b) $\int_S \mathbf{f} \wedge \mathbf{n} \, dS = -\int_V \nabla \wedge \mathbf{f} \, dV$.

- *Hint*: Let $\mathbf{v} = \mathbf{T} \cdot \mathbf{b}$ or $\mathbf{v} = \mathbf{b} \wedge \mathbf{f}$, where $\mathbf{b}$ is an arbitrary constant vector.

Problem 2–9.

(a) Evaluate the surface integrals
$$\int_{\partial D} y_i n_j \, dS \quad \text{and} \quad \int_{\partial D} n_i \frac{\partial n_j}{\partial y_j} \, dS$$
over an arbitrary closed surface ∂D. y_i is the position vector of a point on the surface and n_i is the outer normal to the surface.

(b) Evaluate $\int y_i y_j \, dS$ over the surface of a cube of side length a, where y_i is the position vector of a point relative to an origin at the center of the cube.

Problem 2–10. Prove Equation (2–37). To do this, it will be convenient to show
$$\frac{D}{Dt} \int_{V_m(t)} \rho F \, dV = \int_{V_m(t)} \rho \frac{DF}{Dt} \, dV.$$
In addition, the identity
$$\frac{D\mathbf{x}}{Dt} = \mathbf{u}$$
is necessary, and this should also be demonstrated by means of definition (2–8).

Problem 2–11. Let V be a material volume moving with the fluid and let S be the material surface surrounding V. Assume the fluid is incompressible and inviscid, i.e., no viscosity. As usual, $\mathbf{u}(\mathbf{x}, t)$ is the velocity field. The quantity $\boldsymbol{\omega}(\mathbf{x}, t) \equiv \nabla \wedge \mathbf{u}$ is known as the vorticity field. The vorticity describes the local angular velocity of the fluid. Now, consider the integral of $\mathbf{u} \cdot \boldsymbol{\omega}$ over V,
$$I = \int_V \mathbf{u} \cdot \boldsymbol{\omega} \, dV. \tag{1}$$
If $\mathbf{n} \cdot \boldsymbol{\omega} = 0$ on S, then show ($\mathbf{n}$ is the unit outward normal from)
$$\frac{DI}{Dt} = 0, \tag{2}$$
where D/Dt is the material derivative of the integral (1).

Problems

Hint: At some point you will need to show
$$\mathbf{u} \cdot (\boldsymbol{\omega} \cdot \nabla \mathbf{u}) = \boldsymbol{\omega} \cdot \nabla \left(\frac{\mathbf{u} \cdot \mathbf{u}}{2}\right).$$

Remark: The quantity I is called the helicity of the flow.

Problem 2–12. From Cauchy's equation of motion for the steady flow of an incompressible fluid in the absence of body forces, derive the integral momentum balances for the hydrodynamic force $\mathbf{F}$ and torque $\mathbf{L}$ on an arbitrary body S_B immersed in the fluid:

$$\mathbf{F} = \int_S [\mathbf{T} - \rho \mathbf{u}\mathbf{u}] \cdot \mathbf{n} \, dS,$$

$$\mathbf{L} = \int_S \mathbf{r} \wedge [\mathbf{T} - \rho \mathbf{u}\mathbf{u}] \cdot \mathbf{n} \, dS,$$

where S is an arbitrary closed surface surrounding the body S_B. First recall the definition of force and torque.

Problem 2–13. Derivation of Transport Equations. Consider the arbitrary fluid element depicted in the figure. If we have a flow containing several species that are undergoing reaction (a source/sink per unit volume) and diffusion (a flux of each species *in addition to convection*), derive the equation that governs the conservation of each species. The source of species i that is due to reaction is denoted as R_i (units of mass of i per unit time per unit volume) and the *total* mass flux of species i (diffusion *and* convection) is given by $(\rho_i \mathbf{u} + \mathbf{j}_i)$, in which ρ_i is the mass of species i per unit volume, $\mathbf{u}$ is the total mass average velocity of the fluid, and $\mathbf{j}_i$ is the diffusive flux of species i. Note that both $\mathbf{u}$ and $\mathbf{j}_i$ are vectors. We are *not* using index notation in this problem!

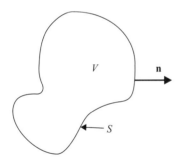

(a) Write down in words the law governing the conservation the mass of species i for the control volume D. Don't forget diffusion and reaction.
(b) Write down the integral relationship equivalent to that in part (a).
(c) Through application of the divergence theorem, obtain a microscopic equation valid at every point in the flow. This will still be in terms of $\mathbf{j}_i$ (we have not specified a constitutive relation here for diffusion, analogous to Newton's law of viscosity or Fourier's law of heat conduction – don't worry about it).

Problem 2–14. Derivation of Basic Equations. Magnetohydrodynamics is the study of the flow of an electrically conducting fluid in the presence of an electromagnetic field. The fundamental equations of magnetohydrodynamics are a combination of electromagnetic field equations and fluid flow equations, modified to include the effect of the interaction between the fluid motion and the electromagnetic field. The usual assumption in magnetohydrodynamics is that the induced current in the fluid will interact with the electromagnetic field in accordance with the classical laws governing electromagnetic interactions.

Basic Principles

For a charged fluid particle moving in an electromagnetic field, the electromagnetic force per unit volume is

$$\mathbf{f} = \mathbf{J} \wedge \mathbf{B},$$

where $\mathbf{J}$ is the current density and $\mathbf{B}$ is the magnetic flux density.

Derive the appropriate form of the Navier–Stokes equation for an electrically conducting Newtonian fluid with constant density and viscosity. This equation is called the magnetohydrodynamic momentum equation. Your derivation should begin from first principles. However, you may assume – but should state in appropriate places – the necessary properties of pressure, viscous stress, and convected time derivatives.

Problem 2–15. Derivation of Transport Equation for a Sedimenting Suspension. There are many parallels among momentum, mass, and energy transport because all three are derived from similar conservation laws. In this problem we derive a microscopic balance describing the concentration distribution $\phi(\mathbf{x}, t)$ of a very dilute suspension of small particles suspended in an incompressible fluid undergoing unsteady flow. [Note: $\phi(\mathbf{x}, t)$ is the local volume fraction of particles in the fluid (i.e. volume of particles/volume of fluid) and hence is dimensionless.]

(a) The flux of particles is the sum of that due to convection, that due to settling $\mathbf{q}_s$ (a vector), and that due to diffusion $\mathbf{q}_D$ (also a vector). These fluxes have units of volume per area per time (e.g., velocity). With this in mind, write an integral balance for the conservation of particles for an arbitrary control volume D. Using the divergence theorem, convert all surface integrals to volume integrals, and so obtain a microscopic equation valid at any point in the flow field.

(b) For small particles of radius a, the flux that is due to settling is given by

$$\mathbf{q}_s = \frac{2}{9} \frac{g a^2 \Delta \rho}{\mu} \phi \, \mathbf{i}_g,$$

and the diffusive flux by

$$\mathbf{q}_D = -\frac{kT}{6\pi \mu a} \nabla \phi.$$

Using these results, obtain an equation in terms of derivatives of the concentration.

(c) Using a characteristic velocity U, characteristic length L, and characteristic time L/U, render the equation dimensionless. What dimensionless parameters appear and what is their physical significance? This equation is the starting point for the study of many problems involving suspensions of particles for which Brownian motion is significant (it is Brownian motion that gives rise to the diffusion term).

Problem 2–16. Fluid Statics (Archimedes' Principle). Assume that we have a spherical coordinate system defined with respect to a Cartesian system according to $x = r \sin\theta \cos\phi$, $y = r \sin\theta \sin\phi$, and $z = r \cos\theta$:

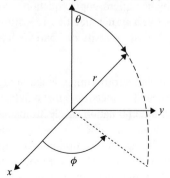

Problems

(a) Obtain an expression for the force on a sphere that is situated at the origin, in terms of the components of the stress **T**,

$$\mathbf{T} = T_{rr}\mathbf{e}_r\mathbf{e}_r + T_{r\theta}\mathbf{e}_r\mathbf{e}_\theta + T_{r\phi}\mathbf{e}_r\mathbf{e}_\phi$$
$$= +T_{\theta r}\mathbf{e}_\theta\mathbf{e}_r + T_{\theta\theta}\mathbf{e}_\theta\mathbf{e}_\theta + T_{\theta\phi}\mathbf{e}_\theta\mathbf{e}_\phi + T_{\phi r}\mathbf{e}_\phi\mathbf{e}_r + T_{\phi\theta}\mathbf{e}_\phi\mathbf{e}_\theta + T_{\phi\phi}\mathbf{e}_\phi\mathbf{e}_\phi.$$

(b) Suppose that the fluid is isothermal and stationary. Further, assume that the fluid and the sphere have the same density ρ. Show that the *net* force on the sphere is zero (that is, prove Archimedes' principle for this case).

Problem 2–17. Fluid Statics. Pool drains can be dangerous things: There was a tragic case recently in this area in which a child was stuck in a drain on the bottom, plugging it, and drowning as a result. Here we look at a somewhat simpler problem. Suppose a ball of radius R is plugging a drain of diameter D at the bottom of a pool of depth h as shown in the figure. Obviously, $R > D/2$ or the ball goes down the drain! Determine the conditions under which the net force on the ball is zero. Assume that the pressure distribution in the drain is just atmospheric pressure, and that in the water is governed by the hydrostatic pressure distribution. If R is 1 ft and D is 6 in, what is the corresponding depth?

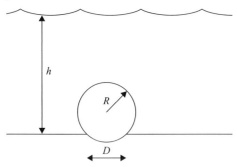

Problem 2–18. Fluid Statics

(a) If air can be treated as an ideal gas, and the temperature of the atmosphere varies linearly with height above the Earth,

$$T(z) = T_0 - \alpha z \quad (T_0, \alpha \equiv \text{constants}),$$

derive an expression for the pressure as a function of z assuming that we can neglect fluid motion in the atmosphere, as well as the Earth's rotation.

(b) A weather balloon, mass m_b, radius R_0, is filled with a mass m_g of some gas. To what altitude will the balloon rise? Assume that the radius of the balloon remains constant. Assume further that $\alpha R_0 \ll 1$ so that we can neglect density variations over the surface of the balloon.

Suppose the balloon is 3 m in diameter and is filled with 3 kg of helium. Further, assume that the mass of the balloon is 4 kg and the diameter is fixed. Finally, assume α is 1°C/100 meters. To what altitude will the balloon rise?

(c) Suppose we have a fluid with the density variation given as

$$\rho(z) = \rho_0(1 + az^2 + bz^4),$$

where $a, b > 0$, and z increases in the direction of g. If a sphere, radius R, density ρ_s, is placed in this fluid, what will be the equilibrium position of the center of the sphere? If the sphere is replaced with a cube of side L, what will the equilibrium position be for the center of the cube?

Basic Principles

Problem 2–19. Fluid Statics. A classic "Honorable Mention" on the Darwin Awards website is the saga of "Lawn Chair Larry" who decided to go flying by attaching 42 helium-filled 60-ft^3 weather balloons to his lawn chair. Instead of leveling off at a height of 30 ft of altitude as he had planned, he wound up at 16,000 ft and actually was cited for violating LAX airspace. Analyze this problem in hydrostatics and determine how many balloons Larry should have used – and if it is possible to control elevation with any precision. One (of many) URLs for the Lawn Chair Larry story is http://www.darwinawards.com/stupid/stupid1997-11.html

Problem 2–20. Fluid Statics. Two fluids are held back by a hinged gate as illustrated in the figure. The lower fluid is of depth h_1 and density ρ_1 and the upper fluid is of depth h_2 and density ρ_2, with $\rho_2 \leq \rho_1$. Determine the moment per unit width about the base of the hinge. Recall that the moment **L** is defined as the cross product of the position vector measured from the point of interest, **x** and the force, i.e., $\mathbf{L} = \int \mathbf{r} \wedge \mathbf{f}\, dS$.

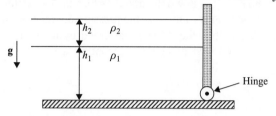

Problem 2–21. Fluid Statics. When a cylindrical tank of liquid is rotated at constant angular velocity ω about its central axis, the free surface attains a steady shape. If the undisturbed (nonrotating) height of the liquid is h, determine the shape of the interface. Be sure to state any and all assumptions. What is the role of interfacial tension? When can it be neglected?

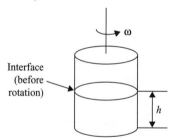

Problem 2–22. Fluid Statics (qualitative questions; assume for these questions that the interfacial tension is zero)

(a) A solid sphere is initially coated with a layer of fluid (sphere radius a, film thickness d). The density of the sphere is ρ_0, the film density is ρ_1, and the exterior fluid density is ρ_2. Suppose $\rho_0 < \rho_1 < \rho_2$. In this case, is the intial configuration consistent with *no motion*? Using what you know of the *hydrostatic* pressure distributions, explain why not (if that is the case), and indicate how the confiugration will initially change.

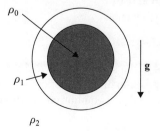

Problems

(b) A liquid layer is in a container with sides parallel to g and bottom perpendicular. The layer of fluid is motionless. At $t = 0$, the container is tilted to an angle ϕ. Discuss the motion of the fluid layer and describe its final steady state (assuming that ϕ is small enough that the fluid all stays in the container).

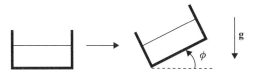

(c) A container of liquid is oscillated back and forth with a velocity $u_{container} = \beta \sin \omega t$. Describe the motion that you would expect of the fluid in the tank. Explain, as well as you can, why the motion occurs, using hydrostatic pressure ideas.

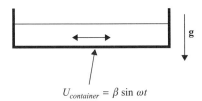

Hint: If the container accerated constantly to the right with acceleration **a**, what shape would you expect the liquid layer to have at steady state?

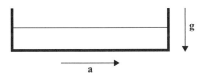

(d) Is the configuration shown below an equilibrium configuration? Consider three cases and explain the expected dynamics (for any case that is not an equilibrium configuration). (1) $\rho_A < \rho_B$; (2) $\rho_A > \rho_B$; and (3) $\rho_A = \rho_B$.

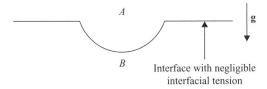

Problem 2–23. Constitutive Equations. As a model of a nonpolar microstructured fluid, consider the material to be described by a single unit vector **p**. Construct the most general relation between the stress tensor **T** and the rate-of-strain tensor **e** that is linear in **e** and depends on **p**. *Note*: when $\mathbf{e} = 0$, the stress is not necessarily isotropic.

Problem 2–24. Polar Fluids. There has been a significant amount of theory developed for so-called *polar fluids*. These are fluids in which it is assumed that the surface couple **r** and the body couple **c** are both nonzero. Show that the stress tensor will no longer, in general, be symmetric for such fluids, but will satisfy the relationship.

$$\boldsymbol{\varepsilon} : \mathbf{T} - \rho \mathbf{c} + \nabla \cdot \mathbf{R} = 0,$$

Basic Principles

where

$$r = n \cdot R.$$

To demonstrate the existence of the surface-couple tensor **R**, you will need to first demonstrate the principle of surface-couple equilibrium,

$$\lim_{l \to 0} \frac{1}{l^2} \int_{S_m(t)} r\, dS \to 0,$$

in analogy with Eq. (2–25).

Problem 2–25. Complex fluids. Consider a suspension consisting of a Newtonian suspending fluid and micrometer-sized particles in which Brownian motion is a factor. The particles are spherical at equilibrium but are made of a Hookean elastic (rubberlike) material. Assume that the suspension is *dilute* (the motion of each particle is independent of the fact that other particles are present). Discuss whether this material can be described at a continuum level as a Newtonian fluid. Does this depend on the magnitude of the elastic modulus? On the shear rate?

Now consider a nondilute version of this suspension in which the motion of each particle is hydrodynamically coupled to the motion of other particles in the suspension. How does your answer change? Suppose the particles have an "extremely large" elastic modulus. Can the suspension then be described as Newtonian? Why not? Or why?

Problem 2–26. Moment equations. The probability density for finding a particle that experiences Brownian and hydrodynamic forces at low Reynolds number satisfies the Smoluchowski equation:

$$\frac{\partial P}{\partial t} + \nabla \cdot \mathbf{j} = 0,$$

where the probability flux **j** is given by

$$\mathbf{j} = (\mathbf{U} - \mathbf{D} \cdot \nabla \ln P)P,$$

and **U** is the imposed velocity.

Directly from the Smoluchowski equation obtain the equations of motion for the mean position of the particle $\langle x \rangle$ and the mean square displacement $\langle (x - \langle x \rangle)(x - \langle x \rangle) \rangle$.

Hint: You can obtain the equations of motion by multiplying the Smoluchowski equation by x and xx and integrating over all space.

Problem 2–27. Surface Divergence Theorem. Prove that

$$\int_C \phi\, \mathbf{t}\, d\ell = \int\int_A \operatorname{grad}_s \phi\, dA - \int\int_A \phi\, \mathbf{n}(\nabla \cdot \mathbf{n})\, dA,$$

where

$$\operatorname{grad}_s \equiv \nabla - \mathbf{n}(\mathbf{n} \cdot \nabla).$$

and **t**, **m**, and **n** are the unit tangents and unit normal vectors, related as shown in the figure by $\mathbf{t} = \mathbf{m} \wedge \mathbf{n}$:

Problems

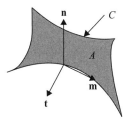

Hint: Start with Stokes' theorem.

Problem 2–28. Interfacial Tension – Capillary Effects. Consider the case of a motionless liquid layer meeting a plane vertical rigid wall as sketched in the figure below. The liquid makes an angle θ with the solid wall, known as the static contact angle, which is a property of the materials involved – the solid, the liquid, and the gas. Derive the relationship between the curvature of the gas-liquid interface and the height of the interface $\zeta(y)$. What are the appropriate boundary conditions for the unknown function $\zeta(y)$? Determine the height h to which the liquid climbs at the rigid wall.

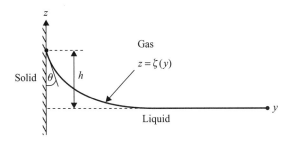

Problem 2–29. Interfacial Tension-Capillary Effects. A rigid sphere of radius a rests on a flat surface, and a small amount of liquid surrounds the point of contact, making a concave–planar lens whose diameter is small compared with a. The angle of contact θ_c with each of the solid surfaces is zero (see Problem 2–28), and the tension in the air–liquid surface is α. Show that there is an adhesive force of magnitude $4\pi a\alpha$ acting on the sphere. (The fact that this adhesive force is independent of the volume of liquid is noteworthy. Note also that the force is repulsive when $\theta_c = \pi$.)

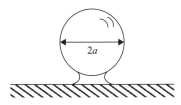

Problem 2–30. Interfacial Tension–Capillary Effects. A straw of radius R is immersed into a liquid through an air–liquid interface. The straw is coated with a material that causes the liquid to contact the surface of the straw with a contact angle θ. The liquid will rise up the straw a distance

$$H = \frac{2\gamma \cos\theta}{\rho g R},$$

where γ is the interfacial tension between the liquid and the air and ρ is the fluid density. Explain why this occurs and derive the formula for H. State all assumptions.

Basic Principles

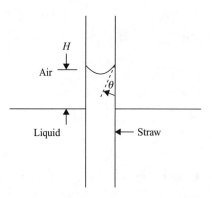

If the coefficient of interfacial tension between air and water is $\gamma = 74$ dyn/cm, what diameter capillary is needed to raise water to a height of 100 m? Does this seem reasonable for, say, redwood trees? If not, what other mechanisms might be at work?

Problem 2–31. Marangoni effects. In a recent fluid mechanics paper, it was claimed that it would take longer for a cooled (initially cold) drop of fluid A to coalesce at an interface separating bulk fluids A and B than if the drop was at the same temperature as the ambient temperature of bulk fluids, $T_A^\infty = T_B^\infty$. You should note that $\rho_A > \rho_B$. In fact, if the density of the drop is only slightly greater than the bulk fluid, the drop may potentially achieve an equilibrium position without ever reaching the interface, where it will remain until it gains too much heat and its temperature approaches that of the bulk fluids. In thinking about this claim, you may *neglect* the change in viscosity or density with temperature.

(a) Do you believe this claim? Explain how you might think about it, noting that the interfacial tension between fluids A and B, σ, is a function of temperature with

$$\frac{\partial \sigma}{\partial T} < 0.$$

Hint: What do you expect the temperature field to look like? Suppose, for example, that the temperature distribution is determined solely by heat conduction.

(b) How would the situation change if $T_{drop} > T_A^\infty = T_B^\infty$?

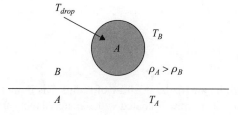

Problem 2–32. Noncoalescence that is Due to Marangoni Effects. In a recent paper [P. Dell'Aversana and G. P. Neitzel, Behavior of noncoalescing and nonwetting drops in stable and marginally stable states, *Exp. Fluids* **36**, 299–308 (2004)], it has been shown that thermocapillary effects involving Marangoni flows can be used to keep a pair of drops from coalescing. One manifestation of this effect concerns the coalescence of a drop at a fluid interface as described in Problem 2–31. Another is concerned with the interactions of two drops that are pushed toward one another along their line of centers by an external force.

Problems

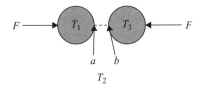

Assume initially (i.e., $t = 0$) that the temperatures in each phase are uniform with $T_1 < T_2 < T_3$. Make a qualitative sketch of the temperature fields at some later time ($t > 0$). Note that we should expect the temperature at point a to be higher than T_1 and the temperature at b to be lower than T_3 because of the transfer of heat between the two drops across the thin film of fluid that separates them. Hence, assuming that

$$\frac{\partial \sigma}{\partial T} < 0,$$

make a sketch of what the Marangoni-driven flow will look like, and discuss how this flow may inhibit coalescence.

3

Unidirectional and One-Dimensional Flow and Heat Transfer Problems

We are now in a position to begin to consider the solution of heat transfer and fluid mechanics problems by using the equations of motion, continuity, and thermal energy, plus the boundary conditions that were given in the preceding chapter. Before embarking on this task, it is worthwhile to examine the nature of the mathematical problems that are inherent in these equations. For this purpose, it is sufficient to consider the case of an incompressible Newtonian fluid, in which the equations simplify to the forms (2–20), (2–88) with the last term set equal to zero, and (2–93).

The first thing to note is that this set of equations is highly nonlinear. This can clearly be seen in the term **u** · grad **u** in (2–88). However, because the material properties such as ρ, C_p, and k are all functions of the temperature θ, and the latter is a function of the velocity **u** through the convected derivative on the left-hand side of (2–93), it can be seen that almost every term of (2–88) and (2–93) involves a product of at least two unknowns either explicitly or implicitly. In contrast, all of the classical analytic methods of solving partial differential equations (PDEs) (for example, eigenfunction expansions by means of separation of variables, or Laplace and Fourier transforms) require that the equation(s) be linear. This is because they rely on the construction of general solutions as sums of simpler, fundamental solutions of the DEs. For nonlinear DEs, however, this superposition principle is not valid, and the analytical methods that have been developed in most introductory engineering mathematics courses cannot be used.

To simplify the problems represented by Eqs. (2–88) and (2–93), one of two possible assumptions is normally introduced. If the bounding surfaces are all at the same temperature, so that the only source of heat is viscous dissipation, it is often a good approximation to assume that the fluid is *isothermal* (i.e., the temperature is a constant, independent of spatial position or time). In this case, we do not need to consider (2–93) at all because it is an equation that is to be used to determine the temperature as a function of position and time. In addition, as noted in Chap. 2, the equation of motion also simplifies when the fluid is isothermal to either the form (2–89) or to (2–91) if the density in the system is also a constant, independent of position. When the temperature at the bounding surfaces is not constant, a different approximation known as the *Boussinesq approximation* is often used to simplify the problem. A detailed discussion of this approximation is postponed to a later point in the book. Here, we simply note the basic idea, which is that the material properties may be approximated as constants, provided that the temperature changes are not too large. In this case, the values of these properties can be evaluated at a representative temperature of the system, such as its mean value. An exception, as we shall see later, is that we sometimes cannot ignore the spatial variations of the density ρ in the body-force term of (2–88) even when the temperature changes are modest. Such density variations can produce motion in the fluid, known as *natural convection*, and this may be very important depending on what other sources of motion exist in the fluid. In any case, the Boussinesq approximation produces a major simplification in Eqs. (2–88) and (2–93) for the analysis of many heat transfer problems, allowing nonlinear terms such as div($2\mu\mathbf{E}$) in (2–88) and div(k grad θ) to be approximated by their linear versions for constant properties.

Even with these simplifications, however, it is rarely possible to obtain analytic solutions for fluid mechanics or heat transfer problems. The Navier–Stokes equation for an isothermal fluid is still nonlinear, as can be seen by examination of either (2–89) or (2–91). The Bousinesq equations involve a coupling between $\mathbf{u}$ and θ, introducing additional nonlinearities. It will be noted, however, that, provided the density can be taken as constant in the body-force term (thus neglecting any natural convection), the fluid mechanics problem is decoupled from the thermal problem in the sense that the equations of motion, (2–89) or (2–91), and continuity, (2–20), do not involve the temperature θ. The thermal energy equation, (2–93), is actually a linear equation in the unknown θ, once the Boussinesq approximation has been introduced. In that case, the only "nonlinear" term is dissipation, but this involves the product $\mathbf{E}:\mathbf{E}$ and can be treated simply as a source term that will be known once Eqs. (2–89) or (2–91) and (2–20) have been solved to determine the velocity. In spite of being linear, however, the velocity $\mathbf{u}$ appears as a coefficient (in the convective derivative term). Even when the form of $\mathbf{u}$ is known (either exactly or approximately), it is normally quite a complicated function, and this makes it extremely difficult to obtain analytic solutions for θ even though the governing equation is linear.

From the discussion of the preceding paragraph, it should be obvious that theoretical studies of fluid mechanics and heat transfer are almost always based on some type of *approximate* solution of the governing equations. There are two basic approaches to obtaining such solutions. One is to introduce a discretized approximation, such as finite-difference, finite-element, or spectral methods, to convert the nonlinear PDEs to a set of nonlinear algebraic equations, and then attempt to solve these equations numerically. Although this approach is not without its own difficulties, it is often possible to obtain solutions over a wide range of parameters. Indeed, in the hands of an experienced and careful practitioner, the use of numerical methods has reached a point that it can often be used as a complement to or even a replacement for laboratory experiments. Of course, it must be remembered that numerical solutions are always approximations of the exact solution, and their interpretation and use must be treated with care. The second class of approximate-solution techniques that has been very effective in the context of fluid mechanics and heat (or mass) transfer are the analytic methods known as asymptotic or perturbation techniques. The basic idea of this approach is to obtain approximate analytic solutions based on a systematic hierarchy of approximations of the basic equations. In addition to the fact that analytic calculation may sometimes be easier or more convenient than numerical methods (which may require very large computational facilities), a primary characteristic of the analytical approach is that it often yields greater and more immediate physical insight. Analytic approximations can also serve as a critical benchmark for numerical methods.

Our main goal in this book is to develop and explain analytical methods for obtaining approximate solutions to (2–20), (2–89) or (2–91), and (2–93), and to use these methods to pursue a detailed and fundamental understanding of fluid mechanics and convective heat transfer (as well as related mass transfer) processes. We do not discuss numerical methods. The interested reader will find many text and reference books that cover that material.[1] We also do not discuss turbulent flows, except in passing. This is not because the governing equations that we have derived in the preceding chapter are not relevant to turbulent flow. Indeed, though turbulent flows can be extremely complex, they are nevertheless governed by the same equations of motion. In fact, a turbulent flow is "just" a particularly nasty

3D time-dependent solution of the Navier–Stokes equations. The main reason we do not discuss these flows here is that the analytical solution techniques that we develop have had relatively little impact on the analysis or understanding of turbulent flows. The most powerful theoretical tools for turbulence research and for the prediction of turbulent flows are currently direct numerical solutions (DNS) of the Navier–Stokes equation, typically by use of spectral techniques for discretization. Again, the interested reader will find many texts and references to modern work on turbulent flows.[2]

Although in the vast majority of this book we consider approximate solutions, there are a few exceptional classes of problems for which an exact analytic solution is possible, and we focus on them in this chapter. This will give us a chance to review analytic solution techniques for linear PDEs. More importantly, it will give us an opportunity to introduce a number of important concepts about scaling and nondimensionalization in a framework of relatively straightforward physical problems and to explore some aspects of the time evolution of steady flows from a state of rest.

A small number of exact solutions of the *nonlinear* Navier–Stokes equations have been discovered more or less by accident. A discussion of some of these solutions can be found in other textbooks.[3] However, they are all "special cases" that do not lead to solutions for a broader class of problems, nor do they generally provide physical insights that can be transferred to other problems. If we approach the question of exact analytic solutions from a more pragmatic or systematic point of view, it is evident that the most important class of problems for which we should expect exact solutions is those for which the nonlinear terms in the equations are identically equal to zero, i.e.,

$$\mathbf{u} \cdot \operatorname{grad} \mathbf{u} \equiv 0.$$

From a mathematical point of view, this requires that the gradients of the velocity $\mathbf{u}$ be orthogonal to $\mathbf{u}$ itself. The obvious question is this: Are there are any physical flows of interest or importance that satisfy this condition? The answer is that there are several. The most important is the class of so-called *unidirectional flows* for which

$$\mathbf{u} = u(q_1, q_2)\mathbf{e}_3,$$

where $\mathbf{e}_3$ is a constant (i.e., nonrotating) unit vector and q_1 and q_2 are spatial coordinates in the plane orthogonal to $\mathbf{e}_3$. We shall discuss a number of flows of this type in this chapter. Examples include flow through a straight tube of arbitrary (but unchanging) cross-sectional shape, or simple shear flow between two plane parallel walls. Closely related to unidirectional flows are 1D flows in which $\mathbf{u}$ follows a single-coordinate direction, and all gradients of the velocity are orthogonal to this direction. An example is the famous *Couette flow* that occurs between two concentric circular cylinders when one (or both) cylinders rotates about its central axis. Here, the flow is in the azimuthal direction, parallel to the motion of the cylinder surfaces, and the velocity gradient is in the radial direction. Not all 1D flows fall into this class. In a later chapter we will discuss the motion produced in a liquid because of the radial expansion or contraction of a spherical bubble. In this case, the flow is strictly radial, but the velocity gradient is also in the radial direction.

We have noted earlier that it can also be very difficult (or impossible) to obtain analytic solutions of (2–93) for the temperature field in a nonisothermal (heat transfer) problem. In this case, the stumbling block is again the term $\mathbf{u} \cdot \operatorname{grad} \theta$. Although this term is linear (provided the Boussinesq approximation is invoked and we neglect natural convection effects), it is typically too complicated to allow analytical solutions to be obtained. We can again consider problems for which an exact solution is possible as being the class in which $\mathbf{u} \cdot \operatorname{grad} \theta \equiv 0$. An example is heat transfer across a fluid between two parallel plane walls that are held at different temperatures, with the flow being unidirectional and thus parallel to the walls. When the offending term in (2–93) is zero, however, the thermal energy balance involves only *pure conduction* as the mode of heat transfer (apart from any effect of the dissipation term, which is usually negligible, as we shall see later). Although pure conduction is generally the only mode of heat transfer in solids, and there are thus many textbooks and papers devoted to analysis or conduction heat transfer in complex geometries, etc.,[4] in most fluids *convection* plays an essential role, as we shall see. Hence, in this book, we focus on heat transfer problems involving convection. In the present chapter, we consider a 1D problem of solidification with a density difference between the two phases. We also consider a pair of heat transfer problems in which the flow is unidirectional, leading to the important phenomena of "Taylor Dispersion" for flow in a straight channel.

A. SIMPLIFICATION OF THE NAVIER–STOKES EQUATIONS FOR UNIDIRECTIONAL FLOWS

We begin by considering the fluid to be Newtonian, incompressible, and isothermal. Furthermore, we consider only a single fluid so that there is no change in density within the domain. Thus the governing equations are (2–20) and (2–91) with μ and ρ given.

The class of *unidirectional* motions that we consider in this chapter is defined simply as a flow in which the direction of motion is independent of position, that is,

$$\mathbf{u}(\mathbf{x}, t) = u(\mathbf{x}, t)\mathbf{e}_3, \quad (3\text{–}1)$$

where $u(\mathbf{x}, t)$ is a scalar function of position and time and $\mathbf{e}_3$ is a constant (nonrotating) unit vector. We do not attempt to identify specific problems at this stage, in which the velocity field actually takes the form (3–1). Instead, we assume that such problems exist and show that the Navier–Stokes and continuity equations are greatly simplified whenever the form (3–1) can be applied. In the process, we derive a linear PDE for the scalar function $u(\mathbf{x}, t)$.

For convenience, we adopt a Cartesian coordinate system with the direction of motion coincident with the z axis. In this system,

$$\mathbf{u} = (0, 0, u), \quad (3\text{–}2)$$

and we can identify $\mathbf{e}_3$ with the Cartesian unit vector in the z direction. We first consider the continuity equation. Because the only nonzero velocity component is in the z direction, this equation becomes simply

$$\frac{\partial u}{\partial z} \equiv 0. \quad (3\text{–}3)$$

It follows that

$$u = u(x, y, t). \quad (3\text{–}4)$$

Hence, as indicated in the introduction to this chapter, u must be independent of position in the flow direction, and it follows from (3–2)–(3–4) that

$$\mathbf{u} \cdot \nabla \mathbf{u} \equiv u \frac{\partial u}{\partial z} \equiv 0. \quad (3\text{–}5)$$

Hence the nonlinear term in the Navier–Stokes equation, (2–91), is identically zero for any flow that satisfies (3–1).

Furthermore, because the x and y components of **u** are zero, it follows from the Navier–Stokes equations that

$$\boxed{\frac{\partial P}{\partial x} = \frac{\partial P}{\partial y} = 0.} \qquad (3\text{--}6)$$

Thus the dynamic pressure is at most a function of z and t:

$$\boxed{P = P(z, t) \text{ only}.} \qquad (3\text{--}7)$$

The only nonzero component of Navier–Stokes equation, (2–91), is the z component that governs $u(x, y, t)$. Taking account of (3–5), we find that this equation is

$$\rho \frac{\partial u}{\partial t} = -\frac{\partial P}{\partial z} + \mu \left(\frac{\partial^2 u}{\partial x^2} + \frac{\partial^2 u}{\partial y^2} \right). \qquad (3\text{--}8)$$

Finally, we may note that the term on the left-hand side and the last two terms on the right-hand side are, at most, functions of x, y, and t. Thus, for any unidirectional flow, $\partial P/\partial z$ cannot depend on z, as suggested by (3–7), but only on t. Indeed, it is convenient to denote $\partial P/\partial z$ as

$$-\frac{\partial P}{\partial z} = G(t), \qquad (3\text{--}9)$$

so that

$$\boxed{\rho \frac{\partial u}{\partial t} = G(t) + \mu \left(\frac{\partial^2 u}{\partial x^2} + \frac{\partial^2 u}{\partial y^2} \right).} \qquad (3\text{--}10)$$

Equation (3–10), which we have derived from the Navier–Stokes equations, governs the unknown scalar velocity function for all unidirectional flows, i.e., for any flow of the form (3–1). However, instead of Cartesian coordinates (x, y, z), it is evident that we could have derived (3–10) by using any cylindrical coordinate system (q_1, q_2, z) with the direction of motion coincident with the axial coordinate z. In this case,

$$u = u(q_1, q_2, t), \qquad (3\text{--}11)$$

and the governing equation for unidirectional flows is

$$\boxed{\rho \frac{\partial u}{\partial t} = G(t) + \mu \left(\nabla_2^2 u \right),} \qquad (3\text{--}12)$$

where

$$\nabla_2^2 \equiv h_1 h_2 \left[\frac{\partial}{\partial q_1} \left(\frac{h_1}{h_2} \frac{\partial}{\partial q_1} \right) + \frac{\partial}{\partial q_2} \left(\frac{h_2}{h_1} \frac{\partial}{\partial q_2} \right) \right]. \qquad (3\text{--}13)$$

Here, h_1 and h_2 are the metrics (or scale factors) of the particular cylindrical coordinate system,[5] defined such that

$$(ds)^2 \equiv \frac{1}{h_1^2}(dq_1)^2 + \frac{1}{h_2^2}(dq_2)^2 + (dz)^2,$$

where ds is the length of an arbitrary differential line element and dq_1, dq_2, and dz represent differential changes in q_1, q_2, and z. Tables with definitions of scale factors and formulae for various vector operations are given in Appendix A for the most common orthogonal coordinate systems.

The introduction of scale factors is necessary because differential changes in q_1 or q_2 of fixed size correspond to line elements of different lengths depending on where we are in the coordinate domain. For example, the length of the arc associated with a differential change in the polar angle $d\theta$ of a circular cylindrical coordinate system depends on proximity to the origin of the coordinate system. The scale factors h_1 and h_2 transform a differential change in q_1 or q_2 to a differential change in length of fixed magnitude, independent of where we are relative to the origin of the coordinate system. Alternatively, we may view them as prescribing the magnitude of the differential change in q_1 or q_2 that is necessary to represent a given differential length in the physical domain. The curvilinear coordinate system that will be of greatest interest here is the circular cylindrical system (r, θ, z). The metrics for this system are $h_1 = 1$ and $h_2 = 1/r$, and the 2D Laplacian, (3–13), in this case is

$$\nabla_2^2 \equiv \frac{1}{r}\frac{\partial}{\partial r}\left(r\frac{\partial}{\partial r}\right) + \frac{1}{r^2}\frac{\partial^2}{\partial \theta^2}. \tag{3–14}$$

For any flow that can be described as *unidirectional*, i.e., as being of the form (3–1), there is a single unknown scalar velocity component to determine, $u(q_1, q_2, t)$.

We have shown that the Navier–Stokes and continuity equations reduce to a governing equation for u of the form (3–12) for any problem in which **u** can be expressed in the form (3–1). We will see that the pressure-gradient function $G(t)$ can always be specified and is considered to be a known function of time. To "solve" a unidirectional flow problem, we must therefore solve Eq. (3–12), with $G(t)$ specified, subject to boundary conditions and initial conditions on u. It is clear from the governing equation that $u(q_1, q_2, t)$ will be nonzero only if either $G(t)$ is nonzero or the value of u is nonzero on one (or more) of the boundaries of the flow domain.

We consider three types of unidirectional flows: steady flows driven either by boundary motion with $G = 0$ or by a nonzero pressure gradient; start-up flows in which a steady boundary motion or a steady pressure gradient is suddenly imposed on a stationary fluid; and transient flows driven by either time-dependent boundary motion or a time-dependent pressure gradient. In addition to changes in the time dependence of either the pressure gradient G or the boundary velocity U, the other thing that can change from one unidirectional flow to another is the geometry of the flow domain. The common feature is that the flow must be in a single direction, and hence any boundaries of the flow domain must be parallel to that direction. However, the cross-sectional geometry of the boundaries is otherwise arbitrary. Two commonly realized cases are two parallel plane walls (of infinite extent) and a circular cylinder, but many other cases are possible.

B. STEADY UNIDIRECTIONAL FLOWS – NONDIMENSIONALIZATION AND CHARACTERISTIC SCALES

We begin with the problem of steady flow between two infinite, parallel plane boundaries. We assume that the pressure gradient G is a nonzero constant and that the upper boundary moves in the same direction as the pressure gradient with a constant velocity U, while the lower boundary is stationary. A sketch of the flow configuration is given in Fig. 3–1. It is conventional to describe the resulting motion of the fluid in terms of Cartesian coordinates x, rather than z, in the direction of motion (parallel to the pressure gradient and the wall

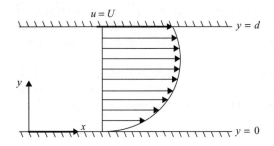

Figure 3–1. A schematic of the steady, unidirectional flow between two infinite parallel plane boundaries, with pressure gradient G acting from left to right in the x direction and the upper plate translating with velocity $u = U$. The arrows represent the magnitude and direction of the fluid velocity at steady state.

velocity), and y, normal to walls. We denote the constant gap width between the two walls as d. Then, starting from (3–10), we see that the equation governing the velocity distribution in the fluid is

$$\boxed{\frac{d^2 u}{dy^2} = -\frac{G}{\mu},} \tag{3–15}$$

which is to be solved subject to the no-slip boundary condition (2–123):

$$\boxed{\begin{aligned} u &= 0 \quad \text{at } y = 0, \\ u &= U \quad \text{at } y = d. \end{aligned}} \tag{3–16}$$

The kinematic boundary condition, (2–118), is satisfied identically because the velocity component normal to the walls is identically equal to zero. Although this problem is extremely simple to solve, the solution is temporarily postponed.

Instead, the procedure of nondimensionalizing the governing DE and boundary conditions is first considered. For now, the motivation for this procedure is not considered, except to say that it will result in a reduction in the number of parameters that characterize the problem (and its solution) from four *dimensional* parameters ($G, \mu, d,$ and U) to a single *dimensionless* parameter. To nondimensionalize, we define a dimensionless velocity,

$$\bar{u} \equiv \frac{u}{u_c}, \tag{3–17}$$

and a dimensionless spatial coordinate,

$$\bar{y} \equiv \frac{y}{\ell_c}, \tag{3–18}$$

where u_c and ℓ_c are known as the *characteristic* velocity and length scale, respectively, and have the same dimensions *as* u and y.

The use of the word "characteristic" in conjunction with the velocity is intended to imply that the magnitude of the velocity, anywhere in the flow, is proportional to u_c. A convenient mathematical representation of this fact is the introduction of the order symbol $u = O(u_c)$, which is stated as u "is of the order u_c" – or, more fully, that the order of magnitude of u is u_c. In the present problem, a reasonable candidate for u_c would seem to be the boundary value U. If this choice is correct, we should expect that doubling the magnitude of U would lead approximately to doubling the magnitude of u everywhere in the fluid. Of course, this choice for u_c is not the only one that could have been made. In particular, another group of

B. Steady Unidirectional Flows – Nondimensionalization and Characteristic Scales

dimensional parameters exists that is proportional to the pressure gradient and has units of velocity is

$$\frac{Gd^2}{\mu}, \qquad (3\text{–}19)$$

and there is no evident reason why this quantity should not have been chosen for u_c, instead of U. For the time being, let us stick with our original choice,

$$u_c = U, \qquad (3\text{–}20)$$

but without forgetting the apparent arbitrariness of this choice relative to the other possibility (3–19).

The physical significance of a characteristic length scale ℓ_c is that it is the distance over which u changes by an amount $O(u_c)$. In the present problem, this would seem to be the distance d between the two parallel plane surfaces,

$$\ell_c = d. \qquad (3\text{–}21)$$

If this choice is correct, then doubling of d should approximately double the distance over which u changes by an amount $O(u_c)$.

With the definitions (3–20) and (3–21) adopted for u_c and ℓ_c, we can express the original problem (3–15) and (3–16) in terms of the dimensionless variables, (3–17) and (3–18), as

$$\boxed{\frac{d^2\bar{u}}{d\bar{y}^2} = -\frac{Gd^2}{\mu U},} \qquad (3\text{–}22)$$

with

$$\boxed{\begin{array}{ll}\bar{u} = 0 & \text{at } \bar{y} = 0, \\ \bar{u} = 1 & \text{at } \bar{y} = 1.\end{array}} \qquad (3\text{–}23)$$

Examining (3–22) and (3–23), we see that the dimensionless problem (and its solution) depends on a single dimensionless parameter,

$$\frac{Gd^2}{\mu U}.$$

This parameter is just the ratio of the two possible velocity scales Gd^2/μ and U, one characterized by the magnitude of the pressure gradient and the other by the magnitude of the boundary velocity. The solution of the problem, in the dimensionless form, (3–22) and (3–23), is

$$\boxed{\bar{u} = -\left(\frac{Gd^2}{\mu U}\right)\left(\frac{\bar{y}^2}{2} - \frac{\bar{y}}{2}\right) + \bar{y}.} \qquad (3\text{–}24)$$

A sketch showing profiles of $\bar{u}$ versus $\bar{y}$ for various values of $Gd^2/\mu U$ is shown in Fig. 3–2.

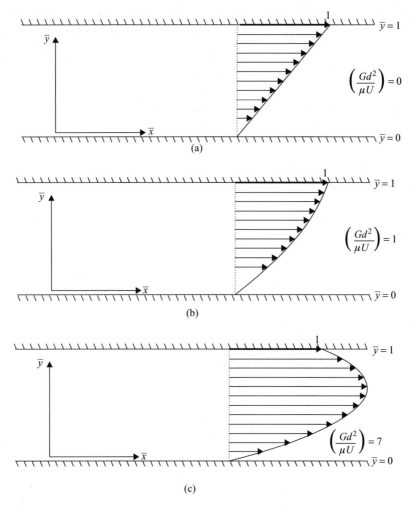

Figure 3–2. Typical velocity profiles for unidirectional flow between infinite parallel plane boundaries: (a) $Gd^2/\mu U = 0$ (simple shear flow), (b) $Gd^2/\mu U = 1$, and (c) $Gd^2/\mu U = 7$. The length of the arrows is proportional to the local dimensionless velocity, with $\bar{u} = 1$ at $\bar{y} = 1$ in all cases. The characteristic velocity scale in this case has been chosen as the velocity of the upper wall, $u_c = U$. The profiles are calculated from Eq. (3–24).

Again, we note that the form of the velocity distribution (usually called the velocity profile) depends on the magnitude of the dimensionless ratio of velocities. When $Gd^2/\mu U \ll 1$, we see from (3–24) that

$$\bar{u} \sim \bar{y}. \qquad (3\text{–}25)$$

In this case, the fluid motion is dominated by the motion of the boundary and the velocity profile reduces to a linear (*simple*) shear flow. When $Gd^2/\mu U \gg 1$, on the other hand, it is apparent from (3–24) that the quadratic contribution dominates the velocity profile; indeed, the quadratic contribution to $\bar{u}$ becomes arbitrarily large as $Gd^2/\mu U$ increases. This increase of $\bar{u}$ does not mean that the actual velocities in the system are blowing up, however. Indeed, we can make $Gd^2/\mu U$ arbitrarily large by simply taking U arbitrarily close to zero while holding Gd^2/μ constant. The increase in $\bar{u}$ simply means that the velocity contribution that is due to the pressure gradient G is increasing *relative* to the boundary

B. Steady Unidirectional Flows – Nondimensionalization and Characteristic Scales

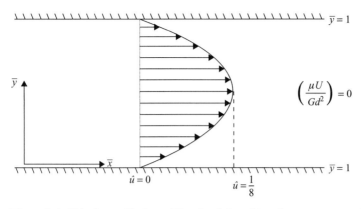

Figure 3–3. Velocity profile for unidirectional flow driven by a pressure gradient between two infinite parallel plane boundaries that are stationary [2D Poiseuille flow, Eq. (3–30)]. The characteristic velocity in this case has been chosen as the centerline velocity divided by 8.

velocity U that was used in the nondimensionalization. The fact is that the boundary velocity U is *not* an appropriate choice for the characteristic velocity when $Gd^2/\mu U \gg U$. In this limit, the magnitude of the velocity in the flow domain is controlled primarily by Gd^2/μ rather than by U. Hence it should be evident by now that

but
$$u_c = U, \quad \text{if} \quad Gd^2/\mu \ll U,$$
$$u_c = Gd^2/\mu, \quad \text{if} \quad Gd^2/\mu \gg U. \tag{3-26}$$

If we consider the governing equations and boundary conditions for $Gd^2/\mu \gg U$, where the second choice for u_c pertains, we find from (3–15) and (3–16) that

$$\frac{d^2 \hat{u}}{d\bar{y}^2} = -1, \tag{3-27}$$

$$\text{with } \hat{u} = 0 \quad \text{at } \bar{y} = 0, \tag{3-28}$$

$$\hat{u} = \frac{U\mu}{Gd^2} \quad \text{at } \bar{y} = 1,$$

where $\hat{u} \equiv u/(Gd^2/\mu)$. In terms of dimensionless variables, the solution in this second case takes the form

$$\hat{u} = -\frac{1}{2}(\bar{y}^2 - \bar{y}) + \frac{U\mu}{Gd^2}\bar{y}. \tag{3-29}$$

Hence, as $U\mu/Gd^2 \to 0$, this solution reduces to

$$\hat{u} = -\frac{1}{2}(\bar{y}^2 - \bar{y}). \tag{3-30}$$

This parabolic velocity profile is characteristic of pressure-gradient-driven flow in the complete absence of boundary motion and is known as a *2D Poiseuille* flow. A sketch of the velocity profile, (3–30), is shown in Fig. 3–3. We see that it is the limiting form of the velocity profile when the motion driven by the pressure gradient dominates the boundary-driven motion.

Unidirectional and One-Dimensional Flow and Heat Transfer Problems

In the intermediate case, in which neither of the limiting cases (3–26) pertains, neither choice for u_c is preferable over the other. In this case, the velocity profile is somewhere between the linear shear flow of $Gd^2/\mu \ll U$ and the parabolic profile of $Gd^2/\mu \gg U$. Either of the dimensionless solution forms, (3–24) or (3–29), is satisfactory. Both show that the form of the velocity profile depends on the dimensionless ratio $Gd^2/\mu U$. Indeed, when converted back to dimensional variables, the two dimensionless solution forms are identical [recall that $\bar{u} = u/U$ in (3–24), but $\hat{u} = u/(Gd^2/\mu)$ in (3–29)]:

$$u = -\frac{Gd^2}{2\mu}\left(\frac{y^2}{d^2} - \frac{y}{d}\right) + U\frac{y}{d}. \qquad (3\text{–}31)$$

This dimensional form of the solution emphasizes the fact that the flow is a linear combination of plane Poiseuille flow and linear shear flow, with the relative magnitude determined by the ratio of Gd^2/μ to U.

Before going on to discuss other unidirectional flow problems, there is one approximation that we have made in deriving the governing equation, (3–10), or (3–12), without justification or discussion, and that is the assumption that the fluid is isothermal. This will often be a very good assumption, but it is never exact because of the presence of *viscous dissipation*. If we return to the full set of governing equations for the motion of an incompressible Newtonian fluid, (2–108)–(2–110), we note that, even for simple unidirectional flows, the viscous dissipation term in Eq. (2–110) is nonzero, indicating that there is kinetic energy of the flow being converted to heat. This indicates that the temperature θ will not be a constant and that the assumption of an "isothermal" fluid is not strictly correct. The important question is to determine whether there are any circumstances when the isothermal fluid assumption can be justified as an approximation. To simplify the discussion, let us consider only the problem of shear flow driven by motion of the boundary at $y = d$ with the pressure gradient $G = 0$.

We assume that the flow will still remain unidirectional, so that u is a function of y only. However, we anticipate that the temperature in the fluid will no longer be a constant, but will also be a function of y because of viscous dissipation. Hence the viscosity μ will also depend on y because it is a function of temperature, and the governing equation for the velocity will then be in the form

$$\frac{d}{dy}\left(\mu \frac{du}{dy}\right) = 0. \qquad (3\text{–}32)$$

Furthermore, to incorporate the influence of viscous dissipation, we must also include the thermal energy equation, (2–110). We seek a steady-state solution, and we assume that the temperature has the same value T_0 at both walls so that any variation of θ with position will be in the y direction. Hence the left-hand side of (2–110) is zero, along with the first term on the right-hand side because p is a constant (this term is generally negligible in any case), and the thermal energy equation thus reduces to the form

$$\frac{d}{dy}\left(k\frac{d\theta}{dy}\right) + \mu\left(\frac{du}{dy}\right)^2 = 0. \qquad (3\text{–}33)$$

To assess the relative importance of the viscous dissipation term, we now introduce characteristic scales, following the guidelines already introduced. Hence we assume that

B. Steady Unidirectional Flows – Nondimensionalization and Characteristic Scales

the characteristic velocity and length scales are U and d [see (3–21) and (3–22)] and define a dimensionless velocity and spatial variable as before. To scale the temperature, we use the wall temperature T_0 as a characteristic temperature, so that a convenient nondimensional temperature is

$$\bar{\theta} = \frac{\theta - T_0}{T_0}.$$

We also note that the viscosity and thermal conductivity now also appear as dependent variables, and so we express them in nondimensional form by using the characteristic values μ_0 and k_0 that correspond to the wall temperature T_0:

$$\bar{\mu} = \frac{\mu}{\mu_0}, \quad \bar{k} = \frac{k}{k_0}.$$

Introducing the nondimensional variables $\bar{u}, \bar{\theta}, \bar{\mu}$, and $\bar{k}$ into (3–32) and (3–33), we can write the resulting equations in the dimensionless form:

$$\boxed{\frac{d}{d\bar{y}}\left(\bar{\mu}\frac{d\bar{u}}{d\bar{y}}\right) = 0,} \tag{3–34}$$

$$\boxed{\frac{d}{d\bar{y}}\left(\bar{k}\frac{d\bar{\theta}}{d\bar{y}}\right) + Br\left[\bar{\mu}\left(\frac{d\bar{u}}{d\bar{y}}\right)^2\right] = 0.} \tag{3–35}$$

It can be seen that the form of the velocity and temperature distributions will depend on the magnitude of a single dimensionless parameter, known as the *Brinkman number*:

$$\boxed{Br \equiv \frac{\mu_0 U^2}{k_0 T_0}.} \tag{3–36}$$

The Brinkman number provides a measure of the magnitude of viscous dissipation relative to the other terms in (3–34) and (3–35). Hence the assumption that viscous dissipation can be neglected is a valid first approximation whenever

$$Br \ll 1.$$

In this case, the solution to (3–35), with the boundary condition that $\bar{\theta} = 0$ at both $\bar{y} = 0$ and 1, is simply $\bar{\theta} \equiv 0$, meaning that the temperature is just a constant T_0 and the assumption that the fluid is isothermal is justified.

In the sections and chapters that follow, we will often make the assumption that viscous dissipation can be neglected. However, we must always keep in mind that this is only an approximation based on the assumption that the Brinkman number is vanishingly small. In the next chapter, we will, however, return to the problem defined by Eqs. (3–34) and (3–35) to consider the effect of viscous dissipation when the Brinkman number is small, but not small enough to allow viscous dissipation to be completely neglected.

Closely related to the pressure-driven unidirectional flow between two parallel plane surfaces is the pressure-driven motion in a straight tube of circular cross section. This is the famous problem studied experimentally as a model for blood flow in the arteries by Poiseuille in 1840.[6] Although this problem could be solved by use of Cartesian coordinates, with z being the axial direction, it is always much simpler to use a coordinate system in which the boundaries of the flow domain are coincident with a line or surface of the coordinate system. In this case, we use (3–12) with ∇_2^2 expressed in terms of the polar coordinates (r, θ) of a circular cylindrical coordinate system (r, θ, z), that is,

$$\frac{1}{r}\frac{\partial}{\partial r}\left(r\frac{\partial u_z}{\partial r}\right) + \frac{1}{r^2}\frac{\partial^2 u_z}{\partial \theta^2} = -\frac{G}{\mu}, \tag{3-37}$$

in which G is the constant pressure gradient. The boundary condition is

$$u_z = 0 \quad \text{at} \quad r = R, \tag{3-38}$$

where R is the radius of the tube. Because the boundary condition is independent of θ, we look for a solution $u_z = u_z(r)$ only, so that

$$\frac{1}{r}\frac{d}{dr}\left(r\frac{du_z}{dr}\right) = -\frac{G}{\mu}. \tag{3-39}$$

In this case, the only source of motion is the axial pressure gradient, and there is therefore only a single choice possible for the characteristic velocity,

$$\bar{u}_z = \frac{u_z}{u_c}, \quad u_c \equiv \frac{GR^2}{\mu},$$

and a single obvious choice for the characteristic length scale,

$$\bar{r} = \frac{r}{\ell_c}, \quad \ell_c \equiv R.$$

In terms of the dimensionless variables $\bar{u}_z$ and $\bar{r}$, the Poiseuille flow problem, (3–38) and (3–39), takes the form

$$\boxed{\frac{1}{\bar{r}}\frac{d}{d\bar{r}}\left(\bar{r}\frac{d\bar{u}_z}{d\bar{r}}\right) = -1, \tag{3-40}}$$

$$\boxed{\text{with} \quad \bar{u}_z = 0 \quad \text{at} \quad \bar{r} = 1. \tag{3-41}}$$

It is noteworthy that the Poiseuille flow problem in this dimensionless form is completely independent of the dimensional parameters G, R, and μ. This means that the form of the velocity profile that we seek is universal and does not depend on the values of these parameters. Integrating (3–41) twice with respect to $\bar{r}$, we obtain

$$\bar{u}_z = -\frac{\bar{r}^2}{4} + A \ln \bar{r} + B, \tag{3-42}$$

where A and B are two constants of integration. The boundary condition, (3–41), gives us one condition from which to determine A and B. A second condition is that $\bar{u}_z$ is bounded for all $\bar{r}$, $0 \leq \bar{r} \leq 1$. This second condition can be satisfied only if $A \equiv 0$, and the solution satisfying the no-slip condition, (3–41), is then

$$\boxed{\bar{u}_z = \frac{1}{4}(1 - \bar{r}^2). \tag{3-43}}$$

If this solution is expressed in terms of dimensional variables, it becomes

$$u_z = \frac{GR^2}{4\mu}\left(1 - \frac{r^2}{R^2}\right). \tag{3-44}$$

B. Steady Unidirectional Flows – Nondimensionalization and Characteristic Scales

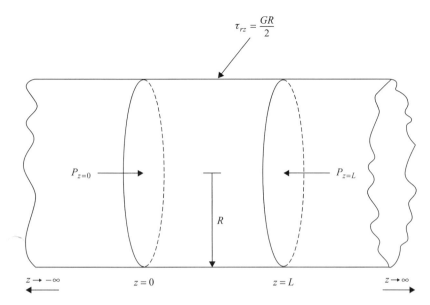

Figure 3–4. A pictorial representation of the force balance on the fluid within an arbitrarily chosen section of a circular tube in steady Poiseuille flow. Pressure forces act on the two cross-sectional surfaces at $z = 0$ and $z = L$, while a viscous stress acts at the cylindrical boundary and exactly balances the net pressure force.

As we noted earlier, the form of the velocity profile is parabolic, independent of the dimensional parameters that serve to determine only the magnitude of u_z. If the velocity profile, (3–44), is integrated over the cross section of the tube, we obtain an equation for the volumetric flux:

$$Q = \int_0^{2\pi} d\theta \int_0^R u_z(r)\, r\, dr \equiv \frac{\pi G R^4}{8\mu}. \tag{3–45}$$

This is the famous *Hagen–Poiseuille law*, which is the basis of capillary viscometry.

In the steady, unidirectional flow problems considered in this section, the acceleration of a fluid element is identically equal to zero. Both the time derivative $\partial u/\partial t$ and the nonlinear "inertial" terms are zero so that $D\mathbf{u}/Dt \equiv 0$. This means that the equation of motion reduces locally to a simple balance between forces associated with the pressure gradient and viscous forces due to the velocity gradient. Because this simple force balance holds at every point in the fluid, it must also hold for the fluid system as a whole. To illustrate this, we use the Poiseuille flow solution. Let us consider the forces acting on a body of fluid in an arbitrary section of the tube, between $z = 0$, say, and a downstream point $z = L$, as illustrated in Fig. 3–4. At the walls of the tube, the only nonzero shear-stress component is τ_{rz}. The normal-stress components at the walls are all just equal to the pressure and produce no *net* contribution to the overall forces that act on the body of fluid that we consider here. The viscous shear stress at the walls is evaluated by use of (3–44),

$$f_z = \mathbf{i}_z \cdot (\mathbf{n} \cdot \mathbf{T}) = -\tau_{rz} = -\mu \frac{du_z}{dr}\bigg|_{r=R} = \frac{GR}{2}, \tag{3–46}$$

and the total viscous force at the walls between $z = 0$ and $z = L$ is

$$F_z = f_z (2\pi R L) = G R^2 \pi L. \tag{3–47}$$

123

However, $G \equiv -\partial P/\partial z$ is a constant [see (3–9)], and so can be expressed in terms of P at $z = 0$ and $z = L$ as

$$G \equiv \frac{1}{L}(P_{z=0} - P_{z=L}). \tag{3–48}$$

Thus, substituting into (3–47), we obtain

$$F_z = (P_{z=0} - P_{z=L})\pi R^2. \tag{3–49}$$

However, the quantity on the right-hand side is just the net pressure force in the z direction acting on the fluid between $z = 0$ and $z = L$: the force $P_{z=0}\pi R^2$ acting on the upstream surface at $z = 0$ minus the opposite-directed force $P_{z=L}\pi R^2$ acting on the downstream surface at $z = L$. Thus we have demonstrated that the net pressure and viscous forces acting on any macroscopic body of fluid in steady, unidirectional Poiseuille flow must be in exact balance. The same conclusion is true for any steady, unidirectional flow.

A reader with some previous experience in fluid mechanics may have noted that some of the results quoted in this section appear to be at odds with observation. The most obvious example of this is the universal form, independent of the flow rate, that we found for the Poiseuille flow velocity profile. Careful experiments on flow through a tube show quite different results. Up to some critical flow rate, a parabolic profile is found in agreement with the Poiseuille solution. For higher flow rates, however, the flow changes dramatically to a highly disordered and time-dependent motion with a time-averaged velocity profile in the z direction that is very nearly a "plug-flow" form – that is, $(u_z)_{avg}$ is independent of r except very near the walls. In addition, if individual velocity components were measured, one would find not only rapid variations with respect to time but also nonzero values for all of the instantaneous components u_r, u_z, and u_θ. These experimental observations are a consequence of the transition from a simple, unidirectional laminar flow to a turbulent flow, which is always fully 3D. We shall see later that the critical flow rate for transition to turbulence is determined by a dimensionless parameter that is known as the Reynolds number. In general, the Reynolds number is defined in terms of the characteristic length and time scales as $Re \equiv u_c \ell_c / \nu$. In the present case, $u_c = (GR^2/\mu)$ and $\ell_c = R$. Experimental studies show that the transition to turbulence in Poiseuille flow through a circular tube is generally observed when

$$Re \equiv \left(\frac{GR^2}{\mu}\right) R/\nu \sim 8000. \tag{3–50}$$

How then are we to interpret solution (3–44), which obviously satisfies the equations of motion and the boundary conditions, but has a universal form independent of the Reynolds number?

Our analysis shows that there is a steady, unidirectional flow solution of the Navier–Stokes equations for the Poiseuille flow problem for all flow rates. However, it says nothing about the stability of this solution. To examine this question, further analysis is necessary.[7]

In principle, the simplest theoretical approach is to examine the stability of the solution to arbitrary "disturbances," that is, arbitrary departures from the laminar solution, (3–44). If these disturbances increase in magnitude with time, the basic laminar motion is said to be unstable, and it is clear that a new form for the velocity field will eventually be realized. If they decrease in magnitude, the flow will revert back to its "undisturbed" laminar form. It is generally possible to carry out such an analysis if the disturbance is assumed to be infinitesimal in magnitude, since then the equations of motion governing the disturbance can be linearized and an arbitrary disturbance analyzed by consideration of the growth or

C. Circular Couette Flow – a One-Dimensional Analog to Unidirectional Flows

decay of its Fourier components (or normal modes). If the disturbance is initially of finite amplitude, however, its evolution in time will be governed by the full nonlinear Navier–Stokes equations (even if the undisturbed motion is unidirectional), and we can generally study the growth or decay of specific disturbances only numerically. Whatever the details of execution, this mathematical determination of instability is reflected in the laboratory by a transition at some critical flow condition from the basic laminar solution to a new, more complicated flow. In some instances this transition may result in turbulence, but in other cases it may lead to only a more complicated laminar flow. Generally, the motion that results from an instability of the basic laminar flow (or a bifurcation from it) is too complicated to be described by means of an analytical solution of the Navier–Stokes equations. (Even if we consider an infinitesimal initial disturbance by means of a linearized stability analysis, the growth of this disturbance will ultimately produce a disturbance of finite amplitude that can be described only by means of the full nonlinear Navier–Stokes equations.) An extreme example is the transition from a steady, unidirectional flow to a fully 3D, time-dependent turbulent velocity field. All that we can say is that the simple, laminar solution can be realized experimentally over only a limited range of flow rates.

In the present book, we focus our attention on the solution of the Navier–Stokes equations for laminar flows, frequently without any attempt to analyze the stability (or experimental realizability) of the resulting solutions. In using these solutions, it is therefore quite apparent that we must always reserve judgment as to the range of parameter values where they will exist in practice. We have already noted that experimental observation shows that Poiseuille flow exists for Reynolds numbers only less than a critical value. A general introduction to hydrodynamic stability theory is given in Chap. 12. It should be noted, however, that the stability of Poiseiulle flow is a very difficult problem, and only a short introductory section that is relevant to this problem is provided.[7]

C. CIRCULAR COUETTE FLOW – A ONE-DIMENSIONAL ANALOG TO UNIDIRECTIONAL FLOWS

We have analyzed several simple examples of steady, unidirectional flows in the preceding section. Returning to the case of flow between two infinite plane walls, the case of simple shear flow is of special significance because it is one of the flows used by rheologists to measure the viscosity and generally characterize the flow behavior of viscous and viscoelastic liquids.[8] The main reason is that the form of the flow between two plane boundaries, one of which is moving and the other stationary is the same, namely

$$u = \frac{U}{d} y, \qquad (3\text{–}51)$$

independent of whether the fluid is Newtonian or non-Newtonian (assuming that the fluid is homogeneous across the gap). Thus the shear rate $\dot{\gamma}$ is uniform across the gap, and its magnitude $\dot{\gamma} = U/d$ is known provided that the boundary velocity and the gap width are known. To see this, we may note that the Cauchy equation of motion, (2–32), for a unidirectional flow in the x direction between two plane walls, and in the absence of a pressure gradient, can be expressed in terms of the deviatoric stress $\boldsymbol{\tau} = \mathbf{T} - p\mathbf{I}$, as

$$\frac{d\tau_{yx}}{dy} = 0. \qquad (3\text{–}52)$$

Hence, the shear stress

$$\tau_{yx} = \text{const.} \qquad (3\text{–}53)$$

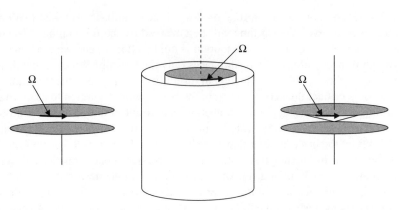

Figure 3–5. Typical rheometer geometries: (a) parallel disk, (b) concentric cylinder (Couette) geometry, (c) cone-and-plate. Either the angular velocity is set and one measures the torque required to produce this rotation rate, or the torque is set and one measures the angular velocity. We analyze the Couette device in this section.

Whatever the nature of the constitutive equation, provided the fluid is homogeneous, this implies that the corresponding component of the rate-of-strain tensor,

$$E_{yx} = \text{const}, \tag{3-54}$$

and the velocity profile must then be linear functions of y and given by (3–51).[9] Linear shear flow is thus an ideal system for rheological studies. To obtain a viscosity, we need only measure the shear stress at one of the boundaries (usually by measuring the total tangential force and dividing by the area), and divide by the shear rate U/d that is known from U and d.

In fact, linear shear flow is rarely used because it is very difficult (nearly impossible) to achieve in the form described. The main problem is that the two plane boundaries are always of finite dimension, and in the process of translating one relative to the other, the common region where the fluid is actually bounded by both boundaries is changing (usually decreasing). This means that the steady shear-flow experiment, in which U is a constant, is limited to a finite time window before "end effects" come strongly into play. As a consequence, most rheological experiments are based on flow devices in which an approximation to simple shear [i.e., the set of conditions (3–51)–(3–54)] is achieved by the rotation of one boundary relative to the other. Typical geometries are the parallel disk device, with two concentric parallel disks, one of which rotates about the common axis; the cone-and-plate, which is very similar except that one of the disks is replaced with a cone with a very shallow cone angle; and the concentric cylinder (or Couette) geometry in which one or both cylinders is rotated about the common axis. These are sketched in Fig. 3–5. The flow in all of them can be analyzed for a Newtonian fluid by an analysis that is very similar to that for unidirectional flows. It must be emphasized, however, that none of these flows is actually unidirectional. On the other hand, they all have a regime of moderate rotation rates in which the velocity field is 1D,[10] with velocity gradients that are orthogonal to the flow direction. This means that the nonlinear terms in the Navier–Stokes are greatly simplified, in a manner similar to unidirectional flows, and it is appropriate to discuss them in this chapter even though they are not unidirectional. The simplest of the three examples just mentioned is the concentric cylinder, Couette flow, and we consider only this one problem in what follows. One initial remark is that the word "circular" is used in the title to this section to distinguish the true Couette flow problem between rotating, concentric cylinders from the idealized

C. Circular Couette Flow – a One-Dimensional Analog to Unidirectional Flows

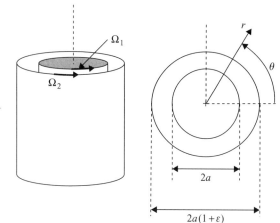

Figure 3–6. The concentric cylinder configuration, including the coordinate axis system.

unidirectional linear shear flow, which is also sometimes (mistakenly) called Couette flow or planar Couette flow.

Now it may appear intuitively evident to the reader, even without any detailed analysis, that the basic velocity profile between two rotating, concentric cylinders will approach the idealized linear shear flow in the limit as the gap width between the cylinders becomes small relative to their radii, and we shall see shortly that this is true. However, the flow is *not unidirectional* regardless of how small the gap width becomes. Nevertheless, provided the rotation rate of the cylinders is not too large, the fluid will follow a circular path between the cylinders. This may at first seem surprising because there is clearly a centripetal acceleration acting on the fluid that is due to curvature of the streamlines, and it is intuitively obvious that this will produce a centrifugal force that has the potential to cause motion that is radially outward across the gap. Indeed, we shall see an example in the next chapter in which an effect of this sort occurs. However, in the Couette flow configuration, the centrifugal force that arises because of centripetal acceleration of fluid elements is balanced exactly by a radial pressure gradient so that the flow remains exactly circular.

To see why this is true, it is simplest to derive a solution for the concentric cylinder problem. For this purpose, it is most convenient to express the governing equations in terms of a circular cylindrical coordinate system (r, θ, z) with the z axis parallel to the central axis of the two cylinders. We denote the radius of the inner cylinder as a and that of the outer cylinder as $a(1 + \varepsilon)$. We assume that either, or both cylinders, may rotate, and denote the angular velocity of the inner cylinder as Ω_1 and that of the outer cylinder as Ω_2. We consider Ω_1 and Ω_2 to be positive when the cylinders rotate in the positive θ direction. A schematic of the concentric cylinder configuration, including the coordinate axis system, is shown in Fig. 3–6. Now, in view of what we have already said, the reader may anticipate that there will be a single nonzero-velocity component u_θ, which will have gradients in the r direction. It may thus be tempting to suppose that we could simply adopt the reduction from the Navier–Stokes to unidirectional flow equations from Section A. However, because the flow is *not* unidirectional, we cannot do this, and we must start again from the full Navier–Stokes equation. For a steady flow, as we assume here, those equations expressed in cylindrical coordinates are the continuity equation

$$\frac{1}{r}\frac{\partial}{\partial r}(r u_r) + \frac{1}{r}\frac{\partial u_\theta}{\partial \theta} + \frac{\partial u_z}{\partial z} = 0, \qquad (3\text{–}55)$$

and the equations of motion

$$\rho\left[u_r\frac{\partial u_r}{\partial r} + \frac{u_\theta}{r}\left(\frac{\partial u_r}{\partial \theta} - u_\theta\right) + u_z\frac{\partial u_r}{\partial z}\right]$$
$$= -\frac{\partial p}{\partial r} + \mu\left[\frac{1}{r}\frac{\partial}{\partial r}\left(r\frac{\partial u_r}{\partial r}\right) + \frac{1}{r^2}\frac{\partial^2 u_r}{\partial \theta^2} + \frac{\partial^2 u_r}{\partial z^2} - \frac{2}{r^2}\frac{\partial u_\theta}{\partial \theta} - \frac{u_r}{r^2}\right], \quad (3\text{--}56)$$

$$\rho\left[u_r\frac{\partial u_\theta}{\partial r} + \frac{u_\theta}{r}\left(\frac{\partial u_\theta}{\partial \theta} + u_r\right) + u_z\frac{\partial u_\theta}{\partial z}\right]$$
$$= -\frac{1}{r}\frac{\partial p}{\partial \theta} + \mu\left[\frac{1}{r}\frac{\partial}{\partial r}\left(r\frac{\partial u_\theta}{\partial r}\right) + \frac{1}{r^2}\frac{\partial^2 u_\theta}{\partial \theta^2} + \frac{\partial^2 u_\theta}{\partial z^2} + \frac{2}{r^2}\frac{\partial u_r}{\partial \theta} - \frac{u_\theta}{r^2}\right], \quad (3\text{--}57)$$

and

$$\rho\left[u_r\frac{\partial u_z}{\partial r} + \frac{u_\theta}{r}\frac{\partial u_z}{\partial \theta} + u_z\frac{\partial u_z}{\partial z}\right]$$
$$= -\frac{\partial p}{\partial z} + \mu\left[\frac{1}{r}\frac{\partial}{\partial r}\left(r\frac{\partial u_z}{\partial r}\right) + \frac{1}{r^2}\frac{\partial^2 u_z}{\partial \theta^2} + \frac{\partial^2 u_z}{\partial z^2}\right]. \quad (3\text{--}58)$$

The boundary conditions are the kinematic condition, (2–113),

$$u_r = 0 \quad \text{at } r = a \text{ and } r = a(1+\varepsilon) \quad (3\text{--}59)$$

and the no-slip condition, (2–122), with the θ component specified by the rotational motion of the cylindrical boundaries:

$$u_z = 0 \quad \text{at } r = a \text{ and } r = a(1+\varepsilon), \quad (3.60\text{a})$$

$$u_\theta = a\Omega_1 \quad \text{at } r = a \text{ and } u_\theta = a(1+\varepsilon)\Omega_2 \quad \text{at } r = a(1+\varepsilon). \quad (3.60\text{b})$$

One immediate comment before discussing the simplification and solution of these equations is that they appear much more complex than the same equations for a Cartesian coordinate system. To cite just one example, we note that the r component of $\nabla^2 \mathbf{u}$ on the right-hand side of (3–56) not only contains second derivatives with respect to r, θ, and z, but a number of additional terms. These arise because the cylindrical coordinate system is *curvilinear*, and indeed are often called "curvature terms" because they arise as a consequence of the curvature of the coordinate lines. The same is true for the $\mathbf{u} \cdot \nabla \mathbf{u}$ terms in which, for example, $(\mathbf{u} \cdot \nabla \mathbf{u})_r \neq \mathbf{u} \cdot \nabla u_r$, again because of curvature terms.

Now, in general, there is no reason to suppose that the solution of (3–55)–(3–58) will be anything but fully 3D, with $u_r = u_r(r, \theta, z)$, $u_\theta = u_\theta(r, \theta, z)$, and $u_z = u_z(r, \theta, z)$. However, for the following analysis, we assume that the cylinders are infinitely long, thus neglecting any z dependence of the flow that is due to end effects that would occur for cylinders of finite length. Of course, in reality, we may make the cylinders very long in comparison with their radius, but not infinite, and the neglect of end effects introduces an approximation into the problem that we do not analyze here because it is extremely complex.[11] With this approximation, we shall assume that the solution takes the simpler form:

$$u_r = u_r(r, \theta), \quad u_\theta = u_\theta(r, \theta), \quad \text{and} \quad u_z = u_z(r, \theta). \quad (3\text{--}61)$$

Then, reexamining the governing equations and boundary conditions, we see that the problem for $u_z(r, \theta)$ is completely homogeneous – that is, the boundary conditions are

C. Circular Couette Flow – a One-Dimensional Analog to Unidirectional Flows

$u_z \equiv 0$ on both cylinder surfaces, and the governing equation is uncoupled from u_r and u_θ. Thus an acceptable solution for u_z is

$$\boxed{u_z(r,\theta) \equiv 0 \quad \text{all } (r,\theta).} \tag{3–62}$$

On the other hand, u_θ is nonzero and *independent* of θ at the boundaries. Thus u_θ must be nonzero within the flow domain, and a plausible assumption for u_θ is that

$$\boxed{u_\theta = G(r) \text{ only.}} \tag{3–63}$$

In this case, it can be seen from the continuity equation, (3–55), that the most general form for u_r is

$$u_r = \frac{F(\theta)}{r}.$$

However, because $u_r = 0$ at both $r = a$ and $r = a(1+\varepsilon)$, it follows that $F(\theta) \equiv 0$ and thus

$$\boxed{u_r(r,\theta) \equiv 0.} \tag{3–64}$$

Finally, the pressure, corresponding to (3–63), must be independent of θ.

In summary, then, we see that an acceptable form for the solution is

$$u_\theta = u_\theta(r), \; p = p(r), \quad \text{and} \quad u_r = u_z = 0. \tag{3–65}$$

The governing equations, (3–56)–(3–58), in this case reduce to the simplified form

$$-\rho \frac{u_\theta^2}{r} = -\frac{dp}{dr}, \tag{3–66}$$

$$0 = \mu \left[\frac{1}{r} \frac{d}{dr}\left(r \frac{du_\theta}{dr}\right) - \frac{u_\theta}{r^2} \right]. \tag{3–67}$$

Equation (3–66) is the balance between centrifugal forces and the radial pressure gradient that is responsible for the fact that $u_r = 0$. Thus, in the Couette flow problem *the acceleration associated with curved fluid pathlines does not necessarily lead to a radial flow*, but may be balanced by a radial pressure gradient.

If we choose a characteristic length scale $\ell_c = a$, a velocity scale $u_c = a\Omega_1$, and a pressure scale $p_c = \rho a^2 \Omega_1^2$, (3–66) and (3–67) can be rewritten in dimensionless form as

$$\boxed{\begin{aligned} \frac{\bar{u}_\theta^2}{\bar{r}} &= \frac{d\bar{p}}{d\bar{r}}, &&(3\text{–}68)\\[4pt] 0 &= \frac{1}{\bar{r}} \frac{d}{d\bar{r}}\left(\bar{r} \frac{d\bar{u}_\theta}{d\bar{r}}\right) - \frac{\bar{u}_\theta}{\bar{r}^2}, &&(3\text{–}69)\\[4pt] &\text{with } \bar{u}_\theta = 1 \quad \text{at } \bar{r} = 1,\\ \bar{u}_\theta &= (1+\varepsilon)\frac{\Omega_2}{\Omega_1} \quad \text{at } \bar{r} = 1+\varepsilon. &&(3\text{–}70) \end{aligned}}$$

It should be noted that these equations are completely characterized by the two dimensionless parameters ε and Ω_2/Ω_1. In particular, the Reynolds number, $Re \equiv u_c \ell_c/\nu = a^2\Omega_1/\nu$, does not appear in spite of the fact that it would have appeared in the full equations, (3–56)–(3–58), if these had been nondimensionalized in the same way. From a mathematical point of view, this is because the viscous terms turned out to be identically equal to zero in (3–56), whereas the inertia and pressure terms were zero in (3–57) – compare (3–66) and (3–67). Thus the form of the velocity and pressure fields in the Couette flow problems is completely independent of Re. In this sense, the Couette flow problem is very similar to a unidirectional flow.

The solution of (3–68) and (3–69) is straightforward. First, if we rewrite (3–69) in the form

$$0 = \bar{r}^2 \frac{d^2 \bar{u}_\theta}{d\bar{r}^2} + \bar{r}\frac{d\bar{u}_\theta}{d\bar{r}} - \bar{u}_\theta, \qquad (3\text{–}71)$$

we see that it is an ordinary differential equation (ODE) of the so-called Euler type, namely, each term is of the form

$$\bar{r}^n \frac{d^n \bar{u}_\theta}{d\bar{r}^n}.$$

The general solution of all such equations is

$$\bar{u}_\theta = \bar{r}^s, \qquad (3\text{–}72)$$

where we obtain a characteristic equation for s by substituting (3–72) into (3–71). The result is

$$(s+1)(s-1) = 0.$$

There are two roots, $s = \pm 1$, corresponding to the two independent solutions

$$\bar{u}_\theta = \bar{r}, \quad \frac{1}{\bar{r}}.$$

A general solution is thus

$$\boxed{\bar{u}_\theta = C_1 \bar{r} + \frac{C_2}{\bar{r}}.} \qquad (3\text{–}73)$$

The constants C_1 and C_2, obtained from application of the boundary conditions, are

$$C_1 = \frac{\frac{\Omega_2}{\Omega_1}(1+\varepsilon)^2 - 1}{(1+\varepsilon)^2 - 1}, \quad C_2 = -\frac{\left(\frac{\Omega_2}{\Omega_1} - 1\right)(1+\varepsilon)^2}{(1+\varepsilon)^2 - 1}. \qquad (3\text{–}74)$$

The profile (3–73) is plotted for several values of ε in Fig. 3–7. If solution (3–73) is now substituted into (3–70), we can integrate to obtain the pressure:

$$\boxed{\bar{p} = C_1^2 \frac{\bar{r}^2}{2} + 2C_1 C_2 \ln \bar{r} - \frac{C_2^2}{2}\frac{1}{\bar{r}^2} + p_0.} \qquad (3\text{–}75)$$

The constant of integration p_0 is arbitrary because the pressure is determinate to within only an arbitrary constant in an incompressible fluid.

We note that, in spite of the superficial similarity to a unidirectional flow, the solution for u_θ is actually quite different from the linear profile of a simple, planar shear flow. This is not at all surprising for an arbitrary ratio of the cylinder radii. However, we should expect that the approximation to a simple shear flow should improve as the gap width becomes

C. Circular Couette Flow – a One-Dimensional Analog to Unidirectional Flows

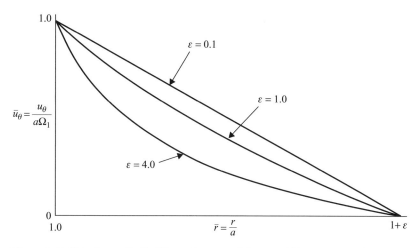

Figure 3–7. Velocity profile (3–73) as a function of the gap width parameter ε for the case $\Omega_2 = 0$.

small compared with the radius of the inner cylinder. Indeed, this is the case. Consider the solutions (3–73) and (3–74) in the thin-gap limit, $\varepsilon \ll 1$. In this case, it can be shown that

$$C_1 = -C_2 = \frac{\Omega_2 - \Omega_1}{2\varepsilon\Omega_1}.$$

Further, to compare with the linear shear solutions obtained earlier, we let

$$\bar{y} = \bar{r} - 1$$

so that $\bar{y} = 0$ at the inner cylinder surface and $\bar{y} = \varepsilon$ at the outer cylinder surface. Hence (3–73) can be written as

$$\bar{u}_\theta = \frac{\Omega_2 - \Omega_1}{2\varepsilon\Omega_1}\left[1 + \bar{y} - \frac{1}{(1+\bar{y})}\right][1 + O(\varepsilon)]$$

$$\approx \frac{\Omega_2 - \Omega_1}{2\varepsilon\Omega_1}[2\bar{y} + O(\bar{y}^2)][1 + O(\varepsilon)] \quad \text{because } 0 \le y \le \varepsilon \ll 1.$$

In writing these results, we have anticipated the notation of asymptotic analysis (which is formally introduced in the next chapter), where the symbol $O(\varepsilon)$ is used to indicate that the magnitude of terms neglected in these equations is proportional to ε, and thus "asymptotically smaller" than the terms that were retained for the limit $\varepsilon \ll 1$. This approximate result is precisely in the form of a simple shear flow provided $\varepsilon \ll 1$, namely

$$\boxed{\bar{u}_\theta = \left(\frac{\Omega_2 - \Omega_1}{\varepsilon\Omega_1}\right)\bar{y}.} \quad (3\text{–}76)$$

Hence, to achieve the best possible approximation to a linear shear flow, the Couette device must have a very thin gap relative to the cylinder radius.

The same conclusion can be reached if we go back and reexamine the simplified governing equation, (3–69). We have nondimensionalized this equation by using the inner cylinder radius as a characteristic length scale for the radial gradients of u_θ. However, for small values of ε, the actual gap width $a\varepsilon$ is much smaller than a. Hence, let us change variables as before,

$$\bar{y} = \bar{r} - 1,$$

Unidirectional and One-Dimensional Flow and Heat Transfer Problems

but then express $\bar{y}$ in the "rescaled" form $\varepsilon \hat{y}$. The advantage is that $\bar{y}$ in the gap varies from only 0 to ε, whereas $\hat{y}$ lies in the range $0 \leq \hat{y} \leq 1$. If we consider the transformation of a typical radial derivative, then

$$\frac{d\bar{u}_\theta}{d\bar{r}} = \frac{d\bar{u}_\theta}{d\bar{y}} = \frac{1}{\varepsilon} \frac{d\bar{u}_\theta}{d\hat{y}}.$$

Whereas, $du_\theta/d\bar{r}$ and $du_\theta/d\bar{y}$ are both very large, $O(1/\varepsilon)$ for the narrow-gap limit, the "rescaled" derivative $du_\theta/d\hat{y}$ is $O(1)$. Now, if we express (3–69) in terms of $\hat{y}$ instead of $\bar{r}$, we obtain

$$(1+\varepsilon\hat{y})^2 \frac{1}{\varepsilon^2} \frac{d^2\bar{u}_\theta}{d\hat{y}^2} + (1+\varepsilon\hat{y})\frac{1}{\varepsilon}\frac{d\bar{u}_\theta}{d\hat{y}} - \bar{u}_\theta = 0. \tag{3–77}$$

Hence $(1 + \varepsilon\hat{y}) \approx 1$, and we see that the first term is the dominant term in (3–69) for the small-gap limit, i.e., we can approximate (3–77) as

$$\frac{d^2\bar{u}_\theta}{d\hat{y}^2} \approx 0. \tag{3–78}$$

We are allowed to do this because we have expressed (3–77) in a form in which the various derivatives are guaranteed to be $O(1)$, i.e., independent of ε. In this regard, we should note that rescaling from $\bar{y}$ (which is scaled by use of a) to $\varepsilon\hat{y}$ is equivalent to nondimensionalizing the radial variable y using the gap width $a\varepsilon$.

A few comments should be made about the use of the Couette flow device for measurements of the viscosity of a fluid. Let us suppose that the fluid is incompressible and Newtonian so that the stress tensor has the form (2–81)

$$\mathbf{T} = -p\mathbf{I} + 2\mu\mathbf{E}.$$

In a rheological experiment, one of the two cylinders is typically rotated with a known angular velocity, and the torque required to produce this motion is measured. Let us suppose that the torque is measured on the inner cylinder. Now, if we ignore the finite length of the Couette device, we have seen that there is a single nonzero component of the velocity $u_\theta(r)$. Hence, if we examine the various components of the rate-of-strain tensor $\mathbf{E}$,

$$\mathbf{E} \equiv \frac{1}{2}(\nabla\mathbf{u} + \nabla\mathbf{u}^T),$$

we see that there is only a single nonzero viscous stress component in the flow, namely,

$$\tau_{r\theta} = 2\mu E_{r\theta} = 2\mu\left(\frac{du_\theta}{dr} - \frac{u_\theta}{r}\right). \tag{3–79}$$

By transforming u_θ from (3–73) and (3–74) back to dimensional variables, we can calculate $\tau_{r\theta}$ to obtain

$$\tau_{r\theta} = -\frac{2C_2 a^2 \Omega_1}{r^2}\mu.$$

Hence, at $r = a$,

$$\tau_{r\theta}|_{r=a} = 2\frac{(\Omega_2 - \Omega_1)(1+\varepsilon)^2}{(1+\varepsilon)^2 - 1}\mu. \tag{3–80}$$

The torque on the inner cylinder is

$$\mathbf{G} = \tau_{r\theta}|_{r=a}(4\pi aL)a\,\mathbf{i}_z.$$

Hence, if we *measure* the torque, we can determine $\tau_{r\theta}|_{r=a}$. Furthermore, if we divide the measured $\tau_{r\theta}|_{r=a}$ by $2E_{r\theta}$, we get exactly the viscosity μ.

C. Circular Couette Flow – a One-Dimensional Analog to Unidirectional Flows

On the other hand, because the circular Couette flow is generally adopted as a convenient substitute for a simple shear flow, it maybe tempting to analyze the experimental data as though we exactly had simple shear between two plane boundaries. This would mean dividing $\tau_{r\theta}|_{r=a}$ with an estimated velocity gradient given by the velocity difference of the two walls divided by the gap width as would be exactly correct for a linear shear flow. The latter is simply

$$\frac{\Delta u}{\text{gap width}} = \frac{a\Omega_2 - a\Omega_1}{a\varepsilon} = \frac{\Omega_2 - \Omega_1}{\varepsilon}. \tag{3-81}$$

On the other hand, the actual velocity gradient at $r = a$ is

$$\left(\frac{\partial u_\theta}{\partial r}\right)_{r=a} = \frac{(2\Omega_2 - \Omega_1)(1+\varepsilon)^2 - \Omega_1}{(1+\varepsilon)^2 - 1}, \tag{3-82}$$

which reduces to the linear profile approximation, (3–81), only in the limit $\varepsilon \ll 1$.

If we were to estimate the viscosity by using only the ratio of the measured shear stress, (3–80), to the actual velocity gradient, (3–82), at $r = a$ (instead of the full strain rate $E_{r\theta}$) as would be valid for a linear shear flow, we would obtain

$$\boxed{\mu^* = \mu \frac{2(\Omega_2 - \Omega_1)(1+\varepsilon)^2}{(2\Omega_2 - \Omega_1)(1+\varepsilon)^2 - \Omega_1}} \tag{3-83}$$

instead of the correct result, μ. If we went one step further and estimated the velocity gradient from (3–81) instead of using the actual (i.e., true) value at $r = a$, we would obtain

$$\boxed{\overline{\mu} = \mu \frac{2\varepsilon(1+\varepsilon)^2}{(1+\varepsilon)^2 - 1}.} \tag{3-84}$$

The values of μ^*/μ and $\overline{\mu}/\mu$ calculated with (3–83) and (3–84) are shown in Table 3–1 for the cases $\Omega_2 = 0$ and $\Omega_1 \neq 0$ for various values of ε. Only in the limit $\varepsilon \ll 1$ does $\mu^* \sim \overline{\mu} \sim \mu$. For $\varepsilon \ll 1$, the gap between cylinders is very thin compared with the cylinder radius, and, in this case, the governing equations (and hence the solutions of these equations) reduce to those for a simple shear flow between parallel plane boundaries. Examining Table 3–1, we see that an error of 10% would be made by using the estimate (3–81) with a gap width that is only 5% of the inner cylinder radius.

One other point about the Couette device for rheological experiments on non-Newtonian materials is that the flow profile is not linear, as can be seen from (3–73). Hence, if the fluid rheology is non-Newtonian, we should generally expect that the flow will change from its form for a Newtonian fluid to something quite different. This means that we do not know the velocity gradients *a priori* in the flow cell, but we must measure them if we are to obtain accurate results for the viscosity μ. In practice, rheometers are built with a thin gap, in which the flow approximates a planar simple shear flow, and it is usually assumed that estimate (3–81) gives the velocity gradient.

We have already seen that this approximate method of data analysis is inaccurate, even for Newtonian fluids, because of the neglect of curvature effects in evaluating **E**. If the fluid is non-Newtonian, there is the further problem that the rate-of-strain component, $E_{r\theta}$, will differ from a constant.

Summarizing the results for Couette flow, we have seen that the fluid moves in circular paths and thus the fluid particles are being accelerated. As a consequence, the inertial terms in the equations of motion are nonzero – clearly the flow is *not* unidirectional. However, this centripetal acceleration does not produce a secondary flow because the nonlinear

Table 3–1. Ratio of inferred-to-true viscosities in a Couette device versus gap width

ε	$\bar{\mu}/\mu$	μ^*/μ
0.05	1.10	1.05
0.1	1.15	1.1
0.2	1.31	1.18
0.4	1.63	1.23
1	2.67	1.6

acceleration terms are exactly balanced by a radial pressure gradient. The general tendency of centripetal acceleration to produce changes in flow direction and thus secondary motions (as will be seen in Chap. 4) does, however, make the Couette flow prone to instabilities when it is subjected to small disturbances. In particular, when

$$\Omega_1 R_1^2 > \Omega_2 R_2^2, \tag{3-85}$$

there is a tendency for centrifugal forces acting on small disturbances of the velocity field to produce a nonzero radial velocity – that is, to break the precise balance between centrifugal force and the force that is due to the pressure gradient. This tendency is counteracted by viscous damping for sufficiently large μ, but detailed analysis as well as careful experiment show that the basic Couette flow becomes unstable at a critical value of the so-called *Taylor number*:

$$T = T_c = \frac{3416}{1 + \frac{\Omega_2}{\Omega_1}} \quad \text{for } 0 \leq \frac{\Omega_2}{\Omega_1} < 1$$

where

$$T_c \equiv \frac{4\Omega_1^2 R_1^4}{\nu^2} \left[\frac{\left(1 - \frac{\Omega_2}{\Omega_1}\right)\left(1 - \frac{R_1^2 \Omega_2}{R_2^2 \Omega_1}\right)}{\left(1 - \frac{R_2^2}{R_1^2}\right)^2} \right]. \tag{3-86}$$

For Taylor numbers exceeding T_c, the flow develops a secondary flow pattern in which u_r and u_z are both nonzero. A sketch of the stability criteria given by (3–86) is shown in Fig. 3–8. The reader who is interested in a detailed description of the stability analysis that leads to the criterion (3–86) is encouraged to consult Chap. 12 or one of the standard textbooks on hydrodynamic stability theory (see Chandrashekhar [1992] for a particularly lucid discussion of the instability of Couette flows).[12]

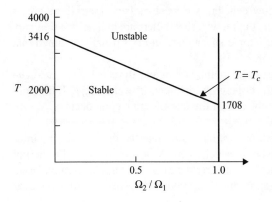

Figure 3–8. Stability diagram for Couette flow with $0 \leq \Omega_2/\Omega_1 < 1$, plotted as Taylor number T versus Ω_2/Ω_1. The flow is stable for $T < T_c$ and unstable for $T > T_c$.

D. START-UP FLOW IN A CIRCULAR TUBE – SOLUTION BY SEPARATION OF VARIABLES

In Section B, we considered several time-independent unidirectional flows. It is evident that the resulting steady velocity profiles can be considered as limiting solutions for large times following the imposition of either a steady boundary motion or a steady pressure gradient. An interesting question that we can study in detail for unidirectional flows is how these steady-state velocity profiles evolve in time if the fluid is initially motionless at the instant of application of the boundary motion or pressure gradient.

In this section, we consider one example of this type of problem, namely, the start-up from rest of the flow in a circular tube when a nonzero pressure gradient is suddenly imposed at some instant (which we shall denote as $t = 0$) and then held at the same constant value thereafter ($t > 0$). In Section G, we consider the start-up of simple shear flow between two infinite plane boundaries. Other relatively simple examples of start-up flows are left to the reader to solve (for example, see Problem 19 at the end of this chapter).

In the present problem of the start-up of pressure-gradient-driven flow in a circular tube, we assume that the tube is very long and focus our attention on only the central region away from the ends so that the motion can be approximated as unidirectional, albeit time dependent. Thus we assume, for $t \geq 0$, that

$$u_z = u_z(r, t), \quad u_r = u_\theta = 0. \tag{3-87}$$

The governing DE, (3–12), is then

$$\rho \frac{\partial u_z}{\partial t} = G + \mu \left[\frac{1}{r} \frac{\partial}{\partial r} \left(r \frac{\partial u_z}{\partial r} \right) \right], \tag{3-88}$$

which is to be solved subject to the no-slip boundary condition,

$$u_z = 0 \quad \text{at} \quad r = R, \quad \text{all } t, \tag{3.89a}$$

the initial condition of no motion

$$u_z = 0 \quad \text{at} \quad t = 0, \quad \text{all } r, \tag{3.89b}$$

and the condition that the solution remains bounded at $r = 0$,

$$u_z \text{ bounded} \quad \text{at} \quad r = 0, \quad \text{all } t. \tag{3.89c}$$

It is clear that a much more complicated analysis would be required for considering the *entry* and *exit* regions near the ends of the tube, because the axial velocity component u_z in that region will be dependent on the distance from the ends of the tube in violation of the basic unidirectional flow assumptions (see Section A).

We begin, again, by nondimensionalizing the governing equations. The characteristic length scale is the tube radius R, and an appropriate choice for the characteristic velocity scale would seem to be GR^2/μ, as this quantity is proportional to the magnitude of the final steady-state velocities [see, e.g., (3–44)]:

$$\ell_c = R, \quad u_c = GR^2/\mu. \tag{3-90}$$

Although other combinations of the parameters (G, μ, ρ, R) can be constructed with dimensions of velocity, the choice (3–90) is the only one in which u_c is proportional to the pressure gradient G. Because the motion is driven solely by the pressure gradient G and the governing equation is linear, it would indeed be surprising if the velocities in the system were not proportional to G.

One important implication of nondimensionalization that we shall use later is that spatial derivatives, such as $\partial \bar{u}_z / \partial \bar{r}$, must satisfy the condition

$$\frac{\partial \bar{u}_z}{\partial \bar{r}} = O(1) \qquad (3\text{--}91)$$

if our choices for ℓ_c and u_c are correct. The symbol $O(A)$ was introduced in the previous section (and will appear throughout this book) as a convenient shorthand notation to indicate that the magnitude of the quantity [for example, $\partial \bar{u}_z / \partial \bar{r}$ in (3–91)] has a numerical value that is *proportional* to A. If this "order-of-magnitude" symbol were applied to the dimensional velocity, for example, we would write

$$u_z = O\left(\frac{GR^2}{\mu}\right),$$

assuming that our choice of the characteristic velocity is correct. When we write $O(1)$, as in (3–83), we imply that the magnitude of $\partial \bar{u}_z / \partial \bar{r}$ is independent of the dimensional parameters of the problem.

To nondimensionalize (3–88) and the boundary conditions (3–89), it is also necessary to identify a characteristic time scale t_c in order to define a dimensionless time $\bar{t} = t/t_c$. The characteristic time scale t_c should be proportional to the time period over which the velocity profile evolves from its initial form, $u_z = 0$, to the final steady state, (3–44). However, it is not immediately obvious how this time scale depends on the dimensional parameters of the problem. Let us leave t_c undefined for the moment and write (3–88) and (3–89) in dimensionless form in terms of t_c, that is,

$$\boxed{\begin{aligned} \frac{R^2}{\nu} \frac{1}{t_c} \left(\frac{\partial \bar{u}_z}{\partial \bar{t}}\right) &= 1 + \frac{1}{\bar{r}} \frac{\partial}{\partial \bar{r}}\left(\bar{r} \frac{\partial \bar{u}_z}{\partial \bar{r}}\right), \\ \text{with } \bar{u}_z &= 0 \quad \text{at } \bar{r} = 1, \text{ all } \bar{t}, \\ \bar{u}_z \text{ bounded} &\quad \text{at } \bar{r} = 0, \text{ all } \bar{t}, \\ \text{and } \bar{u}_z &= 0 \quad \text{at } \bar{t} = 0, \text{ all } \bar{r}. \end{aligned}} \qquad \begin{aligned} (3\text{--}92) \\ \\ (3\text{--}93) \\ \\ \end{aligned}$$

We can convert this problem to a slightly more familiar mathematical form by introducing $\bar{w}(\bar{r}, \bar{t})$ as the difference between $\bar{u}_z(\bar{r}, \bar{t})$ and the final steady-state velocity profile [which is given in dimensionless form in (3–43)], namely,

$$\bar{w}(\bar{r}, \bar{t}) = \bar{u}_z(\bar{r}, \bar{t}) - \frac{1}{4}(1 - \bar{r}^2). \qquad (3\text{--}94)$$

Substituting (3–94) into (3–92) and (3–93), we obtain an equivalent problem for $\bar{w}$ in the form

$$\boxed{\begin{aligned} \frac{R^2}{\nu} \frac{1}{t_c} \left(\frac{\partial \bar{w}}{\partial \bar{t}}\right) &= \frac{1}{\bar{r}} \frac{\partial}{\partial \bar{r}}\left(\bar{r} \frac{\partial \bar{w}}{\partial \bar{r}}\right), \\ \text{with } \bar{w} &= 0 \quad \text{at } \bar{r} = 1, \text{ all } \bar{t}, \\ \bar{w} \text{ bounded} &\quad \text{at } \bar{r} = 0, \text{ all } \bar{t}, \\ \text{and } \bar{w} &= -\frac{1}{4}(1 - \bar{r}^2) \quad \text{at } \bar{t} = 0, \text{ all } \bar{r}. \end{aligned}} \qquad \begin{aligned} (3\text{--}95) \\ \\ (3\text{--}96) \\ \\ \end{aligned}$$

D. Start-Up Flow in a Circular Tube – Solution by Separation of Variables

The DE (3–95) is identical in form to the familiar *heat equation* for radial conduction of heat in a circular cylindrical geometry. Thus we see that the evolution in time of the steady Poiseuille velocity profile is completely analogous to the conduction of heat starting with an initial parabolic temperature profile $-(1-\bar{r}^2)/4$. In our problem, the final steady velocity profile is established by diffusion of momentum from the wall of the tube so that the initial profile for $\bar{w}$ eventually evolves to the asymptotic value $\bar{w} \to 0$ as $\bar{t} \to \infty$. The characteristic time scale for any diffusion process (whether it is molecular diffusion, heat conduction, or the present process) is (ℓ_c^2/diffusivity), where ℓ_c is the characteristic distance over which diffusion occurs. In the present process, $\ell_c = R$ and the kinematic viscosity ν plays the role of the diffusivity so that

$$t_c = \frac{R^2}{\nu}. \tag{3–97}$$

In retrospect, it should perhaps have been evident from (3–95) that this would be the appropriate characteristic time scale.[13] However, without the preceding discussion, the important observation of an analogy between the diffusion of momentum in start-up of a unidirectional flow and the conduction of heat would not have been evident. We shall discuss the nature of this process in more detail after we have solved (3–95) and (3–96) to obtain the time-dependent velocity profile.

Although the problem defined by (3–95) and (3–96) is time dependent, it is linear in $\bar{w}$ and confined to the bounded spatial domain, $0 \leq \bar{r} \leq 1$. Thus it can be solved by the method of separation of variables. In this method we first find a set of *eigensolutions* that satisfy the DE (3–95) and the boundary condition at $\bar{r} = 1$; then we determine the particular sum of those eigensolutions that also satisfies the initial condition at $\bar{t} = 0$. The problem (3–95) and (3–96) comprises one example of the general class of so-called Sturm–Louiville problems for which an extensive theory is available that ensures the existence and uniqueness of solutions constructed by means of eigenfunction expansions by the method of separation of variables.[14] It is assumed that the reader is familiar with the basic technique, and the solution of (3–95) and (3–96) is simply outlined without detailed proofs. We begin with the basic hypothesis that a solution of (3–95) exists in the separable form

$$\bar{w}(\bar{r}, \bar{t}) = R(\bar{r})\Theta(\bar{t}). \tag{3–98}$$

Substituting (3–98) into (3–95), we obtain

$$\frac{1}{\Theta}\frac{d\Theta}{d\bar{t}} = \frac{1}{R}\left[\frac{1}{\bar{r}}\frac{d}{d\bar{r}}\left(\bar{r}\frac{dR}{d\bar{r}}\right)\right]. \tag{3–99}$$

Now, because the left-hand side involves a function of $\bar{t}$ only, whereas the right-hand side involves a function of $\bar{r}$ only, it is evident that each can at most equal a constant. We call this constant $-\lambda^2$. Thence,

$$\frac{1}{\Theta}\frac{d\Theta}{d\bar{t}} = -\lambda^2, \tag{3–100}$$

$$\frac{1}{R}\left[\frac{1}{\bar{r}}\frac{d}{d\bar{r}}\left(\bar{r}\frac{dR}{d\bar{r}}\right)\right] = -\lambda^2. \tag{3–101}$$

The solution of (3–100) for arbitrary λ^2 is

$$\Theta = e^{-\lambda^2 \bar{t}}. \tag{3–102}$$

We can rewrite Eq. (3–101) as

$$\bar{r}^2\frac{d^2 R}{d\bar{r}^2} + \bar{r}\frac{dR}{d\bar{r}} + \lambda^2 \bar{r}^2 R = 0, \tag{3–103}$$

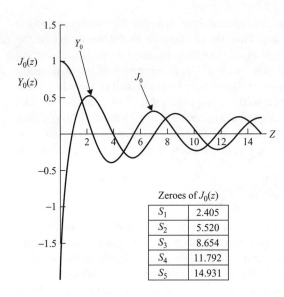

Figure 3–9. The Bessel functions of the first and second kinds of order 0.

Zeroes of $J_0(z)$	
s_1	2.405
s_2	5.520
s_3	8.654
s_4	11.792
s_5	14.931

or, on introducing a change of independent variables,

$$z = \lambda \bar{r}, \tag{3-104}$$

we have

$$z^2 \frac{d^2 R}{dz^2} + z \frac{dR}{dz} + z^2 R = 0. \tag{3-105}$$

This is Bessel's DE of order 0. It has two independent solutions,

$$J_0(z) \quad \text{and} \quad Y_0(z),$$

which are known as Bessel functions of the first and second kinds of order 0.[15] A plot showing the behavior of these two functions is shown in Fig. 3–9. Both oscillate back and forth across zero, but $Y_0(z) \to -\infty$ as $z \to 0$. Hence, the most general solution of the DE (3–95) of the form (3–98) that is bounded at $\bar{r} = 0$ is

$$\bar{w}_n = e^{-\lambda_n^2 \bar{t}} J_0(\lambda_n \bar{r}). \tag{3-106}$$

The subscript n is added in anticipation of the fact that there is an infinite, but discrete, set of values possible for λ such that the general solution, $e^{-\lambda^2 \bar{t}} J_0(\lambda \bar{r})$, satisfies the boundary condition $\bar{w} = 0$ at $\bar{r} = 1$. This set of values of $\lambda = \lambda_n$ is known as the *eigenvalues* of the problem, and the corresponding $\bar{w}_n$ are called the *eigenfunctions*. To determine the eigenvalues λ_n, we apply the boundary condition at $\bar{r} = 1$ to (3–106), that is,

$$\bar{w}_n = e^{-\lambda_n^2 \bar{t}} J_0(\lambda_n \bar{r}) = 0, \quad \text{at } \bar{r} = 1 \text{ for all } \bar{t}. \tag{3-107}$$

This condition requires that

$$J_0(\lambda_n) = 0.$$

Hence the eigenvalues λ_n are equal to the infinite set of zeroes for $J_0(z)$. Referring to Fig. 3–9, we have denoted those zeroes as s_n, with the first crossing for the smallest value of z being s_1, namely,

$$\lambda_n = s_n, \quad n = 1, 2, 3, \ldots, \infty. \tag{3-108}$$

D. Start-Up Flow in a Circular Tube – Solution by Separation of Variables

Thus, because the DE (3–95) and boundary conditions are all linear, the most general form possible for $\bar{w}$ is

$$\bar{w} = \sum_{n=1}^{\infty} A_n \bar{w}_n(\bar{r}, \bar{t}) = \sum_{n=1}^{\infty} A_n e^{-s_n^2 \bar{t}} J_0(s_n \bar{r}), \tag{3–109}$$

where A_n are arbitrary, constant coefficients. This solution satisfies the DE (3–95) and the boundary condition $\bar{w} = 0$ at $\bar{r} = 1$ for any choice of the constant coefficients A_n.

The final step is to choose the A_n so that $\bar{w}(\bar{r}, \bar{t})$ satisfies the initial condition $\bar{w}(\bar{r}, 0) = -(1 - \bar{r}^2)/4$. The general Sturm–Louiville theory[16] guarantees that the eigenfunctions (3–106) form a complete set of orthogonal functions. Thus it is possible to express the smooth initial condition $(1 - \bar{r}^2)$ by means of the Fourier–Bessel series (3–109) with $\bar{t} = 0$, that is,

$$\bar{w}(\bar{r}, 0) = \sum_{n=1}^{\infty} A_n J_0(s_n \bar{r}) = -\frac{1}{4}(1 - \bar{r}^2). \tag{3–110}$$

To determine the A_n, we multiply both sides of (3–110) by $\bar{r} J_0(s_m \bar{r})$ and integrate over $\bar{r}$ from 0 to 1, using the orthogonality properties of J_0:

$$\int_0^1 \bar{r} J_0(s_m \bar{r}) J_0(s_n \bar{r}) d\bar{r} = \begin{cases} 0, & n \neq m \\ \frac{1}{2} J_1^2(s_n), & n = m \end{cases}. \tag{3–111}$$

It follows from this process that

$$A_n = -\frac{\frac{1}{4}\int_0^1 \bar{r}(1 - \bar{r}^2) J_0(s_n \bar{r}) d\bar{r}}{\int_0^1 \bar{r} J_0^2(s_n \bar{r}) d\bar{r}}. \tag{3–112}$$

Finally, using the general properties of the Bessel functions J_0 and J_1, we can show that

$$A_n = -\frac{8}{s_n^3}[J_1(s_n)]^{-1}, \tag{3–113}$$

where J_1 is the Bessel function of the first kind of order 1. This completes the solution of (3–95) and (3–96) for the velocity function $\bar{w}(\bar{r}, \bar{t})$.

Combining (3–113), (3–109), and (3–94) and reverting to dimensional variables, we can express the solution of the full, original problem in terms of the axial velocity profile:

$$\boxed{u_z = \frac{GR^2}{4\mu}\left\{\left[1 - \left(\frac{r}{R}\right)^2\right] - \sum_{n=1}^{\infty} \frac{32}{s_n^3} J_1(s_n)^{-1} e^{-s_n^2 \nu t/R^2} J_0\left(s_n \frac{r}{R}\right)\right\}.} \tag{3–114}$$

Obviously, as $t \to \infty$, this solution reverts to the steady-state Poiseuille profile. To obtain other details of this velocity profile, it is necessary to evaluate the infinite series numerically for each value of t and r. A typical numerical example of the results is shown in Fig. 3–10, where $\bar{u}_z$ has been plotted versus $\bar{r}$ for several values of $\bar{t}$. It can be seen that the initial profile for $\bar{t} = 0.05$ is flat, with $\bar{u}_z$ approximately independent of $\bar{r}$ except for $\bar{r}$ very close to the tube walls. Right at the tube wall, $\bar{u}_z = 0$, and it can be seen that this manifestation of the no-slip condition gradually propagates across the tube by means of the diffusion process discussed earlier. The region in which the wall is felt increases in width at a rate proportional to $\sqrt{\nu t}$ as is typical of diffusion or conduction processes with "diffusivity" ν.

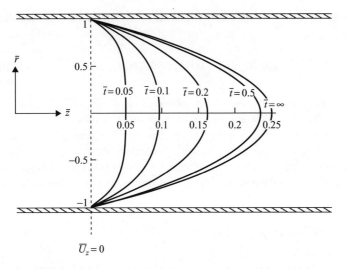

Figure 3–10. Velocity profiles at the different values of $\bar{t}$ depicting the start-up of pressure-gradient-driven flow in a circular tube (far from the ends of the tube). The profiles, from left to right, are for $\bar{t} = 0.05, 0.1, 0.2, 0.5$, and ∞. The velocities in all cases are scaled with the centerline velocity for $\bar{t} = \infty$ divided by 4.

We shall see later that it is often advantageous to consider the temporal or spatial development of a velocity field in terms of diffusion and convection processes. In this framework, the basic concept of diffusion over a distance $\sqrt{\nu t}$ in a time increment t plays a critical role in helping to determine the extent of boundary influence on the flow. It should be recognized, however, that the class of unidirectional flows is a special case in that the direction of diffusion is always at right angles to the direction of motion. In most flows this will not be true, and the influence of the boundary geometry will be propagated by means of momentum transport by both diffusion and convection.

Finally, we return briefly to the problem of choosing appropriate characteristic scales for nondimensionalization. In particular, it is important to recognize the role played by nondimensionalization so that the consequences of an incorrect (or inappropriate) choice of a characteristic scale is clear. The first and most important point is that the introduction of dimensionless variables does not change the problem at all, but simply renames variables. Suppose, for example, that we had used $\ell_c/u_c = \mu/GR$ as a characteristic time scale in (3–93) instead of R^2/ν. Then, instead of

$$\frac{\partial \bar{u}_z}{\partial \bar{t}} = 1 + \frac{1}{\bar{r}}\frac{\partial}{\partial \bar{r}}\left(\bar{r}\frac{\partial \bar{u}_z}{\partial \bar{r}}\right), \tag{3–115}$$

with

$$\bar{t} = t\left(\frac{\nu}{R^2}\right), \tag{3–116}$$

as was actually used earlier in this section, we would have obtained

$$\left(\frac{GR^3\rho}{\mu^2}\right)\frac{\partial \bar{u}_z}{\partial t^*} = 1 + \frac{1}{\bar{r}}\frac{\partial}{\partial \bar{r}}\left(\bar{r}\frac{\partial \bar{u}_z}{\partial \bar{r}}\right), \tag{3–117}$$

in which

$$t^* = t\left(\frac{GR}{\mu}\right). \tag{3–118}$$

The difference in scaling inherent in (3–115) and (3–117) is not significant in this case because the equations are linear. The exact solution must take the form (3–114)

D. Start-Up Flow in a Circular Tube – Solution by Separation of Variables

independently of how (or whether) the DE is nondimensionalized before its solution. Generally speaking, however, the governing equations in fluid mechanics will be nonlinear, and we can seek only approximate solutions obtained by some simplification of the equations. In this case, the relative importance of the various terms in the equations and boundary conditions is frequently determined on the basis of nondimensionalization. To illustrate the general ideas, let us reconsider (3–115) and (3–117) and recall that the basic concept of characteristic scales means that any term that involves only nondimensionalized variables should be $O(1)$ provided the scaling is correct. Hence, assuming that (3–118) is correct, the term on the left-hand side of (3–117) is $O(GR^3\rho/\mu^2)$ whereas the two on the right are $O(1)$. Likewise, if (3–116) is correct, all terms in (3–115) should be $O(1)$. As long as an exact solution of the full equation is achieved, the difference between (3–115) and (3–117) is not important. If we began with the time scaling of (3–118), however, we might be tempted to neglect the left-hand side of (3–117) whenever $GR^3\rho/\mu^2 \ll 1$. It is at this point of approximating the equation before solution that incorrect scaling could cause difficulties if we are not careful in the solution process. It is evident from the DE with correct scaling, (3–115), that the magnitude of the parameter $GR^3\rho/\mu^2$ has *nothing* to do with the relative importance of the various terms. However, suppose we had not previously derived (3–115), but had simply attempted to obtain an approximate solution of the problem for small $GR^3\rho/\mu^2$ by *neglecting* the left-hand side of (3–117). The resulting equation is identical to that for steady flow, (3–40), and the unique solution subject to the condition $\bar{u}_z = 0$ at $\bar{r} = 1$ is just the steady, parabolic Poiseuille profile. However, this solution does not provide a uniformly valid first approximation in time to the actual solution because it does not satisfy the initial condition $\bar{u}_z = 0$ at $t = 0$. Hence the incorrect scaling inherent in (3–117) has resulted in an incorrect approximation to the full equation for $GR^3\rho/\mu^2 \ll 1$ and an approximate solution that does not satisfy all of the boundary and initial conditions of the original problem. Provided that we recognize this last fact, we are forced to conclude that the scaling inherent in (3–117) is not correct – at least for short times, $t^* \ll 1$, or small values of $GR^3\rho/\mu^2$ – and no harm is done provided that we go back and rethink the problem to finally achieve the correct scaling leading to (3–115).

Although we are typically dealing with nonlinear rather than linear DEs, the situation that has been outlined is fairly typical. If we attempt to simplify the governing equations by neglecting terms that appear to be small on the basis of nondimensionalization and if one (or more) of the characteristic scales inherent in the nondimensionalized equations is incorrect, we will always find that it is impossible to satisfy all of the boundary and/or initial conditions of the original problem. Indeed, this should be our signal to reevaluate the choice of characteristic scales. One final point with regard to (3–117) is to explain why the choice of an inappropriate characteristic time scale leads to the false conclusion that the time derivative in (3–88) can be neglected for $GR^3\rho/\mu^2 \ll 1$. The explanation is simple. As long as the nondimensionalization is correct, all terms involving only dimensionless variables are $O(1)$, and the relative magnitude of the various terms can be determined correctly by the magnitude of any dimensionless combination of parameters, such as $GR^3\rho/\mu^2$, that appear. On the other hand, if the characteristic scales are incorrect, then the magnitude of terms involving only dimensionless variables are not necessarily $O(1)$, that is, independent of the dimensional parameters of the problem. Thus, for example, because we know the appropriate scales for the present problem, we can see that as $GR^3\rho/\mu^2$ decreases, the dimensionless time derivative $\partial u_z/\partial t^*$ must be increasing as $(GR^3\rho/\mu^2)^{-1}$, and the conclusion that

$$\left(GR^3\rho/\mu^2\right)\frac{\partial \bar{u}_z}{\partial t^*} \ll 1 \quad \text{for} \quad \left(\frac{GR^3\rho}{\mu^2}\right) \ll 1$$

is incorrect.

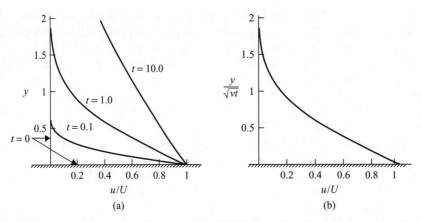

Figure 3–11. (a) The velocity profiles at various increasing times above an infinite flat plate that suddenly begins moving in its own plane at $t = 0$ with a constant velocity U (the Rayleigh problem). (b) The self-similar velocity profile for the same problem obtained by the rescaling of the distance from the plate y with the diffusion length scale $\sqrt{vt}$ for all $t > 0$.

E. THE RAYLEIGH PROBLEM – SOLUTION BY SIMILARITY TRANSFORMATION

In this section, we consider a second example of a transient, unidirectional flow. This is the famous problem, first studied in the 1800s by Lord Rayleigh, in which an initially stationary infinite flat plate is assumed to begin suddenly translating in its own plane with a constant velocity through an initially stationary unbounded fluid.

For convenience, let us adopt a Cartesian coordinate system in which the flat plate is assumed to occupy the xz plane, with the initially stationary fluid occupying the upper half space, $y > 0$. We denote the magnitude of the plate velocity as U so that

$$u_{plate} = \begin{cases} 0, & t < 0 \\ U\mathbf{i}, & t \geq 0 \end{cases}.$$

The problem is to determine the velocity distribution in the fluid as a function of time. In this problem, the fluid motion is due entirely to the motion of the boundary – the only pressure gradient is hydrostatic, and this does not affect the velocity parallel to the plate surface. At the initial instant, the velocity profile appears as a step with magnitude U at the plate surface and magnitude arbitrarily close to zero everywhere else, as sketched in Fig. 3–11. As time increases, however, the effect of the plate motion propagates farther and farther out into the fluid as momentum is transferred normal to the plate by molecular diffusion and a series of velocity profiles is achieved similar to those sketched in Fig. 3–11. In this section, the details of this motion are analyzed, and, in the process, the concept of self-similar solutions that we shall use extensively in later chapters is introduced.

The governing DE for this problem is (3–10) with $G(t) \equiv 0$ and $u = u(y, t)$ only; thus

$$\rho \frac{\partial u}{\partial t} = \mu \frac{\partial^2 u}{\partial y^2}. \tag{3-119}$$

Initially,

$$u(y, 0) = 0, \quad \text{all } y > 0, \tag{3.120a}$$

E. The Rayleigh Problem – Solution by Similarity Transformation

whereas

$$u(0, t) = U, \quad \text{all } t \geq 0, \qquad (3\text{-}120b)$$

and

$$u(y, t) \to 0, \quad \text{as } y \to \infty, \quad t \geq 0. \qquad (3\text{-}120c)$$

The second condition is just the no-slip condition at the plate surface, whereas the third condition arises as a consequence of the assumption that the fluid is unbounded. At any finite $t > 0$, a region will always exist sufficiently far from the plate that no significant momentum transfer will yet have occurred, and in this region the fluid velocity will remain arbitrarily small.

The problem posed by (3–119) with the boundary and initial conditions, (3-120b), is very simple to solve by either Fourier or Laplace transform methods.[14] Further, because it is linear, an exact solution is possible, and nondimensionalization need not play a significant role in the solution process. Nevertheless, we pursue the solution by use of a so-called *similarity transformation*, whose existence is suggested by an attempt to nondimensionalize the equation and boundary conditions. Although it may seem redundant to introduce a new solution technique when standard transform methods could be used, the use of similarity transformations is not limited to linear problems (as are the Fourier and Laplace transform methods), and we shall find the method to be extremely useful in the solution of certain nonlinear DEs later in this book.

Let us begin the solution process in the same way that we have adopted in the preceding sections of this chapter by attempting to nondimensionalize the governing equation and boundary conditions. It is obvious from the problem formulation that an appropriate characteristic velocity for this purpose is

$$u_c = U.$$

However, the geometry of the flow domain is just the upper half-space bounded below by an infinite plane surface, and this does not offer any obvious characteristic length scale. Indeed, it is apparent from the governing equation, (3–119), that the evolution of the velocity profile can be completely described in terms of a 1D diffusion process transporting momentum from the fluid near the plate to the fluid in the rest of the domain, and this suggests that the velocity will vary from U to 0 over a distance that increases continuously with time at a rate proportional to $\sqrt{\nu t}$. Because characteristic scales must be defined in terms of the fixed parameters of a problem, it appears that a physically relevant characteristic length scale simply does not exist in this case. On the other hand, it is obvious that it should be possible to express the solution of any physical problem in a dimensionless form that does not depend on the system of dimensions that is being employed. In most cases, we do this by simply expressing the problem in terms of dimensionless independent and dependent variables, which we obtain from the original dimensional variables by rescaling with respect to characteristic velocity, length, and time scales. In the present case, however, a characteristic length scale does not exist. The only other way in which a solution can be obtained that is independent of the system of dimensions is if it is a function of some *dimensionless combination* of the dimensional independent variables and parameters of the problem, y, t, ν, and U, rather than a function of y and t separately. We call this dimensionless combination a *similarity variable* and denote it as η. The definition of η in terms of y and t is known as the *similarity transformation*:

$$\eta = \eta(y, t, \nu, U). \qquad (3\text{–}121)$$

A solution, such as $u(y, t)$ in the present problem, that depends on a single dimensionless variable η instead of y and t separately is said to be *self-similar*, and this designation is also the source of the names *similarity variable* and *similarity transformation*.[17] The basic idea of self-similarity is that the series of profiles $u(y, t)$ for various fixed times t will collapse into a single, universal form when u is plotted as a function of η rather than as a function of y.

It will be demonstrated later that the generation of a similarity transformation can always be done by a systematic procedure that requires no guesswork or intuition, provided that a similar transformation actually exists. It has been suggested that the absence of any physically significant characteristic length scale in the present problem is a signal that such a transformation must exist, as otherwise it would not be possible to express the solution in a form that does not depend on the arbitrary system of dimensions that has been adopted. We employ two approaches to the generation of a similarity transformation for this case: First, we consider a physical, intuitive approach that is based on our understanding of the diffusion process by which the velocity profile evolves in time and space; second, we obtain the same solution derived by a systematic procedure that requires no intuition or understanding whatsoever, other than the sense to look for a similarity transformation in the first place.

The physical, intuitive approach is very simple. We first recall that, as t increases, the effect of the wall motion is propagated farther and farther out into the fluid as a consequence of the diffusion of momentum normal to the wall. Further, we have seen in the previous section that diffusion will occur across a region of characteristic dimension ℓ_c in a characteristic time increment,

$$t_c = \frac{\ell_c^2}{\nu}.$$

Thus, at any time t^* after the plate begins to move, the momentum generated at its surface will have diffused outward over a distance

$$\ell^* = \sqrt{\nu t^*}.$$

If we now consider two velocity profiles, one at time t_1^* and the other at time t_2^*, with $t_2^* > t_1^*$, we might expect them to reduce to a single "universal" (self-similar) profile if we were to scale the distance from the wall y with the length scale $\ell^* = \ell^*(t^*)$ of the diffusion process, that is,

$$\boxed{\frac{u}{U} = F\left(\frac{y}{\sqrt{\nu t}}\right)} \quad (3\text{--}122)$$

for any y and t. If u/U does depend on only $y/\sqrt{\nu t}$ rather than on y and t independently, we say that u/U is self-similar, and the similarity transformation (3–121) is

$$\eta = \frac{y}{\sqrt{\nu t}}. \quad (3\text{--}123)$$

It will be noted that the similarity variable suggested by our intuitive argument is dimensionless (though y and t are still dimensional).

To determine whether a solution actually exists in the form (3–122), we must first substitute into (3–119) and (3-120b) to see whether this form is consistent with the governing equation and boundary conditions. To do this, we first calculate

$$\frac{\partial u}{\partial t} = U \frac{dF}{d\eta} \frac{\partial \eta}{\partial t} = -U\left(\frac{1}{2t}\right) \eta \frac{dF}{d\eta} \quad (3\text{--}124)$$

E. The Rayleigh Problem – Solution by Similarity Transformation

and

$$\frac{\partial^2 u}{\partial y^2} = U \frac{\partial \eta}{\partial y} \frac{\partial}{\partial \eta}\left(\frac{\partial \eta}{\partial y}\frac{dF}{d\eta}\right) = U \frac{1}{vt}\frac{d^2 F}{d\eta^2}, \quad (3\text{–}125)$$

and then substitute into (3–119). The result is

$$\boxed{\frac{d^2 F}{d\eta^2} + \frac{1}{2}\eta\frac{dF}{d\eta} = 0.} \quad (3\text{–}126)$$

Thus the transformation (3–122) yields an ODE for F with coefficients that are either constant or a function of η only. This equation is, however, only second order in η, and this means that F can satisfy only two boundary conditions. On the other hand, the velocity u in the original problem satisfies three conditions, two boundary conditions and one initial condition. Evidently, if the similarity transformation (3–122) is to be successful, two of these original conditions must reduce to a single condition on F as a function of η. But, in the present case, this clearly happens: The boundary condition on u as $y \to \infty$ and the initial condition for $t = 0$ are both satisfied with the single condition on F:

$$\boxed{F(\eta) \to 0 \quad \text{as} \quad \eta \to \infty.} \quad (3\text{–}127)$$

The boundary condition on u at $y = 0$ requires that

$$\boxed{F(0) = 1.} \quad (3\text{–}128)$$

Thus the Rayleigh problem is reduced to the solution of the ODE, (3–126), subject to the two boundary conditions, (3–127) and (3–128). When a similarity transformation works, this reduction from a PDE to an ODE is the typical outcome. Although this is a definite simplification in the present problem, the original PDE was already linear, and the existence of a similarity transformation is not essential to its solution. When similarity transformations exist for more complicated, nonlinear PDEs, however, the reduction to an ODE is often a critical simplification in the solution process.

Given the equation and boundary conditions, (3–126), (3–127), and (3–128), the solution of the Rayleigh problem is very simple. Integrating (3–126) twice with respect to η, we obtain

$$F(\eta) = A\int_0^{\eta} e^{-s^2/4}dt + B, \quad (3\text{–}129)$$

where s is a dummy variable of integration and A and B are arbitrary constants. Applying the boundary condition, (3–128), we see that

$$B = 1. \quad (3\text{–}130)$$

To apply (3–127), we require the asymptotic value of the integral in (3–129) for $\eta \to \infty$. Instead of evaluating this directly, we first note that the integral is very nearly the error function

$$\text{erf}(z) \equiv \frac{2}{\sqrt{\pi}}\int_0^z e^{-r^2}dr. \quad (3\text{–}131)$$

Using (3–131) and incorporating (3–130), we can express $F(\eta)$ in the form

$$F(\eta) = A\sqrt{\pi}\,\text{erf}\left(\frac{\eta}{2}\right) + 1.$$

Now, it is known[15] that

$$\operatorname{erf}(z) \to 1 \quad \text{as} \quad z \to \infty. \tag{3-132}$$

Thus the boundary condition (3–127) requires that

$$A = -\frac{1}{\sqrt{\pi}}, \tag{3-133}$$

and the solution for u can be expressed in the form

$$\boxed{\frac{u}{U} = 1 - \operatorname{erf}\left(\frac{\eta}{2}\right),} \tag{3-134}$$

where η is defined in (3–123). This velocity profile is plotted in Fig. 3–11 as a function of η and as a function of y for several values of t. This emphasizes the nature of similarity solutions. It has been argued that the lack of a physically meaningful characteristic length scale means that the solution of a problem should be expressible in terms of some dimensionless combination of dimensional, independent variables, such as $y/\sqrt{vt}$ in the present problem. However, this implies that the profiles for different values of the independent variables should collapse to a single, universal form when plotted as a function of the similarity variable rather than as a function of the individual independent variables. In the present case, $u = u(y, t)$ collapses to a single profile for all t when plotted as $u = u(y/\sqrt{vt})$. This internal *similarity* of the profiles for various t is responsible for the name *self-similar*, which is often used to describe solutions of this type.

Our discussion of the Rayleigh problem is essentially complete. Nevertheless, it is useful to briefly reconsider the solution methodology because we have used our intuitive, physical understanding of the problem to simply write the solution form (3–122) and (3–123) that defines the similarity transformation, and it will not always be possible to do this. Indeed, despite the idea that the lack of meaningful characteristic length scales will generally imply the existence of a self-similar solution form, it may still not be clear in any particular case whether a self-similar solution actually exists. For example, it may be possible that a characteristic length scale exists, but we have not identified it. In the Rayleigh problem, we have argued that there is no characteristic length scale. But there is a combination of dimensional parameters with units of length v/U. Although there is no *a priori* reason to expect that v/U is a length scale characteristic of changes of $O(U)$ in the velocity, it would be possible, in principle, that this is the case, and we would not then expect a similarity solution. Fortunately, a straightforward, systematic scheme can be developed from which one can determine both whether a similarity solution exists and, if so, the details of the similarity transformation.

The starting point for this scheme is the assumption of a very general self-similar solution form:

$$\frac{u}{U} = F(\eta) \quad \text{with} \quad \eta = \left[\frac{y}{g(t)}\right]. \tag{3-135}$$

Although we might be tempted to propose even more general forms for η, such as $[y/g(t)]^m$, there is nothing to be gained from doing so in the present problem because $F(\eta)$ is a completely general function that can just as well depend on η^m. Here, $g(t)$ is an as yet unknown function of t to be determined during the course of the solution process. The only thing that we know is that, if a similarity solution exists, the variable η should be dimensionless. From the definition (3–135) of η, we see that $g(t)$ can be interpreted as providing a time-dependent length scale for the region of significant fluid motion. For

E. The Rayleigh Problem – Solution by Similarity Transformation

example, if $F(\eta) = 0.01$ at some value of $\eta = \eta^*$, the value of u is 1% of U at a distance from the moving boundary that depends on $g(t)$ according to $y^* = \eta^* g(t)$.

Necessary conditions for the existence of a self-similar solution are that (1) the governing PDE must reduce to an ODE for F as a function of η alone, and (2) the original boundary and initial conditions must reduce to a number of equivalent conditions for F that are consistent with the order of the ODE. Of course, a proof of *sufficient* conditions for existence of a self-similar solution would require a proof of existence of a solution to the ODE and boundary conditions that are derived for F. In general, however, the problems of interest will be nonlinear, and we shall be content to derive a self-consistent set of equations and boundary conditions and attempt to solve this latter problem numerically rather than seeking a rigorous existence proof. Let us see how the systematic solution scheme based on the general form (3–135) works for the Rayleigh problem.

The first step is always to ask whether the proposed solution form is compatible with the boundary conditions. This step is important because there is no point in spending the time and effort to try to determine $g(t)$ that is compatible with the DE if the solution form could not possibly satisfy the boundary conditions in any case. It is evident that the reduction from a PDE to an ODE by means of the similarity transformation will decrease the number of boundary and initial conditions that can be satisfied by the function $F(\eta)$. Thus two of the original conditions must be expressible in terms of a single condition on F. When (3-120b) is examined, it is obvious that the only possibility is that the conditions for $y \to \infty$ and $t = 0$ collapse into a single condition for F. This means that η must have the same value for $y \to \infty$ as it does for $t = 0$. Evidently, this should be possible with the general form (3–135) provided that the function $g(0) = 0$. Of course, we do not know yet whether we will be able to obtain a function of $g(t)$ that satisfies this condition. All that we know is that the form (3–135) is not *a priori* inconsistent with the boundary and initial conditions.

The second step is to ascertain whether the form (3–135) is compatible with the DE (3–119), and then determine a specific expression for $g(t)$ and verify that we can satisfy the condition $g(0) = 0$. To do this, we differentiate (3–135) and substitute into (3–119). In this case

$$\frac{\partial u}{\partial t} = U \frac{dF}{d\eta} \left(-\frac{\eta}{g} \frac{dg}{dt} \right),$$

$$\frac{\partial^2 u}{\partial y^2} = U \frac{1}{g^2} \frac{d^2 F}{d\eta^2},$$

and (3–119) can be expressed in the form

$$\frac{d^2 F}{d\eta^2} = -\left(\frac{g}{\nu} \frac{dg}{dt} \right) \eta \frac{dF}{d\eta}. \tag{3–136}$$

If a solution of the similarity form actually exists, it is evident that the governing DE for u must yield an ODE for F as a dimensionless function of η only. This means that the coefficients in (3–136) must be either dimensionless constants (independent of η, y, and t) or dimensionless functions of η alone. But we can see that this will be true only if

$$g \frac{dg}{dt} = C\nu, \tag{3–137}$$

for which (3–136) takes the form

$$\frac{d^2 F}{d\eta^2} = C\eta \frac{dF}{d\eta}. \tag{3–138}$$

Condition (3–137) is simply a first-order ODE for $g(t)$. Indeed, if we integrate, we find the general solution

$$g(t) = \sqrt{2\nu Ct} + D \qquad (3\text{–}139)$$

where D is the constant of integration. The condition $g(0) = 0$, which is necessary for a similarity solution to exist, clearly requires $D = 0$.

To this point, nothing has occurred to fix the constant C. We note that if the value is fixed at

$$C = \frac{1}{2}, \qquad (3\text{–}140)$$

Eqs. (3–135) and (3–139) reduce to the forms obtained earlier, namely

$$\frac{d^2 F}{d\eta^2} = -\frac{1}{2}\eta \frac{dF}{d\eta}, \qquad (3\text{–}141)$$

$$\eta = y/\sqrt{\nu t}. \qquad (3\text{–}142)$$

However, the value assigned to the constant C is arbitrary, provided only that $C > 0$. Any choice will yield the same solution. The reader may wish to solve (3–138), leaving C unspecified, to demonstrate this fact. In practice, we would usually choose C to obtain the simplest form for the solution for F. In this case, rather than $C = 1/2$, this suggests $C = 2$, but it is emphasized that this is pure convenience, not a requirement.

In the first analysis of this section, we relied on intuition and some physical insight to guide the choice of form for the similarity variable. However, in this latter derivation, we did not require any real understanding of the physics of the problem, apart from the initial form (3–135), to generate the same similarity transformation.

F. START-UP OF SIMPLE SHEAR FLOW

In the preceding section, we discussed the classical Rayleigh problem of start-up of a unidirectional flow in the unbounded half-plane, $y > 0$, caused by the sudden imposition of tangential motion of an infinite plane wall at $y = 0$. Of course, no real fluid is truly unbounded in extent. Here, we consider the effect of a second stationary plane wall that is parallel to the first one and located at a distance $y = d$ away. It is evident from Section B that the solution at a sufficiently long time following imposition of the boundary motion will be simple shear flow, $u = U(1 - y/d)$. Thus the solution for long times cannot be self-similar. Of course, this should not be a surprise in view of the discussion in the preceding section, as a definite characteristic length scale exists, namely, $\ell_c = d$. On the other hand, for sufficiently short times, $t \ll d^2/\nu$, when the effect of the boundary motion has propagated only a short distance relative to the gap width d, the flow will not be affected significantly by the second wall at $y = d$, and in this case we should expect that the self-similar Rayleigh solution will give a good approximation to the actual velocity profile. In this section, we first derive an exact solution, by means of separation of variables, for the start-up flow between two parallel plane walls that is due to the sudden motion of one of them in its own plane with a constant velocity U. Following this, we explore the form of the solution for $t \ll d^2/\nu$ to show that it does indeed reduce to a self-similar form for sufficiently short times. Last, we see how this same conclusion can be reached (more simply) by considering the governing equations and boundary conditions rescaled for small times, $t = 0(\varepsilon)$, where ε is a small parameter.

The problem is identical to that solved in the preceding section, (3-119) and (3-120b), with the exception that the boundary condition as $y \to \infty$ is replaced with the boundary condition

$$u(d, t) = 0. \qquad (3\text{–}143)$$

F. Start-Up of Simple Shear Flow

However, in this case, it is straightforward to nondimensionalize. The characteristic velocity scale is $u_c = U$, and an appropriate characteristic length scale is $\ell_c = d$. Because the velocity field is established by means of diffusion of momentum, the characteristic time scale is $t_c = d^2/\nu$. Using these characteristic quantities, we find that the problem in dimensionless form is

$$\frac{\partial \bar{u}}{\partial \bar{t}} = \frac{\partial^2 \bar{u}}{\partial \bar{y}^2}, \qquad (3\text{–}144)$$

with $\bar{u}(y, 0) = 0$ all y,

$$\bar{u}(0, \bar{t}) = 1, \quad \text{all } \bar{t} \geq 0, \qquad (3\text{–}145)$$

$$\bar{u}(1, \bar{t}) = 0, \quad \text{all } \bar{t} \geq 0.$$

Because characteristic velocity, length, and time scales all exist, *we do not expect a solution of self-similar form*, but instead seek a solution by using separation of variables.

At steady state, we have already noted that the solution of (3–144) and (3–145) is just

$$\bar{u} = 1 - \bar{y} \quad (\bar{t} \to \infty),$$

and it is convenient, as is usual in start-up problems, to solve for the difference between this steady-state solution and the actual velocity field at any instant:

$$\hat{u} = \bar{u} - (1 - \bar{y}). \qquad (3\text{–}146)$$

Expressing (3–145) and (3–145) in terms of u', we have

$$\frac{\partial \hat{u}}{\partial \bar{t}} = \frac{\partial^2 \hat{u}}{\partial \bar{y}^2}, \qquad (3\text{–}147)$$

with $\hat{u}(y, 0) = -(1 - \bar{y})$,

$$\hat{u}(0, \bar{t}) = 0, \quad \text{all } \bar{t} \geq 0, \qquad (3\text{–}148)$$

$$\hat{u}(1, \bar{t}) = 0, \quad \text{all } \bar{t} \geq 0.$$

Now we seek a solution for (3–147) and (3–148) in the form

$$\hat{u}(\bar{y}, \bar{t}) = F(\bar{y}) H(\bar{t}). \qquad (3\text{–}149)$$

Substituting into (3–147) and solving the resulting ODEs for H and F, we find that the general solution that vanishes as $\bar{t} \to \infty$ is

$$\hat{u} = e^{-\lambda \bar{t}} \left(c_1 \sin \sqrt{\lambda} \bar{y} + c_2 \cos \sqrt{\lambda} \bar{y} \right). \qquad (3\text{–}150)$$

Here, λ is the eigenvalue, which remains to be determined, and c_1 and c_2 are arbitrary constants. It follows from the boundary condition at $\bar{y} = 0$ that $c_2 = 0$, and the condition at $\bar{y} = 1$ yields the set of eigenvalues

$$\lambda = n^2 \pi^2 \quad \text{for } n = 1, 2, \ldots, . \qquad (3\text{–}151)$$

The corresponding eigensolution is then

$$u_n = e^{-n^2 \pi^2 \bar{t}} \sin n \pi \bar{y}, \qquad (3\text{–}152)$$

and the most general solution of (3–147) that satisfies the boundary conditions at $\bar{y} = 0$ and $\bar{y} = 1$ and vanishes at $\bar{t} = \infty$ is

$$\hat{u} = \sum_{n=1}^{\infty} A_n e^{-n^2 \pi^2 \bar{t}} \sin n \pi \bar{y}. \qquad (3\text{–}153)$$

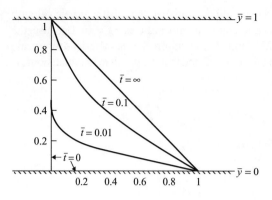

Figure 3–12. The velocity profiles at different times $\bar{t}$ for start-up of simple shear flow between two infinite, parallel plane walls.

All that remains is to determine the coefficients A_n so that the initial condition is satisfied:

$$-(1-\bar{y}) = \sum_{n=1}^{\infty} A_n \sin n\pi\bar{y}. \quad (3\text{–}154)$$

Multiplying both sides of (3–154) by $\sin m\pi y$ and integrating over $\bar{y}$ from 0 to 1, we obtain

$$A_n = \frac{\int_0^1 (\bar{y}-1) \sin n\pi\bar{y}\, d\bar{y}}{\int_0^1 \sin^2 n\pi\bar{y}\, d\bar{y}} \quad (3\text{–}155)$$

or

$$A_n = \frac{2}{n\pi}. \quad (3\text{–}156)$$

Thus,

$$\boxed{\bar{u} = (1-\bar{y}) - \sum_{n=1}^{\infty} \frac{2}{n\pi} e^{-n^2\pi^2 \bar{t}} \sin n\pi\bar{y}.} \quad (3\text{–}157)$$

This solution can also be expressed in terms of dimensional variables:

$$\boxed{\frac{u}{U} = \left(1 - \frac{y}{d}\right) - \sum_{n=1}^{\infty} \frac{2}{n\pi} e^{-n^2\pi^2 \nu t/d^2} \sin n\pi \frac{y}{d}.} \quad (3\text{–}158)$$

The evolution of $\bar{u}$ with respect to $\bar{t}$ is plotted in Fig. 3–12. As in previous examples of start-up flows, we see that the momentum from the lower wall propagates across the gap by means of diffusion in a dimensionless time interval $\bar{t} = O(1)$, or, in dimensional terms, $t = O(d^2/\nu)$.

It is evident, in examining (3–157) or (3–158), that the present exact solution is not self-similar in form. Nevertheless, we expect intuitively that the solution should reduce to the self-similar form for $t \ll O(d^2/\nu)$ [or $\bar{t} \ll O(1)$] because the upper wall at $y = d$ would not yet have any significant impact on the flow. However, it is by no means obvious in examining (3–157) or (3–158) that this will be the case.

Let us then consider the asymptotic form of (3–157) for very short times, $\bar{t} \ll 1$. A convenient way to do this is to introduce an arbitrary small parameter ε and to assume that $\bar{t} = O(\varepsilon)$ with $\varepsilon \ll 1$ in the domain of interest. Thence, introducing the change of variable

$$\bar{t} = \varepsilon \hat{t}, \quad (3\text{–}159)$$

F. Start-Up of Simple Shear Flow

we convert from a limit process involving the independent variable $\bar{t}$ directly to a formal process in which an arbitrarily small time interval in $\bar{t}$ is achieved by taking the asymptotic limit $\varepsilon \to 0$ with the rescaled time $\hat{t} = O(1)$. The main reason for introducing this formal limiting process is that we cannot consider the limit of small $\bar{t}$ in (3–157) without simultaneously considering the limit of small $\bar{y}$. This is a consequence of the fact that momentum diffusion only generates velocities that are significantly different from zero over a dimensionless distance $l/d \sim \sqrt{\bar{t}}$ starting from a stationary fluid at $\bar{t} = 0$. It follows that if we are to establish the asymptotic form of (3–157) for small times by taking the asymptotic limit $\varepsilon \to 0$, we must simultaneously apply the limit $\bar{y} = O(\varepsilon^{1/2}) \to 0$ as $\varepsilon \to 0$. This is most conveniently accomplished by introducing a rescaling of $\bar{y}$,

$$\bar{y} = \varepsilon^{1/2}\hat{y}, \tag{3-160}$$

and then taking the limit $\varepsilon \to 0$ with both $\hat{t}$ and $\hat{y}$ fixed and $O(1)$.

Now let us consider the limiting form of (3–157) for small $\bar{t}$ by introducing (3–159) and (3–160) and taking the limit $\varepsilon \to 0$. On substitution of (3–159) and (3–160) into (3–157), we obtain

$$\bar{u} = 1 - \varepsilon^{1/2}\hat{y} - \sum_{n=1}^{\infty} \frac{2}{n\pi} e^{-n^2\pi^2\varepsilon\hat{t}} \sin n\pi\varepsilon^{1/2}\hat{y}. \tag{3-161}$$

Now, in the limit $\varepsilon \to 0$, the term $\varepsilon^{1/2}\hat{y}$ can be neglected compared with 1, and the infinite sum appears more and more as a continuous variation that can be approximated as an integral. To see this, we define

$$\Delta S = \pi\sqrt{\varepsilon},$$

and rewrite the sum in (3–161) as

$$\sum_{n=1}^{\infty} \frac{2}{n\pi} \frac{1}{\Delta S} e^{-(n\Delta S)^2 \hat{t}} \sin(n\hat{y}\Delta S)\Delta S, \tag{3-162}$$

and consider the limit as ε, and therefore ΔS, goes to zero. In this case,

$$\lim_{\Delta S \to 0} \sum_{n=1}^{\infty} \frac{2}{n\pi} \frac{1}{\Delta S} e^{-(n\Delta S)^2 \hat{t}} \sin(n\hat{y}\Delta S)\Delta S = \frac{2}{\pi} \int_0^{\infty} \frac{1}{S} e^{-S^2\hat{t}} \sin(S\hat{y})dS. \tag{3-163}$$

However, this integral is just

$$\frac{2}{\pi} \int_0^{\infty} \frac{1}{S} e^{-S^2\hat{t}} \sin(S\hat{y})dS = \mathrm{erf}\left(\frac{\hat{y}}{2\sqrt{\hat{t}}}\right). \tag{3-164}$$

Hence we conclude that, for small $\bar{t}$,

$$\bar{u} \sim 1 - \mathrm{erf}\left(\frac{\bar{y}}{2\sqrt{\bar{t}}}\right). \tag{3-165}$$

If $\bar{u}$, $\bar{y}$, and $\bar{t}$ are expressed in terms of dimensional variables, this expression becomes

$$\frac{u}{U} = 1 - \mathrm{erf}\left(\frac{y/d}{2\sqrt{t\frac{\nu}{d^2}}}\right) = 1 - \mathrm{erf}\left(\frac{y}{2\sqrt{\nu t}}\right). \tag{3-166}$$

This is precisely the self-similar solution of the Rayleigh problem, (3–134), which was obtained in the previous section. Notice that the length scale d drops completely out of this limiting form of the solution (3–158). This is consistent with our earlier observation that the presence of the upper boundary should have no influence on the velocity field for sufficiently small times $\bar{t} \ll 1$ (or, equivalently, $t \ll d^2/\nu$).

Thus we see, as expected, that the exact solution (3–157) reduces to a self-similar form for $\bar{t} \ll O(1)$. However, the analysis to demonstrate this fact was somewhat complicated, and we may ask whether a simpler approach might not be possible by direct examination of the governing equation and boundary/initial conditions, (3–144) and (3–145). To determine the asymptotic form for these equations for small $\bar{t}$, we again introduce the small parameter ε and rescale $\bar{y}$ and $\bar{t}$ in (3–144) and (3–145) according to (3–159) and (3–160). The result is

$$\frac{\partial \bar{u}}{\partial \hat{t}} = \frac{\partial^2 \bar{u}}{\partial \hat{y}^2}, \qquad (3\text{–}167)$$

with

$$\bar{u}(\hat{y}, 0) = 0, \quad \text{all } \hat{y} = 0(1),$$
$$\bar{u}(0, \hat{t}) = 1, \quad \text{all } \hat{t} = 0(1), \qquad (3\text{–}168)$$
$$\bar{u}(\varepsilon^{-1/2}\hat{y}, \hat{t}) = 0, \quad \text{all } \hat{t} = 0(1).$$

Thus, in the limit $\varepsilon \to 0$, the present problem reduces to the Rayleigh problem, with governing equation and boundary/initial conditions given by (3–119) and (3-120b).

We have seen that we can obtain the solution for start-up of simple shear flow between two parallel flat plates as an eigenfunction expansion by using the method of separation of variables. The solution is *not* self-similar in form, and this is consistent with the existence of a characteristic length scale d. On the other hand, when we consider the same problem with the upper stationary plate removed (that is, the Rayleigh problem), the solution *is* self-similar, and we have argued that this is to be expected in the absence of a characteristic length scale. Of course, no fluid domain is actually unbounded, and it is important to recognize the role of self-similar solutions in the context of real problems. In particular, we now see that the self-similar Rayleigh solution is the limiting, asymptotic form for small times compared with the time scale for diffusion of momentum across the gap separating the two plates, that is, $t \ll d^2/\nu$. It is important to note that it is much easier to obtain the self-similar solution directly – if one is really interested in only the small time domain – than it is to first derive the exact solution (3–157) or (3–158) and then obtain the self-similar form as a limit. The role of self-similar solutions in general is illustrated by the role of the self-similar solution here; because unbounded domains without characteristic length scales do not occur in reality, we must always expect that self-similar solutions will exist only as an approximation when the flow is sufficiently localized that any characteristic length scales are irrelevant to the problem.

G. SOLIDIFICATION AT A PLANAR INTERFACE

For all of the flows that can be classified as unidirectional, the analysis of Section A shows that the governing equations reduce to solving a single "heat" equation for the magnitude of the scalar velocity component in the flow direction. For heat transfer applications, mathematically analogous problems involve heat transport by *pure conduction*. As noted earlier, there are excellent comprehensive books devoted exclusively to the solution of this class of problems.[4] Here, we consider a related problem, which is chosen because it addresses the physically important coupling of heat transfer in the presence of phase change and also because it is another 1D problem that exhibits a self-similar solution.

We consider a vertically oriented cylindrical container, which may be treated as infinitely long, that is initially filled with a pure, single-component liquid at a temperature θ_1 that exceeds the freezing temperature of the liquid θ_m. Suddenly, at a time that we may denote as $t = 0$, the base of the cylinder is lowered to a temperature θ_0 that is below the freezing point and a thin layer of the liquid instantaneously freezes. The problem that we wish to

G. Solidification at a Planar Interface

analyze is the rate at which the solid–liquid interface propagates up the cylinder, away from the lower boundary. Although the exact problem can be rather complicated, we introduce several simplifying assumptions to make the problem more tractable. First, we assume that the sidewalls of the container are thermally insulated and allow for perfect slip of the fluid. Second, the properties of the liquid- and solid-phase materials are both assumed to be constants, independent of temperature (or other thermodynamic variables). At the moment, we make this latter assumption in a purely *ad hoc* fashion. Later, we shall develop systematic approximation procedures that will allow us to rationalize such assumptions.

The governing equation in each of the two bulk-phase materials is the thermal energy equation, (2–110). In the solid phase, $\mathbf{u} = 0$, and thus

$$\frac{\partial \theta_s}{\partial t} = \kappa_s \nabla^2 \theta_s, \tag{3-169}$$

with $\kappa_s \equiv k_s/\rho_s C_{p_s}$ being the thermal diffusivity of the solid phase. We denote variables for the solid phase with a subscript s. In the liquid, the velocity is generally nonzero. Hence, because we have assumed that the density is independent of temperature, the thermal energy equation in the liquid phase takes the form

$$\rho_L C_{p_L} \left(\frac{\partial \theta_L}{\partial t} + \mathbf{u} \cdot \nabla \theta_L \right) = 2\mu_L \mathbf{E} : \mathbf{E} + k_L \nabla^2 \theta_L. \tag{3-170}$$

We have assumed that the fluid properties are independent of temperature, and hence no natural convection will occur. However, the kinematic condition, (2.114b), at an interface involving a phase transformation from liquid to solid requires that there be a relative velocity in the liquid relative to the velocity of the interface:

$$\mathbf{u} \cdot \mathbf{n} = \left(\frac{\rho_L - \rho_s}{\rho_L} \right) \mathbf{u}^I \cdot \mathbf{n}. \tag{3-171}$$

If $\rho_L = \rho_s$ the only motion in the system will be the motion of the solid–liquid interface as more liquid is solidified, and the problem would consist of heat transfer by conduction in both phases. Otherwise, when $\rho_L \neq \rho_s$, we must retain the convection term in (3–170).

To solve these equations, we require boundary conditions, as well as an expression for the velocity $\mathbf{u}$. Let us denote the position of the bottom of the cylindrical container as $z = 0$. Then, from the assumption of local thermal equilibrium, we require that condition (2–115) be satisfied at $z = 0$,

$$\boxed{\theta_s = \theta_0 \quad \text{at } z = 0} \tag{3-172}$$

and also at the solid–liquid interface, where the temperature of both phases must be equal to the melting/freezing temperature θ_m. We denote the position of this interface as $z_m(t)$. Then (2–115) requires that

$$\boxed{\theta_s = \theta_L = \theta_m \quad \text{at } z = z_m(t).} \tag{3-173}$$

Finally, because we have assumed that the container is infinitely tall, we require that

$$\boxed{\theta_L \to \theta_1 \quad \text{as } z \to \infty.} \tag{3-174}$$

The initial temperature in the container is also θ_1 at all z, except for θ_0 at $z = 0$.

One additional condition is that the walls of the container are perfectly insulated so that

$$\nabla \theta \cdot \mathbf{n} = 0$$

for both the solid and fluid phases. This implies that $\partial \theta_s / \partial r = \partial \theta_L / \partial r = 0$ at the walls of the container. The imposed temperature change at the base is also independent of both r and the polar angle ϕ. Hence, because the fluid velocity $\mathbf{u}$ generated at the interface is also independent of r and ϕ, and we suppose that the container walls are slip surfaces, we can conclude that the whole problem will be 1D with θ_s and θ_L dependent on z and t only, and $\mathbf{u} = u_z \mathbf{i}_z$ in the liquid *independent* of z.

Hence we want to solve (3–169) and (3–170) in the 1D forms,

$$\boxed{\frac{\partial \theta_s}{\partial t} = \kappa_s \frac{\partial^2 \theta_s}{\partial z^2},} \qquad (3\text{–}175)$$

and

$$\boxed{\frac{\partial \theta_L}{\partial t} + \left(\frac{\rho_L - \rho_s}{\rho_L}\right) \frac{dz_m}{dt} \frac{\partial \theta_L}{\partial z} = \kappa_L \frac{\partial^2 \theta_L}{\partial z^2},} \qquad (3\text{–}176)$$

subject to the boundary and initial conditions previously listed. If we count all of the boundary conditions, we find that we have four in total, two each for θ_s and θ_L, and this might seem to be sufficient because the governing equations are each second order in z. However, we need to remember that the pair of conditions, (3–173), has actually introduced a new unknown function into the problem, namely $z_m(t)$. Furthermore, if we think about the problem physically, we realize that we have not completely defined the problem, because the latent heat that is released when the liquid is solidified does not appear anywhere in the equations and boundary conditions that we have written to this point. Although we have utilized a mass balance at the interface in the form of the kinematic condition, (3–171), we have not used the local heat balance in the form of (2–121).

Following the notation of Chap. 2, we denote the latent heat released per unit mass $H_L - H_s$ as L. Then, because $\mathbf{u}$ in the solid phase is zero, condition (2–121a) can be written in the form

$$\boxed{k_L \frac{\partial \theta_L}{\partial z} - k_s \frac{\partial \theta_s}{\partial z} = \rho_s L \frac{dz_m}{dt} \text{ at } S.} \qquad (3\text{–}177)$$

We see that this additional condition now incorporates the latent heat of fusion and also provides an additional equation to determine $z_m(t)$.

The precedent that we have followed in previous problems is to nondimensionalize the governing equations and boundary conditions in order to identify important dimensionless parameters. In this case, it is easy to scale the temperatures so that they vary between 0 and 1, e.g.,

$$\bar{\theta} \equiv \frac{\theta - \theta_0}{\theta_1 - \theta_0}. \qquad (3\text{–}178)$$

However, there is no obvious length scale in the parameters of the problem. The radius of the tube is irrelevant because the temperature fields are independent of r, and the distance from the bottom of the container to the interface, z_m, varies with time. In fact, this is another example in which the solution of the problem exhibits a self-similar form. In view of this, we

G. Solidification at a Planar Interface

simply solve Eqs. (3–175) and (3–176), with boundary and initial conditions in dimensional form.

We may start with Eq. (3–175) for the solid phase, expressed (for convenience) in terms of $\bar{\theta}_s$ as defined as in (3–178). This equation is, in fact, identical to the governing equation, (3–119), for the Rayleigh problem of Section E. The boundary conditions, also expressed in terms of $\bar{\theta}$, are

$$\bar{\theta}_S = 0 \quad \text{at } z = 0, \tag{3-179}$$

$$\bar{\theta}_s = \bar{\theta}_m \equiv (\theta_m - \theta_0)/(\theta_1 - \theta_0) \text{ at } z = z_m(t). \tag{3-180}$$

In the infinitesimal solid layer at $t = 0$, the temperature is also the freezing point θ_m. Now, in view of the analysis of (3–119) for the Rayleigh problem, it is evident that a general solution of (3–175), expressed in self-similar form, is

$$\bar{\theta}_s = c_1 \, \text{erf}(\eta_s) + c_2, \quad \text{with } \eta_s \equiv \frac{z}{2\sqrt{\kappa_s t}}. \tag{3-181}$$

This form for the solution is consistent with the idea, borrowed from the Rayleigh problem, that the temperature imposed at the base of the container will propagate up the cylinder a distance $\sqrt{\kappa_s t}$ by conduction in a time t. Hence this should also be characteristic of the distance that the interface has propagated up into the cylinder in time t.

If we apply the boundary condition, (3–179), $c_2 = 0$. Further, condition (3–180) requires that

$$\bar{\theta}_m = c_1 \, \text{erf}(\eta_{s,m}), \quad \text{with } \eta_{s,m} \equiv \frac{z_m(t)}{2\sqrt{\kappa_s t}}. \tag{3-182}$$

Now, because $\bar{\theta}_m$ and c_1 are constants, $\eta_{s,m}$ must also be a constant independent of t. This implies that

$$\boxed{z_m(t) = \beta\sqrt{4\kappa_s t},} \tag{3-183}$$

where β is a constant still to be determined. Hence,

$$c_1 = \frac{\bar{\theta}_m}{\text{erf}(\beta)},$$

$$\boxed{\bar{\theta}_s = \bar{\theta}_m \frac{\text{erf}(\eta_s)}{\text{erf}(\beta)}.} \tag{3-184}$$

Equation (3–176) for the liquid phase is somewhat more complicated in appearance. Nevertheless, the boundary condition, (3–173), implies that, if a solution exists, it must again be in a self-similar form. In particular,

$$\bar{\theta}_L(z, t) = \bar{\theta}_m \quad \text{at } z = z_m(t) = \beta\sqrt{4\kappa_s t}. \tag{3-185}$$

This can be satisfied only if the solution of (3–176) takes the self-similar form

$$\bar{\theta}_L = \bar{\theta}_L(\eta_L), \quad \text{where } \eta_L \equiv z/\sqrt{4\kappa_L t}. \tag{3-186}$$

The question is whether a solution of this form can satisfy (3–176) and the other boundary and initial conditions of the problem, (3–174), and the initial condition $\bar{\theta}_L = 1$ at $t = 0$. Substituting (3–186) into (3–176), we obtain

$$\frac{d^2 \bar{\theta}_L}{d\eta_L^2} + \left[2\eta_L + 2\beta \left(\frac{\rho_s - \rho_L}{\rho_L} \right) \left(\frac{\kappa_s}{\kappa_L} \right)^{1/2} \right] \frac{d\bar{\theta}_L}{d\eta_L} = 0. \tag{3–187}$$

We see that the governing equation does reduce to a self-similar form. To solve this equation, we can integrate once with respect to η_L to obtain

$$\frac{d\bar{\theta}_L}{d\eta_L} = c_3 \exp\left\{ -\left[\eta_L^2 + 2\beta \left(\frac{\rho_s - \rho_L}{\rho_L} \right) \left(\frac{\kappa_s}{\kappa_L} \right)^{1/2} \eta_L \right] \right\}. \tag{3–188}$$

Integrating again, we can express the solution in the form

$$\bar{\theta}_L = c_3 \int_{\eta_L}^{\infty} \exp\left\{ -\left[u^2 + 2\beta \left(\frac{\rho_s - \rho_L}{\rho_L} \right) \left(\frac{\kappa_s}{\kappa_L} \right)^{1/2} u \right] \right\} du + c_4. \tag{3–189}$$

The boundary condition, (3–174), expressed in terms of the dimensionless temperature $\bar{\theta}_L$, requires that

$$\bar{\theta}_L = 1 \quad \text{as } \eta_L \to \infty \quad \text{and hence } c_4 = 1. \tag{3–190}$$

This condition also encompasses the initial condition that $\bar{\theta}_L = 1$ at $t = 0$. The second boundary condition is

$$\bar{\theta}_L = \bar{\theta}_m \quad \text{at } \eta_{L,m} = z_m / \sqrt{4\kappa_L t} = \beta (\kappa_s / \kappa_L)^{1/2}. \tag{3–191}$$

Hence,

$$c_3 = (\bar{\theta}_m - 1) \bigg/ \int_{\eta_{L,m}}^{\infty} \exp\left\{ -\left[u^2 + 2\beta \left(\frac{\rho_s - \rho_L}{\rho_L} \right) \left(\frac{\kappa_s}{\kappa_L} \right)^{1/2} u \right] \right\} du. \tag{3–192}$$

The only remaining unknown in the solution is the coefficient β that determines the rate of propagation of the interface according to (3–183). To determine β, we must apply the last interface condition, (3–177). To do this, we can evaluate $\partial \bar{\theta}_L / \partial z$ by means of Eq. (3–188), with c_3 and $\eta_{L,m}$ given by (3–191) and (3–192); calculate $\partial \bar{\theta}_s / \partial z$ at $\eta_{s,m} = \beta$ from (3–184); and use (3–183) to calculate the right-hand side of (3–177). The result from Eq. (3–177) is a very complex relationship for β. A somewhat simpler limiting case occurs when $\rho_s = \rho_L$. In this case, the convection velocity in the liquid is zero, and we can express the solution for $\bar{\theta}_L$ in the form

$$\bar{\theta}_L = 1 + (\bar{\theta}_m - 1) \frac{\text{erfc}(\eta_L)}{\text{erfc}(\eta_{L,m})}.$$

The thermal energy balance at the interface then takes the (still complicated) form

$$-\frac{k_L}{k_s} \left(\frac{\kappa_s}{\kappa_L} \right)^{1/2} \frac{(\bar{\theta}_m - 1) \exp[-\beta^2 (\kappa_s / \kappa_L)]}{\text{erfc}\left[\beta (\kappa_s / \kappa_L)^{1/2} \right]} + \frac{\bar{\theta}_m \exp(-\beta^2)}{\text{erf}(\beta)}$$
$$= \frac{\sqrt{\pi} \rho_s L \beta}{\theta_1 - \theta_0} \frac{\kappa_s}{k_s} = \sqrt{\pi} \beta \bar{\theta}_m St. \tag{3–193}$$

H. Heat Transfer in Unidirectional Flows

Here, St denotes the Stefan number:

$$St \equiv \frac{L(\theta_M - \theta_0)}{C_{p,s}},$$

and we see that the propagation rate coefficient β depends on four dimensionless parameters, $\kappa_s/\kappa_L, k_s/k_L, St,$ and $\overline{\theta}_m$.

With the material properties, the latent heat, and the freezing temperature specified, we could solve this equation for β. Rather than provide numerical results, we go one step further in simplifying, so that we can highlight certain important qualitative features without the need for a lot of tedious numerical work. To do this, we assume that all of the material properties are equal in the solid and liquid phases. In this case, (3–193) reduces to the form

$$\frac{\exp(-\beta^2)}{\beta} \left[\frac{\overline{\theta}_m}{\text{erf}(\beta)} - \frac{\overline{\theta}_m - 1}{\text{erfc}(\beta)} \right] = \sqrt{\pi}\, \overline{\theta}_m\, St.$$

One thing that we can see from this expression is that β is reduced as St (or L) is increased. The more latent heat that is released in the transformation from liquid to solid, the slower the interface moves. This is because the process becomes limited by the rate at which heat can be removed from the interface. As more heat is released, the process is slowed down. In contrast, as the specific heat of the solid increases, the rate of freezing increases because more of the heat can be adsorbed with a decreased need to transport the heat away from the interface.

Although we have not made any detailed calculations, we can also see that the presence of convection in the liquid phase will slow the process down. Heat moves *toward* the interface from the liquid side and is removed on the solid side. For the transition of liquid to solid, the convective motion in the liquid is toward the interface. It can be seen from (3–188) that this increases the temperature gradients in the liquid and thus [according to (3–177)] increases the heat flux to the interface. Hence more heat must be removed by the conduction process on the solid side, and the limitation on how fast this can happen slows the rate of freezing. Similarly, if we were to consider the very similar problem of transforming a vapor to a liquid, the convective motion in the vapor is again toward the interface, and this slows the condensation process by increasing the amount of heat that must be transported away from the interface by conduction through the liquid. In contrast, if we were to melt a solid, or vaporize a liquid, the convective motion that is due to the difference in density would enhance the rate (assuming that the liquid is less dense than the solid).

H. HEAT TRANSFER IN UNIDIRECTIONAL FLOWS

We have seen several examples of unidirectional and 1D flows for which the Navier–Stokes equations simplify to a linear form so that exact analytical solutions can be obtained. The closest analogy would be problems for which $\mathbf{u} \cdot \nabla \theta = 0$. Of course, this is just the limit of pure conduction (or pure diffusion) such as the problem considered in the previous section.

In this section, we instead consider two well-known examples of heat transfer in the fully developed, laminar, and unidirectional flow of a Newtonian fluid in a straight circular tube. We begin with a problem in which there is a prescribed heat flux into the fluid at the walls of the tube, so that there is a steady-state temperature distribution in the tube. At the end of the section, we consider the transient evolution of the temperature distribution beginning with an initially sharp temperature jump within the fluid at a fixed position (say $z = 0$), which illustrates an important phenomenon that is known as Taylor dispersion.

Unidirectional and One-Dimensional Flow and Heat Transfer Problems

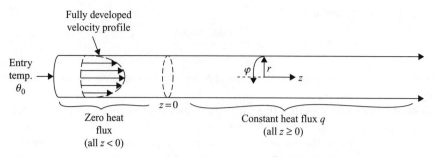

Figure 3–13. A sketch of the configuration for heat transfer in flow through a heated or cooled section of a circular tube.

1. Steady-State Heat Transfer in Fully Developed Flow through a Heated (or Cooled) Section of a Circular Tube

A sketch of the steady-state problem that we will consider is shown in Fig. 3–13. The tube radius is denoted as a, and we utilize the standard cylindrical coordinates (r, ϕ, z).[18] We assume that the wall of the tube is insulated so that the heat flux is zero for positions $z < 0$. On the other hand, beginning at $z = 0$ and for all $z \geq 0$ there is a constant positive heat flux q through the wall of the tube. A reasonable approximation to this condition can be realized if the tube is uniformly wrapped with a heat tape or wire resistance heater beginning at $z = 0$. It may be noted that the problem would be identical from a mathematical point of view if there were a negative heat flux prescribed so that the temperature of the fluid in the tube decreased rather than increased for $z > 0$. We assume that the entry temperature of the fluid into the heated portion of the tube is θ_0 and that viscous dissipation can be neglected.

The governing equation to determine the temperature distribution in the tube is the thermal energy equation, (2–110). To solve this equation, we need to know the form of the velocity distribution in the tube. We have already seen that the steady-state velocity profile for an *isothermal* fluid, far downstream from the entrance to the tube, is the Poiseuille flow solution given by (3–44). In the present problem, however, the temperature must depend on both r and z, and hence the viscosity (which depends on the temperature) will also depend on position. The dependence on z is due to the fact that heat is added for all $z \geq 0$, and thus the temperature must continue to increase with the increase of z. The dependence on r is due to the fact that there must be a nonzero conductive heat flux *in the fluid* at the tube wall to match the prescribed heat flux through the wall, and thus the temperature must have a nonzero r derivative. It follows that the velocity field will generally differ from Poiseuille flow.

A full analysis of the modified velocity field would, in fact, be quite complicated. Because μ must depend on z, as already explained, the exact velocity field is not even unidirectional. For present purposes, we therefore limit our analysis to the simpler limiting situation in which we assume that the changes in temperature are small enough that we can ignore these complicated changes and approximate the velocity field by using the isothermal Poiseuille flow solution:

$$u_z = \frac{GR^2}{4\mu}\left(1 - \frac{r^2}{R^2}\right). \tag{3–44}$$

The most important requirement for approximating the velocity field by means of (3–44) is that the viscosity is (approximately) constant over any cross section of the tube. We shall see that the mean temperature increases in the axial direction over quite large distances compared with the tube radius and we will also neglect the slow decrease in the viscosity and increase

H. Heat Transfer in Unidirectional Flows

in the magnitude of the velocity with z. Although this is offered at the moment as simply an *ad hoc* approximation, we shall see in later chapters that this type of approximation can usually be justified by means of more formal asymptotic arguments.

Finally, use of (3–44) to represent the velocity field also requires that we neglect variations of the density with temperature (and thus with position). If we reexamine the derivation of the governing equations for a steady unidirectional flow, it may not be evident at first that we even need to think about the fluid density because the density appears explicitly only in the time-derivative term of (3–10) or (3–12), and this is zero in a steady flow. However, in using the dynamic pressure (and thus suppressing the body-force term in the Navier–Stokes equation) it was assumed that the density is constant [see Eqs. (2–89)–(2–91)]. If the density is not constant, there will generally be a direct contribution of the body-force term to the motion of the fluid. This contribution is responsible, for example, for "natural convection" flows that occur because of temperature (and density) gradients in a room due to the action of a radiator. A discussion of the conditions for neglect of natural convection effects would take us too far afield for the present. However, it is obvious that they will be negligible if the temperature changes are small enough, and we shall leave it at that for the moment.

Hence, for the analysis in this section, we seek a steady-state solution of (2–110) with velocity approximated as the isothermal Poiseuille flow solution. Because the upstream temperature profile and boundary conditions on the temperature (or temperature gradients) are all independent of the azimuthal angle, we assume that the temperature θ will be a function of only the cylindrical coordinates r and z. Thus Eq. (2–110) can be written in the form

$$2U\left(1 - \frac{r^2}{a^2}\right)\left(\frac{\partial \theta}{\partial z}\right) = \kappa\left[\frac{1}{r}\frac{\partial}{\partial r}\left(r\frac{\partial \theta}{\partial r}\right) + \frac{\partial^2 \theta}{\partial z^2}\right]. \tag{3–194}$$

Here, we have written the Poiseuille solution in terms of the cross-sectional average of the velocity $\langle u_z \rangle \equiv U$. From (3–44) and (3–45), we can see that this is just one-half of the maximum velocity that occurs at the centerline of the tube. The parameter κ that appears is the thermal diffusivity, $\kappa \equiv k/\rho C_p$. We seek to solve (3–194) subject to the conditions that the heat flux at the tube walls is q for $z \geq 0$ and that the temperature is bounded at the centerline, $r = 0$. The condition on the heat flux is applied in the form (2–119). Hence

$$k\frac{\partial \theta}{\partial r} = q \quad \text{at } r = a \quad \text{for all } z \geq 0. \tag{3–195}$$

In writing (3–195) we adopt the convention that q is positive when it represents a heat flux from the tube wall into the fluid, and in this case $(\partial \theta/\partial r)$ is positive.

We begin, as in previous examples, by nondimensionalizing using characteristic scales. We need a characteristic length scale, and for this it seems reasonable to select the tube radius $\ell_c = a$. The other quantity to scale in (3–194) is the temperature θ. However, there is no explicit temperature specified that provides a convenient (or appropriate) choice for scaling θ. Indeed, what is specified is the heat flux q, and if we think about the problem from a qualitative point of view, it is evident that the constant heat flux at the wall will cause the temperature to increase continuously as we move downstream in the tube. Because the increment in θ from its initial value to its value at any fixed z will be proportional to q, we use the quantity $\theta_c = (qa/k)$, which has the same units as θ, to nondimensionalize the temperature. Hence,

$$\bar{\theta} = \frac{\theta - \theta_0}{(qa/k)} \quad \text{with } \hat{z} = z/a \text{ and } \bar{r} = r/a.$$

Using these dimensionless variables, we can write Eq. (3–194) in the nondimensionalized form:

$$2Pe(1 - \bar{r}^2)\frac{\partial \bar{\theta}}{\partial \bar{z}} = \frac{1}{\bar{r}}\frac{\partial}{\partial \bar{r}}\left(\bar{r}\frac{\partial \bar{\theta}}{\partial \bar{r}}\right) + \frac{\partial^2 \bar{\theta}}{\partial \bar{z}^2}. \qquad (3\text{--}196)$$

The one dimensionless parameter that appears in this equation is known as the *Peclet number*:

$$Pe \equiv (Ua/\kappa).$$

The Peclet number represents the ratio of the characteristic magnitude of the convection terms in the thermal energy equation, namely, $\theta_c u_c/\ell_c$, to the conduction terms, $\kappa \theta_c/\ell_c^2$.

Now, assuming that the characteristic scales used to nondimensionalize (3–196) are correct, all of the terms in (3–196) involving dimensionless variables should be independent of the independent parameters of the problem. It follows from this that the relative magnitudes of the various terms should be determined completely by the magnitude of *Pe*. Thus, for *small values of* $Pe \ll 1$, the convection term should be small compared with either of the conduction terms in (3–196), and a first approximation is that the evolution of the temperature field in the heated portion of the tube is dominated by radial and axial conduction.

The more interesting and important situation is that in which $Pe \gg 1$. In this case, the nondimensionalized equation (3–196) suggests that the heat transport process is dominated by convection so that a first approximation is

$$2(1 - \bar{r}^2)\frac{\partial \bar{\theta}}{\partial \bar{z}} \sim 0. \qquad (3\text{--}197)$$

This implies that $\bar{\theta}$ is independent of z, and, because $\bar{\theta} = 0$ at $\bar{z} = 0$, we are led to the conclusion that $\bar{\theta} = 0$ for all $\bar{r}$ and $\bar{z}$. Clearly, however, this "solution" does not satisfy the nonzero heat flux boundary condition at the tube wall, (3–195), and thus it cannot be correct, or at least it cannot be a uniformly valid first approximation to the temperature distribution. Because we derived the approximate form of (3–197) from (3–194) simply by scaling (i.e., nondimensionalizing), using the tube radius as a characteristic length scale, we can only conclude that this choice of characteristic length scale must not be correct, at least for problems in which $Pe \gg 1$. In fact, the assumption that radial and axial changes in the temperature distribution both occur with a characteristic length scale a is not valid in any portion of the tube for large *Pe*.

Near the entry to the heated section of the tube, the heat supplied at the walls can propagate across only a very small fraction of the tube radius. The heat transfer process at the walls is radial conduction, and, because this is a "diffusive" process, we know that it can transfer heat only over a distance $\ell = \sqrt{\kappa t}$ in any time interval t. Therefore, if we assume that the fluid moves downstream with characteristic velocity U, the region of heated fluid at a downstream position $z = a$ will extend only a radial distance in from the wall,

$$\ell = aPe^{-1/2},$$

because the time available for a fluid element to travel from $z = 0$ to this point is only $O(a/U)$ [because $q = O(1)$ this also means that the wall temperature in this region differs only slightly from the inlet temperature, i.e., $[\bar{\theta}_{wall} \sim Pe^{-1/2}]$. Hence, in this "entry" region, the length scale characteristic of radial gradients of the temperature for large *Pe* is clearly not a, as assumed in deriving (3–197) from (3–194), but a much smaller scale that actually depends on *Pe* and gets smaller as *Pe* gets larger. This very thin region near the tube walls

H. Heat Transfer in Unidirectional Flows

for $z = O(a)$ is known as a "thermal boundary layer." The governing equation, (3–197), is not valid in this region because, in taking the limit $Pe \to \infty$ in (3–196), we have failed to notice that the radial derivatives with respect to $\bar{r} = r/a$ actually grow in magnitude with increase of Pe, and thus the term on the left-hand side that involves Pe explicitly is not larger than the radial conduction term on the right-hand side that involves Pe implicitly. In Chaps. 9 and 11, we will discuss the solution of thermal boundary-layer problems by using the methods of asymptotic approximation theory.

Eventually, sufficiently far downstream from $z = 0$, the heating effect of the walls will propagate entirely across the tube, and only then should we expect the length scale characteristic of *radial* gradients of the temperature to be the tube radius. How far downstream do we need to go before this is true? The characteristic time required for the radial conduction process to transport heat a distance equal to the tube radius a is a^2/κ. This requires a distance down the tube of order $U(a^2/\kappa)$. In other words, we must be at a dimensionless distance downstream:

$$\hat{z} = (z/a) \geq O(Ua/\kappa) = O(Pe).$$

To be on the safe side, we consider $\hat{z} \gg O(Pe)$, i.e., $z \gg a\,Pe$. Now, in this regime, the tube radius is an appropriate choice for the characteristic length scale $\ell_c^r = a$ for radial variations of $\bar{\theta}$. However, the length scale characteristic of changes of θ in the z direction must be different from a because having both r and z scaling with a leads to (3–197).

Although the need for a second length scale may be intuitively clear, it is not *a priori* obvious what choice to make for this length scale. We thus follow previous examples in this chapter and introduce the symbol ℓ_c^* to represent this scale, with the hope that an explicit choice for ℓ_c^* will become evident at a later point in the solution process. Hence, introducing

$$\bar{r} = r/a, \quad \bar{z} = z/\ell_c^*, \quad \bar{\theta} = \frac{\theta - \theta_0}{(qa/k)},$$

we can write Eq. (3–194) in the nondimensionalized form:

$$2\varepsilon Pe(1 - \bar{r}^2)\frac{\partial \bar{\theta}}{\partial \bar{z}} = \frac{1}{\bar{r}}\frac{\partial}{\partial \bar{r}}\left(\bar{r}\frac{\partial \bar{\theta}}{\partial \bar{r}}\right) + \varepsilon^2 \frac{\partial^2 \bar{\theta}}{\partial \bar{z}^2}. \tag{3–198}$$

The new dimensionless parameter that appears in (3–198) is the ratio of length scales

$$\varepsilon \equiv (a/\ell_c^*).$$

Although we have not yet specified ℓ_c^* and there is nothing from the nondimensionalized equation, (3–198), that provides a basis at this point to determine what it should be, what we may anticipate from a qualitative point of view is that ℓ_c^* is likely to be much larger than a for large Pe so that the parameter

$$\varepsilon \ll 1.$$

The reason for this is the simple observation that the convection process with a nonuniform velocity profile tends to stretch gradients of temperature in the z direction and thus create an increasing length scale for changes in the z direction as we move farther down the tube.

Hence we shall assume in what follows that the parameter ε is very small. We will have to verify that this assumption is, in fact, true in the region $z \gg aPe$. However, assuming this to be the case, it can be seen from (3–198) that the contribution from axial conduction is very small compared with that of radial conduction, and thus can be neglected, leading to the approximate form

$$2\varepsilon Pe(1-\bar{r}^2)\frac{\partial\bar{\theta}}{\partial\bar{z}} = \frac{1}{\bar{r}}\frac{\partial}{\partial\bar{r}}\left(\bar{r}\frac{\partial\bar{\theta}}{\partial\bar{r}}\right). \qquad (3\text{–}199)$$

It will be noted that we have retained the term involving εPe in spite of the fact that ε is assumed to be very small. However, we need to remember that we have still not specified the axial length scale ℓ_c^*, and because Pe is large, we cannot argue with certainty about whether the product εPe is actually small or large or $O(1)$. From a physical point of view, it is clear that a steady-state temperature distribution can only be established as a balance between radial conduction and either conduction or convection in the axial direction. We have already suggested that axial conduction will play a negligible role when $\varepsilon \ll 1$. In the absence of axial convection, heat would simply enter the fluid at the tube walls and then be conducted radially, so that the temperature at any point in the tube would increase monotonically with time.

Now the question is this: How do we go about specifiying ℓ_c^*? We know from a simple heat balance that the cross-sectionally averaged temperature must continue to increase monotonically with z because we are adding heat to the system at the walls of the tube. In fact, for the regime of $z \gg aPe$, where the effect of the walls has propagated all the way across the tube, we should anticipate that the rate at which θ increases with z (and thus the characteristic length scale for θ changes in the z direction) must be dictated by this rate of heat addition. To see that this is true, we can consider a macroscopic heat balance on a section of the tube between an arbitrarily chosen point $z = z_1 \gg aPe$ and a second point farther along the tube at $z = z_2$ (i.e., $z_2 > z_1$). The total heat flux into the tube from the walls in this section of the tube is simply

$$2\pi aq(z_2 - z_1).$$

On the other hand the net flux of heat out of this section of tube by convection at the two ends is

$$\int_0^{2\pi}\int_0^a \rho C_p(\theta|_{z_2} - \theta|_{z_1})2U\left(1 - \frac{r^2}{a^2}\right)rdrd\varphi.$$

These two terms must be equal at steady state. Hence, introducing the dimensionless forms for r and θ, we can express this condition in the form

$$\Delta z = 2aPe\int_0^1 \Delta\bar{\theta}(1-\bar{r}^2)\bar{r}d\bar{r}, \qquad (3\text{–}200a)$$

where $\Delta z \equiv z_2 - z_1$ and $\Delta\bar{\theta} \equiv \bar{\theta}_2 - \bar{\theta}_1$.

The obvious conclusion from (3–200a) is that a change in the scaled temperature by an amount of order one occurs over a distance along the tube that is proportional to aPe. Hence it follows immediately that this is the appropriate choice for the characteristic length scale ℓ_c^*, i.e.

$$\ell_c^* = aPe$$

Clearly with this choice for ℓ_c^*, the assumption $\varepsilon \ll 1$ is confirmed for $Pe \gg 1$. Furthermore, we see that for this choice for ℓ_c^*, the parameter εPe that appeared in (3–198) is neither

H. Heat Transfer in Unidirectional Flows

large nor small, but actually equal to 1. Indeed, if we return to (3–198) but now with the specific choice $\ell_c^* = aPe$, the equation takes the form

$$2(1 - \bar{r}^2)\frac{\partial \bar{\theta}}{\partial \bar{z}} = \frac{1}{\bar{r}}\frac{\partial}{\partial \bar{r}}\left(\bar{r}\frac{\partial \bar{\theta}}{\partial \bar{r}}\right) + \frac{1}{Pe^2}\frac{\partial^2 \bar{\theta}}{\partial \bar{z}^2}. \tag{3-201}$$

In this form, we can again see that the approximate equation (3–199) is valid only for $Pe \gg 1$ and that the coefficient $\varepsilon Pe = 1$. In the remainder of this section, we use

$$2(1 - \bar{r}^2)\frac{\partial \bar{\theta}}{\partial \bar{z}} = \frac{1}{\bar{r}}\frac{\partial}{\partial \bar{r}}\left(\bar{r}\frac{\partial \bar{\theta}}{\partial \bar{r}}\right) \tag{3-202}$$

in place of (3–199).

To solve (3–202), which is a parabolic equation that is second order in $\bar{r}$ and first order in $\bar{z}$, requires two boundary conditions in $\bar{r}$ and one in $\bar{z}$. The boundary condition (3–195) becomes

$$\frac{\partial \bar{\theta}}{\partial \bar{r}} = 1 \quad \text{at } \bar{r} = 1, \bar{z} > 0. \tag{3-203}$$

We also require

$$\bar{\theta} \text{ bounded at } \bar{r} = 0. \tag{3-204}$$

In the original problem, the solution was also required to satisfy the initial condition $\theta = \theta_0$ (or $\bar{\theta} = 0$) at z (or $\bar{z}) = 0$. Of course, the present approximate formulation is valid only for $z \gg aPe$, and so we cannot apply these initial conditions to the solution of (3–202). However we can see from (3–200a) that an overall integral heat balance from $\bar{z} = 0$ to a large value of $\bar{z}$ (where (3–202) is valid) requires

$$\bar{z} = 2\int_0^1 \bar{\theta}(\bar{z}, \bar{r})(1 - \bar{r}^2)\bar{r}d\bar{r} \tag{3-200b}$$

for arbitrary $\bar{z} \gg 1$. This condition trurns out to be sufficient to obtain a unique solution of (3–202).

Now let us see if we can determine a solution of (3–202), which satisfies the boundary condition (3–203) and (3–204) and the integral constraint (3–200b). The simplest "guess" for $\bar{\theta}$ that can be consistent with the integral constraint (3–200b) is

$$\bar{\theta} = \bar{z}\bar{\theta}_1(\bar{r}) + \bar{\theta}_2(\bar{r}). \tag{3-205}$$

If we substitute this into (3–202), we obtain a pair of coupled equations for the unknown functions $\bar{\theta}_1$ and $\bar{\theta}_2$,

$$\frac{d}{d\bar{r}}\left(\bar{r}\frac{d\bar{\theta}_1}{d\bar{r}}\right) = 0 \tag{3-206}$$

and

$$\frac{1}{\bar{r}}\frac{d}{d\bar{r}}\left(\bar{r}\frac{d\bar{\theta}_2}{d\bar{r}}\right) = 2(1 - \bar{r}^2)\bar{\theta}_1(\bar{r}). \tag{3-207}$$

Integrating (3–206), we find that

$$\bar{\theta}_1 = C_0 \ln(\bar{r}) + C_1. \tag{3-208}$$

Since $\bar{\theta}$ must be bounded at $\bar{r} = 0$, it follows that $C_0 = 0$. Then integrating (3–207), we find

$$\bar{\theta}_2 = 2C_1 \left[\frac{\bar{r}^2}{4} - \frac{\bar{r}^4}{16} \right] + C_2 \ln(\bar{r}) + C_3. \tag{3-209}$$

Once again $C_2 = 0$. Applying the boundary condition (3–203), we find

$$C_1 = 2$$

so that

$$\bar{\theta} = 2\bar{z} + \left(\bar{r}^2 - \frac{\bar{r}^4}{4} \right) + C_3.$$

Finally, we require that the solution satisfy the integral constraint (3–200b). This is sufficient to determine the one remaining constant, which turns out to be $C_3 = -7/24$. Hence,

$$\boxed{\bar{\theta} = 2\bar{z} + \left(\bar{r}^2 - \frac{\bar{r}^4}{4} - \frac{7}{24} \right).} \tag{3-210}$$

In dimensional terms this is

$$\boxed{\theta = \theta_0 + \frac{2qz}{U a \rho C_p} - \frac{qa}{4k}\left[\left(\frac{r}{a}\right)^4 - 4\left(\frac{r}{a}\right)^2 + \frac{7}{6} \right].} \tag{3-211}$$

This is the final form for the first approximation to the temperature field for $z \gg aPe$ and $Pe \gg 1$.

The heat transfer process in the entry zone between $z = 0$ and $z = O(aPe)$ is quite complex. However, far downstream ($z \gg aPe$) the temperature distribution (3–211) has a relatively simple form. The cross-sectionally averaged temperature increases linearly with z

$$\langle \theta \rangle \equiv \frac{1}{\pi a^2} \int_0^{2\pi}\!\!\int_0^a \theta\, r\, dr\, d\varphi = \theta_0 + \frac{2q}{U a \rho C_p} z + \frac{1}{8}\frac{qa}{a}.$$

The radial term in (3–211) adjusts itself to maintain the correct overall heat balance, plus the local balance between conduction of heat in the radial direction and convection in the axial direction. We shall see that the analysis in this section is very similar to that used for the solution of the Taylor dispersion problem, which is discussed in the next section.

Before leaving this section, however, a brief "historical" note is worth consideration. This is concerned with the problem of solving (3–202) subject to the initial condition $\theta = \theta_0$ at $\bar{z} = 0$ so that

$$\boxed{\bar{\theta} = 0 \quad \text{at } \bar{z} = 0.} \tag{3-212}$$

The equation (3–202) with conditions (3–203), (3–204), and (3–212) is known as the *Graetz problem*. The condition (3–212) does not make sense in the present context. Nevertheless, this equation is just the well-known "heat equation" with a variable coefficient, and there is a long history of working on an exact solution of the Graetz problem.

H. Heat Transfer in Unidirectional Flows

In fact, an *exact* solution of the Graetz problem can be achieved by separation of variables.[17] It is convenient to pose the problem in terms of a new unknown temperature function

$$\hat{T} \equiv \bar{\theta} - \bar{\theta}_{approx}, \qquad (3\text{–}213)$$

which is the difference between the exact solution we seek and the approximate solution (3–211) for large $\bar{z}$. The governing equation for $\hat{T}$ is still identical to (3–202), but the boundary and initial conditions become

$$\frac{\partial \hat{T}}{\partial \bar{r}} = 0 \quad \text{at } \bar{r} = 1 \qquad (3\text{–}214)$$

$$\hat{T} = -\bar{\theta}_{approx} \quad \text{at } \bar{z} = 0, \ \bar{r} \leq 1. \qquad (3\text{–}215)$$

The problem for $\hat{T}$ is a classical initial-value problem. If we seek a solution in the form

$$\hat{T}(\bar{r}, \bar{z}) = R(\bar{r}) F(\bar{z}), \qquad (3\text{–}216)$$

the governing equation for $F(\bar{z})$ is

$$\frac{dF}{d\bar{z}} + \frac{\lambda^2}{2} F = 0, \qquad (3\text{–}217)$$

whereas that for $R(\bar{r})$ is

$$\bar{r}^2 \frac{d^2 R}{d\bar{r}^2} + \bar{r} \frac{dR}{d\bar{r}} + \lambda^2 \bar{r}^2 (1 - \bar{r}^2) R = 0. \qquad (3\text{–}218)$$

The solution for F is straightforward:

$$F = C e^{-\frac{\lambda^2}{2} \bar{z}},$$

where C is an arbitrary constant. The solution for R does not correspond to any of the common special functions that usually arise in solutions by separation of variables. However, because we require a solution only in the bounded domain $0 \leq \bar{r} \leq 1$, we can utilize a power series of the form

$$R(\bar{r}; \lambda) = \sum_{i=0}^{\infty} c_i \bar{r}^i.$$

Substituting into (3–218), we can show that

$$c_0 = 1, \quad c_2 = -(\lambda^2/4),$$
$$c_i = -(\lambda/i)^2 (c_{i-2} - c_{i-4}), \quad i \text{ even},$$
$$c_i = 0. \quad i \text{ odd}.$$

Values of the eigenvalues λ can be obtained from the boundary condition, (3–214), and the first several have been evaluated by Siegel et al.[19] A table reproducing these values can be found in the textbook by Slattery.[19] The general solution of the form (3–216) is thus

$$\hat{T} = \sum_{n=0}^{\infty} A_n \exp\left(-\frac{\lambda_n^2}{2} \bar{z}\right) R_n(\bar{r}; \lambda_n). \qquad (3\text{–}219)$$

We obtain the coefficients A_n by applying initial condition (3–215) using the orthogonality properties of the eigenfunctions (as guaranteed by the general Sturm–Liouville theory). Again, numerical values are reported in Slattery.[17]

Unidirectional and One-Dimensional Flow and Heat Transfer Problems

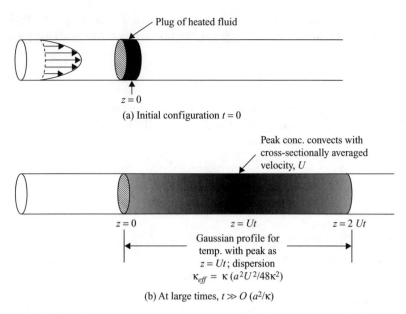

Figure 3–14. A qualitative sketch of the Taylor dispersion problem. Panel (a) represents the initial configuration at $t = 0$. Panel (b) shows a section of the tube at a much later time t^* in which the plug of heated fluid has translated downstream a distance Ut^* and spread symmetrically a distance $O(\sqrt{\kappa_{\mathit{eff}}\, t^*})$ where $\kappa_{\mathit{eff}} = (a^2 U^2 / 48 \kappa)$.

2. Taylor Dispersion in a Circular Tube

In this section we consider a second problem involving heat transfer in Poiseuille flow through a circular tube. In this case, we assume that the fluid in the region $-\delta \le z \le \delta$ is initially heated to a temperature θ_1, while the temperature elsewhere, i.e., $|z| > \delta$, is held at a constant temperature θ_0. The wall of the tube is insulated for all z so that there is no heat loss or gain to the surroundings. Precisely the same problem could be formulated as a mass transfer problem for the redistribution of a solute in a solvent with an initial solute concentration C_1 for $-\delta \le z \le \delta$ and concentration C_0 for $|z| > \delta$, with tube walls that are impermeable to the solute. The only difference is that the thermal diffusivity κ is replaced with the species diffusion coefficient D. However, to make the discussion as straightforward as possible, the analysis in this section is presented as a heat transfer problem.

The famous British physicist and fluid dynamicist G. I. Taylor first analyzed the very same problem,[20] as a mass transfer problem, almost 50 years ago. His analysis pertains to the axial dependence of the cross-sectionally averaged temperature (or concentration) distribution. The transport process that leads to the spread of this cross-sectionally averaged temperature pulse turns out to resemble a pure axial conduction (or diffusion) process and is therefore called *Taylor dispersion*. A qualitative sketch is shown in Fig. 3–14. The use of the word "dispersion" is intended to distinguish this process from a true conduction/diffusion process. The effective thermal diffusivity, usually termed the dispersion coefficient, is not a material constant like κ but depends on the flow and its properties. Although we will derive our results from the viewpoint of an intial radially uniform pulse of heat, we shall see that there are no assumptions in the derivation of the governing equation for the cross-sectionally averaged temperature profile that depend in any essential way on this intial form of the profile at $z = 0$. Hence the resulting equation is equally applicable for analysis of the cross-sectionally averaged temperature for an arbitrararily shaped initial pulse.

H. Heat Transfer in Unidirectional Flows

We can think of the problem just outlined as resulting from one of two initial setups. In the first, we imagine that there is an initial temperature pulse introduced into a tube filled with a stationary fluid, and then set the fluid into motion, at $t = 0$. Both the velocity field and the temperature field begin to evolve in time. However, we can assume that the kinematic viscosity ν is much larger then the thermal diffusivity κ. The velocity field will thus evolve to its steady state very quickly compared with the temperature field. Hence, as an approximation, we can assume that the velocity field is fully developed at all times, i.e.,

$$u_z = 2U\left(1 - \frac{r^2}{a^2}\right). \tag{3-220}$$

This means that we again neglect any change in the fluid viscosity that is due to nonuniformity of temperature in the tube, and we also completely neglect any natural convection effects that are due to the nonuniformity of the fluid density. Alternatively, we might consider that the fluid is moving with the fully developed velocity profile, (3–220), and at some instant (which we can identify as $t = 0$) a pulse of heat is simply injected across the tube. Although we recognize both of these cases as an approximation of what we could actually do in the laboratory, we may anticipate that either will give an adequate basis for thinking about and analyzing the temperature distribution at large times, when the concentration peak has convected far downstream from $z = 0$, which we shall see is the domain of applicability of the Taylor theory.

The Taylor dispersion problem is closely related to that discussed in the previous section, but also differs from it in some important fundamental respects. In the preceding problem, we assumed that the fluid was initially at a constant temperature upstream of $z = 0$ and that there was a constant heat flux into (or out of) the tube for all $z \geq 0$. In that case, the system has a *steady-state* temperature distribution at large times, and it was that steady-state problem that we solved. In the present case, there is no steady state. If the velocity were uniform across the tube instead of having the parabolic form (3–220), the temperature pulse that is initially at $z = 0$ would simply propagate downstream with the uniform velocity of the fluid, gradually spreading in the axial direction because of the action of heat conduction (i.e., the diffusion of heat). After a time t^*, the pulse would have moved downstream by a distance Ut^*, and the temperature pulse would have spread out over a distance of $O(\sqrt{(\kappa t^*)})$. Even in this simple case, there is clearly no steady state. The temperature distribution continues to evolve for all time.[21]

In the problem considered here, however, the velocity (3–220) is nonuniform, and hence the initial temperature pulse is convected downstream at different speeds, with a rate that is maximum at the centerline of the tube and zero at the tube wall. Thus the temperature pulse is rapidly distorted and spread out in the z direction, with the heated fluid initially carried downstream with a maximum velocity $2U$. Nevertheless, if we examine the system at large times, experimental evidence shows that the fluid far enough downstream (i.e., $z \gg 1$) is heated almost uniformly across the complete cross section of the tube, and the temperature pulse has spread symmetrically with respect to a plane cross section of the tube that moves with the *mean* velocity of the fluid, U. This spreading process is qualitatively similar to axial conduction, but with an effective thermal diffusivity that is *much larger* than the molecular thermal diffusivity κ.

The complete heat transfer process from the initial state at $t = 0$ to large times, where the analysis of Taylor applies, is actually quite complex. To discuss it *qualitatively*, it is useful to begin by writing the governing equation in dimensionless form. The starting point is again thermal energy equation (2–110). With viscous dissipation neglected, and assuming that the temperature field will be axisymmetric (i.e., dependent on the cylindrical coordinates r and

Unidirectional and One-Dimensional Flow and Heat Transfer Problems

z, but independent of the polar angle φ), this takes the form (3–194), but now, including the time derivative,

$$\frac{\partial \theta}{\partial t} + 2U\left(1 - \frac{r^2}{a^2}\right)\left(\frac{\partial \theta}{\partial z}\right) = \kappa \left[\frac{1}{r}\frac{\partial}{\partial r}\left(r\frac{\partial \theta}{\partial r}\right) + \frac{\partial^2 \theta}{\partial z^2}\right]. \tag{3–221}$$

We assume that θ is bounded at the centerline $r = 0$ and that the walls of the tube are insulating so that $\partial \theta / \partial r = 0$ at $r = a$ according to (2–219).

To nondimensionalize (3–221) we must identify characteristic scales as demonstrated in previous sections of this chapter. We choose the radius of the tube as a characteristic length scale $\ell_c = a$ and the corresponding ratio a/U as a characteristic time scale t_c. Although we may anticipate from the preceding section that variations in θ with respect to z will occur on a much longer length scale for large times, we are initially concerned with the evolution of θ nearer to t (and z) = 0, and for this purpose the tube radius provides an appropriate length scale for both r and z derivatives. Finally, it is convenient to define a dimensionless temperature, $\hat{\theta} \equiv (\theta - \theta_0)/(\theta_1 - \theta_0)$. The latter is then initially equal to 1 inside the pulse and 0 elsewhere. Because Eq. (3–221) is linear in θ, introduction of a dimensionless temperature does not change anything, and $\hat{\theta}$ can be simply substituted everywhere for θ. We may note that, in addition to the time scale a/U, the ratio a^2/κ is a second combination of the independent parameters of the problem that could have been chosen as a characteristic time scale. We shall come back to the reason why it was not chosen shortly.

If we substitute the dimensionless variables $\hat{z} = z/a$, $\hat{t} = t/(a/U)$, $\bar{r} = r/a$, and $\hat{\theta}$ into (3–221), we obtain

$$Pe\left[\frac{\partial \hat{\theta}}{\partial \hat{t}} + 2(1 - \bar{r}^2)\frac{\partial \hat{\theta}}{\partial \hat{z}}\right] = \frac{\partial^2 \hat{\theta}}{\partial \hat{z}^2} + \frac{1}{\bar{r}}\frac{\partial}{\partial \bar{r}}\left(\bar{r}\frac{\partial \hat{\theta}}{\partial \bar{r}}\right). \tag{3–222}$$

The single dimensionless parameter in this equation is the Peclet number,

$$Pe = Ua/\kappa,$$

as defined in the previous section.

Provided the parameters that we have chosen for characteristic scales are correct, all of the terms in (3–222) other than Pe will be independent of the independent parameters of the problem. Hence, in this case, the magnitude of the Peclet number determines the relative magnitude of the time derivative and convection terms on the left-hand side of (3–222) compared with the conduction terms on the right-hand side. In the Taylor analysis, it is assumed that

$$Pe \gg 1.$$

Because κ is typically very small, this condition on Pe is satisfied for quite modest values of a and U. In any case, the assumption $Pe \gg 1$ means that the terms on the left-hand side of (3–222) are much larger than the conduction terms on the right-hand side in the region where the scaling that we have adopted is correct. As noted earlier, we expect the scaling to be valid when the flow is first initiated and the temperature peak is near $z = 0$. Hence, in this domain, the heat transport process will be dominated by convection for large Pe, and we should be able to neglect conduction altogether. In this case, the initial flat temperature pulse will simply deform from $z = 0$ to positive z by moving along affinely with the fluid. Because the velocity profile is parabolic in shape, the temperature pulse will also become parabolic and extend continuously toward larger values of z everywhere except right at the tube wall where $u_z = 0$.

H. Heat Transfer in Unidirectional Flows

This picture is, however, overly simplified in several respects. Most important is the fact that the convection process will relatively quickly produce an elongated, parabolically shaped temperature front whose dimension in the axial direction becomes very much larger than the tube radius a. The increase of length scale in the axial direction means that the choice of the tube radius a as a characteristic length scale for terms involving z derivatives will be correct only for short times, namely the time for the center of the front to move forward one tube radius, i.e., $t < O(a/U)$ or $\hat{t} < O(1)$. At the same time, this convection process also produces *radial* temperature gradients that did not exist at the initial instant. This suggests that it may not be possible to neglect radial conduction of heat except at short times, $t \leq O(a/U)$. For times that exceed $O(a/U)$, we presumably need a longer length scale to characterize the distance over which the temperature changes in the axial direction. Because the nondimensionalization inherent in (3–222) is not correct except for the relatively short time period up to $t = O(a/U)$, we should not expect the inference that the convection terms dominate the conduction terms for $Pe \gg 1$ to necessarily be valid for larger times. The increase in length scale in the axial direction also suggests that conduction in the axial direction will become less important compared with conduction in the radial direction for large times (as the temperature peak moves to large z), as was also true in the problem discussed in the preceding section. In general, then, it appears that the relative importance of the various terms in the governing equation, (3–222), will be different for short and long times.

Often we will try to obtain approximate solutions of PDEs by approximating the governing equations and boundary conditions by using nondimensionalization to suggest the relative magnitudes of the various terms in order to determine which terms might be neglected in the original equation. For example, in this case, we could use the equation in the form (3–222) to suggest that the conduction terms can be neglected completely for a first approximation to the solution of (3–221), which is valid for "short" times, $t < O(a/U)$. However, we could not say anything about the relative magnitude of the various terms in (3–221) from the nondimensionalized form (3–222) for longer times because the choice of a as a characteristic scale for variations of temperature in the z direction is apparently not correct beyond the initial period. To determine which terms are dominant for large times $t \gg O(a/U)$, we must nondimensionalize again, but now choose a characteristic length scale for changes in the z direction that is appropriate for large times. Generally, this means that a different approximation of (3–221) will be needed for large times than was suggested from (3–222) for $t \leq O(a/U)$.

Approximate solutions generated from approximations of the governing equations for large or small values of characteristic dimensionless parameters like Pe are known as *asymptotic approximations*. When different approximations of the governing equation(s) are found in different parts of the solution domain, it means that different approximations to the solution of the original problem will also be found in these regions. When this occurs, the asymptotic solutions are said to be *singular*. Starting in the next chapter, we will begin a systematic study of asymptotic approximation methods. To obtain a solution of the present problem that is uniformly valid for all times is a complicated calculation that requires the use of a special method, known as the method of "*matched asymptotic expansions*," which is designed for the solution of so-called singular asymptotic problems.

In the present section, however, we follow the lead of Taylor's original work and pursue an approach that is only designed to yield an approximation to the solution for large times and for large values of the Peclet number, $Pe \gg O(1)$. It was stated earlier that the midpoint of the temperature pulse moves with the mean velocity of the fluid for large times where the Taylor analysis is applicable. Although this is a result that we must verify as part of our theoretical work, we shall assume for the time being that it is true. In this case, it is clearly more sensible to develop our theory based on a coordinate frame that also translates with the mean fluid velocity U so that the governing equation, (3–221), can be expressed in the form

$$\frac{D\theta}{Dt} + U\left(1 - 2\frac{r^2}{a^2}\right)\left(\frac{\partial \theta}{\partial \tilde{z}}\right) = \kappa\left[\frac{1}{r}\frac{\partial}{\partial r}\left(r\frac{\partial \theta}{\partial r}\right) + \frac{\partial^2 \theta}{\partial \tilde{z}^2}\right]. \quad (3\text{–}223)$$

Here, the Lagrangian time derivative that appears is

$$\frac{D}{Dt} \equiv \frac{\partial}{\partial t} + U\frac{\partial}{\partial z}$$

and $\tilde{z} = z - Ut$.

The keys to obtaining the long-time Taylor solution are twofold. First, we need to introduce a new length scale that we can again call ℓ_c^*, which is characteristic of the distance over which temperature changes occur in the z direction for large times. The introduction of a new length scale also suggests that we should introduce a new time scale, replacing a/U with a time scale ℓ_c^*/U. Second, we may recognize from the physics inherent in (3–223), and the fact that the convection process creates gradients of θ in the radial direction, that it will be impossible to ignore conduction of heat in the radial direction for large times, even though the parameter Pe is large.

To see the consequences of the first of these two points, let us nondimensionalize again, this time choosing the tube radius as the characteristic length scale for *radial* gradients of θ, but the new longer length scale ℓ_c^* to characterize gradients in the *axial* z direction. As noted earlier, we also introduce the new characteristic time scale ℓ_c^*/U. Hence,

$$\bar{r} = \frac{r}{a}, \quad \bar{z} = \frac{\tilde{z}}{\ell_c^*}, \quad \bar{t} = \frac{t}{(\ell_c^*/U)},$$

The new choice of characteristic length and time scales leads to a modified dimensionless form for (3–223), namely,

$$Pe\frac{a}{\ell_c^*}\left(\frac{D\hat{\theta}}{D\bar{t}} + (1 - 2\bar{r}^2)\frac{\partial \hat{\theta}}{\partial \bar{z}}\right) = \frac{a^2}{\ell_c^{*2}}\frac{\partial^2 \hat{\theta}}{\partial \bar{z}^2} + \frac{1}{\bar{r}}\frac{\partial}{\partial \bar{r}}\left(\bar{r}\frac{\partial \hat{\theta}}{\partial \bar{r}}\right). \quad (3\text{–}224)$$

We have not specified ℓ_c^*, except to say that $\ell_c^* \gg a$. Assuming this to be true, however, we see immediately that conduction of heat in the axial direction can be neglected compared with conduction in the radial direction because $a^2/\ell_c^{*2} \ll O(1)$.

The objective of Taylor's analysis is to predict the evolution of *the cross-sectionally averaged temperature* for large times, where the length scale $\ell_c^* \gg a$. At this point, we thus split $\hat{\theta}$ into a cross-sectionally averaged term $\langle \hat{\theta} \rangle$ and a remainder θ', so that

$$\hat{\theta} \equiv \langle \hat{\theta} \rangle + \theta', \quad (3\text{–}225)$$

where

$$\langle \hat{\theta} \rangle \equiv \frac{1}{2\pi}\int_0^{2\pi}\int_0^1 \hat{\theta}\,\bar{r}\,\partial\bar{r}\,\partial\varphi,$$

and φ is the polar angle, and then use (3–224) to derive governing equations for $\langle \hat{\theta} \rangle$ and θ'. Although it is the cross-sectionally averaged temperature profile that we ultimately seek to predict, we shall see that we must also determine an approximation for θ' to achieve this objective. So, introducing (3–225) into (3–224), we obtain

$$Pe\frac{a}{\ell_c^*}\left[\frac{D\langle \hat{\theta} \rangle}{D\bar{t}} + \frac{D\theta'}{D\bar{t}} + (1 - 2\bar{r}^2)\left(\frac{\partial \langle \hat{\theta} \rangle}{\partial \bar{z}} + \frac{\partial \theta'}{\partial \bar{z}}\right)\right]$$
$$= \frac{a^2}{\ell_c^{*2}}\left(\frac{\partial^2 \langle \hat{\theta} \rangle}{\partial \bar{z}^2} + \frac{\partial^2 \theta'}{\partial \bar{z}^2}\right) + \frac{1}{\bar{r}}\frac{\partial}{\partial \bar{r}}\left(\bar{r}\frac{\partial \theta'}{\partial \bar{r}}\right). \quad (3\text{–}226)$$

H. Heat Transfer in Unidirectional Flows

It will be noted that we have used the fact that $\langle\hat{\theta}\rangle$ is independent of $\bar{r}$.

To obtain an equation for $\langle\hat{\theta}\rangle$, we now average Eq. (3–226) over the cross section of the tube. The result is

$$Pe\frac{a}{\ell_c^*}\left(\frac{D\langle\hat{\theta}\rangle}{D\bar{t}}\right) = \frac{a^2}{\ell_c^{*2}}\left(\frac{\partial^2\langle\hat{\theta}\rangle}{\partial\bar{z}^2}\right) - Pe\frac{a}{\ell_c^*}\left\langle(1-2\bar{r}^2)\frac{\partial\theta'}{\partial\bar{z}}\right\rangle. \quad (3\text{–}227)$$

This approximate equation governs the evolution of the cross-sectionally averaged temperature distribution in the tube. Because $Pe \gg 1$ and $a/\ell_c^* \ll 1$ (this assumption must ultimately be checked), the axial conduction term in (3–227) can be neglected compared with the other two terms. We may also note that the last term in (3–227) can be written in the alternative form

$$Pe\left\langle u'\frac{\partial\theta'}{\partial\bar{z}}\right\rangle \text{ where } u' \equiv \bar{u} - \langle\bar{u}\rangle = 2(1-\bar{r}^2) - 1 \text{ and } \bar{u} \equiv \frac{u}{U}.$$

Hence (3–227) can be written as

$$\boxed{\frac{D\langle\hat{\theta}\rangle}{D\bar{t}} = -\left\langle u'\frac{\partial\theta'}{\partial\bar{z}}\right\rangle.} \quad (3\text{–}228)$$

We can see that *it is the averaged product of the deviation from the mean velocity and the deviation from the mean temperature gradient that is responsible for the Taylor dispersion process*. The anaysis leading to (3–228) is a straightforward manipulation of the exact governing equation, (3–224). However, Taylor's genius was in demonstrating that the right-hand side of Eq. (3–228) is equal to

$$\text{const}\left(\frac{\partial^2\langle\hat{\theta}\rangle}{\partial\bar{z}^2}\right),$$

and thus that (3–228) reduces to a pure conduction equation with respect to the point $\tilde{z} = 0$.

To evaluate the right-hand side of (3–228), we need an expression for θ'. We can obtain an equation for θ' by subtracting (3–227) from (3–226). The result is

$$Pe\frac{a}{\ell_c^*}\left[\frac{D\theta'}{D\bar{t}} + (1-2\bar{r}^2)\left(\frac{\partial\langle\hat{\theta}\rangle}{\partial\bar{z}} + \frac{\partial\theta'}{\partial\bar{z}}\right)\right] = \frac{a^2}{\ell_c^{*2}}\left(\frac{\partial^2\theta'}{\partial\bar{z}^2}\right) + \frac{1}{\bar{r}}\frac{\partial}{\partial\bar{r}}\left(\bar{r}\frac{\partial\theta'}{\partial\bar{r}}\right) + Pe\frac{a}{\ell_c^*}\left\langle u'\frac{\partial\theta'}{\partial\bar{z}}\right\rangle. \quad (3\text{–}229)$$

Now, in principle, to evaluate the Taylor dispersion term in (3–228), we need to solve this equation for θ'. However, if we could solve (3–229) to obtain an exact solution for θ', we could equally well have obtained an exact solution of the original problem, (3–224). Instead, we follow Taylor and seek only *a first approximation* for θ' based on (3–229) for the limit $\bar{t} \gg 1$ and $Pe \gg 1$.

We can begin by examining the terms on the right-hand side of (3–229). Clearly, if $a/\ell_c^* \ll 1$, the axial conduction term is small compared with the other two terms. The dominant term on the right-hand side then depends on the magnitude of $Pe(a/\ell_c^*)$. However, because the length scale ℓ_c^* continues to increase with increase of time, it is clear that, if we consider large-enough times,

$$\ell_c^* \gg a\,Pe, \quad (3\text{–}230)$$

and the middle term will become the largest term on the right-hand side. In spite of the fact that the left-hand side of (3–229) is multiplied by the small parameter $Pe(a/\ell_c^*)$, at least one of the terms on the left-hand side must remain of magnitude comparable with the largest term on the right-hand side. The key assumption that Taylor recognized was that the difference between $\hat{\theta}$ and $\langle\hat{\theta}\rangle$ must become smaller and smaller as $\bar{t}$ is increased, i.e.,

$$\theta' \ll \langle\hat{\theta}\rangle \quad \text{for } \bar{t} \gg 1.$$

This means that the dominant term on the left-hand side of (3–229) is the one involving $\langle\hat{\theta}\rangle$, and thus (3–229) reduces to the following approximate form

$$\frac{1}{\bar{r}}\frac{\partial}{\partial \bar{r}}\left(\bar{r}\frac{\partial \theta'}{\partial \bar{r}}\right) \approx Pe\frac{a}{\ell_c^*}(1 - 2\bar{r}^2)\frac{\partial\langle\hat{\theta}\rangle}{\partial \bar{z}}. \tag{3–231}$$

Condition (3–230) means that the analysis will be valid only for large times. The convection process increases the length in the z direction as $\Delta z \sim 2Ut$. It thus follows that we need a time scale t^* such that $2Ut^* \gg aPe$ or $t^* \gg aPe/2U = O(a^2/\kappa)$.

Given that (3–231) is valid, we can integrate twice with respect to $\bar{r}$ and apply boundary conditions to obtain θ', which we can see will be in the form

$$\theta' = g(\bar{r})\frac{\partial\langle\hat{\theta}\rangle}{\partial \bar{z}}.$$

Hence, substituting into $\langle u'\frac{\partial \theta'}{\partial \bar{z}}\rangle$, we will clearly get a term of the form const $(\partial^2\langle\hat{\theta}\rangle/\partial\bar{z}^2)$, and thus prove that the evolution of the cross-sectionally averaged temperature distribution will occur as an effective conduction process, as asserted by Taylor.

Before carrying out this calculation, however, it may be helpful to consider a more formal justification of Taylor's apparently *ad hoc* approximation of (3–229) in the form (3–231). First, it is necessary that θ' be not only small compared with $\langle\hat{\theta}\rangle$ in a numerical sense, but that it be asymptotically small for $\bar{t} \gg 1$ in the sense that

$$\theta' \equiv Pe^{-m}\tilde{\theta}, \text{ where } \tilde{\theta} \text{ is } O(1), m > 0, \text{ and } Pe \gg 1. \tag{3–232}$$

The question is whether there is a time large enough (or alternatively a position $\bar{z} \gg 1$ corresponding to $\tilde{z} = 0$) for which this scaling (and thus the Taylor dispersion theory) is applicable. We will address this question shortly. For the moment, we simply assume that (3–232) is valid and rewrite (3–229) in terms of $\tilde{\theta}$,

$$\frac{Ua^2}{\kappa\ell_c^*}\left[\frac{1}{Pe^m}\frac{D\tilde{\theta}}{D\bar{t}} + (1-2\bar{r}^2)\left(\frac{\partial\langle\hat{\theta}\rangle}{\partial\bar{z}} + \frac{1}{Pe^m}\frac{\partial\tilde{\theta}}{\partial\bar{z}}\right)\right]$$
$$= \frac{a^2}{\ell_c^{*2}}\frac{1}{Pe^m}\frac{\partial^2\tilde{\theta}}{\partial\bar{z}^2} + \frac{1}{Pe^m}\frac{1}{\bar{r}}\frac{\partial}{\partial\bar{r}}\left(\bar{r}\frac{\partial\tilde{\theta}}{\partial\bar{r}}\right) + \frac{Ua^2}{\kappa\ell_c^*}\frac{1}{Pe^m}\left\langle u'\frac{\partial\tilde{\theta}}{\partial\bar{z}}\right\rangle. \tag{3–233}$$

In this form, we can evaluate the magnitudes of the various terms (term by term). The results are

$$\frac{a}{\ell_c^*}Pe^{1-m}, \quad \frac{a}{\ell_c^*}Pe, \quad \frac{a}{\ell_c^*}Pe^{1-m} = \left(\frac{a}{\ell_c^*}\right)^2 Pe^{-m}, \quad Pe^{-m}, \quad \frac{a}{\ell_c^*}Pe^{1-m}. \tag{3–234}$$

Neglecting the first and third terms on the left-hand side of (3–233) compared with the second, as Taylor did in deriving (3–231), thus requires that

$$Pe \gg 1.$$

If the dominant term on the right-hand side of (3–233) is the radial conduction term, as in (3–231), this requires the condition (3–230). Finally, if the second terms on the two sides of (3–233) are assumed to be of the same order of magnitude, this implies

$$\frac{a}{\ell_c^*}Pe \approx Pe^{-m}.$$

H. Heat Transfer in Unidirectional Flows

We see from this that, in the regime in which the Taylor analysis is valid, the length scale characteristic of changes of $\hat{\theta}$ in the axial direction must be

$$\ell_c^* \approx aPe^{1+m}. \tag{3-235}$$

In this case, Eq. (3–229), is now written in the approximated form

$$\boxed{\frac{1}{\bar{r}}\frac{\partial}{\partial \bar{r}}\left(\bar{r}\frac{\partial \tilde{\theta}}{\partial \bar{r}}\right) = (1 - 2\bar{r}^2)\frac{\partial \langle \hat{\theta} \rangle}{\partial \bar{z}}.} \tag{3-236}$$

The condition (3–235) implies that $\ell_c^* \gg aPe$ for $Pe \gg 1$.

We have seen that the coefficient m must be positive for the Taylor analysis to apply, but it is otherwise undetermined. In a full matched asymptotic analysis, it would be obtained by matching with the solutions for smaller $\bar{t}$, but these solutions are not available here. *Fortunately, the final result for the Taylor dispersion coefficient and the governing equation for $\langle \hat{\theta} \rangle$ are independent of m provided only that it is positive so that the analysis is valid.*

We can now determine θ' by solving either (3–231) or (3–236). We choose to solve Eq. (3–236) for $\tilde{\theta}$. Because $\partial \langle \hat{\theta} \rangle / \partial \bar{z}$ is independent of $\bar{r}$, we can simply integrate twice with respect to $\bar{r}$. The general solution is

$$\tilde{\theta} = \left(\frac{\bar{r}^2}{4} - \frac{\bar{r}^4}{8}\right)\frac{\partial \langle \hat{\theta} \rangle}{\partial \bar{z}} + c \ln \bar{r} + d. \tag{3-237}$$

The conditions on this solution are

$$\tilde{\theta} \text{ bounded at } \bar{r} = 0, \tag{3-238}$$

$$\frac{\partial \tilde{\theta}}{\partial \bar{r}} = 0 \text{ at } \bar{r} = 1, \tag{3-239}$$

$$\int_0^{2\pi} \partial \varphi \int_0^1 \bar{r}\partial \bar{r}[\tilde{\theta}(\bar{r},\bar{t})] = 0. \tag{3-240}$$

The last of these conditions is a consequence of the definition of θ' in Eq. (3–225). The result of applying (3–238)–(3–240) to (3–237) is

$$\boxed{\tilde{\theta} = \left(\frac{\bar{r}^2}{4} - \frac{\bar{r}^4}{8} - \frac{1}{12}\right)\frac{\partial \langle \hat{\theta} \rangle}{\partial \bar{z}}.} \tag{3-241}$$

The approximate solution, (3–241), can now be used to evaluate the Taylor dispersion term in (3–227). Starting from the definition just before (3–228), we obtain

$$\left\langle u'\frac{\partial \theta'}{\partial \bar{z}}\right\rangle = \frac{1}{Pe^m}\left\langle u'\frac{\partial \tilde{\theta}}{\partial \bar{z}}\right\rangle = \frac{1}{Pe^m}\left\langle (1-2\bar{r}^2)\left(\frac{\bar{r}^2}{4} - \frac{\bar{r}^4}{8} - \frac{1}{12}\right)\right\rangle\frac{\partial^2 \langle \hat{\theta} \rangle}{\partial \bar{z}^2} = -\frac{1}{48}\frac{1}{Pe^m}\frac{\partial^2 \langle \hat{\theta} \rangle}{\partial \bar{z}^2}. \tag{3-242}$$

Hence, returning to the governing equation, (3–228), for $\langle \hat{\theta} \rangle$, we obtain

$$\frac{D\langle \hat{\theta} \rangle}{D\bar{t}} = \frac{1}{Pe^m}\left(\frac{1}{Pe^{2+m}} + \frac{1}{48}\right)\frac{\partial^2 \langle \hat{\theta} \rangle}{\partial \bar{z}^2}. \tag{3-243}$$

The term proportional to Pe^{-2-m} is the axial conduction term and is negligible compared with the Taylor dispersion contribution for large Pe. Hence, in the following discussion, we neglect this term.

Unidirectional and One-Dimensional Flow and Heat Transfer Problems

Before commenting on Eq. (3–243), let us convert it back to the dimensional variables

$$t = \frac{\ell_c^*}{U}\bar{t} = \frac{aPe^{m+1}}{U}\bar{t}, \quad \tilde{z} = \ell_c^*\bar{z} = aPe^{m+1}\bar{z},$$

and

$$\theta = \theta_0 + (\theta_1 - \theta_0)\hat{\theta}.$$

This gives

$$\frac{D\langle\theta\rangle}{Dt} = \kappa\left(\frac{Pe^2}{48}\right)\frac{\partial^2\langle\theta\rangle}{\partial\tilde{z}^2}, \quad (3\text{–}244)$$

or, alternatively, in terms of the original Eulerian variables,

$$\frac{\partial\langle\theta\rangle}{\partial t} + U\frac{\partial\langle\theta\rangle}{\partial z} = \kappa\left(\frac{Pe^2}{48}\right)\frac{\partial^2\langle\theta\rangle}{\partial z^2}. \quad (3\text{–}245)$$

A remarkable property of the solution that we have derived, clearly evident in Eq. (3–244) for the cross-sectionally averaged temperature profile, is that the position in the tube where the cross-sectionally averaged temperature is maximum moves as if there were convection downstream *at the mean velocity U*. The temperature pulse also spreads about this plane as though there were axial conduction with an effective thermal diffusivity of

$$\kappa_{e\!f\!f} = \kappa(a^2U^2/48\kappa^2). \quad (3\text{–}246)$$

Here, κ is the thermal diffusivity of the fluid.

We can also calculate the average convective flux in the axial direction, relative to the axes moving at the mean velocity U, according to

$$\langle Q\rangle \equiv \frac{2\rho C_p}{a^2}\int_0^a \theta(r,z)\,u_z(r)\,r\,dr, \quad \text{where } u_z(r) = U\left[1 - 2\left(\frac{r}{a}\right)^2\right]. \quad (3\text{–}247)$$

Substituting for θ and integrating, we obtain

$$\langle Q\rangle = -k\left(\frac{U^2a^2}{48\kappa^2}\right)\frac{\partial\langle\theta\rangle}{\partial z}, \quad (3\text{–}248)$$

where k is the thermal conductivity of the fluid.

However, Eq. (3–248) is just *Fourier's law*, with an effective thermal conductivity $k(a^2U^2/48\kappa^2)$. Further, because there is no net *convection* contribution to the cross-sectionally averaged heat flux calculated with respect to the moving axis $\tilde{z}$, we see that the convective transport rate down the tube is just that due to the *mean* velocity U. Both of these results may at first seem surprising. For example, the maximum rate of transport that is due to convection acting alone is the centerline velocity $2U$, and it is not immediately obvious why the actual "convective" transport rate is slower. In addition, the effective thermal diffusivity is seen in (3–246) to be *inversely* proportional to the molecular thermal diffusivity κ, and this may also seem to be counterintuitive.

Let us consider the second observation first. The result, (3–247), shows that the rate of dispersion of the heated fluid actually decreases if the thermal conductivity increases. This can be explained by consideration of the "competition" between axial convection and radial diffusion as shown in the governing equation, (3–236). The effect of the axial convection is to carry heated fluid downstream with a velocity that ranges from $2U$ at the centerline of the tube to zero at the tube wall. However, the radial gradients produced by axial convection allow for heat to be conducted from the fastest-moving fluid near the center of the tube outward toward the tube wall, where it is convected downstream at a slower rate. The net result is that the midpoint of the spreading front of heated fluid, i.e., the point where $\theta = (\theta_0 + \theta_1)/2$, convects downstream with the mean axial velocity U rather than with the maximum centerline velocity $2U$, whereas the rate of spreading (or dispersion) around this midpoint is driven by the axial convection process. This is *slowed* down by the tendency for diffusion (conduction) to transport heat radially, away from the fastest-moving fluid. Hence the "effective" thermal diffusivity is actually decreased as the molecular diffusivity is increased.

We can now use the Taylor dispersion equation in either of the forms (3–244) or (3–245) to show that the cross-sectionally averaged temperature profile is a Gaussian in the z direction, with the peak concentration remaining at $\tilde{z} = 0$ (i.e., convecting downstream relative to fixed coordinates at the mean velocity U).

The ideas in this section turn out to be extremely valuable, both for estimating dispersion rates and as a basis for determining the (molecular) thermal conductivity by means of measurements of the Taylor dispersion or spreading rates. The latter method is particularly useful for it is difficult to devise other means of measuring the thermal conductivity of a liquid without encountering severe difficulties with nonidealities such as natural convection flows. If we try to make a heat flux measurement in a stationary fluid, the problem is that the thermal conductivity is relatively small, and the convective heat transfer associated with even extremely weak buoyancy-driven motion easily swamps heat transfer by conduction. This problem does not come up in using Taylor dispersion as a way to determine k, however, because the velocities in the tube will typically be very large compared with those of any natural convection flows that would be generated by the small changes in density that are due to the differences in temperature.

The reader who wishes to learn more about Taylor dispersion, and a large number of generalizations to other systems in which dispersion is produced by a coupling between "convection" and "diffusion," may wish to consult the excellent book by Brenner and Edwards.[22]

I. PULSATILE FLOW IN A CIRCULAR TUBE

In the preceding sections of this chapter, we have considered several examples of transient unidirectional flows. In each case, it was assumed that the flow started from rest with the abrupt imposition of either a finite pressure gradient or a finite boundary velocity, and we saw that the flow evolved toward steady state by means of diffusion of momentum with a time scale $t_c = \ell_c^2/v$. Here we consider a final example of a transient unidirectional flow problem in which time-dependent motion is produced in a circular tube by the sudden imposition of a periodic, time-dependent pressure gradient:

$$-\frac{\partial P}{\partial z} = G = G_0(1 + \varepsilon \sin \omega t). \tag{3–249}$$

This problem of pulsatile flow in a circular tube has been studied extensively in the context of model studies of blood flow in the arteries,[23] though it is considerably simpler than the real problem in which the vessel cross section is not circular, the walls are compliant, and the vessels quite short between branch or bifurcation points.[24] In the context of blood flow, and most other applications, we are usually not concerned with the start-up of pulsatile flow from rest, but only with the solution for sufficiently long times after imposition of the pulsatile pressure gradient so that the velocity field is strictly periodic. We shall follow this lead in our present discussion and concentrate primarily on the asymptotic ($t \gg 1$) state in which the velocity is periodic in time. If a more complicated periodic pressure gradient is required, the solution for the velocity field can be achieved by Fourier decomposition of the complex $G(t)$ into a discrete set of Fourier sine modes, with the sinusoidal form, (3–249), then being considered as one typical, but arbitrarily chosen, Fourier component.

In spite of its historical development in biomechanics, however, our main motivation for studying the problem of pulsatile flow in a circular tube is not in this application. Rather, our general goal is to expose the qualitative influence of the fluid's inertia (that is, of acceleration–deceleration effects) when there is an externally imposed time scale $t_c = 1/\omega$ that is characteristic of changes in the flow. In addition, we shall see in the following chapter that the problem provides a very convenient framework for illustrating the concepts of asymptotic analysis as a means of obtaining approximate solutions for large and small frequencies, $\omega \ll 1$ and $\omega \gg 1$.

We begin in this section by deriving the exact solution. Later, in the next chapter, we will reexamine the problem by using asymptotic methods of analysis. The governing DE is just (3–12), with the pressure-gradient function given by (3–249):

$$\boxed{\rho \frac{\partial u_z}{\partial t} = G_0(1 + \varepsilon \sin \omega t) + \mu \left[\frac{1}{r} \frac{\partial}{\partial r} \left(r \frac{\partial u_z}{\partial r} \right) \right].} \qquad (3\text{--}250)$$

The boundary conditions are

$$\boxed{\begin{aligned} u_z &= 0 \quad \text{at} \quad r = R, \\ u_z &\text{ bounded at } r = 0 \end{aligned}} \qquad (3\text{--}251)$$

for all t. In addition, we assume that the fluid begins at rest for $t = 0$, that is,

$$\boxed{u_z = 0 \text{ at } t = 0, \quad \text{all } r (0 \leq r \leq R).} \qquad (3\text{--}252)$$

The first step, as usual, is to introduce characteristic scales and nondimensionalize the problem. In this case, appropriate characteristic scales are

$$u_c = \frac{G_0 R^2}{\mu}, \quad \ell_c = R, \quad \text{and} \quad t_c = \frac{R^2}{\nu}. \qquad (3\text{--}253)$$

The first two are obvious. The time scale R^2/ν is the time for diffusion of momentum across the tube and is certainly relevant to the initial start-up portion of the problem. We recognize the existence of a second time scale in this problem, however, and that is the period $1/\omega$ of the imposed pressure gradient. We may anticipate that this second time scale will be especially relevant for larger times after the initial transients have died out. For the moment, we retain t_c as defined in (3–253). Then, introducing dimensionless variables

$$\bar{u}_z = \frac{u_z}{u_c}, \quad \bar{r} = \frac{r}{\ell_c}, \quad \text{and} \quad \hat{t} = \frac{t}{t_c}$$

I. Pulsatile Flow in a Circular Tube

into the governing equations and boundary conditions, we obtain

$$\frac{\partial \bar{u}_z}{\partial \hat{t}} = 1 + \varepsilon \sin\left(\frac{\omega R^2}{v}\hat{t}\right) + \frac{1}{\bar{r}}\frac{\partial}{\partial \bar{r}}\left(\bar{r}\frac{\partial \bar{u}_z}{\partial \bar{r}}\right), \qquad (3\text{-}254)$$

with $\bar{u}_z = 0$ at $\bar{r} = 1$,

$\bar{u}_z$ bounded at $\bar{r} = 0$, $\qquad (3\text{-}255)$

and $\bar{u}_z = 0$ for $\hat{t} = 0$.

It will be noted that a single dimensionless combination of parameters remains in the problem after nondimensionalization, namely,

$$R_\omega \equiv \frac{\omega R^2}{v}. \qquad (3\text{-}256)$$

In the previous transient unidirectional flow problems that we have considered, the dimensionless governing equations and boundary/initial conditions were completely free of all dimensional parameters. In these problems, however, the only relevant time scale was the diffusion time, ℓ_c^2/v. Here, in contrast, the flow is characterized by a second imposed time scale that is due to the oscillatory pressure gradient, and the form of the resulting flow is predicted by (3-354) to depend upon the ratio of the diffusion time R^2/v to the imposed time scale, $1/\omega$. The dimensionless ratio of time scales, R_ω, can also be considered to be a dimensionless frequency for the flow, and in that context is sometimes called the *Strouhal number*.

Before more is said about the physical significance of R_ω, it is convenient to proceed a few steps farther with the solution of (3-254) and (3-255). Specifically, because the governing equation and boundary conditions are linear, it is clear that the solution can be expressed as the sum of two parts, one representing the response to the steady part of the pressure gradient and the other the response to the transient, oscillatory part. We thus denote $\bar{u}_z$ as

$$\bar{u}_z = u_z^{(0)} + \varepsilon u_z^{(1)}, \qquad (3\text{-}257)$$

where

$$\frac{\partial u_z^{(0)}}{\partial \hat{t}} = 1 + \frac{1}{\bar{r}}\frac{\partial}{\partial \bar{r}}\left(\bar{r}\frac{\partial u_z^{(0)}}{\partial \bar{r}}\right), \qquad (3\text{-}258)$$

with

$u_z^{(0)} = 0$ at $\bar{r} = 1$,

$u_z^{(0)}$ bounded at $\bar{r} = 0$, $\qquad (3\text{-}259)$

$u_z^{(0)} = 0$ at $\hat{t} = 0$,

and

$$\frac{\partial u_z^{(1)}}{\partial \hat{t}} = \sin\left(\frac{\omega R^2}{v}\right)\hat{t} + \frac{1}{\bar{r}}\frac{\partial}{\partial \bar{r}}\left(\bar{r}\frac{\partial u_z^{(1)}}{\partial \bar{r}}\right), \qquad (3\text{-}260)$$

with

$u_z^{(1)} = 0$ at $\bar{r} = 1$,

$u_z^{(1)}$ bounded at $\bar{r} = 0$, $\qquad (3\text{-}261)$

$u_z^{(1)} = 0$ at $\hat{t} = 0$.

Now the problem defined by (3–258) and (3–259) is just the transient start-up flow due to a suddenly imposed constant pressure gradient that was already solved in Section C of this chapter. The final steady-state solution is the steady Poiseuille flow profile:

$$u_z^{(0)} = \frac{1}{4}(1 - \bar{r}^2).$$

Here we focus our attention on the transient pressure-gradient problem defined by (3–260) and (3–261). We have seen in previous transient start-up problems that it is usually advantageous to solve for the difference between the actual instantaneous velocity distribution and the final steady-state profile, as this transforms the problem from a nonhomogeneous boundary-value problem to a homogeneous initial-value problem. In the present case, of course, the solution for large $\hat{t}$ will not be steady but will instead be a strictly periodic transient motion. Nevertheless, we propose adopting the same approach to solving (3–260) and (3–261) that we have used earlier: First, we will obtain the asymptotic solution for $\hat{t} \gg 1$ when all initial transients have died out and the solution has become independent of its initial state, and only then will we show that the complete initial value problem can be solved conveniently by looking for the difference between this periodic, asymptotic solution and the actual instantaneous solution $u_z^{(1)}$.

A key to solving (3–260) and (3–261), either exactly or for the large $\hat{t}$ asymptote, is to note that the velocity field will not generally be in-phase with the oscillating pressure gradient. We can see that this is true by a qualitative examination of the governing equation. It is convenient for this purpose to temporarily rescale time according to

$$\bar{t} = \left(\frac{\omega R^2}{\nu}\right) \hat{t} \qquad (3\text{–}262)$$

so that

$$\left(\frac{\omega R^2}{\nu}\right) \frac{\partial u_z^{(1)}}{\partial \bar{t}} = \sin \bar{t} + \frac{1}{\bar{r}} \frac{\partial}{\partial \bar{r}} \left(\bar{r} \frac{\partial u_z^{(1)}}{\partial \bar{r}}\right). \qquad (3\text{–}263)$$

In this form, the pressure gradient oscillates on a dimensionless time scale of $O(1)$, independent of ω, and we see that the magnitude of R_ω determines the importance of acceleration effects in the fluid relative to the viscous effects (diffusion of momentum) that are represented by the second term on the right-hand side of (3–263). In this role, it is evident that R_ω can be considered as a Reynolds number, based on a characteristic "velocity" ωR. Before proceeding with our discussion, it should be emphasized that the change of variable (3–262) does not alter the problem at all. Indeed, it is precisely equivalent to nondimensionalizing time with respect to $t_c = 1/\omega$ instead of $t_c = R^2/\nu$, as in (3–260). The choice of one characteristic time scale instead of the other is a matter of convenience only, although we may anticipate that the choice $t_c = 1/\omega$ inherent in (3–263) will be more convenient for discussing the long-time asymptotic behavior of $u_z^{(1)}$. In any case, the solution still depends on the ratio of characteristic time scales, R_ω.

Now, examining (3–263), we can see the qualitative nature of the phase relationship between $u_z^{(1)}$ and the pressure gradient by considering the two limiting cases, $R_\omega \gg 1$ and $R_\omega \ll 1$. Let us suppose first that $R_\omega \ll 1$. In this case, once the initial transient development of the flow has been accomplished, it is apparent that the acceleration term in (3–263) can be neglected compared with the viscous term, and this means that, for large times, $u_z^{(1)}$ must be *quasi steady* and completely in-phase with the pressure gradient; that is, if we examine (3–263) with the acceleration terms neglected, it is evident that the solution will be in the form

$$u_z^{(1)} \sim F(\bar{r}) \sin \bar{t} \quad \text{for } \bar{t} \gg 1 \text{ and } R_\omega \ll 1. \qquad (3\text{–}264)$$

I. Pulsatile Flow in a Circular Tube

If $R_\omega \gg 1$, on the other hand, the viscous term in (3–263) is negligible compared with the acceleration term, and

$$u_z^{(1)} \sim -\cos\bar{t}\,\frac{1}{R_\omega}H(\bar{r}) \text{ for } \bar{t} \gg 1 \text{ and } R_\omega \gg 1. \tag{3–265}$$

The restriction to $\bar{t} \gg 1$ is clearly necessary in this case because the asymptotic-solution form (3–265) does not satisfy the initial condition at $\bar{t} = 0$. In (3–265), the velocity lags 90° behind the pressure gradient. In other words, the first peak in G appears at $\bar{t} = \pi/2$, whereas the first peak in $u_z^{(1)}$ does not appear until $\bar{t} = \pi$. It will, perhaps, have been noted that the magnitude of the velocity component $u_z^{(1)}$ appears in (3–265) to be only R_ω^{-1} for $R_\omega \gg 1$. This result has been deduced from the governing equation, (3–263), by the following reasoning. First, the time-dependent velocity field $u_z^{(1)}$ exists only because of the time-dependent pressure-gradient $\sin\bar{t}$. According to (3–263), this pressure gradient must be balanced either by a corresponding acceleration of the fluid or by viscous forces or, of course, some combination. Thus, for any value of R_ω, at least one of the terms corresponding to acceleration or viscous effects must have the same magnitude as the pressure-gradient term, which is $O(1)$. When $R_\omega \ll 1$, it is evident from (3–263) that the balance is between the viscous term and the pressure gradient and that both are $O(1)$ with $u_z^{(1)} = O(1)$. When $R_\omega \gg 1$, on the other hand, the balance must be between acceleration and the pressure gradient, but it appears as though the former is $O(R_\omega)$, whereas the latter is only $O(1)$. This obviously cannot be true. Rather, as R_ω increases, the magnitude of the dimensionless velocity component $u_z^{(1)}$ must decrease as R_ω^{-1} so that the three terms in (3–263) are $O(1)$, $O(1)$, and $O(R_\omega^{-1})$, respectively, as we go from left to right.

The actual solution of problem (3–260) and (3–261) – or the equivalent form (3–263) with $\bar{t}$ used rather than $\hat{t}$ – is straightforward once it is realized that the velocity field will not generally be in-phase with the pressure gradient. A convenient approach that allows a completely general phase relationship between $u_z^{(1)}$ and the pressure gradient $\sin(\omega R^2/\nu)\hat{t}$ is to solve for the complementary *complex* velocity field $\hat{u}_z$ that satisfies

$$\frac{\partial \hat{u}_z}{\partial \hat{t}} = e^{i(\omega R^2/\nu)\hat{t}} + \frac{1}{\bar{r}}\frac{\partial}{\partial \bar{r}}\left(\bar{r}\frac{\partial \hat{u}_z}{\partial \bar{r}}\right), \tag{3–266}$$

with $\hat{u}_z = 0$ at $\bar{r} = 1$,

$\hat{u}_z$ bounded at $\bar{r} = 0$, $\tag{3–267}$

and $\hat{u}_z = 0$ at $\hat{t} = 0$.

When (3–266) and (3–260) are compared, it is evident that

$$u_z^{(1)} = \mathcal{I}m(\hat{u}_z)$$

will satisfy (3–260) and (3–261) if $\hat{u}_z$ is a solution of (3–266) and (3–267). (Note $\mathcal{I}m$ denotes the imaginary part of the quantity in parentheses.)

We solve (3–266)) and (3–267) in two steps. First, we obtain a solution for the periodic velocity field that will exist for $(\omega R^2/\nu)\hat{t} \gg 1$. We then obtain the complete solution, including initial transients, by solving for the difference between $\hat{u}_z$ and this large-time periodic solution. The advantage of solving for the complementary complex velocity field is that the form of the solution for large times, satisfying (3–266) and the first two of the boundary conditions of (3–267), can be guessed *a priori* to be

$$\hat{u}_z = e^{i(wR^2/\nu)\hat{t}}H(\bar{r}). \tag{3–268}$$

Substitution into (3–266)) and the first two conditions of (3–267) shows that this is clearly a solution, provided $H(\bar{r})$ satisfies the ODE and boundary conditions:

$$\frac{d^2 H}{d\bar{r}^2} + \frac{1}{\bar{r}}\frac{dH}{d\bar{r}} - iHR_\omega = -1, \qquad (3\text{–}269)$$

$$H(\bar{r}) = 0 \quad \text{at} \quad \bar{r} = 1,$$

$$\text{and} \quad H(\bar{r}) \quad \text{bounded at} \quad \bar{r} = 0. \qquad (3\text{–}270)$$

Let us denote the periodic solution, (3–268), as $\hat{u}_z^\infty$ to remind us that it is the asymptotic form for $\hat{t}$ large enough that initial transients have died out. To obtain the complete solution of the initial-value problem, (3–266), and (3–267), we then seek a solution for the difference field:

$$\hat{u}_z' = \hat{u}_z - \hat{u}_z^\infty. \qquad (3\text{–}271)$$

Substituting (3–271) into (3–266)), we obtain the governing equation and boundary conditions for $\hat{u}_z'$:

$$\frac{\partial \hat{u}_z'}{\partial \hat{t}} = \frac{1}{\bar{r}}\frac{\partial}{\partial \bar{r}}\left(\bar{r}\frac{\partial \hat{u}_z'}{\partial \bar{r}}\right), \qquad (3\text{–}272)$$

$$\text{with} \quad \hat{u}_z' = 0 \quad \text{at} \quad \bar{r} = 1,$$

$$\hat{u}_z' \quad \text{bounded} \quad \text{at} \quad \bar{r} = 0,$$

$$\text{and} \quad \hat{u}_z' = -H(\bar{r}) \quad \text{for} \quad \hat{t} = 0. \qquad (3\text{–}273)$$

As in the previous transient problems that we have discussed, this problem for the difference between the actual velocity and its large-time form, $\hat{u}_z^\infty$, is reduced to a homogeneous initial-value problem that can be solved by standard methods such as separation of variables. Our interest, in the present problem, is primarily with the large-time behavior of the solution. Hence the details of solving (3–372) and (3–273) are left as an exercise (nontrivial) for the reader, and we concentrate our attention in the remainder of this section on the solution of (3–269) and (3–270).

We begin with the particular solution of (3–269), which we see, by inspection, is just

$$H_p = -\frac{i}{R_\omega}. \qquad (3\text{–}274)$$

The homogeneous solution of (3–269) is a Bessel function. To see this, we note that the general Bessel equation is

$$z^2 \frac{d^2 w}{dz^2} + z\frac{dw}{dz} + (z^2 - v^2)w = 0, \qquad (3\text{–}275)$$

which has Bessel functions of the first and second kind of order v as its solutions. The homogeneous equation for H is easily transformed into the standard form (3–275) by the change of independent variable,

$$\bar{r} = \sqrt{\frac{i}{R_\omega}}R,$$

which yields

$$\frac{d^2 H_h}{dR^2} + \frac{1}{R}\frac{dH_h}{dR} + H_h = 0. \qquad (3\text{–}276)$$

I. Pulsatile Flow in a Circular Tube

Thus the two independent, homogeneous solutions for H are Bessel functions J_0 and Y_0 of order 0 and

$$H = H_p + H_h = -\frac{i}{R_\omega} + A J_0\left[\left(\frac{R_\omega}{i}\right)^{1/2} \bar{r}\right] + B Y_0\left[\left(\frac{R_\omega}{i}\right)^{1/2} \bar{r}\right], \quad (3\text{--}277)$$

where A and B are arbitrary constants.

We have noted previously that $Y_0(z) \to -\infty$ as $z \to 0$. Thus $B = 0$ if H is to be bounded. To obtain A, we apply the condition $H(\bar{r} = 1) = 0$. The resulting solution for $\hat{u}_z^\infty$, which we obtain by substituting the solution for H in Eq. (3–268), is

$$\hat{u}_z^\infty = e^{i\bar{t}}\left[-\frac{i}{R_\omega}\right]\left\{1 - \frac{J_0\left[\left(\frac{R_\omega}{i}\right)^{1/2} \bar{r}\right]}{J_0\left[\left(\frac{R_\omega}{i}\right)^{1/2}\right]}\right\}. \quad (3\text{--}278)$$

The corresponding velocity field $\hat{u}_z^{(1)\infty}$ is

$$\boxed{\hat{u}_z^{(1)\infty} = \mathcal{J}m\left[-\frac{i e^{i\bar{t}}}{R_\omega}\left\{1 - \frac{J_0\left[\left(\frac{R_\omega}{i}\right)^{1/2} \bar{r}\right]}{J_0\left[\left(\frac{R_\omega}{i}\right)^{1/2}\right]}\right\}\right].} \quad (3\text{--}279)$$

We use the superscript ∞ to remind us that this is the long-time periodic form of $u_z^{(1)}$ that pertains once the initial start-up transients have vanished. The result (3–279) is the general exact solution for $\hat{u}_z^{(1)\infty}$, valid for any value of R_ω. Although it has been relatively easy to obtain, it is very complicated to actually evaluate for any arbitrary R_ω. We therefore restrict ourselves here to the case of small $R_\omega \ll 1$.

To evaluate $\hat{u}_z^{(1)\infty}$ for small R_ω from (3–279), we use the small z expansion[13] of $J_0(z)$:

$$J_0(z) \sim 1 - \frac{1}{4}z^2 + \frac{1}{64}z^4 + \cdots +.$$

Let us denote $x \equiv \sqrt{R_\omega}\bar{r}$ and $X \equiv \sqrt{R_\omega}$. Then

$$J_0\left[\left(\frac{1}{i}\right)^{1/2} x\right] \sim 1 + \frac{1}{4}ix^2 - \frac{1}{64}x^4 + \cdots +,$$

and

$$J_0\left[\left(\frac{1}{i}\right)^{1/2} X\right] \sim 1 + i\frac{1}{4}X^2 - \frac{1}{64}X^4 + \cdots +.$$

Thus,

$$u_z^{(1)\infty} = \mathcal{J}m\left[-\frac{i e^{i\bar{t}}}{R_\omega}\left(1 - \frac{1 + i\frac{1}{4}x^2 - \frac{1}{64}x^4 + \cdots +}{1 + i\frac{1}{4}X^2 - \frac{1}{64}X^4 + \cdots +}\right)\right], \quad (3\text{--}280)$$

with an error of $O(R_\omega^3)$ in both the numerator and denominator. The quantity in parenthesis can be reexpressed in the form

$$u_z^{(1)\infty} = \mathcal{J}m \left\{ -\frac{ie^{i\bar{t}}}{R_\omega} \left[\frac{i\frac{1}{4}(X^2 - x^2) - \frac{1}{64}(X^4 - x^4) + \cdots +}{\left(1 - \frac{1}{64}X^4\right) + i\frac{1}{4}X^2 + \cdots +} \right] \right\}. \quad (3\text{–}281)$$

Then, by multiplying both numerator and denominator by the complex conjugate of the denominator and retaining all terms up to $O(x^4, X^4)$, we obtain

$$u_z^{(1)\infty} = \mathcal{J}m \left\{ -\frac{ie^{i\bar{t}}}{R_\omega} \left[i\frac{1}{4}(X^2 - x^2) + \frac{1}{16}X^2(X^2 - x^2) - \frac{1}{64}(X^4 - x^4) \right] \right\}. \quad (3\text{–}282)$$

Finally, because

$$e^{i\bar{t}} = \cos\bar{t} + i\sin\bar{t},$$

we find that

$$u_z^{(1)\infty} \sim -\frac{1}{R_\omega}\left\{-\sin\bar{t}\left[\frac{1}{4}(X^2 - x^2)\right] + \cos\bar{t}\left[\frac{1}{16}X^2(X^2 - x^2) - \frac{1}{64}(X^4 - x^4)\right]\right\} + O(R_\omega^2). \quad (3\text{–}283)$$

Or, substituting for x and X, we obtain

$$\hat{u}_z^{(1)\infty} = \frac{1}{4}(1 - \bar{r}^2)\sin\bar{t} - \frac{1}{16}(R_\omega)\left(\frac{3}{4} - \bar{r}^2 + \frac{\bar{r}^4}{4}\right)\cos\bar{t} + O(R_\omega^2). \quad (3\text{–}284)$$

The error estimate $O(R_\omega^2)$ in (3–283) and (3–284) represents the terms of $O(R_\omega^2)$ that were neglected in the expansions of J_0 in (3–280) and in the simplification of (3–281) to obtain (3–282).

If we now revert to the full dimensional form for the velocity field $\bar{u}_z^\infty$, we have

$$\bar{u}_z^\infty = \bar{u}_z^{(0)\infty} + \varepsilon u_z^{(1)\infty},$$

and thus

$$u_z^\infty = \frac{G_0 R^2}{4\mu}\left\{(1 + \varepsilon \sin\omega t)\left[1 - \left(\frac{r}{a}\right)^2\right]\right.$$
$$\left. - \varepsilon \frac{R_\omega}{4}\left[\frac{3}{4} - \left(\frac{r}{a}\right)^2 + \frac{1}{4}\left(\frac{r}{a}\right)^4\right]\cos\omega t + O(R_\omega^2)\right\}. \quad (3\text{–}285)$$

Examining (3–285), we see that the first term at $O(1)$ is just the quasi-steady Poiseuille solution, which is identical to the steady solution, (3–44), except that the instantaneous pressure gradient,

$$G = G_0(1 + \varepsilon\sin\omega t),$$

appears in place of the steady pressure gradient. This term dominates as $R_\omega \to 0$, and we see that all effects of inertia (or fluid acceleration) are negligible at this level of approximation. The second term is $O(R_\omega)$ and contains the first influence of fluid inertia. It can be seen that inertia causes the velocity field to lag behind the changes in pressure gradient. In addition, the form of the velocity profile is no longer parabolic. It is not surprising that the first influence of inertia is to generate a (small) phase lag between the velocity and the pressure gradient. Indeed, we had anticipated this effect in the approximate limiting forms of the velocity for

Notes

$R_\omega \ll 1$ and $R_\omega \gg 1$ that were given by (3–264) and (3–265). One way of "explaining" the lag between the velocity and pressure for nonzero values of R_ω is simply to note that a finite mass of fluid ($\rho \neq 0$ because $R_\omega \neq 0$) cannot be accelerated instantaneously by the action of a finite force. However, a clearer picture of the significance of large and small values of R_ω emerges if we note that R_ω is the ratio of the time scale for momentum diffusion across the tube, R^2/ν, to the imposed time scale for changes in the flow, $1/\omega$. We have seen in previous unidirectional flow examples that the velocity field always evolves toward a new steady state by diffusion of momentum with a characteristic time scale R^2/ν. Thus, if G is changed at a rate corresponding to a time scale ω^{-1}, which is slow compared with R^2/ν, the velocity field is able to remain arbitrarily close to its steady-state form for each instantaneous value of G. On the other hand, as G changes increasingly rapidly compared with the rate at which the diffusion process takes place (that is, as R_ω increases), the changes in the velocity field will fall further and further behind the changes in the pressure gradient.

We have shown that the first two terms in an asymptotic approximation for $u_z^{(1)\infty}$ can be generated for small R_ω with a modest degree of effort. With sufficient algebra, higher-order terms (that is, terms proportional to R_ω^n with $n > 1$) could be generated in exactly the same way. A much more difficult task is to determine the asymptotic form of (3–279) for $R_\omega \gg 1$. Indeed, we shall not pursue that task in this section. Instead, we shall see in the next chapter that the asymptotic form for both $R_\omega \ll 1$ and $R_\omega \gg 1$ is more easily achieved by direct asymptotic approximation of the governing equation, (3–269), for G, rather than first obtaining an exact solution and then attempting to evaluate it for large and small values for R_ω.

NOTES

1. C. A. Fletcher, *Computational Techniques for Fluid Dynamics* (Springer-Verlag, Berlin, 1988) (in two volumes); R. Peyret and T. D. Taylor, *Computational Methods for Fluid Flows* (Springer-Verlag, Berlin, 1983); P. M. Gresho and R. L. Sani, *Incompressible Flow and the Finite Element Method* (Wiley, New York, 1998) (in two volumes); J. N. Reddy and D. K. Gartling, *The Finite Element Method in Heat Transfer and Fluid Dynamics* (2nd ed.) (CRC Press, Boca Raton, FL, 2000).
2. H. Tennekes and J. L. Lumley, *A First Course in Turbulence* (MIT Press, Cambridge, MA, 1972); S. B. Pope, *Turbulent Flows* (Cambridge University Press, Cambridge, 2000); P. A. Libby, *An Introduction to Turbulence* (Taylor & Francis, London, 1996).
3. H. Schlichting and K. Gersten, *Boundary Layer Theory* (8th ed., revised) (Springer-Verlag, New York, 2000) (see Chap. 5).
4. H. S. Carslaw, Horatio Carslaw, and J. C. Jaeger, 1986, *Conduction of Heat in Solids* (Oxford University Press: Oxford, 1986); M. Ozisik and M. Necati Vzi&scedi Ik, *Heat Conduction* (John Wiley & Sons, Inc, New York, 1993).
5. The metrics (or scale factors) for a large number of orthogonal curvilinear coordinate systems can be found in the appendix by Happel and Brenner (1973): J. Happel and H. Brenner, *Low Reynolds Number Hydrodynamics* (Noordhoff International, Leyden, The Netherlands, 1973).
6. J. L. Poiseuille, "Recherches experimentales sur le movement des liquids dans les tubes de tres petits diameters" *C. R.* **11**, 9612–9617, 1041–8 (1840); **12**, 112–15 (1841); *Mem. Seventes Etrangers* **9**, 433–543 (1846).
7. See Chap. 12, Section I, and references therein.
8. C. W. Macosko, *Rheology: Principles, Measurements and Applications* (Wiley, New York, 1994); K. Walters, *Rheometry* (Chapman & Hall, London, 1975); J. M. Dealy, *Rheometers for Molton Plastics: A Practical Guide to Testing and Property Measurement* (Van Nostrand Rheinhold, New York, 1982); H. A. Barries, J. R. Hutton, and K. Walters, *An Introduction to Rheology* (Elsevier, Amsterdam, 1989).
9. There are two possible exceptions. One is if the fluid is nonhomogeneous. Condition (3–53) remains valid. However, (3–54) is not valid. For example, let us suppose that the fluid is Newtonian,

but nonhomogeneous so that $\mu = \mu(y)$. Then (3–53) does not imply that $du/dy \neq$ const and the velocity profile is not (3–51). A second exception would be if there is slip at the boundaries, as might occur for some polymer/wall material combinations at high shear rate, but in this case the Eqs. (3–52)–(3–54) would still be true.

10. "One-dimensional" indicates that there is only a single nonzero velocity component in an appropriately chosen coordinate system. If we use cylindrical coordinates for the three problems mentioned, the only nonzero velocity component is $u_\theta(r, z)$.

11. A very interesting series of studies of the influence of end effects in the rotating concentric cylinder problem has been published by Mullin and co-workers: T. Mullin, Mutations of steady cellular flows in the Taylor experiment, *J. Fluid Mech.* **121**, 207–18 (1982); T. B. Benjamin and T. Mullin, Notes on the multiplicity of flows in the Taylor experiment, *J. Fluid Mech.* **121**, 219–30 (1982); K. A. Cliff and T. Mullin, A numerical and expwerimental study of anomalous modes in the Taylor experiment, *J. Fluid Mech.* **153**, 243–58 (1985); G. Pfister, H. Schmidt, K. A. Cliffe and T. Mullin, Bifurcation phenomena in Taylor–Couette flow in a very short annulus, *J. Fluid Mech.* **191**, 1–18 (1988); K. A. Cliffe, J. J. Kobine, and T. Mullin, The role of anomalous modes in Taylor–Couette flow, *Proc. R. Soc. London Ser. A* **439**, 341–57 (1992); T. Mullin, Y. Toya, and S. J. Tavener, Symmetry breaking and multiplicity of states in small aspect ratio Taylor–Couette flow, *Phys. Fluids* **14**, 2778–87 (2002).

12. S. Chandrasekhar, *Hydrodynamic and Hydromagnetic stability* (Dover Publications, Inc: New York 1992).

13. Assuming that the characteristic length and velocity scales, (3–90), are correct, the right-hand side of (3–96) is $O(1)$, independent of the dimensional parameters of the problem. Thus the parameter R^2/vt_c must also be $O(1)$. For, if $R^2/vt_c \ll 1$, then the left-hand side would be negligible and the problem reduces back to the steady-flow case. If, on the other hand, $R^2/vt_c \gg 1$, then $\partial \overline{w}/\partial \overline{t} \sim 0$, and $\overline{w}$ should be independent of time, thus adopting its initial form for all $\overline{t}$. However, this is inconsistent with the requirement that $\overline{w} \to 0$ as $\overline{t} \to \infty$ in order that $\overline{u}_z$ achieve its known steady-state form.

14. K. F. Riley, M. P. Hobson, and S. J. Bence, *Mathematical Methods for Physics and Engineering: A Comprehensive Guide* (Cambridge University Press, Cambridge, 1998); J. R. Ockendon, S. Howison, J. Ockendon, A. Lacey, and A. Movchan, *Applied Parital Differential Equations* (Oxford University Press, Oxford, 2003); H. F. Weinberger, *First Course in Partial Differential Equations with Complex Variables and Transform Methods* (Dover, Dover edition: New York, 1995).

15. An excellent source of information about the special functions of mathematical physics is M. Abramowitz and I. A. Stegun, *Handbook of Mathematical Functions* (Dover, New York, 1965).

16. See Chaps. 10 and 11 of Note 14c (Weinberger, *op. cit.*).

17. G. W. Bluman and J. D. Coles, *Similarity Methods for Differential Equations* (Springer-Verlag, New York, 1974); G. I. Barenblatt, *Scaling, Self-Similarity and Intermediate Analysis* (Cambridge University Press, Cambridge, 1994) (originally published in English in 1979 by Consultants Bureau, New York).

18. However, to avoid confusion with the temperature, which we have already denoted as θ, we use φ to represent the polar angle.

19. The problem was solved by R. Siegel, E. M. Sparrow, and T. M. Hallman, Steady laminar heat transfer in a circular tube with a prescribed wall heat flux, *Appl. Sci. Res.* **7**, 386–92 (1958). The description here follows the textbook by J. C. Slattery, *Advanced Transport Phenomena* (Cambridge University Press, Cambridge, 1958).

20. G. I. Taylor, Dispersion of soluble matter in solvent flowing slowly through a tube, *Proc. Roy. Soc. London Ser. A* **219**, 186–203 (1953); G. I. Taylor, Conditions under which dispersion of a solute in a stream of solvent can be used to measure molecular diffusion. *Proc. Roy. Soc. London Ser. A* **225**, 473–7 (1954).

21. In a real system, the tube will, of course, be finite in length and then, at large-enough times, the whole tube will simply be filled with fluid at temperature T_1. Before this, however, the analysis presented in this section will be valid and provide a useful description of the evolution of the temperature distribution.

22. H. Brenner and D. A. Edwards, *Macrotransport Processes* (Butterworth-Heinemann, Boston, 1993).

PROBLEMS

23. J. R. Womerslay, Method for calculation of velocity, rate of flow, and viscous drag in arteries when the pressure gradient is known, *J. Physiol.* **127**, 553–63 (1955).
24. T. J. Pedley, *The Fluid Mechanics of the Large Blood Vessels* (Cambridge University Press, Cambridge, 1980); R. Skalak, N. Ozkaya and T. C. Skalak, Biofluid mechanics, *Annual Rev. Fluid Mech.* **21**, 167–204 (1989).

PROBLEMS

Problem 3–1. Flow in a Cylindrical Tube. Re-solve the start-up of a pressure-driven flow through a tube of radius R, only this time take the pressure gradient to be $G = At^2$ for all $t > 0$, where A is a constant. You may obtain the asymptotic solution at large time by recognizing that $u_\infty = f_1(r)t^2 + f_2(r)t + f_3(r)$. Obtain an integral expression for the coefficients of the decaying solution, but don't evaluate the integral unless you want practice with Bessel functions.

Problem 3–2. Film Flow. A liquid film is flowing down an inclined wall as illustrated in the figure. The channel is configured so that the air in the channel is open to the atmosphere. Assume that the interface is flat and parallel to the walls. Determine the steady-state velocity profile in the liquid film and the volumetric flow rate (per unit width) down the wall when $h_2 \to \infty$. How large must h_2 be in order to neglect the presence of the upper wall on the volumetric flow rate of the liquid? Discuss the limiting cases of the angle $\theta \to 0$ and $\theta \to \pi/2$. In all cases determine the force exerted by the fluid on the lower wall.

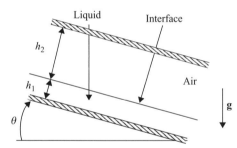

Problem 3–3. Core-Annular Flow

(a) Consider the steady-state flow through a horizontal circular tube of radius R. By what fraction is the volumetric flow rate for a given pressure drop reduced if we place a wire of radius εR at the centerline? Solve the problem for all ε, but calculate the actual fraction for $\varepsilon = 0.1$. The answer may surprise you.

(b) Now, instead of a wire, assume that the central region ($r < \varepsilon R$) consists of a fluid with viscosity μ_1 while the outer region $r > \varepsilon R$ is of viscosity μ_2. Again consider the effects of ε and the fluid viscosities on the volumetric flow rates of the two fluids. This core-annular-flow problem is easy to solve if you remember that shear stress and velocity must be continuous at the interface between the two fluids. Core-annular flow is sometimes used to reduce frictional losses in the pipeline flow of a viscous liquid (e.g., oil) by lubricating the oil with a thin layer (ε close to 1 in this case) of a less-viscous fluid (water).

Problem 3–4. Pressure-Driven Flow in a Rectangular Tube. Obtain the solution for steady-state pressure-driven flow in a rectangular conduit of width $2b$ in the x direction and $2a$ in the y direction. Plot the velocity averaged over the y direction as a function of x for $b/a = 10$. It is the tailing off of the velocity near the sidewalls that dominates dispersion for pressure-driven flow in microfluidic channels. (Hint: Subtract off the velocity profile you would get in the absence of sidewalls to render the PDE homogeneous, and then get a series solution for the remainder. This is fairly easy if you remember to make maximum use of symmetry!)

Problem 3–5. The Homogeneous Flow Apparatus. In an important paper on the structure of suspensions undergoing shear, Graham and Bird [*Ind. Eng. Chem. Fundam.* **23**, 406–10 (1984)] make use of a device they called the Homogeneous Flow Apparatus (HFA). The HFA was designed to produce a simple shear flow between two belts moving in opposite directions (say the z and $-z$ directions). The flow was bounded by sidewalls, which together with the belts formed a rectangle in the x and y directions. The no-slip condition at these bounding walls disturbs the flow and prevents the device from generating a perfect simple shear flow (plane Couette flow). The authors assert that for belts of width 10 cm and a gap between the belts of 5 cm, the wall effects extend only about 1.5 cm into the interior, leaving about 65% of the gap in a homogeneous simple shear flow. They assert that the deviation from simple shear flow in this central region is only 2%. Is this assertion correct? In the drawing below, the flow is unidirectional, and into (or out of) the paper. [Hint: Because the flow is unidirectional (albeit a function of x and y), it satisfies the Laplace equation. To solve this problem, subtract off the simple shear solution (e.g., what happens with no sidewalls), and then solve for the "decaying part" by means of separation of variables – only this time you will get hyperbolics in the x direction rather than pure exponentials, and sines in the y direction.)

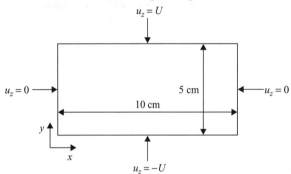

Problem 3–6. Wire Pulling. A wire is pulled through a circular cylinder that connects two large chambers maintained at a constant pressure p_o.

(a) Derive an expression for the steady-state volumetric flow rate between the chambers and an expression for the force required to pull the wire through the cylinder.

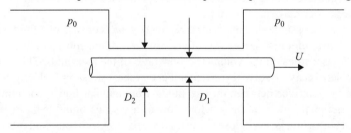

Definition sketch for Problem 3–6(a).

(b) Repeat this calculation when the two chambers are sealed with the total volume equal to that of the fluid. Are the results different? If so, why? Be sure to discuss the limiting case $D_1/D_2 \to 0$.

Problems

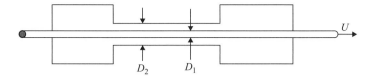

Definition sketch for Problem 3–6(b).

Problem 3–7. Core-Annular Flow. Two fluids are being pumped vertically against gravity in a tube as illustrated in the figure. The heavier fluid B of density ρ_1 and viscosity μ_1 fills the annular space near the walls. The core of the tube contains the lighter fluid A of density ρ_2 and viscosity μ_2. Given the geometry as illustrated and the fluid properties, under what conditions will the heavier fluid adjacent to the walls fall in the same direction as gravity?

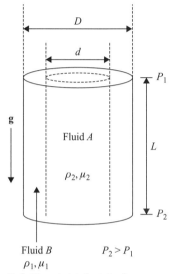

Definition sketch for tube flow.

Problem 3–8. The Dynamics of Capillary Rise in a Tube. In Chap. 2, Problem 2–30, we considered the problem of capillary rise of a liquid up into a soda straw. We assume that the soda straw has an inside diameter R and is placed in a pan of liquid of density ρ. The surface tension between the water and the air is γ, and the water makes a contact angle θ at the water–straw interface. We showed that the liquid will rise up into the straw to an equilibrium height

$$H_0 = \frac{2\gamma \cos\theta}{\rho g R}.$$

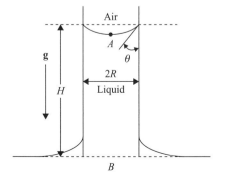

187

(a) Consider the motion of capillary rise from an initial state $H = 0$ to the final equilibrium condition $H = H_0$. Show that the rate at which the capillary fluid will rise is

$$\frac{dH}{dt} = \frac{a^2}{8\mu}\frac{(-\Delta p)}{H},$$

where

$$\Delta p \equiv P_B - P_A,$$

assuming that the velocity profile at any instant of time is given by the *steady* Poiseuille profile. Under what conditions is this assumption valid?

(b) Express the pressure drop $-\Delta p$ in terms of $\alpha, \theta, a, p, g,$ and H, and then show that the dimensionless rate of capillary rise can be expressed in the form

$$\frac{d\overline{H}}{d\bar{t}} = \frac{1-\overline{H}}{\overline{H}},$$

where $\overline{H} \equiv H/H_0$ and $\bar{t} = t/t_c$, where t_c is the characteristic time to attain the equilibrium height H_0. Obtain t_c in terms of the dimensional parameters of the problem.

(c) Obtain an exact solution for $\overline{H}$ as a function of $\bar{t}$.

(d) Show that when $\overline{H} \ll 1$ (that is, for short times),

$$\overline{H} = \sqrt{2}\bar{t}^{1/2}.$$

From this result, demonstrate that the dimensional height of the liquid column, H, increases as the radius of the tube increases, that is, $H \sim a^{1/2}$. Explain the reason for this result in physical terms.

(e) At large times, when $\overline{H} \approx 1$, show that

$$\overline{H} = 1 - e^{-\bar{t}}.$$

This implies that

$$H \sim \frac{1}{a},$$

that is, H decreases as a increases. How can you explain the apparent discrepancy between (d) and (e)?

Problem 3–9. Oscillating Planar Couette Flow. We consider an initially motionless incompressible Newtonian fluid between two infinite solid boundaries, one at $y = 0$ and the other at $y = d$. Beginning at $t = 0$, the lower boundary oscillates back and forth in its own plane with a velocity $u_x = U \sin \omega t$ $(t > 0)$.

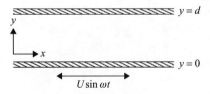

(a) Identify characteristic velocity, length, and time scales, and nondimensionalize the governing DE and boundary conditions. You should note that there are two combinations of dimensional parameters that represent characteristic time scales, ω^{-1} and d^2/ν. What is the physical significance of each? The nature of the solution for the velocity field depends on the magnitude of a single dimensionless parameter. What is it? What is its significance?

Problems

(b) Assume that the boundary motion has been going on for a long period of time so that all initial transients have decayed away ($t \gg d^2/v$) and the velocity field is strictly periodic. Solve for the velocity field in this case. Also, calculate the shear stress at the moving wall. Does fluid inertia increase or decrease the average shear stress?

(c) Determine the limiting form of the solution for $d^2\omega/v \ll 1$ to determine the first effect of inertia. Explain the result in physical terms.

(d) Solve the start-up flow problem, assuming that the fluid is at rest at $t = 0$. Do not evaluate the coefficients, but set up the integral for their calculation.

Problem 3–10. The Rayleigh Problem with Oscillating Boundary Motion. Consider an incompressible, Newtonian fluid that occupies the region above a single, infinite plane boundary. Beginning at $t = 0$, this boundary oscillates back and forth in its own plane with a velocity $u_x = U \sin \omega t (t > 0)$.

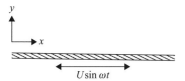

We wish to determine the velocity distribution in the fluid at large times, after any initial transients have decayed, so that the velocity field is strictly periodic. Normally we would proceed as in the related problem, 3–9, by first nondimensionalizing and then solving the problem in dimensionless form. Here, however, there is not an immediately obvious characteristic length scale, so we follow a different procedure.

(a) Solve for the velocity distribution by using the full dimensional equations and boundary conditions. Does the solution exhibit a self-similar form? Should we have expected a self-similar solution? Based on your solution, what is the characteristic length describing the distance the disturbance velocity penetrates into the fluid? Explain this penetration length scale in physical terms.

(b) Based on the characteristic penetration length scale identified in (a), show that the governing equations and boundary conditions can be nondimensionalized to a form that contains no dimensionless parameters at all. This means that the *form* of the solution does not depend on frequency ω, except as a scaling parameter for the independent variables. Does this make sense? Note that this feature in the present problem is fundamentally different from Problem 3–9, where the *form* of the solution was found to depend on the frequency of oscillation ω. This difference may help you to understand the lack of dependence on ω in the present problem.

Problem 3–11. Burial of Pipes to Prevent Frozen Lines. Water pipes are frequently buried in the relative warmer, deep soil to prevent freezing and cracking during very cold weather. It is useful to know how deep to bury the pipes and whether daily, monthly, or seasonal variations in temperatures determine this depth. To model this problem, let us neglect the effects of the pipe and study the unsteady temperature profile in a semi-infinite medium. We will assume that the temperature at the surface of the soil is given by $(T_0 - T_\infty) \sin(\omega t) + T_\infty$; thus we are essentially examining one temperature variation and frequency. The thermal diffusivity of the soil is α, and far away from the surface, the temperature is T_∞, which is greater than the freezing temperature of the fluid in the pipe. Also, we assume that T_0 is less than the freezing temperature.

(a) What are the nondimensional heat conduction equation and boundary conditions for this problem?

- Hint: Let your nondimensional temperature be
$$\theta = (T - T_\infty)/(T_0 - T_\infty),$$
where T is the dimensional temperature.
(b) Determine the long-time temperature profile in the soil and sketch it. Approximately how far does the temperature variation penetrate into the soil?
(c) Which temperature variations are most important in determining the depth of burial, daily, monthly, or seasonal? How deep should the pipe be reasonably buried to prevent freezing?

Problem 3–12. Solidification From a Cylindrical Boundary. A pure liquid is being frozen outside a refrigerated tube, as illustrated in the figure. The bulk liquid is at its freezing temperature T_F and the inner wall of the tube is cooled by convective heat transfer to temperature $T_G < T_F$. The latent heat of fusion is L. Furthermore, we assume that the densities of the liquid and solid phase are equal.

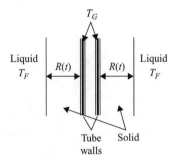

(a) Set up the problem from which you could determine the rate at which the frozen solid advances into the bulk liquid $R_s(t)$. You should carry out this analysis to the point of having obtained a single equation from which you could evaluate the numerical coefficient that relates $R_s(t)$ and time t (hint: should you expect a similarity solution for this problem?)
(b) How does the problem change if, instead of fixing the temperature of the surface of the tube, we assume that heat is transferred to a flowing gas inside the tube with temperature T_G, with a heat transfer coefficient h. You may assume that the heat transfer resistance across the wall of the tube is negligible, that the tube is infinitely long, and that the gas temperature T_G does not change along the length of the tube.
(c) The analysis in part (b) is considerably simplified with the pseudo-steady-state assumption. When is this valid?

Problem 3–13. Reaction and Diffusion at an Interface. Consider a planar interface created by the contact of two miscible fluids. One fluid contains a dilute species A and the other a dilute species B. These diffuse toward one another and react to produce P according to
$$R_P = -R_A = -R_B = -kC_A C_B,$$
where R_i is the rate of production of species i and k is the rate constant. Clearly the reaction occurs near the interface. Far away from the interface, the concentrations of A and B approach C_{A_0} and zero in fluid one, and the concentrations of A and B approach zero and C_{B_0} in fluid two. Let's construct a model for this process and perform a scaling analysis to understand some aspects of the system.

(a) What are the simplified, dimensional species balance equations and boundary/initial conditions that describe the temporal and spatial behavior of this system?

Problems

(b) Non-dimensionalize the equations. What are the characteristic time and length in this system? What measurable aspects of the system do they correspond to?
- Hint: Sketch a figure showing the concentration profiles for A, B, and P to help you understand what is going on here.

(c) Now, let's suppose we can measure the peak concentration of P as a function of time. How do you think P scales with time and the system parameters for short and long times? What do short and long times mean? That is, relative to what?
- Hint: Only some simple algebra is necessary here. You should focus on understanding what is going on in terms of transport mechanisms and time and length scales.

Problem 3–14. Design of a Cross-Flow Filter. In cross-flow filtration, a pressure drop G forces fluid containing neutrally buoyant particles to flow between two porous plates. There is also a transverse flow that forces the particles to collect on one of the plates. A key design question is how one determines the length of the filter.

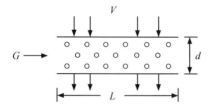

(a) Suppose that the distance between the two plates is d and the fluid has viscosity μ and density ρ. Further, the transverse velocity at the upper porous plate is fixed at V. Assuming the concentration of particles is small and neglecting their effect on the flow, what are the dimensionless equations and boundary conditions that govern the flow field?
(b) What is the average velocity parallel to the plates?
(c) How would you determine the length of the filter, L? Clearly state your reasoning!

Problem 3–15. Sudden Contraction. A horizontal plate is translating at a constant speed U in the x direction, as illustrated in the figure. The plate drags an incompressible Newtonian fluid of constant density and viscosity from a region where the separation between the moving plate and the stationary wall is H into a region where the separation is λH, $\lambda \leq 1$. We wish to analyze the flow structure far upstream from the "sudden contraction" and far downstream after the contraction where the flow is fully developed.

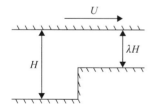

Definition sketch for Problem 3.

(a) Determine the velocity profiles in the for upstream and downstream regions when the net flux (per unit width) through the channel is $(1/2)UH$. Be sure to state any and all assumptions. Does this reduce to the correct answer as $\lambda \to 1$?
(b) Sketch the velocity profiles in the two regions. What is the overall pressure change? Is it positive or negative?

(c) Assuming that the velocity profiles found in part (a) apply right up to and after the contraction, determine both the drag and normal force on the moving plate.

Problem 3–16. Couette Flow with Spatially Stratified Viscosity. Plane Couette flow for an incompressible, Newtonian fluid yields a simple shear flow at steady state, but only if the fluid viscosity is constant. In this problem we examine what happens if there is a linear variation of viscosity with position, e.g., $\mu = \mu_0(1 + Ay/d)$, where A and μ_0 are constants. This could occur, for example, in a bearing in which there is a radial temperature gradient, because viscosity is a function of temperature. Set up the problem of an impulsively started plate at $y = 0$ (and a fixed plate at $y = d$). Solve for the steady-state velocity profile and determine the eigenvalue problem governing the initial transient decaying solution. Obtain the leading-order eigenvalue for this problem. {Hint: Show that the governing equation can be converted to a form of Bessel's equation with solutions $Y_0[2\lambda_n\sqrt{(1 + A\bar{y})/A^2}]$ and $J_0[2\lambda_n\sqrt{(1 + A\bar{y})/A^2}]$, where λ_n are the eigenvalues and $\bar{y} = y/d$.} Plot this eigenvalue as a function of A for both positive and negative values.

Problem 3–17. Plane Couette Flow Driven by Oscillating Shear Stress. Consider an initially motionless incompressible Newtonian fluid confined between two infinite plane boundaries separated by a gap width d as depicted in the figure. Suppose the plane $y = 0$ is oscillated back and forth with an oscillatory *shear stress* $\tau = \tau_0 \sin \omega t$.

(a) Nondimensionalize this problem in a form relevant for large times ($vt/d^2 \gg 1$), and then solve for the velocity distribution in this regime of times. What is the effect of inertia in this case? To consider this last question, you should consider the asymptotic form of the solution for small, but nonzero, values of $d^2\omega/v \ll 1$.

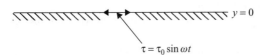

(b) Re-solve this problem for the case in whcih the fluid is initially at rest ($u = 0$ everywhere) and look at the limit $vt/d^2 \ll 1$ and $\omega t \ll 1$. Note that in this limit the shear stress at the wall will be a linearly increasing function of time (i.e., $\tau \approx \tau_0 \omega t$) and the problem will admit a similarity transform. Obtain the similarity transform together with the resulting ODE and boundary conditions, but do not try to obtain an explicit solution. How does the velocity of the plane at $y = 0$ vary with time?

Problem 3–18. Rayleigh Problem with Constant Shear Stress. Consider a semi-infinite body of fluid that is bounded below, at $y = 0$, by an infinite plane, solid boundary. For times $t < 0$, the fluid and the boundary are motionless. However, for $t \geq 0$, the boundary moves in its own plane under the action of a constant force per unit area, F. Assume that the boundary has zero mass so that its inertia can be neglected.

Determine the velocity distribution in the fluid by assuming a solution in the form

$$u = h(t)f(\eta),$$

where $\eta = y/g(t)$. Why does the simpler form $u = f(\eta)$ used in the Rayleigh problem not work in this closely related problem? If you had to program the boundary motion to produce the exact same solution, what would the velocity at the boundary need to be as a function of time? Does this present any difficulties?

Problems

Problem 3–19. Start-Up of Plane Couette Flow Driven by a Constant Shear Stress.
Consider the motion of an incompressible, Newtonian fluid in the region between two parallel plane walls if the lower wall is subjected to a constant force per unit area, F, in its own plane for all $t \geq 0$. The walls are separated by a distance d, there is no pressure gradient, and the fluid is initially at rest.

Determine the velocity distribution in the fluid. Show that your solution reduces for short times to the form $u = h(t)f(\eta)$ obtained in Problem 3–18. What is the criterion for validity of this short-time approximation in this case?

Problem 3–20. Generalization of the Rayleigh Problem. A plate bounding a fluid initially at rest is impulsively started with a velocity given by $u|_{y=0} = U_0 + At$, as depicted in the figure. Determine the shear stress at the wall to within two unknown numerical constants, and show how these may be calculated from simple ODEs. (Hint: Think linearity and break up the problem!)

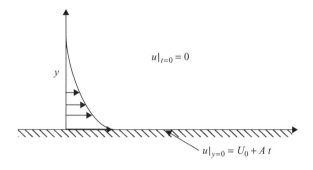

Problem 3–21. The Rayleigh Problem with Prescribed Boundary Acceleration. An infinite plane bounding a quiescent fluid with viscosity μ and density ρ is accelerated from rest with velocity at^2. Set up the equation governing the time-dependent velocity profile in the fluid, and, using a similarity transform, determine how the shear stress at the plane varies with time to within some unknown constant.

Problem 3–22. Rayleigh's Problem for a Non-Newtonian Fluid. Consider an initially quiescent, incompressible, non-Newtonian power-law fluid above an infinite flat plate. The plate is set into motion at $t = 0$ with velocity U. For a power-law fluid, the shear stress is given by

$$\tau_{yx} = k\left(\left|\frac{\partial u}{\partial y}\right|^{n-1} \frac{\partial u}{\partial y}\right),$$

where n is a constant. If $n > 1$, the fluid is shear thickening; if $n < 1$, the fluid is shear thinning; and if $n = 1$ the fluid is Newtonian. Assume a similarity solution of the form

$$u = UF(\eta), \quad \text{where } \eta = Ayt^\alpha.$$

(a) For what values of n can you find a solution by using this particular transform? Give an explanation why this is so using physical and/or mathematical arguments. For such n, what is the general solution? (You needn't evaluate constants, but you should explain how you would.)

Unidirectional and One-Dimensional Flow and Heat Transfer Problems

(b) Using a value of $n = (1/2)$, derive an expression for the unsteady flow field. Plot the flow field and comment on your answer (i.e., Does it make sense to you and why?).

(c) What is the penetration depth as a function of time? How does it compare with that of a Newtonian fluid? Explain the difference.

(d) What is the shear stress acting on the plate?

Problem 3–23. The Controlled Stress Rheometer. A commonly used viscosity measurement tool is the controlled stress rheometer, in which the applied stress is controlled and the resulting motion of a plate is used to calculate the viscosity. A simplified version of such a system is depicted in the figure.

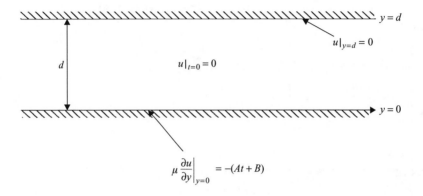

In this problem we are measuring the viscosity by looking at the effect of a linear ramp in the shear stress. The entire system is initially at rest, and at time $t = 0$ a shear stress given by $\tau = -(At + B)$ is applied to the lower wall. The velocity U, of the lower wall is measured as a function of time, with the apparent viscosity being defined as $\mu_{app} = \tau d/U$, where d is the gap width.

(a) If the fluid is actually Newtonian with constant viscosity, how will the apparent viscosity depend on time and the other parameters in the problem for large times? When will this solution be valid?

(b) The solution obtained in part (a) breaks down at short times. Solve for the velocity at these shorter times by using separation of variables. Obtain the eigenfunctions and eigenvalues and show how the constants in the series solution would be obtained.

(c) For very short times the problem may admit a similarity solution. Show that the problem admits such a similarity solution, and give the similarity rule, similarity variable, transformed DE, and boundary conditions. Determine how the apparent viscosity depends on the parameters in the problem to within some unknown constant. When will this solution be valid?

Problem 3–24. Planar Couette Viscometer with Time-Dependent Boundary Motion. It is proposed that the stress–strainrate relationship of a fluid be probed by using a parallel plate shear viscometer in which the imposed velocity of the lower plate is a combination of a linear ramp in velocity and an oscillatory velocity. The resulting problem is depicted in the figure.

Problems

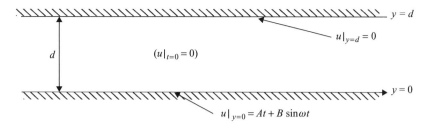

We wish to examine the influence of fluid inertia on the stress measured at the lower plate.

(a) What is the stress at the lower plate at large times? You may leave your answer in terms of complex variables.
(b) How long will we have to wait before the solution obtained in part (a) becomes valid? (Hint: Determine the leading eigenvalue.)

Problem 3–25. Start-Up for a Rotating Cylinder in an Unbounded Fluid. Consider a circular cylinder (infinite length), radius R_1, immersed in a Newtonian fluid. The fluid may be treated as incompressible. The cylinder begins to rotate about its central axis at $t = 0$ with an angular velocity Ω_1 and continues with the same angular velocity for all $t \geq 0$. The fluid is initially motionless.

(a) Find the velocity distribution as a function of r and t (note that $u_\theta \neq 0$, but $u_r = u_z = 0$ – however, the flow field is *not* unidirectional). Is there a similarity solution? How would you calculate the pressure field? What happens in the limit as R_1 becomes arbitrarily small?
(b) Now, consider the same problem, but suppose that there is a second, outer cylinder, of radius $R_2 > R_1$, that is held stationary for all $t \geq 0$. The outer cylinder is assumed to be *concentric* with the inner cylinder. The fluid is initially stationary.
   Determine the solution to the start-up problem in which the inner cylinder suddenly begins (at $t = 0$) to rotate with angular velocity Ω_1. How does this problem differ from (a)? Do the two solutions agree for $t \ll (R_2 - R_1)^2/\nu$?
(c) At some time after the steady flow in part (a) has been established, the cylinder is instantaneously stopped. Calculate the torque required to keep the cylinder from rotating, as a function of time. Notice that the torque vanishes as $t \to \infty$.

Problem 3–26. Spin-Up in a Circular Cylinder. An infinite cylinder is filled with fluid. Initially the fluid is at rest. At time $t = 0$ the cylinder is rotated about its axis with an angular velocity that is linearly increasing in time, e.g., $u_\theta|_{r=R} = At$. In this problem the velocity is in only the θ direction ($u_r = u_z = 0$) and all derivatives with respect to θ and z are zero.

(a) Determine the velocity at large times. The Navier–Stokes equation in the θ direction is given by

$$\rho\left(\frac{\partial u_\theta}{\partial t} + u_r\frac{\partial u_\theta}{\partial r} + \frac{u_\theta}{r}\frac{\partial u_\theta}{\partial \theta} + \frac{u_r u_\theta}{r} + u_z\frac{\partial u_\theta}{\partial z}\right)$$

$$= -\frac{1}{r}\frac{\partial p}{\partial \theta} + \mu\left[\frac{\partial}{\partial r}\left(\frac{1}{r}\frac{\partial}{\partial r}(ru_\theta)\right) + \frac{1}{r^2}\frac{\partial^2 u_\theta}{\partial \theta^2} + \frac{\partial^2 u_\theta}{\partial z^2}\right].$$

(b) Solve the eigenvalue problem (DE and boundary conditions) for the decaying solution by means of separation of variables. This eigenvalue problem is the same you would solve to determine how long it takes to spin up coffee in a coffee cup.

Problem 3–27. Oscillating Cylinder as a Viscometer. Consider a cylinder immersed in a large bath of fluid with kinematic viscosity ν that rotates sinusoidally about its axis with angular velocity $\Omega_z = \Omega \sin(\omega t)$. The cylinder has a radius R, length L, and $L/R \gg 1$. It is proposed to use measurements of the dynamics of the cylinder to determine the viscosity of the fluid.

(a) What are the dimensionless equations and boundary conditions that describe this flow?
(b) What is the long-time oscillating flow field? What is the hydrodynamic torque acting on the cylinder?
(c) Now let us add the dynamics of the cylinder. What is the resonant frequency of the cylinder, assuming it has a density ρ_c?
(d) How can we determine the viscosity of the solution from the resonant frequency of the cylinder? (Presumably it is easier to measure the resonance very accurately, as opposed to the torque.)

Problem 3–28. Rotating Porous Cylinder. A circular cylinder of radius a rotates at a constant angular velocity Ω and generates a flow exterior to the cylinder, i.e., in $a < r < \infty$. The surface of the cylinder is also porous, and fluid is sucked through radially so that the boundary conditions on the cylinder surface are

$$u_r = -U \quad \text{and} \quad u_\theta = \Omega a.$$

Because the domain is unbounded in z and there is axial symmetry, the flow is a function of r only. Obtain the solution for u_r that vanishes at infinity, and derive the following Eulor-type differential equation for u_θ:

$$r^2 \frac{d^2 u_\theta}{dr^2} + (Re + 1)r \frac{du_\theta}{dr} + (Re - 1)u_\theta = 0,$$

$$u_\theta = 1 \quad \text{at } r = 1 \quad \text{and} \quad u_\theta \to 0 \quad \text{as } r \to \infty,$$

where $Re = Ua/\nu$ and ν is the kinematic viscosity of the fluid.

Show that if $Re < 2$ there is only one solution of this equation that has a finite circulation $\Gamma = 2\pi r u_\theta$ at infinity, but that if $Re > 2$ there are many such solutions.

- Hint: An Eulor equation is independent of the units of length and therefore has solutions of the form r^α.

Problem 3–29. Oscillating Flow of a Gas in a Porous Medium. It has been suggested that the permeability of gas-bearing porous rock can be determined by oscillating the pressure at a wellhead and measuring the phase lag in the pressure response at another well. Let us examine this proposal with a simple model.

(a) First, we need to derive an equation that describes the unsteady pressure distribution p in a porous medium. We will assume that the velocity of flow $\mathbf{u}$ follows Darcy's law, namely,

$$\mathbf{u} = -\frac{k}{\mu} \nabla p,$$

as well as conservation of mass in a porous medium,

$$\varepsilon \frac{\partial \rho}{\partial t} + \nabla \cdot (\rho \mathbf{u}) = 0,$$

where ρ and μ are the density and viscosity of the gas, respectively, and k and ε are the permeability and porosity of the rock, respectively.

Problems

Using the ideal gas law, show that
$$\varepsilon \frac{\partial p}{\partial t} = \frac{k}{\mu} \nabla \cdot (p \nabla p).$$

(b) As a first cut, so to speak, let's consider the following *idealized* version of the problem to get some insight into its physics. We will imagine the flow to be 1D. At $x = 0$ (the "wellhead"), the pressure is varied sinusoidally and is given by $p(0, t) = (p_0 - p_\infty) \sin(\omega t) + p_\infty$. The pressure far away from this forcing is p_∞. Show that, for a dimensionless pressure $P = (p - p_\infty)/(p_0 - p_\infty)$, the equation for the pressure can be rewritten as
$$\frac{\partial P}{\partial t} = \frac{k p_\infty}{\varepsilon \mu} \frac{\partial}{\partial x} \left[(1 + \beta P) \frac{\partial P}{\partial x} \right],$$
where $\beta = (p_0 - p_\infty)/p_\infty$. What are the boundary conditions on P? (Don't worry about initial conditions, as we are interested in only the long-time behavior of the pressure.)

(c) If $\beta \ll 1$, then the equation for the pressure becomes linear. Under this condition, what is the long-time behavior of the pressure as a function of x and t?

(d) From the preceding result, how can the phase lag be used to determine the permeability of the rock if a pressure sensor (the other "well") was inserted some distance downstream?
- Hint: If the original forcing is proportional to $e^{i\omega t}$ and the response is proportional to $e^{i(\omega t - \phi)}$, ϕ is the phase lag.

(e) What is a critical limitation of this method?

Problem 3–30. A Model for Cell Migration. Cells proliferate and migrate along surfaces in a variety of processes, including healing and the cultivation of tissue on artificial scaffolds. The following mathematical model has been put forward to describe the 1D, unsteady process:
$$\frac{\partial n}{\partial t} + \frac{\partial}{\partial x}(un) = kn(n_\infty - n),$$
where n is the number density of cells, t is time, and x is position. The velocity $u = [-\mu n(\partial n/\partial x)]$. The boundary conditions are $n = 0$ at $x = \infty$ and $n = n_\infty$ at $x = -\infty$. The initial condition is a step change between the two bounding concentrations at $x = 0$. μ and k are constants.

(a) Explain what transport process is represented by each term in the PDE.
(b) Nondimensionalize this equation and boundary conditions. What are the units of μ and k, and what are your characteristic length and time scales?
(c) Sketch how you think the concentration profile varies with time.
(d) The cell front migrates with some velocity V. Estimate the magnitude of this velocity and the thickness of the front.
(e) After the initial transient dies away, the concentration profile in the frame of reference moving with the front should be time independent. To derive that equation, let $y = x - Vt$ and $\tau - t$. Transform your dimensionless equation from (x, t) to (y, τ) space.
(f) **Extra Credit:** Letting the τ dependence vanish, what is the resulting ODE equation and how might you solve it and determine V?

Problem 3–31. Taylor Dispersion Due to Pressure-Driven Flow in a 2D Channel. Show that the Taylor dispersivity for pressure-driven flow in a 2D channel of width H is
$$D_{eff} = D_m + \frac{2 \langle U \rangle^2 H^2}{105 D_m},$$
where $\langle U \rangle$ is the average velocity in the channel.

Unidirectional and One-Dimensional Flow and Heat Transfer Problems

Problem 3–32. Temperature Distribution in a 2D Channel with a Step Change in Heat Flux. Many analogies exist between energy and momentum transfer. In this problem we look at the temperature distribution acquired by a fluid in a 2D channel as it passes from an insulated region through a region where there is a constant heat flux into the fluid at the walls. Throughout this problem we assume that the velocity distribution is fully developed (parabolic). The DE governing the temperature distribution and the boundary conditions are given in the following equations, which are written in index notation (cf. appendix B),

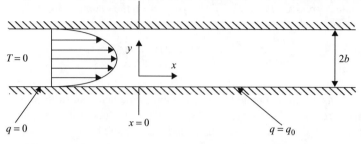

$$u_j \frac{\partial T}{\partial x_j} = \kappa \frac{\partial^2 T}{\partial x_j^2}, \quad -kn_j \frac{\partial T}{\partial x_j}\bigg|_{\substack{y=\pm b \\ x<0}} = 0, \quad -kn_j \frac{\partial T}{\partial x_j}\bigg|_{\substack{y=\pm b \\ x>0}} = q_0.$$

(a) Render the problem and boundary conditions dimensionless. Show, by rendering x dimensionless such that x convection balances y conduction, that for sufficiently high velocities conduction in the x direction is negligible.

(b) Neglecting conduction in the x direction, solve for the asymptotic temperature distribution far down the channel.

(c) Obtain the solution for finite distances down the channel by means of separation of variables. Obtain an explicit equation for the coefficients of this series solution, but do not evaluate the resulting integral.

(d) For small values of x (and values of y close to the wall) it is simpler to solve for the temperature distribution by means of a self-similar solution to the boundary-layer equation. To solve for the temperature profile in this region, let us consider the flow at the upper wall and define a coordinate $s = b - y$. Show that if we neglect conduction in the x direction (as before) and approximate the velocity profile by the linear shear flow $u = \gamma - s$ (its limiting form as $s \to 0$) then the problem will admit a similarity solution. Obtain the similarity transformation and the transformed ODE with corresponding boundary conditions, but do not solve the resulting problem. How does the temperature at the wall and the thickness of the thermal boundary layer vary with x?

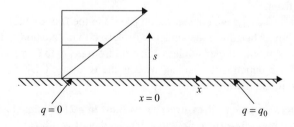

(e) Explicitly determine the domain of validity of the solutions to parts (b)–(d). Under what conditions does the domain of validity of the solution in part (d) entirely vanish?

Problem 3–33. Temperature Distribution in 2D Pressure-Driven Channel Flow with a Step Change in the Wall Temperature. In this problem, we consider the same problem as 3–32, except that the boundary conditions will be different. Suppose we have the channel depicted in the figure:

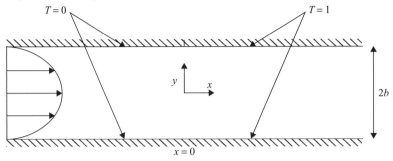

For all $x < 0$ the walls are maintained at a temperature $T = 0$, and for $x > 0$ at a temperature $T = 1$. The fluid velocity at the centerline is U, the width of the channel is $2b$, and the thermal diffusivity is κ? If we assume constant properties everywhere, the problem is governed by the DE:

$$u_j \frac{\partial T}{\partial x_j} = \kappa \frac{\partial^2 T}{\partial x_j^2},$$

where we will assume that the velocity is just unidirectional plane Poiseuille flow at all positions in the channel. Answer the same questions, (a)–(e), for Problem 3–31.

Problem 3–34. Temperature Distribution in Plane Couette Flow Due to a Linearly Increasing Wall Temperature. In this problem you will examine convective heat transfer in plane Couette flow. Suppose we have the channel depicted in the figure:

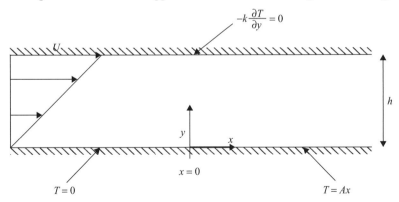

For all $x < 0$ the lower wall is maintained at a temperature $T = 0$, and for $x > 0$ at a linearly increasing temperature $T = Ax$, where A is a constant. The upper wall is insulated, so that there is no heat flux through it. The velocity of the upper wall is U, the width of the channel is h, and the thermal diffusivity is κ. If we assume constant properties everywhere, the problem is governed by the DE:

$$u_j \frac{\partial T}{\partial x_j} = \kappa \frac{\partial^2 T}{\partial x_j^2},$$

where the velocity is just unidirectional plane Couette flow.

Solve the following problems, in each case carefully determining the domain of validity of the solutions:

(a) The asymptotic temperature distribution for large x (this is simple, but not trivial).
(b) The eigenvalue decaying solution. Set up the Sturm–Liouville eigenvalue problem completely, and show how to calculate the coefficients, but don't actually solve the DE for the eigenfunctions. It turns out that the eigenfunctions can be found in terms of Bessel functions of order $1/3$.
(c) The self-similar boundary-layer solution near the entrance to the heated section. Again, set up the problem completely, obtaining the ODE and boundary conditions. How does the heat flux at the wall depend on x?

Problem 3–35. Temperature Distribution in a Combined Plane Couette and Poiseuille Flow with a Linearly Increasing Wall Temperature. In this problem you will examine convective heat transfer in a combined plane Couette and Poiseuille flow. Suppose we have the channel depicted in the figure. The upper wall moves with a velocity U and the lower wall is fixed. In addition to the shear flow there is a pressure-driven backflow resulting in the purely quadratic dependence of velocity on y (e.g., the shear rate at the lower wall is zero).

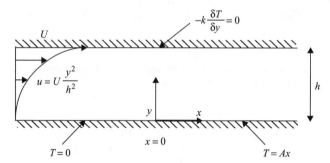

For all $x < 0$ the lower wall is maintained at a temperature $T = 0$, and for $x > 0$ at a linearly increasing temperature $T = Ax$, where A is a constant. The upper wall is insulated, so that there is no heat flux through it. The width of the channel is h, and the thermal diffusivity is κ. If we assume constant properties everywhere, the problem is governed by the DE:

$$u_j \frac{\partial T}{\partial x_j} = \kappa \frac{\partial^2 T}{\partial x_j^2}.$$

Solve the following problems, in each case carefully determining the domain of validity of the solutions:

(a) The pressure gradient required for producing the desired flow.
(b) The asymptotic temperature distribution for large x.
(c) The eigenvalue decaying solution. Set up the Sturm–Liouville eigenvalue problem completely, and show how to calculate the coefficients, but don't actually solve the DE for the eigenfunctions.
(d) The self-similar boundary-layer solution near the entrance to the heated section. Again, set up the problem completely in canonical form, obtaining the ODE and boundary conditions, but don't solve the resulting ODE. In particular determine how the heat flux at the wall and boundary-layer thickness depend on x.

Problem 3–36. Dispersive Transport Due to Oscillatory Flow in a Channel. It is well known that the application of an oscillatory flow to a partially mixed gas in a tube or a channel will lead to a significant enhancement in the rate of transport of a dilute constituent

Problems

in the axial direction from a region of high to low concentration, even though there is no time-averaged mean flow in the axial direction. This phenomena is another manifestation of Taylor-like dispersion, and it plays an important role in such diverse phenomena as pulmonary ventilation and (for liquids) the dispersion of contaminants in tidal estuaries. It can also be used to produce a partial separation of molecular species when there are several components present, because it enhances the differences in transport rate produced by differences in the molecular diffusivity [see, e.g., U. H. Kurzweg and M. J. Jaeger, Diffusional separation of gases by sinusoidal oscillations, *Phys. Fluids* **30**, 1023–5 (1987)].

In this problem, we calculate the time-averaged axial transport rate for the special case of an oscillating pressure-driven flow in a straight wall channel with a cross-sectional shape that we denote as S and a wall that we denote as B. We consider a two-component system consisting of a "carrier" gas and a second dilute species whose concentration we denote as c. The latter is assumed to have a uniform gradient in the axial direction that we denote as β, i.e., $\partial c/\partial z = \beta$. We assume that this second component is sufficiently dilute that the velocity field is wholly determined by the carrier gas, which is modeled as an incompressible, Newtonian fluid with a density ρ and viscosity μ. The pressure gradient is assumed to take the form

$$\frac{\partial p}{\partial z} = -P \cos \omega t.$$

The transport of the diluted species satisfies the species transport equation

$$\frac{\partial c}{\partial t} + w \frac{\partial c}{\partial z} = D \nabla^2 c \text{ in } S,$$

and the walls of the channel are impermeable so that

$$\frac{\partial c}{\partial n} = 0 \text{ on } B,$$

where n denotes the normal variable at the wall.

(a) Assume that the velocity profile $w(x, y, t)$ and the concentration profile $c(x, y, t)$ can be expressed in the forms

$$w = \Re\{f(x, y)e^{i\omega t}\} \text{ and } c = -\beta z + \Re\{\beta g(x, y)e^{i\omega t}\}.$$

Show that the governing equation for the function f can be expressed in the form

$$i\omega f = \frac{P}{\rho} + \nu \nabla^2 f \text{ in } S, \quad \text{with } f = 0 \text{ on } B.$$

Derive a corresponding governing equation and boundary condition for g.

(b) Show that the mean rate of flux in the z direction can be expressed in terms of f and g in the form

$$\text{flux} = D_{\text{eff}} \beta A,$$

where A is the cross-sectional area of S, and

$$D_{\text{eff}} = D(1 + Z) \quad \text{and} \quad Z = \frac{1}{4DA} \int_S \int (f\overline{g} + \overline{f}g) \, dxdy.$$

Here $\overline{f}$ and $\overline{g}$ denote the complex conjugate of f and g. Note that the flux in the absence of any flow would be just $D\beta A$.

(c) Show that the enhancement factor Z can be expressed in the form

$$Z = \frac{1}{2A} \int_S \int \left(\left|\frac{\partial g}{\partial x}\right|^2 + \left|\frac{\partial g}{\partial y}\right|^2 \right) dxdy$$

and is thus positive definite. Hence the rate of mass transfer in the axial direction is always increased compared with what it would be in the absence of any flow.

(d) Now consider the special case of a parallel wall channel of width h. Solve for the functions f and g, and thus obtain an explicit formula for Z. Make a plot of Z versus ω. [For this, you may find it convenient to plot Z versus the dimensionless frequency $h(\omega/\nu)^{1/2}$.]

Problem 3–37. Taylor Dispersion with Streamwise Variations of Mean Velocity. We consider steady, pressure-driven axisymmetric flow in the radial direction between two parallel disks that are separated by a distance $2h$. We assume that the volumetric flow rate in the radial direction is fixed at a value Q and that the Reynolds number is small enough that the Navier–Stokes equations are dominated by the viscous and pressure-gradient terms. Finally, the flow is 1D in the sense that $\mathbf{u} = [u_r(r, z), 0, 0]$. In this problem, we consider flow-induced dispersion of a dilute solute. We follow the precedent set by the classical analysis of Taylor for axial dispersion of a solute in flow through a tube by considering only the concentration profile averaged across the gap, $\langle \cdots \rangle \equiv \frac{1}{2h}\int_{-h}^{h} \cdots dz$.

(a) Determine the velocity profile, with the volume flow rate fixed at Q.

(b) We consider the concentration profile for some dissolved solute, averaged over the azimuthal angle, namely $c(r, z, t) \equiv \frac{1}{2\pi}\int_{0}^{2\pi} c^*(r, z, \phi)d\phi$. The governing equation for c is

$$\frac{\partial c}{\partial t} + u_r(r, z)\frac{\partial c}{\partial r} = D\nabla^2 c.$$

Derive the governing equation for the gap-averaged concentration $\langle c \rangle = \langle c \rangle(r, t)$ in terms of the gap-averaged velocity $\langle u_r \rangle$.

(c) Obtain the governing equation for the concentration perturbation field, $c'(r, z, t) = c(r, z, t) - \langle c \rangle(r, t)$, valid for long times (or large r) compared with the times for diffusion over the distance $2h$, i.e., $\Delta r \gg \langle u_r \rangle h^2/D$.

(d) Solve the equation obtained in (c) for c', and use this solution to evaluate the Reynolds transport term that appears in the equation for $\langle c \rangle$. Hence show that the gap-averaged concentration is governed by the 1D convection–diffusion equation:

$$\frac{\partial \langle c \rangle}{\partial t} + \frac{Q}{4\pi h r}\frac{\partial \langle c \rangle}{\partial r} = \frac{D}{r}\frac{\partial}{\partial r}\left(r\frac{\partial \langle c \rangle}{\partial r}\right) + \frac{2}{105}\left(\frac{Q^2}{16\pi^2 D}\right)\frac{1}{r}\frac{\partial}{\partial r}\left(\frac{1}{r}\frac{\partial \langle c \rangle}{\partial r}\right). \quad (1)$$

Alternatively, this can be written in terms of an "effective diffusivity,"

$$\frac{\partial \langle c \rangle}{\partial t} + \frac{Q}{4\pi h r}\frac{\partial \langle c \rangle}{\partial r} = \frac{1}{r}\frac{\partial}{\partial r}\left(D_{\text{eff}}\, r\frac{\partial \langle c \rangle}{\partial r}\right),$$

where

$$D_{\text{eff}} = D + \frac{2}{105}\left(\frac{Q}{4\pi h r}\right)^2 \frac{h^2}{D}.$$

(e) Suppose that r is not too large, so that the dispersion process is dominated by the second term on the right-hand side of (1). In this case, demonstrate that the average concentration profile can be expressed in the form

$$\langle c \rangle(r, t) = C(s, t),$$

where

$$s \equiv r^2 - \frac{Qt}{2\pi h}$$

Problems

and $C(s, t)$ satisfies a 1D diffusion equation:

$$\frac{\partial C}{\partial t} = \tilde{D}\frac{\partial^2 C}{\partial s^2} \quad \text{with} \quad \tilde{D} = \frac{2}{105}\left(\frac{Q^2}{4\pi^2 D}\right).$$

From this transformation and the solution for C, deduce that the solute concentration distribution corresponding to an initial pulse is both non-Gaussian and subdiffusive. In particular, show that the mean radial position of an initial pulse will increase proportionally to $t^{1/2}$, but that it will spread relative to the mean position only at a rate proportional to $t^{1/4}$.

4

An Introduction to Asymptotic Approximations

Although the full Navier–Stokes equations are nonlinear, we have studied a number of problems in Chap. 3 in which the flow was either unidirectional so that the nonlinear terms $\mathbf{u} \cdot \nabla \mathbf{u}$ were identically equal to zero or else appeared only in an equation for the cross-stream pressure gradient, which was decoupled from the primary *linear* flow equation, as in the 1D analog of circular Couette flow. This class of flow problems is unusual in the sense that exact solutions could be obtained by use of standard methods of analysis for linear PDEs. In virtually all circumstances besides the special class of flows described in Chap. 3, we must utilize the original, nonlinear Navier–Stokes equations. In such cases, the analytic methods of the preceding chapter do not apply because they rely explicitly on the so-called *superposition principle*, according to which a sum of solutions of linear equations is still a solution. In fact, no generally applicable analytic method exists for the exact solution of nonlinear PDEs.

The question then is whether methods exist to achieve approximate solutions for such problems. In fluid mechanics and in convective transport problems there are three possible approaches to obtaining approximate results from the nonlinear Navier–Stokes equations and boundary conditions.

First, we may discretize the DEs and boundary conditions, using such formalisms as finite-difference, finite-element, or related approximations, and thus convert them to a large but finite set of nonlinear algebraic equations that is suitable for attack by means of numerical (or computational) methods. It has become possible, especially with the advent of extremely large and fast computers, to obtain solutions of many flow and transport problems by these techniques. Indeed, the development of suitable numerical methods and their use in solving fluid mechanics problems has become sufficiently widespread and important that it is recognized as an independent subdiscipline known as computational fluid dynamics (CFD). In spite of this, we will not discuss numerical methods in this book. The interested reader will find many reviews of the most current techniques and results in such sources as *The Annual Review of Fluid Mechanics*, and there are also a large number of textbooks devoted exclusively to this subject.[1] For present purposes, we simply note that (1) numerical solutions of the finite-difference or finite-element equations are difficult at large Reynolds numbers; and (2) it can be difficult to understand the physics of a problem from numerical solutions. (For example, the dependence of the solution structure on independent parameters of a problem is difficult to discern from a finite number of numerical solutions at discrete values of these parameters.) Thus numerical solutions offer a very important and useful tool for many problems in fluid mechanics and convective heat or mass transfer, but they are not generally sufficient in themselves, and it is always helpful to supplement numerical analysis with analytic (or experimental) investigations of the same problem.

A. Pulsatile Flow in a Circular Tube Revisited

There are two distinct classes of *analytic approximation* that comprise the second and third approaches that were just mentioned. The first is based on the use of so-called *macroscopic balances*. In this approach, we do not attempt to obtain detailed information about the velocity and pressure fields everywhere in the domain, but only to obtain results that are consistent with the Navier–Stokes equations in an overall (or macroscopic) sense. For example, we might seek results for the volumetric flow rates in and out of a flow system that are consistent with an overall mass or momentum conservation balance but not attempt to determine the detailed form of the velocity profiles. The macroscopic balance approach is described in detail in many undergraduate textbooks.[2] It is often extremely useful for derivation of quantitative relationships among the average inflows, outflows, and forces (or rates of working) within a flow system but is something of a "black-box" approach that provides no detailed information on the velocity, pressure, and stress distributions within the flow domain.

The third and final approach is to seek approximate, *asymptotic* solutions of the governing equations and boundary conditions for very large or very small values of the dimensionless parameters that characterize a particular problem. It is this approach that is the primary focus of the present book.[3] The primary objective is to develop general methods for obtaining approximate solutions of nonlinear DEs by means of perturbation or asymptotic expansions. However, a by-product that is sometimes almost as valuable as the approximate solution is knowledge of the "scaling behavior" of the problem, e.g., the form of the relationship between dependent and independent parameters. We shall see that this level of understanding is achieved in the process of *formulating* the problem in an asymptotic framework, and thus can be attained even when a completely analytic solution of the resulting approximations to the equations and boundary conditions is still not possible.

In the next several chapters, we begin the transition toward more general, approximate-solution methods. We begin, in the present chapter, with the application of asymptotic methods to unidirectional flows, and to some close cousins involving heat transfer and flows that deviate only slightly from unidirectional geometries. These problems introduce the so-called *regular* and *singular* perturbation techniques. Following this, we consider the dynamics of a spherical gas bubble, based on the classical analysis of Rayleigh and Plesset. This very important physical problem allows us to demonstrate several additional asymptotic solution methods, including both *domain perturbations* and *multiple-time-scale methods*.

We begin with a problem that was already solved exactly in the previous chapter, namely, pulsatile flow in a circular tube with a periodic pressure gradient. In particular, we show how asymptotic methods can be applied to obtain both the high- and the low-frequency approximations, $R_\omega \ll 1$ and $R_\omega \gg 1$, for this problem. Although the exact solution and its low-frequency approximation were given in Chap. 3, we shall see that approximate solutions for both the high- and the low-frequency limits can be achieved much more easily by approximation of the *governing equations* (the basis of most asymptotic methods), rather than by first solving the exact equation and then trying to find approximate forms of the exact solution.

A. PULSATILE FLOW IN A CIRCULAR TUBE REVISITED – ASYMPTOTIC SOLUTIONS FOR HIGH AND LOW FREQUENCIES

The analysis leading from the governing equation, (3–254), for pulsatile flow in a circular tube to the exact solution for large times, (3–279), was straightforward, requiring only the recognition of Bessel's equation of order 0, (3–276), and its general solutions $J_0(z)$ and $Y_0(z)$. However, the evaluation of this solution for any arbitrary dimensionless frequency R_ω requires considerable effort, and we thus considered only the limiting case $R_\omega \ll 1$. The asymptotic form for $R_\omega \gg 1$ is very difficult to obtain from the exact solution.

An Introduction to Asymptotic Approximations

In this section we show that asymptotic solutions for $R_\omega \ll 1$ or $R_\omega \gg 1$ at large times can be obtained much more easily if we look for them by directly approximating (3–269), rather than first deriving the exact solution and then trying to deduce its limiting forms. The basis of this direct, asymptotic approximation of the governing equations $R_\omega \ll 1$ ($R_\omega \gg 1$) is to neglect terms that become asymptotically small (compared with other terms in equations) in these limiting cases. For this purpose it is essential to formulate the problem in correct dimensionless form so that we can determine the relative magnitudes of terms by their dependence (or lack of dependence) on R_ω.

In the present case, in which the basic governing equation is linear, the asymptotic analysis serves only to simplify the solution procedure, for example, by avoiding the need to deal with Bessel's equation when $R_\omega \ll 1$. Later, however, we shall see that the same basic methods may often allow approximate analytic solutions to be obtained for nonlinear problems, even when no exact solution is possible.

1. Asymptotic Solution for $R_\omega \ll 1$

Let us begin by considering the case $R_\omega \ll 1$. It is sufficient to consider the problem (3–269) and (3–270) for $H(\bar{r})$, that is,

$$\frac{d^2 H}{d\bar{r}^2} + \frac{1}{\bar{r}} \frac{dH}{d\bar{r}} - iHR_\omega = -1, \qquad (4\text{–}1)$$

$$H(\bar{r}) = 0 \quad \text{at } \bar{r} = 1,$$

$$H \text{ bounded} \quad \text{at } \bar{r} = 0. \qquad (4\text{–}2)$$

In Chap. 3, we derived a general *exact* solution of this problem in terms of Bessel functions J_0 for arbitrary R_ω and then obtained an approximate form for $R_\omega \ll 1$ by approximating this solution. Instead, in the present section, let us suppose from the outset that $R_\omega \ll 1$ and try to seek an approximate solution directly by approximating (4–1) and (4–2).

Assuming that the scaling in (4–1) is correct, terms that do not contain R_ω explicitly are independent of R_ω. Thus, in the limit as $R_\omega \to 0$, Eq. (4–1) reduces to the approximate form

$$\frac{d^2 H}{d\bar{r}^2} + \frac{1}{\bar{r}} \frac{dH}{d\bar{r}} = -1. \qquad (4\text{–}3)$$

The solution of this equation subject to the boundary conditions, (4–2), is just

$$H = \frac{1}{4}(1 - \bar{r}^2), \qquad (4\text{–}4)$$

and if we express this solution in terms of $u_z^{(1)}$ using (3–243), we obtain

$$u_z^{(1)} = \frac{1}{4} \sin \bar{t} (1 - \bar{r}^2). \qquad (4\text{–}5)$$

Thus, in the limit $R_\omega \to 0$, the problem reduces to a quasi-steady Poiseuille flow with an instantaneous pressure gradient $\sin \bar{t}$. In view of the analysis in Chap. 3, this result is not surprising, but we do note that the solution (4–4) was easier to obtain in this case in which we *directly approximated the differential equation* rather than first solving the exact problem and then approximating the solution.

Of course, the limiting case, (4–3), contains no influence of inertia. To determine the effects of inertia for very small, but nonzero, values of R_ω we look for an approximate

A. Pulsatile Flow in a Circular Tube Revisited

solution of (4–1) in the form of an asymptotic expansion in which successive terms are proportional to R_ω^n for $n = 0, 1, 2, \ldots,$

$$H = H_0(\bar{r}) + R_\omega H_1(\bar{r}) + R_\omega^2 H_2(\bar{r}) + O(R_\omega^3). \tag{4–6}$$

Evidently if such a solution exists, the first term, $H_0(\bar{r})$, must be just (4–4). Although it is already known from the exact solution in Chap. 3 that the approximation for small R_ω should take this form, we shall see later that we can determine both the form of asymptotic expansions and the specific functions H_0, H_1, H_2, and so on without any prior knowledge of the exact solution. The form proposed in (4–6) is called a *regular perturbation* (or *asymptotic*) expansion of $H(\bar{r})$ for $R_\omega \ll 1$. We shall discuss the properties of such expansions in more detail in the next section. Here we simply note that the expansion is called *regular* because it is assumed that the same form holds throughout the domain, $0 \leq \bar{r} \leq 1$. We also remark that the convergence of the right-hand side of (4–6) to $H(\bar{r})$ is strictly a consequence of the limit $R_\omega \to 0$ so that each successive term can be made arbitrarily small compared with the terms before it. This includes the error of $O(R_\omega^3)$, which can thus be made arbitrarily small compared with the terms that we have retained in (4–6). *Asymptotic convergence* of this type does not imply convergence to H for some fixed R_ω as the number of terms retained increases, as is true, for example, in power-series approximations of a function. Indeed, for fixed R_ω, we would not necessarily achieve any better approximation to H by the addition of more terms in (4–6), but we shall discuss these questions later.

If we now substitute (4–6) into the governing equation (4–1) for H, we obtain

$$\frac{d^2 H_0}{d\bar{r}^2} + R_\omega \frac{d^2 H_1}{d\bar{r}^2} + R_\omega^2 \frac{d^2 H_2}{d\bar{r}^2} + \frac{1}{\bar{r}} \frac{d H_0}{d\bar{r}}$$
$$+ R_\omega \frac{1}{\bar{r}} \frac{d H_1}{d\bar{r}^2} + R_\omega^2 \frac{1}{\bar{r}} \frac{d H_2}{d\bar{r}} - i H_0 R_\omega - i H_1 R_\omega^2 + O(R_\omega^3) = -1. \tag{4–7}$$

The "order" symbol $O(R_\omega^3)$ represents all terms proportional to R_ω^3 and higher-order terms (i.e., terms proportional to R_ω^m with $m > 3$) that we neglected in writing (4–7) and is known as the *error* term because it represents the largest of the terms we neglected in (4–7) by truncating (4–6) after only three terms. Substituting (4–6) into the boundary conditions (4–2), we obtain

$$H_0(1) + R_\omega H_1(1) + R_\omega^2 H_2(1) + O(R_\omega^3) = 0. \tag{4–8}$$

Now, the parameter R_ω is assumed to be asymptotically small but is otherwise arbitrary; that is, the equalities in (4–7) and (4–8) must be satisfied for any small, but arbitrary, value of R_ω. Thus, if we rewrite (4–7) in the form

$$\left[1 + \frac{1}{\bar{r}} \frac{d}{d\bar{r}}\left(\bar{r} \frac{dH_0}{d\bar{r}}\right)\right] + R_\omega \left[\frac{1}{\bar{r}} \frac{d}{d\bar{r}}\left(\bar{r} \frac{dH_1}{d\bar{r}}\right) - iH_0\right]$$
$$+ R_\omega^2 \left[\frac{1}{\bar{r}} \frac{d}{d\bar{r}}\left(\bar{r} \frac{dH_2}{d\bar{r}}\right) - iH_1\right] + O(R_\omega^3) = 0, \tag{4–9}$$

it is obvious that the terms at each level in R_ω must individually be equal to zero. Similarly, if (4–8) is to hold for arbitrary R_ω it is evident that $H_0(1) = H_1(1) = H_2(1) = 0$. Thus, as we have already noted, the function H_0 satisfies (4–3) plus the condition $H_0(1) = 0$ and

An Introduction to Asymptotic Approximations

is given by (4–4). The functions $H_1(\bar{r})$ and $H_2(\bar{r})$ can be seen from (4–8) and (4–9) to satisfy

$$\frac{1}{\bar{r}}\frac{d}{d\bar{r}}\left(\bar{r}\frac{dH_1}{d\bar{r}}\right) = iH_0 \text{ with } H_1(1) = 0 \text{ and } H_1(0) \text{ bounded,} \qquad (4\text{–}10)$$

and

$$\frac{1}{\bar{r}}\frac{d}{d\bar{r}}\left(\bar{r}\frac{dH_2}{d\bar{r}}\right) = iH_1 \text{ with } H_2(1) = 0 \text{ and } H_2(0) \text{ bounded.} \qquad (4\text{–}11)$$

With $H_0(\bar{r})$ given by (4–4), we easily obtain the solution to (4–10) by integrating twice with respect to $\bar{r}$ and applying boundary conditions. The result is

$$H_1 = \frac{i}{16}\left(\bar{r}^2 - \frac{\bar{r}^4}{4} - \frac{3}{4}\right). \qquad (4\text{–}12)$$

Similarly, with H_1 given by (4–12), the equation and boundary conditions, (4–11), can be solved easily to give

$$H_2 = -\frac{1}{256}\left(\bar{r}^4 - \frac{\bar{r}^6}{9} - 3\bar{r}^2 + \frac{19}{9}\right). \qquad (4\text{–}13)$$

Evidently the same procedure could be used to obtain as many terms as we like in the expansion (4–6). In the present development we stop with H_2. Substituting (4–4) for H_0, plus (4–12) and (4–13) into the asymptotic expansion (4–6), we obtain

$$H = \frac{1}{4}(1 - \bar{r}^2) + R_\omega \frac{i}{16}\left(\bar{r}^2 - \frac{\bar{r}^4}{4} - \frac{3}{4}\right)$$
$$+ R_\omega^2 \frac{1}{256}\left(\frac{\bar{r}^6}{9} - \bar{r}^4 + 3\bar{r}^2 - \frac{19}{9}\right) + O(R_\omega^3). \qquad (4\text{–}14)$$

Now, according to the connection between $\hat{u}_z$ and $u_z^{(1)}$ from Chap. 3,

$$u_z^{(1)} = \mathcal{I}m\left[e^{i\bar{t}}H(\bar{r})\right].$$

Substituting for $H(\bar{r})$ from (4–14), we obtain

$$u_z^{(1)} = \frac{1}{4}\sin\bar{t}(1 - \bar{r}^2) - \frac{1}{16}\cos\bar{t}(R_\omega)\left(\frac{3}{4} - \bar{r}^2 + \frac{\bar{r}^4}{4}\right)$$
$$+ \frac{1}{256}\sin\bar{t}(R_\omega^2)\left(\frac{\bar{r}^6}{9} - \bar{r}^4 + 3\bar{r}^2 - \frac{19}{9}\right) + O(R_\omega^3). \qquad (4\text{–}15)$$

The first two terms in this solution are identical to (3–284), which was obtained from the exact solution by asymptotic expansion of $J_0[(R_\omega/i)^{1/2}\bar{r}]$ for $R_\omega \ll 1$.

There are several important remarks to make concerning the analysis leading to (4–15). First, by introduction of the asymptotic form (4–6) at the beginning and solving

A. Pulsatile Flow in a Circular Tube Revisited

the approximate form of the governing equation (4–9), the analysis has been simplified significantly. For example, at each level of approximation in the asymptotic approach, the ODEs can be solved directly with two integrations with respect to $\bar{r}$, and one requires no knowledge of the Bessel functions $J_0(z)$ or $Y_0(z)$. Second, the solution that was generated (and, indeed, the whole analysis) is precisely equivalent to a solution of (4–1) and (4–2) by the method of successive approximations in which we obtain $H_0(\bar{r})$ by neglecting altogether the term proportional to R_ω in (4–1). This equivalence between successive approximation and asymptotic expansion methods always exists when the asymptotic expansion is regular (that is, valid at all points in the solution domain). It is thus evident that a regular asymptotic expansion will always proceed in increasing powers of the small parameter that appears either in the DEs or the boundary conditions, and this fact allows us to anticipate the form (4–6) without the necessity of having an exact solution available for comparison. Finally, it is important to reiterate the role played by nondimensionalization and the choice of characteristic scales in the asymptotic solution procedure that we have outlined. Specifically, if we return to (4–1), we have implicitly assumed that the term iHR_ω is arbitrarily small in the limit $R_\omega \to 0$ compared with the other terms in the equation for all $\bar{r}$ in the flow domain, $0 \leq \bar{r} \leq 1$. This will clearly be true provided the characteristic scales that are inherent in (4–1) – namely $t_c = 1/\omega$, plus u_c and ℓ_c from (3–253) – provide correct measures of velocity, length, and time scales at all points in the domain. In this case, the magnitude of the terms $d^2H/d\bar{r}^2$ and $(dH/d\bar{r})/\bar{r}$ will be independent of R_ω. Thus, as $R_\omega \to 0$, we can be sure that iHR_ω will become arbitrarily small compared with these terms. If, on the other hand, the characteristic scales were not relevant everywhere in the domain, the magnitude of terms like $d^2H/d\bar{r}^2$ would no longer necessarily be independent of R_ω, and we could make a mistake if we neglected iHR_ω relative to $d^2H/d\bar{r}^2$ even in the limit $R_\omega \to 0$. Fortunately in this case, the scaling and solutions that we have generated are uniformly valid throughout the domain.

Of course, we can never change a physical (or mathematical) problem by simply nondimensionalizing variables, no matter what the choice of scale factors. It is only when we attempt to simplify a problem by neglecting some terms compared with others on the basis of nondimensionalization that the correct choice of characteristic scales becomes essential. Fortunately, as we shall see, an incorrect choice of characteristic scales resulting in incorrect approximations of the equations or boundary conditions will always become apparent by the appearance of some inconsistency in the asymptotic-solution scheme. The main cost of incorrect scaling is therefore lost labor (depending on how far we must go to expose the inconsistency for a particular problem), rather than errors in the solution.

2. Asymptotic Solution for $R_\omega \gg 1$

The next question that we explore is whether the same asymptotic methods that were outlined in the preceding subsection for $R_\omega \ll 1$ could prove equally useful in obtaining explicit analytical results for the asymptotic limit, $R_\omega \to \infty$.

Thus we again consider (4–1) and (4–2) but now for the limit $R_\omega \gg 1$. In this case, we might expect the acceleration term to be larger than the viscous term, which is just the opposite of the limit $R_w \ll 1$. However, it appears as though the acceleration term iHR_ω also becomes larger than the pressure-gradient term in (4–1). But this is clearly impossible! The motion of the fluid exists only because of the pressure gradient. The acceleration *cannot* exceed the pressure gradient. Instead, we should expect that the acceleration and pressure-gradient terms remain in balance even as $R_\omega \to \infty$. Examining (4–1), we see that this can occur only if the magnitude of H decreases as R_ω increases according to

$$H(\bar{r}) = \frac{1}{R_\omega}\overline{H}(\bar{r}).$$

An Introduction to Asymptotic Approximations

In this case, (4–1) takes the form

$$i\overline{H}(\overline{r}) - 1 = \frac{1}{R_\omega}\left[\frac{1}{\overline{r}}\frac{d}{d\overline{r}}\left(\overline{r}\frac{d\overline{H}}{d\overline{r}}\right)\right] \qquad (4\text{–}16)$$

for $R_\omega \gg 1$. Then, in the limit $R_\omega \to \infty$,

$$i\overline{H} = 1 \quad \text{or} \quad \overline{H} = -i, \qquad (4\text{–}17)$$

and this is consistent with the expectation of a balance between acceleration and pressure-gradient terms.

We saw in the previous example for $R_\omega \ll 1$ that the approximate solution, (4–4), which we obtained by taking the limit $R_\omega \to 0$ in the exact equation, (4–1), was just the first term in an asymptotic solution for $R_\omega \ll 1$. Here, we might expect that the approximate equation and solution (4–17) that we obtain by taking the limit $R_\omega \to \infty$ in (4–16) will also be the first term in a formal asymptotic expansion of $\overline{H}$ for $R_\omega \gg 1$. Assuming this expansion is regular, it will take the form

$$\overline{H} = \overline{H}_0(\overline{r}) + \frac{1}{R_\omega}\overline{H}_1(\overline{r}) + O(R_\omega^{-2}). \qquad (4\text{–}18)$$

Now, because $\overline{H}_0 = -i$, it follows from (4–18) that

$$H = -\frac{i}{R_\omega} + O(R_\omega^{-2}), \qquad (4\text{–}19)$$

where the error estimate corresponds to the second term in the expansion (4–18). Thus, referring to (4–17) and (4–18), it follows that

$$u_z^{(1)} = -\frac{1}{R_\omega}\cos\overline{t} + O(R_\omega^{-2}). \qquad (4\text{–}20)$$

Thus Eq. (4–19) yields a uniform (plug) flow that is periodic in time but lags $\pi/2$ radians behind the imposed pressure gradient ($\sim \sin\overline{t}$) as a first approximation to the velocity field for $R_\omega \gg 1$. The fact that the velocity should lag behind the pressure gradient as $R_\omega \to \infty$ is not surprising, as noted in the discussion in Chap. 3, because the balance in (4–1) is increasingly between the pressure-gradient and acceleration terms for large R_ω.

In spite of the fact that some features of the result (4–20) seem reasonable, however, one feature is definitely wrong. The solution (4–19) or (4–20) does *not* satisfy the no-slip boundary conditions $u_z^{(1)} = 0$ at $\overline{r} = 1$. This means that it cannot represent a uniformly valid approximate solution of the original problem, either (3–263) with $u_z^{(1)} = 0$ at $\overline{r} = 1$ or (4–1) and (4–2). What has gone wrong?

The source of the difficulty is seen easily by reexamining the governing DE, (4–16). In this form, the limiting process, $R_\omega \to \infty$, seems to imply that the viscous terms should be negligible everywhere compared with the acceleration and pressure-gradient terms that appear on the left-hand side. However, the neglect of viscous terms reduces (4–16) from a second-order ODE whose solution can be expected to satisfy both the boundary and boundedness conditions on H to an algebraic equation whose solution cannot satisfy these conditions. From a physical point of view, the problem is that the limiting form of (4–16) as $R_\omega \to \infty$ neglects viscous terms, and it is momentum transfer associated with these terms that is responsible for the fact that the fluid does not slip at the boundary (the no-slip

A. Pulsatile Flow in a Circular Tube Revisited

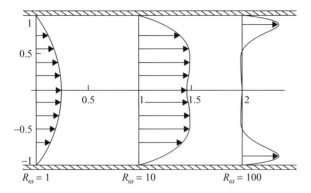

Figure 4–1. Velocity profiles for pulsatile flow in a circular tube for three different values of R_ω all plotted at $\bar{t} = \pi/2$.

condition). It is therefore understandable that the resulting solution does not satisfy the no-slip boundary condition at the tube wall, and this failure indicates that there is something wrong with the limiting process $R_\omega \to \infty$ when applied to (4–16).

The nature of the problem is illustrated in Fig. 4–1, where velocity profiles from the exact solution, (3–279), are plotted for various values of R_ω. We see that, as R_ω increases, the velocity profile does become increasingly blunt in the center of the tube, and the change in $u_z^{(1)}$ with respect to $\bar{r}$ occurs over an increasingly short distance near the tube walls. It thus appears that the problem with (4–16) is that the scaling inherent in its nondimensionalization, namely,

$$\ell_c = R,$$

is not characteristic of the velocity gradients in the vicinity of the tube walls for $R_\omega \gg 1$.

The same conclusion can be reached without recourse to examination of the exact solution for R_ω. For, if the scaling were correct, then the terms $d^2\overline{H}/d\bar{r}^2$ and $\bar{r}^{-1}(d\overline{H}/d\bar{r})$ in (4–16) would be $O(1)$ as actually assumed in the limiting process that led to the solution (4–19) [or (4–20)]. However, this scaling cannot be correct, at least near the walls, because the solution obtained does not satisfy the no-slip condition. Apparently, viscous terms must remain important there even as $R_\omega \to \infty$ because otherwise the no-slip condition cannot be satisfied. The characteristic scale $\ell_c = R$ and the resulting solution (4–19) are perfectly reasonable for most values of $\bar{r}$ in the tube, but neither the solution nor the scaling can be correct near the tube wall.

Let us then reconsider the near-wall region, beginning again with Eq. (4–16). Before proceeding, it is convenient to introduce a change of variables,

$$y = 1 - \bar{r} \quad \text{or} \quad \bar{r} = 1 - y,$$

so that

$$\frac{1}{R_\omega}\left[\frac{d^2\overline{H}}{dy^2} - \frac{1}{(1-y)}\frac{d\overline{H}}{dy}\right] - i\overline{H} = -1 \tag{4–21}$$

and $y = 0$ at the tube wall. Now, all of the preceding discussion indicates that viscous terms must remain important near $y = 0$ even as $R_\omega \to \infty$. Thus,

$$\frac{1}{R_\omega}\left(\frac{d^2\overline{H}}{dy^2}\right) \quad \text{and/or} \quad \frac{1}{R_\omega}\frac{d\overline{H}}{dy}$$

cannot be small compared with unity as suggested by (4–21); that is, derivatives of $\overline{H}$ with respect to y must become large in this region as $R_\omega \to \infty$. However, if derivatives of $\overline{H}$ do become large, it can only mean that $\overline{H}$ must vary over a much smaller length scale than $\ell_c = R$, and, thus, a different nondimensionalization should be introduced into the governing equations in this near-wall region.

Rather than starting with the original dimensional equations and searching for an appropriate characteristic length scale for the near-wall region, we can determine the correct form by simply *rescaling* the previously nondimensionalized equation, (4–21). To do this, let us introduce a new independent spatial variable,

$$Y = y R_\omega^\alpha, \qquad (4\text{--}22)$$

where α is a constant that we will determine later. The motivation for this rescaling can be explained in at least two ways. The simplest idea is that R_ω represents a ratio of the two "natural" length scales of the problem, $\ell_c \equiv R$ and $L \equiv \nu/\omega R$, so that rescaling according to (4–22) simply redefines the radial variable scaled with respect to a new length scale, say, ℓ^*. To see this, we can rewrite (4–22) in the form

$$Y = \left(\frac{y'}{\ell_c}\right)\left(\frac{\ell_c}{L}\right)^\alpha,$$

where y' is the dimensional radial variable. Evidently, if $\alpha = 1$, then

$$Y = \frac{y'}{L},$$

and $L \equiv \nu/\omega R$ would be the new length scale ℓ^* for nondimensionalizing y'. On the other hand, for some other α,

$$Y = \frac{y'}{\ell^*},$$

where

$$\ell^* = (\ell_c)^{1-\alpha} L^\alpha.$$

Thus rescaling in the form (4–22) is precisely equivalent to introducing a new characteristic length scale ℓ^* in the near-wall region. Of course, to completely specify ℓ^* we need to determine α, and we shall see shortly how this can be done. A second way to motivate (4–22) is to simply note that we are *stretching* the variable normal to the wall in such a way that $\overline{H}$ changes from 0 to its mainstream value in an increment of the *rescaled* variable $\Delta Y = O(1)$, whereas in the original nondimensionalized form, $\overline{H}$ apparently changes over a dimensionless distance $\Delta y = O(R_\omega^{-\alpha})$. Thus, in the rescaled system, all spatial derivatives, say, $d^2\overline{H}/dY^2$, will be $O(1)$ and independent of R_ω in the near-wall region, and this will simplify the problem of deciding which terms in the governing equations are to be retained in that part of the domain as $R_\omega \to \infty$.

The coefficient α and the relevant form of the governing equations in the near-wall region are determined by substitution of (4–22) into (4–21), which thus becomes

$$\frac{1}{R_\omega}\left[R_\omega^{2\alpha}\frac{d^2\overline{H}}{dY^2} - R_\omega^\alpha\left(\frac{1}{1 - YR_\omega^{-\alpha}}\right)\frac{d\overline{H}}{dY}\right] - i\overline{H} = -1. \qquad (4\text{--}23)$$

It can be seen that the largest viscous term is $O(R_\omega^{2\alpha-1})$, and thus it follows from (4–23) that

$$2\alpha - 1 = 0$$

A. Pulsatile Flow in a Circular Tube Revisited

if viscous effects are to be equally important as acceleration and pressure-gradient effects in the near-wall region for $R_\omega \to \infty$. Thus

$$\boxed{\alpha = \frac{1}{2},} \tag{4–24}$$

and

$$\boxed{\frac{d^2 \overline{H}}{dY^2} - i\overline{H} + 1 = R_\omega^{-1/2} \frac{d\overline{H}}{dY} + O(R_\omega^{-1}).} \tag{4–25}$$

If we go back to the definition of the rescaling, (4–22), we see that the region $\Delta Y = O(1)$ is actually very thin compared with the radius of the tube. Indeed, in terms of y (which is scaled with respect to R), the near-wall region is only $O(R_\omega^{-1/2})$ in dimension for $R_\omega \gg 1$. This very thin region near the wall where viscous effects are important is called a *boundary layer*. We shall see many other examples of boundary layers in later chapters of this book. Because $d^2\overline{H}/dY^2$ and $d\overline{H}/dY$ are $O(1)$ in the near-wall region, we see that we can obtain a first approximation to the governing equation in the boundary layer by letting $R_\omega \to \infty$ in (4–25).

Before going further with this analysis, it will probably be helpful to recapitulate what we have shown, starting with the original nondimensionalized version of the problem for H, Eq. (4–16) plus the boundary condition $H(1) = 0$. We have seen that the limiting form of (4–16) for $R_\omega \gg 1$ represents a balance between the pressure-gradient and acceleration terms with the viscous terms neglected altogether and has a leading-order solution, (4–19) or (4–20), that does not satisfy boundary conditions at the walls. We surmise from this that the scaling, $\ell_c = R$, inherent in (4–16) is not valid in the vicinity of the tube walls. Thus we search for a new scaling that is consistent with the fact that viscous terms cannot be negligible close to the walls if the no-slip condition is to be satisfied. This new length scale is inherent in the rescaling, (4–22), with $\alpha = 1/2$, and we see from this that the region where viscous effects are important is only

$$\Delta y \sim O(R_\omega^{-1/2}) \quad \text{for} \quad R_\omega \to \infty.$$

In this region, the relevant approximation to the governing equation for H for $R_\omega \gg 1$ is (4–25). Thus, as illustrated in Fig. 4–2, the solution domain, $0 \leq \bar{r} \leq 1$, splits into two parts for $R_\omega \gg 1$. One is the interior region away from the walls, where $\ell_c = R$, and the appropriate nondimensional form of the governing equations is (4–16); the other is a narrow region near the wall of dimension $O(R_\omega^{-1/2})$ relative to the tube radius where the relevant nondimensionalized form of the governing equation is (4–25). If we develop an asymptotic solution of (4–16) for $R_\omega \gg 1$, such as (4–17), we cannot expect it to satisfy boundary conditions at the tube walls because the limiting form of (4–16) is not valid within the region of $O(R_\omega^{-1/2})$ near the tube walls. On the other hand, an asymptotic solution of (4–25) for $R_\omega \gg 1$ will satisfy boundary conditions at the tube wall but cannot ordinarily be expected to give the correct solution outside the narrow, boundary-layer region. Problems in which an asymptotic approximation to the solution requires two (or more) distinct expansions, each valid in different parts of the domain, are called *singular*, and the solutions are known as *singular* or *matched asymptotic expansions*. The word *matched* comes from the observation that two (or more) approximations of the solution to a problem, each valid in different parts of the domain, must have the same functional form in any region of space where their individual domains of validity overlap. We shall see later that this concept of *matching* plays a crucial role in determining solutions of the singular perturbation type.

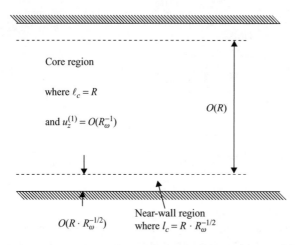

Figure 4–2. A schematic of the flow domain for pulsatile flow in a circular tube at very large values of R_ω. In the core region, the velocity field is characterized by a length scale $\ell_c = R$ and the velocity field is dominated by inertia (acceleration) effects that are due to the time-dependent pressure gradient. In the near-wall region, on the other hand, the characteristic length scale for changes in velocity is much shorter, $O(R R_\omega^{-1/2})$, and viscous effects remain important even for very large values of R_ω.

Let us now return to the solution of our problem for $R_\omega \gg 1$. Although the arguments leading to (4–25) were complex, the resulting equation itself is simple compared with the original Bessel equation. Our objective here is an asymptotic approximation of the solution for the boundary-layer region. In general, we may expect an asymptotic expansion of the form

$$\overline{H}(Y) = \overline{H}_0(Y) + \frac{1}{R_\omega^m}\overline{H}_1(Y) + \cdots.$$

where m is a positive coefficient that must be determined as part of the solution procedure. Here, we seek only the first term in this expansion, corresponding to the fact that we have also obtained only a leading-order approximation to the solution in the interior region away from the walls. Because $d^2\overline{H}/dY^2$ and $d\overline{H}/dY$ are $O(1)$ in the near-wall region, we can obtain the governing equation for the leading-order term in the asymptotic expansion by letting $R_\omega \to \infty$ in (4–25). The result is

$$\frac{d^2\overline{H}_0}{dY^2} - i\overline{H}_0 + 1 = 0,$$

and the general solution of this equation is

$$\overline{H}_0 = -i + A\exp\left(\frac{\sqrt{2}}{2}Y\right)\exp\left(\frac{\sqrt{2}}{2}iY\right) + B\exp\left(-\frac{\sqrt{2}}{2}Y\right)\exp\left(-\frac{\sqrt{2}}{2}iY\right).$$

(4–26)

The boundary condition $\overline{H}_0(0) = 0$ then requires that

$$B = i - A$$

or

$$\overline{H}_0 = i\left[-1 + \exp\left(-\frac{\sqrt{2}}{2}Y\right)\exp\left(-\frac{\sqrt{2}}{2}iY\right)\right]$$
$$+ A\left[\exp\left(\frac{\sqrt{2}}{2}Y\right)\exp\left(\frac{\sqrt{2}}{2}iY\right) - \exp\left(-\frac{\sqrt{2}}{2}Y\right)\exp\left(-\frac{\sqrt{2}}{2}iY\right)\right].$$

(4–27)

A. Pulsatile Flow in a Circular Tube Revisited

The constant A cannot be determined from the boundary condition at the wall but must be obtained from the matching requirement that (4–27) reduce to the form of the core solution (4–17) in the region of overlap between the boundary layer and the interior region. Now, any arbitrarily large, but finite, value of Y will fall within the boundary-layer domain; on the other hand, the corresponding value of y can be made arbitrarily small in the asymptotic limit $R_\omega \to \infty$. Thus the condition of matching is often expressed in the form

$$[\text{B.L.SOLN}]_{Y \gg 1} \Leftrightarrow [\text{INTERIOR SOLN}]_{y \ll 1} \text{ for } R_\omega \to \infty.$$

Although this notation might appear confusing at first, the limiting formulas $Y \gg 1$ and $y \ll 1$, respectively, are intended to serve as a reminder that the region of overlapping validity for $R_\omega \to \infty$ corresponds to large (but finite) values of Y and small (but nonzero) values of y. The implication of the matching formula is that the functional forms of the solution in the different parts of the domain must be the same, to within an error that is asymptotically small in the limit $R_\omega \to \infty$. The magnitude of the error in matching, expressed in the form of an estimate proportional to R_ω^{-n}, depends on the number of terms that have been evaluated in the asymptotic expansions for the solution in the different parts of the domain. In the present case, we have calculated only a single term in both the boundary-layer and interior solutions, and the matching condition can be expressed in the form

$$\frac{1}{R_\omega} \overline{H}_0 \Big|_{Y \gg 1} \Leftrightarrow -\frac{i}{R_\omega} \quad \text{as } R_\omega \to \infty, \tag{4–28}$$

where the right-hand side is the leading-order approximation for H in the interior of the tube and the left-hand side is the boundary-layer solution, (4–27), including the R_ω^{-1} scaling from (4–16). It can be seen from the condition (4–28) that

$$A = 0.$$

It may be noted that the boundary-layer solution with this value for A does not match perfectly with the core solution, $-i/R_\omega$. There is a mismatch in the term

$$\frac{i}{R_\omega} \left[\exp\left(-\frac{\sqrt{2}}{2} Y\right) \exp\left(-\frac{\sqrt{2}}{2} i Y\right) \right].$$

However, this mismatch is asymptotically small in the limit $R_\omega \to \infty$, as we can see by expressing it in terms of the original radial variable r using (4–22), that is,

$$\frac{i}{R_\omega} \left[\exp\left(-\frac{\sqrt{2}}{2} R_\omega y\right) \exp\left(-i \frac{\sqrt{2}}{2} R_\omega y\right) \right] \quad \text{for } R_\omega \to \infty.$$

This small mismatch need not concern us at this stage in the solution scheme because we have so far considered only the first leading-order approximation for $\overline{H}$ in the boundary layer and in the core [see Eq. (4–18)]. With $A = 0$, the final solution for $\overline{H}$ in the boundary-layer region can be expressed in the form

$$\boxed{\overline{H} = i \left[-1 + \exp\left(-\frac{\sqrt{2}}{2} Y\right) \left(\cos \frac{\sqrt{2}}{2} Y - i \sin \frac{\sqrt{2}}{2} Y \right) \right].} \tag{4–29}$$

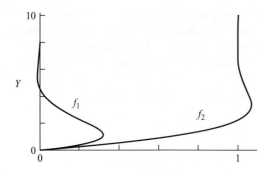

Figure 4–3. The magnitudes of the in-phase and out-of-phase components of the velocity profile, Eq. (4–30), for the axial velocity $u_z^{(1)}$. Note that $u_z^{(1)} R_\omega$ is plotted in order to eliminate the R_ω^{-1} dependence in Eq. (4–30).

Thus, in the boundary-layer region, it follows from (4–1) and (4–2) that

$$u_z^{(1)} = \frac{1}{R_\omega} \left(-\cos \bar{t} \left[1 - \exp\left(-\frac{\sqrt{2}}{2} Y\right) \cos \frac{\sqrt{2}}{2} Y \right] \right. \\ \left. + \sin \bar{t} \left\{ \exp\left(-\frac{\sqrt{2}}{2} Y\right) \sin \frac{\sqrt{2}}{2} Y \right\} \right), \quad (4\text{–}30)$$

where

$$Y = y R_\omega^{1/2} = R_\omega^{1/2}(1 - \bar{r}).$$

In the rest of the tube, we have [Eq. (4–20)]

$$u_z^{(1)} = -\frac{1}{R_\omega} \cos \bar{t}.$$

Thus, near the wall, there is both an *in-phase* and *an out-of-phase* contribution to the velocity field, whereas the approximation to the velocity field in the interior has no in-phase component. Both the in-phase and out-of-phase contributions to (4–30) are modestly complicated in form. Their spatial variations are plotted in Fig. 4–3 for a single, fixed $\bar{t}$.

In the region away from the walls, viscous effects are negligible to the leading order of approximation, and the acceleration of the fluid is in-phase with the pressure gradient (that is, the velocity lags behind the pressure gradient by 90°). No simple explanation for the complicated wavelike character of the solution in the boundary-layer region is possible. In any case, the point in presenting this high-frequency asymptotic solution is primarily as an introduction to the concepts of singular (or matched) asymptotic expansions rather than as a basis for a physical explanation of the flow itself. Readers should not be discouraged if there are detailed questions about the asymptotic techniques that we have discussed in this section that you do not yet fully understand. The main ideas of this section are repeated in the next several sections, in which a more general discussion of asymptotic approximation methods is presented.

B. ASYMPTOTIC EXPANSIONS – GENERAL CONSIDERATIONS

In the previous section we demonstrated the application of asymptotic expansion techniques to obtain the high- and low-frequency limits of the velocity field for flow in a circular tube driven by an oscillatory pressure gradient. In the process, we introduced such fundamental notions as the difference between a *regular* and a *singular* asymptotic expansion and, in the latter case, the concept of *matching* of the asymptotic approximations that are valid in different parts of the domain. However, all of the presentation was *ad hoc*, without the benefit of any formal introduction to the properties of asymptotic expansions. The present section is intended to provide at least a partial remedy for that shortcoming. We note, however,

B. Asymptotic Expansions – General Considerations

that the presentation will still be largely *ad hoc*. It is not our intention to provide a complete mathematical statement of the current status of asymptotic approximation methods; there are a number of comprehensive texts that do a good job of performing that task.[3] In this textbook, the aim is to give students sufficient information so that they can *use* the methods and understand both their advantages and shortcomings. To a large extent, this goal is pursued by means of the many examples of approximate solutions, obtained by means of asymptotic methods, that will be found throughout the following chapters. In this section, a number of necessary and useful general facts about asymptotic expansions are presented.

To do this, let us consider a function, say, $T(\mathbf{x}; \varepsilon)$, that depends on spatial position and on a dimensionless parameter ε that we may assume to be arbitrarily small. In the context of problems to be considered in this text, this function will usually be defined by a DE and boundary conditions, and the parameter ε then appears as a dimensionless parameter in either the equation or boundary conditions. We suppose, for purpose of discussion, that T has an asymptotic expansion for small ε. The general form of such an expansion is

$$T(\mathbf{x}; \varepsilon) = f_0(\varepsilon)T_0(\mathbf{x}) + f_1(\varepsilon)T_1(\mathbf{x}) + f_2(\varepsilon)T_2(\mathbf{x}) + O(f_3(\varepsilon)). \qquad (4\text{--}31)$$

The functions T_0, T_1, and T_2 are independent of ε, and the magnitude of each term is therefore given by the gauge function $f_n(\varepsilon)$. We indicate the size of the error made by truncating (4–31) after three terms with the so-called order symbol, $O()$. Actually, there are two such symbols that could have been used in (4–31). The one known as "big O," which appears in (4–31), means that the magnitude of the largest neglected term in the asymptotic limit $\varepsilon \to 0$ is $f_3(\varepsilon)$; the use of the order-of-magnitude symbol implies that the magnitude of the next term in the expansion is known. This may not always be the case. Then, a second symbol known as "little o," designated as $o()$, is used. In the expansion (4–31) of $T(\mathbf{x}; \varepsilon)$, we would write $o(f_2(\varepsilon))$, meaning that the largest neglected term is *smaller* than $f_2(\varepsilon)$ in the asymptotic limit, $\varepsilon \to 0$. As previously implied, this second symbol is usually used only when we do not know the magnitude of the next largest term in an asymptotic series.

A very important property of an asymptotic expansion is the manner in which it converges to the function that it is intended to represent. Two facts can be stated that relate intimately to the nature of the convergence of an asymptotic expansion. First, if a function such as $T(\mathbf{x}; \varepsilon)$ has an asymptotic expansion for small ε (either for all $\mathbf{x}$ or at least in some subdomain of $\mathbf{x}$), then this expansion is unique (at least in the subdomain). However, second, more than one function T may have the same asymptotic representation through any finite number of terms. The second of these statements implies that one cannot sum an asymptotic expansion to find a unique function $T(\mathbf{x}; \varepsilon)$ as would be possible (in the domain of convergence) with a normal power-series representation of a function. This distinction between an asymptotic and infinite-series representation is reflected in a more formal statement of the convergence properties of both an infinite series and an asymptotic expansion. In the case of an infinite-series representation of some function $T(\mathbf{x}; \varepsilon)$, namely,

$$T(\mathbf{x}; \varepsilon) = \sum_{n=1}^{\infty} a_n T^{(n)}(\mathbf{x}; \varepsilon), \qquad (4\text{--}32)$$

convergence requires that the difference between T and its series representation can be made arbitrarily small by taking a sufficiently large number of terms in the series, that is,

$$\lim_{N \to \infty} \left[\frac{T - \sum_{n=1}^{N} a_n T^{(n)}}{T} \right] \to 0 \qquad (4\text{--}33)$$

An Introduction to Asymptotic Approximations

for fixed ε and $\mathbf{x}$ in the *domain of convergence*. An asymptotic expansion is not necessarily convergent at all in this sense. Instead,

$$\lim_{\varepsilon \to 0} \left[\frac{T - \sum_{n=1}^{N} f_n(\varepsilon) T_n(\mathbf{x})}{T} \right] \to 0 \qquad (4\text{--}34)$$

for *fixed N*. A necessary condition for asymptotic convergence is

$$\lim_{\varepsilon \to 0} \frac{f_{n+1}(\varepsilon)}{f_n(\varepsilon)} \to 0 \quad \text{for all } n. \qquad (4\text{--}35)$$

Thus, according to (4–34), the difference between T and its asymptotic expansion can be made arbitrarily small for any fixed N by taking the limit $\varepsilon \to 0$. It is very important to recognize that asymptotic convergence does not imply that a better approximation will be achieved by taking more terms for any fixed ε, even if ε is small. Indeed, it is possible that the difference between T and its asymptotic expansion may actually diverge as we add more terms while holding ε fixed.

We have already noted that there are two types of asymptotic expansions: regular and singular. If the asymptotic expansion of a function $T(\mathbf{x}; \varepsilon)$ is *regular*, the convergence criterion (4–34) must be satisfied for all $\mathbf{x}$ in the domain of interest. The likelihood of a regular asymptotic expansion is improved if the domain is finite. If the asymptotic expansion is *singular*, on the other hand, the convergence criterion (4–34) is still satisfied but only in some subdomain of the region of interest. For this subdomain, the expansion

$$\sum_{n=1}^{N} f_n(\varepsilon) T_n(\mathbf{x})$$

provides a perfectly adequate representation, but for the rest of the domain one or more additional asymptotic expansions are necessary, each satisfying (4–34) in their subdomain of convergence. Let us suppose, for some particular example, that only one additional expansion is necessary, which we may denote as

$$T = \sum_{n=1}^{N} F_n(\varepsilon) \hat{T}_n(\hat{\mathbf{x}}). \qquad (4\text{--}36)$$

In this second formula, neither the gauge functions nor the spatially dependent coefficients are normally the same as those appearing in the first representation of T. Furthermore, the spatial variable $\hat{\mathbf{x}}$ will frequently be scaled differently from its nondimensionalization in the portion of the domain where

$$\sum_{n=1}^{N} f_n(\varepsilon) T_n(\mathbf{x})$$

is relevant. However, the two asymptotic expansions,

$$T = \sum_{n=1}^{N} f_n(\varepsilon) T_n(\mathbf{x}),$$

and

$$T = \sum_{n=1}^{N} F_n(\varepsilon) \hat{T}_n(\hat{\mathbf{x}}),$$

are both approximations of the same function T, simply valid in different parts of the domain. To provide a uniformly valid asymptotic representation of T for all $\mathbf{x}$, it is necessary that the

C. The Effect of Viscous Dissipation on a Simple Shear Flow

subdomains overlap. In this overlap region, which is a region of common validity for the expansions, the individual expansions must take the same functional form. This is known as the *matching principle*. For two representations of a function to match, it is not sufficient that they have the same numerical value at some fixed **x** (indeed, this property is sometimes associated with the name *patching* to distinguish it from *matching*); rather, they must adopt the same functional form to within an error that can be made arbitrarily small in the limit $\varepsilon \to 0$.

Often the function $T(\mathbf{x}; \varepsilon)$ is defined as the solution of a DE and boundary conditions. In this case, the asymptotic expansion for T will be generated from an asymptotic approximation of the governing equation and boundary conditions for T. We have seen an example of this idea in the preceding section. When the asymptotic expansion for T is singular, different approximations of the equations (and thus of T) will be relevant in the different subdomains of the full solution domain. In this case, the original boundaries of the domain will always be split among the various subdomains, and thus not all of the original boundary conditions can be used for the asymptotic expansion in any given subdomain. This means that *too few* boundary conditions usually will be available to uniquely determine the asymptotic solution of the governing equation in any subdomain. In this case, the concept of matching plays a critical role; the approximate (asymptotic) solutions must match in the regions of overlap between the subdomains, and this provides sufficient additional conditions to obtain unique solutions. Later we shall see many examples of the principles described in this paragraph. Matching was used to determine the high-frequency singular-perturbation solution for pulsatile flow in a circular tube that was discussed in Section A.

A question that is asked frequently is how one can determine, for a particular problem, whether the asymptotic expansion will be *regular* or *singular*. It is, in fact, not always easy to tell *a priori*. One telltale sign of a *singular limit*, corresponding to the necessity for singular or matched asymptotic expansions, occurs when *the limiting process, say, $\varepsilon \to 0$, causes the governing equation to be reduced in order*. An example is the high-frequency limit of the pulsatile flow problem considered in Section A. In this case, if we take the limit $R_\omega^{-1} \to 0$ in governing equation (4–16), the problem is reduced from a second-order ODE equation to an algebraic equation. Whenever the order of the governing equation is reduced, the solution of the resulting equation cannot satisfy all the boundary conditions of the original problem. Hence, as in the pulsatile flow problem, a different approximation of the problem must be required in the vicinity of the boundaries at least, and the asymptotic approximation of the solution for $\varepsilon \ll 1$ will be singular. However, reduction of the order of the governing equation is only one obvious sign of singular asymptotic behavior, and there are many problems that are singular without exhibiting this particular characteristic. Generally speaking, a *singular limit* will be characterized by the necessity for multiple characteristic scales (especially length scales), each valid in different parts of the solution domain.

When in doubt about whether a particular asymptotic limit is regular or singular, it is frequently necessary simply to proceed under the assumption that the asymptotic solution will be *regular*, that is, that the problem can be characterized by a single set of characteristic scales, valid for all **x** in the solution domain. If the limit is, in fact, regular, the asymptotic-solution scheme based on a single nondimensionalization of the governing equations and boundary conditions will proceed successfully at all levels of approximation. Otherwise, an obvious sign of singularity will appear; for example, the solution at some level of approximation either may not satisfy boundary conditions or simply may not exist. In this case, the problem can then be reformulated in a form appropriate to the singular limit.

C. THE EFFECT OF VISCOUS DISSIPATION ON A SIMPLE SHEAR FLOW

We have seen, in the previous chapter, that the effects of viscous dissipation will be small whenever the Brinkman number is small. This is often the case, and it is common practice in theoretical analysis of fluid flow problems to neglect viscous dissipation altogether. There are, however, some instances in which this can lead to important inaccuracies. One example is the neglect of viscous dissipation in the use of simple shear-flow rheometers to measure the viscosity of a fluid. An example of this based on the circular Couette flow was discussed in the preceding chapter.

In this section, we return to the analysis of simple, unidirectional shear flow that was considered in Section B of Chap. 3, but instead of neglecting viscous dissipation altogether, we consider its influence when the Brinkman number is small, but nonzero. The starting point is Eqs. (3–34) and (3–35), which are reproduced here for convenience:

$$\frac{d}{d\bar{y}}\left(\bar{\mu}\frac{d\bar{u}}{d\bar{y}}\right) = 0, \tag{4-37}$$

$$\frac{d}{d\bar{y}}\left(\bar{k}\frac{d\bar{\theta}}{d\bar{y}}\right) + Br\left[\bar{\mu}\left(\frac{d\bar{u}}{d\bar{y}}\right)^2\right] = 0. \tag{4-38}$$

The corresponding boundary conditions are

$$\bar{\theta} = 0 \quad \text{at} \quad \bar{y} = 0, 1, \tag{4-39}$$

$$\bar{u} = 0 \quad \text{at} \quad \bar{y} = 0, \quad \bar{u} = 1 \quad \text{at} \quad \bar{y} = 1.$$

If $Br \equiv 0$, as assumed in Chap. 3, the temperature is constant across the gap (i.e., $\bar{\theta} = 0$ for all $\bar{y}$) and the material coefficients $\bar{\mu}$ and $\bar{k}$ are both equal to one. For small, but nonzero, Br, on the other hand, $\bar{\theta}$ will vary across the gap, and we must take account of the corresponding changes in the material coefficients. For this purpose, it is sufficient to approximate the viscosity and thermal conductivity for small changes in the temperature relative to the value T_0 that exists in the absence of dissipation. Hence, we approximate $\bar{\mu}$ and $\bar{k}$ in the forms

$$\bar{\mu}^{-1} = \left(\frac{\mu}{\mu_0}\right)^{-1} = 1 + \beta_1^*(\theta - T_0) + \beta_2^*(\theta - T_0)^2 + \cdots +,$$

$$\bar{k} \equiv \left(\frac{k}{k_0}\right) = 1 + \alpha_1^*(\theta - T_0) + \alpha_2^*(\theta - T_0)^2 + \cdots +,$$

or

$$\bar{\mu}^{-1} = 1 + \beta_1\bar{\theta} + \beta_2\bar{\theta}^2 + \cdots +, \quad \text{where } \beta_j \equiv \beta_j^* T_0^j \tag{4-40}$$

$$\bar{k} = 1 + \alpha_1\bar{\theta} + \alpha_2\bar{\theta}^2 + \cdots +, \quad \text{where } \alpha_j \equiv \alpha_j^* T_0^j. \tag{4-41}$$

The coefficients α_j^* and β_j^* are material constants for the specific fluid. It will be noted that (4–40) and (4–41) are written in a form suitable for a liquid because it is assumed that the viscosity decreases as the temperature increases.

If we substitute (4–40) and (4–41) into the governing equations (4–37) and (4–38), these can be written in the form

$$\frac{d^2\bar{u}}{d\bar{y}^2} - \frac{\left(\beta_1\frac{d\bar{\theta}}{d\bar{y}} + 2\beta_2\bar{\theta}\frac{d\bar{\theta}}{d\bar{y}} + \cdots +\right)}{(1 + \beta_1\bar{\theta} + \beta_2\bar{\theta}^2 + \cdots +)}\frac{d\bar{u}}{d\bar{y}} = 0, \tag{4-42}$$

$$(1 + \alpha_1\bar{\theta} + \alpha_2\bar{\theta}^2 + \cdots +)(1 + \beta_1\bar{\theta} + \beta_2\bar{\theta}^2 + \cdots +)\frac{d^2\bar{\theta}}{d\bar{y}^2} \tag{4-43}$$

$$+ (1 + \beta_1\bar{\theta} + \beta_2\bar{\theta}^2 + \cdots +)\left(\alpha_1\frac{d\bar{\theta}}{d\bar{y}} + \alpha_2 2\bar{\theta}\frac{d\bar{\theta}}{d\bar{y}} + \cdots +\right)\frac{d\bar{\theta}}{d\bar{y}} + Br\left(\frac{d\bar{u}}{d\bar{y}}\right)^2 = 0.$$

C. The Effect of Viscous Dissipation on a Simple Shear Flow

Clearly, although the physical problem is quite simple, we see that the governing equations are strongly nonlinear, and there is little hope of obtaining anything other than an approximate analytical solution for the limiting case $Br \ll 1$.

To pursue this solution, we follow the prescription of the preceding section and seek a regular asymptotic solution. Thus, we seek a solution in the form

$$\bar{\theta}(\bar{y}; Br) = Br\bar{\theta}_1(\bar{y}) + Br^2\bar{\theta}_2(\bar{y}) + O(Br^3), \quad (4\text{-}44)$$

$$\bar{u}(\bar{y}; Br) = \bar{u}_0(\bar{y}) + Br\,\bar{u}_1(\bar{y}) + Br^2\bar{u}_2(\bar{y}) + O(Br^3). \quad (4\text{-}45)$$

It can be seen from the governing equations (4–37) and (4–38), plus boundary conditions (4–39), that the solution in the limit $Br = 0$ is just $\bar{\theta} = 0$ and $\bar{u}$ equal to $\bar{y}$ (i.e., the simple shear-flow solution). Hence, in writing (4–44), we have set the $O(1)$ term equal to zero. We have also assumed that the gauge functions are simply powers of the small parameter, which in this case is Br. Provided that the asymptotic expansion is regular, we shall see that the gauge functions always turn out to be even powers of the small parameter of the problem, and in fact the solution is equivalent to what would be obtained by means of the method of successive approximations. If the expansion turns out to be singular, it is not possible to anticipate the form of the gauge functions in advance, and then they must be determined as part of the solution process. The symbol $O(Br^3)$ means that the magnitude of the largest term that is not shown explicitly is proportional to Br^3.

To obtain governing equations and boundary conditions for the unknown functions $\bar{\theta}_j(\bar{y})$ and $\bar{u}_j(\bar{y})$ we substitute the expansions (4–44) and (4–45) into the full equations (4–42) and (4–43), as well as into the boundary conditions (4–39). After some algebraic rearrangement, we can write the equations in the form

$$\left(\frac{d^2\bar{u}_0}{d\bar{y}^2}\right) + \left(\frac{d^2\bar{u}_1}{d\bar{y}^2} - \beta_1 \frac{d\bar{\theta}_1}{d\bar{y}}\frac{d\bar{u}_0}{d\bar{y}}\right)Br + \left[\frac{d^2\bar{u}_2}{d\bar{y}^2} - \left(2\beta_2\bar{\theta}_1\frac{d\bar{\theta}_1}{d\bar{y}}\right.\right.$$
$$\left.\left. + \beta_1\frac{d\bar{\theta}_2}{d\bar{y}} - \beta_1^2\bar{\theta}_1\frac{d\bar{\theta}_1}{d\bar{y}}\right)\frac{d\bar{u}_0}{d\bar{y}} - \beta_1\frac{d\bar{\theta}_1}{d\bar{y}}\frac{d\bar{u}_1}{d\bar{y}}\right]Br^2 + O(Br^3) = 0, \quad (4\text{-}46)$$

$$\left[\frac{d^2\bar{\theta}_1}{d\bar{y}^2} + \left(\frac{d\bar{u}_0}{d\bar{y}}\right)^2\right]Br + \left\{\frac{d^2\bar{\theta}_2}{d\bar{y}^2} + \left[\frac{d^2\bar{\theta}_1}{d\bar{y}^2}\bar{\theta}_1(\alpha_1 + \beta_1)\right]\right.$$
$$\left. + \left(\frac{d\bar{\theta}_1}{d\bar{y}}\right)^2\alpha_1 + 2\frac{d\bar{u}_0}{d\bar{y}}\frac{d\bar{u}_1}{d\bar{y}}\right\}Br^2 + O(Br^3) = 0. \quad (4\text{-}47)$$

It may be noted that we retain terms only through $O(Br^2)$. This is because we have retained terms only to this same order in the expansions (4–44) and (4–45). If we retained terms proportional to higher powers of Br in (4–46) or (4–47), they would be incomplete because they would not contain any of the terms involving $\bar{u}_3$ or $\bar{\theta}_3$. Similarly, after substituting into the boundary conditions, we obtain

$$Br\bar{\theta}_1(\bar{y}) + Br^2\bar{\theta}_2(\bar{y}) + O(Br^3) = 0 \quad \text{at } \bar{y} = 0 \text{ and } 1, \quad (4\text{-}48)$$

$$\bar{u}_0(\bar{y}) + Br\,\bar{u}_1(\bar{y}) + Br^2\bar{u}_2(\bar{y}) + O(Br^3) = 0 \quad \text{at } \bar{y} = 0$$
$$= 1 \quad \text{at } y = 1. \quad (4\text{-}49)$$

Now, Eqs. (4–46)–(4–49) are valid for any small but arbitrary value of Br. Because Br is arbitrary, the only way that the sum of the terms on the left-hand side can add up to zero for all allowable values of Br is if each term is equal to zero. An alternative for some *fixed* value of Br might seem to be that some of the terms could be positive and some negative so that they add up to zero. However, all of the coefficients in parentheses are independent of Br, and so if the value of Br were changed, the sum could no longer add up to zero. Hence, each of the coefficients at each order in Br has to be equal to zero, and this condition leads to a hierarchy of equations and boundary conditions for the unknown functions $\bar{u}_j(\bar{y})$ and $\bar{\theta}_j(\bar{y})$. At $O(1)$, we obtain

$$\frac{d^2\bar{u}_0}{d\bar{y}^2} = 0; \quad \bar{u}_0(0) = 0; \quad \bar{u}_0(1) = 1. \tag{4-50}$$

At $O(Br)$, we obtain

$$\frac{d^2\bar{u}_1}{d\bar{y}^2} - \beta_1 \frac{d\bar{\theta}_1}{d\bar{y}} \frac{d\bar{u}_0}{d\bar{y}} = 0; \quad \bar{u}_1(0) = \bar{u}_1(1) = 0. \tag{4-51}$$

$$\frac{d^2\bar{\theta}_1}{d\bar{y}^2} + \left(\frac{d\bar{u}_0}{d\bar{y}}\right)^2 = 0; \quad \bar{\theta}_1(0) = \bar{\theta}_1(1) = 0. \tag{4-52}$$

Finally, at $O(Br^2)$, the result is

$$\frac{d^2\bar{u}_2}{d\bar{y}^2} - \left(2\beta_2\bar{\theta}_1\frac{d\bar{\theta}_1}{d\bar{y}} + \beta_1\frac{d\bar{\theta}_2}{d\bar{y}} - \beta_1^2\bar{\theta}_1\frac{d\bar{\theta}_1}{d\bar{y}}\right)\frac{d\bar{u}_0}{d\bar{y}} - \beta_1\frac{d\bar{\theta}_1}{d\bar{y}}\frac{d\bar{u}_1}{d\bar{y}} = 0;$$
$$\bar{u}_2(0) = \bar{u}_2(1) = 0. \tag{4-53}$$

$$\frac{d^2\bar{\theta}_2}{d\bar{y}^2} + \left[\frac{d^2\bar{\theta}_1}{d\bar{y}^2}\bar{\theta}_1(\alpha_1 + \beta_1)\right] + \left(\frac{d\bar{\theta}_1}{d\bar{y}}\right)^2 \alpha_1 + 2\frac{d\bar{u}_0}{d\bar{y}}\frac{d\bar{u}_1}{d\bar{y}} = 0;$$
$$\bar{\theta}_2(0) = \bar{\theta}_2(1) = 0. \tag{4-54}$$

Before we attempt to solve these equations, a few comments are in order. First, although we have derived a set of DEs and boundary conditions, it should not be assumed that a regular perturbation expansion actually exists. To prove this, we need to show that solutions exist that satisfy these equations and boundary conditions. In the present case, this is relatively straightforward, but we should keep in mind that it is standard practice to begin every problem with the assumption of a regular asymptotic expansion with the understanding that solutions satisfying all of the equations and boundary conditions may not exist, thus signaling the need for a more general ("singular") solution method. The second point is that the governing equations, (4–50)–(4–54), at each level of approximation in the asymptotic expansion scheme are linear in the unknown functions, in contrast to the original equations, which were highly nonlinear. We solve the equations and boundary conditions sequentially, beginning with $\bar{u}_0$, by using the $O(1)$ equations and boundary conditions (4–50), followed by $\bar{\theta}_1$ and $\bar{u}_1$ by means of (4–51) and (4–52) and so on. The nonlinear terms still appear in (4–51)–(4–54), but by the time they appear, they involve only functions that have been evaluated by means of the linear equations at earlier points in the solution scheme. This "linearization" process is typical of many asymptotic problems.

Beginning with (4–50), we see immediately that the solution for $\bar{u}_0$ is

$$\bar{u}_0 = \bar{y}. \tag{4-55}$$

In fact, this is nothing more than the simple, isothermal shear-flow solution obtained earlier, and this is consistent with the fact that $\bar{u}_0$ is the limiting form of (4–37) and (4–38) for $Br \to 0$. Next we turn to the solution of (4–51) and (4–52) for $\bar{\theta}_1$ and $\bar{u}_1$. At this level of approximation, we see the first appearance of nonlinear terms in the equations, as well as

C. The Effect of Viscous Dissipation on a Simple Shear Flow

the first influence of viscous dissipation. If we substitute the known solution, (4–55), the equations become

$$\frac{d^2\bar{u}_1}{d\bar{y}^2} = \beta_1 \frac{d\bar{\theta}_1}{d\bar{y}}, \tag{4-56}$$

$$\frac{d^2\bar{\theta}_1}{d\bar{y}^2} + 1 = 0. \tag{4-57}$$

Integrating (4–57) and applying the boundary conditions, (4–52), we find

$$\bar{\theta}_1 = \frac{1}{2}(\bar{y} - \bar{y}^2). \tag{4-58}$$

Hence, substituting this solution into (4–56), we find that, after integrating and applying boundary conditions, we obtain

$$\bar{u}_1 = \beta_1 \left(-\frac{\bar{y}^3}{6} + \frac{\bar{y}^2}{4} - \frac{\bar{y}}{12} \right). \tag{4-59}$$

Finally, substituting the known solutions (4–55)–(4–59) into (4–53) and (4–54), integrating and applying boundary conditions, we obtain

$$\bar{\theta}_2 = \frac{1}{24}(\bar{y}^2 - \bar{y})[-3\alpha_1(\bar{y}^2 - \bar{y}) + \beta_1(\bar{y}^2 - \bar{y} + 1)], \tag{4-60}$$

$$\bar{u}_2 = -\frac{1}{240}\bar{y}(1 - 3\bar{y} + 2\bar{y}^2)\left[2\beta_2(1 + 3\bar{y} - 3\bar{y}^2) \right. \\ \left. + \beta_1^2(-3 + \bar{y} - \bar{y}^2) + \alpha_1\beta_1.(-1 - 3\bar{y} + 3\bar{y}^2) \right]. \tag{4-61}$$

Examining these solutions, we see that the temperature becomes nonuniform and the velocity profile nonlinear. These results are to be expected from a qualitative point of view. In a sense, the most important conclusion is that the regular asymptotic expansion in terms of the small parameter Br provides a method to obtain an approximate solution of the highly nonlinear boundary-value problem to evaluate the influence of weak dissipation, which can clearly be applied to other problems.

One application of the solutions (4–55)–(4–61) is to evaluate the effect of viscous dissipation in the use of a shear rheometer to measure the viscosity of a Newtonian fluid. In this experiment, we subject the fluid in a thin gap between two plane walls to a shear flow by moving one of the walls in its own plane at a known velocity and then measuring the shear stress produced at either wall (by measuring the total tangential force and dividing by the area). In the absence of viscous dissipation, the velocity profile is linear and the shear rate is simply given by the tangential velocity U divided by the gap width d. Now, the constitutive equation, (2–87), for an incompressible Newtonian fluid applied to this simple flow situation takes the form

$$\tau_{xy} = \mu \frac{du}{dy}. \tag{4-62}$$

Thus, in the isothermal case, we obtain

$$\mu_0 = \frac{\tau_{xy}|_{y=0}}{U/d}, \tag{4-63}$$

where $\tau_{xy}|_{y=0}$ is the measured shear stress at the lower wall (it could also be measured at the upper wall) and μ_0 is the viscosity at the uniform temperature T_0. Now, if there is viscous dissipation, the shear stress at the wall will be different from its value in an isothermal fluid and the shear rate (namely du/dy) will vary with position across the gap. In particular, the shear rate will be different at the two walls, and not equal to U/d at either wall. We assume that the actual shear stress will be measured. Generally, however, the shear rate is not measured but is assumed to be U/d. This will clearly lead to an error in determining the viscosity because the measured shear stress will be divided by an incorrect value for the shear rate.

We can estimate the error for the simplified "rheometer" geometry that we have just analyzed. The dimensionless shear rate can be calculated from the solution, (4–45), with the functions $\bar{u}_j(\bar{y})$ given by (4–55), (4–59), and (4–61). If we assume that the shear stress will be measured at the bottom wall, $\bar{y} = 0$, then we want to evaluate the shear rate at the same location. Thus

$$\left.\frac{d\bar{u}}{d\bar{y}}\right|_{y=0} = 1 + Br\left(-\frac{\beta_1}{12}\right) + Br^2\left(-\frac{\beta_2}{120} + \frac{\beta_1^2}{80} + \frac{\alpha_1\beta_1}{240}\right) + O(Br^3). \quad (4\text{–}64)$$

Therefore, the apparent viscosity, as inferred from the assumed shear rate U/d, is

$$\mu_{app} = \frac{\tau_{xy}|_{y=0}}{U/d}.$$

Although this is identical in appearance to (4–63), in this case, the assumed shear rate U/d is incorrect in the presence of viscous dissipation. The actual viscosity μ_0 at the wall can be calculated as the ratio of the measured shear stress and the corrected shear rate (4–64), which both take into account the influence viscous dissipation:

$$\mu_0 = \frac{\tau_{xy}|_{y=0}}{\dfrac{U}{d}\left(\dfrac{d\bar{u}}{d\bar{y}}\right)_{\bar{y}=0}}.$$

It follows from these two expressions that

$$\mu_{app} = \frac{\mu_0\left(\dfrac{d\bar{u}}{d\bar{y}}\right)_{\bar{y}=0}\dfrac{U}{d}}{\dfrac{U}{d}} = \mu_0\left[1 - \frac{\beta_1}{12}Br + \left(-\frac{\beta_2}{120} + \frac{\beta_1^2}{80} + \frac{\alpha_1\beta_1}{240}\right)Br^2 + O(Br^3)\right].$$

(4–65)

The first correction, assuming that $\beta_1 > 0$, is negative; i.e., the apparent viscosity is slightly smaller than the actual viscosity μ_0. Although the corrections to the viscosity are rather small (for small Br), they provide a warning that viscous dissipation effects could be an important factor in obtaining accurate viscosity measurements for a Newtonian fluid.

D. THE MOTION OF A FLUID THROUGH A SLIGHTLY CURVED TUBE – THE DEAN PROBLEM

We noted in the introduction to Chap. 3 that unidirectional flows occur only when the direction of motion is strictly in a straight-line path. For example, in flow through a circular tube, we assume that the tube axis is straight. An obvious question is whether the analysis and the predicted relationship between volumetric flow rate and the applied pressure gradient will apply if the tube is slightly bent. In this section, we explore this question, closely following an asymptotic analysis that was first published by W. R. Dean in 1927.[4] The problem of flow in curved tubes is relevant to many technological problems such as the design of coiled tube heat exchangers and is also important to the understanding of blood flow in the large arteries.[5] Let us denote the tube radius as a and the coil radius as R. The small parameter in this analysis is the ratio of the tube radius to the radius of curvature of the tube axis, which is assumed to be coiled into a circle, that is, $\varepsilon \equiv a/R \ll 1$. The limiting

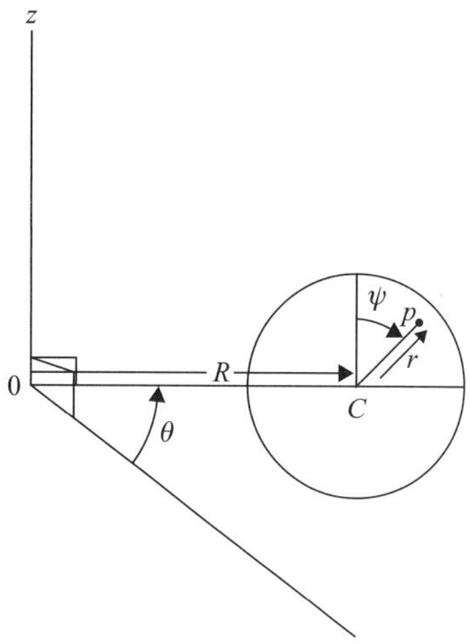

D. The Motion of a Fluid Through a Slightly Curved Tube – The Dean Problem

Figure 4–4. A schematic showing the coordinate system for the motion of a fluid through a slightly curved tube [after Dean[4] (1927)].

case of a perfectly straight tube, for which the unidirectional Poiseuille flow solution applies, corresponds to $\varepsilon = 0$ (that is, $R = \infty$). Of course, a very long tube that is coiled in a circular arc must exhibit a slight pitch orthogonal to the main coil in order to avoid intersecting itself after $360°$ of coil, but we ignore this pitch in the present analysis. The motion of the fluid is due to an imposed pressure gradient along the tube, which we denote as G.

To analyze this problem, a system of coordinates must be specified. We adopt the coordinate system used originally by Dean[4]. (1927) and sketched in Fig. 4–4. The cross section of the tube is circular, with the centerline denoted as C. As stated previously, the centerline traces a circular arc of radius R, and the section of the tube that is shown lies in a plane that makes an angle θ with a fixed axial plane. Within the cross section of the tube, the position of a point P is specified by the polar cylindrical coordinates (r, ψ), defined relative to the centerline of the tube. In this system, the polar angle ψ is measured from the vertical axis that is orthogonal to the OC plane traced by the circular arc of the centerline of the tube. Thus, in general, the point P is specified by the orthogonal coordinates (r, ψ, θ).[6] The components of velocity corresponding to these coordinates are designated as (u, v, w), respectively. It may be noted that distance along the tube, corresponding to the axial variable z in a straight circular tube geometry, is measured by $R\theta$.

In this analysis, we neglect entry and exit effects and concentrate on the *fully developed* flow regime where the motion (u, v, w) is independent of θ. In a straight circular tube, we have already seen that the velocity field takes the form $[(0, 0, w(r)]$. However, in the case of a coiled tube, the tube geometry is no longer unidirectional, and there is no reason to suppose that the velocity field will be so simple. The equations of motion and continuity, specified in dimensional form, for a fully developed, 3D flow are

$$\rho \left(u' \frac{\partial u'}{\partial r'} + \frac{v'}{r'} \frac{\partial u'}{\partial \psi} - \frac{v'^2}{r} - \frac{w'^2 \sin \psi}{R + r \sin \psi} \right)$$
$$= -\frac{\partial p'}{\partial r'} + \mu \left[\left(\frac{1}{r'} \frac{\partial}{\partial \psi} + \frac{\cos \psi}{R' + r' \sin \psi} \right) \left(\frac{\partial v'}{\partial r'} + \frac{v'}{r'} - \frac{1}{r'} \frac{\partial u'}{\partial \psi} \right) \right], \quad (4\text{–}66)$$

$$\rho \left(u' \frac{\partial v'}{\partial r'} + \frac{v'}{r'} \frac{\partial v'}{\partial \psi} + \frac{u'v'}{r'} - \frac{w'^2 \cos \psi}{R' + r' \sin \psi} \right)$$
$$= -\frac{1}{r'} \frac{\partial p'}{\partial \psi} + \mu \left[\left(\frac{\partial}{\partial r'} + \frac{\sin \psi}{R' + r' \sin \psi} \right) \left(\frac{\partial v'}{\partial r'} + \frac{v'}{r'} - \frac{1}{r'} \frac{\partial u'}{\partial \psi} \right) \right], \quad (4\text{–}67)$$

An Introduction to Asymptotic Approximations

$$\rho \left(u' \frac{\partial w'}{\partial r'} + \frac{v'}{r'} \frac{\partial w'}{\partial \psi} + \frac{u'w' \sin \psi}{R' + r' \sin \psi} + \frac{v'w' \cos \psi}{R' + r' \sin \psi} \right)$$

$$= -\frac{1}{R' + r' \sin \psi} \frac{\partial p'}{\partial \theta} + \mu \left[\left(\frac{\partial}{\partial r'} + \frac{1}{r'} \right) \left(\frac{\partial w'}{\partial r'} + \frac{w' \sin \psi}{R' + r' \sin \psi} \right) \right.$$

$$\left. + \frac{1}{r'} \frac{\partial}{\partial \psi} \left(\frac{1}{r'} \frac{\partial w'}{\partial \psi} + \frac{w' \cos \psi}{R' + r' \sin \psi} \right) \right], \qquad (4\text{--}68)$$

and

$$\frac{\partial u'}{\partial r'} + \frac{u'}{r'} + \frac{u' \sin \psi}{R' + r' \sin \psi} + \frac{1}{r'} \frac{\partial v'}{\partial \psi} + \frac{v' \cos \psi}{R' + r' \sin \psi} = 0. \qquad (4\text{--}69)$$

These equations reduce to the equations of motion in cylindrical coordinates if $R^{-1} = 0$ and $R^{-1}(\partial/\partial \theta) = \partial/\partial z$. In the form shown, they apply to a tube of arbitrary radius a and arbitrary radius of curvature R.

To proceed, we nondimensionalize Eqs. (4–66)–(4–69) by using the radius of the tube as a characteristic length scale,

$$\ell_c = a,$$

the axial velocity at the centerline of a *straight* tube with the same axial pressure gradient G as a characteristic velocity scale,

$$u_c = \frac{Ga^2}{4\mu},$$

and a viscous pressure scale,

$$p_c = \frac{\mu u_c}{a}.$$

The result of nondimensionalization for the radial velocity component, (4–66), is

$$Re \left[u \frac{\partial u}{\partial r} + \frac{v}{r} \frac{\partial u}{\partial \psi} - \frac{v^2}{r} - \frac{w^2 \sin \psi}{(R/a) + r \sin \psi} \right]$$

$$= -\frac{\partial p}{\partial r} + \left[\frac{1}{r} \frac{\partial}{\partial \psi} + \frac{\cos \psi}{(R/a) + r \sin \psi} \right] \left(\frac{\partial v}{\partial r} + \frac{v}{r} - \frac{1}{r} \frac{\partial u}{\partial \psi} \right).$$

If we consider the case of a slightly bent tube, where $a/R \ll 1$, this equation can be written in the alternative form:

$$\boxed{\begin{aligned} Re &\left\{ u \frac{\partial u}{\partial r} + \frac{v}{r} \frac{\partial u}{\partial \psi} - \frac{v^2}{r} - w^2 \sin \psi \left(\frac{a}{R} \right) \left[1 - \frac{a}{R} r \sin \psi + O\left(\frac{a}{R} \right)^2 \right] \right\} \\ &= -\frac{\partial p}{\partial r} + \frac{1}{r} \frac{\partial}{\partial \psi} \left(\frac{\partial v}{\partial r} + \frac{v}{r} - \frac{1}{r} \frac{\partial u}{\partial \psi} \right) \\ &\quad + \cos \psi \left(\frac{a}{R} \right) \left[1 - \frac{a}{R} r \sin \psi + O\left(\frac{a}{R} \right)^2 \right] \left(\frac{\partial v}{\partial r} + \frac{v}{r} - \frac{1}{r} \frac{\partial u}{\partial \psi} \right). \end{aligned}} \qquad (4\text{--}70)$$

D. The Motion of a Fluid Through a Slightly Curved Tube – The Dean Problem

Similarly, the nondimensionalized equations for the other velocity components can be written in the form appropriate for small values of a/R:

$$Re\left\{u\frac{\partial v}{\partial r} + \frac{v}{r}\frac{\partial v}{\partial \psi} + \frac{uv}{r} - w^2\cos\psi\left(\frac{a}{R}\right)\left[1 - \frac{a}{R}r\sin\psi + O\left(\frac{a}{R}\right)^2\right]\right\}$$
$$= -\frac{1}{r}\frac{\partial p}{\partial \psi} + \left\{\frac{\partial}{\partial r} + \sin\psi\left(\frac{a}{R}\right)\left[1 - \frac{a}{R}r\sin\psi + O\left(\frac{a}{R}\right)^2\right]\right\}$$
$$\times \left(\frac{\partial v}{\partial r} + \frac{v}{r} - \frac{1}{r}\frac{\partial u}{\partial \psi}\right), \tag{4-71}$$

$$Re\left\{u\frac{\partial w}{\partial r} + \frac{v}{r}\frac{\partial w}{\partial \psi} + (uw\sin\psi + wv\cos\psi)\left(\frac{a}{R}\right)\right.$$
$$\left.\times\left[1 - \frac{a}{R}r\sin\psi + O\left(\frac{a}{R}\right)^2\right]\right\}$$
$$= -\left[1 - \frac{a}{R}r\sin\psi + O\left(\frac{a}{R}\right)^2\right]\frac{1}{r}\frac{\partial p}{\partial \theta}$$
$$+ \left(\frac{\partial}{\partial r} + \frac{1}{r}\right)\left\{\frac{\partial w}{\partial r} + \frac{a}{R}w\sin\psi\left[1 - \frac{a}{R}r\sin\psi + O\left(\frac{a}{R}\right)^2\right]\right\}$$
$$+ \frac{1}{r}\frac{\partial}{\partial \psi}\left\{\frac{1}{r}\frac{\partial w}{\partial \psi} + \frac{a}{R}w\cos\psi\left[1 - \frac{a}{R}r\sin\psi + O\left(\frac{a}{R}\right)^2\right]\right\}. \tag{4-72}$$

Finally, the continuity equation is

$$\frac{\partial u}{\partial r} + \frac{u}{r} + \frac{1}{r}\frac{\partial v}{\partial \psi}$$
$$+ (u\sin\psi + v\cos\psi)\left(\frac{a}{R}\right)\left[1 - \frac{a}{R}r\sin\psi + O\left(\frac{a}{R}\right)^2\right] = 0. \tag{4-73}$$

The general problem, then, is to solve the system of (4–70)–(4–73), subject to the boundary conditions

$$u = v = w = 0 \quad \text{at} \quad r = 1 \tag{4-74}$$

and the requirement that u, v, and w be bounded at the origin $r = 0$. It can be seen, by examination of these equations, that the problem for nonzero values of a/R and Re is highly *nonlinear*. Here, we seek an asymptotic approximation to the solution in the limiting case $a/R \ll 1$.

As a starting point, we recall that the limit $a/R \equiv 0$ corresponds to a straight circular tube, with the flow described by the Poiseuille flow solution $w = (1 - r^2)$, $u = v = 0$. In the present context, we consider small, but *nonzero*, values of a/R, and recognize the Poiseuille flow solution as a first approximation in an asymptotic approximation scheme. In particular, if we assume that a solution exists for **u** in the form of a *regular asymptotic expansion*,

$$\mathbf{u} = \mathbf{u}_0 + \left(\frac{a}{R}\right)\mathbf{u}_1 + \left(\frac{a}{R}\right)^2\mathbf{u}_2 + \cdots +, \tag{4-75}$$

the Poiseuille flow solution corresponds to $\mathbf{u}_0$, and our goal is to obtain additional terms in the expansion, namely, $\mathbf{u}_1$, $\mathbf{u}_2$ ..., and so on. It is, of course, not self-evident that a regular perturbation expansion actually exists for this problem. However, instead of trying to establish this fact in an *a priori* manner, it is common practice initially to seek a solution in the form of a regular expansion, with the proof of existence for such an expansion then residing in the ability to satisfy the resulting DEs and boundary conditions to arbitrary order in a/R. If this can be done, it implies that the scaling inherent in the governing equations is valid throughout the flow domain, and a regular expansion exists.

We assume then that an expansion of the form (4–75) exists for each of the components of $\mathbf{u}$ and for p, namely,

$$
\begin{aligned}
w &= w_0 + \left(\frac{a}{R}\right) w_1 + \left(\frac{a}{R}\right)^2 w_2 + \cdots +, \\
u &= u_0 + \left(\frac{a}{R}\right) u_1 + \left(\frac{a}{R}\right)^2 u_2 + \cdots +, \\
v &= v_0 + \left(\frac{a}{R}\right) v_1 + \left(\frac{a}{R}\right)^2 v_2 + \cdots +, \\
p &= p_0 + \left(\frac{a}{R}\right) p_1 + \left(\frac{a}{R}\right)^2 p_2 + \cdots +.
\end{aligned}
\quad (4\text{–}76)
$$

The coefficients in these expansions, i.e., w_i, u_i, v_i, and p_i, can be functions of (r, ψ) but must be independent of (a/R). To obtain governing equations and boundary conditions for these coefficients, we substitute the expansions (4–76) into the governing equations and boundary conditions (4–70)–(4–74) and collect all terms at each order in a/R. The result is a set of four equations plus boundary conditions, each of which takes the form

$$(fnc_0 \text{ of } r, \psi) + \left(\frac{a}{R}\right)(fnc_1 \text{ of } r, \psi) + \left(\frac{a}{R}\right)^2 (fnc_2 \text{ of } r, \psi) + \cdots + = 0. \quad (4\text{–}77)$$

Because a/R is arbitrary, each of the coefficients labeled "(fnc of r, ψ)" must equal zero, and this yields governing equations for the functions w_i, u_i, v_i, and p_i, in Eqs. (4–76).

At $O(1)$, we obtain a set of coupled equations for the first terms in (4–76), namely

$$Re\left(u_0 \frac{\partial u_0}{\partial r} + \frac{v_0}{r}\frac{\partial u_0}{\partial \psi} - \frac{v_0^2}{r}\right) = -\frac{\partial p_0}{\partial r} + \frac{1}{r}\frac{\partial}{\partial \psi}\left(\frac{\partial v_0}{\partial r} + \frac{v_0}{r} - \frac{1}{r}\frac{\partial u_0}{\partial \psi}\right), \quad (4\text{–}78\text{a})$$

$$Re\left(u_0 \frac{\partial v_0}{\partial r} + \frac{v_0}{r}\frac{\partial v_0}{\partial \psi} + \frac{u_0 v_0}{r}\right) = -\frac{1}{r}\frac{\partial p_0}{\partial \psi} + \frac{\partial}{\partial r}\left(\frac{\partial v_0}{\partial r} + \frac{v_0}{r} - \frac{1}{r}\frac{\partial u_0}{\partial \psi}\right), \quad (4\text{–}78\text{b})$$

$$Re\left(u_0 \frac{\partial w_0}{\partial r} + \frac{v_0}{r}\frac{\partial w_0}{\partial \psi}\right) = -\frac{\partial p_0}{\partial z} + \frac{\partial^2 w_0}{\partial r^2} + \frac{1}{r}\frac{\partial w_0}{\partial r} + \frac{1}{r}\frac{\partial^2 w_0}{\partial \psi^2}, \quad (4\text{–}78\text{c})$$

$$\frac{\partial u_0}{\partial r} + \frac{u_0}{r} + \frac{1}{r}\frac{\partial v_0}{\partial \psi} = 0, \quad (4\text{–}78\text{d})$$

with boundary conditions

$$u_0 = v_0 = w_0 = 0 \quad \text{at} \quad r = 1. \quad (4\text{–}78\text{e})$$

Here, the axial pressure gradient has been expressed in the form $\partial p_0/\partial z$, where $\partial/\partial z \sim R^{-1}(\partial/\partial \theta)$, to emphasize the connection between this problem and the steady motion in a straight tube, which was considered in Chap. 3. We previously indicated that the dimensional

D. The Motion of a Fluid Through a Slightly Curved Tube – The Dean Problem

magnitude of the axial pressure gradient is G. Hence, referring to the definitions of characteristic scales following Eq. (4–69), it follows that

$$-\frac{\partial p_0}{\partial z} = 4,$$

and it is evident from (4–78c) that $w_0 \neq 0$. On the other hand, the equations and boundary conditions for $u_0 = v_0$ are completely homogeneous and independent of w_0, so that an obvious solution is $u_0 = v_0 = 0$, with p_0 being a linear function of z only. In this case, if we assume axisymmetry for w_0, the governing Equation, (4–78), reduces to

$$\frac{1}{r}\frac{d}{dr}\left(r\frac{dw_0}{dr}\right) = -4, \tag{4–79}$$

and the solution, subject to (4–78e), is

$$\boxed{w_0 = 1 - r^2.} \tag{4–80}$$

Of course, this is just the familiar *Poiseuille solution* as expected.

The first influence of the curvature of the tube is felt in the terms of $O(a/R)$ in the expansions (4–76). We obtain the governing equations at $O(a/R)$ by substituting the expansions (4–76) into the dimensionless equations (4–70)–(4–73) and collecting all terms of $O(a/R)$:

$$-Re\, w_0^2 \sin\psi = -\frac{\partial p_1}{\partial r} + \frac{1}{r}\frac{\partial}{\partial \psi}\left(\frac{\partial v_1}{\partial r} + \frac{v_1}{r} - \frac{1}{r}\frac{\partial u_1}{\partial \psi}\right), \tag{4–81a}$$

$$-Re\, w_0^2 \cos\psi = -\frac{1}{r}\frac{\partial p_1}{\partial \psi} + \frac{\partial}{\partial r}\left(\frac{\partial v_1}{\partial r} + \frac{v_1}{r} - \frac{1}{r}\frac{\partial u_1}{\partial \psi}\right), \tag{4–81b}$$

$$Re\left(u_1 \frac{\partial w_0}{\partial r}\right) = \frac{\partial p_0}{\partial z} r \sin\psi - \frac{\partial p_1}{\partial z} + \left(\frac{\partial^2 w_1}{\partial r^2} + \frac{1}{r}\frac{\partial w_1}{\partial r} + \frac{1}{r^2}\frac{\partial^2 w_1}{\partial \psi^2}\right) \tag{4–81c}$$

$$+ \left(\frac{\partial}{\partial r} + \frac{1}{r}\right) w_0 \sin\psi + \frac{1}{r}\frac{\partial}{\partial \psi}(w_0 \cos\psi),$$

$$\frac{\partial u_1}{\partial r} + \frac{u_1}{r} + \frac{1}{r}\frac{\partial v_1}{\partial \psi} = 0. \tag{4–81d}$$

The boundary conditions are

$$\boxed{u_1 = v_1 = w_1 = 0 \quad \text{at} \quad r = 1.} \tag{4–81e}$$

Although the original governing equations are *nonlinear* for arbitrary Re and (a/R), we see that (4–81) (and all higher-order equations) are *linear*. This is typical of regular perturbation solutions of nonlinear PDEs. Equation (4–79) for the leading-order term, w_0, contains only terms that were linear in the original equations. At the next level of approximation, represented by (4–81), nonlinear terms appear, but these are evaluated by use of the leading-order solution that is *now known*. Further examination of (4–81) clearly indicates that the flow at this level of approximation $O(a/R)$ will no longer be 1D. All three velocity components, u_1, v_1 and w_1, must be nonzero because (4–81a)–(4–81c) are nonhomogeneous. Hence we see that a slight departure from the straight tube geometry, $a/R \ll 1$, causes an important qualitative change in the nature of the flow.

To ascertain the details of the changes represented by u_1, v_1, w_1, and p_1, we must solve system of equations (4–81a)–(4–81e). We can begin with the pressure. It follows from (4–81c) that the most general possible form for p_1 is

$$p_1 = H(r, \psi)z + B(r, \psi).$$

However, according to (4–81a) or (4–81b), $\partial p_1/\partial r$ and $\partial p_1/\partial \psi$ are both functions, at most, of r and ψ. Thus we see that

$$H(r, \psi) = 0,$$
$$p_1 = B(r, \psi) \text{ only}. \qquad (4\text{–}82)$$

Hence the pressure contribution at $O(a/R)$ enters, at most, in the pressure-gradient terms in (4–81a) and (4–81b).

To solve (4–81a) and (4–81b), it is convenient to introduce a function $f_1(r, \psi)$, defined such that

$$ru_1 \equiv -\frac{\partial f_1}{\partial \psi} \quad \text{and} \quad v_1 \equiv \frac{\partial f_1}{\partial r}. \qquad (4\text{–}83)$$

If Eqs. (4–81a) and (4–81b) can be satisfied by means of the function f_1, we can see by examining (4–81d) that the continuity equation will be satisfied automatically.[7] Using these forms for u_1 and v_1 and the solution (4–80) for w_0, we can write Eqs. (4–81a) and (4–81b) as

$$-Re(1 - r^2)^2 \sin \psi = -\frac{\partial p_1}{\partial r} - \frac{1}{r}\frac{\partial}{\partial \psi}(\nabla_1^2 f_1), \qquad (4\text{–}84\text{a})$$

$$-Re(1 - r^2)^2 \cos \psi = -\frac{1}{r}\frac{\partial p_1}{\partial \psi} + \frac{\partial}{\partial r}(\nabla_1^2 f_1), \qquad (4\text{–}84\text{b})$$

where

$$\nabla_1^2 \equiv \frac{\partial^2}{\partial r^2} + \frac{1}{r}\frac{\partial}{\partial r} + \frac{1}{r^2}\frac{\partial^2}{\partial \psi^2}. \qquad (4\text{–}85)$$

Hence, eliminating p_1 by cross-differentiating and combining (4–84a) and (4–84b), we obtain a single equation for f_1:

$$r\nabla_1^2(\nabla_1^2 f_1) = 4Re\, r^2(1 - r^2)\cos \psi. \qquad (4\text{–}86)$$

We obtain the governing equation for w_1 by substituting for w_0, as indicated in (4–81c), and noting that $\partial p_0/\partial z \equiv -4$. Because $\partial p_1/\partial z \equiv 0$,[8] the resulting equation takes the form

$$\nabla_1^2 w_1 = 6r \sin \psi - (2ru_1)Re. \qquad (4\text{–}87)$$

Hence we can obtain w_1 only after first solving for u_1 from Eqs. (4–86).

The solution of (4–86) is straightforward. The boundary conditions, (4–81e), expressed in terms of f_1 are

$$\frac{\partial f_1}{\partial r} = \frac{\partial f_1}{\partial \psi} = 0 \quad \text{at} \quad r = 1.$$

The details of solving (4–86) are left to the reader. The solution satisfying the boundary conditions is

$$\boxed{f_1 = \frac{Re}{288}r(1 - r^2)^2(4 - r^2)\cos \psi.} \qquad (4\text{–}88)$$

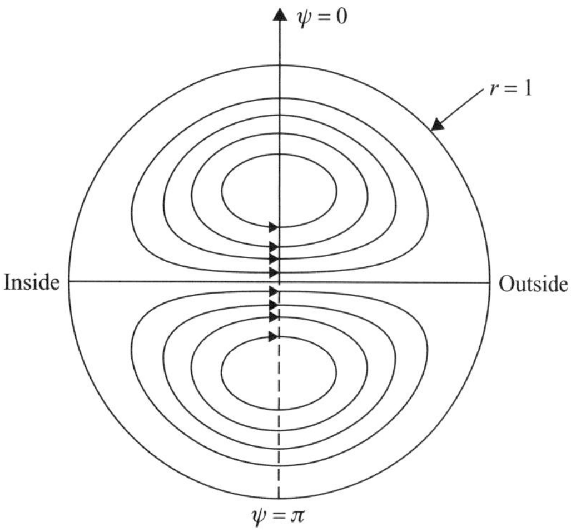

D. The Motion of a Fluid Through a Slightly Curved Tube – The Dean Problem

Figure 4–5. The streamlines for the *secondary* flow of a fluid that is moving through a slightly curved tube. Contour values are plotted in increments of 0.2222.

The corresponding velocity components are

$$u_1 = \frac{Re}{288}(1-r^2)^2(4-r^2)\sin\psi, \qquad (4\text{–}89)$$

$$v_1 = -\frac{Re}{288}(7r^4 - 23r^2 + 4)(1-r^2)\cos\psi. \qquad (4\text{–}90)$$

With u_1 known, (4–87) can now be solved for w_1. Again, the solution is straightforward. The result, after the boundary condition, (4–81e), is applied, is

$$w_1 = -\frac{3}{4}\sin\psi\, r(1-r^2) + Re^2\frac{\sin\psi}{11520}r(1-r^2)(19 - 21r^2 + 9r^4 - r^6). \qquad (4\text{–}91)$$

Thus the presence of a small amount of curvature of the tube axis is seen to produce a significant departure in the nature of the flow even at the leading-order correction, $O(a/R)$. The motion is no longer simply in the axial direction, but there is also motion in the cross-sectional plane. This motion, superposed upon the primary axial flow, is known as a *secondary* flow. A sketch of the streamlines for the secondary flow, given by contours of constant f_1, is shown in Fig. 4–5. The motion at the center of the tube is from the inside to the outside, driven essentially by the centrifugal force exerted on the fluid as it traverses the curved path followed by the tube axis. The return flow nearest the tube walls at the top and bottom is necessary to satisfy continuity.

The path of a typical fluid element is a superposition of the recirculating secondary flow and the primary flow in the axial direction. The result is a helical motion as the fluid element moves through the tube. In addition to the secondary flow, however, the curvature of the tube axis also contributes the modification, (4–91), in the axial profile. When (4–91) and (4–80) are combined in the asymptotic form, (4–76), the axial velocity component takes the form

$$w = (1-r^2) + \left(\frac{a}{R}\right)\left[-\frac{3}{4}r(1-r^2) \right.$$
$$\left. + \frac{Re^2}{11520}r(1-r^2)(19 - 21r^2 + 9r^4 - r^6)\right]\sin\psi + O\left(\frac{a}{R}\right)^2. \qquad (4\text{–}92)$$

Not only is the radial dependence of w changed, but also the axial velocity profile is no longer independent of the polar angle ψ, as shown in Fig. 4–6. Indeed, the contribution to

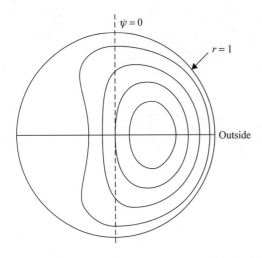

Figure 4–6. Contours of constant axial velocity for pressure-driven flow through a slightly curved tube. Values are plotted in increments of 0.3819.

w at $O(a/R)$ consists of an *asymmetric* profile, with a slightly increased axial flow in the range $0 < \psi < \pi$, which lies on the outer half of the tube, and a slightly decreased flow for $\pi < \psi < 2\pi$, which lies on the inside half of the tube. It should be noted, however, that there is no change at this level of approximation in the total volumetric flow rate through the tube.

Experimentally, it is found that the volumetric flow rate through a curved tube actually decreases relative to the flow rate in a straight tube if all other factors are held constant (namely, the tube length and radius, the pressure gradient, and the fluid viscosity). However, this effect does not appear in the first correction to the Poiseuille flow solution for small a/R. To predict this effect, we must proceed to higher-order corrections in the asymptotic expansions, (4–76). The decrease in volumetric flow rate first appears, in fact, at the next level of approximation, $O(a/R)^2$. However, the algebraic effort necessary to calculate the next terms (u_2, v_2, w_2, p_2) is large, and we do not pursue the detailed calculations here. The physical explanation for decreased flow rates is simply that some of the energy imparted to the fluid through the applied pressure gradient is diverted to driving the secondary flow, which is an added source of dissipation from kinetic energy to heat.

The main conclusions to be drawn from the preceding analysis are twofold: First, the departure from a straight tube to a slightly curved tube destroys the unidirectional nature of the Poiseuille flow, and we may generally expect small departures from strict adherence to a unidirectional geometry to produce similar transitions in the flow for other problems; second, for small a/R, the influence of curvature of the tube is described by a regular perturbation expansion, based on the governing equations in the nondimensionalized forms, (4–70)–(4–73). We should keep in mind that the solution form of (4–76) is valid only in the asymptotic limit $a/R \to 0$. If the tube is more tightly coiled so that the ratio a/R is not small, some other solution technique must be used. Although analytic approximation is possible for some cases, the vast majority of solutions for finite a/R have been carried out numerically. The interested reader may find a summary of additional work on this problem in a review article written by Berger *et al.*[9] More recent studies have focused on numerical investigations at high Dean numbers[10] and on the applications to hemodynamics.[5]

E. FLOW IN A WAVY-WALL CHANNEL – "DOMAIN PERTURBATION METHOD"

In the two preceding sections we considered two problems that could be solved by using a regular perturbation expansion. These problems were similar in that the geometry of the

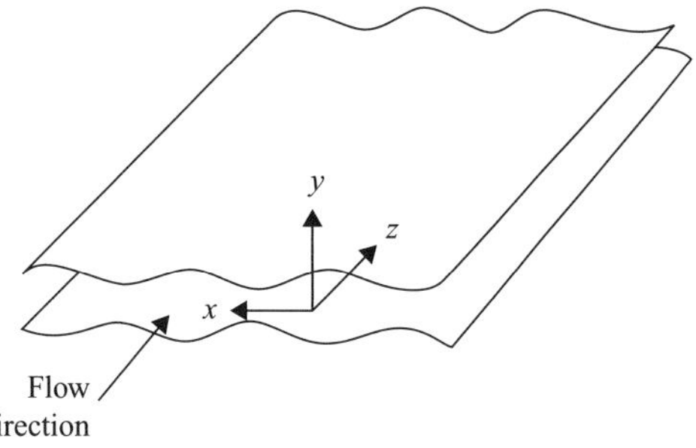

E. Flow in a Wavy-Wall Channel – "Domain Perturbation Method"

Figure 4–7. A sketch of the corrugated flow channel.

flow domain and the boundary conditions were fixed and the small parameter appeared in the DE. Here we consider another type of problem that also allows a regular perturbation solution by means of a method that is usually called the *domain perturbation method*.[11] This method applies to problems in which the geometry of the flow or transport domain is complicated and irregular in the sense that the boundaries do not correspond to coordinate surfaces of any known analytic coordinate system, but are nevertheless "near" to such coordinate surfaces.

Here we consider two generalizations of the problem of pressure-gradient driven unidirectional flow between two infinite parallel plane boundaries. The difference is that, in the present problems, we assume that the boundaries are corrugated rather than flat. In particular, instead of flat walls, we assume that the walls are located at

$$y = \pm h(x) = \pm \frac{d}{2}\left(1 + \varepsilon \sin \frac{2\pi x}{L}\right). \tag{4–93}$$

Although the method we are going to describe can be applied to much more complicated boundary shapes, we assume here that the gap width oscillates in a regular periodic fashion on the length scale $L/2\pi$. In the first example, we assume that the mean-flow direction is parallel to the grooves in the boundaries. In the second, we consider the case in which the flow is perpendicular to the grooves in the boundaries. Both steady and oscillatory flows in corrugated and furrowed channels have received a great deal of study in the research literature because of the potential they have for enhancing heat or mass transfer efficiencies in various types of transport processes.[12]

1. Flow Parallel to the Corrugation Grooves

The first example that we consider is a slight generalization of the unidirectional flow problem of pressure-driven 2D Poiseuille flow between two flat parallel plane boundaries. The only difference is that here we assume that the boundaries are corrugated rather than flat. In this section, we begin with the situation in which the grooves of the corrugated walls are parallel to the pressure gradient. In this case, the flow will still be *unidirectional*. A sketch of the flow domain is shown in Fig. 4–7. We denote the flow direction as z, the direction across the gap as y, and the direction in the plane perpendicular to the flow direction as x.

Although the flow geometry is relatively simple, the boundaries no longer correspond to coordinate surfaces. On the other hand, when the parameter ε is small, they deviate only slightly from the coordinate surfaces $y = \pm d/2$ of the Cartesian coordinate system. We denote the velocity component in the z direction as w. The exact flow problem is to solve the unidirectional flow equation,

$$\mu\left(\frac{\partial^2 w}{\partial x^2} + \frac{\partial^2 w}{\partial y^2}\right) = -G, \tag{4–94}$$

with

$$w = 0 \quad \text{at } y = \pm h(x) = \pm \frac{d}{2}\left(1 + \varepsilon \sin \frac{2\pi x}{L}\right). \tag{4-95}$$

Alternatively, we can require that

$$w = 0 \quad \text{at } y = h(x) = \frac{d}{2}\left(1 + \varepsilon \sin \frac{2\pi x}{L}\right), \tag{4-96}$$

together with the symmetry condition

$$\frac{\partial w}{\partial y} = 0 \text{ at } y = 0. \tag{4-97}$$

Because the boundary geometry depends on x, the unidirectional velocity component is now a function of both x and y.

We could obtain a general solution of Eq. (4–94). However, there is no obvious way to apply the boundary condition at the channel wall, at least in the general form (4–95) or (4–96). The method of domain perturbations provides an approximate way to solve this problem for $\varepsilon \ll 1$. *The basic idea is to replace the exact boundary condition, (4–96), with an approximate boundary condition that is asymptotically equivalent for $\varepsilon \ll 1$ but now applied at the coordinate surface $y = d/2$.* The method of domain perturbations leads to a regular perturbation expansion in the small parameter ε.

To be consistent with our preceding analyses, we begin by nondimensionalizing. For this purpose, we identify characteristic scales. Because the perturbation to the channel shape is assumed to be very small, it is apparent that the relevant velocity scale is

$$u_c \equiv Gd^2/\mu. \tag{4-98a}$$

However, there are clearly two length scales. Gradients across the channel are characterized by the mean channel width d, but gradients in the direction perpendicular to the grooves are characterized by the length scale L. We can denote these length scales as

$$\ell_c^y \equiv d, \ \ell_c^x \equiv L. \tag{4-98b}$$

In terms of dimensionless variables $\bar{u} \equiv u_z/u_c$ and $(\bar{y}, \bar{x}) \equiv (y/d, x/L)$, the problem is now

$$\left(\frac{d^2}{L^2}\right)\frac{\partial^2 \bar{w}}{\partial \bar{x}^2} + \frac{\partial^2 \bar{w}}{\partial \bar{y}^2} = -1, \tag{4-99}$$

with

$$\bar{w} = 0 \quad \text{at } \bar{y} = \bar{h}(\bar{x}) = \frac{1}{2}[1 + \varepsilon \sin(2\pi \bar{x})], \tag{4-100}$$

and

$$\frac{\partial \bar{w}}{\partial \bar{y}} = 0 \quad \text{at } \bar{y} = 0. \tag{4-101}$$

It will be noted that the dimensionless problem is characterized by two dimensionless parameters, the amplitude of the corrugation ε and the ratio of the channel width to the wavelength of the corrugations d/L. The domain perturbation method that we use to solve

E. Flow in a Wavy-Wall Channel – "Domain Perturbation Method"

this problem is based on the assumption that $\varepsilon \ll 1$, as stated already, but we make no assumption about the length-scale ratio.

The key to the domain perturbation method is to transform (4–100) to an asymptotically equivalent boundary condition applied at $\bar{y} = 1/2$. To do this, we first use a Taylor series approximation for $\bar{w}$ at $\bar{y} = \bar{h}$ in terms of $\bar{w}$ and its derivatives evaluated at $\bar{y} = 1/2$. Thus,

$$\bar{w}\mid_{\bar{y}=\bar{h}} = \bar{w}\mid_{\bar{y}=1/2} + \left(\frac{\partial \bar{w}}{\partial \bar{y}}\right)_{\bar{y}=1/2}\left[\frac{\varepsilon}{2}\sin(2\pi\bar{x})\right] + \frac{1}{2}\left(\frac{\partial^2 \bar{w}}{\partial \bar{y}^2}\right)_{\bar{y}=1/2}\left[\frac{\varepsilon}{2}\sin(2\pi\bar{x})\right]^2 + O(\varepsilon^3). \quad (4\text{–}102)$$

Now we seek a solution for $\bar{w}$ in the form of a regular asymptotic expansion:

$$\bar{w} = w_0 + \varepsilon w_1 + \varepsilon^2 w_2 + O(\varepsilon^3). \quad (4\text{–}103)$$

Because $\bar{w}\mid_{\bar{y}=\bar{h}} = 0$ according to (4–100), when we substitute the expansions (4–103) and (4–102) into (4–100), we obtain

$$0 = w_0 + \varepsilon\left[w_1 + \left(\frac{\partial w_0}{\partial \bar{y}}\right)\frac{1}{2}\sin(2\pi\bar{x})\right]$$
$$+ \varepsilon^2\left\{w_2 + \left(\frac{\partial w_1}{\partial \bar{y}}\right)\frac{1}{2}\sin(2\pi\bar{x}) + \frac{1}{2}\left(\frac{\partial^2 w_0}{\partial \bar{y}^2}\right)\left[\frac{1}{2}\sin(2\pi\bar{x})\right]^2\right\} + O(\varepsilon^3) \text{ at } \bar{y}=1/2. \quad (4\text{–}104)$$

Because ε is an arbitrary small parameter, each of the terms in (4–104) must be equal to zero. Thus,

$$\text{at } O(1), \quad w_0 = 0 \quad \text{at } \bar{y} = 1/2; \quad (4\text{–}105\text{a})$$

$$\text{at } O(\varepsilon), \quad w_1 = -\left(\frac{\partial w_0}{\partial \bar{y}}\right)\frac{1}{2}\sin(2\pi\bar{x}) \text{ at } \bar{y} = 1/2; \quad (4\text{–}105\text{b})$$

$$\text{and at } O(\varepsilon^2), \quad w_2 = -\left(\frac{\partial w_1}{\partial \bar{y}}\right)\frac{1}{2}\sin(2\pi\bar{x}) - \frac{1}{2}\left(\frac{\partial^2 w_0}{\partial \bar{y}^2}\right)\left[\frac{1}{2}\sin(2\pi\bar{x})\right]^2 \text{ at } \bar{y} = 1/2. \quad (4\text{–}105\text{c})$$

The original boundary condition, (4–100), at $\bar{y} = \bar{h}(\bar{x})$ is thereby transformed into a hierarchy of asymptotically equivalent boundary conditions for the functions w_0, w_1, and w_2 at $\bar{y} = \frac{1}{2}$. If we first solve for w_0 at $O(1)$ by using (4–105a), then (4–105b) provides a condition for w_1 at $O(\varepsilon)$, and so on to higher-order terms.

Thus, substituting (4–103) into (4–99) and (4–101), we see that the problem at $O(1)$ is

$$\left(\frac{d^2}{L^2}\right)\frac{\partial^2 w_0}{\partial \bar{x}^2} + \frac{\partial^2 w_0}{\partial \bar{y}^2} = -1, \quad (4\text{–}106)$$

with

$$\partial w_0/\partial \bar{y} = 0 \quad \text{at } \bar{y} = 0, \quad (4\text{–}107)$$

plus condition (4–105a). However, the boundary conditions are independent of $\bar{x}$ at this level of approximation, and this is just the standard problem of Poiseuille flow between parallel plane boundaries, in which $w_0 = w_0(\bar{y})$ so that (4–106) becomes

$$\frac{\partial^2 w_0}{\partial \bar{y}^2} = -1. \quad (4\text{–}108)$$

235

The solution, subject to (4–107) and (4–105a), is

$$w_0 = \frac{1}{2}\left(\frac{1}{4} - \bar{y}^2\right). \qquad (4\text{–}109)$$

The problem at $O(\varepsilon)$ is

$$\left(\frac{d^2}{L^2}\right)\frac{\partial^2 w_1}{\partial \bar{x}^2} + \frac{\partial^2 w_1}{\partial \bar{y}^2} = 0, \qquad (4\text{–}110)$$

$$\partial w_1/\partial \bar{y} = 0 \quad \text{at } \bar{y} = 0. \qquad (4\text{–}111)$$

Because $\partial w_0/\partial \bar{y} = -1/2$ at $\bar{y} = 1/2$, the boundary condition (4–105b) gives

$$w_1 = \frac{1}{4}\sin(2\pi\bar{x}) \quad \text{at } \bar{y} = 1/2. \qquad (4\text{–}112)$$

The fact that this boundary condition is nonhomogeneous means that w_1 will be nonzero. This solution at $O(\varepsilon)$ is the first correction for the corrugated geometry of the channel.

To solve this problem, we note that the boundary condition (4–112) imposes a specific $\bar{x}$ dependence on the solution, which will be preserved throughout the domain because the governing equation is linear. This suggests that we should seek a solution in the form

$$w_1 = F(\bar{y})\sin(2\pi\bar{x}). \qquad (4\text{–}113)$$

If we substitute this proposed solution form into (4–110)–(4–112), we find that $F(\bar{y})$ must satisfy the following problem:

$$\frac{d^2 F}{d\bar{y}^2} - \left(\frac{2\pi d}{L}\right)^2 F = 0, \qquad (4\text{–}114)$$

with

$$dF/d\bar{y} = 0 \quad \text{at } \bar{y} = 0, \qquad (4\text{–}115)$$

$$F = 1/4 \quad \text{at } \bar{y} = 1/2. \qquad (4\text{–}116)$$

The general solution of (4–114) is

$$F = A\exp\left(\frac{2\pi d}{L}\bar{y}\right) + B\exp\left(-\frac{2\pi d}{L}\bar{y}\right).$$

When the boundary conditions are applied, this becomes

$$F(\bar{y}) = \frac{1}{4}\frac{\cosh[(2\pi d/L)\bar{y}]}{\cosh(\pi d/L)},$$

and thus

$$w_1 = \frac{1}{4}\frac{\cosh[(2\pi d/L)\bar{y}]}{\cosh(\pi d/L)}\sin(2\pi\bar{x}). \qquad (4\text{–}117)$$

The procedure just outlined can be extended to calculate higher-order terms in the expansion (4–103). We calculate only one more term at $O(\varepsilon^2)$. The governing equation and the symmetry condition for w_2 are identical to (4–110) and (4–111). The boundary condition at $\bar{y} = 1/2$ from the domain perturbation procedure can be obtained from (4–105c):

$$w_2 = \frac{1}{8}\sin^2(2\pi\bar{x})\left[1 - \frac{2\pi d}{L}\tanh\left(\frac{\pi d}{L}\right)\right] \text{at } \bar{y} = 1/2. \qquad (4\text{–}118)$$

E. Flow in a Wavy-Wall Channel – "Domain Perturbation Method"

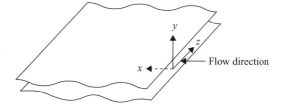

Figure 4–8. A sketch of the flow and wavy-wall channel geometry for motion perpendicular to the grooves in the walls.

Alternatively, this can be written as

$$w_2 = \frac{1}{16}(1 - \cos 4\pi \bar{x})\left[1 - \frac{2\pi d}{L}\tanh\left(\frac{\pi d}{L}\right)\right] \text{ at } \bar{y} = 1/2. \quad (4\text{–}119)$$

The complicated-looking solution for w_2 turns out to be

$$w_2 = \frac{1}{16}\left[1 - \frac{2\pi d}{L}\tanh\left(\frac{\pi d}{L}\right)\right]\left[1 - \frac{\cosh(4\pi d/L)\bar{y}}{\cosh(2\pi d/L)}\cos 4\pi \bar{x}\right]. \quad (4\text{–}120)$$

Given the solutions, (4–109), (4–117), and (4–120), we can calculate the volumetric flux of fluid through the channel (i.e., the volumetric flow rate per unit cross-sectional area):

$$Q = \frac{1}{Ld}\int_0^L \int_{-h(x;\varepsilon)}^{h(x;\varepsilon)} w \, dy \, dx = \frac{Gd^2}{\mu}\int_0^1 \int_{-\bar{h}(\bar{x};\varepsilon)}^{\bar{h}(\bar{x};\varepsilon)} [w_0 + \varepsilon w_1 + \varepsilon^2 w_2 + O(\varepsilon^3)] d\bar{y} d\bar{x}. \quad (4\text{–}121)$$

Substituting for w_0, w_1, w_2 and $\bar{h}$, and evaluating the integral to $O(\varepsilon^2)$, we find

$$Q = \frac{Gd^2}{12\mu}\left[1 + \frac{3}{4}\varepsilon^2 + o(\varepsilon^2)\right]. \quad (4\text{–}122)$$

Here, we use the "little o" symbolism to indicate that the next nonzero term will be smaller than $O(\varepsilon^2)$ but that we do not know its magnitude. We see that the volumetric flux for a given pressure gradient increases compared with that between two flat walls. Note, however, that the first correction occurs only at $O(\varepsilon^2)$, and thus the effect will be small because $\varepsilon \ll 1$.

2. Flow Perpendicular to the Corrugation Grooves

A more complicated problem, which has also received considerable attention in the research literature,[13] occurs when the applied pressure gradient is perpendicular to the grooves. In this case, the cross-sectional area of the channel is varying periodically in the flow direction. A sketch of the flow configuration is shown in Fig. 4–8. In this case, for convenience, we designate the flow direction as x, the cross-channel direction as y, and the corresponding velocity components as u and v. The flow will be 2D. Thus the velocity component in the z direction is zero, and there are also no gradients in the z direction. On the other hand, the flow is no longer unidirectional except in the limit $\varepsilon = 0$. Because the fluid is assumed to be incompressible, the velocity component in the mean-flow (x) direction must increase and decrease as the channel cross section decreases and increases, respectively, in order to preserve a constant-volume flux (i.e., mass conservation). In addition, because $\partial u/\partial x$ is nonzero, $\partial v/\partial y$ must also be nonzero to satisfy continuity, and the velocity component in the orthogonal direction (v) must also be nonzero.

In this case, we must therefore begin with the full Navier–Stokes and continuity equations for a 2D flow, (2–91) and (2–20). In terms of the Cartesian coordinate system described in Fig. 4-8, these are

$$\rho\left(u\frac{\partial u}{\partial x} + v\frac{\partial u}{\partial y}\right) = -\frac{\partial p}{\partial x} + \mu\left(\frac{\partial^2 u}{\partial x^2} + \frac{\partial^2 u}{\partial y^2}\right), \quad (4\text{–}123)$$

$$\rho\left(u\frac{\partial v}{\partial x} + v\frac{\partial v}{\partial y}\right) = -\frac{\partial p}{\partial y} + \mu\left(\frac{\partial^2 v}{\partial x^2} + \frac{\partial^2 v}{\partial y^2}\right), \quad (4\text{–}124)$$

$$\frac{\partial u}{\partial x} + \frac{\partial v}{\partial y} = 0, \quad (4\text{–}125)$$

with the boundary conditions

$$\mathbf{u}\cdot\mathbf{n} = \mathbf{u}\cdot\mathbf{t} = 0 \quad \text{at } y = \pm h(x) = \pm\frac{d}{2}\left(1 + \varepsilon\sin\frac{2\pi x}{L}\right).$$

Here $\mathbf{n}$ and $\mathbf{t}$ are the unit normal and tangent vectors at the channel walls. Because both the normal and the tangential velocity components are zero, the latter is equivalent to saying

$$u = v = 0 \quad \text{at } y = \pm h(x) = \pm\frac{d}{2}\left(1 + \varepsilon\sin\frac{2\pi x}{L}\right). \quad (4\text{–}126)$$

As a consequence of the variations in the channel width with x, the fluid must accelerate and decelerate, and the pressure gradient will therefore be a function of position. However, we can still impose a mean pressure gradient in the x direction, which we denote as $-G$. We can then write

$$\frac{\partial p}{\partial x} = -G + \frac{\partial p'}{\partial x}; \quad \frac{\partial p}{\partial y} = \frac{\partial p'}{\partial y}. \quad (4\text{–}127)$$

In effect, this serves to define the "perturbation" pressure-gradients, $\nabla p'$, which are due to the variation in channel width.

Following our usual custom, we now nondimensionalize. The physically obvious characteristic scales are the length scales for variations of the velocity and perturbation pressure in the x and y directions, and the characteristic magnitude of the velocity in the x direction, namely,

$$\ell_c^y = d, \quad \ell_c^x = L, \quad u_c^x = u_c = Gd^2/\mu. \quad (4\text{–}128)$$

From these scales, we can define the dimensionless variables:

$$\bar{u} = u/u_c, \quad \bar{y} = y/d, \quad \bar{x} = x/L.$$

If we then denote the characteristic velocity in the y direction as u_c^y, we can write the continuity equation in dimensionless form:

$$\frac{Gd^2}{\mu}\left(\frac{d}{L}\right)\frac{\partial \bar{u}}{\partial \bar{x}} + u_c^y\frac{\partial \bar{v}}{\partial \bar{y}} = 0.$$

It is evident from this equation that a convenient choice for u_c^y is

$$u_c^y = u_c(d/L).$$

It may also not be obvious how to scale the pressure. However, if we denote the characteristic pressure as p_c, we can write the x component of the equation of motion as

$$\frac{\rho u_c^2}{L}\left(\bar{u}\frac{\partial \bar{u}}{\partial \bar{x}} + \bar{v}\frac{\partial \bar{u}}{\partial \bar{y}}\right) = G - \frac{p_c}{L}\frac{\partial \bar{p}}{\partial \bar{x}} + G\left(\frac{\partial^2 \bar{u}}{\partial \bar{y}^2} + \frac{d^2}{L^2}\frac{\partial^2 \bar{u}}{\partial \bar{x}^2}\right).$$

E. Flow in a Wavy-Wall Channel – "Domain Perturbation Method"

From this, it is evident that a convenient choice is

$$p_c = GL.$$

Hence the last two dimensionless variables are

$$\bar{v} = v(L/d\,u_c), \quad \bar{p} = p'/GL.$$

We can now write the full problem in dimensionless form:

$$\frac{\rho u_c d}{\mu} \frac{d}{L} \left(\bar{u} \frac{\partial \bar{u}}{\partial \bar{x}} + \bar{v} \frac{\partial \bar{u}}{\partial \bar{y}} \right) = 1 - \frac{\partial \bar{p}}{\partial \bar{x}} + \frac{\partial^2 \bar{u}}{\partial \bar{y}^2} + \frac{d^2}{L^2} \frac{\partial^2 \bar{u}}{\partial \bar{x}^2}, \tag{4–129}$$

$$\frac{\rho u_c d}{\mu} \frac{d}{L} \left(\bar{u} \frac{\partial \bar{v}}{\partial \bar{x}} + \bar{v} \frac{\partial \bar{v}}{\partial \bar{y}} \right) = -\frac{L^2}{d^2} \frac{\partial \bar{p}}{\partial \bar{y}} + \frac{\partial^2 \bar{v}}{\partial \bar{y}^2} + \frac{d^2}{L^2} \frac{\partial^2 \bar{v}}{\partial \bar{x}^2}, \tag{4–130}$$

$$\frac{\partial \bar{u}}{\partial \bar{x}} + \frac{\partial \bar{v}}{\partial \bar{y}} = 0. \tag{4–131}$$

The notation showing u_c explicitly on the left-hand side of (4–129) and (4–130) is preserved to emphasize that this is just the dimensionless parameter known as the Reynolds number, $\rho u_c d/\mu$, multiplied by the geometric parameter d/L. We have noted previously that the Reynolds number provides a measure of the relative importance of the "inertia" or acceleration terms in the Navier–Stokes equations relative to the viscous terms. In the analysis that follows, we make no assumption about the relative magnitudes of d and L. However, we do assume that the Reynolds number is vanishingly small compared even with the geometric amplitude parameter ε, which we shall also assume is small, $\varepsilon \ll 1$, meaning that inertial/acceleration effects can be neglected and that the flow is dominated by viscous and pressure-gradient effects. The resulting problem is still of some practical importance. However, a lot of research has also gone into exploring the flow patterns in a wavy-wall channel when $\rho u_c d/\mu$ is not small, and also the geometry is more extreme, i.e., ε is not small. Although this latter work is beyond the scope of what we consider here, the interested reader can refer to the references listed at the end of the chapter. The reader should be warned, however, that the results reported at finite Reynolds number may be difficult to understand until after we have considered the effects of finite Reynolds number on flow in later chapters.

In this section, we seek a solution for the limiting case $\varepsilon \ll 1$ by means of asymptotic expansions of the form

$$\bar{u} = u_0 + \varepsilon u_1 + O(\varepsilon^2); \bar{v} = v_0 + \varepsilon v_1 + O(\varepsilon^2); \nabla \bar{p} = \varepsilon [G_1(\bar{x},\bar{y})\mathbf{i} + G_2(\bar{x},\bar{y})\mathbf{j}] + O(\varepsilon^2). \tag{4–132}$$

For this limiting case, the boundary conditions, (4–126), can also be approximated in terms of asymptotically equivalent boundary conditions applied at $\bar{y} = \pm 1/2$. The details follow those of the previous subsection entirely and are not repeated here. We again use the symmetry condition at $\bar{y} = 0$ and condition (4–126) at $\bar{y} = 1/2$. The results of applying the domain perturbation analysis to both $\bar{u}$ and $\bar{v}$ are

$$O(1), \quad u_0 = v_0 = 0 \quad \text{at} \quad \bar{y} = 1/2; \tag{4–133}$$

$$O(\varepsilon), \quad u_1 = -\left(\frac{\partial u_0}{\partial \bar{y}}\right) \frac{1}{2} \sin(2\pi \bar{x}); \quad v_1 = -\left(\frac{\partial v_0}{\partial \bar{y}}\right) \frac{1}{2} \sin(2\pi \bar{x}) \quad \text{at } \bar{y} = 1/2. \tag{4–134}$$

An Introduction to Asymptotic Approximations

For the $O(1)$ problem, the boundaries are flat; hence $v_0 = 0$, and the problem is identical to (4–105a), (4–106), and (4–107), with the same solution, (4–109):

$$u_0 = \frac{1}{2}\left(\frac{1}{4} - \bar{y}^2\right). \tag{4–135}$$

At $O(\varepsilon)$,

$$\frac{\partial^2 u_1}{\partial \bar{y}^2} + \frac{d^2}{L^2}\frac{\partial^2 u_1}{\partial \bar{x}^2} = G_1(\bar{x}, \bar{y}), \tag{4–136}$$

$$\frac{\partial^2 v_1}{\partial \bar{y}^2} + \frac{d^2}{L^2}\frac{\partial^2 v_1}{\partial \bar{x}^2} = \frac{L^2}{d^2}G_2(\bar{x}, \bar{y}), \tag{4–137}$$

$$\frac{\partial u_1}{\partial \bar{x}} + \frac{\partial v_1}{\partial \bar{y}} = 0, \tag{4–138}$$

with the boundary conditions

$$v_1 = \partial u_1/\partial \bar{y} = 0 \quad \text{at } \bar{y} = 0, \tag{4–139}$$

and, from (4–134),

$$v_1 = 0; \quad u_1 = \frac{1}{4}\sin(2\pi \bar{x}) \quad \text{at } \bar{y} = 1/2. \tag{4–140}$$

We can solve for the velocity component u_1 by noting that we expect the "symmetry" of the solution in x that is imposed by the boundary condition (4–140) to be preserved for all $\bar{y}$ so that

$$u_1 = F(\bar{y})\sin(2\pi \bar{x}). \tag{4–141}$$

It is then clear from (4–136) that the x component of the pressure-gradient must then have the same dependence on $\bar{x}$,

$$G_1 = C \sin 2\pi \bar{x}, \tag{4–142}$$

where we have assumed that the coefficient C is a constant to be determined. We will, of course, need to verify that the assumption $C = \text{const}$ is sufficiently general by showing that all of the equations and boundary conditions for u_1 and v_1 can be satisfied. The function $F(\bar{y})$ must then satisfy the ODE

$$\frac{d^2 F}{d\bar{y}^2} - \left(\frac{2\pi d}{L}\right)^2 F = C. \tag{4–143}$$

It may be noted that the problem for u_1 appears similar to the problem for w_1 in the preceding section. However, the physics is fundamentally different because $\partial u_1/\partial \bar{x} \ne 0$, whereas $\partial w_1/\partial z \equiv 0$ in the preceding case. This is the reason that $G_1 \ne 0$ in this case.

A general solution of (4–143) that satisfies the symmetry condition (4–139) is

$$F(\bar{y}) = A \cosh[(2\pi d/L)\bar{y}] + C^*, \tag{4–144}$$

where C^* is the constant $-C/(2\pi d/L)^2$. The boundary condition (4–140) then requires that $F = 1/4$ at $\bar{y} = 1/2$. Hence,

$$A \cosh(\pi d/L) + C^* = 1/4, \tag{4–145}$$

and we can express F in the form

$$F(\bar{y}) = (0.25 - C^*)\frac{\cosh(2\pi d \bar{y}/L)}{\cosh(\pi d/L)} + C^*. \tag{4–146}$$

E. Flow in a Wavy-Wall Channel – "Domain Perturbation Method"

This would seem to complete the solution, at least as far as the velocity component u_1 is concerned. However, we should not forget that the constant C^* is still undetermined, and we have not yet determined the velocity component v_1 or the pressure-gradient component G_2.

The constant C^* can be determined from the "mass conservation" constraint that the volume flux must be independent of $\bar{x}$. This means that

$$\int_0^{1/2} [u_0 + u_1 \varepsilon + O(\varepsilon^2)] d\bar{y} = \text{const}, \tag{4–147}$$

i.e., independent of $\bar{x}$. Because $u_1 = F(\bar{y}) \sin[(2\pi d/L)\bar{x}]$, this implies

$$\int_0^{1/2} F(\bar{y}) d\bar{y} = 0. \tag{4–148}$$

To satisfy this condition, we must determine a specific value for C^* such that

$$\int_0^{1/2} \left[(0.25 - C^*) \frac{\cosh(2\pi d \bar{y}/L)}{\cosh(\pi d/L)} + C^* \right] d\bar{y} = 0. \tag{4–149}$$

Hence, carrying out the integration, we find

$$(0.25 - C^*) \left(\frac{L}{\pi d} \right) \tanh\left(\frac{\pi d}{L} \right) + 0.5 C^* = 0, \tag{4–150}$$

and thus

$$C^* = \left\{ \frac{1}{4} \frac{L}{\pi d} \tanh\left(\frac{\pi d}{L} \right) \Big/ \left[\frac{L}{\pi d} \tanh\left(\frac{\pi d}{L} \right) - \frac{1}{2} \right] \right\}. \tag{4–151}$$

This completes the solution for $\bar{u}$ through terms of $O(\varepsilon)$. The only remaining issue is this: Can we determine the $O(\varepsilon)$ contribution to $\bar{v}$ to satisfy the continuity equation, (4–138)?

To solve for v_1, we first note from (4–138) that

$$\frac{\partial v_1}{\partial \bar{y}} = -\frac{\partial u_1}{\partial \bar{x}} = -\left[(0.25 - C^*) \frac{\cosh(2\pi d \bar{y}/L)}{\cosh(\pi d/L)} + C^* \right] 2\pi \cos(2\pi \bar{x}). \tag{4–152}$$

Hence,

$$v_1 = -2\pi \cos(2\pi \bar{x}) \int_0^{\bar{y}} \left[(0.25 - C^*) \frac{\cosh(2\pi dt/L)}{\cosh(\pi d/L)} + C^* \right] dt + D^*, \tag{4–153}$$

where D^* is a constant of integration. Now v_1 must satisfy two boundary conditions. The first is the symmetry condition,

$$v_1 = 0 \quad \text{at } \bar{y} = 0.$$

This implies that $D^* = 0$. The second condition is that

$$v_1 = 0 \quad \text{at } \bar{y} = 1/2.$$

It is clear from (4–149) that this condition is satisfied exactly. Indeed, another way to determine C^* is to use the two boundary conditions on v_1. The main advantage of using (4–147) to determine C^* is that a complete solution can be obtained for u_1 with no need to calculate v_1. Finally, with v_1 known, we can use (4–137) to calculate the $\bar{y}$ component of the pressure-gradient at $O(\varepsilon)$, namely, $G_2(\bar{x}, \bar{y})$.

The solution that we have generated is quite complicated. The fluid velocity in the x direction, $u_0 + \varepsilon u_1$, decreases where the channel expands and increases where it contracts to maintain constant-volume flux along the channel. The pressure increases from the narrow part of the channel to the widest part of the channel, and it is this increasing pressure that slows the fluid down in the x direction. Conversely, from the wide part of the channel to the narrow throats, the pressure decreases in the x direction, and this accelerates the fluid

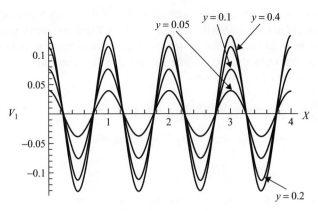

Figure 4–9. An illustration of the predicted flow pattern in a wavy-wall channel. The y-directional velocity, Eq. (4–153), plotted against $\bar{x}$ for various values of $\bar{y}$.

motion along the channel. The motion in the y direction is alternately positive and negative as the fluid moves in and out following the contours of the channel walls. A plot illustrating the predicted motion in the y direction is shown in Fig. 4–9.

We have not chosen to carry out the calculation to higher order in ε. At $O(\varepsilon)$, there is no change in the volumetric flow rate. However, if we examine the boundary condition for u_2 at $\bar{y} = 1/2$ from the domain perturbation expression [(4–104), with u replacing w], we see that there will be two terms, one proportional to $\cos(4\pi\bar{x})$ and the other independent of $\bar{x}$. It is thus clear that the volumetric flow rate will change at $O(\varepsilon^2)$.

F. DIFFUSION IN A SPHERE WITH FAST REACTION – "SINGULAR PERTURBATION THEORY"

In the previous sections we have seen several examples of transport problems that are amenable to analysis by the method of *regular* perturbation theory. As we shall see later in this book, however, most transport problems require the use of singular-perturbation methods. The high-frequency limit of flow in a tube with a periodic pressure oscillation provided one example, which was illustrative of the most common type of singular-perturbation problem involving a boundary layer near the tube wall. Here we consider another example in which there is a boundary-layer structure that we can analyze by using the method of matched asymptotic expansions.

Although this book is almost exclusively about fluid mechanics and heat transfer processes, the problem considered here is the transport of a reactant into a catalyst pellet within which there is a very fast reaction. This is a very convenient problem to illustrate the principles of the matched asymptotic method, and it also has very important practical applications in the field of chemical reaction engineering.

In reality, a typical catalyst pellet will be a porous solid that may be quite complicated or even irregular in shape with a large number of "catalytic" reaction sites distributed throughout. However, to simplify the problem for present purposes, the catalyst pellet will be approximated as being spherical in shape. Furthermore, we will assume that the catalyst pellet is uniform in constitution. Thus we assume that it can be characterized by an effective reaction-rate constant k_1^{eff} that has the same value at every point inside the pellet. In addition, we assume that the transport of reactant within the pellet can be modeled as pure diffusion with a spatially uniform effective diffusivity D_{eff}. To furthor simplify the problem, we assume that the transport of product out of the pellet is decoupled from the transport of reactant into the pellet. Finally, the concentration of reactant in the bulk-phase fluid (usually

F. Diffusion in a Sphere with Fast Reaction – "Singular Perturbation Theory"

a gas) including the surface of the pellet, is assumed to be a constant C_∞ (i.e., independent of position and time). The objective of our calculation is the steady-state concentration distribution of the reactant within the catalyst pellet. We shall see that this concentration distribution is closely related to the "effectiveness" factor for the catalyst pellet, namely,

$$\eta_{cat} \equiv \frac{\text{actual averaged reaction rate within the pellet}}{\text{reaction rate within the pellet if } c = C_\infty \text{ at all points}}.$$

We have not formally derived a *species transport equation*. However, in the present case we have assumed that transport is strictly by an effective diffusion process. Hence the concentration will be spherically symmetric and governed by a radial diffusion equation with a reaction rate R:

$$D_{\text{eff}} \frac{1}{r^2} \frac{\partial}{\partial r}\left(r^2 \frac{\partial c}{\partial r}\right) + R = 0. \tag{4–154}$$

For simplicity we assume that the reaction rate is first order:

$$R = -k_1^{\text{eff}} c. \tag{4–155}$$

We assume that D_{eff} and k_1^{eff} are known. There is a minus sign in (4–155) because the reaction uses up the reactant. As previously noted, the reactant concentration at the outer surface of the pellet is a constant:

$$c = C_\infty \quad \text{at } r = R. \tag{4–156}$$

With a first-order reaction, the governing equation is linear and could thus be solved without any use of scaling or asymptotic methods. However, we could just as easily assume that the reaction rate is second order in c or add other complications that do not so easily allow an *exact* analytic solution. The point here is to illustrate the idea of the asymptotic approximation technique, which is easily generalizable to all of these problems.

The starting point, as usual, is to nondimensionalize the problem. In the present case, this is easily done. We assume that the pellet radius R provides an appropriate characteristic length scale, $\ell_c = R$, and that the concentration can be scaled with the surface concentration

$$\bar{r} = r/R \quad \text{and} \quad \bar{c} = c/C_\infty. \tag{4–157}$$

Hence our problem can be expressed in the dimensionless form

$$\frac{1}{\bar{r}^2} \frac{\partial}{\partial \bar{r}}\left(\bar{r}^2 \frac{\partial \bar{c}}{\partial \bar{r}}\right) - Da(\bar{c}) = 0, \tag{4–158}$$

$$\text{with} \quad \bar{c} = 1 \quad \text{at } \bar{r} = 1. \tag{4–159}$$

There is one dimensionless parameter in this problem, known as the Damköhler number:

$$Da \equiv \frac{k_1^{\text{eff}} R^2}{D_{\text{eff}}}. \tag{4–160}$$

When we say that the reaction is very fast, we mean that the Damköhler number is large, or

$$\varepsilon \equiv Da^{-1} \ll 1. \tag{4–161}$$

Our problem is to determine the concentration distribution within the pellet in the asymptotic limit, $\varepsilon \ll 1$.

As usual in asymptotic problems, we first seek a regular perturbation solution. This means that we seek a solution in the form

$$\bar{c} = \bar{c}_0 + \varepsilon \bar{c}_1 + O(\varepsilon^2). \tag{4-162}$$

We assume that the first term must be $O(1)$, i.e., independent of ε, because the value of $\bar{c} = 1$ at $\bar{r} = 1$. Now we substitute the expansion (4–162) into the governing equation:

$$\bar{c}_0 + \varepsilon \left[\bar{c}_1 - \frac{1}{\bar{r}^2} \frac{\partial}{\partial \bar{r}} \left(\bar{r}^2 \frac{\partial \bar{c}_0}{\partial \bar{r}} \right) \right] + O(\varepsilon^2) = 0. \tag{4-163}$$

Hence the first approximation to this equation is just

$$\bar{c}_0 = 0 \tag{4-164}$$

everywhere within the pellet. This solution satisfies the original DE, and, as a first approximation, it may make sense in the interior of the pellet away from the surface. If the reaction is very fast, the reactant is used up before it can reach the center of the particle. However, the "solution" (4–164) clearly cannot be a uniformly valid approximation because it does not satisfy the boundary condition (4–159) at the surface of the pellet.

We deduce from the analysis of the preceding paragraph that there cannot be a regular asymptotic solution for $\varepsilon \ll 1$. In fact, the problem in the nondimensionalized form, (4–158), displays one of the classic signals that the limit $\varepsilon \equiv Da^{-1} \ll 1$ is singular (see the discussion at the end of Section B). Namely, if we take the limit $\varepsilon \to 0$ in the governing equation (4–158), the order of the DE is reduced. Indeed, in this case, it goes from a second-order ODE to an algebraic equation. Clearly we cannot expect the solution of this algebraic equation to satisfy the boundary conditions of the original problem, and we deduce that the algebraic equation cannot be a valid first approximation of the original problem, at least near the surface of the particle.

The only steps that were taken in deciding that the full (exact) model equation, (4–154), could be approximated by the algebraic equation in the limit as $\varepsilon \to 0$ were to nondimensionalize the problem, and then to assume that the characteristic scales were chosen "correctly" so that the relative magnitude of the two terms is completely determined by the fact that one is multiplied by the small parameter ε whereas the other is not. In fact, because the equation is linear in the concentration, the form of the dimensionless equation will not change regardless of how we decide to nondimensionalize concentration. It is only the choice of characteristic length scale that makes a difference. When we say that the particle radius is the characteristic length scale, we imply that this is the distance over which variations in c of $O(C_\infty)$ occur. In view of the fact that this choice for ℓ_c leads to (4–164) as the first approximation to the governing equation for $\varepsilon \to 0$, we can conclude only that the particle radius is not the correct characteristic length scale for variations in the reactant concentration, at least in the region nearest to the pellet surface for the limit $\varepsilon \to 0$.

In fact, from a mathematical point of view, this problem is very similar to the high-frequency limit of the pulsatile flow problem considered in Section A of this chapter. Away from the pellet surface, the choice $\ell_c = R$ and the resulting dimensionless equation, (4–158), is satisfactory. However, near the pellet surface the diffusion terms *cannot* be negligible relative to the reaction term, even in the limit $\varepsilon \to 0$. Diffusion is the only mechanism for the reactant to get into the particle where it can react. Presumably, just as in the pulsatile flow problem, the appropriate characteristic length scale for variations in the reactant concentration near the pellet surface must be much smaller than the pellet radius (i.e., gradients in the concentration must become increasingly large as $\varepsilon \to 0$). Just as in the previous problem, one way to think about the problem in the limit as $\varepsilon \to 0$ is that a new length scale is developed that has just the "correct" value to ensure that the essential diffusion mechanism is not lost as the reaction goes faster and faster. As a consequence,

F. Diffusion in a Sphere with Fast Reaction – "Singular Perturbation Theory"

there is a so-called "concentration boundary layer" for the reactant near the outer surface of the pellet, with dimensions that decrease with ε at just the rate necessary to retain a balance between the diffusion and reaction processes in the limit as $\varepsilon \to 0$.

To determine the appropriate scaling, and the correct form for the governing equation in this boundary layer region, we use the idea of rescaling that was introduced in Section A. First, for convenience, we follow the analysis from that section, and redefine the dimensionless radial variable as

$$\bar{y} \equiv 1 - \bar{r} \qquad (4\text{--}165)$$

so that the surface of the pellet is $\bar{y} = 0$, and $\bar{y}$ increases as we go toward the center of the pellet. The variable $\bar{y}$ is scaled by use of the pellet radius R, the same as $\bar{r}$. It is also convenient to change the symbolic representation of the dimensionless concentration from $\bar{c}$ to $\bar{C}$. This is simply so that we can distinguish the solution that we will obtain for the concentration in the boundary layer from that in the interior core of the catalyst pellet. The governing equation, written now in terms of $\bar{y}$ and $\bar{C}$, is

$$\varepsilon \frac{1}{(1-\bar{y})^2} \frac{\partial}{\partial \bar{y}} \left[(1-\bar{y})^2 \frac{\partial \bar{C}}{\partial \bar{y}} \right] - \bar{C} = 0. \qquad (4\text{--}166)$$

To determine the appropriate characteristic length scale ℓ^* for the boundary-layer region, we introduce a new dimensionless variable:

$$Y \equiv \bar{y}\varepsilon^{-m}. \qquad (4\text{--}167)$$

The change of variable from $\bar{y}$ to Y is equivalent to scaling the dimensional radial variable $y \equiv R - r$ with the characteristic length scale $\ell^* = R\varepsilon^m$, i.e.,

$$Y = \frac{y}{\ell^*} = \frac{\bar{y}}{\varepsilon^m}.$$

The question now is what value to choose for the scaling coefficient m. To determine this, we substitute the "rescaled" variable $Y \equiv \bar{y}\varepsilon^{-m}$ into (4–166). The result is

$$\varepsilon^{1-2m} \frac{\partial^2 \bar{C}}{\partial Y^2} - \frac{2\varepsilon^{1-m}}{(1-\varepsilon^m Y)} \frac{\partial \bar{C}}{\partial Y} - \bar{C} = 0. \qquad (4\text{--}168)$$

Now, the largest term involving ε is the second derivative term. In the boundary layer, it is essential to retain diffusion, as previously noted, and this term must therefore remain of the same order as the reaction term even as $\varepsilon \to 0$. It follows that

$$m = 1/2. \qquad (4\text{--}169)$$

The characteristic length scale in the boundary layer is therefore

$$\ell^* = \varepsilon^{1/2} R \ll \ell_c \quad \text{as } \varepsilon \to 0.$$

With this choice for ℓ^*, the governing equation in the boundary-layer region takes the form

$$\frac{\partial^2 \bar{C}}{\partial Y^2} - \bar{C} - 2\varepsilon^{1/2}[1 + \varepsilon^{1/2} + O(\varepsilon)] \frac{\partial \bar{C}}{\partial Y} = 0. \qquad (4\text{--}170)$$

The boundary condition at the pellet surface is

$$\bar{C} = 1 \quad \text{at } Y = 0. \qquad (4\text{--}171)$$

An Introduction to Asymptotic Approximations

In addition, the solution in the boundary-layer region must match with the solution of (4–158) that is valid in the interior core of the pellet. This *matching condition* can be expressed in the form

$$\lim_{Y \gg 1} \overline{C} \Leftrightarrow \lim_{y \ll 1} \overline{c} \text{ in the limit as } \varepsilon \to 0. \qquad (4\text{--}172)$$

In terms of the boundary-layer variable Y, the outer edge of the boundary-layer corresponds to a very large value of Y, but, as we can see from (4–167), this corresponds to an arbitrarily small value of the core variable y. The symbolic representation for matching is the double-ended arrow $\Leftrightarrow$. This is intended to distinguish it from a numerical equal sign.

Because $\overline{C}$ and $\overline{c}$ are solutions of the same basic equation, (4–154), just scaled in different ways in the interior core and boundary-layer regions for $\varepsilon \ll 1$, they must take *the same functional form* in the transition from one region to the other. This is the meaning of the matching condition (4–172). We should carefully distinguish between "matching", which requires that the solutions take the same functional form in any region of overlapping validity for the two approximate forms of the solution that will be derived from (4–158) and (4–170), from the notion of "patching," which requires that they have the same numerical value at a fixed point in space. "Matching" is a much more powerful condition, requiring that the value of the function and all of its derivatives should have the same value at any given level of approximation. The procedure for applying the matching condition (4–172) will become clearer when we use it in the course of solving for the concentration profiles in the two regions of the solution domain. For now, we may note that a connection between the solution in the boundary layer and the solution in the interior core region is necessary to uniquely determine both solutions. The governing equation in the boundary layer is a second-order ODE, and thus requires two "boundary" conditions to be uniquely determined. In the original problem, one of these conditions is the value of the concentration at the surface of the pellet, and this still applies to the solution in the boundary-layer region. However, the second condition in the original problem is that the solution remains bounded in the limit as $r \to 0$. This condition can no longer be applied for the boundary-layer solution because the latter is valid in only a thin region near the surface of the pellet. Instead, we require that the boundary-layer solution match the solution in the interior core region according to the condition (4–172).

In the interior region of the pellet, we therefore seek an asymptotic expansion in the form

$$\overline{c} = f_0(\varepsilon)\overline{c}_0 + f_1(\varepsilon)\overline{c}_1 + O(f_2(\varepsilon)). \qquad (4\text{--}173)$$

Unlike the case of a regular perturbation expansion, we cannot know the gauge functions *a priori*, and these must be determined as a part of the solution. To obtain governing equations for the unknown concentration functions $\overline{c}_0$ and $\overline{c}_1$, we substitute (4–173) into the governing equation, (4–158) (remembering that $Da \equiv \varepsilon^{-1}$). Now, recalling from Section B that

$$f_{n+1}(\varepsilon)/f_n(\varepsilon) \to 0 \quad \text{as } \varepsilon \to 0,$$

we see that the governing equation for $\overline{c}_0$ is

$$\overline{c}_0 = 0,$$

regardless of what f_0 turned out to be. In fact, the present problem is quite unusual as all terms in (4–173) turn out to be zero, and we can see that this must be true without having to consider the solution in the boundary-layer region at all. Because the first term in (4–173) is simply 0, the governing equation from (4–158) for the next term is also

$$\overline{c}_1 = 0,$$

F. Diffusion in a Sphere with Fast Reaction – "Singular Perturbation Theory"

independent of f_1, and so on through all terms in the expansion (4–173). This situation is unusual in the sense that one can usually determine the asymptotic expansions in the various parts of the solution domain only one term at a time. The solution at $O(f_1)$, for example, could normally only be determined only after the solution at lower order had been obtained *throughout* the solution domain. In the present case, we can deduce that

$$\bar{c} = 0 \qquad (4\text{–}174)$$

to all orders of approximation in ε. As we shall see, this makes the matching with the solution in the boundary-layer region trivial.

In the boundary-layer region, we also seek a solution in the form of an asymptotic expansion:

$$\overline{C} = \overline{C}_0 + F_1(\varepsilon)\overline{C}_1 + O(F_2). \qquad (4\text{–}175)$$

Here the gauge function for the first term must be independent of ε because $\overline{C} = 1$ at the boundary $Y = 0$. Note again that asymptotic convergence requires that

$$F_{n+1}/F_n \to 0 \quad \text{as } \varepsilon \to 0. \qquad (4\text{–}176)$$

Substituting (4–175) into (4–170), we see that the governing equation for $\overline{C}_0$ is

$$\frac{d^2 \overline{C}_0}{dY^2} - \overline{C}_0 = 0, \qquad (4\text{–}177)$$

which is to be solved with the boundary condition [which we obtain by substituting (4–175) into (4–171)],

$$\overline{C}_0 = 1 \quad \text{at } Y = 0, \qquad (4\text{–}178)$$

and the matching condition (4–172), which, in view of (4–174), becomes

$$\lim_{Y \gg 1} \overline{C} \Leftrightarrow 0 \text{ in the limit as } \varepsilon \to 0 \qquad (4\text{–}179)$$

to all orders of approximation in ε. It will be noted that the diffusion equation reduces, at this level of approximation, to a "Cartesian" form as though the diffusion process were taking place in a plane geometry rather than inside a sphere. This is because the dimension of the boundary layer is vanishingly thin relative to the particle radius in the limit as $\varepsilon \to 0$. Hence the "curvature" terms that appear when the diffusion equation is written in the curvilinear spherical coordinate system drop out in the boundary-layer region at this first level of approximation. Equation (4–177) has the general solution

$$\overline{C}_0 = A_0 e^Y + B_0 e^{-Y}. \qquad (4\text{–}180)$$

In view of the matching condition, (4–179),

$$A_0 = 0,$$

and the boundary condition, (4–178), then leads to the solution

$$\overline{C}_0 = e^{-Y}. \qquad (4\text{–}181)$$

Reexpressed in terms of dimensional variables, this is

$$c_0 = C_\infty \exp\left[-\left(1 - \frac{r}{R}\right)\varepsilon^{-1/2}\right] = C_\infty \exp\left[-\left(1 - \frac{r}{R}\right)\left(\frac{k_1^{eff} R^2}{D_{eff}}\right)^{1/2}\right]. \qquad (4\text{–}182)$$

The reaction is "fast" in the sense that the time scale for the reaction $(k_1^{eff})^{-1}$ is small compared with the time scale for transport into the core of the pellet, R^2/D_{eff}. Hence we see from (4–182) that the reactant is confined to a very thin region near the surface of the pellet. Presumably, if we were going to design a catalyst pellet for this particular reaction, we would want to make it much smaller so that it was more uniformly accessed by the reactant.

Of course, the solution (4–181) is only the first approximation in the asymptotic series (4–175). In writing (4–177), we neglected certain smaller terms in the nondimensionalized equation, (4–170), because they were small compared with the terms that we kept. To obtain the governing equation for the second term in the boundary-layer region, we formally substitute the expansion, (4–175), into the governing equation, (4–170):

$$\frac{\partial^2 \overline{C}_0}{\partial Y^2} - \overline{C}_0 + F_1(\varepsilon)\left(\frac{\partial^2 \overline{C}_1}{\partial Y^2} - \overline{C}_1\right) - 2\varepsilon^{1/2}\frac{\partial \overline{C}_0}{\partial Y} + O(\varepsilon^{1/2} F_1, \varepsilon) = 0. \quad (4\text{–}183)$$

The terms at $O(1)$ lead to the governing equation, (4–177), for $\overline{C}_0$. It can also be seen that $F_1(\varepsilon) = \varepsilon^{1/2}$. If F_1 were larger than $\varepsilon^{1/2}$, then $\overline{C}_1$ would satisfy the same equation, (4–176), as $\overline{C}_0$. However, if we substitute the expansion, (4–175), into the boundary condition at $Y = 0$,

$$\overline{C} = \overline{C}_0 + F_1(\varepsilon)\overline{C}_1 + O(F_2(\varepsilon)) = 1 \quad \text{at } Y = 0,$$

and we see that

$$\overline{C}_1 = 0 \quad \text{at } Y = 0 \quad (4\text{–}184)$$

for any choice for $F_1(\varepsilon)$. Because the matching condition (4–179) also requires that

$$\overline{C}_1 \to 0 \quad \text{for } Y \gg 1, \quad (4\text{–}185)$$

the result for $F_1 > O(\varepsilon^{1/2})$ would be the trivial solution $\overline{C}_1 = 0$. On the other hand, if $F_1 < O(\varepsilon^{1/2})$, Eq. (4–183), would reduce be $\partial \overline{C}_0 / \partial Y = 0$, and this is not satisfied by the solution (4–181) for $\overline{C}_0$. Hence $F_1(\varepsilon) = \varepsilon^{1/2}$ as previously stated, and the governing equation for $\overline{C}_1$ is

$$\left(\frac{\partial^2 \overline{C}_1}{\partial Y^2} - \overline{C}_1\right) - 2\frac{\partial \overline{C}_0}{\partial Y} = 0. \quad (4\text{–}186)$$

Substituting for $\overline{C}_0$, we obtain

$$\frac{\partial^2 \overline{C}_1}{\partial Y^2} - \overline{C}_1 = -2\exp(-Y). \quad (4\text{–}187)$$

The general solution of this equation is

$$\overline{C}_1 = Ye^{-Y} + A_1 e^Y + B_1 e^{-Y}. \quad (4\text{–}188)$$

The matching condition requires that

$$A_1 = 0,$$

and the boundary condition, (4–184), gives

$$B_1 = 0.$$

Hence the boundary-layer solution takes the form

$$\overline{C} = e^{-Y} + \varepsilon^{1/2} Y e^{-Y} + O(\varepsilon) = e^{-Y}[1 + \varepsilon^{1/2} Y + O(\varepsilon)]. \quad (4\text{–}189)$$

F. Diffusion in a Sphere with Fast Reaction – "Singular Perturbation Theory"

The second term in this solution represent the first effect of nonplanar geometry on the transport process. Because the cross-sectional area of the transport path decreases as we move inside the sphere, there is a slight increase in concentration for any fixed $Y > 0$.

Although we could use the same procedure to calculate more terms in the expansion, (4–175), there is little point in doing so, and we truncate the solution at this point.

Now, we noted at the beginning of this section that an important goal of determining the reactant concentration distribution within the pellet is to calculate the "effectiveness" factor η_{cat}. In terms of the reactant concentration, this can be expressed in the form

$$\eta_{cat} \equiv \frac{\int_0^R k_1^{eff} c \, 4\pi r^2 dr}{k_1^{eff} C_\infty (4/3)\pi R^3} = \frac{D_{eff}\left(\frac{dc}{dr}\right)_{r=R} 4\pi R^2}{k_1^{eff} C_\infty . (4/3)\pi R^3}. \qquad (4\text{–}190)$$

The numerator in the last term must equal that in the second term because at steady state the flux of reactant across the particle surface must equal the total rate of consumption of reactant within the pellet. If we convert the expression (4–190) into dimensionless variables, the result is

$$\eta_{cat} = 3\varepsilon(\partial \bar{c}/\partial \bar{r})_{\bar{r}=1}. \qquad (4\text{–}191)$$

Finally, for $\varepsilon \ll 1$, we have shown that

$$\left(\frac{\partial \bar{c}}{\partial \bar{r}}\right)_{\bar{r}=1} = -\left(\frac{\partial \bar{c}}{\partial \bar{y}}\right)_{\bar{y}=0} = -\frac{1}{\varepsilon^{1/2}}\left(\frac{d\bar{C}}{dY}\right)_{Y=0} = \frac{1}{\varepsilon^{1/2}}[1 - \varepsilon^{1/2} + O(\varepsilon)],$$

so that

$$\boxed{\eta_{cat} = 3\varepsilon^{1/2}[1 - \varepsilon^{1/2} + O(\varepsilon)].} \qquad (4\text{–}192)$$

We see that the effectiveness factor decreases as $\varepsilon^{1/2}$ in the limit as $\varepsilon \to 0$. This is an adverse result because it means that, as $\varepsilon \to 0$, a vanishingly small fraction of the catalytically active sites within the pellet are actually involved in the chemical reaction. If the effective reaction time scale is very small compared with the time scale for diffusion across the particle, we waste most of the (expensive) catalytic material within each pellet because the reactant is totally depleted over a very short distance, $O(\varepsilon^{1/2}R)$, and the remainder of the pellet is not actively involved in the conversion of reactant to new products. To improve the effectiveness factor, we must either use much smaller-size catalyst pellets or distribute the catalyst sites in a radially nonuniform manner so that k_1^{eff} becomes a function of r.

Before this section is concluded, there is one key point that should be emphasized. The ultimate goal of the calculation was to calculate the effectiveness factor η_{cat}. In the present case, this came down to solving Eq. (4–170) for the concentration distribution in the boundary-layer near the particle surface. Because of all the assumptions that we made in the original problem setup, Eq. (4–170) is simple and easy to solve. However, even if the equation we achieved in the boundary-layer region had been much more complicated so that we could not solve it, the process of setting up the asymptotic framework, by means of nondimensionalization and rescaling, provides most of the important information about η_{cat}, and would do so even if we had not been able to solve (4–170). If we go back to the definition of η_{cat} in Eq. (4–191), we see that the effectiveness factor depends on $(d\bar{c}/d\bar{r})_{\bar{r}=1}$. However, we see from the rescaling process that

$$\left(\frac{\partial \bar{c}}{\partial \bar{r}}\right)_{\bar{r}=1} = -\left(\frac{\partial \bar{c}}{\partial \bar{y}}\right)_{\bar{y}=0} = -\varepsilon^{-1/2}\left(\frac{\partial \bar{c}}{\partial Y}\right)_{Y=0} = -Da^{1/2}\left(\frac{\partial \bar{c}}{\partial Y}\right)_{Y=0},$$

where $(\partial \bar{c}/\partial Y)_{Y=0}$ is a number that is guaranteed by the scaling to be independent of ε or Da. Hence,

$$\eta_{cat} = 3\varepsilon(\partial \bar{c}/\partial \bar{r})_{\bar{r}=1} = \frac{3}{Da} Da^{1/2}(-\partial \bar{c}/\partial Y)_{Y=0} = \frac{3}{Da^{1/2}}(-\partial \bar{c}/\partial Y)_{Y=0},$$

and we know everything about η_{cat} apart from the numerical coefficient $(-\partial \bar{c}/\partial Y)_{Y=0}$ without solving any DEs. This rather simple example problem thus provides a first glimpse of the power of the asymptotic approach, which frequently allows us to deduce the scaling behavior up to numerical constants without actually needing to solve the problem [i.e., to solve the governing DE(s)].

G. BUBBLE DYNAMICS IN A QUIESCENT FLUID

We conclude this chapter with an important *one dimensional* flow problem, namely the motion induced by changes in the volume of a single bubble that is suspended in an incompressible fluid that is at rest at infinity. If the bubble is *spherical*, as is usually assumed, the motion of the fluid is strictly radial, that is, $u_r \neq 0$, $u_\theta = u_\phi = 0$; hence the flow is 1D but not unidirectional.

This problem has a long history, beginning with important contributions of Lord Rayleigh.[14] In general, contemporary interest in the problem is motivated largely by the sound produced by bubbles that undergo a time-dependent change of volume, particularly in the context of *cavitation bubbles*[15] that are produced in hydromachinery or in the propulsion systems of submarines and other underwater vehicles[16] and by the phenomenon of *sonoluminescence*.[17] The latter is literally the production of a brief flash of light from a collapsing bubble of certain gases that is due to the enormous pressures and temperatures that can be generated by the surface-tension-driven collapse of a bubble. The objective of theoretical analyses for single-bubble motions is usually to predict the bubble radius as a function of time in a prescribed ambient pressure field. Once we know the bubble radius as a function of time, we can determine the velocity and pressure fields throughout the liquid, as well as evaluate the sound produced. If appropriate account is taken of the thermal problem, and a model for sonoluminescence is available, the light produced could also be evaluated.[18] Another important question is whether a real bubble remains spherical or becomes nonspherical as a result of instability of the pulsating spherical bubble.

In this section, we consider these problems in some detail, although with the major simplifications of assuming that the processes are isothermal and that the liquid is incompressible. As we shall see, the governing equations for even this simplified 1D problem are nonlinear, and thus most features can be exposed only by either numerical or asymptotic techniques. In fact, the problem of single-bubble motion in a time-dependent pressure field turns out to be not only practically important, but also an ideal vehicle for illustrating a number of different asymptotic techniques, as well as introducing some concepts of stability theory. It is for this reason that the problem appears in this chapter.

We begin by deriving the well-known *Rayleigh–Plesset* equation, which governs the time dependence of the bubble radius in a prescribed ambient pressure field, subject to the assumption that the bubble remains strictly spherical. We then analyze some of the dynamical behavior predicted by the Rayleigh–Plesset equation. We begin by using a *regular perturbation framework* to consider the *stability* of equilibrium solutions when the bubble is subjected to a perturbation of the pressure. We then consider the response of the bubble to periodic fluctuations of the pressure, including the possibility of resonance when the driving frequency matches the natural oscillation frequency of the spherical bubble. This leads us to introduce a new asymptotic technique known as the method of *multiple-time-scale perturbations* to analyze the bubble response at resonance. Finally, in the last section, we

G. Bubble Dynamics in a Quiescent Fluid

revisit the idea of *domain perturbation techniques* to consider the stability of the oscillating spherical bubble to nonspherical disturbances in its shape.

1. The Rayleigh–Plesset Equation

We consider a single bubble that undergoes a time-dependent change of volume in an incompressible, Newtonian fluid that is at rest at infinity. The bubble may either be a vapor bubble (that is, contains the vapor of the liquid) or it may contain a "contaminant" gas that is insoluble in the liquid (or at least dissolves only very slowly compared with the time scales associated with changes in the bubble volume) or a combination of vapor and contaminant gases. We assume, at the outset, that the bubble remains strictly spherical, and thus that the bubble surface moves only in the radial direction. It follows that the motion induced in the liquid must also be radial, so that

$$u_r = u_r(r, t), \quad u_\theta = u_\phi = 0 \qquad (4\text{–}193)$$

when expressed in terms of velocity components in a spherical coordinate system. Thus, by the assumption of a spherical shape, we implicitly assume that the fluid motion will be 1D.

The assumption of a 1D radial flow leads to a great simplification of the fluid mechanics problem. Indeed, the continuity equation, corresponding to (4–193), takes the very simple form

$$\frac{1}{r^2}\frac{\partial}{\partial r}(r^2 u_r) = 0, \qquad (4\text{–}194)$$

and we see, by integrating, that u_r is completely determined to within a single unknown function of time by the requirement of mass conservation, that is,

$$\boxed{u_r(r, t) = \frac{H(t)}{r^2}.} \qquad (4\text{–}195)$$

From a physical viewpoint, this form for u_r ensures that the total mass flux is the same across any closed surface that encloses the bubble.

The time-dependent function $H(t)$ is determined by the rate of increase or decrease in the bubble volume. The governing equations and boundary conditions that remain to be satisfied are (1) the radial component of the Navier–Stokes equation; (2) the kinematic condition, in the form of Eq. (2–129), at the bubble surface; and (3) the normal-stress balance, (2–135), at the bubble surface with $\hat{\tau} \equiv 0$. Generally, for a gas bubble, the zero-shear-stress condition also must be satisfied at the bubble surface, but $\tau_{r\theta} \equiv 0$, for a purely radial velocity field of the form (4–193), and this condition thus provides no useful information for the present problem.

The relationship between $H(t)$ and the bubble radius $R(t)$ is determined from the kinematic boundary condition. In particular, for a bubble containing only an insoluble gas, the kinematic condition takes the form

$$u_r(R, t) = \frac{dR}{dt}. \qquad (4\text{–}196)$$

For the case of a vapor bubble containing a combination of vapor and insoluble gas, there will be a net mass flux across the interface as the liquid vaporizes or the vapor condenses, and this means that the condition (4–196) is not exactly true. In the case of a pure vapor bubble,

$$u_r(R, t) = \frac{dR}{dt}\left(1 - \frac{\rho_v}{\rho}\right),$$

where ρ_v is the saturation vapor density at the bubble temperature and ρ is the liquid density Hence, because $\rho_v/\rho \ll 1$ in most cases, it follows that the relationship (4–196) can be used as a good approximation for both gas and vapor bubbles. Thus, when (4–195) and (4–196) are combined, it follows that

$$H(t) = R^2 \frac{dR}{dt}. \qquad (4\text{–}197)$$

Finally, to determine $H(t)$ or, equivalently, $R(t)$ as a function of time, we must consider the radial component of the Navier–Stokes equation, that is,

$$\frac{\partial u_r}{\partial t} + u_r \frac{\partial u_r}{\partial r} = -\frac{1}{\rho}\frac{\partial p}{\partial r} + \nu \left[\frac{1}{r^2}\frac{\partial}{\partial r}\left(r^2 \frac{\partial u_r}{\partial r}\right) - \frac{2u_r}{r^2}\right]. \qquad (4\text{–}198)$$

We purposely maintain this equation in dimensional form for the time being. Now, substituting (4–195) into (4–198) and carrying out the indicated differentiation, we obtain a DE for $H(t)$:

$$-\frac{1}{\rho}\frac{\partial p}{\partial r} = \frac{1}{r^2}\frac{\partial H}{\partial t} - \frac{2H^2}{r^5}. \qquad (4\text{–}199)$$

Thus, integrating with respect to r, we obtain

$$\frac{1}{\rho}[p(r,t) - p_\infty(t)] = \frac{1}{r}\frac{\partial H}{\partial t} - \frac{H^2}{2r^4}, \qquad (4\text{–}200)$$

where $p_\infty(t)$ is the ambient pressure in the fluid at a large distance from the bubble. Finally, if we evaluate (4–200) at the bubble surface $r = R(t)$ and express it in terms of R by using (4–197), we obtain

$$R\ddot{R} + \frac{3}{2}(\dot{R})^2 = \frac{1}{\rho}[p(R,t) - p_\infty(t)], \qquad (4\text{–}201)$$

where $p(R,t)$ is the pressure at the bubble surface in the liquid. The over-dots represent time derivates. This gives a direct relationship between the pressure in the liquid and the rate of change of the bubble radius.

A more useful alternative is to express (4–201) in terms of the pressure $p_B(t)$ inside the bubble rather than $p(R,t)$. To do this, we must use the normal-stress balance, (2–135), at the bubble surface, which takes the form

$$-p(R,t) + 2\mu \frac{\partial u_r}{\partial r}\bigg|_{r=R} + p_B(t) - \frac{2\gamma}{R} = 0. \qquad (4\text{–}202)$$

Here, we have assumed that the viscous stress in the gas is negligible and the pressure within the gas is uniform, p_B. The last term in (4–202) is the capillary pressure contribution that is due to surface tension γ. Evaluating $\partial u_r/\partial r|_{r=R}$ by using (4–195) and (4–197), we thus obtain

$$-p(R,t) - \frac{4\mu\dot{R}}{R} + p_B(t) - \frac{2\gamma}{R} = 0. \qquad (4\text{–}203)$$

G. Bubble Dynamics in a Quiescent Fluid

Thus, substituting (4–203) into (4–201), we obtain the famous Rayleigh–Plesset equation:

$$\frac{p_B(t) - p_\infty(t)}{\rho} - \frac{1}{\rho}\left(\frac{2\gamma}{R} + \frac{4\mu \dot{R}}{R}\right) = R\ddot{R} + \frac{3}{2}(\dot{R})^2. \qquad (4\text{–}204)$$

All that remains is a theory to calculate $p_B(t)$ in terms of $H(t)$. For this purpose, we assume

$$p_B = p_v + p_G, \qquad (4\text{–}205)$$

where p_v is the vapor pressure at the bubble temperature and p_G is the pressure contribution for a contaminant gas, which we model here as an ideal gas, so that

$$p_G = G\frac{T}{R^3}. \qquad (4\text{–}206)$$

In general, the temperatures in the gas and liquid both will change with time primarily because of the release (or use) of latent heat at the interface as vapor condenses (or liquid vaporizes). However, for present purposes, we neglect this effect and consider that $T = T_\infty =$ constant everywhere in the system. In terms of practical application, this is a severe assumption, and it is well known that thermal effects can have an important influence on the dynamics of gas/vapor bubbles. However, the isothermal problem retains many of the interesting qualitative features of the full problem and is sufficiently general for our present illustrative purposes. Thus we assume that

$$p_B(t) = p_v(T_\infty) + \frac{\widehat{G}}{R^3(t)}, \qquad (4\text{–}207)$$

where $p_v(T_\infty)$ and $\widehat{G}$ are both constants. The Rayleigh–Plesset equation then takes the form

$$\frac{p_v - p_\infty(t)}{\rho} + \frac{1}{\rho}\left(\frac{\widehat{G}}{R^3} - \frac{2\gamma}{R} - \frac{4\mu \dot{R}}{R}\right) = R\ddot{R} + \frac{3}{2}(\dot{R})^2. \qquad (4\text{–}208)$$

We shall discuss the solution of the Rayleigh–Plesset equation, (4–208), shortly.

Before discussing the solution of the Rayleigh–Plesset equation, a few comments may be useful to put the derivation leading to (4–208) into the context of earlier sections of this book (and this chapter). From a general perspective, the most important point is the distinction between a *unidirectional* and a 1D (but not unidirectional) flow. In particular, 1D flows are often characterized by a strong (or even dominant) contribution from the *inertial* (or nonlinear) terms in the equation of motion. For the special case considered here, the viscous terms in (4–198) are, in fact, identically zero, and the dynamics within the fluid is dominated by *Eulerian acceleration* of fluid elements associated with time-dependent changes in $R(t)$ and by *Lagrangian acceleration* that is due to the *divergence* of streamlines in the radial flow. Lagrangian accelerations are, of course, impossible in a unidirectional flow. In view of the importance of Lagrangian acceleration effects here, it is not surprising that the Rayleigh–Plesset equation is highly *nonlinear*.

As a consequence of this nonlinearity, it is impossible to obtain analytic solutions of the Rayleigh–Plesset equation for most problems of interest, in which $p_\infty(t)$ is specified and the bubble radius $R(t)$ is to be calculated. Indeed, most comprehensive studies of (4–208) have been carried out numerically. These show a richness of dynamic behavior that lies beyond the capabilities of analytic approximation. For example, a typical case might have $p_\infty(t)$ first decrease below $p_\infty(0)$ and then recover its initial value, as illustrated in Fig. 4–10. The bubble radius $R(t)$ first grows up to a maximum (which typically occurs *after* the minimum

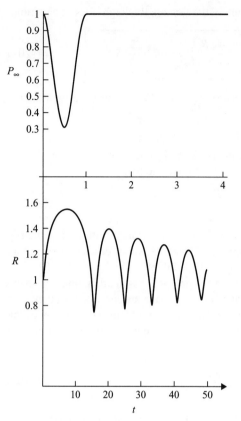

Figure 4–10. The response of a gas bubble to a perturbation of the ambient pressure $p_\infty(t)$. The trace of pressure $p_\infty(t)$ as a function of time is shown in the upper half of the figure, and the bubble radius is shown as a function of time in the lower half. Particularly noteworthy is the time scale for changes in radius compared with that of the pressure change (note that the scale is different in the two parts of the figure).

in p) but then undergoes collapse in a dramatic fashion to a minimum radius, followed by a succession of rebounds and collapses whose amplitude is attenuated by viscosity. It is this latter collapse sequence that is responsible for the sharp crackling sounds that can be heard in cavitating hydromachinery. The reader who is interested in a comprehensive review of the predicted behavior of gas and vapor bubbles in time-dependent pressure fields should consult the excellent paper of Plesset and Prosperetti.[19] Here, we concentrate on results that can be obtained analytically by *regular asymptotic approximations* that correspond to small changes in either bubble *volume* or *shape* relative to some case for which the solution is known (or obtained easily).

Before turning to these asymptotic theories, however, we conclude this subsection by considering one special case in which the Rayleigh–Plesset equation can be integrated exactly. This is the case of a step change in p_∞ from some initial value $p_\infty(0)$ to a different constant value p_∞^*. To integrate (4–208), we first multiply by $2R^2 \dot{R}$. Then every term in the resulting equation is an exact differential (and can be integrated directly) except for the viscous term, which cannot be integrated in this way. We thus consider the approximate case in which $\mu \equiv 0$. In spite of the neglect of viscous effects, the solution provides useful qualitative insight into some of the behavior that is inherent in the Rayleigh–Plesset equation. The result is

$$(\dot{R})^2 = \frac{2}{3} \frac{p_v - p_\infty^*}{\rho} \left(1 - \frac{R_0^3}{R^3}\right) - \frac{2\widehat{G}}{3\rho} \frac{1}{R^3} \ln\left(\frac{R_0^3}{R^3}\right) - \frac{2\gamma}{\rho R} \left(1 - \frac{R_0^2}{R^2}\right), \qquad (4\text{–}209)$$

where we have assumed that

$$R(0) = R_0$$

G. Bubble Dynamics in a Quiescent Fluid

and, for simplicity,

$$\dot{R}(0) = 0.$$

Now we consider the two cases of bubble growth and collapse. For bubble growth, $p_\infty^* < p_\infty(0)$, and the bubble will either reach a new steady equilibrium radius or else continue to grow until $R \gg R_0$. In the latter case,

$$\dot{R} \sim \left[\frac{2}{3} \frac{(p_v - p_\infty^*)}{\rho} \right]^{1/2}, \tag{4-210}$$

and the increase in the bubble radius asymptotes to a constant rate. This corresponds to a smooth increase in the bubble volume at a rate dV/dt, proportional to t^2. For $p_\infty^* > p_\infty(0)$, on the other hand, the bubble collapses, and if $R \ll R_0$, the radius decreases at a rate

$$\dot{R} \sim \left(\frac{R_0}{R} \right)^{3/2} \left[\frac{2}{3} \frac{p_\infty^* - p_v}{\rho} + 2\frac{\gamma}{\rho R_0} - \frac{2\widehat{G}}{\rho R_0^3} \ln \left(\frac{R_0}{R} \right) \right]^{1/2}. \tag{4-211}$$

If the bubble contains mostly insoluble gas, it is unlikely that the bubble will ever satisfy the inequality $R \ll R_0$. In other cases, however, (4–211) shows that the bubble radius will decrease at an ever-increasing rate until the last term finally balances the first two. This occurs at a minimum bubble size:

$$R_{min} = R_0 \exp\left[-\frac{R_0^3}{3\widehat{G}} \left(p_\infty^* - p_v + \frac{3\gamma}{R_0} \right) \right]. \tag{4-212}$$

The relatively mild growth and the violent collapse processes predicted by the asymptotic forms (4–210) and (4–211) are characteristic of the dynamics obtained by more general *numerical* studies of the Rayleigh–Plesset equation, but additional *results* for cases of large volume change are not possible by analytic solution. In the remainder of this section, we consider additional results that can be obtained by asymptotic methods.

2. Equilibrium Solutions and Their Stability

If we refer back to the Rayleigh–Plesset equation, (4–208), it is evident that a bubble in equilibrium must have a radius R_E that satisfies the condition

$$R_E^3 \left[p_v - p_\infty^{(0)} \right] - 2\gamma R_E^2 + \widehat{G} = 0. \tag{4-213}$$

Here, $p_\infty^{(0)}$ is a constant, ambient pressure and p_v is the vapor pressure at the ambient temperature.

Now, if we solve this cubic equation for R_E, we find that there are *three real roots* when $\Delta p \equiv p_v - p_\infty^{(0)}$ is less than a critical value,

$$\boxed{(\Delta p)_{crit} = \left(\frac{32\gamma^3}{27\widehat{G}} \right)^{1/2},} \tag{4-214}$$

but no positive real roots for larger values of Δp. Two of the roots for R_E are positive and one is negative (thus meaningless in this context). Thus, if we plot R_E versus $[p_v - p_\infty^{(0)}]$, we obtain the result sketched in Fig. 4–11. In the language of nonlinear systems, there is a *limit point* at Δp_{crit}, with two steady solutions for smaller values of Δp and no solutions

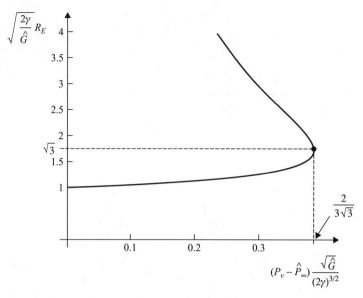

Figure 4–11. Equilibrium radius R_E as a function of the pressure driving force $(p_v - \hat{p}_\infty)$. For $(p_v - \hat{p}_\infty) < \Delta p_{crit}$ [Eq. (4–214)], there are two equilibrium radii possible. However, for $(p_v - \hat{p}_\infty) > \Delta p_{crit}$, no equilibrium radius exists, and the bubble must undergo time-dependent growth.

for $\Delta p > \Delta p_{crit}$. We can find the bubble radius at the limit point, which is denoted in Fig. 4–11 as R_{crit}, directly from (4–213) by calculating the derivative

$$\frac{\partial R_E}{\partial(\Delta p)} = \frac{R_E^2}{(4\gamma - 3R_E \Delta p)}$$

and determining the point where $\partial R/\partial(\Delta p) \to \infty$. Thus,

$$R_E|_{\partial R/\partial(\Delta p)\to\infty} = R_{crit} = \frac{4\gamma}{3\Delta p_{crit}}. \tag{4-215}$$

Substituting this value for R_E into (4–213) leads to the critical value of Δp given by (4–214). Finally, using the expression (4–214), we can express R_{crit} solely in terms of the material parameters $\widehat{G}$ and γ, namely,

$$\boxed{R_{crit} = \sqrt{3\widehat{G}/2\gamma}.} \tag{4-216}$$

Although two steady solutions exist for all $\Delta p \leq \Delta p_{crit}$, an important question is whether either of these solutions is *stable*. Thus, if we were to consider an arbitrarily small change in the bubble volume from one of the predicted equilibrium values, we ask whether the bubble radius will return to the equilibrium value or continue to either grow or collapse.[20] In the former case, the corresponding equilibrium state is said to be *stable* to infinitesimal perturbations in the bubble volume, while it is said to be *unstable* in the latter case. A steady, equilibrium solution that is unstable to infinitesimal perturbations will not be realizable in any real physical system because it is impossible to eliminate disturbances of arbitrarily small magnitude.

Now, the dynamics of changes in bubble radius with time, starting from some initial radius that differs slightly from an equilibrium value, is a problem that is ideally suited to solution by means of a *regular asymptotic approximation*. Of course, the governing equation is still the Rayleigh–Plesset equation. Before beginning our analysis, we follow

G. Bubble Dynamics in a Quiescent Fluid

the precedent of prior sections of this book and express the Rayleigh–Plesset equation in dimensionless form. For this purpose, we scale the bubble radius with the equilibrium value whose stability we seek to establish,

$$R(t) = R_E \tilde{r}(t),$$

and introduce a dimensionless pressure function $\Pi(t)$, ∋

$$p_\infty(t) = p_\infty^{(0)}[1 + \Pi(t)],$$

so that

$$\frac{p_v - p_\infty^{(0)}}{\rho} - \frac{p_\infty^{(0)}}{\rho}\Pi(t) + \frac{1}{\rho}\left(\frac{\hat{G}}{R_E^3 \tilde{r}^3} - \frac{2\gamma}{R_E \tilde{r}} - \frac{4\mu \dot{\tilde{r}}}{\tilde{r}}\right) = R_E^2 \left[\tilde{r}\ddot{\tilde{r}} + \frac{3}{2}(\dot{\tilde{r}})^2\right].$$

Finally, introducing $t = t_c \bar{t}$, we obtain

$$\frac{p_v - p_\infty^{(0)}}{\rho} - \frac{p_\infty^{(0)}}{\rho}\Pi(\bar{t}) + \frac{1}{\rho}\left(\frac{\hat{G}}{R_E^3 \tilde{r}^3} - \frac{2\gamma}{R_E \tilde{r}} - \frac{4\mu}{\tilde{r}}\frac{1}{t_c}\frac{d\tilde{r}}{d\bar{t}}\right) = \frac{R_E^2}{t_c^2}\left[\tilde{r}\frac{d^2\tilde{r}}{d\bar{t}^2} + \frac{3}{2}\left(\frac{d\tilde{r}}{d\bar{t}}\right)^2\right].$$

(4–217)

For this particular problem, in which we will consider a small perturbation in $\tilde{r}$ from the equilibrium value $\tilde{r} = 1$ for a constant far-field pressure $p_\infty^{(0)}$, an appropriate choice for t_c is

$$t_c = R_E \left(\frac{\rho}{\Delta p}\right)^{1/2},$$

where $\Delta p \equiv p_v - p_\infty^{(0)}$. With this choice, the Rayleigh–Plesset equation becomes

$$\boxed{\tilde{r}\frac{d^2\tilde{r}}{d\bar{t}^2} + \frac{3}{2}\left(\frac{d\tilde{r}}{d\bar{t}}\right)^2 = 1 - \alpha\,\Pi(\bar{t}) + \frac{\beta}{\tilde{r}^3} - \frac{2}{We}\frac{1}{\tilde{r}} - \frac{4}{Re}\frac{1}{\tilde{r}}\frac{d\tilde{r}}{d\bar{t}},} \qquad (4\text{–}218)$$

where

$$\alpha \equiv \frac{p_\infty^{(0)}}{\Delta p}, \quad \beta \equiv \frac{\hat{G}}{R_E^3 \Delta p};$$

$$We \equiv \frac{R_E \Delta p}{\gamma} = \frac{\rho R_E u_c}{\gamma} \quad \text{(Weber number)};$$

$$Re \equiv \frac{\rho^{1/2} R_E (\Delta p)^{1/2}}{\mu} = \frac{\rho R_E u_c}{\mu} \quad \text{(Reynolds number)};$$

$$u_c \equiv \frac{R_E}{t_c} = \left(\frac{\Delta p}{\rho}\right)^{1/2}.$$

In the analysis of stability that follows, we seek an approximate asymptotic solution of this equation in which the small parameter is the dimensionless *magnitude* of the initial departure from the equilibrium radius R_E. Hence we seek a solution in the form of a regular perturbation expansion:

$$\boxed{\tilde{r} = 1 + \sum_{n=1}^{N} \varepsilon^n \delta_n(\bar{t}),} \qquad (4\text{–}219)$$

where ε is a small parameter that provides a measure of the magnitude of the initial change in radius from R_E. Here, we consider the asymptotic limit $\varepsilon \to 0$. We can obtain the governing equations for the functions $\delta_n(t)$ by substituting (4–119) into the Rayleigh–Plesset equation, (4–218), and collecting all terms of equal order in ε. The result is

$$1 + \beta - \frac{2}{We} + \varepsilon \left[-\ddot{\delta}_1 - \frac{4}{Re}\dot{\delta}_1 - \left(3\beta - \frac{2}{We}\right)\delta_1 \right]$$
$$+ \varepsilon^2 \left[-\ddot{\delta}_2 - \frac{4}{Re}\dot{\delta}_2 - \left(3\beta - \frac{2}{We}\right)\delta_2 + \left(6\beta - \frac{2}{We}\right)\delta_1^2 \right.$$
$$\left. + \left(\frac{4}{Re} - 1\right)\delta_1\dot{\delta}_1 - \frac{3}{2}\dot{\delta}_1^2 \right] + O(\varepsilon^3) = 0.$$

We note that $\Pi(t) \equiv 0$, by assumption in the present analysis. For arbitrary though small ε, the term at each level in ε is equal to zero, that is,

$$O(1), \quad 1 + \beta - \frac{2}{We} = 0, \tag{4-220a}$$

$$O(\varepsilon), \quad \ddot{\delta}_1 + \frac{4}{Re}\dot{\delta}_1 + \left(3\beta - \frac{2}{We}\right)\delta_1 = 0, \tag{4-220b}$$

$$O(\varepsilon^2), \quad \ddot{\delta}_2 + \frac{4}{Re}\dot{\delta}_2 + \left(3\beta - \frac{2}{We}\right)\delta_2 = \left(6\beta - \frac{2}{We}\right)\delta_1^2 + \left(\frac{4}{Re} - 1\right)\delta_1\dot{\delta}_1 - \frac{3}{2}\dot{\delta}_1^2. \tag{4-220c}$$

We assume that the other parameters that appear are independent of ε. Further, for convenience, we assume that

$$\delta_1(0) = 1, \delta_2(0) = \delta_3(0) = \cdots = \delta_n(0) = 0.$$

The equation at $O(1)$ is just the equilibrium condition, (4–213), now expressed in dimensionless terms. The equation at $O(\varepsilon)$ is

$$\boxed{\ddot{\delta}_1(t) + \frac{4}{Re}\dot{\delta}_1(t) + \left(3\beta - \frac{2}{We}\right)\delta_1(t) = 0.} \tag{4-221}$$

In utilizing the formalism of regular perturbation theory to study stability problems, we normally restrict our attention to determining whether the magnitude of this first term, $\delta_1(t)$, grows or decays from its initial value $\delta_1(0)$ with time. The resulting theory is called *linear stability theory*. This is because the governing equation for $\delta_1(t)$ is identical to the equation that we would obtain by simply assuming that $\tilde{r}(t) = 1 + \varepsilon(t)$ with $[\varepsilon] \ll 1$, and then linearizing (4–219) by neglecting all terms that involve products of $\tilde{r}$ or its time derivatives. Now, to examine the stability question, let us suppose that

$$\delta_1(t) = \delta_0 e^{st}. \tag{4-222}$$

Then the steady solution is stable if the real part of s is negative, but unstable if it is positive. We obtain a characteristic equation for s by substituting (4–222) into (4–221). The result is

$$s^2 + \frac{4}{Re}s + \left(3\beta - \frac{2}{We}\right) = 0. \tag{4-223}$$

and the two roots of this equation are

$$s = -\frac{2}{Re}\left\{1 \pm \left[1 - \frac{Re^2}{4}\left(3\beta - \frac{2}{We}\right)\right]\right\}. \tag{4-224}$$

G. Bubble Dynamics in a Quiescent Fluid

Thus the condition for *instability* is

$$\left(3\beta - \frac{2}{We}\right) < 0,$$

or, on substitution for β and We,

$$\boxed{R_E > \sqrt{3\hat{G}/2\gamma}.} \qquad (4\text{--}225)$$

Referring back to Fig. 4–11 and Eq. (4–216), we see that this limiting value for stability is precisely the same as the critical value R_{crit}. Thus the lower part of the equilibrium solution curve (the lower branch of solutions) for

$$R_E \leq R_{crit}$$

is predicted to be *stable* to infinitesimal changes in the bubble radius, whereas the upper solution branch for

$$R_E > R_{crit}$$

is *unstable*.

Several comments are in order with regard to the preceding analysis. First, because the governing equation at $O(\varepsilon)$ is linear, it pertains equally to the case $\delta_1(0) < 0$ or $\delta_1(0) > 0$. Second, the initial magnitude of $\varepsilon\delta_1(0)$ is small but arbitrary. In particular, for $R_E > R_{crit}$, even an *infinitesimal* perturbation in the bubble volume leads to instability. Third, the governing equation, (4–221), describes only the leading-order (linearized) approximation to the departure from the steady equilibrium radius R_E. Thus, although the first term in the regular perturbation analysis can tell us whether an arbitrarily small change in the bubble radius will initially grow or decay, it cannot determine the ultimate fate of a growing disturbance because the underlying perturbation expansion will break down if the condition $\varepsilon\delta_1(t) \ll 1$ is not maintained. For example, if $\delta_1(0) < 0$ and $R_E > R_{crit}$, the linearized analysis tells us that the bubble radius initially decreases exponentially with time, but it cannot tell us whether the radius ultimately goes to the equilibrium value on the lower solution branch. Similarly, for $\delta_1(0) > 0$, the bubble radius initially increases exponentially from R_E for $R_E > R_{crit}$, but the eventual fate of the bubble cannot be determined from this analysis. Finally, we should point out that the stability condition, (4–225), expressed in terms of the bubble radius R, also provides a useful criterion for the critical pressure difference $(p_v - p_\infty)_{crit}$ that leads to instability. In particular, let us suppose that the bubble is subjected to an environment where the ambient pressure is changing slowly. Then, according to (4–213), as $p_v - \overline{p}_\infty$ increases, the radius of a stable bubble must also increase. However, according to (4–225), there is a maximum that the radius can achieve before the bubble becomes unstable. Thus there is a corresponding maximum in $(p_v - \hat{p}_\infty)$, which is just the critical value $(\Delta p)_{crit}$ given in (4–214):

$$(p_v - \hat{p}_\infty)_{max} = \frac{4}{3}\gamma\left(\frac{2\gamma}{3\hat{G}}\right)^{1/2}.$$

If $\hat{p}_\infty$ decreases below this value, we should expect exponential bubble growth.

For the reader who is knowledgeable about the dynamics of nonlinear systems, it will not be surprising that the stability analysis predicts an exchange of stability for steady solutions of (4–213) as we pass through the limit point from the lower to the upper solution branch. This behavior is expected on general grounds from the theories of nonlinear dynamical systems (see Iooss and Joseph).[21]

An Introduction to Asymptotic Approximations

3. Bubble Oscillations Due to Periodic Pressure Oscillations – Resonance and Multiple-Time-Scale-Analysis

Another important characteristic of the gas bubble is its response to a periodic oscillation of the ambient pressure p_∞. For large-amplitude oscillations of the pressure, or for an initial condition that is not near a stable equilibrium state for the bubble, the response can be very complicated, including the possibility of *chaotic* variations in the bubble radius.[22] However, such features are outside the realm of simple, analytical solutions of the governing equations, and we focus our attention here on the bubble response to asymptotically small oscillations of the ambient pressure, namely,

$$p_\infty(t) = p_\infty^0(1 + \varepsilon \sin \omega t), \qquad (4\text{--}226)$$

where $\varepsilon \ll 1$ and the frequency ω is arbitrary. Furthermore, we restrict our attention to systems that deviate only slightly from a stable equilibrium radius R_E. This implies that the initial state must lie within a small neighborhood of the equilibrium point.

The basis of our analysis is, again, a very simple variation of the regular perturbation technique in which we assume that the bubble radius can be expressed in terms of a regular asymptotic expansion of the form

$$R = R_E\left[1 + \sum_{n=1}^N \varepsilon^n g_n(t)\right]. \qquad (4\text{--}227)$$

Inherent in this form for R is the assumption that the $O(\varepsilon)$ forcing in the ambient pressure leads to an $O(\varepsilon)$ change in the bubble radius, at least at the leading order of approximation.

The governing equations for the coefficient functions $g_n(t)$ are obtained by the standard perturbation procedure of substituting expansion (4–227) into the Rayleigh–Plesset equation in the form (4–208) and then equating the sum of all terms at each order in ε to zero. The result at $O(1)$, expressed in dimensional terms, is

$$p_v + \frac{\widehat{G}}{R_E^3} - p_\infty^0 - \frac{2\gamma}{R_E} = 0.$$

At $O(\varepsilon)$, we obtain the governing equation for g_1, that is,

$$\ddot{g}_1 + \left(\frac{4\mu}{\rho R_E^2}\right)\dot{g}_1 + \frac{1}{\rho}\left(\frac{3\widehat{G}}{R_E^5} - \frac{2\gamma}{R_E^3}\right)g_1 = -\frac{p_\infty^0}{\rho R_E^2}\sin\omega t.$$

This can be expressed in the generic form

$$\ddot{g}_1 + \sigma \dot{g}_1 + \omega_0^2 g_1 = \Delta \sin \omega t \qquad (4\text{--}228)$$

of a *forced harmonic oscillator* with damping factor

$$\sigma \equiv \frac{4\mu}{\rho R_E^2}$$

and a natural frequency

$$\omega_0^2 = \frac{1}{\rho}\left(\frac{3\widehat{G}}{R_E^5} - \frac{2\gamma}{R_E^3}\right).$$

G. Bubble Dynamics in a Quiescent Fluid

The response depends on the relationship between the forcing frequency ω and the natural frequency ω_0.

Before solutions of (4–228) are sought, it is advantageous to rewrite it in dimensionless form. We do this by introducing ω_0^{-1} as a characteristic time scale (note that g_1 is dimensionless by definition). Then,

$$\frac{d^2 g_1}{d\bar{t}^2} + \frac{4}{Re_\omega} \frac{dg_1}{d\bar{t}} + g_1 = -\overline{p}_\infty \sin[(\omega/\omega_0)\bar{t}], \tag{4-229}$$

where $\bar{t} = \omega_0 t$ and

$$\frac{4}{Re_\omega} \equiv \frac{4\mu}{\rho R_E^2 \omega_0} = \frac{\sigma}{\omega_0},$$

$$-\overline{p}_\infty = -\frac{p_\infty^0}{\rho(\omega_0 R_E)^2} = \frac{\Delta}{\omega_0^2}.$$

We begin by considering the solution of (4–229) for the case $\omega \neq \omega_0$. Let us first examine the limiting case of zero damping where $4/Re_\omega \equiv 0$, so that the governing equation is

$$\frac{d^2 g_1}{d\bar{t}^2} + g_1 = -\overline{p}_\infty \sin[(\omega/\omega_0)\bar{t}]. \tag{4-230}$$

We seek a solution of this equation subject to initial conditions

$$g_1(0) = \dot{g}_1(0) = 0, \tag{4-231}$$

which imply that

$$R(0) = R_E, \quad \dot{R}(0) = 0.$$

The general solution of (4–230) is

$$g_1(\bar{t}) = C_1 \sin \bar{t} + C_2 \cos \bar{t} - \frac{\overline{p}_\infty}{1 - (\omega/\omega_0)^2} \sin[(\omega/\omega_0)\bar{t}]. \tag{4-232}$$

Then, applying the boundary conditions, (4–231), we obtain

$$g_1(\bar{t}) = \frac{p_\infty^0}{1 - (\omega/\omega_0)^2} \{(\omega/\omega_0) \sin \bar{t} - \sin[(\omega/\omega_0)\bar{t}]\}. \tag{4-233}$$

Thus, in the absence of viscous damping, we find that the bubble radius oscillates periodically with an amplitude of $O(\varepsilon)$ in response to the oscillating pressure field, provided only that $\omega \neq \omega_0$, as we have assumed. A plot is given in Fig. 4–12 showing the time dependence of g_1/p_∞^0 for several different values of (ω/ω_0). A key point to note about the solution, (4–233), however, is that the magnitude of g_1 becomes *unbounded* in the limit $\omega \to \omega_0$. Indeed, in the limit $\omega = \omega_0$, no bounded solution of the asymptotic form, (4–227), exists. This is a consequence of the *resonant interaction* that occurs when the forcing frequency ω is equal to the natural oscillation frequency of the bubble, ω_0.

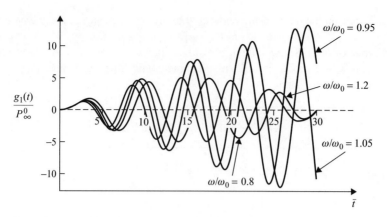

Figure 4–12. Amplitude function for the oscillation of bubble radius in an inviscid fluid that is due to a time-dependent oscillation in the ambient pressure.

A resonant effect is also evident even when the viscous damping term is included in Eq. (4–229). Again, we consider the case $w \ne w_0$. In this case, it is convenient to reformulate the problem in the form

$$\ddot{G}_1 + \frac{4}{Re_\omega}\dot{G}_1 + G_1 = -\overline{p}_\infty e^{i(\omega/\omega_0)\bar{t}}, \qquad (4\text{--}234)$$

where $g_1 = \mathcal{I}m(G_1)$. In this case, we can seek a solution in the form

$$G_1 = Ce^{s\bar{t}}. \qquad (4\text{--}235)$$

Substituting into (4–234) thus yields

$$Ce^{s\bar{t}}\left(s^2 + \frac{4}{Re_\omega}s + 1\right) = -\overline{p}_\infty e^{i(\omega/\omega_0)\bar{t}}. \qquad (4\text{--}236)$$

We thus obtain the particular solution by letting $s = i\omega/\omega_0$; then

$$C = \frac{-\overline{p}_\infty}{-\left(\dfrac{\omega}{\omega_0}\right)^2 + \dfrac{4}{Re_\omega}\left(\dfrac{\omega}{\omega_0}\right)i + 1}. \qquad (4\text{--}237)$$

We obtain homogeneous solutions by letting s be the two roots of the characteristic equation

$$s^2 + \frac{4}{Re_\omega}s + 1 = 0. \qquad (4\text{--}238)$$

These are

$$s = -\frac{2}{Re_\omega}\left[1 \pm \left(1 - \frac{Re_\omega^2}{4}\right)^{1/2}\right]. \qquad (4\text{--}239)$$

Thus the general solution of (4–234) is

$$G_1 = C_1 e^{s_1 \bar{t}} + C_2 e^{s_2 \bar{t}} + \frac{-\overline{p}_\infty}{-\left(\dfrac{\omega}{\omega_0}\right)^2 + \dfrac{4}{Re_\omega}\left(\dfrac{\omega}{\omega_0}\right)i + 1} e^{i(\omega/\omega_0)\bar{t}}, \qquad (4\text{--}240)$$

with the boundary conditions

$$G_1 = \dot{G}_1 = 0 \quad \text{at} \quad t = 0. \qquad (4\text{--}241)$$

There are two cases: When $Re_\omega^2/4 < 1$, both roots s_1 and s_2 are real and negative; when $Re_\omega^2/4 > 1$, s_1 and s_2 are complex conjugates with negative real parts. Hence, in both cases,

G. Bubble Dynamics in a Quiescent Fluid

the homogeneous terms in (4–240) vanish as $\bar{t} \to \infty$ for all nonzero values of σ. For present purposes, we restrict our attention to only the large-time asymptote of (4–240), which thus takes the form

$$G_1 = \frac{-\overline{p}_\infty}{\left[1 - \left(\frac{\omega}{\omega_0}\right)^2\right] + i\frac{4}{Re_\omega}\left(\frac{\omega}{\omega_0}\right)} e^{i(\omega/\omega_0)\bar{t}}. \tag{4–242}$$

The corresponding imaginary part of (G_1) is

$$g_1 = -\frac{\overline{p}_\infty}{\left[1 - \left(\frac{\omega}{\omega_0}\right)^2\right]^2 + \left(\frac{4}{Re_\omega}\frac{\omega}{\omega_0}\right)^2} \left\{\left[1 - \left(\frac{\omega}{\omega_0}\right)^2\right]\sin\frac{\omega}{\omega_0}\bar{t} \right.$$
$$\left. - \frac{4}{Re_\omega}\left(\frac{\omega}{\omega_0}\right)\cos\frac{\omega}{\omega_0}\bar{t}\right\}. \tag{4–243}$$

The solution $g_1/\overline{p}_\infty$ is plotted in Fig. 4–13 for several values of (ω/ω_0) and several values of Re_ω. It will be noted that (4–243) does not reduce to (4–233) in the limit as $Re_\omega^{-1} \to 0$. This is because the roots s_1 and s_2 have negative real parts for all $Re_\omega^{-1} \neq 0$, so that the homogeneous terms in (4–140) vanish for sufficiently large times but are purely imaginary for $Re_\omega^{-1} = 0$ ($s = \pm i$) and thus do not vanish even as $t \to \infty$. Indeed, the second of the two terms in (4–233) is a homogeneous solution, which has no counterpart in (4–243) for $Re_\omega^{-1} \to 0$. In the resonant limit $\omega \to \omega_0$, the solution, (4–243), approaches a maximum magnitude for g_1,

$$(g_1)_{max} \sim \frac{\overline{p}_\infty Re_\omega}{4} \cos\bar{t},$$

though this remains *bounded* for finite $Re_\omega^{-1} \sim O(1)$. In this case, the asymptotic expansion (4–227) remains perfectly well-behaved even as $w \to w_0$. If, on the other hand, $Re_\omega^{-1} \sim O(\varepsilon^p)$ – that is, the viscous damping is small and proportional to the magnitude of the pressure variation to some positive power p – the asymptotic scheme again breaks down in the resonant limit, and some different type of solution must be attempted.

The resonant case $\omega = \omega_0$ in the limit of weak damping is itself quite interesting and also amenable to asymptotic analysis. We have seen, in this case, that an approximation of the form (4–227) is not possible for either $Re_\omega^{-1} = 0$ or $Re_\omega^{-1} \neq O(\varepsilon^p)$ – that is, a response of $O(\varepsilon)$ does not occur in spite of the fact that the pressure variation is $O(\varepsilon)$ and $\varepsilon \ll 1$. An obvious question is whether the amplitude of changes in bubble radius actually remains bounded in the resonant case $\omega = \omega_0$ when $Re_\omega^{-1} \ll 1$. We explore this question briefly in the following pages for the *undamped* case $Re_\omega^{-1} = 0$. The problem for very small, but nonzero, damping is similar, but we shall not consider it here.

If we reexamine the governing equation, (4–230), for $Re_\omega^{-1} \equiv 0$, we see that the mathematical difficulty for $\omega = \omega_0$ is that the forcing function is actually a solution of the homogeneous equation. The forcing function in this case is known as a *secular* term. Obviously, when a secular forcing term is present, a particular solution cannot exist in the form $C \sin \bar{t}$ as is possible for $\omega \neq \omega_0$ – see, for example (4–232) – but must instead take the form $C\bar{t} \cos \bar{t}$. Indeed, substituting this form into the governing equation, (4–230), with $\omega = \omega_0$, we find

$$(g_1)_{particular} = \frac{\overline{p}_\infty}{2}\bar{t}\cos\bar{t} \quad (\omega \equiv \omega_0). \tag{4–244}$$

An Introduction to Asymptotic Approximations

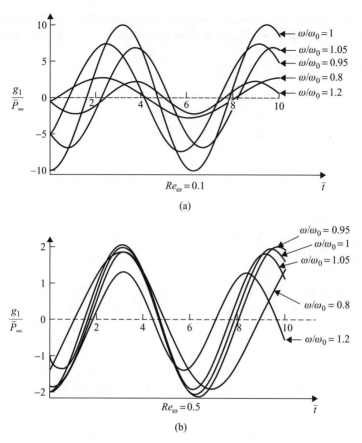

Figure 4–13. Amplitude function for the oscillation of bubble radius in a *viscous* fluid that is due to a time-dependent oscillation in the ambient pressure: (a) $Re_\omega = 0.1$, and (b) $Re_\omega = 0.5$.

Although this solution is perfectly well behaved for $\bar{t} \ll 1$, we see that it grows *without bound* for $\bar{t} \to \infty$. It thus follows that the asymptotic expansion, (4–227), cannot provide a valid asymptotic approximation scheme for $\varepsilon \ll 1$ for all $\bar{t}$, when $\omega = \omega_0$. The fact is that if a bounded solution is to exist for the variation of bubble radius that is due to small pressure oscillations at the resonant frequency, the $O(\varepsilon)$ pressure variation must eventually be balanced by one or more of the *nonlinear* terms that were neglected in deriving (4–230) from the limiting form of the Rayleigh–Plesset equation with $Re_\omega^{-1} \equiv 0$. In the analysis that follows, we show that the resonant response from the Rayleigh–Plesset equation for small ε is that the bubble radius oscillates with an amplitude of $O(\varepsilon^{1/3})$ – that is, it is still asymptotically small but *much larger* than the $O(\varepsilon)$ amplitude of the pressure forcing and much larger than assumed in the expansion, (4–227).

The following analysis for the resonant case is based on the observation that the response of a nonlinear oscillator to weak periodic forcing at a *resonant* frequency is usually characterized by *multiple time scales*, a slow, periodic oscillation superposed on a higher-frequency oscillation that is at (or near) the forcing frequency. A special kind of asymptotic approximation scheme has been developed for this kind of problem, which is known as a *two-timing or two-variable expansion procedure*. A complete description and motivation for the two-timing technique is beyond the scope of this book.[23] Nevertheless, it is perhaps useful to illustrate its application to this specific example problem by proceeding in a purely *formal* manner. Thus we simply note that the basis of the two-timing procedure is to consider the

G. Bubble Dynamics in a Quiescent Fluid

unknown function as depending on two *independent* variables: the regular time $\bar{t}$, which is of $O(1)$, and a second asymptotically longer time-scale variable τ that is related to $\bar{t}$ in the form

$$\tau = \varepsilon^m \bar{t}.$$

Hence any time derivative that appears in the governing equation is now approximated as

$$\frac{\partial}{\partial \bar{t}} \to \frac{\partial}{\partial \bar{t}} + \varepsilon^m \frac{\partial}{\partial \tau}.$$

With this general formalism, the solution for the present problem can be shown to exist in the specific asymptotic form

$$R = R_E \left[1 + \sum_n \varepsilon^{n/3} g_n(\bar{t}, \tau) \right], \qquad (4\text{–}244)$$

$$\text{where } \tau = \varepsilon^{2/3} \bar{t}. \qquad (4\text{–}245)$$

The governing equations for the functions g_n are obtained from the Rayleigh–Plesset equation in the usual manner. We thus substitute (4–244) into the inviscid form of the Rayleigh–Plesset equation (4–230) and collect terms of like powers in ε. The first several functions $g_n(\bar{t}, \tau)$ are found to satisfy the dimensionless equations

$$\ddot{g}_1 + g_1 = 0, \qquad (4\text{–}246)$$

$$\ddot{g}_2 + g_2 = -g_1 \ddot{g}_1 - \frac{3}{2}(\dot{g}_1)^2 + (1 + 3G^*) g_1^2, \qquad (4\text{–}247)$$

$$\ddot{g}_3 + g_3 = -2\frac{\partial^2 g_1}{\partial \bar{t} \partial \tau} - \bar{p}_\infty \sin \bar{t} - (g_1 \ddot{g}_2 + g_2 \ddot{g}_1 + 3\dot{g}_1 \dot{g}_2) \\ + (2 + 6G^*) g_1 g_2 - (1 + 7G^*) g_1^3. \qquad (4\text{–}248)$$

Here the derivatives with respect to $\bar{t}$ are represented as $\dot{g}_n$ and $\ddot{g}_n$, and we also introduce the shorthand notation

$$G^* \equiv \frac{G}{\rho \omega_0^2 R_E^5}.$$

Now, Eq. (4–246) is nothing more than the homogeneous equation obtained in our earlier analysis, and it might appear that nothing has been accomplished because the secular term will now appear in (4–248). However, it is at this point that the analysis departs from the preceding work. For in this case, we consider g_1 to depend not only on the independent variable $\bar{t}$, which appears explicitly in (4–246), but also on the slow time scale τ. Hence a convenient form for the general solution of (4–246) is

$$g_1 = A(\tau) \sin[\bar{t} - \phi(\tau)], \qquad (4\text{–}249)$$

where the slowly varying amplitude and phase functions $A(\tau)$ and $\phi(\tau)$ remain unspecified. However, we shall see that $A(\tau)$ and $\phi(\tau)$ can be chosen such a way that the secular terms that would otherwise appear on the right-hand side of (4–248) are canceled so that g_3 remains bounded for all $\bar{t}$. To see that this is possible, we must proceed with the solution of Eqs. (4–247) and (4–248).

Hence, using g_1 from (4–249), we find that Eq. (4–247) for g_2 takes the form

$$\ddot{g}_2 + g_2 = \frac{A^2}{4}(1 + 6G^*) - \frac{A^2}{4}(7 + 6G^*)\cos\{2[\bar{t} - \phi(\tau)]\}, \qquad (4\text{–}250)$$

and a general solution is

$$\boxed{g_2 = C(\tau)\sin[\bar{t} - \delta(\tau)] + \frac{A^2}{4}(1 + 6G^*) + \frac{A^2}{4}\left(\frac{7}{3} + 2G^*\right)\cos 2[\bar{t} - \phi].} \qquad (4\text{–}251)$$

Again, the long-time-scale functions $C(\tau)$ and $\delta(\tau)$ can be chosen to avoid secular behavior for one of the higher-order terms in the asymptotic expansion.

Finally, we consider the solution of (4–248) for the $O(\varepsilon)$ term in the expansion (4–244). The first thing to notice is that the right-hand side of (4–248) contains a secular term corresponding to the direct pressure forcing at $O(\varepsilon)$, and our immediate reaction may be that it will again be impossible to obtain a bounded solution for g_3. However, in this case, the other terms on the right-hand side of (4–248) involve the as yet undetermined functions $A(\tau)$, $\phi(\tau)$, $C(\tau)$, and $\delta(\tau)$. Clearly, if a bounded solution is to exist in the form (4–244), these functions must be chosen in such a way as to eliminate the secular behavior in the equation for g_3. To see that this is possible, we evaluate the right-hand side of (4–248) by using the general solutions, (4–249) and (4–251), for g_1 and g_2. The result is

$$\ddot{g}_3 + g_3 = -2\left\{\frac{dA}{d\tau}\cos[\bar{t} - \phi(\tau)] + A\frac{d\phi}{d\tau}\sin[\bar{t} - \phi(\tau)]\right\} - \overline{P}_\infty \sin\bar{t}$$
$$+ \left(\frac{A^3}{4}\right)\left[-\frac{7}{6} - 5G^* + 30(G^*)^2\right]\sin[\bar{t} - \phi(\tau)] \qquad (4\text{–}252)$$
$$+ \text{nonsecular terms.}$$

Thus, to avoid secular terms and therefore ensure bounded solutions for g_3, we must choose $A(\tau)$ and $\phi(\tau)$ so that the secular terms on the right-hand side of (4–252) are equal to zero. After some manipulation, this condition leads to the coupled pair of equations

$$\boxed{A\frac{d\phi}{d\tau} = -(\overline{P}_\infty)\frac{\cos\phi}{2} + \left(\frac{A^3}{8}\right)\left[-\frac{7}{6} - 5G^* + 30(G^*)^2\right],} \qquad (4\text{–}253)$$

$$\boxed{\frac{dA}{d\tau} = -(\overline{P}_\infty)\frac{\sin\phi}{2}.} \qquad (4\text{–}254)$$

Hence, provided solutions to these equations exist, we see that

$$\boxed{\begin{aligned} R &= R_E[1 + \varepsilon^{1/3}g_1 + O(\varepsilon^{2/3})], \\ \text{with } g_1 &= A(\tau)\sin[\omega_0\bar{t} - \phi(\tau)], \\ \tau &= \varepsilon^{2/3}\bar{t}, \end{aligned}} \qquad \begin{aligned} (4\text{–}255) \\ (4\text{–}256) \end{aligned}$$

where $A(\tau)$ and $\phi(\tau)$ satisfy (4–253) and (4–254). If we proceed to the next level of approximation to determine g_4, the functions $C(\tau)$ and $\delta(\tau)$ would be chosen to eliminate secular terms in that problem, and so on.

G. Bubble Dynamics in a Quiescent Fluid

To complete the analysis, we must obtain solutions for the amplitude and phase functions $A(\tau)$ and $\phi(\tau)$ from (4–253) and (4–254). We accomplish this by first rescaling time according to

$$t^* = \tau \left(\frac{\overline{p}_\infty}{2}\right),$$

so that (4–253) and (4–254) can be expressed in the simpler form

$$A\frac{d\phi}{dt^*} = -\cos\phi + C\, A^3, \qquad (4\text{--}257)$$

$$\frac{dA}{dt^*} = -\sin\phi, \qquad (4\text{--}258)$$

where

$$C = \frac{1}{8}\left(\frac{2}{\overline{p}_\infty}\right)\left[30(G^*)^2 - 5G^* - \frac{7}{6}\right].$$

Then, when $x \equiv A\cos\phi$ and $y = A\sin\phi$ are introduced, it follows that Eqs. (4–257) and (4–258) can be replaced with the equivalent pair of dynamical equations:

$$\frac{dy}{dt^*} = -1 + C(x^2 + y^2)x, \qquad (4\text{--}259)$$

$$\frac{dx}{dt^*} = -C(x^2 + y^2)y. \qquad (4\text{--}260)$$

The advantage of this latter transformation is that (4–259) and (4–260) can be written in a classical Hamiltonian form,

$$\frac{dy}{dt^*} = \frac{\partial H}{\partial x}, \quad \frac{dx}{dt^*} = -\frac{\partial H}{\partial y}, \qquad (4\text{--}261)$$

where the so-called Hamiltonian function H is defined as

$$H \equiv \frac{C}{4}(x^2 + y^2)^2 - x. \qquad (4\text{--}262)$$

The solution trajectories $x(t^*)$ and $y(t^*)$ of a system of this type correspond to contours of constant values of H.

We can deduce the qualitative significance of this solution of equations (4–259) and (4–260) by plotting the contours of constant H in the x–y plane, as shown in Fig. 4–14 (for two different positive values of C). The relationships between a point on one of these contours and the original functions $A(\tau)$ and $\phi(\tau)$ is indicated in Fig. 4–15. The direction of motion *around* the contours of Fig. 4–14 is counterclockwise, as we can easily verify by examining the sign of dy/dt^* at *various* points along the x axis (that is, $y = 0$), using (4–259). The qualitative nature of the variation of A and ϕ with time depends on the initial conditions $A(0)$ and $\phi(0)$. For each C, there is a single *fixed* point at $\phi = 0, A = 1/\sqrt{C}$, where both ϕ and A are time independent. All other solution trajectories give *oscillatory* values for A. Those trajectories that lie wholly to the *right* of the origin also give oscillatory variations of the phase angle ϕ, whereas all other trajectories yield a monotonic *increase* of ϕ. Hence the predicted response to a time-dependent oscillation of pressure at the *natural* oscillation frequency of the bubble *is* an oscillatory response in bubble volume at the same frequency, with a slow variation in the *amplitude* of this oscillation described by $A(\tau)$ and a slow variation in the phase angle described by $\phi(\tau)$.

An Introduction to Asymptotic Approximations

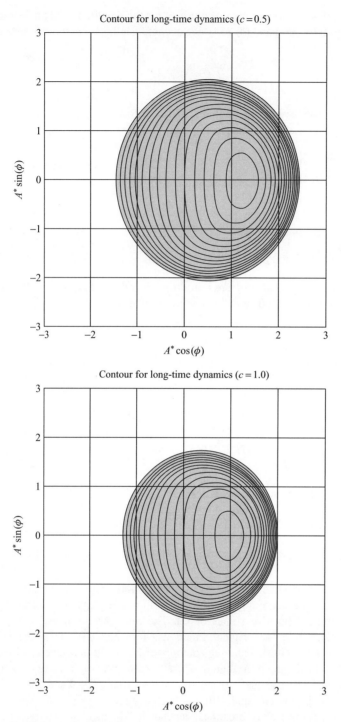

Figure 4–14. The solution trajectories $x(t^*)$ and $y(t^*)$ of the system (4–259) and (4–260), which correspond to contours of constant values of H ($C = 0.5$ and $c = 1.0$).

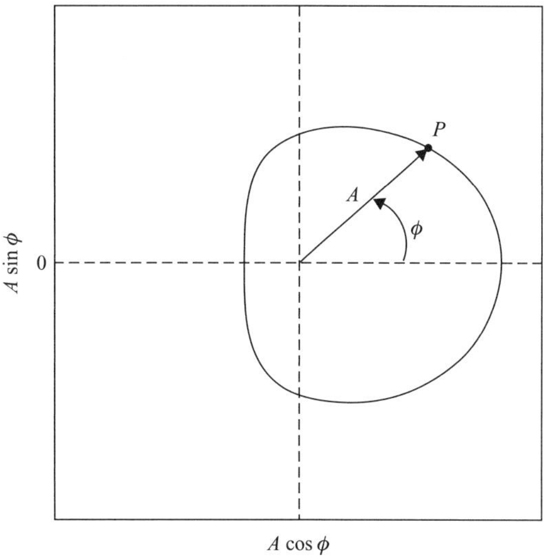

G. Bubble Dynamics in a Quiescent Fluid

Figure 4–15. A schematic showing a single contour from Fig. 4–14 in order to indicate the physical significance of a point on the contour in terms of $A(\tau)$ and $\phi(\tau)$.

A sketch of g_1 as a function of t for $\varepsilon = 0.1$ and $C = 1$ is shown in Fig. 4–16 for two arbitrarily chosen sets of initial values for A and ϕ at $t = 0$.

We see that, even in the resonant case $\omega = \omega_0$ with zero viscous damping, the changes in bubble volume remain small for small-amplitude pressure forcing. Thus, although the analysis is more complicated, it is still possible to obtain approximate solutions for $\varepsilon \ll 1$ by means of an asymptotic perturbation expansion for small changes in the bubble radius. However, it is significant in this case that the $O(\varepsilon)$ forcing produces an $O(\varepsilon^{1/3})$ response in the changes of bubble volume. The solutions of (4–253) and (4–254) describe the slow modulation of amplitude and phase shift in the solutions, as previously discussed.

4. Stability to Nonspherical Disturbances

All of the preceding analysis in this section is predicated on the critical assumption that the bubble remains spherical, so that the flow is *radial*. An important question, which can also be answered by a regular perturbation analysis for small deformations in the bubble *shape*, is whether the spherical surface is stable or whether we should expect deviations to a nonspherical shape as the bubble either expands or contracts. It is well known that the shape of a bubble near a wall, or in a mean flow such as simple shear, will be nonspherical.[24] Here, however, we are concerned with instabilities in the shape of an expanding or collapsing spherical bubble in an unbounded, quiescent fluid. The fact that the interface of a bubble may be unstable in these circumstances is suggested by the early work of Taylor,[25] who showed that an infinite *plane* interface between two fluids of different density in accelerated motion is either stable or unstable depending on whether the acceleration is directed from the heavier to the lighter fluid or conversely. The possible instability of an expanding or collapsing spherical interface may be viewed as a generalization of Taylor's result.[26] The critical difference is that the amplitude of a disturbance on a spherical surface must be *decreased* as the surface expands and *increased* as the surface contracts. Conversely, the wavelength of the disturbance must increase as the surface expands and decrease as it contracts.

In the present case, we consider a bubble whose interface is described in terms of a spherical coordinate system in the asymptotic form

$$r_s = R(t) + \sum_n \varepsilon^n f_n(\theta, \phi, t). \qquad (4\text{–}263)$$

An Introduction to Asymptotic Approximations

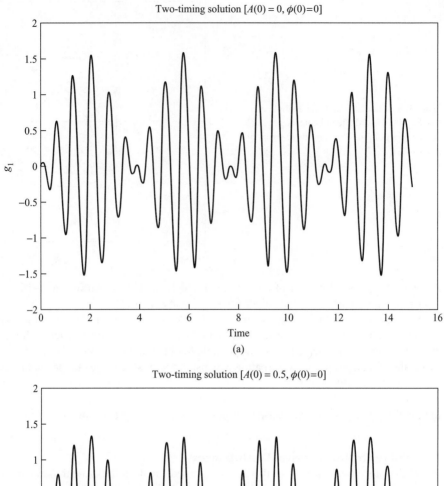

Figure 4–16. The function g_1 defined in Eq. (4–256), for $A(0) = \phi(0) = 0$ and $A(0) = 0.5$, $\phi(0) = 0$. The qualitative difference in these plots is indicative of the sensitivity to initial conditions.

G. Bubble Dynamics in a Quiescent Fluid

Here, ε is a small parameter that will form the basis of an asymptotic approximation for the dynamics of the bubble surface. The question here is whether a bubble with a nonspherical initial shape of small amplitude $O(\varepsilon)$ will return to a sphere – that is, $f_n(\theta, \phi, t) \to 0$ as $t \to \infty$ – or whether the initial disturbance in shape will grow with increase of t.

To analyze this problem, we need to go back to a statement of the problem in general fluid dynamical terms. Thus we begin by restating the governing equations and boundary conditions for an oscillating bubble in a quiescent, incompressible fluid. These are the Navier–Stokes and continuity equations; the three boundary conditions

$$\mathbf{u} \cdot \mathbf{n} = -\frac{1}{|\nabla F|} \frac{\partial F}{\partial t} \quad \text{(kinematic condition)}, \tag{4–264}$$

$$(\boldsymbol{\tau} \cdot \mathbf{n}) \cdot \mathbf{t}_i = 0 \ (i = 1, 2) \quad \text{(zero shear stress)}, \tag{4–265}$$

$$p_B - p + (\boldsymbol{\tau} \cdot \mathbf{n}) \cdot \mathbf{n} = \gamma (\nabla \cdot \mathbf{n}) \quad \text{(normal-stress condition)}, \tag{4–266}$$

that are applied at the surface of the deformed bubble

$$F \equiv r - \left[R(t) + \sum_n \varepsilon^n f_n(\theta, \phi, t) \right] = 0;$$

and the condition

$$\mathbf{u} \to 0 \quad \text{as } |\mathbf{r}| \to \infty. \tag{4–267}$$

Here, p_B is the pressure inside the bubble, γ is the interfacial tension, and $\mathbf{t}_i$ represents one of the pair of orthogonal unit vectors that are tangent at any point to the bubble surface. The problem is to calculate velocity and pressure fields in the fluid, as well as to determine the functions $R(t)$ and $f_n(\theta, \phi, t)$ that describe the geometry of the bubble surface. In the present context, it is the latter part of the problem that is the focus of our interest.

One complication is that the boundary conditions (4–264)–(4–266) must be applied at the bubble surface, which is both unknown [that is, specified in terms of functions $R(t)$ and $f_n(\theta, \phi, t)$ that must be calculated as part of the solution] and *nonspherical*. Further, the normal and tangent unit vectors $\mathbf{n}$ and $\mathbf{t}_i$ that appear in the boundary conditions are also functions of the bubble shape. In this analysis, we use the small-deformation limit $\varepsilon \ll 1$ to simplify the problem by using the *method of domain perturbations* that was introduced earlier in this chapter. First, we note that the unit normal and tangent vectors can be approximated for small ε in the forms

$$\mathbf{n} = \frac{\nabla F}{|\nabla F|} = \mathbf{i}_r - \varepsilon \left(\mathbf{i}_\theta \frac{1}{R} \frac{\partial f_1}{\partial \theta} + \mathbf{i}_\phi \frac{1}{R \sin \theta} \frac{\partial f_1}{\partial \phi} \right) + O(\varepsilon^2), \tag{4–268}$$

$$\mathbf{t}_1 = \mathbf{i}_\theta + \varepsilon \frac{1}{R} \frac{\partial f_1}{\partial \theta} \mathbf{i}_r + O(\varepsilon^2), \tag{4–269}$$

and

$$\mathbf{t}_2 = \mathbf{i}_\phi + \varepsilon \frac{1}{R \sin \theta} \frac{\partial f_1}{\partial \phi} \mathbf{i}_r + O(\varepsilon^2). \tag{4–270}$$

Further, the curvature at the bubble surface becomes

$$\nabla \cdot \mathbf{n} = \frac{2}{R} - \varepsilon \left(\frac{\cos \theta}{\sin \theta} \frac{\partial f_1}{\partial \theta^2} + \frac{\partial^2 f_1}{\partial \theta^2} + \frac{1}{\sin^2 \theta} \frac{\partial^2 f_1}{\partial \phi^2} \right) \frac{1}{R^2} + O(\varepsilon^2). \tag{4–271}$$

An Introduction to Asymptotic Approximations

Thus, at the bubble surface,

$$\mathbf{u} \cdot \mathbf{n} = u_r - \varepsilon \frac{1}{R} \frac{\partial f_1}{\partial \theta} u_\theta - \varepsilon \frac{1}{R \sin\theta} \frac{\partial f_1}{\partial \phi} u_\phi + O(\varepsilon^2), \qquad (4\text{--}272)$$

while

$$(\tau \cdot \mathbf{n}) \cdot \mathbf{t}_1 = \tau_{r\theta} - \varepsilon \left(\frac{1}{R} \frac{\partial f_1}{\partial \theta} \tau_{\theta\theta} + \frac{1}{R\sin\theta} \frac{\partial f_1}{\partial \phi} \tau_{\theta\phi} - \frac{1}{R} \frac{\partial f_1}{\partial \theta} \tau_{rr} \right) + O(\varepsilon^2), \qquad (4\text{--}273)$$

$$(\tau \cdot \mathbf{n}) \cdot \mathbf{t}_2 = \tau_{r\phi} - \varepsilon \left(\frac{1}{R} \frac{\partial f_1}{\partial \theta} \tau_{\theta\phi} + \frac{1}{R\sin\theta} \frac{\partial f_1}{\partial \phi} \tau_{\phi\phi} - \frac{1}{R\sin\theta} \frac{\partial f_1}{\partial \theta} \tau_{rr} \right) + O(\varepsilon^2), \qquad (4\text{--}274)$$

and

$$(\tau \cdot \mathbf{n}) \cdot \mathbf{n} = \tau_{rr} - \varepsilon \left(2\frac{1}{R} \frac{\partial f_1}{\partial \theta} \tau_{r\theta} + 2\frac{1}{R\sin\theta} \frac{\partial f_1}{\partial \phi} \tau_{r\phi} \right) + O(\varepsilon^2). \qquad (4\text{--}275)$$

Finally, we can combine (4–272)–(4–275) with the boundary conditions (4–264)–(4–266).

Of course, the application of these conditions is still complicated by the fact that they must be applied at the slightly deformed surface:

$$r_s = R(t) + \varepsilon f_1(\theta, \phi, t) + O(\varepsilon^2).$$

To avoid this, we use *domain perturbation theory* (see Section E) to transform from the exact boundary conditions applied at $r_s = R + \varepsilon f_1$ to asymptotically equivalent boundary conditions applied at the *spherical* surface $r_s = R(t)$. For example, instead of a condition on u_r at $r = R(t) + \varepsilon f_1$ from the kinematic condition, we can obtain an asymptotically equivalent condition at $r = R$ by means of the Taylor series approximation

$$u_r|_{r=r_s} = u_r|_{r=R} + \left. \frac{\partial u_r}{\partial r} \right|_{r=R} \varepsilon f_1 + O(\varepsilon^2), \qquad (4\text{--}276)$$

and similarly for $\tau_{r\theta}, \tau_{r\phi}$, and τ_{rr}. This simple idea of transforming from boundary conditions at a surface that is not a coordinate surface to asymptotically equivalent boundary conditions on a nearby surface that is a coordinate surface is the essence of the domain perturbation technique. When everything is put together, then, the boundary conditions (4–264)–(4–266) become

$$u_r|_{r=R} + \varepsilon \left(\left. \frac{\partial u_r}{\partial r} \right|_{r=R} f_1 - \frac{1}{R} \frac{\partial f_1}{\partial \theta} u_\theta \bigg|_{r=R} - \frac{1}{R\sin\theta} \frac{\partial f_1}{\partial \phi} u_\phi \bigg|_{r=R} \right) = \frac{dR}{dt} + \varepsilon \frac{\partial f_1}{\partial t} + O(\varepsilon^2), \qquad (4\text{--}277)$$

$$\tau_{r\theta}|_{r=R} + \varepsilon \left(\left. \frac{\partial \tau_{r\theta}}{\partial r} \right|_{r=R} f_1 - \frac{1}{R} \frac{\partial f_1}{\partial \theta} \tau_{\theta\theta} \bigg|_{r=R} - \frac{1}{R\sin\theta} \frac{\partial f_1}{\partial \phi} \tau_{\theta\phi} \bigg|_{r=R} + \frac{1}{R} \frac{\partial f_1}{\partial \theta} \tau_{rr} \bigg|_{r=R} \right)$$
$$+ O(\varepsilon^2) = 0, \qquad (4\text{--}278a)$$

$$\tau_{r\phi}|_{r=R} + \varepsilon \left(\left. \frac{\partial \tau_{r\phi}}{\partial r} \right|_{r=R} f_1 - \frac{1}{R} \frac{\partial f_1}{\partial \theta} \tau_{\theta\phi} \bigg|_{r=R} - \frac{1}{R\sin\theta} \frac{\partial f_1}{\partial \phi} \tau_{\phi\phi} \bigg|_{r=R} + \frac{1}{R\sin\theta} \frac{\partial f_1}{\partial \phi} \tau_{rr} \bigg|_{r=R} \right)$$
$$+ O(\varepsilon^2) = 0, \qquad (4\text{--}278b)$$

G. Bubble Dynamics in a Quiescent Fluid

and

$$(p_B - p_{r=R}) + \frac{\partial}{\partial t}(-p)\bigg|_{r=R} \varepsilon f_1 + \tau_{rr}|_{r=R}$$
$$+ \varepsilon \left(\frac{\partial \tau_{rr}}{\partial r}\bigg|_{r=R} f_1 - \frac{2}{R}\frac{\partial f_1}{\partial \theta}\tau_{r\theta}\bigg|_{r=R} - \frac{2}{R\sin\theta}\frac{\partial f_1}{\partial \phi}\tau_{r\phi}\bigg|_{r=R} \right) + O(\varepsilon^2)$$
$$= \gamma \left[\frac{2}{R} - \varepsilon \left(\frac{\cos\theta}{\sin\theta}\frac{\partial f_1}{\partial \theta} + \frac{\partial^2 f_1}{\partial \theta^2} + \frac{1}{\sin^2\theta}\frac{\partial^2 f_1}{\partial \phi^2} \right)\frac{1}{R^2} + O(\varepsilon^2) \right]. \quad (4\text{--}279)$$

Now the objective is an asymptotic solution for small ε. For simplicity, we consider only the limiting case in which we neglect viscous effects altogether. In the limit $\varepsilon = 0$, considered earlier in deriving the Rayleigh–Plesset equations, the viscous terms in the Navier–Stokes equations were identically equal to zero, and it was only in application of the normal-stress condition that viscosity entered the problem. Here, however, the exact problem is more complicated, and neglect of viscous effects is a significant simplification. It should be noted that the neglect of viscous effects as an approximation for large Reynolds number, $Re \to \infty$, is not as severe an approximation for fluid motion near a gas bubble as for motion near a no-slip surface; in fact, we shall see in Chap. 10 that the limiting form of the Navier–Stokes equation for $\mu = 0$ (that is, the so-called *inviscid equations of motion*) provides a uniformly valid first approximation for the velocity field in this case as $Re \to \infty$. Nevertheless, neglect of viscous effects is treated here strictly as an *ad hoc* approximation. The resulting problem retains much of the essential physics of the full problem, but the reader who wishes to apply the results in a practical context would be well advised to consult with one of the many texts or research papers that discuss viscous effects for this problem.[27]

The form of the asymptotic solution that we seek is

$$\mathbf{u} = \mathbf{u}^0 + \varepsilon \mathbf{u}^1 + O(\varepsilon^2),$$
$$p = p_0 + \varepsilon p_1 + O(\varepsilon^2), \quad (4\text{--}280)$$

and the governing equations in the inviscid approximation are thus

$$\rho \left(\frac{\partial \mathbf{u}^0}{\partial t} + \mathbf{u}^0 \cdot \nabla \mathbf{u}^0 \right) = -\nabla p_0, \quad \nabla \cdot \mathbf{u}^0 = 0, \quad (4\text{--}281)$$

$$\rho \left(\frac{\partial \mathbf{u}^1}{\partial t} + \mathbf{u}^0 \cdot \nabla \mathbf{u}^1 + \mathbf{u}^1 \cdot \nabla \mathbf{u}^0 \right) = -\nabla p_1, \quad \nabla \cdot \mathbf{u}^1 = 0. \quad (4\text{--}282)$$

The boundary conditions, which we obtain by substituting the expansions (4–280) into the conditions (4–277) and (4–279), are, at $O(1)$ and $O(\varepsilon)$,

$$u_r^0|_{r=R} = \frac{dR}{dt}, \quad (4\text{--}283)$$

$$u_r^1|_{r=R} + \left(\frac{\partial u_r^0}{\partial r}\right)_{r=R} f_1 - \frac{1}{r}\frac{\partial f_1}{\partial \theta}u_\theta^0\bigg|_{r=R} - \frac{1}{r\sin\theta}\frac{\partial f_1}{\partial \phi}u_\phi^0\bigg|_{r=R} = \frac{\partial f_1}{\partial t}, \quad (4\text{--}284)$$

and

$$p_B^0 - p^0|_{r=R} = \frac{2\gamma}{R}, \quad (4\text{--}285)$$

$$-p^1|_{r=R} + \frac{\partial}{\partial r}(-p^0)\bigg|_{r=R} f_1 = -\frac{\gamma}{R^2}\left(\frac{\cos\theta}{\sin\theta}\frac{\partial f_1}{\partial \theta} + \frac{\partial^2 f_1}{\partial \theta^2} + \frac{1}{\sin^2\theta}\frac{\partial^2 f_1}{\partial \phi^2} \right). \quad (4\text{--}286)$$

An Introduction to Asymptotic Approximations

The tangential-stress condition is satisfied automatically for an inviscid fluid ($\mu = 0$) because $\tau \equiv 0$ in this case, and there is no viscous contribution to the normal-stress balance.

Thus, examining (4–281), (4–283), and (4–285), which is the full problem at $O(1)$, we see that they are identical to the DEs and boundary conditions that led to the Rayleigh–Plesset equation, except for the neglect of the viscous stress term in (4–285). Thus the solution at $O(1)$ is

$$u_\theta^0 = u_\phi^0 = 0, \quad u_r^0 = \frac{R^2 \dot{R}}{r^2}, \tag{4–287}$$

$$p_0(r,t) = p_\infty(t) + \rho \left[2R(\dot{R})^2 + R^2 \ddot{R} - \frac{1}{2r^4}(R^4 \dot{R}^2) \right], \tag{4–288}$$

where

$$R\ddot{R} + \frac{3}{2}(\dot{R})^2 = \frac{1}{\rho}[p_0(R,t) - p_\infty(t)], \tag{4–289}$$

$$p_0(R,t) = p_B^0(t) - \frac{2\gamma}{R}, \tag{4–290}$$

with

$$p_B^0 = p_v + \frac{\hat{G}}{R^3}. \tag{4–291}$$

The problem at $O(\varepsilon)$ is then to solve (4–282), which can now be expressed in component notation as

$$\begin{aligned}
\rho \frac{\partial u_r^1}{\partial t} + \rho u_r^0 \frac{\partial u_r^1}{\partial r} + \rho u_r^1 \frac{\partial u_r^0}{\partial r} &= -\frac{\partial p_1}{\partial r}, \\
\rho \frac{\partial u_\theta^1}{\partial t} + \rho u_r^0 \frac{\partial u_\theta^1}{\partial r} + \rho \frac{u_r^0 u_\theta^1}{r} &= -\frac{1}{r}\frac{\partial p_1}{\partial \theta}, \\
\rho \frac{\partial u_\phi^1}{\partial t} + \rho u_r^0 \frac{\partial u_\phi^1}{\partial r} + \rho \frac{u_\phi^1 u_r^0}{r} &= -\frac{1}{r \sin\theta}\frac{\partial p_1}{\partial \phi},
\end{aligned} \tag{4–292}$$

$$\frac{1}{r^2}\frac{\partial}{\partial r}(r^2 u_r^1) + \frac{1}{r \sin\theta}\frac{\partial}{\partial \theta}(\sin\theta\, u_\theta^1) + \frac{1}{r \sin\theta}\frac{\partial u_\phi^1}{\partial \phi} = 0, \tag{4–293}$$

with boundary conditions

$$u_r^1|_{r=R} + \left(\frac{\partial u_r^0}{\partial r}\right)_{r=R} f_1 = \frac{\partial f_1}{\partial t}, \tag{4–294}$$

$$-p_1|_{r=R} - \frac{\partial p_0}{\partial r}\bigg|_{r=R} f_1 = -\frac{\gamma}{R^2}\left(\frac{\cos\theta}{\sin\theta}\frac{\partial f_1}{\partial \theta} + \frac{\partial^2 f_1}{\partial \theta^2} + \frac{1}{\sin^2\theta}\frac{\partial^2 f_1}{\partial \phi^2}\right), \tag{4–295}$$

$$u_r^1 \to 0 \quad \text{as} \quad r \to \infty. \tag{4–296}$$

This problem is linear in $\mathbf{u}^1$, p^1, and f_1. Hence, to consider an arbitrary initial bubble shape, we can consider the function f_1 to be a superposition of spherical harmonics, that is,

$$f_1 = \sum_{k,\ell} a_{k\ell}(t) P_k^\ell(\cos\theta) e^{i\ell\phi}. \tag{4–297}$$

G. Bubble Dynamics in a Quiescent Fluid

To examine the stability of some arbitrary initial perturbation in the shape, we must examine the behavior of the time-dependent coefficients $a_{kl}(t)$. Because this problem at $O(\varepsilon)$ is completely linear, the dynamics of the various modes are uncoupled, and it is sufficient to examine any arbitrary single mode for all possible values of k and ℓ:

$$f_1 = a_{k\ell}(t) P_k^\ell (\cos\theta) e^{i\ell\phi}. \tag{4-298}$$

Then the kinematic boundary condition, (4–294), takes the form

$$u_r^1|_{r=R} = \left[\left(\frac{2\dot{R}}{R}\right) a_{k\ell} + \frac{da_{k\ell}}{dt}\right] P_k^\ell(\cos\theta) e^{i\ell\phi}, \tag{4-299}$$

and the normal-stress balance is

$$-p_1|_{r=R} = \left\{ 2\rho \frac{\dot{R}^2}{R} a_{k\ell}(t) P_k^\ell(\cos\theta) e^{i\ell\phi} - \frac{\gamma}{R^2} a_{k\ell}(t) D^2\left[P_k^\ell(\cos\theta)e^{i\ell\phi}\right]\right\}, \tag{4-300}$$

where

$$D^2 \equiv \frac{\cos\theta}{\sin\theta}\frac{\partial}{\partial\theta} + \frac{\partial^2}{\partial\theta^2} + \frac{1}{\sin^2\theta}\frac{\partial^2}{\partial\phi^2}.$$

To solve (4–292), (4–293), (4–296), (4–299), and (4–300), we note that the continuity equation for an inviscid, irrotational flow is satisfied by a velocity field that can be expressed as the gradient of a scalar function, namely,

$$\mathbf{u}^1 = \nabla\phi, \tag{4-301}$$

provided ϕ is a harmonic function, so that

$$\nabla \cdot \mathbf{u}^1 = \nabla \cdot (\nabla\phi) = \nabla^2 \phi = 0. \tag{4-302}$$

However, solutions of Laplace's equation, expressed in terms of spherical coordinates, are just the spherical harmonics:

$$\phi = \begin{cases} r^n P_n^m(\cos\theta) e^{im\phi} \\ r^{-(n+1)} P_n^m(\cos\theta) e^{im\phi} \end{cases}, \quad \text{where } n, m \geq 0. \tag{4-303}$$

In order that $\mathbf{u}^1$ satisfies the far-field condition (4–296), ϕ must be limited to the negative harmonics, that is,

$$\phi = \sum_{n,m} A_{nm} r^{-(n+1)} P_n^m(\cos\theta) e^{im\phi}. \tag{4-304}$$

However, for the distortion mode, (4–298), the kinematic condition, (4–299), requires

$$u_r^1\bigg|_{r=R} \equiv \frac{\partial\phi}{\partial r}\bigg|_{r=R} = \left[\left(\frac{2\dot{R}}{R}\right) a_{kl} + \frac{da_{k\ell}}{dt}\right] P_k^\ell(\cos\theta) e^{i\ell\phi}.$$

Hence,

$$A_{k\ell} = -\frac{R^{k+2}}{k+1}\left(2\frac{\dot{R}}{R} a_{k\ell} + \frac{da_{k\ell}}{dt}\right),$$

and thus

$$\phi = -\frac{R^{k+2}}{k+1}\left(2\frac{\dot{R}}{R}a_{k\ell} + \frac{da_{k\ell}}{dt}\right)\frac{1}{r^{k+1}}P_k^\ell(\cos\theta)e^{i\ell\phi}. \qquad (4\text{--}305)$$

Taken together with (4–301), this result for ϕ is analogous to the result (4–287) for the $O(1)$ problem. All that remains is to obtain the dynamic equation for the coefficients $a_{k\ell}(t)$. For this purpose, we follow the example from the earlier derivation of the Rayleigh–Plesset equation. First, we calculate the pressure $p_1(r, t)$ by means of equations of motion (4–292), and, second, we apply the normal-stress condition, (4–300), to obtain the dynamic equation for the coefficients $a_{k\ell}(t)$. In the interest of brevity all of the details are not displayed here. The result is:

$$\ddot{a}_{k\ell} + 3\frac{\dot{R}}{R}\dot{a}_{k\ell} + (k-1)\left[-\frac{\ddot{R}}{R} + (k+1)(k+2)\frac{\gamma}{\rho R^3}\right]a_{k\ell} = 0. \qquad (4\text{--}306)$$

It will be noted that this result is independent of the azimuthal index l, and it is thus convenient to drop this subscript so that a_{kl} is denoted simply as a_k. Further, the mode $k=1$ is a special case as we can see by writing the expression for the bubble surface for a pure $k=1$ deformation, namely,

$$s = r - R(t) - a_1(t)P_1(\cos\theta).$$

Now, $P_1(\cos\theta) \equiv \cos\theta$. Thus, if $a_1(t) \neq 0$, the *center* of the bubble translates a distance $a_1(t)$ from the origin, but the bubble *shape* is unchanged. To discuss the stability of the spherical shape, we must determine whether the coefficients $a_k(t)$ for $k \geq 2$ increase or decrease with time.

Even without a detailed analysis, there are a few observations and general conclusions that can be made from (4–306). First, for a bubble of *constant volume*,

$$\ddot{a}_k + (k-1)(k+1)(k+2)\frac{\gamma}{\rho R^3}a_k = 0, \qquad (4\text{--}307)$$

and we see that the *natural* oscillation frequency for the k mode is

$$\omega_{0,k}^2 = (k-1)(k+1)(k+2)\frac{\gamma}{\rho R^3}. \qquad (4\text{--}308)$$

This result is well known. More interesting is the coupling between the radial and shape oscillations when $\dot{R} \neq 0$.

Let us recall the properties of a standard linear oscillator equation in the form

$$\ddot{F} + a\dot{F} + bF = 0. \qquad (4\text{--}309)$$

If the second term is zero, $(a=0)$, the solution is

$$F = e^{st}, \quad \text{where } s^2 + b = 0. \qquad (4\text{--}310)$$

Therefore, if $b > 0$, $s = \pm i\sqrt{b}$, and the solution is oscillatory. On the other hand, if $b < 0$, $s = \pm\sqrt{-b}$, and one of the two solutions grows exponentially with time. The second term in (4–309) plays the role of damping if $a > 0$ but is destabilizing (somewhat analogous to

G. Bubble Dynamics in a Quiescent Fluid

a negative viscosity) if $a < 0$. If we now examine (4–306), we see that surface tension is always stabilizing (in the sense that it contributes to $b > 0$), whereas the acceleration term $-\ddot{R}/R$ is stabilizing when $\ddot{R} < 0$ but destabilizing when $\ddot{R} > 0$. The instability that occurs for $\ddot{R}$ sufficiently large and positive is essentially the *Rayleigh–Taylor instability* for this problem (see Chap. 12). Maximum positive values for $\ddot{R}$ tend to occur particularly at the *beginning* of an expansion or near the *minimum* radius when the bubble is collapsing. The term $3(\dot{R}/R)\dot{a}_k$ corresponds to positive damping for $\dot{R} > 0$ but tends to be destabilizing for $\dot{R} < 0$. Indeed, as we shall see shortly, it is possible for the bubble shape to be unstable if $\dot{R} < 0$ even though $\ddot{R} > 0$. The physical origin of the damping or destabilization associated with this term is strictly kinematic in origin. Because of the diverging/converging nature of the streamlines for the $O(1)$ radial motion, the area of the bubble surface is either increased with time or decreased with time, depending on whether $\dot{R} > 0$ or $\dot{R} < 0$, respectively. When the surface area is increased, the magnitude of any surface disturbance is decreased and the wavelength is increased. On the other hand, when the area is decreased, the amplitude of the disturbance is increased and its wavelength is decreased. The increase or decrease in amplitude corresponds to a destabilizing or stabilizing effect.

A more quantitative discussion of the stability is possible. A convenient first step is to introduce the transformation

$$a_k = \frac{1}{R^{3/2}}\alpha_k. \tag{4–311}$$

into (4–306). This leads to the transformed equation

$$\ddot{\alpha}_k - G_k(t)\alpha_k = 0, \tag{4–312}$$

where

$$G_k(t) = \frac{3}{2}\frac{d}{dt}\left(\frac{\dot{R}}{R}\right) + \frac{9}{4}\left(\frac{\dot{R}}{R}\right)^2 + \left\{(k-1)\left[\frac{\ddot{R}}{R} - (k+1)(k+2)\frac{\gamma}{\rho R^3}\right]\right\}. \tag{4–313}$$

Now the time dependence of $G_k(t)$ is determined by the time dependence of $R(t)$, which in turn is dependent on the variation in ambient pressure $p_\infty(t)$. Although the solutions of (4–312) are generally quite complicated, we may hope to learn something that is at least qualitatively useful by examining cases in which the form for G is simple. For example, if $G_k = c$(const), then the solutions of (4–312) are periodic and stable provided $c < 0$.[28] On the other hand, at least one of the solutions of (4–312) is exponentially growing for $c > 0$, and thus the spherical bubble shape is *unstable*. Other forms for $G_k(t)$ are more difficult to analyze in detail, though we may hope that the general "rule" of stability for $G_k < 0$ and instability for $G_k > 0$ might still carry over. In that case, we can achieve some useful insight by determining the factors that control the sign of G_k. For this purpose, it is convenient to rewrite G_k in the form

$$G_k(t) = \frac{3}{4}\left(\frac{\dot{R}}{R}\right)^2 + \left(k + \frac{1}{2}\right)\frac{\ddot{R}}{R} - (k-1)(k+1)(k+2)\frac{\gamma}{\rho R^3}. \tag{4–314}$$

Generally, the surface tension is stabilizing (in the sense that it contributes a negative term to G), and this tendency is enhanced with increase of k.[29] Let us consider the case $\gamma \equiv 0$. Then,

$$G_k(t) = \frac{3}{4}\left(\frac{\dot{R}}{R}\right)^2 + \left(k + \frac{1}{2}\right)\frac{\ddot{R}}{R}. \tag{4–315}$$

An Introduction to Asymptotic Approximations

Clearly, $G_k > 0$ for $\ddot{R} \geq 0$. Instability corresponding to $\ddot{R} > 0$ is the analog of Rayleigh–Taylor instability. For a flat interface, we could have instability *only* for $\ddot{R} \geq 0$. Here, however, we may still have instability even if $\ddot{R} < 0$, provided $\dot{R}/R$ is sufficiently large. To determine the condition for $G > 0$ in terms of controllable parameters, we can substitute for $\ddot{R}/R$ in (4–315) by using the inviscid form of the Rayleigh–Plesset equation, (4–204), with $\gamma \equiv 0$, that is,

$$\frac{\ddot{R}}{R} = -\frac{3}{2}\left(\frac{\dot{R}}{R}\right)^2 + \frac{p_B - p_\infty}{\rho R^2}. \tag{4–316}$$

Then,

$$G_k(t) = k\frac{\ddot{R}}{R} + \frac{p_B - p_\infty}{2\rho R^2}. \tag{4–317}$$

Hence, for $\ddot{R} < 0$, $G(t) > 0$, if

$$\frac{p_B - p_\infty}{\rho} > 2kR|\ddot{R}|. \tag{4–318}$$

Thus, all else being equal, a smaller pressure difference is necessary to induce instability for smaller values of k.

It is important to recognize that the criteria of *instability* for $G_k > 0$ and *stability* for $G_k < 0$ is strictly valid only for G_k independent of t. Indeed, if G_k depends on t, the equation (4–312) is *nonautonomous*, and it is well known that stability can generally be established only by exact integration of the equation. If G_k is "slowly varying," a local analysis based on an instantaneous value of G_k will be qualitatively correct for a finite time interval, but for more general time-dependent forms for G_k, we should not be surprised if the situation turns out to be more complex. To see an example of this, we can consider the special case in which the bubble volume changes periodically with time about some mean value, as may occur in response to an oscillatory variation in p_∞ at a frequency different from the resonant frequency ω_0 (given just prior to Eq. (4–229)). Thus, we let

$$R = R_E(1 + \varepsilon \cos \omega t) \tag{4–319}$$

and evaluate $G_k(t)$. The result is

$$\begin{aligned} G_k(t) = & -\left(k + \frac{1}{2}\right)\varepsilon\omega^2 \cos \omega t \\ & - (k-1)(k+1)(k+2)\frac{\gamma}{\rho R_E^3}(1 - 3\varepsilon \cos \omega t) + O(\varepsilon^2), \end{aligned} \tag{4–320}$$

and thus, letting $\tau = \omega t$, we find that

$$\frac{d^2\alpha_k}{d\tau^2} + \left\{\frac{\omega_{0,k}^2}{\omega^2} + \left[\left(k + \frac{1}{2}\right) - 3\frac{\omega_{0,k}^2}{\omega^2}\right]\varepsilon \cos \tau\right\}\alpha_k = 0, \tag{4–321}$$

where $\omega_{0,k}$ is given by Eq. (4–308).

Equation (4–321) is a special case of a DE that is known as *Mathieu's equation*, for which the standard form is

$$\ddot{y} + (a + 2b\cos t)y = 0. \tag{4–322}$$

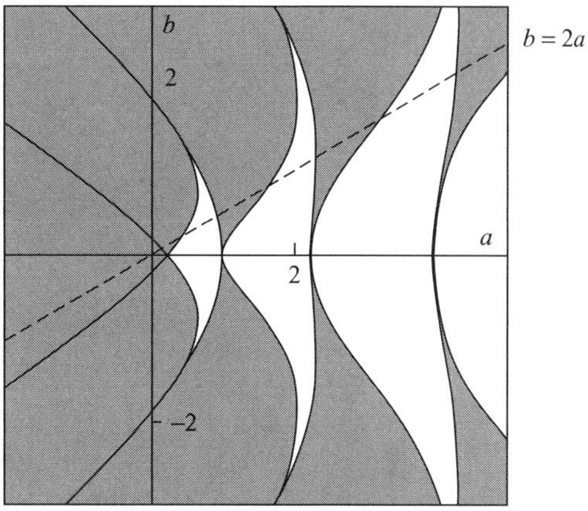

G. Bubble Dynamics in a Quiescent Fluid

Figure 4–17. Stability boundaries for solutions of Mathieu's equation (reproduced from Ref. 30). The shaded regions correspond to unstable regions in the a–b plane and the unshaded are stable regions.

In the present case,

$$a = a_k = \frac{\omega_{0,k}^2}{\omega^2}, \tag{4-323}$$

$$b = b_k = \frac{\varepsilon}{2}\left[\left(k + \frac{1}{2}\right) - 3\frac{\omega_{0,k}^2}{\omega^2}\right]. \tag{4-324}$$

The general theory for linear DEs with periodic coefficients is known as *Floquet theory*. This theory predicts that *unstable* solutions exist for some values of a and b (these solutions consist of modulated oscillations that grow exponentially with increasing t). A plot of the stability boundaries of solutions of Mathieu's equation is reproduced from the book by Bender and Orszag[30] in Fig. 4–17. All solutions are stable in the white regions of the (a,b) plane, whereas unstable solutions exist in the crosshatched region. For much of the (a,b) plane, the existence of *unstable* solutions represents a violation of the general stability rule proposed previously, which suggested that solutions of Eq. (4–312) would tend to be stable for any combination of a and b for which $G_k(t)$ is negative for all t. The condition $G_k < 0$ requires that

$$a_k > 2|b_k|.$$

The straight line corresponding to this condition is superposed on Fig. 4–17. Clearly regions of instability exist in spite of the fact that a_k exceeds $2|b_k|$ so that $G_k < 0$.

Although the behavior of solutions of Mathieu's equation is generally quite complicated, analytic approximations can be obtained by means of an asymptotic expansion for $b \ll 1$. In this case, b is small provided the magnitude of changes in bubble volume is small, that is, $\varepsilon \ll 1$ in (4–319). Therefore, for our problem, the limit $b \ll 1$ corresponds to modulations in bubble shape driven by weak oscillations in the bubble volume, that is, weak oscillations in the ambient pressure $p_\infty(t)$. In this regime, some features of the stability boundary shown in Fig. 4–17 can be predicted analytically. An added bonus of carrying out the necessary analysis is that it exposes the *physical basis* for instability of a bubble in the regime where a_k is asymptotically large compared with b_k so that $G_k < 0$ for all t.

In the remainder of this section, we thus consider the solution of (4–321), expressed in the form (4–322) with $b_k = \varepsilon \bar{b}_k \ll 1|$, namely,

$$\ddot{\alpha}_k + (a_k + 2\varepsilon \bar{b}_k \cos t)\alpha_k = 0. \tag{4-325}$$

Let us suppose, initially, that a solution of (4–325) exists in the form of a regular asymptotic expansion for $\varepsilon \ll 1$, namely,

$$\alpha_k(t) = \alpha_k^{(0)}(t) + \varepsilon \alpha_k^{(1)}(t) + \varepsilon^2 \alpha_k^{(2)}(t) + \cdots +. \tag{4-326}$$

Then, the governing equations for the successive terms in the asymptotic expansion for α can be expressed in the form

$$O(1), \quad \ddot{\alpha}_k^{(0)} + a_k \alpha_k^{(0)} = 0, \tag{4-327a}$$

$$O(\varepsilon), \quad \ddot{\alpha}_k^{(1)} + a_k \alpha_k^{(1)} = (-2\bar{b}_k \cos t) \alpha_k^{(0)}, \tag{4-327b}$$

$$O(\varepsilon^2), \quad \ddot{\alpha}_k^{(2)} + a_k \alpha_k^{(2)} = (-2\bar{b}_k \cos t) \alpha_k^{(1)}, \tag{4-327c}$$

and so on. Now, the general solution of (4–327a) is

$$\alpha_k^{(0)}(t) = A \exp(i\sqrt{a_k}t) + B \exp(-i\sqrt{a_k}t). \tag{4-328}$$

Thus Eq. (4–327b) for $\alpha_k^{(1)}$ becomes

$$\ddot{\alpha}_k^{(1)} + a_k \alpha_k^{(1)} = -2\bar{b}_k A e^{i(1+\sqrt{a_k})t} - 2\bar{b}_k B e^{i(1-\sqrt{a_k})t}. \tag{4-329}$$

Now, however, we can see that if

$$\sqrt{a_k} \pm 1 = \pm\sqrt{a_k}, \tag{4-330}$$

the terms on the right-hand side are *secular* terms and $\alpha_k^{(1)}$ will grow without bound as $t \to \infty$. Condition (4–330) is satisfied only if $a_k = 1/4$. Thus, provided $a_k \neq 1/4$, the solutions for $\alpha_k^{(1)}$ are bounded for all time, but the existence of secular terms for even this one value of a_k in (4–329) already provides significant insight into the physics that underlies the possibility of *unstable* solutions to Mathieu's equation (even though the time-dependent oscillation of the coefficient $a_k + 2\varepsilon \bar{b}_k \cos t$ is very weak for $\varepsilon \ll 1$). The fact is that oscillations of this time-dependent coefficient at the natural frequency of the constant-bubble-volume limit (that is, $b_k \equiv 0$) produce a resonant interaction with the eigenmode for this case [namely, the solution $\alpha_k^{(0)}$ in the present context] and the possibility of growth in the amplitude with time. Subsequent equations in the series have no secular terms provided that $a_k \neq n^2/4$. Thus, provided that $a_k \neq n^2/4$, the solution of Mathieu's equation for $\varepsilon \ll 1$ is bounded for all t. The condition

$$a_k \neq \frac{n^2}{4} \tag{4-331a}$$

also can be written as a condition on the frequency ω:

$$\omega \neq \frac{2}{n}\omega_{0,k}. \tag{4-331b}$$

Thus, for each deformation mode k, there are a discreet set of frequencies for which the bubble shape is unstable. This is consistent with the stability diagram in Fig. 4–17 for the limit $\varepsilon \bar{b} \to 0$.

We can obtain additional insight into the nature of the solutions of Mathieu's equation by extending the asymptotic solution described previously to small, but nonzero, values of ε. Of particular interest is the behavior of solutions near the first resonant instability point $a_k = 1/4$. Referring to Fig. 4–17, we see that there is a finite region around $a_k = 1/4$ where solutions for nonzero values of $\varepsilon \bar{b}_k$ are predicted to be unstable. In this section, we seek an asymptotic expression in terms of ε for the critical values of $a_k \approx 1/4$ that separate the regions of stable and unstable solutions. Thus we suppose that

$$a_k = \frac{1}{4} + \varepsilon a_k^{(1)} + \cdots + \tag{4-332}$$

and again seek stable solutions of (4–325). As is typical of problems that exhibit secular behavior that is due to resonant forcing, the solution for $a \approx 1/4$ exhibits two time scales: a

G. Bubble Dynamics in a Quiescent Fluid

short scale comparable with the time scale of the forcing and a longer time scale representative of the slow modulation of the induced oscillation. Thus we follow the usual two-timing asymptotic procedure and suppose that

$$\alpha_k = \alpha_k(t, \tau) \qquad (4\text{–}333)$$

and

$$\tau = \varepsilon t. \qquad (4\text{–}334)$$

In this case, the governing equation, (4–325), takes the form

$$0 = \frac{d^2\alpha_k}{dt^2} + 2\varepsilon \frac{d^2\alpha_k}{dt\,d\tau} + \varepsilon^2 \frac{d^2\alpha_k}{d\tau^2} + \left(\frac{1}{4} + \varepsilon a_1^{(k)} + \cdots + 2\varepsilon \overline{b}_k \cos t\right)\alpha_k, \qquad (4\text{–}335)$$

and we seek an asymptotic expansion for α_k of the form

$$\alpha_k = \alpha_k^{(0)}(t, \tau) + \varepsilon \alpha_k^{(1)}(t, \tau) + \cdots +. \qquad (4\text{–}336)$$

The governing equations at the first two levels of approximation are then

$$\frac{d^2\alpha_k^{(0)}}{dt^2} + \frac{1}{4}\alpha_k^{(0)} = 0, \qquad (4\text{–}337)$$

$$\frac{d^2\alpha_k^{(1)}}{dt^2} + \frac{1}{4}\alpha_k^{(1)} = -2\frac{d^2\alpha_k^{(0)}}{dt\,d\tau} - a_1^{(k)}\alpha_k^{(0)} - (2\overline{b}_k \cos t)\alpha_k^{(0)}. \qquad (4\text{–}338)$$

The solution of (4–337) is simply

$$\alpha_k^{(0)} = A(\tau)e^{it/2} + A^*(\tau)e^{-it/2}, \qquad (4\text{–}339)$$

where $A(\tau)$ and $A^*(\tau)$ are complex conjugates. Substituting into (4–338), we then obtain the governing equation for $\alpha_k^{(1)}$:

$$\frac{d^2\alpha_k^{(1)}}{dt^2} + \frac{1}{4}\alpha_k^{(1)} = -\left\{\left[i\frac{dA}{d\tau} + a_1^{(k)}A + \overline{b}_k A^*\right]e^{it/2}\right.$$
$$\left. - \left[i\frac{dA^*}{d\tau} - a_1^{(k)}A^* - \overline{b}_k A\right]e^{-it/2} + \overline{b}_k A e^{3it/2} + \overline{b}_k A^* e^{-3it/2}\right\}. \qquad (4\text{–}340)$$

Thus, to eliminate secular terms, we require that

$$i\frac{dA}{d\tau} + a_1^{(k)}A + b_k A^* = 0 \quad \text{and} \quad i\frac{dA^*}{d\tau} - a_1^{(k)}A^* - b_k A = 0, \qquad (4\text{–}341)$$

and this provides dynamic equations for the slow time scale modulation of (4–333). To solve these equations, it is convenient to express A in terms of its *real* and *imaginary* parts:

$$A = B(\tau) + iC(\tau), \quad A^* = B(\tau) - iC(\tau).$$

Then,

$$\frac{dB}{d\tau} + \left[a_1^{(k)} - \overline{b}_k\right]C = 0,$$

$$\frac{dC}{d\tau} - \left[a_1^{(k)} + \overline{b}_k\right]B = 0. \qquad (4\text{–}342)$$

If we eliminate C, we thus obtain a second-order equation for B,

$$\frac{d^2 B}{d\tau^2} = \left[\overline{b}_k^2 - a_1^{(k)2}\right]B, \qquad (4\text{–}343)$$

and the solution of this equation is

$$B = K \exp\left[\pm\sqrt{\bar{b}_k^2 - a_1^{(k)2}}\right]\tau. \tag{4-344}$$

Thus, if $(\bar{b}_k^2 - a_1^{(k)2}) > 0$, solution (4–339) is unstable because the slowly varying amplitude functions grow exponentially in time. On the other hand, for $(\bar{b}_k^2 - a_1^{(k)2}) < 0$, the solution is stable with a two-time scale oscillation in the bubble shape. The stability boundary thus occurs at

$$a_1^{(k)} = \pm \bar{b}_k, \tag{4-345}$$

and so the overall stability boundary for solutions of (4–325) occurs at

$$a_k = \frac{1}{4} \pm \varepsilon \bar{b}_k + O(\varepsilon^2) \tag{4-346}$$

This corresponds to the pair of crossing straight lines in Fig. 4–17 that pass through the point $(a_k = 1/4, \; \varepsilon\bar{b}_k = 0)$.

Hence we can see from this example of a bubble with weak time-periodic oscillations of volume that the qualitative condition for stability $G_k < 0$ does *not* apply unless G_k is constant or only very slowly varying with time (and in the latter case, strictly only for an interval of time that is short compared with the characteristic time scale for variation of G_k). If the coefficient G_k varies at a rate that is comparable with the natural frequency of shape oscillation for a bubble of constant volume, then the bubble may undergo a resonant oscillation of shape to very large amplitudes even though $G_k < 0$ for all times.

NOTES

1. See Ref. 1 in Chap. 3. Typical papers from *Annual Reviews* include A. Leonard, Computing three-dimensional incompressible flows with vortex elements, *Annu. Rev. Fluid Mech.* **17**, 523–59 (1985); M. Y. Hussaini and T. A. Zang, Spectral methods in fluid dynamics, *Annu. Rev. Fluid Mech.* **19**, 339–67 (1987); R. Glowinski and O. Pironneau, Finite element methods for Navier–Stokes equations, *Annu. Rev. Fluid Mech.* **24**, 167–204 (1992); R. Scardovelli and S. Zaleski, Dierect numerical simulation of free-surface and interfacial flow, *Annu. Rev. Fluid Mech.* **31**, 567–603 (1999).
2. S. Whitaker, *Introduction to Fluid Mechanics* (Prentice-Hall, Englewood Cliffs, NJ, 1968); M. M. Denn, *Process Fluid Dynamics* (Prentice-Hall: Englewood Cliffs, NJ, 1980) (paperback edition, Simon & Schuster, New York, 1998); S. Middleman, *Introduction to Fluid Dynamics: Principles of Analysis and Design* (Wiley, New York, 1997).
3. Among the many textbooks written on this topic are A. H. Nayfey, *Perturbation Methods* (Wiley, New York, 200); C. M. Bender and S. A. Orszag, *Advanced Mathematical Methods for Scientists and Engineers: Asymptotic Methods and Perturbation Theory* (Springer-Verlag, New York, 1999); M. Van-Dyke, *Perturbation Methods in Fluid Mechanics* (annotated edition) (Parabolic Press, Stanford, CA, 1975); J. Kevorkian and J. D. Cole, *Perturbation Methods in Applied Mathematics* (Springer-Verlag, New York, 1981); P. A. Lagerstrom, *Matched Asymptotic Expansions, Ideas and Techniques* (Springer-Verlag, New York, 1988); E. J. Hinch, *Perturbation Methods* (Cambridge University Press, Cambridge, 1991).
4. W. R. Dean, Note on the motion of fluid in a curved pipe, *Philos. Mag.* (Series 7) **4**, 208–19 (1927); W. R. Dean, The stream-line motion of fluid in a Curved Pipe, *Philos. Mag.* (Series 7) **5**, 673–95 (1928); W. R. Dean, Fluid in a curved channel, *Proc. R. Soc. London* Ser. A **121**, 402–420 (1928).
5. T. J. Pedley, *The Fluid Mech. of the Large Blood Vessels* (Cambridge University Press, Cambridge, 1980); M. G. Blyth, A. J. Mestel, and L. Zabielski, Arterial bends: The development and decay of helical flows. *Biorheology* **3**, 345–50 (2002); L. Zabielski and A. J. Mestel, Unsteady blood flow

Notes

in a helically symmetric pipe, *J. Fluid Mech.* **370**, 321–45 (1998); L. Zabielski and A. J. Mestel, Steady flow in a helically symmetric pipe, *J. Fluid Mech.* **370**, 297–320 (1998).

6. An alternative would be the toroidal coordinate system, as described by Happel and Brenner (1973), Appendix A-20 (see reference 3, chapter 1).

7. We note, for future reference, that the function f_1 plays a role in the present problem at $O(a/R)$ that is analogous (in the plane) to the *streamfunction* for a general 2D flow. The concept of a streamfunction will be discussed in detail in Chap. 7.

8. In this problem, the pressure-gradient is *fixed* at a dimensionless value of -4, and any changes in the flow may produce changes only in the volumetric flow rate. Hence, $\partial p_n/\partial z \equiv 0$ for all $n \geq 1$. A closely related problem would have the volumetric flow rate fixed, and, in this case, changes in the flow would produce changes in the required pressure-gradient so that $\partial p_n/\partial z \neq 0$ for $n \geq 1$.

9. S. A. Berger, L. Talbot and L. S. Yao, Flow in curved pipes, *Annu. Rev. Fluid. Mech.* **15**, 461–512 (1983).

10. Y. Fan, R. I. Tanner and N. Phan-Thien, Fully developed viscous and viscoelastic flows in curved pipes, *J. Fluid Mech.* **440**, 327–57 (2001); S. C. R. Dennis and N. Riley, On the fully developed flow in a curved pipe at large Dean number, *Proc. R. Soc. London Ser. A.* **434**, 473–8 (1991); W. M. Collins and S. C. R. Dennis, The steady motion of a viscous fluid in a curved tube, *Q. J. Mech. & Appl. Math.* **28**, 133–56 (1975).

11. The method of domain perturbations was used for many years before its formal rationalization by D. D. Joseph: D. D. Joseph, Parameter and domain dependence of eigenvalues of elliptic partial differential equations, *Arch. Ration. Mech. Anal.* **24**, 325–351 (1967). See also Ref. 3f. The method has been used for analysis of a number of different problems in fluid mechanics: A. Beris, R. C. Armstrong and R. A. Brown, Perturbation theory for viscoelastic fluids between eccentric rotating cylinders, *J. Non-Newtonian Fluid Mech.* **13**, 109–48 (1983); R. G. Cox, The deformation of a drop in a general time-dependent fluid flow, *J. Fluid Mech.* **37**, 601–623 (1969); D. D. Joseph, Domain perturbations: The higher order theory of infinitesimal water waves, *Arch. Ration. Mech. Anal.* **51**, 295–303 (1975); D. D. Joseph and R. Fosdick, The free surface on a liquid between cylinders rotating at different speeds, *Arch. Ration. Mech. Anal.* **49**, 321–81 (1973).

12. R. L. Webb, *Principles of Enhanced Heat Transfer* (Wiley, New York, 1994); J. L. Goldstein and E. M. Sparrow, Heat/mass transfer characteristics for flow in a corrugated wall channel, *J. Heat Transfer* **99**, 187–95 (1977); J. E. O'Brien and E. M. Sparrow, Corrugated-duct heat transfer, pressure drop, and flow visualization, *J. Heat Transfer* **104**, 410–16 (1982); G. Wang and S. P. Vanka, Convective heat transfer in periodic wavy passages, *Int. J. Heat Mass Transfer* **38**, 3219–30 (1995).

13. J. C. Burns and T. Parkes, Peristaltic motion, *J. Fluid Mech.* **29**, 405–16 (1967); S. Tsangaris and E. Leiter, Analytical solution for weakly stenosed channels at low Reynolds numbers, in *Proceedings of the First International on Mechanics in Medicine and Biology* (Witzstrock Pub. House, Badon-Badon, West Germany, 1978), pp. 289–92. [also: On laminar steady flow in sinusoidal channels, *J. Eng. Math.* **18**, 89–103 (1984)]; I. J. Sobey, On flow through furrowed channels, Part I. Calculated flow patterns, *J. Fluid Mech.* **96**, 1–26 (1980); S. Selvarajan, E. G. Tulapurkara, and V. Vasanta Ram, A numerical study of flow through wavy-wall channels, *Int. J. Numer. Methods Fluids* **26**, 519–31 (1998).

14. Lord Rayleigh (J. W. Strutt), On the pressure developed in a liquid during the collapse of a spherical cavity, *Philos. Mag. (Series 6)*, **34**, 94–8 (1917).

15. F. R. Young, *Cavitation* (Imperial College Press, London, 2000); C. E. Brennen, *Cavitation and Bubble Dynamics* (Oxford University Press, Oxford, 1994); R. E. Apfel, Acoustic cavitation, in *Methods of Experimental Physics*, edited by P. D. Edmonds (Academic, New York, 1981), Vol. 19, pp. 355–411.

16. T. Clancy, *The Hunt for Red October* (Naval Institute Press, Annapolis, MD, 1984).

17. M. A. Margulis, *Sonochemistry and Cavitation*, translated by G. Leib (Taylor & Francis, London, 1995); L. A. Crum, K. S. Suslick and J. L. Reisse (editors), *Sonochemistry and Sonoluminescence* (Kluwer Acadmic, Dordrecht, The Netherlands, 1998).

18. The mechanism for sonoluminescence is still a matter of considerable controversy. There are a few recent references that illustrate this point: M. P. Brenner, S. Hilgenfeldt and D. Lohse, Single-bubble sonoluminescence, *Rev. Mod. Phys.* **74**, 425–84 (2002); S. Putterman, P. G. Evans,

G. Vazquez and K. Weninger, Is there a simple theory of sonoluminescence? *Nature* (London) **409**, 782–3 (2001); B. D. Storey and A. J. Szeri, Argon rectification and the cause of light emission in single-bubble sonoluminescence, *Phys. Rev. Lett.* **88**, 074301/1–3 (2002).
19. M. S. Plesset and A. Prosperetti, Bubble dynamics and cavitation, *Annu. Rev. Fluid Mech.* **9**, 145–85 (1977). Two more recent papers are Y. Hao and A. Prosperetti, The dynamics of vapor bubbles in acoustic pressure fields, *Phys. Fluids* **11**, 2008–19 (1999); A. Prosperetti, The thermal behavior of oscillating gas bubbles, *J. Fluid Mech.* **222**, 587–615 (1991).
20. An analysis of the dynamics of spherical bubbles that are perturbed away from equilibrium volume by a finite (not arbitrarily small) amount has been reported for the case of a constant differential pressure Δp by H. C. Cheng and L. H. Chen, Growth of a gas bubble in a viscous fluid, *Phys. Fluids* **A3**, 3530–6 (1986).
21. G. Iooss and D. D. Joseph, *Elementary Stability and Bifurcation Theory* (Springer-Verlag, New York, 1997).
22. A. J. Szeri and L. G. Leal, The onset of chaotic oscillations and rapid growth of a gas bubble at subcritical conditions in an incompressible liquid, *Phys. Fluids* **A3**, 551–5 (1991); Z. C. Feng and L. G. Leal, Bifurcation and chaos in shape and volume oscillations of a periodically driven bubble with two-to-one internal resonance, *J. Fluid Mech.* **266**, 209–42 (1994); L. G. Leal and Z. C. Feng, Nonlinear bubble dynamics, *Annu. Rev. Fluid Mech.* **29**, 201–42 (1997).
23. See references to Nayfeh and Hinch in Ref. 3 (this chapter)
24. J. R. Blake and D. C. Gibson, Cavitation bubbles near boundaries, *Annu. Rev. Fluid. Mech.* **19**, 99–123 (1987); E. A. Brujan, G. S. Keen, A. Vogel and J. R. Blake, The final stage of the collapse of a cavitation bubble close to a rigid boundary, *Phys. Fluids* **14**, 85–92 (2002).
25. This type of instability is known as Rayleigh–Taylor instability. It is discussed in Chap. 12.
26. The generalization of the analysis of the Rayleigh–Taylor instability to the acceleration of a spherical interface was independently reported by two authors: M. S. Plesset, On the stability of fluid flows with spherical symmetry, *J. Appl. Phys.* **25**, 96–8 (1954); G. Birkhoff, Note on Taylor instability, *Q. Appl. Math.* **12**, 306–9 (1954); Stability of spherical bubbles, *Q. Appl. Math.* **13**, 451–3 (1956).
27. See, for example, Ref. 14b, or Ref. 18.
28. The amplitude of oscillation remains constant. Thus any dissipation associated with weak viscous effects will lead to a damped oscillation.
29. Note, however, that once G is negative, the effect of increased k, in a system with no viscous damping, is to produce bubble oscillation at higher frequency.
30. C. M. Bender and S. A. Orszag, *Advanced Mathematical Methods for Scientists and Engineers: Asymptotic Methods and Perturbation Theory* (Springer-Verlag, New York, 1999).

PROBLEMS

Problem 4–1. Temperature Distribution in a Cylindrical Rod. We have a long, cylindrical solid rod of thermal conductivity k, density ρ, and heat capacity C_p. The radius of the cylinder is R. We imagine that the rod is immersed in a liquid whose temperature is time-dependent so that the temperature at the surface of the rod is

$$T = T_0(1 + \varepsilon \sin \omega t).$$

(a) Assume that all properties of the rod are temperature independent. What will the temperature be inside the rod for the asymptotic cases of large and small frequencies, assuming that $|\varepsilon| < 1$. (Although an exact solution to this problem is readily derivable, you should nondimensionalize and solve this problem using asymptotic methods. You can assume that $t \gg R^2/\kappa$, where $\kappa = k/\rho C_p$).
(b) Now suppose that the thermal conductivity depends on temperature according to

$$k = k_0 \left[1 + \alpha \frac{(T - T_0)}{T_0} \right]$$

but that ρC_p is still independent of temperature. Again assume that $t \gg R^2/\kappa$.

Problems

   (i) Obtain a nondimensionalized equation and boundary conditions for the temperature in the rod. Can you solve this equation analytically?
   (ii) Assume that $\alpha = O(1)$ and $\varepsilon \ll 1$. Obtain governing equations and boundary conditions from which you could determine the temperature up to and including terms of $O(\varepsilon^2)$ in an asymptotic framework for small ε. [Apart from the leading-order term, you might note that each of the problems you get will be linear, and thus solvable (at least in principle) by analytic means.]
   (iii) Each of the problems obtained in part (ii) still contains the dimensionless frequency parameter, call it X_ω. What is it? Show that solutions of these problems from (ii) can be solved by using asymptotic methods for small frequencies, i.e., $X_\omega \ll 1$. Obtain results including terms to $O(X_\omega)$.

Problem 4–2. Pressure-Driven Flow Through a Slightly Deformed Tube. Consider flow through a straight tube that is due to a constant-pressure gradient, $\partial p/\partial z = -G$. The cross section of the tube is

$$r_S = R(1 + \varepsilon \sin\theta),$$

where (r, θ, z) represent a circular cylindrical coordinate system. Assume that $\varepsilon \ll 1$, and use the method of domain perturbations to determine the steady-state velocity field in the tube to $O(\varepsilon^2)$. Is the volumetric flow rate per unit cross-sectional area larger or smaller than the flow through a circular tube? Discuss the answer to this question from the point of view of a macroscopic force balance.

Problem 4–3. Reaction–Diffusion in a Spherical Catalyst Pellet. Consider a spherical catalyst pellet. Assume that transport of product is by diffusion, with diffusivity D_{eff}. As in the problem discussed in Section F, we assume that the transport of reactant within the pellet is decoupled from the transport of product. Mass transfer in the gas is sufficiently rapid so that the reactant concentration at the catalyst surface is maintained at c_∞.

(a) Suppose that $D_{eff} = D_0$ (independent of position) and the reaction rate is second order with respect to a reactant with concentration c, i.e.,

$$R = -k_1^{eff} c^2.$$

   (i) Determine the concentration distribution of the reactant inside the pellet assuming that the Dahmköhler number is small,

$$Da \ll 1.$$

   Calculate the effectiveness factor η up to terms $O(Da^2)$.
   (ii) Now suppose that $Da \gg 1$. Obtain a first approximation for η in this limit.
   (iii) How long does it take to get to steady state in the above two limiting cases? Under what conditions is it a good approximation to assume that the surface concentration is maintained at c_∞?
(b) Now suppose

$$D_{eff} = D_0(1 - \varepsilon c) \quad \text{and} \quad R = -k_1^{eff} c.$$

Determine c inside the pellet for $\varepsilon \ll 1$ through terms of $O(\varepsilon^2)$, and calculate the effectiveness factor to the same order of approximation.

Problem 4–4. Asymptotic Analysis of Oscillating Planar Couette Flow. When we considered Problem 3–9, we obtained an exact analytic solution (which can then be evaluated to determine the form of the solution for large and small frequencies). Here we consider the same problem, except now we use asymptotic methods to obtain these asymptotic forms for the solution directly.

(a) Use asymptotic methods to obtain an approximate solution for $\omega d^2/\nu \ll 1$ to at least $O(\omega d^2/\nu)^2$. Compare this solution with the limit of the exact solution for $\omega d^2/\nu \ll 1$. Discuss the first two terms in physical terms (quasi-steady? phase lags, etc.).

(b) Use asymptotic approximation methods to obtain a first approximation to the solution for large frequencies, $\omega d^2/\nu \gg 1$.

(c) Determine the velocity profile during the initial transient. Assume that the fluid was at rest at time $t = 0$ and that the frequency of oscillation is small. Does it matter if ω is small?

Problem 4–5. Cross-Flow Filtration – An Asymptotic Approach. Consider the flow between two flat porous plates separated by a distance d. The flow in the direction parallel to the walls is driven by a constant pressure-gradient

$$-\frac{dP}{dx} = G > 0.$$

We assume that the no-slip condition applies at the surface of the plates, but that there is a vertical flow through the porous walls with a normal velocity V.

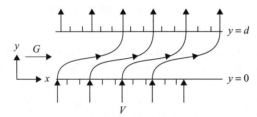

(a) Show that the governing equation for the velocity component parallel to the walls is

$$\rho V \frac{du}{dy} = G + \mu \frac{d^2 u}{dy^2},$$

with $u = 0$ at $y = 0$ and d. Nondimensionalize this equation. The nature of the solution will depend on the single dimensionless parameter, $Re = \rho V d/\mu$. Obtain the exact solution.

(b) Although an exact solution is available, it is instructive to consider asymptotic solutions of this problem for the limiting cases, $Re \ll 1$ and $Re \gg 1$. Derive the first two terms of the asymptotic expansion of the flow field for $Re \ll 1$. Make a plot for $Re = 0.1$ and 1 and compare the exact and asymptotic solutions.

(c) For $Re \gg 1$ it is evident that there exists a boundary-layer in which viscous forces balance the suction effect. Where is the boundary-layer located in the channel? Why must it be located there? Introduce a scaled coordinate $y = Re^\alpha \bar{y}$, where α is a number chosen to balance viscous forces and suction in the governing equation and $\bar{y} = 1 - y$. Why is it necessary to make the change of variables from y to $\bar{y}$ before doing the rescaling? Derive an asymptotic solution for the flow field for $Re \gg 1$. Make a plot for $Re = 1, 10$, and 100 and compare with the exact solution.

Problem 4–6. Parallel Rotating Disks. The flow between a pair of parallel rotating disks is one of the standard geometries used in rheometers, and the flow at higher Reynolds numbers has also been studied in a lot of detail, both because of the complexity of the transition process from one laminar flow to another more complex one as the Reynolds number increases but also because the geometry is relevant to many applications. In all of

Problems

these cases, the radius of the disks is, of course, finite, though often much larger than the gap width between the disks. In this problem, for simplicity, we consider the velocity field for a Newtonian fluid between two *infinite* parallel disks that are separated by a gap H. The lower disk rotates about the central axis with angular velocity Ω_1. Initially we assume that the other disk is stationary.

(a) Write the governing equations and boundary conditions, based on a cylindrical coordinate system with the z axis coincident with the line of centers for the two disks. Nondimensionalize.

(b) Consider the solution for the limiting case in which the Reynolds number

$$Re \equiv \frac{\rho H^2 \Omega}{\mu} = 0.$$

This is called the creeping-flow limit. In this limit, the governing equations are linear, and the solution is expected to exhibit the same spatial symmetry as the imposed behavior at the boundaries of the fluid domain. This suggests that the dimensionless velocity components should take the form $\bar{u}_r = \bar{u}_z = 0$, and $\bar{u}_\theta = \bar{r}F(\bar{z})$, where $u_c = H\Omega$ and $\ell_c = H$. Show that a solution of this form satisfies all of the equations and boundary conditions and determine $F(\bar{z})$.

(c) When the Reynolds number is *not* vanishing, inertia effects will induce secondary currents so that u_z and u_r are nonzero. Determine the inertia-induced changes in the flow, using a regular perturbation scheme for arbitrarily small, but nonzero, Reynolds numbers, including terms to $O(Re^2)$. Sketch the secondary flow patterns that appear at $O(Re)$ in the (r, z) plane, paying special attention to the changes in the flow patterns that occur as the magnitude of the angular velocity Ω is varied.

(d) The problem of flow that is due to rotation of two infinite parallel disks with angular velocities Ω_1 and Ω_2 is one of very few examples in which an exact solution of the full nonlinear Navier–Stokes equations is possible for arbitrary values of the Reynolds number, in the sense that the governing equations can be reduced to a coupled set of ODEs. The form for the velocity component $u_\theta(r, z)$ is dictated by the r dependence of u_θ at the boundaries of the rotating disks to be

$$\bar{u}_\theta = \bar{r}F(\bar{z}),$$

where the nondimensionalization is the same as that done previously. The forms of the velocity components $\bar{u}_r$ and $\bar{u}_z$ and the pressure $\bar{p}$ are then determined from $\bar{u}_\theta$, by use of the r and z components of the Navier–Stokes equations plus the continuity equation, to be

$$\bar{u}_r = \bar{r}H(\bar{z}), \quad \bar{u}_z = G(\bar{z}), \quad \bar{p} = P(\bar{z}).$$

  (i) Derive the coupled set of ordinary ODEs for F, H, G, and P, and obtain corresponding boundary conditions.
  (ii) Demonstrate that the *form* of the solution in the limiting case in which the upper disk is removed ($H \to \infty$, so that we have a single disk rotating in a semi-infinite fluid) is independent of *all* dimensional parameters. Hint: To demonstrate this fact, you need to show that you can rescale all equations and boundary conditions so that none of the dimensional parameters appears explicitly.
  (iii) In the case of two disks, with H finite, it is not possible to eliminate the Reynolds number:

$$Re = \frac{\rho \Omega H^2}{\mu}.$$

Show that the *coupled* set of ODEs for H, G, F, and P can be solved in the limit $Re \ll 1$ by the method of regular perturbation expansions. Sketch the secondary flow patterns that appear at $O(Re)$ in the (r, z) plane, paying special attention to the changes in the flow patterns that occur as the magnitudes and signs of the angular velocities Ω_1 and Ω_2 are varied. Calculate the changes in the shear-stress distributions on the two disks and determine the torque required for maintaining angular velocities Ω_1 and Ω_2. Discuss the sign of the torque components – does it make sense?

Problem 4–7. Planar Channel Flow With an Oscillating Pressure Gradient. We consider the flow of an incompressible, Newtonian fluid between two infinite, plane walls, with a time-dependent pressure-gradient parallel to the walls:

$$-\frac{\partial p}{\partial z} = G_0 \sin \omega t.$$

The walls are separated by a distance d, and the kinematic viscosity of the fluid is ν. This is a time-dependent unidirectional flow that can be solved exactly as in Chap. 3. We are concerned here with the long-term asymptotic behavior only.

(a) Assume ω to be asymptotically small, $\omega \ll 1$. Determine the first three terms in an asymptotic approximation of the exact solution for $(\omega d^2/\nu) \ll 1$.
(b) Determine the first approximation everywhere in the domain for $(\omega d^2/\nu) \gg 1$, again using asymptotic techniques.

Problem 4–8. Pressure-Driven Radial Flow Between Parallel Disks. The flow of a viscous fluid radially outward between two circular disks is a useful model problem for certain types of polymer mold-filling operations, as well as lubrication systems. We consider such a system, as sketched here:

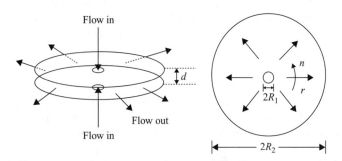

In the figure the flow takes place because of a pressure drop ΔP between the inner and outer radii R_1 and R_2. The two disks are separated by a gap width d.

(a) Assume that the flow is directed radially so that $\mathbf{u} = (u, 0, 0)$ in cylindrical coordinates (r, θ, z). Show that u must then take the form

$$u = \frac{F(z)}{r}.$$

Derive a governing equation for $F(z)$ and show that no solution exists for this equation unless the inertia term involving F^2 is neglected. What is the significance of this result?

(b) In most analyses of this problem, it is assumed, on the basis of the result from part (a), that the relevant Reynolds number is sufficiently small that the creeping-flow approximation can be made so that the velocity field is purely radial. Here, however, we seek

Problems

to understand the role of inertia in this problem by obtaining a solution for small, but nonzero, Reynolds numbers, using the techniques of a regular perturbation expansion.

(i) Obtain a dimensionless form of the governing equations and boundary conditions by assuming that derivatives in the radial direction scale as R_2^{-1} and derivatives with respect to z scale as d^{-1}. Note that the solution to the problem will depend on whether the pressure difference ΔP or the volumetric flow rate Q is held constant (in the former case, the flow rate Q will depend upon the Reynolds number, whereas ΔP will depend on the Reynolds number if Q is held constant). The choice of a characteristic velocity scale will depend on which problem we consider. Eventually, we will consider both cases, so you may wish to consider both possibilities at this point. In any case, you should end up with three dimensionless parameters, $\varepsilon = d/R_2, \delta = R_2/R_1$, and Reynolds number

$$Re \equiv \frac{\rho u_c d^2}{\mu R_2}.$$

(ii) Assume that $Re \ll 1$ and $\varepsilon^2 \ll Re$. Derive governing equations and boundary conditions for the first two terms in an asymptotic expansion of the form

$$\mathbf{u} = \mathbf{u}_0 + Re\mathbf{u}_1 + \cdots +,$$
$$p = p_0 + Re p_1 + \cdots +.$$

It is clear from part (a) that the fluid motion cannot be purely radial except, possibly, in the creeping-flow limit, $Re \to 0$. However, you should be able to demonstrate that the swirl velocity component in the θ direction is zero.

(iii) Obtain the creeping-flow solution for the leading-order terms $\mathbf{u}_0$ and p_0 in the expansion for $Re \ll 1$. Neglect any rearrangement of the flow that occurs in the immediate vicinity of the entry and exit regions at $\bar{r} = \delta$ and $\bar{r} = 1$. You should find that $u_0 \neq 0$, $w_0 = 0$, and $p_0 = p_0(r)$. Derive the relation between volume flow rate Q and pressure drop ΔP that corresponds to this solution, that is,

$$Q = \frac{4\pi d^3 \Delta P}{\ell n(R_2/R_1)}.$$

(iv) Determine the influence of inertia by solving for the second terms in the expansion, namely, $\mathbf{u}_1$ and p_1. As previously noted, the solution will depend on what is held constant. If the pressure difference is maintained at ΔP, independent of Re, the volume flow rate must change. If, on the other hand, the volume flux is maintained at a constant value independent of Re, the required pressure drop ΔP must depend on Re. Consider both cases. In the first case, calculate the effect of inertia on Q. In the second, determine the change in ΔP that is necessary to maintain Q independent of Re.

Problem 4–9. Steady Planar Couette Flow With Porous Boundaries. Newtonian fluid of viscosity μ and density ρ flows between two rigid boundaries $y = 0$ and $y = h$, the lower boundary moving in the x direction with constant speed U, the upper boundary being at rest. The boundaries are porous and there is a vertical velocity $\mathbf{v} = -v_0 \mathbf{e}_y$ with [v_0 a constant at both the upper and lower boundaries, so that there is an imposed downflow across the channel.

(a) Show that the horizontal velocity $\mathbf{u} = u(y)\mathbf{e}_x$ satisfies

$$-\rho v_0 \frac{du}{dy} = \mu \frac{d^2 u}{dy^2},$$

and state the appropriate boundary conditions. By suitable scaling, show that this equation may be written in dimensionless form as

$$\frac{d^2\bar{u}}{d\bar{y}^2} + Re\frac{d\bar{u}}{d\bar{y}} = 0,$$

where $Re = \rho v_0 h/\mu$ is the Reynolds number based on the suction velocity v_0 and the overbars denote dimensionless quantities.

(b) Show by solving the dimensionless DE that

$$\bar{u} = \frac{\exp(-Re\,\bar{y}) - \exp(-Re)}{1 - \exp(-Re)}.$$

A more fundamental way of viewing this problem (and the other unidirectional boundary driven flows from Chap. 3) is in terms of transport of vorticity $\boldsymbol{\omega} = \nabla \wedge \mathbf{u}$. From the solution for $\bar{u}$ show that $\bar{\boldsymbol{\omega}} = \bar{\omega}(\bar{y})\mathbf{e}_z$, where

$$\bar{\omega}(\bar{y}) = \frac{Re\,\exp(-Re\,\bar{y})}{1 - \exp(-Re)}$$

This equation shows that vorticity is a maximum at the moving wall (which acts a source of vorticity) and decays exponentially on a length scale $O(1/Re)$ by the action of viscous forces. Thus for $Re\,\bar{y} \gg 1$ the flow is effectively irrotational. In this respect the problem is atypical as we normally expect vorticity to decay on a diffusive length scale of $O(1/\sqrt{Re})$ (cf. oscillatory Couette flow). The reason for this difference is that in the present problem we have a balance between *convective* inertia and viscous forces as opposed to a balance between *transient* inertia and viscous forces.

(c) Now let us turn our attention to calculating the volumetric flow rate of fluid in the x direction, $\bar{Q} = \int_0^1 \bar{u}(\bar{y})d\bar{y}$. Although we can calculate $\bar{Q}$ exactly, it is instructive to perform perturbation expansions for $Re \ll 1$ and $Re \gg 1$.

(i) For $Re \ll 1$, resolve the problem in the form of a regular expansion of the fluid velocity $\bar{u}(y; Re) = \bar{u}_0 + Re\bar{u}_1 + O(Re^2)$ and hence show that

$$\bar{Q} = \frac{1}{2} - \frac{1}{12}Re + O(Re^2).$$

Does the preceding formula make sense?

(ii) For $Re \gg 1$ it is evident that there exists a boundary layer in which viscous forces balance the suction effect. Where is the boundary-layer located in the channel? Why must it be located there? Introduce a scaled coordinate $Y = Re^\alpha \bar{y}$, where α is a number chosen to balance viscous forces and suction in the governing equation. Deduce that the leading-order velocity in the boundary-layer is $\bar{u} = \exp(-Re\,Y)$ and hence find the corresponding volumetric flow rate $\bar{Q}$ for $Re \gg 1$. Sketch as a function of Re.

Problem 4–10. An Alternative Derivation of the Rayleigh–Plesset Equation. Find the total kinetic energy E_k of the liquid outside a spherical gas bubble that is undergoing time-dependent changes in volume in an unbounded, incompressible, Newtonian fluid. Show that the net rate of working by the pressure inside the bubble $\hat{p}$ at the inner side of the bubble boundary is

$$\dot{W} = 4\pi(\hat{p} - p_\infty)R^2\dot{R},$$

where $R(t)$ is the bubble radius. Hence, use the principle of energy conservation in the macroscopic form

$$\dot{W} = \dot{E}_k + \dot{E}_s + \Phi,$$

where E_S is the surface free energy and Φ is the rate of viscous dissipation in the fluid, to derive the Rayleigh–Plesset equation in the form of Eq. (4–208).

Problem 4–11. The Effects of Weak Inertia for a Cone-and-Plate Flow. A commonly utilized geometry for rheological experiments is the flow between a flat plate and a small-angle cone that is rotating with some angular velocity Ω. The advantage of this geometry relative to either parallel disks or the cylindrical Couette flow is that the shear rate is constant, independent of position in the gap. To determine the velocity field, it is convenient to use a spherical coordinate system (r, θ, ϕ) with its origin coincident with the point where the tip of the cone and the plate "touch" (of course, in reality, a very small gap is retained, but it is convenient to envision the geometry as though the cone and the plate touch). When described using this spherical coordinate system, the problem will be a function of r and θ only. It is generally assumed that the relevant Reynolds number is small enough that the flow is strictly in the azimuthal direction. In this problem, we consider the effects of inertia for small values of the Reynolds number, assuming that the radius of the cone-and-plate is infinite.

Write the governing equations and boundary conditions in dimensionless form. Solve for the velocity profile in the cone-and-plate flow as a regular perturbation problem for small Reynolds numbers. The solution will be very similar to that of the parallel plate problem.

Hint: The angle between the cone and the plate is very small. You will need to rescale the problem for very flat cones to get a tractable problem. It is useful to define the dimensionless angle: $\theta = [(\pi/2) - \alpha\theta^*]$, where θ^* varies between 0 at the flat plate and 1 at the cone. You then take the limit as $\alpha \to 0$ and get rid of a lot of the nasty terms in spherical polar coordinates. You will find that, to do this properly, the perturbation velocities must be rescaled with α, with u_r proportional to α^2 and (from the continuity equation) u_θ proportional to α^3. After the rescaling all the complex stuff is $O(\alpha^2)$ and can be neglected.

Problem 4–12. Consider the flow in a tube of radius a with an accelerating surface velocity as illustrated in the figure.

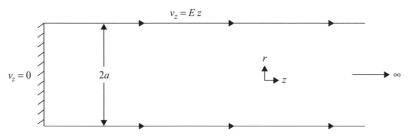

Definition sketch of accelerating tube.

(a) Show that the full Navier–Stokes equations admit an exact solution in which $v_z = Ezf'(r)/r$, where $f'(r)$ is a function of r only. Derive the dimensionless ODE for f and state all boundary conditions.
(b) Obtain the solution to first order in the Reynolds number, $Re = Ea^2/\nu \ll 1$; that is, through the term proportional to Re. Note that Re can be positive or negative depending on the direction of the surface velocity.
(c) Consider now the limit of large Reynolds numbers (both positive and negative). Far from the wall viscosity is not important. Obtain the leading-order (inviscid) solution valid there. Near the accelerating wall a boundary-layer is necessary in order to satisfy the no-slip boundary condition. Determine how the thickness of the boundary layer scales with the Reynolds number and, if possible, construct the leading-order solution that matches your inviscid solution.

Interestingly, for positive Re there is a gap in the Reynolds number, $10.25 < Re < 147$, in which there is *no solution*. Above $Re = 147$ there are two solutions and then above $Re \approx 767$ there are four solutions [see *J. Fluid Mech.* **112**, 127–50 (1981)].

Problem 4–13. Let us reconsider Problem 3–36. In part (d) of this problem, you were asked to obtain an exact solution for the case of a parallel wall channel of width h. In the present problem, we will directly obtain the approximate asymptotic form of this solution for large and small frequencies. The starting point, if you did not do it in solving 3–36, is to nondimensionalize the problem.

(a) For the limit of small dimensionless frequency, you should obtain a solution to third order in the dimensionless frequency. Compare with the exact solution, and comment on the range of frequency for which this asymptotic solution provides a good approximation.

(b) For the limit of large dimensionless frequency, use asymptotic analysis to obtain a uniformly valid first approximation to the exact solution and again compare with the exact solution.

Problem 4–14. Porous Channel Flow. A semi-infinite channel of width $2h$ is closed at one end ($x = 0$) and has porous walls through which fluid is removed at a constant rate V, as illustrated in the figure.

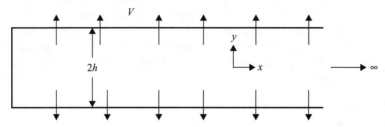

Definition sketch of a porous channel.

(a) Show that the full Navier–Stokes equations admit an *exact* solution in which $v_y = V f(y)$, where $f(y)$ is a nondimensional function of y only. Derive the dimensionless ODE for f and state all boundary conditions. Note that at the porous wall the fluid still satisfies the no-slip boundary condition $v_x = 0$.

(b) Obtain the solution to first order in the Reynolds number, $Re = Vh/v$, assuming that $Re \ll 1$.

(c) Consider now the limit of large Reynolds numbers. Far from the walls, viscosity is not important, and you should obtain the "inviscid" solution valid there. Near the porous walls a boundary layer is necessary to satisfy the no-slip boundary condition. Determine how the thickness of the boundary layer scales with the Reynolds number and construct the leading-order solution that matches your inviscid solution.

Problem 4–15. An Application of Optical Scattering/Fluorescence. A currently active area of research is the application of optical scattering and fluorescence for the noninvasive location of tumors. The patient is injected with a fluorescent dye that preferentially absorbs into a tumor. A portion of the patient's skin located near the suspected tumor is illuminated with a source of photons, such as a laser diode. The wavelength or energy of the photons is such that the absorbed dye fluoresces, and the emitted photons are then detected by a photomultiplier also located on the patient's skin. This optical response is then analyzed to determine the location and extent of the tumor.

Problems

Provided the photons are scattered many times over our length scale of interest, their transport can be fairly well modeled as processes of diffusion and reaction. That is, the number density of photons ϕ is given by

$$\frac{\partial \phi}{\partial t} = D\nabla^2 \phi - \sigma \phi,$$

where D is the effective diffusivity, σ is an absorption rate and t is of course time.

We will study a 1D model problem to understand quantitatively how we might analyze such a response. Assume that the surface of the skin is located at $x = 0$ and that a tumor is located at $x \geq L$. Note we are neglecting any variations parallel to the surface of the skin or tumor. The surface of the skin is exposed to a sinusoidal source of photons such that

$$\phi(0, t) = \phi_0[1 + \sin(\omega t)].$$

To simplify matters we will also assume that most of the absorption of photons by the tumor occurs in a thin region near L. We can then practically model this process as a surface reaction with a rate $r_\phi = -k\phi(L)$, where k is an effective surface rate constant that has units of length/time. Thus the flux of incident photons into the tumor is equal to the rate of consumption by the surface reaction. Note that, for every photon absorbed by the tumor, we assume that a fluorescent photon of a different wavelength is emitted toward the surface of the skin. Further, the number of density of fluorescent photons at the surface of the skin is negligibly small. (The last two boundary conditions are reasonable, simplifying approximations, but in fact not the most accurate. However, they are quite suitable for our purposes.) Our objective is to determine the flux of these emitted photons and how its frequency response can be used to locate the depth of the tumor.

(a) Neglecting the absorption of photons by other than the tumor ($\sigma = 0$), what are the equations and boundary conditions for this problem? (To keep the algebra simple, assume the emitted photons have the same diffusivity D.)
(b) What are the dimensionless equations and boundary conditions? Clearly list all your scalings.
(c) What is the *exact* solution for the flux of emitted photons, $\mathbf{q}_\theta = -D\nabla\theta$, where θ is the number density of fluorescent photons? (This result should be fairly easy to determine, but difficult to evaluate in practice.)
(d) Determine the flux of emitted photons at the surface of the skin for the cases of relatively large and small ω? ω is large and small relative to what?
(e) Practically, we do not know k and D. However, we suspect k is large. Therefore, given measurements of the magnitude and phase shift of the flux, how can we determine the distance L at which the tumor is located?
- Hint: Doubtless you will have solved the complementary complex problem. The phase shift of the flux $\alpha = \arctan[\mathrm{Im}(q)/\mathrm{Re}(q)]$ evaluated at $t = 0$.
(f) To what depth can a tumor be observed with this method?
(g) Qualitatively, what effect does $\alpha > 0$ have on this depth and why?

5

The Thin-Gap Approximation – Lubrication Problems

In Chap. 4 we explored the consequences of a weak departure from strict adherence to the conditions for unidirectional flow; namely, the effect of slight curvature in flow through a circular tube. For that case, the centripetal acceleration associated with the curved path of the primary flow was shown to produce a weak secondary motion in the plane orthogonal to the tube axis. In this chapter we consider another class of deviations from unidirectional flow that occur when the boundaries are slightly nonparallel.

If the boundaries of the flow domain are not parallel, the magnitude of the primary velocity component must vary as a function of distance in the flow direction. This not only introduces a number of new physical phenomena, as we shall see, but it also means that the Navier–Stokes equations cannot be simplified following the unidirectional flow assumptions of Chap. 3, and exact analytical solutions are no longer possible. In this chapter, we thus consider only a special limiting case, known as the "thin-gap" limit, in which the distance between the boundaries is small compared with the lateral gap width. In this case, we shall see that we can obtain approximate analytical solutions by using the asymptotic and scaling techniques that were introduced in the preceding chapter.

The resulting theory, at the leading order of approximation, is applicable to a number of important phenomena. There are two generic classes of problems. When the bounding surfaces do not deform (this usually means that they are solids), the geometry of the gap between the boundaries is known, and we obtain the famous *Reynolds lubrication equations* from which we can calculate the pressure distribution in the gap. We shall see that the combination of relative motion of the boundaries and a thin but nonparallel gap can lead to *very high pressures* that tend to keep the bounding surfaces from coming into contact even when they are subjected to a considerable normal load. This is the basis for *lubrication* between moving surfaces in many mechanical systems.[1] On the other hand, when one boundary (or both) is a fluid interface, the leading-order equations in the thin-gap limit are used to calculate the evolving *shape* of the film. This class of problems is discussed in the next chapter. Again, we shall see that there are many important applications, including such disparate examples as the geological problem of gravitational spreading of a body of viscous fluid on a horizontal surface, or the thinning (and possible rupture) of the thin film between two drops in a coalescence event.[2]

In this chapter, we begin by considering a specific example of a thin-gap problem – namely, flow between two cylinders caused by rotation of the inner cylinder, when the central axes are parallel but offset. In the limit where the gap between the cylinders is small relative to their radii, this is known as the "journal-bearing" problem and is an example of a classical lubrication problem. Following this, we derive governing equations for fluid motion within a completely general 3D thin-gap region. These are then specialized to the appropriate form for the two generic classes of thin-gap problems just mentioned, and the

A. The Eccentric Cylinder Problem

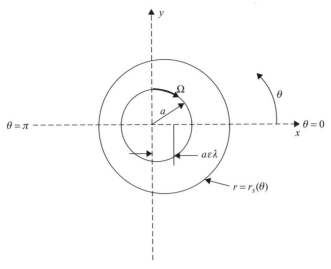

Figure 5–1. The eccentric cylinder geometry demonstrating the cylindrical coordinate system with origin at the center of the inner cylinder.

remainder of this chapter is devoted to a number of examples of lubrication problems. The classical "thin-film" dynamics problems in which the nonlinear DE for the film shape function is solved will then be discussed in the following chapter. We shall see that similarity transformations, first introduced in Chap. 3, will play a key role in solving and understanding this second class of problems.

A. THE ECCENTRIC CYLINDER PROBLEM

As stated, we begin with the special problem of flow between two rotating cylinders whose axes are parallel but offset to produce the eccentric cylinder geometry shown in Fig. 5–1. In the *concentric limit*, this is the famous *Couette flow* problem, which was analyzed in Chap. 3.

To obtain governing equations and boundary conditions, we adopt the circular cylindrical coordinate system that is shown, along with a top view of the eccentric cylinder system, in Fig. 5–1. In this system, the origin of the coordinate system is chosen to be coincident with the central axis of the inner cylinder, which is assumed to have a radius a and an angular velocity Ω in the direction shown. The radius of the outer cylinder is assumed to be $a(1+\varepsilon)$, and its center is offset along the $\theta = 0$ axis relative to the center of the inner cylinder by an amount $a\varepsilon\lambda$. In the journal-bearing problem, the outer cylinder does not rotate ($\Omega = 0$). The surface of the inner cylinder is thus

$$r' = a,$$

and the equation describing the surface of the outer eccentric cylinder is

$$r' = r_s(\theta) = a\varepsilon\lambda \cos\theta + \sqrt{(a\varepsilon\lambda \cos\theta)^2 - a^2\left[\varepsilon^2\lambda^2 - (1+\varepsilon)^2\right]}. \tag{5–1}$$

Note that we use a prime to signify dimensional variables.

The parameter λ determines the degree of eccentricity. If $\lambda = 1$, the two cylinders touch at $\theta = \pi$ whereas they are concentric if $\lambda = 0$. The range of allowable values for λ is thus

$$0 \le \lambda < 1. \tag{5–2}$$

The Thin-Gap Approximation – Lubrication Problems

The gap width between the cylinders varies as a function of θ, being a maximum at $\theta = 0$ and a minimum at $\theta = \pi$. The general expression for the dimensional gap width is

$$h'(\theta) = a(1+\varepsilon)\left[1 - \frac{\varepsilon^2 \lambda^2 \sin^2\theta}{(1+\varepsilon)^2}\right]^{1/2} + a\varepsilon\lambda\cos\theta - a. \qquad (5\text{--}3)$$

The governing equations for arbitrary λ and ε are the Navier–Stokes and continuity equations (3–55)–(3–58). We assume that the cylinders are infinitely long so that all z dependence is eliminated. Further, because there is no motion in the z direction at the boundaries, it can be seen from (3–58) that $u'_z = 0$. Finally, it will be convenient for comparisons between this and later sections of this chapter if we introduce the modified radial variable

$$y' \equiv r' - a \qquad (5\text{--}4)$$

in place of r', so that y' varies from 0 at the surface of the inner cylinder to $y' = h'(\theta)$ on the surface of the outer cylinder. With this change, and the assumptions $u'_z = 0$ and $\partial/\partial z' \equiv 0$, we can write the governing equations in the form

$$\frac{\partial}{\partial y'}[(y'+a)u'_r] + \frac{\partial u'_\theta}{\partial \theta} = 0, \qquad (5\text{--}5)$$

$$\rho\left[u'_r \frac{\partial u'_r}{\partial y'} + \frac{u'_\theta}{y'+a}\left(\frac{\partial u'_r}{\partial \theta} - u'_\theta\right)\right]$$
$$= -\frac{\partial p'}{\partial y'} + \mu\left[\frac{1}{y'+a}\frac{\partial}{\partial y'}\left((y'+a)\frac{\partial u'_r}{\partial y'}\right)\right. \qquad (5\text{--}6)$$
$$\left. + \frac{1}{(y'+a)^2}\frac{\partial^2 u'_r}{\partial \theta^2} - \frac{u'_r}{(y'+a)^2} - \frac{2}{(y'+a)^2}\frac{\partial u'_\theta}{\partial \theta}\right],$$

$$\rho\left[u'_r \frac{\partial u'_\theta}{\partial y'} + \frac{u'_\theta}{y'+a}\left(\frac{\partial u'_\theta}{\partial \theta} + u'_r\right)\right]$$
$$= -\frac{1}{(y'+a)}\frac{\partial p'}{\partial \theta} + \mu\left[\frac{1}{y'+a}\frac{\partial}{\partial y'}\left((y'+a)\frac{\partial u'_\theta}{\partial y'}\right)\right. \qquad (5\text{--}7)$$
$$\left. + \frac{1}{(y'+a)^2}\frac{\partial^2 u'_\theta}{\partial \theta^2} - \frac{u'_\theta}{(y'+a)^2} + \frac{2}{(y'+a)^2}\frac{\partial u'_r}{\partial \theta}\right].$$

The boundary conditions at the cylinder surfaces are just the kinematic and no-slip conditions, namely,

$$u'_r = 0, \quad u'_\theta = -\Omega a \quad \text{at } y' = 0 \qquad (5\text{--}8)$$

and

$$\mathbf{u}' \cdot \mathbf{n} = 0, \quad \mathbf{u}' \cdot \mathbf{t} = 0 \quad \text{at } y' = h'(\theta). \qquad (5\text{--}9)$$

We note that the normal and tangential velocities at the outer cylinder surface are not u'_r and u'_θ because the unit normal and tangent vectors are not equal to $\mathbf{i}_r$ and $\mathbf{i}_\theta$. However, because both $\mathbf{u}' \cdot \mathbf{n}$ and $\mathbf{u}' \cdot \mathbf{t}$ are zero, and each involves a linear combination of u'_r and u'_θ, it is sufficient to apply the conditions

$$u'_r = u'_\theta = 0 \quad \text{at } y' = h'(\theta). \qquad (5\text{--}10)$$

A. The Eccentric Cylinder Problem

1. The Narrow-Gap Limit – Governing Equations and Solutions

Now, the general problem of (5–5)–(5–10) is highly nonlinear and, for an arbitrary occentric cylinder geometry, it can only be solved numerically – i.e., for arbitrary ε and λ in the range $0 < \lambda < 1$.[3] However, for $Re = 0$, an exact analytic solution can be obtained by a coordinate transformation. In addition, for $Re \neq 0$, there are two limiting cases for which we can use asymptotic methods to obtain approximate analytic solutions. These are *slight eccentricity*

$$\lambda \ll 1, \quad \varepsilon = O(1)$$

and the *narrow-gap* case

$$\varepsilon \ll 1, \quad \lambda = O(1).$$

In the present book, we consider only the narrow-gap limit. As noted earlier, this limit is the "journal-bearing" problem, which is an example of the general class of *lubrication problems* in which very large pressures are generated by a relative translational motion of solid "nonparallel" boundaries when the gap between them is very thin.[4] As we shall see later in this section, these very large pressures can be used to support a load on the inner cylinder while maintaining a nonzero gap. The practical utility of the journal bearing derives from the fact that the "cost" in terms of the torque that is required for maintaining rotational motion with angular velocity Ω increases more slowly with decrease of ε than the "load-supporting" pressure force on the inner cylinder increases.

The geometry of the eccentric cylinder configuration in the narrow-gap limit is described by the limiting form as $\varepsilon \to 0$ of the general expressions (5–1) and (5–3). Thus,

$$r'_s(\theta) = a(1+\varepsilon) + a\varepsilon\lambda \cos\theta + O(\varepsilon^2),$$
$$h'(\theta) = a\varepsilon(1 + \lambda \cos\theta) + O(\varepsilon^2).$$

To see how the *thin-gap* approximation $\varepsilon \ll 1$ simplifies this problem, it is again necessary to nondimensionalize the governing equations and boundary conditions. The characteristic velocity for the polar velocity component is clearly

$$u_c = \Omega a. \tag{5–11}$$

Furthermore, the length scale characteristic of velocity gradients in the thin gap is just the characteristic distance across the gap,

$$\ell_c = \varepsilon a, \tag{5–12}$$

and we should expect that the dimensionless derivative $\partial u_\theta/\partial y$ obtained by means of (5–11) and (5–12) is $O(1)$. Gradients along the gap (in the θ direction), on the other hand, are characterized by the length scale a, and this is inherent in the governing equations (5–5)–(5–7). The large difference between the length scale characteristic of changes in velocity along the gap and across the gap – namely, a and $a\varepsilon$, respectively – is typical of thin-gap, lubrication-type geometries.

The only dimensional quantity that remains to be nondimensionalized is the radial velocity component u'_r. To determine a characteristic scale for u'_r, we utilize the continuity equation (5–5) along with the dimensionless, forms, (5–11) and (5–12), for u'_θ and y'. For convenience, let us denote the characteristic magnitude of u'_r as V. Then, substituting $u'_r = Vu_r$, plus (5–11) and (5–12), into (5–5), we obtain

$$\frac{1}{\varepsilon a} \frac{\partial}{\partial y}[(\varepsilon a y + a) V u_r] + a\Omega \frac{\partial u_\theta}{\partial \theta} = 0. \tag{5–13}$$

The Thin-Gap Approximation – Lubrication Problems

Solving for V, this can be expressed in the alternative form

$$V = -\varepsilon\Omega a \left\{ \frac{\partial u_\theta}{\partial \theta} \bigg/ \frac{\partial}{\partial y}[(1+\varepsilon y)u_r] \right\}. \tag{5-14}$$

Because the quantity in curly braces is expressed entirely in terms of dimensionless variables, it is assumed to be $O(1)$. To preserve the balance of terms in (5–13), and thus ensure that mass conservation is respected, it follows that

$$V = O(\varepsilon a \Omega)$$

or

$$u_r' = \varepsilon a \Omega u_r. \tag{5-15}$$

Evidently, if ε were $O(1)$, the scales characteristic of u_r' and u_θ' would be comparable. Here, however, we are concerned with the limit $\varepsilon \ll 1$, and in this case, we see that the characteristic magnitude of the radial velocity component is *very much smaller* than the polar component, u_θ'. We shall see later that it is not all that unusual in fluid mechanics for the velocity components in different directions to exhibit different scales (e.g., boundary layers – Chap. 10). In the present problem this is a consequence of the extreme geometry of the thin-gap limit.

Finally, to nondimensionalize the equations of motion, we require a characteristic pressure scale. However, it is not immediately obvious what choice is appropriate. We temporarily postpone the decision by introducing the formal nondimensionalization

$$p' = \pi p, \tag{5-16}$$

with a characteristic pressure π yet to be determined. With the dimensionless variables, u_r, u_θ, y, and p defined by (5–11), (5–12), (5–15), and (5–16), the continuity and Navier–Stokes equations take the forms

$$\frac{\partial}{\partial y}[(1+\varepsilon y)u_r] + \frac{\partial u_\theta}{\partial \theta} = 0, \tag{5-17}$$

$$\left(\frac{\rho a^2 \Omega}{\mu}\right) \varepsilon \left[\varepsilon \left(u_r \frac{\partial u_r}{\partial y} + \frac{u_\theta}{1+\varepsilon y} \frac{\partial u_r}{\partial \theta} \right) - \frac{u_\theta^2}{(1+\varepsilon y)} \right]$$
$$= -\frac{\pi}{\mu\Omega}\frac{\partial p}{\partial y} + \left\{ \frac{1}{1+\varepsilon y}\frac{\partial}{\partial y}\left[(1+\varepsilon y)\frac{\partial u_r}{\partial y}\right] \right.$$
$$\left. + \frac{\varepsilon^2}{(1+\varepsilon y)^2}\frac{\partial^2 u_r}{\partial \theta^2} - \frac{\varepsilon^2 u_r}{(1+\varepsilon y)^2} - \frac{2\varepsilon}{(1+\varepsilon y)^2}\frac{\partial u_\theta}{\partial \theta} \right\}, \tag{5-18}$$

$$\left(\frac{\rho a^2 \Omega}{\mu}\right) \varepsilon^2 \left[u_r \frac{\partial u_\theta}{\partial y} + \frac{u_\theta}{1+\varepsilon y} \frac{\partial u_\theta}{\partial \theta} + \varepsilon \frac{u_\theta u_r}{(1+\varepsilon y)} \right]$$
$$= -\frac{\pi \varepsilon^2}{\mu\Omega}\frac{1}{(1+\varepsilon y)}\frac{\partial p}{\partial \theta} + \left\{ \frac{1}{1+\varepsilon y}\frac{\partial}{\partial y}\left[(1+\varepsilon y)\frac{\partial u_\theta}{\partial y}\right] \right.$$
$$\left. + \frac{\varepsilon^2}{(1+\varepsilon y)^2}\left(\frac{\partial^2 u_\theta}{\partial \theta^2} - u_\theta\right) + \frac{2\varepsilon^3}{(1+\varepsilon y)^2}\frac{\partial u_r}{\partial \theta} \right\}. \tag{5-19}$$

A. The Eccentric Cylinder Problem

The boundary conditions become

$$u_r = 0, \quad u_\theta = -1 \quad \text{at } y = 0,$$
$$u_r = u_\theta = 0 \quad \text{at } y = h(\theta), \tag{5-20}$$

where

$$h(\theta) = 1 + \lambda \cos\theta \tag{5-21}$$

is the dimensionless gap width.

On examining (5–18) and (5–19), we see that, besides the centrifugal pressure term involving u_θ^2, the largest of the nonlinear, inertial terms is

$$O\left(\varepsilon^2 \frac{\rho a^2 \Omega}{\mu}\right) = O(\varepsilon^2 Re)$$

compared with the *largest* of the viscous terms, which appears to be $O(1)$. The narrow-gap lubrication limit involves neglect of both the $O(\varepsilon)$ and $O(\varepsilon^2)$ contributions to the viscous terms, and the $O(\varepsilon Re)$ and $O(\varepsilon^2 Re)$ inertial terms in (5–18) and (5–19). Indeed, if we seek an asymptotic solution of (5–17)–(5–19) in the form

$$\mathbf{u} = \mathbf{u}^{(0)} + O(\varepsilon, \varepsilon Re) + \cdots +,$$
$$p = p^{(0)} + O(\varepsilon, \varepsilon Re) + \cdots +, \tag{5-22}$$

we can obtain the governing equations for the first term $[\mathbf{u}^{(0)}, p^{(0)}]$ by taking the limit $\varepsilon, \varepsilon Re \to 0$ in the full equations. The result is

$$\frac{\partial u_r^{(0)}}{\partial y} + \frac{\partial u_\theta^{(0)}}{\partial \theta} = 0, \tag{5-23}$$

$$\frac{\partial^2 u_r^{(0)}}{\partial y^2} - \frac{\pi}{\mu \Omega} \frac{\partial p^{(0)}}{\partial y} = 0, \tag{5-24}$$

$$\frac{\partial^2 u_\theta^{(0)}}{\partial y^2} - \frac{\pi \varepsilon^2}{\mu \Omega} \frac{\partial p^{(0)}}{\partial \theta} = 0. \tag{5-25}$$

In contrast to full equations (5–18) and (5–19), these equations are *linear*. Although the parameter ε^2 appears in the pressure-gradient term of (5–25), making it seem at first that this term should have been dropped in the limit $\varepsilon \to 0$, we must remember that the characteristic pressure π has not been specified yet, and it is possible that $\pi \varepsilon^2$ does not become small in the limit $\varepsilon \to 0$. The reader may note that Eq. (5–24) differs fundamentally from the balance (3–68) for the concentric cylinder problem. The acceleration term u_θ^2/r, which is the dominant contribution to the radial pressure gradient in the Couette flow problem, is smaller by $O(\varepsilon)$ than the viscous term shown in (5–24). This is a consequence of the fact that the pressure gradients that are induced in the thin-fluid layer because of the nonuniform gap width are *much larger* than those associated with the centripetal acceleration for $\varepsilon \ll 1$ and $\lambda = O(1)$.

It is important to note that (5–23)–(5–25) are identical in form to the equations that would be obtained if the problem had been formulated from the beginning in terms of a local Cartesian coordinate system, with $\bar{y}$ normal to the surface of the inner cylinder, and

$$x \approx \theta$$

The Thin-Gap Approximation – Lubrication Problems

tangent to that surface, instead of beginning with the cylindrical coordinate system that was actually used. Thus we see that the *curvature* of the boundaries, which occurs on the scale a, can be neglected in the thin-gap limit when formulating the leading-order (lubrication) equations of motion and continuity. All of the terms in (5–17)–(5–19) that existed because the cylindrical coordinate system was curvilinear instead of Cartesian drop out at this leading order of approximation. Although we have demonstrated the reduction to a local Cartesian form only for the specific case of a circular cylindrical boundary geometry, the same result is true for curved boundaries of any shape provided that the thin-film approximation is applicable. We shall use this fact later in this chapter without any added proof.

To complete our derivation of the leading-order equations for flow in the thin gap, we must specify the characteristic pressure π. It is evident in examining (5–23)–(5–25) that one of the pressure-gradient terms must be retained in the limit $\varepsilon \to 0$, as the system of equations would otherwise be overdetermined. Thus, either $\pi = \mu\Omega$ or $\pi = \mu\Omega/\varepsilon^2$. Either choice gives a set of three equations for the three unknowns $u_r^{(0)}$, $u_\theta^{(0)}$, and $p^{(0)}$. However, only the equations corresponding to $\pi = \mu\Omega/\varepsilon^2$ yield solutions that can satisfy all of the boundary conditions (5–20). In particular, if $\pi = \mu\Omega$ it can be shown that the condition $u_r^{(0)} = 0$, at $y = h$, can be satisfied only if $\partial h/\partial \theta = 0$ so that the cylinders are concentric and $u_r^{(0)} \equiv 0$ everywhere. Thus,

$$\pi = \frac{\mu\Omega}{\varepsilon^2}, \qquad (5\text{–}26)$$

and the governing equations reduce to form

$$\frac{\partial u_r^{(0)}}{\partial \bar{y}} + \frac{\partial u_\theta^{(0)}}{\partial \theta} = 0, \qquad (5\text{–}27)$$

$$\frac{\partial^2 u_\theta^{(0)}}{\partial \bar{y}^2} - \frac{\partial p^{(0)}}{\partial \theta} = 0, \qquad (5\text{–}28)$$

$$\frac{\partial p^{(0)}}{\partial y} = 0(\varepsilon^2). \qquad (5\text{–}29)$$

It should be noted that the scaling (5–26) implies that very large pressures, $O(1/\varepsilon^2)$, will build up as the gap width decreases.

According to (5–29), pressure $p^{(0)}$ depends only on θ, and the problem reduces to the solution of (5–27) and (5–28) subject to boundary conditions (5–20). Equations (5–27) and (5–28) are known as the *lubrication (thin-film) equations*. We see that they resemble the equations for unidirectional flow. However, in this case, the boundaries are not required to be parallel. Thus $u_\theta^{(0)}$ can be a function of the streamwise variable θ, and $u_r^{(0)}$ will not be zero in general. Furthermore, because $u_\theta^{(0)}$ is a function of θ, so is $\partial p^{(0)}/\partial \theta$. Finally, whereas the unidirectional flow equations are exact, the lubrication equations are only the leading-order approximation to the exact equations of motion and continuity in the asymptotic, thin-gap limit, $\varepsilon \to 0$.

The solution of (5–27) and (5–28) is straightforward. We integrate (5–28) twice with respect to y, keeping in mind that $\partial p^{(0)}/\partial \theta$ is independent of y according to (5–29). Thus,

$$u_\theta^{(0)} = \frac{dp^{(0)}}{d\theta}\frac{y^2}{2} + ay + b. \qquad (5\text{–}30)$$

A. The Eccentric Cylinder Problem

When the boundary conditions on $u_\theta^{(0)}$ from (5–20) are applied, this becomes

$$u_\theta^{(0)} = -\left(1 - \frac{y}{h}\right) + \frac{dp^{(0)}}{d\theta}\left(\frac{y^2}{2} - \frac{yh}{2}\right). \tag{5-31}$$

It is evident that $u_\theta^{(0)}$ depends both on y and θ, though the velocity field has the same form locally (that is, for fixed θ) as for unidirectional flow between parallel plane boundaries. In this "journal-bearing" problem, the motion is due to the motion of the boundary only. The pressure gradient is a *consequence* of this motion and the fact that the gap width is not constant.

To determine the pressure gradient, we turn to the continuity equation (5–27) and the two boundary conditions on $u_r^{(0)}$ from (5–20). Substituting for $\partial u_\theta^{(0)}/\partial \theta$ from (5–31), we find that the continuity equation becomes a first-order DE for $u_r^{(0)}$:

$$\frac{\partial u_r^{(0)}}{\partial y} = \frac{y}{h^2}\frac{dh}{d\theta} - \frac{d^2 p^{(0)}}{d\theta^2}\left(\frac{y^2}{2} - \frac{yh}{2}\right) + \frac{dp^{(0)}}{d\theta}\left(\frac{y}{2}\frac{\partial h}{\partial \theta}\right).$$

Integrating with respect to y we find that

$$u_r^{(0)} = \frac{y^2}{2h^2}\frac{dh}{d\theta} - \frac{d^2 p^{(0)}}{d\theta^2}\left(\frac{y^3}{6} - \frac{y^2 h}{4}\right) + \frac{dp^{(0)}}{d\theta}\left(\frac{y^2}{4}\frac{dh}{d\theta}\right), \tag{5-32}$$

where the constant of integration has been set equal to zero in order that $u_r^{(0)}$ satisfies the boundary condition

$$u_r^{(0)} = 0 \quad \text{at } y = 0.$$

The second boundary condition on $u_r^{(0)}$ from (5–20) then yields a differential equation for the unknown pressure $p^{(0)}$. Thus, applying

$$u_r^{(0)} = 0 \quad \text{at } y = h$$

to (5–32), we obtain

$$\frac{d}{d\theta}\left[h^3\frac{dp^{(0)}}{d\theta}\right] = -6\frac{dh}{d\theta}. \tag{5-33}$$

This equation is a special case of the famous *Reynolds equation* from general lubrication theory, which will be discussed in the next section.[1]

If we integrate (5–33) once with respect to θ, we find that

$$\frac{dp^{(0)}}{d\theta} = -\frac{6}{h^2} + \frac{c}{h^3}, \tag{5-34}$$

where c is a constant of integration. Because the pressure $p^{(0)}$ must be periodic in θ,

$$\int_0^{2\pi}\left[\frac{dp^{(0)}}{d\theta}\right]d\theta \equiv 0, \tag{5-35}$$

and, thus, integrating (5–34) we find that

$$c = 6\frac{\int_0^{2\pi} d\theta/h^2}{\int_0^{2\pi} d\theta/h^3}. \tag{5-36}$$

The Thin-Gap Approximation – Lubrication Problems

The pressure gradient $dp^{(0)}/d\theta$ can now be expressed in the form

$$\boxed{\frac{dp^{(0)}}{d\theta} = -\frac{6}{h^2} + \frac{6}{h^3}\frac{I_2}{I_3},} \qquad (5\text{–}37)$$

where

$$I_n(\lambda) \equiv \int_0^{2\pi} \frac{d\theta}{(1+\lambda\cos\theta)^n}.$$

Finally, substituting for $dp^{(0)}/d\theta$ in (5–30), we have

$$\boxed{\begin{aligned}u_\theta^{(0)} &= \left[\frac{-3}{(1+\lambda\cos\theta)^2}\right]\left[1 - \frac{I_2}{I_3(1+\lambda\cos\theta)}\right]\\ &\quad\times \left[y^2 - (1+\lambda\cos\theta)y\right] - 1 + \frac{y}{(1+\lambda\cos\theta)}.\end{aligned}} \qquad (5\text{–}38)$$

The analysis of this section is typical of all lubrication problems. First, the equations of motion are solved to obtain a profile for the tangential velocity component, which is always locally similar in form to the profile for unidirectional flow between parallel plane boundaries, but with the streamwise pressure gradient unknown. The continuity equation is then integrated to obtain the normal velocity component, but this requires only one of the two boundary conditions for the normal velocity. The second condition then yields a DE (known as the *Reynolds equation*) that can be used to determine the pressure distribution.

The main results of this subsection are Eqs. (5–37) and (5–38) for $dp^{(0)}/d\theta$ and $u_\theta^{(0)}$ plus the corresponding result for $u_r^{(0)}$, which can be calculated from (5–32). It will be convenient for discussion (and also for comparison with the general results that are presented in the next section) to express these results in dimensional form. Thus, reincorporating dimensional variables from (5–11), (5–12), (5–15), (5–16), and (5–26), we obtain

$$\frac{dp'^{(0)}}{d\theta} = \frac{6\mu\Omega}{\varepsilon^2}\left[-\frac{1}{(1+\lambda\cos\theta)^2} + \frac{1}{(1+\lambda\cos\theta)^3}\frac{I_2}{I_3}\right], \qquad (5\text{–}39)$$

$$\begin{aligned}u_\theta'(0) &= \Omega a\left\{\left[\frac{-3}{(1+\lambda\cos\theta)^2}\right]\left[1 - \frac{I_2}{I_3(1+\lambda\cos\theta)}\right]\right.\\ &\quad\left.\times\left[\left(\frac{y'}{\varepsilon a}\right)^2 - (1+\lambda\cos\theta)\left(\frac{y'}{\varepsilon a}\right)\right] - 1 + \frac{(y'/\varepsilon a)}{(1+\lambda\cos\theta)}\right\},\end{aligned} \qquad (5\text{–}40a)$$

and

$$u_r'(0) = \varepsilon\Omega a u_r^{(0)}. \qquad (5\text{–}40b)$$

Finally, we can evaluate the integrals I_2 and I_3. These are

$$I_2(\lambda) = \frac{2 - 2\lambda^2}{2 + \lambda^2} I_3(\lambda) = \frac{2\pi}{(1-\lambda)^{3/2}(1+\lambda)^{3/2}}.$$

Hence, inserting the ratio I_2/I_3 into (5–39), we obtain

$$\frac{dp'^{(0)}}{d\theta} = \frac{6\mu\Omega}{\varepsilon^2(2+\lambda^2)}\left[\frac{-3\lambda^2 - (2+\lambda^2)\lambda\cos\theta}{(1+\lambda\cos\theta)^3}\right],$$

A. The Eccentric Cylinder Problem

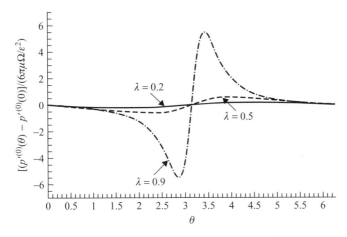

Figure 5–2. The pressure distribution in a journal-bearing flow.

and this can be integrated to give an explicit expression for the pressure:

$$p'^{(0)}(\theta) - p'^{(0)}(0) = -\frac{6\mu\Omega}{\varepsilon^2}\left[\frac{\lambda\sin\theta\,(2+\lambda\cos\theta)}{(2+\lambda^2)(1+\lambda\cos\theta)^2}\right]. \tag{5-41}$$

2. Lubrication Forces

We noted at the beginning of this section that the practical significance of the journal-bearing geometry with $\varepsilon \ll 1$ is that large pressure variations are generated in the gap, which can be used to support a load that is "attached" to the inner cylinder. Now that the pressure and the velocity components $u_r^{(0)}$ and $u_\theta^{(0)}$ are known, we can actually calculate the total force on the inner cylinder.

It is instructive to first consider the problem from a qualitative point of view, beginning with the pressure distribution in the gap as a function of θ. We begin by recalling that $h =$ constant if $\lambda = 0$, and, in this case of two concentric cylinders, $\partial p'^{(0)}/\partial\theta \equiv 0$. It is therefore obvious that a *net* pressure force on the inner cylinder can be generated only if $\lambda \neq 0$. A plot of $p'^{(0)}(\theta) - p'^{(0)}(0)$ as a function of θ for several values of λ, $0 < \lambda < 1$, is given in Fig. 5–2. It is evident that the variation in $p'^{(0)}(\theta)$ increases as λ increases, as expected on qualitative grounds. Referring to Fig. 5–1, we see that there is a strong pressure minimum on the upper half of the cylinder in the quadrant between $\pi/2 < \theta < \pi$, and a corresponding pressure maximum on the bottom half in the quadrant $\pi < \theta < 3\pi/2$.

Thus, with respect to the configuration shown in Fig. 5–1, there is a net pressure *force* exerted on the inner cylinder in the vertical direction as a consequence of its rotation. Note that if the inner cylinder were rotated in the opposite direction (i.e., fluid were forced into the narrowing gap from above instead of below), the sign of the pressure gradient would change, and the force on the inner cylinder would be directed down instead of up. In any case, because $dp^{(0)}/d\theta \sim O(\varepsilon^{-2})$, it is evident that the pressure variations in the gap, and hence the force on the inner cylinder, can be *very large* in the limit $\varepsilon \to 0$. To produce this force, we must rotate the inner cylinder. The torque required for doing this will be proportional to $\partial u_\theta^{(0)}/\partial y'|_{y'=0} = O(\Omega/\varepsilon)$. Thus, by applying a torque of $O(\varepsilon^{-1})$, we achieve a net force of $O(\varepsilon^{-2})$. This distinction between the magnitude of the force produced and the cost in terms of the required force (or in this case torque) to maintain the relative tangential motion across the gap is the basis of the practical applications of lubrication phenomena.

The discussion of the preceding paragraph was mainly qualitative. However, explicit results for the total force acting in the *horizontal* and *vertical* directions (see Fig. 5–1) can

The Thin-Gap Approximation – Lubrication Problems

be calculated from the solutions that we have obtained. To do this, we recall from Chap. 2 that the stress vector (the force per unit area) at any surface in the fluid is given in terms of the (dimensional) stress tensor, $\mathbf{T}'$, as

$$\mathbf{t}'(\mathbf{n}) = \mathbf{T}' \cdot \mathbf{n}.$$

Thus the total force per unit length on the inner cylinder is

$$F'_i = \int_S T'_{ij} n_j dS = -\int_S p' n_i dS + 2\mu \int_S e'_{ij} n_j dS, \tag{5-42}$$

where we have used the general expression (2–87) for the stress in a Newtonian fluid. Let us denote the pressure and viscous contributions to the force as $F_i^{(p)}$ and $F_i^{(v)}$, respectively. The outer unit normal vector $\mathbf{n}$ is given in terms of the Cartesian components in the horizontal (x) and vertical (y) directions as

$$\mathbf{n} = \cos\theta \mathbf{e}_x + \sin\theta \mathbf{e}_y. \tag{5-43}$$

Let us first consider the *pressure* contributions to the force $F_i^{\prime(p)}$, corresponding to the lubrication approximation to the pressure distribution, $p^{\prime(0)}$. The horizontal component $F_x^{\prime(p)}$ can be evaluated immediately. In particular, because $p^{\prime(0)}$ is an odd function of θ,

$$F_x^{\prime(p)} = \int_0^{2\pi} -p^{\prime(0)} \cos\theta a d\theta = 0. \tag{5-44}$$

On the other hand, the contribution in the vertical direction is not zero:

$$F_y^{\prime(p)} = \int_0^{2\pi} -p^{\prime(0)} \sin\theta a d\theta \tag{5-45}$$

If we integrate (5–45) by parts, the right-hand side becomes

$$F_y^{\prime(p)} = a\left[p^{\prime(0)} \cos\theta\right]_0^{2\pi} - \int_0^{2\pi} \cos\theta \left[\frac{dp^{\prime(0)}}{d\theta}\right] a d\theta. \tag{5-46}$$

The first term on the right-hand side is zero. The second term can be evaluated by substituting for $dp^{\prime(0)}/d\theta$ from (5–39). The result is

$$F_y^{\prime(p)} = \frac{6\mu\Omega a}{\varepsilon^2}\left[K_2(\lambda) - \frac{I_2(\lambda)}{I_3(\lambda)} K_3(\lambda)\right], \tag{5-47}$$

where

$$K_n(\lambda) \equiv \int_0^{2\pi} \frac{\cos\theta}{(1+\lambda\cos\theta)^n} d\theta.$$

Because

$$K_n = -\frac{1}{\lambda}(I_n - I_{n-1}),$$

The result (5–47) can also be expressed in the explicit form

$$\boxed{F_y^{\prime(p)} = \frac{12\pi\mu\Omega a}{\varepsilon^2}\left[\frac{\lambda}{(2+\lambda^2)(1-\lambda^2)^{1/2}}\right].} \tag{5-48}$$

It is worth noting that $F_y^{\prime(p)} \to 0$ as $\lambda \to 0$, as expected, because this corresponds to the concentric cylinder case.

A. The Eccentric Cylinder Problem

Now let us consider the *viscous* contributions to the force on the cylinder in the horizontal and vertical directions,

$$F_i'^{(v)} = 2\mu \int_0^{2\pi} e_{ij}'^{(0)} n_j a \, d\theta. \tag{5-49}$$

To evaluate the integrals in (5–49), we would need to determine the components of the rate-of-strain tensor in a Cartesian coordinate system with axes in the horizontal and vertical directions from the velocity components $u_r'^{(0)}$ and $u_\theta'^{(0)}$, given by (5–40a) and (5–40b). This is not a difficult task. However, in the present case, we will be content to show that $F_i^{(v)}$ is $O(\varepsilon^{-1})$ and is thus asymptotically small compared with $F_y^{(p)}$. To see this, we note that

$$\mathbf{e}'^{(0)} = \frac{1}{2}\left(\nabla \mathbf{u}'^{(0)} + \nabla \mathbf{u}'^{(0)T}\right).$$

and the maximum contribution to $\nabla \mathbf{u}'^{(0)}$ comes from $\partial u_\theta'^{(0)}/\partial r'$. However,

$$\frac{\partial u_\theta'^{(0)}}{\partial r'} = O\left(\frac{\Omega}{\varepsilon}\right),$$

and thus, from (5–49),

$$\left|F_i'^{(v)}\right| = O\left(\frac{\mu \Omega a}{\varepsilon}\right). \tag{5-50}$$

In fact, it can be shown by symmetry considerations that $F_x'^{(v)} \equiv 0$, but this is not essential in the present context. The most important conclusion is that the dominant contribution to the force on the inner cylinder is a lift generated by the strong pressure asymmetry in the gap and that viscous contributions to the force and the torque on the inner cylinder are smaller by $O(\varepsilon)$.

It is perhaps worthwhile to dwell briefly on the use of the journal-bearing configuration in practical lubrication applications. In such circumstances, the rotating inner cylinder is normally allowed to "float" to seek a position in which the hydrodynamic force precisely balances the load. We have analyzed the eccentric cylinder problem based on the picture in Fig. 5–1, in which the line of centers between the two cylinders is in the horizontal direction. In that configuration, there is a net vertical force but no horizontal force. Moreover, examination of (5–48) shows that the magnitude of the vertical force for a given pair of cylinders (so that a and ε are fixed) is determined by Ω and λ. The λ dependence is contained in the factor

$$\alpha = \frac{\lambda}{(2+\lambda^2)(1-\lambda^2)^{1/2}}.$$

The dependence of α on λ is shown in Fig. 5–3 for $0 \leq \lambda < 1$. Clearly, as $\lambda \to 1$, so that the geometry becomes increasingly eccentric, the vertically directed force increases. It is obvious, then, that the horizontal configuration of Fig. 5–1 can be an equilibrium configuration if we attach a vertical load to the inner cylinder. If we increase the load on an established journal-bearing system, the configuration will change to a more eccentric geometry (that is, an increased value for λ). The load limit is determined when the value of λ necessary to support the load reaches the maximum possible value, which is determined in practice by surface roughness or other mechanical imperfections that restrict λ to values $\lambda \leq \lambda_{max} < 1$.

One issue that has not been addressed is the transient process that allows λ to change from one value to another when the load is changed, while still maintaining the horizontal configuration as the final steady state. Because there is zero hydrodynamic force in the horizontal direction for any value of λ in the horizontal configuration, the eccentricity cannot change from one steady state to another by simply shifting the inner cylinder in

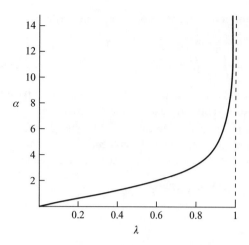

Figure 5–3. The dependence of the magnitude of the hydrodynamic lubrication force on the degree of eccentricity of the journal bearing, $\alpha \equiv F'/(12\pi\mu\Omega a/\varepsilon^2)$.

the horizontal direction. Instead, the response to an increase in the vertical load is that the inner cylinder will initially move down, hence adopting a configuration in which the line of centers is no longer horizontal, but the degree of eccentricity is increased. The horizontal configuration in Fig. 5–1 is convenient for the analysis. However, if we had carried out the analysis with the line of centers between the two cylinders at some other angle besides horizontal, we would have obtained exactly the same result; namely that there is a force perpendicular to the line of centers with magnitude given by (5–48) and zero force along the line of centers. This means that, in the new configuration after the inner cylinder has shifted vertically, the lubrication force on the inner cylinder has a vertical component to support the load and a horizontal component that pulls the inner cylinder back toward the horizontal configuration. In fact, if we neglect transients in the fluid motion [i.e., we adopt the "quasi-steady" assumption that the force on the inner cylinder at each new configuration is just the steady-force components (5–48) perpendicular to the line of centers and zero along the line of centers], we can write the trajectory equation for the position of the inner cylinder as a function of time, beginning with some initial configuration and a change in the load. The details of this are left as a homework problem at the end of the chapter.

B. DERIVATION OF THE BASIC EQUATIONS OF LUBRICATION THEORY

The journal-bearing limit of the eccentric cylinder problem is a specific example of a large class of important problems in which the flow of a viscous fluid in a thin-flow domain of nonuniform thickness, is found to generate large pressure forces that tend to hold the bounding surfaces apart even in the presence of a large normally directed load. It is also representative of an even larger class of problems in which the so-called thin-film approximation of the Navier–Stokes equations can be applied to analyze the flow in at least a part of the flow domain. Some common flow configurations are sketched in Fig. 5–4. In each of these, there is at least a local part of the flow domain between the moving object and a solid wall where we might expect the thin-film approximation to be applicable.

We shall see that there are certain generic features of the thin-film approximation that have already been illustrated by the journal-bearing analysis, which will appear whenever the thin-film approximation is used. *The most significant feature of the journal-bearing analysis, which is shared by all thin-film theories, is the assumption that velocity gradients in the direction "parallel" to the boundaries are asymptotically small compared with those*

B. Derivation of the Basic Equations of Lubrication Theory

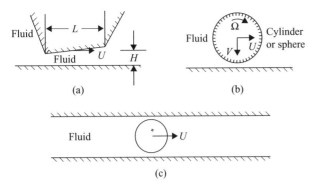

Figure 5–4. Some common lubrication configurations: (a) the slider block, (b) a sphere or cylinder moving in the vicinity of a plane wall, (c) a sphere translating axially in a circular tube.

across the flow domain. This implies that the cross-channel dimension of the flow domain must be asymptotically small compared with either the radius of curvature of the boundaries or with the lateral distance that is characteristic of any change in the gap width. In this sense we may say that the thin-film analysis applies when the shape of the flow domain is only "slowly varying." At the same time, the flow domain is assumed to be "thin" in the sense that the parameter $\varepsilon^2 Re$ is asymptotically small, so that inertial terms in the Navier–Stokes equations can be neglected compared with the viscous and pressure-gradient terms. Subject to these assumptions ($\varepsilon \ll 1$, $\varepsilon^2 Re \ll 1$), we found in the journal-bearing problem that the equations governing the leading-order approximation in the small parameter ε, namely Eqs. (5–27)–(5–29), were identical to the equations for a unidirectional flow, expressed in a form that is identical to the equations for a Cartesian coordinate system. Because of the thin-film assumptions, the curvature terms in the original cylindrical coordinate equations (5–17)–(5–19) were all relegated to the governing equations for higher-order terms in the asymptotic expansion for small ε. In fact, the reduction to a "local" Cartesian coordinate form is a common feature of all thin-film theories. It is important to emphasize that the reduction of the governing equations to the simplified form of (5–27)–(5–29) in the journal-bearing problem, or in any lubrication analysis, does not imply that the flow is unidirectional. The same dominant balance between viscous and pressure-gradient terms is preserved locally, but the slow variation in the gap width, and the dominance of viscous effects because of its small scale, combine to produce a very large variation in the pressure as we move along the thin gap.

Before turning to the development of governing equations for the general class of lubrication problems, based on these "lessons" from the journal-bearing problem, it is perhaps useful to consider any features of the more general class of problems that are fundamentally different. The most obvious is that the whole of the flow domain in the journal-bearing problem actually satisfies the basic assumptions of the thin-film approximation, whereas, in general, the thin-film analysis is applicable in only a portion of the flow domain. The sample problems sketched in Fig. 5–4 illustrate this point. In the journal-bearing problem, the same characteristic length and velocity scales are valid everywhere in the flow domain, and the approximate solution for small $\varepsilon \ll 1$ can be obtained as a *regular* perturbation expansion. For the type of problems shown in Fig. 5–4, however, there are clearly large regions of the flow domain where the thin-film analysis cannot be applied. If we denote the length scale characteristic of the distance across the gap in any of these problems as H and the lateral dimension of the "thin-film region" as L, we can see in all of the problems in the figure that, outside the thin gap between the body and the walls, the scale L is relevant to velocity gradients in all coordinate directions. On the other hand, within the gap, gradients

The Thin-Gap Approximation – Lubrication Problems

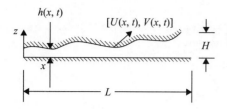

Figure 5–5. A schematic illustration of a 2D thin-film (lubrication) configuration.

across the flow are characterized by the length scale $O(H)$ of the gap width. The fact that different characteristic scales are relevant in different parts of the flow domain means that the approximations of the full governing equations for $\varepsilon \equiv H/L \ll 1$ will be different in different parts of the domain, and thus different asymptotic forms for the solutions will also appear. Alternatively, we may note that the order of the governing equations within the thin film will be reduced in the limit $H/L \ll 1$, e.g.,

$$\nabla^2 u \rightarrow \frac{\partial^2 u}{\partial z^2} + \varepsilon^2 \frac{\partial^2 u}{\partial y^2} \sim \frac{\partial^2 u}{\partial z^2} + O(\varepsilon^2).$$

These are the classic signatures of a *singular* asymptotic limit, for which we must use the method of matched asymptotic expansions to obtain a uniformly valid solution. For purposes of discussion, we denote the region within the thin gap as the *inner* region and the rest of the flow domain as the *outer* region. In principle, our objective is an asymptotic expansion for the solution (in the limit $\varepsilon \to 0$) in both the inner and outer regions, with the two solutions connected in the common region near the ends of the thin gap by *matching* conditions. However, for flow geometries such as those shown in Fig. 5–4, it would be difficult to carry out this procedure. The complexity of the geometry in the outer region would make it extremely unlikely that closed-form analytic solutions could be obtained even for a simplified version of the full Navier–Stokes equations.

Fortunately, as we shall see, a solution for the first term of the asymptotic theory can actually be obtained within the inner region without an explicit calculation of the first term in the outer region. Not only that, but the hydrodynamic contribution to the force or torque on a body in a configuration, like those shown in Fig. 5–4, is dominated by the large pressure forces in the thin-gap (inner) region. For these reasons, analyses of lubrication problems by means of the thin-film asymptotic approach are almost always limited to calculating only the first term in the perturbation expansion for the inner (i.e., thin-film) region. Indeed, this theory, based on only the first term of the asymptotic solution in the thin gap, is what is normally meant by "lubrication theory." We thus begin the process of developing governing equations by focusing on only the "inner" thin-film region. However, we recognize that this thin-film region will generally be only a local part of some more general flow domain, and we will have to return to consider what to do about connecting this "inner" solution with a solution in the difficult "outer" region.

We consider a typical thin-film configuration like that sketched in Fig. 5–5. We have already noted that any curvature of the boundaries will play no role in the leading-order thin-film (or lubrication) approximations provided that the radius of curvature is large compared with the gap width, and there is therefore no loss of generality in assuming one of the two surfaces to be flat. For convenience, we denote the coordinate perpendicular to this surface as z. The other boundary is also a solid surface of fixed shape, with the thickness of the fluid layer between them defined by the function, $h(\mathbf{x}_s, t)$, which can be expressed as a function of the two spatial coordinates that lie in the plane parallel to the lower, flat wall, as well as a function of time. Because the most appropriate coordinate system will depend on the particular problem, it is convenient to maintain the symbolic representation of the

B. Derivation of the Basic Equations of Lubrication Theory

dependence on position in the plane encompassed by $\mathbf{x}_s$ for our subsequent analysis. At least one of the two bounding surfaces will be moving relative to the other one in lubrication problems. The most common configuration is represented by the examples shown in Figs. 5–4(a) and 5–4(c), where one surface moves parallel to the plane of the flat boundary. However, there are also important applications of the theory in which the relative motion is in the normal direction, so as to make the gap either thinner or wider with time. In problems such as those shown in Fig. 5–4, it is often the bounding surface of finite lateral extent (i.e., the slider or the sphere in Fig. 5–4) that is moving. Nevertheless, for purposes of analysis, it is simplest to adopt a coordinate reference frame that is fixed to this moving body so that it is the flat boundary that appears to be moving. In this description, when the relative motion is parallel to the $z = 0$ plane, the geometry will remain fixed, independent of time. It will rarely be the case that the relative motion is a combination of parallel and normal motion. Nevertheless, in order that the analysis of this section can ultimately be applied to both the parallel and normal motions, we initially assume that both $\mathbf{u}'_s$ and u'_z are nonzero at the lower boundary $z' = 0$. It must be noted, however, that h' and u'_z cannot be specified independently as $u'_z \equiv -\partial h'/\partial t'$.

We begin our analysis with the Navier–Stokes and continuity equations for an incompressible fluid, in their most general dimensional form,

$$\nabla' \cdot \mathbf{u}' = 0, \tag{5-51}$$

$$\rho \left(\frac{\partial \mathbf{u}'}{\partial t'} + \mathbf{u}' \cdot \nabla' \mathbf{u}' \right) = -\nabla' p' + \mu \nabla'^2 \mathbf{u}'. \tag{5-52}$$

It will be noted that these equations are written in terms of the dynamic pressure. This is because the hydrostatic pressure plays no role when the boundaries of the flow domain are solid surfaces, as we have already seen in Chap. 2. If we follow the lead of Stone[5] and retain the general coordinate notation already described in connection with the gap function h, these equations can be written in the form

$$\nabla'_S \cdot \mathbf{u}'_S + \frac{\partial u'_z}{\partial z'} = 0, \tag{5-53}$$

$$\rho \left(\frac{\partial u'_z}{\partial t'} + \mathbf{u}'_S \cdot \nabla'_S u'_z + u'_z \frac{\partial u'_z}{\partial z'} \right) = -\frac{\partial p'}{\partial z'} + \mu \left[\nabla'_S \cdot (\nabla'_S u'_z) + \frac{\partial^2 u'_z}{\partial z'^2} \right], \tag{5-54}$$

$$\rho \left(\frac{\partial \mathbf{u}'_S}{\partial t'} + \mathbf{u}'_S \cdot \nabla'_S \mathbf{u}'_S + u'_z \frac{\partial \mathbf{u}'_S}{\partial z'} \right) = -\nabla'_S p' + \mu \left[\nabla'_S \cdot (\nabla'_S \mathbf{u}'_S) + \frac{\partial^2 \mathbf{u}'_S}{\partial z'^2} \right]. \tag{5-55}$$

Here $\mathbf{u}'_S$ represents the components of the velocity vector corresponding to the coordinates $\mathbf{x}_s$ in the plane of the lower (flat) boundary, and ∇'_S is the portion of the gradient operator that involves the same coordinates and therefore contains the metric functions (or scale factors) associated with the geometry.

For purposes of nondimensionalizing, we follow the journal-bearing problem of the preceding section. An appropriate choice for the characteristic velocity when the boundary moves in its own plane is the magnitude of the boundary velocity, $u_c \equiv U$. (We will return to the case in which the boundary moves in the normal direction shortly.) We denote the "horizontal" length scale characteristic of the distance for $O(\langle h \rangle)$ changes in the gap width as ℓ_c. Hence,

$$\mathbf{u}'_S = u_c \mathbf{u}_S, \quad \nabla'_S = \frac{1}{\ell_c} \nabla_S, \tag{5-56}$$

The Thin-Gap Approximation – Lubrication Problems

and the analysis of the preceding section suggests that an appropriate characteristic time scale should be ℓ_c/u_c so that

$$t' = (\ell_c/u_c)\, t. \tag{5-57}$$

We note that the primed variables are all dimensional, whereas the unprimed ones are dimensionless. The ratio of the gap width to the characteristic length scale ℓ_c is denoted as ε, and thus

$$\frac{\partial}{\partial z'} = \frac{1}{\varepsilon \ell_c} \frac{\partial}{\partial z}. \tag{5-58}$$

Finally, because it is necessary to retain the basic form (5–53) of the continuity equation to ensure satisfaction of mass conservation, it follows from (5–56) and (5–58) that

$$u'_z = \varepsilon u_c u_z. \tag{5-59}$$

The preceding description is appropriate for the case of parallel boundary motion. However, if the motion in the lubrication layer results from the relative motion(s) of the boundaries toward (or away from) one another with velocity V, we see from (5–59) that we may simply choose $u_c = V/\varepsilon$. Finally, although the characteristic pressure could also be adapted directly from the analysis of the preceding section, its scaling is a key result of the lubrication theory, and we thus initially adopt the symbolic notation

$$p' = p_c p \tag{5-60}$$

and demonstrate again that the pressure scaling is a natural consequence of the thin-film approximation that is inherent in the length scale and velocity scaling in (5–56)–(5–59), and not a result that is specific to the journal-bearing geometry.

The dimensionless version of the Eqs. (5–53)–(5–55) corresponding to (5–56)–(5–60) is thus

$$\nabla_S \cdot \mathbf{u}_S + \frac{\partial u_z}{\partial z} = 0, \tag{5-61}$$

$$\frac{\rho \varepsilon u_c^2}{\ell_c}\left(\frac{\partial u_z}{\partial t} + \mathbf{u}_S \cdot \nabla_S u_z + u_z \frac{\partial u_z}{\partial z}\right) = -\frac{p_c}{\varepsilon \ell_c}\frac{\partial p}{\partial z} + \frac{\mu u_c}{\varepsilon \ell_c^2}\left[\varepsilon^2 \nabla_S \cdot (\nabla_S u_z) + \frac{\partial^2 u_z}{\partial z^2}\right], \tag{5-62}$$

$$\frac{\rho u_c^2}{\ell_c}\left(\frac{\partial \mathbf{u}_S}{\partial t} + \mathbf{u}_S \cdot \nabla_S \mathbf{u}_S + u_z \frac{\partial \mathbf{u}_S}{\partial z}\right) = -\frac{p_c}{\ell_c}\nabla_S p + \frac{\mu u_c}{\varepsilon^2 \ell_c^2}\left[\varepsilon^2 \nabla_S \cdot (\nabla_S \mathbf{u}_S) + \frac{\partial^2 \mathbf{u}_S}{\partial z^2}\right]. \tag{5-63}$$

In the case of a unidirectional flow, the dominant balance in (5–62) and (5–63) involved only the pressure gradient $\nabla_S p$ and the largest viscous term $\partial^2 \mathbf{u}_S/\partial z^2$. If we adopt the intuitively reasonable assumption that the same two terms should remain in balance in the thin-film approximation, we are led immediately to the result

$$O\left(\frac{p_c}{\ell_c}\right) = O\left(\frac{\mu u_c}{\varepsilon^2 \ell_c^2}\right)$$

from which we see that

$$p_c = \frac{\mu u_c}{\ell_c}\left(\frac{1}{\varepsilon^2}\right). \tag{5-64}$$

We may note that, whatever the scaling might have turned out to be, the pressure must be retained in either (5–62) or (5–63) in the limit $\varepsilon \to 0$. Hence the only other alternative to (5–64) would be to choose p_c to balance the pressure gradient and the largest viscous

B. Derivation of the Basic Equations of Lubrication Theory

term in (5–62). However, this leads to an inconsistency in (5–63), thus indicating that the scaling (5–64) is the only possible choice. We note that *the scaling (5–64) is identical to that found in the journal-bearing problem*. In fact, we now see that very large pressures, $O(\varepsilon^{-2})$, relative to the characteristic viscous pressure, $O(\mu u_c/\ell_c)$, are a general feature of the thin gap-geometry.

With p_c specified in the form (5–64), the Eq. (5–63) can be written in the form

$$\frac{\partial^2 \mathbf{u}_s}{\partial z^2} - \nabla_s p = -\varepsilon^2 \nabla_s \cdot (\nabla_s \mathbf{u}_s) + \varepsilon^2 \, Re \left(\frac{\partial \mathbf{u}_s}{\partial t} + \mathbf{u}_s \cdot \nabla_s \mathbf{u}_s + u_z \frac{\partial \mathbf{u}_s}{\partial z} \right), \quad (5\text{–}65)$$

where

$$Re \equiv \frac{\rho u_c \ell_c}{\mu},$$

and (5–62) becomes

$$\frac{\partial p}{\partial z} = \varepsilon^2 \left\{ \left[\frac{\partial^2 u_z}{\partial z^2} + \varepsilon^2 \nabla_s \cdot (\nabla_s u_z) \right] + \varepsilon^2 \, Re \left(\frac{\partial u_z}{\partial t} + \mathbf{u}_s \cdot \nabla_s u_z + u_z \frac{\partial u_z}{\partial z} \right) \right\}. \quad (5\text{–}66)$$

The boundary conditions at $z = 0$ and $z = h(\mathbf{x}_s)$ are

$$u_z = \mathbf{u}_s = 0 \quad \text{at } z = h(\mathbf{x}_s),$$
$$u_z = -(V'/\varepsilon u_c), \quad \mathbf{u}_s = -(\mathbf{U}'/u_c) \quad \text{at } z = 0. \quad (5\text{–}67)$$

The dimensionless forms of the governing equations emphasize the fact that there are really two dimensionless parameters, ε and $\varepsilon^2 Re$, which must be small for the classical lubrication analysis to apply. It is convenient for present purposes, to assume that Re is fixed independent of ε and thus $\varepsilon^n Re \to 0$ as $\varepsilon \to 0$. Thus in the limit $\varepsilon \to 0$ we may seek an asymptotic solution of (5–61) and (5–65)–(5–67) for the "inner" thin-film part of the flow domain in the form

$$\mathbf{u}_s = \mathbf{u}_s^{(0)} + f_1(\varepsilon) \mathbf{u}_s^{(1)} + \cdots +,$$
$$u_z = u_z^{(0)} + f_1(\varepsilon) u_z^{(1)} + \cdots +, \quad (5\text{–}68)$$
$$p = p^{(0)} + f_1(\varepsilon) p^{(1)} + \cdots +.$$

We note that the gauge function for the first term in these expansions has been immediately set equal to unity because the boundary conditions require that the magnitude of the velocity $\mathbf{u}_s$ be unity (i.e., independent of the small parameter ε).

We can obtain the governing equations for the leading-order term in any asymptotic expansion by letting the small parameter go to zero in the full nondimensionalized equations [in this case by letting $\varepsilon \to 0$ (and $\varepsilon^2 Re \to 0$) in (5–61), (5–65), and (5–66)]. The governing equations for the first approximation to (5–68) in the thin film are thus

$$\boxed{\nabla_s \cdot \mathbf{u}_s^{(0)} + \frac{\partial u_z^{(0)}}{\partial z} = 0,} \quad (5\text{–}69)$$

$$\boxed{\frac{\partial^2 \mathbf{u}_s^{(0)}}{\partial z^2} - \nabla_s p^{(0)} = 0,} \quad (5\text{–}70)$$

The Thin-Gap Approximation – Lubrication Problems

$$\boxed{\frac{\partial p^{(0)}}{\partial z} = 0,} \tag{5-71}$$

and the corresponding boundary conditions are

$$\boxed{\begin{aligned} u_z^{(0)} &= \mathbf{u}_s^{(0)} = 0 \quad \text{at } z = h(\mathbf{x}_s), \\ u_z^{(0)} &= -V, \quad \mathbf{u}_s^{(0)} = -\mathbf{U} \quad \text{at } z = 0. \end{aligned}} \tag{5-72}$$

where we have used the notation $V = (V'/\varepsilon u_c)$ and $\mathbf{U}$ in place of $\mathbf{U}'/u_c$.

These equations, which are known as the *lubrication equations*, are identical in form to those obtained in the journal-bearing problem, and this form is now seen to be *universal* independent of the overall boundary geometry, provided only that the thin-gap approximation, $\varepsilon \ll 1$, is valid. All that really changes from problem to problem in this inner (thin-film) part of the flow domain is the boundary conditions on the upper and lower boundaries and the dependence of the gap width on position and time. We will return to these details shortly and in the context of specific problems. One point to note, however, is that these equations involve derivatives with respect to $\mathbf{x}_s$ as well as z. Thus a unique solution will generally be possible only with the imposition of boundary conditions at the edges of the thin-film region as well as the conditions (5–72) on the upper and lower boundaries. This is the point where the inner and outer domains merge, and we should expect these conditions to derive from the requirement of *matching* between the approximate solutions in the thin gap and in the rest of the flow domain. Before being concerned with the details of matching and of the solution in the outer part of the domain, however, it is useful to proceed a bit further with the analysis in the inner region.

It follows from (5–71) that

$$p^{(0)}(\mathbf{x}_s, z, t) \to p^{(0)}(\mathbf{x}_s, t),$$

and thus (5–70) can be integrated to obtain

$$\mathbf{u}_s^{(0)} = \nabla_s p^{(0)} \frac{z^2}{2} + \mathbf{f}(\mathbf{x}_s, t) z + \mathbf{g}(\mathbf{x}_s, t). \tag{5-73}$$

Applying the boundary conditions (5–72), we obtain

$$\mathbf{u}_s^{(0)} = \nabla_s p^{(0)} \left(\frac{z^2}{2} - \frac{zh}{2} \right) + \mathbf{U} \left(\frac{z}{h} - 1 \right). \tag{5-74}$$

Not surprisingly, in view of the results obtained in Chap. 3, the flow in this case appears as a superposition of a pressure-driven and a boundary-driven shear.

Although the solution (5–74) seems to be complete, the key fact is that the pressure gradient $\nabla_s p^{(0)}$ in the thin gap, and thus $p^{(0)}(\mathbf{x}_s, t)$, is *unknown*. In this sense, the solution (5–74) is fundamentally different from the unidirectional flows considered in Chap. 3, where p varied linearly with position along the flow direction and was thus known completely if p was specified at the ends of the flow domain. The problem considered here is an example of the class of thin-film problems known as "lubrication theory" in which either $h(\mathbf{x}_s)$ and $\mathbf{u}_s$, or $h(\mathbf{x}_s, 0)$ and u_z are prescribed on the boundaries, and *it is the pressure distribution in the thin-fluid layer that is the primary theoretical objective*. The fact that the pressure remains unknown is, of course, not surprising as we have not yet made any use of the continuity equation (5–69) or of the boundary conditions at $z = 0$ and h for the normal velocity component $u_z^{(0)}$.

B. Derivation of the Basic Equations of Lubrication Theory

There are, in fact, two ways to proceed at this point to obtain an eq. for the pressure in the thin gap. One is to substitute for $\mathbf{u}_s^{(0)}$ in Eq. (5–69) and integrate once with respect to z to obtain an expression for $u_z^{(0)}$. This will include a single unspecified "constant" of integration. However, there are two boundary conditions for $u_z^{(0)}$ at $z = 0$ and $z = h(\mathbf{x}_s, t)$. One can be used to determine the constant of integration. The second condition will impose a requirement, from which the only remaining unknown in the problem can be determined, namely, $p^{(0)}(\mathbf{x}_s, t)$.

A second somewhat more "elegant" procedure is to utilize (5–69) to directly obtain an integral constraint that can be used to determine $p^{(0)}(\mathbf{x}_s, t)$. The DE for $p^{(0)}(\mathbf{x}_s, t)$ that results from this constraint is identical to the condition that is obtained following the alternative procedure that was described in the previous paragraph. Most researchers have followed this second approach because it avoids any need to explicitly calculate $u_z^{(0)}$. We will follow this lead, though the first approach is perfectly good and may seem more straightforward to some readers. To derive the required integral constraint, we integrate (5–69) directly with respect to z, i.e.,

$$\int_0^h (\nabla_s \cdot \mathbf{u}_s)\, dz + u_z^{(0)}\Big|_0^h = 0. \tag{5–75}$$

If we apply the boundary conditions for $u_z^{(0)}$ from (5–72), we can write this equation in the form

$$\int_0^h (\nabla_s \cdot \mathbf{u}_s)\, dz = -V = -\frac{\partial h}{\partial t}. \tag{5–76}$$

We can calculate the divergence of $\mathbf{u}_s$ by using (5–74). The result is

$$\nabla_s \cdot \mathbf{u}_s = \nabla_s^2 p^{(0)} \left(\frac{z^2}{2} - \frac{zh}{2}\right) + \nabla_s p^{(0)} \cdot \left(-\frac{z}{2}\nabla_s h\right) + (\nabla_s \cdot \mathbf{U})\left(\frac{z}{h} - 1\right) + \left(-\frac{z}{h^2}\nabla_s h\right) \cdot \mathbf{U}. \tag{5–77}$$

Hence, substituting (5–77) into (5–76) and integrating, we find

$$-\frac{\partial h}{\partial t} = \int_0^h \nabla_s \cdot \mathbf{u}_s\, dz$$
$$= \nabla_s^2 p^{(0)} \left(-\frac{h^3}{12}\right) + \nabla_s p^{(0)} \cdot \left(-\frac{h^2}{4}\nabla_s h\right) + \nabla_s \cdot \mathbf{U}\left(-\frac{h}{2}\right) + \left(-\frac{1}{2}\nabla_s h\right) \cdot \mathbf{U}, \tag{5–78}$$

and this can be rewritten as

$$\boxed{\frac{\partial h}{\partial t} = \nabla_s \cdot \left[\frac{h^3}{12}\nabla_s p^{(0)} + \frac{h}{2}\mathbf{U}\right].} \tag{5–79}$$

In lubrication theory, this equation is known as the *Reynolds equation*. With the boundary motion specified and the shape of the thin film specified, it can be used as a DE for the unknown pressure distribution in the thin film.

To obtain a unique solution of (5–79) for the pressure, the values of $p^{(0)}$ must be specified at the ends of the lubrication layer. The necessary boundary values come from the requirement that the solution within this inner region must match the solution in the outer domain, at the ends of the thin film. Note, however, that this would seem to suggest that the solution within the lubrication layer cannot be determined completely without simultaneously obtaining the leading-order solution in the outer region, including specifically the pressure.

The Thin-Gap Approximation – Lubrication Problems

The asymptotic analysis for small ε in the outer domain proceeds by means of the same steps that we outlined for the inner region. Specifically, we nondimensionalize the full Navier–Stokes and continuity equations by using characteristic scales relevant to the outer region. We then obtain approximate equations for the various terms in an asymptotic expansion of the solution by substituting the expansion into the nondimensionalized equations and requiring that they be satisfied at each order of magnitude in ε. As always, we can obtain governing equations for the first term in the asymptotic series by simply taking the limit $\varepsilon \to 0$ in the full nondimensionalized equations for the outer region. The appropriate characteristic scales for this outer region were anticipated at the beginning of this section. Specifically, the characteristic length scale, ℓ_c, Eqs. (5–56), is now relevant to velocity gradients in all coordinate directions. Thus the same characteristic velocity scale, u_c, is also relevant to both components of the velocity field in the outer domain. If we introduce nondimensional variables

$$u = \frac{u'}{u_c}, \quad v = \frac{v'}{u_c}, \quad x = \frac{x'}{\ell_c}, \quad y = \frac{y'}{\ell_c}, \quad t = \frac{u_c t'}{\ell_c}, \quad p = \frac{p'}{p_c},$$

into the full Navier–Stokes equations, we obtain

$$\frac{\partial u}{\partial t} + u\frac{\partial u}{\partial x} + v\frac{\partial u}{\partial y} = -\frac{p_c}{\rho u_c^2}\frac{\partial p}{\partial x} + \frac{\mu}{\rho u_c \ell_c}\left[\frac{\partial^2 u}{\partial x^2} + \frac{\partial^2 u}{\partial y^2}\right],$$
$$\frac{\partial v}{\partial t} + u\frac{\partial v}{\partial x} + v\frac{\partial v}{\partial y} = -\frac{p_c}{\rho u_c^2}\frac{\partial p}{\partial y} + \frac{\mu}{\rho u_c \ell_c}\left[\frac{\partial^2 v}{\partial x^2} + \frac{\partial^2 v}{\partial y^2}\right]. \quad (5\text{–}80)$$

Although we do not use it here, it is interesting to note that the dimensional equation for the gap,

$$y' = h'(\mathbf{x}'_s, t'),$$

becomes

$$y = \varepsilon h(\mathbf{x}_s, t),$$

when expressed in terms of the outer scaling. Thus, as $\varepsilon \to 0$, the geometric configuration, when seen from the outer point of view, reduces to that of the body "touching" the moving plane boundary with a line source and sink for mass of equal strength located at the two ends of the gap.

As yet unspecified in (5–80) is the characteristic pressure p_c. This is determined from the characteristic pressure in the inner region. The basic idea is that p from the inner and outer domains must match at the entry and exit to and from the gap. It follows from this reasoning that

$$p_c = \frac{\mu u_c}{\ell_c}\left(\frac{1}{\varepsilon^2}\right).$$

Hence Eqs. (5–80) become

$$\nabla p = \varepsilon^2 \left[\nabla^2 \mathbf{u} + \left(\frac{\rho u_c \ell_c}{\mu}\right)\left(\frac{\partial \mathbf{u}}{\partial t} + \mathbf{u}\cdot\nabla\mathbf{u}\right)\right], \quad (5\text{–}81)$$

where $\mathbf{u} = (u, v)$. This shows that the effect of the fluid motion on the pressure in the outer region is negligible relative to the pressure variations that are possible in the inner region provided ε and $\varepsilon^2 Re$ are small. Thus, insofar as the first approximation in the thin gap is concerned, it is only the values of dynamic pressure for a *stationary* fluid that are needed at the ends of the gap. Frequently, the values of the dynamic pressure for a *stationary* fluid will be equal at the two ends of the gap. For example, in the case of a body of finite extent moving in a large body of fluid near a plane wall, the dynamic pressure difference between the two ends of the gap will be zero – i.e., $p^{(0)}(0) = p^{(0)}(L)$ – independent of the orientation

C. Applications of Lubrication Theory

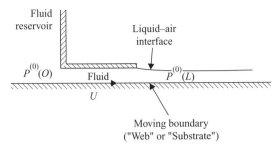

Figure 5–6. A sketch of a film-coating process.

of the plane wall. On the other hand, the dynamic pressure difference will not always be zero. Consider, for example, the configuration depicted in Fig. 5–6, which is typical of film-coating applications. Here, the pressure on the upstream side of the thin gap (the lubrication layer) can be controlled by pressurizing or depressurizing the fluid reservoir, and the static difference in dynamic pressure will not be zero across the gap, that is, $p^{(0)}(0) = p_{reservoir}$ and $p^{(0)}(0) \neq p^{(0)}(L)$ in general.

In any case, the very important conclusion from (5–81) is that the leading-order approximation to the solution in the thin-film region always can be determined completely without the need to determine anything about the velocity field in the rest of the domain where the geometry is much more complicated. This constitutes a very considerable simplification! When the lubrication approximation can be made, we need focus our attention on only the lubrication equations within the thin gap, and in this region the general solutions (5–74) and (5–79) have been worked out already. We shall see in the next section that the dominant contributions to the forces or torques acting on a body near a second boundary always occur in the lubrication layer when $\varepsilon \ll 1$.

C. APPLICATIONS OF LUBRICATION THEORY

In this section we consider the detailed analysis for two applications of lubrication theory: the classic slider-block problem that was depicted in the previous section and the motion of a sphere toward an infinite plane wall when the sphere is very close to the wall. It is the usual practice in lubrication theory to focus directly on the motion in the thin gap using (5–69)–(5–72), or their solutions (5–74) and (5–79), without any mention of the asymptotic nature of the problem or of the fact that these equations (and their solutions) represent only a first approximation to the full solution in the lubrication layer. We adopt the same approach here but with the formal justification of the preceding section.

1. The Slider-Block Problem

We begin by considering the classic slider-block problem that is sketched in Fig. 5–7. Here, a 2D cylindrical body of arbitrary cross section, but with one face that is flat, moves with

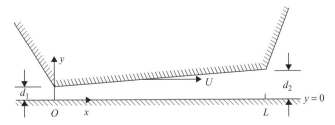

Figure 5–7. The slider-block configuration.

315

The Thin-Gap Approximation – Lubrication Problems

a velocity U parallel to an infinite plane stationary boundary. The plane boundary and the smooth face of the slider block form a lubrication layer that is wider at the front and narrower toward the rear of the body but is everywhere thin in the sense that $d_2/L \ll 1$, where d_2 is the maximum gap width and L is the characteristic length of the lubrication layer. The goals of analysis for this problem are to determine the magnitude of the lift force (normal to the boundary) that is generated by the large pressure and viscous stresses in the lubrication layer, and the magnitude of the tangential force that is necessary to maintain the constant lateral velocity U. The slider-block problem illustrates the key feature of the most common class of lubrication configurations – namely, the generation of very large pressure forces that is due to the relative *tangential* motion of the boundaries that is maintained by a tangential force of smaller magnitude. Provided the relative boundary motion is such that that the fluid is being drawn into the narrowing part of the gap, the pressure will be positive relative to that outside the gap and this force will tend to maintain the separation between two solid boundaries. On the other hand, if the boundary (and fluid) motion is in the opposite direction, the pressure will be reduced relative to the surroundings, thus producing a suction that tends to pull the two surfaces together (or which may create pressure low enough for cavitation to occur if the two surfaces are held apart by some external force).

The analysis of the preceding section and the solutions (5–74) and (5–79) can be applied directly to this problem. The geometry, as pictured in Fig. 5–7, is 2D, and most conveniently described in terms of a Cartesian coordinate system with x in the direction of motion. If we describe the problem with a coordinate system that is fixed with respect to the lower boundary, the geometry is time dependent and the Reynolds equations (5–79) takes the form

$$\frac{\partial h}{\partial t} = \frac{d}{dx}\left(\frac{h^3}{12}\frac{dp^{(0)}}{dx} + \frac{h}{2}U\right). \tag{5–82}$$

However, for reasons explained in the preceding section, it is much simpler to adopt a coordinate reference frame in which the slider remains fixed and the lower boundary moves with velocity $-U$ in the x-direction. In this frame of reference, the geometric configuration for the slider-block problem is independent of time, and the lubrication approximation is intrinsically a steady theory. We need specify only the dimensionless gap width as a function of position, $h(x)$. We formulate the problem by using a Cartesian coordinate system that is fixed with respect to the trailing edge of the slider block, as shown in Fig. 5–7, but with the origin coincident with the lower wall, which is thus $z = 0$, with $x = 0$ and $x = L$ corresponding to the trailing and leading edges of the slider block. Relative to this coordinate system, the upper surface of the lubrication layer is at

$$z' = h'(x') = \frac{d_2 - d_1}{L}x' + d_1. \tag{5–83}$$

If we adopt d_1 as the characteristic gap width $\varepsilon\ell_c$ and L as the characteristic streamwise length scale ℓ_c, the dimensionless gap-width function is

$$\boxed{h(x) = 1 + \left(\frac{d_2}{d_1} - 1\right)x.} \tag{5–84}$$

The relative velocity of the slider block and the lower wall is U, and this is an appropriate choice for the characteristic velocity u_c, which was used to derive Reynolds equation

C. Applications of Lubrication Theory

(5–79). Thus, in dimensionless terms, $U = 1$, and the unknown pressure distribution in the lubrication layer can be obtained from Reynolds equation (5–79), which takes the form

$$\frac{d}{dx}\left[h^3 \frac{dp^{(0)}}{dx}\right] = 6\left(1 - \frac{d_2}{d_1}\right). \tag{5-85}$$

Integrating once with respect to x, we find that

$$\frac{dp^{(0)}}{dx} = 6\left(1 - \frac{d_2}{d_1}\right)\frac{x}{h^3} + \frac{c_1}{h^3}. \tag{5-86}$$

Here, c_1 is a constant of integration. Integrating again, we obtain

$$p^{(0)} = 6\left(1 - \frac{d_2}{d_1}\right)\int_0^x \frac{s\,ds}{h^3(s)} + c_1 \int_0^x \frac{ds}{h^3(s)} + c_2. \tag{5-87}$$

The integration constants c_1 and c_2 are given by the boundary values for $p^{(0)}$ at $x = 0$ and $x = 1$; these come from the dynamic pressure values for a stationary fluid in the region outside the lubrication layer. We assume that the slider block is moving through a large body of fluid that completely surrounds it. In this case, it is obvious that

$$p^{(0)}(0) = p^{(0)}(1) = p_0, \tag{5-88}$$

where p_0 is an arbitrary reference pressure. The fact that p_0 is arbitrary is a consequence of the assumed incompressibility of the fluid. Applying this condition (5–88) to (5–87), we find that

$$c_2 = p_0, \tag{5-89}$$

$$c_1 = 6\left(\frac{d_2}{d_1} - 1\right)\frac{\int_0^1 h^{-3}(s)s\,ds}{\int_0^1 h^{-3}(s)\,ds} = 6\left(\frac{d_2 - d_1}{d_1 + d_2}\right).$$

Hence,

$$\boxed{p^{(0)}(x) - p_0 = 6\left(\frac{d_2 - d_1}{d_1 + d_2}\right)\frac{(x - x^2)}{(1 + bx)^2} \quad \text{with } b \equiv \frac{d_2 - d_1}{d_1},} \tag{5-90}$$

It follows that

$$\boxed{\frac{dp^{(0)}}{dx} = \frac{6(d_2 - d_1)}{d_1 + d_2}\frac{1}{h^3(x)}\left[1 - \left(\frac{d_2}{d_1} + 1\right)x\right],} \tag{5-91}$$

The scaled pressure distribution,

$$p^* - p_0^* \equiv (p^{(0)} - p_0)\bigg/6\left(\frac{d_2 - d_1}{d_1 + d_2}\right),$$

The Thin-Gap Approximation – Lubrication Problems

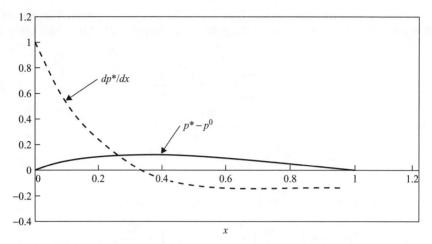

Figure 5–8. The dimensionless pressure and pressure gradient scaled by $6(d_2 - d_1)/(d_2 + d_1)$ plotted as functions of position in the gap.

and the scaled pressure gradient,

$$\frac{dp^*}{dx} \equiv \frac{dp^{(0)}}{dx} \bigg/ 6\left(\frac{d_2 - d_1}{d_1 + d_2}\right),$$

are plotted in Fig. 5–8 for $d_2 = 2$ and $d_1 = 1$. The key point to note is that the pressure is positive in the gap. The magnitude of the dimensionless pressure will increase as the ratio of d_2 to d_1 increases.

If we convert (5–90) and (5–91) to dimensional terms, the pressure distribution becomes

$$p'^{(0)}(x') - p'_0 = \frac{6\mu U}{L}\frac{1}{\varepsilon^2}\left(\frac{d_2 - d_1}{d_1 + d_2}\right)\frac{(Lx' - x'^2)}{(L + bx')^2}, \tag{5–92}$$

and the pressure gradient is

$$\frac{dp'^{(0)}}{dx'} = \frac{6\mu U}{\varepsilon^2}\left(\frac{d_2 - d_1}{d_2 + d_1}\right)\left\{\frac{L - [(d_2/d_1) + 1]x'}{(L + bx')^3}\right\}. \tag{5–93}$$

With dp'/dx' known, the velocity distribution, (5–74), can be specified completely, and the solution for the velocity and pressure fields in the lubrication layer is complete.

Let us now calculate the hydrodynamic forces on the slider block. In general, for any body in a viscous fluid,

$$\mathbf{F}' = \int_A \mathbf{t}'\,dA = \int_A \mathbf{n}\cdot\mathbf{T}'\,dA = -\int_A p'^{(0)}\mathbf{n}\,dA + \int_A \mathbf{n}\cdot\boldsymbol{\tau}'^{(0)}\,dA, \tag{5–94}$$

where A represents the complete exterior surface of the slider block, including both the surface inside and outside the lubrication layer, and p is the dynamic pressure relative to any arbitrary reference pressure (note that a constant reference pressure produces no contribution to the integral involving p). In (5–90), $p'^{(0)}(0)$ was used as the background reference pressure. Thus, to a first approximation,

$$\mathbf{F}' = \int_A -\mathbf{n}\left[p'^{(0)} - p'(0)\right]dA + \int_A \mathbf{n}\cdot\boldsymbol{\tau}'^{(0)}\,dA. \tag{5–95}$$

C. Applications of Lubrication Theory

Let us begin by calculating the forces on the flat surface of the slider block that forms the upper boundary of the lubrication layer. To do this we require an expression for the outer normal $\mathbf{n}$ to this surface. In general, for a surface $z = g(x)$, the unit normal is

$$\mathbf{n} = \pm \frac{\nabla(z-g)}{|\nabla(z-g)|}. \tag{5-96}$$

Here, $g = mx + d_1$, and thus $\nabla(z-g) = \mathbf{e}_z - m\mathbf{e}_x$ and $|\nabla(z-g)| = \sqrt{1+m^2}$, where $m \equiv (d_2 - d_1)/L$. It follows from the definition (5–96) that

$$\mathbf{n} = \pm\left(\frac{-m}{1+m^2}\mathbf{e}_x + \frac{1}{\sqrt{1+m^2}}\mathbf{e}_z\right). \tag{5-97}$$

The choice of the plus or minus sign in (5–97) determines whether the unit normal points into the slider block or points into the gap and is the outer normal as required. It is easy to see which sign to choose by considering the limiting case $m = 0$ in which the surface of the slider block is parallel to the plane wall. In this case

$$\mathbf{n} = \pm \mathbf{e}_z,$$

from which it is evident that the outer normal to the slider plate corresponds to the minus sign in (5–97) and

$$\mathbf{n} = \frac{m}{\sqrt{1+m^2}}\mathbf{e}_x - \frac{1}{\sqrt{1+m^2}}\mathbf{e}_z. \tag{5-98}$$

Thus,

$$\mathbf{n}\cdot\tau'^{(0)} = \frac{m}{\sqrt{1+m^2}}\left[\tau'^{(0)}_{xx}\mathbf{e}_x + \tau'^{(0)}_{xz}\mathbf{e}_z\right] - \frac{m}{\sqrt{1+m^2}}\left[\tau'^{(0)}_{zx}\mathbf{e}_x + \tau'^{(0)}_{zz}\mathbf{e}_z\right]. \tag{5-99}$$

Finally, to calculate the force per unit width on the slider block, we must integrate along s. For this purpose it is convenient to express a differential change along the slider block, ds, in terms of a differential change in x by use of

$$ds = \sqrt{1+(dz'/dx')^2}dx' = \sqrt{1+m^2}dx'. \tag{5-100}$$

Thus the x component of $\mathbf{F}$ on the bottom surface of the slider block is

$$F'_x = \int_0^L -[p'^{(0)}(x) - p'^{(0)}(0)]m\,dx' + \int_0^L [m\tau'^{(0)}_{xx} - \tau'^{(0)}_{xz}]dx'. \tag{5-101}$$

Substituting for $[p'^{(0)}(x) - p'^{(0)}(0)]$ and $[m\tau'^{(0)}_{xx} - \tau'^{(0)}_{xz}]$ by using (5–90) and the corresponding expressions for $\tau'^{(0)}_{xx}$ and $\tau'^{(0)}_{xz}$ calculated from the known velocity field in the lubrication layer, we find that

$$F'_x = -\frac{18\mu UL}{d_2 + d_1}\left(1 - \frac{4}{9}\frac{d_2 + d_1}{d_2 - d_1}\ln\frac{d_2}{d_2}\right). \tag{5-102}$$

Thus, in the limit $\varepsilon \to 0$, we see

$$F'_x \sim O(\varepsilon^{-1}).$$

Thus the force necessary to maintain the constant slider-block velocity U is proportional to ε^{-1} for $\varepsilon \ll 1$, where $\varepsilon \equiv d_1/L$. The normal (or z) component of force on the slider block can be seen from (5–95) and (5–98)–(5–100) to be

$$F'_z = \int_0^L [p'^{(0)}(x') - p'^{(0)}(0)]dx' + \int_0^L [m\tau'^{(0)}_{xz} - \tau'^{(0)}_{zz}]dx'. \tag{5-103}$$

The Thin-Gap Approximation – Lubrication Problems

Noting that $m = O(\varepsilon)$, we can show that the second term in this expression is $O(\varepsilon^{-1})$ but the pressure contribution is $O(\varepsilon^{-2})$. Hence, to a first approximation for small ε,

$$F'_z = \int_0^L [p'^{(0)}(x') - p'^{(0)}(0)]dx'. \tag{5-104}$$

Again, substituting for $[p'^{(0)} - p'^{(0)}(0)]$ and integrating, we find that

$$F'_z \sim \frac{12\mu U L^2}{d_2^2 - d_1^2}\left[1 - \frac{1}{2}\frac{d_1 + d_2}{d_2 - d_1}\ln\left(\frac{d_2}{d_1}\right)\right], \tag{5-105}$$

or, in the limit $\varepsilon \to 0$,

$$F'_z \sim O(\varepsilon^{-2}). \tag{5-106}$$

These results for the slider block are similar to those for the journal bearing, and both are typical of this general class of lubrication problems. The largest force in the lubrication layer is the normal lift force, and this is dominated, for small ε, by the pressure contribution. This force is always $O(\varepsilon^{-2})$ in magnitude, where ε is the dimensionless ratio of gap width to gap length. The tangential force necessary to maintain the relative motion between the boundaries is also large $O(\varepsilon^{-1})$ for small ε, but is smaller than the normal force by $O(\varepsilon)$. Again, this is typical of lubrication problems of this general class.

The only remaining point is the contribution to F'_x and F'_z that is due to pressure and viscous stresses acting on the portions of the slider block that are not included in the lubrication region. At most, however, these contributions are $O(1)$ compared with the $O(\varepsilon^{-1})$ and $O(\varepsilon^{-2})$ contributions from the lubrication layer, and thus they do not need to be calculated explicitly for $\varepsilon \ll 1$. To see this, we note (according to the scaling that we have already done) that the largest viscous stresses in the lubrication layer are proportional to $\partial u'^{(0)}/\partial z'$ and are thus of order $O(\mu(U/\varepsilon L))$. The largest contribution to the pressure in the lubrication layer [relative to $p(0)$] is $O(\mu(U/\varepsilon^2 L))$. Outside the lubrication layer, however, the largest viscous stresses are $O(\mu U/L)$, and the largest pressures in this region are the same magnitude. The ratio of viscous stresses inside the lubrication layer compared with those outside is thus $O(\varepsilon^{-1})$, whereas the ratio of dynamic pressures is $O(\varepsilon^{-2})$. Because the surface area of the slider block is the same order of magnitude inside and outside the lubrication layer, the estimate we made of the relative magnitude of the contributions to the force follows. We thus see that the calculations for F'_x and F'_z from the lubrication layer, (5–102) and (5–105), are, in fact, the correct leading-order approximations (for small ε) to the *total* drag and lift on the slider block. We again note that the dominance of contributions from the lubrication region is typical of problems of this general class. One need not calculate velocity and pressure fields anywhere but in the thin-gap region, for $\varepsilon \to 0$, if the objective is to determine the leading-order approximation to the force and/or torque on the body.

2. The Motion of a Sphere Toward a Solid, Plane Boundary

A second example of the application of lubrication theory is its use in analyzing the motion of a sphere that is pushed by the action of an applied force F toward a solid plane boundary, when the gap between the sphere and the wall at the point of closest approach is small compared with the radius of the sphere.[6] Although this problem is geometrically similar to the slider block in the sense that a body of finite dimensions is moving in the vicinity of an infinite plane boundary, the problem differs in that the motion is normal to the boundary rather than parallel to it and the gap width is time dependent.

The geometry of the problem is sketched in Fig. 5–9. We consider a sphere of radius a, whose center is a distance $a + b$ from an infinite plane solid boundary when measured

C. Applications of Lubrication Theory

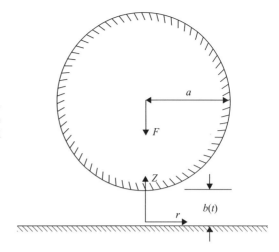

Figure 5–9. A sketch of the geometry and coordinate system for analysis of the motion of a sphere toward a plane wall under the action of an applied force F.

along the normal to the wall that passes through the center of the sphere. The sphere moves toward the wall under the action of an applied force F, and the objective is to calculate its velocity or its position relative to the wall as a function of time. In the present development, we consider the asymptotic solution for $b/a \ll 1$. The flow domain then divides into two parts, the region between the sphere and the wall where the lubrication equations (and solutions) pertain, and the rest of the domain where the lubrication equations cannot be used. As usual, the dominant contributions to the hydrodynamic force on the body come from the thin lubrication layer, and we thus focus our attention on obtaining an approximate solution in that regime, using the general equations from Section B. One question may occur immediately to the reader: If lubrication theory can be applied in this case, where does the lubrication layer end, as it is evident that the gap between the sphere and wall varies from a minimum value of b to a maximum equal to $b + a$ as one traverses parallel to the wall? We shall see that it does not really matter what choice we make. The asymptotic results for $\varepsilon \equiv b/a \to 0$ turn out to be independent of this choice.

Again, we employ the general solutions obtained in Section B to analyze the problem. One difference between this problem and the slider-block problem that was illustrated in the preceding subsection is that the sphere moves normal to the wall rather than parallel to it. Although the governing equations, (5–69)–(5–72), were formulated with the assumption that the relative motions of the boundaries could have both a parallel and normal component, the choice of characteristic velocity u_c was based on the parallel component. In the present case, an obvious choice for the characteristic length scale ℓ_c is the sphere radius a. However, because the boundary motion is strictly in the z direction and the velocity of the sphere will vary with time, it may not be entirely obvious what choice to make for the characteristic velocity scale. A characteristic magnitude for the normal velocity of the sphere can be defined in terms of the applied force F. A combination of dimensional parameters involving F that has units of velocity is $F/\mu a$. Hence, provided F is independent of time, we can use this quantity as the characteristic scale εu_c for V' in the definitions of V and $\mathbf{U}$ in (5–72), with u_c then being $F/\mu a \varepsilon$. If F is time dependent, we can use the value of F at some initial instant of time and then nondimensionalize in a similar fashion.

The value of nondimensionalization is that it allows simplification of the governing equations. The point of the preceding paragraph is that characteristic scales can be determined for the present problem that are consistent with the general analysis of Section B. Hence the same analysis and simplifications can be applied in spite of the qualitative differences

The Thin-Gap Approximation – Lubrication Problems

between this and the classic lubrication problems such as the slider block of the preceding subsection. However, we have already obtained the leading-order (lubrication) equations and solutions for the complete class of thin-film problems in Section B. Thus, rather than being concerned with the details of nondimensionalization for the present problem, we simply utilize the governing equations and solutions from Section B in terms of *dimensional variables*.

The governing equation for the film thickness as a function of time is just the Reynolds equation, (5–79). In the present problem, the horizontal velocity component at the boundary, **U**, is zero, and the last term in (5–79) is therefore neglected. Furthermore, the motion generated in the thin film that is due to the motion of the sphere will be axisymmetric. Hence the most convenient choice of coordinates is cylindrical, with the z axis normal to the wall, as shown in Fig. 5–9. The motion in the thin film will then be independent of the polar angle, and (5–79) can be expressed in terms of dimensional variables in the form

$$12\mu \frac{\partial h}{\partial t} = \frac{1}{r}\frac{\partial}{\partial r}\left(rh^3 \frac{\partial p}{\partial r}\right). \tag{5-107}$$

To utilize (5–107), we must specify the gap-width function $h(r, t)$. For this purpose, we use the cylindrical coordinates previously specified. Then, using elementary geometry, we can write an exact expression for $h(r, t)$:

$$h(r, t) = a + b(t) - (a^2 - r^2)^{1/2}, \tag{5-108}$$

where $b(t)$ is the minimum gap width as shown in Fig. 5–9. However, we expect the thin-film approximation, (5–107), to apply only when $r/a \ll 1$. In this case, the expression (5–108) can be approximated in the form

$$h(r, t) = b(t) + \frac{a}{2}\left(\frac{r^2}{a^2}\right) + O\left(\frac{r^4}{a^4}\right). \tag{5-109}$$

It may be noted that $\partial h/\partial t$ is independent of r. Hence we can integrate (5–107) directly. Integrating once with respect to r, we obtain

$$\frac{\partial p}{\partial r} = 12\mu \frac{\partial h}{\partial t}\left(\frac{r}{2h^3} + \frac{c}{rh^3}\right), \tag{5-110}$$

where c is a constant of integration. We can determine this constant by using the fact, evident from the symmetry of the problem, that $\partial p/\partial r = 0$ at $r = 0$. Thus $c = 0$ and

$$\frac{\partial p}{\partial r} = 6\mu \frac{r}{h^3}\frac{\partial h}{\partial t}.$$

Integrating once more with respect to r we obtain

$$-p(r) + p(R) = 6\mu \frac{\partial h}{\partial t}\int_r^R \frac{s}{h^3(s, t)}ds. \tag{5-111}$$

Here R is a value of r that we assume to be large enough to represent the outermost point within the thin-film region. As noted earlier, the gap width between the sphere and the wall varies monotonically from a minimum of b to a maximum of $a + b$ for $r = a$, and it may seem rather arbitrary to decide what value of r to designate as the largest value that still falls within the thin-film region. This is, in fact, true, but we shall see that R drops out of

C. Applications of Lubrication Theory

the expressions for the pressure (and therefore all quantities related to the pressure) in the limit $\varepsilon \to 0$.

If we now substitute the approximate form for $h(r, t)$ from (5–109) into (5–111) and carry out the integration on the right-hand side, we can finally write the expression for the pressure in the form

$$p(r) - p(R) = -6\mu \frac{db}{dt} \left[\frac{2a^3}{(2ab + r^2)^2} - \frac{2a^3}{(2ab + R^2)^2} \right]. \quad (5\text{--}112)$$

Now, given this form for the pressure distribution, we can calculate the vertical, hydrodynamic force exerted by the fluid on the sphere. As usual in lubrication analyses, the hydrodynamic force on the sphere is dominated by the forces from within the lubrication layer, and, in this region, pressure forces are $O(R/b) \gg 1$ larger than the forces arising because of viscous stresses in the fluid. Thus it follows that the hydrodynamic force exerted by the fluid on the sphere can be expressed symbolically in the form

$$\mathbf{e}_z \cdot \mathbf{F} = -\int_A p(\mathbf{n} \cdot \mathbf{e}_z) dA, \quad (5\text{--}113)$$

where A represents the area of the sphere.

Now we can calculate the unit normal at the sphere surface in terms of our cylindrical coordinates by using the usual expression

$$\mathbf{n} \equiv \nabla f / |\nabla f|,$$

where

$$f \equiv z - h(r, t)$$

is the function that is equal to zero for points on the surface of the sphere. It follows that

$$\mathbf{n} = -\left(1 - \frac{r^2}{a^2}\right)^{1/2} \mathbf{e}_z + \frac{r}{a} \mathbf{e}_r. \quad (5\text{--}114)$$

The differential surface area, expressed in terms of differential changes in the cylindrical coordinates, r and θ, is

$$dA = \left[1 + (\partial z/\partial r)^2\right]^{1/2} r \, dr \, d\theta = \left[1 - (r^2/a^2)\right]^{-1/2} r \, dr \, d\theta. \quad (5\text{--}115)$$

Hence, when (5–113)–(5–115), are combined, it follows that

$$\mathbf{e}_z \cdot \mathbf{F} = 2\pi \int_0^R [p(r) - p(R)] r \, dr. \quad (5\text{--}116)$$

Because $p = p(R)$ everywhere outside the lubrication layer, to a first approximation, it is the pressure difference $p(r) - p(R)$ within the lubrication layer that is responsible for the net lift force on the sphere, and the integration over r is carried out only to $r = R$.

Substituting for $p(r) - p(R)$ from (5–112) and evaluating the integral on the right-hand side of (5–116), we obtain the approximate result for the hydrodynamic force:

$$\mathbf{e}_z \cdot \mathbf{F} = \frac{6\pi \mu \dot{b} a^2}{b} \left[1 - \frac{2ab}{(R^2 + 2ab)} - \frac{R^2(2ab)}{(R^2 + 2ab)^2} \right]. \quad (5\text{--}117)$$

However, within the lubrication layer, $R^2/2ab \gg 1$, and this expression for the force can be further approximated to the form

$$\mathbf{e}_z \cdot \mathbf{F} \sim \frac{6\pi \mu \dot{b} a^2}{b}. \quad (5\text{--}118)$$

The Thin-Gap Approximation – Lubrication Problems

This is the final asymptotic result relating the hydrodynamic force to the motion of the sphere toward the infinite plane wall. We can see from (5–118) that if the sphere were prescribed to move with a constant velocity (i.e., $\dot{b}=$ constant), the hydrodynamic force would blow up as b^{-1}. This implies that the external force applied to the sphere would also be required to go to infinity as b^{-1} to maintain such a motion. Of course, the application of an infinite force is not possible, and for any lesser force the sphere will slow down as it approaches the wall. In the present case, the most straightforward assumption is that the external force on the sphere is a constant F in the $-\mathbf{e}_z$ direction, as indicated in Fig. 5–9.

To determine exactly what will happen, we need to proceed a few more steps with the analysis. If we neglect the inertia of the sphere, the external force F and the hydrodynamic force $\mathbf{e}_z \cdot \mathbf{F}$ must exactly balance, i.e.,

$$-F \sim \frac{6\pi \mu \dot{b} a^2}{b}. \tag{5–119}$$

If F is constant, it follows from (5–119) that db/dt must decrease at just the rate to balance the increase in b^{-1} as the gap width decreases. This means that the sphere velocity rapidly decreases as the sphere approaches the plane wall.

To determine the details of the sphere's motion, we can solve the ODE that is obtained from (5–119):

$$\frac{1}{b}\frac{db}{dt} = -\frac{F}{6\pi \mu a^2}. \tag{5–120}$$

For the case when $F =$ constant, the right-hand side of (5–120) is a constant, independent of time, and we can integrate (5–120) directly to obtain

$$b(t) = b(0) e^{-Ft/6\pi \mu a^2} \tag{5–121}$$

where $b(0)$ represents the gap width at some initial moment $t = 0$. Thus the velocity of the sphere is

$$\frac{db}{dt} = \frac{b(0)F}{6\pi \mu a^2} e^{-Ft/6\pi \mu a^2}. \tag{5–122}$$

It can be seen from either (5–121) or (5–122) that $b(t)$ approaches zero asymptotically only as $t \to \infty$. We therefore conclude that *the sphere will not contact the wall in a finite amount of time* under the action of any finite force.

The result (5–121) is in apparent disagreement with experimental observation, because it is generally agreed that a real sphere will actually contact the wall in a finite time interval when acted on by a finite force that is pushing it toward the wall. It should be remembered, however, that the lubrication analysis just presented assumes that the sphere and the wall are *absolutely smooth*. In reality, neither surface will exactly satisfy this requirement. In normal circumstances, the characteristic length scales of a flow are sufficiently large that the small, microscale imperfections of a solid surface are imperceptible. In the present lubrication theory, however, we are concerned with the motion of the sphere when it is *very close* to the wall, and in this case, such imperfections may become significant if they are not vanishingly small compared with the asymptotically small gap width. Presumably, any microscopic surface roughness will eventually become important as $b \to 0$, and this fact may account for the apparent difference between theory and experiment that was mentioned at the beginning of this paragraph.

D. The Air Hockey Table

Another possible departure from the ideal conditions assumed in the analysis is that the film may become thin enough that the continuum-based hydrodynamic analysis may break down in various ways. First of all, for a sufficiently thin film, either the no-slip conditions or the bulk-phase continuum approximation itself may break down. In either case, the resistance to further decrease in the gap width would be decreased compared with that which underlies the predictions (5–121) and (5–122). However, as we will discuss in more detail in the next chapter, this is not expected to be a factor when the fluid is a small-molecule (Newtonian) liquid until the film thickness is less than about 10–100 Å, and in many cases, the surface roughness discussed in the preceding paragraph will have a longer length scale than this.

The other departure from a pure fluid mechanics phenomenon that occurs in almost every case when the film dimension reaches the order of a few hundred angstroms (still plenty large enough for the fluid motion to be described by the Navier–Stokes equations), is that nonhydrodynamic forces of attraction or repulsion between the two solid surfaces will come into play. The most common of these is van der Waals attraction, but a number of other types of force are also possible depending on the nature of the fluid and the two solid surfaces. When attractive van der Waals forces are present, the attractive force increases as the gap width decreases in proportion to $1/b^3$. This is a strong-enough dependence on the gap width that it can dominate the increased hydrodynamic resistance to further thinning, which as we have seen increases only as $1/b^{3/2}$. Hence, if a particle approaches within approximately 100 Å of a solid surface to which it is attracted by van der Waals interactions, it will come into contact in finite time even if it is perfectly smooth. The presence of very strong forces of nonhydrodynamic origin is a common characteristic of small-scale systems and, when present, can play a dominant role in the dynamics of interactions. In the next chapter, we shall encounter additional examples of problems in which short-range nonhydrodynamic forces can play a crucial role.

D. THE AIR HOCKEY TABLE

The table game of air hockey may be familiar to most readers. It is played by means of the motion of a flat disk that slides seemingly without resistance over the horizontal surface of the playing table. Of course, if the disk were actually in contact with the solid tabletop it could not move freely as it appears to do, and, in reality, it is levitated above the table on a thin layer of air that is blown up through the surface of the table, which is porous. It turns out that the air motion in the gap between the bottom of the disk and the porous tabletop is, in fact, another example to which the thin-film approximation of lubrication theory can be applied. However, the large pressures that are responsible for the lubrication effect do not come from the relative motion of the two solid surfaces, as in the previous lubrication problems, but are due to the normally directed velocity of air through the table surface. The total normal force that is required for lifting the disk is known, namely Mg, where M is the mass of the disk. However, as the air velocity through the porous tabletop is increased above the value necessary to initially suspend the disk, the height d of the lower surface of the disk above the table will increase. It is this height d that we need to calculate in order ultimately to estimate the drag on the levitated disk as it moves horizontally across the tabletop.

In the present section, we consider only the problem of calculating the height of a *stationary* disk as a function of the pressure p'_R in the air reservoir below the table surface. For this purpose, we consider a flat disk of radius R and total mass M, levitated (or suspended) an (unknown) height $d \ll R$ above a horizontal porous surface through which air is blown from a reservoir that is held at a pressure p'_R. A sketch of the configuration is shown in Fig. 5–10. Although the air motion and the air pressure distribution in the thin gap between the disk and the tabletop may be quite complicated, and there may be some significant

The Thin-Gap Approximation – Lubrication Problems

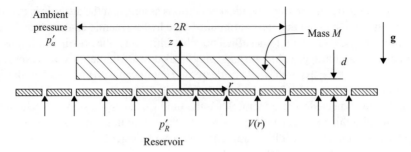

Figure 5–10. The configuration for a "levitated" air hockey puck above the surface of a porous table top through which air is being blown at a superficial velocity V'.

pressure loss in pumping air through the porous tabletop, the minimum reservoir pressure that could possibly produce sufficient force to cause the disk to lift from the table is

$$(p'_R)_{min} - (p'_a) = \frac{Mg}{\pi R^2}. \tag{5-123}$$

Here, p'_a is the ambient pressure, which acts on the upper surface of the disk and opposes p'_R. We thus consider reservoir pressures that exceed this minimum value. It is convenient to parameterize these pressures in terms of a dimensionless pressure p^*_R, ə

$$\boxed{p'_R - p'_a = \frac{Mg}{\pi R^2} p^*_R.} \tag{5-124}$$

Thus we consider cases in which

$$\boxed{p^*_R > 1.} \tag{5-125}$$

Finally, to complete specification of our problem, we need to specify the normal inflow velocity V' across the porous tabletop. For the present purposes, we assume that V' is linearly related to the pressure differences across the porous tabletop:

$$V' = \frac{k}{\mu}(p'_R - p'), \tag{5-126}$$

where k is known as the permeability, μ is the viscosity of air, p'_R is the reservoir pressure, and p' is the local pressure on the other side of the porous plate within the thin gap between the disk and the tabletop. Because p' will vary with radial position r', the normal velocity V' will also be a function of radial position. A simplified approximate problem is to assume that the inflow is *independent* of r'. We shall see that this approximation is appropriate for large reservoir pressures p'_R so that the variations in $(p'_R - p')$ are small compared with the overall pressure difference $(p'_R - p'_a)$ between the reservoir pressure and the ambient pressure p'_a in the air surrounding the disk (but outside the thin gap).

Now, to determine the levitation height of the disk, d, we must first determine the pressure distribution in the thin air gap below the disk and then apply a vertical force balance on the disk. The analysis in the thin gap can be carried out by means of the dimensionless thin-film equations, (5–61), (5–65), and (5–66). These equations are scaled with respect to the characteristic length scale along the gap, ℓ_c, which in this case can be taken as the disk radius $\ell_c = R$; a characteristic velocity scale along the gap u_c; and a characteristic pressure $p_c = [(\mu u_c/R)(1/\varepsilon^2)]$, where the thin-gap parameter $\varepsilon \equiv [(d/R) \ll 1]$.

D. The Air Hockey Table

In our problem, the characteristic air velocities in the thin gap are determined by the normal airflow through the porous lower surface. An appropriate *characteristic* value for this velocity is

$$V_c = \frac{k}{\mu}(p'_R - p'_a). \tag{5-127}$$

The characteristic velocity u_c used in (5–65) and (5–66) is then given in terms of V_c as

$$u_c = \frac{V_c}{\varepsilon}. \tag{5-128}$$

Finally, the Reynolds number that appears in Eqs. (5–65) and (5–66) is $Re \equiv \rho u_c \ell_c/\mu$. A more convenient choice for the present problem is

$$\widetilde{Re} = \frac{\rho V_c d}{\mu}. \tag{5-129}$$

This is related to Re as

$$\widetilde{Re} = \varepsilon^2 \, Re. \tag{5-130}$$

With these definitions, and assuming the flow to be axisymmetric, we can write the governing equations, (5–61), (5–65), and (5–66), using cylindrical coordinates in the form

$$\frac{1}{r}\frac{\partial}{\partial r}(ru) + \frac{\partial w}{\partial z} = 0, \tag{5-131}$$

$$\frac{\partial^2 u}{\partial z^2} - \frac{\partial p}{\partial r} = \widetilde{Re}\left(u\frac{\partial u}{\partial r} + w\frac{\partial u}{\partial z}\right) - \varepsilon^2\left[\frac{1}{r}\frac{\partial}{\partial r}\left(r\frac{\partial u}{\partial r}\right) - \frac{u}{r}\right], \tag{5-132}$$

and

$$\frac{\partial p}{\partial z} = \varepsilon^2\left[\frac{\partial^2 w}{\partial z^2} + \varepsilon^2\frac{1}{r}\frac{\partial}{\partial r}\left(r\frac{\partial w}{\partial r}\right)\right] + \varepsilon^2 \widetilde{Re}\left(u\frac{\partial w}{\partial r} + w\frac{\partial w}{\partial z}\right), \tag{5-133}$$

where u is the radial velocity component and w is the axial component.

In the following discussion, we assume that $\varepsilon \ll 1$ and further that $\varepsilon^2 \widetilde{Re} \ll 1$. Hence, if we seek an asymptotic solution in this thin-film limit, the governing equations for the leading-order approximation are

$$\frac{1}{r}\frac{\partial}{\partial r}(ru) + \frac{\partial w}{\partial z} = 0, \tag{5-134}$$

$$\frac{\partial^2 u}{\partial z^2} - \frac{\partial p}{\partial r} = \widetilde{Re}\left(u\frac{\partial u}{\partial r} + w\frac{\partial u}{\partial z}\right), \tag{5-135}$$

$$\frac{\partial p}{\partial z} = 0. \tag{5-136}$$

The boundary conditions at the tabletop and bottom of the disk are

$$u = w = 0 \quad \text{at } z = 1, \tag{5-137a}$$

$$\left.\begin{array}{l} u = 0 \\ w = \dfrac{p_R - p}{p_R - p_a} \end{array}\right\} \quad \text{at } z = 0, \tag{5-137b}$$

The Thin-Gap Approximation – Lubrication Problems

where p_R and p_a are the reservoir and ambient pressures, now scaled by use of p_c as previously defined. We also assume that the pressure p is equal to the ambient pressure p_a at the edge of the disk, $r = 1$. Although we may expect the thin-film scaling to break down in the immediate vicinity of $r = 1$, we have seen in the previous examples that it is generally sufficient *at leading order* in the thin-film approximation to simply apply an ambient pressure condition at the ends of the film and to ignore any local modifications of either the pressure distribution or the flow in the transition from the thin film to the rest of the flow domain.

Finally, we note again that the height of the disk above the table surface is an unknown, which we must calculate by determining the pressure (and normal viscous stress) distribution on the bottom of the disk within the thin gap and then applying a force balance. At steady state (i.e., with d constant), this force balance can be written in dimensional form as

$$Mg = 2\pi \int_0^R [-(\mathbf{T}' \cdot \mathbf{i}_z) \cdot \mathbf{i}_z \mid_{z'=d} - p'_a] r' dr', \tag{5-138}$$

where

$$\mathbf{T}' = -p'\mathbf{I} + 2\mu \mathbf{E}'.$$

In writing (5–138), we assume that the pressure outside the thin gap is the ambient pressure, which thus produces a downward force on the upper surface of the disk that must be subtracted from the upward force from within the thin film. If we nondimensionalize, we can see that the stress integral is dominated by the pressure contribution, and the force balance, (5–138), can thus be written in the dimensionless form

$$\boxed{Mg = \frac{2\pi \mu V_c R^4}{d^3} \int_0^1 (p \mid_{z=1} - p_a) r \, dr.} \tag{5-139}$$

Clearly, if we can evaluate the integral, this equation can be solved to determine the gap width d as a function of the other parameters.

Although the procedure to determine d is straightforward, the actual implementation is complicated for two reasons. First, for arbitrary $\widetilde{Re}$, the problem is still nonlinear even in the thin-gap limit, and the solution of (5–134)–(5–136) thus generally requires a numerical approach. Second, the blowing velocity through the porous tabletop will generally depend on radial position beneath the disk because of its coupling with the local pressure in the gap, which will in turn depend on radial position.

There are two cases for which approximate analytical solutions can be obtained. In the first, we assume $\widetilde{Re} \ll 1$, so that the leading-order equations for the thin-gap region reduce to the classical lubrication equations. The second case is that in which the reservoir pressure p'_R is large enough that the radial variations in the blowing velocity can be neglected.

1. The Lubrication Limit, $\widetilde{Re} \ll 1$

In this subsection, we begin by considering the limit of the thin-film equations in which $\widetilde{Re} \ll 1$. In this case, the problem reduces to solving a classical lubrication problem, and a slightly modified version of the Reynolds equation, (5–79), can be used to obtain the leading-order approximation to the pressure distribution in the thin gap.

We seek an asymptotic solution of the leading-order thin-film equations, (5–134)–(5–136), in the form

$$\begin{aligned} u &= u_0 + f_1\left(\widetilde{Re}\right) u_1 + \cdots +, \\ w &= w_0 + f_1\left(\widetilde{Re}\right) w_1 + \cdots +, \\ p &= p_0 + f_1\left(\widetilde{Re}\right) p_1 + \cdots +. \end{aligned} \tag{5-140}$$

D. The Air Hockey Table

Thus, at the first approximation, we obtain the classic lubrication equations:

$$\frac{1}{r}\frac{\partial}{\partial r}(ru_0) + \frac{\partial w_0}{\partial z} = 0, \tag{5-141}$$

$$\frac{\partial^2 u_0}{\partial z^2} - \frac{\partial p_0}{\partial r} = 0, \tag{5-142}$$

$$\frac{\partial p_0}{\partial z} = 0, \tag{5-143}$$

with

$$u_0 = w_0 = 0 \quad \text{at } z = 1, \tag{5-144a}$$

$$u_0 = 0 \quad \text{at } z = 0,$$

$$w_0 = \frac{p_R - p_0}{p_R - p_a} \quad \text{at } z = 0. \tag{5-144b}$$

If we review the derivation of the Reynolds equation, (5–79), starting with the governing equations and boundary conditions, (5–69)–(5–72), we see that the present problem differs in that $\partial h/\partial t = \partial d/\partial t \equiv 0$, but there is still a normal velocity at $z = 0$ that is due to blowing of air through the porous tabletop. Hence we can see from (5–75)–(5–77) that the appropriate form of Reynolds equation for the present problem should be

$$-V = \nabla_s \cdot \left(\frac{1}{12}\nabla_s p_0\right), \tag{5-145}$$

where

$$V \equiv (w_0)_{z=0} \equiv \frac{p_R}{p_R - p_a} - \frac{p_0}{p_R - p_a}.$$

In writing (5–145), we have taken account of the fact that the dimensionless film thickness is $h = 1$. If we reexpress this equation in cylindrical coordinates, it can be written in the form

$$r^2\frac{d^2 p_0}{dr^2} + r\frac{dp_0}{dr} - \frac{12}{(p_R - p_a)}r^2 p_0 = -\frac{12 p_R}{p_R - p_a}r^2. \tag{5-146}$$

The homogeneous equation is a slightly disguised form of the modified Bessel's equation of order 0.

We can write it in the "standard form" by introducing a change of the independent variable:

$$\zeta = \left(\frac{12}{p_R - p_a}\right)^{1/2} r = \alpha r. \tag{5-147}$$

This yields

$$\zeta^2 \frac{d^2 p_0}{d\zeta^2} + \zeta \frac{dp_0}{d\zeta} - \zeta^2 p_0 = -p_R \zeta^2. \tag{5-148}$$

There are two independent solutions of the homogeneous equation, known as modified Bessel's functions of order 0, of the first and second kind, $I_0(\zeta)$ and $K_0(\zeta)$. However,

The Thin-Gap Approximation – Lubrication Problems

$K_0(\zeta) \to \infty$ as $\zeta \to 0$, and hence cannot be included in the solution for p_0. The particular solution of (5–148) can be seen to be

$$p_0 = p_R.$$

Hence the general, nonsingular solution for p_0 is

$$p_0 = p_R + c_1 I_0(\zeta). \tag{5–149}$$

The constant c_1 is chosen so that

$$p_0 = p_a \quad \text{at } r = 1. \tag{5–150}$$

This gives

$$c_1 = (p_a - p_R)/I_0(\alpha)$$

so that

$$p_0 = p_R - (p_R - p_a)\frac{I_0(\alpha r)}{I_0(\alpha)}. \tag{5–151}$$

With this leading-order approximation for the pressure in hand, we can obtain a first approximation to the gap height d by applying the force balance in (5–139), namely,

$$Mg = \frac{2\pi \mu V_c R^4}{d^3}(p_R - p_a)\int_0^1 \left[1 - \frac{I_0(\alpha r)}{I_0(\alpha)}\right] r\, dr. \tag{5–152}$$

Carrying out the integration, we can reexpress the result in terms of the *dimensional* pressure difference $p'_R - p'_a$; this gives

$$Mg = \pi R^2 (p'_R - p'_a)\left[1 - \frac{2 I'_0(\alpha)}{\alpha I_0(\alpha)}\right]. \tag{5–153}$$

We can "solve" this equation for the special value of $\alpha = \alpha^*$ that causes the right-hand side to balance Mg. Because

$$\alpha \equiv \left(\frac{12}{p_R - p_a}\right)^{1/2} = \left(\frac{12k}{d^3}\right)^{1/2} R, \tag{5–154}$$

this is tantamount to solving for the unknown gap width d.

The complexity of Eq. (5–153) for α is such that it generally needs to be evaluated numerically. The result for $d^* \equiv d/(12kR^2)^{1/3}$ versus $p_R^* \equiv (p'_R - p'_a)/(Mg/\pi R^2)$ is plotted in Fig. 5–11. Asymptotic approximations can also be obtained from Eq. (5–153) in the two limits $\alpha \ll 1$ and $\alpha \gg 1$. These are

$$d = (3kR^2)^{1/3}\left(\frac{p_R^* - 1}{p_R^*}\right)^{2/3} \quad (\alpha \gg 1), \tag{5–155}$$

$$d = \left(\frac{3}{2}kR^2\right)^{1/3}(p_R^*)^{1/3} \quad (\alpha \ll 1), \tag{5–156}$$

respectively. These asymptotes are also shown in Fig. 5–11. A similar plot that also shows experimental data can be found in the original research paper on this problem by Hinch and Lemaitre.[7]

D. The Air Hockey Table

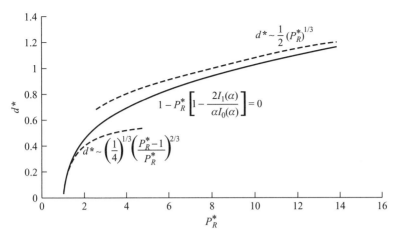

Figure 5–11. A plot of the dimensionless distance of the disk above the table top, $d^* = d/(12kR^2)^{1/3}$, versus the dimensionless pressure of the reservoir relative to ambient, $p_R^* \equiv (p_R' - p_a')/(Mg/\pi R^2)$. Also shown are the two asymptotes, (5–156) and (5–156). Note $d^* = \alpha^{-2/3}$.

We see, as anticipated, that the disk rises to a height above the table that increases with an increase of the dimensionless pressure difference p_R^* between the reservoir and ambient, from the minimum value of $p_r^* = 1$ that is required for lift-off to larger values. If we examine the details of the pressure distribution, Eq. (5–151), plotted in Fig. 5–12 as $\hat{p} \equiv (p_0 - p_a)/(p_a - p_R)$ versus r for various values of α, we find, for $\alpha \gg 1$ (i.e., $p_R^* \approx 1$), that the majority of the pressure drop from p_R' to p_a' occurs near the rim of the disk within a dimensionless distance of $O(\alpha^{-1})$, or a dimensional distance $O(d^3/12k)^{1/2}$. Furthermore, p_0 remains relatively close to the reservoir value, i.e., $\hat{p} \sim 1$ away from the rim.

The analysis in this subsection was carried out under the joint assumptions that $\varepsilon \equiv d/R \ll 1$ and $\widetilde{Re} = \rho V_c d/\mu \ll 1$. If the first of these conditions were to be violated, it would occur for large values of p_R^*, where we should expect the asymptotic formula, (5–156), to hold. In this regime, it can be seen from (5–156) that

$$\frac{d}{R} \equiv \left(\frac{3}{2}\frac{k}{R}\right)^{1/3} \left(p_R^*\right)^{1/3}.$$

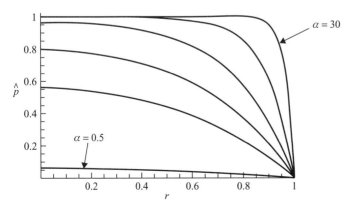

Figure 5–12. Dimensionless pressure distribution, plotted as $\hat{p} \equiv (p_0 - p_a)/(p_a - p_R)$ versus r for various values of $\alpha = 0.5, 2, 3, 5, 10,$ and 30.

The Thin-Gap Approximation – Lubrication Problems

Hence, the thin-film approximation will hold provided that

$$p_R^* \ll \left(\frac{3}{2}\frac{k}{R}\right)^{-1/3} \tag{5-157a}$$

or

$$p_R' - p_a' \ll \frac{Mg}{\pi}\left(\frac{2}{3k}\right)^{1/3} R^{7/3}. \tag{5-157b}$$

Hinch and Lemaitre[7] estimate $k = 2.6 \times 10^{-8}$ m. For disks in the size range from 1 to 10 cm in radius, we see that the condition (5–157) requires

$$p_R^* \ll \sim (10^2).$$

This condition on the reservoir pressure will rarely be violated.

The condition $\widetilde{Re} \ll 1$, on the other hand, requires

$$p_R^* \ll \frac{\mu^{3/2}}{\rho^{3/4} k}\left(\frac{\pi}{Mg}\right)^{3/4}\left(\frac{2}{3}\right)^{1/4} R. \tag{5-158}$$

If we adopt parameters from Hinch and Lemaitre of $k = 2.6 \times 10^{-8}$ m, $g = 9.81$ m s^{-2}, $\rho_{air} = 1.2$ kg m^{-3}, and $\mu_{air} = 1.8 \times 10^{-5}$ kg m^{-1} s^{-1}, we can evaluate the right-hand side of (5–158). The result is

$$p_R^* \ll 0.985 \frac{R}{M^{3/4}}. \tag{5-159}$$

This condition will often be violated for disks of the size and mass used in the air hockey game. For example, for a disk of radius 5 cm and mass of 20 g, the condition (5–159) requires

$$p_R^* \ll 0.926. \tag{5-160}$$

Hence we may expect that the lubrication analysis presented in this subsection will fail when the dimensionless pressure difference becomes much beyond this value.

The obvious question is whether we can say anything about the behavior of the air hockey system when $\widetilde{Re}$ is not small. A direct solution of the governing equations, (5–134)–(5–137), would require a numerical approach because of the nonlinearity of (5–135), if no additional approximation were made. However, as indicated earlier, we may expect that for large-enough values of the pressure difference, $p_R' - p_a'$, it may be possible to neglect the radial variation in the blowing velocity under the disk. Indeed, referring back to the analysis of the present subsection, we see that

$$w_0 = \frac{p_R - p_0}{p_R - p_a} = \frac{I_0(\alpha r)}{I_0(\alpha)}. \tag{5-161}$$

For $\alpha \ll 1$ (i.e., $p_R^* \gg 1$), we see that w_0 is nearly constant. This suggests that we may be able to achieve an adequate approximation to (5–134)–(5–137) by assuming that the blowing velocity at $z = 0$ is completely independent of r, provided we assume $p_R^* \gg 1$

2. The Uniform Blowing Limit, $p_R^* \gg 1$

We begin by adopting the apparently *ad hoc* assumption that

$$\boxed{w = f(z),} \tag{5-162}$$

even for $\widetilde{Re}$ values where the analysis of the preceding section does not apply, i.e., we neglect any radial change in the blowing rate under the disk. Although this assumption is

D. The Air Hockey Table

temporarily made without specifying a corresponding asymptotic limit, we shall see shortly that it may be expected to provide a good approximation to the exact solution of (5–134)–(5–137) whenever $p_R^* \gg 1$, as suggested by the small-$\widetilde{Re}$ analysis of the preceding section. We shall see that the approximate solution that we obtain is actually the first term in an asymptotic solution for p_R^* large.

Assuming that the assumption (5–162) is valid, we can use the continuity equation, (5–134), to determine that the radial velocity component must take the form

$$u = -\frac{r}{2}\frac{\partial f}{\partial z}. \tag{5-163}$$

It then follows that the radial momentum balance, (5–135), becomes

$$-\frac{r}{2}\frac{\partial^3 f}{\partial z^3} - \frac{\partial p}{\partial r} = \widetilde{Re}\left[\frac{r}{4}\left(\frac{\partial f}{\partial z}\right)^2 - \frac{r}{2}f\frac{\partial^2 f}{\partial z^2}\right]. \tag{5-164}$$

Because $p = p(r)$ only according to (5–136), it follows from (5–164) that

$$\frac{\partial p}{\partial r} = -\hat{\beta}r, \tag{5-165}$$

where $\hat{\beta}$ is an unknown that determines the magnitude of the radial pressure gradient, and the function $f(z)$ must then satisfy the ODE

$$-\frac{\widetilde{Re}}{2}\left(\frac{\partial f}{\partial z}\right)^2 + \widetilde{Re}f\frac{\partial^2 f}{\partial z^2} = -2\hat{\beta} + \frac{\partial^3 f}{\partial z^3}. \tag{5-166}$$

If we integrate (5–165), we see that

$$p = -\frac{\hat{\beta}}{2}r^2 + \alpha,$$

and, because $p = p_a$ at $r = 1$, the constant α is

$$\alpha = p_a + \frac{\hat{\beta}}{2}$$

and

$$p = p_a + \frac{\hat{\beta}}{2}(1 - r^2). \tag{5-167}$$

The boundary conditions on $f(z)$ can be obtained from the conditions (5–137) on u and w. These are

$$f(1) = f'(1) = 0,$$
$$f(0) = \frac{p_R - p}{p_R - p_a}, \tag{5-168}$$
$$f'(0) = 0.$$

Now, because Eq. (5–166) for f is third order, it might seem at first that the four boundary conditions (5–168) produce an overdetermined problem. However, the magnitude of the

333

radial pressure gradient $\hat{\beta}$ is also unknown, and we can thus solve (5–166) subject to (5–168) to obtain both $f(z)$ and $\hat{\beta}$.

Before going on to discuss the solution of this problem, we should pause to contemplate whether, or at least under what conditions, the problem as formulated is self-consistent. We have assumed in the first line that w and thus f are functions only of z. On the other hand, the value of $f(0)$ is seen in (5–168) to involve p, and we have found by means of (5–165) that p is a function of r. Clearly, if the solution we are generating is to make any sense, it must be that $\hat{\beta}$ is suitably small that the radially dependent term in (5–167) produces a negligible contribution to either $w(0)$ or $f(0)$. We can determine the necessary condition on $\hat{\beta}$ directly from the expression in (5–168) for $f(0)$, rewritten in the form,

$$f(0) \equiv \frac{p_R - p}{p_R - p_a} = 1 + \frac{1}{p_R - p_a} \frac{\hat{\beta}}{2}(1 - r^2). \tag{5–169}$$

It follows from (5–169) that the radial dependence of the blowing velocity may be neglected provided

$$\frac{\hat{\beta}}{2(p_R - p_a)} \ll 1. \tag{5–170}$$

Because

$$p_R - p_a = \frac{\varepsilon^2 d}{\mu V_c}\left(p'_R - p'_a\right) = \frac{\varepsilon^2 d}{k},$$

we can also express this condition in the form

$$\frac{\hat{\beta} k}{2 d \varepsilon^2} \ll 1. \tag{5–171}$$

In either form (5–170) or (5–171), we see that the uniform blowing analysis requires that p'_R be sufficiently large compared with p'_a [in the case of (5–171) this translates to the condition that d be large enough]. However, to be more specific on the requirements for p'_R, we need to evaluate the constant $\hat{\beta}$. Hence, for the moment, we simply assume that (5–170) or (5–171) is true and consider the necessary physical conditions after we have evaluated $\hat{\beta}$.

a. $\widetilde{Re} \ll 1$. In spite of the simplification inherent in the assumption that the blowing velocity is independent of radial position, the governing equation (5–166) is still nonlinear, and *analytic* solution requires some further approximation. We will see that asymptotic solutions are possible for the two limits in which $\widetilde{Re}$ is either very small or very large. We begin with the first of these cases, namely, $\widetilde{Re} \ll 1$. Of course, we have already used the classical lubrication analysis in section (3a) to obtain an "exact" solution for this case, without the added assumption of spatially uniform blowing beneath the disk, and it is of interest to demonstrate that the solution from this section reduces to that from the lubrication theory when the condition (5–171) is satisfied.

Thus we seek an asymptotic solution of Eq. (5–166), plus boundary conditions (5–168), in the form

$$f = f_0 + \widetilde{Re} f_1 + \cdots +. \tag{5–172a}$$

The pressure-gradient coefficient $\hat{\beta}$ that appears in (5–166) is also unknown, and we should expect a corresponding approximation for $\hat{\beta}$ in the form

$$\hat{\beta} = \hat{\beta}_0 + \widetilde{Re}\hat{\beta}_1 + \cdots +. \tag{5–172b}$$

D. The Air Hockey Table

Following our usual procedure, we substitute the expansions (5–172a) and (5–172b) into (5–166) plus the boundary conditions (5–168) and identify the governing equations and boundary conditions at each level of approximation by collecting terms of equal power in $\widetilde{Re}$.

The governing equation from (5–166) at $O(1)$ is just

$$\frac{d^3 f_0}{dz^3} = 2\hat{\beta}_0, \tag{5-173}$$

which is to be solved subject to the conditions

$$f_0(0) = 1, \quad f_0(1) = 0, \quad f_0'(0) = 0, \quad \text{and} \quad f_0'(1) = 0. \tag{5-174}$$

The first of the conditions (5–174) comes from (5–169), subject to the assumption that the condition (5–170) [or (5–171)] is satisfied.

Integrating (5–173), we obtain the general solution

$$f_0 = \frac{\hat{\beta}_0}{3} z^3 + c_2 z^2 + c_1 z + c_0. \tag{5-175}$$

The four conditions (5–174) are sufficient to determine not only the constants of integration $c_0 - c_2$, but also the first approximation β_0 to β. The result is

$$\boxed{f_0 = 6\left(\frac{z^3}{3} - \frac{z^2}{2}\right) + 1, \\ \beta_0 = 6.} \tag{5-176}$$

The problem at $O(\widetilde{Re})$ provides the first corrections to this result for small but nonzero inertia. At this level, the governing equation for f_1 from (5–166) is

$$\frac{d^3 f_1}{dz^3} - 2\hat{\beta}_1 = -\frac{1}{2}\left(\frac{df_0}{dz}\right)^2 + f_0 \frac{d^2 f_0}{dz^2}, \tag{5-177}$$

and in this case the corresponding boundary conditions are

$$f_1(0) = f_1(1) = f_1'(0) = f_1'(1) = 0. \tag{5-178}$$

Hence, substituting the result (5–176) for f_0 into the right-hand side of (5–177), we find

$$\frac{\partial^3 f_1}{\partial z^3} = 2\hat{\beta}_1 - 6 + 12z - 12z^3 + 6z^4.$$

After integrating and applying the boundary conditions (5–178), the solution at $O(\widetilde{Re})$ is thus

$$\boxed{f_1 = \frac{z^7}{35} - \frac{z^6}{10} + \frac{z^4}{2} - \frac{26}{35}z^3 + \frac{11}{35}z^2, \\ \beta_1 = \frac{27}{35}.} \tag{5-179}$$

Hence, substituting the solutions (5–176) and (5–179) into the original expansions (5–172a) and (5–172b), we see

$$f = 2z^3 - 3z^2 + 1 + \widetilde{Re}\left(\frac{z^7}{35} - \frac{z^6}{10} + \frac{z^4}{2} + \frac{26}{35}z^3 + \frac{11}{35}z^2\right) + O(\widetilde{Re}^2), \tag{5-180a}$$

The Thin-Gap Approximation – Lubrication Problems

with

$$\hat{\beta} = 6 + \frac{27}{35}\widetilde{Re} + O(\widetilde{Re}^2). \tag{5-180b}$$

We also note that the condition (5–171) takes the simple form

$$\frac{3k}{d\varepsilon^2} \ll 1. \tag{5-181}$$

This condition can also be expressed as a condition on the parameter α from the preceding section. Because $\alpha \equiv (12k/d^3)^{1/2}R$, we see that the condition (5–181) implies that the *lubrication* solution should reduce to the first term in this solution when

$$\boxed{\alpha^2 \ll 4.} \tag{5-182}$$

The most straightforward comparison is between the solution (5–167) for the pressure and the solution (5–151) from the lubrication theory with the latter evaluated for small α. Because

$$I_0(x) \cong 1 + \frac{1}{4}x^2 + O(x^4)$$

for $x \ll 1$, we can see that the lubrication solution reduces to the form

$$p_0 = p_a + (p_R - p_a)\frac{\alpha^2}{4}(1 - r^2) + O(\alpha^4)(p_R - p_a), \tag{5-183}$$

or, because $\alpha \equiv [12/(p_R - p_a)]^{1/2}$, this yields

$$p_0 = p_a + 3(1 - r^2) + O[\alpha^4(p_R - p_a)],$$

as expected. Of course, the small-Reynolds-number expansion in this section is easily extended to even higher orders in $\widetilde{Re}$, and the effects of weak inertia evaluated for small α, whereas this is not possible by means of the lubrication theory because it neglects the inertial terms altogether.

b. $\widetilde{Re} \gg 1$. We now turn to the other limit, $\widetilde{Re} \gg 1$, which cannot be addressed by means of standard lubrication theory, even though the thin-film approximation $\varepsilon \ll 1$ is still preserved.

The governing equation is again (5–166), but it is now convenient to write it in the form

$$-\frac{1}{2}\left(\frac{df}{dz}\right)^2 + f\frac{d^2 f}{dz^2} = -\frac{2\hat{\beta}}{\widetilde{Re}} + \frac{1}{\widetilde{Re}}\frac{d^3 f}{dz^3}. \tag{5-184}$$

We have seen, in previous examples, that we can obtain the governing equations for the leading-order approximation of an asymptotic solution by taking the limit of the full governing equation as the small parameter, $\widetilde{Re}^{-1}$, goes to zero. In the present case, this procedure would seem to suggest that the two terms on the right-hand side of (5–184) can be neglected relative to the terms on the left-hand side.

However, the first of the terms on the right-hand side is the pressure gradient that is *responsible* for the radial flow. Clearly, this term *cannot* be smaller than terms on the left-hand side of (5–184), even for $\widetilde{Re} \gg 1$. The reason that it initially appears to be negligible is because we have scaled pressure with $p_c = [(\mu V_c/d)(1/\varepsilon^2)]$, which is appropriate for a flow that is dominated by viscous effects, but is *not* appropriate for $\widetilde{Re} \gg 1$. If we assume that we must retain the radial pressure gradient even for $\widetilde{Re} \gg 1$, this suggests

D. The Air Hockey Table

that $\hat{\beta}/\widetilde{Re}$ must be $O(1)$ to match the magnitude of terms on the left-hand side of (5–184). It is therefore convenient to introduce

$$\beta = \frac{\hat{\beta}}{\widetilde{Re}}, \qquad (5\text{–}185)$$

where $\beta = O(1)$, i.e., it is independent of $\widetilde{Re}$. Referring back to (5–165), we see that the rescaling (5–185) is equivalent to changing the characteristic pressure from $p_c = [(\mu V_c/d)(1/\varepsilon^2)]$ to $p_c^* \equiv p_c \widetilde{Re} = (\rho V_c^2/\varepsilon^2)$. We shall later see that the combination ρu_c^2 is always an appropriate choice as characteristic pressure for high-Reynolds-number flows. The extra ε^{-2} factor appears here because of the thin-film approximation.

With this change in the nondimensionalization of pressure, the governing equation, (5–184), now becomes

$$-\frac{1}{2}\left(\frac{df}{dz}\right)^2 + f\frac{d^2 f}{dz^2} + 2\beta = \frac{1}{\widetilde{Re}}\frac{d^3 f}{dz^3}. \qquad (5\text{–}186)$$

In this case, as $\widetilde{Re} \to \infty$, the viscous term on the right-hand side becomes asymptotically negligible for any point in the flow domain for which the scaling inherent in (5–186) is correct. Indeed, if we introduce asymptotic expansions for f and β of the form

$$\begin{aligned} f &= f_0 + \delta_1(Re^{-1})f_1 + \cdots +, \\ \beta &= \beta_0 + \delta_1(Re^{-1})\beta_1 + \cdots +, \end{aligned} \qquad (5\text{–}187)$$

we see that the governing equation for the leading-order terms is

$$-\frac{1}{2}\left(\frac{df_0}{dz}\right)^2 + f_0\frac{d^2 f_0}{dz^2} = -2\beta_0. \qquad (5\text{–}188)$$

Comparing (5–186) and (5–188), we see that the governing equation is reduced, at the leading order of approximation, from third order in z to second order. This means that $f(z)$ cannot satisfy all four of the boundary conditions (5–168) [and (5–169)], even taking account of the fact that β is still an unknown that can be chosen to satisfy one of these conditions. This implies, as we have seen before, that the asymptotic solution will be singular in this case, with Eq. (5–186) and asymptotic expansions in the forms (5–187) describing the dynamics in a core region between the top of the table and the bottom of the disk, but failing near at least one of these two surfaces because the scaling (nondimensionalization) breaks down at that point. In this latter region, we will find that there must be a boundary layer in which the viscous term remains important even as $\widetilde{Re} \to \infty$. Clearly, this implies that there must be some change in the appropriate characteristic scales in this region. We shall return to this point shortly.

The four boundary conditions for $f_0(z)$ are, again,

$$f_0(1) = f_0'(1) = f_0'(0) = 0; \; f_0(0) = 1, \qquad (5\text{–}189)$$

and the governing equation for f_0 is second order, but with one additional parameter β. This suggests that we should be able to satisfy three of the four boundary conditions with the solution for f_0 and thus that there will be only a single boundary layer at either the bottom of the disk or at the top of the table, i.e., near either $z = 1$ or $z = 0$. Now, the last of the conditions (5–189) represents the blowing flow through the lower wall and must clearly be satisfied whether viscous effects are included or not. Likewise, the first condition $f_0(1) = 0$ represents the statement that there is no flow through the bottom of the disk and must also be satisfied, independent of whether viscous effects are included. The other two conditions on f' represent the no-slip condition at the tabletop and on the bottom of the disk. One may

The Thin-Gap Approximation – Lubrication Problems

anticipate that the solution for f_0, with all viscous terms neglected, may fail to satisfy one or both of the conditions $f_0'(0) = f_0'(1) = 0$. At the tabletop, the fluid comes through the porous surface with *zero* radial momentum, and thus the condition of zero radial velocity is satisfied automatically at this surface, i.e., $f_0'(0) = 0$. Thus it would appear that it is the no-slip condition at the bottom of the disk that will not be satisfied by $f_0(z)$ and that the governing equations in the form (5–186)–(5–188) must fail there.

But we are getting ahead of ourselves. The first step must be to solve the problem posed by (5–186)–(5–189), beginning with the governing equation (5–188) for $f_0(z)$. Although this equation is nonlinear it turns out to have a simple analytic solution in the form of a second-order polynomial in z. Indeed, if we assume that

$$f_0 = a_1 + a_2 z + a_3 z^2$$

and substitute into Eq. (5–188), we see that we have a solution provided that

$$-\frac{1}{2}a_2^2 + 2a_1 a_3 = -2\beta_0.$$

Hence

$$f_0 = -\frac{\beta_0}{a_3} + \frac{a_2^2}{4a_3} + a_2 z + a_3 z^2 \qquad (5\text{--}190)$$

is a general solution, and we can choose the coefficients a_2 and a_3, plus the constant β_0, to satisfy the boundary conditions previously indicated, namely $f_0(0) = 1$, $f_0(1) = f_0'(0) = 0$. The result is

$$\boxed{f_0(z) = 1 - z^2,} \qquad (5\text{--}191)$$

with

$$\boxed{\beta_0 = 1.} \qquad (5\text{--}192)$$

Of course, this is not a uniformly valid first approximation of the solution of the original problem everywhere in the domain $0 \leq z \leq 1$ as it fails to satisfy the no-slip boundary condition at $z = 1$. In particular,

$$\left.\frac{\partial f_0}{\partial z}\right|_{z=1} = -2, \qquad (5\text{--}193)$$

and the asymptotic limit, $\widetilde{Re} \gg 1$, of the solution to (5–184) is *singular*.

In the limit $\widetilde{Re} \gg 1$ there will, in fact, be a boundary layer at the bottom of the disk, as we have already suggested; namely, a region in which the velocity varies over a much shorter length scale in z, such that the viscous term in (5–184) is retained even in the limit $\widetilde{Re} \to \infty$. In other words, the approximation (5–188) of the original problem (5–184) breaks down near $z = 1$ because the length scale, $l_c^z = d$, is not the appropriate choice in that local region of the flow domain. To determine what the appropriate length scale should be, we use the concept of rescaling (as introduced earlier in Chap. 4). For convenience, we first transform from z to $z^* \equiv 1 - z$ so that the boundary of interest is at $z^* = 0$. This is not necessary, but is convenient in the sense that boundary-layer analyses almost always identify the bounding surface as being located where the appropriate coordinate is equal to zero. Thus Eq. (5–184) appears in the slightly modified form

$$-\frac{1}{2}\left(\frac{df}{dz^*}\right)^2 + f\frac{d^2 f}{dz^{*2}} = -2\beta - \frac{1}{\widetilde{Re}}\frac{d^3 f}{dz^{*3}}. \qquad (5\text{--}194)$$

D. The Air Hockey Table

We now seek a new nondimensionalization of the dimensional variable $(z^*)'$ such that the viscous term remains in the problem as $\widetilde{Re} \to \infty$. To do this we introduce a rescaling of z^* by means of the dimensionless parameter $\widetilde{Re}$, i.e.,

$$Z = \widetilde{Re}^m z^*. \tag{5-195}$$

We note that this corresponds to a nondimensional form of $(z^*)'$, with a modified characteristic length scale, i.e.,

$$Z = \frac{z^{*\prime}}{\ell_c^*} \tag{5-196}$$

where we can see from (5–195)

$$\frac{(z^*)'}{\ell_c^*} = \left(\frac{\rho V_c d}{\mu}\right)^m \frac{(z^*)'}{d},$$

and thus

$$\boxed{\ell_c^* = d(\widetilde{Re})^{-m}.} \tag{5-197}$$

Assuming m to be positive, we can see that the characteristic length scale in this boundary-layer region is *much* shorter than d.

To determine m, we use the fact that the boundary-layer scaling must be consistent with retention of the viscous term in the governing equation for f as $\widetilde{Re} \to \infty$. In the present problem, the introduction of a rescaled $(z^*)'$ in (5–196) requires also that the function f be rescaled as well. This is because the radial velocity is related to f in the form

$$u \equiv \frac{r}{2}\frac{df}{dz} \equiv -\frac{r}{2}\frac{df}{dz^*}. \tag{5-197}$$

Because the slip velocity from the core region can be seen from (5–193) to have the value

$$u\,|_{core} = -r \tag{5-198}$$

as we approach the disk, and because we expect the radial velocity to start from this value at the outer edge of the boundary layer and then drop to zero at the actual surface of the disk, it is clear that the radial velocity must be $O(1)$ – i.e., independent of $\widetilde{Re}$, – in the boundary layer, just as it is in the core region.

Going back to the definition (5–197) of u in terms of f, we see that if the rescaling (5–196) is applied to z^*, a corresponding rescaling,

$$F = \widetilde{Re}^m f, \tag{5-199}$$

must be applied to the function f in order that $u = O(1)$ in the boundary layer. Now introducing (5–196) and (5–199) into Eq. (5–194), we obtain

$$-\frac{1}{2}\left(\frac{dF}{dZ}\right)^2 + F\frac{d^2F}{dZ^2} = -2\beta - \widetilde{Re}^{-1+2m}\frac{d^3F}{dZ^3}. \tag{5-200}$$

Once we specify m, this is the nondimensionalized form of the governing equation for f, but now scaled with the length scale ℓ_c^* that is appropriate to the boundary-layer region. The

The Thin-Gap Approximation – Lubrication Problems

choice of m must be made according to the requirement that the viscous term has the same magnitude as the other terms in (5–200) in the limit $\widetilde{Re} \to \infty$. This clearly implies that

$$m = 1/2, \qquad (5\text{–}201)$$

$$\ell_c^* = d\,\widetilde{Re}^{-1/2}. \qquad (5\text{–}202)$$

Thus we see that the function F (and thus the velocity u) varies over a much shorter length scale in this boundary-layer region near the bottom surface of the disk than it does elsewhere in the domain.

In fact, this shorter length scale is imposed on the problem by the physical necessity of retaining viscous terms in order to satisfy the no-slip condition at $z^* = 0$. To say the same thing in another way, an internal length scale is generated by the physics that has just the correct value to maintain viscous effects where they are needed in the domain in order to satisfy all of the physical conditions of the original problem.

With $m = (1/2)$, the governing equation in the boundary-layer region becomes

$$-\frac{1}{2}\left(\frac{dF}{dZ}\right)^2 + F\frac{d^2F}{dZ^2} = -2\beta - \frac{d^3F}{dZ^3}. \qquad (5\text{–}203)$$

Because the radial pressure gradient in the *core* at leading order corresponds to $\beta = 1$ we assume that the same is true in the boundary layer (essentially by matching). The boundary conditions for the function $F(Z)$ are

$$F(0) = F'(0) = 0. \qquad (5\text{–}204)$$

Because β is specified, $F(Z)$ can satisfy only one additional condition. This comes from matching the radial velocity from the boundary-layer solution for large Z with the slip velocity (5–198) from the core solution for $z \to 1$, i.e.,

$$\lim_{Z \gg 1} \frac{dF}{dZ} \Leftrightarrow \lim_{z^* \ll 1} \frac{df}{dz^*} = 2 \text{ in the limit as } \widetilde{Re} \to \infty. \qquad (5\text{–}205)$$

Formally, we then seek an asymptotic solution in the boundary-layer region of the form

$$F = F_0(Z) + \Delta_1\!\left(\widetilde{Re}^{-1}\right)F_1(Z) + \cdots + \qquad (5\text{–}206)$$

for large $\widetilde{Re}$. Here Δ_1 is just the gauge function corresponding to F_1. We have assumed that the gauge function $\Delta_0 = 1$ in the first term of (5–206) as this is required for satisfying the matching condition (5–205). The governing equation and boundary/matching conditions for the leading-order function F_0 are thus (5–203)–(5–205). Because of the nonlinearity of (5–203) no analytic solution has been found, and it is necessary to resort to numerical methods to solve this boundary-layer problem. This was done in the research paper of Hinch and Lemaitre.[7] Two results from their numerical solution are

$$F_0''(0) = 2.624, \qquad (5\text{–}207\text{a})$$

and

$$F \to 2Z - 1.138 \quad \text{as } Z \to \infty. \qquad (5\text{–}207\text{b})$$

D. The Air Hockey Table

The physical interpretation of the last result is that the *normal* velocity is reduced as we approach the bottom of the disk relative to the leading-order approximation of its value from the core solution as $z \to 1$, which completely neglects viscous effects (at the first order of approximation).

One point that may seem strange in the solution procedure just outlined is that we simply assumed that the core solution satisfied the condition of zero normal velocity at the surface of the disk, i.e.,

$$f(z) = 0 \quad \text{at } z = 1 \quad [\text{or } f(z^*) = 0 \quad \text{at } z^* = 0], \tag{5-208}$$

whereas we have just seen that the core problem breaks down as we approach the surface of the disk and is replaced with the boundary-layer approximation. Hence, instead of applying the boundary condition (5–208) at $z = 1$, we should have required that the normal velocity from the core solution as $z \to 1$ match that from the boundary layer for large Z. Strictly speaking, the condition of zero normal velocity at the surface of the disk applies only to the boundary-layer solutions, i.e., $F(0) = 0$. Let us examine the matching condition at the outer edge of the boundary layer which takes the form

$$\lim_{Z \gg 1} \left[\frac{F(Z)}{\widetilde{Re}^{1/2}} \right] \Leftrightarrow \lim_{z^* \ll 1} f(z^*) \quad \text{in the limit as } \widetilde{Re} \to \infty. \tag{5-209}$$

In matching the functions F and f we must remember the $\widetilde{Re}^{1/2}$ scaling from (5–199) in passing from one to the other, as shown. But then we see that, provided $F(Z)$ is well-behaved for $Z \gg 1$, this condition actually implies that $f(z^*) = 0$ as $z^* \to 0$, to *first approximation* in $\widetilde{Re}^{-1}$, as we already assumed in solving the core problem.

By substituting the results obtained for the core and boundary-layer solutions, we can use (5–209) to go one step further with the analysis. In particular, we can write (5–209) in the form

$$\lim_{Z \gg 1} (2Z - 1.138) + \Delta_1(\widetilde{Re}^{-1}) F_1(Z) + \cdots + \\ \Leftrightarrow \lim_{z^* \ll 1} \widetilde{Re}^{1/2} [f_0(z^*) + \delta_1(\widetilde{Re}^{-1}) f_1(z^*)]. \tag{5-210}$$

Now

$$f_0(z) = 1 - z^2$$

according to (5–191), and thus, expressed in terms of z^*,

$$f_0(z^*) = 2z^* - z^{*2}.$$

Substituting into the matching condition (5–210), we obtain

$$\lim_{Z \gg 1} (2Z - 1.138) + \Delta_1(\widetilde{Re}^{-1}) F_1(Z_1) + \cdots + \\ \Leftrightarrow \lim_{z^* \ll 1} \widetilde{Re}^{1/2}(2z^* - z^{*2}) + \widetilde{Re}^{1/2} \delta_1(\widetilde{Re}^{-1}) f_1 + \cdots + \text{ in the asymptotic limit } \widetilde{Re}^{-1} \to 0. \tag{5-211}$$

In considering the matching condition (5–211) it is necessary to either express the core solution in terms of the boundary-layer variables or else express the boundary-layer solution in terms of the core variables. Because both approximations to the solution are valid in the region of overlap where the matching condition applies, either choice is acceptable. However, for present purposes, it is more convenient to introduce the boundary-layer variables into the core solution.

The Thin-Gap Approximation – Lubrication Problems

Hence, because $Re^{1/2}z^* = Z$, we can rewrite the matching condition in the form

$$\lim_{Z \gg 1}(2Z - 1.138) + \Delta_1(\widetilde{Re}^{-1})F_1(Z) + \cdots + \\ \Leftrightarrow 2Z + \widetilde{Re}^{-1/2}Z^2 + Re^{1/2}\delta_1(\widetilde{Re}^{-1})f_1(z^*) + \cdots + \text{as } \widetilde{Re}^{-1} \to 0. \quad (5\text{-}212)$$

We see that the match of the leading-order approximations for the core and boundary-layer solutions is not perfect. First, the term $\widetilde{Re}^{-1/2}Z^2$ does not match with any of the leading-order terms on the left-hand side. However, we cannot expect a match at this level of approximation. This term will not be matched until we consider a term in the boundary-layer region with gauge function equal to $\widetilde{Re}^{-1/2}$, and, in any case it can be made as small as we like relative to the $2Z$ terms by letting $\widetilde{Re}^{-1/2} \to 0$.

The other term that is calculated explicitly but does not match is the constant 1.138 from the leading-order boundary-layer approximation to the normal velocity function. Matching requires a term of $O(1)$, independent of $\widetilde{Re}$, on the right-hand side of (5-212). This can happen only if there is a gauge function $\delta_m(\widetilde{Re}^{-1})$ in the core solution that is equal to $\widetilde{Re}^{-1/2}$. Although we may be tempted to say immediately that this must be the next term in the expansion (5-187), so that

$$\delta_1(\widetilde{Re}^{-1}) \equiv \widetilde{Re}^{-1/2}, \quad (5\text{-}213)$$

we need to be careful. There are two reasons for the higher-order terms in (5-187). One is that such terms may be required by the matching condition, as in the case discussed here. The second is that we need eventually to include terms from the original equation, (5-186), that were neglected at the first approximation. In this case, the missing term can be seen from (5-186) to be $O(\widetilde{Re}^{-1})$. It will thus come into the expansion (5-187) only after the correction required for improving the matching. Thus, as suggested in Eq. (5-213), the matching condition (5-212) requires that the next term in the asymptotic expansion of the core solution be $O(\widetilde{Re}^{-1/2})$ so that

$$\boxed{f = f_0(z) + \widetilde{Re}^{-1/2}f_1(z) + O\left(\delta_2(\widetilde{Re}^{-1})\right).} \quad (5\text{-}214)$$

However, given this information on the gauge function δ_1, we can now write the governing equation for f_1 (and β_1) by substituting (5-214) into (5-186). The result for the $O(\widetilde{Re}^{-1/2})$ problem is

$$-\frac{df_1}{dz}\frac{df_0}{dz} + f_0\frac{d^2f_1}{dz^2} + f_1\frac{d^2f_0}{dz^2} = -2\beta_1. \quad (5\text{-}215)$$

On substituting the known solution for f_0, we find that this becomes

$$\boxed{(1-z^2)\frac{d^2f_1}{dz^2} + 2z\frac{df_1}{dz} - 2f_1 = -2\beta_1.} \quad (5\text{-}216)$$

Referring back to the boundary conditions on f from Eqs. (5-168) and (5-169), we find that f_1 must satisfy the conditions

$$\boxed{f_1(0) = f_1'(0) = 0.} \quad (5\text{-}217)$$

D. The Air Hockey Table

In addition, the matching condition (5–212) requires that

$$\boxed{f_1(1) = -1.138.} \tag{5–218}$$

This completes specification of the problem for f_1.

Before the solution for f_1 is obtained, however, a couple of comments may be worth injecting. The first is that even this second-order approximation for $f(z)$ in the core region contains no direct contribution from the viscous term in (5–186). Yet, this "correction" in (5–214) is proportional to $\widetilde{Re}^{-1/2}$. This may seem strange because the presence of $\widetilde{Re}$ in the solution implies that viscous effects are playing a role. However, the point is that viscosity *directly* enters the *first* approximation in the boundary layer and slows the radial flow relative to what it is predicted by the leading-order core solution in the absence of viscosity. The reduction in radial flux means that less fluid, by $O(\widetilde{Re}^{-1/2})$, is required to be pumped up to the lower surface of the disk than is predicted by the leading approximation to the core solution. It is this fact that is reflected in the mismatch (-1.138) in f between the leading-order approximations in the core and boundary-layer regions. To account for the reduction in the vertical velocity component, *caused by viscous effects in the boundary layer*, we are required to add a correction of comparable magnitude, $O(\widetilde{Re}^{-1/2})$, to the core solution even though the first direct viscous term in the core equation (5–186) appears only at $O(\widetilde{Re}^{-1})$.

In spite of the nonlinearity of the problem for f_1, it turns out that we can obtain an analytic solution of (5–216)–(5–218) for f_1 and β_1. This is because Eq. (5–216) is again satisfied by a second-order polynomial of the form

$$f_1 = b_0 + b_1 z + b_2 z^2, \tag{5–219}$$

provided that only

$$b_0 = \beta_1 + b_2.$$

Then, applying the boundary conditions (5–217) and the matching condition (5–218), we find

$$\boxed{\begin{aligned} f_1(z) &= -1.138 z^2, \\ \beta_1 &= 1.138. \end{aligned}} \tag{5–220}$$

Hence, in the core region,

$$\begin{aligned} \beta &= 1 + 1.138 \widetilde{Re}^{-1/2} + O\left(\delta_2(\widetilde{Re}^{-1})\right) \\ f &= 1 - z^2 + \widetilde{Re}^{-1/2}(-1.138 z^2) + O\left(\delta_2(\widetilde{Re}^{-1})\right). \end{aligned} \tag{5–221}$$

Finally, if we wished to continue to obtain higher-order approximations, the next step would be to seek the second term in the boundary-layer expansion, (5–206). We have now added a second approximation to the core solution. As in the case of the leading-order term $f_0(z)$, we obtained the solution (5–220) by requiring that the normal velocity components (i.e., f) match as we approach the base of the disk, but with no consideration of any boundary condition on the radial velocity component in the core region. Hence, by adding f_1, we will have introduced some additional "slip" that must be matched by a new, second term in the

The Thin-Gap Approximation – Lubrication Problems

boundary-layer expansion (5–206). If we return to the matching condition (5–205) for the radial velocity components, we now have

$$\lim_{Z \gg 1}\left[2+\Delta_1(\widetilde{Re}^{-1})\frac{dF_1}{dZ}\right] \Leftrightarrow \lim_{z^*\ll 1}\left[\frac{df_0}{dz^*}+\widetilde{Re}^{-1/2}\frac{df_1}{dz^*}+O\left(\delta_2(\widetilde{Re}^{-1})\right)\right]$$
$$= 2-2z^*+\widetilde{Re}^{-1/2}[1.138(2-2z^*)]+O(\delta_2) \quad (5\text{–}222)$$
in the limit as $\widetilde{Re}^{-1} \to 0$.

To compare the two expansions for purposes of matching, it is again convenient to express the core solution in terms of boundary-layer variables. Hence we express the condition (5–222) in the form

$$\lim_{Z \gg 1}\left[2+\Delta_1(\widetilde{Re}^{-1})\frac{dF_1}{dZ}\right] \Leftrightarrow 2-2Z\,\widetilde{Re}^{-1/2}+2(1.138)\,\widetilde{Re}^{-1/2}$$
$$-2(1.138)\,Z\widetilde{Re}^{-1}+O\left(\delta_2(\widetilde{Re}^{-1})\right) \text{ as } \widetilde{Re}^{-1}\to 0. \quad (5\text{–}223)$$

We see that the added term in the core region, while improving the match between the z velocity components, has now introduced some additional "slip." Clearly, from (5–223), the largest mismatch between the radial velocity components is $O(\widetilde{Re}^{-1/2})$, and it follows that

$$\Delta_1(\widetilde{Re}^{-1}) = \widetilde{Re}^{-1/2},$$

and

$$\lim_{z \gg 1}\frac{dF_1}{dZ} \sim 2(1.138-Z) \quad (5\text{–}224)$$

in order to satisfy the matching condition (5–223) to the next level of approximation. With $\Delta_1 = \widetilde{Re}^{-1/2}$, we can substitute the boundary-layer expansion; (5–206), into the governing equation, (5–203), for the boundary-layer region and obtain a governing equation for F_1:

$$-\frac{dF_0}{dZ}\frac{dF_1}{dZ}+F_0\frac{d^2F_1}{dZ^2}+F_1\frac{d^2F_0}{dZ^2}=-2(1.138)-\frac{d^3F_1}{dZ^3}. \quad (5\text{–}225)$$

Here, we have again assumed that the radial pressure gradient exactly matches that in the core at the same level of approximation. At the surface of the disk, $F=(dF/dZ)=0$, and thus too

$$F_1 = \frac{\partial F_1}{\partial Z} = 0 \quad \text{at } Z=0. \quad (5\text{–}226)$$

The condition (5–224) provides a third boundary condition through matching with the core solution for the radial velocity. Although (5–224)–(5–226) is a well-posed problem, we shall not solve it here. The solution for F_0 was numerical, and thus the present problem must also be solved numerically. The most efficient approach is to solve the problems for F_0 and F_1 together as this avoids storing the solutions for F_0 and/or interpolating to evaluate the coefficients in (5–225).

The main point here is that the solution procedure for this particular problem of a singular (or matched) asymptotic expansion follows a very generic routine. Given that there are two sub-domains in the solution domain, which overlap so that matching is possible (the sub-domains here are the core and the boundary-layer regions), the solution of a singular perturbation problem usually proceeds sequentially back and forth as we add higher order

D. The Air Hockey Table

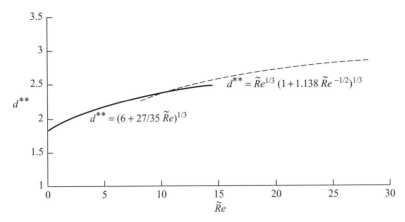

Figure 5–13. Height of disk above the table top for large and small Reynolds number. This plot shows $d^{**} \equiv d/(kR^2 p_R^*/4)^{1/3}$ versus $\tilde{Re}$.

terms. Here, we start with a first approximation in the core, which generates a first approximation in the boundary-layer by matching of the radial velocities, which in turn generates a second approximation in the core due to the mismatch of z-velocity components, and the latter in turn generates a second approximation in the boundary-layer and so on.

c. Lift on the disk. Now that we have generated asymptotic solutions for $\tilde{Re} \ll 1$ and $\tilde{Re} \gg 1$ using the uniform blowing approximation, it is interesting to see what they predict about the height of the disk above the porous tabletop. To do this, we again use the force balance on the disk in the form (5–139):

$$Mg = \frac{2\pi \mu V_c R^4}{d^3} \int_0^1 (p - p_a) r \, dr. \tag{5-227}$$

Now, according to (5–167),

$$p - p_a = \frac{\hat{\beta}}{2}(1 - r^2),$$

and we have shown in the present section that

$$\tilde{Re} \ll 1, \quad \hat{\beta} = 6 + \frac{27}{35}\tilde{Re} + \cdots +,$$
$$\tilde{Re} \gg 1, \quad \hat{\beta} = \tilde{Re}\beta = \tilde{Re}\left[1 + \tilde{Re}^{-1/2}(1.138) + \cdots +\right].$$

Now, if we recall [Eq. (5–124)],

$$p_R' - p_a' = \frac{Mg}{\pi R^2} p_R^*,$$

then we see from Eq. (5–127) that

$$V_c = \frac{k}{\mu} \frac{Mg}{\pi R^2} p_R^*. \tag{5-228}$$

It follows from the force balance (5–227) that

$$\boxed{d^3 = \frac{kR^2 \hat{\beta}}{4} p_R^*.} \tag{5-229}$$

Hence, for $\widetilde{Re} \ll 1$

$$d^3 = \frac{kR^2}{4} p_R^* \left(6 + \frac{27}{35}\widetilde{Re} + \cdots \right). \quad (5\text{–}230)$$

On the other hand, for $\widetilde{Re} \gg 1$,

$$d^3 = \frac{kR^2}{4} p_R^* \widetilde{Re} \left(1 + 1.138\widetilde{Re}^{-1/2} + \cdots \right). \quad (5\text{–}231)$$

Finally, we briefly return to the condition (5–171) for validity of the uniform blowing assumption. If we substitute $\hat{\beta} = \beta(\widetilde{Re})$, this condition becomes

$$\frac{\rho \beta k R^2}{2\mu} \frac{V_c}{d^2} \ll 1.$$

With $V_c = \frac{k}{\mu}(\frac{Mg}{\pi R^2}) p_R^*$ and the expression (5–231) for d^2, this becomes

$$p_R^* \gg 2 \quad \text{for} \quad \widetilde{Re} \gg 1. \quad (5\text{–}232\text{a})$$

Likewise, with $\hat{\beta} = 6$ and d^3 given by (5–226), the condition (5–171) again becomes

$$p_R^* \gg 2 \quad \text{for} \quad \widetilde{Re} \ll 1. \quad (5\text{–}232\text{b})$$

Hence, we expect that the uniform blowing approximation should be valid, provided $p'_R \gg p'_a$.

NOTES

1. A number of textbooks present a general description of lubrication theory, including the Reynolds equation and other aspects. The interested reader may be curious to examine the original paper of Reynolds: O. Reynolds, On the theory of lubrication and its application to Mr. Beauchamp Tower's experiments including an experimental determination of the viscosity of olive oil, *Philos. Trans. R. Soc. London, Ser. A* **177**, 157 (1886). Among the textbooks are Andras Z. Szeri and A. Z. Szeri, *Fluid Film Lubrication: Theory and Design* (Cambridge University Press, Cambridge, 1998); A. Cameron, *Basic Lubrication Theory* (Wiley, New York, 1977).
2. A. Oron, S. H. Davis, and S. G. Bankoff, Long-scale evolution of thin liquid films, *Rev. Mod. Phy.* **69**, 931–80 (1997).
3. The following papers on flow of viscoelastic fluids in the eccentric cylinder geometry also contain examples of calculations for Newtonian fluids: A. N. Beris, R. C. Armstrong, and R. A. Brown, Finite element calculation of viscoelastic flow in a journal bearing, I. Small eccentricities, *J. Non-Newtonian Fluid Mech.* **16**, 141–72 (1984); A. N. Beris, R. C. Armstrong, and R. A. Brown, Finite element calculation of viscoelastic flow in a journal bearing, I. Moderate eccentricity, *J. Non-Newtonian Fluid Mech.* **19**, 323–47 (1986); A. N. Beris, R. C. Armstrong, and R. A. Brown, Spectral/finite element calculations of the flow of a Maxwell fluid between eccentric rotating cylinders, *J. Non-Newtonian Fluid Mech.* **22**, 129–67 (1987).
4. The relevance of the rotating eccentric cylinder geometry to the journal-bearing lubrication system is discussed in many technical papers. As a starting point, the interested reader may wish to refer to A. F. Booker, Design of dynamically loaded journal bearings, in *Fundamentals of the Design of Fluid Film Bearings* (American Society Mechanical Engineers, New York, 1979); K. Walters, Polymers as additives in lubricating oils, *Colloques Internationaux de CNRS*, No. 233, *Polymeres et Lubrification* (Paris), 27–36 (1975); P. K. Goenka, Dynamically loaded journal bearings: Finite element method analysis, *J. Tribol.* **106**, 429–39 (1984); A. S. Lodge, A new method of measuring

multi-grade oil shear elasticity and viscosity at high shear rates, SAE Tech. Paper 872043 (Society of Automotive Engineers, Warrendale, PA, 1987).
5. H. A. Stone, Partial differential equations in thin film flows in fluid dynamics: Spreading droplets and rivulets, in *Nonlinear PDEs in Condensed Matter and Reactive Flows*, edited by H. Berestycki and Y. Pomeau, (springer-verlag, New York, 2000).
6. Many papers have been written that discuss the lubrication analysis of the interaction of a body and a plane wall or two bodies when the separation gap is vanishingly small compared with the length scale of the body. The problem of a sphere near a plane wall in the lubrication limit was considered in the following papers: R. G. Cox and H. Brenner, The slow motion of a sphere through a viscous fluid towards a plane surface, Part II. Small gap widths, including inertial effects, *Chem. Eng. Sci.* **22**, 1753–77(1967); A. J. Goldman, R. G. Cox, and H. Brenner, Slow viscous motion of a sphere parallel to a plane wall, I. Motion through a quiescent fluid, *Chem Eng. Sci.* **22**, 637–51(1967); M. E. O'Neill and K. Stewartson, On the slow motion of a sphere parallel to a nearby wall, *J. Fluid Mech.* **27**, 705–24 (1967). The problem of a parallel cylinder and a plane wall was considered by D. J. Jeffrey and Y. Onishi, The slow motion of a cylinder next to a plane wall, *Q. J. Mech. Appl. Math.* **34**, 129–37 (1981). The motion of two drops was considered by R. H. Davis, J. A. Schonberg, and J. M. Rallison, The lubrication force between two viscous drops, *Phy. Fluids* **A1**, 77–81 (1989). A summary of much of this work may also be found in S. Kim and S. Karrila, *Microhydrodynamics: Principles and Selected Applications* (Butterworth-Heinemann, Boston, 1991).
7. E. J. Hinch and J. Lemaitre, The effect of viscosity on the height of disks floating above an air table, *J. Fluid Mech.* **273**, 313–22 (1994).

PROBLEMS

Problem 5–1. Squeeze Film. Two parallel plane, circular disks of radius R lie one above the other a distance H apart. The space between them is filled with an incompressible Newtonian fluid. One disk approaches the other at constant velocity V, displacing the fluid. The pressure at the edge of the upper disk is atmospheric.

(a) Derive a dimensionless version of the preceding problem from the full Navier–Stokes equations and boundary conditions. What is the Reynolds number for this problem?

(b) Under what conditions can we apply the lubrication approximation to this problem? What is the appropriate choice for the characteristic pressure in the lubrication approximation?

(c) Find the *dynamic* pressure distribution and show that the hydrodynamic force F, resisting the motion in the lubrication limit, is given by

$$F = -\frac{3\pi}{2} \frac{\mu R^4}{H^3} \frac{dH}{dt}.$$

(d) Show that a constant force $-F$ applied to the upper disk will pull it well away from the bottom disk in a time $(3/4)\pi\mu R^4/(H_0^2 F)$, where H_0 is an initial spacing. Note that this time is large when H_0 is small. Give an example for the application of this phenomenon.

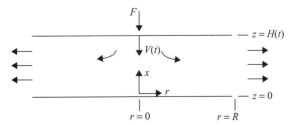

The Thin-Gap Approximation – Lubrication Problems

Problem 5–2. Small-Angle Hinge

(a) Two similar, rigid, plane rectangular plates are joined along an edge of length b by a smooth, oiltight hinge and make a small angle $\alpha(t)$ with one another. This space between

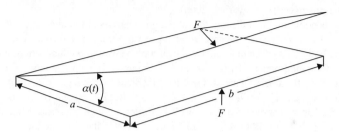

the plates is filled with oil of viscosity μ, which is squeezed out as the plates are pushed together by a force F, perpendicular to each plate, applied at the outer (free) edge. The outer edge is a distance $a \ll b$ from the hinge. Use the fact $\alpha \ll 1$ to obtain

$$F = -\frac{3\mu ab}{2\alpha(t)^3}\frac{d\alpha}{dt},$$

an approximate solution, by means of a lubrication type analysis.

(b) Here we consider a very similar problem, but with a slightly different objective. A plate has a force F applied at the outside edge, pushing it toward the plane. The pressure distribution resulting from the squeeze flow makes the plate want to separate from the plane at the hinge. Using lubrication theory, show that the force F_h exerted by the hinge on the plate for small separation angles is exactly $3F$, e.g., three times the closing force – which is why hinges tend to pop open!

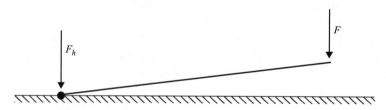

Problem 5–3. Freely Settling Hinge. A plate of density ρm, thickness d, and width W (you may consider the third dimension to be infinite – it is a 2D problem – and $d/W \ll 1$) is hinged at one edge in contact with the bottom of a tank filled with a viscous liquid. The angle separating the plate and the plane is given by α. If we release the plate it will settle toward the plane (e.g., α will decrease in time). In the limit $\alpha \ll 1$, develop an equation governing α as a function of time.

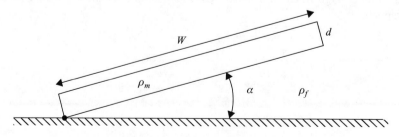

Hint: The net torque on the plate will be zero – the torque that is due to lubrication forces will balance the torque that is due to gravity.

Problems

Problem 5–4. Pressure-Driven Flow Through a Thin Gap. Oil is forced by a constant, small pressure difference Δp through the narrow gap between two cylinders of radius a, with axes parallel and separated by a distance $2a + b$.

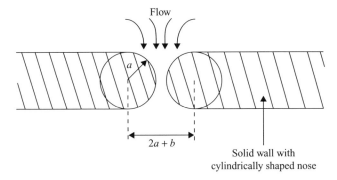

Show that if (b/a) and $(\rho \Delta p / \mu^2)(b^3/a)$ are sufficiently small, then the volume flux per unit width is approximately

$$\frac{2}{9\pi} \frac{\Delta p}{\mu} \left(\frac{b^5}{a} \right)^{1/2}.$$

Problem 5–5. Rotating Cylinders

(a) Two equal cylinders, of radius a, with parallel axes are separated by a distance $2a + b$. The cylinders are rotated with angular velocity Ω and $-\Omega$, respectively. What is the volume flux of fluid/unit length pulled through the gap, what torque is required for maintaining the angular velocity, and how much force must be exerted for maintaining the gap width at a constant value?

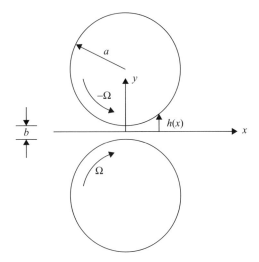

(b) A cylinder of radius a is located with its axis parallel to an infinite, plane interface on which $\partial u / \partial y = 0$ and is separated from it by a distance $a + b$, where $b \ll a$. This cylinder rotates with angular velocity Ω and is attached at its ends to a piece of machinery above it that exerts a net downward force F on the cylinder. Can this load be supported by the high pressure generated in the lubrication layer, and, if so, how fast must the cylinder be rotated? Does the answer change if the plane interface is replaced with an infinite, plane, no-slip boundary?

The Thin-Gap Approximation – Lubrication Problems

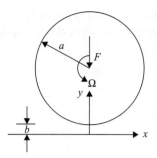

Problem 5–6. Cylinder Moving Toward a Plane Boundary. A cylinder of radius a moves toward an infinite, plane, no-slip boundary under the action of a force F per unit length that is independent of time. The center of the cylinder is located at a distance $b + a$ from the boundary, and we assume $b \ll a$.

(a) Under what conditions is the lubrication approximation valid?
(b) Determine the pressure distribution in the gap between the cylinder and the wall by using lubrication theory.
(c) Calculate the velocity of the cylinder as a function of time.
(d) Determine how long it takes the cylinder to contact the boundary, starting from some initial position $b(0)$.

Repeat the preceding process, assuming that the no-slip boundary is replaced with an *interface* that remains flat and nondeforming and on which the shear stress vanishes. Explain any qualitative differences between the two cases. Can the results of this second calculation be applied to determine the motion of two parallel cylinders of equal radius that are acted on by equal and opposite forces along their line of centers? Explain.

Problem 5–7. Tightly Fitting Sphere in a Cylinder. A sphere is moving along the centerline of a circular cylinder. The sphere radius is a, and the cylinder radius is $a + b$, where $b \ll a$.

(a) Suppose that the cylinder is infinitely long and open at both ends to a large reservoir of the same fluid that is inside the cylinder. If the sphere is to move with a velocity U, what force must be applied? Is there any net volume flux of fluid through the tube? Calculate the relationship between volume flux and U.

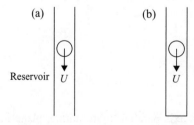

(b) It is proposed to use a cylinder and sphere arrangement like that described in (a) to estimate the viscosity of the fluid. Explain how this could be done. Suppose you have a cylinder with a *closed* end. How will this influence the relationship between the force applied and the sphere velocity? (First describe qualitatively what is going on, and then obtain a quantitative relationship between U and F.)

Problems

Problem 5–8. Converging Flow. Consider converging flow (that is, flow inward toward the vertex) between two infinite plane walls with an included angle 2α, as shown in the sketch. The flow is best described in cylindrical coordinates (r, θ, z), with z being normal to the plane of the flow so that no derivatives exist with respect to z and $v_z = 0$. Also assume that the flow is entirely radial in the sense that fluid moves along rays $\theta =$ constant toward the vertex, that is,

$$v_\theta = 0$$

(obviously, a solution of this form will apply only for large r, away from the slit at the vertex).

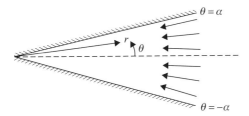

(a) Show that $v_r = f(\theta)/r$.

(b) Obtain the governing differential equation for $f(\theta)$:

$$\left(\frac{2\rho}{\mu}\right) f \frac{df}{d\theta} + \frac{d^3 f}{d\theta^3} + 4 \frac{df}{d\theta} = 0.$$

(c) Assume that no-slip conditions apply at the walls and obtain a sufficient number of boundary conditions or other constraints on f to have a well-posed problem for $f(\theta)$.

(d) Introduce a normalized angle ϕ and a normalized flow variable F, such that

$$\phi = \frac{\theta}{\alpha}, \quad F = \frac{\beta f}{q},$$

where q is the volume flow rate per unit width in the z direction, to obtain the dimensionless equation

$$\operatorname{Re} F \frac{dF}{d\phi} + \frac{d^3 F}{d\phi^3} + 4\alpha^2 \frac{dF}{d\phi} = 0$$

plus boundary conditions and auxiliary conditions on F, where

$$\operatorname{Re} \equiv \frac{2\rho q \alpha}{\mu} \quad \text{(Reynolds number).}$$

(e) Obtain an approximate solution for the creeping-flow limit $\operatorname{Re} \to 0$.

(f) Obtain an approximate solution for $\alpha \to 0$.

(g) Consider the solution for $\operatorname{Re} \to \infty$; show that the solution reduces in this case to

$$F(\phi) = \frac{1}{2}.$$

This solution obviously does not satisfy no-slip conditions on the sidewalls, $\phi = \pm 1$. Explain why. Outline how a solution could be obtained for $\operatorname{Re} \to \infty$ by use of the method of matched asymptotic expansions, which does satisfy boundary conditions on the walls. Be as detailed and explicit as possible, including actually setting up the equations and boundary conditions for the solution in the regions near the walls.

The Thin-Gap Approximation – Lubrication Problems

Problem 5–9. Journal-Bearing Problem. We wish to determine how a journal bearing changes configuration in response to a change in the load. Let us consider a journal bearing that is initially in the configuration shown by Fig. 5–1, with the line of centers being horizontal and eccentricity (λ) at a value λ_0 that is consistent with the support of a vertical load/per unit length of the inner cylinder L_0. Now let us suppose that the load is instantaneously changed to a larger value L_1.

(a) Derive a trajectory equation that describes the motion of the central axis of the inner cylinder with respect to the central axis of the outer cylinder. In doing this, you may make the following assumptions: (1) the inertia of the inner cylinder can be neglected, and its weight is included in the loads L_0 and L_1; (2) that all transients in the fluid velocity field can be neglected or, in other words, that the velocity fields and the corresponding lubrication force are instantaneously equal to their steady-state values for each new configuration.

(b) Integrate these equations, using numerical means if necessary.

(c) Now consider the stability of the steady-state, horizontal configuration assuming that the load is fixed at some constant value L, but that there is a slight perturbation in the horizontal position of the inner cylinder to both more and less eccentric positions.

(d) How would the problem change if we consider the inertial of the inner cylinder, but otherwise assume that the hydrodynamics problem is still quasi-steady? Derive new trajectory equations and discuss how, or whether, the nature of their solutions changes. (Can we get limit cycles?)

Problem 5–10. Drag Due to a Rough Surface. Consider a rough but lubricated surface moving over a smooth surface with velocity U as indicated in the figure. The rough surface is located at $h(y)$ above the smooth surface as given by

$$h(y) = h_0 + h_1 \sin(2\pi y/L),$$

where h_0 is the average height of the rough surface above the smooth surface and h_1 is the size of the sinusoidal asperities, which are distributed with a wavelength of L. We wish to determine the drag on the body with the rough surface, assuming $h_0/L \ll 1$.

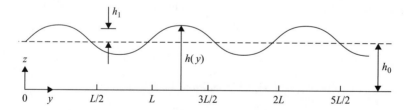

(a) What are the dimensionless governing equations that describe the flow between the two surfaces?

(b) For what conditions is the lubrication theory valid? Is there more than one limit?

(c) Solve the lubrication equations for the flow and pressure fields. Do so for only one limit (your choice) if you think there is more than one limit.

(d) What is the lift and drag on the rough body?

Problem 5–11. Flow Through a Sinusoidal Pipe. Consider the pressure-driven flow of a Newtonian fluid with viscosity μ and density ρ through a sinusoidally varying pipe. The radius R of the pipe is given by

$$R = R_0 [1 + A \sin(2\pi x/\lambda)], \tag{1}$$

Problems

where R_0 is the mean pipe radius, A is a dimensionless amplitude, x is the downstream position, and λ is the wavelength of the variation. The fixed volumetric flow rate through the tube is denoted by Q. Let's consider two asymptotic limits for this flow,

(a) What are the general, dimensionless equations and boundary conditions for this flow, assuming it is axisymmetric and there is no swirl?
(b) Consider the limit of $AR_0/\lambda \ll 1$. What are the zeroth- and first-order perturbation equations and boundary conditions? What is the velocity field of first order?
(c) What is the pressure drop per wavelength for a volumetric flow rate Q?
(d) Now consider the limit of $A \to 1$. What are the equations that govern the flow in the constriction? What is the pressure drop per wavelength for a volumetric flow Q?

Problem 5–12. Flow Through a Porous Tube. Let us consider flow through a cylindrical porous tube, which occurs in many membrane filtration processes. The tube is very long with radius R. At the inlet of the tube, the pressure is P_I. Fluid permeates or leaks out through the wall of the tube with a velocity $k(P - P_s)/\mu$, where P is the local pressure in the fluid, P_s is the pressure on the other side of the membrane, k is a permeation coefficient, and μ is the viscosity of the fluid. We wish to determine how much fluid is filtered as a function of the length of the tube.

(a) Assume steady-state flow of a Newtonian fluid. What are the complete set of equations and boundary conditions that describe the flow?
(b) Nondimensionalize the equations assuming that the characteristic radial and axial velocities are $k(P_I - P_s)/\mu$ and $(P_I - P_s)R^2/L_c\mu$, respectively, where L_c is the characteristic axial direction. What is L_c? What does it represent?
(c) Let $\varepsilon = R/L_C$ and $Re = \rho U_c R/\mu$, where Re is the Reynolds number, U_c is the characteristic axial velocity, and ρ is the density of the field. What are the simplified equations of motions in the limit of ε, $\varepsilon Re \ll 1$. What are the physical conditions that correspond to this limit?
(d) Determine the axial and radial velocity fields and pressure field for these simplified equations.
(e) What is the average axial flow as a function of distance from the entrance?
(f) What is the net flow that has permeated through the tube as a function of distance from the entrance?

Problem 5–13. Two Colliding Spheres. Two spheres of equal radii a suspended in a Newtonian fluid of density ρ and viscosity μ are being pushed toward each other at constant relative velocity U. Calculate the force required for doing so when the separation d between their centers is small, i.e., $(d - 2a)/a \ll 1$.

Problem 5–14. Cylinder Translation Parallel to a Plane Wall. A cylinder of radius a is at a distance h from a plane wall and is being pulled (perpendicular to its central axis) without rotating parallel to the wall at a constant speed U. Determine the force per unit length of the cylinder required for pulling it as a function of the distance h, when $(h - a)/a \ll 1$.

Problem 5–15. Cylinder Rotation Near a Plane Boundary. A cylinder of radius a is at a distance h from a plane wall and is being rotated without translating at a constant speed Ω. Determine the force per unit length of the cylinder required for holding it fixed as a function of the distance h, when $(h - a)/a \ll 1$.

Problem 5–16. Separation of Microscope Slides. You may have observed that it is very difficult to pull apart two glass slides separated by a thin layer of water. To understand the origin of this "adhesive" force, consider two horizontal plates of size $L \times L$ separated by a

The Thin-Gap Approximation – Lubrication Problems

distance H being pulled apart along the normals to the plate surfaces. Calculate the force per unit area required for separating the plates at speed U. Show that the effect of surface tension is small when the glass slides are large, i.e., $H/L \ll 1$. To separate glass slides, one usually slides them parallel to each other. Why?

Problem 5–17. Lubrication in a Cone-and-Plate Geometry. A cone of radius R and angle α is initially in contact with a plane, as depicted in the figure.

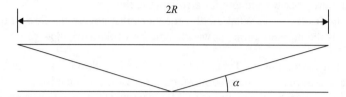

Using lubrication theory, determine the force F necessary to pull the cone off the plane with some velocity V in the limit of small α.

6

The Thin-Gap Approximation – Films with a Free Surface

Although the application of the "thin-film" approximation to analyze lubrication problems is one of its most important successes, there is an even larger body of problems in which the thin-film approximation can still be applied but in which the upper surface (or in some cases both surfaces of the thin film) is an interface. Examples include such diverse applications as gravity currents in geological phenomena, such as the gravitationally driven spread of molten lava; the dynamics of foams and or emulsions for which the thin films between bubbles (or drops) play a critical role in the dynamics; the dynamics of thin films in coating operations, and a variety of other materials processing applications; and thin films in biological systems, such as the coatings of the lung. Not only are the areas of application very diverse, but such films can and do display an astonishing array of complex phenomena, in spite of the limitations inherent in the thin-film assumptions. In part this is a consequence of the wide variety of physical effects that can play a role, including the capillary and Marangoni phenomena associated with surface tension, the possibility of a significant role for nonhydrodynamic effects such as van der Waals forces across the thin film and the possibility of transport processes such as evaporation/condensation.

In this chapter, we derive the governing equations for this class of thin films and show how they can be modified to account for the presence or absence of the various physical phenomena that were mentioned above. We shall see that the result is a dynamical equation for the shape of the film as a function of time, and it is this aspect of the film's behavior that is generally of greatest interest and importance. In even the simplest cases, however, this governing equation is nonlinear and difficult to solve completely. On the other hand, many of the phenomena exhibit self-similar behavior and similarity transformations can often play a very useful role, not only in simplifying the governing equations, but also in identifying the scaling behavior of the solution – for example, the form of the time dependence in gravity or surface tension driven spreading of a liquid "drop." The dynamics of thin films has been a very active area of research in recent years, and this continues to be true. Much of this is based on numerical investigation of the dynamical behavior predicted by the nonlinear thin-film equations. Although this is beyond the scope of the present book, the interested reader may wish to consult the excellent and extensive review of all aspects of these problems by Oron et al.[1]

A. DERIVATION OF THE GOVERNING EQUATIONS

1. The Basic Equations and Boundary Conditions

The problems that we consider in this section are closely related to those in the previous section. As shown in Fig. 6–1, we again consider a thin liquid layer that is bounded below by a solid wall, but with an upper surface that is an interface with an upper fluid that we

The Thin-Gap Approximation – Films with a Free Surface

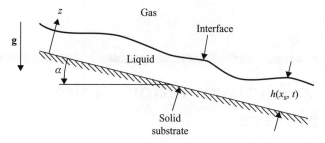

Figure 6–1. A sketch of the generic thin-film configuration for a liquid film on a solid substrate that is tilted to an angle α from the horizontal. The upper surface of the film is an interface.

shall consider to be air or some other gas. Although it is quite possible to consider a thin fluid layer that is bounded above by another liquid, with no change in the nature of the approximations in the thin film itself, the motion in the film is then coupled to the motion in the bounding liquid for which the thin-film approximations are not generally valid. In this class of thin-film problems, it is often of interest to consider gravity-driven flows, and hence we assume that the lower boundary is inclined at an angle α relative to gravity, which is assumed to act in the vertical direction. Because the upper surface is an interface, we know from the discussions at the end of Chap. 2 that the actual pressure will appear in the boundary conditions, and thus it is convenient to retain the gravitational force in the governing equations of motion rather than introducing the dynamic pressure as was done in the previous chapter. Apart from these modest changes and the boundary conditions at the upper interface, the problem statement is basically identical to that given in the previous chapter (Section B), and hence we can utilize almost all of the notation and developments up to the leading order governing equations, which now become

$$\nabla_S \cdot \mathbf{u}_s^{(0)} + \frac{\partial u_z^{(0)}}{\partial z} = 0, \tag{6-1}$$

$$\frac{\partial^2 \mathbf{u}_s^{(0)}}{\partial z^2} - \nabla_S p^{(0)} + \frac{\varepsilon^2 \ell_c^2}{\mu u_c} \rho g \sin\alpha \, \mathbf{e}_x = 0, \tag{6-2}$$

$$\frac{\partial p^{(0)}}{\partial z} + \frac{\varepsilon^3 \ell_c^2 \rho g}{\mu u_c} \cos\alpha = 0. \tag{6-3}$$

The unit vector $\mathbf{e}_x$ is the projection of the gravitational acceleration vector $\mathbf{g}$ onto the plane surface of the lower boundary, i.e., $\mathbf{i}_g - (\mathbf{n} \cdot \mathbf{i}_g)\mathbf{i}_z$. Note that the z axis is always normal to this plane, independent of the angle α. For simplicity, we have assumed that the lower solid boundary is flat. We have already seen that the leading-order approximation of the equations of motion will be unchanged in the thin-film asymptotic limit even if the lower boundary is a curved surface provided only that the radius of curvature is always asymptotically large compared with the film thickness. In the present class of problems, there are two issues to be concerned about if we wish to apply the theory to problems in which the lower boundary is not flat. One is that if gravitational effects are included as previously discussed, we must remember that the angle α is then a function of position $\mathbf{x}_s$. The second, which we shall return to later in this section, is that the overall curvature of the thin film (and thus of the upper interface) may produce a significant contribution to the capillary pressure in the normal-stress boundary condition.

A. Derivation of the Governing Equations

One additional comment regardings Eqs. (6–2) and (6–3) is the fact that we have retained the body-force terms in spite of the fact that they are multiplied by a dimensionless parameter that involves ε to the power of either 2 or 3. However, the magnitude of these terms is not necessarily small because we have not yet specified the characteristic velocity u_c. Unlike the lubrication problems of the preceding chapter, in which a characteristic velocity can be specified in terms of the specified velocity (or force) at the boundaries of the thin film, here the characteristic velocity is dependent on the dominant physical mechanism that is responsible for the evolution of the film with time. One of the possible mechanisms is gravitationally driven motion. If this occurs in a film that is on a surface that has a finite angle of inclination α, then we see from (6–2) that an appropriate choice for u_c would be $\varepsilon^2 \ell_c^2 \rho g/\mu$, and (6–3) would then be approximated as

$$\frac{\partial p^{(0)}}{\partial z} = O(\varepsilon) \approx 0. \tag{6-4}$$

On the other hand, if $\alpha = 0$, and the dynamics of the film is still dominated by body forces, then it appears from (6–3) that $u_c = \varepsilon^3 \ell_c^2 \rho g/\mu$. In other cases, however, gravitational forces may play only a secondary role in the motion of the film, which is instead dominated by capillary forces. Then the appropriate choice for u_c would involve the surface tension rather than either of the choices previously listed and the body-force terms in both (6–2) and (6–3) would be asymptotically small for the limit $\varepsilon \to 0$. This then is a fundamental difference between this class of thin-film problems and the lubrication problems of the previous chapter. Here, the characteristic velocity will depend on the dominant physics, and if we want to derive general equations that can be used for more than one problem, we need to temporarily retain all of the terms that could be responsible for the film motion and only specify u_c (and thus determine which terms are actually large or small) after we have decided which particular problem we wish to analyze.

Apart from the trivial inclusion of the gravitational body-force terms in (6–2) and (6–3), the governing equations, and the analysis leading to them, are identical to the governing equations for the lubrication theory of the previous chapter. The primary difference in the formulation is in the boundary conditions, and the related changes in the physics of the thin-film flows, that arise because the upper surface is now a fluid interface rather than solid surface of known shape. The boundary conditions at the lower bounding surface are:

$$\boxed{u_z^{(0)} = \mathbf{u}_s^{(0)} = 0 \quad \text{at } z = 0.} \tag{6-5}$$

As discussed in Chap. 2, we can specify the position of the upper interface as corresponding to those points $\mathbf{x}'_s$ for which the function $F'(\mathbf{x}', t') \equiv z' - h'(\mathbf{x}'_s, t')$ is identically equal to zero. Then, the usual boundary conditions for a fluid–fluid interface, written in dimensional form, are

$$\mathbf{u}' \cdot \mathbf{n} = \hat{\mathbf{u}}' \cdot \mathbf{n} = -\frac{1}{|\nabla' F'|} \frac{\partial F'}{\partial t'}, \tag{6-6}$$

$$\mathbf{u}' - (\mathbf{u}' \cdot \mathbf{n})\mathbf{n} = \hat{\mathbf{u}}' - (\hat{\mathbf{u}}' \cdot \mathbf{n})\mathbf{n}, \tag{6-7}$$

$$p' - \hat{p}' + 2\mu \left[(\lambda \hat{\mathbf{E}}' - \mathbf{E}') \cdot \mathbf{n} \right] \cdot \mathbf{n} = \gamma (\nabla' \cdot \mathbf{n}), \tag{6-8}$$

$$2\mu[(\mathbf{E}' - \lambda \hat{\mathbf{E}}') \cdot \mathbf{n}] \cdot \mathbf{t}_i - (\nabla'_s \gamma) \cdot \mathbf{t}_i = 0 \quad (i = 1 \text{ or } 2). \tag{6-9}$$

The primes are included to remind the reader that these are dimensional variables. Here, $\mathbf{E}'$ and $\hat{\mathbf{E}}'$ are the rate-of-strain tensors in the fluid on the two sides of the interface as defined following (2–88), λ is the viscosity ratio $\hat{\mu}/\mu$, $\mathbf{t}_i$ stands for one of the two unit tangent vectors ($i = 1$ or 2) in the interfacial surface, and γ is the interfacial tension. In the present analysis, we assume that the second fluid is a gas for which $\lambda \ll 1$. Then, to completely determine the velocity and pressure in the thin film as well as the shape function $h'(\mathbf{x}'_s, t')$, which is an unknown for this class of problems, it is sufficient to apply the kinematic, normal-stress, and shear stress conditions in the forms

$$\mathbf{u}' \cdot \mathbf{n} = -\frac{1}{|\nabla' F'|}\frac{\partial F'}{\partial t'}, \qquad (6\text{--}10)$$

$$p' - p'^* - 2\mu(\mathbf{E}' \cdot \mathbf{n} \cdot \mathbf{n}) = \sigma(\nabla' \cdot \mathbf{n}), \qquad (6\text{--}11)$$

$$2\mu(\mathbf{E}' \cdot \mathbf{n} \cdot \mathbf{t}_i) - (\nabla'_s \sigma) \cdot \mathbf{t}_i = \tau'_s. \qquad (6\text{--}12)$$

All of these boundary conditions are to be applied at the interface $z' = h'(\mathbf{x}'_s, t')$. In writing these conditions, we have used the usual symbol σ to represent the *surface tension* at the gas–liquid interface, and the "ambient" pressure in the gas above the thin film is denoted as p'^*. Finally, τ'_s represents the possibility of an externally applied shear stress on the interface, and will be a useful *ad hoc* way of including the possibility of something like a wind-induced stress at the interface, without actually calculating the velocity profiles and stress in the outer fluid.

It was mentioned earlier that one caveat to the assumption that the lower boundary is flat (i.e., that any overall curvature of the lower boundary can be neglected in deriving the thin-film equations) is that the overall curvature of the thin film (and thus of the interface) may produce a significant contribution to the capillary jump across the interface by means of the normal-stress boundary condition, (6–11). Of course, in the present context, the fluid in the thin film is assumed to be incompressible, and the absolute pressure within the thin film is irrelevant to its dynamics. It is only if there were gradients in the curvature of the solid lower boundary that this capillary pressure jump would be dynamically significant in determining the evolution of the thin film. In the absence of such gradients (for example, if the substrate surface corresponded to a circular cylinder), it is only the variations in the curvature that are due to local changes in the film thickness that will have a dynamical effect on the thin film, and these are included in the dependence of h' on $\mathbf{x}'_s$.

One other limitation in the formulation just presented is the boundary conditions (6–5) at the lower boundary. In some applications, this boundary may actually have a finite velocity. Provided that there is no acceleration associated with this motion, we can simply adopt a frame of reference that is moving with the boundary, and then the conditions (6–5) can be applied. There are two types of acceleration to consider. First, the boundary may be undergoing a purely translational motion with a velocity $\mathbf{U}(t)$ that is a function of time, i.e., $\partial \mathbf{U}/\partial t \ne 0$. In this case, if we adopt a frame of reference that moves with the boundary, we can account for the dynamical effect on the motion of the thin film by simply adding this acceleration to the acceleration that is due to gravity $\mathbf{g}$. Second, the boundary may be rotating in its own plane with angular velocity Ω about some axis of rotation (some specific examples of this type are considered in the Problems section at the end of this chapter). In this case, after some period of time, the fluid in the thin film will also rotate with the same angular velocity, and the fluid elements then experience a centripetal acceleration. If we adopt a coordinate reference frame that rotates with the boundary, the boundary conditions (6–5) still apply, but now the Lagrangian acceleration $D\mathbf{u}/Dt$ that appears in the Navier–Stokes equations is transformed according to

$$D\mathbf{u}/Dt \to (D\mathbf{u}/Dt)_R + \mathbf{\Omega} \wedge \mathbf{\Omega} \wedge \mathbf{r} + 2\mathbf{\Omega} \wedge \mathbf{u}_R.$$

A. Derivation of the Governing Equations

The first term on the right-hand side is the Lagrangian acceleration relative to the rotating reference frame, and $\mathbf{u}_R$ is the velocity of the fluid relative to this same frame of reference. The second term is known as the centrifugal force, and the third term as the Coriolis force. The latter does not play a significant role in thin-film dynamics. The centrifugal force, on the other hand, contributes to the apparent pressure gradient $\nabla'_s p^{(0)}$ by adding a term in the radial direction relative to the axis of rotation. This is because it can be expressed as the gradient of a scalar:

$$\Omega \wedge \Omega \wedge \mathbf{r} = -\nabla \left(\frac{1}{2} \Omega^2 r'^2 \right),$$

where r' is the distance from the axis of rotation. This means that we can replace the pressure with

$$p' - \frac{1}{2} \rho \, \Omega^2 r'^2,$$

and the thin-film problem in the rotating reference frame is then identical in form to that previously outlined. Although we shall not discuss problems involving rotation of the boundary further in the main text of this chapter, as already mentioned, several of the problems at the end of this chapter consider the issue further.

Finally, before going further with the detailed analysis of the thin-film equations and boundary conditions, it is useful to reflect briefly on the qualitative nature of this class of problems relative to the thin-film lubrication problems of the previous chapter. The most important change is that the interface is deformable. Hence, unlike the thin films of lubrication theory, in which the rigid boundaries allowed us to achieve extremely large pressures, in the present class of problems, any tendency toward a local increase in the pressure in the thin film will be compensated for by increased deformation of the interface. At the same time, the deformability of the interface means that all of the boundary conditions, (6–10)–(6–12), are highly nonlinear because the unit normal and tangent vectors depend on the interface shape, which in turn depends on the velocity and pressure distributions in the film. Thus the governing equations for the dynamics of the interface shape function will turn out to also be highly nonlinear, and it is this fact, at least from a mathematical point of view, that is largely responsible for the richness of the collection of possible phenomenon that are known to characterize the motions of thin liquid films (see, e.g., Oron et al[1].).

2. Simplification of the Interface Boundary Conditions for a Thin Film

To apply the boundary conditions, (6–10)–(6–12), to the solution of thin-film problems by using the leading-order governing equations, (6–1)–(6–3), we must simplify, nondimensionalize, and introduce the asymptotic expansions, (5–68). As a starting point, we recall from analytic geometry that we can calculate the unit normal vector from the function F, previously defined, by using the expression

$$\mathbf{n} = \frac{\nabla' F'}{|\nabla' F'|}. \tag{6-13}$$

Now, for a general "shape" function $h'(\mathbf{x}'_s, t')$ we obtain

$$\nabla' F' = \mathbf{i}_z - \nabla'_s h'.$$

Thus,

$$\mathbf{n} = \frac{1}{(1 + |\nabla'_s h'|^2)^{1/2}} \left(\mathbf{i}_z - \nabla'_s h' \right), \tag{6-14}$$

$$\nabla' \cdot \mathbf{n} = -\nabla'_s \cdot \left[\frac{\nabla'_s h'}{\left(1 + |\nabla'_s h'|^2\right)^{1/2}} \right]. \tag{6-15}$$

The Thin-Gap Approximation – Films with a Free Surface

For the present analysis, we are considering a thin film. In this case, it was shown in the previous chapter that $\nabla_s = \ell_c \nabla'_s$ and $h' = \varepsilon \ell_c h$, so that

$$\nabla'_s h' = \varepsilon \nabla_s h. \tag{6–16}$$

Here, we preserve the notation of the previous chapter. The variables without primes are dimensionless.

Utilizing (6–16), we see that rather simpler approximations can be made for the unit normal and related quantities in the thin-film limit, namely,

$$\boxed{\begin{aligned}\mathbf{n} &= [1 + O(\varepsilon^2)]\mathbf{i}_z - \varepsilon \nabla_s h [1 + O(\varepsilon^2)],\\ \mathbf{t}_i &= [1 + O(\varepsilon^2)]\mathbf{i}_i + \varepsilon(\nabla_s h \cdot \mathbf{i}_i)\mathbf{i}_z[1 + O(\varepsilon^2)],\end{aligned}} \tag{6–17}$$

and for the curvature,

$$\boxed{\nabla' \cdot \mathbf{n} = -\frac{\varepsilon}{\ell_c} \nabla_s \cdot (\nabla_s h)[1 + O(\varepsilon^2)].} \tag{6–18}$$

We determine the unit tangent vectors from $\mathbf{n}$ by applying the conditions $\mathbf{n} \cdot \mathbf{t}_i = 0$, $\mathbf{t}_i \cdot \mathbf{t}_j = \delta_{ij}$.

Now, let us consider the various boundary conditions, (6–10)–(6–12). After substituting for $\mathbf{n}$, $\mathbf{t}_i$, and $\nabla \cdot \mathbf{n}$ as appropriate from (6–17) and (6–18), nondimensionalizing by using the characteristic scales (5–56)–(5–60), and substituting the expansions (5–68), we find

$$\boxed{u_z^{(0)} = \frac{\partial h}{\partial t} + \mathbf{u}_s^{(0)} \cdot \nabla_s h \quad \text{at } z = h,} \tag{6–19}$$

$$\boxed{p^{(0)} - p^* + O(\varepsilon^2) = -\frac{\varepsilon^3 \sigma}{\mu u_c} \nabla_s \cdot (\nabla_s h) \quad \text{at } z = h,} \tag{6–20}$$

$$\boxed{\frac{\partial \mathbf{u}_s^{(0)}}{\partial z} - \frac{\varepsilon \sigma_0}{\mu u_c} \nabla_s \hat{\sigma} = \boldsymbol{\tau}_s^* \quad \text{at } z = h.} \tag{6–21}$$

Here p^* is the dimensionless pressure in the gas, i.e., $p'^*/(\mu u_c/\varepsilon^2 \ell_c)$ and $\boldsymbol{\tau}_s^*$ is the dimensionless surface shear stress, $\tau'_s/(\mu u_c/\varepsilon \ell_c)$. In (6–21), we have also introduced the dimensionless surface-tension function $\hat{\sigma}$ such that $\sigma = \sigma_0 \hat{\sigma}$. We denote the value of σ at the ambient temperature (and/or in the absence of surfactants) as σ_0. Hence, if there are no changes in σ from its ambient value, the function $\nabla_s \hat{\sigma} \equiv 0$. Once again, it will be noted that we have retained terms that include the small parameter ε. These are again terms that could be responsible for the motion of the film, and in this case, the appropriate choice for u_c would be either $\varepsilon \sigma_0/\mu$ or $\varepsilon^3 \sigma_0/\mu$, depending on whether the dominant effect is Marangoni or capillary-driven motion, respectively.

3. Derivation of the Dynamical Equation for $h(\mathbf{x}_s, t)$

The problem is now to solve (6–1)–(6–3), subject to the boundary conditions, (6–5) and (6–19)–(6–21). First, we integrate the hydrostatic pressure balance, (6–3), to obtain

$$p^{(0)} = \left(-\frac{\varepsilon^3 \ell_c^2 \rho g}{\mu u_c} \cos \alpha\right) z + \tilde{p}(\mathbf{x}_s, t). \tag{6–22}$$

A. Derivation of the Governing Equations

The unknown function of integration, $\tilde{p}(\mathbf{x}_s, t)$, is determined from the normal-stress boundary condition (6–20) at $z = h$. Specifically, substituting (6–22) into (6–20), we find

$$\tilde{p}(\mathbf{x}_s, t) = p^* + \left(\frac{\varepsilon^3 \ell_c^2 \rho g}{\mu u_c} \cos\alpha\right) h - \left(\frac{\varepsilon^3 \sigma}{\mu u_c}\right) \nabla_s^2 h$$

or

$$p^{(0)} = p^* + \left(-\frac{\varepsilon^3 \ell_c^2 \rho g}{\mu u_c} \cos\alpha\right)(z - h) - \left(\frac{\varepsilon^3 \sigma}{\mu u_c}\right) \nabla_s^2 h. \quad (6\text{--}23)$$

This gives the pressure in the thin film, given the ambient pressure in the gas p^*. Now, because $\nabla_s p^{(0)}$ is independent of z, Eq. (6–21) can be integrated directly to obtain

$$\mathbf{u}_s^{(0)} = \left[\nabla_s p^{(0)} - \left(\frac{\varepsilon^2 \ell_c^2 \rho g}{\mu u_c}\right) \sin\alpha\, \mathbf{e}_x\right]\left(\frac{z^2}{2} + f(\mathbf{x}_s, t)z + g(\mathbf{x}_s, t)\right). \quad (6\text{--}24)$$

The boundary condition (6–5) requires that $g(\mathbf{x}_s, t) \equiv 0$. At the interface,

$$\left.\frac{\partial \mathbf{u}_s^{(0)}}{\partial z}\right|_{z=h} = \frac{\sigma_0}{\mu u_c} \varepsilon \nabla_s \hat{\sigma} + \boldsymbol{\tau}_s^*. \quad (6\text{--}25)$$

Thus

$$\mathbf{u}_s^{(0)} = \left[\nabla_s p^{(0)} - \left(\frac{\varepsilon^2 \ell_c^2 \rho g}{\mu u_c}\right) \sin\alpha\, \mathbf{e}_x\right]\left(\frac{z^2}{2} - hz\right) + \left[\frac{\sigma_0}{\mu u_c} \varepsilon \nabla_s \hat{\sigma} + \boldsymbol{\tau}_s^*\right] z. \quad (6\text{--}26)$$

Although this expression for $\mathbf{u}_s^{(0)}$ is formally complete, it contains the unknown shape function $h(\mathbf{x}_s, t)$. The pressure gradient, on the other hand, is determined from the conditions on pressure at the interface, $z = h$, by means of Eq. (6–23). To obtain a governing equation from which we can determine the unknown shape function, we can follow either of the two paths outlined in the preceding chapter, namely, either integrate the continuity equation, (6–1), to obtain an expression for $u_z^{(0)}$ and apply boundary conditions (6–5) and (6–19), or integrate the continuity equation first to obtain (5–75), and then apply the boundary conditions to evaluate this integral constraint. We follow the latter route.

Substituting the conditions (6–5) and (6–19) into (5–75), we can rewrite the integral constraint as

$$\int_0^h (\nabla_s \cdot \mathbf{u}_s)\,dz + \frac{\partial h}{\partial t} + \mathbf{u}_s^{(0)} \cdot \nabla_s h = 0. \quad (6\text{--}27)$$

Now according to Leibnitz rule from calculus,

$$\nabla_s \cdot \int_0^{h(\mathbf{x}_s,t)} \mathbf{u}_s^{(0)}\,dz = \mathbf{u}_s^{(0)} \cdot \nabla_s h + \int_0^{h(\mathbf{x}_s,t)} (\nabla_s \cdot \mathbf{u}_s^{(0)})\,dz. \quad (6\text{--}28)$$

Hence, we can combine the first and last term of (6–27) to obtain a compact relationship between h and $\mathbf{u}_s^{(0)}$, namely

$$\frac{\partial h}{\partial t} + \nabla_s \cdot \int_0^{h(\mathbf{x}_s,t)} \mathbf{u}_s^{(0)}\,dz = 0. \quad (6\text{--}29)$$

The Thin-Gap Approximation – Films with a Free Surface

Now, if we substitute the expression (6–26) into this integral constraint, we obtain the governing equation for the unknown function h:

$$\boxed{\frac{\partial h}{\partial t} = \nabla_s \cdot \left\{ \frac{h^3}{3} \left[\nabla_s p^{(0)} - \left(\frac{\varepsilon^2 \ell_c^2 \rho g}{\mu u_c} \right) \sin \alpha \mathbf{e}_x \right] - \frac{h^2}{2} \left(\frac{\sigma_0}{\mu u_c} \varepsilon \nabla_s \hat{\sigma} + \tau_s^* \right) \right\}}, \qquad (6\text{–}30)$$

where $p^{(0)}$ is given by Eq. (6–23). We see that (6–30) is a nonlinear, PDE for the film shape function, $h(\mathbf{x}_s, t)$. It has been the subject of many important investigations for the class of problems listed at the beginning of this chapter, but, because of the nonlinearity of (6–30), most are based on numerical solutions (cf. the review paper cited earlier by Oron *et al.*). There are, however, a few problems that can be solved analytically, mainly by means of similarity transformations, and we shall discuss some of these problems in the present chapter (as well as in several of the problems at the end of this chapter).

B. SELF-SIMILAR SOLUTIONS OF NONLINEAR DIFFUSION EQUATIONS[2]

We apply (6–30) for cases in which there is no applied surface stress, the lower substrate is horizontal ($\alpha = 0$), and the surface tension is constant (the latter two assumptions will be relaxed in some of the problems at the end of the chapter). Then, the term proportional to h^2 on the right-hand side of (6–30) is identically equal to zero, and the PDE for h has the generic form

$$\boxed{3 \frac{\partial h}{\partial t} = \nabla_s \cdot \left[h^3 \nabla_s p^{(0)} \right]}. \qquad (6\text{–}31)$$

We shall see that this equation can often be rewritten in a form that is exactly analogous to a 1D diffusion equation in which the diffusivity depends nonlinearly on the "concentration," namely,

$$\frac{\partial c}{\partial t} = \frac{D_0}{c_0^n r^{d-1}} \frac{\partial}{\partial r} \left(c^n r^{d-1} \frac{\partial c}{\partial r} \right), \qquad (6\text{–}32)$$

where $d = 1, 2$, or 3, depending on whether the coordinate system is Cartesian, circular cylindrical, or spherical (generally, either $d = 1$ or $d = 2$ for the dynamics of a thin film), and the diffusivity is

$$D(c) = D_0 (c/c_0)^n. \qquad (6\text{–}33)$$

We shall see that nonlinear diffusion problems, which satisfy (6–32) and (6–33), often have solutions that can be expressed in a self-similar form. Hence, before considering specific examples of the thin-film problems that satisfy (6–31), we take a small detour to consider the form of similarity solutions for the nonlinear diffusion problems, beginning with the classical linear diffusion problem in which $n = 0$.

The classical problem for diffusion of a chemical species in a quiescent fluid is governed by the generic equation

$$\frac{\partial c}{\partial t} = -\nabla \cdot \mathbf{j}. \qquad (6\text{–}34)$$

Here, c is the concentration and $\mathbf{j}$ is the flux of the species in question due to the diffusion process. The species flux vector $\mathbf{j}$ plays the same role in the diffusion problem as the heat flux vector $\mathbf{q}$ plays in the transfer of heat by conduction. In the classical diffusion problem,

B. Self-Similar Solutions of Nonlinear Diffusion Equations

for diffusion in a gas, or in a Newtonian solvent, the relationship between **j** and the gradient of concentration, ∇c, is linear, i.e.,

$$\mathbf{j} = -D\nabla c. \tag{6–35}$$

The result of substituting (6–35) into (6–34) is the classical linear diffusion equation,

$$\frac{\partial c}{\partial t} = D\nabla^2 c. \tag{6–36}$$

A generic problem, which is mathematically analogous to a number of thin-film problems for the shape function h, is the evolution of the "radially" symmetric concentration distribution that evolves at large times from a pulselike initial source. At very large times, the form of the concentration distribution becomes insensitive to its initial "shape," i.e., it is independent of the details of the initial spatial distribution of c and depends on only the "dimension" of the distribution. In d dimensions, the form of the diffusion equation that describes the evolution of this "radially" symmetric concentration distribution is

$$\frac{\partial c}{\partial t} = \frac{D}{r^{d-1}}\frac{\partial}{\partial r}\left(r^{d-1}\frac{\partial c}{\partial r}\right). \tag{6–37}$$

There are three possibilities corresponding to the dimension of the distribution. The first is a 1D concentration distribution ($d = 1$), in which the diffusing species spreads evenly in the $\pm z$ directions from an initial "line" pulse at $z = 0$ on the xz plane. In this case, the variable r in (6–37) is the Cartesian variable z. The second case is a "circularly" symmetric distribution for c ($d = 2$), which evolves by diffusion on a plane from an initial compact "planar" pulse. In this case, r in (6–37) is the radial component of a polar (or cylindrical) coordinate system that lies in the diffusion plane. The third case is a "spherically symmetric" distribution corresponding to $d = 3$, which evolves at long times from a "compact" 3D pulse that diffuses outward into the full 3D space. In this case, r is the radial variable of a spherical coordinate system. To obtain the long-time form of the distribution we must solve (6–37), but subject to the integral constraint that the total amount of the diffusing species is constant, independent of time:

$$Q = \text{const} = \int_V c \, dV \approx \int_r c r^{d-1} dr. \tag{6–38}$$

Here, the integral over V is taken over the whole of the domain containing c; namely, over z from $-\infty$ to $+\infty$ for $d = 1$, from $\theta = 0$ to 2π and $r = 0$ to ∞ for $d = 2$, and from $\theta = 0$ to π, $\phi = 0$ to 2π, and $r = 0$ to ∞ for $d = 3$. The diffusive spread of an initial pulse of material is a classical problem, and the solution of (6–37) subject to (6–38) is well known to be

$$c(r,t) = \frac{Q}{(4\pi Dt)^{d/2}} \exp(-r^2/4Dt), \quad (d = 1, 2, 3). \tag{6–39}$$

As previously indicated, this solution represents the concentration distribution at long times (in which long time means $t \gg L_c^2/D$, where L_c is a length scale characteristic of the initial concentration distribution) and is valid for all three possible values of d.

It will be noted that the solution (6–39) has a self-similar form:

$$c(r,t) = At^\alpha f(Br/t^\beta), \tag{6–40}$$

where A and B are constants. In this particular problem, the governing equation is linear, and we can obtain a closed-form analytic solution, (6–39), for the similarity function $f(Br/t^\beta)$. However, in other cases, the resulting equation for the similarity function may be nonlinear and difficult or impossible to solve analytically. Nevertheless, the fact that the solution can be reduced to a similarity form provides a great deal of the most useful information about the nature or behavior of the solution. We have already encountered some problems of the self-similar type in Chap. 3. We may anticipate from that experience that it will often be possible to deduce the form (6–40) including the numerical values of the scaling parameters α and β from the DE and boundary conditions of the original problem, even when it is not possible to obtain a detailed analytical solution for f. When α and β are known, we know the time dependence of the decay in amplitude, as well as the time dependence of the rate of spread in the c profile. The function f provides the detailed shape of the profile, but in many instances it is the scaling behavior in time that is the most important information in the solution.

It will be useful to review the use of similarity transformations to derive (6–39). We noted in our previous encounter (Chap. 3) that similarity solutions typically exist in problems in which there is no characteristic length scale. Hence, to express the solution of such a problem in a form that is independent of the system of "units," the solution must be "self-similar" in the sense that it depends on a dimensionless combination of the independent variables. The long-time form of the concentration profiles, (6–39), falls into this category. The initial condition, which would dominate the form of the solution at earlier times, is replaced for $t \gg L_c^2/D$ by the integral constraint, (6–38), where there is no explicit dependence on the length scale L_c. If we assume that the form of the distribution for c is dependent on the similarity variable $\eta \equiv Br/t^\beta$, we can see immediately that the integral constraint can be satisfied only if the amplitude of c scales with t to some power. This leads to the suggested similarity form, (6–40). In fact, if we substitute this into the integral constraint, and transform from the independent variables (r,t) to (η,t), we obtain

$$\int_r cr^{d-1}dr = t^{\alpha+\beta d}\int_r \eta^{d-1} f(\eta)d\eta = \text{const (independent of } t\text{)}, \quad (6\text{–}41)$$

where the domain of interest in r is dependent on d, as described earlier. Hence, we see that this condition requires that

$$\alpha + \beta d = 0, \quad (6\text{–}42)$$

so that

$$c = At^{-\beta d}f(\eta), \text{ where } \eta \equiv Br/t^\beta. \quad (6\text{–}43)$$

Now, if this expression for c is substituted into the governing equation, (6–37), the result can be written in the form

$$-\beta d\, f(\eta) - \beta\eta\frac{df}{d\eta} = DB^2 t^{1-2\beta}\eta^{1-d}\frac{d}{d\eta}\left(\eta^{d-1}\frac{df}{d\eta}\right). \quad (6\text{–}44)$$

Now, because f is assumed to be a dimensionless function only of η, the coefficients in this equation can either be constants or functions of η, but must be independent of t. Thus,

$$\beta = 1/2,$$

and in order that f is dimensionless, we can choose $B^2 = 1/2D$ Hence, the governing equation for f is

$$\eta^{1-d}\frac{d}{d\eta}\left(\eta^{d-1}\frac{df}{d\eta}\right) + \eta\frac{df}{d\eta} + df(\eta) = 0. \quad (6\text{–}45)$$

which is to be solved subject to the condition (6–38).

A remarkable feature of Eq. (6–45) is that it has a general solution that is independent of the parameter d, namely

$$f(\eta) = \exp(-\eta^2/2). \quad (6\text{–}46)$$

B. Self-Similar Solutions of Nonlinear Diffusion Equations

Hence the general solution for c is

$$c = At^{-d/2} \exp(-\eta^2/2), \quad \text{with} \quad \eta = r/\sqrt{2Dt}. \tag{6-47}$$

All that remains is to determine the constant A. To do this we apply the integral constraint, (6–38), which takes one of three forms depending on the value of d, namely,

$$\text{for } d = 1, \int_{-\infty}^{\infty} c(r,t) dr = Q,$$

$$\text{for } d = 2, \int_0^{2\pi} \int_0^{\infty} c(r,t) r \, dr \, d\theta = Q, \tag{6-48}$$

$$\text{for } d = 3, \int_0^{2\pi} \int_0^{\pi} \int_0^{\infty} c(r,t) r^2 \sin\theta \, dr \, d\theta \, d\phi = Q.$$

It is left to the reader to show that, in all three cases,

$$A = \frac{Q}{(4\pi D)^{d/2}},$$

and hence that the solution, (6–15), quoted earlier is, in fact, verified.

Stone[2] has summarized a generalization of the solution for this simple linear problem, due to Pattle[3] and Pert[4], which is extremely useful in the analysis of thin-film problems. This is the development of similarity solutions for the d-dimensional symmetric diffusion equation with a diffusivity that depends on the concentration,

$$\boxed{D(c) = D_0(c/c_0)^n,}$$

so that

$$\boxed{\frac{\partial c}{\partial t} = \frac{D_0}{c_0^n r^{d-1}} \frac{\partial}{\partial r} \left(c^n r^{d-1} \frac{\partial c}{\partial r} \right).} \tag{6-49}$$

We again seek a similarity solution for the diffusive spread of the species with concentration c for an initial 1D, 2D, or 3D pulse of finite quantity Q. We can begin with the general similarity form (6–40) for $c(r, t)$, and, in exactly the same fashion as in the previous problem, show that the integral constraint, (6–41), can be satisfied provided the condition (6–42) is satisfied, thus leading to the proposed form (6–43) for c:

$$\boxed{c = At^{-\beta d} f(\eta), \quad \text{where} \quad \eta \equiv Br/t^\beta.} \tag{6-43}$$

To see whether a similarity solution of this proposed form exists for Eq. (6–49), we must determine whether values of β, A, and B exist that reduce (6–49) to an ODE for the dimensionless similarity function $f(\eta)$. To do this, we substitute (6–43) into (6–49). The left-hand side is just

$$\frac{\partial c}{\partial t} = At^{-\beta d - 1} \left[-\beta d f(\eta) - \beta \eta \frac{df}{d\eta} \right]. \tag{6-50}$$

On the other hand, after some algebra, we can show that the right-hand side of (6–49) takes the form

$$\frac{D_0}{c_0^n r^{d-1}} \frac{\partial}{\partial r} \left(c^n r^{d-1} \frac{\partial c}{\partial r} \right) = \frac{D_0 B^2 A^{n+1}}{c_0^n} t^{-2\beta - \beta dn - \beta d} \left[\eta^{1-d} \frac{d}{d\eta} \left(\eta^{d-1} f^n \frac{df}{d\eta} \right) \right]. \tag{6-51}$$

The Thin-Gap Approximation – Films with a Free Surface

The necessary condition for the existence of a similarity solution is that the coefficients in the equation for f [obtained by substituting (6–50) and (6–51) into (6–49)] are either functions of η or constants but do not depend on either t or r independently. Combining the t dependence from the two sides of (6–49), we see that this leads to the condition

$$\beta = \frac{1}{2 + dn}. \qquad (6\text{–}52)$$

We can also choose the coefficient B^2 so that the remaining dimensional coefficient in (6–49) is made dimensionless, and with a numerical value that can be chosen for convenience in solving the resulting equation for $f(\eta)$. We choose

$$B^2 = \frac{\beta c_0^n}{D_0 A^n}\left(\frac{n}{2}\right). \qquad (6\text{–}53)$$

Thus the governing equation for $f(\eta)$ can now be written in the form

$$\frac{n}{2}\left[\eta^{1-d}\frac{d}{d\eta}\left(\eta^{d-1} f^n \frac{df}{d\eta}\right)\right] + \eta\frac{df}{d\eta} + df = 0, \qquad (6\text{–}54)$$

which is to be solved subject to one of the integral constraints, (6–48), depending on the value of d. The self-similarity variable is

$$\boxed{\eta \equiv \left[\frac{nc_0^n}{2D_0(2+dn)A^n}\right]^{1/2} \frac{r}{t^{1/(2+dn)}}.} \qquad (6\text{–}55)$$

Rather than taking a more general approach to the solution of (6–54), we propose a trial form for the solution

$$f = (1 + \delta\eta^2)^\gamma. \qquad (6\text{–}56)$$

Substituting into (6–49), we find that this trial function can provide an exact solution of the problem. Specifically, we find

$$\gamma = 1/n, \; \delta = -1.$$

Thus the solution takes the form

$$\boxed{f(\eta) = (1 - \eta^2)^{1/n}} \qquad (6\text{–}57)$$

This solution is valid for all $n \neq 0$. We note that it is of a fundamentally different form from the solution to the linear diffusion equation previously obtained. One specific point to note is that $f > 0$, so that the solution makes physical sense only for $\eta < 1$. In mathematical terminology, the solution is said to have "compact support" in the sense that f is nonzero only within this region. Thus the value $\eta = 1$ defines the outer boundary of the region in which the pulse of material exists as a function of time.

The only remaining point is to satisfy the appropriate integral constraints, (6–48), and this can be done by choice of the single remaining undetermined coefficient A. In applying these conditions, it will be noted that f, and thus c, is nonzero only out to $\eta = 1$. Thus the conditions (6–48) can be rewritten in the form

$$\text{for } d = 1, \; \int_{-R(t)}^{R(t)} c(r, t)\, dr = Q,$$

$$\text{for } d = 2, \; \int_0^{2\pi}\int_0^{R(t)} c(r, t)\, r\, dr\, d\theta = Q, \qquad (6\text{–}58)$$

C. Films With a Free Surface – Spreading Films $\alpha = 0$

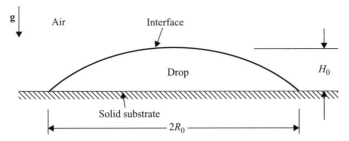

Figure 6–2. A sketch of the initial configuration of spreading drop on a solid substrate. The "drop" may be either a two-dimensional ridge of fluid (with the sketch showing the cross-sectional view), or it may be an axisymmetric drop. The thin film approximation requires that the radial breadth of the drop be much larger than its height. The initial values of these quantities are denoted as R_0 and H_0.

$$\text{for } d = 3, \quad \int_0^{2\pi} \int_0^{\pi} \int_0^{R(t)} c(r,t) r^2 \sin\theta \, dr \, d\theta \, d\phi = Q.$$

The position $R(t)$ corresponds to $\eta = 1$. Hence we can derive a formula for $R(t)$ from (6–55). The net result of applying these conditions is a formula for the last remaining constant in the solution (6–43), namely the amplitude factor A. The result is

$$A^{(2+dn)/2} = Q \left[\frac{nc_0^n}{2D_0(2+dn)} \right]^{d/2} \frac{\Gamma\left(1 + \frac{1}{n} + \frac{d}{2}\right)}{(\pi)^{d/2} \Gamma\left(1 + \frac{1}{n}\right)}. \tag{6–59}$$

The symbol $\Gamma(z)$ represents the gamma function of argument z. Definitions and other properties of the gamma function, including its value for various values of the argument, can be found in the well-known book on special functions by Abramowitz and Stegun.[5]

C. FILMS WITH A FREE SURFACE – SPREADING FILMS $\alpha = 0$

We now consider a number of examples of problems in which the approximate equations for a thin film with a free surface can be used, together with the mathematical theories from the preceding section, to provide solutions.

1. Gravitational Spreading

We begin by considering the problem of the dynamics of 2D and axisymmetric gravity currents over a rigid horizontal surface that is due to the action of gravity, or, more precisely, that is due to the gradient in hydrostatic pressure that is caused by its nonuniform depth. A sketch of the problem is shown in Fig. 6–2. It was originally considered by Huppert[6], and our analysis is based on his theoretical work. As explained qualitatively in Chap. 2, the essential feature of a gravity current is that a fluid with density ρ spreads mainly horizontally into a fluid of different density driven by gravitational forces. If the fluid configuration is like that shown in Fig. 6–2, with the spreading fluid on top of the horizontal boundary, then it will spread outward if its density is larger than that of the exterior fluid. However, if the spreading body of fluid is on the underside of the horizontal surface, it will spread outward if its density is lower than that of the exterior fluid. One geological motivation for interest in this problem in the current thin-film limit, where viscous and gravitational forces dominate the dynamics, might be to describe the rate of spreading of a molten magma. However, the analysis is equally applicable to the spread of a drop of fluid. There are two obvious questions that one may wish to answer: First, how rapidly does the drop spread? And second, what is the shape of the spreading drop as a function of time?

To be specific, we assume that the initial shape of the drop is either a 2D "ridge" of fluid or an axisymmetric "blob," and that the horizontal dimension of this drop is much larger than its maximum height so that the thin-film approximation can be used to obtain a first approximation to its spreading dynamics. We further assume that the upper fluid is a gas so that the applied shear stress at the upper surface can be approximated as zero, and we initially assume that we can neglect the effects of surface tension, i.e., both the Marangoni term in (6–26) and the capillary pressure within the drop, as given by (6–23). It can be seen from this equation that neglect of capillary pressure effects requires

$$Bo^{-1} \equiv \sigma/R_0^2 \rho g \ll 1. \tag{6-60}$$

Here, we have assumed that the characteristic length ℓ_c in the horizontal direction is the initial radius of the drop at the plane of the bounding surface, R_0 (see Fig. 6–2). The initial maximum height of the drop defines the characteristic length scale across the thin film, so that $\varepsilon \equiv H_0/R_0 \ll 1$. We note that the ratio of terms in this condition is the inverse of a dimensionless number that is known as the "Bond number." The Bond number is a ratio of the characteristic magnitude of gravitational forces to capillary forces. With these assumptions, and remembering that $\alpha = 0$, the equation governing the shape function of the drop h can be seen from (6–31) to be

$$\frac{\partial h}{\partial t} = \frac{\varepsilon^3 R_0^2 \rho g}{3\mu u_c} \nabla_s \cdot (h^3 \nabla_s h). \tag{6-61}$$

We have not yet specified the characteristic velocity u_c. However, it appears from (6–61) that a convenient choice would be

$$u_c \equiv \frac{\rho g H_0^3}{\mu R_0}. \tag{6-62}$$

Although this comes from the governing equation as a natural consequence of the nondimensionalization process, it could also be deduced directly from simple scaling arguments. Specifically, the flow in the drop is driven by the horizontal gradient of the hydrostatic pressure, and this has a characteristic magnitude $(\rho g H_0)/R_0$. At steady state, this exactly balances the viscous stress term in the equation of motion, which has a characteristic magnitude $\mu(u_c/H_0^2)$. The balance between these two terms again leads to the estimate (6–62) for the characteristic velocity. The governing equation, (6–61), for the film shape function h is solved subject to the conditions

$$h(\mathbf{x}_s, t) \to 0 \quad \text{as } |\mathbf{x}_s| \to R(t), \tag{6-63}$$

$$\int_{\mathbf{x}_s} h(\mathbf{x}_s, t) dA = q. \tag{6-64}$$

The latter condition is the statement that the total volume of the drop is fixed. The integral is taken over the total area of the drop in the horizontal plane, i.e., for all $|\mathbf{x}_s| \le R(t)$, where $R(t) \equiv R'(t)/R_0$ is the position of the outer edge of the drop in the horizontal plane nondimensionalized by the characteristic length scale R_0, and q is the volume of the drop scaled by either $H_0 R_0^2$ for $d = 2$, or $H_0 R_0$ per unit length for $d = 1$.

C. Films With a Free Surface – Spreading Films $\alpha = 0$

Now, the problem defined by (6–61)–(6–64) is mathematically identical with the nonlinear diffusion problems of the preceding section. Hence this spreading drop problem has similarity solutions that we can read directly from our previous analysis. There are two cases.

For spread of the 2D ridge (which is analogous to the 1D diffusion problem) we have $d = 1$, $n = 3$, and the ratio $c_0^n/D_0 = 3$. Hence the solution is

$$h = \left[q\left(\frac{9}{10}\right)^{1/2} \frac{\Gamma(11/6)}{\sqrt{\pi}\Gamma(4/3)}\right]^{2/5} t^{-1/5} f(\eta), \qquad (6\text{–}65)$$

$$f(\eta) = (1 - \eta^2)^{1/3}, \qquad (6\text{–}66)$$

$$\eta = \frac{(9/10)^{1/5}}{q^{3/5}\left[\dfrac{\Gamma(11/6)}{\sqrt{\pi}\Gamma(4/3)}\right]^{3/5}} \left(\frac{r}{t^{1/5}}\right). \qquad (6\text{–}67)$$

We see from the form of this solution that the height of the spreading drop decreases with time as $t^{-1/5}$ and the outer edge of the drop also spreads outward at a rate proportional to $t^{1/5}$.

For the axisymmetric spread of a drop with a circular base, $d = 2$ and the other parameters are the same. Again the solution can be read directly from the similarity solutions of the preceding section:

$$h = \left[q\left(\frac{9}{16}\right)\frac{\Gamma(7/3)}{\pi\,\Gamma(4/3)}\right]^{1/4} t^{-1/4} f(\eta), \qquad (6\text{–}68)$$

$$f(\eta) = (1 - \eta^2)^{1/3}, \qquad (6\text{–}69)$$

$$\eta = \frac{(9/16)^{1/8}}{\left[q\dfrac{\Gamma(7/3)}{\pi\,\Gamma(4/3)}\right]^{3/8}} \left(\frac{r}{t^{1/8}}\right). \qquad (6\text{–}70)$$

In this case, the height of the spreading drop decreases as $t^{-1/4}$, and the perimeter radius increases as $t^{1/8}$.

A comparison of the axisymmetric solution with experimental data for a spreading drop was given by Huppert.[6] To make this comparison, we need to numerically evaluate the coefficients that appear in (6–68) and (6–70). Because $\Gamma(z+1) = z\Gamma(z)$, we can re-express h and η as

$$h = q^{1/4}\left[\frac{3}{4\pi}\right]^{1/4} t^{-1/4} f(\eta) = 0.699 q^{1/4} t^{-1/4} f(\eta),$$

$$\eta = \frac{(9/16)^{1/8}}{q^{3/8}\left[\dfrac{4}{3\pi}\right]^{3/8}} \left(\frac{r}{t^{1/8}}\right) = 1.592 q^{-3/8} \frac{r}{t^{1/8}}.$$

The Thin-Gap Approximation – Films with a Free Surface

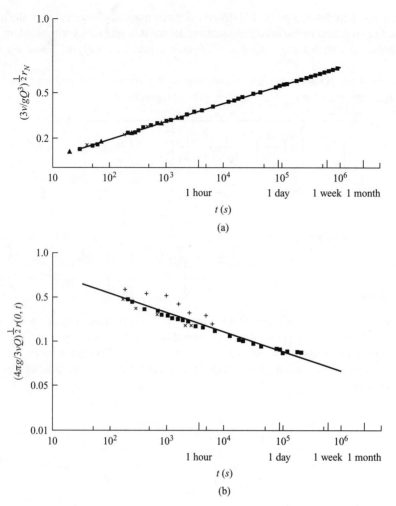

Figure 6–3. Experimental results for the axisymmetric spreading of constant volumes of silicone oils in into air (copied from Fig. 6, Ref. 6.). The straight lines are the best-fit power laws (6–71) and (6–72).

In dimensional terms, the value of h at $r=0$ is thus

$$h'(0,t) = \left[\frac{3\nu q'}{4\pi g}\right]^{1/4} t^{-1/4}, \tag{6-71}$$

and the value of r at the outer rim of the drop (i.e., where $\eta = 1$) is

$$r'_{edge} = \left(\frac{1024}{81\pi^3}\right)^{1/8} \left[\frac{q'^3 g}{3\nu}\right]^{1/8} t'^{1/8}. \tag{6-72}$$

The comparison with experimental results from Huppert's paper is shown in Fig. 6–3. The predictive power of the thin-film theory is obviously very good for this problem.

This may seem surprising if we stop and think a little about the assumptions of the analysis. The theory is based on the thin-film model, in which the flow is nearly a unidirectional flow. The shape function $f(\eta)$ is plotted in Fig. 6–4. Evidently the assumption of nearly unidirectional flow must break down totally at the "front" of the spreading drop. Furthermore, we have completely neglected to say anything about the moving contact line where the drop meets the solid substrate. In spite of these apparent flaws in the theory, the

C. Films With a Free Surface – Spreading Films $\alpha = 0$

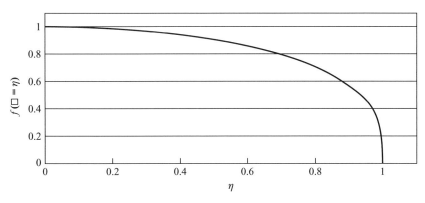

Figure 6–4. The shape function, Eq. (6–69).

predicted and measured results for the spreading rate and for the height of the drop at its center point are very accurate. We can only conclude that the detailed conditions at the front of the drop play, at most, a very minor role in determining its shape and motion. Additional experimental and theoretical work indicates that the insensitivity to conditions at the front is true only if the Reynolds number is very small and the Bond number is large, as is true in our theory and in the experiments that are shown in Fig. 6–3. Gravity currents at higher Reynolds numbers are found to be very strongly influenced by the conditions at the front (see, e.g., Huppert and Simpson[7]). One other point to notice is that the comparison between theory and experimental shapes considered only the height of the drop at the central point. Observations show that the leading edge of a spreading drop does not match well the vertical shape predicted by the theory, even when the spreading process is gravitationally dominated, as in the experiments of Huppert. Instead, the front seems to curl over so that it meets the solid boundary at a much more acute angle, presumably controlled by the dynamics of the contact line region.

2. Capillary Spreading

The preceding analysis was predicated on the assumption shown in (6–60) that the Bond number is very large, i.e., $Bo \equiv (R_0^2 \rho g)/\sigma \gg 1$. In this limiting case, capillary forces play a negligible role in the spreading process. However, for small drops, or in microgravity conditions, this condition will not generally be satisfied. In this section, we consider the opposite limit, $Bo \ll 1$, in which the gravitational effect is negligible and the spreading is totally driven by capillary effects. This problem is significantly more difficult than the preceding case of gravity driven spreading, beginning with the *mechanism* for spreading, which is quite subtle and worthy of discussion before delving into the details of analysis.

Let us start by recalling the discussion of surface tension- or interfacial-tension-driven motions from Chap. 2 for the case of an isothermal, noncontaminated interface. The overriding consideration is that the so-called capillary flows occur because of the presence of spatial gradients in the interface curvature. In the absence of any additional forces or constraints, the *expected* long-time outcome from these flows is that the interface will become spherical, thus representing the surface of constant curvature with the global minimum surface area (and thus the minimum surface free energy) for a given volume of the enclosed fluid. One example from Chap. 2 was the deformed drop, which was driven back to a spherical shape by capillary forces. As explained there, one can think of the internal fluid motion that produces this change of shape as being caused by internal pressure gradients that are due to the gradients in interface curvature along the interface.

The Thin-Gap Approximation – Films with a Free Surface

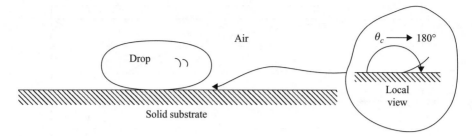

Figure 6–5. A sketch of the configuration of a drop sitting on a solid substrate when the contact angle approaches 180°. The drop will become increasingly spherical in the limit of Bond number, $Bo \ll 1$.

Of course, the tendency of the interface between two immiscible fluids to adopt a *single* spherical shape to achieve the global minimum of surface area may be *prevented* if the system is dominated by forces other than capillary forces or *delayed* to inordinately long times by the "kinetic" barriers of escaping a local free-energy minimum. For example, think of the interface between a high-density fluid layer at the bottom of some container and a less-dense fluid that is above it. With a sufficient density difference, and normal gravity, this is a stable configuration in spite of the fact that the interface is *flat* (rather than spherical) because the configuration is dominated by the gravitational body forces. However, even if the density difference (or gravity) is reduced, this particular configuration remains as a possible equilibrium state. A flat interface is a surface of *constant curvature*, and though the total surface area of the lower fluid (including both the interface and the fluid–solid boundaries) is not minimized by its shape in the container, it is not necessarily true that a configuration with one fluid phase in the shape of a sphere and the second fluid surrounding it in contact with the bottom and sidewalls of the container is a configuration corresponding to a lower overall free energy. Indeed, even if it were, the flat interface separating the two fluids is a local free-energy minimum that is stable to modest perturbations of the interface from this flat shape. Consider, for example, a small-amplitude bulge of the interface upward at some position in the container cross section. This can be thought of as creating a corresponding pressure maximum in the fluid immediately below the bulge that is due to the local curvature and thus capillary pressure jump, which will tend to drive fluid out of the bulge and thus cause the interface to return to its original flat configuration.

If we now think of a drop that is sitting on top of a flat, solid surface, we may ask how these ideas from Chap. 2 apply, and, specifically, whether we should expect the drop to spread as implied in the title to this subsection in the complete absence of any gravitational forces? The answer turns out to depend on the nature of the nonhydrodynamic interactions between the drop and the solid substrate. The most important manifestation of these interactions in the present problem is the equilibrium contact angle, denoted in the sketch of Fig. 2–12 as θ_c. If θ_c is large, say approaching 180°, the drop at its edges must approach the solid surface with this angle, and this means edges of the drop must have a configuration like that sketched in Fig. 6–5. This configuration is not only *inconsistent* (at least locally) with the assumption of a drop shape that satisfies the thin-film approximation, but also one in which there will be a capillary-induced pressure gradient inside the drop that will tend to drive fluid away from the ends of the drop. If the complications of precisely how the contact line moves are ignored it is clear that the net result of this capillary-induced motion will be to produce a shape that is more compact and "spherical," with a final steady state that is a balance between the tendency of the capillary forces to produce a spherical drop and the constraint associated with the fixed contact angle. Therefore, in this case of a large contact angle, the drop clearly does not spread but, in fact, undergoes the opposite type of motion. Indeed if θ_c were exactly 180°, than the drop would end up as exactly spherical, touching

C. Films With a Free Surface – Spreading Films $\alpha = 0$

the solid wall at a single point. The ratio of the drop viscosity to the interfacial tension determines the time scale of this process, but not the ultimate fate of the drop.

Spontaneous spreading under the action of capillary forces occurs only at the other end of the spectra where θ_c is sufficiently small, and the initially flattened drop is constrained by the dual effects of capillary forces and the specified contact angle from the type of motion previously described. If again the details of the contact line motion are ignored, a drop with small contact angle will be driven to spread outward until it achieves a shape that satisfies the dual constraints of constant curvature because of capillary forces and specified contact angle. In the limit $\theta_c \to 0$, the drop will continue to spread indefinitely. However, for any small but finite θ_c, the drop would eventually stop spreading once the balance between fixed contact angle and constant curvature is met. In the preceding section, we saw that there was a "universal" spreading law for the case of gravitationally dominated spreading,

$$R(t) = c(t)^\delta,$$

where the constant δ is $1/5$ for a spreading 2D ridge and is $1/8$ for the axisymmetric drop. We say "universal" in the sense that this result does not depend on the details of the physics near the fluid–solid contact line. When the spreading is due to capillary effects, it has been known since the work of Tanner more than 20 years ago that there is a similar universal spreading law that is apparently independent of the microscale physics at the moving contact line. "Tanner's law" for the axisymmetric case is

$$R(t) = c(t)^{1/10}, \qquad (6\text{--}73)$$

and this is supported by a number of experimental studies. The existence of a universal spreading law in this case is particularly surprising because it is known that there are microscale phenomena near the contact line, such as the presence of a thin "precursor" film that emanates from the edge of the spreading drop, that are apparently present in many but not all cases. Furthermore, the existence of a scaling relationship such as (6–73) is suggestive of the existence of a self-similar solution. This is also surprising for any $\theta_c \neq 0$, because the physics in these cases suggests the existence of a steady solution with $R(t) = \text{const}$. Without considering additional microscale physics, which may maintain the spreading (or "wetting") process beyond the steady-state solution, we are forced to either limit the analysis to $\theta_c = 0$ or to assume that the self-similar analysis is valid for only intermediate times.

In the following discussion, we temporarily ignore the constraint of a fixed contact angle and assume that the drop has achieved a sufficiently flattened configuration that the thin-film assumptions of this section apply. We will try to determine the rate of spreading, much as in the preceding section, but in this case, we will assume that the Bond number is very small so that gravitational forces can be neglected altogether as a first approximation. In this case, the pressure distribution inside the drop, which is given by Eq. (6–23), can be approximated in the form

$$p^{(0)} = p^* - \left(\frac{\varepsilon^3 \sigma}{\mu u_c}\right) \nabla_s^2 h \qquad (6\text{--}74)$$

Hence, because we have assumed that $\alpha = 0$, the governing equation, (6–31), for the shape function h becomes

$$3\frac{\partial h}{\partial t} = -\left(\frac{\varepsilon^3 \sigma}{\mu u_c}\right) \nabla_s \cdot \left[h^3 \left(\nabla_s \nabla_s^2 h\right)\right]. \qquad (6\text{--}75)$$

The Thin-Gap Approximation – Films with a Free Surface

In this case, the characteristic velocity is evidently

$$u_c = \frac{\varepsilon^3 \sigma}{\mu} = \frac{R_0^3 \sigma}{L_0^3 \mu}, \qquad (6\text{–}76)$$

where R_0 and L_0 have the same significance as in the preceding example. Hence, finally,

$$3\frac{\partial h}{\partial t} = -\nabla_s \cdot [h^3 (\nabla_s \nabla_s^2 h)]. \qquad (6\text{–}77)$$

This equation must be solved subject to the constraint, (6–64), of constant volume. We should also expect the solution to satisfy the condition, (6–63), of compact support, but we shall need to return to discuss this point in more detail toward the end of this section.

The first thing to notice about (6–77) is that it cannot be cast into the form of the nonlinear diffusion equation that was discussed in Section B. Hence, unlike the preceding case, here we cannot apply the solution of the diffusion equation from Subsection C.2. However, we can apply the same method of analysis in the hope of finding a self-similar form for the solution of the d-dimensional symmetric version of (6–77). We begin by adopting the notation of (6–37) and/or (6–49) to express (6–77) in the form

$$3\frac{\partial h}{\partial t} = -\frac{1}{r^{d-1}} \frac{\partial}{\partial r} \left\{ r^{d-1} h^3 \frac{\partial}{\partial r} \left[\frac{1}{r^{d-1}} \frac{\partial}{\partial r} \left(r^{d-1} \frac{\partial h}{\partial r} \right) \right] \right\}, \qquad (6\text{–}78)$$

which is to be solved subject again to the constraint of constant volume, (6–64). We seek a similarity solution in the form (6–40)

$$h(r, t) = A t^\alpha H(B r / t^\beta).$$

It was shown in the analysis of subsection B that the constraint of constant volume then requires that

$$\alpha = -\beta d$$

so that [see, e.g., (6–43)]

$$h(r, t) = A t^{-\beta d} H(\eta), \text{ where } \eta \equiv B r / t^\beta.$$

To this point, the analysis is identical to that of the preceding case. However, we now substitute the self-similar form for $h(r,t)$ into the governing equation, (6–78). The result is

$$\left(-\gamma d H - \gamma \eta \frac{dH}{d\eta} \right) \frac{A}{t^{\gamma d + 1}}$$
$$= -\frac{A^4 B^4}{t^{4\gamma(d+1)}} \left(\frac{1}{\eta^{d-1}} \frac{d}{d\eta} \left\{ \eta^{d-1} H^3 \frac{d}{d\eta} \left[\frac{1}{\eta^{d-1}} \frac{d}{d\eta} \left(\eta^{d-1} \frac{dH}{d\eta} \right) \right] \right\} \right). \qquad (6\text{–}79)$$

The condition for reduction of (6–78) to a self-similar form is thus that both sides of this equation should be dependent only on η. The most general way to satisfy this condition is for the power on t on the two sides of the equation to be equal:

$$\gamma d + 1 = 4\gamma(d + 1).$$

C. Films With a Free Surface – Spreading Films $\alpha = 0$

Hence,

$$\gamma = \frac{1}{3d+4}, \qquad (6\text{--}80)$$

and (6–79) is reduced to an ODE for the similarity function $H(\eta)$. In fact, we can simplify Eq. (6–79) significantly even after canceling the t dependence by noting that the remaining term on the left-hand side can be written in the form

$$\frac{\gamma A}{\eta^{d-1}} \frac{d}{d\eta}(\eta^d H).$$

Thus we can integrate both sides of (6–79) once with respect to η to obtain

$$H^2 \frac{d}{d\eta}\left[\frac{1}{\eta^{d-1}} \frac{d}{d\eta}\left(\eta^{d-1} \frac{dH}{d\eta}\right)\right] = -\frac{3\gamma}{A^3 B^4}\eta, \qquad (6\text{--}81)$$

which is a third-order ODE for the unknown function $H(\eta)$. The constraint of constant volume takes the form

$$(2\pi)^{d-1} \int_0^\infty H(\eta)\eta^{d-1} d\eta = B^d A^{-1} q.$$

We can select the constant B such that this condition becomes

$$\int_0^\infty H(\eta)\eta^{d-1} d\eta = 1. \qquad (6\text{--}82)$$

Then, without loss of generality, the constant A can be chosen such that $H(0) = 1$.

We expect the solution of (6–81) to be an even function of η. Hence, in addition to $H(0) = 1$, we expect $H_{\eta\eta}(0) = -\beta$, where β is a positive constant. Because Eq. (6–81) is highly nonlinear, and we cannot find any way to solve it analytically, we are forced to integrate it numerically. It turns out that there is a whole family of solutions for different values of β. However, in view of the fact that the similarity transformation that led to (6–81) is consistent with the experimentally established Tanner's law, a surprising result is that $H(\eta)$ is strictly positive for all η for all values of β. All of the family of solutions exhibits a minimum value of H at some value of $\eta = \eta_0$, but the minimum is always positive. Hence, though the shape function $H(\eta)$ between $\eta = 0$ and $\eta = \eta_0$ looks perfectly reasonable for a spreading drop, none of the solutions has compact support, and thus none actually corresponds to a drop of finite volume. Taken together, these facts suggest that though Eq. (6–81) describes the basic physics away from the point $\eta = \eta_0$, it must be essential to take additional physical effects, such as the presence of van der Waals forces, into account for $\eta \geq \eta_0$.

A very coherent summary of the additional work required for producing a uniformly valid solution of the problem is given in a recent publication by Brenner and Bertozzi,[8] which we briefly recount. There are two main points that should be mentioned. First, if it is assumed that the additional physics required is the inclusion of van der Waals forces, it can be shown that a rigorous asymptotic solution can be constructed[9] in which the similarity solution obtained here is matched to a solution describing a precursor film for $\eta \geq \eta_0$. The precursor film sets the microscopic length scale corresponding to $H(\eta_0)$, and thus the value of β, which in turn has an effect on the prefactor c of the scaling law for the rate of drop spreading [e.g., Eq. (6–73) or the equivalent for the spread of a 1D ridge of fluid by capillarity]. Second,

D. THE DYNAMICS OF A THIN FILM IN THE PRESENCE OF VAN DER WAALS FORCES

There are, of course, many problems involving the dynamics of a thin film of liquid with a free upper surface. We conclude this brief introduction to the topic by considering the dynamics of a film that is thin enough that nonhydrodynamic forces are important. Specifically, we consider a horizontal thin film that is of the order of several tens of nanometers or less across, which is sitting on top of a solid substrate. In this range of film dimensions, van der Waals forces will always be present, though forces of other origins may sometimes be dominant (e.g., hydrophobic forces, Columbic forces, etc.). In the present analysis, we consider the case of constant surface tension, with only a nonretarded London–van der Waals dispersion forces acting across the thin film.

The London–van der Waals force results from the induced dipoles between two separated molecules. Although calculations of the exact quantitative force between bodies requires a detailed understanding of the molecular environment, researchers have used a single parameter, Λ, the London–van der Waals constant, to describe the attraction between particles from induced molecular dipoles. Hamaker[10] used this approach to find the attractive force between two spheres and between a sphere and an infinite plane wall, all in a vacuum. By devising a rule for combining the Λ's for the various materials, Hamaker also solved the problem for bodies separated by a continuous medium. These solutions apply integrated forms of the London–van der Waals point force to discrete bodies, but applications also exist for dispersion forces acting on infinite interfaces between two fluids. Maldarelli and Jain[11] provide a useful review of the governing equations for thin films in the presence of London–van der Waals forces.

Although the dispersion force is a body force in the sense that it depends on the entire volume of the materials involved, it appears only when these materials are separated by at least two interfaces (solid–fluid or fluid–fluid) with either a vacuum or a continuous medium in between, and is therefore often treated as an interfacial effect. Maldarelli and Jain outline two approaches to incorporating London–van der Waals effects in the equations governing the dynamics of a thin film. The first is the "body-force method" in which the dispersion forces are included in the DEs of motion as the gradient of the London–van der Waals potential. An alternative approach, and one that is more convenient for the present application, is termed the "disjoining pressure approach." In this treatment, an additional term, known as the disjoining pressure, is added to the normal-stress boundary condition to account for the London–van der Waals attraction (or repulsion) across the intervening thin film. There is no added force from the interaction of two materials separated by a single interface/boundary because all excess thermodynamic properties between two materials in contact are included in the interfacial tension.

Other than the addition of the disjoining pressure, the derivation of the governing equations is identical to that in Subsection A of the present chapter. Hence the governing equation for the shape function h is given by (6–31) with $\alpha = 0$,

$$3\frac{\partial h}{\partial t} = \nabla_s \cdot \left[h^3 \nabla_s p^{(0)}\right]. \tag{6–83}$$

D. The Dynamics of a Thin Film in the Presence of van der Waals Forces

With the inclusion of the disjoining pressure,

$$\Pi(h) \equiv \frac{A}{6\pi h^3}, \qquad (6\text{--}84)$$

the dimensionless pressure function becomes

$$p^{(0)} = p^* + \left(-\frac{\varepsilon^3 \ell_c^2 \rho g}{\mu u_c}\right)(z - h) - \left(\frac{\varepsilon^3 \sigma}{\mu u_c}\right)\nabla_s^2 h + \left(\frac{A}{6\pi \mu u_c \ell_c^2 \varepsilon}\right)\frac{1}{h^3}. \qquad (6\text{--}85)$$

The dimensional coefficient A that appears in (6–84) is known as the Hamaker constant. Its value depends on the materials involved, but a typical magnitude is 10^{-20} to 10^{-19} J. Generally A is positive, which corresponds to a *positive* disjoining pressure and attraction between the interface and the solid substrate. However, in some circumstances, $A < 0$, and the surfaces repel.

We could apply the equations as written to again consider the spreading problem of the preceding two subsections, now including the effect of van der Waals interactions between the interface of the drop and the wall. However, as indicated earlier, the analysis in this section is of the dynamics of a thin film that is initially of uniform depth h_0 and infinite in lateral extent. The interesting and important problem is to determine under what circumstances the film ruptures, by which we mean that its thickness goes at least locally to zero. An understanding of the conditions for rupture of a liquid film on a solid wall is important in such technologies as the various types of coating processes. It is also qualitatively similar to the rupture process that occurs in the thin film separating two drops during coalescence. Indeed, a similar, if somewhat more complex, analysis to the one we discuss here, has already been developed for that class of problems. In the present section we consider two problems. First, we briefly consider the linear stability of a thin film of uniform thickness subject to infinitesimal perturbations of shape. Second, in circumstances in which the film is unstable, we consider the final stages of the film-rupture process, far into the nonlinear regime, where we will find that there is a self-similar form for the local geometry of the film just before rupture. Between the initial growth of infinitesimal perturbations and the final film-collapse process, we would need to use numerical methods to analyze (6–83) for the film shape versus time. This has been done by Zhang and Lister,[12] following earlier work of other investigators.

One detail of the governing equations, (6–83) and (6–85), may merit some additional discussion before we consider the details of the analysis, and that is the scaling parameters that appear by means of nondimensionalization. In this case, the obvious physical length scale is the unperturbed height of the film, h_0. In the linear stability analysis, we shall consider the dynamics of oscillatory perturbations characterized by a wavelength λ, which thus defines the length scale $\ell_c \equiv \lambda/2\pi$ in the lateral (horizontal) direction. For Eqs. (6–83) and (6–85) to provide a valid approximation of the motion, this wavelength must be large compared with the undisturbed film thickness h_0, i.e.,

$$\varepsilon \equiv \frac{h_0}{\lambda} \ll 1. \qquad (6\text{--}86)$$

There are three dimensionless parameters in addition to ε. These are a parameter that Oron et al.[1] have called the "gravity number,"

$$\overline{G} = \varepsilon G, \quad \text{where } G \equiv \frac{\rho g h_0^2}{\mu u_c}, \qquad (6\text{--}87)$$

a rescaled capillary number

$$\overline{C} = C\varepsilon^{-3}, \quad \text{where } C \equiv \frac{\mu u_c}{\sigma}, \qquad (6\text{--}88)$$

The Thin-Gap Approximation – Films with a Free Surface

and the dimensionless Hamaker constant

$$\overline{A} = \frac{A\varepsilon}{6\pi\mu u_c h_0^2}. \tag{6-89}$$

In writing these definitions, we have used the assumed scaling $\ell_c = h_0/\varepsilon$ to express everything in terms of the physically obvious length scale h_0. The appropriate choice for the velocity scale will depend on which (if any) of the terms in (6–85) is dominant.

1. Linear Stability

In the final section of the preceding chapter, we considered a number of problems in the general area of bubble dynamics that were presented in the framework of analyzing the stability of some particular solution to a perturbation. In the present section, we build on that initial foray into stability theory to discuss the stability of the flat film previously described to infinitesimal perturbations of shape.

Equations (6–83) and (6–85) govern the dynamics of any time-dependent changes in the shape of the thin film, subject of course to the long-wavelength approximation, (6–86). Hence, if we wish to study the stability of the thin film to some perturbation of shape from the uniform h_0, we consider some specified initial shape that we can represent symbolically (in dimensional terms) as

$$h'(\mathbf{x}_s, t) = h_0 + \hat{h}'(\mathbf{x}_s, t). \tag{6-90}$$

We utilize primes to indicate that the corresponding variable is dimensional, and, of course, h_0 is the dimensional unperturbed film thickness. Equations (6–83) and (6–85) are written in dimensionless form with the film thickness h scaled with respect to h_0. Hence the appropriate dimensionless form of (6–90) is

$$h(\mathbf{x}_s, t) = 1 + \hat{h}(\mathbf{x}_s, t). \tag{6-91}$$

Therefore, to study the stability problem, we must solve the governing equations, (6–83) and (6–85), with (6–91) as the initial condition, to determine whether $\hat{h}$ grows or decays in time. The result depends, of course, on the values of the parameters (6–87)–(6–89) and also on the form (and, generally, also the magnitude) of disturbance function. If the disturbance decays so that $h \to 1$ as $t \to \infty$, the system is said to be stable to that particular disturbance.

In a *linear* stability analysis, we consider the stability only to infinitesimal disturbances,

$$\hat{h} \ll 1. \tag{6-92}$$

In this case, we must consider the form of the disturbance in the shape of the film to be arbitrary. Infinitesimal disturbances will be present in any real system at all times, as it is impossible to shield a real system from disturbances that have an arbitrarily small magnitude. At the same time, disturbances of arbitrarily small magnitude will also be present in every conceivable form. However, we shall see that it is not necessary to directly calculate the dynamics of a disturbance with an arbitrary form for the disturbance function $\hat{h}$. Because the condition (6–92) is satisfied, when we substitute (6–91) into (6–83) and (6–85) to obtain a governing equation for the dynamics of $\hat{h}$, we find that we can linearize the equation by neglecting all terms that involve $(\hat{h})^m$, with $m > 1$. Specifically, the governing, linearized equation obtained from (6–83) and (6–85) is

$$3\frac{\partial \hat{h}}{\partial t} = \overline{G}\nabla_s^2 \hat{h} - \overline{C}^{-1}\nabla_s \cdot \left[\nabla_s\left(\nabla_s^2 \hat{h}\right)\right] - 3\overline{A}\nabla_s^2 \hat{h}, \tag{6-93}$$

or, in the case of a 1D disturbance, $\hat{h}(x, t)$,

$$3\frac{\partial \hat{h}}{\partial t} = \overline{G}\frac{\partial^2 \hat{h}}{\partial x^2} - \overline{C}^{-1}\frac{\partial^4 \hat{h}}{\partial x^4} - 3\overline{A}\frac{\partial^2 \hat{h}}{\partial x^2}. \tag{6-94}$$

D. The Dynamics of a Thin Film in the Presence of van der Waals Forces

Now, in view of the linearity of these equations, a superposition of solutions is also a solution, and we can thus represent any *arbitrary* disturbance in terms of a superposition of normal or Fourier modes. In terms of dimensional variables, a typical such mode can be represented in the form

$$\hat{h}' = \beta' \exp[i(k_x x' + k_y y') + s't'], \tag{6–95}$$

with the 1D case represented by setting k_y equal to zero. Here β' is a (arbitrary) constant that specifies the initial amplitude of this mode, k_x and k_y are the wave numbers (inversely related to the wavelengths of this mode in the x and y directions), and s' is the growth-rate factor.

In the present analysis, we have derived the governing equations, (6–93) or (6–94), in terms of *dimensionless* variables,

$$t \equiv \frac{t'}{\ell_c/u_c}, \quad (x, y) \equiv \left(\frac{x'}{\ell_c}, \frac{y'}{\ell_c}\right),$$

where we noted earlier that an appropriate choice for the characteristic length scale in the plane of the thin film is just the wavelength of the disturbance, i.e.,

$$\ell_c \equiv \lambda \equiv k^{-1}.$$

If the wavelengths of the disturbance components in the x and y directions are equal, then

$$k_x = k_y = k.$$

and we can write the dimensionless version of (6–95) in the form

$$\hat{h} = \beta \exp[i(x + y) + st]. \tag{6–96}$$

If k_x and k_y are not equal, we can use either to define the characteristic length scale, and then the ratio would appear in (6–96) as a parameter. For present purposes, it is sufficient to simply consider the 1D case,

$$\hat{h} = \beta \exp(ix + st), \tag{6–97}$$

with the governing equation, (6–94).

By substituting (6–97) into (6–94), we obtain the characteristic equation

$$3s = -\overline{G} - \overline{C}^{-1} + 3\overline{A}, \tag{6–98}$$

from which we can calculate the dimensionless growth-rate factor s corresponding to the disturbance with dimensional wave number k. In the present case, it can be seen from (6–98) that the growth-rate factor s is *real* in all cases. Hence the thin film will be stable to a disturbance of wave number k if $s < 0$, unstable if $s > 0$, and neutrally stable for $s = 0$.

The two simplest cases are those in which $\overline{A}$ and $\overline{C}^{-1}$ are both set equal to zero, so that the only physical effect remaining is the gravitational force, and $\overline{A} = \overline{G} = 0$ so that the only remaining physical effect is the capillary contribution to the motion of the film. In both of these cases, $s < 0$. Because the wavelength that was used for scaling was arbitrary, the film is therefore stable to linear perturbations of all wavelengths. Both the gravitational and capillary effects produce film leveling by physical mechanisms that should be clear from the discussions of Chap. 2.

One exception to the stability in the presence of gravitational and/or capillary effects occurs if the fluid layer is on the underside of the solid surface, rather than on top of it. In that case, the problem is identical except that the gravitational acceleration vector is pointing away from the surface rather than toward it. We can accommodate this in the theory by simply substituting $-g$ for g so that $\overline{G}$ is negative rather than positive. In this case, $s > 0$, provided $\overline{A}$ and $\overline{C}^{-1}$ are both zero, and the film would be unstable to disturbances of any wavelength (i.e., any k). According to the linear stability theory, in this case an initial perturbation of shape will grow exponentially with time at a rate that increases with increasing wave number. Of course, eventually the disturbance will reach a magnitude where the linearization leading to (6–93) or (6–94) is no longer valid, and then the dynamics will have to be described by the full nonlinear equations, (6–83) and (6–85). It will be noted that large values of k correspond to disturbances of small wavelength, and we would expect surface-tension effects to play some role, independent of how small the parameter $\overline{C}^{-1}$ may be.

A more realistic situation is thus one for a sufficiently thick film that van der Waals effects can still be neglected but *both* gravity and capillary effects are present. In this case, $s < 0$ still, if the film is on top of the solid substrate, and in fact the disturbance decays more rapidly than if either of these effects is absent. In fact, according to our linear theory, the leveling rate is proportional to

$$\sim \exp(s't') = \exp\left[-\frac{h_0^3 k^2}{3\mu}(\rho g + k^2 \sigma)t'\right].$$

However, the inverse problem, with the film on the underside of the solid substrate, now features a competition between the stabilizing (leveling) effect of capillary forces against the *destabilizing* effect of gravity. This corresponds to a well-known stability problem, called Rayleigh–Taylor instability, applied to the thin film. In this case,

$$s = \frac{1}{3}\left(|\overline{G}| - \overline{C}^{-1}\right), \quad (6\text{–}99)$$

and the system is linearly *unstable* if

$$|\overline{G}| > \overline{C}^{-1}. \quad (6\text{–}100)$$

If we substitute for $\overline{G}$ and $\overline{C}^{-1}$ from the definitions (6–87) and (6–88), we can show that this condition implies that

$$\boxed{k^2 < k_c^2 \equiv \frac{\rho|g|}{\sigma}.} \quad (6\text{–}101)$$

It is convenient to express this same condition in terms of a *dimensionless* wave number, scaled with respect to the film thickness, i.e., $\overline{k} \equiv kh_0$, namely

$$\boxed{\overline{k}^2 < \overline{k}_c^2 \equiv \frac{\rho|g|h_0^2}{\sigma} \equiv Bo.} \quad (6\text{–}102)$$

The so-called *critical* wave number k_c, equal to the square root of the Bond number, thus separates the long-wavelength disturbances that are unstable from the shorter-wavelength disturbances that are stable because of the influence of capillary effects. We may note that the condition (6–102) shows that the thin-film analysis is valid provided $Bo \ll O(1)$, as this is the condition for the wavelength of the disturbance to be large compared with the film thickness.

Finally, we can consider the problem for a very thin film, $O(100 \text{ Å})$ in width, where van der Waals forces are important, and the characteristic equation is given by (3–220). In this case, if $\overline{A}$ is negative, so that the van der Waals forces are repulsive, then $s < 0$ and the film

D. The Dynamics of a Thin Film in the Presence of van der Waals Forces

is stable to linear perturbations of shape of any wavelength. On the other hand, when $\overline{A}$ is positive and the van der Waals forces are attractive, the film is unstable when

$$3\overline{A} > \overline{G} + \overline{C}^{-1}. \tag{6–103}$$

If we express this condition in dimensional form by again using the definitions of $\overline{A}, \overline{B},$ and $\overline{G}$ we find that it leads to a sufficient condition for instability:

$$\frac{A_H}{2\pi h_0^2} > \rho g h_0^2 + \varepsilon^2 \sigma; \quad \varepsilon = h_0 k. \tag{6–104}$$

In a film of infinite lateral extent, k can range from 0 to ∞, so a necessary condition for instability is that $A_H > 2\pi \rho g h_0^4$. Since all wave numbers are available in a film of infinite extent, we see that this analysis predicts that the thin film will always be unstable, even with the stabilizing influence of surface tension, to disturbances of sufficiently large wavelength when van der Waals forces are present. Similarly, the Rayleigh–Taylor instability that occurs when the film is on the underside of the solid surface will always appear in a film of infinite extent. In reality, of course, the thin film will always be bounded, as by the walls of a container or by the finite extent of the solid substrate. Hence the maximum wavelength of the perturbation of shape is limited to the lateral width, say W, of the film. This corresponds to a minimum possible wave number

$$k_{\min} = W^{-1}.$$

Now it can be seen from (6–104) that this minimum wavenumber must be smaller than a critical value k_c for the system to be unstable

$$\overline{k}_{\min}^2 < \overline{k}_c^2 \equiv \frac{1}{\sigma}\left(\frac{A_h}{2\pi h_0^2} - \rho g h_0^2\right). \tag{6–105}$$

Here, for convenience, we have used values of the wavenumber that are scaled by the undisturbed film width, $\overline{k}_{\min} = h_0 W^{-1}$ and $\overline{k}_c = h_0 k_c$. One way to interpret this condition is that the film height must be less than a critical value for instability

$$h_0^4 < \frac{A}{2\pi(\sigma W^{-2} + \rho g)}. \tag{6–106}$$

According to this result, for any given container dimension, W, there is a film thickness below which the film is unstable. Of course, this latter result is only qualitatively accurate. If we really have a film in a container with bounding lateral boundaries, we must satisfy boundary conditions there as well as at the bottom and top of the film, and we should expect at least quantitative differences from (6–106).

Finally, it may be noted that the viscosity of the thin film has no role in determining stability. Changes in the viscosity are only reflected in the growth (or decay) rate of the disturbance. This is reflected in the dimensional growth rate factor

$$3\mu s' = -\frac{\overline{k}^2}{h_0}\left(\rho g h_0^2 + \sigma \overline{k}^2\right) + \frac{A}{2\pi}\frac{\overline{k}^2}{h_0^3}.$$

2. Similarity Solutions for Film Rupture

The instability in the presence of London–van der Waals attractive forces that was studied in the preceding subsection is indicative of the fact that small perturbations in the film thickness will grow, and this process will eventually cause the film to rupture. Although the growth of a localized disturbance in the film thickness will lead to increased curvature and increased displacement of the film, and thus to an increase in the magnitude of the gravitational and capillary restoring forces, the very strong h^{-3} increase in the van der Waals force is sufficient to drive any existing thin spot to rupture. An important feature of this problem and of a number of other free-surface problems (such as the breakup of a liquid thread by capillary forces), is that the dynamics of the film sufficiently near (in space and time) to the point of rupture is determined solely by the approach to the topological singularity that is rupture. This means that the last phases of the approach to rupture are independent of initial conditions and of the "far-field" conditions away from the immediate point of rupture. In view of what we have already learned about the nature of similarity solutions, it should not therefore be a surprise that the problem exhibits a self-similar solution in this regime.

The overall dynamics of a thin film that is subjected to an initial perturbation in its thickness with a wavelength that is sufficiently long to be unstable according to the analysis of the preceding section is governed by the full nonlinear set of equations, (6–83) and (6–85), for h. It is evident from these equations that, as the film begins to thin so that h decreases and the curvature increases, the gravitational restoring force is quickly dominated by the capillary contribution. In fact, it appears as though the capillary term, in turn, should be dominated by the van der Waals term, thus suggesting that the film dynamics near rupture should be determined completely by a balance of van der Waals forces and viscous dissipation. However, Zhang and Lister[12] have recently carried out a careful numerical analysis of the solutions to (6–83) and (6–85) with $\overline{G} = 0$, which shows that surface tension, van der Waals forces, and viscous dissipation all remain equally important near rupture. Zhang and Lister used the axisymmetric version of the combined equations (6–83) and (6–85) with $\overline{G} = 0$ to study the evolution of an initial disturbance of the form

$$h(r, 0) = \frac{10}{9}\left[1 - \frac{1}{10}\cos\left(\frac{2\pi r}{\lambda}\right)\right].$$

The simulation was carried out on the fixed interval $0 \le r \le \lambda/4$, with boundary conditions $\partial h/\partial r = \partial^3 h/\partial r^3 = 0$ at $r = 0$ and $r = \lambda/4$. The wavelength λ was chosen to be sufficiently long for linear instability, according to (6–104), with the gravitational term neglected. The conditions at $r = 0$ are required for the solution to be regular at that point. The forms of the initial profile and the boundary conditions at $r = \lambda/4$ were chosen for convenience. In any case, it was found that the behavior of the solution near rupture does not depend on these latter choices. A series of the interface profiles, identified by the minimum film thickness, is shown in Fig. 6–6. As expected, there is localized collapse of the film around the point $r = 0$, where the film is initially thinnest. Furthermore, the minimum film thickness and the horizontal length scale of the collapsing region were found to decrease algebraically in time as $(t_R - t)^{-1/5}$ and $(t_R - t)^{-2/5}$, respectively. Here t_R represents the time at actual rupture. This type of algebraic scaling is reminiscent of self-similar behavior, as previously suggested.

An interesting question is whether we can demonstrate that such a self-similar behavior should exist as the film-rupture singularity is approached. We consider the combined equations (6–83) and (6–85) with the gravitational term neglected, as already explained. Using the notation developed earlier in this section, we can write this equation in the form

$$\boxed{3\frac{\partial h}{\partial t} = \nabla_s \cdot \left[h^3 \nabla_s \left(-\overline{C}^{-1}\nabla_s^2 h + \frac{\overline{A}}{h^3}\right)\right].} \qquad (6\text{–}107)$$

D. The Dynamics of a Thin Film in the Presence of van der Waals Forces

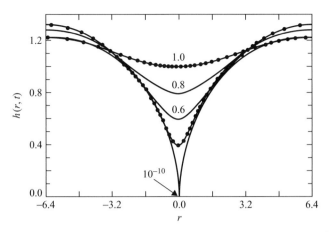

Figure 6–6. Interface profiles at $h_{\min}(t) = 1.0, 0.8, 0.6, 0.4, 0.2$, and 10^{-10} (reproduced from Fig. 1, Ref. 12).

Starting from this equation, we can utilize what we have learned from the numerics and seek a transformation to a self-similar form. The numerical result suggests that we seek a solution of the form

$$h(r, t) = (t_R - t)^{1/5} H(\eta), \quad \text{with } \eta = r/(t_R - t)^{2/5}.$$

However, in case we had not carried out the numerical analysis, we might hope that the form of the proposed self-similar solution might be derivable directly from usual process of reducing the governing equation, (6–107), to a self-similar form, Hence we seek a solution of the same type but with the coefficients unknown, i.e.,

$$\boxed{h(r, t) = (t_R - t)^{\beta} H(\eta) \text{ with } \eta = r/(t_R - t)^{\alpha}.} \quad (6\text{–}108)$$

By substituting (6–108) into (6–107), it is straightforward to show that we can reduce the problem to an ODE for the function $H(\eta)$ and that this process leads to the unique set of values $\beta = 1/5$ and $\alpha = 2/5$ that was suggested by the numerical study by Zhang and Lister of the solutions of (6–107).

Before the details are carried out, however, it is useful to revisit the scaling that led to the nondimensionalized equation, (6–107), because we can show that a slight modification allows us to eliminate the coefficients $\overline{C}^{-1}$ and $\overline{A}$. This involves just two steps. First, we specify the length scale ℓ_c. An appropriate and convenient choice is the wavelength of the disturbance separating stable and unstable solutions in the linear stability analysis. Going back to (6–103), we see that the critical point (i.e., the point where $s = 0$) occurs when

$$3\overline{A} = \overline{C}^{-1}. \quad (6\text{–}109)$$

Substituting for $\overline{A}$ and $\overline{C}^{-1}$ by using their definitions in (6–88) and (6–89), and $\varepsilon = h_0/\lambda$, we find

$$\ell_c \equiv \lambda = h_0^2 (2\pi\sigma/A)^{1/2}. \quad (6\text{–}110)$$

The second step is to change the nondimensionalization of time. We can first convert back to dimensional time t' by using the definition that was built into (6–107),

$$t \equiv \frac{t'}{\ell_c/u_c} = \frac{t'}{\lambda/u_c}, \quad (6\text{–}111)$$

The Thin-Gap Approximation – Films with a Free Surface

and then converting to a new dimensionless time,

$$\hat{t} \equiv \frac{t'}{t_c}, \tag{6–112}$$

with a scaling t_c that remains to be determined. With these changes, (6–107) becomes

$$3\frac{\partial h}{\partial \hat{t}} = \nabla_s \cdot \left[h^3 \nabla_s \left(-\frac{h_0^3 \sigma t_c}{\mu \lambda^4} \nabla_s^2 h + \frac{A t_c}{6\pi \mu \lambda^2 h_0} \frac{1}{h^3} \right) \right]. \tag{6–113}$$

Substituting for λ from (6–110)

$$3\frac{\partial h}{\partial \hat{t}} = \left(\frac{A^2 t_c}{4\pi^2 \mu h_0^5 \sigma} \right) \nabla_s \cdot \left[h^3 \nabla_s \left(-\nabla_s^2 h + \frac{1}{3h^3} \right) \right]. \tag{6–114}$$

Evidently, a convenient choice for t_c is

$$t_c \equiv \frac{4\pi^2 \sigma \mu h_0^5}{A^2}, \tag{6–115}$$

and, in this case, the governing equation for h is reduced to a parameter-free form

$$\boxed{3\frac{\partial h}{\partial \hat{t}} = \nabla_s \cdot \left[h^3 \nabla_s \left(-\nabla_s^2 h + \frac{1}{3h^3} \right) \right].} \tag{6–116}$$

We now seek a self-similar form for the solution by substituting (6–108) into this parameter free form of (6–107). The advantage is that, if we are successful, the governing equation for $H(\eta)$ will be parameter free, and thus can be solved once and for all. The left-hand side of (6–116) takes the form

$$3\frac{\partial h}{\partial \hat{t}} = 3(\hat{t}_R - \hat{t})^{\beta-1} \left[-\beta H + \alpha \eta \frac{dH}{d\eta} \right],$$

and the right-hand side is

$$\nabla_s \cdot [\cdots] = -(\hat{t}_R - \hat{t})^{4\beta - 4\alpha} \frac{d}{d\eta} \left(H^3 \frac{d^3 H}{d\eta^3} \right) - 3(\hat{t}_R - \hat{t})^{-2\alpha} \frac{d}{d\eta} \left(H \frac{dH}{d\eta} \right).$$

The condition for existence of a similarity transformation is that the coefficients of the equation for H must be independent of $\hat{t}$. This condition leads to a pair of equations for α and β,

$$3\beta - 4\alpha + 1 = 0; \quad -\beta - 2\alpha + 1 = 0,$$

which have the unique solution

$$\beta = 1/5, \quad \alpha = 2/5. \tag{6–117}$$

The resulting equation for $H(\eta)$ is

$$\boxed{2\eta \frac{dH}{d\eta} - H = -\frac{5}{3} \frac{d}{d\eta} \left(H^3 \frac{d^3 H}{d\eta^3} \right) - 5 \frac{d}{d\eta} \left(H \frac{dH}{d\eta} \right).} \tag{6–118}$$

Therefore we see that a self-similar form of the governing DE can be obtained. Furthermore, the rate of approach to the rupture singularity, and the rate at which the horizontal scale of the depression in film thickness varies with time to the rupture event, are completely determined by the local balance of capillary and van der Waals forces with viscous dissipation that is inherent in this equation. In particular, those features of the rupture dynamics are independent of initial conditions or of the film dynamics away from the rupture point.

E. Shallow-Cavity Flows

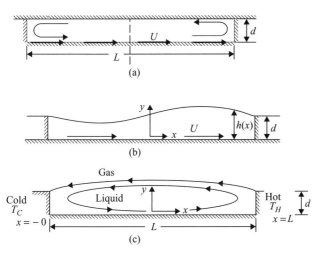

Figure 6–7. Three configurations for the shallow-cavity problem: (a) Four isothermal solid walls with motion driven by tangential motion of the lower wall; (b) the same problem as (a) except, in this case, the upper surface is an interface that may deform because of the flow; (c) the configuration is the same as (b), except, in this case, the lower wall is stationary and the motion in the cavity is assumed to be driven by Marangoni stresses caused by nonuniform interface temperature that is due to the fact that the end walls are at different temperatures.

E. SHALLOW-CAVITY FLOWS

A final class of thin-film flows that we consider in this chapter is that of motions in shallow cavities. In these problems, the fluid is confined to a long thin region bounded below by a horizontal plane boundary of length L that may either be a solid wall or a symmetry plane, and at the two ends by solid vertical boundaries of height $d \ll L$. In the majority of problems of technological interest, the upper boundary is an interface that deforms as a result of the fluid motion within the cavity, but we also consider one problem in which this boundary is assumed to be a solid wall. The motion within the cavity may either be due to imposed motion of the horizontal lower boundary, or again more frequently in problems of technological interest, due to a temperature gradient from one end of the cavity to the other that drives flow either because of buoyancy effects or Marangoni stresses associated with the temperature gradient at the interface, or some combination of these effects.

Three examples of shallow-cavity flows that we consider in this section are sketched in Fig. 6–7. At the top is the case in which all four boundaries are solid walls, the fluid is assumed to be isothermal, and the motion is driven by tangential motion of the lower horizontal boundary. In the middle, a generalization of this problem is sketched in which the fluid is still assumed to be isothermal and driven by motion of the lower horizontal boundary, but the upper boundary is an interface with air that can *deform* in response to the flow within the cavity. Finally, the lower sketch shows the case in which fluid in the shallow cavity is assumed to have an imposed horizontal temperature gradient, produced by holding the end walls at different, constant temperatures, and the motion is then driven by Marangoni stresses on the upper interface. In the latter case, there will also be density gradients that can produce motion that is due to natural convection, but this contribution is neglected here (however, see Problem 6–13.)

The last of the three problems in Fig. 6–7 is qualitatively related to the thermocapillary flows that are important in the processing of single crystals for microelectronics applications. A typical configuration is sketched in Fig. 6–8, in which a cylindrical solid passes through a heating coil (a furnace), is melted, and then resolidifies into a single crystal of high quality.

The Thin-Gap Approximation – Films with a Free Surface

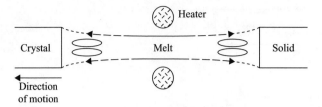

Figure 6–8. A sketch of a crystal growth process from the microelectronics industry.

The neglect of gravitational effects (i.e., buoyancy) is of direct relevance to processing in the microgravity environment of space. Of course, the problem sketched in Fig. 6–7 has a 2D rather than cylindrical geometry. A closer analog of the materials processing configuration would replace the lower solid boundary in Fig. 6–7 with a horizontal symmetry plane, but the analysis of this case is left as a problem (Problem 6–18).

1. The Horizontal, Enclosed Shallow Cavity

We begin by considering the flow within a shallow, horizontal ($\alpha = 0$) cavity as sketched in Fig. 6–7a. We assume that the ratio, d/L, is asymptotically small. We seek only the leading-order approximation within the shallow cavity. Hence the starting point for analysis is the thin-film equations, (6–1)–(6–3). In the present case of a 2D cavity, we can use a Cartesian coordinate system, and, for the present problem, we assume that the fluid is isothermal, so that the body-force term in (6–3) can be incorporated into the dynamic pressure, and hence plays no role in the fluid's motion. In this case, the governing equations become

$$\frac{\partial^2 u^{(0)}}{\partial z^2} - \frac{\partial p^{(0)}}{\partial x} = 0, \qquad (6\text{--}119)$$

$$\frac{\partial u^{(0)}}{\partial x} + \frac{\partial w^{(0)}}{\partial z} = 0, \qquad (6\text{--}120)$$

$$\frac{\partial p^{(0)}}{\partial z} = 0. \qquad (6\text{--}121)$$

The characteristic scales that have been used to produce these nondimensionalized equations are

$$u_c^{(x)} = U, \quad \ell_c^x = L, \quad \ell_c^z = d, \qquad (6\text{--}122\text{a})$$

$$p_c = \frac{\mu U}{L} \frac{1}{\varepsilon^2}, \qquad (6\text{--}122\text{b})$$

with

$$\varepsilon \equiv \frac{d}{L} \ll 1.$$

The variables $u^{(0)}$, $w^{(0)}$, and $p^{(0)}$ are the first terms in an asymptotic expansion [see, e.g., 6–68] for $\varepsilon \to 0$ (with $\varepsilon^2 Re \ll 1$, where $Re = UL/\nu$).

The shallow-cavity *scaling* that produces (6–119)–(6–121) must, of course, break down in the neighborhood of the end walls. The ends of the cavity are impermeable, and thus the flow near the end walls must turn through 180° and return toward the opposite end. We have

E. Shallow-Cavity Flows

previously seen that the vertical (z) velocity component in the interior of a thin-film region, where $\ell_c^x = L$ and $\ell_c^z = d$, is smaller than the horizontal component by $0(\varepsilon)$, i.e.,

$$w_c = \varepsilon U$$

in the present case. However, this clearly does not apply near the end walls. There, because of the turning flow,

$$w_c = u_c = O(U), \qquad (6\text{–}123\text{a})$$

and it follows from this scaling that

$$\ell_c^x = \ell_c^z = d. \qquad (6\text{–}123\text{b})$$

The difference in scaling between the central core of the thin cavity (6–122) and the vicinity of the end walls (6–123) means that the asymptotic solution for $\varepsilon \ll 1$ is singular, and a different set of dimensionless equations and a different form for the asymptotic expansion for $\varepsilon \ll 1$ must be obtained in the end regions. The distinct expansions in the core and end regions are then required to match in the region of overlapping validity.

Although the existence of the end regions, with the essential physical role of turning the fluid through 180°, may make the present problem seem different from the problems we have considered in earlier sections of this and the preceding chapter, this is really not true. In fact, in those problems, the thin-film scaling must also break down at the ends of the thin gap. However, in those problems, we essentially ignored this fact, and, when necessary (as in calculating the pressure gradient along the film in lubrication problems), we simply extended the solution from the thin-film equations all the way to the ends of the gap. In lubrication problems, this is acceptable as an approximation, though it is valid only at the leading-order approximation for $\varepsilon \ll 1$. In the present case, it may not be so obvious that we can determine the solution in the central core without explicitly considering the turning flow at the ends. However, once again, we shall see that it is not necessary to consider the details of flow in the end regions in order to determine the leading-order approximation to the solution in the core. Nevertheless, the end regions do play a crucial role in that the impermeable end walls mean that the net flux of fluid through any vertical plane within the cavity must be identically equal to zero. We will return to the end regions shortly. First, let us consider the problem for the leading-order solution in the core.

We have already noted that the governing equations are (6–119)–(6–121). The boundary conditions at the top and bottom walls are

$$u^{(0)} = w^{(0)} = 0 \quad \text{at } z = 1, \qquad (6\text{–}124\text{a})$$

$$u^{(0)} = 1, \quad w^{(0)} = 0 \quad \text{at } z = 0. \qquad (6\text{–}124\text{b})$$

In addition, as previously noted, the solution must also satisfy the integral constraint,

$$\int_0^1 u^{(0)} dz = 0, \qquad (6\text{–}125)$$

at any position x within the shallow cavity. This latter condition is crucial to the solution in the core region.

The solution of (6–119)–(6–121) largely follows preceding sections. Equation (6–121) tells us that $p^{(0)}$ is a function of x only. Hence, integrating (6–119), we obtain

$$u^{(0)} = \frac{\partial p^{(0)}}{\partial x} \frac{z^2}{2} + a(x)z + b(x).$$

Because $u^{(0)} = 1$ at $z = 0$,

$$b(x) = 1,$$

and the condition $u^{(0)} = 0$ at $z = 1$ then gives

$$a(x) = -\frac{1}{2}\frac{\partial p^{(0)}}{\partial x} - 1,$$

so that

$$u^{(0)} = \frac{1}{2}\frac{\partial p^{(0)}}{\partial x}(z^2 - z) + (1 - z). \qquad (6\text{--}126)$$

Finally, this solution must satisfy the integral constraint, (6–125). The last term in (6–126) represents the motion from left to right in the cavity that is due to the motion of the bottom boundary. The condition of zero net volume flux then requires that there be a contribution to flow in the opposite direction, driven by the pressure gradient $\partial p^{(0)}/\partial x$. Applying (6–125) to the solution (6–126), we obtain

$$\frac{\partial p^{(0)}}{\partial x} = 6. \qquad (6\text{--}127)$$

The solution (6–126) with (6–127), is self-contained. Because of the integral constraint, (6–125), there is no need to consider $w^{(0)}$ in order to obtain an equation for $p^{(0)}$. Indeed, we see that $u^{(0)}$ is completely independent of x. This means that the flow in the core region is *unidirectional* at this leading order of approximation. The continuity equation, (6–120), is reduced to the form

$$\frac{\partial w^{(0)}}{\partial z} = 0,$$

and because $w^{(0)}$ is zero at both $z = 0$ and $z = 1$, it is obvious that

$$w^{(0)} \equiv 0. \qquad (6\text{--}128)$$

Although we have developed a complete solution, it may be confusing to the reader that we have deviated from what has heretofore been a nearly standardized procedure of deriving a "Reynolds equation" by substituting (6–126) into the continuity equation, (6–120), and using it to determine the pressure distribution in thin-film flows. An obvious question is whether this procedure would still work. Hence, instead of immediately applying the constraint (6–125) to the solution (6–126), we could proceed by first differentiating (6–126) with respect to x [if we follow this approach, we must initially assume that $\partial p^{(0)}/\partial x$ is a function of x], and then apply the continuity equation (6–120) plus boundary conditions on w_0 to obtain a Reynolds equation for $p^{(0)}$. Hence, following this procedure, we obtain

$$\frac{\partial w_0^{(0)}}{\partial z} = -\frac{\partial u^{(0)}}{\partial x} = -\frac{\partial^2 p^{(0)}}{\partial x^2}\left(\frac{z^2}{2} - \frac{z}{2}\right),$$

and thus

$$w^{(0)} = -\frac{\partial^2 p^{(0)}}{\partial x^2}\left(\frac{z^3}{6} - \frac{z^2}{4}\right) + c_1.$$

Applying the boundary conditions

$$w_0 = 0 \text{ at } z = 0, 1,$$

E. Shallow-Cavity Flows

we find
$$c_1 = 0,$$
and thus
$$\frac{\partial^2 p^{(0)}}{\partial x^2} = 0.$$

This is Reynolds equation for the present problem. Integrating it, we see that
$$\frac{\partial p^{(0)}}{\partial x} = \alpha = \text{const}.$$

This is, indeed, consistent with (6–127) (as it must be!), but in the absence of the integral constraint, (6–125), there is no way to determine α. Thus the analysis of *this paragraph* is superfluous. It does not hurt us to have done it, but it does not contribute any new insight or information for this particular problem.

At the leading order of approximation represented by (6–126)–(6–128), there is not a turning-flow contribution in the core region. All of this must occur in the end regions. Clearly (6–126) [and thus presumably also (6–127)] is not a uniformly valid first approximation to the solution of the full problem. In spite of the fact that it satisfies the zero-net-flux condition, (6–125), it does not satisfy the impermeability condition

$$u = 0 \text{ at } x = 0, 1 \qquad (6\text{–}129)$$

at the end walls. The flow in the core region is, however, fully determined at this level of approximation by the integral constraint, (6–125), and the dynamics of the turning flow in the end regions is "passive" in the sense that it alters the form of the *core*-region solution only at higher levels of approximation. The flow in the core region is driven by the motion of the lower wall. However, to generate the *reversed* flow in the upper part of the cavity that is necessary to satisfy the zero-flux condition, we *require* a positive pressure gradient that drives the fluid in the direction opposite to the motion of the moving boundary. As previously noted, this is the physics behind the result (6–127).

Before we conclude this section, it is worthwhile to consider the formulation of the problem for flow in the end regions. Instead of obtaining the nondimensionalization and governing equations by *rescaling* the core equations, as we have done in the previous examples of matched asymptotic expansions, here we can directly apply the scaling (6–123) in the form

$$\hat{x} = \frac{x'}{d}, \quad \hat{z} = \frac{z'}{d}, \quad \hat{u} = \frac{u'}{U}, \quad \hat{w} = \frac{w'}{U},$$
$$\hat{p} = p' \Big/ \frac{\mu U_c}{L} \frac{1}{\varepsilon^2}. \qquad (6\text{–}130)$$

The latter comes from the "matching requirement" that the magnitude of the pressure in the end regions must be comparable with that in the core. The definitions (6–130) lead to the following nondimensionalized form of the Navier–Stokes equations for the end regions:

$$\frac{\partial \hat{p}}{\partial \hat{x}} = \varepsilon(\nabla^2 \hat{u}) - \varepsilon^2 Re\left(\hat{u}\frac{\partial \hat{u}}{\partial \hat{x}} + \hat{w}\frac{\partial \hat{u}}{\partial \hat{z}}\right),$$

$$\frac{\partial \hat{p}}{\partial \hat{z}} = \varepsilon(\nabla^2 \hat{w}) - \varepsilon^2 Re\left(\hat{u}\frac{\partial \hat{w}}{\partial \hat{x}} + \hat{w}\frac{\partial \hat{v}}{\partial \hat{z}}\right), \qquad (6\text{–}131)$$

with
$$\frac{\partial \hat{u}}{\partial \hat{x}} + \frac{\partial \hat{w}}{\partial \hat{z}} = 0.$$

The Thin-Gap Approximation – Films with a Free Surface

These equations are to be solved for $\varepsilon \ll 1$ subject to the boundary conditions

$$\hat{u} = \hat{w} = 0 \quad \text{at } \hat{x} = 0 \quad \left(\text{or } \hat{x} = \frac{1}{\varepsilon} \text{ at the other end of the cavity}\right),$$
$$\hat{u} = \hat{w} = 0 \quad \text{at } \hat{z} = 1, \tag{6-132}$$
$$\hat{u} = 1, \ \hat{w} = 0 \quad \text{at } \hat{z} = 0,$$

plus matching with the core solution (here stated only for the end of the cavity at $\hat{x} = 0$):

$$\lim_{\hat{x} \gg 1} \hat{u} \Leftrightarrow \lim_{x \ll 1} u \quad \text{for } \varepsilon \to 0.$$

Because (6–126) and (6–128) show that the core solution takes the form,

$$u = u^{(0)} + o(1) = 6\left(\frac{z^2}{2} - \frac{z}{2}\right) + (1 - z) + o(1),$$

the matching condition becomes

$$\lim_{\hat{x} \gg 1} \hat{u} = 6\left(\frac{\hat{z}^2}{2} - \frac{\hat{z}}{2}\right) + (1 - \hat{z}) + o(1) \quad \text{as } \varepsilon \to 0. \tag{6-133}$$

If we assume an asymptotic expansion for the end regions of the form

$$\hat{u} = \hat{u}_0 + \varepsilon \hat{u}_1 + \cdots +,$$
$$\hat{w} = \hat{w}_0 + \varepsilon \hat{w}_1 + \cdots +, \tag{6-134}$$
$$\hat{p} = \hat{p}_0 + \varepsilon \hat{p}_1 + \cdots +,$$

we see from (6–131) that

$$\frac{\partial \hat{p}_0}{\partial \hat{x}} = \frac{\partial \hat{p}_0}{\partial \hat{z}} = 0,$$

and hence

$$\hat{p}_0 = \text{const}.$$

Thus the governing equations for $\hat{u}_0$ and $\hat{w}_0$ are

$$\nabla^2 \hat{u}_0 - \frac{\partial \hat{p}_1}{\partial \hat{x}} = 0,$$
$$\nabla^2 \hat{w}_0 - \frac{\partial \hat{p}_1}{\partial \hat{z}} = 0, \tag{6-135}$$
$$\frac{\partial \hat{u}_0}{\partial \hat{x}} + \frac{\partial \hat{w}_0}{\partial \hat{z}} = 0.$$

This set of linear equations can be solved, subject to (6–132) and (6–133) with $\hat{u}, \hat{w} \to \hat{u}_0, \hat{w}_0$, to determine the *turning flow* in the end regions. We note that

$$\lim_{\hat{x} \gg 1}(\hat{p}_0 + \varepsilon \hat{p}_1 + \cdots +) \Leftrightarrow \lim_{x \ll 1}\left[p^{(0)} + \cdots +\right] \quad \text{as } \varepsilon \to 0.$$

Now

$$p^{(0)} = \alpha + 6x$$

according to the core solution, where α is an indeterminate constant. Thus, expressing this result from the core solution in terms of the end-region variables, we see

$$\lim_{\hat{x} \gg 1} \hat{p}_0 + \varepsilon \hat{p}_1 + \cdots + \Leftrightarrow \alpha + 6\varepsilon \hat{x} + \cdots + \quad \text{as } \varepsilon \to 0.$$

E. Shallow-Cavity Flows

Hence
$$\lim_{\hat{x} \gg 1} \hat{p}_1 = 6\hat{x}. \tag{6-136}$$

This completes formulation of the problem for a first approximation to flow in the end regions. However, in spite of the fact that this problem can be solved analytically, we do not pursue it here, as it has only a higher-order effect on the velocity and pressure fields in the core, and we do not seek any higher-order approximation beyond the solution (6–126)–(6–128) in the core region.

2. The Horizontal Shallow Cavity with a Free Surface

A problem that is closely related to the *enclosed-cavity* problem of the preceding subsection is that sketched in Fig. 6–7(b). The primary difference is that the upper boundary is now assumed to be an interface that can deform as a consequence of the flow within the cavity. As a consequence, the flow will *not* be unidirectional in the core.

The scaling and governing equations for the leading-order approximation to the core solution for $\varepsilon \ll 1$ are identical to those of the previous section, namely, (6–119)–(6–121), with one notable exception. Because the boundaries in this case include an interface, there is no advantage in suppressing the body-force term by introduction of a dynamic pressure as we did in writing (6–119)–(6–121) from the original thin-film equations, (6–1)–(6–3). If we transform from pressure to dynamic pressure in the governing equations, we must do the same for the pressure that appears in the normal-stress boundery condition and then the body-force term reappears in the interface boundary conditions. Hence, because our shallow cavity is assumed to be horizontal (i.e., normal to **g**) we retain (6–119) and (6–120), but maintain (6–121) in the original form, (6–3), i.e.,

$$\frac{\partial p^{(0)}}{\partial z} = -\varepsilon^3 G, \quad \text{with } G \equiv \frac{L^2 \rho g}{\mu U}. \tag{6-137}$$

The boundary conditions at $z = 0$ are also unchanged and are given by (6–124b). Finally, because of the presence of the impermeable end walls, the zero-flux condition, (6–125), is preserved, though now in the slightly more general form

$$\boxed{\int_0^{h(x)} u^{(0)}(x, z)\, dz = 0,} \tag{6-138}$$

where
$$h(x) \equiv \frac{h'(x)}{d}$$

and h' is the dimensional height of the interface from the plane $z = 0$.

The primary new feature is that the boundary conditions (6–124a) are now replaced with boundary conditions for an interface. We assume that the fluid above the interface is air (or some other gas). Hence these boundary conditions can be adopted from Eqs. (6–9)–(6–21). The kinematic condition, (6–9), (for a steady interface shape) becomes,

$$\boxed{\left[w^{(0)} - \frac{dh}{dx} u^{(0)}\right][1 + O(\varepsilon^2)] = 0 \quad \text{at } z = h.} \tag{6-139}$$

The normal-stress condition, (6–20), is

$$\boxed{p^{(0)} - p^* = -\frac{\varepsilon^3}{Ca} \frac{d^2 h}{dx^2}[1 + O(\varepsilon^2)] + O(\varepsilon^2) = 0} \tag{6-140}$$

The Thin-Gap Approximation – Films with a Free Surface

at $z = h$, where $Ca \equiv \mu U/\sigma$. Finally, the zero-shear-stress condition, suitable for an isothermal and uncontaminated interface with negligible external shear ("wind") stress (i.e., $\nabla_s \sigma = \tau_s^* = 0$) becomes

$$\boxed{\frac{\partial u^{(0)}}{\partial z} + O(\varepsilon^2) = 0 \text{ at } z = h.} \qquad (6\text{–}141)$$

We assume, for simplicity, that the interface is pinned at the corners of the cavity, i.e.,

$$\boxed{h = 1 \text{ at } x = 0, 1,} \qquad (6\text{–}142)$$

and that the cavity is filled exactly to this level in the absence of flow. This determines the total volume of fluid in the shallow cavity, and because the fluid has been assumed to be incompressible, it follows that, whatever shape the interface may take because of deformation in the presence of flow, it must satisfy the additional constraint of constant liquid volume,

$$\boxed{\int_0^1 h(x)dx = 1.} \qquad (6\text{–}143)$$

a. Solution by means of the classical thin-film analysis A solution to this problem [namely, Eqs. (6–119)–(6–120), and (6–137); and the boundary and interface conditions (6–124b), (6–138)–(6–143)] can be obtained following the conventional thin-film analysis as outlined in earlier sections of this, and the preceding, chapter.

Thus we first integrate (6–137) to obtain

$$p^{(0)} = -\varepsilon^3 Gz + \hat{p}(x).$$

Then, applying the normal-stress condition (6–140), we obtain

$$\hat{p}(x) = \varepsilon^3 Gh + p^* - \frac{\varepsilon^3}{Ca}\frac{d^2 h}{dx^2},$$

so that

$$\boxed{p^{(0)} = \varepsilon^3 G(h - z) - \frac{\varepsilon^3}{Ca}\frac{d^2 h}{dx^2} + p^*.} \qquad (6\text{–}144)$$

Because it is clear from this result that $\partial p^{(0)}/\partial x$ is independent of z, we can integrate (6–119) to obtain

$$u^{(0)} = \frac{\partial p^{(0)}}{\partial x}\frac{z^2}{2} + a(x)z + b(x).$$

Applying the boundary condition $u_0 = 1$ at $z = 0$, plus (6–141), we find

$$\boxed{u^{(0)} = \frac{\partial p^{(0)}}{\partial x}\left(\frac{z^2}{2} - hz\right) + 1.} \qquad (6\text{–}145)$$

At this point, we could apply the continuity equation, (6–120), to obtain a version of the Reynolds equation for the pressure. However, as in the preceding subsection, we can

E. Shallow-Cavity Flows

also apply the zero-flux condition, (6–138), which requires that the pressure gradient take the value

$$\boxed{\frac{\partial p^{(0)}}{\partial x} = \frac{3}{h^2}} \qquad (6\text{–}146)$$

at any position x. The reader may wish to demonstrate that application of the continuity equation eventually leads to this same result (see, e.g., Problem 6–10). Finally, combining (6–144) in the differentiated form,

$$\frac{\partial p^{(0)}}{\partial x} = \varepsilon^3 G \frac{dh}{dx} - \frac{\varepsilon^3}{Ca} \frac{d^3 h}{dx^3}, \qquad (6\text{–}147)$$

with (6–146), we obtain a DE for the interface shape function:

$$\boxed{-\frac{\varepsilon^3}{Ca}\frac{d^3 h}{dx^3} + G\varepsilon^3 \frac{dh}{dx} = \frac{3}{h^3}.} \qquad (6\text{–}148)$$

Equation (6–148), plus the boundary conditions (6–142) and the integral constraint (6–143), is *sufficient* to determine $h(x)$. We should note that we do not necessarily expect Eq. (6–148) to hold all the way to the end walls at $x = 0$ and $x = 1$, for it was derived by means of the governing equation, (6–119), (6–120) and (6–137), and these are valid only for the *core* region of the shallow cavity. Nevertheless, we will at least temporarily ignore this fact and integrate (6–148) over the whole domain, with the promise to return to this issue later. Qualitatively, we can see that the interface deformation is determined by a *balance* between the nonuniform pressure associated with the flow in the cavity, e.g., Eq. (6–145), which tends to deform the interface, and the effects of capillary and gravitational forces, both of which tend to maintain the interface in its flat, undeformed state, i.e., $h \equiv 1$.

The governing Eq., (6–148), is highly nonlinear. Although it can be solved numerically, we restrict our considerations to deformations that are a small perturbation from the undeformed state, namely,

$$h = 1 + h_1 \delta + \cdots +, \qquad (6\text{–}149)$$

where δ is a small parameter. In this case, analytic results can be easily derived. Substituting this expression for h into Eq. (6–148), we obtain

$$-\frac{\varepsilon^3}{Ca}\delta \frac{d^3 h_1}{dx^3} + G\varepsilon^3 \delta \frac{dh_1}{dx} = 3(1 - 2h_1\delta + \cdots +). \qquad (6\text{–}150)$$

Clearly, the magnitude of deformation δ is determined by the magnitudes of the parameters Ca/ε^3 and $G\varepsilon^3$. We will assume, for present purposes, that the magnitude is controlled by capillary effects, so that

$$\delta \equiv \frac{Ca}{\varepsilon^3} \ll 1.$$

With this choice for δ, the governing equation for h_1 is

$$-\frac{d^3 h_1}{dx^3} + G Ca \frac{dh_1}{dx} = 3.$$

The Thin-Gap Approximation – Films with a Free Surface

The form of the solution depends on the product, GCa. In keeping with the assumption of a dominant capillary effect, we assume that

$$\text{GCa} \ll 1.$$

Hence we treat the product GCa as an additional small parameter and solve (6–150) by means of a regular perturbation expansion, i.e.,

$$h_1 = h_1^{(0)} + (\text{GCa}) h_1^{(1)} + \cdots + . \tag{6–151}$$

Then, to leading order in GCa,

$$\frac{d^3 h_1^{(0)}}{dx^3} = -3, \tag{6–152}$$

$$h_1^{(0)} = -\frac{x^3}{2} + c_1 x^2 + c_2 x + c_3.$$

The boundary conditions $h_1 = 0$ at $x = 0, 1$ yield

$$h_1^{(0)} = -\frac{1}{2} x^3 + c_1 (x^2 - x) + \frac{1}{2} x,$$

and with the additional constraint (6–143), we find

$$c_1 = \frac{3}{4},$$

so that

$$h_1^{(0)} = -\frac{1}{2} x^3 + \frac{3}{4} x^2 - \frac{1}{4} x. \tag{6–153}$$

The next term in the expansion (6–151) provides an indication of the additional tendency of the body-force contribution to maintain a flat interface. The governing equation is

$$\frac{d^3 h_1^{(1)}}{dx^3} = \frac{dh_1^{(0)}}{dx} = -\frac{3}{2} x^2 + \frac{3}{2} x - \frac{1}{4}. \tag{6–154}$$

The solution of this equation, subject again to the conditions (6–142) and (6–143) is

$$h_1^{(1)} = -\frac{x^5}{40} + \frac{x^4}{16} - \frac{x^3}{24} + \frac{x}{240}. \tag{6–155}$$

The shape functions $h_1^{(0)}$ and $h_1^{(1)}$ are plotted in Fig. 6–9. Although we could proceed to arbitrarily high-order approximations in the small parameter GCa, there is not much to be learned from doing so, and we truncate the analysis at this point.

Although the analysis leading to (6–150) is quite general, we have solved it for only the case in which the gravitational contribution is vanishingly small relative to the capillary contribution. In fact, though we have introduced $\delta \equiv Ca/\varepsilon^3$ as the small parameter in (6–149),

E. Shallow-Cavity Flows

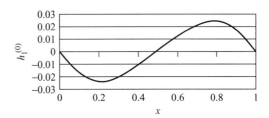
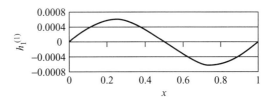

Figure 6–9. The interface shape functions $h_1^{(0)}$ and $h_1^{(1)}$.

an asymptotic expansion of the same form is also relevant for the opposite limit in which the gravitational contribution to (6–148) is dominant over the capillary contribution and

$$\delta \equiv (G\varepsilon^3)^{-1} \ll 1.$$

In that case, the governing equation for h_1 takes the form

$$-\frac{1}{GCa}\frac{d^3 h_1}{dx^3} + \frac{dh_1}{dx} = 3. \tag{6-156}$$

and the small parameter is $(GCa)^{-1}$. However, if we now seek an approximate solution for h_1 in the form of an asymptotic expansion for $(GCa)^{-1} \ll 1$,

$$h_1 = h_1^{(0)} + (GCa)^{-1} h_1^{(1)} + \cdots +, \tag{6-157}$$

the governing equation for the leading term reduces to the form

$$\frac{dh_1^{(0)}}{dx} = 3. \tag{6-158}$$

However, there is no solution of this equation that satisfies the conditions (6–142) at $x = 0$ and $x = 1$, as well as the constant-volume constraint (6–143). This tells us that the gravitational contribution to the interface stress balance is insufficient, by itself, to enforce the condition that the interface is pinned at the ends of the thin gap. In fact, the solution of (6–158) is linear in x and thus can only satisfy the constant-volume constraint, (6–143), by falling below the mean height for $x < (1/2)$ (i.e., $h_1 < 0$) and rising above it for $x > (1/2)$ (i.e., $h_1 > 0$). This makes sense only if the end walls exceed $h' = d$ in height so that the fluid at $x = 1$ remains within the container, and then only if the interface is not required to satisfy the conditions (6–142) or any other condition at the end walls. Typically, however, if the interface is not pinned at the corner as required by (6–142), it is required to satisfy a condition that fixes the "contact angle" between the interface and the end walls, and this condition also cannot be satisfied by the solution of (6–158).

The fact that the leading-order approximation, (6–158), of the governing equation, (6–156), does not produce a solution that can satisfy any of these conditions is telling us that the expansion, (6–157), is not a uniformly valid approximation of the solution of (6–156) for all x. Indeed, from a more general perspective, we see that the limit $(GCa)^{-1} \ll 1$ in (6–156) is *singular*, in the sense that the limit $(GCa)^{-1} \to 0$ produces a lower-order equation (6–157) for h_1, the solution to which cannot possibly satisfy the original two boundary conditions, (6–142), plus the constant-volume constraint, (6–143). Thus the asymptotic solution of (6–156) for $(GCa)^{-1} \ll 1$ must involve a matched asymptotic approximation, with different scaling in the end regions such that capillary effects remain important even as $(GCa)^{-1} \to 0$. This problem is a nice example of a singular-perturbation problem that can be solved by the method of matched asymptotic expansions. However, it is left as an exercise for the reader (Problem 6–11).

b. Solution by means of the method of domain perturbations In the preceding analysis, we have adopted the general approach of a standard thin-film problem. The main results are Eqs. (6–144), (6–145), (6–146), and (6–148) for the pressure $p^{(0)}$, the x component of velocity $u^{(0)}$, and the governing equation for the interface shape function h. Although we did not explicitly calculate the z component of velocity, we could also do that by simply integrating the continuity equation and applying the boundary condition $w^{(0)} = 0$ at $z = 0$. Apart from the inherent assumptions of the thin-film approximation and the conditions (6–142) and (6–143), there was no restriction on the magnitude of the deformation of the interface. In the end, however, the result was the *nonlinear* equation, (6–148), for $h(x)$, and in order to obtain analytic results for h [and thus ultimately also for $u^{(0)}$ and $p^{(0)}$ because these depend on h], we restricted the analysis to small perturbations from the undeformed, flat interface.

It may then occur to the reader that we might have incorporated this small-perturbation assumption from the outset. Instead of deriving the nonlinear equation (6–148) for h, and then linearizing by means of the assumption (6–149), we could have incorporated the small-deformation approximation from the beginning and solved the problem as a regular perturbation in the small parameter, $\delta \ll 1$. Although the end result must ultimately be the same, there is some instructional value in contrasting the two analyses. Thus the regular perturbation approach is summarized in this section. For simplicity, we ultimately restrict our analysis to the case

$$\delta \equiv \frac{Ca}{\varepsilon^3} \ll 1; \quad GCa \ll 1.$$

However, at the outset we can simply carry along δ as a small parameter without specifying its magnitude in terms of either Ca or G.

We begin again with the governing equations, (6–119), (6–120), and (6–137), together with the boundary conditions and integral constraints, (6–124b) and (6–138)–(6–143). However, here we assume from the outset that

$$h = 1 + h_1\delta + h_2\delta^2 + \cdots + . \tag{6–159a}$$

Hence we seek an asymptotic solution for the leading-order terms in the thin-film approximation, i.e., $u^{(0)} p^{(0)} w^{(0)}$, as a regular perturbation in δ, i.e.,

$$\begin{aligned} u^{(0)} &= u_0^{(0)} + u_1^{(0)}\delta + \cdots +, \\ w^{(0)} &= w_0^{(0)} + w_1^{(0)}\delta + \cdots +, \\ p^{(0)} &= p_0^{(0)} + p_1^{(0)}\delta + \cdots + . \end{aligned} \tag{6–159b}$$

E. Shallow-Cavity Flows

The governing equations at $O(1)$ and $O(\delta)$ are obtained from (6–119) and (6–120) plus (6–137). At $O(1)$, we have

$$\frac{\partial^2 u_0^{(0)}}{\partial z^2} - \frac{\partial p_0^{(0)}}{\partial x} = 0, \qquad (6\text{–}160\text{a})$$

$$\frac{\partial u_0^{(0)}}{\partial x} + \frac{\partial w_0^{(0)}}{\partial z} = 0, \qquad (6\text{–}160\text{b})$$

$$\frac{\partial p_0^{(0)}}{\partial z} = -\varepsilon^3 G; \qquad (6\text{–}160\text{c})$$

and at $O(\delta)$,

$$\frac{\partial^2 u_1^{(0)}}{\partial z^2} - \frac{\partial p_1^{(0)}}{\partial x} = 0, \qquad (6\text{–}161\text{a})$$

$$\frac{\partial u_1^{(0)}}{\partial x} + \frac{\partial w_1^{(0)}}{\partial z} = 0, \qquad (6\text{–}161\text{b})$$

$$\frac{\partial p_1^{(0)}}{\partial z} = 0. \qquad (6\text{–}161\text{c})$$

We retain $\varepsilon^3 G$ in (6–160c) because we have not yet specified δ in terms of either Ca or G, and because there is little added complexity in the resulting problem.

The boundary conditions for (6–160) and (6–161) at $z = 0$ are

$$u_0^{(0)} = 1; \quad w_0^{(0)} = u_1^{(0)} = w_1^{(0)} = 0 \text{ at } z = 0.$$

The interface boundary conditions, (6–139)–(6–141), are applied at $z = h$. However, because we now assume that h takes the asymptotic form, (6–159a), it is convenient to use the method of *domain perturbations* to transform these conditions into *asymptotically equivalent boundary conditions* applied at the *undeformed* surface, $z = 1$.

The method of domain perturbations was first introduced in Chap. 4. However, the main ideas are repeated here. We begin with the condition (6–139). Introducing the expansions (6–159b), we find that this becomes

$$w_0^{(0)} + w_1^{(0)}\delta + O(\delta^2) - \frac{dh_1}{dx}\delta\left[u_0^{(0)} + O(\delta)\right] = 0 \text{ at } z = h. \qquad (6\text{–}162)$$

Then, we relate the values of the various dependent variables such as $w_0^{(0)}$ at $z = h = 1 + h_1\delta + O(\delta^2)$ to their value at $z = 1$ by using a simple Taylor series approximation, e.g.,

$$w_0^{(0)}\big|_{z=1+h_1\delta+O(\delta^2)} = w_0^{(0)}\big|_{z=1} + \frac{\partial w_0^{(0)}}{\partial z}\bigg|_{z=1} h_1\delta + O(\delta^2). \qquad (6\text{–}163)$$

We have retained terms only through $O(\delta)$ in both (6–162) and (6–163), though we clearly could carry these expansions to whatever order in δ that we desire. Finally, combining (6–162) and (6–163), we obtain

$$w_0^{(0)}\big|_{z=1} + \frac{\partial w_0^{(0)}}{\partial z}\bigg|_{z=1} h_1\delta + w_1^{(0)}\big|_{z=1}\delta - \delta\frac{dh_1}{dx}u_0^{(0)}\big|_{z=1} + O(\delta^2) = 0. \qquad (6\text{–}164)$$

The Thin-Gap Approximation – Films with a Free Surface

Hence, collecting terms with equal powers of δ, we obtain a hierarchy of boundary conditions at $z = 1$. With our truncated expansions we require only the first two, namely,

$$w_0^{(0)} = 0 \text{ at } z = 1, \qquad (6\text{–}165\text{a})$$

$$w_1^{(0)} = -h_1 \frac{\partial w_0^{(0)}}{\partial z} + \frac{dh_1}{dx} u_0^{(0)} \text{ at } z = 1, \qquad (6\text{–}165\text{b})$$

through terms of $O(\delta)$. These conditions at $z = 1$ are asymptotically equivalent for $\delta \ll 1$ to the original boundary condition, (6–139), applied at $z = h(x)$.

Similarly, the boundary conditions (6–140), and (6–141), can be replaced with asymptotically equivalent boundary conditions at $z = 1$. These are, respectively,

$$\left(p_0^{(0)} - p^*\right) = -\frac{d^2 h_1}{dx^2} \text{ at } z = 1, \qquad (6\text{–}166\text{a})$$

$$p_1^{(0)} + h_1 \frac{\partial p_0^{(0)}}{\partial z} = -\frac{d^2 h_2}{dx^2} \text{ at } z = 1, \qquad (6\text{–}166\text{b})$$

$$\frac{\partial u_0^{(0)}}{\partial z} = 0 \text{ at } z = 1, \qquad (6\text{–}167\text{a})$$

$$\frac{\partial u_1^{(0)}}{\partial z} = -h_1 \frac{\partial^2 u_0^{(0)}}{\partial z^2} \text{ at } z = 1. \qquad (6\text{–}167\text{b})$$

Finally, we transform the integral constraint, (6–143), from its original form as an integral over z from 0 to $h(x)$, i.e.,

$$\int_0^{h(x)} [u_0(z) + u_1(z)\delta + O(\delta^2)] dz = 0$$

for $h = 1 + h_1(x)\delta + O(\delta^2)$, to an integral over $\bar{z}[\equiv z/h(x)]$ from 0 to 1, i.e.,

$$\int_0^1 \left(u_0\{\bar{z}[1 + h_1\delta + O(\delta^2)]\} + u_1(\bar{z})\delta + O(\delta^2) \right) [1 + h_1\delta + O(\delta^2)] d\bar{z} = 0. \qquad (6\text{–}168)$$

Now we are prepared to obtain a solution for $\delta \ll 1$, beginning with the governing equations and boundary conditions at $O(1)$. Equation (6–160c) tells us that $\partial p_0^{(0)}/\partial x$ is independent of z. Hence, integrating (6–160a), we obtain

$$u_0^{(0)} = \frac{\partial p_0^{(0)}}{\partial x} \frac{z^2}{2} + c_1 z + c_2.$$

E. Shallow-Cavity Flows

One boundary condition, as stated above, is $u_0^{(0)} = 1$ at $z = 0$. This implies $c_2 = 1$. A second boundary condition is (6–167a). When this condition is applied, we obtain

$$u_0^{(0)} = \frac{\partial p_0^{(0)}}{\partial x}\left(\frac{z^2}{2} - z\right) + 1. \qquad (6\text{–}169)$$

The pressure gradient is determined by the requirement that the net volume flux be zero at any x. At $O(1)$ this is just the integral constraint (6–168). Substituting (6–169) into this condition yields

$$\frac{\partial p_0^{(0)}}{\partial x} = 3. \qquad (6\text{–}170)$$

Because $\partial p_0^{(0)}/\partial x$ is independent of x, the flow (6–169) is seen to be unidirectional at this level of approximation. This implies $w_0^{(0)} \equiv 0$, as we can see by noting from continuity that $\partial w_0^{(0)}/\partial z = 0$, whereas $w_0^{(0)} = 0$ at $z = 0$ according to the boundary condition (6–165a).

Hence the velocity and pressure distributions at $O(1)$ can be completely determined within the core region (away from the end walls), to within an arbitrary constant for $p_0^{(0)}$. In fact, the flow is a simple unidirectional flow, as is appropriate for traction-driven flow between two plane surfaces. The turning flow that must occur near the ends of the cavity influences the core flow only in the sense that the presence of impermeable end walls requires a pressure gradient in the opposite direction to the boundary motion in order to satisfy the zero-mass-flux constraint. But now, *a remarkable feature* of the domain perturbation procedure is that we can use our knowledge of the *unidirectional* flow that is appropriate for an undeformed interface at $O(1)$ to directly determine the $O(\delta)$ contribution to the interface shape function in (6–159a) without having to determine any other feature of the solution at $O(\delta)$.

We simply use the normal-stress condition, (6–166a). For convenience, because $p_0^{(0)}$ is determined only to the value of its first derivative, we can differentiate (6–166a) and substitute directly to give

$$-\frac{d^3 h_1}{dx^3} = 3. \qquad (6\text{–}171)$$

Hence, integrating, we obtain

$$h_1 = -\frac{x^3}{2} + d_2 x^2 + d_1 x + d_0. \qquad (6\text{–}172)$$

Now, if we assume that this expression can be used all the way to the two end walls, then the boundary conditions (6–142) and the constant-volume constraint (6–143) are sufficient to determine d_0, d_1, and d_2. The result is

$$h_1 = -\frac{1}{2}x^3 + \frac{3}{4}x^2 - \frac{1}{4}x. \qquad (6\text{–}173)$$

This result is precisely the same as the $O(\delta)$ contribution to h_1 [Eq. (6–153)] from the previous "thin-film" analysis.

We proceed one step further to obtain the $O(\delta)$ corrections to the velocity and pressure distributions caused by the $O(\delta)$ perturbation of the interface shape. These solutions at $O(\delta)$ then allow us to calculate the $O(\delta^2)$ perturbation to $h(x)$. The governing equations at $O(\delta)$

The Thin-Gap Approximation – Films with a Free Surface

are (6–161), with $u_1^{(0)} = w_1^{(0)} = 0$ at $z = 0$, and the $O(\delta)$ interface conditions, which are given by (6–165b), (6–166b), and (6–167b), respectively.

Now, we see from (6–161c) that $\partial p_1^{(0)}/\partial x$ is again independent of z, and thus

$$u_1^{(0)} = \frac{\partial p_1^{(0)}}{\partial x}\left(\frac{z^2}{2}\right) + c_1 z + c_2. \tag{6–174}$$

The boundary conditions are $u_1^{(0)} = 0$ at $z = 0$, and, from (6–167b),

$$\frac{\partial u_1^{(0)}}{\partial z} = -3h_1 \text{ at } z = 1.$$

Because h_1 is a function of x, it is clear that $u_1^{(0)}$ must also depend on x, and the flow at this level of approximation is no longer unidirectional. On application of these boundary conditions, the solution (6–174) becomes,

$$\boxed{u_1^{(0)} = \frac{\partial p_1^{(0)}}{\partial x}\left(\frac{z^2}{2} - z\right) - 3h_1(x)z.} \tag{6–175}$$

The integral constraint (6–168) then yields a condition on $\partial p_1^{(0)}/\partial x$. In applying this condition, we first note that

$$u_0\{\bar{z}[1 + h_1\delta + O(\delta^2)]\} \sim u_0(\bar{z}) + 3[(\bar{z}^2 - \bar{z})h_1]\delta + O(\delta^2) \tag{6–176}$$

and that $u_1(\bar{z})$ is just (6–175) with z replaced by $\bar{z}$. Hence the $O(\delta)$ contribution to (6–168) is

$$\int_0^1 \left[3(\bar{z}^2 - \bar{z})h_1 + \frac{\partial p_1^{(0)}}{\partial x}\left(\frac{\bar{z}^2}{2} - \bar{z}\right) - 3h_1\bar{z}\right]d\bar{z} = 0. \tag{6–177}$$

In writing this expression, we have taken account of the fact that $\int_0^1 u_0(\bar{z})d\bar{z} \equiv 0$. Evaluating the integrals, we find

$$\boxed{\frac{\partial p_1^{(0)}}{\partial x} = -6h_1(x).} \tag{6–178}$$

Alternatively, we could have calculated $w_1^{(0)}$ by means of (6–175) and the continuity equation. Together with the boundary condition $w_1^{(0)} = 0$ at $z = 0$, this leads to

$$\boxed{w_1^{(0)} = -\frac{\partial^2 p_1^{(0)}}{\partial x^2}\left(\frac{z^3}{6} - \frac{z^2}{2}\right) + 3\frac{dh_1}{dx}\frac{z^2}{2}.} \tag{6–179}$$

A further condition on (6–179) comes from the kinematic boundary condition, (6–165b),

$$w_1^{(0)} = \frac{dh_1}{dx}u_0^{(0)} \text{ at } z = 1.$$

Substituting for $w_1^{(0)}$ then requires that

$$\frac{1}{3}\frac{\partial^2 p_1^{(0)}}{\partial x^2} = -2\frac{dh_1}{dx}. \tag{6–180}$$

If we integrate once with respect to x this again leads to the condition (6–178).

E. Shallow-Cavity Flows

Now, with $\partial p_1^{(0)}/\partial x$ determined according to (6–178), we can use the normal-stress condition, (6–166b), to determine the $O(\delta^2)$ contribution to the shape function. Again, it is convenient to differentiate (6–166b). The result is

$$\frac{d^3 h_2}{dx^3} = -\frac{\partial p_1^{(0)}}{\partial x} - \frac{dh_1}{dx}\frac{\partial p_0^{(0)}}{\partial z}. \tag{6–181}$$

Now, substituting for $\partial p_1^{(0)}/\partial x$ from (6–178) and recalling from (6–170) that $\partial p_0^{(0)}/\partial z = -\varepsilon^3 G$, we can write this equation in the form

$$\frac{d^3 h_2}{dx^3} = 6h_1(x) + \varepsilon^3 G \frac{dh_1}{dx}, \tag{6–182}$$

with $h_1(x)$ given by Eq. (6–173). Finally, substituting for h_1, we obtain

$$\frac{d^3 h_2}{dx^3} = 6\left(-\frac{x^3}{2} + \frac{3x^2}{4} - \frac{x}{4}\right) + \varepsilon^3 G\left(-\frac{3}{2}x^2 + \frac{3}{2}x - \frac{1}{4}\right). \tag{6–183}$$

If we integrate and apply the conditions (6–142) and (6–143), we obtain the solution for h_2:

$$h_2 = -\left(\frac{x^6}{40} - \frac{3x^5}{40} + \frac{x^4}{16} - \frac{9x^2}{560} + \frac{x}{280}\right) - \varepsilon^3 G\left(\frac{x^5}{40} - \frac{x^4}{16} + \frac{x^3}{24} - \frac{x}{240}\right). \tag{6–184}$$

Finally, introducing the solutions (6–178) and (6–184) into the original expansion, (6–158), and identifying δ as Ca/ε^3 in order to allow comparison with the result obtained in the previous subsection, we obtain

$$\begin{aligned} h = 1 &- \frac{Ca}{\varepsilon^3}\left(\frac{x^3}{2} - \frac{3x^2}{4} + \frac{x}{4}\right) - \left(\frac{Ca^2 G}{\varepsilon^3}\right)\left(\frac{x^5}{40} - \frac{x^4}{16} + \frac{x^3}{24} - \frac{x}{240}\right) \\ &- \left(\frac{Ca}{\varepsilon^3}\right)^2\left(\frac{x^6}{40} - \frac{3x^5}{40} + \frac{x^4}{16} - \frac{9x^2}{560} + \frac{x}{280}\right) + O\left(\frac{Ca}{\varepsilon^3}\right)^3. \end{aligned} \tag{6–185}$$

The first two terms match the result obtained earlier by means of the expansion (6–149) and (6–151) applied to Eq. (6–148). The latter is, however, computed only to terms of $O(Ca/\varepsilon^3)$, and thus does not contain the third term in (6–185). The effort involved in obtaining (6–185) by means of the domain perturbation technique is, however, greater than the analysis to obtain the result (6–148) by means of the thin-film approach, and the latter does not make any *a priori* restriction on the shape function h. These observations suggest that the thin-film approach is both simpler and more powerful for this particular class of problems. It should be emphasized, however, that the domain perturbation technique can sometimes yield results when no other approach will work, and it has proven to be an invaluable tool in obtaining analytic solutions for a wide variety of free-boundary problems, both in fluid mechanics and other subjects.

c. The end regions There is one aspect of the analysis of this section that is worthy of some additional discussion, and that is the fact that the interface shape functions were obtained with the boundary condition, (6–142), and the constant-volume constraint, (6–143), both of which assume that the solutions we have obtained for h are valid all of the way to the end walls at $x = 0$ and $x = 1$. Of course, we know that the analysis leading up to (6–148), for example, is strictly valid only in the "core" of the thin cavity, and breaks down within a distance $O(d)$ from the end walls. This must certainly raise a question about the validity of our procedure of simply ignoring this fact, and extending the core solutions for h right up to $x = 0$ and 1.

The Thin-Gap Approximation – Films with a Free Surface

To show that this procedure is, in fact, acceptable, we need to follow the example of the previous subsection and consider the asymptotic problem in the end regions. For present purposes, it is sufficient to consider the end region at $x = 0$. The scaling (6–130) and the governing equations (6–131) are the same except for the body-force term $-\varepsilon^3 G$ that was retained in this section [see, e.g., Eq. (6–137)]. The boundary conditions at $\hat{x} = 0$ and $\hat{y} = 0$ are also identical to (6–132), but now the upper boundary is the interface. The boundary conditions there can be obtained by nondimensionalizing (6–10)–(6–12) using (6–130). The result is

$$\hat{w} - \hat{u}\frac{\partial h}{\partial \hat{x}} + O(\varepsilon^2) = 0, \tag{6–186a}$$

$$\hat{p} - p^* = \varepsilon\left(2\frac{\partial \hat{w}}{\partial \hat{z}} + O(\varepsilon^3)\right) - \frac{\varepsilon}{Ca}\frac{\partial^2 h}{\partial \hat{x}^2}, \tag{6–186b}$$

$$\left[1 + \left(\frac{\partial h}{\partial \hat{x}}\right)^2\right]\left(\frac{\partial \hat{u}}{\partial \hat{z}} + \frac{\partial \hat{w}}{\partial \hat{x}}\right) - 2\frac{\partial h}{\partial \hat{x}}\left(\frac{\partial \hat{w}}{\partial \hat{z}} + \frac{\partial \hat{u}}{\partial \hat{x}}\right) = 0. \tag{6–186c}$$

Now, we can seek an asymptotic solution for this end region in the form

$$\begin{aligned}h &= 1 + \varepsilon h_1 + \cdots +, \\ (\hat{u},\ \hat{w},\ \hat{p}) &= (\hat{u}_0,\ \hat{w}_0,\ \hat{p}_0) + \varepsilon(\hat{u}_1,\ \hat{w}_1,\ \hat{p}_1) + \cdots +,\end{aligned} \tag{6–187}$$

where the reader is reminded that ε is the aspect ratio of the cavity, $\varepsilon \equiv d/L \ll 1$. Thus, referring to (6–131), we see that the governing equations at $O(1)$ are simply

$$\frac{\partial \hat{p}_0}{\partial \hat{x}} = 0, \quad \frac{\partial \hat{p}_0}{\partial \hat{z}} = -\varepsilon^3 G. \tag{6–188}$$

Hence,

$$\hat{p}_0 = -\varepsilon^3 G \hat{z} + \text{const}, \tag{6–189}$$

where we can obtain the constant in (6–189) by matching this solution with the leading-order expression, (6–144), for the pressure in the core, i.e.,

$$\lim_{\hat{x} \gg 1} \hat{p}_0 + \cdots + \Leftrightarrow \lim_{x \ll 1} p^{(0)} + \cdots + \quad \text{as } \varepsilon \to 0.$$

The governing equations and boundary conditions at $O(\varepsilon)$ define the problem for the $O(1)$ velocity components in the end region. The governing equations are

$$\nabla^2 \hat{u}_0 - \frac{\partial \hat{p}_1}{\partial \hat{x}} = 0, \tag{6–190}$$

$$\nabla^2 \hat{w}_0 - \frac{\partial \hat{p}_1}{\partial \hat{z}} = 0, \tag{6–191}$$

$$\frac{\partial \hat{u}_0}{\partial \hat{x}} + \frac{\partial \hat{w}_0}{\partial \hat{z}} = 0, \tag{6–192}$$

with interface boundary conditions

$$\hat{w}_0 = 0 \quad \text{at } \hat{z} = 1, \tag{6–193a}$$

$$\hat{p}_1 = 2\frac{\partial \hat{w}_0}{\partial \hat{z}} - \frac{\varepsilon^2}{Ca}\frac{\partial^2 h_1}{\partial \hat{x}^2} \quad \text{at } \hat{z} = 1, \tag{6–193b}$$

$$\frac{\partial \hat{u}_0}{\partial \hat{z}} = 0 \quad \text{at } \hat{z} = 1, \tag{6–193c}$$

E. Shallow-Cavity Flows

plus the boundary conditions (6–132) at $\hat{x} = 0$ and $\hat{y} = 0$, and the matching condition

$$\lim_{\hat{x} \gg 1}(\hat{u}_0 + \cdots +) \Leftrightarrow \lim_{x \ll 1}[u^{(0)} + \cdots +].$$

Now, the core solution was developed under the assumption that $Ca/\varepsilon^3 \ll 1$, and, for present purposes, it is convenient to express this condition in terms of the distinguished limit:

$$\overline{Ca} \equiv \frac{Ca}{\varepsilon^{3+m}} = O(1),$$

where m is an arbitrary positive constant. In this case, $Ca/\varepsilon^2 = O(\varepsilon^{1+m})$, and we see from (6–193b) that we obtain a governing equation for h_1 in the limit $\varepsilon \to 0$, which *does not depend on the functions* $\hat{p}_1$ *and* $\hat{w}_0$, namely,

$$\frac{d^2 h_1}{d\hat{x}^2} = 0. \qquad (6\text{–}194)$$

Thus we can consider the leading-order perturbation to h in the end region without having to solve the flow problem that was previously outlined.

Integrating (6–194), we obtain

$$h_1 = c_1 \hat{x} + c_2$$

and, now, the boundary condition (6–142) at $x = 0$ clearly applies, so that

$$c_2 = 0.$$

We must determine the constant c_1 by matching the shape function h with the core solution:

$$\lim_{\hat{x} \gg 1}(1 + \varepsilon c_1 \hat{x} + \cdots +)$$

$$\Leftrightarrow \lim_{x \ll 1}\left(1 + \frac{Ca}{\varepsilon^3}\left\{-\frac{x^3}{2} + [\alpha(z) - p^*]x^2 + d_1 x + d_2\right\} + \cdots +\right) \text{ as } \varepsilon \to 0. \quad (6\text{–}195)$$

It will be noted that we have reverted, on the right hand side, to the general solution obtained for h_1 in the core, without applying the boundary condition, (6–142), or the integral constraint, (6–143), which imply that the core solution for h_1 can be extended all the way to the end walls. If we express the core solution in terms of the end-region variable $\hat{x}$, we see that the matching condition becomes

$$\lim_{\hat{x} \gg 1}(1 + c_1 \hat{x} \varepsilon + \cdots +)$$

$$\Leftrightarrow \lim_{x \ll 1}\left(1 + \frac{Ca}{\varepsilon^3}\left\{-\frac{\varepsilon^3 \hat{x}^3}{2} + [\alpha(z) - p^*] - \varepsilon^2 \frac{\hat{x}^2}{2} + d_1 \varepsilon \hat{x} + d_2\right\}\right) \text{ as } \varepsilon \to 0.$$

Now, $Ca = O(\varepsilon^{3+m})$ in the distinguished limit that we are considering. In this case, the largest term on the right-hand side is the last one, $O(\varepsilon^m)$. The rest are asymptotically smaller in the limit $\varepsilon \to 0$. Hence,

$$1 + c_1 \hat{x} \varepsilon + O(\varepsilon^2) \Leftrightarrow 1 + \frac{Ca}{\varepsilon^3} d_2 + O(\varepsilon^2), \qquad (6\text{–}196)$$

and we see that the two sides will match only if

$$c_1 = 0,$$
$$d_2 = 0.$$

This result is independent of the value of m for $m \leq 1$, including the special case $m = 1$.

403

The value $d_2 = 0$ is exactly what we obtained earlier by simply applying the boundary condition, (6–142), directly to the core solution. We now see that this apparently *ad hoc* procedure works because there is no perturbation to the shape function h through terms of $O(\varepsilon)$ in the end region (i.e., with $c_1 = 0$, $h_1 \equiv 0$). The same result is obtained at the other end of the cavity if we consider the solution there. Hence, both (6–142) and (6–143) can be applied directly to the leading-order terms of the core solution. Of course, the solutions for $\hat{u}_0$, $\hat{w}_0$, and $\hat{p}_1$ will be *nonzero*, and we may expect that higher-order approximations for h will be nontrivial in the end region. Thus it seems unlikely that we could continue to apply the end boundary condition, (6–142), or the integral condition, (6–143), to the core solution as we proceed to higher-order approximations. We would have to determine this by carrying out the details of the solution in both the core and end regions.

3. Thermocapillary Flow in a Thin Cavity

The final example of a shallow-cavity problem that we consider is the flow driven by surface-tension gradients at the upper surface, which is again assumed to be an interface as sketched in Fig. 7(c).[13] The surface-tension gradient is produced when a fixed temperature differential is maintained between the two walls at $x' = 0$ and L.

We assume that the boundaries at $x' = 0$ and L and at $z' = 0$ are all solid walls. The temperature at the left end wall ($x' = 0$) is denoted as T_C and that at the right end wall ($x' = L$) is denoted as T_H, with $T_H > T_C$. The bottom at $z' = 0$ is assumed to be thermally *insulated*, so that

$$\frac{\partial T}{\partial z'} = 0 \quad \text{at } z' = 0. \tag{6–197}$$

Above the interface at $z' = h'(x')$, we assume that there is a gas in which the temperature varies linearly with position from the cold to the hot end of the cell, i.e.,

$$T_g = T_C + \frac{T_H - T_C}{L} x'. \tag{6–198}$$

The thermal boundary condition at the interface is continuity of the heat flux:

$$k_1 (\mathbf{n} \cdot \nabla' T') + k_g (T - T_g) = 0 \quad \text{at } z' = h'(x'). \tag{6–199}$$

Here k_1 is the thermal conductivity of the liquid in the shallow cavity and k_g is a heat transfer coefficient for the gas. Finally, we assume, for simplicity, that the interfacial tension $\sigma(T)$ depends linearly on the interface temperature,

$$\sigma(T) = \sigma_0 - \beta \left[T - \frac{1}{2}(T_H + T_C) \right] = \sigma_0 - \beta(T - \overline{T}). \tag{6–200}$$

Hence, by definition, σ_0 is $\sigma(T)$ at $T = \overline{T}$. The temperature distribution within the liquid in the thin film is determined by means of the thermal energy equation:

$$\rho C_p \left(u' \frac{\partial T}{\partial x'} + w' \frac{\partial T}{\partial z'} \right) = k_1 \left(\frac{\partial^2 T}{\partial x'^2} + \frac{\partial^2 T}{\partial z'^2} \right). \tag{6–201}$$

Now, to proceed further, we need to identify the appropriate additional equations and boundary conditions that govern the fluid motion and then nondimensionalize these equations plus the thermal equations listed above.

To do this, we assume that the temperature difference $\Delta T \equiv T_H - T_C$ is small enough that the Boussinesq approximation, first introduced in Chap. 2, can be employed. By this assumption, we neglect all of the temperature variation of material properties apart from the variation of the interfacial tension, as already indicated by Eq. (6–200). To be specific, we may assume that the other material parameters, such as μ, ρ, k, C_p, etc., are evaluated at

E. Shallow-Cavity Flows

the mean temperature $\overline{T} \equiv [(1/2)(T_C + T_H)]$. The governing equations for the fluid motion are then decoupled from the thermal energy equation and take the same form as for the isothermal flow problems that we have been considering in the prior sections of this and the preceding chapter, namely (6–1)–(6–3). In the present analysis, we assume that buoyancy forces can be neglected as a source of fluid motion relative to Marangoni forces. Assuming that the thin cavity is oriented horizontally [i.e., the angle α that appears in (6–2) and (6–3) is equal to zero], this requires that $(\Delta\rho g \varepsilon^3 \ell_c^2 / \mu u_c) \ll 1$, where $\Delta\rho$ is the magnitude of density changes from the mean density, previously denoted as ρ. To understand the physical significance of this condition, we will have to wait until we have specified ℓ_c and u_c. For convenience, we repeat (6–1)–(6–3) in terms of a 2D Cartesian coordinate system, with z being the coordinate normal to the bottom boundaries:

$$\frac{\partial u}{\partial x} + \frac{\partial w}{\partial z} = 0, \tag{6–202}$$

$$\frac{\partial^2 u}{\partial z^2} - \frac{\partial p}{\partial x} = -\varepsilon^2 \frac{\partial^2 u}{\partial x^2} + \varepsilon \widehat{Re}\left(u \frac{\partial u}{\partial x} + w \frac{\partial u}{\partial z}\right), \tag{6–203}$$

$$\frac{\partial p}{\partial z} = O(\Delta\rho g \varepsilon^3 L^2 / \mu u_c, \; \varepsilon^2, \; \varepsilon^3 \widehat{Re}). \tag{6–204}$$

The scaling inherent in these equations is precisely the same as in the previous thin-film analyses,

$$\frac{\partial}{\partial z'} = \frac{1}{\varepsilon L} \frac{\partial}{\partial z}, \quad \frac{\partial}{\partial x'} = \frac{1}{L} \frac{\partial}{\partial x}, \quad u' = u_c u, \; w' = \varepsilon u_c w,$$

$$p' = \frac{\mu u_c}{L}\left(\frac{1}{\varepsilon^2}\right),$$

where

$$\varepsilon \equiv \frac{d}{L} \ll 1.$$

However, a *difference* from the preceding two subsections is that the characteristic velocity scale u_c is not *imposed* on the problem by a fixed velocity of one of the boundaries, as in the previous examples, but is determined indirectly by the magnitude of tangential stresses from the surface-tension gradient. Rather than trying to deduce this characteristic velocity here, we start by denoting it as u_c, and wait until we consider the interface boundary conditions from which an appropriate choice for u_c will become obvious. One minor notational difference between (6–203) and (5–65) is that we have chosen to express (6–203) in terms of the Reynolds number,

$$\widehat{Re} \equiv u_c d / v,$$

which we define by using the cavity height rather than Re as in (5–65), which we defined by using the cavity length, $Re \equiv u_c L / v$. In this section, we generalize the preceding analyses by including the possibility of inertial effects. In particular, we allow $\widehat{Re} = O(1)$, so that inertial terms are $O(\varepsilon)$.

The fluid dynamical boundary conditions are similar to those applied in the previous problem, with two notable exceptions. First, $u \equiv 0$ at $z = 0$ (i.e., the lower boundary is stationary). Second, the tangential-stress condition is modified to account for the presence of Marangoni stresses that are due to gradients of the interfacial tension at the fluid interface,

The Thin-Gap Approximation – Films with a Free Surface

caused by gradients in the temperature along the interface. In addition, we incorporate the dimensionless form of the thermal energy equation, (6–201), and boundary conditions.

To proceed, it is convenient to introduce a scaled (dimensionless) temperature. We choose

$$\theta \equiv \frac{T - \overline{T}}{\Delta T}, \tag{6-205}$$

where $\Delta T \equiv T_H - T_C$. This makes

$$\boxed{\begin{aligned} \theta &= +1/2 \quad \text{at } x = 1, \\ \theta &= -1/2 \quad \text{at } x = 0. \end{aligned}} \tag{6-206}$$

Furthermore, Eq. (6–198) for the gas temperature, T_g, expressed in terms of θ becomes

$$\boxed{\theta_g \equiv \frac{T_g - \overline{T}}{\Delta T} = -\frac{1}{2} + x,} \tag{6-207}$$

and the interfacial tension from (6–200) is therefore

$$\sigma(\theta) = \sigma_0 - \beta \Delta T \theta. \tag{6-208}$$

The kinematic condition remains in the form (6–19) because it does not involve σ or θ. However, the thermal boundary condition, (6–199) is now

$$\boxed{\frac{\partial \theta}{\partial z} + O(\varepsilon^2) + \frac{k_g d}{k_1}\left(\theta + \frac{1}{2} - x\right) = 0 \quad \text{at } z = h(x),} \tag{6-209}$$

where we have used the thin-film approximations for $\mathbf{n}$ from (6–17) to estimate the magnitudes of the various contributions to the first term in (6–209). The normal-stress condition, (6–20), is essentially unchanged except that we use (6–208) to estimate σ so that

$$\boxed{p - p^* + O(\varepsilon^2) = -\frac{\varepsilon^3 \sigma_0}{\mu u_c}\left(1 - \frac{\beta \Delta T}{\sigma_0}\theta\right)\left(\frac{d^2 h}{dx^2} + O(\varepsilon^2)\right) \quad \text{at } z = h(x).} \tag{6-210}$$

The shear-stress condition requires a bit more care, as indicated earlier. We can revert to the dimensional form, (6–12), with $\tau'_s \equiv 0$ (we assume there are no "wind-driven" shear-stress contributions), and the unit normal and tangent vectors given by the thin-film approximations (6–17). The first term in (6–12) can be shown to take the form

$$2\mu(\mathbf{E}' \cdot \mathbf{n} \cdot \mathbf{t_i}) = \frac{\mu u_c}{\varepsilon L}\left[\frac{\partial u}{\partial z} + O(\varepsilon^2)\right],$$

whereas the surface-tension gradient term is

$$\nabla' \sigma \cdot \mathbf{t_i} = \frac{1}{L}\left(\frac{\partial \sigma}{\partial x} + \frac{\partial h}{\partial x}\frac{\partial \sigma}{\partial z}\right) \quad \text{at } z = h(x).$$

Hence, because

$$\frac{\partial \sigma}{\partial x} = -\beta \Delta T \frac{\partial \theta}{\partial x}, \quad \frac{\partial \sigma}{\partial z} = -\beta \Delta T \frac{\partial \theta}{\partial z},$$

E. Shallow-Cavity Flows

it follows that

$$\frac{\partial u}{\partial z} + O(\varepsilon^2) = -\frac{\varepsilon \beta \Delta T}{\mu u_c}\left(\frac{\partial \theta}{\partial x} + \frac{\partial h}{\partial x}\frac{\partial \theta}{\partial z}\right) \quad \text{at } z = h(x). \tag{6-211}$$

Now, we have used the symbol u_c to represent the characteristic velocity in the x direction within the thin fluid layer. However, unlike the preceding problems, in which a specific velocity scale was imposed at the lower boundary, $z = 0$, here the magnitude of velocities in the thin cavity are determined by the magnitude of Marangoni effects at the upper boundary, by means of (6–211). Indeed, it can be seen from this equation that an appropriate choice for u_c is

$$u_c \equiv \frac{\varepsilon \beta \Delta T}{\mu}. \tag{6-212}$$

The characteristic velocity is determined by the ratio of the characteristic tangential (Marangoni) stress, $O(\beta \Delta T/L)$, which drives this motion to the viscous forces $O(\mu u_c/d)$ that derive from this motion. The definition (6–212) also allows us to return to the condition for neglect of buoyancy forces compared with Marangoni forces as a potential source of fluid motion in the thin cavity. To do this, we introduce the thermal expansion coefficient, which we denote as α, so that the characteristic density difference $\Delta \rho = O(\rho \alpha \Delta T)$. Then the condition $(\Delta \rho g \varepsilon^3 \ell_c^2 / \mu u_c) \ll 1$ can be expressed in the form

$$\left(\frac{\rho g \alpha \Delta T \varepsilon^3 L^2}{\mu}\right)\bigg/\left(\frac{\varepsilon \beta \Delta T}{\mu}\right) \ll 1,$$

which can be viewed as the ratio of a characteristic velocity that is due to buoyancy forces to the characteristic velocity (6–212) that is due to Marangoni stresses. Simplifying this expression, we obtain

$$\rho \frac{\alpha}{\beta} d^2 \ll 1.$$

Thus we see that neglect of buoyancy effects essentially requires that the cavity height be small enough.

The final step in nondimensionalizing is to deal with the thermal energy equation, (6–201). Introducing the standard thin-film scaling for u', w' and spatial derivatives with respect to x' and z', we find that this equation becomes

$$\varepsilon \widehat{Pe}\left(u\frac{\partial \theta}{\partial x} + w\frac{\partial \theta}{\partial z}\right) = \frac{\partial^2 \theta}{\partial z^2} + \varepsilon^2 \frac{\partial^2 \theta}{\partial x^2}, \tag{6-213}$$

where

$$\widehat{Pe} \equiv \frac{u_c d}{\kappa} \quad \text{and} \quad \kappa \equiv \frac{k}{\rho C_p}$$

and $\widehat{Pe}$ is the Peclet number. It is a measure of the relative importance of the convection and conduction contributions to (6–213). In the following analysis, we assume that $\widehat{Pe} = O(1)$.

Finally, we note that the constraint of zero net volume flux must be imposed as in the previous problems of this section, i.e.,

$$\int_0^{(hx)} u\, dz = 0,$$

The Thin-Gap Approximation – Films with a Free Surface

and the same conditions on the function $h(x)$ will be used, namely,

$$h(x) = 1 \quad \text{at} \quad x = 0, 1$$

and the constant-volume constraint

$$\int_0^1 h(x)\partial x = 1.$$

As before, the latter conditions assume that we can extend the solution for $h(x)$ in the *core* region all of the way to the end walls of the cavity. This is a valid assumption at the order of approximation that we consider, but we would need to explicitly consider the solutions in the end regions at higher orders in ε.

a. Solution by means of the classical thin-film analysis We are now in a position to solve the problem of Marangoni-driven circulation in the shallow cavity. We seek a solution in the form of asymptotic expansions in the thin-film limit,[13] $\varepsilon \ll 1$, i.e.,

$$\begin{aligned}
u &= u^{(0)} + \varepsilon u^{(1)} + O(\varepsilon^2), \\
w &= w^{(0)} + \varepsilon w^{(1)} + O(\varepsilon^2), \\
p &= p^{(0)} + \varepsilon p^{(1)} + O(\varepsilon^2), \\
\theta &= \theta^{(0)} + \varepsilon \theta^{(1)} + O(\varepsilon^2).
\end{aligned} \tag{6-214}$$

We retain the superscript notation from the preceding section to designate the order of each variable in terms of the small parameter ε. The governing equations, which we obtained by substituting these expansions into (6–202)–(6–204), plus (6–213), are thus, at $O(1)$,

$$\frac{\partial^2 u^{(0)}}{\partial z^2} - \frac{\partial p^{(0)}}{\partial x} = 0, \tag{6-215a}$$

$$\frac{\partial u^{(0)}}{\partial x} + \frac{\partial w^{(0)}}{\partial z} = 0, \tag{6-215b}$$

$$\frac{\partial p^{(0)}}{\partial z} = 0, \tag{6-215c}$$

$$\frac{\partial^2 \theta^{(0)}}{\partial z^2} = 0, \tag{6-215d}$$

and at $O(\varepsilon)$

$$\frac{\partial^2 u^{(1)}}{\partial z^2} - \frac{\partial p^{(1)}}{\partial x} = \widehat{Re}\left[u^{(0)}\frac{\partial u^{(0)}}{\partial x} + w^{(0)}\frac{\partial u^{(0)}}{\partial z}\right], \tag{6-216a}$$

$$\frac{\partial u^{(1)}}{\partial x} + \frac{\partial w^{(1)}}{\partial z} = 0, \tag{6-216b}$$

$$\frac{\partial p^{(1)}}{\partial z} = 0, \tag{6-216c}$$

$$\frac{\partial^2 \theta^{(1)}}{\partial z^2} = \widehat{Pe}\left[u^{(0)}\frac{\partial \theta^{(0)}}{\partial x} + w^{(0)}\frac{\partial \theta^{(0)}}{\partial z}\right]. \tag{6-216d}$$

E. Shallow-Cavity Flows

We have assumed, in writing (6–216a) and (6–216d), that the Reynolds and Peclet numbers, $\widehat{Re}$ and $\widehat{Pe}$, are *both* $O(1)$.

Finally, substituting (6–214) into the boundary conditions, we have at $O(1)$

$$u^{(0)} = w^{(0)} \quad \text{at } x = 0, 1 \text{ and } z = 0, \tag{6-217a}$$

$$\theta^{(0)} = \frac{1}{2} \quad \text{at } x = 1, \tag{6-217b}$$

$$\theta^{(0)} = -\frac{1}{2} \quad \text{at } x = 0, \tag{6-217c}$$

$$\frac{\partial \theta^{(0)}}{\partial z} = 0 \quad \text{at } z = 0, \tag{6-217d}$$

and at $O(\varepsilon)$

$$u^{(1)} = w^{(1)} = 0 \quad \text{at } x = 0, 1 \text{ and } z = 0, \tag{6-218a}$$

$$\theta^{(1)} = 0 \quad \text{at } x = 0, 1, \tag{6-218b}$$

$$\frac{\partial \theta^{(1)}}{\partial z} = 0 \quad \text{at } z = 0. \tag{6-218c}$$

The interface boundary conditions, (6–209)–(6–211) and (6–19), can also be split into contributions at $O(1)$, $O(\varepsilon)$, and so on. However, to do this we must specify the magnitudes of the several additional, independent parameters that appear in these conditions. First, we assume that the dimensionless heat transfer coefficient that appears in (6–209), which we denote as Bi, is $O(1)$, i.e.,

$$Bi \equiv \frac{k_g d}{k_1} = O(1), \tag{6-219}$$

meaning that it is *independent* of ε. The capillary number, rescaled by ε^3, which appears in (6–210), now takes the form

$$\frac{Ca}{\varepsilon^3} \equiv \left(\frac{\mu u_c}{\varepsilon^3 \sigma_0}\right) = \frac{\beta \Delta T}{\varepsilon^3 \sigma_0}.$$

We denote this parameter as δ,

$$\delta \equiv \frac{Ca}{\varepsilon^3}. \tag{6-220}$$

We may recall from the preceding example that δ determines the magnitude of the interface deformation. We assume, in what follows, that δ is *no larger* than $O(1)$. In fact, we follow the example of the previous section and consider two distinct cases. First, we use the conventional thin-film analysis to obtain the leading-order term (only) in the thin-film parameter ε, but with the possibility of variations in the dimensionless film thickness functions $h(x)$ of $O(1)$, corresponding to $\delta = O(1)$. Second, we employ the domain perturbation technique to consider the case of $\delta = O(\varepsilon) \ll 1$, but in this case, we carry the analysis to second order in ε, thus incorporating the $O(\varepsilon \widehat{Re})$ and $O(\varepsilon \widehat{Pe})$ terms as indicated in (6–216).

The Thin-Gap Approximation – Films with a Free Surface

b. Thin-film solution procedure Let us then begin with the more general situation in which $\delta = O(1)$. In this case, the interface boundary conditions take the form

$$O(1), \quad w^{(0)} - u^{(0)}\frac{dh}{dx} = 0 \tag{6-221a}$$

$$p^{(0)} - p^* = -\left(\frac{d^2h}{dx^2}\right)\frac{1}{\delta} \tag{6-221b}$$

$$\left.\frac{\partial \theta^{(0)}}{\partial z} + Bi\left(\theta^{(0)} + \frac{1}{2} - x\right) = 0 \right\} \text{ at } z = h(x). \tag{6-221c}$$

$$\frac{\partial u^{(0)}}{\partial z} = -\left(\frac{\partial \theta^{(0)}}{\partial x} + \frac{dh}{dx}\frac{\partial \theta^{(0)}}{\partial z}\right) \tag{6-221d}$$

If we were to extend the solution to include corrections at $O(\varepsilon)$, we could use the method of domain perturbations about the $O(1)$ estimate of the interface shape to obtain interface boundary conditions at $O(\varepsilon)$. For example, we can express $h(x)$ in the form of an expansion in ε,

$$h(x) = h^{(0)}(x) + \varepsilon h^{(1)}(x) + \cdots +, \tag{6-222}$$

with $h^0(x)$ denoting the $O(1)$ estimate of the film shape we obtained by using the conditions (6–221) plus Eqs. (6–215) and boundary conditions (6–217). Then, to determine the $O(\varepsilon)$ contribution to the kinematic condition,

$$w - u\frac{\partial h}{\partial x} + O(\varepsilon^2) \quad \text{at } z = h, \tag{6-19}$$

we can introduce a domain perturbation about $z = h^{(0)}(x)$,

$$w|_{z=h} = w|_{z=h^{(0)}(x)} + \frac{\partial w}{\partial z}\bigg|_{z=h^{(0)}(x)}\varepsilon h^{(1)} + O(\varepsilon^2), \tag{6-223}$$

and express (6–19) in the form

$$w^{(0)} - u^{(0)}\frac{dh^{(0)}}{dx} + \varepsilon\left\{w^{(1)} + \frac{\partial w^{(0)}}{\partial z}h^{(1)} - \left[u^{(1)} + \frac{\partial u^{(0)}}{\partial z}h^{(1)}\right]\frac{dh^{(0)}}{dx} - u^{(0)}\frac{dh^{(1)}}{dx}\right\} + O(\varepsilon^2) = 0$$
$$\text{at } z = h^{(0)}(x). \tag{6-224}$$

Hence the boundary condition at $O(\varepsilon)$ is

$$w^{(1)} + \frac{\partial w^{(0)}}{\partial z}h^{(1)} - \left[u^{(1)} + \frac{\partial u^{(0)}}{\partial z}h^{(1)}\right]\frac{dh^{(0)}}{dx} - u^{(0)}\frac{dh^{(1)}}{dx} = 0 \quad \text{at } z = h^{(0)}(x). \tag{6-225}$$

Similarly, we could derive the form of the other interface boundary conditions at $O(\varepsilon)$. However, in this section, we limit our analysis to the solution at $O(1)$.

Let us now seek a solution at $O(1)$. This implies solving (6–215) subject to the $O(1)$ boundary conditions, (6–217), and the $O(1)$ interface conditions, (6–221). We can begin, in this case, by integrating (6–215d) to obtain

$$\theta^{(0)} = d(x)z + e(x). \tag{6-226}$$

The boundary condition (6–217d) requires that

$$d(x) \equiv 0$$

and the thermal condition at $z = \hat{h}$, (6–221c), then requires that

$$e(x) = x - \frac{1}{2}$$

E. Shallow-Cavity Flows

so that

$$\theta^{(0)} = x - \frac{1}{2}. \qquad (6\text{--}227)$$

Now, because $\partial p^{(0)}/\partial x$ is independent of z according to (6–215c), we can integrate (6–215a) to obtain

$$u^{(0)} = \frac{\partial p^{(0)}}{\partial x}\frac{z^2}{2} + a(x)z + b(x). \qquad (6\text{--}228)$$

The boundary condition (6–217a) gives $b(x) = 0$, and the shear-stress condition (6–220d) then yields

$$a(x) = -1 - \frac{\partial p^{(0)}}{\partial x}h(x),$$

so that

$$u^{(0)} = \frac{\partial p^{(0)}}{\partial x}\left[\frac{z^2}{2} - h(x)z\right] - z. \qquad (6\text{--}229)$$

The pressure gradient at $O(1)$ is then determined from the condition of zero volume flux,

$$\int_0^{(x)} u^{(0)} \partial z = 0. \qquad (6\text{--}230)$$

Substituting from (6–229), we find that this gives

$$\frac{\partial p^{(0)}}{\partial x} = -\frac{3}{2h(x)}. \qquad (6\text{--}231)$$

The normal velocity component is obtained from (6–229), by integrating the continuity equation (6–215b). After applying the boundary condition for $w^{(0)}$ at $z = 0$ from (6–217a), this gives

$$w^{(0)} = -\frac{\partial^2 p^{(0)}}{\partial x^2}\frac{z^3}{6} + \frac{\partial}{\partial x}\left(h\frac{\partial p^{(0)}}{\partial x}\right)\frac{z^2}{2}. \qquad (6\text{--}232)$$

We can then use (6–232) and (6–229) in conjunction with the kinematic boundary condition to obtain a differential relationship between $p^{(0)}$ and $h(x)$ (i.e., the Reynolds equation for this problem):

$$-\frac{\partial^2 p^{(0)}}{\partial x^2}\frac{h^3}{6} + \frac{\partial}{\partial x}\left(h\frac{\partial p^{(0)}}{\partial x}\right)\frac{h^2}{2} + \frac{\partial p^{(0)}}{\partial x}\frac{h^2}{2}\frac{dh}{dx} + h\frac{dh}{dx} = 0. \qquad (6\text{--}233)$$

However, as in the previous examples of this thin-cavity section, this relationship does not provide any additional information beyond the result, (6–231). Indeed, if we substitute (6–231) into (6–233), we find that the latter condition is exactly satisfied. This is a consequence, essentially, of the fact that both represent expressions of mass conservation.

Finally, we can differentiate the normal-stress condition (6–221b) to give

$$\frac{\partial p^{(0)}}{\partial x} = -\frac{1}{\delta}\frac{d^3 h}{dx^3}.$$

Hence, substituting for $\partial p^{(0)}/\partial x$ using (6–231), we obtain a DE for the shape function $h(x)$:

$$\frac{d^3 h}{dx^3} = +\frac{3}{2h}\delta. \qquad (6\text{--}234)$$

The Thin-Gap Approximation – Films with a Free Surface

We cannot obtain an analytic solution to this equation for arbitrary δ. However, in the limit $\delta \ll 1$, where the deformation is small, we can obtain an approximate solution in the form

$$h(x) = h_0(x) + h_1(x)\delta + h_2(x)\delta^2 + \cdots +. \tag{6–235}$$

By substitution, (6–234) can be re-written as

$$\frac{d^3 h_0}{dx^3} + \delta\left(\frac{d^3 h_1}{dx^3} - \frac{3}{2h_0}\right) + \delta^2\left(\frac{d^3 h_2}{dx^3} + \frac{3h_1}{2h_0^2}\right) + O(\delta^3) = 0. \tag{6–236}$$

Now, it follows that

$$h_0 = c_1 x^2 + c_2 x + c_3.$$

The boundary conditions, $h_0 = 1$ at $x = 0$ and 1, yield

$$c_3 = 1, \quad c_1 = -c_2.$$

The constant-volume constraint,

$$\int_0^1 h(x)dx = 1,$$

then gives

$$c_1 = c_2 = 0.$$

So, at $0(1)$, we obtain the trivial result

$$h_0 = 1. \tag{6–237}$$

The governing equation for h_1 is then

$$\frac{d^3 h_1}{dx^3} = \frac{3}{2}. \tag{6–238}$$

Integrating and applying the boundary conditions, $h_1 = 0$ at $x = 0, 1$, plus the integral constraint

$$\int_0^1 h(x)dx = 0,$$

yields the solution

$$h_1 = \frac{x^3}{4} - \frac{3x^2}{8} + \frac{x}{8}. \tag{6–239}$$

Finally, we go one step further to solve

$$\frac{d^3 h_2}{dx^3} = -\frac{3}{2}h_1 \tag{6–240}$$

E. Shallow-Cavity Flows

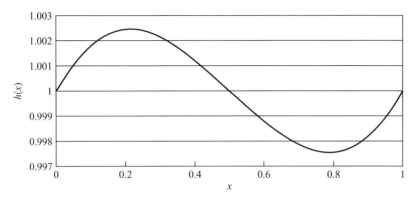

Figure 6–10. The interface shape function, $h(x) = 1 + \delta h_1(x) + O(\delta^2)$, for $\delta = 0.2$.

for $h_2(x)$, subject to the same conditions as h_1. The result is

$$h_2 = -\frac{1}{320}\left(x^6 - 3x^5 + \frac{5}{2}x^4 - \frac{9}{14}x^2 + \frac{1}{7}x\right). \tag{6–241}$$

Although we could continue this process to obtain higher-order terms in the expansions for h, the algebra becomes tiring, and, in any case, we are talking about increasingly small corrections. The basic result of interest is primarily the first perturbation h_1. The first two terms in the shape function $h(x) = 1 + \delta h_1(x)$ for the arbitrarily chosen value of $\delta = 0.2$ are plotted in Fig. 6–10. We see that the interface bulges up in the interval $0 < x < 1/2$ and dips down for $1/2 < x < 1$. The cold end of the cavity is at $x = 0$, and thus the motion at the interface is from $x = 1$ toward $x = 0$. The fact that the interface bulges upward on the left is an indirect consequence of the increased pressure, relative to the mean, that is required for producing the counterflow from $x = 0$ toward $x = 1$ and thus satisfy the constraint (6–230).

c. Solution by the domain perturbation method for $\delta = \varepsilon \ll 1$ In this section, we again consider the small-deformation limit, but this time we use the method of domain perturbations to solve the problem directly. In the process, however, we consider the special limit (i.e., the so-called "distinguished" limit),

$$\delta = \varepsilon \ll 1.$$

This brings the inertial and convective terms, $O(\varepsilon \widehat{Re})$ and $O(\varepsilon \widehat{Pe})$ into the problem at $O(\delta)$ Thus, the governing equations at $O(\delta)$ are (6–216). In the preceding subsection, we obtained a solution for $\delta = O(1)$, but then evaluated the resulting function $h(x)$ in the limit $\delta \ll 1$. In effect, this means that we assume that Eqs. (6–215) are the governing equations to at least the $O(\delta^2)$ term that we actually calculated. This is true provided $O(\delta^2) \gg O(\varepsilon \widehat{Re}, \varepsilon \widehat{Pe})$. Here, we suppose that $O(\delta) = O(\varepsilon \widehat{Re}, \varepsilon \widehat{Pe})$ for $\widehat{Re}, \widehat{Pe} = O(1)$.

The domain perturbation anaylsis largely follows that from the preceding problem. We assume that

$$h = 1 + h_1\delta + h_2\delta^2 + \cdots +, \tag{6–242}$$

with corresponding expansions for u, w, θ, and p of the form

$$\begin{aligned} u &= u_0 + u_1\delta + \cdots +, \\ w &= w_0 + w_1\delta + \cdots +, \\ \theta &= \theta_0 + \theta_1\delta + \cdots +, \\ p &= p_0 + p_1\delta + \cdots +. \end{aligned} \tag{6–243}$$

The Thin-Gap Approximation – Films with a Free Surface

Although we will ultimately consider the special (distinguished) limit where $\delta = \varepsilon$, we initially retain δ as a small parameter that is independent of ε.

The interface boundary conditions at $z = h$ are first converted to asymptotically equivalent boundary conditions (for $\delta \ll 1$) applied at the unperturbed surface at $z = 1$. These conditions are similar to those derived in the preceding section. The kinematic and normal-stress conditions are, in fact, unchanged, and are thus given (in the present notation) as

$$w_0 = 0 \quad (6\text{--}244\text{a})$$

at $z = 1$,

$$w_1 + \frac{\partial w_0}{\partial z} h_1 - u_0 \frac{dh_1}{dx} = 0 \quad (6\text{--}244\text{b})$$

$$p_0 - p^* = -\frac{d^2 h_1}{dx^2} \quad (6\text{--}245\text{a})$$

at $z = 1$,

$$p_1 + h_1 \frac{\partial p_0}{\partial z} = -\frac{d^2 h_2}{dx^2} \quad (6\text{--}245\text{b})$$

respectively. The shear-stress conditions, derived from (6–211), are

$$\frac{\partial u_0}{\partial z} = -\frac{\partial \theta_0}{\partial x} \quad (6\text{--}246\text{a})$$

at $z = 1$,

$$\frac{\partial u_1}{\partial z} + h_1 \frac{\partial^2 u_0}{\partial z^2} = -\left(\frac{\partial \theta_1}{\partial x} + h_1 \frac{\partial}{\partial z}\left(\frac{\partial \theta_0}{\partial x}\right) + \frac{dh_1}{dx}\frac{\partial \theta_0}{\partial z} \right) \quad (6\text{--}246\text{b})$$

and the thermal interface conditions, derived from (6–209), become

$$\frac{\partial \theta_0}{\partial z} = -Bi\left(\theta_0 + \frac{1}{2} - x\right) \quad (6\text{--}247\text{a})$$

at $z = 1$,

$$\frac{\partial \theta_1}{\partial z} + \frac{\partial^2 \theta_0}{\partial z^2} h_1 = -Bi\left(\theta_1 + \frac{\partial \theta_0}{\partial z} h_1\right) \quad (6\text{--}247\text{b})$$

To this point, we have not made any explicit assumption about the relative magnitudes of ε and δ, except for the fact that we have neglected terms of $O(\varepsilon^2)$ in the interface conditions that were used to derive (6–244)–(6–247). If we were to assume that terms of $O(\varepsilon)$ were small compared with terms of $O(\delta^n)$, then the governing equations at all orders in δ would be (6–215), and the domain perturbation method would generate the same solution as that of the small-δ limit of the solution in the previous subsection. If, on the other hand, terms of $O(\varepsilon)$ and $O(\delta^n)$ are of the same magnitude, then at that order in δ, the governing equations would be (6–216) – assuming that $\widehat{Re}$, $\widehat{Pe} = O(1)$, as stated earlier, and the solution at that point would depart from that obtained in the preceding subsection. Here we consider the specific case $n = 1$ and assume $\delta \equiv \varepsilon$. The problem at $O(1)$ is then (6–215), with (6–217) and the first of each pair of conditions (6–244a)–(6–247a). The problem at $O(\delta)$ [or $O(\varepsilon)$, which is the same thing] is then (6–216), (6–218), and the conditions (6–244b)–(6–247b). It should be noted that the distinction between superscripts designating the order in ε and subscripts denoting the order in δ is no longer relevant, and we thus use subscripts for all variables.

E. Shallow-Cavity Flows

We begin with the $O(1)$ problem by integrating (6–215d), and applying the insulating boundary condition (6–217d). This gives

$$\theta_0 = b(x)$$

The thermal condition, (6–247a), then requires that

$$b(x) = x - \frac{1}{2},$$

i.e.,

$$\boxed{\theta_0 = x - \frac{1}{2}.} \quad (6\text{–}248)$$

Next, because $\partial p^{(0)}/\partial x$ is independent of z, according to (6–215c), we can integrate (6–215a) to obtain the usual result:

$$u_0 = \frac{\partial p_0}{\partial x}\left(\frac{z^2}{2}\right) + c(x)z + d(x).$$

The no-slip condition, (6–217a), requires that $d(x) = 0$, and the shear-stress condition at the interface, (6–246a), then requires that

$$\frac{\partial u_0}{\partial z} = -1 \quad \text{at } z = 1$$

or

$$c(x) = -1 - \frac{\partial p_0}{\partial z},$$

so that

$$\boxed{u_0 = \frac{\partial p_0}{\partial x}\left(\frac{z^2}{2} - z\right) - z.} \quad (6\text{–}249)$$

The integral constraint, (6–168), determines the value of $\partial p^{(0)}/\partial x$. At $O(1)$, this takes the form

$$\int_0^1 u_0(\bar{z})\partial\bar{z} = 0.$$

Hence, inserting (6–249) and evaluating integrals, we find

$$\boxed{\frac{\partial p_0}{\partial x} = -\frac{3}{2}.} \quad (6\text{–}250)$$

As in the domain perturbation example of the preceding problem, we see that $\partial p^{(0)}/\partial x$ is independent of x, as one would expect in the absence of interface deformation at this order of approximation, and again

$$\boxed{w_0 \equiv 0.}$$

With the velocity and pressure gradient now evaluated to $O(1)$, we can use the normal-stress balance at $O(1)$, namely (6–245a), to determine the shape function h_1 at $O(\delta)$. If we differentiate (6–245a) and insert $\partial p^{(0)}/\partial x$ from (6–250), we see that

$$\frac{d^3 h_1}{dx^3} = \frac{3}{2}. \quad (6\text{–}251)$$

415

The Thin-Gap Approximation – Films with a Free Surface

This equation can now be integrated twice, and if we assume that the resulting expression is applicable all the way to the end walls at this level of approximation, we can apply the conditions (6–142) and (6–143) to determine h_1. The result is

$$\boxed{h_1 = \frac{x^3}{4} - \frac{3x^2}{8} + \frac{x}{8}.} \qquad (6\text{–}252)$$

As expected, this is identical to the $O(\delta)$ approximation obtained for the small-deformation limit in the preceding part of this subsection, namely Eq. (6–239).

We now consider the $O(\delta)$ (or $O(\varepsilon)$) correction. In the present analysis, we assume that $\widehat{Re}$ and $\widehat{Pe}$ are $O(1)$, and, because $\delta = \varepsilon$, we include the inertial and convective terms as indicated in the governing equations (6–216). As a consequence, the shape function h_2 will differ from the $O(\delta^2)$ term calculated in the earlier analysis [i.e., (6–241)] because the latter is based on the implicit assumption that the classic thin-film equations, (6–215), hold at all orders in the parameter δ.

We begin by evaluating the right-hand side of (6–216a) and (6–216b). The former is actually equal to zero because u_0 is a function of z only and $w_0 = 0$. Hence (6–216a) reduces to the form of the classic thin-film equation, (6–215a). The right-hand side of (6–216d) is, however, nonzero. Substituting for θ_0 and u_0 from (6–248) and (6–249), we have

$$\frac{\partial^2 \theta_1}{\partial z^2} = \widehat{Pe}\left[-\frac{3}{2}\left(\frac{z^2}{2} - z\right) + z\right]. \qquad (6\text{–}253)$$

The general solution for θ_1 thus takes the form

$$\theta_1 = \left(-\frac{z^4}{16} + \frac{5z^3}{12}\right)\widehat{Pe} + a_1(x)z + b_1(x).$$

The insulating boundary condition, (6–218b), requires that $a_1(x) \equiv 0$. The second condition on θ_1 is the thermal condition, (6–247b). Because θ_0 is independent of z, we require that

$$\frac{\partial \theta_1}{\partial z} + Bi\theta_1 = 0 \quad \text{at } z = 1,$$

and thus

$$b_1 = -\frac{(3 + Bi)}{3Bi}\widehat{Pe}.$$

Thus,

$$\boxed{\theta_1 = \left[-\frac{z^4}{16} + \frac{5z^3}{12} - \frac{(3 + Bi)}{3Bi}\right]\widehat{Pe}.} \qquad (6\text{–}254)$$

Next, we turn to the horizontal velocity component u_1. Although the inclusion of convection at this level of approximation has modified the temperature distribution in the core region, this does not play a role in either the velocity fields at $O(\delta)$ or the interface shape at $O(\delta^2)$. This is because the solution of θ turns out to be independent of x and thus does not add anything to the Marangoni-driven tangential stress at this same order of approximation. This means that, in spite of the more general case that we have considered, the result for h_2 should still be the same as obtained earlier as (6–241) under a more restrictive set of assumptions. Because $\partial p_1/\partial x$ is independent of z, we integrate (6–216a) as usual to obtain

$$u_1 = \frac{\partial p_1}{\partial x}\frac{z^2}{2} + c_1(x)z + d_1(x).$$

E. Shallow-Cavity Flows

The no-slip condition, (6–218a), at $z = 0$ requires $d_1(x) \equiv 0$. We evaluate the coefficient $c_1(x)$ by applying the shear-stress condition (6–246b). Because θ_0 is independent of z, and θ_1 is independent of x, this reduces to

$$\frac{\partial u_1}{\partial z} + h_1 \frac{\partial^2 u_0}{\partial z^2} = 0 \quad \text{at } z = 1.$$

Substituting for u_1, h_1, and u_0, we find that this gives

$$c_1(x) = -\frac{\partial p_1}{\partial x} + \frac{3}{2} h_1(x)$$

or

$$u_1 = \frac{\partial p_1}{\partial x}\left(\frac{z^2}{2} - z\right) + \left[\frac{3}{2} h_1(x)\right] z. \qquad (6\text{–}255)$$

The pressure gradient is determined from the $O(\delta)$ contribution to the zero-flux condition, (6–168). To apply this condition, we note that

$$u_0\left\{\bar{z}\left[1 + h_1\delta + O(\delta^2)\right]\right\} = u_0(\bar{z}) - \left[\frac{3}{2}\left(v\bar{z}^2 - \frac{\bar{z}}{3}\right) h_1\right]\delta + O(\delta^2),$$

so that (6–168) becomes at $O(\delta)$

$$\int_0^1 \left[-\frac{3}{2}\left(\bar{z}^2 - \frac{\bar{z}}{3}\right) h_1 + \frac{\partial p_1}{\partial x}\left(\frac{\bar{z}^2}{2} - \bar{z}\right) + \left(\frac{3}{2} h_1\right)\bar{z}\right] d\bar{z} = 0. \qquad (6\text{–}256)$$

Evaluating the various integrals, we find that this gives

$$\frac{\partial p_1}{\partial x} = \frac{3 h_1(x)}{2}. \qquad (6\text{–}257)$$

As usual, we can also derive the same result by first solving for w_1 by means of the continuity equation, (6–216b), plus the impermeability condition, (6–218a), at $z = 0$, and then applying the $O(\delta)$ contribution to the kinematic condition, (6–244b). The result for w_1 is

$$w_1 = -\frac{\partial^2 p_1}{\partial x^2}\left(\frac{z^3}{6} - \frac{z^2}{2}\right) - \frac{3}{4}\frac{dh_1}{dx} z^2, \qquad (6\text{–}258)$$

and the kinematic condition then requires

$$\frac{\partial^2 p_1}{\partial x^2} = \frac{3}{2}\frac{\partial h_1}{\partial x},$$

which is equivalent to (6–257).

Finally, from the normal stress balance,

$$\frac{d^3 h_2}{dx^3} = -\frac{\partial p_1}{\partial x} = -\frac{3}{2} h_1.$$

Clearly, this is identical to (6–240), and the solution has already been calculated as (6–241).

The Thin-Gap Approximation – Films with a Free Surface

NOTES

1. A. Oron, S. H. Davis, and S. G. Bankoff, Long-scale evolution of thin liquid films, *Rev. Mod. Phys.* **69**, 931–80 (1997).
2. The analysis and discussion in this section is adapted from a paper written by H. A. Stone, Partial differential equations in thin film flows in fluid dynamics: spreading droplets and rivulets, which appeared in *Nonlinear PDEs in Condensed Matter and Reactive Flows*, edited by H. Berestycki and Y. Pomeau (Kluwer, Dordrecht, The Netherlands, 2002).
3. R. E. Pattle, Diffusion from an instantaneous point source with a concentration dependent coefficient, *Q. J. Mech. Appl. Math.* **12**, 407–9 (1959).
4. G. J. Pert, A class of similarity solutions of the nonlinear diffusion equation, *J. Phys. A Math. Gen.* **10**, 583–93 (1977).
5. M. Abramowitz and I. A. Stegun, *Handbook of Mathematical Functions* (Dover, New York, 1965).
6. H. E. Huppert, The propagation of two-dimensional and axisymmetric viscous gravity currents over a rigid horizontal surface, *J. Fluid Mech.* **121**, 43–58 (1982).
7. H. E. Huppert and J. E. Simpson, The slumping of gravity currents, *J. Fluid Mech.* **99**, 785–99 (1980).
8. M. Brenner and A. Bertozzi, Spreading of droplets on a solid surface, *Phys. Rev. Lett.* **71**, 593–6 (1993).
9. J. F. Joanny, Dynamics of wetting: Interface profile of a spreading liquid. *J. Mec. Theor. Appl.* Special Issue, 249–71 (1986).
10. H. C. Hamaker, London-van der Waals attraction between spherical particles. *Physica* **4**, 1058–72 (1937).
11. C. Maldarelli and R. K. Jain, The linear, hydrodynamic stability, of an interfacially perturbed, transversely isotropic, thin, planar viscoelastic film. I. General formulation and a derivation of the dispersion equation, *J. Colloid Interface Sci.* **90**, 233–62 (1982).
12. W. W. Zhang and J. R. Lister, Similarity solutions for van der Waals rupture of a thin film on a solid substrate, *Phys. Fluids* **11**, 2454–62 (1999).
13. A. K. Sen and S. H. Davis, Steady thermocapillary flows in two-dimensional slots, *J. Fluid Mech.* **121**, 163–186 (1982).

PROBLEMS

Problem 6–1. Gravity-Driven Flow of a Thin Film. Consider the gravity-driven flow of a thin layer of Newtonian fluid (a so-called gravity current) on an inclined plane of angle θ to the horizontal.

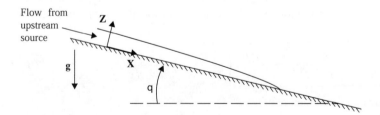

(a) Under the usual assumptions of lubrication theory, derive the following PDE for the thickness $h(x,y,t)$ of the gravity current:

$$\frac{\partial h}{\partial t} + \frac{g \sin\theta}{3\nu} \frac{\partial h^3}{\partial x} = \frac{g \cos\theta}{3\nu} \left[\frac{\partial}{\partial x}\left(h^3 \frac{\partial h}{\partial x} \right) + \frac{\partial}{\partial y}\left(h^3 \frac{\partial h}{\partial y} \right) \right], \qquad (1)$$

where x and y are the downslope and cross-slope coordinates, respectively, ν is the kinematic viscosity of the fluid, and g is the gravitational acceleration constant.

(b) Let us examine the case of a steady flow emanating from a point source of constant-volume flux Q (a crude model of a volcanic eruption). At large distances x downslope

Problems

from the source, the characteristic thickness $\bar{h}(x)$ and the cross-slope extent $\bar{y}(x)$ of the gravity current satisfy $\bar{h}(x) \ll \bar{y}(x) \ll x$. By making suitable approximations and scaling estimates, show that

$$\bar{y}(x) \sim \left(\frac{Qv}{g \sin\theta}\right)^{1/7} (x \cot\theta)^{3/7}, \tag{2}$$

$$\bar{h}(x) \sim \left(\frac{Qv}{g \sin\theta}\right)^{2/7} (x \cot\theta)^{-1/7}, \tag{3}$$

and that the thickness $h(x,y)$ satisfies

$$\frac{\partial h^3}{\partial x} = \cot\theta \frac{\partial}{\partial y}\left(h^3 \frac{\partial h}{\partial y}\right) \tag{4}$$

Here $\sim$ denotes an order-of-magnitude relationship.

(c) Seek a similarity solution to (4) of the form $h(x, y) = \bar{h}(x)H(\eta)$ with $\eta = y/\bar{y}(x)$. In particular, show that

$$H(\eta) = \frac{3}{14}(\bar{\eta}^2 - \eta^2), \tag{5}$$

where $\bar{\eta} = \left(\frac{12005}{36}\right)^{1/7}$. Sketch the function $H(\eta)$.

Problem 6–2. Axisymmetric Gravity Current in a Porous Medium. The spreading of a fluid that is due to buoyancy forces is an important process in porous media. For example, after the release of a large amount of pollutant in an aquifer, one would like to know the extent of the spreading to develop a strategy for remediation. In the process of drilling oil wells, dense drilling fluid leaks into the oil-bearing rock surrounding the well. On completion of drilling, sensors are placed in the well to characterize the properties of the reservoir, but these measurements are affected by the placement of the drilling fluid. The measurements can be corrected provided the extent of the dense drilling fluid is known.

As a model for this transport process, consider the axisymmetric spreading of a fluid of density $\rho + \Delta\rho$ in a porous media containing a fluid of density ρ. Assume the fluid spreads out over an impermeable bottom and that the volume of the dense fluid or gravity current is given by Qt^α, where t is time. The viscosity of the gravity current is μ and the permeability of the porous medium is k. Also, neglect the effects of capillary forces and assume the flow is dominated by a balance between buoyancy and viscous forces. This balance of forces in a porous medium is described by Darcy's equations, which are given by

$$\nabla \cdot \mathbf{u} = 0,$$

$$\mathbf{u} = -\frac{k}{\mu}\nabla p,$$

where $\mathbf{u}$ is the superficial velocity and p is the dynamic pressure. These equations are the conservation of mass and momentum (so-called Darcy's law) for viscous flow in a porous medium.

(a) Without solving any equations, estimate the radius of the gravity current as a function of time and the physical properties of the system. You should show that the radius $r_N(t)$ is given by

$$r_N(t) \sim \left(\frac{g'Qk}{v}\right)^{1/4} t^{(\alpha+1)/4},$$

where v is the kinematic viscosity and $g' = g\Delta\rho/\rho$ is the reduced gravity.

The Thin-Gap Approximation – Films with a Free Surface

(b) Determine the pressure field within the gravity current. Use Darcy's law to find the radial velocity as a function of the height of the gravity current, $h(r,t)$.

(c) Depth average the continuity equation in cylindrical coordinates to show that

$$\frac{\partial h}{\partial t} + \frac{1}{r}\frac{\partial}{\partial r}\int_0^{h(r,t)} ru(r,z)dz = 0.$$

(d) From the integral form of the continuity equation and the velocity field within the current, derive the nonlinear diffusionlike equation for the transport of $h(r,t)$:

$$\frac{\partial h}{\partial t} - \frac{g'k}{\nu}\frac{1}{r}\frac{\partial}{\partial r}\left(rh\frac{\partial h}{\partial r}\right) = 0,$$

where $g' = \Delta\rho/\rho$. What are the boundary conditions on this problem?

(e) Assume that $h(r,t) = Bt^p\phi(\xi)$, where $\xi = r/r_N(t)$ and

$$r_N(t) = \eta_N\left(\frac{g'Qk}{\nu}\right)^{1/4} t^{(\alpha+1)/4}$$

Determine the value of p and B to show that

$$(\xi\phi\phi')' + \frac{1}{4}(\alpha+1)\xi^2\phi' - \frac{1}{2}(\alpha-1)\xi\phi = 0,$$

$$\eta_N = \left[2\pi\int_0^1 \xi\phi(\xi)d\xi\right]^{-1/4}.$$

What are the boundary conditions on $\phi(\xi)$?

(f) What is ϕ for $\alpha = 0$? How fast does the current spread and what is its shape?

(g) For $\alpha > 0$, the equation for ϕ must be solved numerically. What are the two coupled nonlinear firstorder ODEs for this system? What are the initial conditions?

- Hint: Find $\phi'(1)$ from evaluating the governing equation for ϕ at $\xi = 1$.

(h) Solve the coupled ODEs numerically for $\alpha = 0, 1/2, 1, 3/2$, and 2 and plot the results. Note that the case for $\alpha = 0$ provides a check on your numerics. For $\alpha >; 0, \phi \to \infty$ as $\xi \to 0$. Why? (You can use any numerical method that you like, but a fourth-order Runge-Kutta scheme is available on many computer systems or from *Numerical Recipes*.)

(i) What are the values of η_N for the α's considered? How will you integrate ϕ near its singular point?

Problem 6–3. Spin-Coating of a Viscous Fluid (Fixed Volume). Computer disk drives are coated with recording medium by causing a drop of the medium to spread over the surface through rotation (gravity being too slow). To model this process, consider a large drop of a Newtonian fluid of density ρ, viscosity μ, and volume V placed on a rotating horizontal surface as illustrated in the figure. The horizontal surface and drop are rotated rapidly at an angular velocity Ω, which causes the drop to spread. It is desired to estimate the rate $R(t)$ at which the drop spreads in order to know how long we must wait to obtain a coating of a given thickness. You may assume that the drop is axisymmetric and has already spread sufficiently so that it is quite flat. If you have neglected any physical processes, please state why, in appropriate nondimensional terms.

Problems

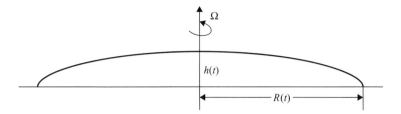

Definition sketch for spreading drop.

Problem 6–4. Spin-Coating (Increasing Volume). In the process of manufacturing silicon ships, wafers are coated with polymeric liquids that are later cured for subsequent masking and etching. The liquids are extremely viscous and do not spread sufficiently fast because of gravitational forces. To accelerate the coating process, the wafer is spun at an angular velocity Ω, which produces a centrifugal force that drives the flow. Assume that the liquid has a density ρ and a Newtonian viscosity μ. The flow may also be assumed to be radially symmetric and the volume of the blob of fluid is given by Qt^α.

(a) Without solving any equations, estimate the radius of this gravity current as a function of time and the physical properties of the system, assuming a balance between viscous and centrifugal forces.

(b) It is observed that as the radius of the current becomes large, instabilities develop at the advancing front of the current. The liquid no longer spreads uniformly in the radial direction, but forms protuberances that spread in a starfishlike pattern. What do you think causes the instability?

Problem 6–5. "Honey on a Spoon". You must have observed that when you dip a spoon into honey (or other viscous fluid) withdraw it and hold the spoon horizontal, you can keep the honey from draining by rotating the spoon. If you rotate the spoon too slowly the honey will drain, but above a critical rotation speed the honey will not drain from the spoon. (Try this with water and it will not work!) We want to analyze this problem and see if we can't predict the critical speed of rotation required to keep the honey on the spoon. The honey forms a thin layer of thickness $h(\theta)$ around the spoon. The spoon is modeled as a cylinder of radius R and rotates about its axis at angular velocity Ω so that the speed of the surface of the cylinder is $U = \Omega R$. The force of gravity acts downward and tries to pull the fluid off the cylinder.

When the layer is thin, $h/R \ll 1$, analyze the flow in the fluid film and derive an equation for the shape $h(\theta)$. Show that there is no solution for the shape when the rotation speed is less than a critical value, and determine this critical value. In film flow problems the characteristic film thickness $|h|$ and the volumetric flux (per unit length of the cylinder) of fluid carried around by the rotating cylinder are related and cannot be specified independently. In this problem it is simplest if you specify the fluid flux carried around by the rotating cylinder as Q per unit length and then determine $|h|$ in terms of Q.

Problem 6–6. Dip Coating. In coating films and other substrates it is necessary to know how thick a fluid film will be after certain time. To model this process, consider a plate that is drawn out of a bath of fluid and held vertically to allow the thin film of liquid to drain off. If the film is initially of uniform thickness h_0, show that

$$h(z, t) = \left(\frac{\nu z}{gt}\right)^{1/2} \quad \text{for } 0 < z < \frac{gh_0^2}{\nu}t$$

$$= h_0 \quad \text{for } z > \frac{gh_0^2}{\nu}t,$$

where z denotes the distance downward from the upper edge of the plate.

The Thin-Gap Approximation – Films with a Free Surface

Problem 6–7. Coating Flows: The Drag-Out Problem A very important model problem in coating theory is sometimes called the "drag-out problem." In this problem, a flat plate is pulled through an interface separating a liquid and a gas at a prescribed velocity U. The primary question is to relate the pull velocity U to the thickness H' of the thin film that is deposited on the moving plate. We consider the simplest case in which the plate is perpendicular to the horizontal interface that exists far from the plate. The density of the liquid is ρ, the viscosity μ, and the surface tension γ.

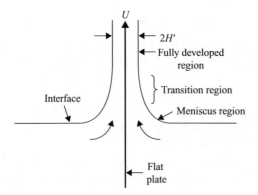

In the case in which the capillary number,

$$Ca \equiv \frac{\mu U}{\gamma},$$

is asymptotically small, this problem can be analyzed by use of the thin-film approximations in this chapter. In particular, we can consider the flow problem as consisting of three regions: a *fully developed flow* region, where the film thickness is constant (H'); a *transition region* where the dominant balance is between viscous and capillary forces and a thin-film analysis can be shown to be appropriate; and a *meniscus region* where the dominant force balance is between capillary and gravitational forces. It may be noted that there are two obvious natural length scales,

$$\ell_1 \equiv \left(\frac{\mu U}{\rho g}\right)^{1/2} \quad \text{and} \quad \ell_2 \equiv \left(\frac{\gamma}{\rho g}\right)^{1/2}.$$

One might logically assume that ℓ_1 should be an appropriate length scale for the fully developed flow region, where the dominant balance is between viscous stresses and gravitational forces, so that $H' \sim \ell_1$. On the other hand, ℓ_2 would seem to be an appropriate length scale for the meniscus region where the balance is between capillary and gravitational forces. We shall see, however, that capillary forces actually play a very important role in the formation of the thin film, and

$$H' \sim \ell_1 Ca^{1/6}$$

in the limit $Ca \ll 1$.

(a) Assume that the thin-film approximation is valid, i.e., $\ell_c^x = \varepsilon^{-1}\ell_c^y$ and $v_c = \varepsilon u_c$, where ε is a small parameter (yet to be determined) in the limit $Ca \to 0$, and ℓ_c^x, ℓ_c^y, u_c, and v_c are the characteristic length and velocity scales in the x and y directions, respectively. Denote the *dimensional* thickness of the "film" as $h'(x')$ and the fully developed thickness of the film (for $x' \to -\infty$) as H'.
Derive the governing equation for $h'(x')$:

$$\frac{1}{3}\frac{\gamma}{\rho}h'^3\frac{d\kappa'}{dx'} + \frac{1}{3}g(h'^3 - H'^3) - \frac{\mu U}{\rho}(h' - H') = 0, \tag{1}$$

Problems

where

$$\kappa' \equiv \frac{d^2 h'/dx'^2}{[1 + (dh'/dx')^2]^{3/2}}. \tag{2}$$

(b) Now consider nondimensionalization in the transition region. In particular, assume that the characteristic length scale in the y direction is

$$\ell_c^y = \delta \ell_1,$$

where δ is a dimensionless number (or parameter) that remains to be determined. Using the fact that the dominant balance is between viscous and capillary forces in this region, show that

$$\varepsilon = Ca^{1/3} \ll 1$$

for $Ca \ll 1$. Hence

$$\ell_c^x = \varepsilon^{-1} \ell_c^y = Ca^{-1/3} \ell_c^y$$

and thus the thin-film approximation is verified in this region.

Write the dimensionless form of the equation for $h'(x')$, assuming that both h' and H' scale as ℓ_c^y. For convenience, denote the dimensionless variables in this region as x, y, h, and H.

(c) It would appear that a different analysis will be necessary in the meniscus region, as it is clear that the thin-film approximation will break down at the bottom of the thin film as we approach the undisturbed interface.

Start with the full 2D equations of motion in the Cartesian coordinate system (x', y'). Assume that ℓ_2 is the appropriate characteristic length scale in both the x and y directions. Hence, in this case, the dimensionless variables are

$$\hat{x} = x'/\ell_2, \quad \hat{y} = y'/\ell_2, \quad \text{and} \quad \hat{h} = h'/\ell_2.$$

Show, in the limit $Ca \ll 1$, that the problem for calculating $\hat{h}$ reduces to a simple balance between the body force and capillary forces:

$$0 = 1 + \frac{d\hat{\kappa}}{d\hat{x}} + O(Ca). \tag{3}$$

Now, show that, if the same scaling $\ell_c = \ell_2 = (\gamma/\rho g)^{1/2}$ is incorporated into the thin-film equation (1) for h', this equation also reduces to Eq. (3) at the leading order of approximation for $Ca \ll 1$. [Hint: h' will scale with ℓ_2 in this region, but the ultimate film thickness H' must still be $O(\delta \ell_1)$.] This means that Eq. (1) could actually have been applied to the whole of the meniscus and transition regions. In the meniscus region, where the thin-film scaling breaks down, we are fortunate in that the dominant capillary and gravity terms are correctly represented provided only that we retain the full exact form (2) for the curvature κ.

(d) The governing equation for $\hat{h}$ can be seen from (3) to be

$$\frac{d}{d\hat{x}} \left[\frac{\hat{h}_{\hat{x}\hat{x}}}{(1 + \hat{h}_{\hat{x}}^2)^{3/2}} \right] + 1 = 0. \tag{4}$$

We may note that this equation is invariant to the position of the origin. Furthermore, the ratio of the length scales in this region to that in the transition region is

$$\frac{\ell_2}{\delta \ell_1} = O\left(\frac{1}{\delta} \left(\frac{\gamma}{\rho g} \right)^{1/2} \left(\frac{\rho g}{\mu U} \right)^{1/2} \right) = O\left(\frac{1}{\delta Ca^{1/2}} \right) \gg 1$$

The Thin-Gap Approximation – Films with a Free Surface

in the limit $Ca \to 0$, provided of course that δ is not large (we shall have to verify this). This means that, as we approach the transition region from the meniscus region, the film thickness will appear from the meniscus point of view to go to zero. We choose to select the origin $\hat{x} = 0$ as the point where the film thickness goes to zero, and thus (4) is subject to the boundary conditions

$$\hat{h} = \hat{h}_{\hat{x}} = 0 \quad \text{at} \quad \hat{x} = 0. \tag{5}$$

At the other end of the meniscus region, which we denote as $\hat{X}_n$

$$\hat{h}_{\hat{x}} \to \infty \quad \text{at} \quad \hat{x} = \hat{X}_n, \tag{6}$$

Now integrate (4) twice with respect to $\hat{x}$. From this, you should be able to show that for $\hat{x} \ll 1$

$$\hat{h} \sim \frac{\hat{X}_n}{2}\hat{x}^2 + O(\hat{x}^3). \tag{7}$$

(e) To determine the shape function h in the transition region, we must solve the governing equation from part (b) subject to *matching* for $x \gg 1$ with the solution (6) for $\hat{x} \ll 1$ in the meniscus region. The dimensionless governing equation for the transition region in the limit $Ca \to 0$ and thus $\delta \to 0$ (this remains to be confirmed) takes the form

$$h^3 \frac{d^3 h}{dx^3} = 3(h - H) + O(\delta^2). \tag{8}$$

Generally, this equation must be solved numerically with $h \to H$, $h_x \to 0$, $h_{xx} \to 0$ as $x \to -\infty$, plus matching with the solution in the meniscus region for $x \to \infty$. Why do you need four boundary conditions? However, you should use Eq. (7) to show that the functional form of the solution for $x \gg 1$ must be

$$h \sim Ax^2 + o(x^2), \tag{9}$$

where A is a constant. (Hint: recall the ratio of length scales $\ell_2/\delta\ell_1 \gg 1$.)

(f) Finally, consider the matching between (7) and (9) in the limit $Ca \to 0$. You should be able to show that $\delta = Ca^{1/6}$ is necessary for matching and that $A = \hat{X}_n/2$. This is the most important result of the whole analysis because it shows that the appropriate scaling of the film width in the transition region is $\ell_1 Ca^{1/6}$. It follows directly that the final film thickness retained on the moving plate is

$$H' = \left(\frac{\mu U}{\rho g}\right)^{1/2} Ca^{1/6} H,$$

where H is an $O(1)$ constant.

Note: To determine numerical values for H and $\hat{X}_n$ we must first numerically solve Eq. (4) subject to the conditions (5) and (6) to obtain $\hat{X}_n$. With A then specified by the matching condition, the numerical solution of (8) subject to the three conditions for $x \to -\infty$ and the matching condition for $x \gg 1$ will allow one to determine the scaled film thickness H.

Problem 6–8. Marangoni-Driven Flow in a Shallow Cavity. In the problem that was considered in Subsection E.3 of this chapter, we described the solution for Marangoni flow in a shallow 2D cavity with a no-slip, thermally insulated lower boundary. In this problem, we consider a generalization of this problem that brings it closer to the case of the crystal growth configuration that was sketched in Fig. 6–8. Re-solve the problem from Subsection 3 assuming that the lower boundary is a nondeforming slip surface (alternatively, we could

Problems

consider the problem as that of a 2D cavity with both the upper and lower surfaces as interfaces and the plane surface at $z = 0$ as a symmetry surface). You may neglect any motions driven by density gradients.

Problem 6–9. Marangoni-Driven Flow in a Capillary Bridge. As a model for investigating the fluid flow associated with the crystal growth configuration in Fig. 6–8, we consider Marangoni-driven flow in an elongated axisymmetric liquid "bridge" that is suspended by interfacial tension effects between two circular disk-shaped boundaries that are held at different temperatures T_0 and T_1. The radius of these disklike end plates is R. We assume that the undeformed (static) shape of the liquid bridge is a circular cylinder of radius R and length L. The liquid bridge is pinned to the edges of the end plates so even with flow its radius is R at $z' = 0$ and L.

(a) By adapting the derivation of the governing equations and boundary conditions for a thin film from Section E, determine the governing equations and boundary conditions that can be used to determine the radius of the liquid bridge as a function of z, assuming that the length of the bridge L is much greater than its undeformed radius R. You may also assume that gravitational effects are negliglible.

(b) Obtain a solution of these equations to determine a first approximation to the flow in the liquid bridge and a first approximation for the modification of the temperature profile from the static distribution.

(c) Based on the results of part (b), obtain a first approximation to the deformation in the shape of the liquid bridge.

Problem 6–10. An Alternative Derivation of (6–146). In Subsection E.2.a, the Eq. (6–146) was derived by application of the zero-flux condition to expression (6–145). Although this is by far the easiest way to achieve (6–146), it must also be true that this equation is equivalent to the Reynolds equation for this problem. Show that one can utilize the same procedure used to derive the Reynolds equation at the beginning of this chapter (Section B) – that is, by using the continuity equation – either directly in the integrated form or indirectly by first deriving an expression for the velocity in the z direction.

Problem 6–11. An Asymptotic Solution of the Thin-Film Profile Equation, (6–156). We have seen in subsection E.2.a that the solution for the film profile equation, (6–156), for weak capillary effects, i.e., $(GCa)^{-1} \ll 1$, is singular in the sense that capillary forces cannot be neglected near the end walls. Set up and obtain the first approximation for the solution of (6–156) in this limit.

Problem 6–12. The Effect of Surfactant on the Capillary-Driven Spread of a Liquid Drop. In considering the spreading of a drop by either gravitational or capillary-driven forces, we neglected the possibility of surface-tension gradients. Here, we reconsider these problems with surface-tension gradients included, caused by the presence of a surfactant. The surfactant is assumed to be insoluble (i.e., it exists only at the interface, so that there is no exchange with the fluid inside the drop). We denote the concentration of surfactant (per unit interface area) as Γ.

(a) Assume that τ_s^* is negligible in (6–12) and that the lower boundary of the thin film is horizontal, namely $\alpha = 0$. Write the (dimensionless) governing equation for the film thickness using the notation of a characteristic velocity u_c and characteristic length scale ℓ_c. Show that the surface transport of surfactant at the drop surface is governed by

$$\frac{\partial \overline{\Gamma}}{\partial t} + \nabla_s \cdot [\mathbf{u}_s^{(0)} \overline{\Gamma}] = \frac{1}{Pe_s} \nabla_s^2 \overline{\Gamma} + O(\varepsilon),$$

where

$$\bar{\Gamma} \equiv \Gamma/\Gamma_0,$$

and Γ_0 is the concentration of surfactant on the drop at $t = 0$, when its radius is R_0 and its maximum height is H_0, assuming that the surfactant is uniformly distributed over the drop surface. Here, the parameter Pe_s is the surface Peclet number,

$$Pe_s \equiv u_c \ell_c / D_s,$$

where D_s is the surface diffusivity for the surfactant along the interface.

For the remainder of this problem, we assume that

$$\sigma = \sigma_0 \hat{\sigma} \quad \text{with} \quad \hat{\sigma} = 1 - \beta \hat{\Gamma}$$

and

$$\hat{\Gamma} \equiv \frac{\Gamma}{\Gamma_0} - 1 = \bar{\Gamma} - 1.$$

(b) Now, let us consider the case of gravitationally driven spreading. Assuming that u_c still scales according to (6–62), what condition must be satisfied in order to neglect surface-tension-gradient-driven flow in the spreading process? Is there any possibility that this effect should be included while still neglecting the capillary pressure contribution to $p^{(0)}$?

(c) Now consider the case in which the spreading process is driven by the capillary pressure contribution to $\nabla_s p^{(0)}$, so that u_c is given by (6–76). Write the governing (dimensionless) equations for h and $\hat{\Gamma}$. Be sure to specify $p^{(0)}$ and $\mathbf{u}_s^{(0)}$. What is the condition for neglect of the gravitational contributions? We wish to consider circumstances in which the Marangoni (surface-tension gradient) contribution is a small correction to the capillary-pressure-driven spreading process.

   (i) For arbitrary Pe_s, what condition must β satisfy in order that the Marangoni effect appears as a correction of $O(\varepsilon^\alpha)$ where $\alpha < 1$? (note that, if $\alpha \geq 1$, additional terms must be considered in the transport equation for $\hat{\Gamma}$ in order to account for the decrease in $\hat{\Gamma}$ that is due to the increase in surface area as the drop spreads.)

   (ii) Now consider the asymptotic limit in which $Pe_s \ll 1$. Derive the governing equations and any necessary boundary conditions to obtain the first, leading order corrections to the spreading rate at $O(Pe_s)$. From a qualitative point of view, will the effect of surfactant concentration gradients be to slow down or to speed up the spreading rate? Explain your answer.

Problem 6–13. Buoyancy-Driven Circulation in a Shallow Cavity. A variation on the problems discussed in Section E is to consider the circulation of fluid within a shallow cavity driven by buoyancy effects when there is a temperature difference between the two ends of the cavity. This has been used as a model problem for the recirculation flows within an estuary.

(a) Set up the problem for the temperature and velocity fields in both the central core region of the cavity and also at the ends. Consider two cases. In one, the upper and lower boundary are imagined as nondeformable no-slip surfaces. In the second, the upper surface is an interface, though in this case we neglect any Marangoni contributions to the recirculation motion.

(b) Solve the equations in the core region to leading order for the temperature and velocity distributions.

Problems

(c) What is the connection with the end regions? Set up the problem for the leading order approximation to the velocity and temperature distributions in the end regions, though you do not have to solve this problem.

Problem 6–14. Surfactant Spreading on a Thin Film. A very important and interesting application of the general theory in this chapter is the spreading of surfactant on the surface of a thin film. One example, which has driven a great deal of the research on this problem, emanates from the use of aerosols that are sprayed into the lungs with the intent of delivering surfactant or other medicines to the thin mucus lining that coats the surfaces of the lungs. The aerosol arrives in the lungs as small droplets that land on the mucus lining, and then must spread rapidly and spontaneously over the surface of the mucus lining to be effective. In this problem, we consider a simple model calculation that provides an estimate of the rate at which an initial 2D or axisymmetric "source" of surfactant spreads over a thin, Newtonian film. The reader interested in following the literature to see more advanced studies can start with the paper "Surfactant spreading on thin viscous films: film thickness evolution and periodic wall stretch," J. L. Bull and J. B. Grotberg, *Exp. Fluids* **34**, 1–15 (2003).

We assume that we have a thin film. A fixed amount of surfactant is deposited onto the surface of this thin film, either as a 2D "strip" or as an axisymmetric drop. We assume that this surfactant is insoluble in the bulk fluid of the thin film. We denote the total (fixed) mass of surfactant as M, which may be related to the surface concentration Γ (measured in units of mass per unit surface area) in a form that encompasses both the 2D and axisymmetric geometries:

$$M = 2\pi^{d-1} \int_0^\infty \Gamma r^{d-1} dr,$$

where $d = 1$ for the case of the 2D strip whose centerline is at $r = 0$ [here we identify r as the Cartesian variable x, say, and the strip extends from $-X(t)$ to $+X(t)$] and $d = 2$ for the axisymmetric drop case (here r is the cylindrical coordinate measured from the axis of symmetry). Far from the origin, we assume that $\Gamma = 0$, and we denote the corresponding surface tension as σ_s. Elsewhere, we assume that there is a linear relationship between σ and Γ of the form

$$\sigma = \sigma_s (1 - \beta \Gamma).$$

The presence of gradients of Γ (and thus of σ) on the interface drives a flow in the thin film that is mainly responsible for the spread of surfactant on the surface of the film.

(a) Obtain dimensionless governing equations for the film thickness h and for the concentration of surfactant Γ, assuming that the film is sitting on a flat, horizontal substrate ($\alpha = 0$). The starting point for the latter is the transport equation for surfactant that was derived in Chap. 2:

$$\frac{\partial \Gamma}{\partial t'} + \nabla_s' \cdot (\Gamma \mathbf{u}_s') + \Gamma (\nabla_s' \cdot \mathbf{n})(\mathbf{u}' \cdot \mathbf{n}) = D_s \nabla_s'^2 \Gamma.$$

Assume that the film is thin, $O(\varepsilon)$, and denote the characteristic velocity in the film as u_c and the horizontal length scale of the spreading surfactant layer as ℓ_c. An appropriate choice for ℓ_c is R_0 (i.e., the initial width of the surfactant film). Show that u_c is

$$u_c = \frac{\varepsilon}{\mu} \sigma_s \beta.$$

In the spread of a surfactant film that is due to Marangoni-driven flow in the thin film, the surface diffusion of surfactant and gravitational and capillary contributions to the motion in the film are often assumed to be negligible. What conditions, in terms of the magnitudes of dimensionless groups, must be satisfied for these assumptions to be valid?

(b) Now simplify the governing equations neglecting the diffusion, gravitational, and capillary pressure terms. You should ultimately find that

$$\frac{\partial h}{\partial t} - \nabla_s \cdot \left(\frac{h^2}{2} \nabla_s \Gamma\right) = 0,$$

$$\frac{\partial \Gamma}{\partial t} - \nabla_s \cdot (\Gamma h \nabla_s \Gamma) = 0.$$

Note that a common form for the second term in each of these equations for the 2D and axisymmetric problems is

$$\nabla_s \cdot (\zeta \nabla_s \psi) = \frac{1}{r^{d-1}} \frac{\partial}{\partial r} \left(\zeta r^{d-1} \frac{\partial \psi}{\partial r}\right).$$

(c) Try to obtain s similarity transformation of the equations from part (b) in the form

$$h(r, t) = H_0(\eta), \quad \Gamma(r, t) = G_0(\eta)/t^b, \quad \text{and} \quad \eta = r/t^a.$$

You should find that this is possible iff

$$2a + b = 1.$$

(d) Now, given that M is independent of t, show that

$$a = 1/3, \quad b = 1/3 \quad \text{for a 2D strip},$$
$$a = 1/4, \quad b = 1/2 \quad \text{for the axisymmetric geometry}.$$

This suggests that a planar strip of surfactant will spread as $t^{1/3}$, whereas an axisymmetric source of surfactant will spread as $t^{1/4}$.

You do not need to attempt to solve the resulting ODEs for H_0 and G_0.

7

Creeping Flows – Two-Dimensional and Axisymmetric Problems

In the preceding chapters, a number of asymptotic methods were introduced for the approximate solution of nonlinear flow problems. In many of the cases considered so far, including all of the problems of the two preceding chapters, the asymptotic limiting process produced a simplification of the full nonlinear problem by restricting the *geometry* of the flow *domain* to one in which certain terms in the equations could be neglected because the length scales in some direction (or directions) become very large compared with the length scales in other directions. In retrospect, even the exact unidirectional flow problems of Chap. 3 can often be regarded as a first approximation of some more general problem in which the geometry reduces to a unidirectional form in the limit as a ratio of two length scales vanishes, e.g., the "Dean" problem of Chap. 4, which reduces to the unidirectional Poiseuille flow problem in the limit as the ratio of the tube radius to the radius of curvature of the tube in the axial direction goes to zero. In some cases, this disparate ratio of length scales was true everywhere in the flow domain, and then the asymptotic solution was found to be "regular"; e.g., the Dean problem or the eccentric cylinder problem of Chap. 5. In others, the region with a small length-scale ratio was restricted to a local part of the overall flow domain, and in these cases, the asymptotic approximation was of the "singular" type, in which the simplified form of the governing equations is valid only locally, and the resulting approximate solution must be "matched" to a solution of the unapproximated equations that are valid elsewhere.

This chapter represents the first step toward a more general class of asymptotic problems, in which we consider approximations that are not restricted to special geometries of the flow domain. We begin with approximations of the Navier–Stokes and continuity equations for isothermal flows of an incompressible fluid. It was shown in Chap. 3 that nondimensionalization reveals the dimensionless combination(s) of independent parameters that control the form of the solution of a DE (or set of equations) that describes some physical process. In general, for isothermal flow of an incompressible, Newtonian fluid in a domain with solid, fixed boundaries, we shall see that there is a single dimensionless group, called the *Reynolds* number, that determines the form of solutions to the Navier–Stokes and continuity equations. When this parameter is very small, the (linear) viscous terms in the equation are dominant over the (nonlinear) inertial or acceleration terms, and a linear approximation of the equations is thus possible in the asymptotic limit as the Reynolds number approaches zero. The class of fluid motions where this approximation can be used is known as *creeping flow*, and this, and the following chapter focuses on detailed analysis for many motions of this type. In subsequent chapters we shall consider flows and convective transport processes for heat or mass transfer under circumstances for which this linearized approximation is not valid. In so doing, we shall see that the creeping-flow solutions from this and the next chapter actually represent the first approximation of an asymptotic solution for the limit of arbitrarily small, but nonvanishing, Reynolds number.

Creeping Flows – Two-Dimensional and Axisymmetric Problems

A. NONDIMENSIONALIZATION AND THE CREEPING-FLOW EQUATIONS

To begin, we restrict our attention to isothermal, laminar flow of an incompressible Newtonian fluid, in which all boundaries are solid surfaces, so that we can use the equations of continuity and motion in the forms (2–20) and (2–91), respectively, that is,

$$\boxed{\rho\left(\frac{\partial \mathbf{u}'}{\partial t'} + \mathbf{u}' \cdot \nabla' \cdot \mathbf{u}'\right) = -\nabla' p' + \mu \nabla'^2 \mathbf{u}',\\ \nabla' \cdot \mathbf{u}' = 0.}$$

These equations contain two independent dimensional parameters, the density ρ and the viscosity μ, which are both constant and assumed to be known. For convenience in what follows, we mark the dimensional variables that appear in these equations with a prime – $\mathbf{u}'$, p', and so on – and, in addition, denote the dimensional gradient operator as ∇'.

To completely characterize the flow, we require, in addition to these equations, the geometry of the flow domain and the boundary conditions that apply at the boundaries of the domain, including the form and magnitudes of the velocity distribution(s) on any inlet to the domain (or the form and magnitude of the velocity field far from any boundaries in an unbounded domain) and the velocities of the bounding surfaces if these are moving in the laboratory frame of reference. From these additional conditions, at least three independent dimensional parameters can normally be specified: a characteristic length scale, ℓ_c; a characteristic velocity scale, U_c; and a characteristic time scale T_c. With these characteristic parameters, plus ρ and μ, it is possible to nondimensionalize (2–20) and (2–91).

Of course, in some instances there may be more than a single length, velocity, or time scale that is evident in a problem. We have encountered this situation previously, for example in Section B of Chap. 3. When this happens, we should endeavor to choose the length, velocity, and time scales for ℓ_c, U_c, and T_c according to the understanding that we have from Chap. 3 of the significance of "characteristic" scales. Specifically, the characteristic length scale ℓ_c is intended to represent the physical distance over which the velocity changes by an amount proportional to U_c. In most cases, this length scale will be evident from the geometry of the flow domain. For example, for translation of a rigid sphere through a large body of liquid, the sphere radius would appear to be an appropriate choice. Similarly, the characteristic velocity scale U_c represents a typical magnitude of fluid velocities in the flow domain and generally will be set by either the boundary velocity or the magnitude of the velocity at large distances from the boundary in an unbounded domain; in the translating sphere problem, for example, an appropriate choice would be the sphere velocity. Finally, the characteristic time scale T_c must provide a measure of the rate of change of velocity as seen by a typical fluid element. Its definition will depend substantially on the details of the particular flow. For example, if either the motion of a boundary or the flow at large distances is time dependent and periodic, then T_c is just proportional to the inverse of the frequency. If the geometry of the flow domain is changing with time, this may provide an appropriate characteristic time scale. Finally, if the flow and geometry are both steady from an Eulerian point of view, the time scale T_c may be just ℓ_c/U_c, that is, the time for a "typical" fluid element to travel a distance ℓ_c.

The reader may again be curious whether there are consequences of making the "wrong" choice for ℓ_c, U_c, or T_c when there is more than one possibility available? We will need to discuss this point in some detail, once we see how we intend to use the nondimensionalized versions of our governing equations (and boundary conditions) in the development of asymptotic approximations. For now, we simply assume that the appropriate choices have been made. Then, for each additional dimensional scale that appears in a particular problem, we get one more dimensionless parameter, in addition to the two that will appear based on

A. Nondimensionalization and the Creeping-Flow Equations

the five independent parameters ρ, μ, ℓ_c, U_c, and T_c. If the added parameter is a second length scale, then the additional dimensionless parameter will be a ratio of this length scale to ℓ_c. Additional velocity or time scales will similarly produce ratios of these quantities to U_c or T_c.

For present purposes, we assume that there is only a single length, velocity, and time scale available for ℓ_c, U_c, and T_c. To express (2–20) and (2–91) in dimensionless form, we introduce dimensionless dependent and independent variables as follows:

$$\mathbf{u} = \frac{\mathbf{u}'}{U_c},$$
$$t = \frac{t'}{T_c},$$
$$\nabla = L_c \nabla', \quad (7\text{–}1)$$
$$p = \frac{p'}{\left(\frac{\mu U_c}{L_c}\right)}.$$

The reader may wish to verify that the dimensionless group $\mu U_c / L_c$ has dimensions of force per unit area and is thus appropriate as a choice for the characteristic pressure. Substituting (7–1) into (2–20) and (2–91), as given at the beginning of this chapter, we find

$$Re\left(\frac{1}{S}\frac{\partial \mathbf{u}}{\partial t} + \mathbf{u} \cdot \nabla \mathbf{u}\right) = -\nabla p + \nabla^2 \mathbf{u}, \quad (7\text{–}2)$$

$$\nabla \cdot \mathbf{u} = 0. \quad (7\text{–}3)$$

Two dimensionless parameters appear in (7–2), known, respectively, as the *Reynolds number*,

$$Re \equiv \frac{\rho U_c L_c}{\mu}, \quad (7\text{–}4)$$

and the *Strouhal number*,

$$S \equiv \frac{T_c}{(L_c/U_c)}. \quad (7\text{–}5)$$

The fact that only two dimensionless combinations of the dimensional parameters appear in (7–2) and (7–3) is one demonstration of the so-called *principle of dynamic similarity*. This principle is nothing more than the observation that the form of the velocity and pressure fields for a typical flow problem will depend on the dimensionless parameters (7–4) and (7–5) rather than on any of the dimensional parameters alone. It may not be obvious without examining the details of specific flow problems that additional dimensionless parameters will not appear in the boundary conditions. However, with the exception of problems that involve multiple scales, as previously described, no additional parameters will appear for flows that involve only solid boundaries.

For steady flows, examination of (7–2) shows that the Reynolds number determines the relative magnitudes of the acceleration terms on the left-hand side and the viscous and pressure-gradient terms on the right. In the case of unsteady flows, the relative magnitude of the Eulerian acceleration is determined by the ratio of the Reynolds number to the Strouhal number, Re/S (note that for steady flows we have already suggested that $T_c = \ell_c/U_c$, and in this case $S = 1$). In particular, in the limit as the Reynolds number becomes small (that is, $Re \ll 1$ and $Re/S \ll 1$), it appears that the acceleration (or inertia) terms in (7–2) will become negligible compared with viscous and pressure-gradient terms. Although we will ultimately see that the situation can be more complicated than it first appears, the resulting linear equations,

$$\nabla^2 \mathbf{u} - \nabla p = 0, \qquad (7\text{–}6\text{a})$$

$$\nabla \cdot \mathbf{u} = 0, \qquad (7\text{–}6\text{b})$$

do represent a valid first approximation to the Navier–Stokes and continuity equations for the limit of very small Reynolds numbers. These equations are known as the *creeping-flow equations*. Clearly, the creeping-flow regime can be reached through a combination of very small velocity or length scales or a very large kinematic viscosity $\nu \equiv \mu/\rho$. These conditions encompass many important flow problems, including chemical, biological, and materials processing applications, as well as many of the very small-scale flows of MEMS and microfluidics systems.[1]

The most important feature of the creeping-flow equations is that they are linear and thus can be solved by a number of well-known methods for linear DEs. In this limited sense, they are similar to the unidirectional flow equations of Chap. 3. However, unlike unidirectional flows, the velocity and pressure fields in the creeping-flow limit can be fully 3D. There is another extremely significant difference between unidirectional and creeping flows. For a flow that is exactly unidirectional, the nonlinear term $\mathbf{u} \cdot \nabla \mathbf{u}$ in the Navier–Stokes equation is identically equal to zero, simply because $\mathbf{u}$ and $\nabla \mathbf{u}$ are orthogonal. On the other hand, in the creeping-flow limit, the nonlinear terms are not identically zero. They are simply neglected as an approximation for small values of R.

The creeping-flow equations involve the dual limit $R \ll 1$ and $R/S \ll 1$ (or simply $R \ll 1$ when the flow is steady and $S = 1$), and the time derivative in the equations of motion is also neglected. For this reason, creeping flows are sometimes called *quasi-steady*. Such a flow can depend on time only as a parameter rather than as a true independent variable. The simplest example is when a boundary of the flow domain has a velocity that depends on time. For example, suppose we were to consider a sphere in an unbounded fluid that is translating back and forth with a velocity that is a periodic function of time, say $U' = U \sin \omega t'$. (In this case, we could define $Re = Ua/\nu$ with $T_c = \omega^{-1}$ so that $S = U/a\omega$). In this case, if $Re \ll 1$ and $Re/S \ll 1$, we would determine the velocity and pressure fields by solving the creeping-flow equations (7–6). Because the velocity of the sphere depends on time, it is obvious that the velocity and pressure fields in the fluid must also depend on time, even though the time derivative $\partial \mathbf{u}/\partial t$ in (7–2) is neglected entirely in the creeping-flow equations. To see why this is true, we can compare two time scales. The first is the time required for the sphere to change velocity by a significant amount, e.g., some fraction of U. This is clearly just $t_1 = \omega^{-1}$. The second is the time scale required for the velocity and pressure fields to achieve a steady-state form following a change in the velocity of the sphere. At low Reynolds numbers the only mechanism for transport of momentum (or vorticity) is diffusion, with the diffusivity ν. To achieve a new steady state, the velocity and pressure fields must be modified over a region of length proportional to the sphere radius a. Hence, the transient process of going from one steady-state flow to another must be the diffusive time scale $t_2 \equiv a^2/\nu$. However, the ratio of these time scales, t_2/t_1, is

$$\frac{t_2}{t_1} = \frac{a^2/\nu}{\omega^{-1}} = \frac{a^2 \omega}{\nu}.$$

A. Nondimensionalization and the Creeping-Flow Equations

But, according to the definitions previously given for Re and S, this ratio is just

$$\frac{Re}{S} \equiv \frac{Ua/\nu}{U/a\omega} = \frac{a^2\omega}{\nu} \ll 1.$$

Hence, $t_2 \ll t_1$, and thus, though the boundary velocity is time dependent, it changes so slowly when the creeping-flow limit is applicable (i.e., $Re/S \ll 1$) that the velocity and pressure fields at each instant are identical to the steady-state velocity and pressure fields for a sphere moving at constant velocity at the current, instantaneous value of $U'(t')$.

The other situation in which the solution of a creeping-flow problem may exhibit a parametric dependence on time is if the geometry of the flow domain is changing. For example, if we consider the motion of a sphere with a constant velocity U toward a planar wall, the configuration clearly changes as a function of time. As the sphere approaches the wall, the ratio of the distance between the sphere and the wall and the sphere radius changes. If the creeping-flow limit is applicable (i.e., $R \equiv Ua/\nu \ll 1$), we would solve Eqs. (7–6) to determine the velocity and pressure distributions in the fluid, and because the solution of these DEs must depend on the geometry of the solution domain (i.e., on the proximity of the sphere to the wall), the velocity and pressure fields at each instant must be different, and thus clearly dependent on time, even though the time derivative $\partial \mathbf{u}/\partial t$ in (7–2) is again neglected entirely in the creeping-flow equations. One way to see why this is true is to compare the time scales required for evolution of the velocity field from one steady state to another, relative to the time scale for a change in the geometry of the flow domain that is due to the relative motion of the sphere and the wall with a characteristic velocity U. Now, a significant change in the geometry of the flow domain occurs when the relative position of the boundaries changes by an amount proportional to the characteristic length scale of the problem, ℓ_c. In the case of the sphere, this is simply the sphere radius, $\ell_c \equiv a$. The characteristic time scale for a change in the geometry of the flow domain is thus $t_1 \equiv \ell_c/U_c = a/U$. On the other hand, at low Reynolds numbers we have already seen that the time scale required for reestablishing a steady-state flow in response to the preceding change in the geometry of the flow domain must be the diffusive time scale, $t_2 \equiv \ell_c^2/\nu$. The ratio of the two time scales, t_2/t_1, for our case of the sphere near a plane wall is thus

$$\frac{t_2}{t_1} = \frac{a^2/\nu}{a/U} = \frac{Ua}{\nu}.$$

However, this is just the Reynolds number that, according to the creeping-motion assumption, is *arbitrarily small*. In addition, in this case, $S = 1$ and $Re/S = Re \ll 1$. It follows that the velocity and pressure fields adjust *instantaneously* relative to the rate at which the geometry of the flow domain changes and therefore always appear to be at steady state with respect to the present configuration. Thus time appears in a creeping-flow solution only as a parameter that characterizes the instantaneous boundary velocity, or boundary geometry, either of which may depend on time.

One final remark should be made about the nondimensionalization that led to (7–2) and (7–3), and thence for $Re \ll 1$ to the creeping-flow equations, (7–6). This concerns the scale factor chosen to nondimensionalize the pressure. Assuming for simplicity that the geometry and boundary conditions lead to a single characteristic length, velocity, and time scale, the nondimensionalization of $\mathbf{u}'$, t', and ∇' in (7–1) is unambiguous. However, even for this simplest case there is a second possible choice for nondimensionalization of p', namely, ρU_c^2, which is another combination of dimensional parameters with dimensions of force per unit area. Let us then suppose that, instead of (7–1), we had used a dimensionless pressure $\widehat{p}$ defined as

$$\boxed{\widehat{p} = \frac{p'}{(\rho U_c^2)}.} \tag{7-7}$$

In this case, instead of (7–2), we would have obtained

$$Re\left(\frac{1}{S}\frac{\partial \mathbf{u}}{\partial t} + \mathbf{u}\cdot\nabla\mathbf{u}\right) = -Re\nabla\widehat{p} + \nabla^2\mathbf{u}, \tag{7-8}$$

and in the limit $Re \to 0$, this would appear to yield

$$\nabla^2\mathbf{u} = 0 \tag{7-9}$$

in place of (7–6a). But the approximation (7–9) cannot be correct. To see this, we need only note that (7–9) and (7–3) now comprise a set of four scalar equations for only three unknowns, that is, the three scalar velocity components. We conclude that the scaling (7–7) is inappropriate, at least for the creeping-flow limit $Re \to 0$. It is important to recognize that the correct scaling will be obtained automatically even if we start with (7–7), assuming only that we require that the pressure be retained in the governing equations. Indeed, examination of (7–8) suggests that, to retain pressure in the limit $Re \to 0$, we should introduce a rescaled pressure in the form

$$p = Re\,\widehat{p}. \tag{7-10}$$

However, this is identical to the original nondimensionalized form in (7–1). To see this, we can rewrite the right-hand side in the form

$$p = \left(\frac{\rho U_c L_c}{\mu}\right)\frac{p'}{\rho U_c^2} = \frac{p'}{\left(\frac{\mu U_c}{L_c}\right)}.$$

In the remainder of this chapter we consider solutions of various flow problems based on the approximate, linear creeping-flow equations.

B. SOME GENERAL CONSEQUENCES OF LINEARITY AND THE CREEPING-FLOW EQUATIONS

A very important consequence of approximating the Navier–Stokes equations by the creeping-flow equations is that the classical methods of linear analysis can be used to obtain exact solutions. Equally important, but less well known, is the fact that many important qualitative conclusions can be reached on the basis of linearity alone, without the necessity of obtaining detailed solutions. This, in fact, will be true of any physical problem that can be represented, or at least approximated, by a system of linear equations. In this section we illustrate some qualitative conclusions that are possible for creeping flows.

1. The Drag on Bodies That Are Mirror Images in the Direction of Motion

Let us consider two solid bodies that translate through a fluid in the creeping-flow regime and are mirror images of one another about a plane orthogonal to the direction of motion. For simplicity in picturing the situation, we may consider the bodies as being held fixed relative to a uniform, undisturbed flow that has a velocity U. A typical example of this situation is illustrated in Fig. 7–1. The two mirror-image configurations are denoted as (a)

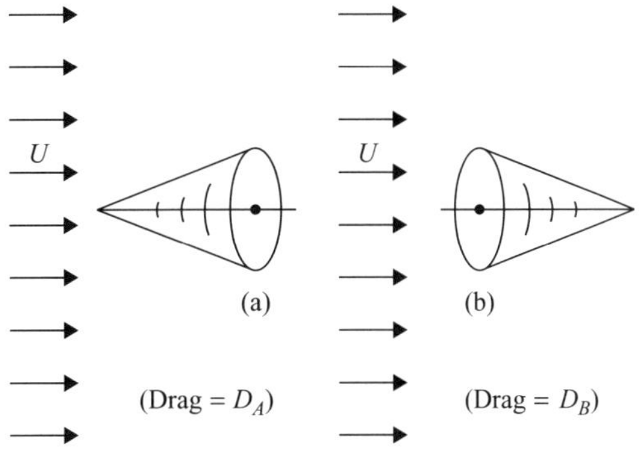

B. Some General Consequences of Linearity and the Creeping-flow Equations

Figure 7–1. A schematic representation of two mirror-image configurations of cone-shaped bodies that are held fixed in a uniform, undisturbed flow. For the limit $Re \equiv 0$, $D_A \equiv D_B$.

and (b). We wish to know whether the drag force D_A for configuration (a) is larger than, smaller than, or equal to the drag force D_B for configuration (b).

To determine D_B, we need to actually solve the creeping-flow and continuity equations,

$$\nabla^2 \mathbf{u} - \nabla p = 0,$$
$$\nabla \cdot \mathbf{u} = 0, \tag{7–11}$$

subject to boundary conditions

$$\mathbf{u} = 0 \text{ on } S(b),$$
$$\mathbf{u} = \mathbf{U} \text{ at } \infty. \tag{7–12}$$

However, once D_B has been determined, we can immediately obtain D_A.

To see that this is true, we simply note that we can obtain the detailed flow problem for configuration (a) directly from (7–11) and (7–12) by simply reversing the sign of the undisturbed flow at infinity. In particular, if the solution for configuration (b) is denoted as $(\mathbf{u}, p)$, the solution for configuration (a) is just

$$\hat{\mathbf{u}} = -\mathbf{u},$$
$$\hat{p} = -p. \tag{7–13}$$

This is a consequence of the linearity of governing equations and boundary conditions (7–11) and (7–12), plus the fact that the geometry of the boundaries are mirror images. It follows that the magnitude of the drag for configuration (a) must be identical (except for a sign change) to the drag for (b), that is,

$$D_A \equiv D_B. \tag{7–14}$$

Clearly, the result (7–14) is a consequence of linearity alone and does not require a solution of the governing equations and boundary conditions. Of course, if we want to determine the actual value of the drag, we will have to solve the problem in either configuration (a) or (b). All that (7–14) tells us is that the drag in the two configurations is equal.

This result may at first seem surprising. For most readers, it might seem that the drag should be lower for the nose-forward configuration (a) as this is more streamlined, and our everyday experience suggests that more streamlined shapes should experience less drag. However, this experience is most likely with the motion of bodies through fluids at relatively large values of the Reynolds number, Re. In fact, for finite Reynolds numbers, e.g., $Re > 1$, the drag in configuration (a) will be lower than the drag in configuration (b). It is difficult to provide an adequate explanation for this result at this point. However, one way to think about it is that the limit $Re = 0$ corresponds to the case in which the inertia (i.e., acceleration) of the fluid is negligible compared with viscous and pressure forces. On the other hand, once

Creeping Flows – Two-Dimensional and Axisymmetric Problems

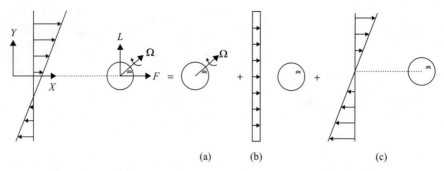

Figure 7–2. Illustration of the decomposition of the problem of a freely rotating sphere in a simple shear flow as the sum of three simpler problems: (a) a sphere rotating in a fluid that is stationary at infinity, (b) a sphere held stationary in a uniform flow, and (c) a nonrotating sphere in a simple shear flow that is zero at the center of the sphere. The angular velocity Ω in (a) is the same as the angular velocity of the sphere in the original problem. The translation velocity in (b) is equal to the undisturbed fluid velocity evaluated at the position of the center of the sphere. The shear rate in (c) is equal to the shear rate in the original problem.

Re is not vanishingly small, the inertia of the fluid will play an increasingly important role with increase of Re. In the creeping-flow case, the fluid with negligible inertia has no more difficulty traversing a path that flows the contours of the body when it is in configuration (a) than (b). Once fluid inertia enters the picture, however, the radial outward flow from the front face of the body in configuration (b) will be much more difficult to deflect back into the main flow direction, and the result will be a larger region of fluid that is disturbed by the presence of the body compared with the more streamlined configuration (a). This results in a larger drag.

2. The Lift on a Sphere That is Rotating in a Simple Shear Flow

As a second example, let us consider a solid sphere rotating with angular velocity Ω about its center, with the center held fixed in space and the fluid "at infinity" undergoing a flow that is a combination of uniform translation with velocity U and linear shear flow with a shear rate γ. The situation is sketched in Fig. 7–2. An analogous problem is the motion of a spinning baseball that is translating through the atmosphere with a velocity $-U$ in the presence of a wind that produces the shear motion because of the no-slip condition at the ground. It is well known to fans of baseball that the ball will follow a curved trajectory with a component of translational motion that is in the direction $\Omega \wedge U$. Hence, with the previously described configuration, the combination of translational and rotational motion leads to a net hydrodynamic force on the baseball that is perpendicular to the U, Ω plane. Such a force, orthogonal to the direction of motion, is usually denoted as a "lift" force. We wish to determine whether there would be any lift on the sphere in the *creeping-flow* limit. Hence, referring to Fig. 7–2, we seek to determine whether there is a force on the sphere in the direction orthogonal to the fluid motion at infinity.

The governing nondimensionalized equations and boundary conditions for this problem are

$$\nabla^2 \mathbf{u} - \nabla p = 0; \quad \nabla \cdot \mathbf{u} = 0, \qquad (7\text{–}15)$$

with

$$\mathbf{u} = \left(\frac{a}{U_c}\right) \Omega \wedge \mathbf{r} \quad \text{on the sphere } r = 1,$$

$$\mathbf{u} = \left(1 + \frac{\gamma a}{U_c} \cdot y\right) \mathbf{i}_z \quad \text{at } \infty.$$

B. Some General Consequences of Linearity and the Creeping-flow Equations

Here, we have utilized the sphere radius as a characteristic length scale and the undisturbed translational velocity evaluated at the sphere center U (see Fig. 7–2) as a characteristic velocity.

Although the problem previously posed may appear to be quite complicated because of the combination of translational, rotational, and shearing motions, a key feature of problems with linear equations and boundary conditions is that the solution can often be expressed as the sum of the solutions to a set of simpler "component" problems. In the present case, instead of directly considering the problem as posed in (7–15), we can use the superposition principle for linear problems to decompose it into a set of three simpler problems, as follows:

(1) A sphere rotating in a fluid that is stationary at infinity, that is,

$$\mathbf{u} = \frac{a}{U_c}(\mathbf{\Omega} \wedge \mathbf{r}) \text{ at } r = 1 \text{ and } \mathbf{u} = 0 \text{ at } \infty;$$

(2) a sphere held stationary in a uniform flow,

$$\mathbf{u} = 0 \text{ at } r = 1 \text{ and } \mathbf{u} = \mathbf{i}_z \text{ at } \infty;$$

(3) a sphere held stationary in a simple shear flow that is zero at the center of the sphere,

$$\mathbf{u} = 0 \text{ at } r = 1 \text{ and } \mathbf{u} = \left(\frac{\gamma a}{U_c}\right) y \mathbf{i}_z \text{ at } \infty.$$

It is a simple matter to see that the lift is zero for each of these component problems. For a sphere rotating in a stationary fluid, it is evident that there is no difference between any of the possible directions in the plane that is orthogonal to $\mathbf{\Omega}$. Hence, because there is no direction in the orthogonal direction that is distinguishable from any other, the force produced by rotation alone must be zero.

The lift on a nonrotating sphere in a uniform flow is also zero because of the symmetry of the problem. Again, because no direction perpendicular to $\mathbf{U}$ can be distinguished from any other, the only possibility is that there can be no force on the sphere in any of these directions. The only force acting on the sphere in this case will be a drag force that is collinear with $\mathbf{U}$.

Finally, the lift on a stationary sphere in simple shear flow can again be shown to be zero because of symmetry. If, for example, the lift were nonzero in the positive y direction (see Fig. 7–2), then it would also have to be nonzero in the negative y direction because the symmetry of the flow problem is such that there is no distinction between $y > 0$ and $y < 0$. Thus the lift must be zero.

Finally, because the lift is zero for each of the component problems and the overall problem is linear, it follows that the lift is zero for the overall problem, (7–15), that is,

$$\boxed{\text{lift} = \sum \text{lift from the component problems} \equiv 0.} \qquad (7\text{–}16)$$

The reader may find the result (7–16) surprising. As already noted, it is well known that a rotating and translating sphere in a stationary fluid will often experience a sideways force (that is, lift) that will cause it to travel in a curved path–think, for example, of a curve ball in baseball or an errant slice or hook in golf. The difference between these familiar examples and the problem previously analyzed is that the Reynolds numbers are not small and the governing equations are the full, nonlinear Navier–Stokes equations rather than the linear creeping-flow approximation. Thus the decomposition to a set of simpler "component" problems cannot be used, and it is not possible to deduce anything about the forces on the

Creeping Flows – Two-Dimensional and Axisymmetric Problems

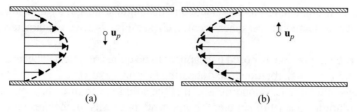

Figure 7–3. A schematic representation of the proof that a spherical particle cannot undergo lateral migration in either 2-D or axisymmetric Poiseuille flow if the disturbance flow is a creeping flow. In (a) we suppose that the undisturbed flow moves from left to right and the sphere migrates inward with velocity $\mathbf{u}_p$. Then, in the creeping-flow limit, if direction of the undisturbed flow is reversed, the signs of all velocities including that of the sphere would also have to be reversed, as shown in (b). Because the problems (a) and (b) are identical other than the direction of the flow through the channel or tube, we conclude that $\mathbf{u}_p = 0$.

body without actually solving the full fluid mechanics problem and calculating the force by integrating the stress vector $\mathbf{n} \cdot \mathbf{T}$ over the sphere surface.

3. Lateral Migration of a Sphere in Poiseuille Flow

One of the best-known experimental results for particle motion in viscous flows is the observation by Segre and Silberberg[2] of lateral migration for a small, neutrally buoyant sphere ($\rho_{sphere} = \rho_{fluid}$) that is immersed in Poiseuille flow through a straight, circular tube or in the pressure-driven parabolic flow (sometimes called 2D Poiseuille flow) between two parallel plane boundaries. The experiments of Segre and Silberberg, and many later investigators, show that a freely suspended sphere in these circumstances will slowly move perpendicular to the main direction of flow until it reaches an equilibrium position that is approximately 60% of the way from the centerline (or central plane) to the wall. Hence a suspension of such spheres flowing in Poiseuille flow through a tube of radius R will tend to accumulate in an annular ring at $r = 0.6R$. Because the Reynolds number for many of the experimental observations was quite small, one might assume that a theoretical explanation could be achieved by using detailed solutions of the creeping-flow equations with suitable boundary conditions. However, in view of the complexity of the geometry (an eccentrically located sphere inside a circular tube), this theoretical problem is extremely complex and difficult to solve, even in the creeping-flow limit. Thus, before actually trying to solve the problem, it is prudent to determine whether lateral migration is possible at all in the creeping-flow limit.

The fact is that a theory based entirely on the creeping-flow approximation will lead to the result that lateral migration is impossible, at least for a single sphere immersed in axisymmetric or two-dimensional Poiseuille flow. To see that this is true, we can refer to Fig. 7–3. Here is a sketch of the hypothetical situation of a sphere that is undergoing lateral migration in Poiseuille flow through a tube. The undisturbed flow in part (a) of Fig. 7–3 is shown moving from left to right, and the sphere is assumed to be migrating radially inward toward the center of the tube. Now, however, if the creeping-motion approximation is valid, the governing equations and boundary conditions are linear in the velocity and pressure, and we can change the signs of all velocities and the pressure and still have a solution of the same problem but with the direction of the undisturbed flow reversed, as shown in Fig. 7–3(b). However, because all the velocities have the opposite sign, the inward migration velocity from configuration (a) must now become an outward migration velocity for configuration (b). But there is now a clear contradiction. The problems in (a) and (b) are clearly indistinguishable in all respects. Thus, if the sphere undergoes a lateral motion, it should be in the same direction in both cases. Because the preceding argument, based on the linearity of the problem, shows that a nonzero migration velocity in case (a) must lead

B. Some General Consequences of Linearity and the Creeping-flow Equations

to a migration velocity of opposite sign in case (b), we can only conclude that the migration velocity must be zero in both cases. In other words, lateral migration cannot occur in the creeping-flow limit.

This conclusion does not, of course, mean that lateral migration cannot occur at all but only that the mechanism is not inherent in the creeping-motion approximation. There have been, in fact, a number of theoretical and experimental studies that show that lateral migration of a rigid sphere in a Newtonian fluid is caused by inertial effects that remain nonzero, though small, for the small Reynolds numbers that were studied experimentally. The reader may find it instructive to think about how the flow-reversal argument of the previous paragraph will fail if the nonlinear, inertial terms are retained in the Navier–Stokes equations. In this case, the solution for case (b) cannot be obtained by simply changing the sign of the solution for case (a). Although the viscous and pressure terms in the Navier–Stokes equations change sign if $(\mathbf{u}, p) \to (-\mathbf{u}, -p)$, the nonlinear term $\mathbf{u} \cdot \nabla \mathbf{u}$ does not! Hence one cannot deduce that the migration velocity must change sign if the direction of flow is reversed. The fact that the flow reversal argument cannot be applied does not prove that migration will necessarily be predicted if the Reynolds number is nonzero. We can only say that there is no simple argument, of the flow-reversal type, that is inconsistent with the existence of lateral migration.

To prove theoretically that a sphere will experience a lateral force, and to evaluate that force, we must solve explicitly for the inertial contributions to the stress $\mathbf{n} \cdot \mathbf{T}$ on the sphere surface. This was done by Ho and Leal[3] and Vasseur and Cox[4] for the case of a sphere in parallel flow between two parallel plane walls at small, but nonzero, Reynolds numbers, and for a sphere in a tube by Ishii and Hasimoto.[5] More recent work can also be found involving both analytical and numerical theory.[6] Researchers have also shown that other departures from the solid-sphere, Newtonian fluid assumptions can lead to lateral migration effects, even for flow at zero Reynolds number. For example, if the fluid is non-Newtonian, both experimental and theoretical studies have shown that lateral migration will occur at zero Reynolds number, though in this case the sphere is found to travel all the way to the centerline or to the wall, depending on the rheological properties of the fluid. For a non-Newtonian fluid, the governing equations are nonlinear in $\mathbf{u}$ because the constitutive equation is nonlinear (see Chap. 2), even if the nonlinear inertial terms are completely negligible. Another class of problems exhibiting lateral migration involves the motion of deformable particles (or drops), where migration can occur even for a Newtonian fluid at zero (or extremely small) Reynolds number. In this case, the shape of the particle is dependent on the flow, and this leads (as we will see later) to nonlinear boundary conditions so that the flow-reversal argument again does not apply. A summary of these various types of migration phenomena can be found in a review paper by Leal,[7] though of course there has been additional research since that time.

The history of many failed theoretical attempts to explain the original observations of Segre and Silberberg using the creeping-flow equations, provides a noteworthy example of the importance of simple arguments based on the linearity of the governing equations and boundary conditions in the creeping-flow limit and a lesson about the need to think creatively at the outset instead of proceeding blindly with detailed analysis.

4. Resistance Matrices for the Force and Torque on a Body in Creeping Flow

Finally, linearity of the creeping-flow equations and boundary conditions allows a great *a priori* simplification in calculations of the force or torque on a body of fixed shape that moves in a Newtonian fluid. To illustrate this assertion, we consider a solid particle of *arbitrary* shape moving with translational velocity $\mathbf{U}(t)$ and angular velocity $\mathbf{\Omega}(t)$ through an unbounded, quiescent viscous fluid in the creeping-flow limit $Re \ll 1$ and $Re/S \ll 1$. The problem of calculating the force or torque on the particle requires a solution of the creeping-flow equations, subject to boundary conditions. In *dimensional terms*, the problem is

$$\mu \nabla^2 \mathbf{u} - \nabla p = 0,$$

$$\nabla \cdot \mathbf{u} = 0, \qquad (7\text{--}17\text{a})$$

with

$$\mathbf{u} = \mathbf{U} + \mathbf{\Omega} \wedge \mathbf{x} \qquad (7\text{--}17\text{b})$$

on the particle surface S and

$$\mathbf{u} \to 0 \quad \text{as} \quad |\mathbf{x}| \to \infty. \qquad (7\text{--}17\text{c})$$

Here, $\mathbf{x}$ is a position vector measured from the center of gravity of the particle. The force on the particle is

$$\mathbf{F} = \int_S (\mathbf{T} \cdot \mathbf{n}) dS, \qquad (7\text{--}18\text{a})$$

and the torque is

$$\mathbf{G} = \int_S (\mathbf{x} \wedge \mathbf{T} \cdot \mathbf{n}) dS, \qquad (7\text{--}18\text{b})$$

where $\mathbf{T}$ is the stress tensor and the integration is over the particle surface.

The critical difficulty with this problem is that the solution depends on the orientations of $\mathbf{U}$ and $\mathbf{\Omega}$ relative to axes fixed in the particle, as well as on the *relative* magnitudes of $\mathbf{U}$ and $\mathbf{\Omega}$. Thus, for every possible orientation of $\mathbf{U}$ and/or $\mathbf{\Omega}$, a new solution appears to be required to calculate $\mathbf{u}, p, \mathbf{F}$, or $\mathbf{G}$. Fortunately, however, the possibility of constructing solutions of a problem as a sum (or superposition) of solutions to a set of simpler problems means that this is not actually necessary in the creeping-flow limit. Rather, to evaluate $\mathbf{u}$, $p, \mathbf{F}$, or $\mathbf{G}$ for any *arbitrary* choice of $\mathbf{U}$ and $\mathbf{\Omega}$, we will show that it is sufficient to obtain detailed solutions for translation in three mutually orthogonal directions (relative to axes fixed in the particle) with unit velocity $\mathbf{U} = \mathbf{e}_i$ and $\mathbf{\Omega} \equiv 0$, and for rotation about three mutually orthogonal axes with unit angular velocity $\mathbf{\Omega} = \mathbf{e}_i$ and $\mathbf{U} \equiv 0$.

To prove the assertion of the previous paragraph, we return to the full problem, (7–17) and (7–18). For convenience, we can consider the problem (7–17) as a superposition of two problems: the first is translation with arbitrary velocity $\mathbf{U}$ and $\mathbf{\Omega} = 0$, and the second is rotation with arbitrary angular velocity $\mathbf{\Omega}$ and $\mathbf{U} = 0$. We can denote the solutions of these two problems as $(\mathbf{u}_1, p_1)$ and $(\mathbf{u}_2, p_2)$, respectively, with the corresponding force and torque being $(\mathbf{F}_1, \mathbf{G}_1)$ and $(\mathbf{F}_2, \mathbf{G}_2)$. The *full* solution of (7–17) is then $\mathbf{u} = \mathbf{u}_1 + \mathbf{u}_2$ and $p = p_1 + p_2$, and the force and torque are $\mathbf{F} = \mathbf{F}_1 + \mathbf{F}_2$ and $\mathbf{G} = \mathbf{G}_1 + \mathbf{G}_2$.

Now, let us consider the translation problem (i.e., $\mathbf{U} \neq 0, \mathbf{\Omega} \equiv 0$), again in dimensional terms,

$$\mu \nabla^2 \mathbf{u}_1 - \nabla p_1 = 0, \quad \nabla \cdot \mathbf{u}_1 = 0, \qquad (7\text{--}19)$$

with

$$\mathbf{u}_1 = \mathbf{U} \text{ on } S,$$

$$\mathbf{u}_1 \to 0 \text{ at } \infty.$$

We see that the problem is linear in $\mathbf{U}$. Thus the solution $(\mathbf{u}_1, p_1)$ can depend only linearly on $\mathbf{U}$, and in view of the relationships (7–18a) and (7–18b), this means that the force and

B. Some General Consequences of Linearity and the Creeping-flow Equations

torque must also be linear functions of **U**. Because $\mathbf{F}_1$ and **U** are true vectors, the most general linear relationship between them is

$$\mathbf{F}_1 = \hat{\mathbf{A}} \cdot \mathbf{U}, \tag{7-20a}$$

where $\hat{\mathbf{A}}$ is a true second-order tensor. Similarly, because $\mathbf{G}_1$ is a *pseudo-vector*, the most general linear relationship between $\mathbf{G}_1$ and **U** is

$$\mathbf{G}_1 = \hat{\mathbf{C}} \cdot \mathbf{U}, \tag{7-20b}$$

where $\hat{\mathbf{C}}$ is a second-order pseudo-tensor.

The components of the tensor $\hat{\mathbf{A}}$ and the pseudo-tensor $\hat{\mathbf{C}}$ have a simple interpretation. For example, the ij component of $\hat{\mathbf{A}}$ is just the i component of the force on the body for translation with unit velocity in the j direction. To see this, we may express (7–20a) in component form:

$$(F_1)_i = A_{ij} U_j.$$

Then if $\mathbf{U} = \mathbf{e}_1$, for example,

$$(F_1)_1 = A_{11},$$
$$(F_1)_2 = A_{21},$$
$$(F_1)_3 = A_{31}.$$

Similarly, the ij component of $\hat{\mathbf{C}}$ is just the i component of the torque produced by translation of the body in the j direction. Thus if we solve the three problems of translation with unit velocity in the three coordinate directions and calculate the components of the force and torque on the body in each case, we completely determine all nine components of $\hat{\mathbf{A}}$ and $\hat{\mathbf{C}}$. But once $\hat{\mathbf{A}}$ and $\hat{\mathbf{C}}$ have been specified, the formulae in (7–20) can be used to evaluate the force and torque for translation of the body in any *arbitrary* direction.

An identical analysis also can be applied to the rotation problem that derives from (7–17), that is, rotation with arbitrary $\boldsymbol{\Omega} \neq 0$ and $\mathbf{U} \equiv 0$. The result for the force and torque in this case is

$$\mathbf{F}_2 = \hat{\mathbf{B}} \cdot \boldsymbol{\Omega},$$
$$\mathbf{G}_2 = \hat{\mathbf{D}} \cdot \boldsymbol{\Omega}. \tag{7-21}$$

Again, specification of the components of $\hat{\mathbf{B}}$ and $\hat{\mathbf{D}}$ requires an evaluation of the three force and torque components for rotation about each of the three coordinate axes.

Thus, combining (7–20) and (7–21), as allowed by the linearity of the governing equations, we find that the force and torque on a particle that moves with *arbitrary* velocities **U** and $\boldsymbol{\Omega}$ through an otherwise quiescent fluid is

$$\mathbf{F} = \mu (\mathbf{A} \cdot \mathbf{U} + \mathbf{B} \cdot \boldsymbol{\Omega}),$$
$$\mathbf{G} = \mu (\mathbf{C} \cdot \mathbf{U} + \mathbf{D} \cdot \boldsymbol{\Omega}). \tag{7-22}$$

Note that the viscosity μ will appear linearly in each of the second-order tensor coefficients in (7–20) and (7–21) and thus has been factored out in (7–22) – that is, $\hat{\mathbf{A}} = \mu \mathbf{A}$, and so on. A most important property of the four remaining *resistance* tensors, **A**, **B**, **C**, and **D**, is that they depend on only the *geometry* of the particle and are independent of all other parameters of the problem including, of course, **U** and **Ω**.

It is important to emphasize the great simplification inherent in (7–22), which results from the linearity of the creeping-flow problem for a body of fixed shape. In the absence of resistance tensors, we could calculate the force or torque on an arbitrary body that translates and/or rotates with arbitrary velocities **U** and **Ω** only by completely resolving the equations of motion for each change in the relative magnitudes of **U** and **Ω** or in their orientation relative to axes that are fixed in the body. Indeed, for *nonzero* Reynolds number, for which the governing equations are the full, nonlinear Navier–Stokes equations, this is precisely what must be done. Once the existence of the resistance tensors is recognized, however, we see that the force and torque for arbitrary **U** and **Ω** can be specified completely by solving a maximum of six *fundamental* problems, corresponding, respectively, to translation in three orthogonal directions with no rotation, and rotation about three orthogonal axes with no translation. These six problems can be solved, once and for all, to determine the components of the tensors **A**, **B**, **C**, and **D**, and these results can then be used for all possible combinations of **U** and **Ω**.

To actually use these results, it is of course necessary to actually calculate the components of the resistance tensors. We have seen that it is necessary to solve only three problems for translation and three problems for rotation in the coordinate directions to specify all of the components of **A**, **B**, **C**, and **D**. It probably does not need to be said that the orientation of the coordinate axes should be chosen to take advantage of any geometric symmetries that can simplify the fluid mechanics problems that must be solved. For example, if we wish to determine the force and/or torque on an ellipsoid for motions of arbitrary magnitude and direction (with respect to the body geometry), we should specify the components of the resistance tensors with respect to axes that are coincident with the principal axes of the ellipsoid, as this choice will simplify the fluid mechanics problems that are necessary to determine these components. If arbitrary velocities **U** and **Ω** are then specified with respect to these same coordinate axes, the Eqs. (7–22) will yield force and torque components in this coordinate system.

Although we will not carry our analysis further, there has been, in fact, significant progress in delineating the properties of the resistance tensors beyond the general formulae (7–22). The most important and general result is the symmetry conditions

$$\boxed{\mathbf{A} = \mathbf{A}^T, \quad \mathbf{D} = \mathbf{D}^T, \quad \text{and} \quad \mathbf{B} = \mathbf{C}^T.} \tag{7–23}$$

It should be noted that there is a direct relationship between the two resistance matrices that represent the coupling between translation and rotation.

In addition to these general symmetry properties, considerable effort has been made to understand the relationships between symmetries in the geometry of the problem and the forms of the resistance tensors. It is beyond our present scope to discuss these relationships in a comprehensive manner; the interested reader can refer to Brenner (1972) or the textbook *Low Reynolds Number Hydrodynamics* by Happel and Brenner (1973) for a detailed discussion of these questions.[8] Here we restrict ourselves to the results for several particularly simple cases. First, if we consider the motion of a body with spherical symmetry in an unbounded fluid, with the origin of coordinates at the geometric center of the body, it can be shown that

$$\boxed{\mathbf{A} = a\mathbf{I}, \quad \mathbf{D} = d\mathbf{I}, \quad \text{and} \quad \mathbf{C} = \mathbf{B} = 0.} \tag{7–24}$$

B. Some General Consequences of Linearity and the Creeping-flow Equations

On the other hand, for an ellipsoid of revolution with the origin of the coordinate system at the geometric center and coordinate axes parallel and perpendicular to the principal axes of the ellipse, it can be shown that

$$\mathbf{A} = \begin{pmatrix} a_\| & 0 & 0 \\ 0 & a_\perp & 0 \\ 0 & 0 & a_\perp \end{pmatrix}, \quad \mathbf{D} = \begin{pmatrix} d_\| & 0 & 0 \\ 0 & d_\perp & 0 \\ 0 & 0 & d_\perp \end{pmatrix}, \quad \text{and} \quad \mathbf{B} = \mathbf{C}^T = \begin{pmatrix} 0 & 0 & 0 \\ 0 & 0 & 0 \\ 0 & 0 & 0 \end{pmatrix}. \tag{7-25}$$

The most obvious example of a body that exhibits *coupling* between translation and rotation is a body with a screwlike structure that clearly *translates* if it is rotated about the screw axis; alternatively, if the particle is to be restrained from translating, a force equal to $-\mu \mathbf{B} \cdot \boldsymbol{\Omega}$ must be applied to *balance* the hydrodynamic force that is generated by its rotation, with

$$\mathbf{B} = \begin{pmatrix} b_1 & 0 & 0 \\ 0 & 0 & 0 \\ 0 & 0 & 0 \end{pmatrix}$$

if the 1 axis is coincident with the screw axis.

Although all the preceding development has been based upon a body moving through an *unbounded* fluid, the general expressions (7–22) for the force and torque in terms of resistance tensors are also applicable to the motion of a spherical particle in a container or in the vicinity of a wall. Of course, the resistance tensors must be modified to account for the presence of the bounding walls. For example, to describe the force or torque on a sphere moving with arbitrary $\mathbf{U}$ and $\boldsymbol{\Omega}$ in the vicinity of an infinite plane wall, the obvious choice is to specify the components of the resistance tensors in a coordinate system whose axes are normal and parallel to the wall. This results in the simplest fluid mechanics problems for determining the components of the resistance tensors – that is, translation with unit velocity parallel and normal to the wall and rotation with unit angular velocity parallel and normal to the wall. We may anticipate intuitively that $\mathbf{A}$ and $\mathbf{D}$ will not be isotropic in this case, nor will $\mathbf{C}$ and $\mathbf{B}$ be zero. The fact that $\mathbf{A}$ and $\mathbf{D}$ are not isotropic is a consequence of the fact that a force or a torque of given magnitude will produce different translational or angular velocities depending on the direction of motion relative to the wall. Thus, for example, the resistance to motion is different for motion parallel and perpendicular to the wall. Note, however, that we can use the arguments of Section A to show that the hydrodynamic force does not depend on the sign of $\mathbf{U}$; that is, the force generated by motion toward the wall is identical to the force on the same body at the same instantaneous position as it moves away from the wall. The fact that $\mathbf{C}$ and $\mathbf{B}$ are nonzero means that translational and rotational motions are coupled. For example, if a force parallel to the wall is applied to the sphere, the sphere not only will translate in that direction but also will rotate because of hydrodynamic interaction with the wall. Indeed, if we express the components of the resistance tensors in a Cartesian coordinate system with axes that are normal and tangential to the wall, it can be shown that

$$\mathbf{A} = \begin{pmatrix} a_\perp & 0 & 0 \\ 0 & a_\| & 0 \\ 0 & 0 & a_\| \end{pmatrix}, \quad \mathbf{D} = \begin{pmatrix} d_\perp & 0 & 0 \\ 0 & d_\| & 0 \\ 0 & 0 & d_\| \end{pmatrix}, \quad \text{and} \quad \mathbf{B} = \mathbf{C}^T = \begin{pmatrix} 0 & 0 & 0 \\ 0 & 0 & b \\ 0 & -b & 0 \end{pmatrix}, \tag{7-26}$$

where we have taken the "1" direction to be normal to the plane wall. The components of $\mathbf{A}$, $\mathbf{D}$, $\mathbf{B}$, and $\mathbf{C}$ will depend on the ratio of the distance between the sphere and the wall to the sphere radius. Even nonspherical bodies in a container or near a wall will be characterized by a linear relationship between the force or torque and the velocities **U** and **Ω**. However, in the latter case, a formulation in terms of resistance tensors will be useful only if the body does not rotate and therefore maintains a fixed orientation relative to the boundaries. The reason is that the force (or torque) on the body generally depends on the orientation and position of the nonspherical body relative to the boundaries, as well as the orientation of **U** and **Ω** relative to the body.

Finally, it is worthwhile to consider briefly the use of the results obtained in this section, and especially Eqs. (7–22), for calculating the motion of a body that is being acted on by a prescribed external force or torque. Equations (7–22) give the hydrodynamic force and torque acting on a body that is moving with a prescribed translational and angular velocity. To calculate the motion of the body in the presence of an external force and/or torque, we must apply Newton's second law to write an equation of motion for the body. However, in the present development, we assume that the acceleration of the body is small compared with the external forces and torques that are acting on it (provided the density of the particle is comparable with that of the fluid, this assumption will always be satisfied in the creeping-flow limit). Hence Newton's second law reduces to the statement that the sum of all forces acting on the particle must add up to zero. Similarly, the sum of all torques acting on the particle must add up to zero. To illustrate the use of (7–22) for the derivation of trajectory equations, we can consider the case of a particle that is subjected to an applied force, $-f$, but zero external torque. Hence, $\mathbf{F} = f$ and $\mathbf{G} = 0$. It follows from (7–22b) that the particle will rotate with angular velocity

$$\boxed{\mathbf{\Omega} = -\mathbf{D}^{-1} \cdot (\mathbf{C} \cdot \mathbf{U}),} \tag{7–27}$$

where $\mathbf{D}^{-1}$ is the inverse of the resistance tensor **D**.[9] Thus the translational velocity, according to (7–22a), can be expressed in the form

$$\mathbf{U} = \mathbf{A}^{-1} \cdot (\mu^{-1} f - \mathbf{B} \cdot \mathbf{\Omega}) = \mathbf{A}^{-1} \cdot \{\mu^{-1} f + \mathbf{B} \cdot [\mathbf{D}^{-1} \cdot (\mathbf{C} \cdot \mathbf{U})]\}.$$

This expression can be solved to obtain a complicated-looking but explicit formula for the translational velocity

$$\boxed{\mathbf{U} = \{\mathbf{I} - \mathbf{A}^{-1} \cdot [\mathbf{B} \cdot (\mathbf{D}^{-1} \cdot \mathbf{C})]\}^{-1} \cdot [\mathbf{A}^{-1} \cdot (\mu^{-1} f)].} \tag{7–28}$$

A particle of general shape with no external torque will rotate as it translates if the coupling tensor **C** is nonzero. Given an applied force in some fixed direction, the particle will translate through the fluid at a rate and in a direction given by (7–28). Generally, the direction will not be collinear with the applied force f, and the magnitude of the velocity will vary as (or if) the particle rotates according to (7–27).

C. REPRESENTATION OF TWO-DIMENSIONAL AND AXISYMMETRIC FLOWS IN TERMS OF THE STREAMFUNCTION

We have seen that the Navier–Stokes and continuity equations reduce, in the creeping-motion limit, to a set of coupled, but linear, PDEs for the velocity and pressure, **u** and p. Because of the linearity of these equations, a number of the classical solution methods can be utilized. In the next three sections we consider the general class of 2D and axisymmetric creeping flows. For this class of flows, it is possible to achieve a considerable simplification of the mathematical problem by combining the creeping-flow and continuity equations to produce a single higher-order DE.

C. Representation of Two-Dimensional and Axisymmetric Flows

The starting point is a general representation theorem from vector calculus that states that any continuously differentiable vector field can be represented by three scalar functions ϕ, ψ, and χ in the form[10]

$$\mathbf{a} = \nabla\phi + \nabla \wedge (\psi \nabla \chi). \tag{7-29}$$

In effect, (7–29) represents a decomposition of the general vector field **a** into an irrotational part, associated with $\nabla\phi$, and a solenoidal (or divergence-free) part, represented by $\nabla \wedge (\psi \nabla \chi)$. It should be noted that general proofs exist that show not only that (7–29) can represent any arbitrary vector field **a** but also that an arbitrary, irrotational vector field can be represented in terms of the gradient of a single scalar function ϕ and that an arbitrary, solenoidal vector field can be represented in the form of the second term of (7–29). Because

$$\nabla \cdot [\nabla \wedge (\psi \nabla \chi)] = \nabla \cdot (\nabla\psi \wedge \nabla\chi) \equiv 0,$$

the representation (7–29) can be given also in terms of a solenoidal vector field **A**, such that

$$\mathbf{a} = \nabla\phi + \nabla \wedge \mathbf{A}, \quad \text{where} \quad \nabla \cdot \mathbf{A} = 0. \tag{7-30}$$

Now, because the velocity field **u** is a continuously differentiable vector field, it can be represented for any Reynolds number by either form (7–29) or (7–30), that is,

$$\mathbf{u} = \nabla\phi + \nabla \wedge (\psi \nabla \chi) \tag{7-31}$$

or

$$\mathbf{u} = \nabla\phi + \nabla \wedge \mathbf{A}, \quad \text{where} \quad \nabla \cdot \mathbf{A} = 0, \tag{7-32}$$

and this is true for an arbitrary 3D motion. The function **A** that appears in (7–32) is known as the vector potential for the vorticity because

$$\nabla^2 \mathbf{A} = -\nabla \wedge \mathbf{u} = -\boldsymbol{\omega}. \tag{7-33}$$

Now the question is whether the representations (7–31) and/or (7–32) lead to any simplification of the mathematical problem for **u**. To answer this question, let us for the moment stick to the creeping-flow limit. For an incompressible fluid, the continuity equation requires that

$$\nabla^2 \phi = 0, \tag{7-34}$$

and the equation of motion reduces to an equation for the vector potential function **A**,

$$\nabla \wedge (\nabla^2 \mathbf{A}) = \nabla p \tag{7-35}$$

or

$$\nabla^4 \mathbf{A} = 0. \tag{7-36}$$

Clearly, **A** must be nonzero in general to satisfy (7–35). On the other hand, the contribution $\nabla\phi$ will only be nonzero if

$$(\mathbf{u} - \nabla \wedge \mathbf{A}) \cdot \mathbf{n} \neq 0,$$

Creeping Flows – Two-Dimensional and Axisymmetric Problems

at the boundaries. Generally, this will *not* be true, and **u** can then be represented simply as

$$\mathbf{u} = \nabla \wedge \mathbf{A}, \tag{7-37}$$

with $\nabla \cdot \mathbf{A} = 0$ and **A** satisfying (7–35) or (7–36). In this case, introduction of the vector potential function **A** allows the problem to be simplified to the extent that the general form (7–37) satisfies continuity for any choice of **A**, and the pressure can be eliminated to obtain a single fourth-order equation for **A**. Nevertheless, for 3D flows, the problem for **A** is not all that much simpler than the original problem because all three components of **A** are generally nonzero.

For 2D and axisymmetric flows, however, the general representation results, (7–31) and (7–32), do lead to a very significant simplification, both for creeping flows and for flows at finite Reynolds numbers where we must retain the full Navier–Stokes equations. The reason for this simplification is that the vector potential **A** can be represented in terms of a single scalar function,

$$\mathbf{A} = h_3 \psi(q_1, q_2)\mathbf{i}_3, \tag{7-38}$$

where $\mathbf{i}_3$ is a unit vector that is orthogonal to the plane of motion for 2D flows and in the azimuthal direction for the axisymmetric case. We shall show shortly that this representation is sufficiently general for arbitrary 2D and axisymmetric flows. First, however, we need to define all of the symbols that appear. The factor h_3 is the scale factor defined for a general, orthogonal curvilinear coordinate system

$$(q_1, q_2, q_3) \tag{7-39}$$

by means of the equation for the length of a differential line element:

$$(ds)^2 \equiv \frac{1}{h_1^2}(dq_1)^2 + \frac{1}{h_2^2}(dq_2)^2 + \frac{1}{h_3^2}(dq_3)^2. \tag{7-40}$$

In general, for 2D flows, q_3 can be identified with a Cartesian variable z, orthogonal to the plane of motion, and $h_3 \equiv 1$. However, for axisymmetric flows, q_3 represents the azimuthal angle ϕ about the axis of symmetry, and h_3 will generally be a function of q_1 and q_2. If we express **A** in terms of spherical coordinates, for example,

$$(q_1, q_2, q_3) \to (r, \theta, \phi),$$

then

$$h_1 = 1, \quad h_2 = \frac{1}{r}, \quad \text{and} \quad h_3 = \frac{1}{r \sin \theta}.$$

The scalar function ψ that appears in (7–38) is known as the *streamfunction*. The physical significance of ψ is best seen through the relationship (7–37). In particular, if we substitute (7–38) into (7–37), we obtain

$$\mathbf{u} = \left(h_2 h_3 \frac{\partial \psi}{\partial q_2}, \; -h_1 h_3 \frac{\partial \psi}{\partial q_1}, \; 0 \right). \tag{7-41}$$

We see, from (7–41), that the form (7–38) is consistent with the existence of a 2D or axisymmetric velocity field and, further, that the magnitudes of the two nonzero-velocity

C. Representation of Two-Dimensional and Axisymmetric Flows

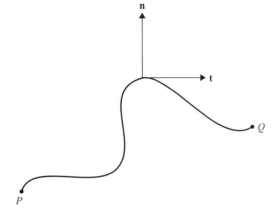

Figure 7–4. An arbitrarily chosen curve between two points, P and Q, with unit normal $\mathbf{n}$ and unit tangent $\mathbf{t}$.

components are directly related to the magnitudes of the spatial derivatives of ψ. In addition, if we calculate the derivative of ψ in the direction of motion, we see that

$$\boxed{\frac{\mathbf{u} \cdot \nabla \psi}{|\mathbf{u}|} \equiv 0.} \tag{7-42}$$

Thus lines of constant ψ are everywhere tangent to the local velocity; that is, curves in space corresponding to constant values of ψ are coincident with the pathlines followed by an element of the fluid.

It is also worth noting that the volume flux of fluid across a curve joining any two arbitrary points, say, P and Q, in the flow domain is directly related to the difference in magnitude of the streamfunction at these two points. For simplicity, let us show that this is true for 2D motions. The axisymmetric case follows in a very similar way. Thus we consider a curve, as shown in Fig. 7–4, that passes through the two points P and Q but is otherwise arbitrary. The volume flux of fluid across this curve, per unit length in the third direction, is

$$J = \int_\ell (\mathbf{u} \cdot \mathbf{n}) d\ell. \tag{7-43}$$

Note that the integral is independent of the particular path between P and Q for an incompressible fluid where $\nabla \cdot \mathbf{u} = 0$. Now, the unit normal is

$$\mathbf{n} = \mathbf{t} \wedge \mathbf{i}_3,$$

where $\mathbf{i}_3$ is the positive unit normal in the third direction. Thus,

$$J = \int_\ell \mathbf{u} \cdot (\mathbf{t} \wedge \mathbf{i}_3) d\ell,$$

where

$$\mathbf{t} d\ell \equiv \left(\frac{1}{h_1} dq_1\right) \mathbf{i}_1 + \left(\frac{1}{h_2} dq_2\right) \mathbf{i}_2,$$

and thus,

$$(\mathbf{t} \wedge \mathbf{i}_3) d\ell \equiv \frac{1}{h_2} dq_2 \mathbf{i}_1 - \frac{1}{h_1} dq_1 \mathbf{i}_2.$$

Hence,

$$J = \int_P^Q \left(\frac{d\psi}{dq_2}dq_2 + \frac{d\psi}{dq_1}dq_1\right);$$

$$J = \int_{\psi_P}^{\psi_Q} d\psi = \psi_Q - \psi_P, \qquad (7\text{--}44)$$

where ψ_Q and ψ_P denote the values of ψ at the points Q and P, respectively.

Introduction of the streamfunction for axisymmetric and 2D problems simplifies their solution in two ways. First, in view of the definitions (7–37) and (7–38), it is evident that the continuity equation for an incompressible fluid,

$$\nabla \cdot \mathbf{u} = 0,$$

will be satisfied automatically for any function ψ that satisfies the other conditions of the problem. Second, the equations of motion are reduced from a coupled pair of equations relating $\mathbf{u}$ and p to a single higher-order equation for the scalar function ψ. In the creeping-flow approximation this equation can be obtained directly from (7–36). Substituting from (7–38), we obtain

$$E^4 \psi = 0, \qquad (7\text{--}45)$$

where

$$E^2 \equiv \frac{h_1 h_2}{h_3}\left[\frac{\partial}{\partial q_1}\left(\frac{h_1 h_3}{h_2}\frac{\partial}{\partial q_1}\right) + \frac{\partial}{\partial q_2}\left(\frac{h_2 h_3}{h_1}\frac{\partial}{\partial q_2}\right)\right].$$

In general, the E^2 operator is not the same as the more familiar $\nabla^2 (\equiv \nabla \cdot \nabla)$ which is defined as

$$\nabla^2 \equiv h_1 h_2 h_3 \left[\frac{\partial}{\partial q_1}\left(\frac{h_1}{h_2 h_3}\frac{\partial}{\partial q_1}\right) + \frac{\partial}{\partial q_2}\left(\frac{h_2}{h_1 h_3}\frac{\partial}{\partial q_2}\right)\right].$$

However, for 2D flows, the third direction corresponds to a Cartesian coordinate direction and

$$h_3 = 1.$$

In this case, E^2 and ∇^2 are identical, and the governing equation for ψ in two dimensions is therefore normally expressed as

$$\nabla^4 \psi = 0, \qquad (7\text{--}46)$$

which is the familiar biharmonic equation in two dimensions.

We shall discuss the solution of (7–45) and (7–46) shortly. First, however, it is worth noting the form of the full Navier–Stokes equations when expressed in terms of ψ. For this purpose, it is convenient first to take the curl of the equations to eliminate the pressure. In the 2D and axisymmetric flow cases considered here, this gives

$$Re\left[\frac{\partial \boldsymbol{\omega}}{\partial t} - \nabla \wedge (\mathbf{u} \wedge \boldsymbol{\omega})\right] + \nabla \wedge (\nabla \wedge \boldsymbol{\omega}) = 0. \qquad (7\text{--}47)$$

Now,

$$\boldsymbol{\omega} = \nabla \wedge [\nabla \wedge (h_3 \psi \mathbf{i}_3)] = -h_3 E^2 \psi \mathbf{i}_3, \qquad (7\text{--}48)$$

D. Two-Dimensional Creeping Flows

according to the definitions of $\boldsymbol{\omega}$ and $\mathbf{u}$ (in terms of ψ). Thus, substituting into (7–47), we obtain

$$Re\left\{\frac{\partial}{\partial t}(E^2\psi) + \frac{h_1 h_2}{h_3}\left[\frac{\partial}{\partial q_1}\left(h_3^2 \frac{\partial \psi}{\partial q_2} E^2\psi\right) - \frac{\partial}{\partial q_2}\left(h_3^2 \frac{\partial \psi}{\partial q_1} E^2\psi\right)\right]\right\} = E^4\psi. \quad (7\text{–}49)$$

Again, for the 2D case, this reduces to

$$Re\left\{\frac{\partial}{\partial t}(\nabla^2\psi) + h_1 h_2\left[\frac{\partial}{\partial q_1}\left(\frac{\partial \psi}{\partial q_2}\nabla^2\psi\right) - \frac{\partial}{\partial q_2}\left(\frac{\partial \psi}{\partial q_1}\nabla^2\psi\right)\right]\right\} = \nabla^4\psi. \quad (7\text{–}50)$$

Clearly, in the limit $Re \to 0$, these equations reduce to the limiting forms (7–45) and (7–46).

D. TWO-DIMENSIONAL CREEPING FLOWS: SOLUTIONS BY MEANS OF EIGENFUNCTION EXPANSIONS (SEPARATION OF VARIABLES)

We saw, in the previous section, that problems of creeping-motion in two dimensions can be reduced to the solution of the biharmonic equation, (7–46), subject to appropriate boundary conditions. To actually obtain a solution, it is convenient to express (7–46) as a coupled pair of second-order PDEs:

$$\nabla^2 \psi = -\omega, \quad (7\text{–}51)$$

$$\nabla^2 \omega = 0. \quad (7\text{–}52)$$

The solution of these equations by means of standard eigenfunction expansions can be carried out for any curvilinear, orthogonal coordinate system for which the Laplacian operator ∇^2 is separable. Of course, the most appropriate coordinate system for a particular application will depend on the boundary geometry. In this section we briefly consider the most common cases for 2D flows of Cartesian and circular cylindrical coordinates.

1. General Eigenfunction Expansions in Cartesian and Cylindrical Coordinates

For Cartesian coordinates, a solution of (7–52) exists in the separable form

$$\omega = X(x)Y(y). \quad (7\text{–}53)$$

Substituting into (7–52), we obtain

$$\frac{X''}{X} = -\frac{Y''}{Y} = \pm m^2, \quad (7\text{–}54)$$

where m is an arbitrary complex number. Hence,

$$X'' \pm m^2 X = 0, \quad Y'' \pm m^2 Y = 0, \quad (7\text{–}55)$$

and from this we deduce that

$$\omega = e^{mx} e^{imy} \quad (7\text{–}56)$$

for arbitrary complex m. Now, to obtain a general solution for ψ, we must solve (7–51) with the right-hand side evaluated using (7–56). Hence,

$$\nabla^2 \psi = -e^{mx} e^{imy} \gamma_m, \quad (7\text{–}57)$$

where γ_m is an arbitrary constant. The solution of (7–57) is the sum of a homogeneous solution of the form (7–56) plus a particular solution to reproduce the right-hand side. After some manipulation, we find

$$\psi_m = \alpha_m e^{mx} e^{imy} + \beta_m x e^{mx} e^{imy} + \delta_m y e^{mx} e^{imy}. \qquad (7\text{–}58)$$

Hence, the most general solution for ψ expressed in Cartesian coordinates is

$$\psi = \sum_m \psi_m \qquad (7\text{–}59)$$

with arbitrary complex values of m.

Starting with (7–51) and (7–52), we can also obtain a general solution for ψ in terms of a circular cylindrical coordinate system, and this solution is more immediately applicable to real problems. The governing equations in polar cylindrical coordinates are

$$\frac{1}{r}\frac{\partial}{\partial r}\left(r\frac{\partial \psi}{\partial r}\right) + \frac{1}{r^2}\frac{\partial^2 \psi}{\partial \theta^2} = -\omega, \qquad (7\text{–}60)$$

$$\frac{1}{r}\frac{\partial}{\partial r}\left(r\frac{\partial \omega}{\partial r}\right) + \frac{1}{r^2}\frac{\partial^2 \omega}{\partial \theta^2} = 0. \qquad (7\text{–}61)$$

We seek a solution of (7–60) and (7–61) in the separable form

$$\omega = R(r)F(\theta) \quad \text{and} \quad \psi = S(r)H(\theta). \qquad (7\text{–}62)$$

Substitution for ω in (7–61) yields

$$\frac{r}{R}\frac{\partial}{\partial r}\left(r\frac{\partial R}{\partial r}\right) = -\frac{F''}{F} = \lambda_n^2,$$

where λ_n is an arbitrary constant, either real or complex. Hence for $\lambda_n \neq 0$ there are two independent solutions for each of the functions R and F, namely,

$$F = \sin \lambda_n \theta, \quad \cos \lambda_n \theta, \quad \text{and} \quad R = r^{\lambda_n}, \, r^{-\lambda_n}. \qquad (7\text{–}63)$$

For $\lambda_n = 0$, on the other hand,

$$F = a_0, \theta \quad \text{and} \quad R = c_0, \ln r. \qquad (7\text{–}64)$$

Hence, the most general solution for ω in the separable form of Eq. (7–62) is

$$\omega = (a_0 + b_0 \theta)(c_0 + d_0 \ln r)$$
$$+ \sum_{n=1}^{\infty} (a_n \cos \lambda_n \theta + b_n \sin \lambda_n \theta)\left(c_n r^{\lambda_n} + d_n r^{-\lambda_n}\right). \qquad (7\text{–}65)$$

To determine ψ, we must solve (7-60) with the general form (7–65) substituted for ω. The general solution for ψ consists of a *homogeneous* solution of the same form as that of (7–65), plus a particular solution. To obtain terms of the particular solution corresponding to the summation in (7–65), we try

$$\psi_p = r^s \sin \lambda_n \theta, \; r^s \cos \lambda_n \theta. \qquad (7\text{–}66)$$

D. Two-Dimensional Creeping Flows

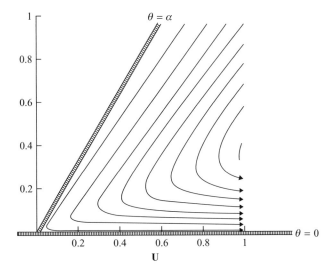

Figure 7–5. 2D flow in a sharp corner, caused by motion of the bottom surface (at $\theta = 0$) with the velocity U. The plot shows streamlines, $\overline{\psi} = \psi/U$, calculated from Eq. (7–79) for $\alpha = \pi/3$. Contour values range from 0 at the walls in increments of 0.02.

Substitution into the left-hand side of (7–60) yields

$$\left(s^2 - \lambda_n^2\right) r^{s-2} \sin \lambda_n \theta, \quad \left(s^2 - \lambda_n^2\right) r^{s-2} \cos \lambda_n \theta. \tag{7–67}$$

Thus, comparing (7–67) and (7–65), we see

$$s = \lambda_n + 2, \quad -\lambda_n + 2, \tag{7–68}$$

except for $\lambda_n = 1$. In this case, the second of the values for s is $s = 1$, and the particular solution form (7–66) reduces to a solution of the homogeneous equation. Thus, in this case, the corresponding particular solutions take the form

$$r^3 \sin \theta, \quad r^3 \cos \theta, \quad r\theta \sin \theta, \quad \text{and } r\theta \cos \theta. \tag{7–69}$$

Finally, for $\lambda_n = 0$, the particular solution is

$$\{(\hat{a}_0 \theta + \hat{b}_0)[\hat{c}_0 r^2 + \hat{d}_0 r^2 (\ln r - 1)]\}. \tag{7–70}$$

Thus, the general solution for ψ is

$$\begin{aligned}\psi &= [c_0 + d_0 \ln r + \hat{c}_0 r^2 + \hat{d}_0 r^2 (\ln r - 1)](a_0 \theta + b_0) \\ &+ (c_1 r + d_1 r^{-1} + \hat{c}_1 r^3)(a_1 \sin \theta + b_1 \cos \theta) + \hat{d}_1 r(\hat{a}_1 \theta \sin \theta + \hat{b}_1 \theta \cos \theta) \\ &+ \sum_{n=2}^{\infty} \left(c_{\lambda_n} r^{\lambda_n} + d_{\lambda_n} r^{-\lambda_n} + \hat{c}_{\lambda_n} r^{\lambda_n + 2} + \hat{d}_{\lambda_n} r^{2 - \lambda_n}\right)(a_{\lambda_n} \sin \lambda_n \theta + b_{\lambda_n} \cos \lambda_n \theta).\end{aligned} \tag{7–71}$$

2. Application to Two-Dimensional Flow near Corners

The general solution (7–71) can be applied to examine 2D flows in the region between two plane boundaries that intersect at a sharp corner. This class of creeping motion problems was considered in a classic paper by Moffatt,[11] and our discussion is similar to that given by Moffatt. A typical configuration is shown in Fig. 7–5 for the case in which one boundary at $\theta = 0$ is moving with constant velocity U in its own plane and the other at $\theta = \alpha$ is stationary.

A noteworthy feature of this configuration is that it lacks a definite physical length scale. One rationalization of this fact is that the corner is generally a localized part of some more complicated global geometry. In any case, an obvious question is the range of validity of the creeping-motion approximation for this situation in which a fixed characteristic length scale (and thus a fixed Reynolds number) does not exist. To answer this question, we need to obtain an estimate for the magnitudes of the inertial and viscous terms in the equations of motion. A starting point is the magnitude of the velocity, corresponding to the form (7–71) for the streamfunction, namely,

$$|\mathbf{u}| = O\left(Ar^{\xi-1}\right), \quad (7\text{–}72)$$

where ξ is the real part of λ and A is a constant with dimension of velocity/(length)$^\xi$. Then, the magnitudes of the inertia and viscous terms in the equations of motion can be estimated as

$$|\mathbf{u} \cdot \nabla \mathbf{u}| = O\left(A^2 r^{2\xi-3}\right),$$
$$|\nu \nabla^2 \mathbf{u}| = O\left(\nu A r^{\xi-3}\right).$$

Hence, the ratio of inertia to viscous terms is

$$\frac{|\mathbf{u} \cdot \nabla \mathbf{u}|}{|\nabla^2 \mathbf{u}|} = O\left(\frac{Ar^\xi}{\nu}\right), \quad (7\text{–}73)$$

and we see that the creeping-motion approximation is valid provided that

$$Re \equiv \frac{Ar^\xi}{\nu} \ll 1, \quad (7\text{–}74)$$

where Re is effectively a Reynolds number based on the distance from the corner. Hence, for $\xi > 0$, inertia is negligible for sufficiently small values of r, whereas for $\xi < 0$, neglect of inertia requires r to be sufficiently large. We focus here on problems in which $\xi > 0$. Because the resulting solution in this case is a *local approximation*, certain features of the flow will generally remain indeterminate. In reality, they are determined by the features of the flow at a large distance from the corner where the creeping-flow approximation breaks down.

The simplest problem of the type considered here is the one sketched in Fig. 7–5, which was originally solved by Taylor.[12] The problem may be considered to be a local approximation for the action of a wiper blade on a solid surface that is completely covered by liquid. The boundary conditions in this case are

$$u_r = U, \quad u_\theta = 0 \quad \text{at} \quad \theta = 0,$$
$$u_r = u_\theta = 0 \quad \text{at} \quad \theta = \alpha, \quad (7\text{–}75)$$

where the r and θ components of velocity are related to the streamfunction by means of the definitions

$$u_r = \frac{1}{r}\frac{\partial \psi}{\partial \theta}, \quad u_\theta = -\frac{\partial \psi}{\partial r}. \quad (7\text{–}76)$$

In view of (7–76), it is clear that the requirement $u_r = U$ (constant) at $\theta = 0$ can be satisfied only by the solution form

$$\psi = rF(\theta).$$

Referring to the general solution (7–71), we find that the terms that are linear in r are

$$\psi = r(A_1 \sin\theta + B_1 \cos\theta + C_1\theta \sin\theta + D_1\theta \cos\theta), \quad (7\text{–}77)$$

D. Two-Dimensional Creeping Flows

where A_1, B_1, C_1, and D_1 are constants. Applying the boundary conditions (7–75) to this solution, we find that

$$U = A_1 + D_1,$$
$$0 = (A_1 + D_1)\cos\alpha + (C_1 - B_1)\sin\alpha + C_1\alpha\cos\alpha - D_1\alpha\sin\alpha,$$
$$0 = B_1,$$
$$0 = A_1\sin\alpha + B_1\cos\alpha + C_1\alpha\sin\alpha + D_1\alpha\cos\alpha.$$

Thus, solving for A_1, B_1, C_1, and D_1, we have

$$B_1 = 0,$$
$$C_1 = \frac{U(\alpha - \sin\alpha\cos\alpha)}{\sin^2\alpha - \alpha^2},$$
$$D_1 = \frac{U\sin^2\alpha}{\sin^2\alpha - \alpha^2},$$
$$A_1 = \frac{U\alpha^2}{\sin^2\alpha - \alpha^2}, \tag{7-78}$$

and

$$\psi = -\frac{Ur}{(\sin^2\alpha - \alpha^2)}[\alpha^2\sin\theta + (\sin\alpha\cos\alpha - \alpha)\theta\sin\theta - (\sin^2\alpha)\theta\cos\theta]. \tag{7-79}$$

The streamlines corresponding to (7–79) are shown in Fig. 7–5. Although the velocity components u_r and u_θ are perfectly well behaved, the shear stress $\tau_{r\theta}$ is *singular* in the limit $r \to 0$. Indeed, if we calculate $\tau_{r\theta}|_{\theta=0}$, we find that

$$\tau_{r\theta}|_{\theta=0} = -\frac{2U\mu}{(\sin^2\alpha - \alpha^2)}\frac{1}{r}(\sin\alpha\cos\alpha - \alpha). \tag{7-80}$$

Clearly, the solution breaks down in the limit $r \to 0$. In fact, according to (7–80), an infinite force is necessary to maintain the plane $\theta = 0$ in motion at a finite velocity U, and this prediction is clearly unrealistic. Presumably, one of the assumptions of the theory breaks down, although a definitive resolution of the difficulty does not exist at the present time. The most plausible explanation is that the no-slip boundary condition is inadequate in regions of extremely high shear stress. However, as discussed in Chapter 2, this issue is still subject to debate.

Closely related to the Taylor problem is the situation sketched in Fig. 7–6, when a flat plate is drawn into a viscous fluid through a free surface (that is, an interface). In reality, of course, the interface will tend to deform as a result of the motion of the plate, but we assume here that the interface remains flat. Then the problem is identical to the previous Taylor problem except for the boundary conditions, which now become

$$u_r = U, \quad u_\theta = 0 \quad \text{at} \quad \theta = -\alpha,$$
$$\tau_{r\theta} \equiv \frac{1}{r}\left(\frac{\partial u_r}{\partial \theta}\right) = 0, \quad u_\theta = 0 \quad \text{at} \quad \theta = 0. \tag{7-81}$$

The solution in this case is

$$\boxed{\psi = Ur(\sin\alpha\cos\alpha - \alpha)^{-1}[\sin\alpha(\theta\cos\theta) - (\alpha\cos\alpha)\sin\theta].} \tag{7-82}$$

Creeping Flows – Two-Dimensional and Axisymmetric Problems

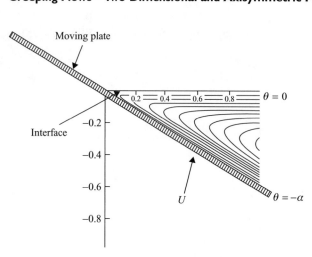

Figure 7–6. 2D flow in a sharp corner created when a flat plate is drawn into a fluid through a flat fluid interface. The plot shows streamlines, $\bar{\psi} = \psi/U$, calculated from Eq. (7–82) for $\alpha = \pi/6$. Contour values range from 0 at the walls in increments of 0.0105.

In this case, it is of interest to calculate the tangential velocity on the free surface,

$$u_r|_{\theta=0} = -U\left[1 - \frac{(\alpha - \sin\alpha)(\cos\alpha + 1)}{-\sin\alpha\cos\alpha + \alpha}\right]. \qquad (7\text{–}83)$$

The term in brackets is positive and independent of r. Hence the velocity on the free surface is constant but smaller than the velocity of the solid plate. The speed of a fluid particle that travels along the interface must therefore increase discontinuously as it reaches the plate and turns the corner – that is, it must undergo an infinite acceleration. This infinite acceleration is produced by an infinite stress and pressure, $O(r^{-1})$, on the plate in the limit $r \to 0$. Again, we conclude that the solution breaks down in the limit $r \to 0$.

A third interesting example of a flow in the vicinity of a corner is the motion of two hinged plates. As sketched in Fig. 7–7, we assume that the plates rotate with angular velocities $-\omega$ and $+\omega$, respectively. Thus, the boundary conditions are

$$u_r = 0, \quad u_\theta = -\omega r \quad \text{at} \quad \theta = +\alpha,$$
$$u_r = 0, \quad u_\theta = \omega r \quad \text{at} \quad \theta = -\alpha. \qquad (7\text{–}84)$$

Because $u_\theta = -\partial\psi/\partial r$, it is evident from the general solution (7–71) and the boundary conditions (7–84) that $\lambda = 2$, so that

$$\psi = r^2(A_2 + B_2\theta + C_2\sin 2\theta + D_2\cos 2\theta). \qquad (7\text{–}85)$$

It is convenient to use the conditions (7–84) at $\theta = \alpha$ and the symmetry conditions $u_\theta = \partial u_r/\partial\theta = 0$ at $\theta = 0$ to determine the constants A_2, B_2, C_2, and D_2. After some manipulation, we find that the solution

$$\boxed{\psi = \omega r^2 \frac{1}{2}(\sin 2\alpha - 2\alpha\cos 2\alpha)^{-1}(\sin 2\theta - 2\theta\cos 2\alpha).} \qquad (7\text{–}86)$$

In this case, both the velocity components and the stress are bounded in the limit $r \to 0$, but the pressure exhibits a $O(\log r)$ singularity.

Finally, it is of interest to consider the nature of the flow near a sharp corner that is induced by an arbitrary "stirring" flow at large distances from the corner. In general,

D. Two-Dimensional Creeping Flows

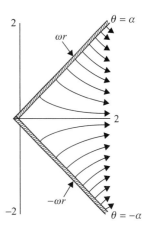

Figure 7–7. 2D flow in the vicinity of the sharp corner between two hinged, plane walls that are rotating toward one another with angular velocity $\omega(-\omega)$. The plot shows streamlines, $\overline{\psi} = \psi/\omega$, calculated from Eq. (7–86) for $\alpha = \pi/44$. Contour values range from 0 at $\theta = 0$ in increments of 0.2105.

there are two fundamental types of flow patterns that can be induced near the corner: an antisymmetric flow, as sketched in Fig. 7–8(a), and a symmetrical flow, as sketched in Fig. 7–8(b). The actual flow near a corner will generally be a mixture of antisymmetrical and symmetrical flow types, but it is permissible in the linear Stokes approximation to consider them separately (the more general flow can then be constructed as a superposition of the simpler fundamental flows). Here we consider only the antisymmetric case, which is the more interesting of the two. Thus we consider the general antisymmetric form for ψ, namely,

$$\psi = \sum_{n=1}^{\infty} r^{\lambda_n} f_{\lambda_n}(\theta), \tag{7–87}$$

where

$$f_{\lambda_n}(\theta) = A_n \cos \lambda_n \theta + C_n \cos(\lambda_n - 2)\theta. \tag{7–88}$$

The boundary conditions at the walls require

$$f(\pm \alpha) = f'(\pm \alpha) = 0. \tag{7–89}$$

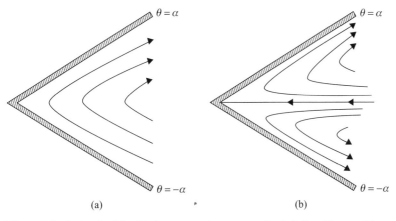

Figure 7–8. A sketch of the 2D flow near a sharp corner that is induced by an arbitrary "stirring" flow at large distances from the corner: (a) antisymmetric, (b) symmetric.

Creeping Flows – Two-Dimensional and Axisymmetric Problems

We focus only on the dominant term in the expansion (7–87) for small r, that is, the term with the largest real part of λ_n,

$$\psi \sim r^{\lambda_1} f_{\lambda_1}(\theta). \tag{7–90}$$

Thus, applying the boundary conditions (7–89) to (7–90), we find that

$$A_1 \cos \lambda_1 \alpha + C_1 \cos(\lambda_1 - 2)\alpha = 0,$$
$$A_1 \lambda_1 \sin \lambda_1 \alpha + C_1(\lambda_1 - 2) \sin(\lambda_1 - 2)\alpha = 0. \tag{7–91}$$

Hence, to obtain a nontrivial solution, λ_1 must satisfy the condition

$$\det \begin{vmatrix} \cos \lambda_1 \alpha & \cos(\lambda_1 - 2)\alpha \\ \lambda_1 \sin \lambda_1 \alpha & (\lambda_1 - 2)\sin(\lambda_1 - 2)\alpha \end{vmatrix} = 0. \tag{7–92}$$

The resulting value of λ_1 is known as the eigenvalue for this problem, and the corresponding function $f_{\lambda_1}(\theta)$ is the eigenfunction. The coefficients A_1 and C_1 corresponding to (7–91) are

$$A_1 = K \cos(\lambda_1 - 2)\alpha,$$
$$C_1 = -K \cos \lambda_1 \alpha,$$

so that

$$\psi = Kr^{\lambda_1}[\cos(\lambda_1 - 2)\alpha \cos \lambda_1 \theta - \cos \lambda_1 \alpha \cos(\lambda_1 - 2)\theta]. \tag{7–93}$$

We cannot determine the coefficient K from the local analysis alone but only by matching the local solution to the stirring flow far from the corner. To obtain λ_1, we express (7–92) in the form

$$(\lambda_1 - 2) \sin(\lambda_1 - 2)\alpha \cos \lambda_1 \alpha = \lambda_1 \sin \lambda_1 \alpha \cos(\lambda_1 - 2)\alpha,$$

or, on rearrangement,

$$-(\lambda_1 - 1) \sin 2\alpha = \sin[2(\lambda_1 - 1)\alpha]. \tag{7–94}$$

This equation has a *real* solution for λ_1 when $2\alpha > 146°$ but no real solutions for $0 \le 2\alpha < 146°$. In this range, λ_1 is complex.

Let us consider this latter case. If we denote $(\lambda - 1)$ as $p + iq$, then we can express (7–94) in the form

$$\sin \xi \cos h\eta = -k\xi,$$
$$\cos \xi \sin h\eta = -k\eta, \tag{7–95}$$

where $\xi = 2\alpha p$, $\eta = 2\alpha q$, and k is the positive parameter $k = \sin 2\alpha / 2\alpha$. Any solution of these equations must satisfy the condition

$$(2n - 1)\pi < \xi_n < \left(2n - \frac{1}{2}\right)\pi,$$

where $\sin \xi_n$ and $\cos \xi_n$ are both negative. The corresponding eigenvalue is

$$\lambda_n = 1 + (2\alpha)^{-1}(\xi_n + i\eta_n). \tag{7–96}$$

The eigenvalue with the least positive real part, which dominates near the corner ($r < 1$), obviously occurs for $n = 1$. Numerical values of ξ_1 and η_1 for $2\alpha < 146°$ were tabulated

D. Two-Dimensional Creeping Flows

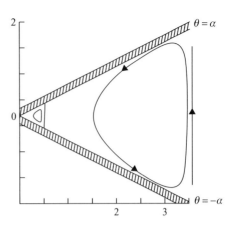

Figure 7–9. The infinite sequence of closed eddies in a sharp corner with an acute angle $2\alpha < 146°$. This sequence of closed streamline flows is commonly known as Moffatt eddies, after the mathematician H. K. Moffatt,[9] who first discovered their existence. Contours shown here are $\psi = 0, -0.005$, and 10, and we see the first two eddies in the sequence. The next eddy, closer to the corner, is too small to see with this resolution.

by Moffatt. The specific values are not significant for present purposes. What is significant is the fact that λ_1 is complex. This feature of the solution implies the existence, for small r, of an infinite *sequence* of closed streamline eddies in the corner, the first two of which are sketched in Fig. 7–9.

To demonstrate that such a sequence of eddies does exist, we can show that there is an infinite sequence of dividing streamlines, $\psi = 0$, for successively smaller values of r. If we introduce (7–96) into (7–93), the dominant streamfunction can be written in the symbolic form

$$\psi_1 = r^{(1+p)}[\cos(q \cdot \ln r)g_1(\theta) - \sin(q \cdot \ln r)g_2(\theta)]. \tag{7–97}$$

As demonstrated by Moffatt, this streamfunction has infinitely many zeroes as r approaches zero, namely

$$q \cdot \ln r = \tan^{-1}\left(\frac{g_1(\theta)}{g_2(\theta)}\right) - \frac{\pi}{2} - n\pi \quad (n = 0, 1, 2, \ldots,). \tag{7–98}$$

Hence antisymmetric flow induced in a corner between two solid boundaries with an acute angle less than 146° may be expected to show an infinite sequence of increasingly small (and weak) eddies.

In this section we have considered the formation of a sequence of eddies near a corner between two plane boundaries with the suggestion that this type of motion could be driven by some external "stirring" motion at large distances from the corner. There have, in fact, been a number of investigations of problems in which the Moffatt eddies occur locally as part of an overall flow structure. Among these are the studies of Wakiya, O'Neill, and others[13] for simple shear flow over a circular cylindrical ridge on a solid, plane surface, as depicted in Fig. 7–10. In this case a sequence of eddies is found in the groove formed at the intersection between the cylinder and the plane wall when the angle of intersection, ϕ, is less than 146.3°. A closely related problem is the 2D motion of two equal, parallel cylinders that are touching, either along the line connecting the centers of the cylinders or perpendicular to it, where a sequence of eddies appears in the neighborhood of the contact point.[14] A summary of these and related problems is available for the interested reader.[15]

Later, in Chap. 9, we shall briefly consider the additional 2D problem of creeping flow past a circular cylinder (that is, a circle) in an unbounded fluid that undergoes a uniform

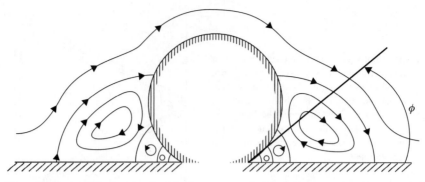

Figure 7–10. A qualitative sketch of the flow over a cylindrical ridge on a plane boundary [adapted from Fig. 1.3.9(a) in Ref. 13].

motion at large distances from the cylinder. For the moment, however, we turn to general solution procedures for other classes of creeping-flow problems.

E. AXISYMMETRIC CREEPING FLOWS: SOLUTION BY MEANS OF EIGENFUNCTION EXPANSIONS IN SPHERICAL COORDINATES (SEPARATION OF VARIABLES)

We saw in Section C that the creeping-motion and continuity equations for axisymmetric, incompressible flow can be reduced to the single fourth-order PDE for the streamfunction,

$$\boxed{E^4 \psi = 0.} \qquad (7\text{–}99)$$

An analytic solution for such a problem can thus be sought as a superposition of separable solutions of this equation in any orthogonal curvilinear coordinate system. The most convenient coordinate system for a particular problem is dictated by the geometry of the boundaries. As a general rule, at least one of the flow boundaries should coincide with a coordinate surface. Thus, if we consider an axisymmetric coordinate system (ξ, η, ϕ), then either $\xi = $ const or $\eta = $ const should correspond to one of the boundaries of the flow domain.

In this section we consider the general solution of (7–99) in the spherical coordinate system (r, θ, φ). This coordinate system is particularly useful for flows in the vicinity of a spherical boundary, but we begin by simply deducing the most general solution of (7–99) that is consistent with the constraint of axisymmetry, namely,

$$u_\theta = 0 \text{ at } \theta = 0, \pi. \qquad (7\text{–}100)$$

Rather than using the polar angle θ as an independent variable, it is more convenient to introduce

$$\eta \equiv \cos\theta$$

so that the coordinate variables are (r, η, ϕ) with $-1 \leq \eta \leq 1$, and the E^2 operator takes the simplified form

$$E^2 \equiv \frac{\partial^2}{\partial r^2} + \frac{(1-\eta^2)}{r^2} \frac{\partial^2}{\partial \eta^2}. \qquad (7\text{–}101)$$

E. Axisymmetric Creeping Flows

The nonzero-velocity components, expressed in terms of the streamfunction, are

$$\boxed{u_r = -\frac{1}{r^2}\frac{\partial \psi}{\partial \eta}, \quad u_\theta \sqrt{1-\eta^2} = -\frac{1}{r}\frac{\partial \psi}{\partial r}.}$$ (7–102)

We shall be interested here in flows that involve an axisymmetric body with the origin of coordinates $r = 0$ inside the body (and at its centre if the body is spherical).
The symmetry condition (7–100) requires that

$$\frac{\partial \psi}{\partial r} = 0 \quad \text{at} \quad \eta = \pm 1 \qquad (7\text{–}103a)$$

If we assume that the surface of the body is impermeable, than $\psi = \text{const}$ on this surface, and it follows from (7–103a) that

$$\psi = \text{const} \quad \text{at} \quad \eta = \pm 1. \qquad (7\text{–}103b)$$

For convenience, we may take const $= 0$ with no loss of generality.

1. General Eigenfunction Expansion

We seek a solution of (7–99) by means of the method of separation of variables. For this purpose, it is more convenient to note that $E^4 \psi = 0$ can be split into two second-order equations,

$$\boxed{E^2 \omega = 0,} \qquad (7\text{–}104)$$

$$\boxed{E^2 \psi = -\omega.} \qquad (7\text{–}105)$$

Obviously, substituting (7–105) into (7–104), we recover (7–99). In the present analysis, we first determine the most general solution for ω by solving (7–104), and then we solve (7–105) for ψ with ω given by this general solution.

To solve (7–104), we assume that

$$\omega = R(r)H(\eta). \qquad (7\text{–}106)$$

Hence, substituting into (7–104), we obtain

$$\frac{r^2}{R}\frac{d^2 R}{dr^2} + \frac{(1-\eta^2)}{H}\frac{d^2 H}{d\eta^2} = 0. \qquad (7\text{–}107)$$

The first term is a function of r only, whereas the second is a function of η. Hence it follows that each must equal a constant. We denote this constant as $n(n+1)$, and thus (7–107) separates into two equations,

$$r^2 \frac{d^2 R}{dr^2} - n(n+1)R = 0, \qquad (7\text{–}108)$$

and

$$(1-\eta^2)\frac{d^2 H}{d\eta^2} + n(n+1)H = 0. \qquad (7\text{–}109)$$

Equation (7–108) is a particular case of Euler's equation, for which a general solution is

$$R = r^s. \qquad (7\text{–}110)$$

Substituting into (7–108), we obtain the "characteristic" equation for s,

$$s(s-1) - n(n+1) = 0. \qquad (7\text{–}111)$$

The two roots of this quadratic equation are

$$s = n+1, \quad s = -n, \qquad (7\text{–}112)$$

Creeping Flows – Two-Dimensional and Axisymmetric Problems

so that the two independent solutions of (7–108) are

$$\boxed{R = r^{n+1}, \ r^{-n}.} \qquad (7\text{–}113)$$

Except for the special case, $n = 0$, a conveniant approach to solving (7–109) is to differentiate the whole equation with respect to η because this transforms it into an equation of well-known form, namely,

$$\frac{d}{d\eta}\left[(1-\eta^2)\frac{dY}{d\eta}\right] + n(n+1)Y = 0, \qquad (7\text{–}114)$$

with

$$Y \equiv \frac{dH}{d\eta}. \qquad (7\text{–}115)$$

Equation (7–114) is called Legendre's equation. In general, it has two independent solutions, known as Legendre's functions of the first and second kind. For the particular choice of the constant $n(n + 1)$, in which n is an integer, the Legendre functions of the first kind are polynomials of degree n, which are known as Legendre polynomials and normally denoted as P_n. The first several Legendre polynomials are

$$P_0 \equiv 1,$$
$$P_1 \equiv \eta, \qquad (7\text{–}116)$$
$$P_2 \equiv \frac{1}{2}(3\eta^2 - 1).$$

For each integer value of n in (7–114), there is a corresponding Legendre polynomial of degree n. The Legendre function of the second kind has a logarithmic singularity at $\eta = \pm 1$. The solution that we seek for Y (and thus for ω) is regular in the domain $-1 \le \eta \le 1$, and so we eliminate this second solution of (7–114) from further consideration. Hence,

$$\boxed{Y(\eta) = P_n(\eta),} \qquad (7\text{–}117)$$

for arbitrary integer values of n.

We have shown that the most general, regular solution for Y is the Legendre polynomial $P_n(\eta)$, for arbitrary integer values of n. However, we require a solution for the function $H(\eta)$ rather than for $Y(\eta)$. To obtain $H(\eta)$ from $Y(\eta)$, we integrate according to (7–115). However, we want the solution for ψ (and thus also the solution for ω) to satisfy the symmetry conditions (7–103) at $\eta = \pm 1$. Because this requires that

$$H(\eta) = 0 \quad \text{at} \ \eta = \pm 1, \qquad (7\text{–}118)$$

a convenient choice for the integral transformation from $Y(\eta)$ to $H(\eta)$ is

$$\boxed{H(\eta) \equiv \int_{-1}^{\eta} Y(\eta) d\eta} \qquad (7\text{–}119)$$

because $H(\eta)$, defined in this way, automatically satisfies the condition (7–118) at both $\eta = \pm 1$ for all n, other than $n = 0$. To see that the condition at $\eta = 1$ is satisfied, we note that the integral in (7–119) can be written in the form

$$\int_{-1}^{\eta} P_0(\eta) P_n(\eta) d\eta$$

E. Axisymmetric Creeping Flows

for any value of n and that the Legendre polynomials are orthogonal functions, that is,

$$\int_{-1}^{1} P_n P_m d\eta = \begin{cases} 0 & \text{for } n \neq m \\ \dfrac{2}{2n+1} & \text{for } n = m \end{cases}. \qquad (7\text{–}120)$$

We denote the polynomials that result from the definition (7–119) as $Q_n(\eta)$,

$$Q_n(\eta) \equiv \int_{-1}^{\eta} P_n(\eta) d\eta, \qquad (7\text{–}121)$$

of which the first several, for $\eta \neq 0$, are

$$Q_1(\eta) \equiv \frac{1}{2}(\eta^2 - 1),$$

$$Q_2(\eta) \equiv \frac{1}{2}(\eta^3 - \eta). \qquad (7\text{–}122)$$

$$Q_3(\eta) = \frac{1}{8}(5\eta^2 - 1)(\eta^2 - 1)$$

These polynomial functions are closely related to the so-called Gegenbauer polynomials[16] and satisfy the orthogonality condition

$$\int_{-1}^{1} \frac{Q_n(\eta) Q_m(\eta)}{(1-\eta^2)} d\eta = \begin{cases} 0 & \text{for } n \neq m \\ \dfrac{2}{n(n+1)(2n+1)} & \text{for } n = m \end{cases}. \qquad (7\text{–}123)$$

Further, as noted earlier, the polynomials $Q_n(\eta)$ are all identically zero at $\eta = \pm 1$, except for $Q_0(\eta)$.

The polynomial functions $Q_n(\eta)$ for $n \geq 1$ are the *eigenfunctions* of (7–109) subject to the symmetry conditions (7–103). The constants $\lambda_n = n(n+1)$ are the corresponding *eigenvalues*. It follows from the preceding discussion that the most general, nonsingular solution of $E^2 \omega = 0$ that satisfies the conditions (7–103) at $\eta = \pm 1$ is

$$\omega = \sum_{n=1}^{\infty} w_n \qquad (7\text{–}124)$$

where

$$\omega_n = \left(\bar{A}_n r^{n+1} + \bar{C}_n r^{-n} \right) Q_n(\eta). \qquad (7\text{–}125)$$

We leave the constant coefficients undetermined (these require imposition of boundary conditions corresponding to some specific problem) and proceed to solve (7–105) for ψ.

Hence, we seek a solution of

$$E^2 \psi = \sum_{n=1}^{\infty} \left(\bar{A}_n r^{n+1} + \hat{C}_n r^{-n} \right) Q_n(\eta) \qquad (7\text{–}126)$$

for arbitrary $\bar{A}_n$ and $\hat{C}_n$. The solution consists of two parts, a homogeneous solution satisfying $E^2 \psi = 0$ and a particular solution that generates the right-hand side of (7–126). The homogeneous solution, consistent with (7–103b), is clearly identical to ω above, that is,

$$\psi_{homog} = \sum_{n=1}^{\infty} \left[B_n r^{n+1} + D_n r^{-n} \right] Q_n(\eta). \qquad (7\text{–}127)$$

Creeping Flows – Two-Dimensional and Axisymmetric Problems

To obtain a particular solution, we try the form

$$\psi_{part} = r^\lambda Q_n(\eta), \tag{7-128}$$

which contains the same η dependence as the right-hand side of (7–126). With this choice for ψ, the left-hand side of (7–126) becomes

$$E^2[r^\lambda Q_n(\eta)] = \alpha r^{\lambda-2} Q_n(\eta), \tag{7-129}$$

where α is a constant that depends on λ and n.

Comparing (7–129) with the right-hand side of (7–126), we see that we will obtain a particular solution of the form (7–128) for each of the terms in (7–126). The term involving r^{n+1} requires $\lambda = n + 3$, whereas the term involving r^{-n} requires $\lambda = 2 - n$. With these values, the particular solution for ψ is

$$\psi_{part} = \sum_{n=1}^{\infty} \left(A_n r^{r+3} + C_n r^{2-n} \right) Q_n(\eta), \tag{7-130}$$

where A_n and C_n are *arbitrary* constants [related to the *arbitrary* constants $\hat{A}_n$ and $\hat{C}_n$ in (7–126)].

Combining (7–127) and (7–130), we can express the general solution for arbitrary, axisymmetric flows, in terms of spherical coordinates, in the form

$$\boxed{\psi = \sum_{n=1}^{\infty} \left(A_n r^{n+3} + B_n r^{n+1} + C_n r^{2-n} + D_n r^{-n} \right) Q_n(\eta).} \tag{7-131}$$

To apply (7–131) to a particular flow problem, we require boundary conditions to determine the four sets of arbitrary constants A_n, B_n, C_n, and D_n.

One extremely useful result can be proven that applies to any problem in which the streamfunction is expressed in the general form (7–131). This result can be formalized in the form of a *theorem*:

For any problem with ψ in the general form (7–131), the force exerted by the fluid on an arbitrary, axisymmetric body with its center of mass at $|\mathbf{x}| = 0$ is generally given by

$$F_z = 4\pi \mu U_c \ell_c C_1, \tag{7-132}$$

where U_c and ℓ_c are the characteristic velocity and length scales and z denotes the direction of the symmetry axis.

The result stated in this theorem is true independent of the values of the other coefficients A_n, B_n, C_n, and D_n. Further, although the numerical value of F_z might appear to depend on the particular velocity and length scale chosen for U_c and ℓ_c, this is not true because the combination $U_c \ell_c C_1$ remains constant for a body of fixed geometry in a particular flow, independent of U_c and ℓ_c. One practical implication of the theorem is that to determine (or estimate) F_z, we require only an accurate numerical value for the coefficient C_1 in (7–131). The accuracy of the other coefficients is irrelevant for purposes of calculating (or estimating) the drag force.

The brute force way to calculate the force on a body is to integrate the surface stress vector over the body surface – in this case,

$$\mathbf{F} = \mu U_c \ell_c \int_S (\mathbf{T} \cdot \mathbf{n}) dA, \tag{7-133}$$

E. Axisymmetric Creeping Flows

where the integral is expressed in terms of dimensionless variables. We could, in principle, simply evaluate (7–133) by using the general solution (7–131) in order to prove (7–132). However, with this approach, we would need to restrict the class of allowable axisymmetric body shapes to ones for which we could actually carry out the integration that is indicated in (7–133). To prove (7–132) for an axisymmetric body of arbitrary shape, we recall that the creeping-motion equation, expressed in terms of the stress tensor, is just

$$\nabla \cdot \mathbf{T} = 0. \tag{7–134}$$

Thus we can apply the divergence theorem to the volume of fluid that is contained between the body surface and any arbitrary closed surface that completely encloses the body to show that

$$0 = \int_V (\nabla \cdot \mathbf{T}) dV = \int_S (\mathbf{T} \cdot \mathbf{n}) dS - \int_{S^*} (\mathbf{T} \cdot \mathbf{n}^*) dS^*. \tag{7–135}$$

Thus the surface integral on S that appears in (7–133) is precisely equal to the surface integral of $\mathbf{T} \cdot \mathbf{n}$ for any closed surface that encloses the body. Here, $\mathbf{n}^*$ denotes the outer normal to the surface S^*, and $\mathbf{n}$ denotes the outer normal to the body surface S. The basic idea of replacing the integral on S (which may be quite complicated for a body of complicated shape) with an integral over some other surface S^*, is to choose S^* so that the integral is as simple as possible. A convenient choice for S^* is a spherical surface that is centered at the center of mass of the body.

Hence, to prove (7–132), we begin with the expression for the z component of $\mathbf{F}$ (that is, the drag force) in terms of a surface integral over a large sphere of radius R^* that is circumscribed about the body and centered at the center of mass, namely,

$$\boxed{F_z = \mathbf{i}_z \cdot \mathbf{F} = \mu U_c \ell_c \int_0^{2\pi} \int_0^{\pi} \mathbf{i}_z \cdot (\mathbf{T} \cdot \mathbf{n}^*)(R^*)^2 \sin\theta \, d\theta \, d\phi.} \tag{7–136}$$

Now, in dimensionless terms,

$$\mathbf{T} = -p\mathbf{I} + \left(\nabla \mathbf{u} + \nabla \mathbf{u}^T\right). \tag{7–137}$$

Thus, for this class of axisymmetric problems, where $T_{r\phi} = 0$,

$$\mathbf{T} \cdot \mathbf{n}^* = \mathbf{T} \cdot \mathbf{i}_r = T_{rr} \mathbf{i}_r + T_{\theta r} \mathbf{i}_\theta, \tag{7–138}$$

where

$$T_{rr} = -p + 2\frac{\partial u_r}{\partial r}, \tag{7–139}$$

$$T_{r\theta} = r\frac{\partial}{\partial r}\left(\frac{u_\theta}{r}\right) + \frac{1}{r}\frac{\partial u_r}{\partial \theta}. \tag{7–140}$$

Now, given the general form (7–131) for the streamfunction ψ, we can obtain corresponding expressions for u_r and u_θ by differentiating ψ using the definitions in (7–102), and thus we can obtain general expressions for T_{rr} and $T_{r\theta}$ that involve all of the coefficients A_n, B_n, C_n, and D_n. Substituting these expressions into (7–136) and carrying out the integration, we find that the result (7–132) follows. It is left to the reader to provide the details of this proof (see Problem 7–11).

Creeping Flows – Two-Dimensional and Axisymmetric Problems

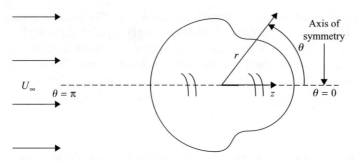

Figure 7–11. A schematic representation of the domain for a uniform flow past an arbitrary, axisymmetric body. For the case of a solid sphere, this is Stokes' problem.

2. Application to Uniform Streaming Flow past an Arbitrary Axisymmetric Body

As an example of the application of (7–131), we consider creeping flow past an arbitrary axisymmetric body with a uniform streaming motion at infinity. For the case of a solid sphere, this is known as *Stokes' problem*. In the present case, we begin by allowing the geometry of the body to be arbitrary (and unspecified) except for the requirement that the symmetry axis be parallel to the direction of the uniform flow at infinity so that the velocity field will be axisymmetric. A sketch of the flow configuration is shown in Fig. 7–11. We measure the polar angle θ from the axis of symmetry on the downstream side of the body. Thus $\eta = 1$ on this axis, and $\eta = -1$ on the axis of symmetry upstream of the body.

The asymptotic form of the streamfunction for $r \to \infty$ can be obtained from the uniform streaming condition at infinity, that is,

$$\mathbf{u} = \mathbf{i}_z \quad \text{as } r \to \infty. \tag{7–141}$$

Because

$$\mathbf{i}_z = \mathbf{i}_r \cos\theta - \mathbf{i}_\theta \sin\theta = \mathbf{i}_r \eta - \mathbf{i}_\theta (1-\eta^2)^{1/2}, \tag{7–142}$$

it follows from (7–141) that

$$\boxed{u_r = \eta, \quad u_\theta = -(1-\eta^2)^{1/2} \quad \text{for } r \to \infty.} \tag{7–143}$$

Thus, from the definitions (7–102), we see

$$\frac{\partial \psi}{\partial r} = r(1-\eta^2), \quad \frac{\partial \psi}{\partial \eta} = -\eta r^2 \quad \text{as } r \to \infty, \tag{7–144}$$

or, integrating these expressions, we obtain

$$\psi \to \frac{r^2}{2}(1-\eta^2) + \text{const} \quad \text{for } r \to \infty. \tag{7–145}$$

The constant that appears in (7–145) is completely arbitrary. Notice from (7–102) that we can always add an *arbitrary* constant to ψ without changing the velocities at all. Because the constant is arbitrary, we set it equal to zero, so that

$$\psi \to \frac{r^2}{2}(1-\eta^2) \quad \text{as } r \to \infty. \tag{7–146}$$

E. Axisymmetric Creeping Flows

Note that the choice const $= 0$ here also requires that the constant in (7–103) be zero, so that

$$\psi = 0 \text{ at } \eta = \pm 1 \quad (\text{all } r). \tag{7–147}$$

Because the general solution (7–131) is expressed in terms of the modified Gegenbauer polynomials $Q_n(\eta)$, it is convenient to also express (7–146) in terms of these functions. Referring to definitions (7–132), we see that (7–146) can also be expressed in the form

$$\psi \to -r^2 Q_1(\eta) \text{ as } r \to \infty. \tag{7–147}$$

Thus, regardless of the details of the body geometry, or the form of the boundary conditions at the body surface, the streamfunction must exhibit the asymptotic form (7–147) if there is a uniform streaming flow at large distances from the body. Comparing (7–147) and (7–131), we see that many of the coefficients A_n and B_n in the general axisymmetric solution must be zero for this case. In particular, the asymptotic condition (7–147) requires that

$$\begin{aligned} A_n &= 0, \quad \text{all } n, \\ B_n &= 0, \quad n \geq 2, \\ B_1 &= -1, \end{aligned} \tag{7–148}$$

and thus the general solution (7–131) becomes

$$\psi = -r^2 Q_1 + \sum_{1}^{\infty} \left(C_n r^{2-n} + D_n r^{-n} \right) Q_n(\eta). \tag{7–149}$$

Again, it is emphasized that though the form of this solution does depend on the fact that we have used eigenfunctions for a spherical coordinate system, it is strictly independent of the body geometry – this will come into play only when we try to apply boundary conditions at the body surface to determine C_n and D_n. For now we simply leave these constants unspecified.

By expressing the streamfunction in the form (7–149), we have essentially split the corresponding velocity field into two parts: the free-stream (or undisturbed) flow and an as yet unspecified *disturbance flow* that is due to the presence of the body in the flow. Although the constants C_n and D_n remain to be determined, there are several general features of the disturbance flow that are worth mentioning here. First, it is clear from the theorem, (7–132), that C_1 must be nonzero for any case in which the net force on the body is nonzero. For the problem considered here of uniform flow past a stationary body of axisymmetric, though otherwise arbitrary, geometry, the net force will always be nonzero, and thus $C_1 \neq 0$. Only the magnitude of C_1 will change from case to case. Second, because $C_1 \neq 0$, the largest possible contribution of the particle to the *far-field* (that is, $r \gg 1$) form of the disturbance flow is the term

$$\psi' = C_1 r Q_1(\eta). \tag{7–150}$$

Creeping Flows – Two-Dimensional and Axisymmetric Problems

In terms of velocity components, this contribution to the disturbance flow is

$$\boxed{u'_r = -\frac{C_1}{r}P_1(\eta); \quad u'_\theta\sqrt{1-\eta^2} = -\frac{C_1}{r}Q_1(\eta).} \qquad (7\text{–}151)$$

The most important feature of (7–151) is that the disturbance velocities fall off only *linearly* with distance from the body. Thus, for a body that exerts a net force on the fluid, the disturbance produced is extremely *long range* in a low-Reynolds-number flow. One important implication is that the motion of such a body will be extremely sensitive to the presence of another body, or other boundaries, even when these are a considerable distance away. For example, the velocity of a sphere moving through a quiescent fluid toward a plane wall under the action of buoyancy is reduced by approximately 12% when the sphere is 10 radii from the wall, and by 35% when it is 4 radii away.

Now, we have expressed the general streamfunction, (7–149), and the disturbance flow contribution in (7–150) and (7–151), in terms of spherical coordinates. However, we have not yet specified a body shape. Thus the linear decrease of the disturbance flow with distance from the body must clearly represent a property of creeping-flows that has nothing to do with specific coordinate systems. Indeed, this is the case, and the velocity field (7–151) plays a very special and fundamental role in creeping-flow theory. It is commonly known as the *Stokeslet velocity field* and represents the motion induced in a fluid at $Re = 0$ by *a point force at the origin* (expressed here in spherical coordinates).[17] We shall see later that the Stokeslet solution plays an important role in many aspects of creeping-flow theory.

F. UNIFORM STREAMING FLOW PAST A SOLID SPHERE – STOKES' LAW

The analysis of the preceding section was carried out by use of a spherical coordinate system, but the majority of the results are valid for an axisymmetric body of arbitrary shape. The necessity to specify a particular particle geometry occurs only when we apply boundary conditions on the particle surface (that is, when we evaluate the coefficients C_n and D_n in the spherical coordinate form of the solution). For this purpose, an exact solution requires that the body surface be a coordinate surface in the coordinate system that is used, and this effectively restricts the application of (7–149) to streaming flow past spherical bodies, which may be solid, as subsequently considered, or spherical bubbles or drops, as considered in section H.

Here, we consider *Stokes' problem* of uniform, streaming motion in the positive z direction, past a stationary solid sphere. The problem corresponds to the schematic representation shown in Fig. 7–11 when the body is spherical. This problem may also be viewed as that of a solid spherical particle that is translating in the negative z direction through an unbounded stationary fluid under the action of some external force. From a frame of reference whose origin is fixed at the center of the sphere, the latter problem is clearly identical with the problem pictured in Fig. 7–11. Because we have already derived the form for the streamfunction under the assumption of a uniform flow at infinity, we adopt the latter frame of reference. The problem then reduces to applying boundary conditions at the surface of the sphere to determine the constants C_n and D_n in the general equation (7–149). The boundary conditions on the surface of a solid sphere are the kinematic condition and the no-slip condition,

$$u_r = u_\theta = 0 \quad \text{for } r = 1. \qquad (7\text{–}152)$$

F. Uniform Streaming Flow past a Solid Sphere – Stokes' Law

In terms of the streamfunction, these conditions require that

$$\frac{\partial \psi}{\partial \eta} = \frac{\partial \psi}{\partial r} = 0 \quad \text{at } r = 1. \tag{7–153}$$

It follows from the first of these conditions that $\psi = \text{const}$ on the sphere surface, and, because $\psi = 0$ at $\eta = \pm 1$ for all r (because of symmetry), we can see that the appropriate version of the first condition in (7–153) is

$$\psi = 0 \quad \text{at } r = 1. \tag{7–154}$$

Hence we solve for C_n and D_n by using the second of the conditions in (7–153), plus the condition (7–154).

The two boundary conditions $\psi = (\partial \psi/\partial r) = 0$ at $r = 1$ lead to a pair of algebraic equations for C_n and D_n, namely,

$$0 = -Q_1 + \sum_{n=1}^{\infty} [C_n + D_n] Q_n(\eta), \tag{7–155}$$

$$0 = -2Q_1 + \sum_{n=1}^{\infty} [C_n(2-n) - n D_n] Q_n(\eta). \tag{7–156}$$

Because the $Q_n(\eta)$ are a complete set of orthogonal functions, those equations can be solved for the constant coefficients C_n and D_n. The result is that

$$\begin{aligned} C_n = D_n = 0, \; n \geq 2, \\ C_1 = \frac{3}{2}, \; D_1 = -\frac{1}{2}. \end{aligned} \tag{7–157}$$

Hence, in dimensionless terms,

$$\boxed{\psi = -\left(r^2 - \frac{3r}{2} + \frac{1}{2r}\right) Q_1(\eta).} \tag{7–158}$$

We can obtain the corresponding expressions for the velocity components u_r and u_θ by substituting this solution for ψ into the definitions (7–102). If we denote the relative translational velocity between the sphere and the surrounding fluid as U_∞, we can write the resulting expressions in dimensional form:

$$\boxed{\begin{aligned} u'_r &= U_\infty \left[1 - \frac{3}{2}\left(\frac{a}{r'}\right) + \frac{1}{2}\left(\frac{a}{r'}\right)^3\right] \cos\theta, \\ u'_\theta &= -U_\infty \left[1 - \frac{3}{4}\left(\frac{a}{r'}\right) - \frac{1}{4}\left(\frac{a}{r'}\right)^3\right] \sin\theta. \end{aligned}} \tag{7–159}$$

Furthermore, because $\nabla p' = \mu \nabla^2 \mathbf{u}'$, we can use these expressions for the velocity components to obtain the (dimensional) pressure

$$\boxed{p' = P_0 - \frac{3\mu U_\infty}{2a}\left(\frac{a}{r'}\right)^2 \cos\theta.} \tag{7–160}$$

Creeping Flows – Two-Dimensional and Axisymmetric Problems

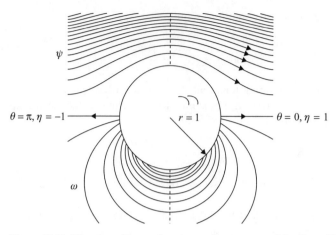

Figure 7–12. The streamlines and contours of constant vorticity for uniform streaming flow past a solid sphere (Stokes' problem). The streamfunction and vorticity values are calculated from Eqs. (7–158) and (7–162). Contour values plotted for the streamfunction are in increments of 1/16, starting from zero at the sphere surface, whereas the vorticity is plotted at equal increments equal to 0.04125.

Finally, because $C_1 = 3/2$, the theorem (7–132) yields the force law

$$F_z = 6\pi \mu a U_\infty. \tag{7–161}$$

This result is known as *Stokes' law*.

A plot showing lines of constant ψ in a plane that is parallel to the free stream and passing through the center of the sphere is shown in Fig. 7–12. One striking feature is that the flow is completely symmetric about $\eta = 0$ (that is, $\theta = \pi/2$). Apart from the arrows that have been added to the figure, it would not be possible from the shape of the streamlines to tell whether the fluid is moving from right to left or left to right.

Another feature of the flow that we can calculate from ψ by using the relationship (7–48) is the vorticity:

$$\boldsymbol{\omega} = -h_3 E^2 \psi \mathbf{i}_\phi.$$

It may be recalled that the vorticity, defined as $\boldsymbol{\omega} \equiv \nabla \wedge \mathbf{u}$, provides a measure of the local rate of rotation of fluid elements. In the present problem, the vorticity becomes

$$\boldsymbol{\omega} = -\frac{1}{2r^2}(1 - \eta^2)^{1/2} \mathbf{e}_\phi \mathbf{i}_\phi. \tag{7–162}$$

A plot showing lines of constant ω ($\equiv |\boldsymbol{\omega}|$) is also shown in Fig. 7–12. Again, the fore–aft symmetry of the flow is clearly evident. The fact that the constant-vorticity lines all terminate on the body surface is indicative of the fact that the source of vorticity in the flow is the sphere surface. Indeed, the governing equation for vorticity in the creeping-flow limit can be shown from (7–6) to be[18]

$$\nabla^2 \boldsymbol{\omega} = 0.$$

Further, there is no vorticity in the undisturbed flow. Hence vorticity is created at the body surface and then transported into the fluid by means of diffusion (we shall see later that it is the no-slip condition that is responsible for production of vorticity). In the creeping-flow limit, there is no convection of vorticity. We shall see later that it is the convection of vorticity

F. Uniform Streaming Flow past a Solid Sphere – Stokes' Law

from upstream toward the downstream direction that is responsible for asymmetry in the flow at finite Reynolds numbers, including the development of strong disturbances in the downstream wake. Because vorticity only diffuses in the creeping-flow limit, the disturbance produced by the sphere is fore–aft symmetric. There is, in fact, a strong similarity between the transport of vorticity in 2D and axisymmetric flows and the transport of heat from a heated body. Not surprisingly, constant-vorticity lines are often similar in appearance to isotherms in the thermal problem. In the simplest thermal problem, however, the surface of the body would be assumed to be at constant temperature so that the temperature field in the case of pure conduction from a sphere would be spherically symmetric. The sphere surface is *not* a uniform source of vorticity. In regions where the velocity gradients are largest, the surface vorticity is highest. As a consequence, even for creeping flow with a spherical body, the vorticity distribution is not spherically symmetric, and this is evident in Fig. 7–12.

Finally, before leaving this topic, we return briefly to the result (7–161) for the hydrodynamic drag on the sphere that is due to its motion relative to the surrounding fluid. The theorem (7–132) provides a convenient way to evaluate the drag, once a solution for the streamfunction has been obtained in the form (7–149). In this case, however, the body geometry is simple and we can also evaluate the drag directly by applying the formula (2–29) from Chap. 2 for the force per unit area on the surface in terms of the stress tensor, and then integrating over the sphere surface as shown in (7–133). The result for the (dimensional) drag force is

$$F_{drag} = \int_A \mathbf{i}_z \cdot (\mathbf{i}_r \cdot \mathbf{T}')dA, \qquad (7\text{–}163)$$

or substituting for $\mathbf{i}_z$ and $\mathbf{T}'$ in terms of their components in a spherical coordinate system, we have

$$F_{drag} = \int_0^{2\pi}\int_0^{\pi} \left[(\cos\theta)T'_{rr} - (\sin\theta)T'_{r\theta}\right]\Big|_{r=a}\, a^2 \sin\theta\, d\theta\, d\phi. \qquad (7\text{–}164)$$

Now, according to the constitutive equation (2–81) for an incompressible, Newtonian fluid, the T'_{rr} component is

$$T'_{rr} = -p' + 2\mu\left(\frac{\partial u'_r}{\partial r'}\right),$$

and the $T'_{r\theta}$ component is

$$T'_{r\theta} = \mu\left[r'\frac{\partial}{\partial r'}\left(\frac{u'_\theta}{r'}\right) + \frac{1}{r'}\frac{\partial u'_r}{\partial \theta}\right].$$

However, at the sphere surface, we see from (7–159) that

$$\left(\frac{\partial u'_r}{\partial r'}\right)_{r=a} = \left(\frac{\partial u'_r}{\partial \theta}\right)_{r=a} = 0. \qquad (7\text{–}165)$$

Hence

$$T'_{rr}\Big|_{r=a} = -P_0 + \frac{3\mu U_\infty}{2a}\cos\theta \qquad (7\text{–}166)$$

and

$$T'_{r\theta}\Big|_{r=a} = -\frac{3\mu U_\infty}{2a}\sin\theta. \qquad (7\text{–}167)$$

Substituting into (7–164), we thus have

$$F_{drag} = \int_0^{2\pi}\int_0^{\pi}\left[-P_0\cos\theta + \frac{3\mu U_\infty}{2a}(\cos^2\theta + \sin^2\theta)\right]a^2 \sin\theta\, d\theta\, d\phi, \qquad (7\text{–}168)$$

and integrating, we obtain

$$F_{drag} = 2\pi \mu a U_\infty + 4\pi \mu a U_\infty = 6\pi \mu a U_\infty. \qquad (7\text{--}169)$$

Of course, the result is again *Stokes' law*. Although $(\cos^2\theta + \sin^2\theta) = 1$, we display this term explicitly in (7–168) because it provides a basis for identifying the source of the two contributions to Stokes' law. Indeed, when (7–164)–(7–167) are compared with (7–168), it is evident that the $\cos^2\theta$ contribution, which leads to the term $2\pi \mu a U_\infty$ in (7–169), is due to the dynamic pressure distribution on the sphere surface, whereas the $\sin^2\theta$ term corresponds to the viscous tangential-stress contribution. It is common practice in fluid mechanics to identify the pressure contributions to the force on a body as the *form drag* and the viscous stress contributions as the *friction drag*. It is noteworthy that the form drag and friction drag contributions to (7–169) are of comparable magnitude for this low-Reynolds-number flow.

Stokes' law can be used in conjunction with a force balance on the sphere to obtain an estimate for the sphere velocity for gravity-driven motion at steady-state, where acceleration effects are not present. For this purpose, we need to know that the net buoyancy force on the sphere is

$$F_{buoyancy} = \frac{4}{3}\pi a^3 (\rho_s - \rho) g. \qquad (7\text{--}170)$$

This buoyancy force consists of two distinct contributions: $(4/3)\pi a^3 \rho_s g$ is the gravitational body force on a sphere of density ρ_s, and $(4/3)\pi a^3 \rho g$ is the net force contribution due to the hydrostatic pressure distribution at the sphere surface. At steady-state, the two forces, (7–169) and (7–170), must balance, that is,

$$6\pi \mu a U_\infty = \frac{4}{3}\pi a^3 (\rho_s - \rho) g. \qquad (7\text{--}171)$$

Thus, solving for U_∞, we obtain

$$\boxed{U_\infty = \frac{2}{9}\frac{a^2(\rho_s - \rho)g}{\mu},} \qquad (7\text{--}172)$$

which is known as the *terminal velocity* for a sedimenting sphere in a quiescent fluid. We emphasize that the results obtained in this section are valid only for cases in which the Reynolds number, $Re \equiv \rho U_\infty a/\mu$, is very small, $Re \ll 1$. Given the expression (7–172) for the terminal velocity, we can see that this condition is typically satisfied for small particles or highly viscous fluids.

G. A RIGID SPHERE IN AXISYMMETRIC, EXTENSIONAL FLOW

1. The Flow Field

Stokes' problem of a rigid sphere in a steady, uniform flow is relevant to the steady translation of the sphere through an unbounded quiescent fluid that is due to the action of a body force on the sphere. The most common example of such a problem is the sedimentation of a sphere that is due to the action of gravity, when the sphere density is not equal to the density of the suspending fluid. For many applications, however, we would like to understand the fluid motion in the vicinity of a body when the fluid at infinity is undergoing some nonuniform motion. For purposes of studying such a problem, it is convenient to consider a body whose density is equal to the density of the fluid so that the only motion in the fluid is due to the flow at infinity and its interaction with the body. Such bodies are called *neutrally buoyant*, for obvious reasons. In creeping-flow, the solution for a nonneutrally buoyant body in some general flow at infinity can be constructed by superposition from the

G. A Rigid Sphere in Axisymmetric, Extensional Flow

solution for a neutrally buoyant body in the same undisturbed flow and Stokes' solution for sedimentation/translation through a quiescent fluid. Of course, it will generally be difficult to consider the fluid motion in the vicinity of a body for some completely arbitrary flow at infinity. For creeping flow, for which solutions can be constructed by superposition, a reasonable alternative is to consider a hierarchy of nonuniform undisturbed flows beginning with flows in which the magnitude of the velocity varies *linearly* with spatial position.

A more direct motivation for studying linear flows is that we are frequently interested in applications of creeping-flow results for particles that are very small compared with the length scale L that is characteristic of changes in the undisturbed velocity gradient for a general flow. In this case, we may approximate the undisturbed velocity field in the vicinity of the particle by means of a Taylor series approximation, namely,

$$\mathbf{u}'_\infty = \mathbf{u}'_0 + (\nabla'\mathbf{u}')_0 \cdot \mathbf{x}' + \cdots +,$$

where $(\nabla'\mathbf{u}')_0$ is the velocity gradient evaluated at the position of the center of mass of the particle, $\mathbf{x}' = 0$. The next term involving the gradient of $\nabla'\mathbf{u}'$ is smaller by a factor proportional to the ratio of the length scale of the body ℓ to the macroscopic scale L. We shall see later that the net hydrodynamic force on a sphere in a linear undisturbed flow is proportional to the difference between the velocity of the center of mass of the sphere and the velocity of the undisturbed flow evaluated at the position occupied by the center of mass. Thus, because the external force on the sphere is zero (recall that it is neutrally buoyant), it follows that

$$\mathbf{u}_p = \mathbf{u}'_0, \tag{7-173}$$

where $\mathbf{u}_p$ is the velocity of the center of mass of the body.

Thus the undisturbed flow that is seen in a frame of reference that translates with the particle is just

$$\mathbf{u}'_\infty = (\nabla'\mathbf{u}')_0 \cdot \mathbf{x}' + \cdots +. \tag{7-174}$$

Now, however, this can be expressed as the sum of a linear straining flow and a purely rotational flow, that is,

$$\mathbf{u}'_\infty = \mathbf{E}'_0 \cdot \mathbf{x}' + \mathbf{\Omega}'_0 \cdot \mathbf{x}' + \cdots +, \tag{7-175}$$

where $\mathbf{E}'_0$ and $\mathbf{\Omega}'_0$ are the rate-of-strain and vorticity tensors defined in terms of $(\nabla'\mathbf{u}')_0$, and we see that the motion of a small particle in a general linear flow can be described as a superposition of its motion in a pure straining flow and a purely rotational flow. In general, a particle in the purely rotational flow will simply rotate with the vorticity of the undisturbed flow, though we shall not prove this result here. The response of the fluid and particle to a pure straining flow is generally more complicated. In this section, we consider the special case of a rigid sphere in an axisymmetric pure straining flow,

$$\mathbf{u}'_\infty = -E(x'\mathbf{i} + y'\mathbf{j} - 2z'\mathbf{k}), \tag{7-176}$$

where

$$\mathbf{E} = E \begin{pmatrix} 1 & 0 & 0 \\ 0 & 1 & 0 \\ 0 & 0 & -2 \end{pmatrix}.$$

A plot showing streamlines for this flow with a spherical body at the origin (calculated later in this section) is shown in Fig. 7–13. For $E > 0$, there is flow outward away from the sphere along the axis of symmetry and flow inward in the plane orthogonal to this axis. This flow is called *uniaxial extensional flow*. For $E < 0$, the direction of fluid motion is reversed and the undisturbed flow is known as *biaxial extensional flow*. In either case, with an axially

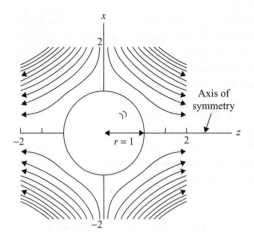

Figure 7–13. The streamlines for axisymmetric flow in the vicinity of a solid sphere with *uniaxial* extensional flow at infinity. When the direction of motion is reversed at infinity, the undisturbed flow is known as *biaxial* extensional flow. The stream-function values are calculated from Eq. (7–185). Contour values are plotted in equal increments equal to 0.5.

symmetric body at the origin $\mathbf{x} = 0$, with its symmetry axis parallel to the symmetry axis of the flow, it is evident that the complete velocity field will be axisymmetric. Hence the general eigenfunction expansion (7–131) should be applicable in creeping-flow, and if the body is spherical as we assume here, it should be possible to determine all of the coefficients A_n, B_n, C_n, and D_n from the form of the undisturbed flow and the boundary conditions on the sphere surface.

To do this, we must first determine the form of streamfunction at large distances from the sphere when the velocity field has the asymptotic form (7–176). Specifically, if we transform (7–176) to spherical coordinates by using the general relationships

$$\mathbf{i} = \mathbf{i}_r \sin\theta \cos\phi + \mathbf{i}_\theta \cos\theta \cos\phi - \mathbf{i}_\phi \sin\phi,$$

$$\mathbf{j} = \mathbf{i}_r \sin\theta \sin\phi + \mathbf{i}_\theta \cos\theta \sin\phi + \mathbf{i}_\phi \cos\phi, \qquad (7\text{–}177)$$

$$\mathbf{k} = \mathbf{i}_r \cos\theta - \mathbf{i}_\theta \sin\theta,$$

and

$$x' = r' \sin\theta \cos\phi,$$
$$y' = r' \sin\theta \sin\phi, \qquad (7\text{–}178)$$
$$z' = r' \cos\theta,$$

we find that

$$\mathbf{u}'_\infty = -Er'(1 - 3\cos^2\theta)\mathbf{i}_r - Er'(3\cos\theta \sin\theta)\mathbf{i}_\theta. \qquad (7\text{–}179)$$

A relevant characteristic velocity in this case is thus

$$u_c = Ea, \qquad (7\text{–}180)$$

where a is the sphere radius, and in dimensionless terms we therefore have

$$\mathbf{u}_\infty = -r(1 - 3\cos^2\theta)\mathbf{i}_r - r(3\sin\theta \cos\theta)\mathbf{i}_\theta, \qquad (7\text{–}181)$$

where r is now the dimensionless radial variable that is equal to unity ($r = 1$) at the sphere surface.

From the form (7–181) and the definitions (7–102) of the streamfunction in terms of the axisymmetric velocity components u_r and u_θ, we can easily obtain the asymptotic form for ψ. The result is

$$\boxed{\psi = r^3 \eta(1 - \eta^2) = -2r^3 Q_2(\eta) \quad \text{for } r \to \infty.} \qquad (7\text{–}182)$$

G. A Rigid Sphere in Axisymmetric, Extensional Flow

Thus, if we impose this far-field condition on the general axisymmetric solution (7–131), we must require that

$$\begin{aligned} A_n &= 0, \quad n \geq 1, \\ B_n &= 0, \quad n \geq 3, \\ B_2 &= -2, \\ B_1 &= 0. \end{aligned} \quad (7\text{--}183)$$

The latter condition simply avoids the presence of a uniform streaming flow in the far field.

To complete the solution, we must apply boundary conditions at the surface of the solid sphere. Thus, again, we require that

$$u_r = u_\theta = 0 \quad \text{at} \ r = 1,$$

and this translates to the two conditions on ψ

$$\psi = \frac{\partial \psi}{\partial r} = 0 \quad \text{at} \ r = 1.$$

Hence, because the $Q_n(\eta)$ are a complete set of orthogonal eigenfunctions, it follows from (7–131) and (7–183) that

$$C_2 = 5, \quad D_2 = -3, \quad (7\text{--}184)$$
$$C_n = D_n = 0 \ \text{for} \ n \neq 2.$$

Hence, the solution for axisymmetric straining flow past a solid sphere is

$$\boxed{\psi = 2\left(-r^3 + \frac{5}{2} - \frac{3}{2}\frac{1}{r^2}\right) Q_2(\eta).} \quad (7\text{--}185)$$

A sketch of the streamlines corresponding to this solution is given in Fig. 7–13.

There are a few brief, but useful, observations about the form (7–185) and about the solution in general. First, it may be noted that $C_1 = 0$. Hence it can be shown that the net hydrodynamic force on the sphere is zero. Second, the sign of E plays no role in solution of this problem. Hence, for a solid sphere, the solution (7–185) is valid for both uniaxial and biaxial flows and the form of the velocity field is unchanged by a change from $E > 0$ to $E < 0$, though, of course, the direction of motion does reverse.

2. Dilute Suspension Rheology – The Einstein Viscosity Formula

The velocity field given by (7–185) will be used later to estimate heat transfer rates for spherical particles in a straining flow. Here, we focus on a different application of (7–185), namely, its use in predicting the effective viscosity of a dilute suspension of solid spheres. To carry out the calculation, it is first necessary to briefly discuss the properties of a suspension in a more general framework.

Let us then consider a suspension of identical, neutrally buoyant solid spheres of radius a. We are interested in circumstances in which the length scale of the suspension at the particle scale (that is, the particle radius) is very small compared with the characteristic dimension L of the flow domain so that the suspension can be modeled as a continuum with properties that differ from the suspending fluid because of the presence of the particles. Our goal is to obtain an *a priori* prediction of the macroscopic rheological properties when the suspension is extremely dilute, a problem first considered by Einstein (1905) as part of his Ph.D. dissertation.[19] A suspension is called *dilute* if the volume fraction of particles is small enough that each particle moves independently of all the others, with no hydrodynamic interactions possible.

To discuss the macroscopic (or "bulk") properties of a suspension, it is necessary to specify the connection between local variables at the particle scale and macroscopic variables at the scale L. One plausible choice, in view of the relationship between continuum and molecular variables in Chap. 2, is to assume that the macroscopic variables are just volume averages of the local variables. In particular, we assume in the discussion that follows that the macroscopic (or bulk) stress can be defined as a volume average of the local stress in the suspension, namely,

$$\langle \sigma_{ij} \rangle \equiv \frac{1}{V} \int_V \sigma_{ij} dV, \qquad (7\text{--}186)$$

where σ_{ij} may pertain either to a point in the suspending fluid or to a point inside one of the particles.[18] The volume V is an averaging volume whose linear dimensions are large compared with the characteristic microscale of the suspension (so that V will contain a statistically significant number of spheres) but still arbitrarily small compared with the dimension L of the flow domain. Similarly, we define a bulk or macroscopic strain rate as

$$\langle e_{ij} \rangle \equiv \frac{1}{V} \int_V e_{ij} dV, \qquad (7\text{--}187)$$

and it follows that

$$\langle \sigma_{ij} \rangle - 2\mu \langle e_{ij} \rangle \equiv \frac{1}{V} \int_V (\sigma_{ij} - 2\mu e_{ij}) dV. \qquad (7\text{--}188)$$

Now, in the solid particles $e_{ij} = 0$, and within the suspending fluid σ_{ij} differs from $2\mu e_{ij}$ by the pressure multiplied by the unit tensor δ_{ij}. Thus,

$$\langle \sigma_{ij} \rangle - 2\mu \langle e_{ij} \rangle = -p^* \delta_{ij} + \frac{1}{V} \sum \int_{V_p} \sigma_{ij} dV, \qquad (7\text{--}189)$$

where $p^* \delta_{ij}$ is an isotropic term that is related to the bulk pressure and the summation is over all of the particles within V.

The expression (7–189) for the bulk stress is, of course, valid for arbitrary concentrations of particles, but the volume integrals over V_p are exceedingly difficult to evaluate in general because the value for a particular particle depends on the complete configuration of particles in the suspension. For a *dilute* suspension of identical particles, on the other hand, the problem simplifies immensely, because the integral over V_p is *exactly the same* for all particles, and the expression (7–189) can be replaced with

$$\langle \sigma_{ij} \rangle - 2\mu \langle e_{ij} \rangle = -p^* \delta_{ij} + n \int_{V_p} \sigma_{ij} dV, \qquad (7\text{--}190)$$

where n is the number of particles per unit volume and the integration is now over the volume of a single sphere.

G. A Rigid Sphere in Axisymmetric, Extensional Flow

To evaluate the integral over V_p, it is convenient to convert it to an integral over the *surface* of the sphere by means of the divergence theorem. To do this, we note that

$$\frac{\partial \sigma_{ij}}{\partial x_j} \equiv 0 \qquad (7\text{--}191)$$

within a freely suspended sphere, and thus

$$\sigma_{ij} = \frac{\partial}{\partial x_k}(x_j \sigma_{ik}). \qquad (7\text{--}192)$$

It follows that

$$\int_{V_p} \sigma_{ij} dV = \int_{S_p} x_j \sigma_{ik} n_k dS, \qquad (7\text{--}193)$$

where S_p is the sphere surface and **n** is the unit outer normal to S_p. Hence, to evaluate the bulk stress for a dilute suspension of identical spheres, we must evaluate the surface integral, (7–193), for some specified flow.

In the present circumstances, we consider a *dilute* suspension that is undergoing an axisymmetric pure straining flow, $\langle e_{ij}\rangle x_j$. For this case, the velocity field in the vicinity of each of the spherical particles is given by (7–185), namely,

$$\begin{aligned}
u_r &= \frac{1}{r^2}\frac{\partial \psi}{\partial \eta} = \left(r - \frac{5}{2r^2} + \frac{3}{2r^5}\right)(3\cos^2\theta - 1), \\
u_\theta &= \frac{1}{r}\frac{1}{\sqrt{1-\eta^2}}\frac{\partial \psi}{\partial r} = -3\left(r - \frac{1}{r^4}\right)\sin\theta\cos\theta,
\end{aligned} \qquad (7\text{--}194)$$

and the integral on S_p can be evaluated for the nonzero components of $\langle e_{ij}\rangle$. The result, after some manipulation, is

$$\int_{S_p} x_j \sigma_{ik} n_k dA = \frac{4\pi}{3} a^3 5 \langle e_{ij}\rangle \mu. \qquad (7\text{--}195)$$

Because the volume fraction of spheres is just

$$C = \frac{4\pi a^3}{3} n,$$

it follows that

$$\langle \sigma_{ij}\rangle = -p^* \delta_{ij} + 2\mu\left(1 + \frac{5}{2}C\right)\langle e_{ij}\rangle. \qquad (7\text{--}196)$$

It can, in fact, be proven that a dilute suspension of rigid spheres *will always be Newtonian* at the first $O(C)$ correction to the bulk stress, with an effective viscosity given by (7–196) of the form

$$\mu^* = \mu\left(1 + \frac{5}{2}C\right). \qquad (7\text{--}197)$$

This is the famous result obtained by Einstein for the viscosity of a dilute suspension of spheres.[20] Although the integral (7–195) leading to this result was evaluated by use of the

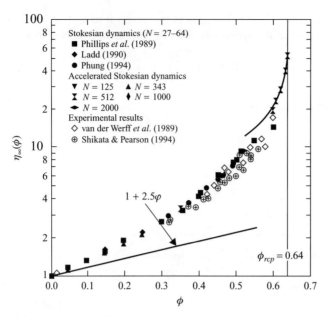

Figure 7–14. A comparison of experimental and simulation data for the viscosity of a suspension of spherical particles versus the Einstein prediction (7–197). The maximum volume fraction for random cubic packing of equal-size spheres is 0.64 (figure courtesy of J. F. Brady).

velocity field (7–194) for the specific case of an axisymmetric extensional flow, the same result, in fact, is obtained for any arbitrary linear flow, including simple shear flow.

It is obvious from (7–196) that the effective viscosity of a dilute suspension of rigid spheres exceeds the viscosity of the suspending fluid by an amount that is proportional to the volume fraction of the particles. The increase in effective viscosity can be understood to be a consequence of the fact that the rigid particles cannot deform in response to the straining motion in the fluid and thus generate a disturbance flow that increases the rate of viscous dissipation relative to that which would occur in the absence of the particles. In this sense, the particles produce a macroscopic effect that is equivalent to an increase of the viscosity of the suspension when it is considered as a homogeneous continuum. It may be noted from (7–196) that the bulk stress for a given bulk strain rate, $\langle e_{ij} \rangle$, is completely independent of the addition of vorticity to the flow. This is because the particle response to vorticity in the linear creeping-motion approximation is to simply rotate with the local angular velocity of the fluid with no additional disturbance of the fluid's motion and hence no additional contribution to the rate of viscous dissipation.

The prediction (7–197) is compared with experimental data for a suspension of monodisperse spherical particles in Fig. 7–14. The result (7–197) is actually independent of whether the suspension is monodisperse or polydisperse as the contribution of each particle is independent of the presence of other particles. However, we see that the result (7–197) holds only for extremely small volume fractions and then the data begin to increase more rapidly with volume fraction. When this happens, particle interactions begin to be important, both hydrodynamic and also those that are due to any interaction forces that may exist between particles. Then the size distribution, the nature of hydrodynamic forces, and other factors such as the importance of Brownian motion begins to play a role.

A general discussion of the rheology of suspensions and other materials that are *heterogeneous* at the microscale is beyond the scope of the present book. However, the interested reader may wish to refer to the original paper of Batchelor[21] or to the textbooks of Schowalter, Russel et al., or Larson for a more comprehensive presentation of this material.[22]

H. TRANSLATION OF A DROP THROUGH A QUIESCENT FLUID AT LOW Re

A second problem, closely related to Stokes' problem, is the steady, buoyancy-driven motion of a bubble or drop through a quiescent fluid. There are many circumstances in which the buoyancy-driven motions of bubbles or drops are of special concern to chemical engineers. Of course, bubble and drop motions may occur over a broad spectrum of Reynolds numbers, not only the creeping-flow limit that is the focus of this chapter. Nevertheless, many problems involving small bubbles or drops in viscous fluids do fall into this class.[23]

As we have noted previously, problems involving the motions of two contiguous fluids that are separated by an interface are especially difficult to solve because the shape of the interface is not known *a priori* and will generally change in response to the motions of the fluids. This general difficulty is true of the motions of bubbles or drops. Indeed, it is well known that the deformation of shape produced in some flows can lead to breakup.[24] On the other hand, in the absence of flow, the equilibrium shape of a drop or bubble is spherical because of the tendency for systems with nonzero interfacial tension to adopt a configuration that minimizes surface area, and thereby minimizes the free energy. We may expect, then, that flow conditions will exist in which the deformation from this equilibrium state is small, and for these cases we may be able to obtain analytical results by means of the method of *domain perturbations* as introduced for a different class of bubble motions in Chap. 4. On the other hand, when the deformations are not small, analytical results are not generally possible even in the creeping-flow limit because the boundary conditions are highly nonlinear. In fact, highly efficient numerical methods have been developed in recent years to deal with this class of free-boundary problem, especially in the creeping-flow regime in which the linearity of the governing DEs has led to methods, known as boundary integral methods, based on the use of fundamental solutions. We will discuss the foundations of these methods in the next chapter.

In this section, we begin by considering the buoyancy-driven motion of a single gas bubble or drop through an otherwise stationary viscous fluid, under the assumption that the bubble or drop shape is nearly spherical. We denote the viscosities and densities of the two fluids as μ, $\hat{\mu}$, ρ, and $\hat{\rho}$ with the variables with carets corresponding to the fluid inside the drop. In this section, we also assume that the interfacial tension, which we denote as γ, is uniform at the drop surface, and that the Reynolds numbers for both the interior and exterior flows are sufficiently small that the creeping-motion approximation can be applied for both fluids. Under these circumstances, experimental evidence shows (and we will assume) that the drop or bubble will translate with a constant velocity **U**. In addition, though we must consider the shape of the drop to be unknown, we may also anticipate that it will be axisymmetric about an axis that is collinear with the velocity vector **U**.

The governing equations are thus the creeping-flow equations for both the inner and outer fluids. As in previous problems, it is important to write both the DEs and the boundary conditions in dimensionless form. A convenient choice for nondimensionalization is to use the radius of the undeformed drop as a characteristic length scale, $\ell_c = a$; the (as yet unknown) translational velocity of the drop as a characteristic velocity scale, $u_c = U$; and characteristic pressures, $p_c \equiv \mu U/a$ and $\hat{p}_c \equiv \hat{\mu} U/a$, as we have seen earlier, are appropriate in the creeping-flow limit. In dimensionless terms, the governing equations are thus

$$\nabla^2 \mathbf{u} - \nabla p = 0,$$
$$\nabla \cdot \mathbf{u} = 0, \quad (7\text{–}198)$$

$$\nabla^2 \hat{\mathbf{u}} - \nabla \hat{p} = 0,$$
$$\nabla \cdot \hat{\mathbf{u}} = 0. \quad (7\text{–}199)$$

Creeping Flows – Two-Dimensional and Axisymmetric Problems

We adopt a coordinate reference frame that is *fixed at the center of mass of the drop*, as shown in Fig 7–11. Note, however, that we re-orient so that the z-axis is vertical and $\mathbf{g} = -g\,\mathbf{i}_z$. Then, the far-field boundary condition takes the form

$$\boxed{\mathbf{u} \to \mathbf{i}_z \text{ as } |\mathbf{x}| \to \infty.} \qquad (7\text{–}200)$$

Inside the drop, we require that the velocity and pressure fields be bounded at the origin [which is a singular point for the spherical coordinate system that we will use to solve (7–199)]. Finally, at the drop surface, we must apply the general boundary conditions at a fluid interface from Section L of Chap. 2. However, a complication in using these boundary conditions is that the drop shape is actually unknown (and, thus, so too are the unit normal and tangent vectors $\mathbf{n}$ and $\mathbf{t}$ and the interface curvature $\nabla \cdot \mathbf{n}$). As already noted, we can expect to solve this problem analytically only in circumstances when the shape of the drop is approximately (or exactly) spherical, and, in this case, we can use the method of *domain perturbations* that was first introduced in Chap. 4. In this procedure, we assume that the shape is nearly spherical, and develop an asymptotic solution that has the solution for a sphere as the first approximation. An obvious question in this case is this: When may we expect the shape to actually be approximately spherical?

To answer this question, we must first go back and reexamine the boundary conditions that apply at the interface. These are the continuity of the tangential velocity (no slip between the two liquids), which is given as Eq. (2–122); the kinematic condition on the normal velocity components, which is given by (2–117) plus (2–129); the tangential-stress condition (2–141) with $\text{grad}_s\,\gamma = 0$; and the normal-stress condition (2–135). We see that all of these conditions involve the shape of the drop, at least implicitly by means of the unit normal and tangent vectors [see, e.g., Eq. (2–128)]. Thus, because the shape depends on the flow, the problem is nonlinear even though the creeping-flow equations (7–198) and (7–199) are linear. As shown in Chap. 2, we represent the shape of the drop in terms of the points $\mathbf{x}$ where a function

$$F'(\mathbf{x}) \equiv r' - [a + f'(\mathbf{x})] \qquad (7\text{–}201)$$

is equal to zero. The primes on r', F' and f' indicate that these functions are dimensional. Clearly, if the function $f' = 0$, then F' equals zero for $r' = a$ and the shape is a sphere of radius a.

Now, the drop shape will be non-spherical if this is necessary to satisfy the boundary conditions at its surface, and, specifically, to satisfy the normal-stress balance, (2–135). To determine the condition that leads to small deformations, we can nondimensionalize the boundary conditions using the same characteristic scales that were used for the governing equations, (7–198) and (7–199). The result for the normal-stress balance is

$$\boxed{\lambda\hat{p} - p + [(\boldsymbol{\tau} - \lambda\hat{\boldsymbol{\tau}})\cdot\mathbf{n}]\cdot\mathbf{n} = \frac{1}{Ca}(\nabla\cdot\mathbf{n}) \text{ at } F = 0.} \qquad (7\text{–}202)$$

It may be noted here that $\boldsymbol{\tau} = 2\mathbf{e}$ and $\hat{\boldsymbol{\tau}} = 2\hat{\mathbf{e}}$, where $\mathbf{e}$ and $\hat{\mathbf{e}}$ are the dimensionless rate-of-strain tensors and $\lambda = \hat{\mu}/\mu$ is the viscosity ratio. The dimensionless parameter that appears in (7–202) is the capillary number:

$$Ca \equiv \frac{\mu U}{\gamma}.$$

A qualitative prediction of the role of the capillary number in drop deformation can be obtained from the normal-stress balance. Now, if the shape is spherical, the capillary term on the right-hand side is a constant, and thus the viscous pressure and stress contributions

H. Translation of a Drop Through a Quiescent Fluid at Low *Re*

on the left-hand side must also produce a constant value (corresponding to the pressure jump across the interface that is due to surface tension) if the normal-stress balance is to be satisfied. In general, however, the pressure and stress differences will not reduce to this very simple form, but will instead vary as a function of position on the surface of the drop. Such a nonuniform distribution of pressure and stress will tend to deform the drop. In fact, in this case, the normal-stress balance can be satisfied only if the drop deforms to a shape where the interface curvature ($\nabla \cdot \mathbf{n}$) varies in precisely the same way as the surface pressure and stress difference. The *magnitude* of the change in shape that is required to satisfy the normal-stress balance can be seen from (7–202) to depend on Ca. For $Ca \ll 1$, very small deviations from a spherical shape (where $\nabla \cdot \mathbf{n} =$ const) can produce sufficient variation in $(1/Ca)(\nabla \cdot \mathbf{n})$ to balance the pressure or stress variations over the drop surface. Hence, for $Ca \ll 1$, a drop will remain approximately spherical in shape. This is consistent with the interpretation of Ca as the ratio of characteristic viscous (and pressure) forces that tend to deform the drop, to capillary forces that tend to maintain the drop in a spherical shape. The limit $Ca \ll 1$ corresponds to dominant interfacial-tension effects, and this accounts for the tendency of the drop to remain almost spherical in this case. On the other hand, for $Ca \gg 1$, pressure and stress variations in (7–202) can be balanced only by large deformations of shape to produce large variations in the curvature term $(1/Ca)(\nabla \cdot \mathbf{n})$.

In the present section, we consider the translational motion of a drop through a quiescent fluid for $Ca \ll 1$. As we have just discussed, any deformation in this limit is expected to remain very small, and this facilitates the derivation of an approximate analytic solution by the method of domain perturbations.

For convenience in what follows, we express the drop shape in the form

$$F(\mathbf{x}) \equiv r - \left[1 + Ca f_0(\mathbf{x}) + O(Ca^2)\right] = 0. \tag{7–203}$$

This expression is consistent with the fact that a first approximation to the drop shape is just a sphere,

$$r = 1,$$

and that the dimensionless version of the shape function can be expressed in the form of an asymptotic expansion

$$f(\mathbf{x}) = Ca \sum_{n=0}^{N} f_n(\mathbf{x}) \, Ca^n. \tag{7–204}$$

The first approximation of the deviation from sphericity is contained in the function $f_0(\mathbf{x}, t)$, and we have anticipated, from the preceding arguments, that the magnitude of term is proportional to Ca. The presumption that the expansion for $f(\mathbf{x})$ proceeds in whose powers of Ca is based on the fact that domain perturbation techniques almost always lead to a regular (rather than singular) asymptotic structure.

Thus, to obtain a general solution for $Ca \ll 1$, we seek a regular asymptotic approximation for all of the unknowns, $\mathbf{u}, \hat{\mathbf{u}}, p, \hat{p}$, and f. Hence,

$$\mathbf{u}(\mathbf{x}) = \sum_{n=0}^{N} \mathbf{u}^{(n)}(\mathbf{x}) \, Ca^n + O(Ca^{N+1}), \tag{7–205}$$

with similar expansions for $\hat{\mathbf{u}}, p$, and $\hat{p}$. The governing equations for the first approximations, i.e., $\mathbf{u}^{(0)}, \hat{\mathbf{u}}^{(0)}, p^{(0)}$, and $\hat{p}^{(0)}$, are still just the creeping-flow equations, (7–198) and (7–199), and the leading-order velocity field in the outer fluid still satisfies the far-field condition, (7–200). To determine the boundary conditions at the drop surface for the various terms in the expansions like (7–205) for $\mathbf{u}$, $\hat{\mathbf{u}}$, p, and $\hat{p}$, we must substitute these expansions plus the approximations for the unit normal and tangent vectors, $\mathbf{n}$ and $\mathbf{t}$, corresponding to the expansion (7–204) into the exact conditions previously listed. For this purpose, we may note that the first two terms in the approximation for the unit normal can be expressed in the form

$$\mathbf{n} \equiv \frac{\nabla F}{|\nabla F|} = \mathbf{i}_r - Ca \nabla f_0 + O(Ca^2). \tag{7–206}$$

It can be seen from (7–206) that the first approximation to the unit normal to the drop surface is just

$$\mathbf{n} = \mathbf{i}_r,$$

as expected, because the first approximation to the shape is a sphere. Further, because we seek an axisymmetric solution for the velocity and pressure fields, the only relevant unit tangent vector at this level of approximation is

$$\mathbf{t} = \mathbf{i}_\theta.$$

It follows that the leading-order approximations to the boundary conditions (2–122), (2–117), (2–129), (2–135), and (2–141) at the interface of the drop take the simple forms

$$u_\theta^{(0)} = \hat{u}_\theta^{(0)} \quad \text{at } r = 1, \tag{7–207}$$

$$u_r^{(0)} = \hat{u}_r^{(0)} = 0 \quad \text{at } r = 1, \tag{7–208}$$

$$e_{r\theta}^{(0)} = \lambda \hat{e}_{r\theta}^{(0)} \quad \text{at } r = 1, \tag{7–209}$$

$$\hat{p}_{tot}^{(0)} - p_{tot}^{(0)} + 2\left[e_{rr}^{(0)} - \lambda \hat{e}_{rr}^{(0)}\right] = \frac{2}{Ca} - \nabla^2 f_0 \quad \text{at } r = 1. \tag{7–210}$$

Here, e_{ij} is the ij component of the rate-of-strain tensor. Because the leading-order approximation of the shape is a sphere, these conditions with $f_0 = 0$ are just the exact interface boundary conditions applied at the surface of a spherical drop. The only possible confusion with these conditions is that all terms appear to be $O(1)$ except for the Ca^{-1} term in (7–210). It should be noted, however, that this is just the dimensionless form of the capillary pressure jump for a spherical drop, i.e.,

$$\frac{(\gamma/2a)}{p_c} = \frac{2}{Ca}.$$

This term is thus balanced by a constant difference in the dimensionless pressures on the left-hand side of (7–210).

In summary then, the leading-order problem is just the translation of a spherical drop through a quiescent fluid. The solution of this problem is straightforward and can again be approached by means of the eigenfunction expansion for the Stokes' equations in spherical coordinates that was used in section F to solve Stokes' problem. Because the flow both inside and outside the drop will be axisymmetric, we can employ the equations of motion and continuity, (7–198) and (7–199), in terms of the streamfunctions $\psi^{(0)}$ and $\hat{\psi}^{(0)}$, that is,

$$E^4 \psi^{(0)} = 0 \quad \text{and} \quad E^4 \hat{\psi}^{(0)} = 0. \tag{7–211}$$

H. Translation of a Drop Through a Quiescent Fluid at Low Re

Thus, in both the inner and outer fluids, the general solution of (7–206) in spherical coordinates has the form derived earlier in this section:

$$\psi = \sum_{n=1}^{\infty} \left(A_n r^{n+3} + B_n r^{n+1} + C_n r^{2-n} + D_n r^{-n} \right) Q_n(\eta),$$

$$\hat{\psi} = \sum_{n=1}^{\infty} \left(\hat{A}_n r^{n+3} + \hat{B}_n r^{n+1} + \hat{C}_n r^{2-n} + \hat{D}_n r^{-n} \right) Q_n(\eta). \quad (7\text{–}212)$$

The set of eight constants, A_n, B_n, C_n, D_n, $\hat{A}_n$, $\hat{B}_n$, $\hat{C}_n$, and $\hat{D}_n$, must be determined from the boundary conditions.

As in the case of streaming flow past a solid sphere, the far-field condition (7–200) requires that

$$\psi^{(0)} \to -r^2 Q_1(\eta) \quad \text{as } r \to \infty,$$

and thus $A_n = B_n = 0$ for all n, except $B_1 = -1$, and

$$\psi = -r^2 Q_1 + \sum_{n=1}^{\infty} \left[C_n r^{2-n} + D_n r^{-n} \right] Q_n(\eta). \quad (7\text{–}213)$$

Inside the drop, on the other hand, we require that the velocity components $u_r^{(0)}$ and $u_\theta^{(0)}$ be bounded at $r = 0$. Because

$$u_r = -\frac{1}{r^2} \frac{\partial \psi}{\partial \eta} \quad \text{and} \quad u_\theta \sqrt{1-\eta^2} = -\frac{1}{r} \frac{\partial \psi}{\partial r}$$

according to (7–102), it follows that

$$\hat{C}_n = \hat{D}_n = 0 \quad \text{all } n, \quad (7\text{–}214)$$

and thus

$$\hat{\psi} = \sum_{n=1}^{\infty} \left(\hat{A}_n r^{n+3} + \hat{B}_n r^{n+1} \right) Q_n(\eta). \quad (7\text{–}215)$$

It remains to determine the four sets of constants, $\hat{A}_n$, $\hat{B}_n$, C_n, and D_n, and the function f_0 that describes the $O(Ca)$ correction to the shape of the drop. For this, we still have the five independent boundary conditions, (7–207)–(7–210). It can be shown that the conditions (7–207), (7–208), and (7–209) are sufficient to completely determine the unknown coefficients in (7–213) and (7–215). Indeed, *for any given (or prescribed) drop shape*, the four conditions of no-slip, tangential-stress continuity and the kinematic condition are sufficient along with the far-field condition to completely determine the velocity and pressure fields in the two fluids. The normal-stress condition, (7–210), can then be used to determine the leading-order shape function f_0. Specifically, we can use the now known solutions for the leading-order approximations for the velocity components and the pressure to evaluate the left-hand side of (7–210), which then becomes a second-order PDE for the function f_0. The important point to note is that we can determine the $O(Ca)$ contribution to the unknown shape knowing only the $O(1)$ contributions to the velocities and the pressures. This illustrates a universal feature of the domain perturbation technique for this class of problems. If we solve for the $O(Ca^m)$ contributions to the velocity and pressure, we can

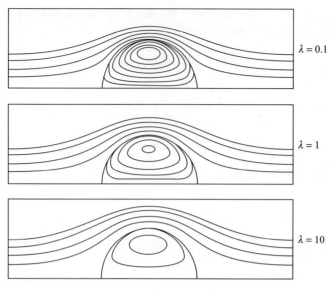

Figure 7–15. Streamlines for the Hadamard–Rybczynski solution for translation of a spherical drop through a quiescent fluid (plotted in a frame of reference that is fixed to the drop).

use the normal-stress balance to determine the $O(Ca^{m+1})$ approximation to the interface shape.

It is left to the reader to provide the details of the algebra necessary to determine the constants in (7–213) and (7–215) from the boundary conditions (7–207)–(7–209). The result is

$$C_n = D_n = \hat{A}_n = \hat{B}_n = 0, \quad n \neq 1, \tag{7-216}$$

$$C_1 = \frac{3\lambda + 2}{2(\lambda + 1)}, \quad D_1 = -\frac{\lambda}{2(\lambda + 1)}, \quad -\hat{A}_1 = \hat{B}_1 = \frac{1}{2(\lambda + 1)}, \tag{7-217}$$

where

$$\lambda \equiv \frac{\hat{\mu}}{\mu}.$$

Thus,

$$\psi = -Q_1 \left[r^2 - \frac{3\lambda + 2}{2(\lambda + 1)} r + \frac{\lambda}{2(\lambda + 1)} \frac{1}{r} \right], \tag{7-218}$$

$$\hat{\psi} = \frac{Q_1}{2} \frac{r^2 - r^4}{(\lambda + 1)}. \tag{7-219}$$

This result is known as the *Hadamard–Rybczynski* solution. It describes the velocity field for a spherical drop that is translating through an otherwise motionless fluid. A plot of the streamlines is shown in Fig. 7–15 for several different ratios of the internal to external viscosity. Obviously, for $\lambda = 10$, the interior fluid is moving slowly, and this is reflected in the small number of streamlines inside the drop. In addition, the exterior fluid is required to come almost to a stop on the drop surface, and this results in rather large velocity gradients (the streamlines are close together). As λ is decreased, on the other hand, the interior fluid moves more freely, and the velocity gradients in the exterior fluid are reduced.

H. Translation of a Drop Through a Quiescent Fluid at Low Re

The drag on a *spherical drop* can be predicted from (7–132) and the result (7–217) for C_1, namely,

$$\boxed{\text{drag} = 4\pi a \mu U C_1 = 4\pi a \mu U \left[\frac{3\lambda + 2}{2(\lambda + 1)}\right].} \qquad (7\text{–}220)$$

We see that as the viscosity ratio is decreased there is a monotonic decrease in the drag, with the maximum value occurring in the limiting case, $\lambda \to \infty$, where (7–220) becomes

$$\boxed{D = 6\pi \mu a U,} \qquad (7\text{–}221)$$

which is simply Stokes law for the drag on a rigid, no-slip sphere that moves with relative velocity U through a quiescent fluid. The fact that the drag is decreased with decrease of λ is qualitatively consistent with the decrease in the magnitude of the velocity gradients in the exterior fluid. We will discuss some other aspects of the result (7–220) at the end of this section. First, however, we wish to complete the solution we began earlier.

The remaining step is to obtain the shape function f_0. The first step is to evaluate the left-hand side of the normal-stress balance (7–210),

$$\left[\hat{p}_{tot}^{(0)} - p_{tot}^{(0)} + 2\frac{\partial u_r^{(0)}}{\partial r} - 2\lambda \frac{\partial \hat{u}_r^{(0)}}{\partial r}\right]_{r=1}, \qquad (7\text{–}222)$$

by use of the Hadamard–Rybczynski solution. Here $\hat{p}_{tot}^{(0)}$ and $p_{tot}^{(0)}$ represent the sum of hydrostatic and dynamic (flow-induced) pressure variations in the fluid.

To calculate $\partial u_r^{(0)}/\partial r|_{r=1}$, we must first calculate $u_r^{(0)}$ by using the definition

$$u_r^{(0)} \equiv -\frac{1}{r^2}\frac{\partial \psi^{(0)}}{\partial \eta}$$

and (7–218), and then differentiate with respect to r. Similarly, we can evaluate $\partial \hat{u}_r^{(0)}/\partial r$ from (7–219). The pressure in the exterior fluid, on the other hand, is

$$p_{tot}^{(0)} = -\sum_{n=1}^{\infty} \frac{2(2n-1)}{n+1} \frac{P_n(\eta)}{r^{n+1}} C_n + p_{hydrostatic}. \qquad (7\text{–}223)$$

To evaluate the first term on the right-hand side, we use the numerical value for C_1 from (7–217). The hydrostatic pressure contribution to (7–223) can be written in nondimensionalized form as

$$p_{hydrostatic} = \left(\frac{\mu U}{a}\right)^{-1}[-\rho g(za) + c], \qquad (7\text{–}224)$$

where c is an arbitrary constant. Because

$$z = r\eta = r P_1(\eta),$$

it follows that

$$p_{hydrostatic} = -\frac{\rho g a^2}{\mu U} P_1(\eta) + C^* \quad \text{at } r = 1. \qquad (7\text{–}225)$$

Hence, at the sphere surface ($r = 1$),

$$p_{tot}^{(0)} = -\frac{P_1}{2}\left(\frac{3\lambda + 2}{\lambda + 1}\right) - \frac{\rho g a^2}{\mu U} P_1(\eta) + C^*. \qquad (7\text{–}226)$$

Finally, the pressure inside the drop $\hat{p}_{tot}^{(0)}$ can be evaluated in a similar manner. When the results for $\partial u_r^{(0)}/\partial r$, $\partial \hat{u}_r^{(0)}/\partial r$, $p_{tot}^{(0)}$, and $\hat{p}_{tot}^{(0)}$ are combined in the form (7–222), we obtain

$$\frac{3}{2}\left(\frac{3\lambda+2}{\lambda+1}\right) P_1(\eta) - \frac{\rho g a^2}{\mu U}(\kappa-1)P_1(\eta) + \pi = \frac{2}{Ca} - \nabla^2 f_0, \qquad (7\text{–}227)$$

where $\kappa = \hat{\rho}/\rho$ and $\pi \equiv C^* - C$ is a constant that can be equated to the pressure difference between the internal and external fluid that is due to the capillary pressure jump, $2/Ca$. What remains are the two terms on the left-hand side that are proportional to $P_1(\eta)$ and the term $-\nabla^2 f_0$ on the right-hand side. Normally, this condition would lead to a nonzero result for f_0, meaning that the drop has deformed in the flow by an amount proportional to Ca. In the present problem, however, it turns out that $f_0 \equiv 0$, meaning that, in spite of the flow, the drop remains spherical to at least the $O(Ca)$ level of approximation. The fact of the matter is that the two terms involving $P_1(\eta)$ in (7–227) actually cancel identically. To see this, we need to remind ourselves that the velocity U, which represents the translational velocity of the drop, has not yet been determined. For this purpose, we need to consider a force balance on the spherical drop. At steady-state, this requires that the net buoyancy force be exactly balanced by the hydrodynamic drag, given at this level of approximation by (7–220). Thus,

$$\frac{4}{3}\pi a^3 g \rho (\kappa - 1) = 4\pi a \mu U \left[\frac{3\lambda+2}{2(\lambda+1)}\right], \qquad (7\text{–}228)$$

and, rearranging, we obtain

$$U = \frac{2}{3}\left(\frac{\lambda+1}{3\lambda+2}\right)\frac{a^2 g \rho (\kappa - 1)}{\mu}. \qquad (7\text{–}229)$$

Comparing (7–229) with the left-hand side of (7–227), we see that the two terms involving $P_1(\eta)$ in the normal-stress balance cancel exactly, and (7–227) reduces to the simple form

$$\pi = \frac{2}{Ca} - \nabla^2 f_0. \qquad (7\text{–}230)$$

Thus the normal-stress balance is precisely satisfied with $f_0 = 0$. The capillary (or interfacial-tension) contribution is simply to produce a jump in pressure equal to $\pi = 2/Ca$ across the drop surface. This pressure jump is, in fact, precisely the result (2–138) that was derived earlier, simply written in dimensionless terms based on a characteristic pressure, $p_c = \mu U/a$.

The fact that the normal-stress balance is satisfied with $f_0 = 0$ proves that the sphere is an equilibrium shape for a drop that is translating at low Reynolds number through a quiescent fluid. This may, at first, seem to be a rather unremarkable result. After all, the equilibrium shape for a stationary drop would clearly be spherical as a consequence of the tendency to minimize interface area (surface free energy) for any immiscible system where $\gamma \neq 0$. However, in the result that we have obtained, there is an exact balance between viscous and pressure forces for a drop of spherical shape, and thus no tendency for the drop to deform at any arbitrary capillary number, including $Ca = 0$ where $\gamma \equiv 0$. This is, in fact, a unique result! For virtually any other problem involving the motion of bubbles or drops in a viscous fluid, the shape would be nonspherical even for $Ca \ll 1$. Even for the buoyancy-driven motion of a bubble or drop in a quiescent fluid, the shape becomes nonspherical if the Reynolds number is nonzero. That particular problem, for small, but nonzero, Reynolds number, was solved originally by Taylor and Acrivos.[25] In the creeping-flow solution however, the sphere is an equilibrium shape and the magnitude of the capillary number (interfacial tension) determines only the difference in pressure between the inside and outside of the drop.

H. Translation of a Drop Through a Quiescent Fluid at Low Re

One point that has not been emphasized is that all of the preceding analysis and discussion pertains only to the steady-state problem. From this type of analysis, we cannot deduce anything about the *stability* of the spherical (Hadamard–Rybczynski) shape. In particular, if a drop or bubble is initially nonspherical or is perturbed to a nonspherical shape, we cannot ascertain whether the drop will evolve toward a steady, spherical shape. The answer to this question requires additional analysis that is not given here. The result of this analysis[26] is that *the spherical shape is stable to infinitesimal perturbations of shape for all finite capillary numbers but is unstable in the limit $Ca = \infty$ ($\gamma = 0$)*. In the latter case, a drop that is initially elongated in the direction of motion is predicted to develop a tail. A drop that is initially flattened in the direction of motion, on the other hand, is predicted to develop an indentation at the rear. Further analysis is required to determine whether the magnitude of the shape perturbation is a factor in the stability of the spherical shape for arbitrary, finite Ca.[27] Again, the details are not presented here. The result is that finite deformation can lead to instability even for finite Ca. Once unstable, the drop behavior for finite Ca is qualitatively similar to that predicted for infinitesimal perturbations of shape at $Ca = \infty$; that is, oblate drops form an indentation at the rear, and prolate drops form a tail.

Before concluding this section, let us return to the result (7–220) for the drag on the drop. We noted that this expression reduces to Stokes' law in the limit $\lambda \to \infty$. On other hand, in the limit $\lambda \to 0$, the expression (7–220) becomes

$$\boxed{D = 4\pi \mu a U} \qquad (7\text{–}231)$$

which is the drag on a *spherical bubble* at $Re \ll 1$. It is interesting that the drag on a solid sphere exceeds the drag on a spherical bubble by only a factor of $3/2$. *This result is illustrative of a general observation that the drag on a body of fixed shape at low Reynolds number is remarkably insensitive to the boundary conditions at the body surface.*

Although the limit $\lambda \to 0$ in (7–218)–(7–220) may at first seem straightforward, it is not obvious that we can take this limit without violating the condition

$$\hat{Re} \equiv \frac{\hat{\rho} U a}{\hat{\mu}} \ll 1.$$

Surprisingly, however, the solution of (7–218)–(7–220) yields the correct result for $\lambda \to 0$, without regard to whether or not $\hat{Re}$ remains small. The reason is that, as $\lambda \to 0$, the motion of the fluid outside the drop becomes increasingly insensitive to the motion inside. Thus, provided that we focus on the motion of the external fluid and on results like (7–231) that depend on this motion, the details of the inside flow are unimportant. One clear way to see that this is true is to examine the boundary conditions at the drop surface for $\lambda \to 0$. In particular, the boundary condition (7–209) can be written in the form

$$r \frac{\partial}{\partial r} \left(\frac{u_\theta}{r} \right) = \lambda \left[r \frac{\partial}{\partial r} \left(\frac{\hat{u}_\theta}{r} \right) \right] \quad \text{at } r = 1. \qquad (7\text{–}232)$$

Thus, in the limit $\lambda \to 0$, this reduces to

$$r \frac{\partial}{\partial r} \left(\frac{u_\theta}{r} \right) = \tau_{r\theta} = 0 \quad \text{at } r = 1. \qquad (7\text{–}233)$$

But insofar as the flow in the *exterior* fluid is concerned, this condition

$$\boxed{\tau_{r\theta} = 0 \quad \text{at } r = 1} \qquad (7\text{–}234)$$

plus the kinematic condition

$$\boxed{u_r = 0 \quad \text{at } r = 1} \qquad (7\text{–}235)$$

is *completely sufficient to determine the exterior flow* (that is, the constants C_n and D_n) without any need to consider the fluid motion inside the drop! Indeed, theories of bubble motion in viscous liquids are almost always obtained by applying the free-shear condition (7–234) and the kinematic condition (7–235) (or their generalizations for bodies of nonspherical geometry) as boundary conditions to directly determine the velocity and pressure fields in the liquid phase, rather than first solving the full flow problem for a drop with arbitrary λ and then letting $\lambda \to 0$.

I. MARANGONI EFFECTS IN THE MOTION OF BUBBLES AND DROPS

The analysis of the preceding section is based on the assumption that the interfacial tension at the drop surface is uniform so that grad $\gamma \equiv 0$ and the condition (2–141) reduces to the requirement that the tangential components of stress in the two fluids are equal at the drop interface. However, the interfacial tension at any point on the interface depends on the thermodynamic state (p and T) as well as the concentrations of any solute molecules that are adsorbed at the interface. Hence, in a real flow system, we must often expect γ to vary from point to point on the interface, and it is important to consider how gradients of γ may influence the flow.

In this section, we turn our attention to an example of Marangoni effects on the motion of a bubble (or drop), in which the source of the nonuniform interfacial tension at the drop surface is a nonuniform temperature. This phenomena was first illustrated experimentally by Young, Goldstein and Block[28] more than 40 years ago, and was discussed qualitatively in Section M of Chap. 2. In this example, a bubble is placed in a body of liquid in which a vertical temperature gradient is developed by some external means (for example, heating the fluid from below), so that the temperature decreases with increasing height. In this system, the temperature gradient in the fluid produces a surface-tension gradient at the bubble surface. Because the surface tension is found to decrease as the temperature increases [see, e.g., Eq. (2–145)], that is,

$$\frac{\partial \gamma}{\partial \theta} = -\beta < 0, \qquad (7\text{–}236)$$

the surface tension will increase from the bottom toward the top of the bubble. Hence a tensile force is developed in the bubble surface that tends to produce a fluid motion from the bottom toward the top of the bubble. If this Marangoni effect is strong enough, it may completely counteract the tendency of the bubble to rise in the liquid. The mechanism that makes this possible is most easily visualized by analogy to the downward motion produced by a swimmer in response to a sweeping upward motion of his arms. This was illustrated qualitatively in Fig. 2–20. Experimental observation, in fact, demonstrates that the motion of the bubble can be arrested if the temperature gradient becomes sufficiently large. Surface-tension-induced flows generated by temperature gradients are usually called *thermocapillary motions*. Interest in this class of fluid motions has been particularly strong in recent years as a consequence of the potential for materials processing applications in the microgravity environment of space flight, in which buoyancy-driven motions are nonexistent and thermocapillary phenomena can play a dominant role. One application of thermocapillary-driven motions is to provide a mechanism for removal of small gas bubbles from specialty glasses and other space-processed materials.[29]

In the remainder of this section an approximate analysis is provided of the thermocapillary problem of motion of a bubble in a gravitational field.[28] Thus we suppose that we have

I. Marangoni Effects in the Motion of Bubbles and Drops

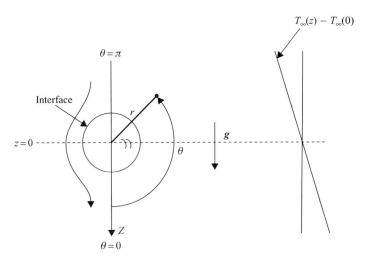

Figure 7–16. Schematic diagram of the coordinate systems and undisturbed temperature distribution for thermocapillary-driven motion of a gas bubble.

a spherical bubble, which translates vertically through a clean stationary fluid, in which we impose a temperature distribution that decreases linearly with height at large distances from the bubble, that is,

$$T_\infty(z) = T_\infty(0) + \alpha z, \quad (7\text{–}237)$$

where $T_\infty(0)$ is a reference value at $z = 0$, which we take for convenience to correspond to a horizontal plane that coincides with the bubble center. A sketch of the problem is shown in Fig. 7–16. We note that we have maintained the same convention adopted earlier in which the fluid far from the drop moves in the positive-z direction. Hence z decreases from the bottom to the top of Fig. 7–16 (i.e., in the opposite direction to **g**). With the linear temperature field at large distances from the bubble surface, an exact analysis of the problem would require a solution of the thermal energy equation in order to determine the temperature distribution in the fluid, and hence the temperature distribution at the bubble surface. For example, if convection of heat could be neglected compared with conduction, and viscous dissipation also were negligible, we should solve $\nabla^2 T = 0$ with (7–237) as the asymptotic boundary condition at infinity, and $\partial T / \partial r = 0$ at the bubble surface (if we assume $\kappa_{bubble}/\kappa_{fluid} \ll 1$ where $\kappa \equiv k/\rho C_p$). However, for present purposes, we adopt an *ad hoc* approximate approach and simply assume that the temperature distribution (7–237) is maintained throughout the fluid, including the fluid immediately adjacent to the bubble surface. This approximation to the temperature field would be exact if heat transfer were dominated by conduction ($\nabla^2 T = 0$ in the fluid) and the thermal conductivities of the liquid and gas were equal ($\kappa_{bubble}/\kappa_{fluid} = 1$). However, we expect, in any case, that it is qualitatively correct in the sense that the temperature decreases monotonically on the bubble surface as we move from the bottom toward the top.

The problem, then, is to solve the creeping-flow equations in the fluid, subject to the condition

$$\boxed{\mathbf{u} \to \mathbf{i}_z \quad \text{for} \quad |\mathbf{r}| \to \infty,} \quad (7\text{–}238)$$

plus boundary conditions at the bubble surface. Here, we have adopted a frame of reference that is fixed at the center of the bubble and nondimensionalized with respect to the (as yet unknown) translational velocity U. The conditions at the bubble surface are the kinematic condition (2–130),

$$u_r = 0 \quad \text{at } r = 1, \tag{7-239}$$

and the tangential stress balance (2–141), combined with (2–146), which can be written in the dimensionless form

$$\tau_{r\theta} - \frac{(1-\eta^2)^{1/2}}{\mu U} \frac{\partial \gamma}{\partial \eta} = 0 \quad \text{at } r = 1. \tag{7-240}$$

As usual, $\eta \equiv \cos\theta$. To evaluate $\partial \gamma / \partial \eta$ in (7–240), we use the temperature distribution (7–237), which is now assumed to hold throughout the fluid, and the relationship (7–236) between the surface tension and temperature. The temperature distribution can be expressed in terms of spherical coordinates as

$$\theta_\infty(z) = \theta_\infty(0) + \alpha r \eta. \tag{7-241}$$

Hence, at $r = a$,

$$\frac{\partial \gamma}{\partial \eta} = \frac{\partial \gamma}{\partial \theta}\frac{\partial \theta}{\partial \eta} = -\alpha\beta a \tag{7-242}$$

and the tangential-stress condition (in nondimensionalized form) becomes

$$\tau_{r\theta} = -\frac{\alpha\beta a(1-\eta^2)^{1/2}}{\mu U} \quad \text{at } r = 1. \tag{7-243}$$

Because the problem is axisymmetric about the z axis, we can solve the creeping-flow equation in terms of the streamfunction ψ. Hence, after applying the asymptotic condition (7–238), we can express the solution in the form

$$\psi = -r^2 Q_1(\eta) + \sum_{n=1}^{\infty} \left(C_n r^{2-n} + D_n r^{-n}\right) Q_n(\eta). \tag{7-244}$$

The coefficients C_n and D_n must be determined by application of the boundary conditions (7–239) and (7–243) at the bubble surface $r = 1$. The first of these requires that $\psi = 0$ at $r = 1$. Hence,

$$0 = -Q_1(\eta) + \sum_{n=1}^{\infty} [C_n + D_n] Q_n(\eta). \tag{7-245}$$

Hence $D_1 = 1 - C_1$ and $D_n = -C_n (n \geq 2)$.

The second condition requires that

$$r\frac{\partial}{\partial r}\left(\frac{u_\theta}{r}\right) = -\frac{\alpha\beta a}{\mu U}(1-\eta^2)^{1/2} \quad \text{at } r = 1, \tag{7-246}$$

where

$$u_\theta \equiv -\frac{1}{\sqrt{1-\eta^2}}\frac{1}{r}\frac{\partial \psi}{\partial r}.$$

I. Marangoni Effects in the Motion of Bubbles and Drops

Hence,

$$6Q_1 + \sum_{n=1}^{\infty} C_n(-4n-2)Q_n = -2\frac{\alpha\beta a}{\mu U}Q_1. \tag{7-247}$$

Solving (7–247), we find that

$$C_1 = 1 + \frac{1}{3}\frac{\alpha\beta a}{\mu U},$$

$$D_1 = -\frac{1}{3}\frac{\alpha\beta a}{\mu U}, \tag{7-248}$$

$$C_n = D_n = 0 \text{ for } n \geq 2.$$

It follows from (7–132) that the hydrodynamic drag on the bubble is

$$F = 4\pi\mu a U\left(1 + \frac{1}{3}\frac{\alpha\beta a}{\mu U}\right). \tag{7-249}$$

Thus the hydrodynamic force is larger than it would be in the absence of the thermocapillary (Marangoni) contribution to the shear stress at the bubble surface. As a consequence, the bubble moves slower. Indeed, at steady-state, the bubble velocity can be calculated from the overall force balance:

$$\frac{4}{3}\pi a^3 \rho g = 4\pi\mu a U\left(1 + \frac{1}{3}\frac{\alpha\beta a}{\mu U}\right). \tag{7-250}$$

The result is

$$U = \frac{1}{3}\frac{a^2 \rho g}{\mu}\left(1 - \frac{\alpha\beta}{a\rho g}\right). \tag{7-251}$$

It would appear from (7–251) that the critical value of α to completely arrest the bubble motion is

$$\alpha_{crit} = \frac{a\rho g}{\beta}. \tag{7-252}$$

The correct result, if we determine the actual temperature distribution at the bubble surface by solving $\nabla^2 T = 0$, as outlined earlier, was shown by Young et al.[28] to be

$$\alpha_{crit} = \frac{2}{3}\frac{a\rho g}{\beta}. \tag{7-253}$$

Clearly the simplification of the problem that we have used does not change this result except in the quantitative sense.

Several brief comments should be made regarding the solution just outlined. First, the derivation assumes, implicitly, that $U \neq 0$; remembering the far-field condition (7–238) and the form of the streamfunction (7–244), the reader may well wonder about the efficacy of trying to use this solution to deduce conditions when $U = 0$. The simplest response is

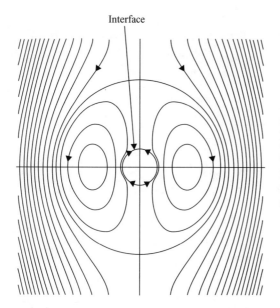

Figure 7–17. Streamlines for thermocapillary motion of a gas bubble for $\alpha = 0.8$ $(a^2 \rho g/\beta)$, where the bubble velocity is reduced to 20% of its value in the absence of thermocapillary effects. The stream-function values are calculated from Eq. (7–244) with coefficients C_n and D_n from Eqs. (7–248). Contour values are plotted in increments of 0.7681.

to note that the analysis we have given is clearly appropriate for any $U \neq 0$, no matter how small, and we rely on this fact to hypothesize that the limiting process $U \to 0$ is nonsingular. However, it is a useful exercise to go back and reformulate the problem starting with the presumption $U = 0$, and this is suggested as a problem at the end of this chapter. Second, the reader is reminded that it is only the bubble that is not moving when $U = 0$. The fluid is, in fact, undergoing a surface-tension-driven recirculating flow of the form shown in Fig. 7–17. Indeed, it is the downward thrust on the bubble that results from this flow that is responsible for balancing the buoyancy force and maintaining the bubble in a stationary position.

J. SURFACTANT EFFECTS ON THE BUOYANCY-DRIVEN MOTION OF A DROP

A more common source of Marangoni effects in systems of interest to chemical engineers is surfactants, as discussed in Chap. 2. This is particularly pertinent to the motion of gas bubbles (or drops) in water, or in any liquid that has a large surface tension (the surface tension of a pure air–water interface is approximately 70 dyn/cm). Experiments on the motion of gas bubbles in water at low Reynolds numbers show the perplexing result illustrated in Fig. 7–18. For bubbles larger than about 1 mm millimeter in diameter, the translation velocity is approximately equal to the predicted value for a spherical bubble with zero shear stress at the interface, that is,

$$U = \left(\frac{\Delta \rho g a^2}{3\mu} \right).$$

However, for smaller bubbles the translation velocity decreases until it is approximately two-thirds of this value; that is, the small bubbles move through the liquid with a velocity that is consistent with a no-slip boundary condition at the interface (hence, yielding Stokes' result $6\pi \mu a U$ for the drag) rather than a free-surface condition [leading to the value $4\pi \mu a U$, according to (7–206)].

A qualitative explanation for this observation was discovered many years ago by Frumkin and Levich,[30] who carried out experiments on buoyancy-driven motion of gas bubbles in

J. Surfactant Effects on the Buoyancy-Driven Motion of a Drop

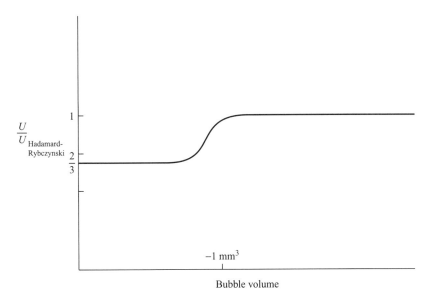

Figure 7–18. A qualitative sketch representing the terminal velocity of a gas bubble in water at low Reynolds number.

very highly purified liquids under a close approximation to isothermal conditions. These experiments led to two main conclusions. First, as the liquid became increasingly purified, the transition from the theoretically predicted bubble motion to motion of the bubble as a no-slip body occurred for smaller and smaller bubbles (or drops). Second, experiments in water showed extreme sensitivity to even minute concentrations of contaminant (even 1 ppm by weight of some contaminants led to an observable effect on bubble motion).

The explanation for these observations and for the existence of a range of bubble sizes in which the bubble seems to act as a no-slip body is that real fluid systems almost invariably contain small quantities of contaminant, and this material can modify boundary conditions at *the bubble (or drop) surface*. This is especially true if the contaminant is a surfactant – that is, a solute that is preferentially adsorbed at the interface, as this means that infinitesimal amounts of surfactant in the bulk-phase fluids can yield very significant concentrations of surfactant at the interface. For air–water or oil–water systems, the chemical structure of a typical surfactant will be made of two distinct moieties: One part is polar (and thus hydrophilic), and the other is nonpolar (and thus hydrophobic). In general, a surfactant can modify the physical character of the interface in two quite distinct ways. When sufficient concentrations of several different surfactants are present, the surfactant mixture can form a semipermanent film at the interface, with a definite 2D structure. For example, this is typical of surfactant mixtures that are used as emulsifiers designed to stabilize an emulsion by creating a physical barrier to droplet coalescence. On the other hand, for low surfactant concentrations or when a single surfactant species is present, the adsorbed surfactant can remain mobile on the interface and it then influences dynamical behavior by producing spatial variations in the interfacial tension. The dynamic effects of interfacial-tension gradients are known as Marangoni effects. These latter effects were first introduced at the end of Chap. 2 and were also discussed in section E on shallow cavity flows in Chap. 6.

If the concentration of surfactant is uniform at the interface, the effect is a uniform decrease in surface tension, which may generally change the bubble shape (because of an increase in the capillary number) but otherwise has no influence on the bubble dynamics or on the hydrodynamic drag. The fact is, however, that the surfactant concentration at

Creeping Flows – Two-Dimensional and Axisymmetric Problems

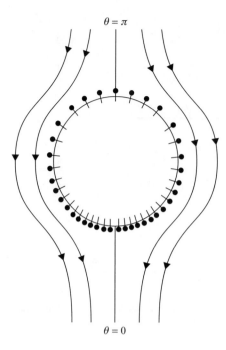

Figure 7–19. A pictorial representation of the distribution of surfactant on the surface of a rising gas bubble. The small "sticks" at the bubble interface are intended to represent surfactant, which adsorbs preferentially at the gas–liquid interface. The fluid motion, from the top of the bubble toward the bottom, convects surfactant toward the rear of the bubble where it tends to accumulate. This tendency is counteracted to some extent by diffusion that tends toward a uniform surfactant distribution.

a bubble surface does not remain uniform as the bubble translates through the liquid. Although the transport processes for surfactants are quite complicated, the essential feature is that convection of surfactant molecules sweeps them from the front toward the rear of the bubble, and a gradient of surfactant concentration is established as a balance between this convective effect and the combination of bulk-phase mass transfer and surface diffusion that drives the system back toward a uniform surfactant concentration. But a nonuniform surfactant concentration implies that the surface tension must also vary over the bubble surface. Because surface tension goes down as the concentration of surfactant goes up, the net result is the development of a gradient of surface tension at the bubble surface, from a maximum value at the front of the bubble ($\theta = \pi$) where the surfactant concentration is lowest, to a minimum value at the rear ($\theta = 0$) where surfactant concentration is largest. Hence, referring to Fig. 7–19, we expect that

$$\frac{d\gamma}{d\theta} > 0. \qquad (7\text{–}254)$$

However, such a gradient of γ modifies the boundary condition at the bubble (or drop) surface, from a zero-shear-stress condition ($\tau_{r\theta} = 0$) to the condition (2–141), which becomes

$$\tau_{r\theta} = -\frac{1}{a}\frac{d\gamma}{d\theta} \qquad (7\text{–}255)$$

for this axisymmetric problem. Here, a is the bubble radius. Clearly, the existence of a gradient of γ is equivalent to a tensile force in the interface, which acts in the direction of increasing γ (in the direction to oppose the external flow). Thus, as $d\gamma/d\theta$ increases, the bubble surface becomes increasingly resistant to tangential motions and the bubble acts increasingly as a no-slip body. The limit to this process is complete immobilization of the interface. As this limit is approached, the convection of surfactant becomes weaker because of the decrease in interfacial velocity. Examination of (7–255) shows that, for a given gradient of γ, the effect on the shear stress increases as the bubble radius decreases.

J. Surfactant Effects on the Buoyancy-Driven Motion of a Drop

This is in qualitative accord with the experimental observation that the bubble reverts to no-slip behavior for small radii below some critical threshold value. This effect of nonuniform surfactant concentrations is one of the most easily observed manifestations of Marangoni effects in the motion of bubbles or drops.

1. Governing Equations and Boundary Conditions for a Translating Drop with Surfactant Adsorbed at the Interface

In the present subsection, we consider the analysis of these effects for a translating drop (which also reduces to the case of a bubble when the internal viscosity is small compared with that of the suspending fluid). We begin by stating the governing equations and boundary conditions based on the discussion of boundary conditions at a fluid interface with surfactant present in Chap. 2. To simplify the analysis, and thus allow us to concentrate our efforts on understanding the Marangoni effects, we introduce a number of assumptions. We assume that the creeping-motion assumption can be used for the motion both inside and outside the drop, and we assume for simplicity that the drop remains *spherical*. We have seen that a drop rising under the action of buoyancy through a viscous fluid with a uniform surface tension at the interface will remain spherical for arbitrary values of the capillary number Ca. This does not imply that the same result will be true if the interfacial tension is nonuniform. Indeed, the modifications of the flow caused by the Marangoni stresses, as well as the nonuniform interfacial tension, will *together* tend to produce modifications of the drop or bubble shape. Here, we assume that the capillary number is small enough to minimize these effects, and we consider only a spherical drop.

Even with these assumptions, it is not generally possible to obtain analytic solutions of the resulting problem because of the complexity of the surfactant transport problem and the coupling between surfactant transport and fluid motion.[31] Numerical methods have been used to consider general cases. Here, we consider several limiting cases in which analytic results can be obtained. Although reasonably limited in terms of cases of practical interest, these are quite useful in verifying the mechanisms described previously, as well as illustrating some of the details of the problem.

The majority of the governing equations and boundary conditions have been given already in previous sections of this Chapter. The governing equations of motion and continuity, assuming that the creeping-motion assumption is valid, are (7–198) and (7–199), with the boundary condition at infinity again given by (7–200). It should be noted, of course, that the equations and boundary conditions have been scaled with the steady, buoyancy-driven velocity of the drop, U, and this remains unknown other than the fact that it is nonzero and must be determined at the end by means of a balance between the buoyancy and drag forces on the drop. In all of the problems that we consider, the flow field is axisymmetric. Hence (7–198) and (7–199) can be replaced with the governing equations in terms of the streamfunction (7–211), with general solutions (7–212). The specific forms that satisfy the boundary conditions at infinity and the condition of boundedness at the center of the drop are again precisely the forms given earlier, (7–213) and (7–215), i.e.,

$$\psi = -r^2 Q_1 + \sum_{n=1}^{\infty} \left(C_n r^{2-n} + D_n r^{-n} \right) Q_n(\eta), \qquad (7\text{–}213)$$

$$\hat{\psi} = \sum_{n=1}^{\infty} \left(\hat{A}_n r^{n+3} + \hat{B}_n r^{n+1} \right) Q_n(\eta). \qquad (7\text{–}215)$$

All that remains is to apply the boundary conditions at the interface of the drop to determine the remaining coefficients in (7–213) and (7–215).

Two of these boundary conditions are straightforward, namely, continuity of velocity and the kinematic conditions. Because the drop shape is assumed to be spherical, these conditions simply reduce to

$$u_r = \hat{u}_r = 0 \quad \text{at } r = 1 \qquad (7\text{–}256)$$

and continuity of the tangential velocity component,

$$u_\theta = \hat{u}_\theta \quad \text{at } r = 1. \qquad (7\text{–}257)$$

Because the shape is spherical, the tangential-stress condition takes the form

$$\tau_{r\theta} - \lambda \hat{\tau}_{r\theta} = -\frac{1}{Ca} \frac{\partial (\gamma/\gamma_s)}{\partial \theta} \quad \text{at } r = 1. \qquad (7\text{–}258)$$

In this expression, $Ca \equiv \mu U/\gamma_s$ is the capillary number based on the clean interface value of the interfacial tension. To proceed beyond this, we need to specify the gradient of the interfacial tension at the surface of the drop. For this purpose, we return to Section N of Chap. 2, where we discuss the boundary conditions at an interface with adsorbed surfactant.

We formulate the problem in general terms first, and then consider several simplifying approximations in the subsequent subsections. Hence we begin by assuming that the surfactant may be soluble in both of the bulk fluids, that it adsorbs to and from the bulk fluids following Langmuir kinetics, and that the interfacial tension is related to the interface concentration of surfactant by means of the Langmuir limit of (2–155),

$$\gamma = \gamma_s + RT\Gamma_\infty \ln[1 - (\Gamma'/\Gamma_\infty)].$$

The reader is reminded that γ_s is the interfacial tension of the clean fluid interface and Γ_∞ is the maximum surfactant concentration that can be accommodated on the interface. It will be convenient to scale the interface concentrations with Γ_∞. In this case, we replace the preceding dimensional expression with

$$\gamma/\gamma_s = 1 + (RT\Gamma_\infty/\gamma_s) \ln(1 - \Gamma), \qquad (7\text{–}259)$$

where we now understand that $\Gamma = \Gamma'/\Gamma_\infty$, both here and in subsequent equations, stands for the *dimensionless* interface concentration.

At steady-state, the transport of surfactant in the two bulk fluids is governed by convection diffusion equations [these are analgous to convection and diffusion (or conduction) in the equations for heat transfer – (2–110)]:

$$Pe \, \nabla \cdot (c\mathbf{u}) = \nabla^2 c, \qquad (7\text{–}260\text{a})$$

$$\hat{Pe} \, \nabla \cdot (\hat{c}\hat{\mathbf{u}}) = \nabla^2 \hat{c}, \qquad (7\text{–}260\text{b})$$

where c and $\hat{c}$ are both scaled with respect to c_∞, $u_c = U$, and $\ell_c = a$, and the Peclet numbers are

$$Pe = Ua/D \quad \text{and} \quad \hat{Pe} = Ua/\hat{D},$$

with D and $\hat{D}$ being the bulk-phase diffusion coefficient.

J. Surfactant Effects on the Buoyancy-Driven Motion of a Drop

The transport of surfactant on the interface is governed by the sequential process of diffusion to the interface, followed by adsorption–desorption, and then transport of the adsorbed surfactant on the interface by a combination of convection and diffusion. The latter process is described by Eq. (2–164), but now for steady-state and for a fixed shape. We express the flux to the interface in terms of the net diffusive flux:

$$\nabla_s \cdot (\mathbf{u}_s \Gamma) - \frac{1}{Pe_s} \nabla_s^2 \Gamma = \frac{c_\infty a}{\Gamma_\infty} \left(\frac{1}{Pe} \mathbf{n} \cdot \nabla c - \frac{1}{\hat{Pe}} \mathbf{n} \cdot \nabla \hat{c} \right)_s. \qquad (7\text{–}261)$$

Here Pe_s is the "interface" Peclet number defined in terms of the interface diffusivity D_s. The diffusive flux on the right-hand side is equal to the net rate of adsorption (minus desorption). Because we have adopted Langmuir kinetics according to (2–150), the dimensionless form of the balance between diffusion to the interface and adsorption–desorption takes the form

$$\frac{c_\infty a}{\Gamma_\infty} \left(\frac{1}{Pe} \mathbf{n} \cdot \nabla c - \frac{1}{\hat{Pe}} \mathbf{n} \cdot \nabla \hat{c} \right)_s$$
$$= Bi \left\{ k c_s (1 - \Gamma) - \Gamma + \frac{\hat{\alpha}}{\alpha} [\hat{k} \hat{c}_s (1 - \Gamma) - \Gamma] \right\}. \qquad (7\text{–}262)$$

Here, the Biot number $Bi \equiv \alpha a / U$ provides a measure of the relative importance of kinetic desorption relative to interfacial convection, c_s and $\hat{c}_s$ are the bulk-phase concentrations evaluated in the limit as we approach the surface of the drop, $k \equiv \delta c_\infty / \alpha$ is a measure of the ratio of adsorption to desorption to the exterior fluid and $\hat{k} \equiv \hat{\delta} c_\infty / \alpha$ is the same quantity for the drop. We recall that δ and α are the adsorption and desorption rate constants defined following (2–150).

Now, returning to (7–258) with the expression for the interfacial tension, (7–259), we find that

$$\tau_{r\theta} - \lambda \hat{\tau}_{r\theta} = \frac{1}{Ca} \left[\frac{RT\Gamma_\infty}{\gamma_s} \left(\frac{1}{1 - \Gamma} \right) \frac{\partial \Gamma}{\partial \theta} \right] \quad \text{at } r = 1. \qquad (7\text{–}263)$$

The parameter

$$E = \frac{RT\Gamma_\infty}{\gamma_s}$$

is called the "elasticity parameter." The ratio of $Ma \equiv E/Ca$ is also called the Marangoni number. Hence (7–263) can be written as

$$\tau_{r\theta} - \lambda \hat{\tau}_{r\theta} = \frac{E}{Ca} \left(\frac{1}{1 - \Gamma} \right) \frac{\partial \Gamma}{\partial \theta} = Ma \left(\frac{1}{1 - \Gamma} \right) \frac{\partial \Gamma}{\partial \theta} \quad \text{at } r = 1. \qquad (7\text{–}264)$$

This completes our derivation of the governing equations and boundary conditions. Generally, the boundary conditions and the associated equations for transport of surfactant produce a strongly *nonlinear* problem for which numerical methods provide the best approach. At the end of this section, references are provided for additional numerical studies.[32]

In the remainder of this section, we solve these equations for several cases in which the problem allows for an analytic approximation. We do this in the context of a somewhat simpler class of problems, in which we assume that the surfactant is insoluble inside the drop. This simplifies Eq. (7–262) to the form

$$\frac{c_\infty a}{\Gamma_\infty}\left(\frac{1}{Pe}\mathbf{n}\cdot\nabla c\right)_s = Bi[kc_s(1-\Gamma)-\Gamma], \qquad (7\text{–}265)$$

and (7–261) becomes

$$\nabla_s\cdot(\mathbf{u_s}\Gamma) - \frac{1}{Pe_s}\nabla_s^2\Gamma = \frac{c_\infty a}{\Gamma_\infty}\left(\frac{1}{Pe}\mathbf{n}\cdot\nabla c\right)_s \qquad (7\text{–}266)$$

Although this approximation is not essential to the solution procedures, or to the qualitative physics, it does simplify the analysis by eliminating the need to solve (7–260) for $\hat{c}$. We may also note that an additional simplification occurs if $\Gamma'/\Gamma_\infty \ll 1$. In this case, the Langmuir relationship for the interfacial tension, (7–259), can be *linearized*:

$$\gamma \sim \gamma_s - RT\Gamma[1 + O(\Gamma'/\Gamma_\infty)], \qquad (7\text{–}267)$$

$$\tau_{r\theta} - \lambda\hat{\tau}_{r\theta} = \frac{E}{Ca}\nabla_s\Gamma = Ma\nabla_s\Gamma \quad \text{at } r=1. \qquad (7\text{–}268)$$

In fact, for cases in which $\Gamma'/\Gamma_\infty \ll 1$, we do not need to start with the linear expressions (7–267) and (7–268). They will evolve from the full nonlinear problem for any case in which the limit $\Gamma'/\Gamma_\infty \ll 1$ is relevant. We may note, however, that many research papers use the linear expressions as the basis of their analysis from the beginning. Although this is perfectly acceptable, it must be remembered that the resulting analysis will be valid only provided that the surfactant concentrations are low enough to justify the linearization of (7–259) to the form (7–267).

Finally, before any details of analysis are considered, it is useful to review the physics associated with various limiting values of the dimensionless parameters. Qualitatively, the convective transport of surfactant on the interface leads to concentration gradients, with the highest surfactant concentration at the back of the drop and the lowest concentration at the front. This produces an interfacial-tension gradient, with the highest value at the front and the lowest at the back, which tends to inhibit the interfacial velocity and thus slow the translational motion of the drop. These phenomena have been previously described in preceding chapters as Marangoni effects. Now, the tendency to increase the concentration at the back of the drop will be inhibited by the existence of a *maximum* concentration, Γ_∞, and the associated increase in the sensitivity of the interfacial tension to changes in concentration as $\Gamma' \to \Gamma_\infty$. In addition, the presence of diffusion on the interface also tends to maintain a uniform concentration, as is evident from (7–261). However, the Peclet number Pe_s is generally quite large (the interfacial diffusivity is quite small) and this effect is often not a major factor. A more important effect is *often* the possibility of adsorption and desorption to and from the interface, which also tends to diminish the interfacial concentration gradients. At the back of the drop, as Γ increases relative to Γ_{eq}, the rate of desorption increases, whereas the adsorption rate increases at the front of the drop as Γ decreases relative to Γ_{eq}. The details of these surfactant processes depend on which of the three basic transport mechanisms is limiting the process; interfacial convection and diffusion, adsorption–desorption from the bulk fluid, or diffusion from the bulk fluid to the vicinity of the interface.

One important *limiting case* that can be studied analytically occurs when convection at the interface is intrinsically fast compared with either bulk diffusion or the adsorption–desorption kinetics,

$$\left(\frac{c_\infty a}{\Gamma_\infty}\right)\frac{1}{Pe} \ll O(1) \quad \text{or } Bi \ll O(1). \qquad (7\text{–}269)$$

J. Surfactant Effects on the Buoyancy-Driven Motion of a Drop

This case is the so-called "insoluble" limit, in the sense that *the distribution of surfactant on the interface is decoupled to a first approximation from the bulk-phase mass transfer processes.* When interface diffusion is also small compared with convection, $Pe_s \gg O(1)$, the combination of Eqs. (7–265) and (7–266) leads to the conclusion that

$$\nabla_s \cdot (\mathbf{u}_s \Gamma), \approx 0, \tag{7–270}$$

and thus either $\mathbf{u}_s$ or Γ must be zero at the interface. This leads to a so-called "spherical-cap" configuration in which $\Gamma = 0$ over the front portion of the drop while $\mathbf{u}_s = 0$ over a "cap" at the rear. This is a major simplification of the basic problem because it suggests that the details of the surfactant mass transport process are not an important factor, except for determining the point of transition from $\Gamma = 0$ to $\mathbf{u}_s = 0$. We shall discuss this case shortly.

A second class of problems occurs when the intrinsic rate of kinetic and diffusive exchange of surfactant is of the same order of magnitude as interface convection, so that the bulk-phase mass transfer problem is a key factor in determining the interfacial distribution of surfactant. One problem that resembles the "insoluble" surfactant limit is that of a so-called "incompressible" surfactant, where

$$Ma \gg O(1). \tag{7–271}$$

In this case, very tiny changes in the interface concentration produce a large contribution to the shear-stress balance, as can be seen from (7–268). As a result, because the shear stress is $O(1)$ in the dimensionless form of (7–268), it follows that Γ is always very close to Γ_{eq}, and the leading-order approximation is that of a drop with no-slip boundary conditions on the interface. It is because of the latter condition that the "incompressible" limit is said to resemble the "insoluble" surfactant limit. However, in this case $\Gamma \cong \Gamma_{eq}$, not 1(i.e. $\Gamma' \sim \Gamma_\infty$), and bulk-phase mass transport will play an important role in determining the departure from this limiting case. This case is formulated as Problem 7–21 at the end of this chapter.

Finally, a third limiting case that can be analyzed occurs if the kinetic rate is fast compared with interface convection,

$$Bi \gg \frac{c_\infty a}{\Gamma_\infty Pe}, \tag{7–272}$$

so that the interface concentration will be in local equilibrium with the bulk concentration, according to (7–265),

$$\Gamma = \frac{kc_s}{1 + c_s}. \tag{7–273}$$

In this case, *the interfacial gradients of surfactant occur because of interfacial gradients in the bulk concentration driven by bulk convection rather than interfacial convection.* Although generally requiring numerical analysis, an analytic approximation can be obtained in the limit $k \ll O(1)$, where the interface concentration is small. Again, we shall consider this case shortly.

2. The Spherical-Cap Limit
[$c_\infty a/\Gamma_\infty Pe \ll O(1)$ or $Bi \ll O(1)$; and $Pe_s \gg O(1)$]

We begin with the limiting case of a spherical cap bubble. This problem has been studied by many investigators, beginning most likely with Savic[33] and continuing with an important investigation by Davis and Acrivos.[34] Here we largely follow the analysis of Sadhal and Johnson.[35] Although there at least two small parameters in the problem, we seek only the leading-order approximation.

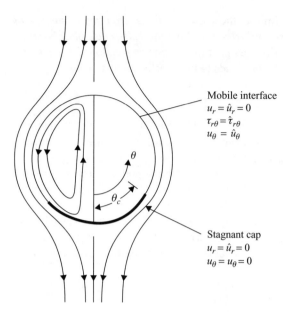

Figure 7–20. The spherical-cap model (from Ref. 34).

As already stated, the limiting form of the governing mass transfer problem for this limit of "insoluble" surfactant is (7–270). Thus, in this case, we do not need to consider either the bulk transport or surfactant adsorption–desorption processes and the problem is greatly simplified. The governing equation (7–270) requires that either $\mathbf{u}_s$ or Γ be zero at every point on the drop interface. To verify this fact, we may note that the surfactant interface concentration is axisymmetric so that the solution of (7–270) reduces to the form

$$u_s \Gamma = \frac{A}{\sin \theta}, \tag{7-274}$$

where A is a constant of integration. Hence, to have a bounded solution at $\theta = 0, \pi$, we must choose $A = 0$. Thus, wherever surfactant is present, the interface must be completely immobilized,

$$\boxed{u_s = 0 \quad \text{for} \quad 0 \leq \theta < \theta_c,} \tag{7-275}$$

or, otherwise, the interface must be completely devoid of surfactant

$$\boxed{\Gamma = 0 \quad \text{for} \quad \theta_c < \theta \leq \pi.} \tag{7-276}$$

Note that the polar angle is again defined as $\theta = \pi$ at the front stagnation point and $\theta = 0$ at the back. We thus see that either the interface velocity is zero or that $\Gamma = 0$ and the tangential-stress must be continuous across the interface [see, e.g., (7–264)]. This physical picture has been called the spherical-cap model. It is sketched in Fig. 7–20. The problem then is to determine the velocity fields in the two fluids and determine the critical angle θ_c.

Up to the point of applying the boundary conditions at the surface of the drop, this problem is actually identical to the problem of a rising drop under the action of buoyancy with a clean fluid interface. The solutions, satisfying axisymmetry, and the uniform flow at infinity were given previously as (7–213) and (7–215). Now, at the drop interface, the normal velocity vanishes:

$$\boxed{\psi = \hat{\psi} = 0 \quad \text{at} \quad r = 1, \text{ all } \theta.} \tag{7-277}$$

J. Surfactant Effects on the Buoyancy-Driven Motion of a Drop

Incorporating this condition, we can now write the general solutions in the form

$$\psi = (-r^2 + r^{-1})Q_1(\eta) + \sum_{n=1}^{\infty} C_n(r^{2-n} - r^{-n})Q_n(\eta), \tag{7-278}$$

$$\hat{\psi} = \sum_{n=1}^{\infty} \hat{A}_n(r^{n+3} - r^{n+1})Q_n(\eta). \tag{7-279}$$

The additional boundary conditions are continuity of the tangential velocity,

$$\boxed{\frac{\partial \psi}{\partial r} = \frac{\partial \hat{\psi}}{\partial r} \quad \text{at } r = 1 \text{ and all } \eta;} \tag{7-280}$$

no slip at the rear of the drop,

$$\boxed{\frac{\partial \psi}{\partial r} = \frac{\partial \hat{\psi}}{\partial r} = 0 \quad \text{at } r = 1 \text{ and } \cos\theta_c < \eta \leq 1;} \tag{7-281}$$

and continuity of the tangential-stress along the "clean" portion of the interface,

$$\boxed{\frac{\partial}{\partial r}\left(\frac{1}{r^2}\frac{\partial \psi}{\partial r}\right) = \frac{\hat{\mu}}{\mu}\frac{\partial}{\partial r}\left(\frac{1}{r^2}\frac{\partial \hat{\psi}}{\partial r}\right) \quad \text{at } r = 1 \text{ and } -1 \leq \eta < \cos\theta_c.} \tag{7-282}$$

The first of these conditions yields

$$\hat{A}_1 = C_1 - 3/2 \quad \text{and} \quad \hat{A}_n = C_n \text{ for } n \geq 2. \tag{7-283}$$

The noslip condition then requires that

$$(2C_1 - 3)Q_1(\eta) + \sum_{n=2}^{\infty} 2C_n Q_n(\eta) = 0 \quad \text{for } \cos\theta_c < \eta \leq 1 \tag{7-284}$$

and the continuity of shear stress requires that

$$\left[\frac{6\mu + 9\hat{\mu}}{2(\mu + \hat{\mu})}\right] Q_1(\eta) = \sum_{n=1}^{\infty} C_n(2n+1)Q_n(\eta) \quad \text{for } -1 \leq \eta < \cos\theta_c. \tag{7-285}$$

Although it remains "only" to determine the coefficients C_n to satisfy the last two conditions, it will not be obvious to most readers how this can be done. Because the method is special, we follow the outline of Sadhal and Johnson,[35] who used a method "invented" by Collins[36] to solve a similar mathematical problem in electrostatics.

For this purpose, we note that the polynomials $Q_n(\eta)$ can be expressed in terms of functions $T_k^m(\cos\theta)$ that are known as the associated Legendre functions:

$$Q_n(\eta) = -\sin\theta \, T_n^{-1}(\eta). \tag{7-286}$$

We also modify the coefficients from C_n to C_n^* according to the definitions

$$C_1 = C_1^* + \frac{2\mu + 3\hat{\mu}}{2(\mu + \hat{\mu})} \quad \text{and} \quad C_n = C_n^*. \, (n \geq 2) \tag{7-287}$$

Creeping Flows – Two-Dimensional and Axisymmetric Problems

Hence, the conditions (7–284) and (7–285) can be written in a form suitable for application of Collins' method:

$$\sum_{n=1}^{\infty} C_n^*(2n+1)T_n^{-1}(\eta) = 0 \quad \text{for } -1 \leq \eta < \cos\theta_c,$$

$$\sum_{n=1}^{\infty} C_n^* T_n^{-1}(\eta) = \frac{1}{4}\frac{\mu}{\mu+\hat{\mu}}\sin\theta \quad \text{for } \cos\theta_c < \eta \leq 1.$$

The interested reader may wish to consult Sadhal and Johnson for the details of solving those two equations by using the method of Collins. For present purposes, it is not essential to follow the remainder of the analysis. What is important is that a closed-form analytic solution can be obtained with the constants

$$C_1 = \left[\frac{\mu}{4\pi(\mu+\hat{\mu})}\left(2\theta_c + \sin\theta_c - \sin 2\theta_c - \frac{1}{3}\sin 3\theta_c\right) + \frac{2\mu+3\hat{\mu}}{2(\mu+\hat{\mu})}\right], \qquad (7\text{–}288)$$

$$C_n = -\frac{\mu}{4\pi(\mu+\hat{\mu})}\left\{\sin(n+2)\theta_c - \sin n\theta_c + \sin(n+1)\theta_c - \sin(n-1)\theta_c \right.$$
$$\left. -2\left[\frac{\sin(n+2)\theta_c}{n+2} + \frac{\sin(n-1)\theta_c}{n-1}\right]\right\}(n \geq 2). \qquad (7\text{–}289)$$

Although complete, one factor has been left unspecified, and that is the "cap angle"; namely, the angle at which the boundary conditions change from a clean interface at the front of the drop to a noslip interface at the back.

If we think about the physics of establishing the immobile cap at the rear of the bubble, it is necessary that the local Marangoni stress balance the net shear stress on the interface. In effect, this means that the condition (7–258) must be satisfied at all points on the surface where the cap exists,

$$(1-\eta^2)^{1/2}\frac{\partial(\gamma/\gamma_s)}{\partial\eta} = Ca(\tau_{r\theta} - \lambda\hat{\tau}_{r\theta}) \quad \text{at } r = 1, \qquad (7\text{–}290)$$

but with the shear stress components being the values for a no-slip surface at the rear of the drop. Because there is a maximum change possible in the interfacial tension from the clean interface value γ_s to the value γ_{min} obtained from (7–259) for $\Gamma' = \Gamma_\infty$, there will typically be only a finite-size region that can be immobilized before $\Gamma' \to \Gamma_\infty$ at $\theta = 0$. Hence, if we can determine $\gamma_s - \gamma_{min}$ experimentally, it should be possible to relate this quantity to the cap angle.

To calculate this angle, we first express the right-hand side of (7–290) in terms of the streamfunction that was previously calculated:

$$(1-\eta^2)\frac{\partial(\gamma/\gamma_s)}{\partial\eta} = Ca\left[\frac{\partial}{\partial r}\left(\frac{1}{r^2}\frac{\partial\psi}{\partial r}\right) - \frac{\hat{\mu}}{\mu}\frac{\partial}{\partial r}\left(\frac{1}{r^2}\frac{\partial\hat{\psi}}{\partial r}\right)\right] \quad \text{at } r = 1. \quad (7\text{–}291)$$

Then evaluating the right-hand side using (7–278) and (7–279) we find

$$\frac{(1-\eta^2)^{1/2}}{Ca}\frac{d(\gamma/\gamma_s)}{d\eta} = -\left(6 + 9\frac{\hat{\mu}}{\mu}\right)T_1^{-1}(\eta) + \left(1 + \frac{\hat{\mu}}{\mu}\right)\sum_{n=1}^{\infty}(4n+2)C_n T_n^{-1}(\eta),$$

J. Surfactant Effects on the Buoyancy-Driven Motion of a Drop

or, after substituting for C_n according to (7–287), we have

$$\frac{(1-\eta^2)^{1/2}}{Ca}\frac{d(\gamma/\gamma_s)}{d\eta} = \left(1 + \frac{\hat{\mu}}{\mu}\right) \sum_{n=1}^{\infty} (4n+2)C_n^* T_n^{-1}(\eta). \tag{7–292}$$

Finally, integrating, we obtain

$$\frac{1}{Ca}\left(\frac{\mu}{\mu+\hat{\mu}}\right)\left(1 - \frac{\gamma_{min}}{\gamma_s}\right) = \int_{\eta_c}^{1} \left[\sum_{n=1}^{\infty}(4n+2)C_n^* T_n^{-1}(\eta)\right] \frac{d\eta}{(1-\eta^2)^{1/2}}. \tag{7–293}$$

After substituting for C_n^* and integrating the right-hand side of this expression, Sadhal and Johnson[35] show that it can be expressed as a function of θ_c and (7–293) written as

$$\boxed{\frac{1}{Ca}\left(1 - \frac{\gamma_{min}}{\gamma_s}\right) = \frac{1}{\pi}[3\theta_c + 3\sin\theta_c - \theta_c(1+\cos\theta_c)].} \tag{7–294}$$

This expression provides a direct relationship between the minimum and maximum interfacial tensions and the cap angle for any given capillary number. Equation (7–294) is plotted in Fig. 7–20. If we are not careful to think about what we have done, the result (7–294) may seem rather strange as it does not seem to contain any indication of the drop size. This clearly cannot be correct. The fact is that the left-hand side of (7–294) still contains one unknown, namely the translational velocity of the drop U that appears in the capillary number, and this does depend on the drop size. Hence, to complete the calculation, we need to obtain one additional result, namely the drag on the bubble and, hence, by means of a balance with buoyancy, the translational velocity as a function of θ_c.

To calculate the drag, we can use the formula (7–132),

$$F = 4\pi\mu a U C_1,$$

with the constant obtained from (7–288). The result for the force is

$$\boxed{F = 4\pi\mu aU\left[\frac{\mu}{4\pi(\mu+\hat{\mu})}\left(2\theta_c + \sin\theta_c - \sin 2\theta_c - \frac{1}{3}\sin 3\theta_c\right)\right.\\ \left. + \frac{2\mu + 3\hat{\mu}}{2(\mu+\hat{\mu})}\right].} \tag{7–295}$$

One immediately obvious limiting case is *no surfactant* (i.e., $\theta_c = 0$) for which we recover the Hadamard–Rybczynski result

$$F = 4\pi\mu aU \frac{2\mu + 3\hat{\mu}}{2(\mu+\hat{\mu})}.$$

In the limit $\hat{\mu}/\mu \gg 1$, this reduces to Stokes' law, as it should. We also note that the limiting case of a *completely immobilized interface* ($\theta_c = \pi$) reduces to Stokes' law. The terminal velocity corresponding to (7–295) is

$$U = \frac{1}{3}\frac{\Delta\rho g a^2}{\mu}\frac{1}{C_1}. \tag{7–296}$$

Creeping Flows – Two-Dimensional and Axisymmetric Problems

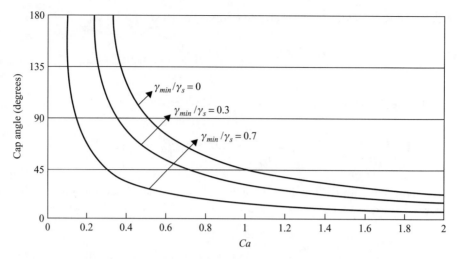

Figure 7–21. The cap angle as a function of Ca.

Finally, substituting (7–296) back into (7–294), now rewritten in the form

$$3\left(\frac{\gamma_s - \gamma_{min}}{\Delta\rho g a^2}\right) = \frac{1}{\pi}[3\theta_c + 3\sin\theta_c - \theta_c(1 + \cos\theta_c)]\frac{1}{C_1(\lambda, \theta_c)}. \qquad (7\text{–}297)$$

we obtain a complicated but explicit expression for the cap angle in terms of the viscosity ratio λ, and the ratio of terms on the left-hand side of (7–297). This relationship is shown in Fig. 7–22.

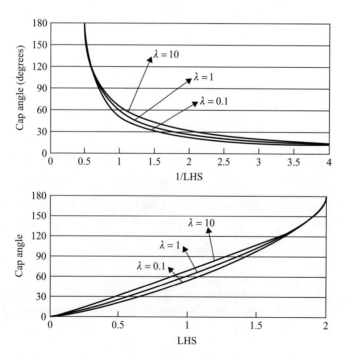

Figure 7–22. The cap angle as a function of $(\gamma_s - \gamma_{min})/\Delta\rho g a^2$ for several values of the viscosity ratio λ. LHS: Lefthand side.

J. Surfactant Effects on the Buoyancy-Driven Motion of a Drop

3. The Limit of Fast Adsorption Kinetics ($Bi \gg 1$)

We now turn to a case in which the interface concentration is determined by the mass transfer process to and from the bulk fluids. We begin with the case in which the fastest of the surfactant transport processes is the adsorption–desorption of surfactant between the exterior bulk fluid where it is assumed to be soluble and the interface, $Bi \gg 1$. In particular, we assume that $Bi \gg (c_\infty a/\Gamma_\infty k)Pe^{-1}$, so that, according to (7–265),

$$\Gamma = \frac{kc_s}{1+kc_s}, \qquad (7\text{–}298)$$

where c_s is the bulk concentration of surfactant in the limit as $r \to 1$. Clearly, in this case, if there is to be a concentration gradient along the interface of the drop, it must be because there is a corresponding gradient of the bulk concentration along this surface produced by convection in the bulk fluid.

For the case of a drop rising by means of buoyancy through a quiescent Newtonian fluid, the velocity and concentration fields will be axisymmetric. Hence, because the Reynolds numbers are assumed to be low, the governing equations of motion are again

$$E^4 \psi = 0, \quad \text{and} \quad E^4 \hat{\psi} = 0, \qquad (7\text{–}211)$$

with $\psi = \hat{\psi} = 0$ at $\eta = \pm 1$. The boundary conditions are zero normal velocity at the interface [Eq. (7–277)],

$$\psi = \hat{\psi} = 0 \quad \text{at } r = 1;$$

uniform flow at large distances from the drop [Eq. (7–147)],

$$\psi \to -r^2 Q_1(\eta) \quad \text{as } r \to \infty;$$

and boundedness for $\hat{\psi}$ at $r = 0$. Finally, we must satisfy continuity of the tangential velocity components at $r = 1$ [Eq. (7–280)],

$$\frac{\partial \psi}{\partial r} = \frac{\partial \hat{\psi}}{\partial r} \quad \text{at } r = 1 \text{ and all } \eta;$$

and continuity of the shear stress, also at $r = 1$,

$$\frac{\partial}{\partial r}\left(\frac{1}{r^2}\frac{\partial \psi}{\partial r}\right) - \frac{\hat{\mu}}{\mu}\frac{\partial}{\partial r}\left(\frac{1}{r^2}\frac{\partial \hat{\psi}}{\partial r}\right) = -Ma\frac{1-\eta^2}{1-\Gamma}\frac{\partial \Gamma}{\partial \eta} \quad \text{at } r = 1. \qquad (7\text{–}299)$$

For the mass transfer problem, we have the governing equation (7–260):

$$\nabla^2 c = Pe \, \nabla \cdot (c \mathbf{u}).$$

The interfacial mass transport equation is (7–266). For convenience, we write this equation in the form

$$\nabla_s \cdot (\mathbf{u}_s \Gamma) - \frac{1}{Pe_s}\nabla_s^2 \Gamma = \frac{\chi k}{Pe}(\mathbf{n} \cdot \nabla c)_s \quad \text{at } r = 1, \qquad (7\text{–}300)$$

Creeping Flows – Two-Dimensional and Axisymmetric Problems

using the definitions

$$\frac{c_\infty a}{\Gamma_\infty} \equiv k\chi \equiv \left(\frac{\beta c_\infty}{\alpha}\right)\left(\frac{a\alpha}{\beta\Gamma_\infty}\right). \tag{7-301}$$

The convenience of the form (7–300) will become evident shortly. In addition, we assume that the dimensionless bulk concentration

$$c = 1 \quad \text{as} \quad r \to \infty \tag{7-302}$$

and that Γ is given at $r = 1$ by Eq. (7–298). Finally, we suppose that $Pe_s \gg 1$ so that interface diffusion is neglected altogether.

We consider a solution of this problem only for the limiting case of small interface concentrations,

$$k \ll 1.$$

The problem for arbitrary k was solved recently, by use of numerical methods by Wang et al.,[37] and the small-k expansion followed here is similar to an analysis published in the appendix of this same paper. In this solution, we assume that the interfacial concentration is given by (7–298), and thus, by reference to (7–265), that

$$Bi \gg \chi Pe^{-1}.$$

To obtain a nonuniform concentration distribution, we need to retain convection in the preceding bulk transport equation, (7–260). On the other hand, for present illustrative purposes, we do not want to try to solve the full convection–diffusion equation. This will be the subject of later chapters of this book. Hence we assume that Pe is small but not vanishingly small. A convenient way to incorporate this is to consider the distinguished limit, in which $Pe = Zk$, where Z is $O(1)$. Finally, we assume that Ma is arbitrary, though finite.

Formally we expand the dependent variables in powers of k,

$$\boxed{\begin{aligned}
\mathbf{u} &= \mathbf{u}_0 + \mathbf{u}_1 k + \mathbf{u}_2 k^2 + O(k^3), \\
\Gamma &= \Gamma_0 + \Gamma_1 k + \Gamma_2 k^2 + O(k^3), \\
c &= c_0 + c_1 k + c_2 k^2 + O(k^3), \\
\psi &= \psi_0 + \psi_1 k + \psi_2 k^2 + O(k^3), \\
\hat{\psi} &= \hat{\psi}_0 + \hat{\psi}_1 k + \hat{\psi}_2 k^2 + O(k^3),
\end{aligned}} \tag{7-303}$$

and substitute into the various equations and boundary conditions previously listed.

At $O(1)$, we see from (7–298) that

$$\boxed{\Gamma_0 = 0.}$$

This means that there are no surfactant concentration gradients at this level of approximation, and the Marangoni stress condition, (7–299), is reduced to continuity of the viscous shear stress from the two fluids:

$$\frac{\partial}{\partial r}\left(\frac{1}{r^2}\frac{\partial \psi_0}{\partial r}\right) - \frac{\hat{\mu}}{\mu}\frac{\partial}{\partial r}\left(\frac{1}{r^2}\frac{\partial \hat{\psi}_0}{\partial r}\right) = 0 \quad \text{at } r = 1. \tag{7-304}$$

J. Surfactant Effects on the Buoyancy-Driven Motion of a Drop

Hence, with the governing equations (7–211), plus additional boundary conditions (7–277), (7–147) and (7–280), all of which apply directly to ψ_0 and $\hat{\psi}_0$, we see that the $O(1)$ problem reduces to the Hadamard–Rybczynski problem. The solution can be written in the form

$$\psi_0 = \left(-r^2 + \frac{6\mu + 9\hat{\mu}}{6\mu + 6\hat{\mu}} r - \frac{3\hat{\mu}}{6\mu + 6\hat{\mu}} \frac{1}{r}\right) Q_1(\eta), \qquad (7\text{–}305a)$$

$$\hat{\psi}_0 = -\frac{\mu}{2\mu + 2\hat{\mu}} (r^4 - r^2) Q_1(\eta). \qquad (7\text{–}305b)$$

For the bulk-phase concentration of surfactant, we solve the $O(1)$ problem

$$\nabla^2 c_0 = 0, \qquad (7\text{–}306)$$

with

$$c_0 \to 1 \quad \text{as } r \to \infty, \qquad (7\text{–}307)$$

and because $\Gamma_0 = 0$ it follows from (7–300) that

$$\left.\frac{\partial c_0}{\partial r}\right|_{r=1} = 0. \qquad (7\text{–}308)$$

The general axisymmetric solution of (7–306) is

$$c_0 = \sum_{n=0}^{\infty} \left[a_n r^n + b_n r^{-(n+1)}\right] P_n(\eta). \qquad (7\text{–}309)$$

Applying the two boundary conditions gives the trivial solution

$$c_0 = 1. \qquad (7\text{–}310)$$

The next order problem is $O(k)$. At this level, the interface concentration is still independent of position, as we can see by applying the expansions (7–303) to (7–298), from which we see that

$$\Gamma_1 = 1. \qquad (7\text{–}311)$$

The $O(k)$ term from the interface shear-stress condition (7–299) is, in general,

$$\frac{\partial}{\partial r}\left(\frac{1}{r^2}\frac{\partial \psi_1}{\partial r}\right) - \frac{\hat{\mu}}{\mu}\frac{\partial}{\partial r}\left(\frac{1}{r^2}\frac{\partial \hat{\psi}_1}{\partial r}\right) = -Ma\,(1-\eta^2)\frac{\partial \Gamma_1}{\partial \eta} \quad \text{at } r = 1,$$

but, in view of (7–311), this again reduces to continuity of the viscous shear-stress contributions. Now, however, the boundary condition at infinity, which was nonzero at $O(1)$, is homogeneous at $O(k)$, and the solution of (7–211) with completely homogeneous boundary conditions is just

$$\psi_1 = \hat{\psi}_1 = 0. \qquad (7\text{–}312)$$

The surfactant mass transfer problem, on the other hand, has a nontrivial solution at this level of approximation. The equations and boundary conditions at $O(k)$ are

$$\nabla^2 c_1 = 0, \qquad (7\text{–}313)$$

Creeping Flows – Two-Dimensional and Axisymmetric Problems

with

$$c_1 \to 0 \quad \text{as } r \to \infty \tag{7-314}$$

and the $O(k)$ contribution from (7–300)

$$\frac{\chi}{Z} \frac{\partial c_1}{\partial r} = -\frac{\partial}{\partial \eta}[(1-\eta^2)^{1/2}(u_\theta)_0 \Gamma_1] \quad \text{at } r = 1. \tag{7-315}$$

If we evaluate the $O(1)$ approximation to the velocity from (7–305a),

$$(u_\theta)_0|_{r=1} = \frac{\mu}{2(\mu+\hat{\mu})}(1-\eta^2)^{1/2},$$

we can write the condition (7–315) in the form

$$\left.\frac{\partial c_1}{\partial r}\right|_{r=1} = \frac{Z}{\chi}\left(\frac{\mu}{\mu+\hat{\mu}}\right)\eta. \tag{7-316}$$

Again, the general axisymmetric solution of (7–313), which vanishes at infinity, is

$$c_1 = \sum_{n=0}^{\infty} b_n r^{-(n+1)} P_n(\eta). \tag{7-317}$$

Applying the boundary condition (7–316), we have

$$\boxed{c_1 = -\frac{Z}{2\chi}\left(\frac{\mu}{\mu+\hat{\mu}}\right)\frac{P_1(\eta)}{r^2}.} \tag{7-318}$$

At this stage, we see that convection in the interface produces a bulk-phase concentration field that has a nonzero gradient at $r = 1$, though there has not yet been an effect on the flow because of the assumption of very small concentrations.

The next approximation is $O(k^2)$. Now, because the bulk concentration at $r = 1$ is dependent on position, it follows from (7–298) and the expansion (7–303), that the surface concentration of surfactant will also depend on spatial position. Indeed,

$$\boxed{\Gamma_2 = c_1|_{r=1} - \Gamma_1 \, c_0|_{r=1} = -\frac{Z}{2\chi}\left(\frac{\mu}{\mu+\hat{\mu}}\right)P_1(\eta) - 1.} \tag{7-319}$$

As expected, we see that this interface contribution produces a gradient from a minimum concentration at the front of the drop (where $\eta = 1$) to a maximum concentration at the back of the drop (where $\eta = -1$). Now, the Marangoni stress term on the right-hand side of the shear-stress condition, (7–299), at $O(k^2)$ is nonzero:

$$\left. \begin{aligned} \frac{\partial}{\partial r}\left(\frac{1}{r^2}\frac{\partial \psi_2}{\partial r}\right) - \frac{\hat{\mu}}{\mu}\frac{\partial}{\partial r}\left(\frac{1}{r^2}\frac{\partial \hat{\psi}_2}{\partial r}\right) &= -Ma(1-\eta^2)\frac{\partial \Gamma_2}{\partial \eta} \\ &= \frac{MaZ}{2\chi}\left(\frac{\mu}{\mu+\hat{\mu}}\right)(1-\eta^2) \end{aligned} \right\} \quad \text{at } r = 1, \tag{7-320}$$

and there will be a direct Marangoni contribution to the velocity field.

The general solutions of the governing equations (7–211) for ψ_2 and $\hat{\psi}_2$, which satisfy (7–277), (7–147) and (7–280), are

$$\psi_2 = \sum_{n=1}^{\infty} C_n(r^{2-n} - r^{-n})Q_n(\eta); \quad \hat{\psi}_2 = \sum_{n=1}^{\infty} C_n(r^{n+3} - r^{n+1})Q_n(\eta). \tag{7-321}$$

J. Surfactant Effects on the Buoyancy-Driven Motion of a Drop

Hence, substituting these solutions into (7–320), we obtain

$$\left(\frac{\mu}{\mu+\hat{\mu}}\right)\sum_{n=1}^{\infty}(4n+2)C_n Q_n(\eta) = \frac{MaZ}{\chi}\left(\frac{\mu}{\mu+\hat{\mu}}\right)Q_1(\eta). \tag{7-322}$$

It follows that

$$C_n = 0, \quad n \geq 2, \quad C_1 = \frac{MaZ}{6\chi},$$

and thus

$$\psi_2 = \frac{MaZ}{6\chi}(r - r^{-1})Q_1(\eta), \tag{7-323a}$$

with

$$\hat{\psi}_2 = \frac{MaZ}{6\chi}(r^4 - r^2)Q_1(\eta). \tag{7-323b}$$

If we combine (7–323) with the zero-order solution, (7–305), we see, as expected, that the presence of surfactant on the interface retards the flow. This is consistent with our qualitative expectations based on the fact that the surfactant concentration increases as we move from the front to the back of the drop. However, one surprising feature of the solution (7–323) is that there is no dependence on the viscosity ratio. This flow is established as a consequence of the shear-stress balance, (7–320). Clearly, the shear-stress difference [the left-hand side of (7–320)] does depend on the viscosity ratio; however we see from (3–322) that the Marangoni stress that drives the flow also depends on viscosity ratio in precisely the same form.

Now, with Γ_2, ψ_2, and $\hat{\psi}_2$ all known, we can solve the mass transfer problem for the corresponding bulk-phase concentration at $O(k^2)$. The governing equation for c_2 (recall that $Pe = Zk$) is

$$\nabla^2 c_2 = -Z\left(\frac{1}{r^2}\frac{\partial \psi_0}{\partial \eta}\frac{\partial c_1}{\partial r} - \frac{1}{r^2}\frac{\partial \psi_0}{\partial r}\frac{\partial c_1}{\partial \eta}\right). \tag{7-324}$$

We seek a solution such that $c_2 \to 0$ at $r = \infty$, which also satisfies the $O(k^2)$ condition from the interface transport equation,

$$\frac{\chi}{Z}\frac{\partial c_2}{\partial r} = -\frac{\partial}{\partial \eta}\left[(1-\eta^2)^{1/2}(u_\theta)_0 \Gamma_2\right] \quad \text{at } r = 1. \tag{7-325}$$

Substituting on the right-hand side of (7–324) yields

$$\nabla^2 c_2 = \frac{Z^2}{\chi}\left(\frac{\mu}{\mu+\hat{\mu}}\right)\left[\left(-\frac{1}{r^3}+\frac{A}{r^4}-\frac{B}{r^6}\right)\eta^2 + \left(\frac{1}{2r^3}-\frac{A}{4r^4}-\frac{B}{4r^6}\right)(1-\eta^2)\right]. \tag{7-326}$$

To simplify the notation, we have used

$$A \equiv \frac{6\mu+9\hat{\mu}}{6\mu+6\hat{\mu}} \quad \text{and} \quad B \equiv \frac{3\hat{\mu}}{6\mu+6\hat{\mu}}.$$

Substituting for $(u_\theta)_0$ and Γ_2 in the boundary condition (7–325) gives

$$\left.\frac{\partial c_2}{\partial r}\right|_{r=1} = -\frac{Z}{2\chi}\left(\frac{\mu}{\mu+\hat{\mu}}\right)\left[2\eta - (1-3\eta^2)\frac{Z}{2\chi}\left(\frac{\mu}{\mu+\hat{\mu}}\right)\right]. \tag{7-327}$$

Creeping Flows – Two-Dimensional and Axisymmetric Problems

To solve (7–326), it is useful to note that the right-hand side can be expressed in the form

$$\frac{1}{r^2}\left[\frac{\partial}{\partial r}\left(r^2\frac{\partial c_2}{\partial r}\right) + \frac{\partial}{\partial \eta}\left((1-\eta^2)\frac{\partial c_2}{\partial \eta}\right)\right]$$
$$= \frac{Z^2}{\chi}\left(\frac{\mu}{\mu+\hat{\mu}}\right)\left[\left(-\frac{1}{r^3} + \frac{5A}{6r^4} - \frac{B}{2r^6}\right)P_2(\eta) - \left(\frac{A}{6r^4} + \frac{B}{2r^6}\right)P_0(\eta)\right]. \tag{7-328}$$

To obtain particular solutions, the easiest approach is to assume solutions of the general form $c_2 = \text{const} \cdot r^m P_2(\eta)$ or $\text{const} \cdot r^m P_0$ and choose m and const so that, when substituted into the left-hand side of (7–328) one of the terms on the right is generated. When this is done on a term-by-term basis, and the homogeneous solution is added, the general solution of (7–328) that vanishes at infinity is

$$\boxed{\begin{aligned} c_2 &= \frac{Z^2}{\chi}\left(\frac{\mu}{\mu+\hat{\mu}}\right)\left[\left(\frac{1}{6r} - \frac{5A}{24r^2} - \frac{B}{12r^4}\right)P_2(\eta) - \left(\frac{A}{12r^2} + \frac{B}{24r^4}\right)\right] \\ &\quad + \sum_{n=0}^{\infty} b_n r^{-(n+1)} P_n(\eta). \end{aligned}} \tag{7-329}$$

The boundary condition (7–327) then gives the constants b_n:

$$b_0 = \frac{Z^2}{6\chi}\left(\frac{\mu}{\mu+\hat{\mu}}\right)(A+B), \tag{7-330a}$$

$$b_1 = \frac{Z}{2\chi}\left(\frac{\mu}{\mu+\hat{\mu}}\right), \tag{7-330b}$$

$$b_2 = \frac{2}{3}\left(\frac{Z}{2\chi}\right)^2\left(\frac{\mu}{\mu+\hat{\mu}}\right)^2 + \left(\frac{Z^2}{3\chi}\right)\left(\frac{\mu}{\mu+\hat{\mu}}\right)\left(-\frac{1}{6} + \frac{5A}{12} + \frac{B}{3}\right), \tag{7-330c}$$

$$b_n = 0, \quad n \geq 3. \tag{7-330d}$$

We proceed one more step with the expansion in k because we are interested in calculating the stress to $O(k^3)$. For this purpose, we are required to solve only for the C_1 coefficient in the solution for ψ_3. The governing equations for ψ_3 and $\hat{\psi}_3$ are again

$$E^4\psi_3 = 0 \quad \text{and} \quad E^4\hat{\psi}_3 = 0.$$

The boundary condition (7–299) at $O(k^3)$ is

$$\frac{\partial}{\partial r}\left(\frac{1}{r^2}\frac{\partial \psi_3}{\partial r}\right) - \frac{\hat{\mu}}{\mu}\frac{\partial}{\partial r}\left(\frac{1}{r^2}\frac{\partial \hat{\psi}_3}{\partial r}\right) = -Ma(1-\eta^2)\left(\frac{\partial \Gamma_2}{\partial \eta} + \frac{\partial \Gamma_3}{\partial \eta}\right). \tag{7-331}$$

To evaluate the right-hand side, we need Γ_3. Substituting the expansions (7–303) into (7–298), we see that

$$\Gamma_3 = (c_2 - \Gamma_1 c_1 - \Gamma_2 c_0)|_{r=1}. \tag{7-332}$$

Evaluating the right-hand side of (7–331), we find

$$\frac{\partial}{\partial r}\left(\frac{1}{r^2}\frac{\partial \psi_3}{\partial r}\right) - \frac{\hat{\mu}}{\mu}\frac{\partial}{\partial r}\left(\frac{1}{r^2}\frac{\partial \hat{\psi}_3}{\partial r}\right) = 2Ma\left[\frac{Z}{\chi}\left(\frac{\mu}{\mu+\hat{\mu}}\right)Q_1(\eta) + \cdots +\right]. \tag{7-333}$$

J. Surfactant Effects on the Buoyancy-Driven Motion of a Drop

We have displayed only the term proportional to Q_1 because it is only this term that will make a C_1 contribution to ψ_3, and hence it is only this term that will make a nonzero contribution to the force on the drop.

The analysis follows that previously given for ψ_2. Hence the general solution satisfying all but the shear-stress condition is

$$\psi_3 = \sum_{n=1}^{\infty} C_n(r^{2-n} - r^{-n})Q_n(\eta); \quad \hat{\psi}_3 = \sum_{n=1}^{\infty} C_n(r^{n+3} - r^{n+1})Q_n(\eta) \quad (7\text{–}334)$$

and, when substituted into (7–333), yields

$$\left(\frac{\mu}{\mu+\hat{\mu}}\right)\sum_{n=1}^{\infty}(4n+2)C_n Q_n(\eta) = -\frac{2MaZ}{\chi}\left(\frac{\mu}{\mu+\hat{\mu}}\right)Q_1(\eta) + \cdots + . \quad (7\text{–}335)$$

Hence

$$C_1 = -\frac{MaZ}{3\chi}, \quad (7\text{–}336)$$

$$\psi_3 = -\frac{MaZ}{3\chi}(r - r^{-1})Q_1(\eta) + \sum_{n=2}^{\infty} C_n(r^{2-n} - r^{-n})Q_n(\eta). \quad (7\text{–}337)$$

We do not require the other terms in (7–337) for calculating the drag on the drop. For this we can again use the formula (7–132),

$$F = 4\pi\mu a U C_1.$$

The result is

$$F = 4\pi\mu a U \left[\frac{6\mu + 9\hat{\mu}}{6\mu + 6\hat{\mu}} + k^2\left(\frac{MaZ}{6\chi}\right) - k^3\left(\frac{MaZ}{3\chi}\right) + O(k^4)\right]. \quad (7\text{–}338)$$

We note that the first Marangoni contribution increases the drag for a given U, but the second term actually goes in the opposite direction. The drag increases because the flow around the drop produces a gradient of surfactant concentration on the drop surface from a minimum at the front of the drop to a maximum at the rear. This produces a Marangoni contribution to the shear-stress balance that tends to immobilize the interface and thus increase the drag. The fact that the next correction goes in the opposite direction simply indicates that the leading-order contribution to the surfactant concentration gradient overestimates its magnitude for small but finite k.

The most interesting feature of the expression (7–338), however, is that the Marangoni corrections do not depend on the viscosity ratio. The reason for this can be seen if we trace the derivation of the $O(k^2)$ and $O(k^3)$ terms starting from (7–320) to (7–322). On the right-hand side of (7–320), we can see that the Marangoni contribution to the stress depends on the ratio of μ to $\hat{\mu}$, as we would expect. However, when we get to (7–322), we see that the viscous stress has the exact same dependence on these quantities, and the solution that results (7–323) is independent of the *viscosity ratio*. The same thing happens at the next level. It is not obvious that it will continue, however, as it seems rather remarkable that the correction to the drag, which itself depends on the viscosity ratio, should be independent of that ratio.

NOTES

1. M. Gad-El Hak, *The MEMS Handbook* (Mechanical Engineering) (CRC, Boca Raton, FL, 2001); N.-T. Nguyen, S. Werely, and S. T. Wereley, *Fundamentals and Applications of Microfluidics* (Artech House, Norwood, MA, 2002); T. M. Squires and S. R. Quake, Microfluidics: Fluid physics at the nanoliter scale, *Rev. Mod. Phys.* **77**, 977–1026 (2005); D. J. Beebe, G. A. Mensing, and G. M. Walker, Physics and applications of microfluidics in biology, *Annu. Rev. Biomed. Eng.* **4**, 261–286 (2002); J. P. Brody, P. Yager, R. E. Goldstein, and R. H. Austin, Biotechnology at low Reynolds numbers, *Biophy. J.* **71**, 3430–41 (1996); H. A. Stone, A. D. Stroock, and A. Adjari, Engineering flows in small devices: Microfluidics toward a lab-on-a-chip, *Annu. Rev. Fluid Mech.* **36**, 381–411 (2004).
2. G. Segre and A. Silberberg, Behavior of macroscopic rigid spheres in Poiseuille flow, Part 1, Determination of local concentration by statistical analysis of particle passages through crossed light beams, *J. Fluid Mech.* **14**, 115–36 (1962); Part 2, Experimental results and interpretation, *J. Fluid Mech.* **14**, 137–57 (1962).
3. B. P. Ho and L. G. Leal, Inertial migration of rigid spheres in two-dimensional unidirectional flows, *J. Fluid Mech.* **65**, 365–400 (1974).
4. P. Vasseur and R. G. Cox, The lateral migration of a spherical particle in two-dimensional shear flows, *J. Fluid Mech.* **78**, 385–413 (1976).
5. K. Ishii and H. Hasimoto, Lateral migration of a spherical particle in flows in a circular tube, *J. Phys. Soc. Jpn.* **48**, 2144–53 (1980).
6. J. A. Schonberg and E. J. Hinch, Inertial migration of a sphere in Poiseuille flow, *J. Fluid Mech.* **203**, 517–524 (1989); E. S. Asmolov, The inertial lift on a small particle in a weak-shear parabolic flow, *Phys. of Fluids* **14**, 15–28 (2002); P. Cherukat, J. B. McLaughlin, and D. S. Dandy, A computational study of the inertial lift on a sphere in a linear shear flow field, *Int. J. of Multiphase Flow*, **25**, 15–33 (1999).
7. L. G. Leal, Particle motion in a viscous fluid, *Annu. Rev. of Fluid Mech.* **12**, 435–76 (1980).
8. H. Brenner, Dynamics of neutrally buoyant particles in low Reynolds number flows, *Prog. Heat Mass Transfer* **6**, 509–74 (1972); J. Happel and H. Brenner, *Low Reynolds Number Hydrodynamics* (Noordhoff, Leyden, The Netherlands, 1973).
9. The inverse of any second-order tensor $\mathbf{M}$ is defined such that $\mathbf{M}^{-1} \cdot \mathbf{M} = \mathbf{M} \cdot \mathbf{M}^{-1} = \mathbf{I}$, where $\mathbf{I}$ is the unit tensor.
10. An excellent sourcebook for vector and tensor calculus is R. Aris, *Vectors, Tensors and the Basic Equations of Fluid Mechanics* (Prentice-Hall, Englewood Cliffs, NJ, 1962).
11. H. K. Moffatt, Viscous and resistive eddies near a sharp corner, *J. Fluid Mech.* **18**, 1–18 (1964).
12. G. I. Taylor, Deposition of viscous fluid in a plane surface, *J. Fluid Mech.* **9**, 218–224 (1960).
13. S. Wakiya, Application of bipolar co-ordinates to the two-dimensional creeping-motion of a liquid. I. Flow over a projection or a depression on a wall, *J. Phys. Soc. Jpn.* **39**, 1113–20 (1975); M. E. O'Neill, On the separation of a slow linear shear flow from a cylindrical ridge or trough in a plane, Z. Angew. Math. Phys. **28**, 438–48 (1977); A. M. J. Davis and M. E. O'Neill, Separation in a Stokes flow past a phase with a cylindrical ridge or trough, *Q. J. Mech. Appl. Math.* **30**, 355–68 (1977).
14. J. M. Dorreapaal and M. E. O'Neill, The existence of free eddies in a streaming Stokes flow, *Q. J. Mech. Appl. Math.* **32**, 95–107 (1979).
15. M. E. O'Neill and K. B. Ranger, Particle–fluid interaction, in *Handbook of Multiphase Flow*, G. Hetsroni (ed.) (Hemisphere, New York, 1982), pp. 1–96.
16. M. Abramowitz and I. A Stegun, *Handbook of Mathematical Functions* (Dover, New York, 1965).
17. H. Hasimoto and O. Sano, Stokeslets and eddies in creeping-flow, *Annu. Rev. Fluid Mech.* **12**, 335–63 (1980).
18. To obtain this result, we simply take the curl of both terms in (7–6) and use the well-known result $\nabla \wedge (\nabla \phi) = 0$ for any scalar ϕ.
19. A. Einstein, *Investigations on the Theory of the Brownian Movement* (Dover, New York, 1956).
20. It is interesting to note that Einstein's original published result was incorrect, with the coefficient 3/2 instead of 5/2. Although Einstein quickly published a correction to the result, there is perhaps

Notes

some "comfort" for the student in realizing that even brilliant researchers can make published mistakes.

21. G. K. Batchelor, The stress system in a suspension of force-free particles, *J. Fluid Mech.* **41**, 545–70 (1970).
22. W. R. Schowalter, *Mechanics of Non-Newtonian Fluids* (Pergamon, New York, 1978); W. B. Russell, D. A. Saville, and W. R. Schowlater, *Colloidal Dispersions* (Cambridge University Press (Cambridge, 1989); R. G. Larson, *The Structure and Rheology of Complex Fluids* (Oxford University Press, New York, 1999).
23. An excellent review of work through 1972 is J. F. Harper, The motion of bubbles and drops through liquids, *Adv. Appl. Mech.* **12**, 59–129 (1972).
24. H. A. Stone, Dynamics of drop deformation and breakup in viscous fluids, *Annu. Rev. Fluid Mech.* **26**, 65–102 (1994).
25. T. D. Taylor and A. Acrivos, On the deformation and drag of a falling viscous drop at low Reynolds number, *J. Fluid Mech.* **18**, 466–76 (1964).
26. M. Kojima, E. J. Hinch, and A. Acrivos, The formation and expansion of a toroidal drop moving in a viscous fluid, *Phys. Fluids* **27**, 19–32 (1984).
27. C. J. Koh and L. G. Leal, The stability of drop shapes for translation at zero Reynolds number through a quiescent fluid, *Phys. Fluids A* **1**, 1309–13 (1989).
28. N. O. Young, J. S. Goldstein and M. J. Block, The motion of bubbles in a vertical temperature gradient, *J. Fluid Mech.* **6**, 350–6 (1959).
29. A useful compilation of research on bubble and drop dynamics in the context of space experiments and space processing of materials can be found in the proceedings of a colloquium sponsored by NASA; T. G. Wang (ed.), *Proceedings of the Third International Colloquium on Drops and Bubbles*, AIP Conference Proceedings, **197** (American Institute of Physics, Melville, NY, 1989).

    Additional papers that contain reference to materials processing applications in low-gravity environments (space) include these: R. S. Subramanian, Thermocapillary migration of bubbles and droplets, *Adv, Space Res.* **3**, 145 (1983); K. D. Barton and R. S. Subramanian, The migration of liquid drops in a vertical temperature gradient, *J. Colloid Interface Sci.* **133**, 211–22 (1989); R. M. Merritt and R. S. Subramanian, Migration of a gas bubble normal to a plane horizontal surface in a vertical temperature gradient, *J. Colloid. Interface Sci.* **131**, 514–25 (1989); R. S. Subramanian, The motion of bubbles and drops in reduced gravity, in *Transport Processes in Bubbles, Drops and Particles*, R. P. Chhabra and D. DeKee (eds.) (Hemisphere, New York, 1990); G. Wozniak, J. Siekmann, and J. Srulijies, Thermocapillary bubble and drop dynamics under reduced gravity – Survey and prospects, *Z. Flugwiss. Weltraumforsch.* **12** 137–144 (1988).
30. A. N. Frumkin and V. G. Levich, *Zh. Fiz. Khim.* **21**, 1183 (1947) (in Russian). This work, as well as related research on the motion of drops and bubbles in fluids, is summarized in the textbook (translated from the Russian); V. G. Levich, *Physicochemical Hydrodynamics* (Prentice-Hall, Englewood Cliffs, NJ, 1962).
31. B. Cuenot, J. Magnaudet, and B. Spennato, The effect of slightly soluble surfactants on the flow around a spherical bubble, *J. Fluid Mech.* **399**, 25–53 (1996); Y. Wang, D. T. Papageorgiou, and C. Maldarelli, Increased mobility of a surfactant-retarded bubble at high bulk concentrations, *J. Fluid Mech.* **390**, 251–70 (1999).
32. M. D. LeVan and J. A. Holbrook, Motion of a droplet containing surfactant, *J. Colloid Interface Sci.* **131**, 242–51 (1989); J. A. Holbrook and M. D. LeVan, Retardation of droplet motion by surfactant, *Chem. Eng. Commun.* **20**, 273–90 (1983).
33. P. Savic, Circulation and distortion of liquid drops falling through a viscous medium. *Res. Counc. Can. Div. Mech. Eng. Rep.* MT-22.
34. R. E. Davis and A. Acrivos, The influence of surfactants on the creeping motion of small bubbles, *Chem. Eng. Sci.* **21**, 681–5 (1966).
35. S. Sadhal and R. E. Johnson, Stokes flow past bubbles and drops partially coated with thin films. Part 1. Stagnant cap of surfactant film – Exact solution, *J. Fluid Mech.* **126**, 237–50 (1982).
36. W. D. Collins, On some dual series equations and their application to electrostatic problems for spheroidal caps, *Proc. Cambridge Philos. Soc.* **57**, 367 (1961).
37. See the second entry of Ref. 31.

Creeping Flows – Two-Dimensional and Axisymmetric Problems

PROBLEMS

Problem 7–1. General Consequences of the Linearity and Quasi-Steady Nature of Creeping Flow.

(a) A jellyfish swims by slowly drawing fluid into its umbrella-like body (see the sketch) and then ejecting it with high velocity, propelling itself forward. Would this still work if the jellyfish were of microscopic dimensions (consider the dimensionless velocity of the jellyfish here)? Briefly explain your answer.

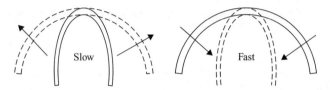

(b) A straight, slender rodlike particle sediments through a quiescent fluid under the action of gravity ($\rho_{particle} \neq \rho_{fluid}$). The Reynolds number is very small so that the creeping-motion approximation is relevant. Show that the particle will fall vertically if it is oriented with $\alpha = 0$ or $\alpha = \pi/2$. What can you say for $\alpha \neq 0, \pi/2$? Justify your conclusions.

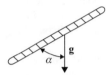

(c) A neutrally buoyant drop (that is, $\rho_{drop} = \rho_{suspending\ fluid}$) in simple shear flow near a plane wall is observed to migrate in the direction normal to the wall. Is this possible in creeping-flow? Discuss. (Note that the shape of the drop will not generally remain spherical.)

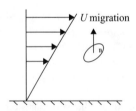

Problem 7–2. Linearity of Creeping Flow; Force on a Sphere Near a Plane Wall. A sphere is a distance h from a plane and is acted on by a force F_i. We perform two experiments. In the first experiment the force F is applied perpendicular to the plane, and the sphere is observed to move with a velocity U_a. In the second, the force is applied parallel to the plane, and the observed velocity is U_b. Using this, determine the velocity for an arbitrary applied force.

Hint: The velocity will be a function of both F_i and n_i, the unit normal to the plane.

Problem 7–3. Linearity of Creeping Flow. In the laboratory you are looking at the sedimentation of rods under creeping-flow conditions (zero Reynolds number). These may be regarded as bodies of revolution characterized by an orientation vector (director) p_i. In a simple experiment you measure the sedimentation velocity when the director is parallel to gravity (e.g., point down) to be 0.03 cm/s and when it is perpendicular to gravity to

Problems

be 0.01 cm/s. In a very viscous fluid the velocity is proportional to the force. What is the maximum distance the rod can travel laterally (perpendicular to gravity) during the time it takes to fall 10 cm vertically? For what angle between the director and gravity does this occur?

Problem 7–4. Linearity of Creeping Flow; Drag on a Cubical Body. Using simple arguments based on the linearity of the creeping-flow equations, prove that the drag on a cube under creeping-flow conditions is the same regardless of its orientation relative to the direction of motion.

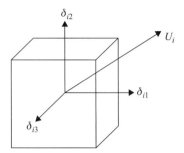

Problem 7–5. Linearity of Creeping Flow; Drag on a Translating Body. We are interested in solving for the drag of an object moving with some constant velocity through a viscous fluid (creeping flow) as a function of the orientation of the object. How many problems will we have to solve if the object is

(a) isotropic,
(b) a body of revolution (the orientation may be characterized by a single unit vector),
(c) an arbitrary body (the orientation may be characterized by two **orthogonal** unit vectors).

Determine the functional form of the force vs. orientation relationship for each case.

Problem 7–6. Linearity and Superposition. Determine the flow between two concentric spheres of radii a and b that are rotating about different axes with angular velocities Ω_i^a and Ω_i^b, respectively, assuming that the Reynolds number for the motion is small. Also determine the torque exerted by the fluid on each sphere.

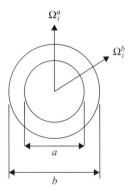

Definition sketch for rotating spheres.

Problem 7–7. Accelerating Motion of a Sphere in a Quiescent Fluid. Consider a rigid sphere of radius a, which executes a rectilinear oscillatory motion with velocity

$$U(t) = U \sin \omega t$$

Creeping Flows – Two-Dimensional and Axisymmetric Problems

in an unbounded, quiescent body of fluid. We consider the asymptotic limit

$$\frac{Ua}{\nu} \ll 1, \quad \frac{a^2\omega}{\nu} \sim O(1).$$

In this case, nonlinear inertia terms in the Navier–Stokes equation can be neglected to leading-order of approximation, but the acceleration term, $\partial \mathbf{u}/\partial t$, must be retained.

(a) Derive dimensionless forms for the governing equations and boundary conditions in the limit just described for the case in which t is sufficiently large that initial transients associated with start-up flow at $t = 0$ can be neglected. For this purpose, it is convenient to adopt a spherical coordinate system whose origin is fixed at the position initially occupied by the center of the sphere. Show that the displacement of the sphere can be neglected at the first level of approximation so that boundary conditions can be applied at $r = a$ for all times t.

(b) Solve for the velocity and pressure fields for the problem previously defined. *Hints:* Do not assume that the motion of the fluid is in-phase with the motion of the sphere. In addition, this problem is an example of a case in which we can anticipate the form of the solution based on the symmetry of the boundary conditions. In particular, if the motion of the sphere is along the z axis, as depicted here, and the spherical coordinate system is set up with the polar angle θ measured from the z axis, you may wish to try

$$u_r = e^{-i\bar{t}} \overline{f}(\bar{r}) \cos\theta,$$

$$u_\theta = e^{-i\bar{t}} g(\bar{r}) \sin\theta,$$

where, for algebraic convenience,

$$\overline{f}(\bar{r}) = -\frac{2}{r}\frac{df}{dr}.$$

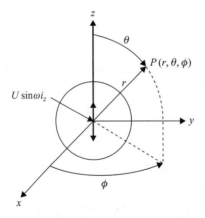

You should find that the governing equation takes the form

$$\frac{d}{dr}\left[\nabla_r^2\left(\nabla_r^2 f\right) + i\left(\frac{\omega a^2}{\nu}\right)\nabla_r^2 f\right] = 0,$$

where

$$\nabla_r^2 \equiv \frac{1}{r^2}\frac{d}{dr}\left(r^2\frac{d}{dr}\right).$$

Hence, because $\mathbf{u}$ and all its derivatives vanish as $r \to \infty$, it follows that

$$\nabla_r^2\left(\nabla_r^2 f\right) + i\left(\frac{\omega a^2}{\nu}\right)\nabla_r^2 f = 0.$$

Problems

(c) Solve the equation from part (b) for f and calculate the drag on the sphere. In dimensional terms, you should find that

$$F = 6\pi\mu a\left[1 + \left(\frac{a^2\omega}{2\nu}\right)^{1/2}\right]Ue^{-i\omega t}$$
$$+ 3\sqrt{2}\pi a^2 \sqrt{\frac{\mu\rho}{\omega}}\left[1 + \frac{2}{9}\left(\frac{a^2\omega}{2\nu}\right)^{1/2}\right](-i\omega Ue^{-i\omega t}).$$

Discuss the limit $\omega \to 0$. Show that there is a contribution to F for $\omega \gg 1$ that is independent of the viscosity. This term is known as the added-mass contribution and is identical to the force on an accelerating sphere in an inviscid fluid (that is, a fluid with $\mu = 0$). It predicts that there is a contribution to the force that has the effect of adding an additional mass to the sphere that is equal to $1/2$ of the sphere volume times the fluid density.

(d) Given the result in part (c) for the force on an oscillating sphere, we can obtain the corresponding result for the force on an accelerating sphere with velocity $u_s(t)$. The simplest way is to note that $u_s(t)$ can be represented in terms of a Fourier transform,

$$u_s(t) = \int_{-\infty}^{\infty} u_\omega e^{-i\omega t}\,d\omega,$$

where

$$u_\omega = \frac{1}{2\pi}\int_{-\infty}^{\infty} u_s(\tau)e^{i\omega\tau}\,d\tau,$$

so that $F \equiv F_\omega$ from part (c) can be interpreted as the force on a sphere with velocity $u_\omega e^{-i\omega t}$. Then $F(t)$ corresponding to $u_s(t)$ is just

$$F = \int_{-\infty}^{\infty} F_\omega\,d\omega.$$

Show that

$$F = 2\pi\rho a^3\left(\frac{1}{3}\frac{du_s}{dt} + \frac{3\nu u_s}{a^2} + \frac{3}{a}\sqrt{\frac{\nu}{\pi}}\int_{-\infty}^{t}\frac{du_s}{d\tau}\frac{d\tau}{\sqrt{t-\tau}}\right).$$

Here, the first term is the added mass contribution, the second is Stokes' law, and the third is known as the Basset memory integral contribution. Evaluate this expression for

$$u_s = \begin{cases} 0; & t < 0 \\ u_\infty; & t \geq 0 \end{cases}.$$

Problem 7–8. Sphere in a Linear Flow. A rigid sphere is translating with velocity $\mathbf{U}$ and rotating with angular velocity $\boldsymbol{\Omega}$ in an unbounded, incompressible Newtonian fluid. The position of the sphere center is denoted as $\mathbf{x}_\rho$ (that is, $\mathbf{x}_\rho$ is the position vector). At large distances from the sphere, the fluid is undergoing a simple shear flow (this is the undisturbed velocity field). We may denote this flow in the form

$$\mathbf{u}_\infty = \boldsymbol{\Gamma}\cdot\mathbf{x},$$

where

$$\boldsymbol{\Gamma} = \gamma\begin{bmatrix} 0 & 1 & 0 \\ 0 & 0 & 0 \\ 0 & 0 & 0 \end{bmatrix},$$

and $\mathbf{x}$ is the general position vector associated with any arbitrary point in the fluid. (Note that $\mathbf{x}_\rho \neq 0$.) If the appropriate Reynolds number is very small, so that the creeping-motion approximation is valid, it can be shown that the hydrodynamic force and torque acting on the body are expressible in the forms

$$\mathbf{F} = a\mu(\mathbf{A}\cdot\mathbf{U} + \mathbf{B}\cdot\mathbf{\Omega} + \mathbf{D}:\mathbf{\Gamma}),$$

$$\mathbf{T} = a\mu(\mathbf{C}\cdot\mathbf{U} + \mathbf{E}\cdot\mathbf{\Omega} + \mathbf{G}:\mathbf{\Gamma}),$$

where $\mathbf{A}$, $\mathbf{B}$, $\mathbf{C}$, $\mathbf{E}$, $\mathbf{D}$, and $\mathbf{G}$ are constant tensors. Here, a is the particle radius and μ is the fluid viscosity.

(a) Demonstrate that the preceding result for $\mathbf{F}$ and $\mathbf{T}$ is true.
(b) Indicate the forms for $\mathbf{D}$ and $\mathbf{G}$ (that is, what terms are zero or nonzero and what can we say beyond that?).
(c) Demonstrate the validity of the general formula for $\mathbf{F}$ and $\mathbf{G}$ for any linear flow of the type

$$\mathbf{u}_\infty = \mathbf{\Gamma}\cdot\mathbf{x}.$$

Problem 7–9. Motion of a Force- and Torque-Free Axisymmetric Particle in a General Linear Flow. We consider a force- and torque-free axisymmetric particle whose geometry can be characterized by a single vector $\mathbf{d}$ immersed in a general linear flow, which takes the form far from the particle: $\mathbf{v}^\infty(r) = \mathbf{U}^\infty + \mathbf{r}\wedge\mathbf{\Omega}^\infty + \mathbf{r}\cdot\mathbf{E}^\infty$, where $\mathbf{U}^\infty$, $\mathbf{\Omega}^\infty$, and $\mathbf{E}^\infty$ are constants. Note that $\mathbf{E}^\infty$ is the symmetric rate-of-strain tensor and $\mathbf{\Omega}^\infty$ is the vorticity vector, both defined in terms of the undisturbed flow. The Reynolds number for the particle motion is small so that the creeping-motion approximation can be applied.

(a) Show that the translational velocity of the particle must take the form

$$\mathbf{U} = \mathbf{U}^\infty + \frac{X^G}{X^A}\mathbf{d}(\mathbf{d}\cdot\mathbf{E}^\infty\cdot\mathbf{d}) + 2\frac{Y^G}{Y^A}[\mathbf{d}\cdot\mathbf{E}^\infty - \mathbf{d}(\mathbf{d}\cdot\mathbf{E}^\infty\cdot\mathbf{d})],$$

where the scalar functions X^A, Y^A, X^G, and Y^G are defined in the resistance tensors that relate the force to the relative translational velocity $\mathbf{U}_\infty - \mathbf{U}$ and to the rate of strain $\mathbf{E}^\infty$.

$$R_{ij}^{FU} = X^A d_i d_j + Y^A(\delta_{ij} - d_i d_j),$$

$$R_{kij}^{FE} = X^G\left(d_i d_j - \frac{1}{3}\delta_{ij}\right)d_k + Y^G(d_i\delta_{jk} + d_j\delta_{ik} - 2d_i d_j d_k).$$

It is also assumed that there is no translational/rotational coupling, i.e., the particle has no chirality.

(b) Show that the angular velocity of the particle is given by

$$\mathbf{\Omega} = \mathbf{\Omega}^\infty + \frac{Y^H}{Y^C}\mathbf{d}\wedge(\mathbf{E}^\infty\cdot\mathbf{d})$$

where Y^C is the analog of Y^A in the expression for $R_{ij}^{T\Omega}$ and Y^H is the scalar function giving the torque that is due to the straining motion:

$$R_{kij}^{TE} = Y^H(\varepsilon_{ikl}d_j + \varepsilon_{jkl}d_i)d_l.$$

(c) For a body possessing fore–aft symmetry (i.e., $\mathbf{d}$ and $-\mathbf{d}$ are equally valid descriptions of the geometry) what simplifications to the results in parts (a) and (b) are possible?

Problem 7–10. Translation of a Bubble Through a Quiescent Fluid. A spherical gas bubble rises through a stationary liquid that is sufficiently viscous that the creeping-flow approximation is valid. Assume that the boundary conditions for the liquid motion at the bubble surface are

$$\mathbf{u}\cdot\mathbf{e}_r = 0, \qquad \mathbf{e}_\theta\cdot(\mathbf{e}_r\cdot\mathbf{T}) = 0,$$

Problems

where $\mathbf{e}_r$ and $\mathbf{e}_\theta$ are unit tangent vectors normal and parallel to the bubble surface. Further, assume that the gas inside the bubble has negligible viscosity and density relative to the liquid.

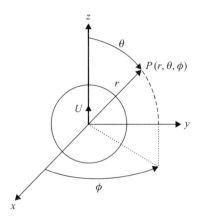

Determine the velocity and pressure fields in the liquid as well as the velocity of the bubble by means of a full eigenfunction expansion for ψ in spherical coordinates. Does this solution satisfy the normal-stress condition on the bubble surface?

Problem 7–11. Force on a Body in Axisymmetric Flow.

(a) Prove that the hydrodynamic force exerted by the fluid on a solid sphere in *any* axisymmetric creeping-flow is

$$F_z = 4\pi\mu U a C_1,$$

where

$$\psi = \sum_{n=1}^{\infty} \left(A_n r^{n+3} + B_n r^{n+1} + C_n r^{2-n} + D_n r^{-n} \right) Q_n(\eta).$$

(b) Use the divergence theorem to demonstrate that the force on an *arbitrarily shaped* axisymmetric body in creeping-flow is proportional to C_1.

Problem 7–12. Bubble in an Axisymmetric Flow. A gas bubble is immersed in a viscous Newtonian fluid that is undergoing an axisymmetric extensional flow. The fluid is viscous enough that the relevant Reynolds number is small so that the creeping-motion approximation can be applied. The capillary number based on the extension rate, E, and the surface tension, σ, is small, i.e.,

$$Ca = \frac{\mu E a}{\sigma} \ll 1.$$

Here, a is the radius of the undeformed bubble and μ is the viscosity of the liquid. Assume that the boundary conditions for the liquid motion at the bubble surface are

$$\mathbf{u} \cdot \mathbf{n} = 0, \quad \mathbf{t} \cdot (\mathbf{n} \cdot \mathbf{T}) = 0, \quad \mathbf{n} \cdot \mathbf{n} \cdot \mathbf{T} = \sigma \nabla \cdot \mathbf{n},$$

where $\mathbf{n}$ and $\mathbf{t}$ are unit tangent vectors normal and parallel to the bubble surface.

(a) Nondimensionalize and determine the governing equations and boundary conditions for the first two terms in an asymptotic approximation for $Ca \ll 1$. To do this, you will need to use the method of domain perturbations
(b) Determine the first approximation for the streamfunction, and use this to determine the first approximations for the velocity and pressure fields for $Ca \ll 1$.

(c) Using the solution from (b), determine the shape of the bubble at $O(Ca)$.
(d) Solve for the streamfunction at $O(Ca)$.

Problem 7–13. The Viscosity of a Dilute Suspension of Spherical Bubbles. Consider again the Problem 7–12. This time, assume that the bubble remains spherical. If you have already solved 7–12, the solution for part (b) provides the velocity and pressure fields for a spherical bubble. If you have not previously solved 7–12, you will need to determine the solution for a spherical bubble in an axisymmetric extensional flow, using the boundary conditions

$$\mathbf{u} \cdot \mathbf{e}_r = 0, \quad \mathbf{e}_\theta \cdot (\mathbf{e}_r \cdot \mathbf{T}) = 0.$$

Using this solution, determine the effective viscosity for a dilute "suspension" of spherical bubbles. Discuss the fact that the effective viscosity is smaller than for a dilute suspension of solid spherical particles.

Problem 7–14. The Young–Goldstein–Block Problem Revisited. Let us reconsider the Young–Goldstein–Block problem, but in this case we directly seek the solution for the case in which the temperature gradient has the value that causes the velocity of the bubble, **U**, to be exactly equal to zero. Starting with the governing Stokes' equations and boundary conditions, nondimensionalize and re-solve for this particular case. It may be useful to remember that this solution is valid when the hydrodynamic force on the bubble is exactly equal to its buoyant force.

Problem 7–15. Creeping Fows Near a Cylindrical Body. Although there is no solution for a 2D cylinder translating in Stokes' flow (Stokes' paradox), there is a solution for a cylinder rotating or in a pure straining motion. Obtain these solutions. Determine the hydrodynamic torque per unit length exerted on a cylinder of radius a rotating about its centerline at angular velocity Ω in a quiescent fluid at zero Reynolds number.

Problem 7–16. Two-Dimensional Creeping Flow. Consider the 2D problem depicted in the figure. Two belts separated by an angle 2α are in motion, such that each has a velocity of $+U$ in the radial direction. We wish to determine the velocity profile using a stream-function formulation. The velocities are given by

$$u_\theta = -\frac{\partial \psi}{\partial r}; \quad u_r = \frac{1}{r}\frac{\partial \psi}{\partial \theta}.$$

Recall that the general expression for a separable streamfunction in the cylindrical geometry is given by

$$\psi_\lambda = r^\lambda f_\lambda(\theta),$$

where, in general, we have

$$f_\lambda(\theta) = A_\lambda \sin(\lambda\theta) + B_\lambda \cos(\lambda\theta) + C_\lambda \sin[(\lambda - 2)\theta] + D_\lambda \cos[(\lambda - 2)\theta].$$

We also have the repeated root special cases:

$$f_0(\theta) = A_0 + B_0\theta + C_0 \sin(2\theta) + D_0 \cos(2\theta),$$
$$f_1(\theta) = A_1 \sin(\theta) + B_1 \cos(2\theta) + C_1\theta \sin(\theta) + D_1\theta \cos(\theta),$$
$$f_2(\theta) = A_2 + B_2\theta + C_2 \sin(2\theta) + D_2 \cos(2\theta).$$

Using the preceding relations, solve for the streamfunction and velocity distribution. *Hint*: Think about how the radial velocity and streamfunction have to depend on r and about symmetry relations for u_r and ψ in θ.

Problems

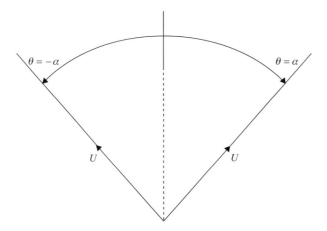

Problem 7–17. Two-Dimensional Creeping flow. Consider the two-dimensional problem depicted in this figure. Fluid drains from a trough with interior angle 2α at a rate Q/W, where W is the extension of the trough in the third dimension. We wish to determine the velocity profile by using a stream-function formulation. The velocities are given by

$$u_\theta = -\frac{\partial \psi}{\partial r}; \quad u_r = \frac{1}{r}\frac{\partial \psi}{\partial \theta}.$$

Recall that the general expression for a separable streamfunction in the cylindrical geometry is given by

$$\psi_\lambda = r^\lambda f_\lambda(\theta),$$

where, in general, we have

$$f_\lambda(\theta) = A_\lambda \sin(\lambda\theta) + B_\lambda \cos(\lambda\theta) + C_\lambda \sin[(\lambda-2)\theta] + D_\lambda \cos[(\lambda-2)\theta].$$

We also have the repeated root special cases:

$$f_0(\theta) = A_0 + B_0\theta + C_0 \sin(2\theta) + D_0 \cos(2\theta),$$
$$f_1(\theta) = A_1 \sin(\theta) + B_1 \cos(2\theta) + C_1\theta \sin(\theta) + D_1\theta \cos(\theta),$$
$$f_2(\theta) = A_2 + B_2\theta + C_2 \sin(2\theta) + D_2 \cos(2\theta).$$

Using the preceding relations, solve for the streamfunction and velocity distribution. *Hint*: Think about how the radial velocity and streamfunction have to depend on r, and about symmetry relations for u_r and ψ in θ.

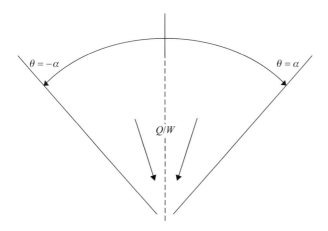

Problem 7–18. Two-Dimensional Creeping Flow

(a) Consider the 2D flow of a ribbon of fluid confined between two plates moving with opposite velocity:

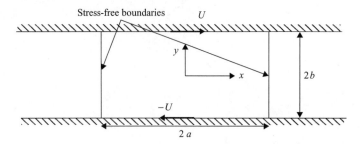

If the boundaries at $x = \pm a$ are stress free, write down the general series solution to the problem making maximum use of symmetry. Determine all of the eigenvalues and all but the last set of coefficients explicitly (e.g., leave the solution in terms of some unknown coefficients A_m). Show how you would obtain these final coefficients, but don't evaluate the integrals.

(b) Now consider the same geometry but assume that the horizontal boundaries at $y = \pm b$ are moving in the same direction. This profile is the solution to slug flow through a channel, where the slug is of length $2a$ and the channel is of width $2b$.

Plot the resulting streamlines for $a/b = 2$, keeping a reasonable number of eigenfunctions in the expansion. Note that the streamlines are closer together in regions of higher shear rate (the shear rate is inversely proportional to the streamline spacing). [*Hint*: You will find the Matlab command "contour" extremely helpful!! In plotting it, calculate the value of the streamfunction for an array of values of x and y, and put it in some array z. You can then plot the contours (constant streamfunction) by just using the contour(x, y, z) command – Matlab does all the interpolation for you!]

Problem 7–19. Two-Dimensional Creeping flow in a Vertical Slot. Consider creeping-flow in a slot of infinite length as depicted in the figure. The flow is driven by a lid moving with velocity **U**, and the walls of the slot ($x = \pm a$) may be considered to be stress free. There is no flow through the boundaries of the slot, so the walls are a streamline. Solve for the velocity profile in terms of the streamfunction. *Hint*: The velocity and the streamfunction should die away as $y \to \infty$. This should suggest what functional representation you should take for the y dependence of the streamfunction (e.g., trigonometric, hyperbolic, or exponential).

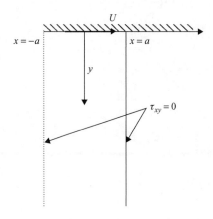

Problems

Problem 7–20. Sphere in a Parabolic Flow. Use the general eigenfunction expansion for axisymmetric creeping-flows, in spherical coordinates, to determine the velocity and pressure fields for a solid sphere of radius a that is held fixed at the central axis of symmetry of an unbounded parabolic velocity field,

$$\mathbf{u}_\infty = u_z(r)\mathbf{i}_z,$$

where

$$u_z(r) = U\left(1 - \frac{r^2}{d^2}\right).$$

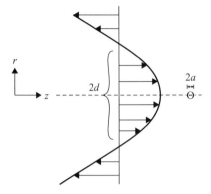

Use the result from Problem 7–11 to calculate the drag force on the sphere. You should find that

$$F_z = 6\pi\mu a U\left(1 - \frac{2a^2}{3d}\right).$$

This result should be checked against *Faxen's law*, which will be proven in Chap. 8. Faxen's law for a solid sphere holds that the force and torque that are due to an *arbitrary undisturbed flow*, $\mathbf{u}_\infty = \mathbf{u}(r)$, that satisfies the creeping-motion equations can be calculated from the following formula:

$$\mathbf{F} = 6\pi\mu a\left[\mathbf{u}_\infty(r) + \frac{a^2}{6}\nabla^2\mathbf{u}_\infty(r)\right]_{r=0},$$

and

$$\mathbf{T} = 4\pi\mu a^3\left[\nabla \wedge \mathbf{u}_\infty(r)\right]_{r=0},$$

where the subscript zero means that the indicated quantities are to be evaluated at the point in space occupied by the center of the sphere.

Problem 7–21. The Motion of a Spherical Bubble with Surfactant for $Ma \gg 1$. Consider the axisymmetric, buoyancy-driven translational motion of a gas bubble through an unbounded fluid. Assume that the capillary number is small enough that the bubble can be approximated as spherical. The Reynolds number is small enough to utilize the creeping-flow approximation. Far from the bubble, the bulk concentration of surfactant is denoted as c_∞. Calculate the solution for the steady-state motion of this bubble in the limit for asymptotically large values of the Marangoni number, through terms of $O(Ma^{-1})$. At the leading-order you should find that interface of the bubble is completely immobilized and $c = c_\infty$ throughout the bulk fluid.

Problem 7–22. The Viscosity of a Multicomponent Membrane. An interesting generalization of the Einstein calculation of the effective viscosity of a dilute suspension of spheres is to consider the same problem in two dimensions. This is relevant to the effective viscosities of some types of multicomponent membranes. Obtain the Einstein viscosity correction at small Reynolds number for a dilute suspension of cylinders of radii a whose axes are all aligned. Although there is no solution to Stokes' equations for a translating cylinder, there is a solution for a force- and torque-free cylinder in a 2D straining flow. The result is

$$\mu_{eff} = \mu(1 + 2\phi),$$

where $\phi = \pi a^2 n$ is the area fraction of cylinders. [Reference: J. F. Brady, The Einstein viscosity correction in n dimensions, *Int. J. Multiphase Flow* **10**, 113–14 (1984)].

Problem 7–23. Diffusivity of a Protein in a Porous Medium. The diffusion of proteins in a porous medium, such as gels or chromatographic columns, is important for many processes in biotechnology. These include separations of valuable proteins, mass transfer in bioreactors, and biosensing. We wish to determine how the diffusivity of the protein scales with its size and the nature of the porous medium.

(a) We can make very reasonable estimate of the diffusion by making use of the Stokes–Einstein relationship, according to which $D = kT$ (hydrodynamic mobility), and modeling the protein as a sphere in the so-called Brinkman model for the porous medium with a permeability α^{-2}, which depends on the volume fraction of solid phase. The hydrodynamic mobility is just the ratio of the drag to the velocity, F/U. Using this model, derive an estimate for the diffusion coefficient of a protein in a porous medium
  - "Recall" that the Brinkman's model corresponds to the pair of governing equations

$$\nabla \cdot \mathbf{u} = 0, \quad \mu \nabla^2 \mathbf{u} - \nabla p = \mu \alpha^2 \mathbf{u},$$

  where $\mathbf{u}$ is velocity, p is pressure, and μ is viscosity.

(b) How does the diffusion coefficient scale with volume fraction if we assume a dilute suspension of spheres for the estimate of α?

Problem 7–24. Sedimentation of a Colloidal Aggregate. Colloidal particles often aggregate because of London–van der Waals or other attractive interparticle forces unless measures are taken to stabilize them. The aggregation kinetics are such that the aggregate formed has a fractal dimension D_f, which is often less than the spatial dimension. The fractal dimension measures the amount of mass in a sphere of radius R, i.e., mass $\sim R^{D_f}$. For a fractal aggregate composed of N primary particles of radius a_p with mass m_p, estimate the sedimentation velocity of the aggregate when the Reynolds number for the motion is small. What is the appropriate Reynolds number?

Problem 7–25. Sedimentation and Growth of an Aggregate of Yeast Cells. Certain yeast cells have been genetically modified to cause them to aggregate when in solution. This has been accomplished by causing certain cells to grow flagellum and others not. The "hairy" ones tend to aggregate, whereas the "bald" ones do not. By using this preferential aggregation we can design a separations process for a mixed batch of hairy and bald cells. Let's model each cell as a sphere of, say, 1 μm suspended in water at room temperature. The cells are slightly heavier than water, $\Delta\rho/\rho = 0.001$. They are small enough that they are subject to Brownian motion, which can be characterized by a diffusivity D.

(a) Estimate the sedimentation rate of a single cell (i.e., neglecting Brownian motion).
(b) If we have a vertical column of suspension of height H such that there is no flux of cells out the top or bottom, determine the concentration profile of the cells in the column.

Problems

You may assume that the concentration of cells is so dilute that they do not interact with one another.

(c) The hairy cells will aggregate forming a "fractal" object – that is, the characteristic size R of the aggregate grows as the number of particles N to some power: $R \sim N^\alpha$, where α is not necessarily 3 as it would be for a nonfractal object. Estimate the sedimentation velocity of these fractal aggregates.

(d) The aggregation process occurs because of Brownian motion of the suspended cells. In the limit where the sticking is fast (define what fast means) estimate the rate of growth of the fractal aggregates.

Problem 7–26. An Alternative Derivation of the Solution for Stokes' Flow. In Subsection B.4, we showed that the force acting on a sphere that translates through a fluid at low Reynolds number can be expressed in terms of a "resistance tensor" $\mathbf{A}$ in the form $\mathbf{F} = \mathbf{A} \cdot \mathbf{U}$. A generalization of this idea is that the pressure and velocity fields around the sphere must also be a linear function of the vector $\mathbf{U}$, and thus expressible in terms of a "vector pressure" and a "tensor velocity" in the form

$$p = \mathbf{P}_U \cdot \mathbf{U} \quad \text{and} \quad \mathbf{u} = \mathbf{V}_U \cdot \mathbf{U}.$$

Show that the pressure and velocity functions defined by

$$\mathbf{P}_U = f(r)\mathbf{r}, \quad \mathbf{V}_U = \frac{1}{2\mu} f(r)\mathbf{rr} + g^H(r)\mathbf{rr} + h^H(r)\mathbf{I}$$

must satisfy the following "equidimensional" equations:

$$\frac{d^2 f}{dr^2} + \frac{4}{r}\frac{df}{dr} = 0, \quad \frac{d^2 g^H}{dr^2} + \frac{6}{r}\frac{dg^H}{dr} = 0, \quad \frac{d^2 h^H}{dr^2} + \frac{2}{r}\frac{dh^H}{dr} = -2g^H.$$

Hence show that the force density on the surface of the sphere is given by

$$\sigma \cdot \mathbf{n}|_{r=a} = -\frac{3}{2}\frac{\mu}{a}\mathbf{U}.$$

8

Creeping Flows – Three-Dimensional Problems

Although the solution methods of the preceding chapter are very useful and allowed us to derive a number of interesting results, they were all based on transforming the creeping-flow and continuity equations into a single higher-order differential for the streamfunction so that the classical methods of eigenfunction expansions can be used to obtain a general representation of the solution. This approach will work whenever the geometry of the boundaries and the form of any imposed flow are consistent with either a 2D or axisymmetric velocity and pressure field. We applied it to some representative 2D problems, as well as to problems involving the motions of spherical or near-spherical particles, bubbles, and drops in axisymmetric applied flows. There are, of course, other problems involving axisymmetric bodies in an axisymmetric flow – for example, any body of rotation in which the rotation axis is parallel to the axis of symmetry of the undisturbed flow. These problems can all be formulated in the same way, and, provided the geometry of the body is coincident with a coordinate surface in some orthogonal coordinate system, the same procedures can be followed.

In spite of the fact that there are actually quite a large number of axisymmetric problems, however, there are many important and apparently simple-sounding problems that are not axisymmetric. For example, we could obtain a solution for the sedimentation of any axisymmetric body in the direction parallel to its axis of symmetry, but we could not solve for the translational motion in any other direction (e.g., an ellipsoid of revolution that is oriented so that its axis of rotational symmetry is oriented perpendicular to the direction of motion). Another example is the motion of a sphere in a simple linear shear flow. Although the undisturbed flow is 2D and the body is axisymmetric, the resulting flow field is fully 3D. Clearly, it is extremely important to develop a more general solution procedure that can be applied to fully 3D creeping-flow problems.

Generally speaking, there are two approaches that have been used successfully to solve general 3D creeping-flow problems. The first is a natural extension of the procedures of the two preceding sections, in the sense that it is based on an eigenfunction expansion obtained by use of a general solution of the creeping-flow equations in terms of harmonic functions, which was originally due to Lamb.[1] A detailed description of Lamb's solution and its application to the solution of creeping-flow problems can be found in *Low Reynolds Number Hydrodynamics* by Happel and Brenner,[2] among a number of other available sources. However, in spite of the fact that Lamb's general solution does provide an effective procedure for solving creeping-flow problems, we do not pursue it here.

Instead, we describe more recently developed approaches that are more general and powerful, as well as some very useful techniques that can be used to obtain more limited information but without actually solving any detailed flow problems.

A. SOLUTIONS BY MEANS OF SUPERPOSITION OF VECTOR HARMONIC FUNCTIONS

We begin with a powerful solution method that can be applied for general 3D flows whenever the boundaries of the domain can be expressed as a coordinate surface for some orthogonal coordinate system. In this case, we can use an invariant vector representation of the velocity and pressure fields to simultaneously represent (solve) the solutions for a complete class of related problems by using so-called vector harmonic functions, rather than solving one specific problem at a time, as is necessary when we are using standard eigenfunction expansion techniques.

For this purpose, we must begin by discussing some important (preliminary) concepts.

1. Preliminary Concepts

a. Vector "equality" – pseudo-vectors The first important concept is the significance of the apparently trivial equality

$$A = B \qquad (8\text{--}1)$$

when used in a general vector analysis, where A and B may represent scalars, vectors, or tensorial quantities of any order. In particular, if (8–1) derives from some physical law or principle, it implies "equality" on a number of different levels:

(1) A and B must have the same physical dimensions.
(2) A and B must have the same tensorial rank (that is, if A is a scalar, then B must be a scalar; if A is a vector, then B must be a vector; and so on).
(3) A and B must have the same parity (that is, if A is a pseudo-vector, then B is a pseudo-vector; if A is a true vector, then B is a true vector; and so on).
(4) A and B have the same tensorial symmetry.

We shall subsequently see that the second and third requirements play a critical role in the representation of solutions to the creeping-flow equations.

A general discussion of the concept of pseudo-vectors (in general, pseudo-scalars, pseudo-tensors, and so on) can be found in many books on vector analysis. However, generally speaking, a pseudo-vector differs from a true vector because it changes sign on inversion of the coordinate axes (alternatively, we may think of inversion as changing from a right- to a left-handed coordinate reference frame). Thus, if we consider an orthogonal transformation, specified by the transformation matrix $\mathbf{L}$, a pseudo-vector transforms according to the rule

$$\mathbf{B} = (\det \mathbf{L})\mathbf{L} \cdot \overline{\mathbf{B}},$$

whereas a true vector transforms according to

$$\mathbf{A} = \mathbf{L} \cdot \overline{\mathbf{A}},$$

as we have already seen in Chap. 2. Similarly, a second-order pseudo-tensor transforms according to

$$\mathbf{D} = (\det \mathbf{L})\mathbf{L} \cdot \overline{\mathbf{D}} \cdot \mathbf{L}^{\mathrm{T}},$$

whereas a regular second-order tensor transforms according to

$$\mathbf{C} = \mathbf{L} \cdot \overline{\mathbf{C}} \cdot \mathbf{L}^{\mathrm{T}}.$$

Because $\det \mathbf{L}$ can have only two values for an orthogonal transformation, $+1$ or -1, it follows that the difference between pseudo-vectors or pseudo-tensors and normal vectors and tensors is, at most, only in the sign, and even this difference disappears if the transformation is a rotation rather than an inversion.

Common examples of pseudo-vectors that will be relevant later include the angular velocity vector $\boldsymbol{\Omega}$, the torque $\mathbf{T}$, the vorticity vector $\boldsymbol{\omega}$ (or the curl of any true vector), and the cross product of two vectors. The inner scaler product of a vector and a pseudo-tensor or a pseudo-vector and a regular tensor will both produce a pseudo-vector. It will also be useful to extend the notion of a pseudo-vector to scalars that are formed as the product of a vector and a pseudo-vector. The third-order, alternating tensor $\boldsymbol{\varepsilon}$ is a pseudo-tensor of third order as may be verified by reviewing its definition

$$\varepsilon_{ijk} \equiv \begin{cases} +1 & \text{if } (ijk) \text{ is an even permutation of } (123) \\ -1 & \text{if } (ijk) \text{ is an odd permutation of } (123). \\ 0 & \text{otherwise (some or all subscripts are equal)} \end{cases}$$

A second-order pseudo-tensor can be formed as the inner product of $\boldsymbol{\varepsilon}$ and any regular vector including the general position vector $\mathbf{x}$; namely, $\boldsymbol{\varepsilon} \cdot \mathbf{x}$, but no pseudo-vector can be formed by means of any standard vector operation that is linear in $\mathbf{x}$. We may also note that quantities that involve "products" of two (or any even number of) pseudo-quantities will generally become true scalars, vectors, or tensors. For example, if $\mathbf{v} = \mathbf{V} \cdot \mathbf{A}$, where $\mathbf{V}$ is a pseudo-tensor and $\mathbf{A}$ is a pseudo-vector, then $\mathbf{v}$ is a true vector. Another true vector is the curl of a pseudo-vector, $\nabla \wedge \boldsymbol{\omega}$, say, or the vector product of a true vector and a pseudo-vector $\mathbf{v} \wedge \boldsymbol{\omega}$. Finally, we note that, whereas the vector vorticity $\boldsymbol{\omega} \equiv \nabla \wedge \mathbf{u}$ is clearly a pseudo-vector, as already stated, the vorticity tensor defined as $\boldsymbol{\Omega} \equiv [(1/2)(\nabla \mathbf{u} - \nabla \mathbf{u}^T)]$ is a regular tensor. One way to see this is to note that $\boldsymbol{\Omega} = -[(1/2)(\boldsymbol{\varepsilon} \cdot \boldsymbol{\omega})]$. Because $\boldsymbol{\varepsilon}$ is a third-order pseudo-tensor and $\boldsymbol{\omega}$ is pseudo-vector, their inner product is a regular second-order tensor. A simple test to determine whether a scalar, vector, or tensor formed as a product of vectors and tensors is a true- or pseudo-quantity is to see whether its components change sign upon transformation from a right- to left-handed coordinate system. Thus, for example, $\mathbf{v} \wedge \boldsymbol{\omega}$ is a true vector because $\boldsymbol{\omega}$ and the vector product both change sign upon coordinate transformation so that the product $\mathbf{v} \wedge \boldsymbol{\omega}$ is invariant.

b. Representation theorem for solution of the creeping-flow equations Another necessary preliminary involves a few words about the solution of the creeping-flow equations (7–6). However, it is convenient for some problems involving bubbles or drops (to be considered later) to formulate the equations in terms of the disturbance pressure $p' = p - p_\infty$, where p_∞ is a constant, arbitrary reference pressure. One important fact is that the disturbance pressure (and thus also the pressure itself) is a harmonic function. To see this, we can simply take the divergence of the creeping-flow equation, (6–6a), expressed in terms of p'. Since $\mathbf{u}$ is a solenoidal vector function, according to the continuity equation, it follows from

$$\nabla \cdot (\mu \nabla^2 \mathbf{u} - \nabla p') = 0,$$

that

$$\nabla^2 p' = 0. \tag{8-2}$$

A second, even more important, fact is that a general solution of the creeping- (dimensional) flow equations is

$$\boxed{\mathbf{u} = \frac{\mathbf{x}}{2\mu} p' + \mathbf{u}^{(H)},} \tag{8-3}$$

A. Solutions by Means of Superposition of Vector Harmonic Functions

where $\mathbf{u}^{(H)}$ is also a harmonic function, that is,
$$\nabla^2 \mathbf{u}^{(H)} = 0.$$

We can demonstrate this easily by substituting (8–3) into the creeping-flow equations. To satisfy the continuity equation, the harmonic function $\mathbf{u}^{(H)}$ is required to satisfy the condition

$$\nabla \cdot \mathbf{u}^{(H)} = -\frac{1}{2\mu}(3p' + \mathbf{x} \cdot \nabla p'). \tag{8-4}$$

Although it is not essential, we shall see that it is convenient in the developments that follow to maintain the general solution (8–3) in dimensional form.

c. Vector harmonic functions Finally, to develop a general representation procedure for solutions of the creeping-flow equations, we require a brief introduction to harmonic functions, expressed in a coordinate-independent vector form by using the general position vector $\mathbf{x}$. It is convenient to group the harmonic functions into two categories: *decaying* harmonics, whose magnitude decreases with increase in $r \equiv |\mathbf{x}|$, and *growing* harmonics, whose magnitudes increase with r. The decaying harmonics are conveniently represented by means of higher-order derivatives of $1/r$, which is the fundamental solution of Laplace's equation. The first decaying harmonic beyond $1/r$ is thus

$$\nabla\left(\frac{1}{r}\right) \to \frac{x_i}{r^3},$$

the second is

$$\nabla\left(\frac{\mathbf{x}}{r^3}\right) \to \left(\frac{x_i x_j}{r^5} - \frac{\delta_{ij}}{3r^3}\right),$$

and so forth. The first several *decaying* harmonics are thus

$$\frac{1}{r}, \frac{x_i}{r^3}, \left(\frac{x_i x_j}{r^5} - \frac{\delta_{ij}}{3r^3}\right), \left(\frac{x_i x_j x_k}{r^7} - \frac{x_i \delta_{jk} + x_j \delta_{ki} + x_k \delta_{ij}}{5r^5}\right),$$
$$\left(\frac{x_i x_j x_k x_l}{r^9} - \frac{\delta_{ij} x_k x_l + \delta_{ik} x_j x_l + \delta_{il} x_j x_k + \delta_{jk} x_i x_l + \delta_{jl} x_i x_k + \delta_{kl} x_i x_j}{7r^7} \right. \tag{8-5}$$
$$\left. + \frac{\delta_{ij}\delta_{kl} + \delta_{ik}\delta_{jl} + \delta_{il}\delta_{jk}}{35r^5}\right), \text{ etc.,}$$

or

$$\phi_{-(n+1)} \equiv \frac{(-1)^n}{1 \cdot 3 \cdot 5 \cdots (2n-1)} \frac{\partial^{n-1}\left(\frac{1}{r}\right)}{\partial x_i \partial x_j \partial x_k \cdots}, \quad n = 0, 1, 2, \ldots, .$$

The growing harmonics, on the other hand, are just

$$r^{2n+1}\phi_{-(n+1)}$$

or

$$1, x_i, \left(x_i x_j - \frac{r^2}{3}\delta_{ij}\right), \left[x_i x_j x_k - \frac{r^2}{5}(x_i \delta_{jk} + x_j \delta_{ki} + x_k \delta_{ij})\right],$$
$$\left[x_i x_j x_k x_l - \frac{r^2}{7}(\delta_{ij} x_k x_l + \delta_{ik} x_j x_l + \delta_{il} x_j x_k + \delta_{jk} x_i x_l + \delta_{jl} x_i x_k + \delta_{kl} x_i x_j) \right. \tag{8-6}$$
$$\left. + \frac{r^4}{35}(\delta_{ij}\delta_{kl} + \delta_{ik}\delta_{jl} + \delta_{il}\delta_{jk})\right], \text{ etc.}$$

It is important to note that these *vector harmonic functions* involve only the general position vector $\mathbf{x}$ and its magnitude $r \equiv |\mathbf{x}|$ and thus can be represented in any coordinate system that is convenient for a particular problem. Another property of these tensorial harmonic functions is that they are *irreducible*. This means that the harmonics of second and higher order are symmetric with respect to the interchange of any two indices, as well as being traceless with respect to the contraction of any two indices. We shall see that this property makes a difference in the formulation of the problem.

Now, let us see how the preliminary concepts in this section can be put together to achieve a general representation procedure for the solution of general classes of creeping-flow problems. We begin our discussion with the simplest case of spherical particles.

2. The Rotating Sphere in a Quiescent Fluid

We begin with some simple problems to illustrate the basic ideas. The simplest is the flow induced by rotation of a sphere of radius a with angular velocity Ω in an unbounded, quiescent fluid. Although we could formulate this problem in dimensionless terms by using Ωa as a characteristic velocity (where $\Omega \equiv |\Omega|$) and the sphere radius as a characteristic length scale, it is more convenient because the solution representation was carried out in terms of dimensional variables to simply solve it in dimensional form.

The starting point is the pressure p'. The pressure is a harmonic function and a true scalar. Further, it must be a decaying function of distance from the origin because the flow decays at infinity. Finally, a key observation in the present problem is that the pressure must be a linear function of Ω. This is because the governing equations are linear and Ω appears linearly in the boundary condition. Hence both p' and $\mathbf{u}$ must be linear functions of Ω.

Now, the solution for p' must be "constructed" solely from Ω and the general position vector $\mathbf{x}$ in the form of the vector harmonic functions. Now, there is only a single scalar function that can be constructed from Ω and the decaying harmonics, (8–5), that is linear in Ω, namely,

$$c_1 \, \Omega \cdot \frac{\mathbf{x}}{r^3}, \tag{8-7}$$

where c_1 is an arbitrary constant. The inner product of Ω with the higher-order harmonics produces either a vector or a tensor, whereas all vector products of Ω and the vector harmonic functions produce at least a vector. Thus the only candidate for p' is (8–7). However, p' cannot be of this form because p' is a true scalar, while $\Omega \cdot \mathbf{x}$ is a pseudo-scalar. To see this, we simply note that Ω is a pseudo-vector, so that the product $\Omega \cdot \mathbf{x}$ changes sign upon inversion from a right- to a left-handed system. Thus the only possibility consistent with all of the conditions on p' is $c_1 = 0$, so that

$$p' \equiv 0. \tag{8-8}$$

It follows, therefore, from (8–3) that $\mathbf{u} = \mathbf{u}^{(H)}$ must be a decaying harmonic function, linear in Ω and a true vector. The only combination of Ω and the vector harmonic functions that satisfies these conditions is

$$\mathbf{u} = c_2 \Omega \wedge \left(\frac{\mathbf{x}}{r^3}\right). \tag{8-9}$$

Again, c_2 is an arbitrary constant.

To determine c_2, and thus complete our solution of the rotating sphere problem, we apply the boundary condition

$$\mathbf{u} = \Omega \wedge (a\mathbf{e}_x) \quad \text{at } r = a. \tag{8-10}$$

Here, we denote the position vector $\mathbf{x}$ at the sphere surface as $a\mathbf{e}_x$ where $\mathbf{e}_x$ is a unit vector in the direction of $\mathbf{x}$ and $r = a$ is the sphere radius. Now, evaluating (8–9) at the sphere surface, we find

$$\mathbf{u} = c_2 \Omega \wedge \left(\frac{a\mathbf{e}_x}{a^3}\right), \tag{8-11}$$

A. Solutions by Means of Superposition of Vector Harmonic Functions

and we see by comparing (8–10) and (8–11) that

$$c_2 = a^3.$$

Hence, the final solution for the flow produced by a rotating sphere is

$$\boxed{\mathbf{u} = \boldsymbol{\Omega} \wedge \mathbf{x}\left(\frac{a^3}{r^3}\right).} \tag{8–12}$$

3. Uniform Flow past a Sphere

A second straightforward example, solved previously by other means, is Stokes' original problem of uniform flow past a stationary sphere. To apply the methods of the preceding subsection to this problem, it is convenient to transform to the *disturbance flow* problem,

$$\mathbf{u}' = \mathbf{u} - \mathbf{U}, \tag{8–13}$$

where $\mathbf{U}$ is the undisturbed, uniform fluid velocity and $\mathbf{u}'$ satisfies

$$\boxed{\nabla^2 \mathbf{u}' - \nabla p' = 0, \quad \nabla \cdot \mathbf{u}' = 0,} \tag{8–14}$$

with

$$\boxed{\begin{aligned}\mathbf{u}' &= -\mathbf{U} \text{ on the sphere } r = a, \\ \mathbf{u}' &\to 0 \text{ at } \infty.\end{aligned}} \tag{8–15}$$

Thus $\mathbf{u}'$ and p' are decaying functions of r.

To construct a solution for $(\mathbf{u}', p')$, we again begin with the pressure p'. In this case, p' is a decaying harmonic function, linear in $\mathbf{U}$ and a true scalar, that must be constructed solely from $\mathbf{U}$ and the vector harmonic functions. Examination of the decaying harmonics, (8–5), shows that there is a single combination of $\mathbf{U}$ and the vector harmonics that satisfy those conditions, namely,

$$\boxed{p' = c\frac{\mathbf{U} \cdot \mathbf{x}}{r^3}.} \tag{8–16}$$

Again, c is an arbitrary constant, and in this case the quantity on the right-hand side is a true scalar and thus has the same parity as p'.

Now, with the form (8–16) for p', we see from (8–3) that

$$\mathbf{u}' = \frac{\mathbf{x}}{2\mu}\left(c\frac{\mathbf{U} \cdot \mathbf{x}}{r^3}\right) + \mathbf{u}^{(H)}. \tag{8–17}$$

The function $\mathbf{u}^{(H)}$ is harmonic, decaying, linear in $\mathbf{U}$ and a true vector. Again, examining the decaying vector harmonic functions (8–5), we find that there are only two products of $\mathbf{U}$ and the decaying harmonics that are true vectors. Thus the most general form for $\mathbf{u}^{(H)}$ is

$$\mathbf{u}^{(H)} = \alpha\left(\frac{\mathbf{xx}}{r^5} - \frac{\mathbf{I}}{3r^3}\right) \cdot \mathbf{U} + \beta\frac{\mathbf{U}}{r}, \tag{8–18}$$

where α and β are arbitrary constants. The continuity condition imposes one constraint on the constants c, α and β. Substituting (8–18) into (8–17) and calculating $\nabla \cdot \mathbf{u}'$ [or using the condition (8–4) directly], we find that

$$\beta = \frac{c}{2\mu}. \tag{8–19}$$

Creeping Flows – Three-Dimensional Problems

Thus the general form for $\mathbf{u}'$ that satisfies continuity is

$$\mathbf{u}' = \frac{\mathbf{x}(\mathbf{U}\cdot\mathbf{x})}{r^3}\left(\frac{c}{2\mu}+\frac{\alpha}{r^2}\right)+\frac{\mathbf{U}}{r}\left(\frac{c}{2\mu}-\frac{\alpha}{3r^2}\right). \qquad (8\text{--}20)$$

The boundary condition in this case requires that

$$\mathbf{u}' = -\mathbf{U} \quad \text{at } r = a. \qquad (8\text{--}21)$$

Hence

$$-\mathbf{U} = \frac{\mathbf{e}_x(\mathbf{U}\cdot\mathbf{e}_x)}{a}\left(\frac{c}{2\mu}+\frac{\alpha}{a^2}\right)+\frac{\mathbf{U}}{a}\left(\frac{c}{2\mu}-\frac{\alpha}{3a^2}\right). \qquad (8\text{--}22)$$

The two types of terms appearing in this relation are $\mathbf{U}\cdot\text{const}$ and $\mathbf{e}_x(\mathbf{U}\cdot\mathbf{e}_x)\cdot\text{const}$. Clearly, these terms exhibit a different dependence on position on the sphere surface. Thus, to satisfy (8–22), we must require that the coefficients c and α satisfy

$$1+\frac{c}{2a\mu}-\frac{\alpha}{3a^3}=0,$$

$$\frac{c}{2a\mu}+\frac{\alpha}{a^3}=0.$$

Hence,

$$\alpha = \frac{3a^3}{4}, \quad c = -\frac{3a\mu}{2}, \qquad (8\text{--}23)$$

and thus substituting back into (8–16) and (8–20), we have

$$\mathbf{u} = \mathbf{U}-\left(\frac{3}{4}\frac{a}{r}+\frac{1}{4}\frac{a^3}{r^3}\right)\mathbf{U}-\left(\frac{3a}{4r^3}-\frac{3a^3}{4r^5}\right)\mathbf{x}(\mathbf{U}\cdot\mathbf{x}), \qquad (8\text{--}24)$$

$$p = -\frac{3a\mu}{2}\frac{\mathbf{U}\cdot\mathbf{x}}{r^3}. \qquad (8\text{--}25)$$

This is nothing more than Stokes' solution, obtained earlier in Chap. 7 by other (more cumbersome) techniques.

B. A SPHERE IN A GENERAL LINEAR FLOW

In the preceding two subsections, we demonstrated the construction of solutions of the creeping-flow equations by means of the superposition of vector harmonic functions. However, in both cases, the problem could have been solved by the techniques described earlier, and though the present methodology requires less effort, its very important fundamental advantages have not really been exposed. A third problem from Chap. 7 was the motion in the vicinity of a sphere with the flow at large distances being an axisymmetric extensional flow. This is one specific example of the motion of a small, spherical particle in a general flow, which we may denote as $\mathbf{u}_\infty(\mathbf{x})$. If the particle is small relative to the length-scale characterizing u_∞, we can approximate the flow field in the neighborhood of the particle, relative to a point fixed at the center of the sphere, by means of a Taylor series approximation:

$$\mathbf{u}_\infty(\mathbf{x}) = \mathbf{U}_\infty + \boldsymbol{\Gamma}\cdot\mathbf{x}+\mathbf{K}:\mathbf{xx}+\cdots+. \qquad (8\text{--}26a)$$

B. A Sphere in a General Linear Flow

Here, $\mathbf{U}_\infty$ is the translational velocity of the undisturbed flow evaluated at the particle center, $\boldsymbol{\Gamma}$ is the (second-order) velocity gradient tensor evaluated at that point $(\nabla \mathbf{u}_\infty)$, and $\mathbf{K}$ is the third-order tensor representing the second derivative $\frac{1}{2}\nabla\nabla\mathbf{u}_\infty$. We assume that the flow in the neighborhood of the sphere is a creeping flow, and thus that (8–26a) satisfies the creeping-flow and continuity equations. This means that $\boldsymbol{\Gamma}$ is traceless. The third order tensor $\mathbf{K}$ is clearly symmetric with respect to its last two indices, i.e., $K_{ijk} = K_{ikj}$, and it must also be traceless with respect to contraction of its first two indices, $K_{mmk} = 0$, in order to satisfy $\nabla \cdot \mathbf{u}_\infty(\mathbf{x}) = 0$. Finally, for (8–26a) to satisfy the creeping-flow equations, there must be a corresponding Taylor series for the undisturbed pressure $p_\infty(\mathbf{x})$. The first nonzero term in this expansion corresponds to the quadratic term in (8–26a),

$$p_\infty(\mathbf{x}) = 2\mu \mathbf{x} \cdot (\mathbf{K} : \mathbf{I}) + \cdots + . \tag{8-26b}$$

Now, because the creeping-flow equations are linear, we can consider a sphere in each of the component flows of (8–26) separately, and then add the solutions together to solve the full problem. The first term, namely a sphere in a fluid that is translating relative to it with a uniform velocity was just solved in the preceding section. If we have a freely suspended sphere that sediments through the suspending fluid, this will be the first approximation for the flow in the neighborhood of the sphere. If the sphere is neutrally buoyant, on the other hand, it will translate along with the fluid and $\mathbf{U}_\infty$ is zero, at least if the fluid motion far from the sphere is uniform. A special case of a sphere in an axisymmetric flow that varies *linearly* with position [i.e., a special case of the second term in (8–26a)] was considered in Chap. 7. One characteristic of the solution was that there was no contribution to the hydrodynamic force on the sphere. We shall see that this is a characteristic of the solution for a sphere at the origin of any linear flow. Hence, if the sphere is neutrally buoyant, it will still translate with the undisturbed uniform velocity of the fluid, and again $\mathbf{U}_\infty = 0$. If we consider the quadratic terms in (8–26a), however, there will generally be a nonzero contribution to the hydrodynamic force on the sphere. Then a neutrally buoyant sphere will translate with a velocity such that the net hydrodynamic force on the sphere is zero. Then $\mathbf{U}_\infty$ will be nonzero, even for a neutrally buoyant particle.

To proceed, we again formulate the problem in terms of the disturbance velocity and pressure fields $(\mathbf{u}', p')$, namely,

$$\begin{aligned} \mathbf{u}' &= \mathbf{u} - \mathbf{u}_\infty(\mathbf{x}) = \mathbf{u} - (\mathbf{U}_\infty + \boldsymbol{\Gamma} \cdot \mathbf{x} + \mathbf{K} : \mathbf{xx} + \cdots +), \\ p' &= p - p_\infty(\mathbf{x}) = p - (2\mu \mathbf{x} \cdot (\mathbf{K} : \mathbf{I}) + \cdots +), \end{aligned} \tag{8-27}$$

which satisfies (8–14), but with boundary conditions

$$\boxed{\begin{aligned} \mathbf{u}' &= -\mathbf{U}_\infty - \boldsymbol{\Gamma} \cdot (a\mathbf{e}_x) - \mathbf{K} : (a^2 \mathbf{e}_x \mathbf{e}_x) + \cdots + \quad \text{on} \quad r = a, \\ \mathbf{u}' &\to 0 \quad \text{as } r \to \infty. \end{aligned}} \tag{8-28}$$

There is one new factor that needs to be considered for this problem. The vector harmonic functions are *irreducible* tensors. Hence, for any sum of such terms to satisfy the boundary conditions, (8–28), we should also express the tensors, $\boldsymbol{\Gamma}$ and $\mathbf{K}$, in terms of *irreducible* tensors. Now, we have seen previously that $\boldsymbol{\Gamma}$ can be expressed in terms of the rate-of-strain and vorticity tensors:

$$\boldsymbol{\Gamma} = \mathbf{E} + \boldsymbol{\Omega}.$$

Although the rate-of-strain tensor is *irreducible*, the antisymmetric vorticity tensor is not. A well-known result from tensor calculus[3] is that an arbitrary second-order tensor $\mathbf{M}$ can be written as the sum of three terms, each of which involves only irreducible components of **M**,

$$M_{ij} = H_{ij} + \varepsilon_{ijk} N_k + S \delta_{ij}. \tag{8-29}$$

Here $H_{ij} = \frac{1}{2}(M_{ij} + M_{ji}) - \frac{1}{3}(M_{kk})\delta_{ij}$ is symmetric and traceless (hence irreducible), whereas N_k and S are a vector and scalar, respectively, and thus are automatically irreducible. The vector is related to **M** in the form

$$N_k = \varepsilon_{kmn} M_{nm}^a \quad \text{and} \quad M_{nm}^a = \frac{1}{2}(M_{nm} - M_{mn}), \tag{8-30a}$$

whereas

$$S = \frac{1}{3}(M_{kk})\delta_{ij}. \tag{8-30b}$$

We may note also that $M_{mn}^a = -\frac{1}{2}\varepsilon_{mnl} N_l$. In our system, the tensor **M** is the velocity gradient tensor, **Γ**. Hence, tr **M** = tr **Γ** = 0, so that $S = 0$, and **H** is just the rate-of-strain tensor **E**. The vector **N** is the vorticity vector **ω**. Hence, expressing **Γ** in terms of irreducible tensors reduces to writing it in terms of **E** and **ω**:

$$\mathbf{\Gamma} = \mathbf{E} + \boldsymbol{\varepsilon} \cdot \boldsymbol{\omega} \quad \text{or} \quad \Gamma_{ij} = E_{ij} + \varepsilon_{ijk}\omega_k, \tag{8-31}$$

so that

$$\mathbf{\Gamma} \cdot \mathbf{x} = \mathbf{E} \cdot \mathbf{x} + \boldsymbol{\omega} \wedge \mathbf{x}. \tag{8-32}$$

The rate-of-strain tensor is symmetric and traceless (hence irreducible), whereas the vorticity is a vector and is thus automatically irreducible.

The task of expressing **K** in terms of *irreducible* tensors is somewhat more involved. The general problem of reducing tensors of any order into irreducible components is discussed by Coope et al.,[3] which was referred to earlier. These authors show that a general third-order tensor can always be expressed in the form

$$Q_{ijk} = \gamma_{ijk} + \varepsilon_{ijp} D_{pk}^1 + \varepsilon_{ikp} D_{pj}^2 + \varepsilon_{kjp} D_{pi}^3 + \delta_{ij} v_k^1 + \delta_{ik} v_j^2 + \delta_{kj} v_i^3 + \varepsilon_{ijk} S. \tag{8-33}$$

Here, **γ** is the unique symmetric and traceless (irreducible) part of **Q**, whereas $\mathbf{D}^{(i)}$ are symmetric and traceless second-order (irreducible) tensors, and $\mathbf{v}^{(i)}$ and S are vectors and a scalar. It may be noted that, if we write a general third-order tensor in the form (8–33), there is no loss of generality in assuming that the second-order tensor components are symmetric. The antisymmetric part of any second-order tensor can always be represented by a vector. For example, if $\mathbf{D} = \mathbf{D}^s + \mathbf{D}^a$ then the antisymmetric part can always be written as $\mathbf{D}^a = \boldsymbol{\varepsilon} \cdot \mathbf{d}$ where $\mathbf{d} = -\frac{1}{2}\boldsymbol{\varepsilon} : \mathbf{D}^a$ and included in the vector terms of (8–33).

The specific case of the third-order tensor $\mathbf{K} \equiv \frac{1}{2}\nabla\nabla\mathbf{u}_\infty$ was considered by Nadim and Stone.[4] In index notation, the irreducible description of **K** is

$$K_{ijk} = \gamma_{ijk} - \frac{1}{6}(\varepsilon_{ijl}\delta_{mk} + \varepsilon_{ikl}\delta_{mj})\theta_{lm} + \frac{1}{10}(4\delta_{ij}\delta_{kl} - \delta_{kj}\delta_{il} - \delta_{ki}\delta_{jl})\tau_l. \tag{8-34}$$

Here **γ** is the symmetric and traceless (i.e., irreducible) part of **K**,

$$\gamma_{ijk} \equiv \frac{1}{3}(K_{ijk} + K_{kij} + K_{jki}) - \frac{1}{15}(\delta_{ij}K_{kmm} + \delta_{ik}K_{jmm} + \delta_{jk}K_{imm}), \tag{8-35a}$$

θ is a symmetric and traceless second-order tensor,

$$\theta_{lm} \equiv \varepsilon_{pql} K_{qpm} + \varepsilon_{pqm} K_{qpl}, \tag{8-35b}$$

and **τ** is a vector (automatically irreducible)

$$\tau_l \equiv K_{lmm}. \tag{8-35c}$$

B. A Sphere in a General Linear Flow

With this decomposition of **K** into irreducible form, the solution for a spherical particle in the quadratic flow part of (8–26) can be represented as the sum of the terms that involve γ, θ, and τ combined with the vector harmonic functions (8–5) to produce a true scalar for p' and a true vector for $\mathbf{u}'$. This problem is posed in the problems section at the end of the chapter.

In this section, we consider the problem of a nonrotating sphere in a general linear flow of an unbounded fluid, namely,

$$\mathbf{u}_\infty = \mathbf{E} \cdot \mathbf{x} + \boldsymbol{\omega} \wedge \mathbf{x}. \tag{8–36}$$

In other words, we consider the first departure in (8-26) from a uniform flow. As previously noted, we solved one example of this type in the preceding chapter, namely axisymmetric pure straining flow, where

$$\boldsymbol{\omega} \equiv 0 \quad \text{and} \quad E_{ij} = \pm \begin{cases} E & \text{if } i = j = 1, 2 \\ -2E & \text{if } i = j = 3 \\ 0 & \text{if } i \neq j \end{cases}.$$

However, this case is exceptional in the sense that for all other choices of **E** and $\boldsymbol{\omega}$, the velocity and pressure fields will be fully 3D and the eigenfunction expansion of the preceding chapter cannot be used. In addition, instead of focusing on a specific example of the general class of undisturbed flows (8–36), as would be required by any of the methods of solution discussed in the previous chapter, we show that the present method allows the solution to be completed for *arbitrary* **E** and $\boldsymbol{\omega}$. Hence we *simultaneously* determine the solution for all possible flows of the general class (8–36).

In the present case, the boundary conditions (8–28) at $r = a$ become

$$\mathbf{u}' = -\mathbf{E} \cdot (a\mathbf{e}_x) - \frac{1}{2}[\boldsymbol{\omega} \wedge (a\mathbf{e}_x)] \quad \text{at } r = a. \tag{8–37}$$

In view of the linearity of the Stokes' equations, we can construct a solution to this problem as the sum of two solutions, one in which the external flow is the purely extensional part of (8–36) characterized by **E**, and the other in which the external flow is just the rotational motion associated with the vorticity $\boldsymbol{\omega}$. The solution for the disturbance velocity for the latter problem is identical to the solution for the rotating sphere in the. Here we seek a solution for the extensional component of (8–36).

To construct this solution, we begin, as in the previous examples, with the pressure p'. In this case, p' is a decaying harmonic, which is linear in **E** and a true scalar. The only true scalar that can be formed by **E** and the decaying harmonic functions (8–5) is

$$p' = c_1 \mathbf{E} : \left(\frac{\mathbf{xx}}{r^5} - \frac{\mathbf{I}}{3r^3} \right). \tag{8–38}$$

Because tr **E** = 0,

$$\boxed{p' = c_1 \frac{\mathbf{x} \cdot \mathbf{E} \cdot \mathbf{x}}{r^5}.} \tag{8–39}$$

The constant c_1 will be determined shortly by means of the boundary condition (8–37).

A general form for the homogeneous contribution $\mathbf{u}^{(H)}$ to the velocity can be obtained in a similar manner. In this case, $\mathbf{u}^{(H)}$ must consist of a sum of decaying harmonics, all of which are linear in **E** and are true vectors since $\mathbf{u}^{(H)}$ is a true vector. The most general form for $\mathbf{u}^{(H)}$, satisfying these constraints, is

$$\mathbf{u}^{(H)} = c_2 \mathbf{E} \cdot \frac{\mathbf{x}}{r^3} + c_3 \mathbf{E} : \left(\frac{\mathbf{xxx}}{r^7} - \frac{2\mathbf{xI}}{5r^5} \right). \tag{8–40}$$

The constants c_2 and c_3 are again arbitrary, apart from the continuity constraint (8–4). To obtain a complete form for the disturbance velocity corresponding to (8–36), we combine (8–39) and (8–40) in the form (8–3), and add the solution for the rotational component of the undisturbed flow:

$$\mathbf{u}' = c_2 \mathbf{E} + \mathbf{x}(\mathbf{x} \cdot \mathbf{E} \cdot \mathbf{x}) \left(\frac{c_1}{2r^5 \mu} + \frac{c_3}{r^7} \right) + \mathbf{E} \cdot \mathbf{x} \left(-\frac{2c_3}{5r^5} \right) + c_4 \boldsymbol{\omega} \wedge \frac{\mathbf{x}}{r^3}. \tag{8-41}$$

In this case, the continuity condition (8–4) yields $c_2 = 0$.

We obtain the remaining constants c_1, c_3, and c_4 by applying the boundary condition (8–37). The result is

$$-\mathbf{E} \cdot (a\mathbf{e}_x) - \frac{1}{2} \boldsymbol{\omega} \wedge (a\mathbf{e}_x) = \mathbf{e}_x (\mathbf{e}_x \cdot \mathbf{E} \cdot \mathbf{e}_x) \left(\frac{c_1}{2a^2 \mu} + \frac{c_3}{a^4} \right)$$

$$+ \mathbf{E} \cdot \mathbf{e}_x \left(-\frac{2c_3}{5a^4} \right) + c_4 \left(\frac{\boldsymbol{\omega} \wedge \mathbf{e}_x}{a^2} \right), \tag{8-42}$$

and thus

$$c_3 = \frac{5a^5}{2}, \quad c_4 = -\frac{a^3}{2}, \quad \text{and} \quad c_1 = -5a^3 \mu. \tag{8-43}$$

Substituting these values as well as $c_2 = 0$ into (8–39) and (8–41), and adding the undisturbed flow, we therefore obtain

$$\mathbf{u} = \mathbf{E} \cdot \mathbf{x} \left(1 - \frac{a^5}{r^5} \right) + \frac{1}{2} (\boldsymbol{\omega} \wedge \mathbf{x}) \left(1 - \frac{a^3}{r^3} \right) - \mathbf{x}(\mathbf{x} \cdot \mathbf{E} \cdot \mathbf{x}) \left(\frac{5}{2} \frac{a^3}{r^5} - \frac{5a^5}{2r^7} \right). \tag{8-44}$$

The corresponding pressure distribution is

$$p = -5\mu (\mathbf{x} \cdot \mathbf{E} \cdot \mathbf{x}) \left(\frac{a^3}{r^5} \right) + p_\infty. \tag{8-45}$$

The expressions (8–44) and (8–45) represent a complete, exact solution of the creeping-flow equations for a completely *arbitrary* linear flow. Among the linear flows of special interest are axisymmetric pure strain, which was solved by means of the eigenfunction expansion for axisymmetric flows in the previous chapter, and simple shear flow, for which

$$\mathbf{u} = Gx_2 \mathbf{i}_1, \text{ that is, } \mathbf{E} = \begin{pmatrix} 0 & G/2 & 0 \\ G/2 & 0 & 0 \\ 0 & 0 & 0 \end{pmatrix} \text{ and } \boldsymbol{\omega} = -G\mathbf{i}_3. \tag{8-46}$$

It is of interest, in this latter case, to express the general solution (8–44) in terms of components in a spherical coordinate system. For this purpose, we note that the general position vector can be expressed in the form

$$\mathbf{x} = x_1 \mathbf{e}_1 + x_2 \mathbf{e}_2 + x_3 \mathbf{e}_3, \tag{8-47}$$

with

$$\begin{aligned} \mathbf{e}_1 &= \mathbf{i}_r \sin\theta \cos\phi + \mathbf{i}_\theta \cos\theta \cos\phi - \mathbf{i}_\phi \sin\phi, \\ \mathbf{e}_2 &= \mathbf{i}_r \sin\theta \sin\phi + \mathbf{i}_\theta \cos\theta \sin\phi + \mathbf{i}_\phi \cos\phi, \\ \mathbf{e}_3 &= \mathbf{i}_r \cos\theta - \mathbf{i}_\theta \sin\theta, \end{aligned} \tag{8-48}$$

B. A Sphere in a General Linear Flow

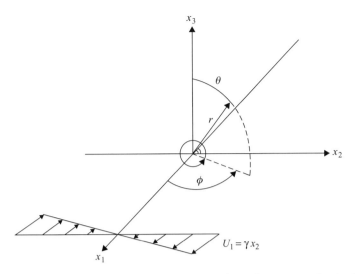

Figure 8–1. The undisturbed flow and spherical coordinate system for simple shear flow, $u = \gamma y$, in the vicinity of a sphere.

and

$$x_1 = r \sin\theta \cos\phi,$$
$$x_2 = r \sin\theta \sin\phi, \quad (8\text{--}49)$$
$$x_3 = r \cos\theta.$$

Note that the spherical coordinates (r, θ, ϕ) are defined, as sketched in Fig. 8–1, with the polar angle θ measured from the x_3 axis and the azimuthal angle ϕ measured around the x_3 axis with $\phi = 0$ on the $x_1 x_3$ plane. For the specific case of simple shear flow, given by (8–46), the general solution (8–44) takes the form

$$\boxed{\begin{aligned}\mathbf{u} &= \frac{G}{2}(x_2\mathbf{e}_1 + x_1\mathbf{e}_2)\left(1 - \frac{a^5}{r^5}\right) - \left(\frac{G}{2}x_1\mathbf{e}_2 - \frac{G}{2}x_2\mathbf{e}_1\right)\left(1 - \frac{a^3}{r^3}\right) \\ &\quad - \frac{5}{2}Gx_1 x_2 (x_1\mathbf{e}_1 + x_2\mathbf{e}_2 + x_3\mathbf{e}_3)\left(\frac{a^3}{r^5} - \frac{a^5}{r^7}\right).\end{aligned}} \quad (8\text{--}50)$$

Thus, if we substitute (8–48) and (8–49) into (8–50) and collect all of the terms corresponding to each vector component, we find that

$$\boxed{\begin{aligned}\mathbf{u} &= \mathbf{i}_r \left\{ Gr\sin^2\theta \sin\phi\cos\phi \left[1 - \frac{5}{2}\left(\frac{a^3}{r^3}\right) + \frac{3}{2}\left(\frac{a^5}{r^5}\right)\right]\right\} \\ &\quad + \mathbf{i}_\theta \left[Gr\sin\theta\cos\theta\sin\phi\cos\phi\left(1 - \frac{a^5}{r^5}\right)\right] \\ &\quad + \mathbf{i}_\phi \left[-\frac{G}{2}r\sin\theta + \frac{Gr}{2}\sin\theta\frac{a^3}{r^3} + \frac{Gr}{2}(\sin\theta\cos 2\phi)\left(1 - \frac{a^5}{r^5}\right)\right].\end{aligned}} \quad (8\text{--}51)$$

This is the velocity field for a stationary, nonrotating sphere in a simple shear flow. A sketch of the fluid pathlines in the $x_1 x_2$ plane for this case is reproduced in Fig. 8–2(a). If the sphere is allowed to rotate with some angular velocity $\mathbf{\Omega}$, the corresponding velocity field

Creeping Flows – Three-Dimensional Problems

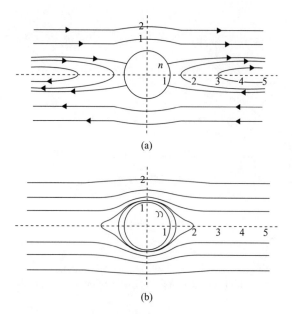

Figure 8–2. Fluid pathlines in the x_1x_2 plane (see Fig. 8–1) for simple shear flow past (a) nonrotating and (b) rotating spheres. The nonrotating case was obtained from Eq. (8–51), and the rotating case was obtained by adding (8–54) to (8–51).

can be obtained as a direct superposition of the present solution for a nonrotating sphere in simple shear and the solution (8–12) for rotation with velocity Ω in a quiescent fluid. If the sphere rotates as a consequence of the shear flow, it is evident that Ω must be in the same direction as the vorticity vector $\boldsymbol{\omega}$, that is,

$$\Omega = -\Omega \mathbf{i}_3. \tag{8–52}$$

In that case, the velocity field (8–12) becomes

$$\mathbf{u} = (-\Omega x_1 \mathbf{e}_2 + \Omega x_2 \mathbf{e}_1)\left(\frac{a^3}{r^3}\right), \tag{8–53}$$

and when this is expressed in terms of spherical coordinates, it simplifies dramatically to the form

$$\mathbf{u} = -\Omega \left(\frac{a}{r}\right)^3 r \sin\theta \, \mathbf{i}_\phi. \tag{8–54}$$

The composite velocity field for a sphere rotating with the angular velocity (8–52) in a simple shear flow, Eq. (8–46), is thus the sum of (8–51) and (8–54). The fluid pathlines in the x_1x_2 plane for this case are shown in Fig. 8–2(b). The corresponding torque on the sphere is

$$\mathbf{T} = -8\pi a^3 \mu \left(\frac{G}{2} - \Omega\right) \mathbf{i}_3. \tag{8–55}$$

To determine the angular velocity Ω, we must balance the hydrodynamic torque against the magnitude of any externally applied torque on the sphere. If the sphere is free to rotate – that is, $\mathbf{T} = 0$ – it follows that

$$\Omega = \frac{G}{2}. \tag{8–56}$$

Thus the sphere rotates with an angular velocity that is just $1/2$ the vorticity of the undisturbed flow. The velocity field for a freely rotating sphere in a simple shear flow is identical to Eq. (8–51), except that

$$u_\phi = -\frac{G}{2} r \sin\theta + \frac{Gr}{2} \sin\theta \cos 2\phi \left(1 - \frac{a^5}{r^5}\right). \tag{8–57}$$

C. Deformation of a Drop in a General Linear Flow

Clearly, for either a nonrotating or freely rotating sphere, the velocity field in a simple shear flow is fully 3D, so that the solution methods of earlier sections would necessarily fail.

The technique outlined in this section is extremely simple and powerful, allowing solutions for a complete class of undisturbed flows to be obtained at the same time. The reader should note that all the eigenfunction-based methods of the preceding sections are included within the present methodology, which may thus seem to supersede them. To some degree, this is in fact true. However, the earlier methods are still frequently used, often because of the preference of the researcher or engineer for the older, more classical techniques. It should also be noted that instances will arise, in using the present techniques, in which application of the boundary conditions to determine constants is extremely complicated if it is done (as previously) for the most general form of the undisturbed flow. However, because the *general* form of the solution is the same for all members of a particular family of flows, it is often possible to use the detailed solution for one specific flow to determine the constants in the general solution form. In this case, the solution for the specific flow problem might be determined by the methods of previous sections (or by the generalization using Lamb's general solution that was mentioned at the beginning of this section).

C. DEFORMATION OF A DROP IN A GENERAL LINEAR FLOW

We noted in chapter 7 that a drop will deform in almost any viscous flow, other than in steady translation through a stationary fluid. In all other cases, at zero Reynolds number, the magnitude of the deformation for a given flow primarily depends on the capillary number (and to a lesser extent upon the ratio of internal to external viscosity $\lambda \equiv \hat{\mu}/\mu$). The type of motion also influences the degree of deformation. We can illustrate the qualitative idea by thinking of two limiting cases. In the first, we imagine a drop immersed in a fluid that is undergoing a purely rotational 2D flow, and in the second, the fluid undergoes a 2D pure straining flow with the same magnitude for the velocity gradient. It is intuitively obvious that the drop will not deform in the pure rotational flow (it will simply rotate) but will deform substantially in the extensional flow. If we have a 2D flow that is somewhere between these extremes, the degree of deformation will be decreased by the addition of vorticity (holding $|\nabla \mathbf{u}| = $ const).

A qualitative prediction of the role of the capillary number in drop deformation can be obtained from the normal-stress balance, (2–135), written in a dimensionless form that is suitable to the low-Reynolds-number limit that we consider here. Thus,

$$\hat{p}_{tot} - p_{tot} + [(\boldsymbol{\tau} - \lambda\hat{\boldsymbol{\tau}}) \cdot \mathbf{n}] \cdot \mathbf{n} = \frac{1}{Ca}(\nabla \cdot \mathbf{n}). \qquad (8\text{–}58)$$

Here, the characteristic pressure and deviatoric stress is μG. The capillary number is $Ca \equiv \mu G a/\gamma$, and λ is the viscosity ratio, $\lambda \equiv \hat{\mu}/\mu$.

Now, if the shape is spherical, the capillary term on the right-hand side is a constant, and thus the viscous pressure and stress contributions on the left must also produce a constant value that corresponds to the jump in pressure across the interface that is due to surface tension. In general, however, the pressure and stress differences will not reduce to this very simple form, but will instead vary as a function of position on the surface of the drop. Such a *nonuniform* distribution of pressure and stress will tend to deform the drop. In fact, in this case, the normal-stress balance can be satisfied only if the drop deforms to a shape where the interface curvature ($\nabla \cdot \mathbf{n}$) varies in precisely the same way as the surface pressure and stress difference.

The *magnitude* of the change in shape that is required for satisfying the normal-stress balance can be seen from (8–58) to depend on Ca. For $Ca \ll 1$, very small deviations from a spherical shape (where $\nabla \cdot \mathbf{n} = $ const) can produce sufficient variation in $(1/Ca)(\nabla \cdot \mathbf{n})$

to balance the pressure or stress variations over the drop surface. Hence, for $Ca \ll 1$, a drop will remain approximately spherical in shape. This is consistent with the interpretation of Ca as the ratio of characteristic viscous (and pressure) forces to capillary forces. Thus the limit $Ca \ll 1$ corresponds to dominant surface-tension effects, and this accounts for the tendency of the drop to remain almost spherical. On the other hand, for $Ca \gg 1$, pressure and stress variations in (8–58) can be balanced only by large deformations of shape to produce large variations in the curvature term $(1/Ca)(\nabla \cdot \mathbf{n})$. In fact, experimental observation shows that steady shapes are not always possible for very large values of Ca. Essentially, in this case, no steady shape exists that allows a balance between the two sides of (8–58), and the drop must undergo a time-dependent change of shape, leading possibly to breakup into two or more pieces. Such breakup effects are extremely important in processing equipment such as mixers, or blenders, that are designed to disperse one continuous fluid in another, but an analysis of them is beyond the scope of this book.

Instead, in this section, we consider the deformation of a drop in a general linear flow, for $Ca \ll 1$, where the shape remains approximately spherical. We assume that the density of the drop is equal to that of the suspending fluid and that the surface tension is constant on the drop surface – that is, there are no thermal gradients and no surfactants present. The fact that the drop is neutrally buoyant means that it does not translate relative to the surrounding fluid. Thus, at large distances from the drop, we can assume that the fluid undergoes a general linear flow of the form

$$\mathbf{u}'_\infty \to G\left(\mathbf{E} \cdot \mathbf{x}' + \frac{1}{2}\boldsymbol{\omega} \wedge \mathbf{x}'\right), \quad |r'| \to \infty,$$

where G represents the magnitude of $\nabla \mathbf{u}$, $\mathbf{E}$ is the dimensionless rate-of-strain tensor and $\boldsymbol{\omega}$ is the dimensionless vorticity vector. Hence, in dimensionless terms, with $u_c = Ga$,

$$\boxed{\mathbf{u}_\infty \to \left(\mathbf{E} \cdot \mathbf{x} + \frac{1}{2}\boldsymbol{\omega} \wedge \mathbf{x}\right), \quad |r| \to \infty.} \quad (8\text{--}59)$$

We have already seen that this general linear flow includes simple shear, uniaxial extension, and other flows of interest.

In the following analysis, we assume that the Reynolds numbers for the flows inside and outside the drop are both extremely small, that is,

$$Re \equiv \frac{\rho G a^2}{\mu}, \quad \widehat{Re} \equiv \frac{\hat{\rho} G a^2}{\hat{\mu}} \ll 1,$$

so that the governing equations are again the creeping-flow equations, namely, (7–11). The far-field boundary condition in this case is (8–59). In addition, the boundary conditions (2–112), (2–122), and (2–141) with grad $\gamma \equiv 0$ must be applied at the surface of the drop, which we represent as the set of points $\mathbf{x}$ where the function $F(\mathbf{x}, t) \equiv 0$. The normal-stress balance, given earlier as (8–58), is also applied at $F = 0$.

In general, the problem just defined is nonlinear, in spite of the fact that the governing, creeping-flow equations are linear. This is because the drop shape is unknown and dependent on the pressure and stresses, which in turn, depend on the flow. Thus $\mathbf{n}$ and F are also unknown functions of the flow field, and the boundary conditions (2–112), (2–122), (2–141), and (8–58) are therefore nonlinear. Thus, for arbitrary Ca, for which the deformation may be quite significant, the problem can be solved only numerically. Later in this chapter, we briefly discuss a method, known as the boundary Integral method, that may be used to carry out such numerical calculations. Here, however, we consider the limiting case

C. Deformation of a Drop in a General Linear Flow

$Ca \ll 1$. As we have discussed already, the deformation in this limit is very small, and we shall see that we can obtain an approximate analytic solution using the asymptotic *method of domain perturbations*.

Thus, for convenience, we express the drop shape in terms of those values of the radial variable r such that a function $F = 0$. In the present case, we can write F in the form

$$F(\mathbf{x}, t) \equiv r - [1 + Ca f(\mathbf{x}, t)] \equiv 0, \qquad (8\text{–}60)$$

which is consistent with the fact that a first approximation to the drop shape is just a sphere,

$$r = 1.$$

The deviation from sphericity is contained in the function $f(\mathbf{x}, t)$, and we have anticipated, from the arguments at the beginning of this subsection, that the magnitude of the deviation from sphericity is proportional to Ca. Thus, for small Ca, a first approximation to the unit normal vector $\mathbf{n}$, expressed in terms of a spherical coordinate system, is just the unit vector in the radial direction, $\mathbf{i}_r$. A second approximation to $\mathbf{n}$ can be obtained by application of the definition of $\mathbf{n}$ in terms of F, namely,

$$\boxed{\mathbf{n} \equiv \frac{\nabla F}{|\nabla F|} = \mathbf{i}_r - Ca \nabla_s f + O(Ca^2).} \qquad (8\text{–}61)$$

The symbol $O(Ca^2)$ in (8–61) is the usual asymptotic order symbol and signifies that the next term in the approximation for $\mathbf{n}$ has a magnitude of order Ca^2 for $Ca \ll 1$. Substituting (8–60) and (8–61) into the boundary conditions (2–112), (2–122), (2–141), and (8–58), we find that

$$\boxed{\mathbf{u} = \hat{\mathbf{u}} \quad \text{at } r \approx 1,} \qquad (8\text{–}62)$$

$$\boxed{u_r = Ca \frac{\partial f}{\partial t} + O(Ca^2) \quad \text{at } r \approx 1,} \qquad (8\text{–}63)$$

$$\boxed{[(\hat{p} - p)\mathbf{I} + \boldsymbol{\tau} - \lambda \hat{\boldsymbol{\tau}}] \cdot \mathbf{i}_r = \frac{\mathbf{i}_r}{Ca}\left[2 - Ca \nabla_s^2 f + O(Ca^2)\right] \quad \text{at } r \approx 1.} \qquad (8\text{–}64)$$

The condition (8–64) incorporates both (2–141) and (8–58). Again, there are additional corrections that are proportional to Ca^2 or smaller for $Ca \ll 1$. Note that we do not assume that $Ca\, (\partial f / \partial t)$ is necessarily small, and it is not necessary to make this assumption to solve the problem. Examination of (8–61)–(8–64) shows that the boundary conditions at this level of approximation are linear in the unknowns $\mathbf{u}$, $\hat{\mathbf{u}}$, p, $\hat{p}$, and f. Thus the first approximation of the overall problem for $Ca \ll 1$ is linear.

To obtain a solution for the *complete class* of flows given by (8–59), we again construct a solution of the creeping-motion equations by means of the superposition of vector harmonic functions. The development of a general form for the pressure and velocity fields in the fluid exterior to the drop follows exactly the arguments of the preceding solution for a solid sphere in a linear flow, and the solution therefore takes the same general form [see Eq. (8–39) and (8–41)], that is,

$$\boxed{p' = c_1 \frac{\mathbf{x} \cdot \mathbf{E} \cdot \mathbf{x}}{r^5},} \qquad (8\text{–}65)$$

$$\boxed{\begin{aligned}\mathbf{u} = \mathbf{E}\cdot\mathbf{x} &+ \frac{1}{2}(\boldsymbol{\omega}\wedge\mathbf{x})+\mathbf{x}(\mathbf{x}\cdot\mathbf{E}\cdot\mathbf{x})\left(\frac{c_1}{2r^5}+\frac{c_3}{r^7}\right)\\ &-\frac{2c_3}{5r^5}(\mathbf{E}\cdot\mathbf{x})+c_4\left(\frac{\boldsymbol{\omega}\wedge\mathbf{x}}{r^3}\right).\end{aligned}} \qquad (8\text{-}66)$$

Similarly, a general solution can be obtained for the fluid inside the drop (note, however, that we must use growing harmonics to avoid any singularity at the origin of the coordinate system). The results are

$$\boxed{\hat{p} = \mathbf{E}:\left(\mathbf{xx}-\frac{r^2}{3}\mathbf{I}\right)d_2,} \qquad (8\text{-}67)$$

$$\boxed{\hat{\mathbf{u}} = d_3(\mathbf{E}\cdot\mathbf{x}) + d_4(\boldsymbol{\omega}\wedge\mathbf{x}) - \frac{2}{21}d_2\mathbf{x}(\mathbf{x}\cdot\mathbf{E}\cdot\mathbf{x}) + \frac{5}{21}d_2 r^2 \mathbf{E}\cdot\mathbf{x}.} \qquad (8\text{-}68)$$

Finally, because the shape function f is linearly related to the dynamic variables $\mathbf{u}$, $\hat{\mathbf{u}}$, and $(\mathbf{T}-\lambda\hat{\mathbf{T}})\cdot\mathbf{n}$, it follows that f must be expressible in invariant form as a linear function of $\mathbf{E}$ or $\boldsymbol{\omega}$. Because f is a true scalar, it follows that

$$\boxed{f(\mathbf{x},t) = b_1(\mathbf{x}\cdot\mathbf{E}\cdot\mathbf{x}).} \qquad (8\text{-}69)$$

To complete the solution, we must apply the boundary condition (8–63) at the interface to determine the coefficients c_i, d_i, and b_i, in (8–65) – (8–69). The first condition, (8–62), requires that the velocity components match at the drop surface, which at first approximation is just $r=1$. Thus, the normal component of (8–62) yields

$$\mathbf{n}\cdot\left[1+\left(\frac{c_1}{2}+c_3\right)-\frac{2c_3}{5}\right]\mathbf{E}\cdot\mathbf{n} = \mathbf{n}\cdot\left[d_3\mathbf{E}+\frac{3}{21}d_2\mathbf{E}\right]\cdot\mathbf{n}$$

or

$$1+\left(\frac{c_1}{2}+\frac{2c_3}{5}\right) = \left(d_3+\frac{3}{21}d_2\right). \qquad (8\text{-}70)$$

Similarly, the tangential component of (8–62) gives

$$\begin{aligned}1-\frac{2}{5}c_3 &= d_3+\frac{5}{21}d_2,\\ \frac{1}{2}+c_4 &= d_4.\end{aligned} \qquad (8\text{-}71)$$

To apply the stress balance, (8–64), we must first calculate the viscous stresses associated with $\mathbf{u}$, $\hat{\mathbf{u}}$, p, $\hat{p}$. For the fluid exterior to the drop, the stress field at $r=1$ is found to be

$$\mathbf{T}\cdot\mathbf{n}|_{r=1} = -p\mathbf{n} + 2\Big\{\mathbf{E}\cdot\mathbf{n}+\frac{c_1}{2}[\mathbf{E}\cdot\mathbf{n}-4\mathbf{n}(\mathbf{n}\cdot\mathbf{E}\cdot\mathbf{n})]$$
$$-\frac{c_3}{5}[-8\mathbf{E}\cdot\mathbf{n}+20\mathbf{n}(\mathbf{n}\cdot\mathbf{E}\cdot\mathbf{n})]-3c_4(\boldsymbol{\omega}\wedge\mathbf{x})\Big\},$$

C. Deformation of a Drop in a General Linear Flow

and the stress field inside the drop is

$$\hat{\mathbf{T}} \cdot \mathbf{n}|_{r=1} = -\hat{p}\mathbf{n} + 2\left\{d_3 \mathbf{E} \cdot \mathbf{n} + \frac{d_2}{21}\left[8\mathbf{E} \cdot \mathbf{n} - \frac{19}{2}(\mathbf{n} \cdot \mathbf{E} \cdot \mathbf{n})\mathbf{n}\right]\right\}.$$

Hence, matching the *tangential* components of stress, we find that

$$c_4 = 0, \tag{8-72}$$

$$1 + \frac{c_1}{2} + \frac{8c_3}{5} = \lambda\left(d_3 + \frac{8}{21}d_2\right). \tag{8-73}$$

To evaluate the *normal* component of the stress balance, we must first evaluate $\nabla \cdot \mathbf{n}$. This is slightly more subtle than it may seem. Because we are calculating everything in general vector/tensor form, we note that the surface gradient operator in (8–61) can be expressed in the form

$$\nabla_s f = (\mathbf{I} - \mathbf{n}\mathbf{n}) \cdot \nabla f = \left(\mathbf{I} - \frac{\mathbf{x}\,\mathbf{x}}{r\,r}\right) \cdot \nabla[b_1(\mathbf{x} \cdot \mathbf{E} \cdot \mathbf{x})].$$

Thus the unit normal $\mathbf{n}$ can be expressed in the form

$$\mathbf{n} = \frac{\mathbf{x}}{r} + \left(2\frac{\mathbf{x} \cdot \mathbf{E} \cdot \mathbf{x}}{r^2}\mathbf{x} - \mathbf{E} \cdot \mathbf{x} - \mathbf{x} \cdot \mathbf{E}\right) b_1 Ca,$$

and therefore

$$\nabla \cdot \mathbf{n} = 2 + 4(\mathbf{x} \cdot \mathbf{E} \cdot \mathbf{x})b_1 Ca + 0(Ca^2).$$

It follows from the normal component of Eq. (8–64) that

$$\hat{p}_0 - p_0 = \frac{2}{Ca},$$

$$\left[1 - 3\left(\frac{c_1}{2}\right) - \frac{12c_3}{5}\right] - \lambda\left(d_3 - \frac{1}{14}d_2\right) = 2b_1. \tag{8-74}$$

We thus have six equations, (8–70)–(8–74), that can be solved for the unknowns $c_1, c_3, c_4, d_2, d_3, d_4$ in terms of b_1. The results are

$$c_4 = 0, \tag{8-75}$$

$$d_4 = \frac{1}{2}, \tag{8-76}$$

$$c_1 = -\frac{5(\lambda - 1)}{2\lambda + 3}(2) - \frac{8}{2\lambda + 3}b_1, \tag{8-77}$$

$$c_3 = 5\left(\frac{\lambda - 1}{2\lambda + 3}\right) + \frac{20(3\lambda + 2)}{(2\lambda + 3)(19\lambda + 16)}b_1, \tag{8-78}$$

$$d_2 = \frac{84}{19\lambda + 16}b_1, \tag{8-79}$$

$$d_3 = \frac{5}{2\lambda + 3} - \frac{4(16\lambda + 19)}{(2\lambda + 3)(19\lambda + 16)}b_1. \tag{8-80}$$

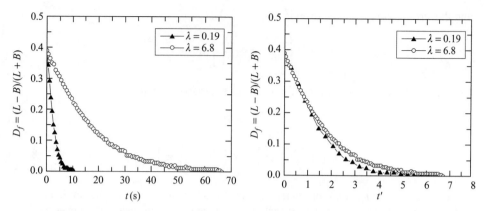

Figure 8–3. Drop deformation versus time for relaxation of the pair of drops shown in Fig. 2.15. The figure on the right-hand side shows the relaxation data as a function of the actual time. The data on the right-hand side are shown plotted versus a dimensionless time scale, $t^* = t/t_c^*$, where $t_c^* = \frac{(2\lambda+3)(19\lambda+16)}{40(\lambda+1)}\frac{\mu R}{\sigma}$.

The rate of change of b_1 describes the response of the interface to any rate of change of the flow. To obtain an equation for b_1, we apply the kinematic condition, (8–63). The result is

$$Ca\frac{db_1}{dt} = \frac{5}{2\lambda+3} - \frac{40(\lambda+1)}{(2\lambda+3)(19\lambda+16)}b_1. \qquad (8\text{--}81)$$

Although $Ca \ll 1$, we do not neglect $Ca\,(db_1/dt)$, which may, in fact, be $O(1)$. In the absence of flow, this result shows that the relaxation from a deformed elliptic shape will occur on a dimensionless time-scale

$$t = \frac{(2\lambda+3)(19\lambda+16)}{40(\lambda+1)Ca}.$$

Because time was nondimensionalized in this problem by the inverse shear rate, G^{-1}, this means that the characteristic time scale for relaxation in the absence of flow is

$$t_c^* = \frac{(2\lambda+3)(19\lambda+16)}{40(\lambda+1)}\frac{\mu R}{\sigma}.$$

We have already shown visual evidence in Chap. 2 (Fig. 2–15), that this is the correct time scale for the relaxation process. Further quantitative evidence is offered in Fig. 8–3. In this figure is plotted the measured deformation from Fig. 2–15, specified in terms of the dimensionless parameter $D_f \equiv (L-B)/(L+B)$ where L is the length of the drop and B is its breadth at the equator. It can be seen that when time is scaled with t_c^*, the deformation curves for the two different viscosity ratio fluids are reduced approximately to a single curve.

The solution of (8–81) for b_1 can be obtained easily. The two terms on the right-hand side of this equation represent the stretching of the drop that is due to the strain rate $\mathbf{E}$, and the restoring force that is due to interfacial tension. The dimensionless relaxation rate, for return to a spherical shape in the absence of a flow, is $(5/6)(1/Ca)$ when the viscosity ratio is small, and $(20/19\lambda)(1/Ca)$ for $\lambda \gg 1$. Hence, for a relatively inviscid drop, the relaxation rate is determined by the external fluid viscosity, whereas the relaxation rate for a very viscous drop is determined by the internal fluid viscosity. The steady-state solution of (8–81) is

$$b_1 = \frac{19\lambda+16}{8(\lambda+1)}, \qquad (8\text{--}82)$$

C. Deformation of a Drop in a General Linear Flow

and the corresponding drop shape is

$$r_s = 1 + Ca \left[\frac{19\lambda + 16}{8(\lambda + 1)} \right] (\mathbf{x} \cdot \mathbf{E} \cdot \mathbf{x}). \tag{8–83}$$

Thus, for a simple shear flow ($u_1 = x_2$),

$$\mathbf{E} = \begin{pmatrix} 0 & 1/2 & 0 \\ 1/2 & 0 & 0 \\ 0 & 0 & 0 \end{pmatrix}, \quad \text{and} \quad r_s = 1 + Ca \left[\frac{19\lambda + 16}{8(\lambda + 1)} \right] (x_1 x_2),$$

whereas for a pure uniaxial elongation ($u_3 = x_3, u_1 = -\frac{1}{2}x_1, u_2 = -\frac{1}{2}x_2$),

$$\mathbf{E} = \begin{pmatrix} -1/2 & 0 & 0 \\ 0 & -1/2 & 0 \\ 0 & 0 & 1 \end{pmatrix}, \quad \text{and} \quad r_s = 1 + Ca \left[\frac{19\lambda + 16}{8(\lambda + 1)} \right] \left[x_3^2 - \frac{1}{2}(x_1^2 + x_2^2) \right].$$

The corresponding drop shapes are sketched in Fig. 8–4. If we express these equations for drop shape in terms of spherical coordinates as shown in Fig. 8–5 ($x_1 = r \sin\theta \cos\phi, x_2 = r \sin\theta \sin\phi, x_3 = r \cos\theta$), we find that

$$r_s = 1 + Ca \left(\frac{19\lambda + 16}{8(\lambda + 1)} \right) \sin^2\theta \sin\phi \cos\phi \tag{8–84}$$

for simple shear flow, and

$$r_s = 1 + Ca \left[\frac{19\lambda + 16}{8(\lambda + 1)} \right] \left(\frac{3}{2} \cos^2\theta - \frac{1}{2} \right) \tag{8–85}$$

for uniaxial extension. The major axis of deformation for simple shear flow lies in the x_1, x_2 plane at 45° from the x_1 axis, and the *ratio* of the longest to the shortest axis in this case is

$$\alpha = \frac{1 + \frac{1}{2}Ca\left[\frac{19\lambda + 16}{8(\lambda + 1)}\right]}{1 - \frac{1}{2}Ca\left[\frac{19\lambda + 16}{8(\lambda + 1)}\right]}. \tag{8–86}$$

In the extensional flow, the principal axis of deformation is x_1 and the shape is axisymmetric about this axis. The ratio of major to minor axis in this case is

$$\alpha = \frac{1 + Ca\left[\frac{19\lambda + 16}{8(\lambda + 1)}\right]}{1 - \frac{1}{2}Ca\left[\frac{19\lambda + 16}{8(\lambda + 1)}\right]}. \tag{8–87}$$

The drop in the extensional flow is slightly more elongated than in the simple shear flow. Although the deformation is small in all cases for the limit $Ca \ll 1$, the slight difference in shape indicated by (8–86) and (8–87) is illustrative of the general principle that extensional flows are more efficient at stretching deformable bodies than flows that contain vorticity, such as the simple shear flow.

Creeping Flows – Three-Dimensional Problems

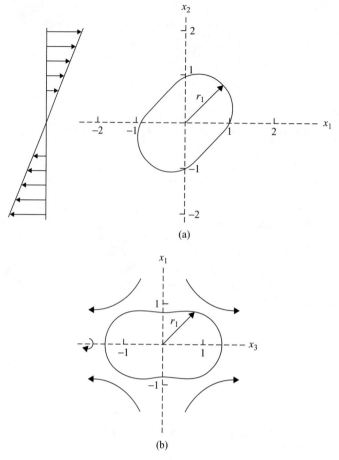

Figure 8–4. Shape of a viscous drop in (a) simple, linear shear flow and (b) uniaxial extension for $\lambda = 1$ and $Ca = 0.2$.

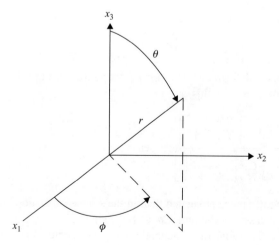

Figure 8–5. Relationship between Cartesian coordinates (x_1, x_2, x_3), and spherical coordinates (r, θ, ϕ).

D. FUNDAMENTAL SOLUTIONS OF THE CREEPING-FLOW EQUATIONS

All of the techniques for solving creeping-flow problems to this point (Chap. 7 and the previous sections of this chapter) were based on the representation of solutions in terms of either eigenfunction expansions and/or vector harmonic functions. These techniques, and their generalization to other coordinate systems besides spherical, provide a powerful arsenal of methods to attack problems in which the geometry of the solid boundaries or interfaces either coincides with a coordinate surface(s) in some orthogonal coordinate system[5] or else lies close to such a surface, as in the deforming drop problem of the previous section. However, the linearity of the creeping-flow equations allows another class of solution methods that is more readily applied to problems with *complicated boundary geometries*. These methods for solving Stokes' equations are based on *fundamental* solutions, corresponding to the flow produced by a point force in a fluid.[6] For problems involving *solid* boundaries, a general solution of the creeping-flow equations can be obtained in integral form, corresponding to a *distribution of point forces* over all of the boundaries. For problems involving a fluid interface, a similar integral formulation is obtained easily by generalization to a distribution of singularities at the interface.

The solution of a specific problem is then transformed to determining the distribution of surface forces that is necessary to satisfy boundary conditions. If the boundary (or interface) shapes are complex, or at least not "close" to coordinate surfaces in some coordinate system, the solution of this distribution problem can only generally be accomplished numerically, and this leads to so-called *boundary-integral techniques* that have been used widely in the research literature[7] and are particularly well suited to free-boundary problems involving large steady or transient deformations of an interface.[8] We discuss the basic principles of these boundary-integral methods at the end of this section, though we do not discuss the actual numerical implementation. If the boundary shapes are simpler, an *analytic* approximation for the distribution of point forces can sometimes be achieved by means of equivalent, *internal* distributions of force and force multipole singularities. One particularly important example is the solution for very elongated, slender bodies for which the integral representation can be reduced, at least approximately, to a line distribution of point forces along the centerline of the body; the resulting theory is known as *slender-body theory* for creeping flows.[9]

This section of the chapter is organized along the following lines: First, we discuss the fundamental solution of the creeping-flow equations for a point force in an unbounded fluid and show that it leads to a general integral representation for the solution of the creeping-flow equations; we then consider the special cases, including slender bodies, when the integral equation that results from applying boundary conditions can be solved analytically; finally, we return to the full integral equation and discuss its use in the context of the numerical, boundary-integral technique.

1. The "Stokeslet": A Fundamental Solution for the Creeping-Flow Equations

We begin by considering the solution of the creeping-flow equations for a point force **f** in an unbounded fluid. This solution is a *fundamental solution* in the sense that it can be obtained formally from the governing equations

$$\nabla \cdot \mathbf{u} = 0, \quad (8\text{--}88\text{a})$$

$$\nabla^2 \mathbf{u} - \nabla p = \delta(\mathbf{x})\mathbf{f}, \quad (8\text{--}88\text{b})$$

for a point force **f** applied to the fluid at $\mathbf{x} = 0$. Rather than solving (8–88) directly, by means of Fourier transforms or some similar technique, we resort to the method of superposition of harmonic functions as described in Section A of this chapter.

Creeping Flows – Three-Dimensional Problems

This method is particularly simple to apply in the case of a point force **f**. The (disturbance) pressure must be a decaying harmonic function that is linear in **f** and a true scalar. Hence,

$$\boxed{p = c_1 \frac{\mathbf{f} \cdot \mathbf{x}}{r^3}.} \tag{8–89}$$

The (disturbance) velocity is thus

$$\mathbf{u} = \frac{\mathbf{x}}{2}\left(\frac{c_1 \mathbf{f} \cdot \mathbf{x}}{r^3}\right) + \mathbf{u}^H, \tag{8–90}$$

where $\mathbf{u}^H$ is harmonic, decaying, linear in **f** and a true vector, that is,

$$\mathbf{u}^H = \alpha\left(\frac{\mathbf{xx}}{r^5} - \frac{\mathbf{I}}{3r^3}\right)\cdot\mathbf{f} + \beta\frac{\mathbf{f}}{r}, \tag{8–91}$$

where α and β are coefficients that must still be determined.

If we apply the continuity equation, (8–88a), we find that

$$\frac{c_1}{2} = \beta$$

so that

$$\mathbf{u} = \frac{c_1}{2}\left[\frac{\mathbf{f}}{r} + \frac{\mathbf{x}(\mathbf{f}\cdot\mathbf{x})}{r^3}\right] + \alpha\left(\frac{\mathbf{xx}}{r^5} - \frac{\mathbf{I}}{3r^3}\right)\cdot\mathbf{f}. \tag{8–92}$$

To this point, the solution is essentially identical to that obtained for uniform flow past a sphere in Section A. Beyond this, however, the two differ. To satisfy boundary conditions at the surface of a sphere, a nonzero value of α is necessary. However, for a point force, $\alpha = 0$. The simplest way to see this is to note that c_1 is dimensionless, whereas α must have dimensions of (length)^2. But a point force has no characteristic length scale, so $\alpha = 0$. Thus,

$$\boxed{\mathbf{u} = \frac{c_1}{2}\left[\frac{\mathbf{f}}{r} + \frac{\mathbf{x}(\mathbf{f}\cdot\mathbf{x})}{r^3}\right].} \tag{8–93}$$

To determine the scalar constant c_1, we make use of an integral relationship between the force **f** and the resultant force exerted on the fluid outside a control surface S.

First, consider the fluid region V between a body of arbitrary geometry ∂D and control surface S of arbitrary shape that completely encloses it. Because the creeping-motion approximation is being used,

$$\nabla \cdot \mathbf{T} = 0 \tag{8–94}$$

everywhere in the fluid region V, and thus

$$\int_V \nabla \cdot \mathbf{T}\, dV = 0. \tag{8–95}$$

Thus, by applying the divergence theorem to (8–95), we see that

$$\int_{\partial D} \mathbf{T}\cdot\mathbf{n}\, dA = \int_S \mathbf{T}\cdot\mathbf{n}_s\, dA. \tag{8–96}$$

Here the integral on the left-hand side is taken over the surface of the body ∂D and **n** is the *outer* normal *from the body into the fluid*. The integral on the right-hand side is taken over the control surface S with outer normal $\mathbf{n}_S$. However, the integral on the left-hand side

D. Fundamental Solutions of the Creeping-Flow Equations

is just the hydrodynamic force acting on the body (that is, the negative of the force acting from the body onto the fluid). It thus follows that

$$\mathbf{F} = \int_S \mathbf{T} \cdot \mathbf{n}_s dA. \tag{8–97}$$

In the present case, we have a point force at the origin, and the same expression holds but with $-\mathbf{f}$ replacing $\mathbf{F}$.

To obtain c_1, we apply (8–97) to the solution (8–89) and (8–93), with S chosen for convenience as a *sphere* centered at the origin. The surface stress, evaluated with (8–89) and (8–93), is

$$\mathbf{T} \cdot \mathbf{n} = [-p\mathbf{I} + (\nabla \mathbf{u} + \nabla \mathbf{u}^T)] \cdot \mathbf{n}$$
$$= -3c_1 \frac{\mathbf{xx} \cdot \mathbf{f}}{r^4}. \tag{8–98}$$

Hence, integrating over the spherical surface S (arbitrary radius), we obtain from (8–97)

$$\mathbf{f} = 4\pi c_1 \mathbf{f}. \tag{8–99}$$

Hence,

$$c_1 = \frac{1}{4\pi} \tag{8–100}$$

so that

$$\boxed{\mathbf{u} = \frac{1}{8\pi}\left[\frac{\mathbf{f}}{r} + \frac{\mathbf{x}(\mathbf{f} \cdot \mathbf{x})}{r^3}\right],} \tag{8–101}$$

$$\boxed{p = \frac{1}{4\pi}\left(\frac{\mathbf{f} \cdot \mathbf{x}}{r^3}\right).} \tag{8–102}$$

The associated expression for the stress tensor $\mathbf{T}$ is

$$\boxed{\mathbf{T} = -\frac{3}{4\pi}\frac{\mathbf{xx}(\mathbf{x} \cdot \mathbf{f})}{r^5}.} \tag{8–103}$$

The solutions (8–100) and (8–102) are called the *stokeslet* solution, apparently a name coined by Hancock.[10] It is customary to choose

$$\mathbf{f} = 8\pi \alpha \delta(\mathbf{x}) \tag{8–104}$$

so that the point-force singularity is characterized in magnitude and direction by α.

We shall see that the stokeslet solution plays a fundamental role in creeping flow theory. We have already seen in Section E of Chap. 7 that it describes the disturbance velocity far away from a body of any shape that exerts a nonzero force on an unbounded fluid. Indeed, when nondimensionalized and expressed in spherical coordinates, it is identical to the velocity field, (7–151). In the next section we use the stokeslet solution to derive a general integral representation for solutions of the creeping-flow equations.

2. An Integral Representation for Solutions of the Creeping-Flow Equations that is due to Ladyzhenskaya

To obtain a general integral representation for solutions of the creeping-flow equations, it is necessary first to derive a general integral theorem reminiscent of the Green's theorem from vector calculus.

Creeping Flows – Three-Dimensional Problems

Let us then consider the space outside a closed surface ∂D, and let $\mathbf{u}$ and $\hat{\mathbf{u}}$ be any smooth vector fields (such as velocity fields) in this exterior domain that satisfy the conditions

$$\nabla \cdot \mathbf{u} = 0, \ \nabla \cdot \hat{\mathbf{u}} = 0 \tag{8-105}$$

(such vector fields are called *solenoidal*) and vanish at infinity as r^{-1} or faster. In addition to $\mathbf{u}$ and $\hat{\mathbf{u}}$, we also define the two tensor (stress) functions

$$\mathbf{T} \equiv -p\mathbf{I} + (\nabla \mathbf{u} + \nabla \mathbf{u}^T), \tag{8-106}$$

$$\hat{\mathbf{T}} \equiv -\hat{p}\mathbf{I} + (\nabla \hat{\mathbf{u}} + \nabla \hat{\mathbf{u}}^T), \tag{8-107}$$

where p and $\hat{p}$ are any smooth scalar fields that vanish as r^{-2} or faster. Then a general vector identity between these functions is

$$\hat{\mathbf{u}} \cdot \nabla \cdot \mathbf{T} - \mathbf{u} \cdot \nabla \cdot \hat{\mathbf{T}} = \nabla \cdot (\hat{\mathbf{u}} \cdot \mathbf{T} - \mathbf{u} \cdot \hat{\mathbf{T}}) + (\nabla \mathbf{u} : \hat{\mathbf{T}} - \nabla \hat{\mathbf{u}} : \mathbf{T}). \tag{8-108}$$

However, if we utilize (8–106) and (8–107), then it can be shown that the last term in (8–108) is identically equal to zero. Hence, integrating over V, we obtain

$$\int_V [\hat{\mathbf{u}} \cdot (\nabla \cdot \mathbf{T}) - \mathbf{u} \cdot (\nabla \cdot \hat{\mathbf{T}})] dV = \int_V \nabla \cdot [\hat{\mathbf{u}} \cdot \mathbf{T} - \mathbf{u} \cdot \hat{\mathbf{T}}] dV, \tag{8-109}$$

and applying the divergence theorem, we obtain

$$\int_V [\hat{\mathbf{u}} \cdot (\nabla \cdot \mathbf{T}) - \mathbf{u} \cdot (\nabla \cdot \hat{\mathbf{T}})] dV = \int_S [\hat{\mathbf{u}} \cdot \mathbf{T} - \mathbf{u} \cdot \hat{\mathbf{T}}] \cdot \mathbf{n} dA - \int_{\partial D} [\hat{\mathbf{u}} \cdot \mathbf{T} - \mathbf{u} \cdot \hat{\mathbf{T}}] \cdot \mathbf{n} dA, \tag{8-110}$$

where S is any surface enclosing ∂D. Finally, if we let the surface $S \to \infty$, we see that the integral over S is $O(r^{-1})$ and thus vanishes. It follows that

$$\boxed{\int_V [\hat{\mathbf{u}} \cdot (\nabla \cdot \mathbf{T}) - \mathbf{u} \cdot (\nabla \cdot \hat{\mathbf{T}})] dV = -\int_{\partial D} \mathbf{n} \cdot [\hat{\mathbf{u}} \cdot \mathbf{T} - \mathbf{u} \cdot \hat{\mathbf{T}}] dA,} \tag{8-111}$$

where V is now the whole space exterior to ∂D.

The integral theorem (8–111), leads directly to a general integral representation for solutions of the creeping-flow equations. To see this, let $\mathbf{u}, \mathbf{T}$ represent a solution of the creeping-flow equations – arbitrary except that they must be $O(r^{-1})$ and $O(r^{-2})$, respectively, for $r \to \infty$, as we assumed in deriving (8–111). Further, let $\hat{\mathbf{u}}, \hat{\mathbf{T}}$ be the fundamental solution of the creeping-flow equations for a point force at a point $\boldsymbol{\xi}$, that is, referring to (8–100),

$$\hat{\mathbf{u}} = \frac{1}{8\pi} \left[\frac{(\mathbf{x} - \boldsymbol{\xi})(\mathbf{x} - \boldsymbol{\xi})}{R^3} + \frac{\mathbf{I}}{R} \right] \cdot \mathbf{e}, \tag{8-112}$$

where the point force is assumed to be of unit magnitude $\mathbf{f} = \mathbf{e}$, and $R \equiv |\mathbf{x} - \boldsymbol{\xi}|$ is the distance between a point $\mathbf{x}$ and the point of application of the force $\boldsymbol{\xi}$. The corresponding stress $\hat{\mathbf{T}}$ is

$$\hat{\mathbf{T}} = -\frac{3}{4\pi} \frac{(\mathbf{x} - \boldsymbol{\xi})(\mathbf{x} - \boldsymbol{\xi})(\mathbf{x} - \boldsymbol{\xi})}{R^5} \cdot \mathbf{e}. \tag{8-113}$$

D. Fundamental Solutions of the Creeping-Flow Equations

Now, substituting (8–112) and (8–113) into (8–111) and noting that

$$\nabla \cdot \mathbf{T} \equiv 0, \tag{8-114a}$$

$$\nabla \cdot \hat{\mathbf{T}} = \delta(\mathbf{x} - \boldsymbol{\xi})\mathbf{e}, \tag{8-114b}$$

we obtain

$$\int_V [-\delta(\mathbf{x} - \boldsymbol{\xi})\mathbf{u} \cdot \mathbf{e}] dV_\xi = -\int_{\partial D} \mathbf{n} \cdot [(\hat{\mathbf{U}} \cdot \mathbf{e}) \cdot \mathbf{T} - \mathbf{u} \cdot (\hat{\boldsymbol{\Sigma}} \cdot \mathbf{e})] dA_\xi, \tag{8-115}$$

where we have introduced the shorthand notation

$$\hat{\mathbf{u}} = \hat{\mathbf{U}} \cdot \mathbf{e}, \quad \hat{\mathbf{T}} = \hat{\boldsymbol{\Sigma}} \cdot \mathbf{e} \tag{8-116}$$

in place of (8–112) and (8–113). The symbols dV_ξ and dA_ξ indicate that the variable of integration is ξ, and the fixed point is thus $\mathbf{x}$. The tensor component of $\hat{\mathbf{U}}$ (i.e., U_{ik}) is the i component of the velocity field generated by a point force of unit magnitude in the k direction. Similarly, the component Σ_{ijk} is the ij component of the stress tensor corresponding to the flow produced by a point force in the k direction.

To simplify (8–115), we can factor out the constant unit vector $\mathbf{e}$ from all terms and use the integral property of the delta function to evaluate the term on the left-hand side. The result, after substituting for $\hat{\mathbf{U}}$ and $\hat{\boldsymbol{\Sigma}}$ from (8–112), (8–113), and (8–116), is

$$\boxed{\begin{aligned}\mathbf{u}(\mathbf{x}) = &-\frac{3}{4\pi} \int_{\partial D} \left[\frac{(\mathbf{x} - \boldsymbol{\xi})(\mathbf{x} - \boldsymbol{\xi})(\mathbf{x} - \boldsymbol{\xi})}{R^5} \cdot \mathbf{u}(\boldsymbol{\xi})\right] \cdot \mathbf{n} \, dA_\xi \\ &+ \frac{1}{8\pi} \int_{\partial D} \left[\frac{\mathbf{I}}{R} + \frac{(\mathbf{x} - \boldsymbol{\xi})(\mathbf{x} - \boldsymbol{\xi})}{R^3}\right] \cdot \mathbf{T}(\boldsymbol{\xi}) \cdot \mathbf{n} \, dA_\xi.\end{aligned}} \tag{8-117}$$

The corresponding form for the pressure is

$$\boxed{\begin{aligned}p(\mathbf{x}) = &\frac{1}{4\pi} \int_{\partial D} \frac{(\mathbf{x} - \boldsymbol{\xi})}{R^3} \cdot \mathbf{T}(\boldsymbol{\xi}) \cdot \mathbf{n} \, dA_\xi \\ &+ \frac{1}{2\pi} \int_{\partial D} \left[\frac{\mathbf{I}}{R} - \frac{3(\mathbf{x} - \boldsymbol{\xi})(\mathbf{x} - \boldsymbol{\xi})}{R^3}\right] \cdot \mathbf{u}(\boldsymbol{\xi}) \cdot \mathbf{n} \, dA_\xi.\end{aligned}} \tag{8-118}$$

This is the famous integral representation for the solution of the creeping-flow equations that is usually attributed to Ladyzhenskaya.[11] Because the derivation requires $\mathbf{u} \sim r^{-1}$ for large r, we recognize that $\mathbf{u}$ must be interpreted as the disturbance velocity field if we wish to apply (8–117) to a problem that involves an undisturbed velocity field $\mathbf{u}^\infty(\mathbf{x})$ (which is a solution of the creeping-flow equations) at large distances from the body (or boundary) that is denoted as ∂D. To apply (8–117) directly to the actual velocity field, we let $\mathbf{u} = \mathbf{u}'$ in (8–117), where $\mathbf{u}' = \mathbf{u} - \mathbf{u}^\infty$; $\mathbf{u}$ is now the true velocity, and $\mathbf{u}^\infty$ is the undisturbed velocity. This gives

$$\boxed{\begin{aligned}\mathbf{u}(\mathbf{x}) = \mathbf{u}^\infty(\mathbf{x}) &- \frac{3}{4\pi} \int_{\partial D} \left[\frac{(\mathbf{x} - \boldsymbol{\xi})(\mathbf{x} - \boldsymbol{\xi})(\mathbf{x} - \boldsymbol{\xi})}{R^5} \cdot \mathbf{u}\right] \cdot \mathbf{n} \, dA_\xi \\ &+ \frac{1}{8\pi} \int_{\partial D} \left[\frac{\mathbf{I}}{R} + \frac{(\mathbf{x} - \boldsymbol{\xi})(\mathbf{x} - \boldsymbol{\xi})}{R^3}\right] \cdot \mathbf{T} \cdot \mathbf{n} \, dA_\xi.\end{aligned}} \tag{8-119}$$

Creeping Flows – Three-Dimensional Problems

The term

$$\int_{\partial D} \mathbf{n} \cdot [\hat{\mathbf{u}} \cdot \mathbf{T}^\infty - \mathbf{u}^\infty \cdot \hat{\boldsymbol{\Sigma}}] dA_\xi$$

is identically equal to zero, as the reader may wish to verify.

The formula (8–117) [or (8–119)] provides a formal solution of the creeping-motion equations in a compact form. The first integral on the right-hand side is denoted as the *double-layer potential* and has a density function that is just the velocity $\mathbf{u}$ on the boundaries ∂D of the flow domain. The second integral on the right-hand side is termed the *single-layer potential*, and its density function is the surface-stress vector $\mathbf{t} = \mathbf{T} \cdot \mathbf{n}$. Of course, (8–117) and (8–119) do not provide a solution for any specific problem until the density functions $\mathbf{u}$ and $\mathbf{T} \cdot \mathbf{n}$ are specified on ∂D. *In fact, all that we really have done is to obtain an integral formula for $\mathbf{u}$ that is equivalent to the differential form of the creeping-flow equation,* (7–6). To obtain a solution for any particular problem, we must determine the density functions so that the velocity field $\mathbf{u}$ satisfies the boundary conditions on ∂D. In general, this requires numerical solution of the integral equations that result from applying boundary conditions to (8–117) or (8–119). In fact, this is the essence of the so-called *boundary-integral method* for solution of creeping-flow equations; this technique has been used widely in research and is especially suitable for free-surface and other Stokes' flow problems with complicated boundary geometries.[12] At the end of this section, we discuss some principles of the boundary-integral technique. First, however, we discuss some alternative techniques for the *analytic* solution of problems involving flow exterior to a solid body.

E. SOLUTIONS FOR SOLID BODIES BY MEANS OF INTERNAL DISTRIBUTIONS OF SINGULARITIES

One important class of Stokes' flow problems involves motion past a stationary solid surface. In this case, the no-slip boundary condition is

$$\mathbf{u} = 0 \quad \text{for all } \mathbf{x} \text{ on } \partial D, \tag{8–120}$$

and the integral formula (8–119) can be applied for $\mathbf{x}$ on ∂D to obtain

$$\boxed{\mathbf{u}^\infty(\mathbf{x}) = -\frac{1}{8\pi} \int_{\partial D} \left[\frac{\mathbf{I}}{R} + \frac{(\mathbf{x} - \boldsymbol{\xi})(\mathbf{x} - \boldsymbol{\xi})}{R^3} \right] \cdot \mathbf{t}(\boldsymbol{\xi}) dA_\xi.} \tag{8–121}$$

The solution of this integral equation gives us the unknown surface-stress vector $\mathbf{t}(\boldsymbol{\xi}) = \mathbf{T}(\boldsymbol{\xi}) \cdot \mathbf{n}$. Then the general solution of the creeping-flow equations is

$$\boxed{\mathbf{u}(\mathbf{x}) = \mathbf{u}^\infty(\mathbf{x}) + \frac{1}{8\pi} \int_{\partial D} \left[\frac{\mathbf{I}}{R} + \frac{(\mathbf{x} - \boldsymbol{\xi})(\mathbf{x} - \boldsymbol{\xi})}{R^3} \right] \cdot \mathbf{t}(\boldsymbol{\xi}) dA_\xi,} \tag{8–122}$$

where $\mathbf{x}$ is now a fixed point in the flow field and $\mathbf{t}(\mathbf{x})$ is the distribution of surface stress on the boundary ∂D that we obtain by solving (8–121). We do not discuss the solution of the integral equation (8–121) here. The main objective in deriving (8–122) is to show that a general solution of the creeping-flow equations for flow past stationary solid surfaces can be expressed completely as a superposition of surface forces (stokeslets) at the boundaries ∂D. In fact, a solution of the creeping-flow equations can always be written solely as a distribution of stokeslets over the bounding surfaces, even if these are not solid and stationary, but the simple identity of the stokeslet density function with the actual surface stress is valid only

E. Solutions for Solid Bodies by Means of Internal Distributions of Singularities

for this special case. To retain the simple physical interpretation of the density functions for other kinds of boundaries, we must use the more general form (8–119).

1. Fundamental Solutions for a Force Dipole and Other Higher-Order Singularities

In this section, we pursue the basic idea that the solution for creeping flow past a body, in terms of a *surface distribution* of point forces (stokeslets) on the surface of that body, can sometimes be replaced with an *internal distribution* of point forces and higher-order singularities. This is based on two key points. First, if we begin with a solution of the creeping-flow equations, then a derivative of any order of that solution is still a solution of the creeping-flow equations. In particular, if we start with the stokeslet solutions, (8–101), (8–102), and (8–104), which we denote here as $(\mathbf{u}_s, p_s)$,

$$\mathbf{u}_s(\mathbf{x}; \boldsymbol{\alpha}) = \frac{\boldsymbol{\alpha}}{r} + \frac{(\boldsymbol{\alpha} + \mathbf{x})\mathbf{x}}{r^3},$$

$$p_s(\mathbf{x}; \boldsymbol{\alpha}) = 2\frac{\boldsymbol{\alpha} \cdot \mathbf{x}}{r^3}, \tag{8-123}$$

then a derivative of any order of $\mathbf{u}_s$ and p_s is also a solution of the creeping-flow equations, corresponding to a point singularity that is a derivative of the same order of a point force $\mathbf{f}$. Second, a stokeslet solution for a point force at $\mathbf{x} = \boldsymbol{\xi}$ can be expressed in terms of a formal *multipole* expansion in the form of a generalized Taylor series about $\mathbf{x}$,

$$\mathbf{u}_s(\mathbf{x} - \boldsymbol{\xi}) = \mathbf{u}_s(\mathbf{x}) - (\boldsymbol{\xi} \cdot \nabla)\mathbf{u}_s(\mathbf{x}) + \frac{1}{2}(\boldsymbol{\xi} \cdot \nabla)^2 \mathbf{u}_s(\mathbf{x}) + \cdots +, \tag{8-124}$$

with a similar series expression for p_s. However $(\boldsymbol{\xi} \cdot \nabla)\mathbf{u}_s$ and $(\boldsymbol{\xi} \cdot \nabla)^2 \mathbf{u}_s$ are the velocity fields generated by a *force dipole*, $(\boldsymbol{\xi} \cdot \nabla)\mathbf{f}_s$, and a *force quadrapole*, $(\boldsymbol{\xi} \cdot \nabla)^2 \mathbf{f}_s$, respectively. Thus, for flow past a solid body, we always can use the generalized Taylor series to replace the surface distribution of stokeslets in the solution (8–122) with an equivalent internal distribution of stokeslets and higher-order singularities (inside the body).

The obvious question is whether there is any advantage to be gained, especially in view of the fact that we must replace a surface distribution of stokeslets only with a whole hierarchy of higher-order singularities inside. One possibility is that we may be able to replace the surface distribution with an internal distribution of lower spatial order. Thus, for example, for axisymmetric bodies, it may be possible to express the solution in terms of a *line distribution* of singularities along the axis of symmetry of the body. In this case, the 2D integral equation that arises from application of boundary conditions [Eq. (8–121)] would be replaced with a 1D, though more complicated, integral equation. Another possibility is that we may be able to replace a stokeslet distribution on a body that has a very complicated surface (for example, lots of "bumps") with an internal distribution of singularities on a nearby surface that has a much simpler geometry. In any case, however, the use of internal distributions of singularities will be an advantage only if the number of terms in the multipole expansion can be limited to a relatively small set. In other words, it will be an advantage only if the Taylor series, (8–124), can be truncated after a finite number of terms. Intuitively, for an exact solution, this will require bodies of simple geometry in relatively simple flows. Alternatively, if the internal surface (or line) is close enough to the surface of the body, it should be possible to *approximate* the multipole expansion by a small number of terms (or even one term), as the higher-order terms will decrease rapidly in magnitude. Beyond these generalities, we cannot offer more definitive criteria for recognizing problems for which internal distributions of singularities offer an advantage over the solutions in terms of a surface distribution of stokeslets [Eq. (8–122)]. This is, in fact, a subject of current research.

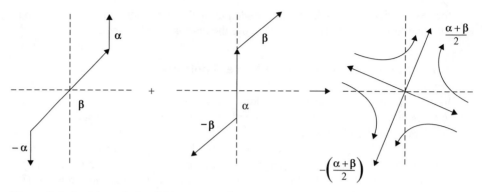

Figure 8–6. Stresslet velocity field constructed as a sum of the Stokes' dipoles $\mathbf{u}_{SS}(\mathbf{x}; \alpha, \beta)$ and $\mathbf{u}_{SD}(\mathbf{x}; \beta, \alpha)$.

In the remainder of this section, we outline results for two types of problems for which internal distributions of singularities have been used to advantage. The first, originally pursued by Chwang and Wu,[13] considers bodies of very simple shape – spheres, prolate ellipsoids of revolution (spheroids), and similar cases for which *exact* solutions can be obtained for some flows either by a point or line distribution involving only a few singularities. The second class of problems is for very slender bodies for which an *approximate* solution can be obtained by means of a distribution of stokeslets along the particle centerline.[14]

To pursue the use of internal singularities for bodies of simple geometry, we begin by discussing the solutions of the creeping-flow equations for force dipoles and higher-order multipole singularities. These may be obtained formally by differentiation of the basic stokeslet solution. For example, the so-called *Stokes dipole* solution is

$$\mathbf{u}_{SD}(\mathbf{x}; \alpha, \beta) \equiv -(\beta \cdot \nabla)\mathbf{u}_s(\mathbf{x}; \alpha) = \frac{(\beta \wedge \alpha) \wedge \mathbf{x}}{r^3}$$
$$- \left[\frac{(\alpha \cdot \beta)\mathbf{x}}{r^3} - \frac{3(\alpha \cdot \mathbf{x})(\beta \cdot \mathbf{x})\mathbf{x}}{r^5}\right] \quad (8\text{–}125\mathrm{a})$$

$$p_{SD}(\mathbf{x}; \alpha, \beta) \equiv -(\beta \cdot \nabla)p_s(\mathbf{x}; \alpha) = 2\mu\left[-\frac{\alpha \cdot \beta}{r^3} + 3\frac{(\alpha \cdot \mathbf{x})(\beta \cdot \mathbf{x})}{r^5}\right]. \quad (8\text{–}125\mathrm{b})$$

This corresponds to the velocity and pressure fields generated by a pair of forces, one at $\mathbf{x} = \beta/2$ with strength α and the other at $\mathbf{x} = -\beta/2$ with strength at $-\alpha$, in the limit as $|\beta| \to 0$ (see, for example, the sketch in Fig. 8-6). Similarly, the *Stokes quadrapole* solution is

$$\mathbf{u}_{S4}(\mathbf{x}; \alpha, \beta, \delta) = (\delta \cdot \nabla)(\beta \cdot \nabla)\mathbf{u}_s(\mathbf{x}; \alpha) \quad (8\text{–}126\mathrm{a})$$
$$p_{S4}(\mathbf{x}; \alpha, \beta, \delta) = (\delta \cdot \nabla)(\beta \cdot \nabla)p_s(\mathbf{x}; \alpha), \quad (8\text{–}126\mathrm{b})$$

and we can obtain solutions for even higher-order singularities by continued differentiation as indicated in (8–125) and (8–126).

It is useful to express the Stokes' dipole solution as the sum of two component parts, because each has a clear physical significance. These components are defined in a manner that is analogous to the symmetric and antisymmetric components of a second-order tensor, that is,

$$\mathbf{u}_{SD}(\mathbf{x}; \alpha, \beta) = \mathbf{u}_{SS}(\mathbf{x}; \alpha, \beta) + \mathbf{u}_R(\mathbf{x}; \alpha, \beta), \quad (8\text{–}127)$$

E. Solutions for Solid Bodies by Means of Internal Distributions of Singularities

where

$$\mathbf{u}_{SS}(\mathbf{x}; \boldsymbol{\alpha}, \boldsymbol{\beta}) \equiv \frac{1}{2}[\mathbf{u}_{SD}(\mathbf{x}; \boldsymbol{\alpha}, \boldsymbol{\beta}) + \mathbf{u}_{SD}(\mathbf{x}; \boldsymbol{\beta}, \boldsymbol{\alpha})], \tag{8–128}$$

$$\mathbf{u}_R(\mathbf{x}; \boldsymbol{\alpha}, \boldsymbol{\beta}) \equiv \frac{1}{2}[\mathbf{u}_{SD}(\mathbf{x}; \boldsymbol{\alpha}, \boldsymbol{\beta}) - \mathbf{u}_{SD}(\mathbf{x}; \boldsymbol{\beta}, \boldsymbol{\alpha})]. \tag{8–129}$$

The symmetric part, given by (8–128), is known as the *stresslet* solution, and the antisymmetric part, (8–129), is known as the *rotlet* solution.

We can obtain the detailed form of the rotlet easily by substituting the Stokes' dipole solution, (8–125a), into the definition (8–129). The result is

$$\frac{1}{2}[\mathbf{u}_{SD}(\mathbf{x}; \boldsymbol{\alpha}, \boldsymbol{\beta}) - \mathbf{u}_{SD}(\mathbf{x}; \boldsymbol{\beta}, \boldsymbol{\alpha})] = \frac{(\boldsymbol{\beta} \wedge \boldsymbol{\alpha}) \wedge \mathbf{x}}{r^3}. \tag{8–130}$$

Hence, if we define $\boldsymbol{\gamma} \equiv \boldsymbol{\beta} \wedge \boldsymbol{\alpha}$, the rotlet velocity field is

$$\boxed{\mathbf{u}_r(\mathbf{x}; \boldsymbol{\gamma}) = \frac{\boldsymbol{\gamma} \wedge \mathbf{x}}{r^3},} \tag{8–131a}$$

and the corresponding pressure is

$$\boxed{p_r(\mathbf{x}; \boldsymbol{\gamma}) = 0.} \tag{8–131b}$$

The physical significance of the rotlet solution is that it is the flow that is due to a singular, point torque at the origin. To see this, we can calculate the moment exerted on the fluid outside an arbitrary control surface S centered at the origin by a rotlet of strength $\boldsymbol{\gamma}$.

Thus,

$$\mathbf{M} = \int_S \mathbf{x} \wedge (\mathbf{T}_R \cdot \mathbf{n}) \, dA, \tag{8–132}$$

and the result after some straightforward manipulation is

$$\mathbf{M}_R = 8\pi \mu \boldsymbol{\gamma}. \tag{8–133}$$

The stresslet solution, (8–128), takes the form

$$\boxed{\mathbf{u}_{SS}(\mathbf{x}; \boldsymbol{\alpha}, \boldsymbol{\beta}) = \left[-\frac{\boldsymbol{\alpha} \cdot \boldsymbol{\beta}}{r^3} + \frac{3(\boldsymbol{\alpha} \cdot \mathbf{x})(\boldsymbol{\beta} \cdot \mathbf{x})}{r^5}\right] \mathbf{x},} \tag{8–134a}$$

$$\boxed{p_{SS}(\mathbf{x}; \boldsymbol{\alpha}, \boldsymbol{\beta}) = 2\mu \left[-\frac{\boldsymbol{\alpha} \cdot \boldsymbol{\beta}}{r^3} + \frac{3(\boldsymbol{\alpha} \cdot \mathbf{x})(\boldsymbol{\beta} \cdot \mathbf{x})}{r^5}\right].} \tag{8–134b}$$

The stresslet exerts zero net force or torque on the fluid, that is,

$$\mathbf{F}_{SS} = 0 \quad \text{and} \quad \mathbf{M}_{SS} = 0, \tag{8–135}$$

and can be thought of as a straining motion of the fluid that is symmetric about the $\boldsymbol{\alpha}, \boldsymbol{\beta}$ plane with principal axes of strain $\boldsymbol{\alpha} + \boldsymbol{\beta}$, $\boldsymbol{\alpha} - \boldsymbol{\beta}$, and $\boldsymbol{\alpha} \wedge \boldsymbol{\beta}$; see Fig. 8–6.

The Stokes' quadrapole solutions are more complicated; however, one component turns out to be particularly useful in the solution of Stokes' flow problems, and that is the *potential dipole* solution, namely,

$$\mathbf{u}_D(\mathbf{x}; \boldsymbol{\delta}) \equiv -\frac{1}{2} \nabla^2 \mathbf{u}_s(\mathbf{x}; \boldsymbol{\delta}). \tag{8–136}$$

Creeping Flows – Three-Dimensional Problems

This solution has the simple form

$$\mathbf{u}_D = -\frac{\boldsymbol{\delta}}{r^3} + \frac{3(\boldsymbol{\delta}\cdot\mathbf{x})\mathbf{x}}{r^5},$$

$$p_D = 0.$$
(8–137)

Again, the potential dipole exerts zero net force on the fluid. Physically, it corresponds to a mass dipole at the origin – that is, the flow generated by a mass source at $\mathbf{x} = \boldsymbol{\delta}/2$ and a mass sink at $\mathbf{x} = -\boldsymbol{\delta}/2$ in the limit $|\boldsymbol{\delta}| \to 0$.

Unfortunately, the use of fundamental solutions to solve Stokes' flow problems by means of internal distributions of singularities has not yet become completely systematized. Many (and perhaps all) problems involving flow in the vicinity of a solid sphere can be solved by a superposition of point-force and/or higher-order singularities at the center of the sphere. Some problems involving prolate axisymmetric ellipsoids (spheroids) can be solved by a superposition of point-force and higher-order singularities on the symmetry axis between the two foci of the ellipse. However, general rules that are guaranteed to guide the choice of singularities or the choice of internal surface geometry for all possible situations cannot be offered. If there is a net force on the body, the leading-order term must be a stokeslet or stokeslet distribution, and experience suggests that the stokeslet is always accompanied by the potential dipole. If there is a net torque on the body, we require a rotlet or rotlet distribution, and this will be the leading term if the net force is zero. Finally, if the net force and torque are both zero, the leading term often (though not always) involves the stresslet. Several examples may illustrate these points.

2. Translation of a Sphere in a Quiescent Fluid (Stokes' Solution)

The translation of a sphere through a quiescent fluid produces a net force on the sphere. Thus, to construct a solution by means of internal singularities we require a stokeslet of strength α located at the sphere center. However, by itself, the stokeslet flow does not satisfy the no-slip and kinematic boundary conditions at the surface of the sphere. In particular, if we assume that velocities have been nondimensionalized with the magnitude of the sphere's velocity U, the boundary condition at the sphere surface, $r = 1$, is

$$\mathbf{u} = \mathbf{i}_u \quad \text{at } r = 1,$$
(8–138)

where spatial variables are nondimensionalized by use of the sphere radius a. The stokeslet velocity field at $r = 1$ takes the form

$$\mathbf{u}_s = \boldsymbol{\alpha} + (\boldsymbol{\alpha}\cdot\mathbf{i}_r)\mathbf{i}_r.$$
(8–139)

Hence it is clear that the stokeslet solution alone cannot satisfy (8–138). However, it has already been suggested that the stokeslet is most often accompanied by the potential dipole $\mathbf{u}_D$. At the sphere surface, the potential dipole field becomes

$$\mathbf{u}_D = -\boldsymbol{\delta} + 3(\boldsymbol{\delta}\cdot\mathbf{i}_r)\mathbf{i}_r.$$
(8–140)

Hence when (8–139) and (8–140) are compared, it is obvious that a superposition of the stokeslet and potential dipole solutions will satisfy the creeping-flow equations and also the boundary condition (8–138). Specifically,

$$\mathbf{u} = \mathbf{u}_S(\mathbf{x}; \boldsymbol{\alpha}) + \mathbf{u}_D(\mathbf{x}; \boldsymbol{\delta}),$$
(8–141)

where

$$\boldsymbol{\alpha} - \boldsymbol{\delta} = \mathbf{i}_u,$$
(8–142)

$$\boldsymbol{\alpha} + 3\boldsymbol{\delta} = 0.$$
(8–143)

E. Solutions for Solid Bodies by Means of Internal Distributions of Singularities

Thus,

$$\boxed{\delta = -\frac{\mathbf{i}_u}{4}, \quad \alpha = \frac{3}{4}\mathbf{i}_u.} \qquad (8\text{–}144)$$

It may be noted that the force on the sphere in dimensionless terms, as given by (8–104), is $\mathbf{f} = 6\pi \mathbf{i}_u$. In dimensional terms this becomes $\mathbf{f}_D = (6\pi \mu U a)\mathbf{i}_u$. Not surprisingly, we recover Stokes's law.

3. Sphere in Linear Flows: Axisymmetric Extensional Flow and Simple Shear

We have previously obtained solutions by other techniques for the problems of a rigid sphere immersed in axisymmetric straining or simple shear flow of an unbounded fluid. In this subsection, it is shown that those two problems also can be solved very simply by means of a superposition of fundamental singularities at the center of the sphere.

We begin with the problem of a solid sphere centered at the stagnation point of an axisymmetric straining flow in which the undisturbed velocity at infinity takes the dimensionless form

$$\mathbf{u}^\infty = \mathbf{E} \cdot \mathbf{x}, \qquad (8\text{–}145)$$

with

$$\mathbf{E} = \begin{bmatrix} -1 & 0 & 0 \\ 0 & -1 & 0 \\ 0 & 0 & +2 \end{bmatrix}.$$

In this technique, we seek the *disturbance* velocity field,

$$\mathbf{u}' = \mathbf{u} - \mathbf{u}^\infty, \qquad (8\text{–}146)$$

so that

$$\mathbf{u}' \to 0 \quad \text{as } |\mathbf{x}| \to \infty, \qquad (8\text{–}147)$$

$$\mathbf{u}' = -\mathbf{E} \cdot \mathbf{x} \quad \text{for } \mathbf{x} \in \mathbf{x}_s \text{ (sphere surface)}. \qquad (8\text{–}148)$$

We have already noted that a systematic procedure does not yet exist for choosing the singularities necessary to solve any particular problem. Thus we rely on a combination of intuition and a knowledge of the properties of the basic singularities to make an educated guess. In the present case, it is clear from symmetry considerations alone that there is no net force or torque on the sphere. Thus the solution cannot involve either a stokeslet or a rotlet. Furthermore, we have already seen that the flow field associated with the stresslet is an axisymmetric extensional motion, and the form of the disturbance flow at the sphere surface is also an axisymmetric extension. Thus an educated guess in this case is that the dominant singularity at the sphere center should be a stresslet. Then, because the stokeslet is always accompanied by a potential dipole and the stresslet is just a derivative of the stokeslet, it seems reasonable to suppose that the stresslet should be accompanied by a derivative of the potential dipole – namely, the potential quadrapole.

Thus we suggest trying to construct a solution that is a superposition of a stresslet and a potential quadrapole $\mathbf{u}_{D4}$ both located at the center of the sphere, that is,

$$\boxed{\mathbf{u}' = \mathbf{u}_{SS}(\mathbf{x}; \alpha, \beta) + \mathbf{u}_{D4}(\mathbf{x}; \delta, \gamma),} \qquad (8\text{–}149)$$

Creeping Flows – Three-Dimensional Problems

where

$$\boxed{\mathbf{u}_{D4}(\mathbf{x}; \boldsymbol{\delta}, \boldsymbol{\gamma}) \equiv \boldsymbol{\gamma} \cdot \nabla[\mathbf{u}_D(\mathbf{x}; \boldsymbol{\delta})] = \frac{6(\boldsymbol{\delta} \cdot \boldsymbol{\gamma})\mathbf{x}}{r^5} - \frac{15(\boldsymbol{\delta} \cdot \mathbf{x})(\boldsymbol{\gamma} \cdot \mathbf{x})\mathbf{x}}{r^7} + \frac{3(\boldsymbol{\delta} \cdot \mathbf{x})\boldsymbol{\gamma}}{r^5}.} \qquad (8\text{–}150)$$

Hence, at the surface of the sphere

$$\mathbf{u}' = [-\boldsymbol{\alpha} \cdot \boldsymbol{\beta} + 6(\boldsymbol{\delta} \cdot \boldsymbol{\gamma})]\mathbf{i}_r + 3(\boldsymbol{\delta} \cdot \mathbf{i}_r)\boldsymbol{\gamma} + [3(\boldsymbol{\alpha} \cdot \mathbf{i}_r)(\boldsymbol{\beta} \cdot \mathbf{i}_r) - 15(\boldsymbol{\delta} \cdot \mathbf{i}_r)(\boldsymbol{\gamma} \cdot \mathbf{i}_r)]\mathbf{i}_r. \qquad (8\text{–}151)$$

On the other hand, we have previously shown that the boundary condition (8–148) can be expressed in terms of spherical coordinates in the form (7–181), namely,

$$\mathbf{u}' = (1 - 3\cos^2\theta)\mathbf{i}_r + 3\sin\theta\cos\theta\,\mathbf{i}_\theta. \qquad (8\text{–}152)$$

It follows, by equating (8–151) and (8–152), that

$$\boldsymbol{\alpha} = \alpha\mathbf{i}_z, \quad \boldsymbol{\beta} = \beta\mathbf{i}_z, \quad \boldsymbol{\gamma} = \gamma\mathbf{i}_z, \quad \text{and} \quad \boldsymbol{\delta} = \delta\mathbf{i}_z, \qquad (8\text{–}153)$$

where

$$\beta\gamma = -1 \quad \text{and} \quad \alpha\beta = -5. \qquad (8\text{–}154)$$

Hence, we can write the solution (8–149) in the form

$$\boxed{\mathbf{u}' = -5\mathbf{u}_{SS}(\mathbf{x}; \mathbf{i}_z, \mathbf{i}_z) - \mathbf{u}_{D4}(\mathbf{x}; \mathbf{i}_z, \mathbf{i}_z),} \qquad (8\text{–}155)$$

$$\boxed{\mathbf{u} = \mathbf{u}^\infty + \mathbf{u}'.} \qquad (8\text{–}156)$$

The closely related problem of a rigid sphere in a linear shear flow is very easy to solve now by analogy to the solution for an axisymmetric straining flow. We consider the problem in the form

$$\mathbf{u}^\infty = \boldsymbol{\Gamma} \cdot \mathbf{x}, \qquad (8\text{–}157)$$

with

$$\boldsymbol{\Gamma} = \begin{bmatrix} 0 & 0 & 1 \\ 0 & 0 & 0 \\ 0 & 0 & 0 \end{bmatrix}.$$

Thus,

$$\mathbf{u}^\infty = z\mathbf{i}_1. \qquad (8\text{–}158)$$

The boundary condition at the sphere surface for the disturbance flow,

$$\mathbf{u}' \equiv \mathbf{u} - \mathbf{u}^\infty, \qquad (8\text{–}159)$$

is thus

$$\mathbf{u}' = -\boldsymbol{\Gamma} \cdot \mathbf{x} \quad \text{for } \mathbf{x} \in \mathbf{x}_s \text{ (sphere surface).} \qquad (8\text{–}160)$$

In this case, if the sphere does not rotate (that is, $\mathbf{u} = 0$ for $\mathbf{x} \in \mathbf{x}_s$), there will be a torque in the direction of the 2 axis, and the solution must therefore involve a rotlet, as well as a stresslet and a potential quadrapole. Indeed, if we derive the expression for surface velocity as a superposition of these three singularity types located at the sphere center and compare

E. Solutions for Solid Bodies by Means of Internal Distributions of Singularities

it with the boundary condition (8–160) expressed in terms of spherical coordinates, it is not difficult to show that the solution can be expressed in the form

$$\mathbf{u}' = -\frac{5}{6}\mathbf{u}_{SS}(\mathbf{x}; \mathbf{i}_x, \mathbf{i}_z) - \frac{1}{2}\mathbf{u}_R(\mathbf{x}; \mathbf{i}_y) - \frac{1}{6}\mathbf{u}_{D4}(\mathbf{x}; \mathbf{i}_x, \mathbf{i}_z),\quad (8\text{–}161)$$

with

$$\mathbf{u} = \mathbf{u}^\infty + \mathbf{u}'.$$

This solution is identical to (8–50).

4. Uniform Flow past a Prolate Spheroid

We have seen in the preceding two subsections that solutions for creeping flow in the vicinity of a spherical body can be generated by means of a superposition of fundamental singularities at the center of the sphere. Although this technique, in some instances, may be simpler to apply or more convenient than the methods discussed earlier, the reader may nevertheless wonder at the need to have introduced yet another new technique. For solutions involving spheres only, this skepticism would be well founded. However, the use of internal distributions of singularities applies equally well to certain creeping-flow problems involving nonspherical bodies. Although these problems could also be solved by generalization of the methods introduced earlier, these methods would require the use of elliptical or other nonspherical coordinates. This is not true of the use of internal distributions of singularities.

The solution of Stokes' flow problems by internal distributions of singularities for nonspherical bodies has been discussed in a series of papers by Chwang and Wu[13] and more recently by Kim.[14] Here, as an example to demonstrate basic principles, we consider the relatively simple problem of uniform flow past a prolate spheroid

$$\frac{x'^2}{a^2} + \frac{r'^2}{b^2} = 1 \quad (r'^2 \equiv y'^2 + z'^2,\ a \geq b), \quad (8\text{–}162)$$

with the free-stream velocity given by

$$\mathbf{u}'^\infty = U_1 \mathbf{i}_x + U_2 \mathbf{i}_y \quad \text{as } |\mathbf{x}| \to \infty, \quad (8\text{–}163)$$

where $\mathbf{i}_x$ and $\mathbf{i}_y$ are unit vectors. In dimensionless terms, with a characteristic length scale a and a characteristic velocity scale, say, U_1, the prolate spheroid is specified as

$$x^2 + \left(\frac{a^2}{b^2}\right) r^2 = 1 \quad (8\text{–}164)$$

with the free-stream velocity given by

$$\mathbf{u}^\infty = \mathbf{i}_x + \frac{U_2}{U_1}\mathbf{i}_y \quad \text{as } |\mathbf{x}| \to \infty. \quad (8\text{–}165)$$

For bodies of nonspherical shape, it is generally necessary to use a *distribution* of singularities over an interior *surface*, but this does not offer much advantage over a solution in terms of a distribution of stokeslets on the body surface itself, by means of (8–121) and (8–122). However, for prolate spheroids, we shall see that it is sufficient to utilize a *line* distribution of singularities along the central axis of revolution, between the two foci. In this case, the original problem of determining a *surface* distribution of stokeslets is reduced to the *one-dimensional* problem of determining the distribution of stokeslets and higher-order singularities on a line.

Although there are no firm rules to determine the *type* of singularities that should be used, we attempt to use the qualitative insight of the preceding sections. Thus, because the net force on the body will be nonzero, we try to construct a solution by using a line distribution of stokeslets, between the foci of the ellipsoid, $x = -c$ and $+c$ (dimensionless), where

$$c = \left(1 - \frac{b^2}{a^2}\right)^{1/2} = e, \qquad (8\text{–}166)$$

and e is also known as the eccentricity $0 \le e < 1$. In addition, because we have found earlier that the stokeslet is always accompanied by a potential dipole, we also include a line distribution of potential dipoles along the centerline from $x = -c$ to c. Hence we try to construct a solution for the disturbance flow in the form

$$\mathbf{u}' = -\int_{-c}^{c} [\alpha_1 \mathbf{u}_s(\mathbf{x} - \boldsymbol{\xi}; \mathbf{i}_x) + \alpha_2 \mathbf{u}_s(\mathbf{x} - \boldsymbol{\xi}; \mathbf{i}_y)] \, d\xi$$
$$+ \int_{-c}^{c} (c^2 - \xi^2)[\beta_1 \mathbf{u}_D(\mathbf{x} - \boldsymbol{\xi}; \mathbf{i}_x) + \beta_2 \mathbf{u}_D(\mathbf{x} - \boldsymbol{\xi}; \mathbf{i}_y)] \, d\xi, \qquad (8\text{–}167)$$

$$p' = -\int_{-c}^{c} [\alpha_1 p_s(\mathbf{x} - \boldsymbol{\xi}; \mathbf{i}_x) + \alpha_2 p_s(\mathbf{x} - \boldsymbol{\xi}; \mathbf{i}_y)] \, d\xi, \qquad (8\text{–}168)$$

where $\boldsymbol{\xi} = \xi \mathbf{i}_x$.

The proposed form, (8–167) and (8–168), clearly satisfies the creeping-flow equations (7–6) and also satisfies the free-stream boundary condition, namely,

$$\mathbf{u}' \to 0 \quad \text{as } |\mathbf{x}| \to \infty. \qquad (8\text{–}169)$$

The question is whether the no-slip condition at the spheroid surface can be satisfied, that is,

$$\mathbf{u}' = -\mathbf{i}_x - \frac{U_2}{U_1} \mathbf{i}_y \quad \text{for } \mathbf{x} \in \mathbf{x}_s \text{ (spheroid surface)}, \qquad (8\text{–}170)$$

through the choice of constants α_1, α_2, β_1, and β_2. If the suggested solution form is correct, then this should be possible. However, it is also possible that we may need additional singularities or that the singularities should be distributed with different weighting functions or over some other portion of the centerline, and in this case no choice of α_1, α_2, β_1, and β_2 will work. To see whether we can satisfy the no-slip conditions, we make use of the integrated form of $\mathbf{u}'$, which can be written as

$$\mathbf{u}' = -(2\alpha_1 \mathbf{i}_x + \alpha_2 \mathbf{i}_y) B_{1,0} - (\alpha_1 r \mathbf{i}_r + \alpha_2 y \mathbf{i}_x)\left(\frac{1}{R_2} - \frac{1}{R_1}\right)$$
$$+ (\alpha_1 r \mathbf{i}_x - \alpha_2 y \mathbf{i}_r) r B_{3,0} + \nabla\left[-2\beta_1 B_{1,1} + \beta_2 y\left(\frac{x-c}{r^2} R_1 - \frac{x+c}{r^2} R_2 + B_{1,0}\right)\right], \qquad (8\text{–}171)$$

where $\mathbf{e}_r = (y\mathbf{e}_y + z\mathbf{e}_z)/r$ is the radial unit vector in the yz plane and

$$R_1 = [(x+c)^2 + r^2]^{1/2},$$
$$R_2 = [(x-c)^2 + r^2]^{1/2},$$
$$B_{1,0} = \log \frac{R_2 - (x-c)}{R_1 - (x+c)},$$
$$B_{1,1} = R_2 - R_1 + x B_{1,0},$$
$$B_{3,0} = \frac{1}{r^2}\left(\frac{x+c}{R_1} - \frac{x-c}{R_2}\right),$$
$$B_{3,1} = \left(\frac{1}{R_1} - \frac{1}{R_2}\right) + x B_{3,0}.$$

E. Solutions for Solid Bodies by Means of Internal Distributions of Singularities

On the surface of the spheroid,

$$r^2 = (1-e^2)(1-x^2), \quad R_1 = 1+ex, \quad R_2 = 1-ex,$$

$$B_{1,0} = \log\frac{1+e}{1-e} = L_e, \quad B_{3,0} = \frac{2e}{(1-e^2)(1-e^2x^2)}.$$

Hence, after some manipulation, the velocity *at the spheroid surface* can be expressed in the form

$$\mathbf{u}'_s = \left[\frac{2\alpha_1}{e} - 2(\alpha_1 + \beta_1)L_e\right]\mathbf{i}_x - \frac{2}{e}\left(\alpha_1 - \frac{2e^2\beta_1}{1-e^2}\right)\frac{b^2\mathbf{i}_x + e^2 x r \mathbf{i}_r}{1-e^2 x^2}$$

$$- \left[\frac{2e^2\beta_2}{1-e^2} + (\alpha_2 - \beta_2)L_e\right]\mathbf{i}_y - 2e\left(\alpha_2 - \frac{2e^2\beta_2}{1-e^2}\right) y \frac{\frac{b^2}{a^2}x\mathbf{i}_x + r\mathbf{i}_r}{\frac{b^2}{a^2}(1-e^2 x^2)}. \quad (8\text{--}172)$$

Thus, the no-slip condition, (8–170), is satisfied if

$$\boxed{\begin{aligned}\alpha_1 &= \frac{2\beta_1 e^2}{(1-e^2)} = e^2[-2e + (1+e^2)L_e]^{-1}, & (8\text{--}173)\\ \alpha_2 &= \frac{2\beta_2 e^2}{(1-e^2)} = 2\frac{U_2}{U_1}e^2[2e + (3e^2 - 1)L_e]^{-1}. & (8\text{--}174)\end{aligned}}$$

Although this completes the formal solution and demonstrates that the distribution of stokeslets and potential dipoles proposed in (8–167) is sufficient to solve the problem, it is of interest to calculate the force on the body.

In view of the properties of the stokeslet and potential dipole solutions, the force acting on the spheroid is simply represented by the accumulative strength of the stokeslet distribution along the centerline of the spheroid. Thus, in dimensionless terms,

$$\mathbf{F} = -8\pi \int_{-c}^{c} (-\alpha_1 \mathbf{i}_x - \alpha_2 \mathbf{i}_y) dx. \quad (8\text{--}175)$$

Hence, evaluating the integral by using the expressions (8–173) and (8–174) for α_1 and α_2, we obtain

$$\boxed{\mathbf{F} = 6\pi\left(C_{F1}\mathbf{i}_x + \frac{U_2}{U_1}C_{F2}\mathbf{i}_y\right),} \quad (8\text{--}176)$$

where

$$\boxed{\begin{aligned}C_{F1} &= \frac{8}{3}e^3\left[-2e + (1+e^2)\log\frac{1+e}{1-e}\right]^{-1}, & (8\text{--}177\text{a})\\ C_{F2} &= \frac{16}{3}e^3\left[2e + (3e^2-1)\log\frac{1+e}{1-e}\right]^{-1}. & (8\text{--}177\text{b})\end{aligned}}$$

In dimensional terms, then, the total force is

$$\boxed{\mathbf{F}_D = 6\pi\mu a(U_1 C_{F1}\mathbf{i}_x + U_2 C_{F2}\mathbf{i}_y).} \quad (8\text{--}178)$$

Creeping Flows – Three-Dimensional Problems

For the limiting case of a sphere, $e \to 0$, and
$$C_{F1} = C_{F2} = 1,$$
as expected. On the other hand, for very slender spheroids, with $b/a = (1 - e^2)^{1/2} \ll 1$, the force coefficients take the limiting form

$$\boxed{C_{F1} \approx \frac{2}{3} \frac{1}{\log(2a/b) - 1/2}, \quad C_{F2} \approx \frac{4}{3} \frac{1}{\log(2a/b) + 1/2},} \quad (8\text{–}179)$$

so that
$$\frac{C_{F1}}{C_{F2}} \to \frac{1}{2} \quad \text{as} \quad \frac{b}{a} \to 0. \quad (8\text{–}180)$$

This latter result also can be obtained by means of the approximate *slender-body* procedure outlined in the next section. The fact that the ratio of the force coefficients C_{F1}/C_{F2} approaches 1/2 only for an extremely elongated, needlelike shape is of qualitative interest because it reflects a general property of creeping flows; namely, the resistance to motion of a body is remarkably insensitive to the geometric configuration. In particular, the ratio 1/2 means that an elongated, needlelike body will translate at a speed when moving parallel to its symmetry axis that is only a factor of 2 faster than the same body would translate under the action of the same force when moving perpendicular to its symmetry axis. The insensitivity of the resistance to the particle geometry is also evident in the fact that the force law (8–178) for a long slender body differs from that for a sphere of radius a primarily in the logarithmic factor $\log(2a/b)$, which varies only *slowly* with the axis ratio. In particular, the drag on a slender rod aligned parallel to its direction of motion is only $2/3 \log(2a/b)$ times the drag on a sphere having the same diameter as the length of the rod. Although this ratio vanishes as $b/a \to 0$, it is only 1/12 for $b/2a \sim 10^{-4}$.

5. Approximate Solutions of the Creeping-Flow Equations by Means of Slender-Body Theory

In the preceding subsection, we saw that exact solutions for creeping flow past a solid body can be obtained by means of internal distributions of fundamental singularities and that this can lead to significant simplification when the geometry of the body is either spherical or a prolate, axisymmetric ellipsoid (other body shapes that have been studied include oblate spheroids[15]). In the prolate ellipsoid case, instead of solving for the strength of a surface distribution of stokeslets, as suggested by the integral formulation (8–121), we need determine only the distribution of stokeslets and higher-order singularities on the centerline of the body (and this reduces to a single point at the center for a sphere). Hence, as noted earlier, the problem of solving creeping-flow problems is reduced to the solution of 1D integral equations for the singularity distribution functions. Unfortunately, however, for complicated flows it is not clear what particular singularities are needed, nor what the functional forms of the density functions should be, and this currently reduces the practical usefulness of the singularity superposition technique.

In one case, however, the problem is greatly simplified. This is the creeping motion of a fluid relative to a *very slender body* – namely, a body in which the typical cross-sectional dimension is very small compared with the body's length.[9] In this case, an approximate solution can be obtained for bodies of *arbitrary* cross-sectional shape by use of only a stokeslet distribution along the centerline of the body. Intuitively, it is clear that this must be the case. An exact solution can always be obtained, at least in principle, by means of a *surface* distribution of stokeslets, and this solution, in turn, can be expressed in terms of an equivalent *internal* distribution of stokeslets, Stokes dipoles, and higher-order singularities by means of the generalized Taylor series approximation that was initially given as

E. Solutions for Solid Bodies by Means of Internal Distributions of Singularities

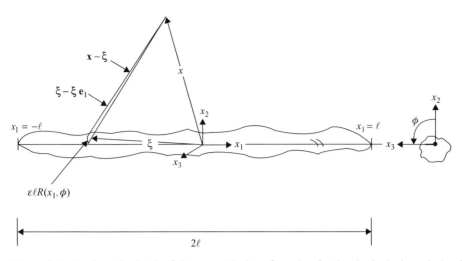

Figure 8–7. A schematic sketch of the geometrical configuration for the slender-body analysis of an elongated body of arbitrary cross-sectional shape [derivation of Eq. (8–181)].

Eq. (8–124). However, for a slender body, the magnitude of the position vector that connects points at the body surface and a point on the centerline is very small (for the same x_1), and thus the stokeslet distribution at the body surface can be accurately approximated by use of only a stokeslet distribution on the centerline. This qualitative discussion can be formalized in a straightforward manner.

Let us then consider the elongated solid particle that is sketched in Fig. 8–7. The length of the particle is 2ℓ, and the surface of the particle can be denoted in the form

$$\sqrt{x_2'^2 + x_3'^2} = \varepsilon \ell R(\mathbf{x}_1, \phi),$$

where the x_1 axis is assumed to be parallel to the particle centerline, ϕ is the azimuthal angle about the x_1 axis, R is an $O(1)$ quantity, and ε is the slenderness ratio. In dimensionless terms, the equation for the body surface becomes

$$\sqrt{x_2^2 + x_3^2} = \varepsilon R(\mathbf{x}_1, \phi),$$

where we have taken the half-length of the body ℓ as a characteristic length scale, and the body now lies in the range $-1 \le x_1 \le 1$. For the slender-body approximation that we shall seek here, it is assumed that $\varepsilon \ll 1$.

In the case of an axisymmetric body, the cross-sectional shape at any point is a circle (R independent of ϕ), and for the special case of a spheroid, $\varepsilon \equiv a/b$, where a and b are the semi-minor and semi-major axis lengths.

We have already seen in Eq. (8–122) that a general solution of the creeping-flow equations for flow past a solid body is

$$\mathbf{u}(\mathbf{x}) = \mathbf{u}^\infty(\mathbf{x}) + \frac{1}{8\pi} \int_{\partial D} \left[\frac{\mathbf{I}}{R} + \frac{(\mathbf{x} - \boldsymbol{\xi})(\mathbf{x} - \boldsymbol{\xi})}{R^3} \right] \cdot \mathbf{t}(\boldsymbol{\xi}) dA_\xi,$$

where $\mathbf{t}(\boldsymbol{\xi}) = \mathbf{n} \cdot \mathbf{T}$. Now, $\mathbf{x}$ is the position vector for a fixed point in the flow domain, and $\boldsymbol{\xi}$ is the position vector for a variable point of integration on the body surface, ∂D. However, for a slender body, the vector $\mathbf{x} - \boldsymbol{\xi}$ can be approximated by the vector $\mathbf{x} - \xi \mathbf{e}_1$ as suggested

Creeping Flows – Three-Dimensional Problems

by the sketch in Fig. 8–7, where $-1 \le \xi \le 1$. Thus the general solution (8–122) can be approximated in the form of a line integral along the centerline of the particle:

$$u_i(\mathbf{x}) = u_i^\infty(\mathbf{x}) + \frac{1}{8\pi} \int_{-1}^{1} \left[\frac{\delta_{ij} r + (x_i - \xi \delta_{i1})(x_j - \xi \delta_{j1})}{r^3} \right] f_j(\xi) d\xi, \qquad (8\text{–}181)$$

where

$$r \equiv \sqrt{(x_1 - \xi)^2 + \rho^2}, \qquad \rho^2 \equiv x_2^2 + x_3^2,$$

and $f_i(\xi)$ is the integral of the local surface-stress vector $t_j(\xi)$ around the closed perimeter of the particle at each point $x_1 = \xi$. For convenience, we have reverted to a component form for (8–181). The error in approximating the exact solution (8–122) by the approximate form (8–181) generally depends on the geometry of the ends of the body. For a particle with rounded ends, the error is of order $\varepsilon^2 \ln \varepsilon$.

To apply (8–181) to obtain the solution for a specific problem, we must determine the weighting function $f_j(\xi)$ by applying the boundary condition $\mathbf{u} = 0$ at ∂D. Thus proceeding formally, we obtain

$$u_i^\infty(\mathbf{x}) = -\frac{1}{8\pi} \int_{-1}^{1} \left[\frac{\delta_{ij}}{r} + \frac{(x_i - \xi \delta_{i1})(x_j - \xi \delta_{j1})}{r^3} \right] f_j(\xi) d\xi \qquad (8\text{–}182)$$

for $\mathbf{x} \in \partial D$. Although the reader may question whether the approximation (8–182) remains valid for $\mathbf{x} \in \partial D$, as the approximation $\mathbf{x} - \boldsymbol{\xi} \approx \mathbf{x} - \xi \mathbf{e}_1$ is clearly in error when $\boldsymbol{\xi}$ is in the neighborhood of $\mathbf{x}$, it can be shown rigorously that (8–181) remains valid (basically because the approximation referred to previously fails over a portion of ∂D that is very small). Thus the solution of creeping-flow problems for slender bodies is reduced to solving an integral equation of the first kind, (8–182), for the stokeslet density function $f_j(\xi)$.

To this end, let us consider the integral terms in (8–182), beginning with the first term. With $\rho = \varepsilon R(\mathbf{x})$, the integral is

$$\int_{-1}^{1} \frac{f_i(\xi) d\xi}{\sqrt{(x - \xi)^2 + \varepsilon^2 R^2}}. \qquad (8\text{–}183)$$

Because this integral is singular at $x = \xi$ in the limit $\varepsilon \to 0$, we rewrite it in the form

$$\int_{-1}^{1} \frac{f_i(\xi) d\xi}{\sqrt{(x - \xi)^2 + \varepsilon^2 R^2}} \equiv f_i(x) \int_{-1}^{1} \frac{d\xi}{\sqrt{(x - \xi)^2 + \varepsilon^2 R^2}} + \int_{-1}^{1} \frac{f_i(\xi) - f_i(x)}{\sqrt{(x - \xi)^2 + \varepsilon^2 R^2}} d\xi.$$

$$(8\text{–}184)$$

In the limit $\varepsilon \to 0$, the right-hand side of (8–184) is dominated by the first term. Specifically,

$$I \equiv \int_{-1}^{1} \frac{d\xi}{\sqrt{(x - \xi)^2 + \varepsilon^2 R^2}} = \ln\left\{ (x + 1) + [(x + 1)^2 + \varepsilon^2 R^2]^{1/2} \right\}$$
$$- \ln\left\{ (x - 1) + [(x - 1)^2 + \varepsilon^2 R^2] \right\} \qquad (8\text{–}185)$$
$$\cong -2 \ln \varepsilon + \ln \frac{4(1 - x^2)}{R^2} + \cdots + = -2 \ln \varepsilon + O(1).$$

E. Solutions for Solid Bodies by Means of Internal Distributions of Singularities

The symbol $O(1)$ means that the error in approximating the integral as $-2 \ln \varepsilon$ is independent of ε as $\varepsilon \to 0$, and thus is much smaller than $-2 \ln \varepsilon$. The second integral in (8–184) also represents an error of $O(1)$. Similarly,

$$\int_{-1}^{1} \frac{(x-\xi)^2 d\xi}{[(x-\xi)^2 + \varepsilon^2 R^2]^{3/2}} \to I - 2 + O(\varepsilon^2) = -2\ln\varepsilon + O(1). \qquad (8\text{–}186)$$

Hence, utilizing (8–185) and (8–186), we can approximate the integral equation (8–182) in the form

$$u_i^\infty(x_1, 0, 0) = \frac{\ln \varepsilon}{4\pi}[f_i(x_1) + \delta_{i1} f_1(x_1)] + O(1). \qquad (8\text{–}187)$$

With $\mathbf{u}^\infty(\mathbf{x})$ specified, the algebraic equation (8–187) can thus be solved to determine a first approximation to the stokeslet density distribution in the slender-body limit, $\varepsilon \to 0$.

In general, it can be seen from (8–187) that

$$f_1(x_1) \approx \frac{2\pi u_1^\infty(x_1, 0, 0)}{\ln \varepsilon} \quad \text{and} \quad f_i \approx \frac{4\pi u_i^\infty(x_1, 0, 0)}{\ln \varepsilon} \quad \text{for } i = 1, 2. \qquad (8\text{–}188)$$

Thus, for $\mathbf{u}^\infty = \mathbf{e}_1$ (that is, uniform flow parallel to the slender-body axis, with $\mathbf{u}^\infty$ nondimensionalized by the magnitude of the velocity, U),

$$f_1 \approx \frac{2\pi}{\ln \varepsilon}, \quad f_2 = f_3 = 0, \qquad (8\text{–}189)$$

and the dimensionless drag force on the body is therefore

$$F_1 = -\int_{-1}^{1} f_1 dx_1 = \frac{4\pi}{\ln(1/\varepsilon)}. \qquad (8\text{–}190)$$

The corresponding dimensional drag force is thus

$$(F_D)_1 = \frac{4\pi \mu \ell U}{\ln(1/\varepsilon)}. \qquad (8\text{–}191)$$

For $\mathbf{u}^\infty = \mathbf{e}_2$, on the other hand,

$$f_2 \cong \frac{4\pi}{\ln \varepsilon}, \quad f_1 = f_3 = 0, \qquad (8\text{–}192)$$

and the dimensional force is

$$(F_D)_2 = -\mu U \ell \int_{-1}^{1} f_2 dx_1 = \frac{8\pi \mu \ell U}{\ln(1/\varepsilon)}. \qquad (8\text{–}193)$$

Comparing the slender-body results (8–191) and (8–193) with the limiting form of the exact solution for prolate ellipsoids (spheroids) [Eqs. (8–176) and (8–179)], we see that the two are identical. Again, the force on a very elongated body is seen to be approximately proportional to its length. Furthermore, the ratio of the force for motion parallel and perpendicular to the particle axis, respectively, is just $(F_D)_2/(F_D)_1 = 2$. This result provides further evidence of the relative insensitivity of the drag to the body geometry in the creeping-flow limit.

F. THE BOUNDARY-INTEGRAL METHOD

We have previously obtained the general integral representation of the velocity and pressure fields that is known as the Ladyzhenskaya solution, (8–117)–(8–119). It should be emphasized that this "solution" does not actually solve any creeping-flow problem, but simply recasts the creeping-flow equations into an integral form. However, this integral form of the creeping-flow equations, (8–117) or (8–119), is the basis of a powerful solution method known as the boundary-integral method.

In this method, the boundary conditions are used to specify either the velocity or stress at the boundaries, thus converting (8–117) or (8–119) into an *integral equation* for whichever of the surface velocity or stress vector is not specified at the boundaries by the boundary conditions. Once this integral equation is solved, we know both $\mathbf{u} \cdot \mathbf{n}$ and $\mathbf{T} \cdot \mathbf{n}$ at the boundaries and then (8–117) or (8–119) can be used to calculate the velocity at any interior point in the solution domain. A general advantage of the boundary-integral formulation is that it reduces the dimensionality of the problem that must be solved. In the most general case, in which the fluid motion is fully 3D, it is obvious that the integral equation derived from (8–117) or (8–119) is only 2D. (If $\mathbf{x} \to \mathbf{x}_s \in D$, these equations are completely defined on the 2D surface ∂D.) If the flow problem is axisymmetric, as in the example of a sphere in a uniaxial extensional flow or an ellipsoid of revolution translating parallel to its axis of revolution, the velocity and the surface stress will be independent of the azimuthal angle, and we can reduce the integral equation derived from either (8–117) or (8–119) to a 1D problem by first evaluating the integrals with respect to the azimuthal coordinate.

The solution of the integral equations must be done *numerically* in virtually all cases. In this book, we do not discuss the strictly numerical issues of solving these equations. There have been excellent books and review papers written that cover this topic at a level of depth that could not be emulated here (see, for example, the excellent book of Pozrikidis[7]). Instead, the focus is on explaining the principles of using (8–117) or (8–119) to formulate the *boundary-integral equations*.

The key to using (8–117) or (8–119) as the basis for derivation of the boundary-integral equations is that they must be applied at the boundaries of the flow domain where boundary conditions are specified. For this purpose, we need to understand the behavior of the single- and double-layer terms as we approach a boundary, ∂D. The single-layer terms are continuous in the entire fluid domain, including boundaries, provided only that the latter are sufficiently smooth. However, the double-layer terms are not continuous at ∂D, but suffer a jump.[11] In particular, let us define a function $\mathbf{W}(\mathbf{x})$

$$\mathbf{W}(\mathbf{x}) \equiv -\frac{3}{4\pi} \int_{\partial D} \frac{(\mathbf{x} - \boldsymbol{\xi})(\mathbf{x} - \boldsymbol{\xi})(\mathbf{x} - \boldsymbol{\xi})}{R^5} \cdot \mathbf{u} \cdot \mathbf{n} dA_\xi, \qquad (8\text{–}194)$$

in which $\mathbf{n}$ is the unit normal pointing out of the fluid domain. Then the jump condition for the double-layer term can be expressed in the form

$$\mathbf{W}_i(\mathbf{x}_s) = \frac{1}{2}\mathbf{u}(\mathbf{x}_s) + \mathbf{W}_s(\mathbf{x}_s), \qquad (8\text{–}195\text{a})$$

$$\mathbf{W}_e(\mathbf{x}_s) = -\frac{1}{2}\mathbf{u}(\mathbf{x}_s) + \mathbf{W}_s(\mathbf{x}_s). \qquad (8\text{–}195\text{b})$$

Here, $\mathbf{W}_i(\mathbf{x}_s)$ and $\mathbf{W}_e(\mathbf{x}_s)$ denote, respectively, the limiting values of $\mathbf{W}(\mathbf{x})$ as $\mathbf{x}$ approaches $\mathbf{x}_s \in \partial D$ from inside and outside the fluid domain. The function $\mathbf{W}_s(\mathbf{x}_s)$ is $\mathbf{W}(\mathbf{x})$ evaluated exactly at the point $\mathbf{x} = \mathbf{x}_s \in \partial D$. The reader may well be puzzled why we should care about the value of $\mathbf{W}(\mathbf{x})$ for points $\mathbf{x}$ that lie out of the fluid domain. This will become evident when we consider the case of applying (8–117) or (8–119) at a liquid–liquid interface.

F. The Boundary-Integral Method

1. A Rigid Body in an Unbounded Domain

The most straightforward class of problems that can be solved by use of the boundary-integral technique is the motion of a solid body in an unbounded fluid that is undergoing a motion $\mathbf{u}_\infty(\mathbf{x})$ that is a solution of the creeping-flow equations far from the body. The general representation (8–119) provides an expression for the velocity in the fluid domain in terms of the single- and double-layer potentials. On the surface of a rigid particle that is undergoing a general translational and rotational motion, the velocity is

$$\mathbf{u}(\mathbf{x}_s) = \mathbf{U} + \mathbf{\Omega} \wedge \mathbf{x}_s. \tag{8-196}$$

Now, if $\mathbf{u}(\boldsymbol{\xi})$ is a constant, as in the present case, it can be shown that the integral of the double-layer potential is zero (the proof of this statement is left to the reader). Thus (8–119) reduces to

$$\mathbf{u}(\mathbf{x}) = \mathbf{u}_\infty(\mathbf{x}) + \frac{1}{8\pi} \int_{\partial D} \left[\frac{\mathbf{I}}{R} + \frac{(\mathbf{x} - \boldsymbol{\xi})(\mathbf{x} - \boldsymbol{\xi})}{R^3} \right] \cdot (\mathbf{T}(\boldsymbol{\xi}) \cdot \mathbf{n}) dA_\xi, \tag{8-197}$$

which shows that flows past a rigid particle can be represented with just the single-layer potential.

Now, generally, it is the motion of the particle that is specified, and the surface stress vector $\mathbf{T}(\boldsymbol{\xi}) \cdot \mathbf{n}$ is unknown. However, if we evaluate (8–197) at points $\mathbf{x}_s$ on the boundary of the particle, it is converted into an integral equation from which the surface-stress distribution can be determined. Specifically, taking account of the condition (8–195), we obtain from (8–197)

$$\frac{1}{2}\mathbf{u}(\mathbf{x}_s) = \mathbf{u}_\infty(\mathbf{x}_s) + \frac{1}{8\pi} \int_{\partial D} \left[\frac{\mathbf{I}}{R} + \frac{(\mathbf{x}_s - \boldsymbol{\xi})(\mathbf{x}_s - \boldsymbol{\xi})}{R^3} \right] \cdot (\mathbf{T}(\boldsymbol{\xi}) \cdot \mathbf{n}) dA_\xi. \tag{8-198}$$

Hence, knowledge of the boundary velocity $\mathbf{u}(\mathbf{x}_s)$ and the form of the undisturbed flow evaluated at the body surface $\mathbf{u}_\infty(\mathbf{x}_s)$ allows a direct calculation of the surface-force vector $\mathbf{T} \cdot \mathbf{n}$ by means of a solution of the integral equation, (8–198). It is emphasized that we do not actually address the numerical problem of solving (8–198). We note, however, that it is an integral equation of the *first kind*, and it is known that there can be numerical difficulties with the solution of this class of integral equations. The reader who wishes to learn more about the details of numerical solution should consult one of the general reference books that were listed in the introduction to this section.

The formulation (8–198) was used by Youngren and Acrivos[16] to calculate the force on solid particles of different shapes translating through an unbounded *stationary* fluid, $\mathbf{u}_\infty(\mathbf{x}_s) = 0$, in what was likely the first application of the boundary-integral method to creeping-flow problems. Many subsequent investigators have used it to calculate forces on bodies of complicated shape, in a variety of undisturbed flows.[17]

2. Problems Involving a Fluid Interface

The value of the boundary-integral method is particularly evident if we consider problems in which one or more of the boundaries is a fluid interface. Here, for simplicity, we consider the generic problem of a drop in an unbounded fluid that is undergoing some mean motion that causes the drop to deform in shape. This type of problem is particularly difficult because the shape of the interface is unknown and is often changing with time. We shall see that the boundary-integral formulation provides a powerful basis to attack this class of problems, and, in fact, is largely responsible for much of the considerable theoretical progress that has been made in recent years in understanding phenomena such as the deformation and breakup of a viscous drop in various types of linear shear flow in the creeping-motion regime.[8]

For convenience, we begin by restating the governing equations and boundary conditions in nondimensional form:

$$\nabla^2 \mathbf{u} - \nabla p = 0, \quad \nabla \cdot \mathbf{u} = 0; \tag{8-199a}$$

$$\nabla^2 \hat{\mathbf{u}} - \nabla \hat{p} = 0, \quad \nabla \cdot \hat{\mathbf{u}} = 0; \tag{8-199b}$$

with

$$\mathbf{u} \to \mathbf{u}_\infty \quad \text{as } |\mathbf{x}| \to \infty, \tag{8-200}$$

$$\left.\begin{array}{r} \mathbf{u} = \hat{\mathbf{u}} \\[4pt] \dfrac{1}{|\nabla F|}\dfrac{\partial F}{\partial t} + \mathbf{u}\cdot\mathbf{n} = 0 \\[6pt] (\mathbf{T} - \lambda\hat{\mathbf{T}}) + \dfrac{1}{Ca}[\mathrm{grad}_s\, \gamma' - (\nabla\cdot\mathbf{n})\mathbf{n}] = 0 \end{array}\right\} \text{ at } \mathbf{x} = \mathbf{x}_s.$$

(8-201a)
(8-201b)
(8-201c)

Here,

$$Ca \equiv \frac{\mu u_c}{\gamma_s} \quad \text{and} \quad \lambda \equiv \frac{\hat{\mu}}{\mu}.$$

We denote the characteristic length and velocity scales as ℓ_c and u_c, the characteristic time scale as ℓ_c/u_c, and the characteristic stress and pressure scales as $\mu u_c/\ell_c$ and $\lambda\mu u_c/\ell_c$ for the fluid outside and inside the drop, respectively. We express the interfacial tension in the form

$$\gamma = \gamma_s(1 + \gamma') \tag{8-202}$$

where γ_s is the equilibrium value of the interfacial tension at the ambient temperature and with uniform coverage of surfactant (if any is present). In the preceding equations, the unit normal points from the drop outward into the suspending fluid. The interface is defined in terms of a function F such that

$$F(\mathbf{x}, t) = 0 \quad \text{for } \mathbf{x} \in \mathbf{x}_s. \tag{8-203}$$

The function F is unknown and must be calculated as part of the solution of the problem.

To approach this problem by means of the boundary-integral technique, we first note the general expression for the velocity in the exterior fluid:

$$\mathbf{u}(\mathbf{x}) = \mathbf{u}^\infty(\mathbf{x}) + \frac{3}{4\pi}\int_{\partial D}\left[\frac{(\mathbf{x}-\boldsymbol{\xi})(\mathbf{x}-\boldsymbol{\xi})(\mathbf{x}-\boldsymbol{\xi})}{R^5}\cdot\mathbf{u}\right]\cdot\mathbf{n}\,dA_\xi \\ - \frac{1}{8\pi}\int_{\partial D}\left[\frac{\mathbf{I}}{R} + \frac{(\mathbf{x}-\boldsymbol{\xi})(\mathbf{x}-\boldsymbol{\xi})}{R^3}\right]\cdot\mathbf{T}\cdot\mathbf{n}\,dA_\xi, \tag{8-204}$$

where ∂D represents the surface of the drop. It will be noted that the signs are opposite those in the original expression, (8–119), because the direction of the normal vector is into rather than out of the fluid as assumed in (8–119). The velocity inside the drop is

$$\hat{\mathbf{u}}(\mathbf{x}) = -\frac{3}{4\pi}\int_{\partial D}\left[\frac{(\mathbf{x}-\boldsymbol{\xi})(\mathbf{x}-\boldsymbol{\xi})(\mathbf{x}-\boldsymbol{\xi})}{R^5}\cdot\hat{\mathbf{u}}(\boldsymbol{\xi})\right]\cdot\mathbf{n}\,dA_\xi \\ + \frac{1}{8\pi}\int_{\partial D}\left[\frac{\mathbf{I}}{R} + \frac{(\mathbf{x}-\boldsymbol{\xi})(\mathbf{x}-\boldsymbol{\xi})}{R^3}\right]\cdot\hat{\mathbf{T}}(\boldsymbol{\xi})\cdot\mathbf{n}\,dA_\xi. \tag{8-205}$$

F. The Boundary-Integral Method

Now, we combine these two expressions, taking account of the boundary conditions (8–201a) and (8–201c). First, we evaluate (8–204) and (8–205) at the surface of the drop. Taking account of the jump conditions (8–195a) and (8–195b), we obtain

$$\frac{1}{2}\mathbf{u}(\mathbf{x}_s) = \mathbf{u}^\infty(\mathbf{x}_s) + \frac{3}{4\pi}\int_{\partial D}\left[\frac{(\mathbf{x}_s-\boldsymbol{\xi})(\mathbf{x}_s-\boldsymbol{\xi})(\mathbf{x}_s-\boldsymbol{\xi})}{R^5}\cdot\mathbf{u}(\boldsymbol{\xi})\right]\cdot\mathbf{n}\,dA_\xi$$
$$-\frac{1}{8\pi}\int_{\partial D}\left[\frac{\mathbf{I}}{R}+\frac{(\mathbf{x}_s-\boldsymbol{\xi})(\mathbf{x}_s-\boldsymbol{\xi})}{R^3}\right]\cdot\mathbf{T}(\boldsymbol{\xi})\cdot\mathbf{n}\,dA_\xi, \quad (8\text{--}206)$$

$$\frac{1}{2}\hat{\mathbf{u}}(\mathbf{x}_s) = -\frac{3}{4\pi}\int_{\partial D}\left[\frac{(\mathbf{x}_s-\boldsymbol{\xi})(\mathbf{x}_s-\boldsymbol{\xi})(\mathbf{x}_s-\boldsymbol{\xi})}{R^5}\cdot\hat{\mathbf{u}}(\boldsymbol{\xi})\right]\cdot\mathbf{n}\,dA_\xi$$
$$+\frac{1}{8\pi}\int_{\partial D}\left[\frac{\mathbf{I}}{R}+\frac{(\mathbf{x}_s-\boldsymbol{\xi})(\mathbf{x}_s-\boldsymbol{\xi})}{R^3}\right]\cdot\hat{\mathbf{T}}(\boldsymbol{\xi})\cdot\mathbf{n}\,dA_\xi. \quad (8\text{--}207)$$

Now if we multiply (8–207) by the viscosity ratio λ and add it to (8–206), taking into account the condition (201a),

$$\mathbf{u}(\mathbf{x}_s) = \hat{\mathbf{u}}(\mathbf{x}_s),$$

we obtain

$$\frac{1}{2}(1+\lambda)\mathbf{u}(\mathbf{x}_s) = \mathbf{u}_\infty(\mathbf{x}_s) + (1-\lambda)\frac{3}{4\pi}\int_{\partial D}\left[\frac{(\mathbf{x}_s-\boldsymbol{\xi})(\mathbf{x}_s-\boldsymbol{\xi})(\mathbf{x}_s-\boldsymbol{\xi})}{R^5}\cdot\mathbf{u}(\boldsymbol{\xi})\right]\cdot\mathbf{n}\,dA_\xi$$
$$-\frac{1}{8\pi}\int_{\partial D}\left[\frac{\mathbf{I}}{R}+\frac{(\mathbf{x}_s-\boldsymbol{\xi})(\mathbf{x}_s-\boldsymbol{\xi})}{R^3}\right]\cdot[\mathbf{T}(\boldsymbol{\xi})-\lambda\hat{\mathbf{T}}(\boldsymbol{\xi})]\cdot\mathbf{n}\,dA_\xi. \quad (8\text{--}208)$$

Finally, we can substitute for the stress difference by using (8–201c):

$$\frac{1}{2}(1+\lambda)\mathbf{u}(\mathbf{x}_s) = \mathbf{u}_\infty(\mathbf{x}_s) + (1-\lambda)\frac{3}{4\pi}$$
$$\times\int_{\partial D}\left[\frac{(\mathbf{x}_s-\boldsymbol{\xi})(\mathbf{x}_s-\boldsymbol{\xi})(\mathbf{x}_s-\boldsymbol{\xi})}{R^5}\cdot\mathbf{u}(\boldsymbol{\xi})\right]\cdot\mathbf{n}\,dA_\xi$$
$$-\frac{1}{8\pi}\int_{\partial D}\left[\frac{\mathbf{I}}{R}+\frac{(\mathbf{x}_s-\boldsymbol{\xi})(\mathbf{x}_s-\boldsymbol{\xi})}{R^3}\right]$$
$$\times\left\{-\frac{1}{Ca}[\mathrm{grad}_s\gamma'-(\nabla\cdot\mathbf{n})\mathbf{n}]\right\}\cdot\mathbf{n}\,dA_\xi. \quad (8\text{--}209)$$

This is the key result for application of the boundary-integral technique to interface dynamics problems. Let us first consider a case in which the interfacial tension is constant, i.e., $\mathrm{grad}_s\gamma' = 0$. In this case, if the shape of the interface is specified, then $\nabla\cdot\mathbf{n}$ is known, and (8–209) is an integral equation for the interface velocity, $\mathbf{u}(\mathbf{x}_s)$. Hence the problem defined by (8–199)–(8–203) can be solved as follows for a specified undisturbed flow, $\mathbf{u}\to\mathbf{u}_\infty$ as $|\mathbf{x}|\to\infty$. The drop shape is initially specified (usually as a sphere). The integral equation (8–209) is then solved to obtain the interface velocity, $\mathbf{u}(\mathbf{x}_s)$. Then, with $\mathbf{u}(\mathbf{x}_s)$ known, we can use a discretized form of the kinematic condition, (8–201b), to increment the drop shape forward one step in time. We then return to (8–209) with this new drop shape, and again solve for $\mathbf{u}(\mathbf{x}_s)$, and so on. This process continues as long as the interface shape continues to evolve. If there is a steady-state solution, and our numerical scheme is working properly, we should find that

$$\mathbf{u}(\mathbf{x}_s)\cdot\mathbf{n}\to 0$$

at large time. If, on the other hand, no steady-state shape exists for the particular undisturbed flow, initial shape, and specified value of Ca, the shape will continue to evolve with time. It should be noted that the preceding description did not involve calculating the fluid velocity except at the interface. Clearly, once (8–209) is solved to determine $\mathbf{u}(\mathbf{x}_s)$ at any point in time (i.e., corresponding to the instantaneous shape of the drop), we can use (8–204) and (8–205) to determine the velocity at any point $\mathbf{x}$ either inside or outside the drop. However, to calculate the time-dependent shape of the drop, we do not need to calculate the velocity or pressure in the fluids. In cases in which it is the evolution of the drop shape that is the primary result of interest, the fact that we get this without the necessity of calculating velocity or pressure fields is a major advantage of the boundary-integral formulation. Furthermore, as in the example of the previous subsection, we also benefit from the fact that the boundary-integral formulation reduces the spatial dimension of the problem by one.

If the interfacial tension is not constant, there is an additional Marangoni stress term in (8–209). If $\text{grad}_s \gamma'$ is known, the calculation can proceed as previously described. However, generally the interfacial tension is not constant because the temperature of the interface is not constant or because there are surfactants present with a nonuniform concentration at the interface. In either case, the nonuniformity of temperature or surfactant concentration is generally affected by the motion of the fluids inside and outside the drop. In principle, we can use the known velocity fields in the thermal energy equations for the two fluids (or in the surfactant transport equations) to determine the corresponding temperature or concentration distributions. However, from a practical point of view, if we must calculate the velocity fields at points other than at the interface, the primary computational advantage of the boundary-integral formulation is lost. Two cases can be considered for which this is not the case. In the first case, there is the possibility that the transport process in the bulk fluids is dominated by diffusion (i.e., by conduction in the thermal problem), so that we can calculate the temperature or surfactant concentration fields in the bulk fluids without needing the velocity fields. However, as we shall see in the next chapter, this requires the Peclet number for the bulk fluid transport process to be small. From a practical point of view, however, it is unlikely that this condition will be satisfied while still maintaining convection as an important transport mechanism within the interface. If the interface transport process is also dominated by diffusion, the temperature (or concentration) will be uniform and $\text{grad}_s \gamma' = 0$.

The second case is of greater practical importance. This is the problem of an *insoluble* surfactant—namely, a surfactant that does not exchange significantly with the bulk fluids. In this case, the surfactant concentration distribution at the interface is determined by a balance between diffusion and convection on the interface, together with "dilution" if the area of the interface changes with deformation of shape (the latter occurs because the total quantity of an insoluble surfactant is fixed), but there is no coupling with the external flow except as this influences the velocity distribution in the interface. The governing equation for the interface concentration for this case was given in Chap. 2 as (2–162). For this equation, all that we need to know is the interface velocity $\mathbf{u}_s \equiv \mathbf{u}(\mathbf{x}_s)$, and this comes automatically in the boundary-integral formulation, without the need for any additional calculations. The reader who is interested in surfactant effects may wish to refer to Stone and Leal, or to other more recent references such as Eggleton *et al.*, or Li and Pozrikidis, all of which are listed in Ref. 8.

3. Problems in a Bounded Domain

Finally, we briefly consider the case in which the flow domain is bounded. A typical problem is the motion of a particle or drop in the vicinity of a plane wall or within a circular tube. We might also be interested in systems in which there are multiple particles. One approach to this

F. The Boundary-Integral Method

class of problems is to simply proceed by "brute force," using (8–117) or (8–119) with ∂D recognized as the sum of all the surfaces in the problem; i.e., the particle/drop plus the surface of the wall or tube, or the sum of all of the particle surfaces in the multiparticle problem. For example, in the case of a solid particle near a solid wall with the velocity of both the particle and the wall specified, we can apply (8–198) at both the particle surface and on the wall to calculate the surface-force vector $\mathbf{T} \cdot \mathbf{n}$ on both surfaces. From this, we could evaluate the effect of the wall on the hydrodynamic force experienced by the particle. However, though the wall may be unbounded in extent, we will only be able to discretize its surface out to some finite distance, and this will introduce an unwanted approximation into the calculation, in addition to those already inherent in solving the integral equation numerically.

A more efficient approach is to base the boundary-integral formulation on a fundamental solution (or more accurately a *Green's function*) that incorporates the relevant boundary conditions at one or more of the surfaces. In the case of a particle or drop moving near an infinite plane wall, this means finding a solution for a point force that exactly satisfies the no-slip and kinematic boundary conditions at the wall. If we were to consider the motion of a particle or drop in a tube, it would be useful to have the solution for a point force satisfying the same conditions on the tube walls.

The solution for a point force near a solid, plane wall of infinite extent was originally obtained by Blake.[18] For this relatively simple geometry, we can construct the solution by considering an image system of point singularities applied at a point on the opposite side of the wall (i.e., out of the actual fluid domain). To see that this might work, we could begin with a simpler problem. Let us suppose that we have a point force of strength F applied at a point $x_3 = h$ normal to a plane-bounding surface at $x_3 = 0$. If we were to suppose that there were an oppositely oriented point force applied in the same fluid but on the opposite side of the boundary at $x_3 = -h$, then the sum of the two solutions would be a velocity field in the fluid for $x_3 \geq 0$ that satisfies the kinematic condition $\mathbf{u} \cdot \mathbf{n} = 0$ at $x_3 = 0$. Of course, the solution we seek is also supposed to satisfy the no-slip condition at $x_3 = 0$. To satisfy this condition, we require additional singularities at the image point, but the strength of these singularities must decrease as h is increased. Clearly, as the point $x_3 = h$ moves farther from the wall at $x_3 = 0$, the tangential velocity components generated by it and its image stokeslet at $x_3 = -h$ will get smaller and smaller, and thus require weaker contributions from whatever other singularities we need at the image point.

Blake has shown that the image system required for satisfying both the kinematic and no-slip conditions at the boundary $x_3 = 0$ for a point force of strength F at $x_3 = \xi_3 = h$ consists of a stokeslet (point force) of the same strength but oppositely directed, plus a Stokes dipole [Eqs. (8–125)] of strength $2hF$ and a potential dipole of strength $2h^2 F$. The sense of these additional singularities for a point force perpendicular and parallel to the boundary is shown in Fig. 8–8. Utilizing the same notation that was originally used in the derivation of the Ladyzhenskaya solution, (8–112) and (8–113), but here in component form, the velocity component in the i direction for a point force applied at (ξ_1, ξ_2, ξ_3) in the j direction is

$$U_i^j = \frac{1}{8\pi} \left\{ \left(\frac{1}{r} - \frac{1}{R}\right) \delta_{ij} + \frac{r_i r_j}{r^3} - \frac{R_i R_j}{R^3} \right. \\ \left. + 2\xi_3 (\delta_{j\alpha}\delta_{\alpha l} - \delta_{j3}\delta_{3l}) \frac{\partial}{\partial R_l} \left[\frac{\xi_3 R_i}{R^3} - \left(\frac{\delta_{i3}}{R} + \frac{R_i R_3}{R^3}\right) \right] \right\}, \quad (8\text{–}210)$$

the pressure is

$$Q^j = \frac{1}{4\pi} \left[\frac{r_j}{r^3} - \frac{R_j}{R^3} - 2\xi_3 (\delta_{j\alpha}\delta_{\alpha l} - \delta_{j3}\delta_{3l}) \frac{\partial}{\partial R_l} \left(\frac{R_3}{R^3}\right) \right], \quad (8\text{–}211)$$

Creeping Flows – Three-Dimensional Problems

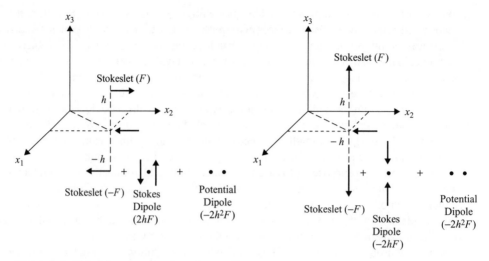

Figure 8–8. Blake's image system for a plane wall.

and the stress is

$$\Sigma_{ik}^{j} = \frac{3}{4\pi} \left\{ \frac{r_i r_j r_k}{r^5} - \frac{R_i R_j R_k}{R^5} - 2\xi_3 (\delta_{j\alpha}\delta_{\alpha l} - \delta_{j3}\delta_{3l}) \left[-\frac{\xi_3}{R^5} \delta_{ik} R_l \right. \right.$$
$$\left. \left. + \frac{x_3}{R^5}(R_i \delta_{lk} + \delta_{il} R_k) + \frac{R_i \delta_{3l} R_k}{R^5} - \frac{5 R_i R_l R_k x_3}{R^7} \right] \right\},$$
(8–212)

where the point force is located at (ξ_1, ξ_2, ξ_3), $r \equiv [(x_1 - \xi_1)^2 + (x_2 - \xi_2)^2 + (x_3 - \xi_3)^2]^{1/2}$, and $R \equiv [(x_1 - \xi_1)^2 + (x_2 - \xi_2)^2 + (x_3 + \xi_3)^2]^{1/2}$.

Now, for example, if we wish to consider the motion of a solid particle moving in the vicinity of a plane wall, we can use these solutions in (8–115) to derive a solution that is analogous to (8–117). The result is the same as before,

$$\mathbf{u}(\mathbf{x}) = -\int_{\partial D} \mathbf{n} \cdot [(\mathbf{U} \cdot \mathbf{e}) \cdot \mathbf{T} - \mathbf{u} \cdot (\Sigma \cdot \mathbf{e})] \, dA_\xi,$$

except of course that the fundamental solutions are (8–210) and (8–212) instead of (8–112) and (8–113). The result is a more complicated solution form, but the advantage is that the surface integral on ∂D involves only the particle surface. The boundary conditions at the wall are satisfied automatically. An implementation of this solution to the motion of a solid particle near a planar wall can be found in Ascoli et al. (1989),[19] but there are a number of applications of this basic idea of implementing a Green's function that satisfies boundary conditions on one of the boundaries of the flow domain as a way of reducing the size of the surface that must be discretized in the boundary integral technique.

G. FURTHER TOPICS IN CREEPING-FLOW THEORY

The preceding sections have been concerned primarily with direct solution techniques for problems in creeping-flow theory. Here, we discuss several general topics that evolve directly from these developments. The first two involve application of the so-called *reciprocal theorem* of low-Reynolds-number hydrodynamics.

G. Further Topics in Creeping-Flow Theory

1. The Reciprocal Theorem

The reciprocal theorem is derived directly from the general integral formula, (8–111). For this purpose, we identify **u** and **û**, as well as **T** and **T̂**, as the solutions of two creeping-flow problems for flow past the same body but with different boundary conditions on the body surface ∂D.

Then, because

$$\nabla \cdot \mathbf{T} = 0, \quad \nabla \cdot \hat{\mathbf{T}} = 0,$$

(8–111) can be written in the form

$$\int_{\partial D} \hat{\mathbf{u}} \cdot \mathbf{f} \, dA = \int_{\partial D} \mathbf{u} \cdot \hat{\mathbf{f}} \cdot dA, \tag{8–213}$$

where

$$\mathbf{f} = \mathbf{T} \cdot \mathbf{n} \quad \text{and} \quad \hat{\mathbf{f}} = \hat{\mathbf{T}} \cdot \mathbf{n}.$$

The relationship (8–213) is the famous *reciprocal theorem* of Lorentz.[20] As we shall see shortly, it is an extremely useful result that often leads to the impression of "getting something for nothing."[21]

2. Faxen's Law for a Body in an Unbounded Fluid

As a first application of the reciprocal theorem, we consider its use in calculating the hydrodynamic force on a body in an undisturbed flow, $\mathbf{u}^\infty(\mathbf{x})$.

In particular, let us suppose that we have obtained the solution of the creeping-motion equations for uniform flow **U** past a *stationary* body ∂D. Equivalently, we may consider the case of a particle that translates with velocity $-\mathbf{U}$ in a fluid at rest at infinity. We denote the solution of this problem as **u** and the corresponding surface-force vector on ∂D as **f**. Then, on applying the reciprocal theorem, we find that

$$\mathbf{U} \cdot \int_{\partial D} \hat{\mathbf{f}} \, dA = \mathbf{U} \cdot \hat{\mathbf{F}} = \int_{\partial D} \hat{\mathbf{u}} \cdot \mathbf{f} \, dA. \tag{8–214}$$

But this simple formula provides an example of the idea of something for nothing. For if we have actually solved the uniform-flow problem, we can immediately deduce the force on the same body held fixed in any undisturbed flow $\mathbf{u}^\infty(\mathbf{x})$ that satisfies the creeping-flow equations. In particular,

$$\boxed{\mathbf{U} \cdot \hat{\mathbf{F}} = \int_{\partial D} \mathbf{u}^\infty \cdot \mathbf{f} \, dA.} \tag{8–215}$$

Hence we can obtain the force $\hat{\mathbf{F}}$ for the undisturbed flow $\mathbf{u}^\infty(\mathbf{x})$ by means of a simple integration of the right-hand side of (8–215), with **f** known from the solution of the uniform-flow problem but without any need to actually solve the creeping-flow problem involving $\mathbf{u}^\infty(\mathbf{x})$. To obtain all three components of $\hat{\mathbf{F}}$, we require a solution for uniform flow past the same body along any three mutually perpendicular directions.

An especially powerful result is obtained if we apply the formula (8–215) to a stationary solid sphere of radius a in the undisturbed flow $\mathbf{u}^\infty(\mathbf{x})$. In this case, the solution for uniform flow past a stationary solid sphere yields the general result

$$\mathbf{f} = \frac{3}{2a} \mu \mathbf{U}. \tag{8–216}$$

Hence, from (8–215),

$$\mathbf{U} \cdot \hat{\mathbf{F}} = \int_{\partial D} \mathbf{u}^\infty(\mathbf{x}) \cdot \left(\frac{3}{2a}\mu \mathbf{U}\right) dA.$$

Because $\mathbf{U}$ is arbitrary,

$$\boxed{\hat{\mathbf{F}} = \frac{3}{2a}\mu \int_{\partial D} \mathbf{u}^\infty(\mathbf{x}) \, dA.} \qquad (8\text{–}217)$$

Finally, supposing the origin to be at the center of the sphere, we can expand $\mathbf{u}^{(\infty)}$ at ∂D in a multipole representation in terms of the values of $\mathbf{u}^{(\infty)}$ and higher-order derivatives of $\mathbf{u}^{(\infty)}$ at the position occupied by the center of the sphere:

$$\mathbf{u}^{(\infty)}(\mathbf{x}) = \mathbf{u}^{(\infty)}(0) + \mathbf{x} \cdot \left[\nabla \mathbf{u}^{(\infty)}\right]_0 + \frac{\mathbf{x}\mathbf{x}}{2!} : \left\{\nabla\left[\nabla \mathbf{u}^{(\infty)}\right]\right\}_0 + \cdots + . \qquad (8\text{–}218)$$

Inserting (8–218) into (8–217), we can show that

$$\hat{\mathbf{F}} = 6\pi\mu a \left\{\mathbf{u}^{(\infty)}(0) + \frac{a^2}{6}\left[\nabla^2 \mathbf{u}^{(\infty)}\right]_0 + \text{const}\left[\nabla^4 \mathbf{u}^{(\infty)}\right]_0 + \cdots + \right\}. \qquad (8\text{–}219)$$

In deriving (8–219) from (8–217), it may be noted that the integrals of all the odd-number terms in (8–218) vanish because ∂D is a sphere. However, because

$$\nabla^2 \mathbf{u} - \frac{1}{\mu}\nabla p = 0$$

we see that $\nabla^4 \mathbf{u} = 0$, as well as all other terms, $\nabla^{2n} \mathbf{u}$, for $n \geq 2$. Hence (8–219) reduces to the exact result

$$\boxed{\hat{\mathbf{F}} = 6\pi\mu a \left\{\mathbf{u}^{(\infty)}(0) + \frac{a^2}{6}\left[\nabla^2 \mathbf{u}^{(\infty)}\right]_0\right\}.} \qquad (8\text{–}220)$$

This important result is known as *Faxen's law*.[22] According to this law, if we specify the undisturbed velocity $\mathbf{u}^\infty(\mathbf{x})$, then the force on a sphere can be calculated directly from the formula (8–220), without any need to actually solve the flow problem corresponding to the free-stream velocity $\mathbf{u}^\infty(\mathbf{x})$.

A number of interesting results can be obtained from the Faxen formula, (8–220). The first is to note that the force on a stationary sphere in an arbitrary linear flow,

$$\mathbf{u}^\infty = \mathbf{U}_\infty + \mathbf{\Gamma}_\infty \cdot \mathbf{x}, \qquad (8\text{–}221)$$

where $\mathbf{U}_\infty$ and $\mathbf{\Gamma}_\infty$ are independent of $\mathbf{x}$, depends only on the translational velocity of the undisturbed fluid evaluated at the position occupied by the body center $\mathbf{x}_p$ and is otherwise independent of the velocity gradient tensor $\mathbf{\Gamma}_\infty$. If the sphere is translating with a velocity $\mathbf{U}_p$, then the force depends on the *relative* translational velocity between the body and the undisturbed fluid evaluated at the position occupied by the center of the sphere,

$$\boxed{\hat{\mathbf{F}} = 6\pi\mu a(\mathbf{U}_\infty + \mathbf{\Gamma}_\infty \cdot \mathbf{x}_p - \mathbf{U}_p).} \qquad (8\text{–}222)$$

G. Further Topics in Creeping-Flow Theory

Thus, in the absence of an external force on the sphere (in the presence of gravity, we assume $\rho_p = \rho_{fluid}$), it must translate at steady state with the undisturbed velocity of the fluid evaluated at the position occupied by its center point, that is,

$$\boxed{\mathbf{U}_p = \mathbf{U}_\infty + \mathbf{\Gamma}_\infty \cdot \mathbf{x}_p.} \qquad (8\text{–}223)$$

It follows that a neutrally buoyant sphere in any linear flow can be treated as though the origin of coordinates for the undisturbed flow is coincident with the center of the sphere, namely,

$$\mathbf{u}^\infty = \mathbf{\Gamma}_\infty \cdot (\mathbf{x} - \mathbf{x}_p). \qquad (8\text{–}224)$$

In a flow with *quadratic* dependence on spatial position, on the other hand, the hydrodynamic force on a sphere will not simply equal Stokes' drag. For example, let us suppose that we have a sphere in a 2D, Poiseuille flow:

$$\mathbf{u}^\infty = \frac{G}{2\mu}(dy - y^2)\mathbf{i}_x, \qquad (8\text{–}225)$$

where G is the negative of the pressure gradient and d is the distance between the two walls. Then, according to Faxen's law, the hydrodynamic force is

$$\boxed{\hat{\mathbf{F}} = 6\pi\mu a \left[\frac{G}{2\mu}\left(dy_p - y_p^2\right)\mathbf{i}_x - \mathbf{U}_p - \frac{Ga^2}{6\mu}\mathbf{i}_x \right].} \qquad (8\text{–}226)$$

Again, suppose that there is no external force acting on the sphere (suppose it is neutrally buoyant). Then, according to (8–226),

$$\boxed{\mathbf{U}_p = \frac{Gd^2}{2\mu}\left[\bar{y}_p - \bar{y}_p^2 - \frac{1}{3}\left(\frac{a}{d}\right)^2\right]\mathbf{i}_x,} \qquad (8\text{–}227)$$

where $\bar{y}_p = y_p/d$. Thus the particle translates slower by an amount $(1/3)(a/d)^2$ than it would do if the center of the particle simply translated with the undisturbed velocity of the fluid, evaluated at the position occupied by the particle center. It should be noted, however, that we obtained the results (5–226) and (5–227) without taking any account of the hydrodynamic interaction between the sphere and the channel walls; that is, we calculated the force on the sphere by implicitly assuming that the only effect of the channel walls is to create the parabolic profile (8–225), by means of the no-slip condition. This should be a reasonable approximation for $a/d \ll 1$, provided the particle is not too near either of the walls, that is, $\bar{y}_p \neq 0, 1$. In this case, however, the correction in (8–227) will be very small. This suggests that it would be valuable to try to include the direct hydrodynamic effect of the walls, but this is beyond our present scope (notice, however, that we will briefly consider the interaction with a single wall later in this section).

3. Inertial and Non-Newtonian Corrections to the Force on a Body

Another consequence of the integral theorem (8–111) is that we can calculate *inertial and non-Newtonian corrections* to the force on a body directly from the creeping-flow solution. Let us begin by considering inertial corrections for a Newtonian fluid. In particular, let us recall that the creeping-flow equations are an approximation to the full Navier–Stokes equations we obtained by taking the limit $Re \to 0$. Thus, if we start with the full equations of motion for a steady flow in the form

$$\nabla \cdot \mathbf{T} = Re(\mathbf{u} \cdot \nabla \mathbf{u}), \qquad (8\text{–}228)$$

then the creeping-flow equation is obtained as the limit $Re \to 0$. Alternatively, however, it is intuitively obvious that the creeping flow solution also can be obtained as the first term in a series approximation of the form

$$\mathbf{u} = \mathbf{u}_0 + Re\mathbf{u}_1 + \cdots +,$$
$$\mathbf{T} = \mathbf{T}_0 + Re\mathbf{T}_1 + \cdots +, \qquad (8\text{--}229)$$

for $Re \ll 1$. This type of approximation scheme will be discussed at length in Chap. 9. For now, we simply note that substitution of (8–229) into (8–228) yields the approximation

$$\nabla \cdot \mathbf{T}_0 + Re(\nabla \cdot \mathbf{T}_1 - \mathbf{u}_0 \cdot \nabla \mathbf{u}_0) + \cdots + = 0. \qquad (8\text{--}230)$$

Hence, if this approximation is to hold for small, but arbitrary, values of the Reynolds number, it is obvious that each term in (8–230) must individually be equal to zero, that is,

$$\nabla \cdot \mathbf{T}_0 = 0, \qquad (8\text{--}231)$$
$$\nabla \cdot \mathbf{T}_1 - \mathbf{u}_0 \cdot \nabla \mathbf{u}_0 = 0, \qquad (8\text{--}232)$$

and so on. The first of these equations is just the creeping-flow equation. Thus, as previously suggested, the creeping-flow equation is the first approximation to the full equations of motion for small values of the Reynolds number, $Re \ll 1$. The advantage of deriving it by means of Eq. (8–229) is that we also obtain a governing equation for the second- and higher-order approximations in the series. It should be noted that the derivation of (8–231) and (8–232) does not guarantee that an approximate solution of the form (8–229) actually exists. We can establish this only by showing that solutions to (8–231), (8–232), etc., exist that satisfy the boundary conditions at each order of approximation.

For present purposes, we assume that the approximation (8–229) is valid. Then, if we wished to obtain a correction to the solution of the creeping-flow equation, (8–231), to account for the first influence of the acceleration or inertial contributions to the equations of motion, we would solve (8–232) subject to appropriate boundary conditions with $\mathbf{u}_0$ being the creeping-flow solution. Hence, to calculate a first correction to the force on ∂D, we would expect to solve (8–232) for $\mathbf{u}_1$ and only then to calculate $\mathbf{F}_1$:

$$\mathbf{F} = \mathbf{F}_0 + Re\mathbf{F}_1 + \cdots +, \qquad (8\text{--}233)$$

where

$$\mathbf{F}_1 = \int_{\partial D} \mathbf{T}_1 \cdot \mathbf{n} dA. \qquad (8\text{--}234)$$

Surprisingly, however, it is not actually necessary to solve for the velocity and pressure fields, $\mathbf{u}_1$ and p_1, in order to determine the first correction to the force $\mathbf{F}_1$. Instead, by manipulating the integral theorem (8–111), we can determine $\mathbf{F}_1$ based on the solution of creeping-flow problems only.

To see that this is true, let us identify the velocity field $\mathbf{u}$ in (8–111) as the creeping-flow solution for translation of the body ∂D with unit velocity in the $\mathbf{e}$ direction. Then,

$$\int_V \mathbf{u} \cdot (\nabla \cdot \hat{\mathbf{T}}) dV = \int_{\partial D} \hat{\mathbf{u}} \cdot (\mathbf{T} \cdot \mathbf{n}) dA - \mathbf{e} \cdot \int_{\partial D} \hat{\mathbf{T}} \cdot \mathbf{n} dA. \qquad (8\text{--}235)$$

Now let $\hat{\mathbf{u}}, \hat{\mathbf{T}}$ stand for the solution $(\mathbf{u}_1, \mathbf{T}_1)$ of (8–232), subject to the boundary conditions,

$$\mathbf{u}_1 = 0 \quad \text{on} \quad \partial D. \qquad (8\text{--}236)$$

G. Further Topics in Creeping-Flow Theory

This condition derives from the fact that the boundary condition for the disturbance flow is

$$\hat{\mathbf{u}} = -\mathbf{u}^{(\infty)}(\mathbf{x}) \quad \text{on } \partial D.$$

Hence, for arbitrary Re,

$$\mathbf{u}_0 = -\mathbf{u}^{(\infty)}(\mathbf{x}) \quad \text{on } \partial D,$$

and

$$\mathbf{u}_1 = 0 \quad \text{on } \partial D.$$

Now, if we carefully examine (8–235), the first integral on the right-hand side is zero because of (8–236), whereas the second integral is just $\mathbf{e} \cdot \mathbf{F}_1$. Hence,

$$\boxed{\mathbf{e} \cdot \mathbf{F}_1 = -\int_V \mathbf{u} \cdot (\mathbf{u}_0 \cdot \nabla \mathbf{u}_0) dV.} \qquad (8\text{–}237)$$

To calculate the component of $\mathbf{F}_1$ in the $\mathbf{e}$ direction, we therefore require the solution of the original Stokes' flow problem to obtain $\mathbf{u}_0$, plus the solution of a second Stokes' flow problem for translation through a quiescent fluid to obtain $\mathbf{u}$. However, we do *not* have to determine $\mathbf{u}_1, p_1$ to calculate $\mathbf{F}_1$, and this represents a substantial simplification of the problem.

The formulation leading to (8–237) has been used to great advantage in calculating inertial corrections to the motion of rigid bodies in flows with small, but nonzero, Reynolds number.[23] It must be remembered that the force $\mathbf{F}_1$ represents only a small correction to the actual force on the body. Nevertheless, the configurations of particles or bodies in Stokes' flows are often "indeterminate" in the sense that they depend solely on the initial configuration, and thus corrections to the particle trajectory that are very small at any instant may have a major *accumulative* effect on the system. One example of this is the radial position of a solid spherical particle that is being carried through a circular tube in a Newtonian, incompressible fluid. In this case, as we have already seen in Subsection B.3 of Chap. 7, the particle in creeping flow remains precisely at the radial position that was set by the initial configuration – that is, lateral or cross-stream motions (migrations) do not occur. However, if we take fluid inertia into account, there is no reason why the particle may not translate sideways, and the formula (8–237) provides a means to evaluate the lateral force based on only a pair of *creeping*-flow solutions; one for the original problem, and the other for translation normal to the channel walls through an otherwise quiescent fluid. Indeed, when the formula for the inertial contribution $\mathbf{F}_1$ to the force is evaluated, it is found that a lateral force does exist that drives particles away from the walls and away from the centerline, toward an intermediate, equilibrium position that is 60% of the way from the centerline to the walls.[24] This same result has been observed experimentally, beginning with the famous papers of Segre and Silberberg.[25]

The reader may note that the preceding developments actually provide a basis for calculating corrections to the force on bodies, ∂D, because of a number of different kinds of weak departures from the creeping-flow limit for Newtonian fluids. One additional example is the effects of weak non-Newtonian contributions to the motion of a body. In this case, for vanishingly small Reynolds numbers, we can symbolically write the equation of motion in the form

$$\boxed{\nabla \cdot \mathbf{T} = De[\mathbf{f}(\mathbf{u})],} \qquad (8\text{–}238)$$

where the left-hand side is the basic Stokes' operator and the right-hand side is the *nonlinear* correction that is due to the non-Newtonian fluid properties. The parameter De is known

575

as the *Deborah number* (see Section J in Chap. 2) and provides a measure of the relative magnitudes of the Newtonian and non-Newtonian contributions to the fluid's behavior. Clearly, a development that mirrors the preceding analysis can be used to obtain an equation from which the first correction, $De\,\hat{\mathbf{F}}_1$, to the force on a body can be calculated from the solution of the original creeping-flow problem only, plus one additional creeping flow solution for translation through a quiescent fluid. The result, analogous to Eq. (8–237), is

$$\mathbf{e} \cdot \hat{\mathbf{F}}_1 = -\int_V \mathbf{u} \cdot \mathbf{f}(\mathbf{u}_0)dV. \qquad (8\text{--}239)$$

As before, $\mathbf{u}_0$ is the creeping-flow solution of the same problem and $\mathbf{u}$ is the creeping flow solution for translation with velocity $\mathbf{e}$ through a quiescent fluid. The interested reader may find examples of the application of the formula (8–239) in published papers.[26]

4. Hydrodynamic Interactions Between Widely Separated Particles – The Method of Reflections

We have seen in Chap. 7 that the presence of a rigid body in a Stokes' flow produces a disturbance in the velocity field that decays only *algebraically* with increasing distance from the body. The rate of decay depends on the type of disturbance: If there is a net hydrodynamic force on the body, the far-field disturbance flow is dominated by the stokeslet velocity field (8–123) with a strength α that depends on the net force, and the velocity disturbance decays as $|r|^{-1}$; if there is no net force on the body, but the undisturbed flow is linear so that the dominant disturbance mode is a rotlet or stresslet, the disturbance decays as $|r|^{-2}$, and so forth. As a consequence of the long-range perturbation to the velocity field in such a case, we may expect significant hydrodynamic interactions when other bodies or boundaries are present, even when the separation distance is relatively large. Thus, for example, to achieve a dilute suspension in which each particle is hydrodynamically isolated from the others, we require an extremely small volume fraction of particles (<1%). Further, the sedimentation velocity of a solid particle is still significantly influenced by walls or other particles when separation distance is more than 10 particle radii away.

Clearly the analysis of particle–particle and particle–wall interactions is very important. Unfortunately, most of the classical analytical techniques are very difficult (or impossible) to apply. Most analytical methods rely on the existence of a coordinate system in which the particle and boundary surfaces are all coincident with coordinate surfaces, and there are very few particle–particle or particle–wall geometries that fall into this niche. The exceptions are two spheres, or a sphere and a plane wall, for which *bispherical* coordinates may be employed to obtain exact eigenfunction expansions for solution of the creeping-flow equations. In fact, this coordinate system has been used to solve for the translation of two spheres along their line of centers and for translation of a sphere near an infinite plane wall or interface. However, even for these particular geometries, for which the bispherical coordinate system can be used, the analysis is quite complicated (and tedious) to use, and the results for such entities as the force or torque are generally obtained in the form of an infinite series that converges very slowly when the separation distance between the surfaces is small. Thus it is essential to have other solution procedures that do not rely on a specific coordinate system. One is the boundary-integral method described in the preceding section F, but ultimately this requires numerical methods for solving the boundary-integral equations.

The best available *analytic* technique is known as the *method of reflections*. It is an iterative scheme that applies when the separation distance between the particles or between the particle and wall is large relative to the characteristic dimension of the particle(s).[27] The basic idea is to approximate the solution as a series of terms that satisfy the creeping-flow

G. Further Topics in Creeping-Flow Theory

equations at each level, but only alternatively satisfy the boundary conditions at the solid surfaces. To illustrate the idea, let us suppose that we have two particles, A and B, that move in an unbounded fluid with velocities $\mathbf{U}_A$ and $\mathbf{U}_B$, We denote a length scale characteristic of the particles as a and denote the distance between the particles as d. The leading-order approximation to the velocity field in the vicinity of particle A is thus the creeping-flow solution, $\mathbf{u}_1^A$, which vanishes at infinity and satisfies the boundary condition

$$\mathbf{u}_1^A = \mathbf{U}_A \quad \text{on } \partial D_A. \tag{8–240}$$

Similarly, the leading-order approximation in the vicinity of B is the Stokes' velocity field that satisfies the boundary condition

$$\mathbf{u}_1^B = \mathbf{U}_B \quad \text{on } \partial D_B. \tag{8–241}$$

Now, neither $\mathbf{u}_1^A$ nor $\mathbf{u}_1^B$ is exact, because neither satisfies the no-slip condition on the surface of the other particle. For example, the motion of particle A produces a disturbance velocity in the vicinity of B of order $|\mathbf{U}_A| \cdot a/d$ and the motion of B similarly produces a disturbance velocity of the fluid in the vicinity of A. Thus, to improve the approximation of the velocity field near particle A, we add to $\mathbf{u}_1^A$ a correction $\mathbf{u}_2^A$ that is a solution of the creeping-flow equations that vanishes at infinity and satisfies the condition

$$\mathbf{u}_2^A = -\mathbf{u}_1^B \quad \text{on } \partial D_A. \tag{8–242}$$

Near particle B, we add a correction $\mathbf{u}_2^B$ that satisfies the creeping-flow equations, vanishes at infinity, and satisfies the condition

$$\mathbf{u}_2^B = -\mathbf{u}_1^A \quad \text{on } \partial D_B. \tag{8–243}$$

Now, however, each of the correction fields $\mathbf{u}_2^A$ and $\mathbf{u}_2^B$ produces an additional disturbance motion near the *other* particle that is smaller again by an amount proportional to (a/d). Thus, to improve the approximation near each particle, we add velocity fields $\mathbf{u}_3^A$ and $\mathbf{u}_3^B$, which satisfy the boundary conditions

$$\mathbf{u}_3^A = -\mathbf{u}_2^B \quad \text{on } \partial D_A \quad \text{and} \quad \mathbf{u}_3^B = -\mathbf{u}_2^A \quad \text{on } \partial D_B. \tag{8–244}$$

Clearly this procedure can be continued indefinitely. At each successive level of approximation, the magnitude of the correction decreases in proportion to (a/d). In the vicinity of ∂D_A the velocity field is approximated as

$$\mathbf{u}^A = \mathbf{u}_1^A + \mathbf{u}_2^A + \mathbf{u}_3^A + \cdots +, \tag{8–245}$$

and the hydrodynamic force is similarly approximated as

$$\mathbf{F}^A = \mathbf{F}_1^A + \mathbf{F}_2^A + \mathbf{F}_3^A. \tag{8–246}$$

Similar expressions hold for the particle B.

We can now apply the preceding general analysis to the specific case of *two identical spheres* of radius a, the centers of which are separated by a distance d, as illustrated in Fig. 8–9. The line connecting the particle centers is assumed to make an angle θ with the horizontal. Thus, at $\theta = 0$, the line of centers is horizontal, whereas for $\theta = \pi/2$ the line of centers is vertical. In the absence of any particle interaction, we assume that the spheres would translate vertically. At the leading order, the solution for both spheres is thus Stokes' solution

$$\mathbf{u}_1^A = \mathbf{u}_S\left(\mathbf{x}; \frac{3}{4}\mathbf{U}_A^0\right) + \mathbf{u}_D\left(\mathbf{x}; -\frac{1}{4}\mathbf{U}_A^0\right) \tag{8–247a}$$

$$\mathbf{u}_1^B = \mathbf{u}_S\left(\mathbf{x}; \frac{3}{4}\mathbf{U}_B^0\right) + \mathbf{u}_D\left(\mathbf{x}; -\frac{1}{4}\mathbf{U}_B^0\right). \tag{8–247b}$$

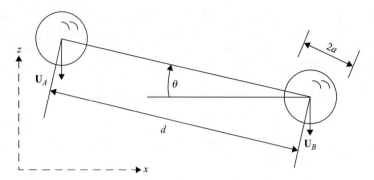

Figure 8–9. Two identical spheres of radius a, separated by a distance d, that are translating through a viscous fluid with velocity $\mathbf{U}_A = \mathbf{U}_B$. The line of centers makes an angle θ with the horizontal. For $0 < \theta < \pi/2$, the pair of spheres drifts toward the right. For $\pi/2 < \theta < \pi$, it drifts to the left.

Here, the superscript zero signifies that $\mathbf{U}_A^0$ and $\mathbf{U}_B^0$ are the velocities that the spheres would have in the absence of any hydrodynamic interaction between them. Because the particles are *identical*,

$$\mathbf{U}_A^0 = \mathbf{U}_B^0 = -U^0 \mathbf{i}_z. \tag{8–248}$$

Hence the velocity induced at particle A that is due to the motion of B and the velocity at particle B that is due to the motion of A are identical, that is,

$$\mathbf{u}_1^A(B) = \mathbf{u}_1^B(A) = -\frac{3}{4}\left(\frac{a}{d}\right)U^0(1 + \sin^2\theta)\mathbf{i}_z + \frac{3}{4}\left(\frac{a}{d}\right)U^0 \sin\theta\cos\theta\, \mathbf{i}_x + O\left[\left(\frac{a}{d}\right)^3 U^0\right]. \tag{8–249}$$

But each of these disturbance velocities is just the superposition of a pair of uniform streaming flows.

Now, we can consider two possible consequences of the disturbance flow that is induced in the vicinity of each particle by the motion of the other particle. If the particle *velocity* is assumed to be *fixed*, in the form of Eq. (8–248), there will be a change in the hydrodynamic force acting on the body [that is, the external force required for producing the specified motion, (8–248), will be changed]. If, on the other hand, the *external* (and thus the hydrodynamic) *force* is held *constant*, the velocity of the particles will change, and (8–248) is then just the first approximation to the particle velocity.

In the analysis outlined earlier in this section, we essentially assumed that the velocities of the particles were fixed, and calculated the resulting hydrodynamic forces. Thus we apply the boundary conditions (8–242) and (8–243) to determine $\mathbf{u}_2^A$ and $\mathbf{u}_2^B$, and we see that $\mathbf{u}_2^A$ and $\mathbf{u}_2^B$ are just the (creeping-flow) disturbance velocities for simple translation of spheres A and B with velocities $-\mathbf{u}_1^B(A)$ and $-\mathbf{u}_1^A(B)$ through an unbounded fluid. The solutions for $\mathbf{u}_2^A$ and $\mathbf{u}_2^B$ therefore can be expressed in the same form as (8–247a) and (8–247b), namely,

$$\mathbf{u}_2^A = \mathbf{u}_S\left(\mathbf{x}; -\frac{3}{4}\mathbf{u}_1^B(A)\right) + \mathbf{u}_D\left[\mathbf{x}; \frac{1}{4}\mathbf{u}_1^B(A)\right] \tag{8–250a}$$

$$\mathbf{u}_2^B = \mathbf{u}_S\left(\mathbf{x}; -\frac{3}{4}\mathbf{u}_1^A(B)\right) + \mathbf{u}_D\left[\mathbf{x}; \frac{1}{4}\mathbf{u}_1^A(B)\right]. \tag{8–250b}$$

G. Further Topics in Creeping-Flow Theory

The corresponding hydrodynamic force on the particles is

$$\boxed{\begin{aligned}\mathbf{F} = {}& 6\pi\mu a U^0 \left[1 - \frac{3}{4}\left(\frac{a}{d}\right)(1 + \sin^2\theta)\right]\mathbf{i}_z \\ & + 6\pi\mu a U^0 \left(\frac{a}{d}\right)\frac{3}{4}\sin\theta\cos\theta\,\mathbf{i}_x + O\left(\frac{a}{d}\right)^2.\end{aligned}} \tag{8–251}$$

Hence, in order to maintain the constant velocity $-U^0\mathbf{i}_z$, it would be necessary to apply extra forces to both spheres, one with a vertical component of magnitude,

$$6\pi\mu a U_0 \left(\frac{3}{4}\right)\left(\frac{a}{d}\right)(1 + \sin^2\theta),$$

and a horizontal component, from right to left, of magnitude,

$$6\pi\mu a U_0 \left(\frac{3}{4}\right)\left(\frac{a}{d}\right)\sin\theta\cos\theta.$$

The other approach to analysis, as indicated earlier, is to hold the force on the body constant and allow its velocity to vary. In this case, we assume that

$$\begin{aligned}\mathbf{U}_A &= \mathbf{U}_A^0 + \left(\frac{a}{d}\right)\mathbf{U}_A^1 + \cdots +, \\ \mathbf{U}_B &= \mathbf{U}_B^0 + \left(\frac{a}{d}\right)\mathbf{U}_B^1 + \cdots +,\end{aligned} \tag{8–252}$$

and the boundary conditions (8–242) and (8–243) are replaced with

$$\mathbf{u}_2^B = -\mathbf{u}_1^A(B) + \mathbf{U}_B^1\left(\frac{a}{d}\right) \quad \text{on } \partial D_B, \tag{8–253}$$

$$\mathbf{u}_2^A = -\mathbf{u}_1^B(A) + \mathbf{U}_A^1\left(\frac{a}{d}\right) \quad \text{on } \partial D_A, \tag{8–254}$$

Then, in place of the solutions (8–250), we find

$$\begin{aligned}\mathbf{u}_2^A &= \mathbf{u}_S\left\{\mathbf{x}; -\frac{3}{4}\left[\mathbf{u}_1^B(A) - \mathbf{U}_A^1\left(\frac{a}{d}\right)\right]\right\} + \mathbf{u}_D\left\{\mathbf{x}; \frac{1}{4}\left[\mathbf{u}_1^B(A) - \mathbf{U}_A^1\left(\frac{a}{d}\right)\right]\right\}, \\ \mathbf{u}_2^B &= \mathbf{u}_S\left\{\mathbf{x}; -\frac{3}{4}\left[\mathbf{u}_1^A(B) - \mathbf{U}_B^1\left(\frac{a}{d}\right)\right]\right\} + \mathbf{u}_D\left\{\mathbf{x}; \frac{1}{4}\left[\mathbf{u}_1^A(B) - \mathbf{U}_B^1\left(\frac{a}{d}\right)\right]\right\},\end{aligned} \tag{8–255}$$

and the hydrodynamic force on the spheres is

$$\begin{aligned}\mathbf{F}_B = {}& \left\{6\pi\mu a U^0\left[1 - \frac{3}{4}\left(\frac{a}{d}\right)(1 + \sin^2\theta)\right]\right\}\mathbf{i}_z + 6\pi\mu a \mathbf{U}_B^1\left(\frac{a}{d}\right) \\ & - 6\pi\mu a U^0 \frac{3}{4}\cdot\left(\frac{a}{d}\right)\sin\theta\cos\theta\,\mathbf{i}_x,\end{aligned} \tag{8–256}$$

$$\begin{aligned}\mathbf{F}_A = {}& -\left\{6\pi\mu a U^0\left[1 - \frac{3}{4}\left(\frac{a}{d}\right)(1 + \sin^2\theta)\right]\right\}\mathbf{i}_z + 6\pi\mu a \mathbf{U}_A^1\left(\frac{a}{d}\right) \\ & - 6\pi\mu a U^0 \frac{3}{4}\cdot\left(\frac{a}{d}\right)\sin\theta\cos\theta\,\mathbf{i}_x.\end{aligned}$$

Now, if the net force on the spheres is assumed to be fixed (thus equal to the hydrodynamic force in the absence of interactions, $-6\pi\mu a U^0\mathbf{i}_z$), it follows that

$$\mathbf{U}_A^1 = \mathbf{U}_B^1 = U^0\left[-\frac{3}{4}(1 + \sin^2\theta)\mathbf{i}_z + \frac{3}{4}\sin\theta\cos\theta\,\mathbf{i}_x\right]. \tag{8–257}$$

Thus, the overall motion is

$$\mathbf{U}_A = \mathbf{U}_B = U^0 \left(-1 - \frac{3}{4}\left(\frac{a}{d}\right)(1 + \sin^2 \theta) \right) \mathbf{i}_z + \frac{3}{4}\left(\frac{a}{d}\right) U^0 \sin\theta \cos\theta \mathbf{i}_x. \qquad (8\text{--}258)$$

The two spheres move at the *same* velocity to this order of approximation, but the configuration shown in Fig. 8–9 drifts toward the *right* as it sediments vertically downward. The exceptional cases are $\theta = 0, \pi/2$. For $\theta = 0$, the two spheres lie in the horizontal plane and they sediment vertically downward with a speed

$$U^0 \left[1 + \frac{3}{4}\left(\frac{a}{d}\right) \right]. \qquad (8\text{--}259)$$

On the other hand, for $\theta = \pi/2$, the line of centers is vertical and the two spheres again sediment vertically downward with a speed

$$U^0 \left[1 + \frac{3}{2}\left(\frac{a}{d}\right) \right]. \qquad (8\text{--}260)$$

The reader may note that the far-field hydrodynamic interactions cause the vertical configuration to sediment slightly faster than the horizontal configuration. This is qualitatively similar to the fact that a vertical rod falls faster than the same rod in a horizontal orientation [see equations (8–191) and (8–193)].

NOTES

1. H. Lamb, *Hydrodynamics* (6th ed.) (Dover, New York, 1945, reprint of Cambridge University Press edition, 1932).
2. J. Happel and H. Brenner, *Low Reynolds Number Hydrodynamics* (Noordhoff, Leyden, The Netherlands, 1973).
3. J. A. R. Coope, R. F. Snider, and F. R. McCourt, Irreducible Cartesian tensors, *J. Chem. Phys.* **43**, 2269–75 (1965).
4. A. Nadim and H. A. Stone, The motion of small particles and droplets in quadratic flows, *Stud. Appl. Math.* **85**, 53–73 (1991).
5. Many such coordinate systems exist. Besides cylindrical, spherical, and Cartesian, the most common are probably elliptical, elliptical cylindrical, bispherical, and bicylindrical. A very useful compilation of the properties of vectors and vector operators, in a large number of orthogonal coordinate systems, may be found in an appendix to the book by Happel and Brenner, *op. cit.*
6. S. Kim and S. J. Karrila, *Microhydrodynamics: Principles and Selected Applications* (Butterworth-Heinemann, Stoneham, MA, 1991).
7. C. Pozrikidis, *Boundary Integral and Singularity Methods for Linearized Viscous Flow* (Cambridge, 1992).
8. A few examples of recent papers involving the use of boundary-integral methods for problems involving a deforming fluid interface are as follows: C. Pozrikidis, Interfacial dynamics for Stokes flow, *J. Comput. Phys.* **169**, 250–301 (2001); J. M. Rallison, A numerical study of the deformation and burst of a viscous drop in general shear flows, *J. Fluid Mech.* **109**, 465–82 (1981); H. A. Stone and L. G. Leal, Relaxation and breakup of an initially extended drop in an otherwise quiescent fluid, *J. Fluid Mech.* **198**, 399–427 (1989); H. A. Stone and L. G. Leal, The effects of surfactants on drop deformation and breakup, *J. Fluid Mech.* **220**, 161–86 (1990); J. M. Rallison and A. Acrivos, A numerical study of the deformation and burst of a viscous drop in an extensional flow, *J. Fluid Mech.* **89**, 191–200 (1978); M. Loewenberg and E. J. Hinch, Numerical simulation of a concentrated emulsion in shear flow, *J. Fluid Mech.* **321**, 395–419 (1996); M. Loewenberg and E. J. Hinch, Collision of two deformable drops in shear flow, *J. Fluid Mech.* **338**, 299–315 (1997); Z. Zinchenko, M. A. Rother, and R. H. Davis, A novel boundary-integral algorithm for viscous interaction of deformable drops, *Phys. Fluids* **9**, 1493–511 (1997); Z. Zinchenko and R. H. Davis,

Notes

Large-scale simulations of concentrated emulsion flows, *Philos. Trans. R. Soc. London Ser. A* **361**, 813–45 (2003); C. D. Eggleton, T. M. Tsai, and K. J. Stebe, Tip streaming from a drop in the presence of surfactants, *Phy. Rev. Lett.* **87**, 048302/1–4 (2001); X. Li and C. Pozrikidis, Effect of surfactants on drop deformation and on the rheology of dilute emulsions in Stokes flow, *J. Fluid Mech.* **341**, 165–94 (1997).

9. G. K. Batchelor, Slender-body theory for particles of arbitrary cross-section in Stokes flow, *J. Fluid Mech.* **44**, 419–40 (1970); R. G. Cox, The motion of long slender bodies in a viscous fluid, Part 1, General theory, *J. Fluid Mech.* **44**, 791–810, (1970); Part 2, Shear Flow, *J. Fluid Mech.* **45**, 625–657 (1971); J. B. Keller and S. T. Rubinow, Slender-body theory for slow viscous flow, *J. Fluid Mech.* **75**, 705–14 (1976); R. E. Johnson, An improved slender-body theory for Stokes flow, *J. Fluid Mech.* **99**, 411–31 (1980); A. Sellier, Stokes flow past a slender particle, *Proc. R. Soc. London Ser. A* **455**, 2975–3002 (1999).

10. G. I. Hancock, The self-propulsion of microscopic organisms through liquids, *Proc. R. Soc. London Ser A*, **217**, 96–121 (1953).

11. O. A. Ladyzhenskaya, *The Mathematical Theory of Viscous Incompressible Flow* (Gordon & Breach, New York, 1963).

12. See Refs. 7 and 8 of this chapter.

13. A. T. Chwang and T. Y. Wu, Hydromechanics of low-Reynolds number flow, Part 1, Rotation of axisymmetric prolate bodies, *J. Fluid Mech.* **63**, 607–22 (1974); A. T. Chwang and T. Y. Wu, Hydromechanics of low-Reynolds number flow, Part 2, Singularity method for Stokes flows, *J. Fluid Mech.* **67**, 787–815 (1975); A. T. Chwang, Hydromechanics of low-Reynolds number flow, Part 3, Motion of a spheroidal particle in quadratic flows, *J. Fluid Mech.* **72**, 17–34 (1975); A. T. Chwang and T. Y. Wu, Hydromechanics of low-Reynolds number flow, Part 4, Translation of spheroids, *J. Fluid Mech.* **75**, 677–89 (1975).

14. S. Kim, Singularity solutions for ellipsoids in low-Reynolds number flows: With applications to the calculation of hydrodynamic interactions in suspensions of ellipsoids, *Int. J. Multiphase Flow* **12**, 469–91 (1986).

15. See Chap. 3 of Ref. 6.

16. G. K. Youngren and A. Acrivos, Stokes flow past a particle of arbitrary shape: A numerical method of solution, *J. Fluid Mech.* **69**, 377–403 (1975).

17. In addition to the recent book by Pozrikidis (Ref. 7), a good general reference to the boundary-integral method is S. Weinbaum, P. Ganatos, and Z. Y. Yan, Numerical multipole and boundary integral equation techniques in stokes flow, *Annu. Rev. Fluid Mech.* **22**, 275–316 (1990).

18. J. R. Blake, A note on the image system for a stokeslet in a no-slip boundary, *Proc. Cambridge Philos. Soc.* 70, 303–10 (1971); D. F. Katz, J. R. Blake, and S. L. Paveri-Fontana, On the movement of slender bodies near plane boundaries at low Reynolds number (applied to micro-organisms), *J. Fluid Mech.* **72**, 529–40 (1975).

19. E. P. Ascoli, D. S. Dandy, and L. G. Leal, Low Reynolds number hydrodynamic interaction of a solid particle with a planar wall, *Int. J. Numer. Methods Fluids* **9**, 651–88 (1989); E. P. Ascoli, D. S. Dandy, and L. G. Leal, Buoyancy-driven motion of a deformable drop toward a planar wall at low Reynolds number, *J. Fluid Mech.* **213**, 287–311 (1990).

20. H. A. Lorentz, Ein Aligemeiner Satz, die Bewegung einer Reibendon Fliissigkeit Betreffend, Nebst Einigen Anwendungen Desselben (a general theorem concerning the motion of a viscous fluid and a few consequences derived from it), *Versl. Kon. Akad. Wetensch. Amsterdam* **5**, 168–74 (1896).

21. I attribute this observation about the reciprocal theorem, and other related results, to a lecture given by H. Brenner at Caltech in the 1970s.

22. The original reference to this work was H. Faxen's Ph.D. thesis, Uppsala University, Uppsala, Sweden (1921).

23. L. G. Leal, Particle motion in a viscous fluid, *Annu. Rev. Fluid Mech.* **12**, 435–76 (1980).

24. B. P. Ho and L. G. Leal, Inertial migration of rigid spheres in two-dimensional unidirectional flows, *J. Fluid Mech.* **65**, 365–400 (1974); P. Vasseur and R. G. Cox, The lateral migration of a spherical particle in two-dimensional shear flows, *J. Fluid Mech.* **78**, 385–413 (1976); K. Ishii and H. Hasimoto, Lateral migration of a spherical particle in flows in a circular tube, *J. Phys. Soc. Jpn.* **48**, 2144–53 (1980).

For the reader interested in this phenomenon, additional work in more recent years includes the following papers:

J. B. McLaughlin, Inertial migration of a small sphere in linear shear flows, *J. Fluid Mech.* **224**, 261–74 (1991); J. A. Schonberg and E. J. Hinch, Inertial migration of a sphere in Poiseuille flow, *J. Fluid Mech.* **203**, 517–24 (1989); A. J. Hogg, The inertial migration of non-neutrally buoyant spherical particles in two-dimensional shear flows, *J. Fluid Mech.* **272**, 285–315 (1994); E. S. Asmolov, The inertial lift on a spherical particle in a plane Poiseuille flow at large channel Reynolds number, *J. Fluid Mech.* **381**, 63–87 (1999); P. Cherukat, J. B. McLaughlin, and D. S. Dandy, A computational study of the inertial lift on a sphere in a linear shear flow field, *Int. J. Multiphase Flow* **25**, 15–33 (1999); A. M. Leshansky and J. F. Brady, Force on a sphere via the generalized reciprocal theorem, *Phys. Fluids* **16**, 843–4 (2004); J. Magnaudet, Small inertial effects on a spherical bubble, drop or particle moving near a wall in a time-dependent linear flow, *J. Fluid Mech.* **485**, 115–42 (2003).

25. G. Segre and A. Silberberg, Behavior of macroscopic rigid sphere in Poiseuille flow, Part 1, Determination of local concentration by statistical analysis of particle passages through crossed light beams, *J. Fluid Mech.* **14**, 115–136 (1962); Part 2, Experimental results and interpretation, *J. Fluid Mech.* **14**, 137–57 (1962).

26. L. G. Leal, The slow motion of slender rod-like particles in a second-order fluid, *J. Fluid Mech.* **69**, 305–37 (1975); B. P. Ho and L. G. Leal, Migration of rigid spheres in a two-dimensional unidirectional shear flow of a second-order fluid, *J. Fluid Mech.* **76**, 783–99 (1976); P. C. H. Chan and L. G. Leal, The motion of a deformable drop in a second-order fluid, *J. Fluid Mech.* **92**, 131–70 (1979); L. G. Leal, The motion of small particles in non-Newtonian fluids, *J. Non-Newtonian Fluid Mech.* **5**, 33–78 (1979); R. J. Phillips, Dynamic simulation of hydrodynamically interacting spheres in a quiescent second-order fluid, *J. Fluid Mech.* **315**, 345–65 (1996).

27. An alternative description of the application of the method of reflections can be found in the textbook by Happel and Brenner, (see Ref. 2).

PROBLEMS

Problem 8–1. Prove that $u_i^{(p)} = [(x_i/2\mu)p]$ with p being the (harmonic) pressure, is a particular solution to the Stokes' flow equations.

Problem 8–2. Derive the formula for the first five spherical harmonics from the formulae $\phi_{n+1} = \frac{\partial^n (r^{-1})}{\partial x_i \partial x_j \ldots}$ for the decaying functions and $r^{2n+1}\phi_{n+1}$ for the growing functions.

Problem 8–3. Consider the flow that is due to a point source of fluid, such that the fluid emanates from the origin at a rate Q (vol/time – a scalar). We may assume that this flow is a creeping motion. Under what conditions will this be true? What is the resulting pressure and velocity distribution? There is obviously more than one way to solve this problem, but the objective here is to use the singularity methods of this chapter.

Problem 8–4. Consider a sphere suspended in the general linear shear flow $u_i = \Gamma_{ij} x_j$ under creeping-flow conditions. If the sphere is free to rotate, what is its angular velocity Ω_i?

Problem 8–5. The annular region between two concentric rigid spheres of radii a and λa (with $\lambda > 1$) is filled with Newtonian fluid of viscosity μ and density ρ. The outer sphere is held stationary whereas the inner sphere is made to rotate with angular velocity Ω. Assume that inertia is negligible so that the fluid is in the Stokes' flow regime.

(a) Without any detailed calculation, explain why the pressure field in the fluid must be identically zero. Show that this implies that continuity is trivially satisfied.

Problems

(b) By solving the Stokes' equations, show that the fluid velocity is

$$\mathbf{u} = \frac{1}{1-\lambda^3}\left[1 - \left(\frac{\lambda a}{r}\right)^3\right]\boldsymbol{\Omega}\wedge\mathbf{r},$$

where $\mathbf{r}$ is the position vector and r is the magnitude of $\mathbf{r}$.

(c) Calculate the stress field in the fluid. Deduce that the couple $\mathbf{L}$ exerted on the inner sphere by the fluid is

$$\mathbf{L} = -8\pi\mu a^3\left(\frac{\lambda^3}{\lambda^3-1}\right)\boldsymbol{\Omega}.$$

Briefly comment on the limit $\lambda \gg 1$. For $\lambda \to 1$, show that the couple is to leading order

$$\mathbf{L} = -\frac{8}{3}\pi\mu a^3 \frac{1}{\lambda-1}\boldsymbol{\Omega},$$

and give an physical interpretation of this result.

(d) What is the condition that inertia is indeed negligible for the case $\lambda \gg 1$? Explain why this is not the appropriate condition for the case $\lambda \to 1$.

Problem 8–6. Sphere in a Quadratic Flow. Consider a solid sphere of radius a in a general quadratic undisturbed flow,

$$\mathbf{u}_\infty = \mathbf{U} + \mathbf{K}:\mathbf{xx},$$

where $\mathbf{U}$ is a constant vector and $\mathbf{K}$ is a constant third-order tensor.

(a) Determine the velocity and pressure field in terms of $\mathbf{U}$ and $\mathbf{K}$, assuming that the sphere is stationary (i.e., it neither translates nor rotates) and its center lies at the origin of the coordinate reference system (that is, $\mathbf{x}$ is a general position vector defined with respect to the same origin of coordinates). In doing this problem, the decomposition of $\mathbf{K}$ into irreducible components, as discussed in Section B, is essential.

(b) Show that problem (7–20) is a special case of this problem (what form do you need to assume for $\mathbf{K}$?). See if the solutions match if the general solution here is specialized to Problem 7–20.

(c) What is the force and torque on the sphere?

Problem 8–7. The Hele–Shaw Cell. The Hele–Shaw cell is perhaps one of the simplest constructions of a nearly unidirectional flow; however, as we shall see in this problem the flow inside the cell has some remarkable properties. A Hele–Shaw cell consists of two flat parallel rigid walls of vertical separation h and horizontal extent l with $h/l \ll 1$. The gap between the fluids is occupied with Newtonian fluid and contains obstacles in the form of cylinders with generators normal to the walls. The fluid is being driven through the cell by a steady horizontal pressure gradient $\nabla p = \mathbf{G}$ applied between the ends of the cell.

(a) Use lubrication theory to show that the horizontal velocity defined by means of $\mathbf{u} = u(z)\mathbf{e}_x + v(z)\mathbf{e}_y$ is given by

$$\mathbf{u} = -\frac{1}{2\mu}z(h-z)\nabla p, \tag{1}$$

where $\nabla = \mathbf{e}_x\frac{\partial}{\partial x} + \mathbf{e}_y\frac{\partial}{\partial y}$ is the 2D horizontal gradient operator and p is the local pressure. Note that the ratio $v(z)/u(z)$ is independent of z so that the streamline pattern is also independent of z.

(b) Deduce that the horizontal volumetric flow rate $\mathbf{q} = \int_0^h \mathbf{u}\,dz$ is given by

$$\mathbf{q} = -\frac{h^3}{12\mu}\nabla p. \tag{2}$$

583

Conservation of mass dictates that $\nabla \cdot \mathbf{q} = 0$. Show that this implies that the pressure is a harmonic function, $\nabla^2 p = 0$. In this respect the flow in a Hele–Shaw cell is analogous to the flow in an isotopic porous medium described by Darcy's law.

(c) Suppose there is one circular cylinder of radius a contained in the cell. Our objective is to find the pressure distribution around the cylinder. The pressure must be linear in $\mathbf{G}$ (by linearity of the Laplace equation) and the corresponding solution to $\nabla^2 p = 0$ is

$$p = p_0 + \alpha \mathbf{G} \wedge \mathbf{x} + \beta \frac{\mathbf{G} \wedge \mathbf{x}}{r^2}, \tag{3}$$

where α and β are constants, p_0 is an arbitrary pressure, and $\mathbf{x}$ is the position vector. The appropriate boundary conditions are zero flux of fluid at the surface of the cylinder and $\nabla p \to \mathbf{G}$ at large distances from the cylinder. Apply these conditions to deduce that

$$p = p_0 + \left(1 - \frac{a^2}{r^2}\right) \mathbf{G} \wedge \mathbf{x}. \tag{4}$$

Show that the resulting velocity profile is

$$\mathbf{u} = -\frac{1}{2\mu} z(h-z) \left[\left(1 - \frac{a^2}{r^2}\right) \mathbf{G} + \frac{2a^2}{r^4} \mathbf{G} \cdot \mathbf{xx} \right]. \tag{5}$$

(d) Show that, by taking the curl of (1),

$$\operatorname{curl} \mathbf{u} = \nabla \wedge \mathbf{u} = \frac{\partial v}{\partial x} - \frac{\partial u}{\partial y} = 0. \tag{6}$$

This result implies that (at any fixed z) the flow past a cylindrical obstacle will correspond to the 2D *potential* flow past that obstacle. For the Hele–Shaw cell the pressure p plays the role of the velocity potential ϕ. In fact, (4) is nothing but the solution for the velocity potential outside a circular cylinder of radius a. There is, however, one important caveat. Let us consider the circulation Γ defined by

$$\Gamma = \int_C \mathbf{u} \cdot d\mathbf{x}, \tag{7}$$

where C is any closed contour (perhaps including the obstacle) lying in a horizontal plane. Show that

$$\Gamma = -\frac{1}{2\mu} z(h-z) [p]_c, \tag{8}$$

where $[p]_c$ is the change in pressure on traversal of the contour. However, the pressure is a single-valued function; hence this jump must be identically zero and thus $\Gamma = 0$. This shows that the flow in a Hele–Shaw cell corresponds to 2D potential flow with *zero circulation* and hence (by the Kutta–Joukowski theorem) a zero lift force.

Some excellent photographs of flow past obstacles in a Hele–Shaw cell are contained in *An Album of Fluid Motion* by M. Van Dyke (Parabolic Press, Stanford, CA, 1982).

Problem 8–8. Reciprocal Theorem for the Navier–Stokes Equations. Derive the reciprocal theorem for the Navier–Stokes equations, not the Stokes equations. You should have a volume integral remaining.

Problem 8–9. Torque on a Sphere in a General Stokes Flow. Use the reciprocal theorem for Stokes equations to derive the following expression for the torque exerted on a sphere of radius a that is held fixed in the Stokes flow $\mathbf{u}^\infty(x)$:

$$\mathbf{T} = -8\pi \mu a^3 \nabla \wedge \mathbf{u}^\infty(x)|_{x=0}.$$

Problems

Problem 8–10. Symmetry of the Grand Resistance Tensor. Use the reciprocal theorem to show that the "grand resistance tensor" is symmetric. The grand resistance tensor relates the hydrodynamic force/torque on a particle to its velocity/angular velocity:

$$\begin{pmatrix} \mathbf{F} \\ \mathbf{T} \end{pmatrix} = \begin{pmatrix} \hat{\mathbf{A}} & \hat{\mathbf{C}} \\ \hat{\mathbf{B}} & \hat{\mathbf{D}} \end{pmatrix} \cdot \begin{pmatrix} \mathbf{U} \\ \mathbf{\Omega} \end{pmatrix},$$

and symmetry implies that $\hat{\mathbf{A}} = \hat{\mathbf{A}}^T$, $\hat{\mathbf{D}} = \hat{\mathbf{D}}^T$, and $\hat{\mathbf{B}} = \hat{\mathbf{C}}^T$.

Problem 8–11. The Effect of Surfactant on Drop Deformation in a General Linear Flow. We wish to consider the effect of interface contamination by surfactant on the deformation of a drop in a steady linear flow at very low Reynolds number, that is, $\mathbf{u}_\infty = G[(1/2)\boldsymbol{\omega} \wedge \mathbf{x} + \mathbf{E} \cdot \mathbf{x}]$. The local interface concentration of surfactant is denoted as Γ^*, specified in units of mass per unit of interfacial area. We assume that Γ^* is small so that a linear relationship exists between Γ^* and interfacial tension, σ,

$$\sigma_S - \sigma = \Gamma^* RT,$$

where R is the gas constant, T is the absolute temperature, and σ_S is the interfacial tension of the clean interface without surfactant ($\Gamma^* = 0$). In the absence of motion, the surfactant concentration at the interface is uniform, Γ_0.

(a) Derive *dimensionless* forms of the governing creeping-flow equations and boundary conditions for the problem, assuming that the surfactant can be approximated as *insoluble* in the bulk fluids, so that the distribution of surfactant at the interface is governed by a balance between advection with the surface velocity $\mathbf{u}_s$, diffusion with interface diffusivity D_S, and changes in the local concentration that are due to stretching and distortion of the interface; namely, in dimensional terms (see Chap. 2)

$$\frac{\partial \Gamma^*}{\partial t'} + \nabla'_s \cdot (\Gamma^* \mathbf{u}'_s - D_S \nabla' \Gamma^*)$$
$$+ \Gamma^* (\nabla'_s \cdot \mathbf{n})(\mathbf{u}' \cdot \mathbf{n}) = 0.$$

To define a dimensionless surface concentration, use $\Gamma = \Gamma^*/\Gamma_0$. The dimensionless parameters that should appear on nondimensionalization are

$$\lambda = \frac{\hat{\mu}}{\mu}, \quad Ca_S = \frac{\mu G a}{\sigma_2}, \quad \beta = \frac{\Gamma_0 RT}{\sigma_S},$$

$$Pe_S = \frac{Ga^2}{D_S}.$$

The latter is know as the interface Peclet number and provides a measure of the relative importance of convection and diffusion in determining the surfactant distribution on the interface.

Note: In dimensionless terms, the uniform equilibrium concentration of surfactant is $\Gamma = 1$, and the corresponding interfacial tension is

$$\sigma^* = \sigma_S(1 - \beta).$$

For some purposes, it is more convenient to compare results obtained for the full problem with results for the uniformly contaminated surface σ^* rather than for the clean interface value σ_S. If we choose this route, we can express all results in terms of the capillary number Ca^*, based on σ^*, by noting that $Ca^* = Ca_S/(1 - \beta)$.

(b) Let us now consider the changes in drop shape produced by increases of the local shear rate G. For this purpose, it is convenient to define

$$\gamma \equiv \frac{\sigma^* a}{\mu D_S}$$

585

so that $Pe_S \equiv \gamma Ca^*$, where γ is a parameter that depends on only material properties. We consider the limit

$$Ca^* \ll 1 \quad \text{and} \quad \gamma = O(1) (\text{i.e.}, Pe_S \ll 1)$$

so that both the drop shape and the surfactant concentration distribution are only slightly perturbed from their equilibrium values. For the general linear flow considered here, the first *corrections* to the description of the surface shape and the surfactant distribution are expected to take the forms

$$r = 1 + Ca^* b_r(t) \frac{\mathbf{x} \cdot \mathbf{E} \cdot \mathbf{x}}{r^2},$$

$$\Gamma = 1 + \gamma Ca^* b_\Gamma(t) \frac{\mathbf{x} \cdot \mathbf{E} \cdot \mathbf{x}}{r^2}.$$

Use the general representation of solutions for creeping flows in terms of vector harmonic functions to solve for the velocity and pressure fields in the two fluids, as well as the deformation and surfactant concentration distribution functions, *at steady state*. You should find

$$b_r = \frac{5}{4} \frac{(16 + 19\lambda) + 4\beta\gamma/(1-\beta)}{10(1+\lambda) + 2\beta\gamma/(1-\beta)},$$

plus a comparable expression for b_Γ.

(c) Discuss your solution in physical terms. The classical limit of a drop in the absence of surfactant, Eq. (8–83), is obtained for both $\beta \to 0$ and $\gamma \to 0$. Explain this. Is the deformation increased or decreased in the presence of surfactant, at a fixed dimensionless shear rate Ca^*? Explain.

Problem 8–12. Electrophoretic Mobility. Under the action of an external electric field, a charged colloidal particle will move at a velocity determined by the force that is due to the electric field and the hydrodynamic drag or mobility, the latter being essentially the inverse of the drag (i.e., the velocity that is due to a unit force, rather than the drag that is due to a unit velocity). In many systems, however, the problem is complicated by the fact that there are ionizable species in the liquid. In this case, the charge on the colloidal particle attracts ions of opposite charge and these form a local cloud of counterions near the particle that is known as the electrical double layer. The thickness of this double layer is determined by a balance between the attractive Coulombic force and diffusion of the charged ions away from the particle. The problem in this case is to calculate the effect of the double layer on the mobility of the particle, which is known as the electrophoretic mobility.

In general, this is a moderately difficult problem. An exception occurs, however, when the double layer is thin relative to the dimension of the colloidal particles. In this case, to determine the electrophoretic mobility of the particle we need to solve Stokes equations

$$0 = -\nabla p + \mu \nabla^2 \mathbf{u}, \quad \nabla \cdot \mathbf{u} = 0,$$

subject to the boundary conditions

$$\mathbf{u} \sim -\mathbf{U}, \quad p \sim 0 \quad \text{as } r \to \infty,$$

$$\mathbf{u} = \mathbf{u}_s = \frac{\varepsilon}{\mu} \zeta \nabla \psi \quad \text{at } r = a,$$

along with the requirement that the phoretic motion be force free:

$$\mathbf{F}^h = \int_{S_p} \sigma^h \cdot \mathbf{n} \, dS = 0,$$

where $\sigma^h = -p\mathbf{I} + 2\mu\mathbf{e}$ is the hydrodynamic stress in the fluid, $\mathbf{n}$ is the outer normal to the surface, and the integral is over the particle surface at $r = a$. In these equations, $\mathbf{U}$ is the electrophoretic velocity we seek, μ is the viscosity of the fluid, ε is the electrical permittivity of the fluid, ζ is the constant zeta potential of the particle and ψ is the electrostatic potential that can be shown to satisfy the following problem:

$$\nabla^2 \psi = 0,$$

$$\nabla \psi \sim -\mathbf{E}^\infty \quad \text{as } r \to \infty,$$

$$\mathbf{n} \cdot \nabla \psi = 0 \quad \text{at } r = a,$$

where $\mathbf{E}^\infty$ is the constant externally imposed electric field.

From the solutions of the preceding electrostatic and Stokes' flow problems, show that the electrophoretic mobility μ_e defined by

$$\mathbf{U} = \mu_e \mathbf{E}^\infty$$

is given by the Smoluchowski formula

$$\mu_e = \frac{\varepsilon}{\mu} \zeta.$$

(*Note*: In solving the preceding problems, be sure to exploit the linearity of the Stokes problem in $\mathbf{U}$ and the electrostatic problem in $\mathbf{E}^\infty$.)

Problem 8–13. Lateral Migration of a Deformable Drop in 2D Poiseuille Flow. A viscous drop of viscosity $\hat{\mu}$ and density $\hat{\rho}$ is carried along in the unidirectional motion of an incompressible, Newtonian fluid of viscosity μ and density $\rho \equiv \hat{\rho}$ between two infinite plane walls. The radius of the undeformed drop is denoted as a, and the distance between the walls is d. We assume that the capillary number, $Ca \equiv a\mu G/\sigma$, is small so that the drop deformation is also small. Here, σ is the interfacial tension, and G is the mean shear rate of the undisturbed flow.

The flow configuration is sketched in the figure. The coordinate direction normal to the plane walls is specified as x_3, and the velocity of the center of mass is denoted as $\mathbf{U}_S$. Further, the undisturbed velocity, pressure, and stress are denoted, respectively, as $\mathbf{V}, \mathbf{Q}, \mathbf{T}$. In the most general case,

$$\mathbf{V} = \left(\alpha + \beta x_3 + \gamma x_3^2\right) \mathbf{e}_1 - \mathbf{U}_S,$$

where we have adopted a coordinate frame that moves with the drop velocity $\mathbf{U}_S$.

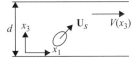

Now, if the Reynolds number of the flow is sufficiently small for the creeping-motion approximation to apply, it can be shown by the arguments of Subsection B.3 in Chap. 7 that no lateral motion of the drop is possible unless the drop deforms. In other words, $\mathbf{U}_S = U_S \mathbf{e}_1$ in this case, though, of course, $\mathbf{U}_S$ is not generally equal to the undisturbed velocity of the fluid evaluated at the x_3 position of the drop center. The drop may either lag or lead (in principle) because of a combination of interaction with the walls and the hydrodynamic effect of the quadratic form of the undisturbed velocity profile – see Faxen's law. Because the drop deforms, however, lateral migration can occur even in the complete absence of inertia (or non-Newtonian) effects. In this problem, our goal is to formulate two possible ways to evaluate the lateral velocity of the drop. In the first, we seek to evaluate the lateral velocity by means of a full perturbation solution of the problem to $O(\delta)$, where δ represents the magnitude of the drop deformation. In the second, we use the reciprocal theorem to show that the lateral velocity can be evaluated at $O(\delta)$ using only velocity and stress fields for motion of the *undeformed* drop.

(a) Let us denote the velocity, pressure, and stress fields outside and inside the drop as $\mathbf{U}$, P, and $\mathbf{S}$ and $\tilde{\mathbf{U}}$, $\tilde{P}$, $\tilde{\mathbf{S}}$, respectively. Further, let us denote the drop surface in terms of F:
$$F \equiv r - 1 - \delta f = 0.$$
Suppose that $\delta \ll 1$, so that all of the velocity, pressure, and stress fields can be expressed in terms of an asymptotic expansion of the form
$$\mathbf{U} = \mathbf{U}_0 + \delta \mathbf{U}_1 + \cdots +,$$
$$\tilde{\mathbf{U}} = \tilde{\mathbf{U}}_0 + \delta \tilde{\mathbf{U}}_1 + \cdots +,$$
$$F \equiv r - 1 - \delta f_1 + \cdots + = 0.$$
Derive dimensionless equations and boundary conditions whose solution would be sufficient to determine the drop velocity (and shape) to $O(\delta)$. Use the method of domain perturbations to express all boundary conditions at the deformed drop interface in terms of equivalent conditions at the spherical surface of the undeformed drop. Show that $\delta \equiv Ca$.

(b) Instead of solving part (a) directly to terms of $O(\delta)$, the option is to apply the reciprocal theorem. To apply the reciprocal theorem to obtain $\mathbf{U}_S^{(1)}$, it is necessary to solve only the $O(1)$ problem from part (a) and the "complementary" creeping-flow problem of a drop translating perpendicular to the walls through a quiescent Newtonian fluid. Write the governing equations and boundary conditions for translation of a *spherical* drop with velocity $\mathbf{e}_3$ normal to the two plane walls. The velocity, pressure, and stress fields for this complementary problem should be denoted as $\mathbf{u}$, q, $\mathbf{t}$ and $\tilde{\mathbf{u}}$, $\tilde{q}$, and $\tilde{\mathbf{t}}$ for the fluid outside and inside the drop, respectively.

Note: To apply the reciprocal theorem, the shape of the drop for the complementary problem generally would have to be exactly the *same* as the shape in the problem of interest. However, because the original problem can be reduced by means of domain perturbations to an equivalent problem with the boundary conditions applied at the spherical surface, $r = 1$, we may also conveniently choose the drop to be spherical for this complementary problem.

(c) Beginning with the reciprocal theorem for the disturbance velocity fields in the fluid volume outside the undeformed drop surface,
$$\int_{V_f} [(\nabla \cdot \mathbf{S} - \nabla \cdot \mathbf{T}) \cdot (\mathbf{u} + \mathbf{e}_S) - \nabla \cdot \mathbf{t} \cdot (\mathbf{U} - \mathbf{V})] dV = 0,$$
and the corresponding expression for the fluid volume inside the underdeformed drop, show that
$$\int_{A_d} [(\mathbf{S} - \kappa \tilde{\mathbf{S}}) \cdot \mathbf{u} - (\mathbf{t} - \kappa \tilde{\mathbf{t}}) \cdot \mathbf{U} - \kappa \tilde{\mathbf{t}} \cdot (\mathbf{U} - \tilde{\mathbf{U}}) - \mathbf{T} \cdot \mathbf{u} + \mathbf{t} \cdot \mathbf{V}] \cdot \mathbf{n} dA,$$
where A_d is the spherical surface of the undeformed drop and κ is the viscosity ratio $\kappa \equiv \hat{\mu}/\mu$. Derive an expression for the drop velocity normal to the plane walls at $O(\delta)$, and express this formula entirely in terms of the stress and velocity fields $\mathbf{S}$, $\tilde{\mathbf{S}}$, $\mathbf{U}$, $\tilde{\mathbf{U}}$ at $O(1)$, plus the shape function $f^{(1)}$ at $O(\delta)$. Show that $f^{(1)}$ can be calculated from the velocity and stress fields at $O(1)$. [Reference: P. C. H Chan and L. G. Leal, *J. Fluid Mech.*, **92**, 131–70 (1979).]

Problems

Problem 8–14. Slender-Body Dynamics

(a) A long solid cylinder of length 2ℓ and radius a is held stationary in a pure straining flow, the undisturbed velocity of which is given by $u_1 = Gx_1$, $u_2 = -Gx_2$, and $u_3 = 0$. The center of the cylinder is at $x_1 = \ell/4$, $x_2 = 0$, $x_3 = 0$. The slenderness ratio, $\varepsilon \equiv a/\ell$, is asymptotically small, $\varepsilon \ll 1$. Obtain an expression for the force on the cylinder.

(b) Determine the hydrodynamic torque [up to $O(\ln\varepsilon)^{-2}$] exerted on a slender particle when the long axis is rotating about the center of the particle with a constant angular velocity Ω through a quiescent Newtonian fluid.

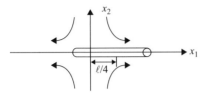

(c) We consider the motion of a slender torus through a quiescent Newtonian fluid. The radius of the torus is R, and its cross-sectional radius is a. We assume $a/R \ll 1$. Apply slender-body analysis to show that the ratio in force for motion parallel to the torus (say, x_1) and broadside (say, x_3) is

$$\frac{F_3}{F_1} = \frac{4}{3} + O(\ln\varepsilon)^{-1}.$$

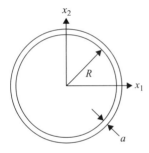

Problem 8–15. Sphere in a Parabolic Flow. Consider a solid spherical particle that is held stationary in a parabolic shear flow. In dimensionless terms,

$$\mathbf{u}_\infty = (x^2 + y^2)\mathbf{i}_z,$$

where the spatial variables x and y are nondimensionalized by use of the sphere radius. Solve for the velocity field in the creeping-flow limit by means of a superposition of a stokeslet, a potential dipole, a Stokes quadrapole, and a potential octupole, all located at the center of the sphere. What is the force on the sphere? Is this consistent with the solution of Problem 7–20 in Chap. 7?

Problem 8–16. Axisymmetric Prolate Spheroid Held Fixed in a Shear Flow. Let us consider an axisymmetric, prolate ellipsoid whose surface is given by

$$\frac{x^2}{a^2} + \frac{r^2}{b^2} = 1, \quad \text{where } r^2 = y^2 + z^2 \text{ and } a \geq b.$$

The foci of the ellipsoid are located at $x = \pm c$, where

$$c = (a^2 - b^2)^{1/2}.$$

Suppose that the spheroid is held fixed in a simple shear flow,
$$\mathbf{u}_\infty = Gy\mathbf{e}_x.$$
Show that the disturbance velocity field necessary to satisfy no-slip boundary conditions at the surface of the ellipsoid can be expressed in terms of a line distribution of stresset, rotlet, and potential quadrapole singularities of the form

$$\mathbf{u} = \int_{-c}^{c} (c^2 - \xi^2)[\alpha \mathbf{U}_{SS}(\mathbf{x} - \xi; \mathbf{e}_x, \mathbf{e}_y) + \gamma \mathbf{U}_R(\mathbf{x} - \xi; \mathbf{e}_x; \mathbf{e}_a)]d\xi$$
$$+ \beta \int_{-c}^{c} (c^2 - \xi^2)\frac{\partial}{\partial y}\mathbf{U}_D(\mathbf{x} - \xi; \mathbf{e}_x)d\xi,$$

where α, γ, and β are constants. What is the torque on the ellipsoid?

Note: The integrals can all be expressed in terms of the function $B_{m,n}(\mathbf{x})$, defined by

$$B_{m,n}(\mathbf{x}) \equiv \int_{-c}^{c} \frac{\xi^n d\xi}{|\mathbf{x} - \xi|^m} (n = 0, 1, 2, \ldots; m = -1, 1, 3, 5, \ldots,).$$

A recurrence formula that proves useful is

$$B_{m,n} = -\frac{C^{n-1}}{m-2}\left[\frac{1}{R_2^{m-2}} + \frac{(-1)^n}{R_1^{m-2}}\right] \quad (n \geq 2)$$
$$+ \frac{n-1}{m-2} B_{m-2,n-2} + x B_{m,n-1},$$

where

$$R_1 = [(x+c)^2 + r^2]^{1/2}, \quad R_2 = [(x-c)^2 + r^2]^{1/2},$$
$$B_{1,0} = \log\frac{R_2 - (x-c)}{R_1 - (x+c)}, \quad B_{1,1} = R_2 - R_1 + xB_{1,0},$$
$$B_{3,0} = \frac{1}{r^2}\left(\frac{x+c}{R_1} - \frac{x-c}{R_2}\right), \quad B_{3,1} = \left(\frac{1}{R_1} - \frac{1}{R_2}\right) + xB_{3,0}.$$

[Reference: A. T. Chwang and T. Y. Wu, *J. Fluid Mech.* **67**, 787–815 (1975).]

Problem 8–17. Selective Withdrawal From a Porous Medium. Consider a porous medium that is bounded on one side by a vertical impermeable wall. We assume that Darcy's law applies so that

$$\mathbf{u} = -\frac{k}{\mu}\nabla p$$

where k is the permeability of the porous medium. Here $\mathbf{u}$ is the superficial (or spatially averaged) velocity. Two fluids fill the pores of the medium, one over the other: a fluid of density ρ_1 for $z > 0$ and a heavier fluid with density ρ_2 for $z < 0$, where z is the vertical coordinate. The upper fluid is to be withdrawn from a point sink at a volumetric flow rate Q. The sink is located a distance H above the undisturbed interface of the fluids and a distance L from the vertical wall.

(a) Using a domain perturbation, determine the leading-order correction to the shape of the interface for a small rate of withdrawal (quantify "small" nondimensionally).
  - *Hint*: You will need to add three additional image sinks to model the zeroth-order pressure field.

Problems

(b) Does the presence of the wall increase or decrease the deformation of the interface? Why?

Problem 8–18. Brownian Diffusion of Slender Rigid Rods. Consider a slender rigid rod with length 2ℓ and a typical aspect ratio $\varepsilon = R/\ell$, where R is a typical radius of the rod. We wish to quantify some of the diffusional aspects of the particle.

(a) From the resistance/mobility relationships derived in Subsection E.5 for slender bodies, what is the translational diffusivity tensor and the *average* translational diffusivity of a rigid rod?
(b) What is the rotational diffusivity about the minor axes of a slender rigid rod?
  - *Hint*: Rotational Brownian diffusivity is the manifestation of random walks of the orientation of the rod. By analogy with translational diffusion, the rotational diffusivity $\mathbf{D}_r = kT\mathbf{M}_{\Omega G}$, where $\mathbf{M}_{\Omega G}$ is the mobility tensor relating angular velocity and torque.

Problem 8–19. Stresslet Drift. In this problem we consider the drift of a stresslet that is due to the presence of a plane boundary. As we have noted (Eq. 8–128), the stresslet is specified by a second-order physical tensor S_{ij} that is symmetric ($S_{ij} = S_{ji}$) and traceless ($S_{ij}\delta_{ij} = 0$). The drift velocity u_i is proportional to S_{ij} for creeping flows, and will be a function of the unit normal n_i describing the orientation of the plane.

(a) Show that the drift velocity of an arbitrary stresslet is characterized by only two constants, and
(b) prove that if $n_i = \delta_{i3}$ then the drift normal to the plane is proportional to only a single element of the S_{ij} tensor. You may find the following list of third-order tensors useful:

$$\delta_{ij}n_k,\ n_in_jn_k,\ \delta_{ik}n_j,\ \varepsilon_{ijk},\ \delta_{jk}n_i,\ \varepsilon_{jkl}n_ln_i$$

Problem 8–20. Unsteady Gas Flow in Porous Media. The unsteady flow of a gas in a porous medium is important in several economically valuable geophysical systems, such as natural gas and geothermal reservoirs. To develop some insight into these systems we will develop a simple model describing the flow. We may assume that Darcy's law applies or that

$$\mathbf{u} = -\frac{k}{\mu}\nabla p,$$

where k is the permeability of the porous medium, which is partially determined by the accessible porosity of the medium. (The microstructure of the porous rock is also important). Here $\mathbf{u}$ is the superficial (or spatially averaged) velocity.

(a) Using the equation of continuity for flow in a porous material and assuming the ideal gas law, show that the pressure distribution for isothermal gas flow is given by

$$\frac{\partial p}{\partial t} = \frac{k}{\mu\phi}\nabla\cdot(p\nabla p),$$

where ϕ is the porosity (or volume fraction of interstitial space) of the rock.
(b) The preceding equation is a nonlinear diffusion equation. Why?
(c) Let's consider the following ideal problem to assess the importance of the nonlinearity. Gas is withdrawn at a constant pressure p_i along an infinite plane into a porous medium. Initially and far away from the plane of withdrawal, the gas pressure is p_∞. Show that the PDE can be transformed to the ODE,

$$[(1+\alpha f)f']' + 2\eta f' = 0,$$

where $\eta = \frac{1}{2}(\mu\phi k p_\infty)^{1/2} x t^{-1/2}$, $\alpha = (p_i - p_\infty) p_\infty$, $(prime) = (d/d\eta)$, and x and t are the spatial and temporal coordinates. What are the boundary conditions for this equation?

(d) An analytic solution exists for only $\alpha = 0$. (What is it?) Determine the numerical solution for a range of α's. Plot the dimensionless pressure and velocity profiles.
  - *Hint*: A useful approach to solving this problem is the shooting method.

(e) Is the penetration depth of the pressure still proportional to the square root of time? How does the rate of withdrawal vary with time?

(f) Based on your numerical results, how does the penetration depth of the pressure quantitatively depend on the physical parameters, α and time?

(g) What is your assessment of the importance of the nonlinearity in this system?

9

Convection Effects in Low-Reynolds-Number Flows

In the preceding chapters, we focused mainly on fluid dynamics problems, with only an occasional problem involving heat or mass transfer. In this chapter, we change our focus to problems of heat (or single-solute mass) transfer. Specifically, we address the problem of heat (or mass) transfer from a finite body to a surrounding fluid that is moving relative to the body. In this chapter, we concentrate on problems in which the fluid motion is viscous in nature, and thus is "known" (or can be calculated) from creeping-flow theory. Later, after we have considered flows at nonzero Reynolds number, we will also consider heat (or mass) transfer for this situation.

In all of the fluid mechanics problems that we have considered until now, the nonlinear inertia terms in the equations of motion were either identically zero or small compared with the viscous terms. We begin this chapter by considering the corresponding heat (or mass) transfer problem, in which the fluid motion is "slow" in a sense to be described shortly, so that convection effects are weak and the transport process is dominated by conduction. When convection terms in the thermal energy equation can be neglected altogether, the resulting pure conduction problem is mathematically and physically analogous to the creeping flows that we have been studying in the preceding two chapters. The transport of heat is purely "diffusive" in this limit, i.e., conduction, just as the transport of momentum (or vorticity) in a creeping flow is also "diffusive." However, when the heat transfer medium is a fluid, it is almost impossible to completely eliminate convection effects. Even apparently tiny contributions that are due to convection seem always to produce a measurable effect on the heat transfer rate. (The difficulty of completely eliminating convection effects is one of the reasons why it is relatively difficult to actually measure the thermal conductivity of a liquid). Hence the focus of our discussion in this chapter is convection. In the limit of heat transfer dominated by conduction, we are concerned primarily with a prediction of the role played by *weak-convection* effects.

Following this initial topic, we then consider the opposite limit in which the transport process is dominated by convection. In this case, the concept of a thermal boundary layer plays a critical role.

A. FORCED CONVECTION HEAT TRANSFER – INTRODUCTION

We begin by considering the general problem of heat transfer from a hot or cold body to an adjacent flowing fluid. In many instances, we could also pose the same problems as mass transfer of a single, dilute solute that is being released from the surface of the same body. For convenience of presentation, all problems are described from the heat transfer perspective. However, at the end of this introductory section, a brief comment is made on the analogy between the heat and mass transfer problems, and when it is appropriate we

Convection Effects in Low-Reynolds-Number Flows

will also indicate when the detailed analysis in subsequent sections may be applied to the corresponding mass transfer problem.

1. General Considerations

The starting point for any analysis of heat transfer effects in a flowing fluid is the thermal energy equation, (2–114). The goal is to solve this equation to determine the temperature distribution in the fluid, given appropriate boundary conditions and the velocity field **u**. In the problems of fluid motion considered in preceding sections of this book, we have generally assumed the fluid to be isothermal. One resultant simplification was that the fluid properties, such as density and viscosity, which depend on temperature, were treated as known constants, independent of position in the flow domain. When the fluid temperature is not uniform, however, all of the material properties, μ, ρ, C_p, and k, will vary from point to point throughout the domain. As a consequence, the problem of obtaining exact solutions for the velocity and temperature fields becomes highly coupled; to ascertain the values of μ, ρ, C_p, and k at any point, we must know the temperature T at that point. However, to determine T, we must know not only ρ, C_p, and k, which appear directly in the thermal energy equation, but also the velocity field **u**. To obtain **u**, we must know μ and ρ, and for this we must have T. Evidently, an exact solution would require that the equations of continuity, motion, and thermal energy be solved simultaneously.

As a consequence of the complexity of the full nonisothermal problem, as outlined in the preceding paragraph, it is necessary to introduce some rather severe approximations to make any progress. The most common approach is to completely neglect the spatial dependence of the material properties, approximating them as constant at the value appropriate to the ambient temperature of the fluid. The reader should recognize that this is a very strong approximation. At minimum, it requires that the actual variations in temperature be asymptotically small. Even then, the treatment of all material properties as constant is too harsh an assumption in some cases. For example, it is well known that a fluid near a heated (or cooled) body may move, even if there is no motion imposed on the fluid at large distances from the body. Such motions, called *natural convection*, are driven by the differences in density of the fluid that occur when the temperature of one part is different from that of another. Evidently, if we wish to analyze natural convection motions, we must retain at least the variation of density ρ with position in the nonisothermal fluid. Even if the fluid moves relative to a heated body for other reasons (for example, it may be "pumped"), there will always be some superposed *natural convection* present when the fluid is nonisothermal. However, if the density variations in the fluid are not too large, natural convection may represent only a small correction to the motion that would be present in the absence of temperature variations, and it may be acceptable, as an approximation, to neglect natural convection altogether. This is tantamount to approximating the density as constant.

The approximation of all fluid properties as constant, including the complete neglect of any natural convection, is known as the *forced convection* approximation. We shall adopt it for this chapter, as well as for Chap. 11. Unlike most approximations that are introduced in this book, the forced convection approximation is adopted initially on an *ad hoc* basis, without a rigorous asymptotic justification.[1] We may be reassured that the resultant analysis is relevant to many conditions of practical interest by the fact that it has been adopted almost universally for analysis of heat transfer problems, in which the fluid motion is not due *solely* to natural convection.

To focus our subsequent discussion, let us begin by considering the problem of heat removal from a body of arbitrary shape that is immersed in a fluid that is undergoing a uniform motion of magnitude U_∞ relative to the body. The situation is illustrated in Fig. 9–1. We shall suppose that the fluid has an ambient temperature T_∞ far from the body. Furthermore, we assume, for simplicity, that the body is heated in such a way that its

A. Forced Convection Heat Transfer – Introduction

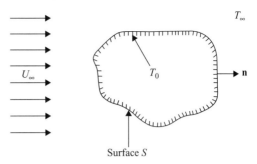

Figure 9–1. A schematic representation of a body of arbitrary shape and constant surface temperature, T_0, immersed in a fluid of constant properties, ρ, μ, C_p, k, and ambient temperature T_∞ that undergoes a uniform motion of magnitude U_∞ relative to the body. The body surface is denoted as S, and the unit outer normal to the surface is **n**.

surface is maintained at a constant, uniform temperature T_0. Finally, we adopt the forced convection approximation and treat the fluid properties ρ, μ, C_p, and k as constants with values corresponding to the ambient temperature T_∞. With all of these approximations, the thermal energy equation, (2–93), becomes

$$\rho C_p \left(\frac{\partial T}{\partial t} + \mathbf{u} \cdot \nabla T \right) = k \nabla^2 T + 2\mu (\mathbf{E} : \mathbf{E}). \tag{9–1}$$

This is to be solved subject to boundary conditions

$$T = T_0 \quad \text{on } S, \tag{9–2a}$$

where S represents the body surface and

$$T \to T_\infty \quad \text{at } |\mathbf{r}| \to \infty, \tag{9–2b}$$

where **r** is a position vector with origin inside S. The last term in (9–1) represents the irreversible dissipation of kinetic energy (associated with **u**) to heat.

In the forced convection approximation, with ρ and μ fixed, the Navier–Stokes and continuity equations can be solved (at least in principle) to determine the velocity field **u**, and this solution is completely independent of the temperature distribution in the fluid. Once the velocity **u** is known, the thermal problem, represented by (9–1) and (9–2), can then be solved (again, in principle) to determine the temperature field T. Because the boundary values of T are assumed to be constant, we may anticipate that the temperature distribution throughout the fluid will be independent of time (with the exception of some initial period after the heated body is first introduced into the moving fluid). It is the steady-state temperature field that is most often our goal. For this reason, the time derivative in (9–1) will be dropped in subsequent developments.

Among the various assumptions that have been introduced, the one that may seem to be the most problematical is that the surface temperature is a constant independent of time or position on the body surface. Of particular concern is the latter condition. Frequently, instead of the surface temperature being specified, it is the heat flux at the surface that is fixed, and then the surface temperature will be unknown and generally a function of position on the surface. Often, however, the methodology developed for the constant-temperature case can be utilized with the heat flux specified. Further, because the thermal energy equation is linear in the temperature, there are also methods of converting the constant-temperature results into methods for calculating the surface temperature when the heat flux is specified. We shall revisit these issues especially in Chap. 11 after we have considered the constant-surface-temperature problem at both high and low Reynolds numbers.

2. Scaling and the Dimensionless Parameters for Convective Heat Transfer

Certain features of the thermal problem can be ascertained by means of nondimensionalization alone. An appropriate characteristic velocity is U_∞. Furthermore, it is convenient to introduce a "dimensionless" (i.e., rescaled) temperature,

$$\theta \equiv \frac{T - T_\infty}{T_0 - T_\infty}, \qquad (9\text{--}3)$$

which varies between one at the body surface and zero at a large distance from the body. Finally, we suppose that variations in θ of $O(1)$ occur over a characteristic length scale that is proportional to a typical linear dimension of the body, ℓ. Then, from the examples of previous chapters, the steady-state nondimensionalized version of (9–1) becomes

$$\boxed{Pe(\mathbf{u} \cdot \nabla \theta) = \nabla^2 \theta + 2Br(\mathbf{E} : \mathbf{E}).} \qquad (9\text{--}4)$$

Here, $\mathbf{u}$ and $\mathbf{E}$ represent the dimensionless velocity and rate-of-strain fields, and

$$Pe \equiv \frac{U_\infty \ell}{\kappa}, \quad \kappa \equiv \frac{k}{\rho C_p}, \quad \text{and} \quad Br \equiv \frac{\mu U_\infty^2}{k(T_0 - T_\infty)}.$$

The dimensionless Peclet number Pe provides a measure of the relative magnitude of the convection terms in (9–4) compared with conduction. It is sometimes useful to note that

$$Pe = RePr, \qquad (9\text{--}5)$$

where Re is the Reynolds number and

$$Pr \equiv \frac{\mu C_p}{k} = \frac{\nu}{\kappa} \qquad (9\text{--}6)$$

is the *Prandtl number*. The Prandtl number depends on fluid properties only and is the ratio of the diffusivity for momentum (the kinematic viscosity) compared with the thermal diffusivity κ.

The relative magnitude of heat generation by means of viscous dissipation is determined by the *Brinkman number*, Br, as noted already in Chap. 4. Although viscous dissipation can be significant for highly viscous fluids, the Brinkman number is often small, and we consider only the limiting case $Br \to 0$, for which the effects of viscous dissipation can be neglected completely.[2] In this case, (9–4) takes the simpler limiting form

$$\boxed{Pe(\mathbf{u} \cdot \nabla \theta) = \nabla^2 \theta.} \qquad (9\text{--}7)$$

The dimensionless boundary conditions corresponding to expressions (9–2a) and (9–2b) are

$$\boxed{\begin{aligned} \theta &= 1 \quad \text{on } S, \\ \theta &\to 0 \quad \text{at } |r| \to \infty. \end{aligned}} \qquad \begin{aligned} (9\text{--}8\text{a}) \\ (9\text{--}8\text{b}) \end{aligned}$$

It is clear from (9–7) and (9–8) that the form of the temperature distribution will depend on the geometry of the heated body (represented by S) and the magnitude of the Peclet number, Pe. In addition, because the form of the velocity field depends on the Reynolds number, that is,

$$\mathbf{u} = \mathbf{u}(\mathbf{r}; Re),$$

A. Forced Convection Heat Transfer – Introduction

we see from (9–7) and (9–8) that the temperature distribution for a body of specified shape also will depend on Re, that is,

$$\theta = \theta(\mathbf{r}; Re, Pe) \tag{9-9a}$$

or, equivalently,

$$\theta = \theta(\mathbf{r}; Re, Pr). \tag{9-9b}$$

A primary objective of most calculations of the temperature field near a heated or cooled body is the overall rate of heat transfer between the body and the fluid. For the case of a body of fixed shape, the heat flux at the surface is due solely to heat conduction. Thus the total rate of heat transfer is simply

$$\boxed{Q = \int_A -k(\nabla T \cdot \mathbf{n})\bigg|_s dA.} \tag{9-10}$$

Here, k is the thermal conductivity, S stands for the body surface, and $\mathbf{n}$ is the outer unit normal to the surface, as shown in Fig. 9–1. The integration is carried out over the complete body surface A.

The *dimensionless* total rate of heat transfer is known as the *Nusselt number Nu* and is normally defined as

$$Nu \equiv \frac{2Q}{(\text{surface area})k(T_0 - T_\infty)/\ell}, \tag{9-11}$$

where ℓ is the characteristic linear dimension of the body. Combining the definitions (9–10) and (9–11) and expressing the integral in terms of dimensionless variables ($dA = \ell^2 \, d\bar{A}$), we obtain

$$Nu \equiv \frac{-2\ell^2}{(\text{area})} \int_{\bar{A}} (\nabla \theta \cdot \mathbf{n})_{\bar{s}} \, d\bar{A}, \tag{9-12}$$

and thus, in view of (9–9), we see that

$$\boxed{\begin{aligned} Nu &= Nu(Re, Pe), \\ Nu &= Nu(Re, Pr). \end{aligned}} \tag{9-13}$$

Hence, from dimensional analysis alone, we see that a correlation should be expected relating Nu to the independent dimensionless parameters Re and Pe (or Re and Pr).

It was noted earlier that the dependence of θ and/or Nu on the Reynolds number is a consequence of the dependence of the velocity field on Re, as indicated just before (9–9). In some circumstances, however, the form of $\mathbf{u}$ is independent of Re. Two examples are the velocity profiles in unidirectional flows (Chap. 3) and the creeping-flow approximation to the velocity fields for $Re \ll 1$. In such cases

$$\mathbf{u} = \mathbf{u}(\mathbf{r}) \text{ only,}$$

and it follows that

$$\boxed{\theta = \theta(Pe), \quad Nu = Nu(Pe).} \tag{9-14}$$

To evaluate heat transfer rates by means of the definition (9–12) or (9–10), it is necessary to solve (9–7) and (9–8). Although this problem is linear in θ, the coefficient $\mathbf{u}$ in (9–7) is almost always too complex to allow exact analytic solutions for θ. The best that we can do, analytically, is to solve (9–7) approximately for limiting values of Re and Pe.

Convection Effects in Low-Reynolds-Number Flows

We begin in the next several sections with the limit $Pe \ll 1$, in which convection effects are weak. Following that we consider the opposite limit $Pe \gg 1$, where convection effects are extremely important to the heat transfer process.

3. The Analogy with Single-Solute Mass Transfer

We may note that the problem defined by (9–7) and (9–8) is identical mathematically to the corresponding single-solute *mass transfer* problem, provided the conditions at the body surface and at infinity are such that we can specify the solute concentration as known constants. In this case, we can substitute concentration c (measured as mass fraction of solute) for the temperature T in the definition (9–3) of θ. Then the boundary conditions are identical to (9–8), and the governing equation for solute transport is also the same as (9–7), with the exception that the Peclet number must now be defined in terms of the diffusivity D for the solute in the solvent, rather than the thermal diffusivity κ. Hence, in this case

$$Pe \equiv \frac{U_\infty \ell}{D},$$

and instead of the Peclet number being the product of the Reynolds number times the Prandtl number, it is the Reynolds number times the Schmidt number:

$$Pe \equiv ReSc, \quad \text{where} \quad Sc \equiv \frac{v}{D}.$$

The condition of constant concentration c at the body surface is again not always applicable. Common examples of mass transfer problems involving a single solute are the dissolution of a solid into an ambient liquid or gas; mass transfer involving a chemical reaction at a surface in which a reactant is transported to the surface but no product is released (e.g., $2Mg + O_2 \rightarrow 2MgO$ at high temperature, where gaseous oxygen diffuses to the surface, reacts, but no gaseous products are produced); or the transport of a solute across a fluid interface. In the latter case, the concentration will rarely be independent of position at the interface because the transport rate will vary with position due to the influence of flow. However, if the intrinsic dissolution or reaction rate is fast enough so that the actual rate of reaction or dissolution is controlled by the rate of mass transfer to (or from) the surface, then the concentration in the fluid very near the surface will be constant. Again, it is of interest to analyze more general surface boundary conditions, and we will briefly discuss this topic in Chap. 11. Of course, many mass transfer problems involve multiple components, and in these cases a more specialized theoretical framework is required.[3]

Another important practical difference between the heat and mass transfer problems is the magnitude of the Prandtl number and the Schmidt numbers. In the present chapter, these parameters appear in only the *relationship* between the Peclet number and the Reynolds number. However, we shall see in Chap. 11 that this is not always true. In any case, the magnitude of the Schmidt number is typically larger than the magnitude of the Peclet number. For liquids, typical values of the Schmidt number are

$$Sc > O(10^3),$$

whereas values of the Prandtl number for typical liquids, apart from a few very viscous oils and greases, are

$$Pr \leq O(5).$$

On the other hand, for gases,

$$Sc = O(1),$$
$$Pr = O(0.01-0.1).$$

A. Forced Convection Heat Transfer – Introduction

Finally, one *fundamental difference* between heat and mass transport is that mass transfer may modify the fluid flow and/or produce a convective contribution to the transport rate to (or from) a surface, even when natural convection effects are completely negligible, because the mass transfer process produces a nonzero normal velocity at the transport boundary. To see that this is the case, we must carefully reconsider the mass balance at the boundary. First, let us introduce some convenient notation. If we denote the solvent as A and the solute as B, then by c we mean the concentration c_B. Because c_A and $c_B (\equiv c)$ are specified as mass fractions, the sum $c_A + c_B = c_A + c = 1$, and hence $\partial c_A/\partial n = -\partial c/\partial n$, where we can denote the surface as $n=0$ and derivatives normal to the surface as $\partial c/\partial n$. The key to the correct mass balance is to note that there can be no net flux of the (inert) solvent at the mass transfer surface. Hence there must be a normal velocity away from the surface to balance the diffusive flux of solvent toward the surface because of the gradient in the solvent concentration, $\partial c_A/\partial n$:

$$c_A v' - D\frac{\partial c_A}{\partial n'} = 0 \quad \text{at } n'=0.$$

Here we include the primes on v' and n' to remind us that these are dimensional quantities. It follows that

$$v' = \frac{D}{c_A}\frac{\partial c_A}{\partial n'} \quad \text{at } n'=0.$$

Now because $c_A + c = 1$ we can express the normal velocity in terms of the concentration of the component whose distribution and mass flux rate we wish to calculate:

$$v' = -\frac{D}{1-c_0}\frac{\partial c}{\partial n'}\bigg|_{n'=0}.$$

The minus sign implies that if c increases with n', the induced motion will be toward the surface (this is often associated with the name "suction"), whereas the opposite will be true if c decreases with n' (termed "blowing"). If we introduce $\theta \equiv (c - c_\infty)/(c_0 - c_\infty)$, and the characteristic velocity and length scales used to derive (9–7), we can write this in dimensionless form as

$$v = -\frac{B}{Pe}\frac{\partial \theta}{\partial n} \quad \text{at } n=0.$$

Here, B is a dimensionless parameter that is often called the "blowing" number:

$$B \equiv \frac{c_0 - c_\infty}{1 - c_0}.$$

However, as noted earlier, the direction of this induced velocity depends on the sign of B. Clearly, we can neglect the change in the *flow* that is due to the normal velocity at the boundary only if $B/Pe \ll 1$.

It may also be noted that the dimensionless overall rate of mass transfer to (or from) a body is usually known in the mass transfer literature as the *Sherwood number*, rather than the Nusselt number. In general, the overall rate of mass transfer is

$$J = -\int_A (D\nabla' c \cdot \mathbf{n})\bigg|_S dA + \int_A v'c \bigg|_S dA.$$

The Sherwood number is normally defined in complete analogy with the Nusselt number,

$$Sh \equiv \frac{2J}{(\text{surface area})D(c_0 - c_\infty)/\ell}.$$

Convection Effects in Low-Reynolds-Number Flows

Expressing J in terms of dimensionless variables, we obtain

$$Sh = \frac{-2\ell^2}{(area)} \left[\int_A (\nabla\theta \cdot \mathbf{n})_{\bar{s}} \, d\bar{A} + B \int_A (\nabla\theta \cdot \mathbf{n}) \left(\theta + \frac{c_\infty}{c_0 - c_\infty}\right)\Big|_{\bar{s}} d\bar{A} \right].$$

In the limit $B \ll 1$, this reduces to the same form as the expression for the Nusselt number:

$$Sh = \frac{-2\ell^2}{(area)} \left[\int_A (\nabla\theta \cdot \mathbf{n})_{\bar{s}} \, d\bar{A} \right].$$

It must be noted, however, that the flow, and therefore $(\nabla\theta \cdot \mathbf{n})_{\bar{s}}$, will still be affected by blowing/suction unless the *ratio B/Pe* is also vanishingly small.

B. HEAT TRANSFER BY CONDUCTION ($Pe \to 0$)

When the dimensionless form of the thermal energy equation is compared with the dimensionless Navier–Stokes equation, it is clear that the Peclet number plays a role for heat transfer that is analogous to the Reynolds number for fluid motion. Thus it is natural to seek approximate solutions for asymptotically small values of the Peclet number, analogous to the low-Reynolds-number approximation of Chaps. 7 and 8.

Let us suppose, then, that an asymptotic solution exists for $Pe \ll 1$ in the form of a regular perturbation expansion:

$$\theta = \theta_0(\mathbf{x}; Re) + Pe\theta_1(\mathbf{x}; Re) + O(Pe^2). \tag{9-15}$$

If we substitute (9–15) into (9–7) and (9–8) and remember that $\mathbf{u}$ is independent of Pe, we see that the leading term satisfies

$$\nabla^2 \theta_0 = 0, \tag{9-16}$$

with boundary conditions

$$\begin{aligned} \theta_0 &= 1 \quad \text{on } S, \\ \theta_0 &\to 0 \quad \text{at } |r| \to \infty. \end{aligned} \tag{9-17}$$

Equation (9–16) is known as the steady-state *heat conduction equation* and is completely analogous to the creeping-motion equation of Chaps. 7 and 8. It can be seen that convection plays no role in the heat transfer process described by (9–16) and (9–17). Thus the form of the velocity field is not relevant, and in spite of the initial assumption (9–15), there is no dependence of θ_0 on the Reynolds number of the flow. The solution of (9–16) and (9–17) depends on only the geometry of the body surface, represented in (9–17) by S.

A simple example for which problems (9–16) and (9–17) can be solved easily is the case of a heated sphere. In this case, we may choose the sphere radius as the characteristic length scale for nondimensionalization, and the problem is to solve (9–16) subject to the boundary condition $\theta = 1$ at $r = 1$. This can be done easily. We may first note that a general solution of Laplace's equation in spherical coordinates is

$$\theta_0 = \sum_{n=0}^{\infty} \left[A_n r^n + B_n r^{-(n+1)} \right] P_n(\eta). \tag{9-18}$$

B. Heat Transfer by Conduction ($Pe \to 0$)

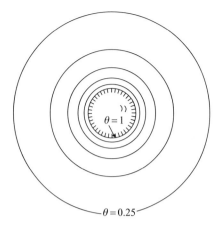

Figure 9–2. Contours of constant temperature for pure conduction heat transfer from a heated (or cooled) sphere. For constant surface temperature, $\theta = 1$, the isotherms are spherically symmetric, as indicated by Eq. (9–19).

Here $\eta \equiv \cos\theta$, and A_n and B_n are coefficients to be determined from the boundary conditions. The second of the conditions (9–17) requires that

$$A_n = 0 \quad \text{all } n.$$

The boundary condition at $r = 1$ is then satisfied provided that

$$B_n = 0, \quad n \geq 1, \quad \text{and} \quad B_0 = 1.$$

Thus, the solution of (9–16) and (9–17) for a heated sphere is simply

$$\boxed{\theta_0 = \frac{1}{r}.} \tag{9–19}$$

According to this solution, which is plotted in Fig. 9–2, the temperature falls off inversely with distance from the sphere in all directions. This very simple solution is the equivalent of Stokes' solution for creeping motion past a sphere. Indeed, it may be recalled from Chap. 7 that the disturbance to the velocity field that is due to a sphere in Stokes' flow also decreased as r^{-1} for large r.

With the approximate result (9–19) for the temperature field, we can easily evaluate the overall heat transfer rate, that is, the Nusselt number for this pure conduction limit. For a sphere, the general expression (9–12) for the Nusselt number becomes

$$Nu = -\frac{1}{2\pi} \int_0^{2\pi} \int_0^\pi \left(\frac{\partial \theta}{\partial r}\right)_{r=1} \sin\theta \, d\theta \, d\phi, \tag{9–20}$$

where we have used $\ell = a$. Hence, substituting for $(\partial\theta/\partial r)_{r=1}$ from (9–19), we obtain

$$\boxed{Nu_0 = 2.} \tag{9–21}$$

The subscript 0 in this case is intended to serve as a reminder that this result is only a first approximation to the dimensionless rate of heat transfer for $Pe \ll 1$.

Clearly, in view of what was stated at the end of the previous section, the analysis in this section could equally well be viewed as *pure diffusion* of a *solute* in circumstances in which the solute concentration can be specified at the surface of the sphere.

C. HEAT TRANSFER FROM A SOLID SPHERE IN A UNIFORM STREAMING FLOW AT SMALL, BUT NONZERO, PECLET NUMBERS

1. Introduction – Whitehead's Paradox

A natural question is to what extent the overall rate of heat transfer is modified by convection for small, but nonzero, values of the Peclet number. To obtain a more accurate estimate of Nu for this case, it would appear, from what has been stated thus far, that we must calculate added terms in the regular asymptotic expansion (9–15) for θ. To attempt this, we must substitute (9–15) into (9–7) and (9–8) to obtain governing equations and boundary conditions for the subsequent terms θ_n. In this section, we consider only the second approximation, θ_1. The governing equation and boundary conditions derived from (9–7), (9–8), and (9–15) are

$$\nabla^2 \theta_1 = \mathbf{u} \cdot \nabla \theta_0, \tag{9-22}$$

$$\theta_1 = 0 \quad \text{on} \quad S, \tag{9-23a}$$

$$\theta_1 \to 0 \quad \text{at} \quad |r| \to \infty. \tag{9-23b}$$

To complete the specification of the problem for θ_1, we must specify a particular velocity field $\mathbf{u}$. In the case of $Re \ll 1$, we can use the creeping-flow solutions of Chaps. 7 and 8, and it is again convenient to focus our attention on the case of a sphere in a uniform streaming flow, in which a first approximation to the velocity field is given by the Stokes' solution, Eq. (7–158), from which we can calculate the velocity components by means of (7–102).

In this case, the full thermal energy equation, (9–7), takes the form

$$\frac{1}{r^2}\left\{\frac{\partial}{\partial r}\left(r^2 \frac{\partial \theta}{\partial r}\right) + \frac{\partial}{\partial \eta}\left[(1-\eta^2)\frac{\partial \theta}{\partial \eta}\right]\right\}$$
$$= Pe\left[\left(1 - \frac{3}{2r} + \frac{1}{2r^3}\right)\eta\frac{\partial \theta}{\partial r} + \frac{(1-\eta^2)}{r}\left(1 - \frac{3}{4r} - \frac{1}{4r^3}\right)\frac{\partial \theta}{\partial \eta}\right]. \tag{9-24}$$

Hence the expansion (9–15) apparently can be written in the slightly simplified form [see Eqs. (9–14)]

$$\theta = \theta_0(\mathbf{x}) + Pe\,\theta_1(\mathbf{x}) + O(Pe^2), \tag{9-25}$$

and the governing equation, (9–22), for the second term, θ_1, becomes

$$\frac{1}{r^2}\left\{\frac{\partial}{\partial r}\left(r^2 \frac{\partial \theta_1}{\partial r}\right) + \frac{\partial}{\partial \eta}\left[(1-\eta^2)\frac{\partial \theta_1}{\partial \eta}\right]\right\} = -\frac{1}{r^2}\left(1 - \frac{3}{2r} + \frac{1}{2r^3}\right)P_1(\eta). \tag{9-26}$$

The homogeneous solution of this equation is just (9–18). A particular solution, corresponding to each of the terms on the right-hand side of (9–26) can be obtained in the form $r^s P_1(\eta)$. Combining the homogeneous and particular solutions, we can express the general solution of (9–26) as

$$\theta_1 = \sum_{n=0}^{\infty}\left[A_n r^n + B_n r^{-(n+1)}\right]P_n(\eta) + \left(\frac{1}{2} - \frac{3}{4r} - \frac{1}{8r^3}\right)P_1(\eta). \tag{9-27}$$

To this point, the regular perturbation scheme (9–15) [or (9–25)] has seemed to proceed in a completely straightforward manner. Now, however, when we attempt to determine values

C. Heat Transfer From a Solid Sphere in a Uniform Streaming Flow

for A_n and B_n by application of the boundary conditions (9–23), we surprisingly find that no combination is satisfactory. The term $P_1(\eta)/2$, which appears in the particular solution, does not vanish as $r \to \infty$, and there is no corresponding term in the homogeneous solution to cancel it so as to satisfy the condition (9–23b). The unavoidable, but seemingly paradoxical, conclusion is that no solution to (9–26) exists that will satisfy the boundary condition (9–23b) at infinity. Thus an asymptotic solution of (9–24) with the boundary conditions (9–23) apparently does not exist in the form (9–25) or (9–15) for $Pe \ll 1$. This result is sometimes called *Whitehead's paradox* for convective heat transfer. The important implication is that *the pure conduction solution (9–19) is not a uniformly valid first approximation to the temperature field, even in the limit as* $Pe \to 0$. To those readers who have studied all or parts of the previous two chapters, this may seem quite alarming if we recognize that the creeping-flow solutions of the preceding two chapters were obtained from the full Navier–Stokes equations by precisely the same type of analysis for $Re \ll 1$. In effect, to obtain the creeping-flow approximation (7–6a) from the full Navier–Stokes equations in the dimensionless form (7–2), we assumed that a uniformly valid perturbation expansion (i.e, a *regular* expansion) exists for the velocity and pressure in the limit $Re \to 0$ with the creeping-flow solution as the first term. The fact that this same type of assumption does not work for the corresponding thermal problem should cause some concern that it is not valid for the fluid mechanics problem either, and thus that the creeping-flow solutions that we have generated may also not be a uniformly valid first approximation to the solution of the Navier–Stokes equations for $Re \ll 1$. We shall return to this question at the end of this section. First, let us consider the thermal problem, which is the main focus of this chapter.

Although the conclusion of the preceding paragraph is clear, the reader may well be puzzled for an explanation of what has gone wrong. To ascertain the difficulty, we must carefully retrace our steps. In particular, the only two approximations that were made in the analysis leading to (9–26) were the neglect of viscous dissipation and the assumption inherent in (9–24) and (9–25) that convection terms in the thermal energy equation are negligible compared with conduction terms everywhere in the domain for $Pe \ll 1$, so that a uniformly valid *first* approximation to the temperature field in the limit $Pe \to 0$ can be obtained from the pure conduction equation, (9–16). Indeed, if both of these assumptions were correct, the approximate equation (9–16) would have to be a *uniformly* valid first approximation of the full thermal equation (9–1). The neglect of viscous dissipation is not a critical assumption; even if viscous dissipation were retained, the same difficulty would occur in attempting to obtain a solution as a regular perturbation of the form (9–25). The fact that such a solution does *not* exist can therefore indicate *only* that the neglect of convection everywhere in the fluid domain compared with conduction is incorrect, even in the limit as $Pe \to 0$. But this assumption is inherent in the nondimensionalized form of the thermal energy equation, (9–7) or (9–24). In particular, at any point where the nondimensionalization is correct, the terms $\mathbf{u} \cdot \nabla\theta$ and $\nabla^2\theta$ are $O(1)$ – that is, *independent* of Pe – and the convection terms can be made arbitrarily small compared with conduction by simply letting $Pe \to 0$.

The unavoidable conclusion is that the characteristic scales $\ell_c = \ell$ and $u_c = U_\infty$ used to obtain Eqs. (9–7) or (9–24) cannot be relevant everywhere in the domain. For purposes of simply writing the thermal energy equation in nondimensional form, this is not a serious problem. Indeed, in the transition between the exact dimensional and nondimensional forms – (9–1)–(9–4) – we are simply making a formal change of variables. If the original exact dimensional equation is correct, then the exact nondimensionalized equation is correct, independent of the length or velocity scales used in nondimensionalization. It is only when we attempt to approximate the nondimensionalized equation by estimating the magnitude of the various terms through the magnitude of dimensionless parameters such as Pe that we

Convection Effects in Low-Reynolds-Number Flows

will make mistakes unless we have used the physically relevant characteristic scales in the nondimensionalization process. Because the problem in solving (9–26) arises in attempting to satisfy boundary conditions at infinity, we may surmise that Whitehead's paradox is a consequence of the fact that the radius of the sphere is not an appropriate characteristic length scale far from the sphere. Hence the asymptotic approximation (9–25), based on (9–24), with (9–19) as the first term, is *not* valid in this region.

To corroborate this suggestion, it is instructive to consider the magnitudes of a typical conduction term in (9–25), $\partial^2\theta/\partial r^2$, and the largest convection term, $Pe\,\eta(\partial\theta/\partial r)$. If we evaluate the order of magnitude of these terms by using the pure conduction approximation for θ, we find that

$$\frac{\partial^2\theta_0}{\partial r^2} \sim O\left(\frac{1}{r^3}\right) \quad \text{and} \quad Pe\,\eta\frac{\partial\theta_0}{\partial r} \sim O\left(\frac{Pe}{r^2}\right) \quad \text{for } r \gg 1. \qquad (9\text{–}28)$$

The nondimensionalization leading to (9–7) and/or (9–24) has led us to the conclusion that convection terms should be negligible compared with conduction terms anywhere in the domain if Pe is sufficiently small. However, the estimates (9–28) show clearly that this is not true. Specifically, for any arbitrarily small Pe, we can always find a value of $r \sim O(Pe^{-1})$ such that the conduction and convection terms are of equal importance. We conclude that the nondimensionalization leading to (9–7) can be valid only in the part of the domain within a distance $r < O(Pe^{-1})$ from the sphere. It follows that a regular (uniformly valid) asymptotic expansion cannot exist for $Pe \ll 1$.

We may recall that the case just described bears a considerable resemblance to several problems of preceding chapters. Specifically, we may recall pulsatile flow at high frequencies, which was described in Chap. 4, diffusion in a sphere with fast reaction (again in Chap. 4), or the air hockey problem at large Re in Chap. 5. It is recommended that the reader review the discussion of the physical significance of nondimensionalization and the manner in which it may lead to incorrect approximations of governing equations when limits are applied to the dimensionless parameters that appear. In the pulsatile flow problem, we found that the obvious physical length scale, the tube radius, was not characteristic of velocity gradients in the region very near the tube wall for $Re_\omega \gg 1$. Hence, a regular perturbation solution could not be found, and it was necessary to use the method of matched asymptotic expansions with different approximations to the governing equations valid in different parts of the domain, corresponding to the regions where the solution was characterized by different length scales. It may seem at first that a fundamental difference exists between the present problem and the pulsatile flow problem, as no solution satisfying boundary conditions could be obtained in the latter case, even for the first term in the asymptotic expansion, whereas in our problem we seemingly obtain a perfectly good solution for θ_0 but cannot find a solution for the second term, θ_1. However, this distinction is not fundamentally significant.

To illustrate this fact, we may consider the 2D heat transfer problem of uniform flow past a heated circular cylinder with uniform surface temperature. In this case, if we look for a solution in the asymptotic form (9–15) for low Peclet numbers, the nondimensional governing equation and boundary conditions for θ_0 are again (9–16) and (9–17), but this time are expressed in cylindrical coordinates, namely,

$$\frac{1}{r}\frac{\partial}{\partial r}\left(r\frac{\partial\theta_0}{\partial r}\right) = 0, \qquad (9\text{–}29)$$

$$\theta_0 = 1 \quad \text{at} \quad r = 1, \qquad (9\text{–}30a)$$

$$\theta_0 \to 0 \quad \text{at} \quad r \to \infty. \qquad (9\text{–}30b)$$

C. Heat Transfer From a Solid Sphere in a Uniform Streaming Flow

A general solution of (9–29) that is independent of position on the cylinder surface and at infinity, in accord with the boundary conditions (9–30), is

$$\theta_0 = C_1 \ln r + C_2, \tag{9-31}$$

where C_1 and C_2 are coefficients that must be determined from the boundary conditions. On examination, however, it is evident that no choice for these coefficients will actually satisfy the boundary conditions. In particular, to satisfy (9–30), we require that $C_2 = 1$. However, the resulting function cannot be made to vanish at infinity for any choice of C_1. Hence, for this 2D heat transfer problem, *no solution exists even for the first term* in a regular perturbation expansion for $Pe \ll 1$. This is analogous to the behavior found earlier for the pulsatile flow problem. In fact, it is the heated sphere problem that is unusual in the sense that a solution of (9–16) exists satisfying both (9–17a) and (9–17b) even though (9–16) is not a uniformly valid first approximation to (9–7), as we have already seen.

To obtain a valid approximate solution for heat transfer from a sphere in a uniform streaming flow at small, but nonzero, Peclet numbers, we must resort to the method of matched (or singular) asymptotic expansions.[4] In this method, as we have already seen in Chap. 4, two (or more) asymptotic approximations are proposed for the temperature field at $Pe \ll 1$, each valid in different portions of the domain but linked in a so-called overlap or *matching* region where it is required that the two approximations reduce to the same functional form. The approximate forms of (9–1), from which these matched expansions are derived, can be obtained by nondimensionalization by use of characteristic length scales that are *appropriate to each subdomain*.

2. Expansion in the Inner Region

Let us suppose, based upon the estimates of (9–28), that the dimensionless form (9–24) of the thermal energy equation is valid within the region $1 \le r \le O(Pe^{-1})$. In other words, we suppose that within this so-called *inner* region, the sphere radius is an appropriate characteristic length scale as we have assumed in the nondimensionalization leading up to (9–24). Hence, within this region, the dimensionless temperature field can be represented in the form (9–25) with $\theta_0 = 1/r$, that is,

$$\theta(r, \eta, Pe) = \frac{1}{r} + \sum_{n=1}^{N} f_n(Pe)\theta_n(r, \eta), \tag{9-32}$$

where the gauge functions $f_n(Pe)$ are, for the moment, unspecified except for the requirements that

$$f_1 \to 0 \quad \text{and} \quad \frac{f_{n+1}}{f_n} \to 0 \quad \text{as } Pe \to 0. \tag{9-33}$$

If $f_1(Pe) = Pe$, as assumed in the regular perturbation expansion (9–25), then, of course, the governing equation for θ_1 is still (9–26), and the general solution for θ_1 is still (9–27). However, we do not know *a priori* what form f_1 and the other $f_n(Pe)$ in (9–32) should take; this must be determined as part of the solution of the problem.

The fact that the first term in (9–32) is still $1/r$ is a consequence of the fact that θ_0 must satisfy Laplace's equation, the boundary condition $\theta_0 = 1$ at $r = 1$, and also be a decreasing function of r in order that the total heat flux is conserved for increasing values of r. The only solution of $\nabla^2 \theta_0 = 0$ with the latter two properties is $\theta_0 = 1/r$.

It should be noted, however, that the inner region does not extend to $r \to \infty$. Hence boundary conditions cannot be imposed on any of the terms of (9–32) in the limit $r \to \infty$, and it is more or less accidental that the first term of (9–32) is consistent with the original boundary condition (9–8b) for $r \to \infty$. The most that we can require of the approximate solution (9–32) is that it matches the corresponding small-Pe approximation that is valid in the so-called outer region where $r \geq O(Pe^{-1})$.

We shall return shortly to attempt to obtain a second approximation for θ, corresponding to the θ_1 term in (9–32). Before doing this, however, it is necessary to consider a first approximation to θ for the outer region.

3. Expansion in the Outer Region

In the outer part of the domain, the preceding discussion indicates that conduction and convection terms in the thermal energy equation will remain the same order of magnitude, even as $Pe \to 0$. Obviously, in the nondimensionalized form (9–7), this does not appear to be the case because both $(\mathbf{u} \cdot \nabla \theta)$ and $\nabla^2 \theta$ should be $O(1)$ (that is, independent of Pe), if the choice of characteristic scales were correct. Thus, we conclude, as already stated, that the choice $\ell_c = a$ as a characteristic length scale must break down if we are too far from the sphere. Certainly, on intuitive grounds, this would seem to make sense. One would be surprised if any detailed feature of the temperature field at very large distances from a heated body were to remain sensitive to the body geometry – either its shape or size – except insofar as these factors control the total heat flux from the body to the fluid. Indeed, if we are sufficiently far from the sphere, we should anticipate that it will simply appear as a point source of heat in the uniform streaming flow. The question is, if the sphere radius is not appropriate as a characteristic length scale in this outer part of the domain, then what choice is appropriate and how shall we find it?

One approach is simply to guess until we (hopefully) hit the correct choice. An obvious possibility in the present problem is κ/U_∞, as this is another combination of independent dimensional parameters, independent of a, which has dimensions of length. However, this choice is by no means unique. There are, in fact, an infinite number of other possibilities, for example, $a^n(\kappa/U_\infty)^{1-n}$ for any n, $0 \leq n < 1$. In some singular-perturbation problems, the "obvious" guess turns out to be correct, but in others it does not. It is therefore imperative to develop a *systematic* strategy for determining the characteristic length scale for those parts of the domain for which the physical dimension of the boundaries does not provide a characteristic length scale.

The most effective approach, as we have already seen in previous problems, is *rescaling*. In this case, we *rescale* the original nondimensionalized radial variable r in the very general form

$$\boxed{\rho = Pe^m r.} \tag{9–34}$$

For any nonzero m, this newly rescaled radial variable ρ is just the original *dimensional* radial variable, r', nondimensionalized with respect to the new length scale

$$\ell^* = a^{1-m}\left(\frac{\kappa}{U_\infty}\right)^m.$$

In the present problem, the rescaling parameter m must be chosen in such a way that both conduction and convection terms are retained in the limit $Pe \to 0$ when the thermal energy equation is expressed in terms of the outer variable ρ. Before actually attempting to determine m, however, the reader may wish to know why the radial variable is rescaled but not η or θ. In the case of θ, the governing equation is linear, and thus rescaling θ serves no purpose in spite of the fact that comparison with the leading term in the inner solution, expressed in terms of the outer variable ρ (as is appropriate in the overlap or matching region between the two solutions) shows clearly that the first approximation to θ in the outer region will be $O(Pe^m)$. On the other hand, the independent variable η is bounded, $-1 \leq \eta \leq 1$, and there is simply no plausible basis to suggest that thermal gradients with respect to η will

C. Heat Transfer From a Solid Sphere in a Uniform Streaming Flow

depend on Pe in some part of the domain. Thus the only rescaling is (9–34). If (9–34) is now substituted into the dimensionless thermal energy equation, (9–24), scaled with respect to a, we obtain the same equation scaled with respect to ℓ^*:

$$\frac{Pe^{2m}}{\rho^2}\left\{\frac{\partial}{\partial\rho}\left(\rho^2\frac{\partial\oplus}{\partial\rho}\right)+\frac{\partial}{\partial\eta}\left[(1-\eta^2)\frac{\partial\oplus}{\partial\eta}\right]\right\}$$
$$=Pe^{1+m}\left[\left(1-\frac{3Pe^m}{2\rho}+\frac{Pe^{3m}}{2\rho^3}\right)\eta\frac{\partial\oplus}{\partial\rho}+\frac{(1-\eta^2)}{\rho}\left(1-\frac{3Pe^m}{4\rho}-\frac{Pe^{3m}}{4\rho^3}\right)\frac{\partial\oplus}{\partial\eta}\right], \quad (9\text{–}35)$$

where we have denoted the dimensionless temperature as $\oplus$ in order to be able to distinguish it from the temperature approximation in the inner region, which we have denoted previously as θ.

All that remains to completely specify the form of the governing equation in this outer region is to determine the parameter m and thus specify the characteristic length scale ℓ^* for this region. To do this, we recall that conduction and convection terms are expected to remain in balance in this outer portion of the domain as $Pe \to 0$. We see from (9–35) that this implies that

$$2m = 1 + m$$

or, simply,

$$m = 1. \qquad (9\text{–}36)$$

In this case,

$$\boxed{\rho = Pe\, r = \left(\frac{U_\infty a}{\kappa}\right)\left(\frac{r'}{a}\right).}$$

Thus the appropriate characteristic length scale ℓ^* in this outer portion of the domain is just κ/U_∞. Although this choice was recognized immediately in this problem as an obvious possibility (and thus could have been guessed *a priori*), we shall see eventually that the rescaling of variables is not usually so obvious.

With $m = 1$, the thermal energy equation, (9–35), can now be rewritten in the dimensionless form appropriate to the outer region of the domain:

$$\boxed{\begin{aligned}\nabla_\rho^2\oplus &= \left[\eta\frac{\partial\oplus}{\partial\rho}+\frac{(1-\eta^2)}{\rho}\frac{\partial\oplus}{\partial\eta}\right]+Pe\left[-\frac{3\eta}{2\rho}\frac{\partial\oplus}{\partial\rho}-\frac{3}{4\rho^2}(1-\eta^2)\frac{\partial\oplus}{\partial\eta}\right]\\ &\quad +Pe^3\left[\frac{\eta}{2\rho^3}\frac{\partial\oplus}{\partial\rho}-\frac{(1-\eta^2)}{4\rho^4}\frac{\partial\oplus}{\partial\eta}\right].\end{aligned}} \qquad (9\text{–}37)$$

If we examine the terms on the right-hand side, we see that the largest convection contribution for $Pe \to 0$ is that due to the undisturbed, uniform streaming flow. The corrections to the rate of convection that are due to the modifications of the velocity field by the sphere are $O(Pe)$ and $O(Pe^3)$, respectively. According to (9–8b), the solution of (9–37) is required to vanish as $\rho \to \infty$. The boundary condition (9–8a) does not apply to the outer solution because the sphere surface is not a part of the outer region. Instead, the solution of (9–37) is required to match in the limit $Pe \to 0$ with the inner approximation to the solution of the full problem, (9–32), for $\rho \ll 1$. A schematic of the solution domain summarizing the scaling, governing equations and boundary conditions is shown in Fig. 9–3.

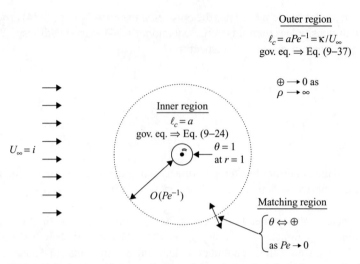

Figure 9–3. A schematic representation of the solution domain for heat transfer from a sphere in a uniform flow in the limit of small Peclet numbers, that is, $Pe \to 0$.

We now suppose that the outer solution for $Pe \to 0$ can be represented by an asymptotic expansion of the form

$$\oplus(\rho, \eta, Pe) = \sum_{n=0}^{N} F_n(Pe) \oplus_n (\rho, \eta), \qquad (9\text{--}38)$$

where the gauge functions $F_n(Pe)$ are unknown but must again satisfy the requirement that

$$\frac{F_{n+1}}{F_n} \to 0 \quad \text{as } Pe \to 0.$$

On substituting (9–38) into (9–37), we find that, to leading order,

$$\nabla_\rho^2 \oplus_0 = \eta \frac{\partial \oplus_0}{\partial \rho} + \frac{(1-\eta^2)}{\rho} \frac{\partial \oplus_0}{\partial \eta}, \qquad (9\text{--}39)$$

which, as we have already noted, is nothing more than the thermal energy equation with the velocity **u** replaced with the nondimensionalized, free-stream velocity, $\mathbf{i}_\infty$. To solve (9–39), it is convenient to introduce the transformation, first discovered by Goldstein,[5]

$$\oplus_0 = e^{\rho \eta / 2} \Phi, \qquad (9\text{--}40)$$

with Φ now considered as the independent unknown. Substituting (9–40) into (9–39), we obtain

$$\nabla_\rho^2 \Phi - \frac{1}{4}\Phi = 0. \qquad (9\text{--}41)$$

A general solution of this equation, which is bounded at $\eta = \pm 1$, can be obtained easily by separation of variables in terms of modified Bessel functions of the first and second kind,

C. Heat Transfer From a Solid Sphere in a Uniform Streaming Flow

$I_{n+1/2}(\rho/2)$ and $K_{n+1/2}(\rho/2)$. However, $I_{n+1/2}(\rho/2) \to \infty$ *exponentially* as $\rho \to \infty$. Thus, in view of the condition $\oplus \to 0$ as $\rho \to \infty$, the most general allowable solution for Φ is

$$\Phi = \sum_0^\infty A_n \sqrt{\frac{\pi}{\rho}} K_{n+1/2}\left(\frac{\rho}{2}\right) P_n(\eta). \tag{9-42}$$

After substituting back into (9–40), we have

$$\boxed{\oplus_0 = \sqrt{\frac{\pi}{\rho}} e^{\rho\eta/2} \sum_{k=0}^\infty A_k K_{k+1/2}\left(\frac{\rho}{2}\right) P_k(\eta),} \tag{9-43}$$

where $K_{k+1/2}$ is the Bessel function of the second kind and half order and $P_k(\eta)$ is the Legendre polynomial of order k.

The coefficients A_k must be determined so that the solution (9–43) matches, at leading order, the first (pure conduction) approximation for the inner solution θ, that is,

$$F_0(Pe) \oplus_0 |_{\rho \ll 1} \Leftrightarrow \frac{1}{r} \quad \text{as } Pe \to 0. \tag{9-44}$$

To carry out this matching, we first note that

$$K_{k+1/2}\left(\frac{\rho}{2}\right) = \sqrt{\frac{\pi}{\rho}} e^{-\rho/2} \sum_{m=0}^k \frac{(k+m)!}{(k-m)! m!} \frac{1}{\rho^m} \tag{9-45}$$

and then express the inner approximation $1/r$ in terms of outer variables so that the two sides of (9–44) can be compared directly, that is,

$$\boxed{F_0(Pe) \oplus_0 (\rho, \eta)|_{\rho \ll 1} \Leftrightarrow \frac{Pe}{\rho} \quad \text{as } Pe \to 0.} \tag{9-46}$$

It is immediately obvious that the first approximation in the inner and outer regions can match only if

$$F_0(Pe) = Pe. \tag{9-47}$$

To complete the matching, we may note that the matching region between inner and outer expansions corresponds to $\rho \ll 1$ and/or $r \gg 1$. Indeed, for any arbitrarily large, but finite, value of r, we can make the corresponding value of ρ arbitrarily small, though nonzero, by taking the limit $Pe \to 0$. Thus, for purposes of matching, we can approximate the left-hand side of (9–46) by expanding the exponential function as a power series in ρ. At leading order, $\exp[\rho(n-1)/2] \sim 1 + O(\rho)$. Thus, the matching condition (9–46) can be expressed in the form

$$Pe \left\{ \pi \sum_{k=0}^\infty A_k P_k(\eta)[1 + O(\rho)] \sum_{m=0}^k \frac{(k+m)!}{(k-m)! m!} \frac{1}{\rho^{m+1}} \right\} \Leftrightarrow \frac{Pe}{\rho} \tag{9-48}$$

as $Pe \to 0$. When the two sides of (9–48) are compared, it is evident that the inner and outer approximations can match only if

$$A_0 = \frac{1}{\pi}, \quad A_k = 0 \ (k \geq 1). \tag{9-49}$$

With $A_0 = 1/\pi$, it is obvious that there is a term, $1/\rho$, on the left-hand side that matches precisely with the inner solution. The fact that all other $A_k = 0$ may require some additional explanation.

Let us suppose, for example, that $A_1 \ne 0$. Then there would be an additional term in (9–48) from the outer solution of the form

$$Pe\left\{\pi A_1 P_1(\eta)[1 + O(\rho)]\left(\frac{1}{\rho} + \frac{2}{\rho^2}\right)\right\}.$$

But such a term could not be matched with the solution in the inner region, either in the present form or even after we have added additional terms to the inner expansion. For example, to obtain a term from the inner solution that would equal Pe/ρ^2 when expressed in terms of outer variables, it would be necessary to start with a term in the inner solution of the form

$$\frac{1}{Pe}\left(\frac{1}{r^2}\right).$$

But this would imply the existence of a term $O(Pe^{-1})$ in the inner approximation, and such a term would be asymptotically large compared with the largest term in the inner expansion (9–32), which we have already proven to be $O(1)$, because of the boundary value $\theta = 1$ at $r = 1$. It is therefore evident that no such term can exist and $A_1 = 0$. Similarly, it can be shown that the other $A_k \equiv 0$, as these would require terms of $O(Pe^{-k})$ in the inner expansion for matching.

Finally, let us examine the matching condition (9–48) once more with the coefficients A_k evaluated according to (9–49). The result is

$$(Pe)\frac{1}{\rho}[1 + O(\rho)] \Leftrightarrow \frac{1}{\rho}(Pe) \quad \text{for } Pe \to 0. \tag{9–50}$$

The $O(\rho)$ term on the left-hand side is just the second term of the power-series approximation of $\exp[\rho(\eta - 1)/2]$. We see that the first term on the left-hand side matches precisely with the first approximation in the inner region. However, there is a *mismatch* that is due to the second term from the exponential, which has no counterpart in the first approximation for the inner solution. However, this term is $O(Pe)$ and independent of ρ. A term of this form could be generated only in the inner region from a term that is $O(Pe)$ and independent of r. However, we have so far considered only the first term in the inner region, which is $O(1)$. The mismatch evident in (9–50) at $O(Pe)$ can be made arbitrarily small by taking the asymptotic limit $Pe \to 0$. It can be removed only by considering additional terms in the expansion (9–32) for the inner region. We will return to this task shortly.

First, however, it is worthwhile to consider briefly the qualitative features of the first approximation to the temperature field in the outer region, which we have just shown to be

$$\oplus(\rho, \eta, Pe) = \frac{Pe}{\rho}e^{-\rho(1-\eta)/2} + O(F_1(Pe)), \tag{9–51}$$

where $F_1(Pe)/Pe \to 0$ as $Pe \to 0$. This is, in fact, nothing more than the fundamental solution for a point source of heat at $\rho = 0$ in a uniform streaming flow. Contours of constant temperature, corresponding to this solution, are shown in Fig. 9–4. In contrast to the temperature field for pure conduction, which we showed to decrease linearly with distance from the sphere, we see from (9–51) that the presence of convection in the outer region causes $\oplus$ to diminish exponentially with distance from the sphere, except in a region arbitrarily near $\eta = 1$, where it continues to decrease algebraically as $1/\rho$. Thus, in spite of

C. Heat Transfer From a Solid Sphere in a Uniform Streaming Flow

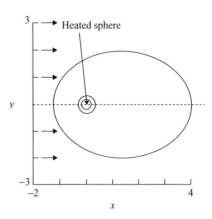

Figure 9–4. Contours of constant temperature (isotherms) for heat transfer from a sphere in a uniform flow at low Peclet numbers according to Eq. (9–51). Note that the sphere appears in this representation, in which $\ell_c = \kappa/U_\infty$, as a *point* source (sink) of heat (the radius of the sphere, a, is vanishingly small compared with κ/U_∞ in the limit as $Pe \to 0$). In the inner region, near to the sphere, the isotherms at leading order of approximation are still spherical as illustrated in Fig. 9–2. The three contours plotted are $\oplus_0 = 0.25, 2.75$, and 5.25.

the fact that the pure conduction solution appeared to satisfy the original boundary condition $\theta = 0$ as $r \to \infty$, and thus to provide a uniformly valid first approximation to the temperature field for $Pe \ll 1$, we see now that it did not actually have the correct functional dependence on r for $r \to \infty$. The symmetry axis $\eta = 1$, where $\oplus$ continues to decrease algebraically, lies downstream of the sphere, and the solution (9–51) is thus seen to exhibit a *thermal wake* structure. Heat released from the sphere is, at very large distances, mainly concentrated very near the symmetry axis $\eta = 1$, where it is transported by convection with the uniform streaming motion of the fluid. Unlike the first approximation in the inner region, which is radially symmetric, this first approximation in the outer region contains a contribution that is due to convection, and the fore–aft symmetry of the inner (conduction-dominated) temperature field is lost as the motion of the fluid carries heat in the downstream direction.

4. A Second Approximation in the Inner Region

The first approximation to the temperature field in the outer region, Eq. (9–51), has been constructed in such a way that it asymptotically matches the first approximation in the inner region, $\theta = 1/r$, in the region of overlap. However, we have already noted that the match between these two leading-order approximations is not perfect. Rather, there is an $O(Pe)$ mismatch, which can be made arbitrarily small in the limit $Pe \to 0$ but is nonetheless nonzero. To determine the precise form of this mismatch, we investigate the approximate form of (9–51) for $\rho \ll 1$. In particular, if we expand the exponential and retain the first two terms, we find that

$$\oplus(\rho, \eta, Pe) \to \frac{Pe}{\rho} + \frac{Pe}{2}(P_1 - P_0) + O(Pe\rho) + O(F_1(Pe)).$$

When written in terms of the inner variable $r = \rho/Pe$, this becomes

$$\oplus(\rho, \eta, Pe) \to \frac{1}{r} + \left(\frac{Pe}{2}\right)(P_1 - P_0) + O(Pe^2 r) + O(F_1(Pe)). \quad (9\text{–}52)$$

The first term matches identically with the pure conduction, leading-order approximation in the inner region, $1/r$, but the second term clearly represents an $O(Pe)$ mismatch between the first approximations in the two regions. The third term from the power-series approximation of the exponential is $O(Pe\rho)$, but we see from (9–52) that this leads to a mismatch that is $O(Pe)^2$ when expressed in terms of inner variables. A term of $O(Pe^2)$ cannot appear until at least the third approximation in the inner region.

We conclude from (9–52) that the largest mismatch between the first terms in the asymptotic expansions for the inner and outer regions is $O(Pe)$. In addition, we may note that the terms that were neglected in the governing equation for the inner region at the leading order of approximation are also $O(Pe)$. Hence the magnitude of the second term in the inner expansion must be $O(Pe)$ – that is, $f_1(Pe) \equiv Pe$ – and the solution for θ_1 will both include the first $O(Pe)$ convection contributions in the governing equation (9–24) for the inner region and also match the $O(Pe)$ term in (9–52) in order to remove the largest mismatch between the inner and outer approximations. With $f_1(Pe) \equiv Pe$, the governing equation for θ_1 is precisely the same, Eq. (9–26), that led to Whitehead's paradox. Hence the general solution is exactly (9–27). Now, however, we recognize θ_1 as the second term in an expansion that holds only in the inner portion of the domain, rather than the second term in a regular expansion that was assumed valid throughout the domain. Thus, unlike the analysis that led to Whitehead's paradox, we do not attempt to apply boundary conditions on θ_1 for $r \to \infty$, but instead require that this solution match at $O(Pe)$ with the first approximation in the outer region.

Hence, to determine the correct form for θ_1 in the inner region, we first apply the boundary condition

$$\theta_1 = 0 \quad \text{at} \quad r = 1 \tag{9–53}$$

to (9–27). The result is

$$A_n = -B_n, \quad n = 0, 2, 3, \ldots,$$

and

$$A_1 = \frac{3}{8} - B_1. \tag{9–54}$$

Thus,

$$\theta_1 = \sum_{n=0}^{\infty} A_n \left[r^n - r^{-(n+1)} \right] P_n(\eta) + \left(\frac{1}{2} - \frac{3}{4r} + \frac{3}{8r^2} - \frac{1}{8r^3} \right) P_1(\eta), \tag{9–55}$$

and we are left with the task of determining the coefficients A_n. To do this, let us consider the matching between the outer solution for small ρ, expression (9–52), and the first two terms of the inner solution, namely,

$$\theta_0 + Pe\theta_1 + O(f_2(Pe))|_{r \gg 1} \Leftrightarrow Pe\,\Theta_0 + O(F_1(Pe))|_{\rho \ll 1} \quad \text{as} \quad Pe \to 0. \tag{9–56}$$

Using (9–55) for large r and (9–52), this matching condition reduces to

$$\frac{1}{r} + Pe \left[\sum_{n=0}^{\infty} A_n r^n P_n(\eta) + \frac{1}{2} P_1 + O\left(\frac{1}{r}\right) \right] + O(f_2(Pe))$$
$$\Leftrightarrow \frac{1}{r} + \frac{Pe}{2}(P_1 - P_0) + O(Pe^2 r) + O(F_1(Pe)) \quad \text{as} \quad Pe \to 0. \tag{9–57}$$

Hence we see that the two expressions are identical up to and including all terms of $O(Pe)$ if

$$A_0 = -\frac{1}{2} \quad \text{and} \quad A_n = 0 \quad \text{for} \quad n \geq 1. \tag{9–58}$$

We also see that the term $Pe P_1/2$, which led to Whitehead's paradox, matches automatically with a term in the outer solution. As suggested earlier, it is now evident that Whitehead's

C. Heat Transfer From a Solid Sphere in a Uniform Streaming Flow

paradox resulted from an attempt to apply boundary conditions on the inner solution for $r \to \infty$, where it is not a valid approximation, instead of matching asymptotically with the outer solution for large r in the limit $Pe \to 0$.

Before proceeding, it is perhaps useful to return momentarily to the matching condition (9–57) to ensure that we understand why terms like $O(Pe/r)$ in the inner solution and $O(Pe^2 r)$ in the outer solution can be neglected in determining A_n. The latter is neglected at the present level of approximation because we have so far considered only the $O(1)$ and $O(Pe)$ terms in the inner expansion. If we proceed to a term of $O(Pe^2)$ in the inner expansion, however, the $Pe^2 r$, term from $F_0 \oplus_0$ would clearly have to be included in the matching. In the inner expansion on the left-hand side of (9–57), we have neglected $O((1/r)Pe)$, as well as smaller terms $O((1/r^n)Pe)$. This is because such terms can be matched only to higher-order terms in the outer expansion. To see this, we can express the term Pe/r in outer variables:

$$O\left(\frac{1}{r}Pe\right) \to \left(\frac{1}{\rho}Pe^2\right).$$

Hence, we see that it corresponds to a contribution of $O(Pe^2)$ in the outer expansion and thus again need not be considered in the matching at the present level of approximation where we have included only the first $O(Pe)$ term in the outer region.

With the coefficients A_n determined according to (9–58), the complete inner solution to $O(Pe)$ is

$$\theta = \frac{1}{r} + Pe\left[\frac{1}{2}\left(\frac{1}{r}-1\right) + \eta\left(\frac{1}{2} - \frac{3}{4r} + \frac{3}{8r^2} - \frac{1}{8r^3}\right)\right] + O(f_2(Pe)). \quad (9\text{–}59)$$

Although terms like $(1/r)Pe$ can be neglected in the matching process, they are not generally negligible in the inner region and thus must be included in (9–59).

We noted at the beginning of this chapter that a primary objective in heat transfer calculations is the overall rate of heat transfer. In dimensionless terms this is the Nusselt number [see Eq. (9–20)]. With the first two terms of the inner solution in hand, we are now in a position to extend our previous result for pure conduction from a sphere, (9–21), to include the first $O(Pe)$ contribution of convection for small Pe. In particular, substituting the approximation (9–59) into (9–20) and carrying out the integration, we find the following correlation between Nu and Pe:

$$Nu = 2\left[1 + \frac{1}{2}Pe + O(f_2(Pe))\right]. \quad (9\text{–}60)$$

The influence of convection at small Pe is a small increase in the overall rate of heat transfer above its value for pure conduction.

5. Higher-Order Approximations

A great potential advantage of the matched asymptotic expansion procedure is that we can generate as many terms as we wish in the asymptotic expansion by simply extending the procedure described above. Of course, the effort required to do this increases as we proceed because of the increased algebraic complexity of the problem at each level of approximation. One way around this difficulty is to use so-called symbolic manipulation packages to carry out the algebra on a computer, and this has actually been done in a few instances. However, we do not discuss this approach here. Part of the reason is that the addition of large numbers of additional terms to a result like (9–60) often does not significantly extend the range of the small parameter where the asymptotic expression gives accurate results.

In spite of the limited practical value of calculating many terms in the asymptotic expansion for small Pe, it is nevertheless instructive to consider briefly the addition of the next term, for it illustrates a fundamental difference between regular and singular asymptotic expansions. When there is a regular asymptotic solution for some DE and boundary conditions, the expansion must always proceed sequentially in increasing powers of the small parameter that appears in the problem. Thus, if the parameter is ε, the expansion will be $(1, \varepsilon, \varepsilon^2, \varepsilon^3, \ldots,)$. The fact that this is true can be seen easily if we note that the sequence of equations generated by a regular perturbation scheme is identical to that obtained in a solution by the method of successive approximations. Returning to the heated sphere problem, it is tempting to assume that the third term in the inner expansion (9–32) should be $O(Pe^2)$, that is,

$$\theta = \frac{1}{r} + Pe\theta_1 + Pe^2\theta_2 + \cdots + . \tag{9-61}$$

However, Acrivos and Taylor[4] have shown that the particular solution to the equation for θ_2 contains a term $-1/2 \ln r$ [see their Eq. (31)] that, when multiplied by Pe^2 and expressed in the outer variable, $\rho = rPe$, generates a term $-(Pe^2/2)\ln Pe$ that must be matched with a corresponding term in the outer expansion. But this is not possible because the next term in the outer solution which matches (9–59) is $O(Pe^2)$, and the outer expansion thus does not contain a term $O(Pe^2 \ln Pe)$. Consequently, we must backtrack and assume that $f_2(Pe)$ in (9–32) is $(Pe^2 \ln Pe)$ instead of Pe^2. In this case

$$\nabla_r^2 \theta_2 = 0, \tag{9-62}$$

and it can be shown that

$$\theta_2 = \frac{1}{2}\left(\frac{1}{r} - 1\right). \tag{9-63}$$

Acrivos and Taylor actually evaluated two additional terms in the inner expansion. Their expression for the correlation between Nusselt number and Peclet number is

$$Nu = 2\left[1 + \frac{1}{2}Pe + \frac{1}{2}Pe^2 \ln Pe + 0.41465 Pe^2 + \frac{1}{4}Pe^3 \ln Pe + O(Pe^3)\right]. \tag{9-64}$$

Unlike the regular perturbation expansion discussed earlier, the method of matched asymptotic expansions often leads to a sequence of gauge functions that contain terms like $Pe^2 \ln Pe$ or $Pe^3 \ln Pe$ that are intermediate to simple powers of Pe. Thus, unlike the regular perturbation case, for which the form of the sequence of gauge functions can be anticipated in advance, this is not generally possible when the asymptotic limit is singular. In the latter case, the sequence of gauge functions must be determined as a part of the matched asymptotic-solution procedure.

One final question that is worth brief comment is the applicability of the analysis in this section to the corresponding single-component mass transfer problem. Assuming that the condition of constant surface concentration is applicable, the basic question has to do with the importance of blowing or suction at the surface (see Subsection A.3 of this chapter). In this case, we can neglect this effect only when $B \ll Pe \ll 1$. This is a rather severe

C. Heat Transfer From a Solid Sphere in a Uniform Streaming Flow

condition, suggesting that it is difficult to find conditions in real applications in which the result (9–64) can be used to estimate the Sherwood number for the mass transfer rate.

6. Specified Heat Flux

One feature of the preceding analysis that may seem unduly restrictive is that we assume the actual temperature is known both in the far field ("the ambient value") and at the surface of the body. Often, in reality, it is not the temperature that is specified by the physics of the problem, but rather the heat flux, i.e., for this problem

$$-k\frac{\partial T}{\partial r} = q \quad \text{(constant)} \quad \text{at } r = a \qquad (9\text{–}65)$$

In this case, the overall heat flux is known (specified), and there is no point in calculating a Nusselt number. Instead, it is the temperature at the surface of the body that is unknown. In addition, the boundary condition (9–2a) is now replaced with (9–65), and we may ask how the analysis of the preceding sections is changed. The job of answering this question is partly left to the problems section at the end of this chapter. However, it is worthwhile to consider some of the issues here.

First, if we go back to Section A, we see that some aspects of the nondimensionalization will need to be changed. The definition of a dimensionless temperature according to (9–3) no longer makes sense, because the temperature at the surface of the body is no longer known. Instead, we can see from the form of (9–65) that an appropriate way to define a dimensionless temperature is

$$\theta \equiv \frac{T - T_\infty}{(qa/k)}. \qquad (9\text{–}66)$$

Alternatively, we can think of an appropriate generalized form of the dimensionless temperature as

$$\theta \equiv \frac{T - T_\infty}{T_c}, \qquad (9\text{–}67)$$

with

$$T_c \equiv qa/k \qquad (9\text{–}68)$$

for the specified heat flux problem, and

$$T_c \equiv T_0 - T_\infty \qquad (9\text{–}69)$$

for the preceding specified temperature problem.

If we introduce the definition (9–67) into (9–1), the governing equation reduces to almost the same dimensionless form as before, namely (9–4), except now with the generalized definition of the Brinkman number as

$$Br \equiv \frac{\mu U_\infty^2}{kT_c} \qquad (9\text{–}70)$$

Thus, when the Brinkman number is small, the governing equation is again precisely (9–7), but now with boundary conditions

$$-\frac{\partial \theta}{\partial r} = 1 \quad \text{at} \quad r = 1 \text{ (i.e., on } S\text{)},$$
$$\theta \to 0 \quad \text{for} \quad |r| \to \infty. \qquad (9\text{–}71)$$

The temperature at the surface of the sphere is unknown.

We shall see later that the solutions of the constant-surface-temperature and constant-heat-flux problems can often be quite different. However, the formal solutions of these two problems in the asymptotic limit $Pe \ll 1$ are actually very similar. The problem for the constant-heat-flux case is still singular, and the scaling and nondimensionalized equations for the inner and outer regimes are identical to what we obtained already for the constant-temperature problem. The details of showing this, plus the analysis of solutions, are left to the reader by means of Problem 9–7 at the end of this chapter. We simply note here that the leading-order approximation in the inner region is still the pure conduction solution

$$\theta_0 = 1/r,$$

and the first approximation in the outer region is still precisely (9–51). Only at the second- and higher-order approximations does the change in boundary conditions at the surface of the sphere influence the solution, and even the second approximation in the inner region is changed very little, becoming

$$\theta = \frac{1}{r} + Pe\left[-\frac{1}{2} + \eta\left(\frac{1}{2} - \frac{3}{4r} + \frac{9}{16r^2} - \frac{1}{8r^3}\right)\right] + O(f_2(Pe)) \qquad (9\text{–}72)$$

instead of (9–59). Thus, in dimensionless terms, the change of boundary conditions produces a very modest change in either the mathematical problem or its solution.

One important difference to remember, however, is that it is the surface temperature that is unknown in this problem, rather than the heat flux in the previous case. If we express (9–72) in dimensional form

$$T = T_\infty + \left(\frac{qa}{k}\right)\left\{\frac{a}{r'} + Pe\left[-\frac{1}{2} + \eta\left(\frac{1}{2} - \frac{3a}{4r'} + \frac{9a^2}{16r'^2} - \frac{a^3}{8r'^3}\right)\right] + O(f_2(Pe))\right\}. \qquad (9\text{–}73)$$

Hence the surface temperature is

$$T_s = T_\infty + \left(\frac{qa}{k}\right)\left[1 + \left(-\frac{1}{2} + \frac{3}{16}\eta\right)Pe + O(f_2(Pe))\right]. \qquad (9\text{–}74)$$

We see that the T_s depends on the rate of heat release q and that the overall effect of convection is to reduce the surface temperature and make it nonuniform, with the surface temperature being highest at $\eta = 1$ (i.e., the downstream stagnation point of the sphere) and lower at the front, $\eta = -1$. The asymmetry is due to the fact that the radial temperature gradient is slightly increased at the front relative to the back and thus requires a slightly lower surface temperature to sustain the heat flux q, compared with the surface temperature that is required at the back.

D. UNIFORM FLOW PAST A SOLID SPHERE AT SMALL, BUT NONZERO, REYNOLDS NUMBER

Now, let us return to the fluid mechanics problem of streaming flow past a solid sphere at small, but nonzero, Reynolds numbers. The objective is to see how the steady-state creeping-flow solution is modified if we consider the nonlinear terms in (7–2) for $Re \ll 1$. As we shall see, there is essentially the same problem of trying to obtain a solution of (7–2) in the form of a regular perturbation expansion as was encountered when we tried to obtain such a solution of the thermal problem for $Pe \ll 1$. Indeed, this problem and its resolution are very

D. Uniform Flow Past a Solid Sphere at Small, but Nonzero, Reynolds Number

similar to the problem of convective heat transfer at low Pe that we have just considered. It is only that the details are somewhat more complicated.

The problem of obtaining an inertial correction to Stokes' solution was first investigated in 1889 by the (then) young English mathematician Whitehead,[6] who was attempting to obtain corrections to Stokes' formula for the drag force on a sphere because the latter had been found, experimentally, to provide accurate predictions *only* for very small Reynolds numbers, $Re < O(0.1)$. Our experience from the previous section should alert us to the fact that the creeping-flow solution of Stokes is only the first approximation in a *singular* asymptotic expansion for $Re \to 0$. However, Whitehead did not have the benefit of any knowledge of asymptotic methods, which were not developed until much later, and he attempted to obtain a solution for small, but nonzero, Reynolds numbers by the method of successive approximation. As pointed out in Chapt. 4, this is equivalent for small Re to a regular perturbation scheme, and we should anticipate that it will fail. Nevertheless, it is instructive to consider Whitehead's original analysis, expressed in the language of asymptotic analysis.

Thus we begin by considering the full Navier–Stokes equation expressed in terms of the streamfunction ψ and nondimensionalized by use of the undisturbed velocity of the fluid relative to the sphere U_∞ as a characteristic velocity and the sphere radius a as a characteristic length scale. Using spherical coordinates, with $\eta = \cos\theta$, this equation is

$$E^4\psi = Re\left\{\frac{1}{r^2}\left[\frac{\partial\psi}{\partial r}\frac{\partial}{\partial\eta}(E^2\psi)\right] + \frac{2}{r^2}E^2\psi\left(\frac{\partial\psi}{\partial r}\frac{\eta}{1-\eta^2} + \frac{1}{r}\frac{\partial\psi}{\partial\eta}\right)\right\}. \quad (9\text{--}75)$$

The boundary conditions are (see Chap. 7, Section E)

$$\frac{\partial\psi}{\partial r} = \psi = 0 \quad \text{at } r = 1,$$
$$\psi = 0 \quad \eta = \pm 1, \quad (9\text{--}76)$$
$$\psi \to -r^2 Q_1(\eta) \quad \text{as } r \to \infty.$$

Following Whitehead's approach, we seek a solution of (9–75) and (9–76) by means of a regular perturbation expansion of the form

$$\boxed{\psi(r,\eta;Re) = \psi_0(r,\eta) + Re\,\psi_1(r,\eta) + \cdots.} \quad (9\text{--}77)$$

Substituting (9–77) into (9–75) and (9–76), we obtain

$$E^4\psi_0 + Re\,E^4\psi_1 + O(Re^2) = Re\left\{\frac{1}{r^2}\left[\frac{\partial\psi_0}{\partial r}\frac{\partial}{\partial\eta}(E^2\psi_0) - \frac{\partial\psi_0}{\partial\eta}\frac{\partial}{\partial r}(E^2\psi_0)\right]\right.$$
$$\left. + \frac{2}{r^2}E^2\psi_0\left(\frac{\partial\psi_0}{\partial r}\frac{\eta}{1-\eta^2} + \frac{1}{r}\frac{\partial\psi_0}{\partial\eta}\right)\right\} + O(Re^2), \quad (9\text{--}78)$$

with

$$\frac{\partial\psi_0}{\partial r} + Re\frac{\partial\psi_1}{\partial r} + O(Re^2) = \psi_0 + Re\,\psi_1 + O(Re^2) = 0, \quad r = 1,$$
$$\psi_0 + Re\,\psi_1 + O(Re^2) = 0, \quad \eta = \pm 1, \quad (9\text{--}79)$$
$$\psi_0 + Re\,\psi_1 + O(Re^2) \to -r^2 Q_1(\eta) \quad \text{as } r \to \infty.$$

Convection Effects in Low-Reynolds-Number Flows

Thus, because Re is an arbitrary small parameter,

$$E^4 \psi_0 = 0, \tag{9-80}$$

$$\frac{\partial \psi_0}{\partial r} = \psi_0 = 0 \quad \text{at} \quad r = 1,$$

$$\psi_0 = 0 \quad \text{at} \quad \eta = \pm 1, \tag{9-81}$$

$$\psi_0 \to -r^2 Q_1(\eta) \quad \text{as} \quad r \to \infty,$$

and, at $O(Re)$,

$$E^4 \psi_1 = \frac{1}{r^2} \left[\frac{\partial \psi_0}{\partial r} \frac{\partial}{\partial \eta} (E^2 \psi_0) - \frac{\partial \psi_0}{\partial \eta} \frac{\partial}{\partial r} (E^2 \psi_0) \right] \\ + \frac{2}{r^2} E^2 \psi_0 \left(\frac{\partial \psi_0}{\partial r} \frac{\eta}{1 - \eta^2} + \frac{1}{r} \frac{\partial \psi_0}{\partial \eta} \right), \tag{9-82}$$

with

$$\frac{\partial \psi_1}{\partial r} = \psi_1 = 0 \quad \text{at} \quad r = 1,$$

$$\psi_1 = 0 \quad \text{at} \quad \eta = \pm 1, \tag{9-83}$$

$$\frac{1}{r} \frac{\partial \psi_1}{\partial r}, \quad \frac{1}{r^2} \frac{\partial \psi_1}{\partial \eta} \to 0 \quad \text{as} \quad r \to \infty.$$

The last condition follows from the fact that the first approximation, ψ_0, satisfies the nonhomogeneous boundary condition $\psi \to -r^2 Q_1(\eta)$ as $r \to \infty$ (that is, $\mathbf{u}_0 \to \mathbf{i}$ as $r \to \infty$), so that $\mathbf{u}_1 \to 0$ as $r \to \infty$.

The solution of (9–80) and (9–81) is Stokes' solution, Eq. (7–158), derived in Chap. 7. In view of our experience with the analogous heat transfer problem at low Peclet numbers, it should come as no surprise that a solution for the next term, ψ_1, satisfying (9–82) and (9–83), is not possible. In particular, we obtain the governing equation for ψ_1 by substituting the known solution for ψ_0 into the right-hand side of (9–82). This gives

$$E^4 \psi_1 = \frac{9}{2} \left(\frac{2}{r^2} - \frac{3}{r^3} + \frac{1}{r^5} \right) Q_2(\eta), \tag{9-84}$$

and a general solution of this equation is

$$\psi_1 = \sum_{1}^{\infty} \left(A_n r^{n+3} + B_n r^{n+1} + C_n r^{2-n} + D_n r^{-n} \right) Q_n(\eta) \\ + \frac{3}{16} \left(2r^2 - 3r - \frac{1}{r} \right) Q_2(\eta), \tag{9-85}$$

with A_n, B_n, C_n, and D_n being arbitrary constants that should, in principle, be determined from the boundary conditions (9–83). However, if we examine (9–85), we find that there

D. Uniform Flow Past a Solid Sphere at Small, but Nonzero, Reynolds Number

is no choice of the constants that will satisfy the boundary condition on ψ_1 as $r \to \infty$. In particular, there is no term in the homogeneous part of the solution that can be used to cancel the term $3/8r^2 Q_2(\eta)$ that appears in the particular solution and clearly violates the requirement that $\psi \sim O(r^n)$ with $n < 2$. We conclude that *no solution exists* at $O(Re)$ in the regular expansion (9–77) that satisfies the DE and boundary conditions in Eqs. (9–82) and (9–83). This is the original version of *Whitehead's paradox*. It has been suggested that the inability to generate a second approximation for small Reynolds numbers, based on Stokes' solution ψ_0 as the first approximation, so discouraged Whitehead that he "gave up" his career as a mathematician and turned to philosophy.[7]

Of course, our experience with the analogous thermal problem at low Peclet numbers suggests that it is the assumption of a *regular* perturbation expansion, valid for all r, that leads to Whitehead's paradox. However, as we have noted, Whitehead did not have the benefit of any knowledge of singular perturbations or the method of matched asymptotic expansions, which came into being only some 80 years later, with the pioneering publications of Kaplan[8] and Proudman and Pearson.[9] Indeed, even an explanation for the Whitehead paradox did not appear until 1911, when the German mathematician C.W. Oseen published a landmark paper that not only explained the difficulty but also provided a method of circumventing the problem.[10] Oseen's explanation was based on a comparison of the magnitude of viscous terms and inertial terms in (9–75), using Stokes' solution. In deriving Stokes' solution, it is assumed by means of the expansion (9–77) that viscous terms are dominant over inertia terms *everywhere* in the solution domain ($1 \leq r \leq \infty$) provided only that the Reynolds number is small enough. However, as $r \to \infty$, the magnitude of the largest term on the right-hand side of (9–75) can be estimated by use of Stokes' solution to be

$$Re\left(u_r \frac{\partial u_r}{\partial r}\right) \sim O\left(\frac{Re}{r^2}\right). \tag{9-86}$$

A typical viscous term, on the other hand, is

$$\frac{\partial^2 u_r}{\partial r^2} \sim O\left(\frac{1}{r^3}\right). \tag{9-87}$$

Comparing (9–86) and (9–87), we clearly see that inertia terms are small compared with viscous terms only for

$$r < O\left(\frac{1}{Re}\right). \tag{9-88}$$

Hence, as Oseen noted, we cannot expect the Stokes' solution to provide a uniformly valid first approximation to the solution of (9–75), but instead expect that it will break down for large values of $r \geq O(Re^{-1})$. Thus Whitehead's attempt to evaluate the second term in the expansion (9–77) was unsuccessful for large r. Indeed, as noted earlier in conjunction with the thermal problem, it is not so much a surprise that we cannot obtain a solution for ψ_1 as it is a surprise to be able to solve for ψ_0 by using the boundary condition (9–81) for $r \to \infty$, in spite of the fact that the governing equation (9–80) is not a valid first approximation to the full Navier–Stokes equation except for $r < O(1/Re)$.

If we examine the corresponding problem of streaming flow past a circular cylinder for $Re \to 0$, we find, in fact, that a solution of Stokes' equation cannot be obtained to

Convection Effects in Low-Reynolds-Number Flows

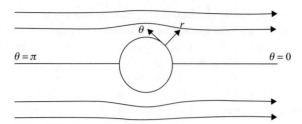

Figure 9–5. A sketch showing the cylindrical coordinate system used for analysis of streaming flow past a circular cylinder.

satisfy boundary conditions at infinity. In particular, for a 2D flow, the creeping-motion approximation to the Navier–Stokes equation can be expressed in the form

$$\nabla^4 \psi_0 = 0, \qquad (9\text{--}89)$$

where ∇^4 is the biharmonic operator ($\nabla^2 \cdot \nabla^2$). If we consider streaming flow past a circular cylinder, the uniform velocity condition at infinity is most conveniently expressed in terms of a circular cylindrical coordinate system,

$$\psi_0 \to r \sin\theta \quad \text{as } r \to \infty, \qquad (9\text{--}90)$$

where $\theta = 0$ and π are the downstream and upstream symmetry axes, respectively. A sketch of the problem, including the cylindrical coordinate system, is shown in Fig. 9–5. In addition to (9–90), the solution must satisfy boundary conditions $u_r = u_\theta = 0$ at the cylinder surface, plus the symmetry condition $u_\theta = 0$ at $\theta = 0, \pi$. In terms of the streamfunction ψ_0, these conditions take the form

$$\psi_0 = \frac{\partial \psi_0}{\partial r} = 0 \quad \text{at } r = 1,$$
$$\psi_0 = 0 \quad \text{at } \theta = 0, \pi. \qquad (9\text{--}91)$$

It is left to the reader to supply the details of solving (9–89). However, the most general solution that also satisfies the boundary conditions (9–91) is

$$\psi_0 = C\left(\frac{1}{r} - r + r \ln r\right) \sin\theta. \qquad (9\text{--}92)$$

The coefficient C in (9–82) is arbitrary, and the condition (9–90) for $r \to \infty$ has not been applied in deriving (9–92). Thus it would seem that C should be chosen to satisfy this condition. However, it is obvious in examining (9–92) that there is no value of C that will work. For any nonzero value of C, the right-hand side of (9–92) increases for large r in proportion to $r \ln r$, and this term is arbitrarily large compared with the required form (9–90). Thus, in the case of streaming flow past a circular cylinder, we must conclude that *no solution exists* for the creeping-flow equations that satisfies boundary conditions for $r \to \infty$. In fact, the same conclusion is true for streaming flow past a 2D cylinder of

D. Uniform Flow Past a Solid Sphere at Small, but Nonzero, Reynolds Number

arbitrary cross-sectional shape. The fact that no creeping-flow solution exists for streaming flow in an unbounded 2D flow was actually discovered by Stokes and is known as *Stokes' paradox*. A very similar result was shown in Section C for heat conduction from a heated cylinder in an unbounded 2D domain.

Of course, neither the Stokes' nor Whitehead's paradox is a true paradox, for we now understand both why they occur and how they can be avoided by use of the method of matched asymptotic expansions. In the present section, we do not pursue all of the details of this analysis. The reader who wishes to see the detailed analysis for both a sphere and a cylinder may refer to the landmark paper of Proudman and Pearson.[9] Here we consider only a few more general introductory issues, and enough of the analysis to provide a justification for the creeping-flow solutions of the previous two chapters as a useful first approximation for $Re \ll 1$, in the same sense that the pure conduction solution was found to be a useful first approximation of the thermal problem for $Pe \ll 1$. One point relevant to both problems is that the singular nature of the low-Reynolds-number approximation (necessitating use of the matched asymptotic expansion technique) is a consequence of the assumption of an *unbounded* domain. Specifically, as long as r remains smaller than $O(Re^{-1})$, Stokes' equation (9–80), and hence the solution of the equation, will provide a valid first approximation of the complete problem (9–75) and (9–76) for small Reynolds numbers [and the same is true of the conduction solution of the thermal problem for $r < O(Pe^{-1})$]. In a *bounded* domain, all points can be held within the region of validity of Stokes' solution provided that we consider sufficiently small Reynolds numbers so that $r_{max} < O(Pe^{-1})$. For an *unbounded* domain, on the other hand, there *always* will be a region where the Stokes' approximation breaks down.

Let us now return to the problem of streaming flow past a sphere in an unbounded domain at small Reynolds numbers. It is presumably obvious, by analogy to the low-Peclet-number heat transfer problem, that the source of Whitehead's difficulty in obtaining a regular perturbation solution of (9–75) is the fact that the sphere radius used to nondimensionalize (9–75) is appropriate as a characteristic length scale only over the region $r < O(Re^{-1})$. Far from the sphere, there must exist a second characteristic length scale, hence a second nondimensional form of the Navier–Stokes equation [replacing (9–75)] and a second asymptotic expansion [replacing (9–77)]. We have seen this situation now several times: When two different length scales are relevant in different parts of the domain, two different approximations to the governing equations will result in the limit $Re \to 0$, and this requires two different asymptotic approximations to the solution. The asymptotic limit is *singular*, and we can attack the problem by using the method of matched asymptotic expansions. The situation is illustrated schematically in terms of the solution domain for $Re \ll 1$ in Fig. 9–6.

In the so-called inner region nearest the sphere, the sphere radius is appropriate as the characteristic length scale, and thus equation (9–75) is applicable. To solve this equation for $Re \ll 1$, we again assume the existence of an asymptotic expansion, but unlike for (9–77), we do not presuppose any knowledge of the gauge functions, so that

$$\psi = \sum_{n=0}^{N} f_n(Re)\psi_n(r, \eta), \qquad (9\text{–}93)$$

subject only to the convergence condition

$$\lim_{Re \to 0} \frac{f_{n+1}(Re)}{f_n(Re)} \to 0. \qquad (9\text{–}94)$$

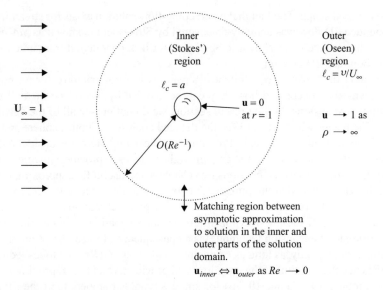

Figure 9–6. A schematic of the asymptotic solution domain for *low* Reynolds number streaming flow past a sphere, that is, $Re \ll 0$.

The terms in (9–93) are expected to satisfy the symmetry conditions at $\eta = \pm 1$ and boundary conditions at the sphere surface, namely,

$$\psi_n(r, \eta) = 0, \quad \eta = \pm 1, \tag{9-95a}$$

$$\psi_n(r, \eta) = \frac{\partial \psi_n}{\partial r} = 0 \quad \text{at } r = 1. \tag{9-95b}$$

However, the boundary condition at infinity is no longer applied to the inner expansion because $r \to \infty$ lies outside the inner region. Instead, we will require that the inner expansion match the outer expansion in the region of overlap between the two approximations.

In the outer region, we must first determine the appropriate characteristic length scale. To do this, we follow the example of the small Peclet number heat transfer problem of the previous section, and introduce a rescaling,

$$\boxed{\rho = r \, Re^m .} \tag{9-96}$$

As before, we recognize this rescaling as defining a new nondimensionalized radial variable, with a new characteristic length scale

$$\ell = a(Re)^{-m} \tag{9-97}$$

depending on m. To determine m, we must substitute the rescaled variable into (9–75). First, however, we note that the rescaling (9–96) in this case requires a corresponding rescaling of the streamfunction ψ. This is because velocities, nondimensionalized with respect to U_∞, are expected to be $O(1)$ in both the inner and outer regions. However, in terms of the streamfunction,

$$u_r = -\frac{1}{r^2} \frac{\partial \psi}{\partial \eta}, \quad u_\eta = -\frac{1}{r\sqrt{1-\eta^2}} \frac{\partial \psi}{\partial r}. \tag{9-98}$$

D. Uniform Flow Past a Solid Sphere at Small, but Nonzero, Reynolds Number

Thus, if we introduce the rescaling (9–96) into these definitions, it is evident that ψ should be rescaled in the outer region according to

$$\Psi = Re^{2m}\psi. \tag{9–99}$$

Now we determine the scaling constant m by substituting (9–96) and (9–99) into the Navier–Stokes equation (9–75) and requiring that the limiting form of the resulting equation for $Re \to 0$ contain viscous terms and at least one inertia term, as suggested by Oseen's argument. On substitution of outer variables, (9–75) becomes

$$Re^{2m} E_\rho^4 \Psi = Re^{m+1} \left\{ \frac{1}{\rho^2} \left[\frac{\partial \Psi}{\partial \rho} \frac{\partial}{\partial \eta}(E_\rho^2 \Psi) - \frac{\partial \Psi}{\partial \eta} \frac{\partial}{\partial \rho}(E_\rho^2 \Psi) \right] \right. \\ \left. + \frac{2}{\rho^2} E_\rho^2 \Psi \left(\frac{\partial \Psi}{\partial \rho} \frac{\eta}{1-\eta^2} + \frac{1}{\rho} \frac{\partial \Psi}{\partial \eta} \right) \right\}, \tag{9–100}$$

where

$$E_\rho^2 \equiv \frac{\partial^2}{\partial \rho^2} + \frac{1-\eta^2}{\rho^2} \frac{\partial^2}{\partial \eta^2}. \tag{9–101}$$

Thus, because inertial terms and viscous terms are of equal importance in the outer region for $Re \to 0$, it follows that

$$2m = m+1 \quad \text{or} \quad m = 1. \tag{9–102}$$

Thus,

$$\rho = rRe, \quad \Psi = Re^2 \psi, \tag{9–103}$$

and the characteristic length scale in this outer region has been shown to be

$$\ell = a(Re)^{-1} = \frac{v}{U_\infty}. \tag{9–104}$$

The governing equation in the outer region is just (9–100) with $m = 1$.

We assume, following our previous analyses, that there exists an asymptotic expansion for Ψ in the form

$$\Psi = \sum_{n=0}^{N} F_n(Re) \Psi_n(\rho, \eta), \tag{9–105}$$

with

$$\lim_{Re \to 0} \frac{F_{n+1}}{F_n} \to 0. \tag{9–106}$$

In addition to the governing equation (9–100), the solution in this outer region must satisfy

$$\Psi(\rho, \pm 1) = 0, \tag{9–107a}$$

$$\Psi \to -\rho^2 Q_1 \quad \text{as} \quad \rho \to \infty. \tag{9–107b}$$

The latter condition is the uniform velocity at infinity, expressed in terms of outer variables. Of course, the outer region solution is not required to satisfy boundary conditions at the

Convection Effects in Low-Reynolds-Number Flows

sphere surface as this is not part of the outer domain. Instead, we require matching between the solutions in the two parts of the domain, that is,

$$Re^2 \psi|_{r \gg 1} \Leftrightarrow \Psi_{\rho \ll 1} \quad \text{as } Re \to 0. \tag{9–108}$$

In the limit as $Re \to 0$, the outer portion of the inner region, $r \gg 1$, corresponds to arbitrarily small values of the outer variable, $\rho \equiv Re\, r \ll 1$.

In the present problem, our analysis begins with a first approximation to the solution in the outer region. We obtain this by simply noting that the asymptotic form (9–107b) for large ρ is actually an exact solution of the full governing equation (9–100). Thus we find that

$$\boxed{\Psi_0 = -\rho^2 Q_1, \quad F_0(Re) = 1.} \tag{9–109}$$

Because the boundary condition at infinity provides a valid first approximation to the solution in the whole of the outer domain, we see why it was possible to obtain Stokes' solution by applying the free-stream boundary condition for $r \to \infty$.

However, rather than simply accepting Stokes' solution as the first approximation for $Re \ll 1$ in the inner region, we will show how it is obtained in the present framework of matched asymptotic expansions. To do this, we note that the governing equation for ψ_0 in the expansion (9–93) is simply the Stokes' equation for any choice of the gauge function $f_0(Re)$, namely,

$$E^4 \psi_0 = 0, \tag{9–110}$$

which has the general solution (see Chap. 7, Section E), satisfying the symmetry condition (9–95a),

$$\psi_0 = \sum_{n=1}^{\infty} \left(A_n r^{n+3} + B_n r^{n+1} + C_n r^{2-n} + D_n r^{-n} \right) Q_n(\eta). \tag{9–111}$$

In our earlier derivation of Stokes' solution, (7–158), we applied the boundary condition (9–95b), plus the free-stream condition for $r \to \infty$. Here, however, the boundary condition at infinity is formally replaced with the matching condition (9–108) using the first approximation (9–109) to the solution in the outer region. To carry out the matching, we express ψ_0 in terms of outer variables by multiplying by Re^2, as indicated in (9–108), and converting from r to ρ. Then the matching condition becomes

$$-\rho^2 Q_1 \Leftrightarrow f_0(Re) \left[\sum_{n=1}^{\infty} \left(A_n \rho^{n+3} \frac{1}{Re^{n+1}} + B_n \rho^{n+1} \frac{1}{Re^{n-1}} \right. \right. \tag{9–112}$$
$$\left. \left. + C_n \rho^{2-n} Re^n + D_n \rho^{-n} Re^{n+2} \right) Q_n(\eta) \right].$$

Here, the left-hand side is just the first approximation (9–109) in the outer region. Let us examine the various terms on the right-hand side. There is only one term that exhibits ρ^2 dependence, namely,

$$f_0(Re) B_1 \rho^2 Q_1(\eta).$$

Hence, if the two sides of (9–112) are to match, it is evident that

$$\boxed{f_0(Re) = 1,}$$
$$B_1 = -1. \tag{9–113}$$

D. Uniform Flow Past a Solid Sphere at Small, but Nonzero, Reynolds Number

The terms involving coefficients A_n are all $O(Re^{-2})$ *or larger*. Because the biggest term in the outer region is $O(1)$, it is clear that matching requires that

$$A_n \equiv 0, \quad \text{all } n \geq 1. \tag{9-114}$$

Similarly, all of the terms involving B_n for $n \geq 2$ are $O(Re^{-1})$ or larger, and hence

$$B_n \equiv 0, \quad \text{all } n \geq 2. \tag{9-115}$$

Thus we obtain values for all the coefficients A_n and B_n in the first inner approximation (9–111) from the matching condition (9–112). Because the first approximation in the outer region is just the free-stream boundary condition, expressed in outer variables, it should not be surprising that the coefficients A_n and B_n by means of matching are found to have the same values that were originally found by application of the boundary condition (9–81) for $r \to \infty$. The other coefficients, C_n and D_n, cannot be determined by matching. For example, the terms in (9–112) involving C_n are all $O(Re)$ or smaller, when expressed in outer variables, and therefore cannot be required to match in (9–112) at the present level of approximation, as we have considered only the leading $O(1)$ term so far in the outer region. Similarly, the terms involving D_n are $O(Re^3)$ or smaller and cannot be required to match at this leading level of approximation. Instead, to determine C_n and D_n, we must apply the boundary condition (9–95b) at $r = 1$. The values we obtain are just those of Stokes' solution, namely,

$$\begin{aligned} C_n &= D_n = 0, \\ C_1 &= \frac{3}{2}, \quad D_1 = -\frac{1}{2}, \end{aligned} \tag{9-116}$$

and thus

$$\boxed{\psi_0 = -\left(r^2 - \frac{3}{2}r + \frac{1}{2r}\right) Q_1(\eta).} \tag{9-117}$$

So far, we have just reproduced Stokes' solution, though we now recognize it as the first approximation in the inner region only, rather than a universally valid first approximation everywhere in the domain. It is perhaps worth reiterating that Stokes' solution matches the first approximation in the outer region to $O(1)$ only. The terms involving C_1 and D_1 cause a mismatch between the leading-order approximate solutions in the inner and outer regions of $O(Re)$ and $O(Re^3)$, respectively. To see this, we simply reexpress (9–117) in terms of outer variables,

$$Re^2 \psi_0 = -\left(\rho^2 - \frac{3}{2}\rho Re + \frac{1}{2\rho} Re^3\right) Q_1(\eta), \tag{9-118}$$

and compare with the left-hand side of (9–112).

The next step is to seek a second approximation in the outer expansion (9–105) that matches with the $O(Re)$ mismatch in (9–118). However, we do not pursue the derivation of higher-order terms in the expansions (9–93) and (9–105) here. As mentioned earlier, the interested reader can find all of the details in the paper by Proudman and Pearson[9]. We simply give the results. The first two terms in the asymptotic expansion for Ψ in the outer region are

$$\boxed{\Psi \sim -\rho^2 Q_1 - Re \left\{\frac{3}{2}(1+\eta)\left[1 - e^{-\rho(1-\eta)/2}\right]\right\} + O(F_2(Re)).} \tag{9-119}$$

Convection Effects in Low-Reynolds-Number Flows

We may note that Ψ behaves quite differently for large ρ than the Stokes' solution ψ_0 for large r. In particular, ψ_0 is symmetric about $\eta = 0$, whereas Ψ is asymmetric at $O(Re)$. It is at this level of approximation that the first effects of convection appear in the asymptotic solution for $Re \to 0$.

The term at $O(Re)$ in (9–119) represents a wakelike solution for the vorticity, which is analogous to the thermal wake in Eq. (9–51). To see this, we can calculate the vorticity associated with (9–119) by means of the definition

$$\omega \equiv E_\rho^2 \Psi = \frac{3}{2} Re Q_1 e^{-\rho(1-\eta)/2}. \tag{9-120}$$

Except for a small neighborhood $\eta = 1$ (the downstream axis of symmetry), the vorticity, which is generated by the no-slip condition at the sphere surface, decreases exponentially with increase of ρ. Very near $\eta = 1$, on the other hand, the vorticity decreases only algebraically with ρ. This is a consequence of the fact that the vorticity generated at the sphere surface is swept downstream by the flow into a wakelike region. For comparison, it may be noted that the vorticity in the inner region, which is wholly transported by diffusion at $O(1)$, only decreases as $1/r$. A difference between (9–120) and the thermal wake solution for $\oplus_0$ is that $\omega \equiv 0$ at $\eta \equiv \pm 1$, whereas $\oplus_0$ was maximum at $\eta = 1$. The vorticity generated on opposite sides of the sphere has opposite signs, and thus cancels at $\eta = 1$.

The $O(Re)$ term in (9–119) also can be associated with the existence of a "defect" in the flux of momentum downstream of the sphere, at $\eta = 1$, relative to that approaching the sphere from upstream. This *momentum defect* is proportional in magnitude to the drag force on the sphere.

The corresponding inner solution, through terms of $O(Re)$, is

$$\psi = \left(1 + \frac{3}{8} Re\right) \psi_0 + Re \frac{3}{16} Q_2 \left(2r^2 - 3r + 1 - \frac{1}{r} + \frac{1}{r^2}\right). \tag{9-121}$$

We see that there are two terms at $O(Re)$, one that is proportional to ψ_0 and the other proportional to Q_2, which is an odd function of η. Thus, again the symmetry of the Stokes' solution is broken, and there is a distinction between the upstream and downstream flows. With the first two terms of the inner expansion now known, we can use (7–136) to calculate the drag to $O(Re)$. The result is

$$\boxed{drag = 6\pi a\mu U \left[1 + \frac{3}{8} Re + o(Re)\right].} \tag{9-122}$$

The effect of inertia is to slightly increase the drag relative to Stokes' drag.

A large number of additional terms have actually been calculated in the present problem. The next term in the inner expansion and in the expression for the drag was found by Proudman and Pearson[9] to be $O(Re^2 \ln Re)$,

$$\boxed{drag = 6\pi a\mu U \left[1 + \frac{3}{8} Re + \frac{9}{40} Re^2 \ln Re + O(Re^2)\right].} \tag{9-123}$$

Higher-order terms to $O(Re^3)$ were obtained by Chester and Breach.[11] It is disappointing, but fairly typical of asymptotic approximations, that the calculation of many terms achieves a relatively small increase in the range of Reynolds number in which the drag can be evaluated accurately compared with (9–122). We may recall that asymptotic convergence is achieved by taking the limit $Re \to 0$ for a fixed number of terms in the expansion, rather than an increasing number of terms for some fixed value of Re.

E. HEAT TRANSFER FROM A BODY OF ARBITRARY SHAPE IN A UNIFORM STREAMING FLOW AT SMALL, BUT NONZERO, PECLET NUMBERS

We have invested considerable effort to analyze the rate of heat transfer from a heated sphere in a uniform streaming flow at low Peclet number. Now that we understand the asymptotic structure of the low-Peclet-number limit, however, we can very easily extend our result for the first two terms in the expression for Nu, (9–60), for the same flow to bodies of *arbitrary* shape, which may be either *solid* or *fluid*, and to *arbitrary values of the Reynolds number* provided only that $Pe \ll 1$. This extension was first demonstrated by Brenner,[12] and our discussion largely follows his original analysis.

Let us first consider the general formulation of the problem. If we adopt the linear dimension of the body as a characteristic length scale, we have seen that the governing equations and boundary conditions for forced convection heat transfer can be written in the forms (9–7) and (9–8), namely,

$$Pe(\mathbf{u} \cdot \nabla \theta) = \nabla^2 \theta, \tag{9-124}$$

$$\theta = 1 \quad \text{on} \quad S, \tag{9-125}$$

$$\theta \to 0 \quad \text{as} \quad |\mathbf{r}| \to \infty, \tag{9-126}$$

and this is true independent of the body geometry. Furthermore, for small values of Pe, the solution of (9–124) for a body of arbitrary shape must take the form of a matched asymptotic expansion, with the solution in the previous section for a solid sphere being a special case. Thus there exists an inner region, within a dimensionless distance $|\mathbf{r}| < O(Pe^{-1})$ from the body, where

$$\theta = \theta_0 + Pe\,\theta_1 + o(Pe); \tag{9-127}$$

and the governing equations and boundary conditions are

$$\nabla^2 \theta_0 = 0,$$
$$\theta_0 = 1 \quad \text{on} \quad S, \tag{9-128}$$

$$\nabla^2 \theta_1 = \mathbf{u} \cdot \nabla \theta_0,$$
$$\theta_1 = 0 \quad \text{on} \quad S. \tag{9-129}$$

There must also be an outer region with

$$\tilde{x} = Pe\,x, \quad \tilde{y} = Pe\,y, \quad \tilde{z} = Pe\,z, \tag{9-130}$$

and a governing equation and boundary condition

$$\tilde{\nabla}^2 \oplus = \mathbf{V} \cdot \tilde{\nabla} \oplus, \quad \text{with}$$
$$\oplus \to 0 \quad \text{as} \quad \tilde{r} = (\tilde{x}^2 + \tilde{y}^2 + \tilde{z}^2)^{1/2} \to \infty. \tag{9-131}$$

Here

$$\mathbf{V} = \mathbf{u}\left(x = \frac{\tilde{x}}{Pe}, \quad y = \frac{\tilde{y}}{Pe}, \quad z = \frac{\tilde{z}}{Pe}\right), \tag{9-132}$$

and thus, in the limit $Pe \to 0$, with $\tilde{x}$, $\tilde{y}$, and $\tilde{z}$ fixed, which is applicable to the outer region,

$$\mathbf{V} = \mathbf{i}, \tag{9-133}$$

where it is assumed that the undisturbed streaming flow is parallel to the x axis. From the analysis for the special case of a sphere, we assume that there exists an asymptotic expansion in this outer region of the form

$$\oplus = Pe\oplus_0 + Pe^2\oplus_1 + o(Pe^2), \tag{9-134}$$

and thus, substituting (9–134) into (9–131), we have

$$\tilde{\nabla}^2 \oplus_0 = \mathbf{i} \cdot \tilde{\nabla} \oplus_0 = \frac{\partial \oplus_0}{\partial \tilde{x}}, \quad \text{with} \tag{9-135}$$

$$\oplus_0 \to 0 \quad \text{as} \quad \tilde{r} \to \infty.$$

Finally, we note that the inner and outer expansions must match in the region of overlap between the two regions.

So far, we have done no more than restate the governing equations and boundary conditions that were derived in Section C for the heated sphere in a uniform streaming flow, but without specifying the geometry of the body that we wish to consider. Clearly, if we wish to determine the details of the temperature distribution in the fluid, it will be necessary to restrict ourselves to a specific body geometry, both because of the need to apply boundary conditions on the body surface in (9–128) and (9–129) and because the velocity field that appears in (9–129) (and higher-order approximations in both the inner and outer regions) depends on the body geometry. However, if our objective is a prediction of the correlation between Nu and Pe for small Pe, we can determine the first convective contribution in a form similar to Eq. (9–60) without the need to restrict ourselves to a particular body geometry.

To be more specific, if we use the definition of the Nusselt number

$$Nu = \frac{Q'}{2\pi k \ell_c (T_0 - T_\infty)}, \quad \text{with} \quad Q' = \iint_S \mathbf{q}' \cdot \mathbf{n} \, dS', \tag{9-136}$$

where Q' is the overall dimensional rate of heat transfer, ℓ_c is a characteristic length scale chosen so that the surface area is $4\pi \ell_c^2$, and $\mathbf{q}'$ is the local dimensional heat flux vector,

$$\mathbf{q}' = -k\nabla T, \tag{9-137}$$

then, corresponding to (9–127), we have

$$Nu = Nu_0 + Pe\,Nu_1 + o(Pe), \tag{9-138}$$

where

$$Nu_0 = \frac{1}{2\pi} \iint_S \mathbf{q}_0 \cdot \mathbf{n} \, dS, \quad \mathbf{q}_0 = -\nabla \theta_0; \tag{9-139}$$

$$Nu_1 \equiv \frac{1}{2\pi} \iint_S \mathbf{q}_1 \cdot \mathbf{n} \, dS, \quad \mathbf{q}_1 = -\nabla \theta_1. \tag{9-140}$$

E. Heat Transfer From a Body of Arbitrary Shape in a Uniform Streaming Flow

The key result that we shall demonstrate in this section is that

$$\boxed{Nu_1 = \frac{1}{4}(Nu_0)^2.} \tag{9-141}$$

It is clear that Nu_0, which is the dimensionless overall heat flux in the pure conduction limit, will depend on the geometry of the body. However, once Nu_0 is known (either by theoretical calculation or, perhaps, by experiment), we can calculate the first convective contribution in (9–138) by means of (9–141), with no extra work, so that

$$\boxed{Nu = Nu_0 + \frac{1}{4}(Nu_0)^2 Pe + o(Pe),} \tag{9-142}$$

and we shall demonstrate that this is true independent of the body shape, the Reynolds number of the flow, or whether the dynamic boundary condition at the body surface is the no-slip condition (as is appropriate for a solid body) or continuity of velocity and stress (as is appropriate for a drop). The only requirements are that the flow far from the body be a uniform streaming motion,

$$\mathbf{u} \to \mathbf{i} \quad \text{as } r \to \infty, \tag{9-143}$$

as already assumed in (9–133) and that

$$\mathbf{u} \cdot \mathbf{n} = 0 \quad \text{on } S, \tag{9-144}$$

which implies that the body shape is fixed with respect to axes fixed in the undisturbed flow and that the boundary S is impermeable.

The key to obtaining the result (9–142) is the fact that Nu_0 and Nu_1, which would seem to require a specific body geometry in view of the definitions (9–139) and (9–140), can both be calculated by means of equivalent integral expressions over a spherical surface at some arbitrarily large distance from the body. The reader may anticipate that the form of the temperature field at a large distance from the body is not strongly dependent on the body shape. However, the more important fact is that the details of the *far-field* temperature distribution that are necessary to calculate Nu_0 and Nu_1 can be determined completely from a knowledge of the asymptotic structure of the problem without the need to specify a particular body geometry.

To demonstrate these facts, and hence prove (9–141), let us begin by deriving formulas from which Nu_0 and Nu_1 can be calculated on the basis of the far-field temperature distribution. First, for Nu_0, the desired result is obtained easily from the governing equation (9–128) for θ_0 in the inner region, rewritten in the form

$$\nabla \cdot \mathbf{q}_0 = 0. \tag{9-145}$$

Now, let σ denote the surface of a sphere that circumscribes the arbitrarily shaped body S and is constrained to lie within the inner region where (9–145) holds, but is otherwise assumed to have an arbitrarily large radius [that is, the radius $r_\sigma \gg 1$ but subject to $r_\sigma < O(Pe^{-1})$ as $Pe \to 0$]. Then,

$$\int_V \nabla \cdot \mathbf{q}_0 \, dV \equiv 0, \tag{9-146}$$

where V denotes the portion of the fluid domain between the body surface S and the spherical surface σ. Now, applying the divergence theorem to (9–146), we find that

$$-\int_S \mathbf{q}_0 \cdot \mathbf{n} dS + \int_\sigma \mathbf{q}_0 \cdot \mathbf{n}_\sigma dS = 0. \tag{9–147}$$

Here, $\mathbf{n}$ is the outer normal for the arbitrarily shaped body and $\mathbf{n}_\sigma$ is the outer normal for the circumscribed sphere at $r_\sigma [\equiv (x^2 + y^2 + z^2)^{1/2}]$. However, according to the definition (9–139), the first term in (9–147) is just $2\pi Nu_0$, and it therefore follows that

$$\boxed{Nu_0 = -\frac{1}{2\pi}\int_\sigma \mathbf{q}_0 \cdot \mathbf{n}_\sigma dS.} \tag{9–148}$$

Hence, instead of calculating Nu_0 from (9–139) applied at the body surface, S, we see that it can be obtained by evaluating the integral (9–148), taken over the surface of a circumscribed sphere at a large distance from the body.

A similar expression to (9–148) also can be obtained for Nu_1. As before, we begin with the governing equation for θ_1 in the form

$$\nabla \cdot (\mathbf{q}_1 + \mathbf{u}\theta_0) = 0. \tag{9–149}$$

Now, this equation can be integrated over V, and the divergence theorem again applied to obtain

$$\boxed{Nu_1 = -\frac{1}{2\pi}\int_\sigma \mathbf{q}_1 \cdot \mathbf{n}_\sigma dS.} \tag{9–150}$$

However, this expression does not prove to be convenient for calculating Nu_1, as we shall see shortly. Instead, we derive an alternative formula for Nu_1, obtained by multiplying (9–149) by θ_0 and rewriting the result in the form

$$\nabla \cdot \left(\theta_0 \mathbf{q}_1 - \theta_1 \mathbf{q}_0 + \frac{1}{2}\theta_0^2 \mathbf{u}\right) = 0. \tag{9–151}$$

Here we have used the fact that

$$\mathbf{q}_1 \cdot \nabla \theta_0 = \mathbf{q}_0 \cdot \nabla \theta_1 = \nabla \cdot (\theta_1 \mathbf{q}_0)$$

because $\nabla \cdot \mathbf{q}_0 = 0$. Now we can apply the divergence theorem after integrating (9–151) over the domain V between the body and the large sphere of radius r_σ. Taking account of the facts that $\mathbf{u} \cdot \mathbf{n} = 0$, $\theta_1 = 0$, and $\theta_0 = 1$ on the body surface S, we obtain

$$-\int_S \mathbf{q}_1 \cdot \mathbf{n} dS + \int_\sigma \theta_0 \mathbf{q}_1 \cdot \mathbf{n}_\sigma dS - \int_\sigma \theta_1 \mathbf{q}_0 \cdot \mathbf{n}_\sigma dS + \frac{1}{2}\int_\sigma \theta_0^2 \cdot \mathbf{u} \cdot \mathbf{n}_\sigma dS = 0. \tag{9–152}$$

However, the first term in (9–152) is just

$$2\pi Nu_1 \equiv \int_S \mathbf{q}_1 \cdot \mathbf{n} dS,$$

so that we can calculate Nu_1, at least in principle, by evaluating the remaining terms in (9–152), all of which are integrals over the surface of the circumscribed sphere at r_σ. Of course, it is not evident at this stage why it should be advantageous to evaluate these three integrals instead of the single integral in (9–150). We shall return to this point shortly.

E. Heat Transfer From a Body of Arbitrary Shape in a Uniform Streaming Flow

Let us now show that the desired results (9–141) and (9–142) can be obtained from (9–148) and (9–152) under the very general conditions listed just after (9–142). To do this, let us begin with (9–141) and the general observation that Nu_0, and thus the integral in (9–148), cannot depend on the radius of the circumscribed sphere (r_σ) because it is arbitrary. Furthermore, Nu_0 must be bounded. The only contribution to $\mathbf{q}_0$ that satisfies these requirements is the one that varies as $O(r^{-2})$ for $r \gg 1$. Thus, to calculate Nu_0 from (9–148), it is obvious that we need determine only the $1/r^2$ dependence of $\mathbf{q}_0$. Given the definition (9–139) for $\mathbf{q}_0$ in terms of θ_0, it is clear that we need only evaluate terms of θ_0 that vary like $O(r^{-1})$, for large $r \gg 1$.

If we now examine solutions of Laplace's equation (9–128) for θ_0, we find that there is only one term that exhibits an r^{-1} dependence, namely, C/r, where C is an arbitrary constant. Furthermore, Brenner[12] has shown that the next term in an expression for θ_0 for large $r \gg 1$ will be a solid spherical harmonic of $O(r^{-3})$, provided only that the origin of coordinates is chosen at the proper point inside the body, and this is always true independent of the body shape. Thus, in general,

$$\theta_0 = \frac{C}{r} + O(r^{-3}) \tag{9-153}$$

for large values, $r \gg 1$. If we combine (9–153) with (9–148), the constant C can be evaluated in terms of Nu_0, and the result is

$$\boxed{\theta_0 = \frac{Nu_0}{2r} + O(r^{-3}).} \tag{9-154}$$

Again, it is emphasized that this far-field *form* for θ_0 at large r is independent of the body shape. Only the constant Nu_0 and the term at $O(r^{-3})$ will depend on the geometry of the body.

Now, we may note, by analogy to the analysis of Section C for the specific case of a solid sphere, that the asymptotic form (9–154) of the first term in the inner expansion for $r \gg 1$ is all that we need to obtain a *complete* solution for the first term, $\oplus_0$, in the expansion for the outer region. Indeed, if we examine the governing equation and boundary condition (9–135) and the matching condition derived from (9–154),

$$\lim_{\tilde{r} \ll 1} \oplus_0 \Leftrightarrow \frac{Nu_0}{2\tilde{r}} + O\left(\frac{Pe^2}{\tilde{r}^3}\right), \tag{9-155}$$

we see that the leading-order outer problem is *identical* to that for the solid sphere, and this is true independent of dynamic boundary conditions on the body surface, the Reynolds number of the flow, or the body shape. The solution is therefore

$$\boxed{\oplus_0 = \frac{Nu_0}{2\tilde{r}} \exp\left[-\frac{1}{2}(\tilde{r} - \tilde{x})\right].} \tag{9-156}$$

But now we can turn to the second term, $O(Pe)$, in the asymptotic expansion for the inner region. We can obtain matching conditions for this $O(Pe)$ inner problem by expanding (9–156) for small $\tilde{r}$ and expressing the result in terms of inner variables. The result is

$$\theta_1|_{r \gg 1} \Leftrightarrow -\frac{Nu_0}{4} + \frac{Nu_0 x}{4r}. \tag{9-157}$$

Together with the governing equation and boundary condition, (9–129), the matching condition (9–157) yields a well-posed problem that could be solved, in principle, to obtain

θ_1. Note, however, that the solution will depend on the body geometry – both through the boundary condition on S and through the form of the velocity field $\mathbf{u}$ near the body, which now enters the problem for the first time through the governing equation for θ_1. Thus, if we really require a knowledge of θ_1 in the vicinity of the body, we cannot avoid an explicit dependence on body geometry. If, on the other hand, we require only the contribution of θ_1 to the overall heat transfer rate (Nu_1), this can be determined directly without the need to solve for θ_1 or specify the boundary geometry.

To do this, we return to (9–152). We have seen already that the first term is just $2\pi Nu_1$. We can evaluate the other terms knowing only the far-field form (9–157) for θ_1 obtained from matching. We note, first, that all three terms must be independent of the radius of the spherical surface, r_σ for $r_\sigma \gg 1$. This means that the integrands must be $O(r^{-2})$ or smaller for $r \gg 1$, and in the limit $r_\sigma \gg 1$, we need evaluate only the $O(r^{-2})$ contributions to the integrands to evaluate Nu_1. Let us begin with the second term. Because $\theta_0 \sim O(r^{-1})$ for large r, we require $\mathbf{q}_1$ only to $O(r^{-1})$, or, equivalently, θ_1 to $O(1)$ for $r \gg 1$. But the asymptotic form for θ_1 at $O(1)$ is known from the matching condition (9–147),

$$\theta_1 = -\frac{Nu_0}{4} + \frac{Nu_0 x}{4r} + O(r^{-1}).$$

However, when we substitute this expression for $\nabla \theta_1$ (or $\mathbf{q}_1$) in the second integral of (9–152), we find that

$$\int_\sigma \theta_0 \mathbf{q}_1 \cdot \mathbf{n}_\sigma dS \equiv 0.$$

Similarly, if we examine the last term in (9–152), we know that $\theta_0^2 \sim O(r^{-2})$, and thus we require $\mathbf{u}$ only to $O(1)$ to evaluate the integral. But again, this is known from the far-field boundary condition on $\mathbf{u}$, namely,

$$\mathbf{u} \sim \mathbf{i}.$$

Substituting this form for $\mathbf{u}$ and the expression (9–154) for θ_0 into the last integral in (9–152), we again find that

$$\int_\sigma \theta_0^2 \mathbf{u} \cdot \mathbf{n}_\sigma dS \equiv 0$$

Finally, let us consider the third term in (9–152). Here, we require θ_1 to $O(1)$ – the asymptotic form (9–157) – and $\mathbf{q}_0$ to $O(r^{-2})$ – that is, $-\nabla \theta_0$ with θ_0 evaluated from (9–154). In this case, however, when we evaluate the integral, we do not get zero, but

$$\int_\sigma \theta_1 \mathbf{q}_0 \cdot \mathbf{n}_\sigma dS \equiv -\pi Nu_0^2/2.$$

Hence we find from (9–152) that

$$Nu_1 = \frac{Nu_0^2}{4},$$

as already given in Eq. (9–141). The intriguing feature of the preceding analysis is that the result (9–141) for *bodies of arbitrary shape* is obtained for "free," once the asymptotic analysis has been carried out for one specific case (the sphere) so that that solutions (9–154) and (9–156) are known, and the asymptotic structure of the problem is understood.

One comment before going further is that Nu_1 could not have been evaluated by means of a similar analysis starting from (9–150). The problem with (9–150) is that it requires $\mathbf{q}_1$ to $O(r^{-2})$ and thus θ_1 to $O(r^{-1})$. However, from the far-field matching condition (9–157), we only know θ_1 to $O(1)$. Hence, to evaluate Nu_1 through (9–150), we would have to actually

F. Heat Transfer From a Sphere in Simple Shear Flow at Low Peclet Numbers

solve the inner problem to get θ_1 throughout the inner region. Anticipation of this difficulty with (9–150) was the motivation for derivation of (9–152).

In reviewing our analysis of (9–152) leading to (9–141), we may note that we have used the conditions (9–143) and (9–144) only on the velocity field. Thus, as stated earlier, the result (9–141) or (9–142) is valid in streaming flow for any heated body with a uniform surface temperature provided that these conditions are satisfied and that $Pe \ll 1$. Higher-order terms in (9–142) will depend on the details of the flow, and thus on the Reynolds number Re, as well as the shape and orientation of the body relative to the free stream. However, in the creeping-flow limit, Brenner was able to extend (9–142) to one additional term for a particle of *arbitrary* shape,

$$\frac{Nu}{Nu_0} = 1 + \frac{Nu_0}{4} Pe + \frac{Nu_0}{4} f Pe^2 \ln Pe + O(Pe^2), \tag{9–158}$$

where f is the magnitude of the force exerted by the fluid on the particle, divided by $6\pi \mu \ell_c U$.

F. HEAT TRANSFER FROM A SPHERE IN SIMPLE SHEAR FLOW AT LOW PECLET NUMBERS

In previous sections of this chapter, we have considered forced convection heat transfer at low Peclet number from particles in a uniform streaming flow. The results are applicable if the density of the particle is different from that of the fluid, so that the particle is subject to gravitational and/or inertial forces that give it a *translational* motion relative to the fluid. In this case, we have seen that the relationship between Nu and Pe takes a common form,

$$Nu = Nu_0 + PeNu_1 + O(Pe^2 \ln Pe), \tag{9–159}$$

for bodies of arbitrary fixed shape, which may be either fluid or solid, and the first two terms hold even if the Reynolds number is unrestricted provided the flow remains laminar and steady.

In many physical processes, however, the particle has a density very nearly equal to that of the fluid, and the particle then has no (or negligible) translational velocity relative to the fluid. Provided that the linear dimensions of the particle are small compared with distances over which the velocity gradient in the ambient flow field changes significantly, the flow near the particle is then effectively due to a force-free particle in an ambient velocity field that can be approximated as varying linearly with position. An important question is whether the rate of heat transfer from a heated sphere in such a flow is still given by a correlation of the form (9–159) for $Pe \ll 1$. More generally, of course, we may ask whether the type of flow that exists far from a heated particle has any effect on the rate of heat transfer from the particle for arbitrary Pe. Although an analysis for $Pe \ll 1$ cannot provide a complete answer to this more general question, it can serve as a useful initial test case.

Thus in this section, we consider the special case of heat transfer from a rigid sphere when the undisturbed fluid motion, relative to axes that translate with the sphere, is a simple, linear shear flow,

$$\mathbf{u}_\infty = \gamma y \mathbf{i}. \tag{9–160}$$

The sphere is heated, as before, in such a way that its surface temperature is constant. We assume that the sphere is free of any external couples so that it can rotate with the fluid. Thus the angular velocity of the sphere is just $\Omega = \gamma/2$. The situation is sketched in Fig. 9–7. It is convenient to solve the problem by use of a spherical coordinate system (r, θ, ϕ), where the azimuthal angle ϕ is measured from the yz plane around the z axis and the polar angle θ

Convection Effects in Low-Reynolds-Number Flows

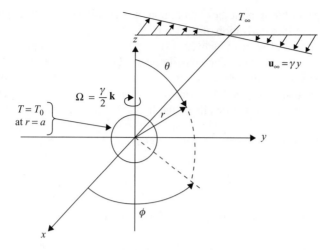

Figure 9–7. A schematic representation of a sphere of radius a and a constant surface temperature T_0 in a fluid of ambient temperature T_∞ that is undergoing a simple shear flow $\mathbf{u}_\infty = \gamma y \mathbf{i}$. Because the sphere is assumed to rotate freely in the ambient flow, it rotates with an angular velocity $\mathbf{\Omega} = (\gamma/2)\mathbf{k}$ about the z axis.

is measured from the z axis. The relationship between spherical and Cartesian coordinates was illustrated in Fig. 9–5.

The governing DE and boundary conditions for the temperature field are again (9–1) and (9–2), and the dimensionless equation and boundary conditions are (9–7) and (9–8), with the same definition for the dimensionless temperature (9–3) and the sphere radius as a characteristic length scale. Only the form of the velocity field, $\mathbf{u}$, and the choice of characteristic velocity are different for this case of a linear shear flow far from the sphere. The appropriate choice for the characteristic velocity is

$$u_c = \gamma a. \tag{9–161}$$

Hence, the Peclet number that appears in (9–7) is defined as

$$\boxed{Pe \equiv \frac{a^2 \gamma \rho C_p}{k}.} \tag{9–162}$$

To specify the velocity field $\mathbf{u}$, we must solve the Navier–Stokes equations subject to the boundary condition (9–160) at infinity. For present purposes, we follow the example of Section C and assume that the Reynolds number, defined here as $Re \equiv a^2 \gamma \rho / \mu$, is very small so that the creeping-flow solution for a sphere in shear flow (obtained in Chap. 8, Section B) can be applied throughout the domain in which θ differs significantly from unity. Hence, from (8–51) and (8–57), we have

$$\boxed{\begin{aligned} u_r &= \frac{1}{2}\left(r - \frac{5}{2}r^{-2} + \frac{3}{2}r^{-4}\right)\sin^2\theta \sin 2\phi, & (9\text{–}163\text{a}) \\[4pt] u_\theta &= \frac{1}{2}(r - r^{-4})\sin\theta \cos\theta \sin 2\phi, & (9\text{–}163\text{b}) \\[4pt] u_\phi &= -\frac{1}{2}r\sin\theta + \frac{1}{2}(r - r^{-4})\sin\theta \cos 2\phi, & (9\text{–}163\text{c}) \end{aligned}}$$

F. Heat Transfer From a Sphere in Simple Shear Flow at Low Peclet Numbers

where these velocity components also have been nondimensionalized with respect to the characteristic velocity scale (9–161).

We now seek a solution of (9–7) and (9–8) for small values of the Peclet number, $Pe \ll 1$, by using the matched asymptotic expansion procedure that was detailed for uniform flow past a sphere in Section C. Although the reader may not immediately see that the derivation of an asymptotic solution for this new problem necessitates use of the matched asymptotic expansion technique, an attempt to develop a regular expansion for θ for $Pe \ll 1$ leads to a Whitehead-type paradox similar to that encountered for the uniform-flow problem.

As in the uniform-flow problem, the solution domain divides into two parts. In the so-called inner region, the sphere diameter provides an appropriate characteristic length scale, and the dimensionless form of (9–7) is applicable, along with the boundary condition (9–8a) at the sphere surface. We assume, in this region, that an asymptotic expansion for θ exists in the form

$$\theta = \theta_0 + \sum_n f_n(Pe)\,\theta_n, \tag{9-164}$$

with boundary conditions on θ_n being

$$\theta_0 = 1, \quad \theta_n = 0 \ (n \geq 1) \quad \text{at } r = 1. \tag{9-165}$$

Obviously, for $Pe \ll 1$, the governing equation for θ_0 is again Laplace's equation for pure conduction,

$$\nabla^2 \theta_0 = 0, \tag{9-166}$$

and we have already seen that the solution of this equation, satisfying the condition (9–165) on the sphere surface is

$$\boxed{\theta_0 = \frac{1}{r}.} \tag{9-167}$$

The governing equation at the next level in the expansion (9–164) depends, of course, on the gauge function $f_1(Pe)$. Although we might be tempted to guess that $f_1(Pe)$ will be equal to Pe, analogous to the streaming-flow problem, we have seen that the gauge functions do not necessarily follow in the simple sequence $(1, Pe, Pe^2, \ldots,)$, which would appear if the asymptotic solution were regular. Thus it is prudent to be cautious at this stage and solve for the first approximation in the outer domain before we attempt to anticipate the form for the second approximation in the inner region.

To obtain the governing equation for the outer domain, we introduce the general rescaling

$$\rho = rPe^m \tag{9-168}$$

for the radial variable, following the analysis of Section C and recall that the rescaling parameter m must be chosen in such a way that both conduction and convection terms are retained when the thermal energy equation is expressed in terms of ρ and the limit $Pe \to 0$ is applied with $\rho = O(1)$. To distinguish the outer temperature field from the inner temperature field, it is also convenient to change symbols for the temperature from θ to $\oplus$.

The rescaled governing equation, expressed in terms of ρ and $\oplus$, then takes the form

$$Pe^{2m} \left\{ \frac{1}{\rho^2} \left[\frac{\partial}{\partial \rho} \left(\rho^2 \frac{\partial \oplus}{\partial \rho} \right) + \frac{1}{\sin \theta} \frac{\partial}{\partial \theta} \left(\sin \theta \frac{\partial \oplus}{\partial \theta} \right) + \frac{1}{\sin^2 \theta} \frac{\partial^2 \oplus}{\partial \phi^2} \right] \right\}$$
$$= Pe \left\{ \left[\left(\frac{1}{2} \frac{\rho}{Pe^m} + O(Pe^{2m}) \right) \sin^2 \theta \sin 2\phi \right] Pe^m \frac{\partial \oplus}{\partial \rho} \right.$$
$$+ \left\{ \frac{1}{2} \left[\frac{\rho}{Pe^m} - O(Pe^{4m}) \right] \sin \theta \cos \theta \sin 2\phi \right\} \frac{Pe^m}{\rho} \frac{\partial \oplus}{\partial \theta} \qquad (9\text{--}169)$$
$$+ \left. \left\{ -\frac{1}{2} \frac{\rho}{Pe^m} \sin \theta + \frac{1}{2} \left[\frac{\rho}{Pe^m} - O(Pe^{4m}) \right] \sin \theta \cos 2\phi \right\} \frac{Pe^m}{\rho \sin \theta} \frac{\partial \oplus}{\partial \phi} \right\}.$$

Examination of the right-hand side shows that, in this outer region, the largest convection terms derive from the linear $O(r)$ contributions to the velocity field (9–163). But this is not at all surprising. These terms are just the dimensionless undisturbed linear shear flow $\mathbf{u}_\infty = y\mathbf{i}$ expressed in spherical coordinates, and the rescaled equation is simply reflecting the obvious result that the dominant convection contribution, sufficiently far from the sphere, is just that associated with the undisturbed linear shear flow. A very similar result was also obtained in the streaming-flow problem in which it was shown that the leading-order approximation to the convection terms in the outer region was due to the undisturbed *uniform* velocity, $\mathbf{u}_\infty = \mathbf{i}$.

If we retain only the largest terms on the right-hand side of (9–169), the equation takes the form

$$Pe^{2m} \left(\nabla_\rho^2 \oplus \right) = Pe \left[\frac{1}{2} \rho \sin^2 \theta \sin 2\phi \frac{\partial \oplus}{\partial \rho} + \frac{1}{2} \sin \theta \cos \theta \sin 2\phi \frac{\partial \oplus}{\partial \theta} \right.$$
$$+ \left. \left(-\frac{1}{2} \sin \theta + \frac{1}{2} \sin \theta \cos 2\phi \right) \frac{1}{\sin \theta} \frac{\partial \oplus}{\partial \phi} \right] + O(Pe^{3m+1}). \qquad (9\text{--}170)$$

Following the reasoning outlined in Section C, the rescaling parameter m is now chosen by the requirement that Eq. (9–170) preserve a balance between conduction and convection terms in the limit as $Pe \to 0$. Thus

$$2m = 1 \quad \text{or} \quad m = \frac{1}{2}, \qquad (9\text{--}171)$$

and the rescaling equation, (9–168), becomes

$$\boxed{\rho = rPe^{1/2}.} \qquad (9\text{--}172)$$

Further, with $m = 1/2$, the governing equation (9–170) can be rewritten as

$$\boxed{\nabla_\rho^2 \oplus - \hat{y} \frac{\partial \oplus}{\partial \hat{x}} = O(Pe^{3/2}).} \qquad (9\text{--}173)$$

Here, $\hat{x} = xPe^{1/2}$, $\hat{y} = yPe^{1/2}$, and $\rho = (\hat{x}^2 + \hat{y}^2 + \hat{z}^2)^{1/2}$.

Before discussing the solution of (9–173), it is perhaps useful to review once more the rationale for rescaling and the choice $m = 1/2$ for the rescaling parameter. The change of variables (9–168) is just one way of changing the nondimensionalization of the radial variable r' from the sphere radius a to the "new" length scale,

$$\ell = a \left(\frac{a^2 \gamma \rho C_p}{k} \right)^{-m},$$

F. Heat Transfer From a Sphere in Simple Shear Flow at Low Peclet Numbers

that is appropriate to the region far from the sphere. If ℓ is chosen properly, it should reflect the distance over which $\oplus$ changes in this region, and then derivatives of $\oplus$ with respect to ρ will be $O(1)$. The only choice possible for m that is consistent with the idea that convection should be as important as conduction in the outer region is (9–171).

Let us now turn to the solution for the temperature distribution in the outer region, which we assume has an asymptotic expansion for small Pe of the form

$$\oplus = \sum_n F_n(Pe)\oplus_n, \tag{9-174}$$

where, as usual,

$$\lim_{Pe \to 0} \frac{F_{n+1}}{F_n} \to 0.$$

The governing equation for the first term in this expansion is the limit of (9–173) for $Pe \to 0$,

$$\nabla_\rho^2 \oplus_0 - \hat{y}\frac{\partial \oplus_0}{\partial \hat{x}} = 0. \tag{9-175}$$

This equation is to be solved subject to the boundary condition

$$\oplus_0 \to 0 \quad \text{as} \quad \rho \to \infty. \tag{9-176}$$

Furthermore, in the limit $Pe \to 0$, the solution should match the first approximation for θ in the inner domain, that is,

$$F_0(Pe) \oplus_0 (\rho, \theta, \phi)|_{\rho \ll 1} \Leftrightarrow Pe^{1/2}\frac{1}{\rho} \quad \text{as} \quad Pe \to 0, \tag{9-177}$$

where the right-hand side is just the inner solution, $1/r$, expressed in terms of the outer variable. It is evident from (9–177) that

$$F_0(Pe) = Pe^{1/2}. \tag{9-178}$$

A general solution to Eq. (9–175) that satisfies the boundary condition (9–176) was obtained some years ago by Elrick, who used Fourier transforms.[13] Elrick was seeking temperature distribution produced by a point source of heat in a simple shear flow. It is not surprising that his solution should be relevant to the present problem. When viewed with the scale of resolution,

$$\ell = a\left(\frac{a^2\gamma\rho C_p}{k}\right)^{-1/2},$$

which is characteristic of the outer region for $Pe \to 0$, the sphere of radius a appears only as a point source of heat. Elrick's solution has the form

$$\oplus_0 = \frac{A}{2\sqrt{\pi}} \int_0^\infty \frac{ds}{\left(1+\frac{1}{12}s^2\right)^{1/2} s^{3/2}} \exp\left[-\frac{\left(\hat{x}-\frac{1}{2}\hat{y}s\right)^2}{4s\left(1+\frac{1}{12}s^2\right)} - \frac{\hat{y}^2+\hat{z}^2}{4s}\right], \tag{9-179}$$

in which A is an arbitrary constant related to the strength of the point heat source and s is a dummy variable of integration. In the present application, the constant A must be determined from the matching condition (9–177). To apply this condition, we require the limiting form of (9–179) for $\rho \ll 1$. It is straightforward to show that

$$\lim_{\rho \to 0} \Theta_0 \sim A \left[\frac{1}{\rho} + \frac{\Gamma}{2\sqrt{\pi}} + O(\rho) \right], \qquad (9\text{–}180)$$

where

$$\Gamma \equiv \int_0^\infty \frac{ds}{s^{3/2}} \left[\left(1 + \frac{1}{12}s^2\right)^{-1/2} - 1 \right] = -0.9104.$$

Now, applying the matching condition (9–177), we obtain

$$Pe^{1/2} A \left[\frac{1}{\rho} + \frac{\Gamma}{2\sqrt{\pi}} + O(\rho) \right] \Leftrightarrow Pe^{1/2} \frac{1}{\rho} \quad \text{as} \quad Pe \to 0. \qquad (9\text{–}181)$$

It is evident that $A = 1$. There is a mismatch in ρ of $O(1)$ for $\rho \ll 1$. However, this mismatch need not concern us at the present level of approximation. A corresponding term, independent of r (because $\Gamma/2\sqrt{\pi}$ is independent of ρ), will be generated only by a term in the inner expansion (9–164) that has a gauge function $O(Pe^{1/2})$. At the present leading order of approximation, such terms have not yet been considered.

Now, with A determined, our solution for the first term in the outer expansion is completed, and we can turn to the problem of obtaining a second term in the asymptotic expansion for θ in the inner region. In view of the fact that the mismatch between the first terms in the inner and outer expansions has been shown in the previous paragraph to be $O(Pe^{1/2})$, it is clear that the gauge function for the second term in the inner expansion (9–164) must be

$$f_1(Pe) = Pe^{1/2}. \qquad (9\text{–}182)$$

With (9–182), the inner expansion has the form

$$\theta = \frac{1}{r} + Pe^{1/2} \theta_1 + \sum_{n=2} f_n(Pe)\theta_n,$$

and the governing equation for θ_1 is easily seen from (9–7) to again be

$$\boxed{\nabla^2 \theta_1 = 0,} \qquad (9\text{–}183)$$

with the boundary condition (9–165)

$$\boxed{\theta_1 = 0 \quad \text{at} \quad r = 1.} \qquad (9\text{–}184)$$

In addition to (9–183) and (9–184), θ_1 must satisfy the matching condition

$$Pe^{1/2} \left[\frac{1}{\rho} + \frac{\Gamma}{2\sqrt{\pi}} + O(\rho) \right]_{\rho \ll 1} \Leftrightarrow \left(\frac{Pe^{1/2}}{\rho} + Pe^{1/2}\theta_1 \right)_{r \gg 1} \quad \text{as} \quad Pe \to 0, \qquad (9\text{–}185)$$

that is,

$$\theta_1 \to \frac{\Gamma}{2\sqrt{\pi}} \quad \text{for} \quad r \gg 1. \qquad (9\text{–}186)$$

The reader may have noted that the governing DE at this level of approximation is still the steady-state conduction equation. Thus, even at this second level of approximation, convection plays no direct role in the heat transfer process in the inner region. Indeed, the

F. Heat Transfer From a Sphere in Simple Shear Flow at Low Peclet Numbers

governing equation (9–183) and the boundary condition (9–184) are both homogeneous. Thus the temperature distribution in the inner region changes at $O(Pe^{1/2})$ only because of the effect of convection on the solution in the outer region and the resultant mismatch (9–185) at $O(Pe^{1/2})$, which leads to the nonhomogeneous matching condition (9–186). This is quite different from the uniform streaming problem, analyzed in Section C, where the first correction to the pure conduction solution in the inner region occurred at $O(Re)$ and contained a direct contribution that was due to convection.

The general solution of (9–183) for θ_1, which satisfies (9–184) and is a function of r only [as suggested by (9–186)], is

$$\theta_1 = C\left(1 - \frac{1}{r}\right). \tag{9–187}$$

The matching condition (9–186) then requires

$$C = \frac{\Gamma}{2\sqrt{\pi}}. \tag{9–188}$$

As usual, even with the condition (9–186) satisfied, there is still a mismatch between the solutions in the two parts of the domain. To see that this mismatch can be neglected at this level of approximation, let us reexamine the full matching condition (9–185) with θ_1 incorporated. The condition is

$$Pe^{1/2}\left[\frac{1}{\rho} + \frac{\Gamma}{2\sqrt{\pi}} + O(\rho)\right]_{\rho \ll 1} + O(F_1(Pe))$$
$$\Leftrightarrow \left\{\frac{Pe^{1/2}}{\rho} + Pe^{1/2}\left[\frac{\Gamma}{2\sqrt{\pi}}\left(1 - \frac{Pe^{1/2}}{\rho}\right)\right]\right\}_{\rho \gg 1} + O(f_2(Pe)) \quad \text{as} \quad Pe \to 0. \tag{9–189}$$

The right-hand side is the first two terms of the inner solution $\theta_0 + Pe^{1/2}\theta_1$, expressed in terms of outer variables. The first two terms on the two sides of (9–189) match exactly. However, there is a mismatch at $O(Pe)$. In particular, on the right-hand side, the term $\Gamma/2\sqrt{\pi}(-Pe/\rho)$ is not yet matched with a corresponding term in the outer solution. Of course, this is to be expected because we have so far considered only the leading-order solution of $O(Pe^{1/2})$ in the outer region. Likewise, the term $(O(\rho Pe^{1/2}))$ in the outer solution has not yet been matched with any term in the inner expansion, but this is, again, to be expected because a term of this form could derive only from a term $O(rPe)$ in the inner solution, and we have not yet considered the $O(Pe)$ approximation in the inner expansion (9–164).

Evidently it would be a relatively straightforward matter to proceed to higher-order approximations in both the inner and outer portions of the domain. The next step would be to seek a second term in the outer solution, and the form of the mismatch in (9–189) suggests that this term should be $O(Pe)$, that is, $F_1(Pe) = Pe$. However, we shall not pursue further correction terms here but rather content ourselves with calculating the first two terms in the approximation to the Nusselt number for small Pe, corresponding to the first two terms in the inner expansion for the temperature distribution, that is,

$$\theta \sim \frac{1}{r} + Pe^{1/2}\left[\frac{\Gamma}{2\sqrt{\pi}}\left(1 - \frac{1}{r}\right)\right] + o(Pe^{1/2}). \tag{9–190}$$

Convection Effects in Low-Reynolds-Number Flows

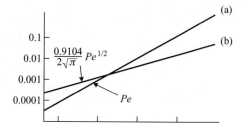

Figure 9–8. Correlation between the Nusselt number and Peclet number for heat transfer at low Peclet number from a sphere with constant surface temperature in (a) uniform streaming flow and (b) simple shear flow.

Substituting (9–190) into the definition (9–20) for the Nusselt number,

$$Nu = -\frac{1}{2\pi} \int_0^{2\pi} \int_0^{\pi} \left(\frac{\partial \theta}{\partial r}\right)_{r=1} \sin\theta \, d\theta \, d\phi,$$

we obtain

$$Nu = 2 + \frac{0.9104}{\sqrt{\pi}} Pe^{1/2} + o(Pe^{1/2}). \qquad (9\text{–}191)$$

The first term in this expression is, of course, the familiar result for pure conduction from a heated sphere and is the same for all flows. The second term represents the first contribution of convection and should be compared with the second term in Eq. (9–60), which is the Nusselt number for forced convection heat transfer at low Pe when the flow is uniform streaming. The most important observation is that the dependence on Pe is different in the two cases, being $O(Pe)$ in the uniform streaming flow and $O(Pe^{1/2})$ in simple shear flow. The two results, (9–60) and (9–191), are plotted in Fig. 9–8. Evidently, the Nusselt number for simple shear flow exceeds the value for uniform streaming flow for $Pe \ll 1$ where the two asymptotic predictions are valid. Although the numerical difference between the two results is small, the most important conclusion from the analysis is not the numerical magnitude of corrections to the conduction heat flux but rather the fact that the asymptotic form of the convection contribution clearly depends on flow type. In general, *heat transfer correlations developed for one type of flow will not carry over to some other type of flow.*

Before we leave the present problem, the reader's attention is called to several generalizations of the predicted relationship (9–161) between Nu and Pe for $Pe \ll 1$. These generalizations clearly illustrate the power of the asymptotic method to provide insight into the form of correlations between dimensionless parameters, with a minimum of detailed analysis. The first is due to Batchelor[14] and Acrivos,[15] who showed that the correlation (9–191), first derived for a sphere in linear shear flow, could be generalized easily and extended to the much more general case of a rigid, heated sphere in an arbitrary linear flow

$$\mathbf{u} = \mathbf{\Gamma} \cdot \mathbf{x}. \qquad (9\text{–}192)$$

Here, $\mathbf{\Gamma}$ is a constant second-order tensor (the velocity gradient tensor) that is arbitrary except for the requirement $\mathrm{tr}\,\mathbf{\Gamma} = 0$ so that $\nabla \cdot \mathbf{u} \equiv 0$.

The class of general linear flows represented by (9–192) includes simple shear flow as a special case,

$$\mathbf{\Gamma} = \begin{pmatrix} 0 & \gamma & 0 \\ 0 & 0 & 0 \\ 0 & 0 & 0 \end{pmatrix}, \qquad (9\text{–}193)$$

F. Heat Transfer From a Sphere in Simple Shear Flow at Low Peclet Numbers

and a number of commonly studied extensional flows such as

$$\text{uniaxial extension} \quad \Gamma = \begin{pmatrix} 2 & 0 & 0 \\ 0 & -1 & 0 \\ 0 & 0 & -1 \end{pmatrix} E, \tag{9-194}$$

$$\text{biaxial extension} \quad \Gamma = \begin{pmatrix} -2 & 0 & 0 \\ 0 & 1 & 0 \\ 0 & 0 & 1 \end{pmatrix} E, \tag{9-195}$$

$$\text{hyperbolic extension} \quad \Gamma = \begin{pmatrix} 1 & 0 & 0 \\ 0 & -1 & 0 \\ 0 & 0 & 0 \end{pmatrix} E. \tag{9-196}$$

These flows are sketched in Fig. 9–9. Another interesting subset of the general case, (9–192), is 2D linear flows. In this case, the complete set of possible motions can be represented as a one-parameter family of the form (9–192) with

$$\Gamma = \begin{pmatrix} 1+\lambda & 1-\lambda & 0 \\ -(1-\lambda) & -(1+\lambda) & 0 \\ 0 & 0 & 0 \end{pmatrix} \frac{E}{2} \tag{9-197}$$

and $-1 \leq \lambda \leq 1$. The case $\lambda = 1$ is just the hyperbolic flow listed as Eq. (9–196); $\lambda = -1$ is a purely rotational flow with streamlines that are circles; and $\lambda = 0$ is simple shear flow, expressed in terms of axes that are rotated 45° from the direction of motion. The parameter E that appears in the various expressions for Γ is usually expressed in units of inverse seconds and is often called the shear rate.

In spite of the apparent generality of the class of motions considered, Batchelor[14] was able to obtain the following correlation for the Nusselt number as a function of the Peclet number,

$$\boxed{\frac{Nu}{Nu_0} = 1 + \alpha Nu_0 Pe^{1/2} + o(Pe^{1/2})} \tag{9-198a}$$

for a rigid spherical particle in any of the class of linear flows represented by (9–192). Shortly thereafter, Acrivos[15] showed that the correlation (9–198a) could actually be extended for spherical particles to include two more terms with a minimum of additional effort. Acrivos' result is

$$\boxed{\frac{Nu}{Nu_0} = 1 + \alpha Nu_0 Pe^{1/2} + \alpha^2 Nu_0^2 Pe + \alpha^3 Nu_0^3 Pe^{3/2} + o(Pe^{3/2}).} \tag{9-198b}$$

In both (9–198a) and (9–198b), $Nu_0 = 2$, and α is a constant coefficient that depends on the type of linear flow that we consider, that is, on the form of Γ. A general formula for computing α from the specified form for Γ is contained in Batchelor's original paper.

The essence of Batchelor's generalization can be summarized as follows. First, the leading approximation in the outer region for all flows of the general class, (9–192), satisfies an equation of the same form,

$$\nabla_\rho^2 \oplus_0 - (G_{ij}\widehat{x}_j) \cdot \frac{\partial \oplus_0}{\partial \widehat{x}_i} = 0, \tag{9-199}$$

representing a balance between conduction and convection with the undisturbed velocity field. Second, in view of the common form (9–199), it is clear that the rescaling (9–172),

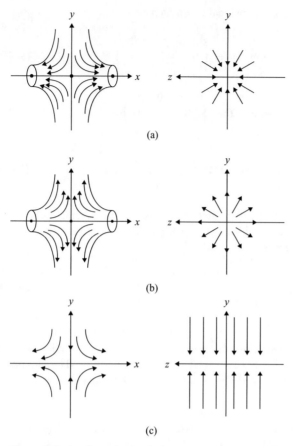

Figure 9–9. A schematic sketch showing the flow patterns for (a) uniaxial extensional flow, (b) biaxial extensional flow, and (c) hyperbolic (or 2D) extensional flow. Each part of the figure shows the flow from two perspectives: one along the z axis toward the xy plane and the other along the x axis toward the yz plane. In the first two cases, the flow is axisymmetric, with the x axis being the symmetry axis. In the 2D case, the flow is invariant in the z direction.

originally derived for the special case of simple shear flow, actually applies to all problems in which the ambient flow is a linear function of spatial position. Third, the fundamental solution of (9–199) will always reduce to the asymptotic form

$$\oplus_0 \sim \frac{A_1}{\rho} - \alpha \quad \text{for } \rho \ll 1 \tag{9-200}$$

(that is, for all G_{jj}), in which A_1 and α are constants of $O(1)$ that depend on the form of G_{jj}. Hence the mismatch between the pure conduction solution in the inner region and the solution of (9–199) in the outer region will be $O(Pe^{1/2})$, and the second term in the inner expansion will thus satisfy (9–183) and (9–184) and be of the general form (9–187). Only the flow-dependent constant $C = \alpha$ will vary from case to case for the general class of linear flows, and Batchelor's result (9–198a) thus follows directly from the analysis for simple shear flow. Acrivos' generalization is based on the fact that (9–199) and (9–200) actually hold in the outer region up to the terms of $O(Pe^{3/2})$, when the first neglected convection terms in (9–173) come into play.

Acrivos[15] also employed arguments very similar to those of Brenner[12] and Batchelor[14], previously outlined, to show that (9–198a) was actually valid for particles of *arbitrary*

G. Strong Convection Effects in Heat and Mass Transfer at Low Reynolds Number

shape in an arbitrary linear flow. For such particles, the only change in the arguments of the preceding section is a generalization of Laplace's equation for conduction to the general form (9–154) and the observation that the coefficients A_1 and α in (9–200) will depend on particle geometry as well as flow type.

G. STRONG CONVECTION EFFECTS IN HEAT AND MASS TRANSFER AT LOW REYNOLDS NUMBER – AN INTRODUCTION

In the preceding sections of this chapter, we considered weak convection effects in the transfer of heat from a hot or cold body. In the remainder of this chapter, we continue our study of convection effects in heat (or mass) transfer problems, but now we focus on the limit $Pe \gg 1$, where the convection of heat appears dominant in the governing equation (9–7) relative to conduction. We actually consider the limit $Pe \gg 1$ in two steps. In this chapter, we consider problems in which $Re \ll 1$. Then, following Chap. 10, in which we consider flow problems in the limit $Re \gg 1$, we return to consider the high-Re and high-Pe regimes in Chap. 11.

Because the class of problems that we can currently study is restricted to flows in which the Reynolds number is very small, $Re \ll 1$, and because $Pe \equiv Re\,Pr$, an obvious question is whether the combination $Re \ll 1$ and $Pe \gg 1$ is achievable in real systems. The key is to remember that the Prandtl number is an independent material parameter. For gases, for which $Pr < O(1)$, and for relatively inviscid liquids such as water that have $Pr \sim O(1)$ or slightly larger, Pe will always be small when Re is small. On the other hand, viscous oils and greases can have $Pr \sim 10^3 - 10^6$, and for these fluids Pe may be large, even though Re is small. This provides one clear motivation for studying heat transfer for the dual limit $Re \ll O(1)$ and $Pe \gg O(1)$.

Another motivation is provided by the analogous mass transfer problem, as discussed in Subsection A.3. The main point to note here is that the Schmidt number is $O(10^3)$ or larger for almost all solute/solvent systems in the liquid state and is $O(1)$ only for systems consisting of two gases (the one present at lowest concentration being designated as the solute). Thus, even for small Reynolds numbers, the Peclet number defined for mass transfer, i.e., $Pe = ReSc$, is very often larger than $O(1)$. Furthermore, for large Peclet number, the conditions for neglecting the effects of blowing/suction are much less severe than in the low-Pe limit.

It should be emphasized that the motivation for seeking asymptotic results for large and small values of the Peclet number is not based entirely on the presence of systems that have large or small values of this parameter. In terms of the Nusselt (or equivalently the Sherwood) number, although the two limiting solutions strictly pertain only for $Pe \gg 1$ or $Pe \ll 1$, if we plot the two asymptotes on a log–log plot as Nu (or Sh) versus Pe, they intersect for $Pe = O(1)$, and in cases in which an exact solution has been obtained numerically, or for which experimental data are available, it is typically found that the exact results never differ from the closest asymptote by more than 20%–30%. Hence a quite reasonable engineering estimate of the rate of transport can usually be obtained by a smooth interpolation between the two limiting curves over a range of about one decade on either side of the intersection point.

The forced convection heat transfer problem [Eq. (9–7) plus boundary conditions] is *linear* in θ, but it still cannot be solved exactly (except for special cases) for $Pe \geq O(1)$ because of the complexity of the coefficient **u**. What may appear surprising at first is that simplifications arise in the limit $Pe \gg 1$, which allow an approximate solution even though no analytic solutions (exact or approximate) are possible for intermediate values of Pe. This is surprising because the importance of the troublesome convection term, which is the source of the complicated coefficient **u** in Eq. (9–7), would appear to increase as Pe is increased.

To see how the problems simplify in the limit $Pe \gg 1$, it is necessary to consider the solution of some specific prototype problems in detail. These solutions in the *strong-convection* limit are presented in the next several sections of this chapter. It will not surprise the reader who has come this far that we approach the limit $Pe \to \infty$ by means of the methods of asymptotic approximation theory. However, the problems we pursue here, as well as in Chaps. 10 and 11, provide the first real insight into the full power and generality of the methods of asymptotic approximation and dimensional analysis. The asymptotic analyses of the preceding sections of this chapter were important in elucidating the role of dimensional analysis in the solution of nonlinear transport problems and especially in demonstrating how to deal with the fact that the temperature fields generally need not be characterized everywhere in the domain by the apparently obvious length scale imposed by the boundary geometry.

In practical terms, however, the preceding results of this chapter contribute only *small corrections* to the dominant pure conduction correlations between the Nusselt number and the independent dimensionless parameters Pe or Re. In contrast, the asymptotic analysis for strong convection plays a critical role in achieving *the leading-order approximation* for correlations between independent and dependent dimensionless groups. Furthermore, we shall see that it is possible to deduce the complete functional form of this relationship, under extremely general conditions, by nothing more than a straightforward combination of nondimensionalization and an asymptotic *formulation* for the problem. For example, in Section H, we consider the problem of heat transfer from a solid sphere in uniform streaming flow at low Reynolds number but high Peclet number. In this case, we can show that

$$Nu = cPe^{1/3} \tag{9-201}$$

to leading order of approximation by simply formulating the asymptotic problem but without solving any differential equations! The asymptotic formulation of the problem guarantees that the constant c is independent of Pe and has a numerical value of $O(1)$, which depends on only the geometry of the body. It is only if we wish to obtain a precise value for c that we must solve the leading-order set of DEs. Although we would get different numerical values of c for different body shapes, we shall see that the general form of the correlation between Nu and Pe remains the same for bodies of arbitrary geometry at this leading order of approximation. Knowledge of the general form of a correlation like (9–201) is almost always more significant than the value of c. For example, with the general form known, a rather small number of experiments could presumably be used to determine c for a quite arbitrarily shaped body. Without this knowledge, on the other hand, a much more extensive experimental program would be required for determining the functional dependence of Nu on Pe, and it would not presumably be evident without much more experimentation that the same form should hold for solid bodies of many different shapes.

But before the virtues of the results and the approach are extolled, the method must be described in detail. Let us therefore return to a systematic development of the ideas necessary to solve transport (heat or mass transfer) problems (and ultimately also fluid flow problems) in the strong-convection limit. To do this, we begin again with the already-familiar problem of heat transfer from a solid sphere in a uniform streaming flow at sufficiently low Reynolds number that the velocity field in the domain of interest can be approximated adequately by Stokes' solution of the creeping-flow problem. In the present case we consider the limit $Pe \gg 1$. The resulting analysis will introduce us to the main ideas of thermal (or mass transfer) *boundary-layer theory*.

H. HEAT TRANSFER FROM A SOLID SPHERE IN UNIFORM FLOW FOR $Re \ll 1$ AND $Pe \gg 1$

We begin our analysis of strong convection effects by returning to the transfer of heat from a uniformly heated sphere in a uniform streaming flow at low Reynolds number, when the velocity field can be approximated by Stokes' solution. This problem was first considered by Acrivos and Goddard.[16] In dimensionless form, the problem we aim to solve is still given by Eqs. (9–7) and (9–8). In these equations and boundary conditions, we have assumed that the characteristic length scale for variations in the temperature field is the sphere radius $\ell_c = a$. After substituting Stokes' solution for $\mathbf{u}$ and expressing (9–7) in spherical coordinates, we obtain (9–24), as before. In the present case, however, we consider the limit in which $Pe \gg 1$. To do this, we propose an asymptotic expansion for θ in the form

$$\theta = \theta_0 + \sum_{n=1}^{N} F_n(Pe^{-1})\theta_n. \qquad (9\text{--}202)$$

As usual, the governing equation for the first term in this expansion is simply generated by taking the limit $Pe \to \infty$ in the full equation, (9–7) or (9–24), namely,

$$\boxed{\mathbf{u} \cdot \nabla \theta_0 = 0.} \qquad (9\text{--}203)$$

We see that (9–203) involves a complete neglect of conduction relative to convection.

Equation (9–203) can be solved quite easily. The physical significance of (9–203) is that θ_0 must be constant along lines (or surfaces) parallel to $\mathbf{u}$. In other words, the projection of $\nabla \theta_0$ in the direction of $\mathbf{u}$ is zero. We may also note that the streamfunction ψ is, by *definition*, constant along lines parallel to $\mathbf{u}$, that is,

$$\mathbf{u} \cdot \nabla \psi = 0. \qquad (9\text{--}204)$$

Comparing (9–203) and (9–204), we conclude that

$$\boxed{\theta_0 = \theta_0(\psi),} \qquad (9\text{--}205)$$

that is, θ is constant on streamlines of the flow. We can obtain the relationship (9–205) in a more formal way by changing independent variables in (9–203) from (r, η) to (r, ψ), with the result

$$u_r \left(\frac{\partial \theta_0}{\partial r}\right)_\psi \equiv 0. \qquad (9\text{--}206)$$

Integrating (9–206), we again get the result (9–205). In Stokes' solution for streaming flow past a sphere, all streamlines begin "at infinity" in front of the sphere and terminate "at infinity" downstream, as we can see from the streamline sketch in Fig. 7–12. When streamlines begin and end at infinity, we say that the streamlines are *open*. In contrast, if the fluid is recirculating in some part of the domain, the streamlines in that region are said to be *closed*.

Equation (9–205) is the general solution of (9–203), but in this case, in which all streamlines are open, we can go one step further because the boundary condition (9–8) for $r \to \infty$ implies that $\theta = 0$ sufficiently far upstream (or downstream) on every streamline. Hence, in combination with (9–205), we conclude that

$$\boxed{\theta_0 = 0} \qquad (9\text{--}207)$$

everywhere in the fluid domain. This "solution" satisfies (9–203) and the boundary condition for $r \to \infty$. However, it clearly cannot be uniformly valid throughout the domain because it does not satisfy the boundary condition $\theta = 1$ at the sphere surface. An alternative way to say this is that a regular perturbation expansion of the form (9–202) does not exist, and this is reflected in the fact that we cannot obtain even the first term that satisfies both of the boundary conditions (9–8).

The problem lies in the scaling inherent in (9–7) or (9–24); we assume that the sphere radius is an appropriate characteristic length scale for θ throughout the fluid domain. The resulting nondimensionalized form of the thermal energy balance leads to the conclusion that conduction terms can be neglected everywhere relative to convection for $Pe \to \infty$, and hence to the solution (9–207). However, it is clear from a physical point of view that conduction terms *cannot* be negligible sufficiently near the surface of the sphere. Because $\mathbf{u} = 0$ at the sphere surface, the only mechanism for heat transfer from the sphere to the fluid is conduction. This is true *independent* of the magnitude of the Peclet number and is inherent in the formula (9–10) for the total heat flux. With conduction completely neglected, as in (9–203), it is not surprising that we end up with the solution (9–207). The dimensionless temperature upstream is $\theta = 0$, and we have thrown out the only possible mechanism for transfer of heat from the sphere (where $\theta = 1$) to the fluid. The only possible conclusion is that conduction *cannot* be negligible near the sphere surface, and thus that the sphere radius is *not* an appropriate characteristic length scale in this region. This may be contrasted with the $Pe \ll 1$ limit for the same problem, where we found that the sphere radius did not serve as an appropriate length scale at *large* distances from the sphere.

The fact that the temperature field apparently must be characterized by two different length scales (the sphere radius for the majority of the fluid domain and some other scale very near to the sphere surface) means that the asymptotic solution for $Pe \gg 1$ must be singular, as explained in preceding chapters (as well as earlier sections of this chapter). For this particular problem, the limiting process $Pe \to \infty$ applied to Eq. (9–7) or (9–24) provides a classic indicator of singular asymptotic behavior. Specifically, in the limit $Pe \to \infty$, *the highest-order derivatives are lost*, and this means that the corresponding solution cannot possibly satisfy all of the boundary conditions of the original problem. Thus it will not be a valid first approximation in the whole domain. This is precisely the same type of behavior that we first observed in the high-frequency limit of flow in a tube with a sinusoidal variation of the pressure gradient in time, as well as in other examples of boundary-layer theory in previous chapters.

We have anticipated that the characteristic length scale $\ell_c = a$ breaks down in the neighborhood of the sphere surface. In the language of matched asymptotic expansions, then, the temperature field must be approximated by two asymptotic expansions for $Pe \gg 1$: one (termed *outer*) in the region within a distance of $O(a)$ from the sphere where $\ell_c = a$ and Eqs. (9–202)–(9–207) are valid, and the other (termed *inner*) in some region much closer to the sphere surface where the details of the scaling and solution remain to be defined. Before attempting to determine an appropriate characteristic length scale and other features of the temperature field in the inner region by formal analysis of the governing equation and boundary condition, it is useful to discuss the expected qualitative behavior in this region.

Beginning right at the sphere surface, heat is transferred radially outward by conduction. However, because Pe is large, very small convection velocities can overwhelm thermal conduction, and the heat transfer process becomes convection dominated at a very short distance from the sphere surface. Thus, before the heat released from the sphere can propagate very far outward in the radial direction, it is swept around the sphere and downstream into a wake. Very near the sphere surface where convection comes into play, the dominant velocity component is in the tangential direction. As a result, the region of heated fluid

H. Heat Transfer From a Solid Sphere in Uniform Flow for $Re \ll 1$ and $Pe \gg 1$

adjacent to the sphere is very thin. In fact, this region is known as the *thermal boundary layer*.

We can continue this discussion a little further to see how the dimension of this thin boundary layer should be expected to depend on the dimensional parameters of the problem. Specifically, the time scale characteristic of the *conduction* process (conduction is simply the diffusion of heat) is

$$t^* = O(L^2/\kappa), \tag{9-208}$$

where L is a length scale over which diffusion takes place and κ is the thermal diffusivity $k/\rho C_p$. On the other hand, the time scale characteristic of the motion of a fluid element completely around the sphere is

$$\hat{t} = O(a/U_\infty). \tag{9-209}$$

Actually, $\hat{t}$ is a lower bound on the so-called convective time scale because the actual velocity of fluid elements very close to the sphere surface will be a small fraction of U_∞. Nevertheless, $\hat{t}$ provides an estimate of the time scale *available* for heating of a fluid element by conduction. Thus, substituting $\hat{t}$ for t^* in (9–208), we obtain an indication of the radial distance from the sphere over which we may expect the fluid to be heated by conduction, namely,

$$L^2 \sim \frac{a\kappa}{U_\infty}. \tag{9-210}$$

Thus, the distance nondimensionalized with respect to a is

$$\ell^2 \sim \left(\frac{\kappa}{aU_\infty}\right) = Pe^{-1}. \tag{9-211}$$

We cannot take this estimate too seriously because of uncertainty about the magnitude of the velocity in the region where the heat transfer process occurs and the greatly simplified picture that is presumed to describe the heat transfer process (essentially a decoupled, sequential process of diffusion followed by convection). Nevertheless, (9–211) provides a strong indication that the thickness of the thermal boundary layer should decrease as Pe increases, i.e., as U_∞ increases or κ decreases. Furthermore, this estimate illustrates strongly the source of the difficulty in using the sphere radius everywhere in the domain as a characteristic length scale for nondimensionalization. In particular, as $Pe \to \infty$, the preceding qualitative argument indicates that we should expect the temperature to decrease from $\theta = 1$ at the sphere surface (to $\theta = 0$) in a region near the sphere surface that gets thinner as the Peclet number increases. As a consequence,

$$\frac{\partial \theta}{\partial r} \to \infty \quad \text{as} \quad Pe \to \infty, \tag{9-212}$$

and it is not at all evident that conduction terms in (9–24), such as

$$\frac{1}{Pe} \frac{\partial^2 \theta}{\partial r^2}$$

will be particularly small as $Pe \to \infty$. On the other hand, in deriving (9–203) by letting $Pe \to \infty$ in (9–24), it was intrinsically presumed that $(1/Pe)\nabla^2 \theta \to 0$, that is, that $\nabla^2 \theta$ is $O(1)$ independent of Pe. The fact that $\partial \theta/\partial r$, with r nondimensionalized by a, increases in magnitude with Pe means that the radius is not an appropriate length scale for the thermal boundary-layer region. Thus (9–24) and the corresponding expansion (9–202) with leading term $\theta_0 = 0$ are not applicable to this part of the domain, as stated already.

From a physical point of view, we can say that the physics introduces a new "intrinsic" length scale for $Pe \gg 1$, which is precisely what is needed to maintain a balance between

Convection Effects in Low-Reynolds-Number Flows

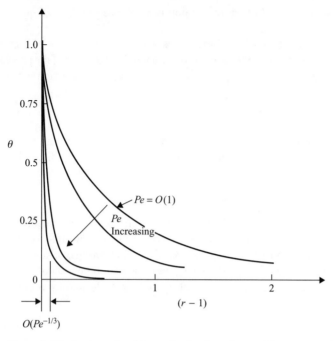

Figure 9–10. A schematic of the qualitative dependence of the temperature distribution on radial distance from the surface of a heated solid body ($\theta_s = 1.0$) for $Re \ll 1$ and increasing Peclet number from $O(1)$ to asymptotically large values, $Pe \gg 1$.

convection and conduction near the sphere surface *even* as $Pe \to \infty$. As suggested by the sketch in Fig. 9–10, this occurs by having the temperature drop occur in an ever-narrowing region as Pe increases; this region is what we have called the *thermal boundary layer*.

1. Governing Equations and Rescaling in the Thermal Boundary-Layer Region

We thus see from the preceding discussion that the limit $Pe \to \infty$ must be analyzed as a singular-perturbation problem with the length scale a and Eq. (9–202) and the boundary condition (9–8b) holding in an outer region and the as yet to be determined thermal boundary-layer equations and scaling holding in an inner region immediately adjacent to the sphere surface.

To obtain the nondimensionalized form of the thermal energy equations appropriate to the inner boundary-layer region, we must determine a length scale characteristic of the distance over which θ decreases by $O(1)$ from its surface value $\theta = 1$. As before, we might accomplish the process of determining the appropriate length scale by starting with the dimensional form of the thermal energy equation and simply looking for a combination of dimensional parameters with units of length that produce a dimensionless form of this equation in which at least some of the conduction terms are retained compared with convection terms in the limit as $Pe \to \infty$. In the weak-convection problems considered in earlier sections of this chapter, this would, in fact, be a feasible proposition because the new characteristic length scale turned out to equal the only combination of dimensional parameters, not involving a, that had units of length (namely, κ/U_∞). However, it has been pointed out that any combination of the form $(\kappa/U_\infty)^m a^{1-m}$ is an equally feasible candidate as a characteristic length scale, and this suggests the need to use the formal *rescaling* procedure to determine the length scale that is characteristic of the thermal boundary layer. This procedure begins with the previously nondimensionalized equation, (9–24).

H. Heat Transfer From a Solid Sphere in Uniform Flow for $Re \ll 1$ and $Pe \gg 1$

For the purposes of rescaling (9–24), it is convenient to redefine variables in the form

$$r - 1 = y \tag{9–213}$$

so that y is a radial variable that is zero at the sphere surface. Then the rescaling to a new nondimensionalized radial variable appropriate to the boundary-layer region can be rewritten as

$$\boxed{y = Pe^{-m}Y,} \tag{9–214}$$

where m remains to be determined. The rescaled variable, in this case Y, is simply the dimensional radial variable y' nondimensionalized with the length scale

$$\ell^* = a^{1-m}\left(\frac{\kappa}{U_\infty}\right)^m. \tag{9–215}$$

With the correct choice for m, this is the length scale characteristic of the inner (or boundary-layer) region. In the rescaled variables, the change in θ from $\theta = 1$ to approximately the free-stream value $\theta = 0$ will occur over an increment $\Delta Y = O(1)$ so that $\partial\theta/\partial Y = O(1)$ *independent* of Pe. For $m > 0$, an increment $\Delta Y = O(1)$ clearly corresponds to a very small increment in the radial distance Δy, scaled with respect to a. In particular, the thickness of the so-called thermal boundary layer is only $O(Pe^{-m})$ relative to the sphere radius a.

To determine m, we substitute the rescaling (9–214) into the governing equation (9–24) and choose m so that both convection and the largest conduction term (or terms) are retained in the inner region, even as $Pe \to \infty$. The only subtlety in carrying out this scheme is to remember that we are focusing on a very thin region $O(Pe^{-m})$, and in this region the magnitude of the velocity components will depend on Pe. To see that this must be true, and to facilitate the rescaling process, it is convenient to consider a Taylor series approximation of the velocity components u_r and u_θ for $y \ll 1$. This is the region of interest for the boundary-layer analysis. Thus, starting with

$$u_r = \left(1 - \frac{3}{2r} + \frac{1}{2r^3}\right)\eta, \tag{9–216}$$

we find that the Taylor series representation for $(r-1) = y \ll 1$ is

$$u_r = u_r|_{r=1} + \left(\frac{\partial u_r}{\partial r}\right)_{r=1}(r-1) + \left(\frac{\partial^2 u_r}{\partial r^2}\right)_{r=1}\frac{(r-1)^2}{2} + \cdots +, \tag{9–217}$$

where, according to (9–216),

$$u_r|_{r=1} = 0,$$

$$\left(\frac{\partial u_r}{\partial r}\right)_{r=1} = 0,$$

$$\left(\frac{\partial^2 u_r}{\partial r^2}\right)_{r=1} = 3\eta.$$

Thus, for the region near the sphere's surface,

$$\boxed{u_r \sim \frac{3}{2}(r-1)^2\eta + O((r-1)^3).} \tag{9–218}$$

Convection Effects in Low-Reynolds-Number Flows

Similarly, we can show that

$$u_\theta \sim -\frac{3}{2}(r-1)(1-\eta^2)^{1/2} + O((r-1)^2). \tag{9-219}$$

Clearly, if we restrict our attention to a region $(r-1) = O(Pe^{-m})$, the magnitudes of both u_r and u_θ in that region will depend on Pe. Indeed, if we introduce the rescaling (9–214) into (9–218) and (9–219) and restrict our attention to $Y = O(1)$, we confirm this fact:

$$u_r \sim \frac{3}{2}Y^2\eta\left(\frac{1}{Pe^{2m}}\right) + O(Pe^{-3m}), \tag{9-220}$$

$$u_\theta \sim -\frac{3}{2}Y(1-\eta^2)^{1/2}\left(\frac{1}{Pe^m}\right) + O(Pe^{-2m}). \tag{9-221}$$

Evidently, to leading order in Pe for $Pe \gg 1$, we can neglect all but the first term in the Taylor series approximation for u_r and u_θ.

Let us now consider rescaling the full thermal energy equation (9–24) by using the results (9–220) and (9–221). When expressed in terms of the boundary-layer variable Y, we obtain

$$\frac{1}{Pe}\left\{\frac{\partial^2\theta}{\partial Y^2}Pe^{2m} + \frac{2Pe^m}{(1+Pe^{-m}Y)}\frac{\partial\theta}{\partial Y} + \frac{1}{(1+Pe^{-m}Y)^2}\frac{\partial}{\partial\eta}\left[(1-\eta^2)\frac{\partial\theta}{\partial\eta}\right]\right\}$$
$$= \left[\frac{3}{2}Y^2Pe^{-2m} + O(Pe^{-3m}Y^3)\right]\eta Pe^m\frac{\partial\theta}{\partial Y}$$
$$+ \frac{1-\eta^2}{(1+Pe^{-m}Y)}\left[\frac{3}{2}YPe^{-m} + O(Pe^{-2m}Y^2)\right]\frac{\partial\theta}{\partial\eta}. \tag{9-222}$$

Clearly, for $Pe \gg 1$, the dominant conduction term is $\partial^2\theta/\partial Y^2$. Thus, retaining the largest terms on the two sides, we see that

$$Pe^{2m-1}\frac{\partial^2\theta}{\partial Y^2} + O(Pe^{m-1}) = \left[\frac{3}{2}Y^2\eta\frac{\partial\theta}{\partial Y} + \frac{3}{2}(1-\eta^2)Y\frac{\partial\theta}{\partial\eta}\right]Pe^{-m} + O(Pe^{-2m}). \tag{9-223}$$

Hence, if conduction and convection are both to be retained in the boundary-layer region for $Pe \to \infty$, we see that

$$2m - 1 = -m$$

or

$$m = \frac{1}{3}. \tag{9-224}$$

Rewriting (9–223), we thus obtain

$$\frac{\partial^2\theta}{\partial Y^2} = \frac{3}{2}Y^2\eta\frac{\partial\theta}{\partial Y} + \frac{3}{2}(1-\eta^2)Y\frac{\partial\theta}{\partial\eta} + O(Pe^{-1/3}). \tag{9-225}$$

H. Heat Transfer From a Solid Sphere in Uniform Flow for $Re \ll 1$ and $Pe \gg 1$

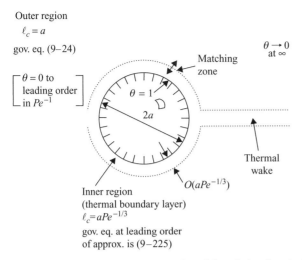

Figure 9–11. A schematic representation of the solution domain for forced convection heat transfer that is due to uniform flow past a solid sphere with a uniform surface temperature for $Re \ll 1$ but $Pe \gg 1$.

This is known as the *thermal boundary-layer equation* for this problem. Because we have obtained it by taking the limit $Pe \to \infty$ in the full thermal energy equation (9–222) with $m = 1/3$, we recognize that it governs only the first term in an asymptotic expansion similar to (9–202) for this inner region.

One interesting feature is that the operator $\nabla^2 \theta$, which is expressed on the left-hand side of (9–222) in terms of rescaled spherical coordinate variables, takes a form in the limiting approximation (9–225) that appears to be just the normal derivative term in $\nabla^2 \theta$ for a Cartesian coordinate system. In fact, we shall see that boundary-layer equations always can be expressed in terms of a *local* Cartesian coordinate system, with one variable normal to the body surface at each point (Y in this case) and the others tangent to it. This reduction of the equations in the boundary layer to a local, Cartesian form is due to the fact that the dimension of the boundary layer is so small relative to that of the body that surface curvature effects play no role.

To complete a solution for the leading-order approximation to the temperature field for $Pe \gg 1$, we need only solve (9–225) subject to appropriate boundary conditions. At the sphere surface, the condition

$$\boxed{\theta = 1 \quad \text{at} \quad Y = 0} \tag{9-226}$$

still holds. However, the second of the two conditions (9–8) cannot be applied directly because $r \to \infty$ is outside the domain of validity for this inner region. We should not forget, on the other hand, that the leading-order approximation in the outer region was already shown to be $\theta_0 = 0$. Thus, though we cannot directly apply the boundary condition $\theta \to 0$ as $r \to \infty$ to the solution (9–225), the matching condition with the first term in the outer region still produces the condition

$$\boxed{\theta \to 0 \quad \text{for} \quad Y \gg 1 \quad \text{as} \quad Pe \to \infty,} \tag{9-227}$$

which is equivalent. A summary sketch showing a qualitative representation of the solution domain for this problem in the limit of large Pe is shown in Fig. 9–11.

We shall very shortly consider the solution of (9–225) subject to the conditions (9–226) and (9–227). First, however, there is a very important result that we can draw from the

Convection Effects in Low-Reynolds-Number Flows

work we have done so far. This is the form of the relationship between the overall rate of heat transfer and the independent dimensional parameters of the system, which can now be determined in spite of the fact that we have not yet solved any DEs other than the trivial one, (9–203). In the case of heat transfer for $Re \ll 1$, the only independent dimensionless parameter is the Peclet number Pe, and in dimensionless terms the objective is thus the relationship

$$Nu = Nu(Pe),$$

where, for the present spherical geometry,

$$\boxed{Nu \equiv \int_{-1}^{1} -\left(\frac{\partial \theta}{\partial r}\right)_{r=1} d\eta,} \qquad (9\text{–}228)$$

as shown in (9–20). However, the formulation of the boundary-layer equations has shown that

$$\left(\frac{\partial \theta}{\partial r}\right)_{r=1} = Pe^{1/3} \left(\frac{\partial \theta}{\partial Y}\right)_{Y=0}, \qquad (9\text{–}229)$$

where $(\partial \theta / \partial Y)_{Y=0}$ is a function of η that is $O(1)$ in magnitude and completely independent of Pe. It follows from this and (9–228) that the correlation between Nu and Pe must take the form

$$\boxed{Nu = cPe^{1/3},} \qquad (9\text{–}230)$$

where

$$c \equiv \int_{-1}^{1} -\left(\frac{\partial \theta}{\partial Y}\right)_{Y=0} d\eta$$

is a numerical coefficient of magnitude $O(1)$.

The correlation (9–230) is the most important result of the analysis of this section. To determine the coefficient c, we must solve (9–225)–(9–227). We will do this in the next subsection. However, we are guaranteed that c will be of $O(1)$, and we may not always wish to proceed further in an analysis of this kind. The fact that a result like (9–230) can be obtained without the necessity of solving any DEs apart from the trivial equation [(9–203)] illustrates part of the power of the asymptotic method that we have been developing in the last several chapters.

2. Solution of the Thermal Boundary-Layer Equation

The asymptotic formulation of the previous subsection has led not only to the important result given by (9–230) but also to a very considerable simplification in the structure of the governing equation in the thermal boundary-layer region. As a consequence, it is now possible to obtain an analytic approximation for θ.

To do this we introduce a similarity transformation of the form

$$\boxed{\theta = \theta(\zeta), \quad \text{where} \quad \zeta = \frac{Y}{g(\eta)}} \qquad (9\text{–}231)$$

into Eq. (9–225). The function $g(\eta)$ determines the dependence of the boundary-layer thickness on η. In particular, assuming for the moment that a solution of the form (9–231)

H. Heat Transfer From a Solid Sphere in Uniform Flow for $Re \ll 1$ and $Pe \gg 1$

exists, we can associate the edge of the boundary layer with the value of $\zeta = \zeta^*$, where θ has some arbitrary small value, say, 0.01. Then, in terms of Y, the edge will be located at $\zeta^* g(\eta)$. The derivatives of θ required in (9–225) are

$$\frac{\partial \theta}{\partial Y} = \frac{1}{g}\frac{d\theta}{d\zeta},$$

$$\frac{\partial^2 \theta}{\partial Y^2} = \frac{1}{g^2}\frac{d^2\theta}{d\zeta^2},$$

$$\frac{\partial \theta}{\partial \eta} = -\frac{\zeta}{g}\frac{dg}{d\eta}\left(\frac{d\theta}{d\zeta}\right).$$

Substituting into (9–225), we thus obtain

$$\boxed{\frac{d^2\theta}{d\zeta^2} + \frac{3}{2}\zeta^2 \frac{d\theta}{d\zeta}\left[\frac{(1-\eta^2)}{3}\frac{dg^3}{d\eta} - g^3\eta\right] = 0} \qquad (9\text{–}232)$$

after multiplying through by g^2 to make the coefficient of the highest-order derivative equal to unity.

If a similarity solution exists, the coefficients of (9–232) must be either a constant or a function of ζ only. This means that

$$\boxed{\frac{(1-\eta^2)}{3}\frac{dg^3}{d\eta} - g^3\eta = \text{const} = 2.} \qquad (9\text{–}233)$$

The numerical value of the constant in (9–233) is arbitrary (provided it is nonzero) but is conveniently chosen as 2. However, the solution for θ would be unchanged if any other nonzero value were chosen. The constant cannot be zero because the resulting equation for θ does not have a solution that can satisfy both of the boundary conditions (9–226) and (9–227). Corresponding to (9–233), Eq. (9–232) for θ now becomes

$$\boxed{\frac{d^2\theta}{d\zeta^2} + 3\zeta^2 \frac{d\theta}{d\zeta} = 0,} \qquad (9\text{–}234)$$

and the problem has been reduced to the solution of (9–233) and (9–234) subject to appropriate boundary conditions.

Let us begin with (9–233). For this purpose, it is convenient to rewrite (9–233) in the form

$$\frac{dg^3}{d\eta} + \frac{3}{2}g^3 \frac{d\ln(1-\eta^2)}{d\eta} = \frac{6}{1-\eta^2}. \qquad (9\text{–}235)$$

Then the left-hand side can be integrated directly to obtain the homogeneous solution $g^3 = c_1(1-\eta^2)^{-3/2}$. It is left to the reader to verify that the general solution of (9–235) is

$$\boxed{g^3 = \frac{c_1}{(1-\eta^2)^{3/2}} + \frac{6}{(1-\eta^2)^{3/2}}\int_{-1}^{\eta}(1-t^2)^{1/2}dt.} \qquad (9\text{–}236)$$

The homogeneous solution is singular at both $\eta = 1$ and -1. The particular solution, on the other hand, blows up at $\eta = 1$ (that is, at the axis of symmetry downstream of the sphere) but is finite at $\eta = -1$. In view of the physical interpretation of $g(\eta)$ as representing the η dependence of the boundary-layer thickness, we expect that g must be finite at $\eta = -1$ (the upstream symmetry axis). Thus we require that

$$c_1 = 0,$$

and

$$g(\eta) = \frac{6^{1/3}}{(1-\eta^2)^{1/2}} \left[\int_{-1}^{\eta} (1-t^2)^{1/2} dt \right]^{1/3}. \tag{9-237}$$

Even with this choice, $g(\eta) \to \infty$ as $\eta \to 1$. This means that the boundary-layer approximation breaks down in the limit as we approach the downstream axis of symmetry. We do not pursue this point here, except to comment that it is not very surprising that the boundary-layer solution should break down in this region. This is because the physical meaning of the rescaling used in deriving (9–225), as well as the qualitative discussion leading up to that rescaling, is that derivatives of θ in the radial direction are asymptotically large compared with derivatives along the sphere surface, that is,

$$\frac{\partial \theta}{\partial y} = Pe^{1/3} \frac{\partial \theta}{\partial Y} \gg \frac{\partial \theta}{\partial \eta}.$$

Near the rear-stagnation point, however, this assumption cannot remain valid because the flow turns the corner and carries the thermal layer into a "wake" along the axis of symmetry. Mathematically, the fact that $g \to \infty$ means that the boundary-layer scaling is breaking down. Thus, if we were to completely analyze the temperature distribution in the whole fluid domain, it would be necessary to include at least two additional asymptotic regions: one along the downstream symmetry axis to analyze the structure of the thermal wake for $Pe \to \infty$ and at least one more within some small neighborhood of the rear-stagnation point to "connect" the boundary layer on the sphere with the downstream wake.

We do not pursue solutions for these additional regions here. Careful research[17] has shown that we can evaluate the coefficient c in (9–230) with negligible error for $Pe \gg 1$ by evaluating the integral in (9–230) using the boundary-layer result for $(\partial \theta / \partial Y)_{Y=0}$ for all η in the range $-1 \leq \eta \leq 1$. In particular, the breakdown in the boundary-layer solution at $\eta = 1$ contributes an error in c_1 of only $O(Pe^{-1})$, which is asymptotically small for $Pe \to \infty$.

To complete the present solution, we need only solve (9–234) subject to the conditions

$$\theta(0) = 1, \tag{9-238}$$

$$\theta(\infty) \to 0, \tag{9-239}$$

which are derived from (9–226) and (9–227). Integrating (9–234) once with respect to ζ, we obtain

$$\frac{d\theta}{d\zeta} = c_1 e^{-\zeta^3},$$

and after two integrations we get the general solution

$$\theta = c_2 + c_1 \int_0^{\zeta} e^{-t^3} dt.$$

H. Heat Transfer From a Solid Sphere in Uniform Flow for $Re \ll 1$ and $Pe \gg 1$

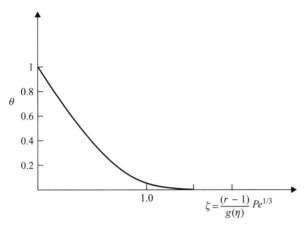

Figure 9–12. The self-similar temperature profile given by Eq. (9–240) for forced convection heat transfer from a heated (or cooled) solid sphere in a uniform velocity field at small Re and large Pe. The function $g(\eta)$ represents the dependence of the thermal boundary-layer thickness on η and is given by (9–237).

Applying the boundary conditions (9–238) and (9–239) to determine c_1 and c_2, we find that

$$\theta = 1 - \frac{\int_0^\zeta e^{-t^3} dt}{\int_0^\infty e^{-t^3} dt}. \tag{9-240}$$

This solution for the temperature distribution is plotted in Fig. 9–12.

The denominator of (9–240) is the well-known gamma function $\Gamma(4/3)$:

$$\int_0^\infty e^{-t^3} dt = \int_0^\infty \frac{1}{3} e^{-u} u^{-2/3} du = \frac{1}{3}\Gamma(1/3) = \Gamma(4/3) = 0.8930.$$

The final step in the present analysis is to use the solutions (9–240) and (9–237) to evaluate c in (9–230). After some straightforward manipulation, we find that

$$c = \frac{3}{2(6)^{1/3}\Gamma(4/3)} \left(\frac{\pi}{2}\right)^{2/3} = 1.249. \tag{9-241}$$

Thus,

$$Nu = 1.249 Pe^{1/3} + o(Pe^{1/3}). \tag{9-242}$$

It should be remembered that (9–225) and its solution (9–240) represent only the first term in an asymptotic expansion for $Pe \to \infty$ in the boundary-layer regime, and (9–203) and its solution (9–207) is the leading-order term in a corresponding expansion for the outer region. To obtain the next level of approximation, it is necessary to calculate an additional term in both of these expansions. We do not pursue this calculation here because of the complexity of the details. Nevertheless, it is worth recording the result obtained originally by Acrivos and Goddard[16]:

$$\boxed{\frac{Nu}{Pe^{1/3}} = 1.249 + .922 Pe^{-1/3} + o(Pe^{-1/3}).} \qquad (9\text{--}243)$$

I. THERMAL BOUNDARY-LAYER THEORY FOR SOLID BODIES OF NONSPHERICAL SHAPE IN UNIFORM STREAMING FLOW

It was stated that the most important result of the previous section is the correlation, at the leading order of approximation, between the Nusselt and Peclet numbers, namely, equation (9–230):

$$Nu = cPe^{1/3}.$$

However, all of the preceding analysis was for the special case of a solid, spherical body in uniform, steady streaming flow or, equivalently, for translation of a sphere, without rotation, through an otherwise quiescent fluid. What makes (9–230) especially significant is that it is valid for a much more general class of problems. In this section, we consider solid bodies of *arbitrary* shape for a uniform streaming flow in the creeping-flow limit (or, equivalently, we consider translation of the body along a rectilinear path through an otherwise quiescent fluid). For this class of problems, the correlation (9–230) applies to all cases that involve smooth bodies without either sharp corners or regions of extreme curvature that generate regions of closed streamlines[18] for $Re \ll 1$. We shall shortly return to discuss the reasons for these limitations on allowable body shapes. It should be noted, however, that the restriction is rather mild. Almost all smooth particles will fall within the allowed class. In this case, only the coefficient c in (9–230) varies with the geometry of the body, and even then it is always guaranteed to be $O(1)$ in magnitude.

To understand the generality claimed for the correlation (9–230), it is well to remind ourselves that the general form (9–230) was deduced completely from the *asymptotic structure* of the heat transfer problem for the sphere at $Pe \gg 1$. The fact that (9–230) can be applied for a much wider class of particle geometries is equivalent to stating that the asymptotic structure for $Pe \gg 1$ is invariant to geometry within the limitations on shape that were previously stated. The ease with which we can generalize the formulation to a body of arbitrary shape is a graphic demonstration of the power of the asymptotic method. We need only backtrack through the analysis for a sphere to see that the same asymptotic formulation and thus the result (9–230) are still valid.

We begin by noting the obvious fact that the dimensionless form of the thermal energy equation (9–7) is preserved. Only the form of the velocity vector **u** will vary with the geometry of the body. Thus the analysis for the outer region also will be preserved, leading to the solution (9–207), provided only that no regions of closed streamline or recirculating flow exist in the fluid domain. Because a fluid without inertia will tend to follow faithfully the contours of a body surface, one's intuition might suggest that recirculating flows would never occur for streaming motion at $Re \ll 1$. However, it is necessary to be slightly cautious on this point, as recent research results have demonstrated the existence of counterexamples.[19] For example, some body shapes have been found,[20] such as a hollow hemisphere moving along its symmetry axis, as sketched in Fig. 9–13(a), that exhibit closed streamline "wakes" in zero Reynolds number flow. In addition, the region between two spheres that are translating

I. Thermal Boundary-Layer Theory for Solid Bodies of Nonspherical Shape

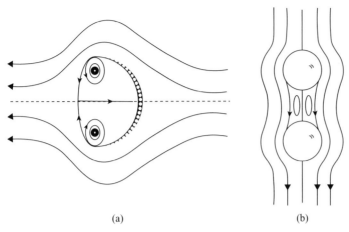

Figure 9–13. Creeping flows with regions of closed streamline motion: (a) streaming flow past a hollow hemispherical cap; (b) two spheres translating along their line of centers.

with equal velocity along their line of centers has been found to contain closed streamlines if the distance between the spheres is smaller than one diameter, as sketched in Fig. 9–13(b). Nevertheless, for a single body in a zero-Reynolds-number streaming flow, the likelihood of a recirculating region is reasonably remote and does *not* occur for single bodies of "simple" shape, such as ellipsoids of arbitrary axis ratio. In all such cases, the high-Peclet-number heat transfer problem reduces to the analysis of a thermal boundary-layer region near the surface of the body.

Now, a key feature of the boundary-layer analysis in the case of a sphere was the fact that it was only the form of the velocity field very close to the sphere surface that determined the critical scaling parameter m, and thus the form of the correlation (9–230). At the level of approximations (9–218) and (9–219), we found that the velocity component tangent to the sphere surface varied linearly with distance from the surface y, whereas the velocity component normal to the surface varied quadratically as y^2. But it is a simple matter to show that this dependence of the tangential and normal velocity components on distance from the surface is actually typical of all bodies on which the no-slip condition is applied. To simplify the details, we restrict attention to axisymmetric or 2D bodies (with the streaming motion parallel to the symmetry axis or perpendicular to the cylindrical axis, respectively). Furthermore, we assume that the geometry of the body is one for which there is a known analytic orthogonal coordinate system with the surface of the body corresponding to a constant value of one of the spatial coordinates. Although it will be intuitively obvious that the principles we will outline below must also hold for other body geometries (3D shapes, for example), an analytic "proof" would be much more complicated for these other cases. Nevertheless, the general scaling results that have their origin in the linear dependence of the tangential velocity on distance from a no-slip surface will clearly carry over to any (reasonably smooth) solid body.

To be specific, we denote the three coordinate directions as (q_1, q_2, q_3), with q_3 being either the axial coordinate z in the 2D case or the azimuthal angle ϕ for axisymmetric geometries. We require a set of scale factors (h_1, h_2, h_3), corresponding to q_i, defined such that the length of a differential line element can be expressed in terms of increments in the q_i:

$$ds^2 = \frac{(dq_1)^2}{h_1^2} + \frac{(dq_2)^2}{h_2^2} + \frac{(dq_3)^2}{h_3^2}. \tag{9–244}$$

Convection Effects in Low-Reynolds-Number Flows

For a 2D or axisymmetric body, the surface corresponds to either q_1 or $q_2 = $ const. For convenience, we assume that it is q_1 (though the required changes are completely trivial if it is q_2 instead). If we denote the velocity vector as

$$\mathbf{u} = u_1 \mathbf{i}_1 + u_2 \mathbf{i}_2 = (u_1, u_2, 0), \qquad (9\text{--}245)$$

then it is the component u_2 that is parallel to the body surface and u_1 that is perpendicular to this surface. The two nonzero-velocity components are related by means of the continuity equation, which becomes

$$\frac{\partial}{\partial q_1}\left(\frac{u_1}{h_2 h_3}\right) + \frac{\partial}{\partial q_2}\left(\frac{u_2}{h_1 h_3}\right) = 0 \qquad (9\text{--}246)$$

when expressed in component form. The thermal energy equation in dimensionless form is

$$h_1 h_2 h_3 \left[\frac{\partial}{\partial q_1}\left(\frac{h_1}{h_2 h_3}\frac{\partial \theta}{\partial q_1}\right) + \frac{\partial}{\partial q_2}\left(\frac{h_2}{h_1 h_3}\frac{\partial \theta}{\partial q_2}\right)\right] = Pe\left[u_1 h_1 \frac{\partial \theta}{\partial q_1} + u_2 h_2 \frac{\partial \theta}{\partial q_2}\right]. \qquad (9\text{--}247)$$

In the limit $Pe \gg 1$, we have seen that the thermal boundary layer equations hold for a region very close to the surface of the body, and thus that we require only the limiting form of the velocity components as we approach the body surface. We can again use a Taylor series approximation to deduce the appropriate form for the velocity components,

$$\mathbf{u}(\mathbf{x}) = \mathbf{u}(\mathbf{x}_s) + (\mathbf{x} - \mathbf{x}_s) \cdot \nabla \mathbf{u} + O|\mathbf{x} - \mathbf{x}_s|^2 \qquad (9\text{--}248)$$

where

$$(\mathbf{x} - \mathbf{x}_s) \cdot \nabla \mathbf{u} = \mathbf{i}_1 h_1 (q_1 - q_1^s)\left[\frac{\partial u_1}{\partial q_1} + h_2 u_2 \frac{\partial}{\partial q_2}\left(\frac{1}{h_1}\right)\right]$$
$$+ \mathbf{i}_2 h_1 (q_1 - q_1^s)\left[\frac{\partial u_2}{\partial q_1} - h_2 u_1 \frac{\partial}{\partial q_2}\left(\frac{1}{h_1}\right)\right]$$
$$+ \mathbf{i}_1 h_2 q_2 \left[\frac{\partial u_1}{\partial q_2} - h_1 u_2 \frac{\partial}{\partial q_1}\left(\frac{1}{h_2}\right)\right]$$
$$+ \mathbf{i}_2 h_2 q_2 \left[\frac{\partial u_2}{\partial q_2} + h_1 u_1 \frac{\partial}{\partial q_1}\left(\frac{1}{h_2}\right)\right]$$

for the 2D and axisymmetric cases that are being considered. Now, the no-slip and kinematic boundary conditions require $u_1 = u_2 = 0$, and thus also $\partial u_1/\partial q_2 = \partial u_2/\partial q_2 = 0$. We then see from continuity that $\partial u_1/\partial q_1 = 0$. Hence, from (9–248), we see that

$$u_2 \approx h_1 \left(\frac{\partial u_2}{\partial q_1}\right)_{q_1 = q_1^s}(q_1 - q_1^s) = h_1 \alpha(q_2)(q_1 - q_1^s), \quad \text{where} \quad \alpha(q_2) \equiv \left(\frac{\partial u_2}{\partial q_1}\right)_{q_1 = q_1^s} \qquad (9\text{--}249)$$

The continuity equation (9–246) can then be used to determine u_1:

$$u_1 \approx -h_2 h_3 \frac{\partial}{\partial q_2}\left(\frac{\alpha(q_2)}{h_3}\right)\frac{(q_1 - q_1^s)^2}{2}. \qquad (9\text{--}250)$$

Hence, for small distances from the no-slip surface, the tangential velocity varies linearly with distance, whereas the normal velocity varies quadratically.

I. Thermal Boundary-Layer Theory for Solid Bodies of Nonspherical Shape

We can now find the appropriate form for the thermal boundary-layer equation by substituting the approximate forms for the two velocity components into the thermal equation (9–247) and rescaling according to

$$(q_1 - q_1^s) = Y Pe^{-m}. \tag{9-251}$$

The result is

$$Pe^{2m} h_1^2 \frac{\partial^2 \theta}{\partial Y^2} + O(Pe^m) = Pe\left[-h_1 h_2 h_3 \frac{\partial}{\partial q_2}\left(\frac{\alpha}{h_3}\right) \frac{Y^2}{2Pe^m} \frac{\partial \theta}{\partial Y} + h_1 h_2 \alpha(q_1) \frac{Y}{Pe^m} \frac{\partial \theta}{\partial q_2}\right], \tag{9-252}$$

and we see again that

$$m = 1/3, \tag{9-253}$$

and thus

$$\frac{\partial^2 \theta}{\partial Y^2} + O(Pe^{-1/3}) = \left(\frac{h_2}{h_1}\right)\left[\alpha(q_2) Y \frac{\partial \theta}{\partial q_2} - h_3 \frac{\partial}{\partial q_2}\left(\frac{\alpha}{h_3}\right) \frac{Y^2}{2} \frac{\partial \theta}{\partial Y}\right]. \tag{9-254}$$

This is the generalization of the thermal boundary-layer equation (9–225).

In view of the general scaling result, (9–253), it is clear that the general form of the correlation between Nusselt number and Peclet number,

$$Nu = cPe^{1/3}, \tag{9-255}$$

is preserved for arbitrary solid, no-slip bodies of axisymmetric or 2D shape, subject only to the assumption that the streamlines of the flow are open (that is, there exist no regions of recirculating motion). In fact, very similar arguments can be made without the restriction to axisymmetric or 2D shapes, thus demonstrating that (9–30) is actually applicable to bodies of arbitrary shape (provided again that there is no region of recirculating flow and that the boundary is smooth and contains no sharp corners or regions of extreme curvature[21]). For the 2D or axisymmetric case, the coefficient c in equation (9–255) is given by

$$c \equiv \int_S -\left(\frac{\partial \theta}{\partial Y}\right)_0 dS, \tag{9-256}$$

where S denotes the body surface, and the differential surface element is

$$dS \equiv \frac{dq_2 dY}{h_1 h_2}.$$

1. Two-Dimensional Bodies

For 2D body shapes, $h_3 \equiv 1$. In addition either $h_2/h_1 \equiv 1$ (for example elliptical cylindrical, bipolar, or parabolic cylindrical coordinate[22]), or $h_2/h_1 \equiv 1 + O(Pe^{-1/3})$ (for circular cylindrical coordinates assuming that $r = 1$ is the surface of the cylinder). Hence (9–254) simplifies to the universal form, at least for all 2D geometries for which there is a known analytic coordinate system:

$$\frac{\partial^2 \theta}{\partial Y^2} + O(Pe^{-1/3}) = \alpha(q_2) Y \frac{\partial \theta}{\partial q_2} - \left(\frac{\partial \alpha}{\partial q_2}\right) \frac{Y^2}{2} \frac{\partial \theta}{\partial Y}. \tag{9-257}$$

Convection Effects in Low-Reynolds-Number Flows

In the case of a solid sphere, considered in the preceding section, we solved the thermal boundary-layer equation analytically by using a similarity transformation. An obvious question is whether we may also solve (9–257) by means of the same approach. To see whether a similarity solution exists, we apply a similarity transformation of the form

$$\theta = \theta(\eta), \quad \eta \equiv \frac{Y}{g(q_2)}, \tag{9–258}$$

to Eq. (9–257). The result is

$$\theta'' + \left(\alpha g^2 g' + \frac{1}{2}\alpha' g^3\right)\eta^2 \theta' = 0.$$

Thus, if a similarity solution exists,

$$\boxed{\alpha g^2 g' + \frac{1}{2}\alpha' g^3 = \text{const} = 3,} \tag{9–259}$$

$$\boxed{\theta'' + 3\eta^2 \theta' = 0.} \tag{9–260}$$

The constant coefficient that appears in (9–259) and (9–260) is arbitrary. The value 3 is chosen for convenience. To determine whether a similarity solution exists, we must determine whether solutions of (9–259) and (9–260) can be found that satisfy appropriate boundary conditions. Specifically, to satisfy boundary condition on θ for $Y = 0$ and the matching condition as $Y \to \infty$, we require that the similarity function satisfies

$$\boxed{\begin{array}{l}\theta = 1 \quad \text{at} \quad \eta = 0, \\ \theta \to 0 \quad \text{as} \quad \eta \to \infty.\end{array}} \tag{9–261}$$

In addition, $g(q_2)$ is required to be *finite* except possibly at values of x corresponding to thermal wakes (for example, at the downstream stagnation point in the case of a sphere) for which the assumption of a thin thermal layer is no longer valid.

Now, Eq. (9–260) is identical to equation (9–234), which was found earlier for the sphere, and we have already seen that it can be solved subject to the conditions (9–261). The solution for θ is given in (9–240). The existence of a similarity solution to (9–257) thus rests with Eq. (9–259). Specifically, for a similarity solution to exist, it must be possible to obtain a solution of (9–259) for $g(q_2)$, which remains finite for all q_2 except possibly at a stagnation point where $\alpha = 0$, from which a thermal wake may emanate.

To seek a solution for (9–259), it is convenient to rewrite it in the form

$$\frac{\alpha}{3}\frac{dg^3}{dq_2} + \frac{1}{2}\frac{\partial \alpha}{\partial q_2} g^3 = 3, \tag{9–262}$$

of a first-order linear ODE for g^3. In this form, it is straightforward to write a general solution

$$g^3(q_2) = \frac{c}{\alpha^{3/2}} + \frac{9}{\alpha^{3/2}} \int_{q_{2_0}}^{q_2} [\alpha(t)]^{1/2} dt,$$

I. Thermal Boundary-Layer Theory for Solid Bodies of Nonspherical Shape

where q_{2_0} denotes the front stagnation point on the body surface (where $\alpha = 0$). Because q_{2_0} is the upstream stagnation point, we must choose c such that g remains finite as $q_2 \to q_{2_0}$. Thus we require that $c = 0$ and

$$g^3(q_2) = \frac{9}{\alpha^{3/2}} \int_{q_{2_0}}^{q_2} [\alpha(t)]^{1/2} dt. \qquad (9\text{--}263)$$

With this definition, $g(q_2)$ remains finite for all q_2 (other than the rear-stagnation point), and we claim to have constructed a similarity solution for the complete class of smooth 2D solid bodies (with no closed-streamlines).

To verify this claim, we first note that the absence of a closed-streamline region means that only two zeros are present in the function $\alpha(q_2)$ on the surface: one at the front-stagnation point $q_2 = q_{2_0}$ and the other at the rear-stagnation point where the downstream symmetry axis of the flow and the body surface intersect. Thus the solution (9–263), which was constructed to remain bounded at q_{2_0}, is now seen to remain bounded at all positions on the body surface (because $\alpha \neq 0$), except at the rear-stagnation point where $\alpha \to 0$ and $g \to \infty$. However, as we have already noted in the case of the sphere, the rear-stagnation point is the point of departure for the thermal wake, in which heat is carried downstream by convection along the symmetry axis, and the boundary-layer assumptions clearly fail at this point because the thermal layer is not thin. Thus, with the exception of the rear-stagnation point where the whole analysis breaks down, $g(q_2)$, as defined in (9–263), remains finite for all q_2 and thus satisfies the critical requirement for existence of a similarity solution. Furthermore, this is true for any 2D body so long as flow is characterized everywhere by open streamlines.

2. Axisymmetric Bodies

The thermal boundary-layer equation, (9–257), also applies for axisymmetric bodies. One example that we have already considered is a sphere. However, we can consider the thermal boundary layer on any body of revolution. A number of orthogonal coordinate systems have been developed that have the surface of a body of revolution as a coordinate surface. Among these are prolate spheroidal (for a prolate ellipsoid of revolution), oblate spheroidal (for an oblate ellipsoid of revolution), bipolar, toroidal, paraboloidal, and others.[22] These are all characterized by having $h_3 = h_3(q_1, q_2)$, and either $h_2/h_1 \equiv 1$ or $h_2/h_1 \equiv 1 + O(Pe^{-1/3})$ (assuming that the surface of the body corresponds to $q_2 = 1$). Hence the thermal boundary-layer equation takes the form

$$\frac{\partial^2 \theta}{\partial Y^2} + O(Pe^{-1/3}) = \left[\alpha(q_2) Y \frac{\partial \theta}{\partial q_2} - h_3 \frac{\partial}{\partial q_2}\left(\frac{\alpha}{h_3}\right) \frac{Y^2}{2} \frac{\partial \theta}{\partial Y} \right]. \qquad (9\text{--}264)$$

We know that a similarity solution of this equation exists for the special case of a sphere, and we thus seek again to solve (9–264) by means of similarity transformation for the general case. If we simply follow the prescription of the preceding section, we obtain the same DE and boundary conditions for the similarity function $\theta(\eta)$, but now the equation for $g(q_2)$ takes a slightly more general form,

$$\frac{dg^3}{dq_2} + \frac{3}{2} \frac{h_3}{\alpha} \frac{\partial}{\partial q_2}\left(\frac{\alpha}{h_3}\right) g^3 = \frac{9}{\alpha}. \qquad (9\text{--}265)$$

Provided a solution of this equation exists such that g remains finite as $q_2 \to q_{2_0}$, the similarity transformation is successful. This has been shown to be the case for the case of a sphere, where $q_1 = r$, $q_2 = \theta$, $\alpha = -3\sin\theta/2$, and $h_3 = (\sin\theta)^{-1}[1 + O(Pe^{-1/3})]$. Solutions can also be obtained for other axisymmetric bodies, but to go further, we would

Convection Effects in Low-Reynolds-Number Flows

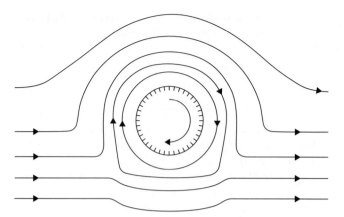

Figure 9–14. A qualitative sketch of the flow pattern for uniform, creeping flow past a rotating circular cylinder.

need to solve the corresponding creeping-flow problem to obtain $\alpha(q_2)$, and that has not been done.

3. Problems with Closed Streamlines (or Stream Surfaces)

The reader may well ask whether anything can be said for bodies that generate regions of recirculating motion in uniform streaming flow. To answer this question, it is advantageous to consider two possible configurations of this general class. In the first configuration, an example of which is sketched in Fig. 9–14, the body is completely surrounded by closed streamlines (or stream surfaces), as would happen if the body were rotating with an axis of rotation that is normal to the direction of the undisturbed flow; in the second configuration, a region of closed streamlines or recirculating flow is generated near the body (usually downstream) but it does not surround the body and is in contact with only a part of the body surface. Two examples of this case are sketched in Fig. 9–13. In the first case, the boundary-layer analysis of this section is not at all relevant, and we must discuss the problem on a completely different basis, which will be described in Section L of this chapter. For streaming-flows (or equivalently, simple rectilinear translation through a quiescent fluid), the need to defer discussion of this case is not much of a limitation, because very few bodies rotate as a consequence of translation in the absence of an external torque. The second class of closed-streamline (or recirculating) flows has a higher probability of occurring, though a precise statement of particle geometries that lead to steady, closed-streamline (or stream surface) flows at low Reynolds number is not known at present. In any case, the boundary-layer results of this section still can be applied, at least qualitatively, to this second class of streaming-flow problems even though they involve regions of recirculating motion adjacent to the body.

To do this, we must anticipate one result from the general discussion of Section L on high-Peclet-number heat transfer in regions of closed-streamlines: The dimensionless temperature gradient in such regions is determined primarily by its size. For example, in a closed-streamline region of $O(1)$ in extent, dimensionless, steady-state temperature gradients will also be $O(1)$.

Suppose we consider a hypothetical situation in which a region of closed-streamline flow, with a linear dimension of $O(1)$ relative to the body, is generated downstream of some point A on the body surface. Upstream of A, the streamlines adjacent to the body surface are all open and the boundary-layer scaling is still relevant so that $\partial\theta/\partial q_1 \sim O(Pe^{1/3})$.

J. Boundary-Layer Analysis of Heat Transfer From a Solid Sphere

Downstream of A, on the other hand, we have already noted that temperature gradients reflect the physical dimensions of the recirculating region. If we assume that these dimensions are $O(1)$, or, indeed, any size larger than $O(Pe^{-1/3})$, the temperature gradient at the body surface downstream of A will be asymptotically small compared with that upstream, and, as a consequence, almost all the heat transfer will occur in the upstream region. In this case, to leading order in Pe, the total heat flux will still be $O(Pe^{1/3})$, as it was in the absence of a closed-streamline region, but the coefficient c in the correlation (9–230) will reflect the fact that the dominant heat transfer takes place on only the fraction of the body surface that is upstream of A. The total heat flux downstream of A, assuming the closed-streamline region to be $O(1)$ in size, will be only $O(1)$ – insignificant compared with that upstream of A for $Pe \gg 1$.

J. BOUNDARY-LAYER ANALYSIS OF HEAT TRANSFER FROM A SOLID SPHERE IN GENERALIZED SHEAR FLOWS AT LOW REYNOLDS NUMBER

It is perhaps timely to stop and reflect upon the nature of the thermal boundary-layer analysis to determine whether other generalizations of the basic result (9–230) may be possible. In particular, heat transfer from solid bodies occurs frequently when the fluid motion seen by the body cannot be approximated as a uniform streaming flow, and the reader may ask whether the correlation (9–230) can be applied in these cases with a proper choice for the characteristic velocity that appears in Pe. It is especially interesting, in this regard, to compare the present analysis with the corresponding low-Peclet-number problem that appeared earlier in this chapter.

One immediately evident contrast between the limits $Pe \to 0$ and $Pe \to \infty$ is the nature of the dependence of the temperature field on the velocity field. In the low-Peclet-number limit, the temperature field near the body is dominated by conduction and is thus relatively insensitive to the details of the fluid motion. When convection effects do come into the low-Peclet-number problem, it is primarily the form of the velocity field at *large* distances from the body that determines the temperature field and the dependence of the Nusselt number on the Peclet number. On the other hand, at high-Peclet-number, the heat transfer process is confined by convective effects to a very thin region near the body surface, and it is only the local form of the velocity distribution in this region that matters. As a consequence one might suppose that the qualitative nature of the high-Peclet-number process should be invariant to changes in the form of the flow at large distances from the body, in contrast with the low-Peclet-number case. Indeed, whatever the nature of the motion elsewhere in the fluid, the local forms (9–249) and (9–250) must be preserved in the immediate vicinity of the body surface, provided we describe the problem in a frame of reference that is fixed at the center of the body and rotates with it so that $u = v = 0$ on the body surface.

In view of these facts, it is tempting to suppose that the correlation

$$Nu \sim Pe^{1/3}$$

should hold for solid bodies, even when the fluid far from the body undergoes a more or less arbitrary motion. If true, this would be an extremely important result, because solid particles are very frequently subjected to motions that are much more complicated than a simple, uniform streaming flow.

Unfortunately, however, there are a large number of different types of flow conditions for which the boundary-layer form of the heat transfer correlation (9–255) is *not* applicable. This applies, basically, to any flow configuration in which the body is completely surrounded by a region of closed streamlines (or pathlines, if the flow is not 2D or axisymmetric). We will discuss high-Peclet-number heat transfer in such cases in Section L. Here, we consider

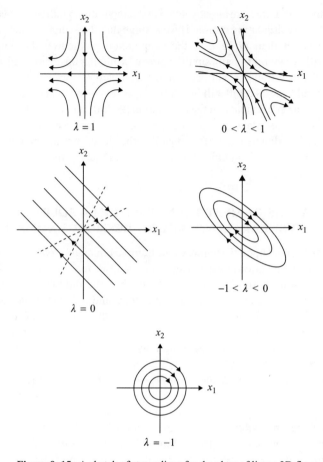

Figure 9–15. A sketch of streamlines for the class of linear 2D flows for $-1 \leq \lambda \leq 1$.

only cases for which the boundary-layer analysis can be used. In general, this requires that

(1) the streamlines (or fluid pathlines) of the undisturbed flow at infinity must be open, as seen in a fixed, laboratory reference frame, and
(2) the particle, when viewed in the same reference frame, cannot rotate.

A general discussion of flow types and particle shapes for which these conditions are satisfied is beyond the scope of this book. As an alternative, at least a qualitative sense may be imparted for the issues involved by consideration of the special case of a *spherical* particle and the class of linear 2D flows described by

$$\mathbf{u} = \mathbf{\Gamma} \cdot \mathbf{x} = (\mathbf{E} + \mathbf{\Omega}) \cdot \mathbf{x}, \qquad (9\text{–}266)$$

where

$$\mathbf{\Gamma} = \frac{1}{2} \begin{bmatrix} 1+\lambda & 1+\lambda & 0 \\ -1+\lambda & -(1+\lambda) & 0 \\ 0 & 0 & 0 \end{bmatrix}$$

and $-1 \leq \lambda \leq 1$. A sketch of the streamlines for these flows is given in Fig. 9–15. The limit $\lambda = 1$ is plane hyperbolic (extensional) flow, $\lambda = -1$ is a purely rotational flow with

J. Boundary-Layer Analysis of Heat Transfer From a Solid Sphere

circular streamlines, and $\lambda = 0$ is simple shear flow. Viewed from another standpoint, flows with $\lambda > 0$ have a strain rate that exceeds the vorticity – that is, $|E| > |\mathbf{\Omega}|$ – culminating at $\lambda = 1$ in a flow with no vorticity. On the other hand, flows with $\lambda < 0$ have a strain rate that is smaller than the vorticity, $|E| < |\mathbf{\Omega}|$, culminating at $\lambda = -1$ in a flow that is purely rotational. For simple shear flow, $\lambda = 0$ and $|E| = |\mathbf{\Omega}|$. A special feature of simple shear flow is that it separates those flows with open streamlines (hyperbola) from flows with closed streamlines (ellipses). For the problem at hand, we assume that the sphere is neutrally buoyant and subject to no body forces. Hence, because $Re \ll 1$, it will translate in a linear flow with its center moving as an element of the fluid, and there is no loss of generality in assuming that $\mathbf{x} = 0$ corresponds to the center of the sphere.

The question at hand is whether circumstances exist for this rather simple situation in which the conditions (1) and (2) are satisfied so that boundary-layer analysis can be applied. So far as the first condition is concerned, the only flows of (9–266) that have open streamlines are those with $\lambda \geq 0$ (which includes simple shear flow). On the other hand, there is a nonzero hydrodynamic torque on the sphere that causes it to rotate for all flows in this subgroup except $\lambda = 1$. Thus, for a sphere in the general linear 2D flow, given by (9–266), there are only two cases that satisfy the conditions for applicability of boundary-layer theory:

(1) *The sphere in pure extensional flow* ($\lambda = 1$). In this case, the streamlines are open (hyperbola at infinity), and the net hydrodynamic torque on the sphere is zero. Thus, if no external torque is applied to the sphere, it will not rotate.
(2) *The sphere in one of the open-streamline flows*, $0 \leq \lambda \leq 1$, but with an external torque applied so that the sphere cannot rotate. Note that the sphere would rotate, in the absence of an external torque, at the angular velocity that causes the hydrodynamic torque to vanish.

These are the only cases for a sphere in a linear velocity field for which a boundary-layer analysis is expected to apply. In other cases, the sphere is surrounded by a zone of closed streamlines.

Similarly, if we consider the case of a circular cylinder with its axis at $\mathbf{x} = 0$ and oriented normal to the plane of the 2D flow, (9–266), the requirement of open-streamline flow is satisfied only if $\lambda = 1$ or if the cylinder is constrained from rotation and $0 \leq \lambda < 1$.

Let us consider, then, the situation in which the streamlines near a heated body are open and a boundary-layer structure is expected to hold for $Pe \gg 1$. How will the problem differ from that analyzed in Sections H and I? Because the leading-order approximation to θ is determined entirely by the form of the velocity field near the body surface, and this structure is invariant to changes both in body geometry and in the nature of the velocity field away from the surface, it is evident that there will be no qualitative change at all if the problem is one for which the boundary-layer structure can be expected. Indeed, if we examine the analysis in Section I, it should become apparent that it is only the functional form of the coefficients $\alpha(q_2) \equiv (\partial u_2/\partial q_1)_{q_1=q_1^s}$ and α' that should change from case to case and the general similarity solution should still apply.

The only changes required in these solutions are due to the fact that $\alpha(q_2)$ may be more complex than for uniform streaming flows. For example, a qualitative sketch of the flow structure for a nonrotating cylinder in simple shear flow at low Reynolds number is shown in Fig. 9–16.[23] It is evident in this case that there are four stagnation points on the cylinder surface rather than two, as in the streaming-flow problem. Two of the streamlines that lead to the stagnation points A and C are lines of inflow, and two from B and D are lines of outflow, where we should expect a thin thermal wake. In the limit as these outflow points are approached, we thus expect a breakdown of the similarity solution with $g \to \infty$. At the inflow stagnation points, on the other hand, we require that g be finite. To accommodate

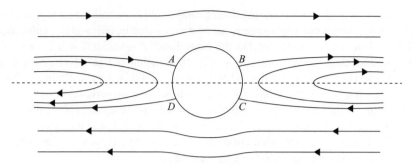

Figure 9–16. A qualitative sketch of the streamlines near a nonrotating circular cylinder in a simple 2D shear flow.

these changes in the form of α, we can define $q_2 = q_2^0 = 0$ in $g(q_2)$ as being coincident with point A in Fig. 9–16. Then, g is defined by (9–266) for the interval between A and B with $q_2 > 0$, and as

$$g^3 = -\frac{9}{(\alpha)^{3/2}} \int_0^{q_2} (\alpha)^{(1/2)} dt \quad \text{with} \quad \alpha \equiv -\frac{\partial u_2}{\partial q_1} \qquad (9\text{–}267)$$

in the interval A to D with $q_2 < 0$. At B and D, $g \to \infty$. The other regions of thermal boundary-layer structure between B and C and C and D do not need to be treated explicitly because of the symmetry of the problem. Among the problems at the end of this chapter, additional examples are posed involving nonuniform flows for which the thermal boundary-layer analysis can be applied.

K. HEAT (OR MASS) TRANSFER ACROSS A FLUID INTERFACE FOR LARGE PECLET NUMBERS

The correlation (9–230) was shown to be valid for heat transfer from solid bodies of arbitrary shape in a variety of arbitrary undisturbed flows, subject only to the condition that the body neither rotate nor be placed in an undisturbed flow that has closed streamlines at infinity.

Another question that the reader may ask is whether the restriction to solid bodies is really necessary for the validity of (9–230). In particular, an important problem closely related to the results in the preceding sections is the transport of heat (or a solute) across a fluid–fluid interface in the regime of large Peclet numbers. We consider this question in two stages. First, we consider some general principles, and second we discuss the specific problem of transport from a rising bubble or drop, which is closely related to the problem discussed in Section H.

1. General Principles

Let us begin by discussing the problem of heat or mass transport across a fluid interface from a general point of view. The basic message from the preceding several sections is that the transport process at large Peclet number is dominated by the nature of the velocity field in the immediate vicinity of the surface. When we consider a fluid interface, the boundary conditions change from a no-slip condition in which the tangential velocity component goes to zero in a frame of reference that moves with the surface (i.e., with the body) to the conditions of continuous tangential velocity and continuous tangential stress. With the exception of interfaces in which tangential motion is completely inhibited by the presence of Marangoni stresses, this means that the tangential velocity at the interface is nonzero. This means that the local approximations for the velocity components (9–249) and (9–250)

K. Heat (or Mass) Transfer Across a Fluid Interface for Large Peclet Numbers

must be modified, and, in view of the role that these local approximations play in the scaling process that leads to the thermal boundary-layer equations, we must expect both the scaling (9–253) and the resultant correlation for the transport rate (9–255) to change.

The consequences of this change can be explored first in rather general terms without the need for reference to a specific problem. To see the general situation, it is sufficient to think in terms of a local 2D Cartesian coordinate system. The resulting analysis will be applied directly only to a 2D problem. However, as we have seen in the preceding sections, the same qualitative result will be obtained for axisymmetric or even fully 3D problems. In the simplest view, the only difference between transport across a fluid interface and previous problems is in the Taylor series approximations for the velocity components (u, v). For convenience we assume that the local coordinate system is defined so the interface corresponds to $y = 0$. Because the first nonzero term for the tangential component is the slip velocity, the Taylor series approximation then takes the form,

$$u \sim u|_{y=0} + O(y). \qquad (9\text{--}268)$$

If we consider transport in the liquid at a gas–liquid interface (for example for transport from a bubble to a surrounding liquid), the shear stress is actually zero in the liquid and the first nonzero correction will be $O(y^2)$ rather than $O(y)$.

It then follows from (9–268) and the continuity equation that the normal velocity component must be linear in y, that is,

$$v \sim -\frac{\partial}{\partial x}(u_{y=0})y + O(y^2). \qquad (9\text{--}269)$$

Again, at a liquid–gas interface, the first nonzero correction in (9–269) will be $O(y^3)$ rather than $O(y^2)$. These changes in the local form of the velocity components, relative to their form at a no-slip boundary, have a profound influence on the structure of the thermal boundary layer.

To see this, we may first express the thermal energy equation (9–7) in terms of the local coordinates (x, y) and introduce the approximate forms, (9–268) and (9–269), for u and v at small values of y:

$$[u|_{y=0} + O(y)]\frac{\partial \theta}{\partial x} + \left[-\frac{\partial}{\partial x}(u_{y=0})y + O(y^2)\right]\frac{\partial \theta}{\partial y} = \left(\frac{\partial^2 \theta}{\partial x^2} + \frac{\partial^2 \theta}{\partial y^2}\right)\frac{1}{Pe}. \qquad (9\text{--}270)$$

Then, introducing a rescaling of the general form

$$Y = yPe^m, \qquad (9\text{--}271)$$

we obtain

$$u_s(x)\frac{\partial \theta}{\partial x} - \frac{du_s}{dx}Y\frac{\partial \theta}{\partial Y} = Pe^{2m-1}\frac{\partial^2 \theta}{\partial Y^2} + O(Pe^{-m}, Pe^{-1}), \qquad (9\text{--}272)$$

where we have denoted $u|_{y=0}$ as $u_s(x)$. Hence, in this case, the balance between conduction and convection in the limit $Pe \to \infty$ requires that

$$m = \frac{1}{2}. \qquad (9\text{--}273)$$

It follows that the thickness of the thermal boundary layer is $O(Pe^{-1/2})$ rather than $O(Pe^{-1/3})$, and the correlation between Nu and Pe takes the general form

$$Nu \sim cPe^{1/2}. \qquad (9\text{-}274)$$

We may note that the thermal boundary layer in this case is asymptotically thin relative to the boundary layer for a solid body. This is a consequence of the fact that the tangential velocity near the surface is larger, and hence convection is relatively more efficient. From a simplistic point of view, the larger velocity means that convection parallel to the surface is more efficient, and hence the time available for conduction (or diffusion) normal to the surface is reduced. Thus, the dimension of the fluid region that is heated (or within which solute resides) is also reduced. Indeed, if we define Pe by using a characteristic length scale ℓ_c and a characteristic velocity scale u_c, heat can be conducted a distance

$$\ell^* = \sqrt{\kappa t} = \ell_c Pe^{-1/2} \qquad (9\text{-}275)$$

in a time period ℓ_c/u_c (or solute can be diffused over a distance $\ell^* = \sqrt{Dt} = \ell_c Pe^{-1/2}$, where D is the diffusivity and $Pe \equiv ReSc$ in the case of mass transfer)

The critical point is that the same scaling and the same correlation (9–274) will apply for any system in which there is heat or mass transfer across an interface at high Pe. The only thing that will change with the geometry of the boundary or with the viscosity ratio and/or the ratio of thermal or solute diffusivities is the $O(1)$ coefficient c.

2. Mass Transfer from a Rising Bubble or Drop in a Quiescent Fluid

An example for which the analysis of the preceding section will apply qualitatively is to the heat or mass transfer from a bubble or drop that rises (or falls) through a quiescent fluid at low Re and large Pe. Especially for the case of a gas bubble, the most natural problem to solve in this case is the mass transfer of a *solute* to the surrounding fluid.

We have already noted that *mass transfer* in a liquid is almost always characterized by large values of the Peclet number (the Peclet number for mass transfer involves the product of the Schmidt number and Reynolds number instead of the Prandtl number and Reynolds number) and that the dimensionless form of the convection–diffusion equation governing transport of a single solute through a solvent is still (9–7), with θ now being a dimensionless solute concentration. For transfer of a solute from a bubble or drop into a liquid that previously contained no solute, the concentration θ at large distances from the bubble or drop will satisfy the condition

$$\theta \to 0 \quad \text{as} \quad |\mathbf{r}| \to \infty. \qquad (9\text{-}276)$$

The condition at the bubble or drop surface (that is, within the liquid in the limit as we approach the surface) can be quite complicated as was already pointed out in Section A of this chapter. Here, for simplicity, we initially confine our attention to a bubble and suppose that the solute concentration inside the bubble remains constant (though this assumption will clearly break down eventually if we assume that there is no source for production of additional solute inside the bubble). We will further assume that the bubble volume remains constant in the time interval of interest (again, the corresponding condition $\mathbf{u} \cdot \mathbf{n} = 0$ at the surface will never be precisely true, as we have seen in Section A). Finally, we assume that the solute concentration on the liquid side of the interface is a constant c_0 that can, in principle, be related to the constant concentration inside the bubble by means of the condition of

K. Heat (or Mass) Transfer Across a Fluid Interface for Large Peclet Numbers

thermodynamic equilibrium (by means of Henry's law) that is generally satisfied locally across the interface. Hence, in this case,

$$\theta \equiv \frac{c}{c_0}, \tag{9-277}$$

and the mass transfer problem has the same dimensionless form

$$\mathbf{u} \cdot \nabla \theta = \frac{1}{Pe} \nabla^2 \theta, \tag{9-278}$$

$$\theta = 1 \quad \text{at } S,$$

$$\theta \to 0 \quad \text{at } \infty, \tag{9-279}$$

as solved in Section H for heat transfer from solid bodies.

Of course the feature that differs in this case is the form of the velocity field $\mathbf{u}$. For simple translation of a bubble through a quiescent fluid (that is, the uniform-flow problem) at zero Reynolds number, this is solely a consequence of the change from no-slip conditions for a solid body to the condition of vanishing tangential stress at the surface of a clean bubble (recall that the shape remains spherical for $Re \ll 1$; see Chap. 7, Section H). It is this change that leads to the form of the correlation (9–274).

In the remainder of this section, we evaluate the coefficient c in (9–274) for the specific case of a translating gas bubble. In this case, the full creeping-flow solution for the velocity field is

$$u_r = \left(1 - \frac{1}{r}\right) \cos\theta, \quad u_\theta = -\left(1 - \frac{1}{2r}\right) \sin\theta, \quad u_\phi = 0. \tag{9-280}$$

Thus, transforming from r to y according to (9–213) and expanding u_r and u_θ about $y = 0$, we find that

$$u_r \sim (\cos\theta)y + 0(y^2) = (\cos x)y + 0(y^2), \tag{9-281}$$

$$u_\theta \sim -\frac{1}{2}\sin\theta + 0(y) = -\frac{1}{2}\sin x + 0(y), \tag{9-282}$$

where $x = \theta$. It may be noted that the relationship between u_r and u_θ is not in accord with (9–268) and (9–269) because the geometry in this case is axisymmetric rather than 2D. With the approximate forms (9–281) and (9–282), the thermal boundary-layer equation takes the specific form

$$(\cos x) Y \frac{\partial \theta}{\partial Y} - \frac{1}{2} \sin x \frac{\partial \theta}{\partial x} = \frac{\partial^2 \theta}{\partial Y^2}. \tag{9-283}$$

To solve this equation, it is convenient to introduce the transformation $\eta = \cos x$:

$$\eta Y \frac{\partial \theta}{\partial Y} + \frac{1}{2}(1 - \eta^2) \frac{\partial \theta}{\partial \eta} = \frac{\partial^2 \theta}{\partial Y^2}. \tag{9-284}$$

Convection Effects in Low-Reynolds-Number Flows

Then, following the example of previous problems in this chapter, we seek a similarity solution of the form

$$\theta = \theta(\xi), \quad \text{with} \quad \xi \equiv \frac{Y}{g(\eta)}. \tag{9-285}$$

Introducing (9–285) into (9–284), we find that

$$\left[\eta g^2 - \frac{1}{2}(1-\eta^2) gg'\right] \xi \frac{d\theta}{d\xi} = \frac{d^2\theta}{d\xi^2}. \tag{9-286}$$

Thus, if a similarity solution exists, the coefficient in square brackets must be a nonzero constant that we choose, for convenience, as -2. It follows that

$$\boxed{\frac{d^2\theta}{d\xi^2} + 2\xi \frac{d\theta}{d\xi} = 0,} \tag{9-287}$$

with the boundary conditions

$$\begin{aligned}\theta = 1 & \quad \text{at} \quad \xi = 0, \\ \theta \to 0 & \quad \text{as} \quad \xi \to \infty.\end{aligned} \tag{9-288}$$

A solution also must exist for $g(\eta)$ satisfying

$$\boxed{\frac{1}{4}(1-\eta^2) \frac{dg^2}{d\eta} - \eta g^2 = 2,} \tag{9-289}$$

with $g(\eta)$ being finite for all η except possibly $\eta = 1$, where a thermal wake may invalidate the boundary-layer assumptions.

It is a simple matter to solve (9–287), subject to (9–288). The result is

$$\boxed{\theta = 1 - \frac{\int_0^{\xi} e^{-t^2} dt}{\int_0^{\infty} e^{-t^2} dt} = 1 - \frac{2}{\sqrt{\pi}} \int_0^{\xi} e^{-t^2} dt.} \tag{9-290}$$

A general solution of (9–289) is also straightforward, namely,

$$g = \frac{c_1}{1-\eta^2} + \frac{2\sqrt{2}}{(1-\eta^2)} \left[\int_{-1}^{\eta} (1-s^2) ds\right]^{1/2}. \tag{9-291}$$

This solution is finite for all η, except for the rear-stagnation point at $\eta = 1$, provided $c_1 = 0$. Thus,

$$\boxed{g(\eta) = \frac{2\sqrt{2}}{(1-\eta^2)} \left[\int_{-1}^{\eta} (1-s^2) ds\right]^{1/2}.} \tag{9-292}$$

Equation (9–274) can now be interpreted as a relationship between the Peclet number and the Sherwood number, and the constant c in (9–274) can be calculated from the definition

$$c \equiv \int_{-1}^{1} -\left(\frac{\partial \theta}{\partial Y}\right)_0 d\eta, \tag{9-293}$$

L. Heat Transfer at High Peclet Number Across Regions of Closed-Streamline Flow

where

$$\left(\frac{\partial \theta}{\partial Y}\right)_0 = -\frac{2}{\sqrt{\pi}} \frac{1}{g(\eta)}. \tag{9-294}$$

When the indicated integration, is carried out, the result is

$$c = \frac{4}{\sqrt{6\pi}} = 0.9213. \tag{9-295}$$

L. HEAT TRANSFER AT HIGH PECLET NUMBER ACROSS REGIONS OF CLOSED-STREAMLINE FLOW

1. General Principles

In the five preceding sections of this chapter we considered heat (or mass) transfer at high Peclet number and low Reynolds number for a variety of circumstances in which the heated fluid region is confined to a thin thermal boundary layer within $O(Pe^{-1/3})$ or $O(Pe^{-1/2})$ of the body surface. This limited radial extent of heated fluid is a consequence of the fact that convection is an intrinsically more efficient process than conduction for large Pe, in conjunction with the open streamline structure of the flow, which means that heat, initially transferred to the fluid at the body surface by conduction, can progress only a very short distance out from the surface before it is swept downstream around the body by convection and into the thermal wake. Because the thermal region is so thin in these cases, the temperature gradients at the body surface are large and the Nusselt number increases as $Pe^{1/3}$ or $Pe^{1/2}$.

It was already indicated in the preceding sections that this thermal boundary-layer structure does *not* occur when a particle (or body) is entirely surrounded by closed streamlines (or closed stream surfaces). In this case, the convection process near the body can no longer transfer heat directly to the streaming flow where it is carried into the wake, but instead circulates it only in a closed path around the body. Thus the heat transfer process is fundamentally altered, because heat can escape from the body only by *diffusing* slowly across the region of closed streamlines (or stream surfaces). Because the size of this region is independent of Pe, the steady-state temperature gradients will be $O(1)$, and we expect that

$$\boxed{Nu \sim O(1) \quad \text{as} \quad Pe \to \infty.} \tag{9-296}$$

Assuming that this result is correct, it is very important because comparison with (9-230) or (9-274) shows that the rate of heat transfer from a particle (or heated body) to a surrounding fluid at high Peclet numbers depends critically on whether the streamlines (or stream surfaces) near the heated surface are open or closed.

In this section, we shall show how the temperature distribution can be calculated for heat transfer at high Pe across regions of closed-streamline motion. However, the most important goal for the reader is to have a clear, qualitative understanding of the fundamental difference between open- and closed-streamline flows for convective heat transfer at high Peclet number. In the preceding section, we gave a "derivation" of the fact that the dimension of the thermal boundary layer will be $O(Pe^{-1/2})$ if the convection process is characterized by the macroscopic characteristic velocity scale u_c [Eq. (9-275)]. For a solid body, on the other hand, the actual velocity in the region of interest is smaller,

$$O\left(u_c \left(\frac{\ell^*}{\ell_c}\right)\right),$$

where ℓ^* is the (dimensional) thermal boundary-layer thickness. As a consequence, the characteristic time available for conduction is longer than for an interface where the relevant velocity is u_c,

$$t^* = \frac{\ell_c^2}{u_c \ell^*}, \qquad (9\text{-}297)$$

and the predicted thickness of the thermal layer, nondimensionalized by the particle scale a, becomes

$$\frac{\ell^*}{\ell_c} \sim (Pe)^{-1/3}, \qquad (9\text{-}298)$$

that is, the thermal boundary-layer thickness is increased relative to (9-275) because the available time for conduction of heat outward from the body surface is increased.

Now, we may consider how these time scale arguments change when heat is transferred across a region of *closed* streamlines or stream surfaces, still at large Peclet number. In this case, the convection process circulates heat only around a closed path and thus does not carry the heated fluid away from the body surface. As a consequence, the restriction on available time for conduction of heat outward from the body is no longer relevant, and the conduction process, though relatively inefficient for $Pe \gg 1$, has an indefinite time to reach a steady-state configuration in which the whole of the closed-streamline (or stream surface) region is heated. Assuming that the size of this region is $O(1)$ relative to the heated body, it follows that the dimensionless temperature gradient at steady state must be $O(1)$, and the Nusselt number will approach an $O(1)$ value for $Pe \to \infty$, as indicated in Eq. (9-296). Indeed, at steady state, the *whole* of the temperature decrease from the surface value $\theta = 1$ to the free-stream value $\theta = 0$ must occur across the recirculating flow region. If this were not true, the temperature on the outermost closed streamline (or stream surface) would be $O(1)$, though obviously smaller than 1, and there would need to be a thermal boundary layer on this surface of the type described in the preceding section. But then, the heat flux from the closed-streamline region to the bulk fluid stream across this thermal boundary layer would be $O(Pe^{1/2})$, and this obviously could not be consistent with a temperature gradient and heat flux of $O(1)$ across the closed-streamline region. Thus we conclude that $\theta \to 0$ on the outermost closed streamline (or stream surface) of the recirculating zone. The qualitative arguments presented here clearly indicate that, for $Pe \gg 1$, the rate of heat transfer from a particle or heated body to a surrounding fluid depends critically on whether the streamlines near the heated surface are open or closed.

2. Heat Transfer from a Rotating Cylinder in Simple Shear Flow

In the remainder of this section, we consider the details of solution for a particular problem of a heated body surrounded by a region of closed streamlines. The problem that is closest to that analyzed in previous sections of this chapter is a heated sphere in a flow where the sphere rotates [such as the general linear flows of Eq. (9-256) with $\lambda \neq 1$] and is thus surrounded by a closed region of recirculating fluid. In fact, however, the only problem of this general type that has been worked out in detail is the sphere in a simple shear flow ($\lambda = 0$) at low Reynolds number,[24] and even in that case, the flow is fully 3D, the geometry of the recirculating region is complicated, and the analysis is quite difficult in detail. As an alternative, we consider a problem that retains the essential features of the heated sphere but is simpler to solve. This is the 2D analogue of a freely rotating, heated circular cylinder in a simple shear flow.[25] Following the detailed analysis for the circular cylinder, we shall briefly return to the results for a sphere in shear flow.

L. Heat Transfer at High Peclet Number Across Regions of Closed-Streamline Flow

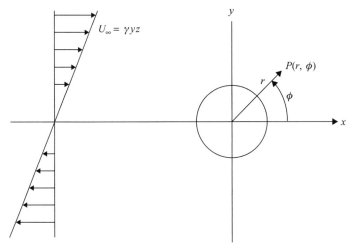

Figure 9–17. The coordinate axes and undisturbed flow for a cylinder in an unbounded simple shear flow.

To begin our analysis of the cylinder problem, it is necessary to consider briefly the velocity field. The fluid mechanics solution for a circular cylinder rotating with an imposed angular velocity Ω in a simple shear flow was given originally by Bretherton[26] for the creeping-flow limit. In this case, the velocity field can be specified in terms of a streamfunction (see Chap. 7) such that

$$u_r \equiv \frac{1}{r}\frac{\partial \psi}{\partial \phi}, \quad u_\theta \equiv -\frac{\partial \psi}{\partial r}. \tag{9-299}$$

A sketch of the physical problem and a definition of the cylindrical polar coordinates (r, ϕ) are given in Fig. 9–17. For the creeping-flow limit, the governing equations and boundary conditions are

$$\nabla^4 \psi = 0, \tag{9-300}$$

with

$$\psi = 0, \quad \frac{\partial \psi}{\partial r} = \Omega \quad \text{at} \quad r = 1,$$

$$\psi \to \frac{1}{2}y^2 = \frac{1}{2}r^2 \sin^2 \phi \quad \text{as} \quad r \to \infty. \tag{9-301}$$

Here, the velocities are nondimensionalized with respect to Ga, where G is the shear rate of the undisturbed flow and a is the cylinder radius. The Bretherton solution of (9–300) and (9–301) is

$$\psi = \frac{1}{2}y^2 - \frac{1}{4}\left[2(1 - 2\Omega)\ln r + 1 + \left(\frac{1}{r^2} - 2\right)\cos 2\phi\right], \tag{9-302}$$

and the corresponding torque on the cylinder is

$$T = -2\pi \left(\frac{v}{Ga^2}\right)(1 - 2\Omega). \tag{9-303}$$

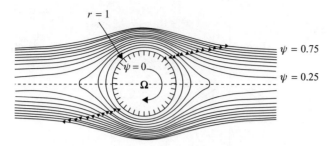

Figure 9–18. Streamlines for a freely rotating circular cylinder in simple, linear shear flow (9–306). Contours 0 to 0.75 in increments of 1/16.

In the present case, there is no externally applied torque on the cylinder (i.e., it is free to rotate in the flow). Thus the hydrodynamic torque must also vanish at steady state and

$$\Omega = \frac{1}{2}. \qquad (9\text{--}304)$$

It follows that

$$\psi = \frac{1}{2}y^2 - \frac{1}{4}\left[1 + \left(\frac{1}{r^2} - 2\right)\cos 2\phi\right]. \qquad (9\text{--}305)$$

Alternatively,

$$\psi = \frac{1}{4}(r^2 - 1) - \frac{1}{4}(r^2 - 2 + r^{-2})\cos 2\phi. \qquad (9\text{--}306)$$

A full asymptotic analysis by Robertson and Acrivos[23] has shown that this solution of the creeping-motion equations is a uniformly valid first approximation for $Re \ll 1$.

A plot of the streamlines (9–306) is shown in Fig. 9–18. It is evident that all streamlines for $\psi < 1/4$ are closed, whereas those for $\psi > 1/4$ are open. The dividing streamline $\psi = 1/4$ can be thought of as "closing at infinity." The fact that $\psi = 1/4$ is the critical value can be seen by means of the following argument. First, if we evaluate ψ at $\phi = 0$ or π, we find that

$$\psi|_{\phi=0,\pi} = \frac{1}{4}\left(1 - \frac{1}{r^2}\right). \qquad (9\text{--}307)$$

Thus, $\psi \to 1/4$ as $r \to \infty$. On the other hand, for $\phi = \pi/2$,

$$\psi|_{\pi/2} = \frac{r^2}{2} - \frac{3}{4} + \frac{1}{4r^2}, \qquad (9\text{--}308)$$

and on this axis, $\psi = 1/4$ at $r \sim 1.32$. Thus streamlines for $\psi > 1/4$ cross the y axis but do not cross the x axis, whereas those for $\psi < 1/4$ (including $\psi = 0$) cross both axes.

Now let us turn again to the thermal energy equation,

$$\mathbf{u} \cdot \nabla \theta = \frac{1}{Pe}\nabla^2 \theta, \qquad (9\text{--}309)$$

L. Heat Transfer at High Peclet Number Across Regions of Closed-Streamline Flow

which we must solve to determine the temperature distribution in the fluid. The boundary conditions are

$$\boxed{\begin{aligned} \theta &= 1 \quad \text{at } r = 1, \\ \theta &\to 0 \quad \text{at } r \to \infty. \end{aligned}} \qquad (9\text{--}310)$$

In writing (9–309), we have used Ga as the characteristic velocity, so that the Peclet number is

$$Pe \equiv \frac{Ga^2}{\kappa}.$$

We seek an approximate solution in the form of an asymptotic expansion for the limit $Pe \to \infty$.

To obtain the governing equation for the first term in this expansion, we take the limit $Pe \to \infty$ in (9–309). The result is

$$\mathbf{u} \cdot \nabla \theta_0 = 0, \qquad (9\text{--}311)$$

and we have already seen that a general solution of this equation is

$$\theta_0 = \theta_0(\psi). \qquad (9\text{--}312)$$

In an open-streamline flow, the functional dependence of temperature on ψ can be obtained from the known dependence at ∞. This can be used, in the present case, to see that

$$\theta_0 = 0 \qquad (9\text{--}313)$$

for $\psi \geq 1/4$. However, for the closed-streamline region, $\psi < 1/4$, (9–312) simply tells us that θ_0 is constant on any streamline but provides no basis to determine which value applies to which streamline. Generally, this apparent *indeterminacy* in the function $\theta_0(\psi)$ is characteristic of all problems involving closed streamlines or closed stream surfaces – we can say only that θ_0 is constant on streamlines for $Pe \gg 1$ but cannot determine the actual dependence of θ_0 on ψ without further analysis of some kind.

If we consider the qualitative description of heat transfer in closed-streamline regions presented in the preceding section, we may recognize that, to obtain the temperature distribution, we need to somehow include conduction effects because this is the primary mode of heat transfer *across* streamlines for $Pe \gg 1$, when the streamlines are closed. In the boundary-layer problems of the preceding sections of this chapter, the necessity for conduction signaled the need for rescaling to a different characteristic length scale and the development of a second distinct expansion with a new leading-order approximation. Here, on the other hand, the scaling and the general solution (9–312) are perfectly valid, and a different approach is necessary to see how the *very slow and inefficient* conduction process (acting over a very long period of time) can completely control the distribution of θ_0 within the closed-streamline region. For this purpose let us return to the full thermal energy equation in the time-dependent form that applies during the slow evolution of the temperature distribution toward a steady state,

$$\frac{\partial \theta}{\partial t} + \mathbf{u} \cdot \nabla \theta = \frac{1}{Pe} \nabla^2 \theta. \qquad (9\text{--}314)$$

Convection Effects in Low-Reynolds-Number Flows

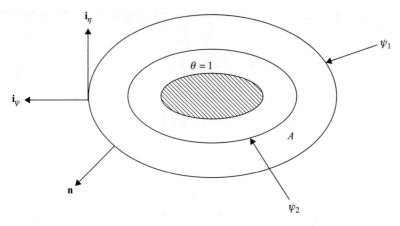

Figure 9–19. A sketch of a typical 2D closed-streamline configuration. Here, ψ_1 and ψ_2 are any two arbitrarily chosen closed streamlines. The area of the region between these streamlines is denoted as A. The unit normal is denoted as $\mathbf{n}$, and $\mathbf{i}_\psi$ and $\mathbf{i}_\eta$ are a pair of orthogonal unit vectors that are normal and tangential to the streamline (note that $\mathbf{i}_\psi \equiv \mathbf{n}$).

In writing (9–314), we have incorporated a characteristic time scale $t_c = a/U = G^{-1}$. This equation also can be written in the form

$$\frac{\partial \theta}{\partial t} = \operatorname{div}\left(-\mathbf{u}\theta + \frac{1}{Pe}\nabla\theta\right) \tag{9–315}$$

because div $\mathbf{u} \equiv 0$. Now suppose that we integrate (9–315) over area A between any two arbitrarily chosen closed streamlines, say, ψ_1 and ψ_2. A sketch of a typical configuration is shown in Fig. 9–19. Formally, this gives

$$\int_A \left(\frac{\partial \theta}{\partial t}\right) dA = \int_A \operatorname{div}\left(-\mathbf{u}\theta + \frac{1}{Pe}\nabla\theta\right) dA, \tag{9–316}$$

which is nothing more than a macroscopic heat balance for the fluid region bounded by ψ_1 and ψ_2. The significance of the right-hand side of this balance is best exposed by application of the divergence theorem. This gives

$$\int_A \left(\frac{\partial \theta}{\partial t}\right) dA = \int_{\psi_1} \left(-\mathbf{u}\theta + \frac{1}{Pe}\nabla\theta\right) \cdot \mathbf{n}\, dl - \int_{\psi_2} \left(-\mathbf{u}\theta + \frac{1}{Pe}\nabla\theta\right) \cdot \mathbf{n}\, dl \tag{9–317}$$

where

$$\mathbf{n} = \mathbf{i}_\psi$$

is a unit vector that is orthogonal to the streamlines. However, by definition,

$$\mathbf{u} \cdot \mathbf{n} = 0 \tag{9–318}$$

on any streamline. Hence (9–317) reduces to

$$\int_A \left(\frac{\partial \theta}{\partial t}\right) dA = \int_{\psi_1} \frac{1}{Pe}\nabla\theta \cdot \mathbf{n}\, dl - \int_{\psi_2} \frac{1}{Pe}\nabla\theta \cdot \mathbf{n}\, dl. \tag{9–319}$$

We see that, in the transient period before establishing a steady state, heat accumulates in the region between any two arbitrarily chosen closed streamlines solely because of conduction of heat across the streamlines. In addition, in the limit $Pe \gg 1$, it is evident that the rate of change of the temperature distribution is very slow. Once $|\nabla\theta| \sim O(1)$, the left-hand side

L. Heat Transfer at High Peclet Number Across Regions of Closed-Streamline Flow

of (9–319) changes on a dimensionless time scale $\Delta t \sim O(Pe)$ only. Eventually, however, steady state will be established, and it is this steady-state temperature distribution that we seek.

Clearly, whatever the final steady-state solution may be, it must satisfy the macroscopic condition (9–319), with the left-hand side set equal to zero. Because ψ_1 and ψ_2 are completely arbitrary, the resulting steady-state condition can be expressed in the form

$$\frac{1}{Pe} \int_\psi \nabla\theta \cdot \mathbf{n}\, dl = C, \tag{9–320}$$

where C is a constant, independent of ψ. Physically, this condition states that the total heat flux across any streamline that is due to conduction must be independent of the particular streamline that we choose. Because one possible choice for a closed streamline in the present case is $\psi = 0$ coincident with the cylinder surface, we see that

$$C = \frac{1}{Pe} \int_0^{2\pi} \frac{\partial \theta}{\partial r}\bigg|_{r=1} d\phi. \tag{9–321}$$

But this integral is just the negative of the dimensionless heat flux per unit length of the cylinder, namely,

$$Nu = -\frac{1}{\pi} \int_0^{2\pi} \frac{\partial \theta}{\partial r}\bigg|_{r=1} d\phi. \tag{9–322}$$

Thus

$$\boxed{C = -\frac{Nu\,\pi}{Pe},} \tag{9–323}$$

and we can immediately determine the Nusselt number once C is known.

To see whether the condition (9–320) provides the added input that is required for determining $\theta_0(\psi)$ in the closed-streamline region, it is convenient to express it in terms of the independent variables ψ and η (which we choose as lines orthogonal to ψ) instead of cylindrical coordinates (r, ϕ). To do this, we need only introduce the appropriate scale or metric factors for the (ψ, η) coordinate system, h_ψ and h_η. The reader is reminded that these scale factors are defined such that the length of a differential line element, expressed in the (ξ, η) system, takes the form

$$(ds)^2 \equiv h_\psi^{-2}(d\psi)^2 + h_\eta^{-2}(d\eta)^2. \tag{9–324}$$

Equivalently,

$$h_\psi^{-2} \equiv \left(\frac{\partial x}{\partial \psi}\right)^2 + \left(\frac{\partial y}{\partial \psi}\right)^2, \quad h_\eta^{-2} \equiv \left(\frac{\partial x}{\partial \eta}\right)^2 + \left(\frac{\partial y}{\partial \eta}\right)^2. \tag{9–325}$$

Now,

$$\nabla\theta \cdot \mathbf{n} = h_\psi \frac{\partial \theta}{\partial \psi}, \tag{9–326}$$

$$d\ell = \frac{1}{h_\eta} d\eta. \tag{9–327}$$

Thus (9–320) can be written in the form

$$\boxed{\frac{1}{Pe} \int_\psi \frac{h_\psi}{h_\eta} \frac{\partial \theta}{\partial \psi} d\eta = C.} \tag{9–328}$$

Convection Effects in Low-Reynolds-Number Flows

However, according to (9–312), $\theta \sim \theta_0(\psi)$ is a function of ψ only. Thus,

$$\frac{1}{Pe}\left(\int_\psi \frac{h_\psi}{h_\eta}d\eta\right)\frac{d\theta_0}{d\psi} = C,$$

where the coefficient in parenthesis depends on ψ only. Let us denote this coefficient as $\Gamma(\psi)$. Then

$$\boxed{\frac{1}{Pe}\Gamma(\psi)\frac{d\theta_0}{d\psi} = C.} \qquad (9\text{–}329)$$

On integration and applying the boundary condition $\theta_0 = 1$ at the cylinder surface, $\psi = 0$, we find that

$$\theta_0 = 1 + PeC\int_0^\psi \frac{ds}{\Gamma(s)}. \qquad (9\text{–}330)$$

The function $\Gamma(\psi)$ depends on the form of the stream-function distribution and can be calculated once the flow is specified.

The solution (9–330) applies within the closed-streamline region. If we assume that it applies right up to the separating streamline $\psi = 1/4$ (as is true at the leading order of approximation), then the constant C can be determined from the second condition

$$\theta_0 = 0 \quad \text{at} \quad \psi = 1/4. \qquad (9\text{–}331)$$

In this case,

$$C = -\frac{1}{Pe\int_0^{1/4}\frac{ds}{\Gamma(s)}}, \qquad (9\text{–}332)$$

$$\boxed{\theta_0 = 1 - \frac{\int_0^\psi \frac{ds}{\Gamma(s)}}{\int_0^{1/4}\frac{ds}{\Gamma(s)}}.} \qquad (9\text{–}333)$$

In addition, we see from (9–323) that

$$\boxed{Nu_0 = \frac{1}{\pi\int_0^{1/4}\frac{ds}{\Gamma(s)}}.} \qquad (9\text{–}334)$$

Numerical evaluation of the integral in (9–334), using the solution (9–306) for ψ and the definitions (9–325) for h_ψ, and h_η to obtain $\Gamma(\psi)$, yields

$$\boxed{Nu_0 = 5.73.} \qquad (9\text{–}335)$$

L. Heat Transfer at High Peclet Number Across Regions of Closed-Streamline Flow

This is the primary result of the present calculation and confirms the general prediction that

$$Nu \to O(1) \quad \text{as} \quad Pe \to \infty$$

for heat transfer from a body that is embedded within a region of closed-streamline flow.

The reader may note that the solution (9–333) for θ_0 approaches the same numerical value for $\psi \to 1/4$ as does the first-order approximation (9–313) for θ_0 in the region $\psi > 1/4$. It thus may seem that we have constructed a uniformly valid solution for the first term in a regular perturbation expansion for $Pe \gg 1$. Unfortunately, this is not the case. The governing equation for θ_0, (9–311), assumes that convection is dominant over conduction. But a careful analysis of the magnitude of conduction and convection terms, based on the leading-order solution (9–333) for θ_0, indicates that there is a region arbitrarily close to $\psi = 1/4$ where conduction is not small compared with convection as assumed. Thus the limit $Pe \to \infty$ is singular, and a complete solution for $Pe \gg 1$ would require use of the method of matched asymptotic expansions. However, the "correction" necessary in the vicinity of $\psi = 1/4$ does not change (9–333) in the closed-streamline region nor the resulting estimate for the Nusselt number Nu_0. Indeed, the asymptotic solution in this case is analogous to Stokes' solution for flow past a sphere in the sense that the singular nature of the limit $Pe \gg 1$ does not appear explicitly in the leading-order approximation for θ but only when we attempt to generate higher-order approximations to the solution. We are content here with the leading-order estimates (9–313) and (9–333).

Before concluding the discussion of high-Peclet-number heat transfer in low-Reynolds-number flows across regions of closed streamlines (or stream surfaces), let us return briefly to the problem of heat transfer from a *sphere* in simple shear flow. This problem is qualitatively similar to the 2D problem that we have just analyzed, and the physical phenomena are essentially identical. However, the details are much more complicated. The problem has been solved by Acrivos,[24] and the interested reader may wish to refer to his paper for a complete description of the analysis. Here, only the solution and a few comments are offered. The primary difficulty is that an integral condition, similar to (9–320), which can be derived for the net heat transfer across an arbitrary isothermal stream surface, does not lead to any useful quantitative results for the temperature distribution because, in contrast with the 2D case in which the isotherms correspond to streamlines, the location of these stream surfaces is *a priori* unknown. To resolve this problem, Acrivos shows that the more general steady-state condition,

$$\int_A \nabla^2 \theta \, dA = 0, \qquad (9\text{–}336)$$

must be used, which is obtained from (9–316) by application of the divergence theorem to only the first term on the right-hand side. After considerable effort, beginning with (9–312) and (9–336), Acrivos obtained the result

$$\boxed{Nu = 8.9 \quad \text{for} \quad Pe \to \infty.}$$

This corresponds qualitatively to the 2D result (9–335).

It is important to recognize that the analysis presented in this section is generally applicable to any high-Peclet-number heat transfer process that takes place across a region of closed-streamline flow. In particular, the limitation to small Reynolds number is not an intrinsic requirement for any of the development from Eq. (9–309) to Eq. (9–334). It is only in the specification of a particular form for the function $\Gamma(\psi)$ that we require an analytic solution for ψ and thus restrict our attention to the creeping-flow limit. Indeed, all of (9–309)–(9–334) apply for any closed-streamline flow at any Reynolds number, provided only that $Pe \gg 1$. The same is true of the qualitative discussion of the heat transfer mechanism for closed-streamline (or stream surface) flows at the beginning of the section.

NOTES

1. A detailed discussion of this approximation and the asymptotic conditions required for it to be accepted may be found in Chap. 12, Section D, of the predecessor to this book: L. G. Leal, *Laminar Flow and Convective Transport Processes* (Butterworth-Heinemann, Boston, 1992).
   In general, it is sufficient that the Rayleigh number is asymptotically small, where the Rayleigh number is defined in terms of the characteristic velocity of the external flow, u_c, and the characteristic length scale of the flow, ℓ_c, as
   $$Ra \equiv \frac{\ell_c \, \beta g |T_0 - T_\infty|}{u_c^2} \ll 1.$$
   Here g is the gravitational acceleration constant, and β is the thermal expansion coefficient of the liquid.
2. Viscous dissipation effects can be important in a number of technological applications, either because velocity gradients are large, the fluid viscosity is large, or a combination of these. One class of problems of particular significance to chemical engineers occurs in many polymer processing systems. The interested reader may consult textbooks such as S. Middleman, *Fundamentals of Polymer Processing* (McGraw-Hill, New York, 1977); J. R. A. Pearson, *Mechanics of Polymer Processing* (McGraw-Hill [Hemisphere], Washington, 1983); D. G. Baird and D. I. Collias, *Polymer Processing: Principles and Design* (Wiley, New York, 1998).
3. R. B. Bird, W. E. Stewart, and E. N. Lightfoot, *Transport Phenomena* (2nd ed.) (Wiley, New York, 2001).
4. The original analysis of this problem was published by A. Acrivos and T. E. Taylor, Heat and mass transfer from single spheres in Stokes flow, *Phys. Fluids* **5**, 387–94 (1962).
5. S. Goldstein, The steady flow of viscous fluid past a fixed obstacle at small Reynolds numbers, *Proc. R. Soc. Ser. A*, **123**, 225–35 (1929).
6. A. N. Whitehead, Second approximation to viscous fluid motion, *Q. J. Math.* **23**, 143–52 (1889).
7. This oft-related story is easy to accept, as it required several decades before the German mathematician Oseen finally explained the Whitehead paradox and more than 60 years before the "modern" resolution was developed by means of the techniques of matched asymptotic expansions; it is clearly a rather drastic oversimplification. The young mathematician Whitehead is, in fact, Alfred North Whitehead, who became one of the most famous and accomplished British mathematicians and philosophers of his time. Together with his former student, Bertrand Russell, he authored the landmark three-volume *Principia Mathematica* (1910, 1912, 1913), and he contributed significantly to twentieth-century logic and metaphysics. He was elected a Fellow of the Royal Society in 1903 for his work on universal algebra, he was appointed to the professorial chair in applied mathematics at the Imperial College of Science and Technology in 1914, and then served as professor of philosophy at Harvard University from 1924 until he retired in 1937. He was awarded the Order of Merit in 1945.
8. S. Kaplan, Low Reynolds number flow past a circular cylinder, *J. Math. Mech.* **6**, 595–603 (1957); S. Kaplan and P. A. Lagerstrom, Asymptotic expansions of Navier–Stokes solutions for small Reynolds numbers, *J. Math. Mech.* **6**, 585–93 (1957). These and other published and previously unpublished works of Kaplan are reproduced in the following book: S. Kaplan, *Fluid Mechanics and Singular Perturbations* P. A. Lagerstrom, L. N. Howard, and C. S. Lin (eds.). (Academic, New York, 1957).
9. I. Proudman and J. R. A. Pearson, Expansions at small Reynolds numbers for the flow past a sphere and a circular cylinder, *J. Fluid Mech.* **2**, 237–62 (1957).
10. A reproduction of the work in this paper can be found in the textbook by C. W. Oseen, *Hydrodynamik* (Akademie Verlagsgellschatt, Leipzig, 1927).
11. W. Chester and D. R. Breach, On the flow past a sphere at low Reynolds number, *J. Fluid Mech.* **37**, 751–60 (1969).
12. H. Brenner, Forced convection heat and mass transfer at small Peclet numbers from a particle of arbitrary shape, *Chem. Eng. Sci.* **18**, 109–22 (1963).

Problems

13. D. E. Elrick, Source functions for diffusion in uniform shear flow, *Aus. J. Phys.* **15**, 283–88 (1962).
14. G. K. Batchelor, Mass transfer from a particle suspended in fluid with a steady linear ambient velocity distribution, *J. Fluid Mech.* **95**, 369–400 (1979).
15. A. Acrivos, A note on the rate of heat or mass transfer from a small particle freely suspended in a linear shear field, *J. Fluid Mech.* **98**, 299–304 (1980).
16. A. Acrivos and J. D. Goddard, Asymptotic expansions for laminar convection heat and mass transfer, *J. Fluid Mech.* **23**, 273–91 (1965).
17. See Appendix 2 in Ref. 16.
18. *Streamline* implies the existence of a streamfunction, which we have seen to be true only for axisymmetric and 2D flows. In three dimensions, recirculating flows are associated with regions of closed *stream surfaces*, or *pathlines*.
19. Among references that discuss closed-streamline patterns and eddies in low-Reynolds-number flows, the reader may wish to refer to Ref. 13, Chap. 7, and D. J. Jeffrey and J. D. Sherwood, Streamline patterns and eddies in low-Reynolds-number flow, *J. Fluid Mech.* **96**, 315–34 (1980); A. M. J. Davis and M. B. O'Neill, The development of viscous wakes in a Stokes flow when a particle is near a large obstacle, *Chem. Eng. Sci.* **32**, 899–906 (1977); A. M. J. Davis and M. B. O'Neill, Separation in a slow linear shear flow past a cylinder and a plane, *J. Fluid Mech.*, **81**, 551–64 (1977).
20. See Ref. 14, Chap. 7.
21. This latter restriction is necessary to preserve the ordering inherent in boundary-layer rescaling that requires $\partial/\partial q_1 \gg \partial/\partial q_2$.
22. J. Happel and H. Brenner, *Low Reynolds Number Hydrodynamics* (Noordhoff, Leyden, The Netherlands, 1973), appendix.
23. A general study of the streamlines for a circular cylinder in simple shear flow can be found in the following papers: C. R. Robertson and A. Acrivos, Low Reynolds number shear flow past a rotating circular cylinder, Part I, Momentum Transfer, *J. Fluid Mech.* **40**, 685–704 (1970); R. G. Cox, I. Y. Z. Zia, and S. G. Mason, Particle motions in sheared suspensions, 15. Streamlines around cylinders and spheres, *J. Colloid Interface Sci.* **27**, 7–18 (1968).
24. A. Acrivos, Heat transfer at high Peclet number from a small sphere freely rotating in a simple shear flow, *J. Fluid Mech.* **46**, 233–40 (1971).
25. N. A. Frankel and A. Acrivos, Heat and mass transfer from small spheres and cylinders freely suspended in shear flow, *Phys. Fluids* **11**, 1913–18 (1968).
26. F. P. Bretherton, Slow viscous motion around a cylinder in simple shear, *J. Fluid Mech.* **12**, 591–613 (1962).

PROBLEMS

Problem 9–1. Heat Conduction From a Sphere. We consider heat transfer from a spherical body into a surrounding fluid that is completely motionless. If the temperature is denoted as T, then the equation governing the steady-state temperature distribution is

$$\nabla^2 T = 0.$$

(a) Suppose the surface temperature of the sphere is constant with a value T_0 and the temperature of the fluid far from the sphere is T_∞ (also constant).

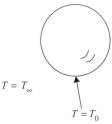

Show that $T = T_\infty + \frac{(T_0 - T_\infty)}{r} a$, where a is the sphere radius.

Convection Effects in Low-Reynolds-Number Flows

(b) Now suppose that the surface of the sphere is insulating so that

$$\frac{\partial T}{\partial r} = 0 \text{ as } r = a$$

and that there is a linear variation in the temperature of the fluid at infinity, that is,

$$T = T_\infty + \boldsymbol{\beta} \cdot \mathbf{x} \text{ as } r \to \infty,$$

where $\mathbf{x}$ is a general position vector with origin at the sphere center. Determine the temperature distribution in the fluid for an arbitrary vector $\boldsymbol{\beta}$.

(c) You should find that the disturbance produced by the sphere decreases as r^{-1} for case (a) and r^{-2} for case (b). Why does this happen?

(d) Finally, suppose that the sphere has a thermal conductivity λk in a medium with thermal conductivity k. In this case, determine the temperature field both outside and inside the particle when the temperature at infinity is $T = T_\infty + \boldsymbol{\beta} \cdot \mathbf{x}$. Also calculate the total heat flux from the sphere Q, where $Q = \int_{S_p} \mathbf{q} \cdot \mathbf{n} dS$.

Problem 9–2. Heat Conduction From a Cylinder to a Plane Wall Across a Thin Gap. Consider a heated cylinder whose centerline lies parallel to a cooled, infinite plane wall. The distance between the centerline of the cylinder and the wall is $a + h$, where a is the cylinder radius. We suppose that the surface of the cylinder is held at a constant prescribed temperature T_1 and the wall temperature is T_0, with $T_0 < T_1$. The fluid around the cylinder is assumed to remain motionless so that all heat transfer between the cylinder and wall occurs by conduction.

(a) Assume that $h \ll a$, and obtain an approximation by using lubrication ideas for the heat flux between the cylinder and wall.

(b) Repeat the calculation for a heated sphere, instead of a heated cylinder, and comment (or explain) any difference between the results for a cylinder and for a sphere.

Problem 9–3. Heat Conduction in a 2D Fin. Heat transfer from fins can often be analyzed by approximate transport equations. The idea is to exploit the geometry of the problem to reduce the calculational effort. Consider the 2D problem sketched in the figure.

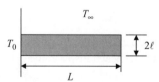

Definition sketch for fin problem.

(a) When the aspect ratio is large, $L/2\ell \gg 1$, derive the following 1D transport equation for the "lumped" temperature $\overline{T}(x)$:

$$\frac{d}{dx}\left(k\frac{d\overline{T}}{dx}\right) - \frac{h}{\ell}(\overline{T} - T_\infty) = 0, \tag{1}$$

$$\overline{T}(0) = T_0, \quad \overline{T}(L) = T_\infty. \tag{2}$$

You should derive this equation by performing a heat balance over a differential element of width Δx and height 2ℓ. Here, k is the thermal conductivity of the fin and h is a heat transfer coefficient between the fin and the surrounding fluid.

(b) An alternate derivation of the 1D model, which shows clearly the nature of the approximation, is to integrate the DE for the temperature over the height 2ℓ. By doing this, show that the same 1D model equation is obtained. How does the temperature $\overline{T}$ in (1) relate

Problems

to the temperature distribution in the differential form of the energy balance? Explain the appearance of the heat source term $h/\ell(\overline{T} - T_\infty)$ in (1).

(c) Obtain the solution to (1). Under what conditions do you expect the 1D lumped approximation to be valid? State your reasoning in terms of the dimensionless parameters hL/k and $2\ell/L$. [It may help to use the results of part (d).]

(d) Solve for the exact temperature field $T(x,y)$ by separation of variables. Discuss the limits of small and large Biot numbers ($Bi = hL/k$).

Problem 9–4. Matched Asymptotic Expansion. The following equation is a 1D analog of the Navier–Stokes equation:

$$\left(D^2 - \varepsilon^2 \frac{1}{r}\frac{\partial \psi}{\partial r}\right) D^2 \psi = 0,$$

where

$$D^2 \equiv \frac{\partial^2}{\partial r^2} - \frac{2}{r^2},$$

$$1 \le r < \infty.$$

Obtain an asymptotic solution of this equation for $\varepsilon \ll 1$, including terms of $O(\varepsilon^2)$, with the boundary conditions

$$\psi = \frac{\partial \psi}{\partial r} = 0 \quad \text{at } r = 1,$$

$$\psi - \frac{1}{2}r^2 \to 0 \quad \text{as } r \to \infty.$$

Note: The equation $(D^2 - \varepsilon^2)\phi$ has solutions

$$e^{-\varepsilon r}\left(1 + \frac{1}{\varepsilon r}\right) \quad \text{and} \quad e^{\varepsilon r}\left(1 - \frac{1}{\varepsilon r}\right).$$

Problem 9–5. Heat Transfer From a Freely Rotating Circular Cylinder in Shear Flow for $Pe \ll 1$. Let us consider the rate of heat transfer from a heated cylinder, whose central axis is perpendicular to the plane of a simple shear flow. The ambient temperature of the fluid is denoted as T_∞, and the surface of the cylinder is T_0. You may also assume that the velocity field near the cylinder is given in terms of the streamfunction as

$$\psi = \frac{1}{4}(r^2 - 1) - \frac{1}{4}(r^2 - 2 + r^{-2})\cos 2\phi$$

when the cylinder is freely rotating.

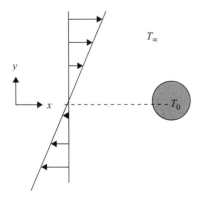

Here,
$$u_r \equiv \frac{1}{r}\frac{\partial \psi}{\partial \phi}, \quad u_\phi = -\frac{\partial \psi}{\partial r}$$

$$\mathbf{u}^\infty = y\mathbf{i}.$$

(a) Solve the problem of heat transfer from the cylinder for small $Pe \ll 1$, and prove
$$Nu = 2\left(1.372 - \frac{1}{2}\log Pe\right)^{-1}.$$

What is the magnitude of the error in this expression for the Nusselt number?

Note: As part of the analysis for this problem, it is useful to have a solution in cylindrical coordinates of
$$\nabla^2 w - y\frac{\partial w}{\partial x} = 0$$

for a point source of w at the origin. This solution is

$$w = \int_0^\infty \frac{dS}{2S\left(1 + \frac{1}{12}S^2\right)^{1/2}} \exp\left\{-\left[\frac{\left(x - \frac{1}{2}yS\right)^2}{4S\left(1 + \frac{1}{12}S^2\right)} + \frac{y^2}{4S}\right]\right\}.$$

You should derive this solution as part of the analysis of the overall problem.

(b) Suppose there is a nonzero torque applied to the cylinder. How does this change the problem? Consider carefully, starting with a recalculation of the stream-function field. Carry out calculations for enough cases to illustrate any qualitative trends in the heat transfer problem. (To be specific, you may wish to begin by considering the case in which the applied torque completely inhibits the rotation of the cylinder.)

Problem 9–6. Inertial Effects on the Motion of a Gas Bubble for $Re \ll 1$. A gas bubble rises through an infinite body of fluid under the action of buoyancy. The Reynolds number associated with this motion is very small but nonzero. Assume that the bubble remains *spherical*, and use the method of matched asymptotic expansions to calculate the drag on the bubble, including the first correction that is due to inertia at $O(Re)$. You may assume that the viscosity and density of the gas are negligible compared with those of the liquid so that you can apply the boundary conditions

$$u_r = 0 \quad \text{and} \quad \tau_{r\theta} = 0 \quad \text{at } r = a,$$

where a is the bubble radius. The motion of the gas may thus be ignored in the calculation. Would the bubble actually remain spherical? If not, calculate the bubble shape through terms of $O(Re)$ by assuming that the shape is only slightly perturbed from spherical and by using the method of domain perturbations. What condition (or conditions) must be satisfied to ensure that the near-sphere approximation is valid at $O(Re)$?

Problem 9–7. Heat Transfer from a Sphere at $Pe \ll 1$, with Specified Heat Flux. Consider heat transfer from a solid sphere of radius a that is immersed in an incompressible Newtonian fluid. Far from the sphere this fluid has an ambient temperature T_∞ and is undergoing a uniform flow with a velocity U_∞. At the surface of the sphere the heat flux is independent of position with a specified value q. Determine the temperature distribution on the surface of the sphere, assuming that the Peclet number is small.

Problems

Problem 9–8. Inertial Corrections on the Rotation of a Sphere in a Quiescent Fluid. A rigid sphere rotates with a constant angular velocity Ω in an infinite, quiescent fluid. It is desired to calculate the torque on the sphere to include any inertial correction at $O(Re)$, where

$$Re = \frac{\Omega a^2 \rho}{\mu} \ll 1.$$

Can this be done by a regular perturbation solution, or is it necessary to use the method of matched asymptotic expansions? Calculate the torque to include the first inertial correction.

Problem 9–9. Solid Sphere in a Uniaxial Extensional Flow at $Re \ll 1$. We consider uniaxial extensional flow exterior to a solid sphere whose center is located at the origin so that

$$\mathbf{u}_\infty = \begin{pmatrix} -E & 0 & 0 \\ 0 & -E & 0 \\ 0 & 0 & 2E \end{pmatrix} \cdot \mathbf{x}.$$

Here, $\mathbf{x}$ is zero at the sphere center. Let us suppose that the Reynolds number,

$$Re = \frac{\rho E a^2}{\mu} \ll 1,$$

is very small but nonzero. We desire to determine the first *inertial* contribution to the flow, beyond the classical creeping-flow solution.

(a) By considering the solution of the problem arising in a regular perturbation scheme at $O(Re)$, show that a singular-perturbation solution is necessary.
(b) Develop appropriately scaled governing equations and boundary conditions for both the inner region nearest the sphere and the outer region in the far field. Using these equations and boundary conditions, determine the leading-order approximation to the solution in the outer region, and use this solution, plus matching, to demonstrate that the creeping-flow solution (obtained by assuming that the creeping-flow approximation is uniformly valid) is the proper leading-order approximation in the inner region.
(c) Demonstrate that the second term for the asymptotic approximation in the outer region is $O(Re^{3/2})$ in magnitude, and derive a governing equation for the stream-function contribution at this order, namely, Ψ_1. This governing equation can be reduced to a simpler form by means of the substitution

$$E_\rho^2 \Psi_1 = \exp\left[\frac{1}{4}\rho^2 P_2(\eta)\right] \Phi(\rho, \eta),$$

but, so far as we know, this outer problem at $O(Re^{3/2})$ has not been solved analytically.
(d) Derive governing equations and boundary conditions also for the second term in the expansion of ψ for the inner region by assuming that $f_1(Re) = Re$, that is,

$$\psi = \psi_0 + f_1(Re)\psi_1 \ldots.$$

Obtain a solution for ψ_1, and demonstrate that the assumption $f_1(Re) = Re$ is required for matching. In fact, you should find that ψ_1 can be determined to within a single constant even though the lack of an analytic solution for Ψ_1 means that a detailed match of inner and outer solutions cannot be carried out.

Problem 9–10. Mass Transfer From a Spherical Gas Bubble for $Re \ll 1, Pe \ll 1$. Consider a *spherical* gas bubble and suppose that it contains a gaseous species A that is soluble in the liquid that surrounds the bubble and a second dominant species B that is insoluble in the liquid. We wish to calculate the rate of mass transfer in the creeping-flow limit when the bubble is rising through an unbounded body of liquid because of buoyancy. The governing equation for transport of a single solute A in the liquid is

$$\frac{\partial C}{\partial t} + \mathbf{u} \cdot \nabla C = D \nabla^2 C,$$

where D is the mutual diffusion coefficient for the species A in the liquid and C is the concentration of species A in the liquid. You may assume that $C \to 0$ far from the bubble. Obtain an expression for the rate of mass transfer in the limits $Re \ll 1$ and $Pe \ll 1$, where the Peclet number for mass transfer is defined as

$$Pe \equiv \frac{Ua}{D},$$

with U being the rise velocity and a being the bubble radius. You may assume that any change in the volume of the bubble is negligible, and also neglect any change in the concentration of A inside the bubble. Hence, at equilibrium, the concentration of A on the liquid side of the gas–liquid interface is constant at a value C_0 that is given by Henry's law.

For this problem, you should assume that the mass transfer process occurs within the region surrounding the bubble where the Stokes' approximation to the velocity field can be used. Thus the solution will be valid provided that $Re \ll Pe$. Explain the reason for this condition. Your calculation of mass transfer rate should be carried out to include the first correction because of convention.

Problem 9–11. The Probability Density Function for Microstructured Fluids at $Pe \ll 1$.
In describing the mechanical response of microstructured fluids, e.g., polymers, emulsions, colloidal dispersions, etc., one needs to determine the pair distribution function – the probability density $P(\mathbf{r})$ for finding a particle at a position $\mathbf{r}$ given a particle at the origin in suspensions, or the probability density of the end-to-end vector in polymers, or a measure of the deformation of drops in an emulsion. This probability density satisfies an advection–diffusion or Smoluchowski equation of the following (when suitable approximations have been made) form:

$$\frac{\partial P}{\partial t} + \nabla \cdot \mathbf{u} P - D \nabla^2 P = 0. \qquad (1)$$

Here D is the diffusivity and $\mathbf{u}$ is the advective velocity. The characteristic length is taken to be a. The boundary conditions to be satisfied by P are

$$P \to 1 \quad \text{as} \quad r \to \infty, \qquad (2)$$

$$D \mathbf{n} \cdot \nabla P = \mathbf{n} \cdot \mathbf{u} \quad \text{at} \quad r = 2a. \qquad (3)$$

For the mechanical response one subjects the material to an oscillatory shearing motion with the imposed velocity

$$\mathbf{u} = \mathbf{r} \cdot \mathbf{E} e^{i\omega t}.$$

Here $\mathbf{E}$ is the symmetric part of the velocity gradient tensor and is independent of r with a magnitude $|E| = \dot{\gamma}$ and ω is the frequency of the oscillation.

(a) Nondimensionalize the equations to define an appropriate Peclet number, Pe. Then show that at small Peclet numbers the pair distribution function P can be written as

$$P = 1 + Pe f(r) \frac{\mathbf{r} \cdot \mathbf{E} \cdot \mathbf{r}}{r^2} e^{it} + \cdots + \qquad (4)$$

Problems

and that the equation for the function $f(r)$ is, to leading order in Pe,

$$\frac{1}{r^2}\frac{d}{dr}r^2\frac{df}{dr} - \frac{6}{r^2}f - i\alpha f = 0, \tag{5}$$

with

$$f \to 0 \quad \text{as} \quad r \to \infty \tag{6}$$

$$\frac{df}{dr} = 2 \quad \text{at} \quad r = 2. \tag{7}$$

In Eqs. (4)–(7), the variables such as **E**, r, and t are understood to be dimensionless. Furthermore, in (5), $\alpha = \omega a^2/D$.

(b) Obtain the solution for f in the limits of small and large ω.

(c) Now consider the case of steady shearing ($\omega = 0$) and discuss the nature of the perturbation expansion for small Pe. Is it a regular perturbation expansion? What is the next term in the expansion for P [see, Eq. (4)] and why? Be careful here to estimate the size of the various terms you expect to appear.

Problem 9–12. Thermal Boundary Layer for a Circular Cylinder at $Re \ll 1$ and $Pe \gg 1$. For flow past a circular at low Reynolds number, $Re \ll 1$, a first (inner) approximation to the velocity field is

$$\psi = -\frac{1}{2\ln Re}\left(\frac{1}{r} - r + 2r\ln r\right)\sin\theta.$$

Suppose the cylinder is heated and $Pe \gg 1$. Find an appropriate thermal boundary-layer equation, and deduce the general form of the relationship between Nu and Pe.

Problem 9–13. Heat Transfer From a Circular Cylinder in an Inviscid Fluid. Determine the dimensionless rate of heat transfer from a circular cylinder of radius a immersed in a uniform flow of an *inviscid* fluid. Because there is no vorticity in the free stream, the velocity field can be written in terms of a potential, $\mathbf{u} = \nabla\phi$, with $\nabla^2\phi = 0$. The boundary conditions are $\mathbf{u} \sim \mathbf{U}$ as $r \to \infty$, and $\mathbf{u}\cdot\mathbf{n} = 0$ at $r = a$. The cylinder is maintained at temperature T_1, while the fluid far from the cylinder is at T_0.

(a) First, assume the Peclet number for heat transfer is small. (The solution for a line source of heat in a uniform flow can be obtained by use of Fourier transforms and involves the modified Bessel function K_0). Determine the solution and heat transfer rate through terms of $O(Pe)$.

(b) Now, assume that the Peclet number is large. Obtain the first leading-order approximation for the heat transfer rate.

Problem 9–14. Mass Transfer From a Sphere With a First-Order Reaction at Its Surface. A small spherical particle of radius a is immersed in a uniform flow with velocity **U** far from the fixed particle. On the surface of the particle a chemical reaction takes place in which a solute c in the fluid is consumed according to a first-order reaction, $r = -kc$. Estimate the net rate of consumption, R, of c, in dimensionless form, when the fluid far from the particle has solute concentration c_0. The molecular diffusivity of the solute is a constant D. R depends on two parameters, a Peclet number and a Damköhler number; $R(Pe, Da)$.

(a) Determine the form of R in the various limiting cases when the Peclet number is small. What are these limiting cases?

(b) Repeat this problem, but now assume that the Peclet number is large.

Problem 9–15. Evaporating Spherical Particle for $Pe \ll 1$. A spherical particle at temperature T_s is evaporating into a gas whose temperature far from the particle is T_∞. The total mass flux from the particle due to evaporation is $\dot{m}$. Although the particle shrinks in size as it evaporates, the density of the vapor exceeds that of the liquid, and the evaporation generates a net motion in the surrounding gas. The mass flux is related to the rate of evaporation, which is in turn dependent on the net heat flux to the drop. Recall that the boundary condition at the drop surface is

$$[\mathbf{n} \cdot \mathbf{q}] = h_{vap} \frac{da}{dt},$$

where [ ] denotes the jump in heat flux, h_{vap} is the latent heat of vaporization, and a is the radius of the drop. The mass flux is related to the rate at which the interface moves by

$$\dot{m} = (\rho_g - \rho_\ell) 4\pi a^2 \frac{da}{dt}$$

where ρ_g and ρ_ℓ are the densities of the gas and liquid, respectively. In a quasi-steady approximation we may neglect the unsteady terms $(\partial/\partial t)$ in the energy and momentum equations.

(a) First, neglecting the mass flux from the drop surface (the appropriate Peclet number is small) determine the temperature profile in the gas and the rate of evaporation $\dot{m}$.
(b) Now include the fluid motion induced by the evaporation on the temperature profile and thus the correction to the evaporation rate for small Peclet number.

Problem 9–16. Evaporating Spherical Particle for $Pe \gg 1$. Do Problem 9–15 for large Peclet number.

Problem 9–17. Heat Transfer From an Ellipsoid of Revolution at $Pe \gg 1$. In a classic paper, Payne and Pell [*J. Fluid Mech.* **7**, 529(1960)] presented a general solution scheme for axisymmetric creeping-flow problems. Among the specific examples that they considered was the uniform, axisymmetric flow past prolate and oblate ellipsoids of revolution (spheroids). This solution was obtained with prolate and oblate ellipsoidal coordinate systems, respectively.

In *the prolate case*, the ellipsoidal coordinates (ξ, η) are related to circular cylindrical coordinates (r, z) by the transformation $z = c \cosh \xi \cos y$ and $r - c \sinh \xi \sin \eta$. The coordinate surface $\xi = \xi_0$ (constant) defines the surface of a prolate spheroid, with major and minor semiaxes a_0 and b_0 given by

$$a_0 = c \cosh \xi_0, \quad b_0 = c \sinh \xi_0.$$

The line $\xi_0 = 0$ is a degenerate ellipse that reduces to a line segment $-c \leq z \leq c$ along the z axis.

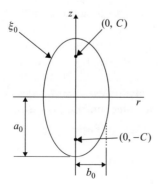

Problems

In these terms, the solution of Payne and Pells for uniform streaming flow with undisturbed velocity U past a prolate spheroid with semimajor and semiminor axes a_0 and b_0 is

$$\psi = \frac{1}{2}Ur^2\left[1 - A\ln\frac{s+1}{s-1} + B\left(\frac{s}{s^2-1}\right)\right],$$

where

$$A = -\frac{\frac{1}{2}(s_0^2+1)}{s_0 - \frac{1}{2}(s_0^2+1)\ln\frac{s_0+1}{s_0-1}},$$

$$B = \frac{s_0^2 - 1}{s_0 - \frac{1}{2}(s_0^2+1)\ln\frac{s_0+1}{s_0-1}},$$

and $s = \cosh\xi$, $s_0 = \cosh\xi_0$, with ξ_0 given in terms of a_0 and b_0 as indicated.

In *the oblate case*, the coordinate transformation is

$$z = c\sinh\xi \cos\eta,$$
$$r = c\cosh\xi \sin\eta$$

with $\xi = \xi_0$ corresponding to the surface of an oblate ellipsoid of revolution (spheroid) with major and minor axes:

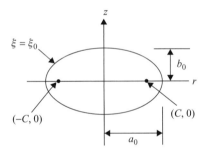

$$a_0 = c\cosh\xi_0,$$
$$b_0 = c\sinh\xi_0,$$

as before. In this case, the solution of Payne and Pells for the uniform streaming flow problem is

$$\psi = \frac{1}{2}Ur^2\left(1 - C\frac{\tau}{1+\tau^2} - D\cot^{-1}\tau\right),$$

with

$$A \equiv \frac{(1+\tau_0^2)}{\tau_0 + (1-\tau_0^2)\cot^{-1}\tau_0},$$

$$B \equiv -\frac{(1-\tau_0^2)}{\tau_0 + (1-\tau_0^2)\cot^{-1}\tau_0},$$

where $\tau = \sinh\xi$ and $\tau_0 = \sinh\xi_0$. One interesting special case is flow past a flat, circular disk, where $\tau_0 = 0$, and

$$\psi = \frac{1}{2}Ur^2\left[1 - \frac{2}{\pi}\left(\cot^{-1}\tau + \frac{\tau}{1+\tau^2}\right)\right].$$

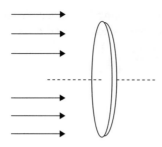

(a) With the preceding solution given, develop thermal boundary-layer equations for a heated ellipsoid ($Pe \gg 1$), and determine the relationship between Nu and Pe for each of the two cases of a prolate and oblate ellipsoid.

(b) Discuss the influence of the ellipsoid geometry on the rate of heat transfer. To provide a fixed frame of reference, assume that a_0 and b_0 are varied in such a way that the aspect ratio of the ellipsoid changes but the *volume* is held fixed. Thus, as you vary the body shape, the surface area will change, as well as the heat flux distribution on the body surface. (You may find it convenient to define the Peclet number in terms of the length scale of a sphere of equal volume, so that its value remains fixed for fixed U/k as the geometry is varied). Include the limiting case of a circular disk in your discussion.

Problem 9–18. Heat Transfer From a Nonrotating Cylinder in a Simple Shear Flow.
A heated circular cylinder is suspended in a fluid that is undergoing a simple shear flow. Assume that the cylinder does not rotate so that the streamfunction representing flow in the (inner) region near the cylinder is

$$\psi = \frac{1}{2}y^2 - \frac{1}{4}\left[2\ln r + 1 + \left(\frac{1}{r^2} - 2\right)\cos 2\phi\right]$$

in dimensionless terms [see Eq. (9–302)]. The temperature at the surface of the cylinder is held constant at a value T_0, and the ambient temperature of the fluid is T_1. Calculate the first approximation to the rate of heat transfer from the cylinder (per unit length) to the fluid, assuming that the Peclet number is large,

$$Pe = \frac{Ga^2}{k} \gg 1.$$

Problem 9–19. Heat Transfer From a Rotating Cylinder in Simple Shear Flow. We consider the same situation as in Problem 9–18, but the torque acting on the cylinder is now assumed to be too small to keep the cylinder from rotating. Hence it rotates with an angular velocity Ω, which, in dimensionless terms, is in the range,

$$0 < \Omega < \frac{1}{2}$$

between free rotation and no rotation. Thus the velocity field is now given, in the region near the cylinder, by

$$\psi = \frac{1}{2}y^2 - \frac{1}{4}\left[2(1 - 2\Omega)\ln r + 1 + \left(\frac{1}{r^2} - 2\right)\cos 2\phi\right]$$

[that is, Eq. (9–302) with $\Omega \neq 0$ or $\Omega \neq 1/2$]. Obtain the relationship between Nu and Ω, assuming that $Pe \gg 1$.

Problem 9–20. Mass Transfer From a Spherical Bubble in an Extensional Flow. Consider a *spherical* gas bubble that is suspended in a liquid that undergoes an axisymmetric extensional flow. The bubble contains a component A that is soluble in the exterior liquid

Problems

and a major component B that is insoluble. Assume that the concentration of A is constant inside the bubble and that the bubble volume is fixed. Determine the concentration distribution in the liquid for species A and the mass transfer rate from the gas to the liquid (the Nusselt number), assuming $Pe \gg 1$.

Note: In dimensional terms, the steady-state distribution of A in the liquid is governed by

$$\mathbf{u} \cdot \nabla C_A = D \nabla^2 C_A,$$

where D is the mutual diffusion coefficient for species A and C_A is the concentration of A in the liquid.

Problem 9–21. Nonlinear Microrheology. A common way to measure the viscosity of a liquid is to drop a heavy spherical ball into a large tank of the fluid and measure the steady speed with which the ball descends. When the ball is only slightly more dense than the liquid, the Reynolds number based on the ball's fall speed will be small, and the viscosity can be determined from the Stokes' settling velocity: $U_{Stokes} = [(2/9)\Delta \rho a^2 g/\mu]$, where $\Delta \rho$ is the density difference between the ball and the fluid, a is the ball radius, g is the acceleration of gravity, and μ is the fluid viscosity.

When the fluid is a colloidal dispersion of small submicrometer-sized particles, rather than a simple liquid, the same approach can be used to measure the "effective viscosity" of the colloidal dispersion. Of interest is to relate the measured viscosity to the disturbance of the microstructure of the colloidal dispersion created by the falling ball. As the ball falls, it pushes the background colloidal particles out of its way and creates a variation in the local concentration of the colloidal particles. Because of their Brownian motion the colloidal particles diffuse down the concentration gradient and this entropic restoring force increases the drag on the falling ball. Thus, to determine the increased drag caused by the colloidal particles, we need to determine the disturbed microstructure and how this affects the drag.

The simplest model of this process leads to the following Smoluchowski equation to determine the concentration distribution of the background colloidal particles:

$$\mathbf{U} \cdot \nabla g(r) = D \nabla^2 g(r), \tag{1}$$

where $\mathbf{U}$ is the constant velocity of the falling ball, D is the relative diffusivity of the background colloidal particles to the falling ball, and $g(r)$ is the probability density of finding a background colloidal particle relative to the falling ball. The boundary conditions for (1) are

$$g(r) \sim n \quad \text{as } r \to \infty \quad \text{and} \quad \hat{\mathbf{r}} \cdot (D \nabla g - \mathbf{U} g) = 0 \quad \text{at } r = a+b, \tag{2}$$

where n is the number density of colloidal particles and $\hat{\mathbf{r}}$ is a unit vector from the falling ball surface directed into the colloidal dispersion. The falling ball has radius a and the background particles have radius $b(<a)$ and the origin is at the center of the falling ball.

There are three forces that are acting on the falling ball: gravity, the viscous drag of the solvent, and the entropic force that is due to the disturbed microstructure of the colloidal particles. A force balance on the falling ball gives

$$0 = -6\pi \mu a \mathbf{U} + \frac{4}{3}\pi a^3 \Delta \rho \mathbf{g} - kTn \oint_{r=a+b} \hat{\mathbf{r}} g(r) dS, \tag{3}$$

where kT is the thermal energy and the integral is over a spherical surface of radius $a+b$ corresponding to the distance of closest approach of the background colloidal particles to the falling ball.

Question: Determine the "effective viscosity" μ_{eff} from the force balance (3) by relating U to the external force of gravity from the Stokes' settling velocity

$$\mathbf{U} = \frac{2}{9}\Delta\rho a^2 \mathbf{g}/\mu_{eff}, \tag{4}$$

in the two limiting cases of (i) small Peclet number and (ii) large Peclet number, where $Pe = U(a+b)/D$ and the diffusivity from the Stokes–Einstein equation is given by $D = kT/6\pi\mu(a+b)$. Express your answer as the ratio μ_{eff}/μ and in terms of the volume fraction of the background colloidal particles, $\phi_b = 4\pi b^3 n/3$.

[Note that $g(r)$ is a function of the constant fall speed U from the Smoluchowski equation (1), and that at small Pe the $O(Pe)$ problem must be solved to get the correction to the fall speed.]

Problem 9–22. Flow in a Brinkman Medium. Fluid flow in a packed bed or porous medium can be modeled as flow in a Brinkman medium, which we may envision as a bed of spherical particles. Each particle in the bed (there are n particles per unit volume) exerts a drag force on the fluid proportional to fluid velocity relative to the particle given by Stokes' law, i.e., ($\sim -6\pi\mu a \mathbf{u}$, where a is the characteristic size of a bed particle). Thus the equations describing the fluid motion on an averaged scale (averaged over many bed particles, for example) are

$$\rho\frac{D\mathbf{u}}{Dt} = -\mu\alpha^2\mathbf{u} - \nabla p + \mu\nabla^2\mathbf{u}, \tag{1}$$

$$\nabla \cdot \mathbf{u} = 0, \tag{2}$$

where $\alpha^2 = [(9/2)\phi a^{-2}]$, and $\phi = 4\pi a^3 n/3$ is the volume fraction of particles of size a. The individual particle's drag forces have been converted to a body force on the averaged scale.

The Brinkman equation (1) transitions properly from Stokes' flow (in the absence of inertia) to Darcy flow in a porous medium:

$$\nabla p = -\mu\alpha^2 \mathbf{u},$$

with its balance between a pressure gradient and viscous drag. We can now model flow problems in porous media by assuming a particle is immersed in a Brinkman medium rather than a viscous fluid.

In the dilute limit $\phi \to 0$ the viscous drag force introduces a new length $a\phi^{-1/2}$. On scales of $O(a)$ the solution to (1) is the same as for Stokes' flow, but on scales of $O(a\phi^{-1/2})$ the body force exerted by the n fixed particles changes the nature of the velocity disturbance caused by a particle; the intervening particles screen the velocity disturbance, resulting in a decay faster than for Stokes' flow. In particular, for a point force immersed in a Brinkman medium at zero particle Reynolds number, $Re = \rho U a/\mu \ll 1$, with a uniform flow $\mathbf{U}$ at infinity, the velocity field is given by

$$\mathbf{u} = \mathbf{U} - \frac{3}{4}\mathbf{U} \cdot \left\{ \mathbf{I}\frac{2}{\alpha^2 r^3}[(1+\alpha r + \alpha^2 r^2)e^{-\alpha r} - 1] \right.$$
$$\left. + \mathbf{xx}\frac{2}{\alpha^2 r^5}[3 - (3+3\alpha r + \alpha^2 r^2)e^{-\alpha r}] \right\}, \tag{3}$$

which shows quite clearly the screening. The velocity $\mathbf{u}$ behaves as Ua/r for $r \sim a$ and as $Ua^3\phi^{-1}r^3$ for $r \sim a\phi^{-1/2}$.

Problems

(a) The problem is to estimate the correction to the drag exerted on a particle in uniform flow in a Brinkman medium that is due to inertial effects. There is a dual limit of $Re \to 0$ and $\phi \to 0$. You should state whether the asymptotic expansion for small Re and small ϕ will be regular or singular, and why. You should obtain the scaling of the next correction to the drag and the well-posed boundary-value problem, which includes the boundary conditions, that needs to be solved to obtain this next term. To get started, the drag force on the particle at zero particle Reynolds number from (3) is

$$\mathbf{F} = -6\pi \mu a \mathbf{U}\left[1 + \alpha a + \frac{1}{3}(\alpha a)^2\right].$$

You now need to determine corrections to this expression for small Re.

(b) If we now make the particle hot relative to the surrounding fluid, how does the Nusselt number depend on the Peclet number $Pe = Ua/\alpha$ in the limit $Pe \to 0$ for the cases previously considered?

Problem 9–23. Orientational Distribution of Brownian Magnetic Rods. Consider a very dilute (no interparticle interactions) suspension of slender rigid rods with length 2ℓ and a typical aspect ratio $\varepsilon = R/\ell$, where R is a typical radius of a rod. Suppose the rigid rods have a permanent magnetic moment $\mathbf{p}$ along their major axis, as is the case for liquid crystals and ferrofluids. On applying a magnetic field $\mathbf{B}$, the rods will tend to orient so as to align their moment with the field. However, rotational Brownian diffusion will disperse the particles from perfect alignment. Thus the average orientation of the particles is determined from a balance between magnetic and thermal forces. One may in fact write an equation for the probability distribution, P, for the orientation of the particles, which, in consideration of the symmetry of the problem here, is given by,

$$\frac{d}{d\theta}[\Omega(\theta)P(\theta)\sin\theta] = \frac{D}{\ell^2}\frac{d}{d\theta}\left[\sin\theta\frac{dP(\theta)}{d\theta}\right],$$

where D is the rotational diffusivity of the rod and $\theta \in [0, \pi]$ is the angle of orientation of the rod measured from the axis of the magnetic field $[\mathbf{B}\cdot\mathbf{p} = pB\cos(\theta)]$. Here $\Omega(\theta)$ is the rotational velocity of the particles. Note that the right-hand side of the equation describes the dispersion of the particle orientations that is due to rotational Brownian diffusion and the left-hand side describes the convection of the orientations by forced rotation. We wish to determine how the magnetic field affects the orientation of the particles.

(a) What is the relationship between Ω and the applied magnetic field?
- *Hint*: The magnetic field will induce a torque on the rods given by $\ell\mathbf{p}\wedge\mathbf{B}$.

(b) What is the probability distribution, $P(\theta)$? If you nondimensionalize, there will be a "Peclet" number for this problem, which can be thought about as a dimensionless magnetic field, and you can consider the asymptotic limits in which $Pe \ll 1$ and $Pe \gg 1$.

(c) Sketch the distribution in these limiting cases and explain its features.
- *Hint*:

$$\int_0^{2\pi}\int_0^{\pi} P(\theta)\sin\theta\, d\theta\, d\phi = 1.$$

Problem 9–24. Mass Transfer from a Sheet in Extension. Consider a polymer-solvent sheet in extension such that it is stretched as it is pulled horizontally. The dimensional tangential velocity of the sheet is given by $u = u_0 E^{x/L}$, where u_0 is the initial velocity, E is a constant draw ratio, x is the distance downstream, and L is the extent of the draw zone. We wish to determine the mass transfer of solvent from the polymer. The concentration of solvent far from the sheet is zero, and on the surface of the sheet, it is C_0. The Peclet number

Convection Effects in Low-Reynolds-Number Flows

for this process, $Pe = u_0 L/D \gg 1$, where D is the diffusivity of the solvent, and so we can expect a boundary layer.

(a) Use the continuity equation to determine the approximate velocity field near the surface of the polymer film.
(b) Rescaling as necessary, what is the resultant boundary-layer equation for concentration?
(c) What is the thickness of the boundary layer?
(d) Look for a solution to the boundary-layer equation of the form $c(x, Y) = f(\eta)$, where Y is the rescaled coordinate normal to the polymer film and similarity variable $\eta = Y/g(x)$. What is $g(x)$?
(e) What is $f(\eta)$?
(f) What is the effective mass transfer coefficient as a function of the physical parameters of the problem? How does mass transfer rate scale with E?

Problem 9–25. Mass Transfer From a Gas Bubble in a Bioreactor. Gas bubbles are injected through a sparger into a bioreactor to oxygenate a liquid growth medium. The liquid is sufficiently viscous that the bubbles flow at low Reynolds number. You have been asked to model the mass transfer of oxygen in the reactor.

(a) Show that the mass transfer coefficient for a single gas bubble, k_b, is given by

$$\frac{k_b L}{D} = C \left(\frac{UL}{D} \right)^{1/2},$$

if $UL/D \gg 1$ and where U is rise velocity of the bubble, L is some characteristic length of the bubble, D is the diffusivity of oxygen in the liquid, and C is some $O(1)$ constant.
(b) Compute C for a dilute dispersion of spherical bubbles.
(c) Qualitatively, how does C change for a nondilute dispersion of bubbles? (Does it increase or decrease?) Make your best guess of how C scales with the volume fraction of bubbles, clearly stating your reasoning.
(d) Suppose the bioreactor is a well-stirred tank and that several experiments were performed with different spargers. Each sparger produces bubbles of a different size for the same pressure drop. In the experiments the concentration of oxygen was measured as a function of time for each sparger. Explain in detail how you would analyze and correlate the data to evaluate the spargers.

Problem 9–26. The Effective Thermal Conductivity of a Dilute Suspension of Spheres. Consider a medium of thermal conductivity k_1 containing identical spherical inclusions of radius a and thermal conductivity k_2. Our aim is to derive an expression for the effective thermal conductivity of the suspension (defined as the medium plus inclusions) correct to first order in the volume fraction of the inclusions.

The temperature field $T(\mathbf{r})$ in the suspension satisfies Laplace's equation. We define the average temperature gradient $\mathbf{G}$ as

$$\mathbf{G} = \frac{1}{V} \int_V \nabla T \, dV,$$

where V is the volume of the suspension. The average temperature gradient $\mathbf{G}$ generates an average heat flux $-\mathbf{F}$ where

$$\mathbf{F} = \frac{1}{V} \int_V k \nabla T \, dV$$

(a) The suspension is sufficiently dilute so that one can neglect interactions between the inclusions. Hence show that

$$\mathbf{F} = k_1 \mathbf{G} + n \mathbf{S},$$

Problems

where n is the number density of the inclusions, $\mathbf{S} = (k_2 - k_1) \int_{V_s} \nabla T \, dV$, and V_s is the volume of a single inclusion. What is the physical meaning of $\mathbf{S}$?

(b) To compute $\mathbf{S}$ one must solve for the temperature field in the suspension subject to an ambient temperature gradient $\mathbf{G}$. By solving Laplace's equation for the temperature field, show that

$$T(\mathbf{r}) = \left(1 - \frac{k_2 - k_1}{k_2 + 2k_1} \frac{a^3}{r^3}\right) \mathbf{G} \cdot \mathbf{r} \quad \text{for} \quad r > a,$$

$$T(\mathbf{r}) = \frac{3k_1}{k_2 + 2k_1} \mathbf{G} \cdot \mathbf{r} \quad \text{for} \quad r \leq a.$$

(c) Hence show that the effective conductivity of the suspension k^*, defined by means of $\mathbf{F} = k^* \mathbf{G}$ is

$$\frac{k^*}{k_1} = 1 + 3 \frac{\alpha - 1}{\alpha + 2} \phi,$$

where ϕ is the volume fraction of inclusions and $\alpha = k_2/k_1$. Give physical interpretations of the cases $\alpha < 1$, $\alpha = 1$ and $\alpha > 1$.

Problem 9–27. Effective Conductivity of a Dense Suspension of Spheres. Consider a suspension of nearly closely packed, perfectly conductive spheres imbedded in a matrix of conductivity λ. We would like to determine the effective conductivity of this two-phase material as function of the distribution of particles, or microstructure, and the separation between spheres, when this distance is small.

(a) Assuming the spheres reside in a simple cubic lattice; what is the effective conductivity? Why is your answer reasonable?
(b) Suppose the spheres resided in a body-centered cubic lattice; what then would be the effective conductivity?

Hints: For a general framework, see Problem 9–26, though be careful not to accidentally incorporate any results that require the particle concentration to be small. Note that it is sufficient to determine the temperature field in the thin-gaps, and for this purpose we can apply the thin-gap approximation to this thermal problem.

Problem 9–29. Effective Conductivity of a Dilute Suspension of Ordered Rods. Consider a very dilute (no interparticle interactions) suspension of slender rigid rods with length 2ℓ and aspect ratio $\varepsilon = R/\ell$, where R is the radius of a rod. The rods are perfectly conducting and are aligned in a medium of conductivity λ. We would like to determine the conductivity along the direction of the rods.

To determine the effective conductivity, we will use the relationship between the average heat flux and the particle dipole, $\mathbf{S}$, discussed in Problem 9–26. To compute the dipole, we must determine the local heat flux on the surface of the particle.

For a perfectly conducting particle, there is an integral relationship between the temperature field and the heat flux given by,

$$T(\mathbf{x}) = T^\infty(\mathbf{x}) + \frac{1}{4\pi \lambda} \int_{\partial \Omega} \mathbf{q} \cdot \mathbf{n} \frac{1}{r} dS,$$

where T is the temperature on the surface of the particle, T^∞ is the imposed temperature field at infinity, $\mathbf{q}$ is the heat flux, $\mathbf{n}$ is the outward normal on the surface of the particle, and $r = \|\mathbf{x} - \mathbf{y}\|$. The points $\mathbf{x}$ and $\mathbf{y}$ are on the surface of the particle and the surface integral is with respect to the variable $\mathbf{y}$.

Convection Effects in Low-Reynolds-Number Flows

(a) Derive the integral formula for T (*note*: this step can be skipped at the discretion of the instructor because the necessary result was already previously given).

(b) Simplify the preceding integral equation for a slender body.

(c) Show that, for a rod immersed in a linear temperature field such that $T^\infty = Gx_1$, the heat flow per unit length of rod, $Q_z(x_1) = (\mathbf{q} \cdot \mathbf{n})P = 2\pi\lambda G x_1/\ln\varepsilon$, where x_1 is measured along the length of a rod and $P = 2\pi R$ is the perimeter of a rod.

(d) Show that the effective conductivity along the direction of the rods is

$$\lambda_{\text{eff}} = \lambda\left(1 + \frac{2}{3}\frac{\phi}{\varepsilon^2 \ln\varepsilon^{-1}}\right),$$

where ϕ is the volume fraction of particles. How does this compare with the effective conductivity for a dilute suspension of spheres?

10

Laminar Boundary-Layer Theory

In Chap. 9 we considered strong-convection effects in heat (or mass) transfer problems at low Reynolds numbers. The most important findings were the existence of a thermal boundary layer for open-streamline flows at high Peclet numbers and the fundamental distinction between open- and closed-streamline flows for heat or mass transfer processes at high Peclet numbers. An important conclusion in each of these cases is that conduction (or diffusion) plays a critical role in the transport process, even though $Pe \to \infty$. In open-streamline flows, this occurs because the temperature field develops increasingly large gradients near the body surface as $Pe \to \infty$. For closed-streamline flows, on the other hand, the temperature gradients are $O(1)$ – except possibly during some initial transient period – and conduction is important because it has an indefinite time to act.

In this chapter we continue the development of these ideas by considering their application to the approximate solution of fluid mechanics problems in the asymptotic limit $Re \to \infty$, with a particular emphasis on problems in which boundary layers play a key role. Before embarking on this program, however, it is useful to highlight the expected goals and limitations of the analysis in which we formally require $Re \to \infty$ but still assume that the flow remains laminar. In practice, of course, most flows will become unstable at a large, but finite, value of Reynolds number and eventually undergo a transition to turbulence, and this is the flow we will see in the lab. However, the existence of an instability leading to a branch of unsteady and complicated solutions of the Navier – Stokes equations at some Reynolds number (e.g., a turbulent velocity field) does not preclude the existence of one or more steady, laminar solutions at the same Reynolds numbers. It is these latter solutions that we seek. In part, the practical significance of high-Reynolds-number laminar solutions is that they provide considerable insight and often even accurate results for large, but *finite*, Reynolds numbers below the instability limit. In addition, the high-Reynolds-number laminar-flow solutions provide approximate base solutions whose stability can be studied to understand the mechanisms and critical conditions for instability.

As already indicated, our approach to the construction of approximate solutions for $Re \gg 1$ will be to use the method of matched asymptotic expansions. In view of this, it is worth noting that the earliest use of boundary-layer theory for fluid mechanics problems at high Reynolds number predated considerably the first *formal* use of asymptotic expansion procedures to solve PDEs. The initial applications of the method of matched asymptotic expansions appeared in the mid-1950s,[1] whereas the essentials of boundary-layer theory had already been presented in 1904 by Prandtl[2] in a paper that revolutionized fluid mechanics. Prandtl's approach, as well as that of many subsequent investigators in the intervening 35–40 years, was an *ad hoc*, but physically motivated, simplification of the Navier–Stokes equations for large Reynolds number. The interested reader may wish to refer to the classic textbook by Schlichting[3] for a derivation of boundary-layer theory based on this physical approach.

Laminar Boundary-Layer Theory

Prandtl's theory was revolutionary because it provided for the first time a theoretical understanding of the critical role played by viscous effects in determining fluid motions in the limit $Re \to \infty$. Before Prandtl, *inviscid* flow theory had dominated attempts to describe fluid motions for $Re \gg 1$ but was in serious disagreement with experimental observation[4] because it could not deal in any way with *separation phenomena* – namely, the existence of recirculating wakes and large *form* drag in flow past nonstreamlined bodies. It was only with the advent of Prandtl's boundary-layer theory that a theoretical basis existed to predict whether fluid motion past a body of given shape would remain attached (with correspondingly "small" drag) or would "separate" (with a large form drag). The importance of this advance in the development of rational aerodynamic design of airfoil shapes with large lift/drag ratios cannot be overemphasized.[5] We shall return to the problem of predicting separation phenomena later in this chapter.

A potential advantage of the "physical" approach to boundary-layer theory is that it forces an emphasis on the underlying physical description of the flow. However, unlike the asymptotic approach presented here, the physically derived theory provides no obvious means to improve the solution beyond the first level of approximation. Provided that the physical picture underlying the analysis is properly emphasized, the asymptotic approach can incorporate the principal positive aspect of the earlier theories within a rational framework for systematic improvement of the approximation scheme.

A. POTENTIAL-FLOW THEORY

The first part of this chapter is concerned largely with a specific prototype problem in which a stationary solid body is immersed in an unbounded, incompressible fluid that is undergoing a steady, uniform translational motion at large distances from the body ("at infinity"). For simplicity, we shall assume in most instances that the body is 2D; namely, that it extends indefinitely in the third direction, z, without change of shape so that its geometry can be specified completely by its cross sectional shape in the $x\,y$ plane. The streaming motion at infinity is then assumed to be parallel to the $x\,y$ plane.

To analyze streaming flow at high Reynolds number past a 2D body, the starting point is the full, steady-state Navier–Stokes and continuity equations, nondimensionalized by use of the streaming velocity U_∞ as a characteristic velocity scale and a scalar length of the body in the $x\,y$ plane, say, a, as the characteristic length scale, namely,

$$\mathbf{u}\cdot\nabla\mathbf{u} = -\nabla p + \frac{1}{Re}\nabla^2\mathbf{u}, \qquad (10\text{--}1)$$

$$\nabla\cdot\mathbf{u} = 0. \qquad (10\text{--}2)$$

Pressure has been nondimensionalized by use of ρU_∞^2, as is appropriate for flow at large Reynolds number.

A convenient way to discuss some aspects of flow at high Reynolds number is in terms of the *transport of vorticity* rather than directly in terms of velocity and pressure. We recall that the vorticity is defined as the curl of the velocity,

$$\boldsymbol{\omega} \equiv \nabla \wedge \mathbf{u}. \qquad (10\text{--}3)$$

Hence, physically, it represents the local rotational motion of the fluid. We obtain an equation for transport of vorticity directly by taking the curl of all terms in Eq. (10–1). Because

$$\nabla \wedge (\nabla \phi) \equiv 0 \qquad (10\text{--}4)$$

A. Potential-Flow Theory

for any scalar ϕ, we see that the pressure is eliminated, and we have remaining

$$\mathbf{u} \cdot \nabla \boldsymbol{\omega} = \boldsymbol{\omega} \cdot \nabla \mathbf{u} + \frac{1}{Re} \nabla^2 \boldsymbol{\omega}. \qquad (10\text{--}5)$$

The left-hand side represents the advection (or convection) of vorticity by the velocity $\mathbf{u}$, and the second term on the right-hand side represents the transport of vorticity by diffusion (with diffusivity = the kinematic viscosity ν). These two terms are "familiar" in the sense that they resemble the convection and diffusion terms appearing in the transport equation for any passive scalar. A counterpart to the second term does not appear in these transport equations, however. Known as the *production* term, it is associated with the intensification of vorticity that is due to stretching of vortex lines. It is not a true production term, however, because it cannot produce vorticity where none exists. Indeed, because (10–5) contains $\boldsymbol{\omega}$ linearly in every term, it is clear that vorticity can be neither created nor destroyed in the interior of an isothermal, incompressible fluid: It can only be convected, diffused, or changed in magnitude once it is already present.[6]

The main value of the vorticity transport equation, in the present context, is that a direct analogy exists for 2D motions between this equation and the thermal energy equation of Chap. 9. Specifically, for a 2D flow,

$$\boldsymbol{\omega} = \omega \mathbf{k},$$

where $\mathbf{k}$ is a unit vector normal to the plane of flow, so that

$$\boldsymbol{\omega} \cdot \nabla u \equiv 0$$

and the general equation (10–5) then reduces to the form

$$\mathbf{u} \cdot \nabla \omega = \frac{1}{Re} \nabla^2 \omega, \qquad (10\text{--}6)$$

where ω is the scalar magnitude of $\boldsymbol{\omega}$. If we compare (10–6) with the thermal energy equation (9–7), we see that the forms are identical, and Re plays the same role as Pe. In particular, in a 2D motion, vorticity, like heat, is convected and diffused and it is produced only at the boundaries of the flow. The main differences between the thermal and vorticity transport problems are that θ and $\mathbf{u}$ are truly independent variables in the forced convection approximation, whereas ω and $\mathbf{u}$ are intimately connected by means of (10–3). In addition, the magnitude of the vorticity cannot be specified on the boundaries because $\mathbf{u}$ is already given, and it is almost never independent of position on the surface as we assumed for θ. Nevertheless, there is considerable similarity between heat transfer at high Pe and vorticity transport at high Re, which we can use to advantage now that we have solved the thermal problem for $Pe \gg 1$.

In particular, let us start with the nondimensionalized vorticity transport equation (10–6) and attempt to obtain an approximate solution for $Re \gg 1$. We expect an asymptotic expansion with Re^{-1} as the small parameter, but we restrict our attention here to the leading-order term in this expansion, which we can obtain by solving the limiting form of (10–6) for $Re \to \infty$, namely,

$$\mathbf{u} \cdot \nabla \omega = 0. \qquad (10\text{--}7)$$

However, this is precisely the same form as Eq. (9–3), which we have already shown to have the solution

$$\omega = \omega(\psi), \qquad (10\text{--}8)$$

Laminar Boundary-Layer Theory

that is, the vorticity is constant along a streamline. Hence, if the flow is characterized by open streamlines, it follows that

$$\omega \equiv 0 \tag{10-9}$$

on any streamline that begins upstream in the uniform-flow region (where $\omega \equiv 0$). Flows in which $\omega = 0$ are known as *irrotational*.

Although $\omega = 0$ is a valid, leading-order approximation for the solution of (10–6) as $Re \to \infty$, we cannot tell without further analysis if it corresponds to a uniformly valid solution of the original problems (10–1) and (10–2) for **u**. For this, we need to determine the corresponding velocity fields. We do this most easily by expressing (10–9) in terms of **u**,

$$\nabla \wedge \mathbf{u} = 0. \tag{10-10}$$

Then, introducing the streamfunction, by means of the usual definition (see Chap. 7, Section C),

$$\mathbf{u} \equiv \nabla \wedge (\psi \mathbf{k}), \tag{10-11}$$

and combining (10–10) and (10–11), we find that

$$\nabla^2 \psi = 0. \tag{10-12}$$

Thus, we see that ψ in a 2D, irrotational flow satisfies Laplace's equation. Such motions are therefore also known as *potential flows*.

If we compare (10–12) with the full 2D Navier–Stokes equation, expressed in terms of the streamfunction, we note that the latter is fourth order (the viscous terms generate $\nabla^4 \psi$), whereas (10–12) is only second order. As a result, it is clear that the velocity field obtained from (10–12) will, at most, be able to satisfy only one of the boundary conditions of the original problem at the body surface. Intuitively, we may anticipate that the kinematic condition on the normal component of velocity,

$$\mathbf{u} \cdot \mathbf{n} = 0 \quad \text{on } S, \tag{10-13}$$

is the more crucial of the two original boundary conditions (the second being the no-slip condition) because it ensures that the predicted flow field moves *around* the body and not through it. From a physical viewpoint, no-slip at the wall can be established only when the fluid viscosity is nonzero. Hence, for an inviscid fluid, we also should expect that the no-slip condition must be abandoned. These intuitive arguments are, in fact, correct. It is generally possible to solve (10–12) subject to the kinematic condition (10–13), but it is *not* possible to simultaneously satisfy the no-slip condition. Thus the approximate solution (10–9) is clearly not valid in the immediate vicinity of the body surface.

In spite of this, we shall see that *potential-flow theory* plays an important role in the development of asymptotic solutions for $Re \gg 1$. Indeed, if we compare the assumptions and analysis leading to (10–9) and then to (10–12) with the early steps in analysis of heat transfer at high Peclet number, it is clear that the solution $\omega = 0$ is a valid first approximation for $Re \gg 1$ everywhere except in the immediate vicinity of the body surface. There the body dimension, a, that was used to nondimensionalize (10–1) is not a relevant characteristic length scale. In this region, we shall see that the flow develops a boundary layer in which viscous forces remain important even as $Re \gg 1$, and this allows the no-slip condition to be satisfied.

Before discussing the flow structure in the boundary-layer region, we digress briefly to say a few words about the historical development of boundary-layer theory. The early years of theoretical studies of high-Reynolds-number flows were, in fact, dominated by attempts to obtain solutions for the so-called potential-flow problem satisfying (10–12) and (10–13).[7] The reader may well wonder how early researchers could overlook the obvious

A. Potential-Flow Theory

flaw of not satisfying the no-slip condition. However, before being too critical, it should be remembered (see the discussion in Subsection L.3 in Chap. 2) that the no-slip condition is not a consequence of a general conservation principle such as Cauchy's equations of motion but is similar to constitutive equations in representing a mathematical hypothesis consistent with observed behavior. To early workers in fluid mechanics, the necessity of satisfying the no-slip condition was not so obvious, especially for very large values of the Reynolds number (or, equivalently, small viscosities). Indications that the potential-flow theory was incomplete, even as an approximation for $Re \gg 1$, were mainly macroscopic and based on comparisons between experimental observations and predictions of the potential-flow theory. The most dramatic evidence of a fundamental flaw in the theory was known as D'Alembert's paradox. This paradox was simply that the drag on a 2D body of arbitrary shape was predicted to be zero by the potential-flow theory but was found to increase monotonically with Re in experimental observations.

A second, equally dramatic difference between the potential-flow theory and experimental observation was that the flow patterns were often completely different. In the case of streaming flow past a circular cylinder, for example, the potential-flow solution is fore–aft symmetric with no indication of a wake downstream of the body. To show this we simply solve the potential-flow equation in cylindrical coordinates,

$$\frac{1}{r}\frac{\partial}{\partial r}\left(r\frac{\partial \psi}{\partial r}\right) + \frac{1}{r^2}\frac{\partial^2 \psi}{\partial \theta^2} = 0, \tag{10--14}$$

subject to the boundary conditions

$$\psi \to r\sin\theta \quad \text{as } r \to \infty, \tag{10--15}$$

$$\psi = 0 \quad \text{at } r = 1. \tag{10--16}$$

The condition (10–16) is just the kinematic condition (10–13) expressed in terms of the streamfunction, whereas (10–15) requires that the velocity field approach a uniform streaming motion at large distances from the cylinder. Equation (10–14), subject to (10–15) and (10–16), is solved easily by means of separation of variables, or other standard transform methods, with the resulting solution

$$\boxed{\psi = r\sin\theta\left(1 - \frac{1}{r^2}\right).} \tag{10--17}$$

This solution is clearly symmetric with respect to the fore–aft (upstream and downstream) directions. A sketch of streamlines (10–17) is shown in Fig. 10–1. For comparison we show streakline photographs of flow past a cylinder at two values of Reynolds number, $Re = 13.1$ and 26, in Fig. 10–2. The most obvious difference between the photographs and the predicted streamlines is the asymmetric flow pattern, with the large region of recirculation immediately to the rear of the cylinder. Of course, the theory is a limiting result for $Re = \infty$, whereas the photographs are for *finite* Reynolds number. The main reason for the moderate experimental Reynolds number is, in fact, that the flow pictured in Fig. 10–2 becomes unsteady at somewhat larger Reynolds numbers, and there is a complex series of transitions until eventually the motion becomes highly turbulent. Thus it is not necessarily true that the photographs are indicative of what the laminar flow would look like if we could take the limit $Re \to \infty$ without encountering instabilities. Nevertheless, there is nothing to suggest that the flow pattern would revert to one resembling (10–17). In fact, the unsteady flows encountered experimentally at large Reynolds numbers continue to show a strong degree of asymmetry and a clear remnant of the closed-streamline pattern in

Laminar Boundary-Layer Theory

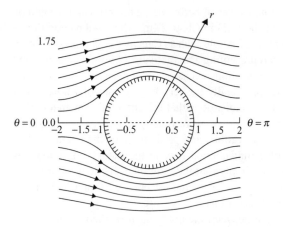

Figure 10–1. Streamlines for uniform streaming flow past a circular cylinder in the potential-flow limit, according to (10–17). Contours shown range from $\psi = 0$ to $\psi = 1.75$ in increments of 0.25.

the form of a downstream pattern of vortices that is known as the Karman vortex street. All this input suggests strongly that the potential-flow pattern, (10–17) and Fig. 10–1, is highly unrealistic for $Re \gg 1$.

One way to think of the difference between (10–17) and the actual velocity field near a cylinder for $Re \gg 1$, which allows us to take advantage of the close analogy to the heat transfer problem for high Peclet numbers discussed in the preceding chapter, is that the potential-flow solution lacks the vorticity characteristic of the motion of real fluids. It is particularly instructive to think of the transient physical process of establishing a steady-state velocity (or vorticity) field, beginning with the fluid initially at rest relative to the body and then suddenly imposing a uniform motion on the fluid at $t = 0$. Initially, $\boldsymbol{\omega} \equiv 0$ everywhere, and we have already seen that vorticity is not created within the fluid, so that the motion must remain irrotational [and precisely of the form (10–17) if the body is a circular cylinder] unless vorticity is produced at the bounding surfaces of the fluid and transported into the fluid by means of diffusion or convection according to (10–6). The mechanism for vorticity production at a *solid* bounding surface is the rotational motion imparted by the no-slip condition. Initially, in a real, viscous fluid, the motion will be very close to the irrotational potential flow except in the immediate vicinity of the body surface where the no-slip condition produces a large gradient in the tangential velocity component.[8] If we think in terms of vorticity, this corresponds to $\boldsymbol{\omega} = 0$ almost everywhere, but with a very strong source at the body surface. In the real fluid, this vorticity is transported radially outward into the fluid by diffusion, until eventually a steady-state configuration is achieved when production is balanced by convection and diffusion. From this viewpoint, it is obvious that the deficiency in the approximation (10–7) – or, equivalently, the potential-flow solution – is that it neglects vorticity diffusion completely and thus allows no mechanism for vorticity to "escape" from the body surface. As a consequence, $\boldsymbol{\omega} \equiv 0$ everywhere in the fluid, according to (10–9).

The conclusion to be drawn from the preceding discussion is that the potential-flow theory (10–9) [or, equivalently, (10–12) and (10–13)] does *not* provide a uniformly valid first approximation to the solution of the Navier–Stokes and continuity equations (10–1) and (10–2) for $Re \gg 1$. Furthermore, our experience in Chap. 9 with the thermal boundary-layer structure for large Peclet number would lead us to believe that this is because the velocity field near the body surface is characterized by a length scale $O(a Re^{-n})$, instead of the body dimension a that was used to nondimensionalize (10–2). As a consequence, the terms $\nabla^2 \omega$ and $\mathbf{u} \cdot \nabla \omega$, in (10–6), which are nondimensionalized by use of a, are not $O(1)$ and independent of Re everywhere in the domain, as was assumed in deriving (10–7), but instead are increasing functions of Re in the region very close to the body surface. Thus in

A. Potential-Flow Theory

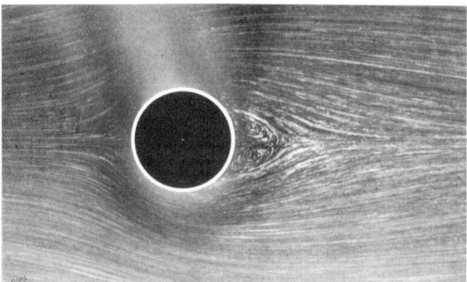

Figure 10–2. Streamline visualization of uniform streaming flow of a Newtonian fluid past a circular cylinder at Reynolds numbers $Re = 13.1$ (bottom) and $Re = 26$ (top). These photos were taken using a long exposure time to record the pathways of very small tracer particles in the flow. [Source: with M. D. Van Dyke, *An Album of Fluid Motion* Parabolic Press, Stanford, CA, 1982; original photos were taken by S. Taneda. The photo depicting $Re = 26$ was originally published by Taneda in 1956 in the *Journal of the Physical Society of Japan* **11**, 302–7.]

this region it is *not* true that $(1/Re)\nabla^2\omega$ becomes negligible compared with $\mathbf{u} \cdot \nabla\omega$, in the limit $Re \to \infty$, as Eq. (10–6) suggests.

We see then that (10–7) – and potential-flow theory – provides a leading-order approximation for $Re \gg 1$ in an outer region away from the body but is *not* valid very near the body surface. *Physically, to retain the essential effects of viscosity in the vicinity of the body, there is an internal readjustment in the flow that produces very large gradients in ω (or $\mathbf{u}$)*

Laminar Boundary-Layer Theory

near the surface. This is not reflected in Eqs. (10–1), (10–2), or (10–6), which are scaled with respect to the length scale of the body. To determine the form of the Navier–Stokes equation that holds near the body, we must rescale (10–1) and (10–2), or (10–6), to introduce a characteristic length scale that is appropriate to this inner boundary-layer region, and then employ the method of matched asymptotic expansions to obtain a uniformly valid approximation to the solution for $Re \gg 1$.

B. THE BOUNDARY-LAYER EQUATIONS

The discussion in the previous section leads to the following qualitative description of the flow domain for $Re \gg 1$:

(1) *Outer region*, where variations of velocity are characterized by the length scale a of the body and *potential-flow theory* provides a valid first approximation in an asymptotic expansion of the solution for $Re \to \infty$.
(2) *Inner region*, a boundary layer, of thickness $O(a\, Re^{-\alpha})$ near the body surface (with α yet to be determined), where *viscous effects* must be included, even in the limit $Re \to \infty$.

Because this inner boundary-layer region is infinitesimal in thickness relative to a, all curvature terms that appear when the equations of motion are expressed in curvilinear coordinates will drop out to first order in Re^{-1}, thus leaving boundary-layer equations that are effectively expressed in terms of a local Cartesian coordinate system.

To illustrate this latter point, we first derive equations for the inner boundary-layer region for the specific problem of streaming flow past a circular cylinder, starting from the equations of motion expressed in a cylindrical coordinate system. These are

$$\left(u_r \frac{\partial u_\theta}{\partial r} + \frac{u_\theta}{r} \frac{\partial u_\theta}{\partial \theta} + \frac{u_r u_\theta}{r} \right) = -\frac{1}{r} \frac{\partial p}{\partial \theta} + \frac{1}{Re} \left[\frac{\partial^2 u_\theta}{\partial r^2} + \frac{\partial}{\partial r} \left(\frac{u_\theta}{r} \right) + \frac{1}{r^2} \frac{\partial^2 u_\theta}{\partial \theta^2} + \frac{2}{r^2} \frac{\partial u_r}{\partial \theta} \right], \tag{10–18}$$

$$\left(u_r \frac{\partial u_r}{\partial r} + \frac{u_\theta}{r} \frac{\partial u_r}{\partial \theta} - \frac{u_\theta^2}{r} \right) = -\frac{\partial p}{\partial r} + \frac{1}{Re} \left[\frac{\partial^2 u_r}{\partial r^2} + \frac{\partial}{\partial r} \left(\frac{u_r}{r} \right) + \frac{1}{r^2} \frac{\partial^2 u_r}{\partial \theta^2} - \frac{2}{r^2} \frac{\partial u_\theta}{\partial \theta} \right], \tag{10–19}$$

$$\frac{1}{r} \frac{\partial}{\partial r}(r u_r) + \frac{1}{r} \frac{\partial u_\theta}{\partial \theta} = 0. \tag{10–20}$$

For convenience, we change from r to y,

$$r - 1 \equiv y,$$

so that y is a radial variable (normal to the body surface) that is *zero* at the surface, and we also change from θ to x,

$$\theta \equiv x.$$

Then, rewriting (10–18)–(10–20), we have

$$\left(v \frac{\partial u}{\partial y} + \frac{u}{1+y} \frac{\partial u}{\partial x} + \frac{uv}{1+y} \right)$$

$$= -\frac{1}{1+y} \frac{\partial p}{\partial x} + \frac{1}{Re} \left[\frac{\partial^2 u}{\partial y^2} + \frac{\partial}{\partial y} \left(\frac{u}{1+y} \right) + \frac{1}{(1+y)^2} \frac{\partial^2 u}{\partial x^2} + \frac{2}{(1+y)^2} \frac{\partial v}{\partial x} \right], \tag{10–21}$$

B. The Boundary-Layer Equations

$$\left(v\frac{\partial v}{\partial y} + \frac{u}{1+y}\frac{\partial v}{\partial x} - \frac{u^2}{1+y}\right)$$
$$= -\frac{\partial p}{\partial y} + \frac{1}{\text{Re}}\left[\frac{\partial^2 v}{\partial y^2} + \frac{\partial}{\partial y}\left(\frac{v}{1+y}\right) + \frac{1}{(1+y)^2}\frac{\partial^2 v}{\partial x^2} - \frac{2}{(1+y)^2}\frac{\partial u}{\partial x}\right], \quad (10\text{–}22)$$

$$\frac{1}{1+y}\frac{\partial}{\partial y}[(1+y)v] + \frac{1}{1+y}\frac{\partial u}{\partial x} = 0, \quad (10\text{–}23)$$

where we have denoted $u_r \equiv v$ and $u_\theta \equiv u$ in anticipation of a reduction, locally, to a Cartesian form for these equations in the boundary layer.

To obtain the appropriate form of the equations of motion for the inner, boundary-layer region, we must rescale with a new characteristic length scale. Following the precedent of previous analyses, we do this by rescaling the radial variable y in the equations of motion and continuity, (10–21)–(10–23), according to

$$\boxed{Y = \text{Re}^\alpha y,} \quad (10\text{–}24)$$

as it is the direction normal to the surface that is expected to exhibit a length scale much smaller than a and dependent on Re. If we introduce (10–24) into the continuity equation (10–23), we find that

$$\frac{1}{1+(Y/\text{Re}^\alpha)}\text{Re}^\alpha\frac{\partial}{\partial Y}\left[\left(1+\frac{Y}{\text{Re}^\alpha}\right)v\right] + \frac{1}{1+(Y/\text{Re}^\alpha)}\frac{\partial u}{\partial x} = 0. \quad (10\text{–}25)$$

Thus, approximating $1 + Y/\text{Re}^\alpha$ as 1, as is appropriate at the leading order of approximation, we have

$$\text{Re}^\alpha\frac{\partial v}{\partial Y} + \frac{\partial u}{\partial x} = 0. \quad (10\text{–}26)$$

Clearly, the magnitude of either u or v must depend on Re in this region if the form of the continuity equation is to be preserved in the limit $Re \to \infty$, and this condition is necessary to ensure that the boundary-layer solution is consistent with conservation of mass. The tangential velocity component u will be required to *match* the tangential velocity component from the potential-flow solution in the region of overlapping validity between the inner (boundary-layer) and outer (potential-flow) regions. But the potential-flow solution, say, (10–17) for the circular cylinder, clearly yields a tangential velocity in the vicinity of the body that is $O(1)$, and this implies that $u = O(1)$ in the inner boundary-layer domain if the two solutions are to match. The normal velocity component, on the other hand, approaches zero in the potential-flow approximation as the body surface is approached. Thus a normal velocity component of $O(Re^{-\beta})$ in the boundary-layer region can be matched to within an arbitrarily small error for $Re \to \infty$ with the potential-flow solution. This suggests that the rescaling

$$\boxed{V = \text{Re}^\beta v} \quad (10\text{–}27)$$

is allowable, and we see immediately from (10–26) that $\alpha = \beta$ so that

$$\boxed{v\frac{\partial u}{\partial x} + \frac{\partial V}{\partial Y} = 0} \quad (10\text{–}28)$$

in terms of the rescaled inner variables Y and V. The reader should recall that α is to be chosen so that derivatives with respect to Y are $O(1)$ – independent of Re – in this inner region. The dimension of the inner region is $O(a\,Re^{-\alpha})$ according to (10–24), and, provided that $\alpha > 0$, as we have assumed, this is extremely small for $Re \gg 1$. Accordingly, it is not surprising that the continuity equation reduces to a local Cartesian form (10–28). If we go back to Eq. (10–25), we see that the "curvature" terms are $O(Re^{-\alpha})$ compared with the terms that we have retained. It must be remembered, however, that Eq. (10–28) governs only the first term in an asymptotic expansion of the full solution for the inner region. If we wished to calculate higher-order contributions to this expansion, we would have to retain the curvature terms in Eq. (10–25) as they would appear in the governing equations at $O(\mathrm{Re}^{-\alpha})$. In the present analysis, we will be content with obtaining only the first term in both the inner and outer regions (the latter being the potential-flow solution).

To determine α and the appropriate form of the equations of motion, (10–21) and (10–22), in the inner region, we substitute the rescaled variables (10–24) and (10–27) into these equations. The result for the tangential component (10–21) is

$$\left\{ V \frac{\partial u}{\partial Y} + \frac{u}{[1+(Y/Re^\alpha)]} \frac{\partial u}{\partial x} + \frac{1}{Re^\alpha} \frac{uV}{[1+(Y/Re^\alpha)]} \right\}$$

$$= -\frac{1}{[1+(Y/Re^\alpha)]} \frac{\partial p}{\partial x} + Re^{2\alpha-1} \frac{\partial^2 u}{\partial Y^2} + \frac{Re^\alpha}{Re} \frac{\partial}{\partial Y} \left[\frac{u}{1+(Y/Re^\alpha)} \right]$$

$$+ \frac{1}{Re[1+(Y/Re^\alpha)]^2} \left(\frac{\partial^2 u}{\partial x^2} + \frac{2}{Re^\alpha} \frac{\partial V}{\partial x} \right). \tag{10–29}$$

We can obtain the governing equation for the first approximation to u and V by taking the limit $Re \to \infty$,

$$u\frac{\partial u}{\partial x} + V\frac{\partial u}{\partial Y} + O(Re^{-\alpha}) = -\frac{\partial p}{\partial x} + Re^{2\alpha-1}\left[\frac{\partial^2 u}{\partial Y^2} + O(Re^{-\alpha})\right] + O(Re^{-1}). \tag{10–30}$$

Hence, the largest viscous term is $O(Re^{2\alpha-1})$ and it is clear that

$$\alpha = \frac{1}{2} \tag{10–31}$$

if viscous effects are to be retained in the limit $Re \to \infty$. It follows from (10–31) that the dimensionless boundary-layer thickness is $O(Re^{-1/2})$.

The reader may note that we have thus far considered only the tangential component of the equations of motion. To see whether any additional information can be obtained from the normal component equation (10–22), we substitute (10–24) and (10–27) with $\beta = \alpha = 1/2$. The result is

$$\frac{\partial p}{\partial Y} = O(Re^{-1/2}), \tag{10–32}$$

from which we conclude that the pressure is constant across the boundary layer to a first approximation; that is, p is a function of x only. Equations (10–28), (10–30), and (10–32) are sufficient to determine the three unknowns in the boundary-layer region, u, V, and p. We shall return shortly to the method of solution.

First, however, it is important to recognize that the *form* of equations (10–28), (10–30), and (10–32) is *independent* of the geometry of the body (i.e., independent of the cross-sectional shape for any 2D body). Although we started our analysis with the specific problem of flow past a circular cylinder, and thus with the equations of motion in cylindrical coordinates, the equations for the leading-order approximation in the inner (boundary-layer) region reduce to a local, Cartesian form with Y being normal to the body surface and x

B. The Boundary-Layer Equations

tangential. The geometry of the body does not directly enter these so-called boundary-layer equations at all. Indeed, we shall see shortly that it is only through the requirement of matching with the outer potential-flow solution that the boundary-layer solution depends on the body geometry.

Let us then consider an *arbitrary* 2D body. In all cases, the governing equations at leading order in the boundary layer are

$$\frac{\partial u}{\partial x} + \frac{\partial V}{\partial Y} = 0, \tag{10--33}$$

$$u\frac{\partial u}{\partial x} + V\frac{\partial u}{\partial Y} = -\frac{\partial p}{\partial x} + \frac{\partial^2 u}{\partial Y^2}, \tag{10--34}$$

$$\frac{\partial p}{\partial Y} = 0. \tag{10--35}$$

These are known as the *boundary-layer equations*. There is one further simplification that we can always introduce. In particular, if we integrate (10–35), we see that the pressure distribution in the boundary layer is a function of x only. Thus the pressure gradient in the boundary layer ($dp/\partial x$) is also independent of Y and must have the same form as the pressure gradient in the outer potential flow, evaluated in the limit as we approach the body surface, namely,

$$\left(\frac{\partial p}{\partial x}\right)_{\text{boundary layer}} = \lim_{y \to 0} \left(\frac{\partial p}{\partial x}\right)_{\text{potential flow}} \tag{10--36}$$

In other words, the pressure distribution in the boundary-layer is completely determined at this level of approximation by the limiting form of the pressure distribution impressed at its outer edge by the potential flow. It is convenient to express this distribution in terms of the potential-flow velocity distribution. In particular, let us define the tangential velocity function $u_e(x)$ as

$$u_e(x) \equiv (\mathbf{u} \cdot \mathbf{t})|_{y \to 0}, \tag{10--37}$$

where $y = 0$ denotes the body surface and $\mathbf{t}$ is a unit tangent vector (that is, $\mathbf{n} \cdot \mathbf{t} \equiv 0$). In terms of the velocity and pressure, the leading-order equation of motion in the potential-flow region is

$$\mathbf{u} \cdot \nabla \mathbf{u} = -\nabla p. \tag{10--38}$$

However, the normal velocity component is zero at the surface of the body, and (10–38) is reduced in the limit $y \to 0$ to an equation for the pressure gradient that is required in (10–36),

$$\frac{\partial p}{\partial x}\bigg|_{y \to 0} = -u_e \frac{du_e}{\partial x}. \tag{10--39}$$

This is a form of the equation that is often referred to as *Bernoulli's equation*. Once the potential-flow velocity field is known, it provides a means to determine $\partial p/\partial x$ at the outer edge of the boundary-layer region and hence, by means of (10–36), the pressure gradient $\partial p/\partial x$ in the boundary layer itself.

Laminar Boundary-Layer Theory

Thus, insofar as the boundary layer is concerned, the pressure gradient may be regarded as "known" – it is actually obtained by solving the outer potential-flow problem – and the boundary-layer equations can be written as

$$u\frac{\partial u}{\partial x} + V\frac{\partial u}{\partial Y} = u_e(x)\frac{du_e}{dx} + \frac{\partial^2 u}{\partial Y^2}, \qquad (10\text{--}40)$$

$$\frac{\partial u}{\partial x} + \frac{\partial V}{\partial Y} = 0, \qquad (10\text{--}41)$$

with u and V being unknown. To demonstrate that these equations and the potential-flow equation (10–12) actually provide a basis to determine a uniformly valid approximation to the solution of the equations of motion for $Re \to \infty$, we must consider boundary conditions and matching conditions for the solutions in the two regions.

The boundary conditions are straightforward. The outer potential-flow solution is required to take the form of the undisturbed velocity field at infinity; that is, in the case of uniform streaming motion,

$$\mathbf{u} \to \mathbf{e} \quad \text{as } |\mathbf{r}| \to \infty, \qquad (10\text{--}42)$$

where $\mathbf{e}$ is a unit vector in the direction of the free-stream motion. At the body surface, the no-slip and kinematic conditions give boundary conditions for the inner fields:

$$u = V = 0 \quad \text{at } Y = 0. \qquad (10\text{--}43)$$

However, the conditions (10–42) and (10–43) are *not* sufficient to obtain unique solutions in either the inner or outer domains.

The boundary-layer equations are third order with respect to Y and first order with respect to x [this can be seen clearly if (10–40) and (10–41) are expressed in terms of the streamfunction]. Hence, to specify completely the velocity profiles in the boundary layer, we require one additional boundary condition in Y and an initial profile at the leading edge of the boundary layer (x is usually defined so that this point corresponds to $x = 0$). In addition, the potential-flow equations are second order and thus require at least one boundary condition in addition to (10–42). In Section A, it was suggested that an appropriate condition was

$$\mathbf{u} \cdot \mathbf{n} = 0 \quad \text{on the body surface } S_B. \qquad (10\text{--}44)$$

However, this was predicated on the assumption that the potential-flow solution was to be applied in the whole of the domain, whereas we now see that it should apply only up to a distance of $O(Re^{-1/2})$ from S_B, but not right at the surface itself.

To determine the additional conditions that must be imposed to obtain a well-posed problem for the velocity fields in the inner and outer domains, we must examine the *matching conditions* in the region of overlap between them. These matching conditions can be written in general form as

$$\lim_{y \ll 1} u(x, y) \Leftrightarrow \lim_{Y \gg 1} u(x, Y) \quad \text{as } Re \to \infty, \qquad (10\text{--}45)$$

$$\lim_{y \ll 1} v(x, y) \Leftrightarrow \lim_{Y \gg 1} \frac{1}{\sqrt{Re}} V(x, Y) \quad \text{as } Re \to \infty, \qquad (10\text{--}46)$$

B. The Boundary-Layer Equations

where $y = 0$ is still used to denote the body surface in outer variables. The condition (10–46) includes $\sqrt{Re}$ so that both sides can be expressed in equivalent terms according to (10–27) with $\beta = 1/2$.

Let us now examine the matching conditions in more detail. Turning first to (10–46), we see that the right-hand side is arbitrarily small, $O(Re^{-1/2})$ for $Re \to \infty$, relative to the left-hand side, provided only that $V(x,\infty)$ is finite. Thus, subject to a *posteriori* verification of this latter fact, we see that the matching condition (10–46) is precisely equivalent to the boundary condition (10–44), which can thus be used for the first (i.e., potential-flow) approximation for the normal velocity component in the outer solution. But this condition, together with (10–42), is sufficient to completely determine the outer-flow solution, which we see is just the potential-flow solution of the previous section. The only difference is that (10–44) is now seen as a *matching* condition, valid only as a leading-order approximation that is applied at the outer edge of the boundary layer, rather than an exact condition that is applied right at the body surface, as before.

Let us suppose that we have solved the potential-flow problem. Then the left-hand side of (10–45) becomes a *known* function of x, which we have denoted previously as $u_e(x)$, and this matching condition becomes a boundary condition for the boundary-layer solution

$$\lim_{Y \gg 1} u(x, Y) = u_e(x). \qquad (10\text{--}47)$$

Together with (10–43) and an initial profile at $x = 0$, this condition is sufficient to completely specify the solution of (10–40) and (10–41).

It would thus appear that the potential-flow and boundary-layer equations (10–12), (10–40), and (10–41), plus the boundary conditions (10–42) and (10–43) and the matching conditions (10–44) and (10–47), provide a self-contained basis to determine the *leading-order approximation* to the velocity and pressure fields for steady, laminar streaming flows past 2D bodies of arbitrary shape. First, we solve the potential-flow problem by using (10–42) and (10–44). Then, with $u_e(x)$ known, we calculate the pressure gradient in the boundary layer by using (10–39) and solve the boundary-layer equations (10–40) and (10–41) with the boundary conditions (10–43) and (10–47), plus an initial condition at $x = 0$. This prescription works for a wide variety of *streamlined* body shapes. However, there is one intrinsic approximation in the formulation that can invalidate the procedure (or at least decrease its domain of validity) for bodies of *nonstreamlined* shape (such as the circular cylinder that we originally began to analyze). This is the assumption, inherent in (10–44) and the corresponding matching condition (10–46), that all regions of nonzero vorticity remain confined to a very thin layer, with a thickness of $O(Re^{-1/2})$, immediately adjacent to the body surface. For the horizontal flat plate and streamlined airfoil shapes that we consider in the next section, this assumption is actually satisfied. However, assuming that the photographs in Fig. 10–2 for large, but finite, Reynolds number are at least qualitatively indicative of the asymptotic flow structure for $Re \to \infty$, it is patently obvious that the assumption is drastically violated for the circular cylinder (and all nonstreamlined bodies) because of the large, recirculating region of flow immediately downstream. In such cases, a boundary layer still exists over the front part of the body, but at some point (called the *separation point*), this boundary layer departs radically from the surface, leading to the formation of the recirculating region (of nonzero vorticity) downstream.

The reader may wonder why this change in flow structure invalidates the boundary-layer theory. Upstream of the separation point, the flow still divides into a boundary layer of the usual kind on the surface and an outer, potential flow. However, the potential-flow approximation applies only outside any region of nonzero vorticity. When these regions all remain attached to the body and are of $O(Re^{-1/2})$ in thickness, it is adequate for at least the first term in an asymptotic solution to approximate this region as encompassing the whole of the domain external to the body itself. When the boundary layer separates, however, it almost always leads to the formation of a large region of recirculating flow downstream in which the vorticity is nonzero. Thus, in these cases, the potential-flow solution should be applied to the region outside both the body and this attached recirculating wake, with (10–44) applied to this "composite" body rather than the solid body alone. Furthermore, because the potential-flow equations are elliptic in type, we should expect the solution everywhere in the domain to be influenced by the modification in boundary shape to include the recirculating wake. Thus, to obtain the pressure distribution in the boundary layer before the separation point, it is necessary, in principle, to solve the potential-flow problem including any recirculating flow regions with dimensions of $O(1)$ as part of the "effective" body. But now we encounter a severe difficulty.

The shape of the recirculating wake region is *a priori* unknown, and it can be calculated only in the context of a complete asymptotic theory in the whole domain – appropriate scaling and governing equations in the recirculating flow region and any boundary layers adjacent to this region, plus, in all likelihood, a wake region downstream of the recirculating zone. At present, and in spite of intensive effort for many years, such a theory does not exist.[9] One problem is that the wake region becomes unstable at quite modest Reynolds numbers [$\sim O(40)$ for the circular cylinder], and thus experimental studies do not provide a clear indication of the scale of the recirculating region or the magnitude of velocities within it in a range of Reynolds number that can be extrapolated to $Re \sim \infty$. In the absence of this kind of input, it has been extremely difficult to develop an asymptotic theory for the recirculating wake. Recently, numerical solutions have been obtained for steady flow past a circular cylinder up to $Re \sim 600$.[10] These appear to be suggestive of the asymptotic behavior for $Re \to \infty$ and may eventually lead to a satisfactory asymptotic solution for laminar, separated flows at $Re \gg 1$. For the moment, however, the available proposals for the flow structure in the limit $Re \to \infty$ remain controversial, and the steady, separated-flow problem is still regarded as one of the two most important unsolved problems in fluid mechanics (the other being a satisfactory theoretical description of turbulence by some means other than direct numerical simulation[11]).

What is amazing about boundary-layer theory is that it still has been incredibly useful for precisely the class of problems in which separation occurs. Although it does *not* provide a uniformly valid description of velocity and pressure fields in such cases, it can be used to determine when (that is, for what body geometries) and where the boundary layer will separate. Beyond the separation point, the boundary-layer assumptions break down. For example, the dimension of the region of nonzero vorticity is no longer $O(Re^{-1/2})$. Hence a prediction of separation is tantamount to the statement that the boundary-layer theory is capable of providing an accurate prediction of the point on the body surface where it becomes invalid! We obtain this prediction, as we shall see, by calculating the pressure distribution for potential flow around the body *without* taking into account the possibility of a closed-streamline wake and using this pressure distribution in the boundary-layer equations. The existence of a separation point is signaled if $\partial u/\partial Y|_{Y=0} \to 0$ at some point on the body surface, and when this occurs, the boundary-layer solution develops a singularity.[12] This scheme, though clearly an *ad hoc* application of the asymptotic, boundary-layer approach, works because the *actual* pressure distribution in the presence of the recirculating wake and the predicted pressure distribution from potential-flow theory in the absence of a wake are reasonably similar almost up to the point of actual separation. This is illustrated in Fig. 10–3, where we show a comparison between an experimental pressure distribution on a circular cylinder for relatively small Reynolds numbers,

B. The Boundary-Layer Equations

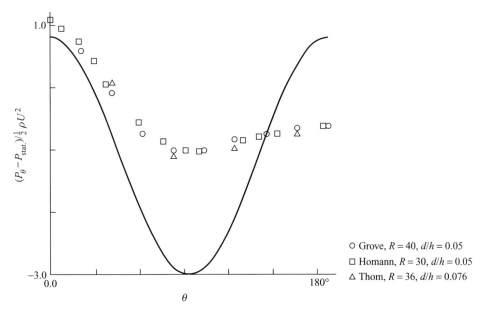

Figure 10–3. A comparison of the pressure distribution on the surface of a circular cylinder according to potential-flow theory (———) and as measured experimentally for $30 \leq Re \leq 40$. The experimental data are taken from A. S. Grove et al., Steady separated flow past a circular cylinder, *J. Fluid Mech.* **19**, 60–80 (1964), where original references are given to Homann's and Thorn's papers. The offset in stagnation point pressure at $\theta - 0$ is a result of the finite Reynolds number, as shown in Grove et al. (1964). The difference in the pressure between $\theta = 0$ and the point of separation (which occurs at $\theta \sim 135°$ at this relatively low Reynolds number) is partly due to finite Reynolds number effects within the boundary layer and partly due to the influence of the recirculating wake on the external flow, which is felt significantly upstream of the separation point.

$30 \leq Re \leq 40$, and the predicted pressure distribution from potential-flow theory. The ability of the boundary-layer theory to predict separation is probably its most important characteristic.

It has been stated repeatedly that the boundary-layer and potential-flow equations apply to only the leading term in an asymptotic expansion of the solution for $Re \gg 1$. This is clear from the fact that we derived both in their respective domains of validity by simply taking the limit $Re \to \infty$ in the appropriately nondimensionalized Navier–Stokes equations. Frequently, in the analysis of laminar flow at high Reynolds number, we do not proceed beyond these leading-order approximations because they already contain the most important information: a prediction of whether or not the flow will separate and, if not, an analytic approximation for the drag. Nevertheless, the reader may be interested in how we would proceed to the next level of approximation, and this is described briefly in the remainder of this section.[13]

At the next level, we first calculate the second term in the outer solution. If we examine the equation of motion for the outer region,

$$\mathbf{u} \cdot \nabla \mathbf{u} + \nabla p = \frac{1}{Re} \nabla^2 \mathbf{u}, \quad \nabla \cdot \mathbf{u} = 0, \qquad (10\text{–}48)$$

we see that the first neglected term in the preceding analysis was $O(Re^{-1})$. On the other hand, examination of the matching conditions (10–45) and (10–46) shows that there is a

Laminar Boundary-Layer Theory

mismatch between the first term in the outer region that satisfies the boundary condition (10–44) and the boundary-layer solution that takes a value

$$\lim_{Y \gg 1} \frac{1}{\sqrt{Re}} V(x, Y) = O(Re^{-1/2}) \tag{10–49}$$

at the outer edge of the boundary-layer. Taken together, (10–48) and (10–49) imply that the next term in the asymptotic expansion for the outer region must be $O(Re^{-1/2})$, that is,

$$\mathbf{u} = \mathbf{u}_0 + \frac{1}{\sqrt{Re}} \mathbf{u}_1 + o(Re^{-1/2}), \tag{10–50}$$

$$p = p_0 + \frac{1}{\sqrt{Re}} p_1 + o(Re^{-1/2}). \tag{10–51}$$

To obtain the governing equations for the first and second terms in this expansion, we substitute (10–50) and (10–51) into (10–48):

$$\mathbf{u}_0 \cdot \nabla \mathbf{u}_0 + \nabla p_0 + \frac{1}{\sqrt{Re}} (\mathbf{u}_0 \cdot \nabla \mathbf{u}_1 + \mathbf{u}_1 \cdot \nabla \mathbf{u}_0 + \nabla p_1) = O(Re^{-1}), \tag{10–52}$$

$$\nabla \cdot \mathbf{u}_0 + \frac{1}{\sqrt{Re}} (\nabla \cdot \mathbf{u}_1) + o(Re^{-1/2}) = 0. \tag{10–53}$$

Hence, at $O(1)$,

$$\mathbf{u}_0 \cdot \nabla \mathbf{u}_0 + \nabla p_0 = 0, \quad \nabla \cdot \mathbf{u}_0 = 0, \tag{10–54}$$

which, together with the boundary conditions (10–42) and (10–44), is just the potential-flow problem we discussed in Section A of this chapter. At $O(Re^{-1/2})$,

$$\mathbf{u}_0 \cdot \nabla \mathbf{u}_1 + \mathbf{u}_1 \cdot \nabla \mathbf{u}_0 + \nabla p_1 = 0, \quad \nabla \cdot \mathbf{u}_1 = 0, \tag{10–55}$$

and, at this level of approximation, the boundary condition at infinity is

$$\mathbf{u}_1 \to 0 \quad \text{as } |\mathbf{x}| \to \infty, \tag{10–56}$$

and the matching condition (10–46), together with (10–49), yields

$$\mathbf{u}_1 \cdot \mathbf{n} = V(x, \infty) \quad \text{at } y = 0. \tag{10–57}$$

At this second level, the outer solution is modified to account for the weak outward displacement of the potential flow that is caused by the presence of the boundary layer at the body surface. It can be seen in (10–57) that this displacement manifests itself in the outer flow as a weak "blowing" velocity normal to the body surface.

In the inner boundary-layer region, there are neglected terms of $O(Re^{-1/2})$ in the first-order equation we obtain from (10–29) by taking the limit $Re \to \infty$. In addition, the $O(Re^{-1/2})$ velocity field $\mathbf{u}_1$ in the outer region requires an $O(Re^{-1/2})$ correction in the inner region because of both the matching condition (10–45) and the modification of the pressure gradient imposed on the boundary layer from the outer flow. To obtain governing equations for the higher-order terms in the boundary-layer regime, it follows that we would again use an expansion of the form

$$u = u_0(x, Y) + Re^{-1/2} u_1(x, Y) + o(Re^{-1/2}),$$
$$V = V_0(x, Y) + Re^{-1/2} V_1(x, Y) + o(Re^{-1/2}) \tag{10–58}$$

in (10–29) with $\alpha = 1/2$.

C. Streaming Flow Past a Horizontal Flat Plate – The Blasius Solution

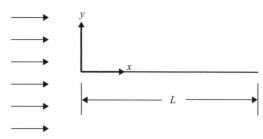

Figure 10–4. A schematic of the geometric configuration for streaming flow past a horizontal flat plate of finite length L.

C. STREAMING FLOW PAST A HORIZONTAL FLAT PLATE – THE BLASIUS SOLUTION

In this section we apply the general boundary-layer analysis of the previous sections to the problem of laminar streaming flow past a horizontal flat plate at high Reynolds number. We assume, at the outset, that the plate length is finite (denoted as L) and that the flow at infinity is parallel to the plate surface. A schematic representation of the flow configuration is given in Fig. 10–4. In this case, we define the Reynolds number as $Re \equiv U_\infty L/\nu$. Later, we shall see that the plate can be either finite or semi-infinite in length with no change in the solution structure.

Following the prescription for solving boundary-layer problems outlined in the previous section, we begin with the outer potential-flow problem. For this purpose, it is convenient to adopt a fixed system of Cartesian coordinates (x, y), with y normal to the body surface and x tangent to the surface with $x = 0$ denoting the upstream (or *leading*) edge of the plate. Then the potential-flow equation is just

$$\frac{\partial^2 \psi}{\partial x^2} + \frac{\partial^2 \psi}{\partial y^2} = 0, \qquad (10\text{–}59)$$

which is to be solved subject to the free-stream boundary condition

$$\psi \to y \quad \text{as } y \to \infty \qquad (10\text{–}60)$$

and the matching condition (10–44) at $y = 0$, which becomes just

$$\psi = 0 \quad \text{at } y = 0 \qquad (10\text{–}61)$$

when expressed in terms of the streamfunction ψ. The solution of (10–59)–(10–61) is trivial, namely,

$$\boxed{\psi = y, \quad \text{for all } x \text{ and } y.} \qquad (10\text{–}62)$$

This solution simply implies that the velocity everywhere in the outer region is just the undisturbed free-stream velocity. The flat plate is idealized mathematically as having zero thickness, and, because the no-slip condition does not apply to the potential-flow regime, the fluid moves past the plate with no disturbance whatsoever.

The next step is to solve the boundary-layer equations (10–40) and (10–41) with $\partial p/\partial x$ calculated from (10–39). However, in this case,

$$u_e = 1,$$

and thus

$$\partial p/\partial x = 0 \qquad (10\text{–}63)$$

Laminar Boundary-Layer Theory

to leading order in the boundary layer. It follows that the boundary-layer problem reduces to

$$u\frac{\partial u}{\partial x} + V\frac{\partial u}{\partial Y} = \frac{\partial^2 u}{\partial Y^2}, \quad \frac{\partial u}{\partial x} + \frac{\partial V}{\partial Y} = 0, \qquad (10\text{–}64)$$

with boundary conditions from (10–43) and (10–47),

$$u = V = 0 \quad \text{at } Y = 0, \ x > 0, \qquad (10\text{–}65)$$

$$u \to 1 \quad \text{as } Y \to \infty, \ x > 0, \qquad (10\text{–}66)$$

and an initial profile

$$u = 1 \quad \text{at } x = 0. \qquad (10\text{–}67)$$

The latter condition corresponds to the physical fact that the fluid is not disturbed at all by the plate until it actually encounters the no-slip condition for $x \geq 0$.

The problem (10–64)–(10–67) was first solved by Blasius, in his doctoral thesis (1908), using the similarity transformation[14]

$$u \equiv f'(\eta) \quad \text{with} \quad \eta \equiv \frac{Y}{g(x)}. \qquad (10\text{–}68)$$

It is convenient to define f so that $u = \partial f/\partial \eta$, because otherwise V must involve an integral of f. As it is, we see from the continuity equation and the boundary condition on V at $Y = 0$ that

$$V = -\int_0^Y \frac{\partial u}{\partial x} dY = g'(x) \int_0^\eta t f''(t) dt.$$

Integrating by parts, we obtain

$$V \equiv g'(x)(\eta f' - f), \qquad (10\text{–}69)$$

where $f(0) = 0$ so that $V = 0$ at $Y = \eta = 0$ and $f'(0) = 0$ so that $u = 0$ at $\eta = 0$. Substituting (10–68) and (10–69) into (10–64) yields

$$f''' + gg' ff'' = 0. \qquad (10\text{–}70)$$

Boundary conditions on f can be derived from (10–65) and (10–66):

$$f = f' = 0 \quad \text{at } \eta = 0, \qquad (10\text{–}71)$$

$$f' = 1 \quad \text{as } \eta \to \infty. \qquad (10\text{–}72)$$

Because the DE for f is third order, these are all the boundary conditions that we can impose.

Thus, if a similarity solution of the form (10–68) is to be possible, the condition (10–72) must represent both the boundary condition for $Y \to \infty$ and the initial condition at $x = 0$. This is possible, in the present problem, provided that

$$g(0) = 0, \qquad (10\text{–}73)$$

and this is a necessary condition for the existence of a similarity solution.

C. Streaming Flow Past a Horizontal Flat Plate – The Blasius Solution

A second necessary condition is that the coefficient gg' in Eq. (10–70) must be either zero or a nonzero constant. However, it cannot be zero because no solution of $f''' = 0$ exists to satisfy (10–71) and (10–72). Because $gg' \neq 0$, we choose

$$\boxed{gg' = 1} \tag{10-74}$$

for convenience, and then

$$\boxed{f''' + ff'' = 0.} \tag{10-75}$$

The latter is known as the *Blasius equation*. To complete the similarity solution, we must solve (10–74) subject to (10–73), and (10–75) subject to (10–71) and (10–72).

The solution for $g(x)$ is trivial. Equation (10–74) can be written in the alternative form

$$\frac{dg^2}{dx} = 2. \tag{10-76}$$

Hence, integrating once and applying the boundary condition $g(0) = 0$, we obtain

$$\boxed{g = \sqrt{2x}.} \tag{10-77}$$

The solution for $f(\eta)$ cannot be obtained analytically. Although the similarity transformation has reduced the set of PDEs, (10–64), to a single ODE, (10–75), the latter is still *nonlinear*. In fact, Blasius originally solved (10–75) by using a numerical method, but with the algebra carried out by hand! Fortunately, today accurate numerical solutions can be obtained with a computer. The main difficulty in solving (10–75) numerically is that most methods for solving ODEs are set up for initial-value problems.

To convert a nonlinear boundary-value problem like (10–71), (10–72), and (10–75) to an initial-value form suitable for solution by means of standard numerical methods, the most common approach is to use a "shooting" method. In this method, we first guess a value of f'' at $\eta = 0$, say,

$$f''(0) = A,$$

and solve (10–75) as an initial-value problem, with $f(0) = f'(0) = 0$ and $f''(0) = A$. Of course, it is highly unlikely that $f'(\infty)$ will be equal to 1 for an arbitrary choice of A. However, if A is not too far from the correct value, we may find that

$$f'(\infty) \to B \quad \text{as } \eta \to \infty.$$

Then, if the problem is solved again for $f''(0) = A + \varepsilon$, we should find $f'(\eta) \to B + \delta$ as $\eta \to \infty$. Provided $\delta \ll 1$ when $\varepsilon \ll 1$, a new guess for $f''(0)$ can be obtained because we have effectively calculated

$$\Delta = \frac{\Delta f'(\infty)}{\Delta f''(0)}.$$

For example, a new guess for $f''(0)$ could be

$$f''(0) = A + \left[\frac{\Delta f''(0)}{\Delta f'(\infty)}\right](1 - B) = A + \frac{1 - B}{\Delta}.$$

Although "shooting" works well for the Blasius equation, an unusual feature of this particular equation is that we can transform directly from the first solution for $f''(0) = A$ to the exact solution once we have calculated $f'(\infty) = B$. To see how this works, let us define an auxiliary function $F(z)$ that satisfies the following equations and boundary conditions:

$$\begin{aligned} F''' + FF'' &= 0, \\ F(0) &= 0, \\ F'(0) &= 0, \\ F''(0) &= A, \end{aligned} \quad (10\text{–}78)$$

where the independent variable in this case has been denoted as z. However, this is just the initial-value problem that was solved as first step in the "shooting" method. Thus the solution is known to yield

$$F'(z) = B \quad \text{as } z \to \infty. \quad (10\text{–}79)$$

But now let us introduce a new function $f(\eta)$, where

$$f = \frac{1}{\sqrt{B}} F \quad \text{and} \quad \eta = \sqrt{B} z. \quad (10\text{–}80)$$

Then, transforming from $F(z)$ to $f(\eta)$ in problem (10–78), using (10–80), we find that

$$\begin{aligned} f''' + ff'' &= 0, \\ f(0) &= 0, \\ f'(0) &= 0, \\ f''(0) &= A/B^{3/2}. \end{aligned}$$

Furthermore, because the function $F(z)$ satisfies (10–79), the new function f must satisfy

$$f'(\infty) = 1.$$

But this is just the *Blasius* problem, and $f(\eta)$ is therefore the Blasius function. Clearly we need solve only a *single* trial problem with initial value $f''(0) = A$ to obtain the asymptotic value $f'(\infty) = B$, and we can transform directly to the Blasius solution, which is obtained as an initial-value problem with $f''(0) = A/B^{3/2}$.

A plot for $f'(\eta)$ versus η was obtained by numerical solution of the Blasius equation and is shown in Fig. 10–5. We find that

$$\boxed{f''(0) = 0.469} \quad (10\text{–}81)$$

and that $f'(\eta) \sim 0.99$ for $\eta \sim 3.6$. The latter result provides a qualitative estimate of the width of the boundary layer. In terms of the original variables, nondimensionalized by the plate length, $\eta = 3.6$ corresponds to

$$y \sim 3.6 \sqrt{\frac{2x}{Re}}. \quad (10\text{–}82)$$

We recognize that the numerical coefficient 3.6 is arbitrary and dependent on the value of $f'(\eta) = u$, that we choose to represent an effective "edge" of the boundary layer. However, the dependence on x and Re are independent of this choice and are thus of considerable physical interest. In particular, the dependence of the boundary-layer thickness on $\sqrt{x}$

C. Streaming Flow Past a Horizontal Flat Plate – The Blasius Solution

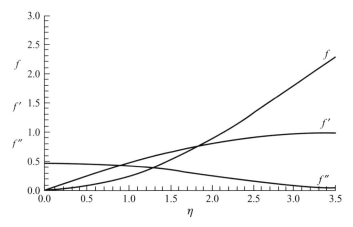

Figure 10–5. The Blasius function and its derivatives as a function of the similarity variable η.

indicates a singularity in the solution in the limit as $x \to 0$. This shows up most directly in the shear stress at the plate surface,

$$\tau_{xy} = \mu \left(\frac{\partial u}{\partial y}\right)_{y=0} = \frac{\mu U_\infty}{L}\sqrt{\frac{Re}{2x}} f''(0). \qquad (10\text{–}83)$$

Obviously, $\tau_{xy} \to \infty$ as $x^{-1/2}$ for $x \to 0$. We may note also that the normal velocity component diverges as $x^{-1/2}$ for all η other than $\eta = 0$, where $V \equiv 0$ for all x.

This *singularity* in the solution as $x \to 0$ indicates that the boundary-layer approximation breaks down as we approach the leading edge of the plate. This is not surprising. The boundary-layer approximation is based on the assumption that derivatives with respect to y exceed those with respect to x by a large amount, proportional to $\sqrt{Re}$. However, this assumption clearly breaks down near $x = 0$, where there is a discontinuity in boundary conditions on the axis $y = 0$. Fortunately, the error inherent in the boundary-layer solution for $x \to 0$ does not have a serious effect on the solution for other values of x, nor does it introduce an appreciable error in the predicted total force on the plate because τ_{xy} is still integrable for $x \to 0$. The reader interested in the correct flow structure for $x \ll 1$ may wish to refer to a paper by Carrier and Lin.[15] Here we simply ignore the singularity in the boundary-layer solution at $x = 0$ because it does not affect the calculated force on the plate. We obtain the latter by integrating the local shear stress (10–83) over the top and bottom surfaces of the plate. The result, per unit length of the plate in the third direction z, is

$$D = 2\mu U_\infty \sqrt{\frac{Re}{2}} f''(0) \int_0^1 \frac{dx}{\sqrt{x}},$$

$$D = 4\mu U_\infty \sqrt{\frac{Re}{2}} f''(0). \qquad (10\text{–}84)$$

It is common practice to report the dimensionless drag coefficient (in analogy to Nu) rather than D, that is,

$$C_D \equiv \frac{D}{\rho U_\infty^2 L} = \frac{4}{\sqrt{2}} \frac{f''(0)}{\sqrt{Re}}. \qquad (10\text{–}85)$$

Laminar Boundary-Layer Theory

The functional dependence of C_D on $Re^{-1/2}$ is a feature common to all streamlined bodies for which the boundary-layer analysis is valid over the complete body surface (i.e., the flow does not separate). The correlation

$$C_D = \frac{\text{const}}{\sqrt{Re}} \qquad (10\text{--}86)$$

is, in this sense, analogous to the correlations between Nu and Pe that were derived in Chap. 9.

Two features of the Blasius solution are worth mentioning here. The first is the asymptotic behavior of the normal velocity component for large Y. It was noted earlier that the matching condition between the inner and outer regions was satisfied to $O(Re^{-1/2})$ by the boundary-layer and potential-flow solutions, provided conditions (10–44) and (10–47) were satisfied and the normal velocity component $V(x, Y)$ remained finite for $Y \to \infty$. The latter condition is necessary because the boundary-layer equations are only third order in Y, and thus the behavior of V at large Y cannot be specified in advance, while retaining the boundary conditions (10–43) and the matching condition (10–47). By examining the numerical solution of the Blasius equation and/or developing a series expansion of the solution for large η (see Schlichting, 1968[3]), it can be shown that

$$f(\eta) \to \eta - 1.217 \quad \text{for } \eta \to \infty.$$

Hence,

$$V = \frac{1}{\sqrt{2x}}(\eta f' - f) \to \frac{1}{\sqrt{2x}}(1.217) \quad \text{for } \eta \to \infty.$$

Except in the limit $x \to 0$, which we have already discussed, we see that V is finite as required. If we transform back to outer variables, we find that

$$v = \frac{V}{\sqrt{Re}} = \frac{1}{\sqrt{2}}\left(\frac{1.217}{\sqrt{x Re}}\right). \qquad (10\text{--}87)$$

Thus, as far as the flow in the outer region is concerned, the presence of the boundary layer acts like a weak vertical flow at the plate surface, which tends to displace streamlines outward. This feature would appear in the solution for the outer region at $O(Re^{-1/2})$, as a correction to the $O(1)$ solution given by (10–62). The physical reason for the outward displacement of streamlines in the outer flow is the deceleration of fluid that occurs in the boundary layer because of the no-slip condition at the plate surface.

A second property of the Blasius solution is worth mentioning because it reflects a general property of the boundary-layer equations. The fact is that the Blasius solution is actually independent of the length of the plate. This may seem strange or even incorrect at first because we have obviously used the plate length L to nondimensionalize the equations leading to (10–59), and the plate length L also appears in the Reynolds number that was used to rescale the boundary-layer equations (10–64). However, we have shown that

$$u = f'(\eta),$$

where

$$\eta = \frac{Y}{\sqrt{2x}},$$

and we can show that η is completely independent of L. To see this, we can rewrite η in terms of the original dimensional variables (x', y'):

$$\eta = \frac{1}{\sqrt{2}}\frac{Y}{\sqrt{x}} = \frac{1}{\sqrt{2}}\frac{y\sqrt{Re}}{\sqrt{x}} = \frac{1}{\sqrt{2}}\frac{(y'/L)}{(x'/L)^{1/2}}\left(\frac{U_\infty L}{\nu}\right)^{1/2} = \frac{y'}{\sqrt{2x'}}\sqrt{U_\infty/\nu}.$$

D. Streaming Flow Past a Semi-Infinite Wedge – The Falkner–Skan Solution

It is in fact the distance from the leading edge of the plate that provides the relevant measure of the relative importance of viscous and inertia terms in the equation

$$Re_x = \frac{U_\infty x'}{\nu}. \tag{10–88}$$

Indeed, we have already seen that the solution breaks down as $x' \to 0$.

The fact that the solution is independent of L is a reflection of the fact that the boundary-layer equations are *parabolic*, with characteristics that proceed in the direction of increasing x. This means that the solution of the boundary-layer equations at a given position x^* depends only on conditions within the boundary layer for $x < x^*$ (upstream) but is not at all influenced directly by conditions downstream. This is a general characteristic of the boundary-layer equations (10–40) and (10–41). In the problem of flow past a flat plate, it is reflected by the fact that the solution at any point $x' < L$ is completely unaltered by the proximity of x' to the end of the plate. Indeed, the solution for $x' < L$ is precisely the same as if the plate were *semi-infinite* in extent. We already have seen that the boundary-layer equations are local in form, with no direct dependence on the geometry of the body except through the form of the pressure gradient. In other words, the solution in the boundary-layer depends upon the geometry of the body only because the solution in the outer region depends on the geometry. In the case of a flat plate, however, the first approximation in the outer region is completely independent of the length of the plate. In fact, the free-stream motion is not changed by the plate at all. Thus the boundary-layer solution is also independent of L. There is, however, one caveat to this discussion, and that is that the time required for establishing a steady-state solution on a semi-infinite flat plate would be infinity. Suppose that we consider the impulsive start-up problem in which the fluid is assumed to go from rest to a uniform velocity U_∞ at some initial instant $t = 0$ or, alternatively, the plate begins to translate with velocity $-U_\infty$. Far from the leading edge of the plate this problem looks like (i.e., can be approximated, as) the Rayleigh problem from Chap. 3. On the other hand, the leading edge of the plate will be felt downstream by means of the diffusion and convection of vorticity a distance of $O(U_\infty t)$, beyond which the velocity profile will still be the evolving Rayleigh profile.[16]

D. STREAMING FLOW PAST A SEMI-INFINITE WEDGE – THE FALKNER–SKAN SOLUTION

We saw in the previous section and in Chap. 9 that the boundary-layer equations very frequently have similarity solutions. In fact, in every case that we have studied, we have been able to find a self-similar solution. In Chap. 3, where the concept of self-similar solutions we first introduced, it was suggested that they should exist only for problems that are not characterized by a physical length scale. The local nature of the boundary-layer equations might lead to the assumption that the physical length scale of the problem, namely, the length scale of the body, would never be relevant and thus that similarity solutions should always be possible for these equations. However, this is not true. The form of the *momentum* boundary-layer equations depends on the form of the pressure gradient $-\partial p/\partial x$, which is imposed from the outer, potential flow around the body, and the potential-flow solution will generally introduce a length scale into the problem. Thus, in general, we should not expect to be able to obtain similarity solutions of the momentum boundary-layer equations. The flat plate problem of the preceding section is a special case in that the potential-flow solution introduces no geometric parameters. In heat transfer problems, the geometry of the body enters through the velocity field, but the thermal boundary-layer equations depend on only the local form of the velocity components at the body surface. These local approximations

Laminar Boundary-Layer Theory

are independent of the length scale of the body, and thus it is not surprising that similarity solutions can be obtained for *arbitrarily* shaped bodies, as shown in Chap. 9.

The conclusion from the previous paragraph is that similarity solutions of the momentum boundary-layer equations should not generally be expected. An interesting question is whether similarity solutions can be obtained in any case other than the flat plate problem in the previous section. To answer this question, we start with the boundary-layer equations in their most general form:

$$u\frac{\partial u}{\partial x} + V\frac{\partial u}{\partial Y} = u_e\frac{du_e}{dx} + \frac{\partial^2 u}{\partial Y^2}, \tag{10-89}$$

$$\frac{\partial u}{\partial x} + \frac{\partial V}{\partial Y} = 0, \tag{10-90}$$

with

$$u = V = 0 \quad \text{at } Y = 0,$$
$$u \to u_e(x) \quad \text{as } Y \to \infty. \tag{10-91}$$

The only distinction between bodies of different geometry is the function $u_e(x)$. Thus the question we wish to answer is whether similarity solutions exist for functions $u_e(x)$ other than $u_e(x) = $ const (the flat plate problem).

To answer this question, let us attempt a similarity solution of the general form

$$u = u_e(x)f'(\eta) \quad \text{with } \eta = \frac{Y}{g(x)}. \tag{10-92}$$

The function $u_e(x)$ must be included in (10–92); otherwise $f'(\eta)$ would have to be a function of x as $\eta \to \infty$ [see (10–91)]. The form for V corresponding to (10–92) can be obtained by use of (10–90) and the boundary condition $V = 0$ at $Y = 0$, that is,

$$V = -\int_0^Y \left(\frac{\partial u}{\partial x}\right)dY = -\int_0^\eta g\left(\frac{\partial u}{\partial x}\right)d\eta. \tag{10-93}$$

The result of substituting for $(\partial u/\partial x)$ from (10–92) and carrying out the integration is

$$V = -\frac{d}{dx}(gu_e)f + g'u_e\eta f', \tag{10-94}$$

where $f(0) = 0$. If we now substitute for u and V in the boundary-layer equation (10–89), we obtain

$$f''' + \left[g\frac{d}{dx}(gu_e)\right]ff'' + g^2u_e'\left[1 - (f')^2\right] = 0. \tag{10-95}$$

The boundary conditions (10–91) become

$$f = f' = 0 \quad \text{at } \eta = 0, \tag{10-96}$$
$$f' \to 1 \quad \text{as } \eta \to \infty. \tag{10-97}$$

D. Streaming Flow Past a Semi-Infinite Wedge – The Falkner–Skan Solution

In general, (10–89) requires an initial profile at $x = 0$ (corresponding to the most upstream point in the boundary layer). Evidently, if a similarity solution does exist, the boundary condition (10–97) must represent this initial condition, as well as the boundary condition for $Y \to \infty$. This implies that $g(0) = 0$; that is, the boundary-layer thickness must go to zero as $x \to 0$. This can occur only for a body that has a *pointed* leading edge (such as the flat plate).

If we examine (10–95), the existence of a similarity solution requires that the coefficients involving $g(x)$ and $u_e(x)$ must be either zero or a nonzero constant. Let us denote these constants as α and β, that is,

$$g\frac{d}{dx}(u_e g) = \alpha, \tag{10–98}$$

$$g^2 u'_e = \beta. \tag{10–99}$$

We suppose that both of these constants are nonzero. Equations (10–98) and (10–99) should then lead to a unique combination of functions $u_e(x)$ and $g(x)$ for which similarity solutions exist. Of course, in the approach that we have adopted, the problem of determining a function $u_e(x)$ that allows similarity solutions is a purely mathematical question. The related physical problem is to determine whether the resulting forms for $u_e(x)$ correspond to any realizable body shape.

Let us first resolve the mathematical question by solving (10–98) and (10–99). To do this, we substitute (10–99) into (10–98). This gives

$$2\beta + u_e \frac{dg^2}{dx} = 2\alpha. \tag{10–100}$$

Now, adding (10–99) and (10–100), we obtain

$$\frac{d}{dx}(g^2 u_e) = 2\alpha - \beta \tag{10–101}$$

and, integrating this equation, we obtain

$$g^2 u_e = (2\alpha - \beta)x + c, \tag{10–102}$$

where c is an arbitrary constant of integration. Now, substituting for g^2 from (10–99), we find that

$$\frac{1}{u_e}\frac{du_e}{dx} = \frac{\beta}{[(2\alpha - \beta)x + c]}, \tag{10–103}$$

and integration of this equation gives

$$u_e = c_1[(2\alpha - \beta)x + c]^{\beta/(2\alpha - \beta)}, \tag{10–104}$$

where c_1 is another arbitrary constant. Finally, with u_e known, we obtain g^2 from (10–102). The result is

$$g^2 = \frac{1}{c_1}[(2\alpha - \beta)x + c]^{(2\alpha - 2\beta)/(2\alpha - \beta)}. \tag{10–105}$$

Evidently, $(2\alpha - \beta)$ cannot be zero, but we shall return to this point momentarily.

Now, the remaining questions are whether the requirements for similarity impose additional constraints on the constants α, β, c, and c_1, and then whether the resulting form for u_e corresponds to any physically realizable geometry for the body. Let us first consider the

Laminar Boundary-Layer Theory

case $\alpha, \beta \neq 0$, which turns out to be the most interesting. In this case, either α or β can be set equal to unity with no loss of generality. We choose $\alpha = 1$. Thus,

$$u_e = c_1[(2-\beta)x + c]^{\beta/(2-\beta)}, \tag{10-106}$$

$$g^2 = \frac{1}{c_1}[(2-\beta)x + c]^{(2-2\beta)/(2-\beta)}. \tag{10-107}$$

One of the remaining conditions for existence of a similarity solution is that $g(0) = 0$. Thus we can see from (10–107) that

$$c = 0.$$

What remains for u_e is

$$u_e = c_1[(2-\beta)x]^{\beta/(2-\beta)}.$$

However, the constant $c_1(2-\beta)^{\beta/(2-\beta)}$ is superfluous because it always can be made equal to unity by a proper choice of characteristic velocity U_c. Thus, finally, we can write

$$\boxed{u_e = x^{\beta/(2-\beta)} = x^m,} \tag{10-108}$$

where

$$m = \frac{\beta}{2-\beta} \quad \text{or} \quad \beta = \frac{2m}{m+1}, \tag{10-109}$$

$$c_1 = (2-\beta)^{-\beta/(2-\beta)}. \tag{10-110}$$

Substituting for β and c_1 in (10–107), we see that the corresponding form for g is

$$\boxed{g = \left(\frac{2}{m+1}x^{1-m}\right)^{1/2}.} \tag{10-111}$$

Thus a *necessary* condition for the existence of similarity solutions for $\alpha, \beta \neq 0$ is that $u_e = x^m$. The *sufficient* condition is that $u_e = x^m$ and a solution $f(\eta)$ exists that satisfies (10–95) in the form

$$\boxed{f''' + ff'' + \beta\left[1 - (f')^2\right] = 0} \tag{10-112}$$

and the boundary conditions (10–96) and (10–97). Equation (10–112) is known as the *Falkner–Skan* equation.

Assuming that (10–112) can be solved subject to (10–96) and (10–97), the question is whether $u_e = x^m$ corresponds to any physically realizable body shapes. To answer from first principles, we would have to solve an *inverse problem* in potential-flow theory. We do not propose to do that here. Rather, we simply state the result, which is that

$$u_e = x^m$$

for streaming flow past a semi-infinite wedge of included angle $\pi\beta$, with the motion at infinity parallel to the bisector of the wedge, or flow past a corner with either a slip boundary or a negligible boundary layer on the upstream surface.[17] These flow configurations are sketched in Fig. 10–6. A special case of (a) is $\beta = 0$, when the definition of g and the governing Falkner–Skan equation reduce to the Blasius equations for flow past a parallel flat plate.

D. Streaming Flow Past a Semi-Infinite Wedge – The Falkner–Skan Solution

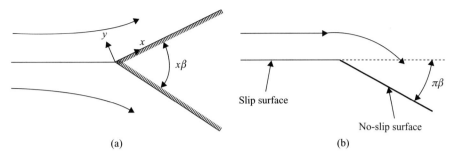

Figure 10–6. A sketch of flow configurations that correspond to the Falkner–Skan equation: (a) The body is a semi-infinite wedge with included angle $\pi\beta$. The boundary-layer coordinates (x, Y) are shown for the upper surface, with $x = 0$ corresponding to the leading edge of the wedge. (b) Flow around a corner, in this case corresponding to negative β. The upstream surface can be considered as a slip surface, or else a short-enough no-slip surface that one can neglect the thickness of any boundary layer at the corner.

When $0 < \beta < 1$, we have the wedge geometry. A plot of the solutions for $f'(\eta)$ [$\equiv u/u_e(x)$] is given for various β in Fig. 10–7. The constant m increases monotonically in the range $0 < m < 1$ as β increases. Further, because $u_e = x^m$ and $dp/dx = -u_e(du_e/\partial x)$, it follows that the magnitude of the pressure gradient in the streamwise direction,

$$\frac{\partial p}{\partial x} = -mx^{2m-1} \qquad (10\text{--}113)$$

increases as the wedge angle increases. We will refer to $\partial p/\partial x < 0$ as a *favorable* pressure gradient because the pressure decreases in the direction of flow, and this tends to accelerate the fluid in the boundary layer relative to the velocity it would have in the absence of the pressure gradient (i.e., for the flat plate $m = 0$). This is reflected in the velocity profiles in Fig. 10–7.

The case of $\beta = 1$ ($m = 1$) deserves special mention. In this case, the geometry reduces to flow directly toward a perpendicular flat plate. The position $x = 0$ is chosen in this instance to be the midpoint, so that flow near the surface is diverted either left or right as it approaches the plate from $x < 0$ or $x > 0$. The resulting motion is known as the *two-dimensional stagnation flow* and corresponds to the potential flow $u = x$, $v = -y$. However, if we recall that the leading edge of the body should be sharp in order for the condition $g(0) = 0$

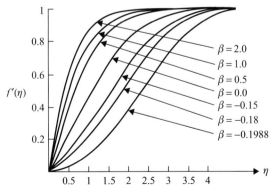

Figure 10–7. Solutions of the Falkner–Skan equation, plotted as f' versus η for several different values of β. The values of β that correspond to the infinite wedge configuration of Fig. 10–6 are $0 \leq \beta \leq 1$. The solutions for $\beta > 1$ and $\beta < 0$ are discussed in the text.

Laminar Boundary-Layer Theory

to make sense, we may be skeptical that a solution of similarity form should exist for this problem. Indeed, if we examine the function $g(x)$ for $\beta = m = 1$, we find that

$$g = 1, \tag{10–114}$$

and thus the general similarity form (10–92) reduces to

$$\boxed{u = xf'(Y),} \tag{10–115}$$

which is not a similarity solution at all. The transformation (10–115) reduces the laminar boundary-layer equation to the same ODE that is obtained from the Falkner–Skan equation for $\beta = 1$, namely,

$$\boxed{f''' + ff'' + (1 - f'^2) = 0,} \tag{10–116}$$

but in this case f is a function of Y alone. The reader may recall that the function g provides a measure of the boundary-layer thickness as a function of x. In this limiting case, it reverts to a constant. Physically, the tendency of vorticity to diffuse out from the body surface (and thus to increase the thickness of the boundary layer as we move in the downstream direction) is precisely compensated for in this instance by the normal motion of fluid toward the boundary, which convects vorticity *toward* the boundary at a rate that just balances diffusion in the opposite direction.

Hartree[18] also obtained a family of solutions for β between 0 and -0.1988 that were physically acceptable in the sense that $f' \to 1$ *from below* as $\eta \to \infty$. Several such profiles are sketched in Fig. 10–7. These correspond to the boundary layer downstream of the corner in Fig. 10–6(b) (assuming that the upstream surface is either a slip surface or is short enough that one can neglect any boundary layer that forms on this surface). It should be noted that solutions of the Falkner–Skan equation exist for $\beta < -0.1988$, but these are unacceptable on the physical ground that $f' \to 1$ *from above* as $\eta \to \infty$, and this would correspond to velocities within the boundary layer that *exceed* the outer potential-flow value at the same streamwise position, x. It may be noted from Fig. 10–7 that the shear stress at the surface ($\eta = 0$) decreases monotonically as β is decreased from 0. Finally, at $\beta = -0.1988$, the shear stress is exactly equal to zero, i.e., $f''(0) = 0$. It will be noted from (10–113) that the pressure gradient

$$\frac{\partial p}{\partial x} > 0$$

for $\beta < 0$. This means that the pressure increases in the direction of motion, and the resulting pressure gradient tends to *decelerate* the fluid in the boundary layer relative to its velocity in the absence of a pressure gradient. A pressure gradient acting in this sense is known as an *adverse* pressure gradient. This deceleration is reflected by the fact that $f'(\eta)$ is smaller for any $\eta \neq 0$ for $m < 0$ than it is for $m = 0$. Furthermore, we see that, when the pressure gradient increases only a little above zero, corresponding to $\beta = -0.1988$, the solution exhibits zero shear stress, $f''(0) = 0$. For more negative values of β, the boundary layer does not exist.

We have shown from a mathematical viewpoint that similarity solutions of the boundary-layer equations exist whenever

$$u_e = x^m$$

for *arbitrary m*. From this infinity of possible cases, we have identified two problems for $-0.0908 \leq m \leq 1$ (i.e., $-0.1998 \leq \beta \leq 1$) where the functional dependence of u_e on x^m corresponds to a realizable, physical problem. However, this limitation on m does not mean that solutions of the Falkner–Skan equations may not exist for m outside this range. Indeed, such solutions have been found numerically by Hartree and Stewartson.[18] As m increases above 1.0, the solution of the Falkner–Skan equation smoothly continues the trend indicated in Fig. 10–7 for increasing values of m. However, for $m > 1.0$, it has been pointed out by Hartree and Stewartson that these solutions do not lead to real values of u and V and thus are not of any physical significance or interest. For $\beta < -0.1988$, we have already pointed out that the solutions of the Falkner–Skan equation are unacceptable on physical grounds because $f' \to 1$ *from above* as $\eta \to \infty$. The reader may well ask why we bother to even mention these solutions of the Falkner–Skan equation for m outside the range $-0.0908 \leq m \leq 1$, as there is evidently no corresponding physical problem. The main justification is the *qualitative* connection between solutions for $m < 0$ and boundary-layer *separation*.

Stewartson[19] has suggested (as a qualitative model) that the Falkner–Skan equation might be viewed as describing the velocity profile in a boundary layer in which the constant β is treated as a parameter that determines the *local* pressure gradient. Thus, for example, if we consider a circular cylinder, the pressure gradient is negative over the front half of the cylinder, which would be interpreted in this qualitative view as $\beta > 0$, with an initial value of unity at the front-stagnation point. β then decreases monotonically as we pass around the body, and finally passes through zero at the equator. In this framework, the point of separation would occur at the point where the local pressure gradient gives $\beta = -0.1988$. Of course, this qualitative interpretation by means of the Falkner–Skan solution is a gross oversimplification. The primary message, which does turn out to be true, is that separation may be expected when a boundary layer encounters a region with only a very weakly increasing pressure (i.e., a small adverse pressure gradient). We shall see in the next section that this is, in fact, the case and we discuss the "mechanism" of separation in detail.

E. STREAMING FLOW PAST CYLINDRICAL BODIES – BOUNDARY-LAYER SEPARATION

In the previous two sections we studied boundary-layer problems that yield similarity solutions. Here, we turn to a general class of problems that cannot be solved by means of similarity transforms, namely, flow past cylindrical bodies, with the circular cylinder serving as a convenient prototype. This class of problems is interesting not only because it requires a new approach to obtain solutions but also because many cylindrically shaped bodies exhibit boundary-layer separation. Experimentally, steady separated flows are generally identified by the presence of a pair of large recirculating eddies downstream of the body. Another indication of separation that is more important from a practical point of view is that the drag on the body is much larger than it is for the flat plate or other streamlined bodies that do not lead to a separated flow.

The cause of large drag in the case of a body like a circular cylinder is the asymmetry in the velocity and pressure distributions at the cylinder surface that results from separation. All bodies in laminar streaming flow at large Reynolds number are subjected to viscous stresses that boundary-layer analysis shows must be

$$\tau_{xy} \sim \frac{\partial u}{\partial y} \sim \sqrt{Re}\left(\frac{\partial u}{\partial Y}\right) \sim O\left(\sqrt{Re}\right). \qquad (10\text{–}117)$$

These viscous stresses lead to a contribution to the drag coefficient (that is, the dimensionless drag per unit length for a 2D body):

$$\boxed{C_D \equiv \frac{drag}{\rho U_\infty^2 \ell_c} \sim O(Re^{-1/2}) \quad \text{for} \quad Re \to \infty.} \qquad (10\text{–}118)$$

Laminar Boundary-Layer Theory

In addition, there are pressure forces at the body surface. We have seen from boundary-layer theory that the pressure does not vary to leading order across the boundary layer. Hence, the pressure on the body surface can be approximated by the pressure distribution imposed on the boundary layer by the outer potential flow. The pressure variations in this outer region according to (10–1) are $O(\rho U_\infty^2/\ell_c)$. However, so long as the boundary layer does not separate, the *net* contribution to the drag that is due to pressure variations over the body surface will be zero. This is, in fact, nothing more than D'Alembert's paradox revisited. Even though there are large pressure variations over the body surface, the downstream force that is due to pressure forces acting over the front portion of the body is precisely balanced by an equal and opposite upstream force that is due to pressure forces on the rear of the body. On the other hand, when the boundary layer separates, the pressure at the rear of the body is sharply reduced and the pressure distribution becomes very asymmetric.[20] Thus there is a net force on the body that is due to the imbalance between the pressure forces on the front and rear of the body. Because the pressure variations are $O(\rho U^2/\ell_c)$, the net force on the body will be $O(\rho U^2 \ell_c)$, and the corresponding contribution to the drag coefficient will be

$$\boxed{C_D \sim O(1) \quad \text{as} \quad Re \to \infty.} \tag{10–119}$$

The drag contribution associated with an asymmetric pressure distribution is sometimes known as *form* drag, whereas that due to the viscous stresses is called *friction* drag. Comparing (10–118) and (10–119), we see that the form drag on a body in separated flow is asymptotically large compared with the drag on a streamlined body without separation. Indeed, the very large drag penalty associated with separation explains the emphasis in aerodynamics on the design of streamlined cross-sectional shapes for aircraft wings that achieve maximum lift without inducing separation. Regardless of the specific application, the relative magnitudes of the form and friction contributions to the drag show why it is critically important to have an *a priori* method to predict when separation can be expected.

Let us now turn to the application of boundary-layer theory to streaming flow past cylindrical bodies by using the circular cylinder to illustrate the details of the calculation. We have already noted that experimental studies show the existence of separated flow in the case of a circular cylinder. Thus this should provide an interesting test case for prediction of separation by means of boundary-layer theory. As we noted earlier, we carry out the analysis initially ignoring the possibility that the flow may separate. In particular, we calculate the pressure gradient in the boundary layer by using potential-flow theory for streaming flow around the cylinder alone (that is, without any attempt to incorporate the recirculating wake as part of a composite body). The presence of a separation point is then signaled in the boundary-layer solution by the existence of a point on the body surface where the velocity gradient $\partial u/\partial Y$ vanishes, namely,

$$\boxed{\left.\frac{\partial u}{\partial Y}\right|_{Y=0} \Rightarrow 0 \quad \text{as} \quad x \to x_s,} \tag{10–120}$$

where x_s is used to represent the position of the separation point.

The potential-flow solution for streaming motion past a circular cylinder was obtained earlier and given in terms of the streamfunction in (10–17). To calculate the pressure gradient in the boundary layer, we first determine the tangential velocity function, u_e, as defined in (10–37):

$$u_e = \lim_{r \to 1}\left(\frac{\partial \psi}{\partial r}\right). \tag{10–121}$$

E. Streaming Flow Past Cylindrical Bodies – Boundary-Layer Separation

Adopting the local Cartesian coordinates appropriate to the boundary-layer equations, we can express the result as

$$u_e = 2\sin x. \tag{10-122}$$

The corresponding pressure gradient according to (10–39) is

$$\frac{\partial p}{\partial x} = -2\sin 2x. \tag{10-123}$$

The velocity function $u_e(x)$ and the pressure gradient dp/dx are sufficient to completely specify the boundary-layer problem, but it is of interest to compare the predicted pressure distribution on the cylinder surface with experimentally measured results, and for this purpose it is convenient to proceed one step beyond (10–123) to calculate p. To do this, we integrate with respect to x,

$$p = c + \cos 2x, \tag{10-124}$$

where c is a constant of integration that is determined by the choice of a reference pressure. For experimental work it is convenient to use the pressure in the free stream as the reference value: Let us denote this reference pressure as p_∞. To relate p_∞ to p on the body surface, we can integrate (10–19) with u_r evaluated from the potential-flow solution:

$$u_r = -\left(1 - \frac{1}{r^2}\right)\cos\theta \tag{10-125}$$

along the front symmetry axis $\theta = 0$ from $r = 1$ to $r \to \infty$. The result is

$$p|_{r=1,\theta=0} = p_\infty + \frac{1}{2}. \tag{10-126}$$

Hence, comparing (10–126) and (10–124) at $x = 0$, we see that

$$c = p_\infty - \frac{1}{2}, \tag{10-127}$$

and we can rewrite (10–124) as

$$p - p_\infty = -\frac{1}{2} + \cos 2x. \tag{10-128}$$

This quantity was plotted in Fig. 10–3.[21] Also shown is the measured pressure distribution for Re in the range 30 to 40. It is evident that the measured and predicted distributions are in close agreement over the front portion of the cylinder. Thus it is not surprising that boundary-layer theory, based on the potential-flow pressure distribution, should be quite accurate up to the vicinity of the separation point. It is this fact that explains the ability of boundary-layer theory to provide a reasonable estimate of the onset point for separation, as we shall demonstrate shortly.

It will be noted that the pressure gradient over the front half of the cylinder is *favorable* – that is, the pressure decreases in the direction of motion. Beyond the halfway point ($\theta = x = \pi/2$), on the other hand, the pressure begins to increase; in this region the pressure gradient is *adverse*. Our experience with the solutions of the Falkner–Skan equation suggests that this latter region is a candidate for boundary-layer separation. Indeed, experimental observation corroborates this fact. If we stop for a moment and think qualitatively about cylindrical bodies of other cross-sectional shapes, it should be clear that the potential-flow pressure distribution will *always* have the same *qualitative* appearance as shown in Fig. 10–3. At the front-stagnation point, there is always a pressure maximum. Furthermore, as the fluid moves around the cylinder, its velocity tangent to the surface will increase, and thus the pressure will decrease until the point of maximum body width is reached. Beyond this point, the tangential velocity in the potential-flow region will decrease, and the pressure will then increase. Thus, for every cylindrical body, the boundary layer will experience a region of increasing pressure (an adverse pressure gradient) and thus have the potential to separate. Experience has shown that we can avoid separation only by reducing the magnitude of this adverse pressure gradient by streamlining the body shape beyond the point of maximum width. Not surprisingly, the resulting cross-sectional profile resembles an airfoil. Obviously, to distinguish between cases that will and will not separate, it is extremely important to have a method to solve the boundary-layer equations for cylindrical bodies of quite general shapes. One such method is based on the so-called *Blasius series* and is presented in the following discussion.

The boundary-layer problem for the specific case of a circular cylinder is (10–40), (10–41), (10–43), and (10–47), with u_e and $\partial p/\partial x$ given by (10–122) and (10–123). The first point to note is that a similarity solution does not exist for this problem. Furthermore, in view of the qualitative similarity of the pressure distributions for cylinders of arbitrary shape, it is obvious that similarity solutions do not exist for any problems of this general class. The Blasius series solution developed here is nothing more than a power-series approximation of the boundary-layer solution about $x = 0$.

We begin by noting that the velocity function $u_e(x)$ can be expressed, for any cylindrical body that has a stagnation point at $x = 0$ and is symmetrical about an axis parallel to the free stream, in the general form

$$u_e(x) = u_1 x + u_3 x^3 + u_5 x^5 + \cdots +, \tag{10–129}$$

Here, u_1, u_3, u_5, and so forth are the numerical coefficients of a power-series expansion of $u_e(x)$ and are assumed to be known. In the case of a circular cylinder, the first few are

$$u_1 = 2, \tag{10–130}$$

$$u_3 = -\frac{1}{3}, \tag{10–131}$$

$$u_5 = +\frac{1}{60}, \tag{10–132}$$

$$u_n = -(-1)^{(n+1)/2} \left(\frac{2}{n!}\right). \tag{10–133}$$

It may be noted that the radius of convergence for this power-series representation of $\sin x$ is $|x| < \infty$. In general, the number of terms required in the series (10–129) to give a good approximation of $u_e(x)$ will vary depending on the form of $u_e(x)$ and thus the geometry of the body. Asymmetric shapes resembling airfoils with long tapering rear sections will generally require many more terms than a symmetric shape such as a circular or elliptic cylinder.

Because the boundary-layer velocity u must match $u_e(x)$ for large Y, it seems reasonable to assume that u can be represented throughout the boundary layer by a similar power series

E. Streaming Flow Past Cylindrical Bodies – Boundary-Layer Separation

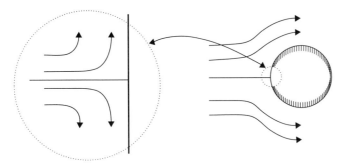

Figure 10–8. A sketch showing streaming flow past a circular cylinder. On the right is the flow as seen with a course level of resolution. On the left is the flow in the immediate vicinity of the front-stagnation point, as seen from a much finer resolution. It is evident that the *local* flow examined in a region close enough to the stagnation point reduces to a classic stagnation point flow as described by the Falkner–Skan equation for $\beta = 1$.

in x but with coefficients that depend on Y because $u = 0$ at $Y = 0$. Thus we propose an approximation:

$$u = F_1'(Y)x + F_3'(Y)x^3 + F_5'(Y)x^5 + \cdots +, \qquad (10\text{–}134)$$

with

$$F_1'(0) = F_3'(0) = F_5'(0) = F_n'(0) = 0, \qquad (10\text{–}135)$$

$$F_n'(Y) = u_n \quad \text{as } Y \to \infty \quad \text{for all } n. \qquad (10\text{–}136)$$

However, with this form for u, it is evident from the continuity equation that a power-series representation also must exist for V,

$$V = -\int_0^Y \frac{\partial u}{\partial x} dY = -\left(F_1 + 3F_3 x^2 + 5F_5 x^4 + \cdots +\right), \qquad (10\text{–}137)$$

where

$$F_1(0) = F_3(0) = F_5(0) = F_n(0) = 0. \qquad (10\text{–}138)$$

If we substitute for u, V, and u_e in the boundary-layer equation (10–40) by using the power-series approximations in (10–129), (10–134), and (10–137), and collect all terms of like power in x, we obtain a set of ODEs for the coefficients $F_1(Y)$, $F_3(Y)$, and so forth:

$$F_1''' + F_1 F_1'' + \left(u_1^2 - F_1'^2\right) = 0, \qquad (10\text{–}139)$$

$$F_3''' + F_1 F_3'' + 4\left(u_1 u_3 - F_1' F_3'\right) + 3F_1'' F_3 = 0, \qquad (10\text{–}140)$$

$$F_5''' + F_1 F_5'' + 6\left(u_1 u_5 + \frac{1}{2}u_3^2 - F_1' F_5'\right) + 5F_1'' F_5 = 3\left[(F_3')^2 - F_3 F_3''\right]. \qquad (10\text{–}141)$$

These ODEs are to be solved subject to (10–135), (10–136), and (10–138). The first is just the Falkner–Skan equation for a 2D stagnation flow. This is not at all surprising because the geometry of a cylinder near $x = 0$ is locally just the stagnation point geometry, as sketched in Fig. 10–8. Subsequent equations are linear in F_3, F_5, and so forth. There is therefore

Laminar Boundary-Layer Theory

no doubt that solutions exist for the functions $F_n(Y)$, though the equations will have to be solved numerically in view of the fact that (10–139) is nonlinear.

The only difficulty with the approached outlined here is that the ODEs, (10–139)–(10–141), as well as the equations for the higher-order functions $F_n(Y)$, all depend explicitly on the numerical coefficients u_1, u_3, and u_5 from the expansion (10–129). This is not surprising. These coefficients are the only source of dependence of the boundary-layer equations on the geometry of the cylinder. Thus it would seem inevitable that they should appear explicitly in the analysis. However, this means that the solution for every new problem, that is, every new cylinder geometry, would require a new numerical integration of the set of ODEs for $F_n(Y)$. The astonishing fact, however, is that a slight reformulation of the power-series approximation yields a set of ODEs and boundary conditions that do not contain the coefficients u_i at all. This transformation was originally discovered by the English mathematician Leslie Howarth.[22]

Howarth proposed to replace the power series (10–134) with a power series of the form

$$u = u_1 x f_1'(Y\sqrt{u_1}) + 4u_3 x^3 f_3'(Y\sqrt{u_1}) + 6u_5 x^5 f_5'(Y\sqrt{u_1}) + \cdots +, \quad (10\text{–}142)$$

where

$$u_5 f_5 = u_5 g_5 + \frac{u_3^2}{u_1} h_5,$$

$$u_7 f_7 = u_7 g_7 + \frac{u_3 u_5}{u_1} h_7 + \frac{u_3^3}{u_1^2} k_7,$$

and so on. This is known as the *Blasius series*. With this form for u, we obtain in place of (10–139)–(10–141) the alternative set

$$f_1'^2 - f_1 f_1'' = 1 + f_1''', \quad (10\text{–}143)$$

$$4 f_1' f_3' - 3 f_1'' f_3 - f_1 f_3'' = 1 + f_3''', \quad (10\text{–}144)$$

$$6 f_1' g_5' - 5 f_1'' g_5 - f_1 g_5'' = 1 + g_5''', \quad (10\text{–}145)$$

$$6 f_1' h_5' - 5 f_1'' h_5 - f_1 h_5'' = \frac{1}{2} + h_5''' - 8(f_3'^2 - f_3 f_3''), \quad (10\text{–}146)$$

and so on. The boundary conditions for these equations are

$$\begin{aligned} f_1(0) &= f_1'(0) = 0, \quad f_1'(\infty) = 1, \\ f_3(0) &= f_3''(0) = 0, \quad f_3'(\infty) = \frac{1}{4}, \\ g_5(0) &= g_5'(0) = 0, \quad g_5'(\infty) = \frac{1}{6}, \\ h_5(0) &= h_5'(0) = 0, \quad h_5'(\infty) = 0. \end{aligned} \quad (10\text{–}147)$$

The key point to note is that (10–143)–(10–147) are completely independent of the geometric coefficients u_1, u_3, u_5, and so on. Similarly, if we continued the series to higher-order terms, the corresponding equations and boundary conditions would all be independent of the u_i. This means that the functions $f_i(Y\sqrt{u_1})$ are *universal functions* that apply to cylindrical bodies of arbitrary cross-sectional shape or to any boundary-layer problem for which the

E. Streaming Flow Past Cylindrical Bodies – Boundary-Layer Separation

function $u_e(x)$ can be expressed as either a power series or polynomial in odd powers of x. Thus we can integrate (10–143)–(10–146) once and for all and tabulate the results. Application to a specific problem then involves only the algebra of inserting the coefficients u_n into the series for u, V, or any other quantity that we may wish to calculate. Tables of values for the functions $f'_n(z)$ for $n \leq 11$ are given in books on boundary-layer theory,[23] such as that by Schlichting, as well as in the paper by Howarth.

It is of primary interest in many cases to calculate $\partial u/\partial Y$ at $Y = 0$. This quantity is proportional to the shear stress and can be used in the absence of separation to calculate the frictional force on a body. More importantly, however, condition (10–120) indicates that separation will occur if $\partial u/\partial Y \to 0$ at any point x on the body surface (other than the stagnation point $x = 0$). Furthermore, the point x where this occurs should provide an estimate of the position of the separation point. To calculate $(\partial u/\partial Y)_{Y=0}$ from the Blasius series solution, we require numerical values for the second derivative of $f_n(Y)$ at $Y = 0$, namely,

$$\boxed{\begin{aligned}\frac{1}{\sqrt{u_1}}\left(\frac{\partial u}{\partial Y}\right)_{Y=0} &= u_1 x f''_1(0) + 4u_3 x^3 f''_3(0) + 6x^5 \left[u_5 g''_5(0) + \frac{u_3^2}{u_1} h''_5(0)\right] \\ &\quad + 8x^7 \left[u_7 g''_7(0) + \frac{u_3 u_5}{u_1} h''_7(0) + \frac{u_3^3}{u_1^2} k''_7(0)\right] + O(x^9).\end{aligned}} \quad (10\text{–}148)$$

The numerical values of the coefficients required for evaluating (10–148) are

$$\boxed{\begin{aligned}f''_1(0) &= 1.2326, \\ f''_3(0) &= 0.7244, \\ g''_5(0) &= 0.6347, \\ h''_5(0) &= 0.1192, \\ g''_7(0) &= 0.5792, \\ h''_7(0) &= 0.1829, \\ k''_7(0) &= 0.0076.\end{aligned}} \quad (10\text{–}149)$$

At the beginning of this section, we started out to analyze the boundary-layer solution for the specific case of a circular cylinder. Thus it is of interest to apply (10–148) and (10–149) to that problem by using the u_n coefficients given by (10–133). The result is

$$\frac{1}{\sqrt{u_1}}\left(\frac{\partial u}{\partial Y}\right)_{Y=0} = 2x(1.2326) - \frac{4}{3}x^3(0.7244) + 6x^5\left(\frac{0.6347}{60} + \frac{0.1192}{18}\right) + O(x^7). \quad (10\text{–}150)$$

A plot of the right-hand side of (10–150) is shown in Fig. 10–9 for the two-term approximation in which we neglect the contribution of the x^5 term, and for the four-term approximation truncated after x^7, as indicated in (10–150). At both levels of approximation, $\partial u/\partial Y \to 0$ for $x > \pi/2$. With the first two terms only, (10–150) yields $x_s \sim 1.597$ (or $\theta_s = 91.5°$), which is barely beyond the point $x = \pi/2 = 1.57$ where the pressure begins to increase. If we were to retain all terms in (10–148) up to order x^{11}, we would find a predicted separation point at

$$\theta_s = 109.6°.$$

Laminar Boundary-Layer Theory

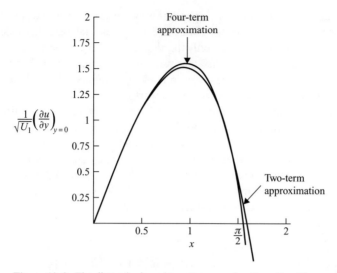

Figure 10–9. The dimensionless shear stress as a function of position on the surface of a circular cylinder as calculated with the approximate Blasius series solution. Note that x is measured in radians from the front-stagnation point. The predicted point of boundary-layer separation corresponds to the second zero of $\partial u/\partial Y|_0$. and is predicted to occur just beyond the minimum pressure point at $x = \pi/2$.

Thus the boundary-layer solution, obtained by means of the Blasius series, indicates that separation should be expected at about $\theta_s \sim 110°$. Careful *numerical* solution of the boundary-layer equations in the form (10–41) with u_e given by (10–122) yields a predicted separation point at

$$\theta_s = 104.5°.$$

Experimental observations of the flow past a circular cylinder show that separation does indeed occur, with a separation point at $\theta_s \sim 110°$. It should be noted, however, that steady recirculating wakes can be achieved, even with artificial stabilization,[24] only up to $Re \sim 200$, and it is not clear that the separation angle has yet achieved an asymptotic ($Re \to \infty$) value at this large, but finite, Reynolds number. In any case, we should not expect the separation point to be predicted too accurately because it is based on the pressure distribution for an unseparated potential flow, and this becomes increasingly inaccurate as the separation point is approached. The important fact is that the boundary-layer analysis does provide a method to predict whether separation should be expected for a body of specified shape. This is a major accomplishment, as has already been pointed out.

Before we turn to other topics, it is worth considering the physical events that lead to separation. There are two plausible ways to explain the phenomenon. A common feature of these two mechanisms is that viscous effects play a critical role. Experimentally, we find that separation (or at least the downstream recirculating wake that we associate with separation) usually occurs for Reynolds numbers larger than some critical value, and we infer from this that separation is basically a large-Reynolds-number phenomenon. Indeed, this point of view is consistent with the fact that separation can be predicted by boundary-layer theory, which is an asymptotic theory for $Re \to \infty$. In spite of this, separation is a phenomenon that we can explain only by considering the consequences of *viscous* contributions to the motion of the fluid.

The most commonly accepted explanation for separation can be paraphrased in the following way. In a potential flow with no viscosity, there is an exact interchange of kinetic energy and work that is due to the action of the pressure gradient for a fluid element that is near the body surface. On the front portion of the body, the pressure gradient is favorable and

F. Streaming Flow Past Axisymmetric Bodies – A Generalization of the Blasius Series

the fluid element gains kinetic energy. On the back side, it encounters an adverse pressure gradient causing it to slow down, but the acceleration process on the front side has provided just enough kinetic energy for the fluid element to reach the rear-stagnation point before it is brought to rest by the pressure forces at the back. The presence of viscous effects within the boundary layer alters this picture in a profound way. Some of the work done by the favorable pressure gradient at the front is lost to viscous dissipation, and the kinetic energy of the fluid element is decreased. Thus, when it encounters the adverse pressure gradient at the rear of the body, it no longer has enough kinetic energy to make it to the rear-stagnation point, and its tangential motion is completely arrested at some intermediate point. Because fluid mass cannot simply accumulate at this point, the only possibility is for the fluid element to depart from the surface; that is, the boundary layer separates from the body. In this view of separation, it is the local dynamics within the boundary-layer that triggers the separation process, and the large recirculating region of high vorticity downstream of the body is seen to exist as a consequence of the separation of the boundary layer.

An alternative point of view is that vorticity accumulates at the rear of the body, which leads to a large recirculating eddy structure, and as a consequence, the flow in the vicinity of the body surface is forced to detach from the surface. This is quite a different mechanism from the first one because it assumes that the primary process leading to separation is the production and accumulation of vorticity rather than the local dynamics within the boundary layer.[25] However, viscosity still plays a critical role for a solid body in the *production* of vorticity. In fact, for any finite Reynolds number, there is probably some element of truth in both explanations. Furthermore, it is unlikely that experimental evidence (or evidence based on numerical solutions of the complete Navier–Stokes equations) will be able to distinguish between these ideas, because such evidence for steady flows will inevitably be limited to moderate Reynolds numbers.

F. STREAMING FLOW PAST AXISYMMETRIC BODIES – A GENERALIZATION OF THE BLASIUS SERIES

Essentially all of the preceding developments of this chapter have assumed that the geometry of the flow can be approximated as 2D. Although we have suggested that the same general principles will apply for bodies of arbitrary, but smooth, shape, it is worth considering at least the generalization to axisymmetric flows, as this includes the important practical problem of streaming flow past a sphere. We do not need to reiterate every detail of the analysis, as some parts are virtually identical to the 2D problems discussed earlier. Instead, we concentrate on those features for which it is not clear that the same general analysis should apply.

For convenience, we focus our discussion initially on streaming flow past a sphere. The configuration is sketched in Fig. 10–10, where we also indicate the direction of motion relative to the spherical coordinates (r, θ). The dimensionless equations of motion for the outer region, in which $\ell_c = a$ (the sphere radius), are identical to (7–50), and it is clear that the leading-order approximation for $Re \gg 1$ is thus

$$E^2 \psi = 0, \qquad (10\text{–}151)$$

with

$$\psi = 0 \quad \text{at} \quad r = 1, \qquad (10\text{–}152)$$

$$\psi \to -\frac{1}{2} r^2 \sin^2 \theta \quad \text{for} \quad r \to \infty. \qquad (10\text{–}153)$$

Laminar Boundary-Layer Theory

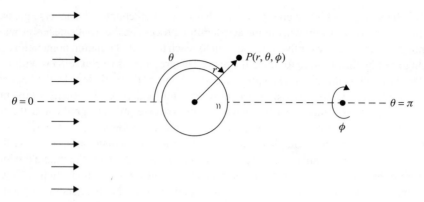

Figure 10–10. A schematic representation for streaming flow past a sphere. The flow is axisymmetric and thus is independent of the azimuthal angle ϕ. The polar angle θ is measured from the symmetry axis and is equal to zero on the upstream axis.

This is equivalent to the potential-flow problem in two dimensions. The solution is

$$\psi = -\frac{1}{2}\sin^2\theta\left(r^2 - \frac{1}{r}\right), \qquad (10\text{--}154)$$

with corresponding velocity components

$$u_\theta = \sin\theta\left(1 + \frac{1}{2r^3}\right) \qquad (10\text{--}155)$$

$$u_r = -\cos\theta\left(1 - \frac{1}{r^3}\right). \qquad (10\text{--}156)$$

It will be noted that this solution again exhibits fore-aft symmetry and satisfies the condition $u_r = 0$ at the sphere surface $r = 1$. However, like the potential-flow solutions in two dimensions, this first outer solution does not satisfy the no-slip condition at $r = 1$, and this points clearly to the necessity for a thin boundary-layer near the body surface where viscous effects are important.

Thus, in a manner entirely equivalent to the two-dimensional analysis, we seek rescaled equations in the inner (boundary-layer) region very near to the sphere surface within which the tangential velocity goes from the potential-flow value $(3/2)\sin\theta$ to 0 at the body surface. The only difference from the previous analysis is in the detailed form of the Navier–Stokes and continuity equations for axisymmetric geometries. When expressed in terms of spherical coordinates, these equations are

$$\frac{1}{r^2}\frac{\partial}{\partial r}(r^2 u_r) + \frac{1}{r\sin\theta}\frac{\partial}{\partial \theta}(u_\theta \sin\theta) = 0, \qquad (10\text{--}157)$$

$$\frac{1}{Re}\left[\frac{1}{r^2}\frac{\partial}{\partial r}\left(r^2\frac{\partial u_r}{\partial r}\right) + \frac{1}{r^2\sin\theta}\frac{\partial}{\partial \theta}\left(\sin\theta\frac{\partial u_r}{\partial \theta}\right) - \frac{2u_r}{r^2} - \frac{2}{r^2}\frac{\partial u_\theta}{\partial \theta} - \frac{2u_\theta \cos\theta}{r^2}\right]$$
$$= \frac{\partial p}{\partial r} + u_r\frac{\partial u_r}{\partial r} + \frac{u_\theta}{r}\frac{\partial u_r}{\partial \theta} - \frac{u_\theta^2}{r}, \qquad (10\text{--}158)$$

F. Streaming Flow Past Axisymmetric Bodies – A Generalization of the Blasius Series

and

$$\frac{1}{Re}\left[\frac{1}{r^2}\frac{\partial}{\partial r}\left(r^2\frac{\partial u_\theta}{\partial r}\right) + \frac{1}{r^2\sin\theta}\frac{\partial}{\partial\theta}\left(\sin\theta\frac{\partial u_\theta}{\partial\theta}\right) + \frac{2}{r^2}\frac{\partial u_r}{\partial\theta} - \frac{u_\theta}{r^2\sin^2\theta}\right]$$
$$= \frac{1}{r}\frac{\partial p}{\partial\theta} + u_r\frac{\partial u_\theta}{\partial r} + \frac{u_\theta}{r}\frac{\partial u_\theta}{\partial\theta} + \frac{u_\theta u_r}{r}. \tag{10–159}$$

Thus, when we apply the usual boundary-layer rescaling to the continuity equation,

$$\boxed{u_r = \frac{1}{\sqrt{Re}}V, \quad r - 1 = \frac{Y}{\sqrt{Re}},} \tag{10–160}$$

we obtain

$$\frac{1}{[1+(Y/\sqrt{Re})]^2}\frac{\partial}{\partial Y}\left\{[(1+(Y/\sqrt{Re})]^2 V\right\} + \frac{1}{[1+(Y/\sqrt{Re})]\sin\theta}\frac{\partial}{\partial\theta}(u\cdot\sin\theta) = 0, \tag{10–161}$$

where the tangential component u_θ has been denoted as u. Hence, in the limit $Re \to \infty$,

$$\boxed{\frac{\partial V}{\partial Y} + \frac{1}{\sin\theta}\frac{\partial}{\partial\theta}(u\cdot\sin\theta) = O(Re^{-1/2}) \to 0.} \tag{10–162}$$

Similarly, the tangential component of the momentum equations becomes

$$\frac{\partial^2 u}{\partial Y^2} + \frac{2}{\sqrt{Re}}\frac{\partial u}{\partial Y} + O(Re^{-1})$$
$$= \frac{\partial p}{\partial\theta} + V\frac{\partial u}{\partial Y} + u\frac{\partial u}{\partial\theta} + \frac{1}{\sqrt{Re}}uV - \frac{Y}{\sqrt{Re}}\frac{\partial p}{\partial\theta} - \frac{Y}{\sqrt{Re}}u\frac{\partial u}{\partial\theta} + O(Re^{-1}), \tag{10–163}$$

so that, in the limit $Re \to \infty$,

$$\boxed{\frac{d^2 u}{\partial Y^2} = \frac{\partial p}{\partial\theta} + V\frac{\partial u}{\partial Y} + u\frac{\partial u}{\partial\theta}.} \tag{10–164}$$

The normal component gives

$$\frac{\partial p}{\partial Y} = +\frac{1}{\sqrt{Re}}u^2 + O(Re^{-1}), \tag{10–165}$$

so that

$$\boxed{\frac{\partial p}{\partial Y} = 0} \tag{10–166}$$

for $Re \to \infty$.

As usual, the governing equations for the leading-order approximation in the inner boundary-layer region are the limit of the full rescaled equations for $Re \to \infty$, namely,

Laminar Boundary-Layer Theory

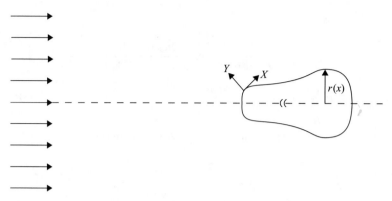

Figure 10–11. A schematic representation for streaming flow past an arbitrary axisymmetric body. The body geometry is specified by the function $r(x)$. which measures the distance from the symmetry axis to the body surface as a function of the position x.

(10–162), (10–164), and (10–166). It is important to note that the same rescaling (10–160), as deduced originally for 2D flows, still applies for this axisymmetric problem (and indeed would apply to any axisymmetric geometry). Furthermore, with the exception of the continuity equation, the form of the boundary-layer equations at leading order is identical to the 2D case.

To obtain a solution for the leading approximation in the inner region, we need to solve (10–162), (10–164), and (10–166) subject to the boundary conditions

$$u = V = 0 \quad \text{at} \quad Y = 0 \tag{10–167}$$

and the matching condition

$$u \to \frac{3}{2}\sin\theta \quad \text{as} \quad Y \to \infty. \tag{10–168}$$

The latter, taken with (10–166) and the Bernoulli equation in the outer region, implies that

$$\frac{\partial p}{\partial \theta} = \frac{9}{8}\sin 2\theta \tag{10–169}$$

in the boundary layer.

Although the preceding governing equations have been derived for the specific case of a sphere, Schlichting[23] has shown that the same general form would apply to any axisymmetric body. In particular, if we define the geometry in terms of a function $r(x)$, as shown in Fig. 10–11, the governing boundary-layer equations for a general axisymmetric body can be shown to take the form

$$\frac{\partial (ru)}{\partial x} + \frac{\partial (rV)}{\partial Y} = 0, \tag{10–170}$$

$$u\frac{\partial u}{\partial x} + V\frac{\partial u}{\partial Y} = -\frac{\partial p}{\partial x} + \frac{\partial^2 u}{\partial Y^2}, \tag{10–171}$$

$$\frac{\partial p}{\partial Y} = 0, \tag{10–172}$$

$$u = V = 0 \quad \text{at} \quad Y = 0, \tag{10–173}$$

$$u \to u_e(x) \quad \text{as} \quad Y \to \infty. \tag{10–174}$$

F. Streaming Flow Past Axisymmetric Bodies – A Generalization of the Blasius Series

The governing equations for the sphere are a special case, with

$$r(x) = \sin\theta = \sin x, \qquad (10\text{--}175)$$

$$u_e = \frac{3}{2}\sin\theta = \frac{3}{2}\sin x. \qquad (10\text{--}176)$$

The general problem (10–170)–(10–174) can be solved by a generalization of the Blasius series approximation from Section E. Let us suppose that the two functions $r(x)$ and $u_e(x)$, which differentiate between bodies with different geometries, can be approximated in the form

$$r = r_1 x + r_3 x^3 + r_5 x^5 + \cdots +, \qquad (10\text{--}177)$$

$$u_e = u_1 x + u_3 x^3 + u_5 x^5 + \cdots +, \qquad (10\text{--}178)$$

Then a power-series approximation also can be obtained for u within the boundary layer of the form

$$u = u_1 x f_1'(\eta) + 2u_3 x^3 f_3'(\eta) + 3u_5 x^5 f_5'(\eta) + O(x^7). \qquad (10\text{--}179)$$

Furthermore, Froessling[26] followed the same general approach pioneered by Howarth (1935) to show that the governing equations and boundary conditions for the functional coefficients in (10–179) can be made independent of the coefficients u_i and r_i from the expansions (10–177) and (10–178). This requires the following substitutions:

$$f_3 = g_3 + \frac{r_3 u_1}{r_1 u_3} h_3, \qquad (10\text{--}180)$$

$$f_5 = g_5 + \frac{r_5 u_1}{r_1 u_5} h_5 + \frac{u_3^2}{u_1 u_5} k_5 + \frac{r_3 u_3}{r_1 u_5} j_5 + \frac{r_3^2 u_1}{r_1^2 u_5} m_5, \qquad (10\text{--}181)$$

and so forth. The resulting equations for the coefficients at $O(x)$ and $O(x^3)$ are

$$f_1''' + f_1 f_1'' - \frac{1}{2}(f_1'^2 - 1) = 0, \qquad (10\text{--}182)$$

$$g_3''' + f_1 g_3'' = 2 f_1' g_3' - 2 f_1'' g_3 - 1, \qquad (10\text{--}183)$$

$$h_3''' + f_1 h_3'' = 2 f_1' h_3' - 2 f_1'' h_3 - \frac{1}{2} f_1 f_1'', \qquad (10\text{--}184)$$

with boundary conditions

$$f_1(0) = g_3(0) = h_3(0) = 0,$$
$$f_1'(\infty) \to 1, \quad g_3'(\infty) \to \frac{1}{2}, \quad h_3'(\infty) \to 0, \qquad (10\text{--}185)$$
$$f_1'(0) = g_3'(0) = h_3'(0) = 0.$$

Numerical solutions for all of the coefficients through $O(x^5)$ in (10–181) were actually obtained originally by Froessling, and later Scholkemeier[27] solved for the 10 additional

Laminar Boundary-Layer Theory

functions required for completely specifying (10–181) to $O(x^7)$. The results from Froessling have been reproduced in the book by Schlichting.[23]

Most important of these are the numerical values

$$\begin{aligned} f_1''(0) &= 0.9277, \\ g_3''(0) &= 1.0475, \\ h_3''(0) &= 0.0448, \\ g_5''(0) &= 0.9054, \\ h_5''(0) &= 0.0506, \\ k_5''(0) &= 0.1768, \\ j_5''(0) &= 0.0291, \\ m_5''(0) &= -0.0244, \end{aligned} \qquad (10\text{–}186)$$

because these can be used to determine the shear stress at the body surface and thus test for the existence of a separation point. In particular,

$$\left.\frac{\partial u}{\partial Y}\right|_{Y=0} = u_1 x f_1''(0)\sqrt{2u_1} + 2u_3 x^3 \sqrt{2u_1}\, f_3''(0)\ldots. \qquad (10\text{–}187)$$

Now, for the case of a sphere,

$$r_1 = 1, \quad r_3 = -\frac{1}{6}, \ldots,$$
$$u_1 = \frac{3}{2}, \quad u_3 = -\frac{1}{4}, \ldots, \qquad (10\text{–}188)$$

If we truncate after only two terms in (10–187), we therefore find that

$$\left.\frac{\partial u}{\partial Y}\right|_{Y=0} = \frac{3\sqrt{3}}{2}(0.9277)[1 - 0.3925 x^2 + \cdots +]. \qquad (10\text{–}189)$$

Thus a separation point exists, according to this approximation, at

$$x_s = 1.596 \quad (\theta_s \sim 91.5°). \qquad (10\text{–}190)$$

This is very close to the result for the similar two-term approximation for a circular cylinder. According to Schlichting, if we retain all terms through $O(x^7)$, we find that

$$x_s = 1.913 \quad (\theta_s \sim 109.6°). \qquad (10\text{–}191)$$

Again, this is almost identical to the result for the circular cylinder.

The most important point of the preceding analysis is that the same scaling laws and the same asymptotic solution structure apply for both axisymmetric and 2D geometries. Indeed, though we do not pursue the subject here, the same general solution scheme applies for arbitrary 3D geometries. The only complication in solving the general equations for *axisymmetric* bodies is the shape function $r(x)$ that appears in the continuity equation, and we have seen that a slight generalization of the Blasius series yields a universal power-series solution for arbitrary $r(x)$.

G. THE BOUNDARY LAYER ON A SPHERICAL BUBBLE

By now the reader may have come to associate the momentum boundary layer as a thin region of $O(Re^{-1/2})$ adjacent to a body surface across which the dimensionless tangential velocity changes by $O(1)$ in order to satisfy the no-slip condition at the body surface. To be sure, many boundary layers do exhibit this structure, but it is overly restrictive as a general description. A more accurate and generally applicable description is that the boundary layer is a thin region of $O(Re^{-1/2})$ adjacent to a body surface inside of which the vorticity generated at the body surface is confined.

In this regard, it is of interest to contrast the two problems of the streaming motion of a fluid at large Reynolds number past a solid sphere and a spherical bubble. In the case of a solid sphere, the potential-flow solution (10–155)–(10–156) does not satisfy the no-slip condition at the sphere surface, and the necessity for a boundary layer in which viscous forces are important is transparent. For the spherical bubble, on the other hand, the no-slip condition is replaced with the condition of zero tangential stress, $\tau_{r\theta} = 0$, and it may not be immediately obvious that a boundary layer is needed. However, in this case, the potential-flow solution does not satisfy the zero-tangential-stress condition (as we shall see shortly), and a boundary-layer in which viscous forces are important still must exist. We shall see that the detailed features of the boundary layer are different from those of a no-slip, solid body. However, in both cases, the surface of the body acts as a source of vorticity, and this vorticity is confined at high Reynolds number to a thin $O(Re^{-1/2})$ region near the surface.

The source of vorticity at a solid, no-slip surface is the velocity gradient that is generated in satisfying the no-slip condition. This mechanism yields vorticity of $O(Re^{1/2})$ at the body surface. At an interface where the tangential velocity is *not* zero, on the other hand, vorticity is produced by rotation of fluid elements caused by the surface curvature. This latter mechanism generates vorticity of magnitude proportional to the local curvature of the surface in the direction of the motion of the fluid. As an example of vorticity production in the latter case, we may consider the condition of zero tangential stress at the surface of a bubble whose shape we assume, for simplicity, to be spherical. In this case, for an axisymmetric motion,

$$\tau_{r\theta} = \frac{\partial u_\theta}{\partial r} - \frac{u_\theta}{r} = 0, \qquad (10\text{–}192)$$

and if this condition is expressed in terms of vorticity,

$$\omega \equiv \frac{1}{r}\left[\frac{\partial}{\partial r}(ru_\theta) - \frac{\partial u_r}{\partial \theta}\right], \qquad (10\text{–}193)$$

we find that

$$\boxed{\omega|_{r=1} = \frac{2u_\theta}{r}\Big|_{r=1} = 2u_\theta.} \qquad (10\text{–}194)$$

The corresponding condition for motion past an axisymmetric bubble of arbitrary shape is

$$\boxed{\omega|_{r=1} = 2\kappa u_s,} \qquad (10\text{–}195)$$

that is, the vorticity is equal to twice the product of the tangential velocity u_s and the local curvature in the direction of motion κ.

Regardless of the source of the vorticity, it remains confined near the body surface for $Re \gg 1$ because the time scale for radial diffusion is limited by convection around the body

Laminar Boundary-Layer Theory

to $O(L/U_\infty)$, where L is a characteristic linear dimension of the body. This restricted time scale for radial diffusion then leads to the same estimate of the thickness of the boundary layer, $O(LRe^{-1/2})$, in all cases. However, the *strength* of the boundary layer, as indicated either by the magnitude of the vorticity or the magnitude of velocity gradients, differs dramatically between the cases of no-slip and slip boundaries. When the body surface is solid, the velocity must change by $O(U_\infty)$ across this layer, and this requires that the velocity gradients be $O((U_\infty/L)Re^{1/2})$, as we have already seen in the preceding sections of this chapter. For the bubble, on the other hand, it is the gradient of tangential velocity normal to the surface that must change by $O(U_\infty/L)$; that is, the dimensionless velocity gradient must change by $O(1)$ in order for the shear stress to go to zero at the bubble surface. Thus the tangential velocity must change by $O((U_\infty/L)\cdot L(Re)^{-1/2}) = O(U_\infty Re^{-1/2})$. In other words, the change in the velocity across the boundary layer is *very small* compared with the characteristic velocity in the potential-flow region. This suggests that the velocity distribution in the boundary layer will be the potential flow profile plus an $O(Re^{-1/2})$ correction to satisfy the zero-shear-stress condition at the bubble surface.

To verify these predictions of dimensional reasoning, we now turn to a detailed analysis for a bubble moving through a quiescent fluid with velocity U_∞. An alternative description of the problem from the point of view of a reference system fixed at the center of the bubble is that the fluid undergoes a uniform motion with velocity U_∞ past a stationary bubble. We will adopt this latter viewpoint for our analysis because it is the same as that used in the earlier sections of this chapter. At the beginning, we assume that the bubble is *spherical*, though we may anticipate, by direct observation or by examination of the pressure distribution imposed on a bubble in a potential flow, that a real bubble will tend to deform to a flattened, axisymmetric shape.[28] We shall return at the end of this section to consider the necessary conditions for a bubble to remain approximately spherical when the Reynolds number is not small. In addition, we will actually calculate the bubble shape when the deformation is small. However, the primary goal here is to examine the change in boundary-layer structure for a curved surface when the surface boundary condition is changed from no-slip to zero shear stress. For this purpose it is adequate to concentrate on the limiting case of a perfectly spherical bubble.

As usual, we begin by nondimensionalizing the equations of motion by using the bubble radius as a characteristic length scale. Thus we obtain (10–157)–(10–159), which are to be solved subject to the boundary conditions

$$u_r = 0 \quad \text{at} \quad r = 1, \qquad (10\text{--}196)$$

$$\tau_{r\theta} = r\frac{\partial}{\partial r}\left(\frac{u_\theta}{r}\right) = 0 \quad \text{at} \quad r = 1, \qquad (10\text{--}197)$$

$$u_r \to -\cos\theta, \quad u_\theta \to \sin\theta \quad \text{as} \quad r \to \infty. \qquad (10\text{--}198)$$

Clearly, in the limit $Re \gg 1$, the leading-order approximation for the solution to this problem is identical to the inviscid flow problem for a solid sphere. Although the no-slip boundary condition has been replaced in the present problem with the zero-shear-stress condition, (10–197), this has no influence on the leading-order inviscid flow approximation because the potential-flow solution can, in any case, only satisfy the kinematic condition $\mathbf{u}\cdot\mathbf{n} = 0$ at $r = 1$. Hence the first approximation in the outer part of the domain where the bubble radius is an appropriate characteristic length scale is precisely the same as for the no-slip sphere, namely, (10–155) and (10–156). However, this solution does *not* satisfy the zero-shear-stress condition (10–197) at the bubble surface, and thus it is clear that the inviscid flow equations do not provide a uniformly valid approximation to the Navier–Stokes

G. The Boundary Layer on a Spherical Bubble

equations even for $Re \gg 1$. The problem is the implicit assumption, inherent in the nondimensionalized equations (10–151)–(10–153), that viscous terms can be neglected everywhere, including the region near the bubble surface. Both the source of this incorrect assumption (that is, the use of the body radius as a characteristic length scale everywhere in the domain) and its remedy are the same as in the preceding sections of this chapter, though the details are different, as we shall see.

Thus, to obtain governing equations appropriate to the inner region near the surface, we introduce the usual boundary-layer rescaling,

$$u_r = \frac{1}{\sqrt{Re}} V, \quad r - 1 = \frac{Y}{\sqrt{Re}}, \tag{10–199}$$

into (10–157)–(10–159). The resulting equations are (10–161), (10–163), and (10–165), in which we retain terms of both $O(1)$ and $O(Re^{-1/2})$. The boundary conditions (10–196) and (10–197) in terms of inner variables are

$$V = 0 \quad \text{at} \quad Y = 0, \tag{10–200}$$

$$\frac{\partial u}{\partial Y} - \frac{u}{\sqrt{Re}} + O(Re^{-1}) = 0 \quad \text{at} \quad Y = 0. \tag{10–201}$$

As in the previous section we denote the tangential velocity component within the boundary layer as u to distinguish it from the tangential component u_θ in the outer, inviscid flow region. The boundary-layer solution is also required to match the outer inviscid solution, namely,

$$\lim_{Y \gg 1} u \Leftrightarrow \frac{3}{2} \sin\theta - \frac{3}{2} Y \sin\theta \left(\frac{1}{\sqrt{Re}}\right) + O\left(Re^{-1}\right), \tag{10–202}$$

$$\lim_{Y \gg 1} \frac{1}{\sqrt{Re}} V \Leftrightarrow \lim_{(r-1) \ll 1} (u_r) \tag{10–203}$$

(both as $Re \to \infty$). The second of these produces the boundary condition $u_r = 0$ at $r = 1$, which was already used for the leading-order approximation in the inviscid flow region (10–155) and (10–156). The matching condition (10–202) incorporates the limiting form of (10–155) for $y \equiv (r-1) \ll 1$ on the right-hand side and provides a boundary condition for the leading approximation (at least) to u in the boundary layer,

$$u \to \frac{3}{2} \sin\theta \quad \text{for} \quad Y \to \infty. \tag{10–204}$$

Now, let us suppose, following the scaling arguments at the beginning of this section, that the first two terms in the solution for the inner boundary-layer region can be expressed in the form

$$\begin{aligned}
u &= u_0 + \frac{1}{\sqrt{Re}} u_1 + \cdots +, \\
V &= V_0 + \frac{1}{\sqrt{Re}} V_1 + \cdots +, \\
p &= p_0 + \frac{1}{\sqrt{Re}} p_1 + \cdots +.
\end{aligned} \tag{10–205}$$

Laminar Boundary-Layer Theory

Thus, substituting into (10–161), (10–163), and (10–165), we find the governing equations at $O(1)$ for the first terms in (10–205), namely,

$$\frac{\partial V_0}{\partial Y} + \frac{1}{\sin\theta}\frac{\partial}{\partial\theta}(\sin\theta u_0) = 0, \tag{10-206}$$

$$\frac{\partial^2 u_0}{\partial Y^2} = \frac{\partial p_0}{\partial\theta} + V_0\frac{\partial u_0}{\partial Y} + u_0\frac{\partial u_0}{\partial\theta}, \tag{10-207}$$

$$\frac{\partial p_0}{\partial Y} = 0. \tag{10-208}$$

The corresponding boundary conditions from (10–200)–(10–202) are

$$\left.\begin{aligned} V_0 &= 0 \\ \frac{\partial u_0}{\partial Y} &= 0, \end{aligned}\right\} \quad \text{at } Y = 0, \tag{10-209}$$

$$u_0 \to \frac{3}{2}\sin\theta \quad \text{as } Y \to \infty. \tag{10-210}$$

Not surprisingly, these equations and boundary conditions are just the limiting forms of (10–161), (10–163), (10–165), and (10–200)–(10–202) for $Re \to \infty$.

The solution for (10–206)–(10–210) is just

$$u_0 = \frac{3}{2}\sin\theta \quad \text{and} \quad V_0 = -3Y\cos\theta, \tag{10-211}$$

that is, the boundary (matching) condition for $Y \gg 1$ is a uniformly valid solution. This is *not* a surprising result. The scaling arguments at the beginning of this section suggest that the change in the velocity profile in the boundary layer that is necessary to satisfy the zero-shear-stress boundary condition is only an $O(Re^{-1/2})$ correction to the potential-flow solution. The $O(1)$ solution in the boundary layer, (10–211), is nothing but the largest term in an expansion of the potential-flow solution for $y \equiv (r-1) \ll 1$. Indeed, if we simply express the potential-flow solution in terms of inner variables, we find that

$$u_\theta \to \frac{3}{2}\sin\theta - \frac{3}{2}Y\sin\theta\frac{1}{\sqrt{Re}} + O(Re^{-1}), \tag{10-212}$$

$$u_r\sqrt{Re} = V \to -3Y\cos\theta + 6Y^2\cos\theta\frac{1}{\sqrt{Re}} + O(Re^{-1}), \tag{10-213}$$

and the first terms are identical to (10–211).

Normally, the next step in the method of matched asymptotic expansions would be to seek a second approximation in the outer region, followed by a second approximation in the inner region and so on. However, if we examine the governing equations (10–157)–(10–159) for the outer region, we see that the largest neglected term is $O(Re^{-1})$. Furthermore, because the first approximation to the velocity distribution in the inner region turned out to be just the first term of the outer solution for $(r-1) \ll 1$ and $Re \to \infty$, there is no mismatch at this level to generate a nonzero second approximation in the outer region. Instead, the next term is the $O(Re^{-1/2})$ correction in the expansion (10–205) for the inner boundary-layer region.

G. The Boundary Layer on a Spherical Bubble

To determine the form of this term, we must solve the governing equations and boundary conditions that derive from (10–161), (10–163), (10–165), and (10–200)–(10–202) at $O(Re^{-1/2})$. These are

$$\frac{\partial V_1}{\partial Y} + \frac{1}{\sin\theta}\frac{\partial}{\partial \theta}(u_1 \sin\theta) = -2V_0 - Y\frac{\partial V_0}{\partial Y} \qquad (10\text{–}214)$$

$$\frac{\partial^2 u_1}{\partial Y^2} = \frac{\partial p_1}{\partial \theta} + V_0 \frac{\partial u_1}{\partial Y} + V_1 \frac{\partial u_0}{\partial Y} + u_1 \frac{\partial u_0}{\partial \theta} + u_0 \frac{\partial u_1}{\partial \theta} + u_0 V_0 - Y\frac{\partial p_0}{\partial \theta} - Yu_0 \frac{\partial u_0}{\partial \theta} - 2\frac{\partial u_0}{\partial Y}, \qquad (10\text{–}215)$$

$$\frac{\partial p_1}{\partial Y} = u_0^2, \qquad (10\text{–}216)$$

with boundary conditions

$$V_1 = 0, \quad \text{at } Y = 0, \qquad (10\text{–}217)$$

$$\frac{\partial u_1}{\partial Y} - u_0 = 0 \quad \text{at } Y = 0, \qquad (10\text{–}218)$$

$$u_1 \to -\frac{3}{2}Y \sin\theta \quad \text{as } Y \to \infty \text{ (match)}. \qquad (10\text{–}219)$$

The latter condition requires that the boundary-layer solution match $O(Re^{-1/2})$ with the inner limit of the outer solution (10–212).

Considerable simplification occurs when we substitute for u_0 and V_0 from (10–211). First, the right-hand side of (10–214) is identically zero, and we recover the usual form of the continuity equation for (u_1, V_1), namely,

$$\frac{\partial V_1}{\partial Y} + \frac{1}{\sin\theta}\frac{\partial}{\partial \theta}(u_1 \sin\theta) = 9Y \cos\theta. \qquad (10\text{–}220)$$

Similarly, the right-hand side of (10–215) also simplifies. First, note that

$$\frac{\partial p_0}{\partial \theta} = -u_0 \frac{\partial u_0}{\partial \theta}$$

because u_0 is independent of Y. Thus,

$$\frac{\partial^2 u_1}{\partial Y^2} = \frac{\partial p_1}{\partial \theta} + V_0 \frac{\partial u_1}{\partial Y} + V_1 \frac{\partial u_0}{\partial Y} + u_1 \frac{\partial u_0}{\partial \theta} + u_0 \frac{\partial u_1}{\partial \theta} + u_0 V_0.$$

However, according to (10–216) and the fact that u_0 is a function of θ only

$$\frac{\partial p_1}{\partial \theta} = 2u_0 \frac{\partial u_0}{\partial \theta} Y.$$

Hence, substituting for u_0 from (10–211), we obtain

$$\frac{\partial p_1}{\partial \theta} = \frac{9}{2} Y \sin\theta \cos\theta. \qquad (10\text{–}221)$$

But this is just $-u_0 V_0$, and so finally we obtain

$$\frac{\partial^2 u_1}{\partial Y^2} = V_0 \frac{\partial u_1}{\partial Y} + V_1 \frac{\partial u_0}{\partial Y} + u_1 \frac{\partial u_0}{\partial \theta} + u_0 \frac{\partial u_1}{\partial \theta}. \qquad (10\text{–}222)$$

Laminar Boundary-Layer Theory

Thus the problem at $O(Re^{-1/2})$ is to solve the linear equations (10–220) and (10–222), subject to the boundary and matching conditions (10–217)–(10–219). We may note that (10–222) is actually independent of V_1 because $(\partial u_0/\partial Y) \equiv 0$. Hence we can solve (10–222) first by using the boundary conditions (10–218) and (10–219), and then we can obtain V_1 from the continuity equation (10–220) and the boundary condition (10–217).

The solution of (10–222), (10–218), and (10–219) can be obtained by similarity transformation. For this purpose, however, it is convenient to formulate the problem in terms of the *difference* velocity,

$$\hat{u}_1 \equiv u_1 - \left(-\frac{3}{2} Y \sin\theta\right). \tag{10–223}$$

Substituting (10–223) into (10–218), (10–219), and (10–222), and using the $O(1)$ solution (10–211) to evaluate the coefficients, we obtain

$$\frac{\partial^2 \hat{u}_1}{\partial Y^2} = -3Y \cos\theta \frac{\partial \hat{u}_1}{\partial Y} + \frac{3}{2} \cos\theta \hat{u}_1 + \frac{3}{2} \sin\theta \frac{\partial \hat{u}_1}{\partial \theta}, \tag{10–224}$$

$$\frac{\partial \hat{u}_1}{\partial Y} = 3 \sin\theta \quad \text{at } Y = 0, \tag{10–225a}$$

$$\hat{u}_1 \to 0 \quad \text{as } Y \to \infty. \tag{10–225b}$$

To solve (10–224) and (10–225), a similarity transformation is proposed of the general form

$$\hat{u}_1 = h(\theta) f(\eta), \quad \text{where } \eta = \frac{Y}{g(\theta)}. \tag{10–226}$$

It is necessary to include the multiplying coefficient $h(\theta)$ because of the form of the boundary condition (10–225a). In particular, if we substitute (10–226) into (10–225a), we find that

$$h(\theta) \frac{1}{g(\theta)} f'(0) = 3 \sin\theta, \tag{10–227}$$

so that

$$\frac{h(\theta)}{g(\theta)} = 3 \sin\theta. \tag{10–228}$$

Although it might at first seem that (10–227) could be satisfied with $h = \text{const}$ by proper choice of $g(\theta)$, the resulting similarity transform would not reduce (10–224) to an ODE. With the present formulation, $g(\theta)$ still can be chosen to reduce (10–224) to a similarity form. To see that this is possible, we substitute for $\hat{u}_1$ in the form

$$\hat{u}_1 = 3 \sin\theta g(\theta) f(\eta), \quad \text{where } \eta = \frac{Y}{g(\theta)}, \tag{10–229}$$

into the governing equation (10–224) and boundary conditions (10–225). The result is

$$\frac{d^2 f}{d\eta^2} = \left(-3g^2 \cos\theta - \frac{3}{2} gg' \sin\theta\right) \eta \frac{df}{d\eta} + \left(3g^2 \cos\theta + \frac{3}{2} \sin\theta gg'\right) f, \tag{10–230}$$

G. The Boundary Layer on a Spherical Bubble

with
$$f'(0) = 1,$$
$$f(\eta) \to 0 \quad \text{as} \quad \eta \to \infty. \tag{10–231}$$

However, the two coefficients involving g in (10–230) are identical, and the condition for existence of a similarity solution thus reduces to a single, first-order ODE for $g(\theta)$,

$$\frac{3}{2} g g' \sin\theta + 3g^2 \cos\theta = 2, \tag{10–232}$$

where the constant 2 is arbitrary (provided only that the constant is nonzero). Hence the problem reduces to solving (10–232) for g, with the requirement that g remains bounded for all θ (except possibly $\theta = \pi$, where a wake may be expected), plus the ODE

$$\frac{d^2 f}{d\eta^2} + 2\eta \frac{df}{d\eta} - 2f = 0, \tag{10–233}$$

with boundary conditions (10–231) for f.

Solving (10–233) is straightforward. We let

$$s = f/\eta, \tag{10–234}$$

and (10–233) is transformed to

$$\frac{d^2 s}{d\eta^2} + 2\left(\frac{1+\eta^2}{\eta}\right) \frac{ds}{d\eta} = 0. \tag{10–235}$$

Integrating twice with respect to η, we obtain

$$s(\eta) = k_1 \int_\eta^\infty \frac{1}{t^2} e^{-t^2} dt + k_2 \tag{10–236}$$

or

$$f(\eta) = k_1 \eta \int_\eta^\infty \frac{1}{t^2} e^{-t^2} dt + k_2 \eta. \tag{10–237}$$

The two integration constants are evaluated by application of the boundary conditions (10–231). The condition

$$f(\eta) \to 0 \quad \text{as} \quad \eta \to \infty$$

requires $k_2 = 0$. (The reader may wish to verify that the first term goes to zero as $\eta \to \infty$ for arbitrary k_1.) We facilitate application of the second condition by expressing $f(\eta)$ in the alternative form

$$f(\eta) = k_1 \left(e^{-\eta^2} - 2\eta \int_\eta^\infty e^{-t^2} dt \right), \tag{10–238}$$

which we obtained from (10–237) by integrating by parts. It follows from (10–238) that

$$\frac{df}{d\eta} = -2k_1 \int_\eta^\infty e^{-t^2} dt. \tag{10–239}$$

Therefore, applying (10–231), we obtain

$$1 = -2k_1 \int_0^\infty e^{-t^2} dt = -2k_1 \sqrt{\pi}/2$$

or

$$k_1 = -\frac{1}{\sqrt{\pi}}. \tag{10-240}$$

Thus, finally,

$$\boxed{f(\eta) = -\frac{1}{\sqrt{\pi}} e^{-\eta^2} + \eta \operatorname{erfc}(\eta),} \tag{10-241}$$

where

$$\operatorname{erfc}(\eta) \equiv \frac{2}{\sqrt{\pi}} \int_\eta^\infty e^{-t^2} dt \tag{10-242}$$

is the complementary error function, whose values are tabulated in most mathematics handbooks (see Abramowitz and Stegun[29]).

To solve for $g(\theta)$, we can multiply (10–232) by $\sin^3\theta$ and then rewrite the resulting equation in the form

$$\frac{\partial}{\partial \theta}\left(\frac{3}{4}\sin^4\theta g^2\right) = 2\sin^3\theta. \tag{10-243}$$

Integrating and applying the condition that $g(\theta)$ is finite at $\theta = 0$, it can be shown easily that

$$\boxed{g(\theta) = \frac{\sqrt{8/3}}{\sin^2\theta}\left(\frac{2}{3} - \cos\theta + \frac{\cos^3\theta}{3}\right)^{1/2}.} \tag{10-244}$$

Hence, combining the results of our analysis, we find that

$$\boxed{u_1 = -\frac{3}{2} Y \sin\theta + 3g(\theta)\sin\theta f(\eta),} \tag{10-245}$$

where $f(\eta)$ and $g(\theta)$ are given in (10–241) and (10–244), respectively. Thus the general solution for the inner boundary-layer region through terms of $O(Re^{-1/2})$ is

$$\boxed{u = \frac{3}{2}\sin\theta + \frac{1}{\sqrt{Re}}\left[-\frac{3}{2}Y\sin\theta + 3g(\theta)\sin\theta f(\eta)\right] + O(Re^{-1}).} \tag{10-246}$$

From (10–245) and the continuity equation (10–220), the normal velocity component V_1 can be obtained, if so desired.

This completes the solution to $O(Re^{-1/2})$. It should be noted that the first two terms in (10–246) are, in fact, nothing but the first two terms in the inviscid solution, evaluated in the inner region, namely, (10–213). Thus, to $O(Re^{-1/2})$, we see that the solution in the complete domain consists of the inviscid solution (10–155) and (10–156), with an $O(Re^{-1/2})$ viscous correction in the inner boundary-layer region to satisfy the zero-shear-stress boundary condition at the bubble surface. Because the viscous correction in the inner region is only $O(Re^{-1/2})$, the governing equation for it is *linear*. Hence, unlike the no-slip boundary layers considered earlier in this chapter, it is possible to obtain an analytic solution for the leading-order departure from the inviscid flow solution.

G. The Boundary Layer on a Spherical Bubble

It is of particular interest to calculate the force acting on the bubble as a complement to the drag law obtained earlier for the force on a spherical bubble in *creeping flow*. One way to determine the force is to calculate the normal stress and pressure forces on the bubble surface and integrate in the usual manner. However, this approach turns out to be somewhat complicated because it is necessary to include the leading-order departures in u, V, and p from the inviscid flow solution to obtain the first nonzero contribution to the drag (note that the inviscid flow solution alone gives zero drag when used in this manner). Here, we follow an alternative approach pioneered by Levich,[30] based on an overall mechanical energy balance, in the general form

$$\boxed{\mathbf{U} \cdot \mathbf{F}_{drag} = \int_V \Phi dV,} \qquad (10\text{--}247)$$

where $\Phi \equiv \nabla \mathbf{u} : \mathbf{T}$ is the rate of viscous dissipation in the fluid. Physically, the condition (10–247) states that there is a precise balance at steady state between the rate of working on the fluid that is due to the bubble motion and the rate of dissipation of mechanical energy in the fluid to heat. This macroscopic balance applies to an arbitrary body undergoing an arbitrary (though steady) motion, but it is particularly useful, as we shall see, for the special case of bubble motion at high *Re*.

Before considering the application of (10–247), it is instructive to outline its derivation. To do this, we consider a so-called "control volume" of fluid lying within an arbitrarily large spherical surface in the fluid that circumscribes the particle or body of interest and is fixed with respect to the center of the particle, so that $\mathbf{u} \to \mathbf{U}_\infty$ at its surface as the radius is increased to very large values. We begin with the equation of motion in the form

$$\rho \frac{D\mathbf{u}}{Dt} = \nabla \cdot \mathbf{T}, \qquad (10\text{--}248)$$

which applies at all points within the fluid. We can obtain differential mechanical energy balance directly from (10–248) by taking its inner (scalar) product with $\mathbf{u}$ (see Chap. 2, Section **D**), and we can write it in the form

$$\frac{\rho}{2} \frac{D}{Dt}(\mathbf{u} \cdot \mathbf{u}) = \nabla \cdot (\mathbf{u} \cdot \mathbf{T}) - \nabla \mathbf{u} : \mathbf{T}. \qquad (10\text{--}249)$$

Hence, at steady state,

$$\frac{\rho}{2} \nabla \cdot [(\mathbf{u} \cdot \mathbf{u})\mathbf{u}] = \nabla \cdot (\mathbf{u} \cdot \mathbf{T}) - \nabla \mathbf{u} : \mathbf{T}, \qquad (10\text{--}250)$$

where we have utilized the continuity condition $\nabla \cdot \mathbf{u} = 0$ to rewrite the left-hand side. This mechanical energy equation is satisfied at every point in the fluid. Hence it also applies in the integral sense:

$$\int_V \nabla \cdot \left[\frac{\rho}{2}(\mathbf{u} \cdot \mathbf{u})\mathbf{u}\right] dV = \int_V \nabla \cdot (\mathbf{u} \cdot \mathbf{T}) dV - \int_V (\nabla \mathbf{u} : \mathbf{T}) dV, \qquad (10\text{--}251)$$

where V represents the whole of the volumetric region within the control volume. But now the divergence theorem can be applied to both the left-hand side of (10–251) and to the first term on the right-hand side. The integral on the left-hand side is equal to zero, that is,

$$\int_A \frac{\rho}{2}(\mathbf{u} \cdot \mathbf{n})(\mathbf{u} \cdot \mathbf{u}) dA \equiv -\int_{A_{particle}} \frac{\rho}{2}(\mathbf{u} \cdot \mathbf{n})(\mathbf{u} \cdot \mathbf{u}) dA + \int_{A_\infty} \frac{\rho}{2} U_\infty^3 \cos\theta \sin\theta d\theta d\phi \equiv 0, \qquad (10\text{--}252)$$

Laminar Boundary-Layer Theory

where the symbol A_∞ is used to denote the surface of the large circumscribed sphere in the fluid. The integral at the particle surface is zero because $\mathbf{u} \cdot \mathbf{n} = 0$ at $A_{particle}$. The first integral on the right-hand side of (10–251) becomes

$$\int_A \mathbf{n} \cdot (\mathbf{u} \cdot \mathbf{T}) dA = \int_A \mathbf{u} \cdot (\mathbf{n} \cdot \mathbf{T}) dA = \mathbf{U}_\infty \cdot \int_{A_\infty} \mathbf{T} \cdot \mathbf{n} dA - \int_{A_{particle}} \mathbf{u} \cdot (\mathbf{n} \cdot \mathbf{T}) dA.$$
(10–253)

For the spherical bubble, the second term on the right hand side is zero because the integrand is proportional to the shear stress $\tau_{r\theta}$, which is zero at the bubble surface. The first term on the right is just

$$\mathbf{U}_\infty \cdot \int_{A_\infty} \mathbf{T} \cdot \mathbf{n} dA = \mathbf{U}_\infty \cdot \int_{A_{particle}} \mathbf{T} \cdot \mathbf{n} dA = \mathbf{U}_\infty \cdot \mathbf{F}_{drag}.$$
(10–254)

The equality of the integral of $\mathbf{T} \cdot \mathbf{n}$ over A_∞ and $A_{particle}$ is true only to $O(Re^{-1/2})$ because it does not account for the deficit in the momentum flux in the wake region behind the bubble. However, this approximation is adequate for present purposes. Combining (10–251)–(10–254), we obtain (10–247).

Let us now apply (10–247) to calculate the drag on the spherical bubble, following the analysis of Levich.[30] First, it is convenient to nondimensionalize:

$$U_\infty F_{drag} = \mu U_\infty^2 a \int_V (\nabla \mathbf{u} : \mathbf{T}) dV.$$
(10–255)

Thus, expressing the drag in terms of the dimensionless drag coefficient,

$$C_D \equiv \frac{F_{drag}}{\pi a^2 \left(\frac{1}{2}\rho U_\infty^2\right)},$$

we find that

$$C_D = \frac{4}{Re}\left[\int_0^\pi \int_1^\infty r^2 (\nabla \mathbf{u} : \mathbf{T}) \sin\theta d\theta dr\right],$$
(10–256)

where

$$\Phi \equiv \nabla \mathbf{u} : \mathbf{T} = 2\left(\frac{\partial u_r}{\partial r}\right)^2 + 2\left(\frac{1}{r}\frac{\partial u_\theta}{\partial \theta} + \frac{u_r}{r}\right)^2 + 2\left(\frac{u_r}{r} + \frac{u_\theta \cos\theta}{r}\right)^2$$
$$+ \left[r\frac{\partial}{\partial r}\left(\frac{u_\theta}{r}\right) + \frac{1}{r}\frac{\partial u_r}{\partial \theta}\right]^2.$$
(10–257)

But now we can obtain a first approximation to the integral term in (10–256) by using the *inviscid* flow solution. In particular, we have shown that

$$u_r = u_r^p + O(Re^{-1/2}),$$
$$u_\theta = u_\theta^p + O(Re^{-1/2}),$$
(10–258)

throughout the domain $1 \leq r \to \infty$, where u_r^p and u_θ^p are the inviscid velocity components given in (10–155) and (10–156), and the $O(Re^{-1/2})$ terms correspond to the viscous correction within the boundary layer. We have not, of course, proven that the boundary layer is the only region where $O(Re^{-1/2})$ corrections exist to the inviscid flow solution. In fact, Moore[31] has demonstrated that velocity corrections of this order also must exist in a thin wake of width $O(Re^{-1/4})$ around the axis of symmetry.

G. The Boundary Layer on a Spherical Bubble

Now, substituting (10–258) into (10–257) and integrating according to (10–256), we find that

$$\int_V \Phi dV = 6 + O(Re^{-1/2}). \tag{10--259}$$

Thus the drag coefficient is

$$\boxed{C_D = \frac{24}{Re}[1 + O(Re^{-1/2})].} \tag{10--260}$$

Although we have determined the $O(Re^{-1/2})$ velocity corrections within the boundary layer, we cannot determine the numerical coefficient of the $O(Re^{-1/2})$ correction for C_D without a detailed analysis of the downstream wake. This is beyond the scope of the present work, but Moore carried out the necessary analysis to show that

$$C_D = \frac{24}{Re}\left[1 - \frac{2.2}{Re^{1/2}} + O(Re^{-1})\right]. \tag{10--261}$$

The amazing feature of (10–260) is that it is obtained entirely from the inviscid flow solution – the boundary-layer analysis does play an important role in demonstrating that the volume integral of Φ, based on the inviscid velocity field, will provide a valid first approximation to the total viscous dissipation but does not enter directly.

An obvious question that may occur to the reader is why the very simple method of integrating the viscous dissipation function has not been used earlier for calculation of the force on a solid body. The answer is that the method provides no real advantage except for the motion of a shear-stress-free bubble because the easily attained inviscid or potential-flow solution does not generally yield a correct first approximation to the dissipation. For the bubble, $\nabla \mathbf{u} : \mathbf{T} = O(1)$ everywhere to leading order, including the viscous boundary layer where the deviation from the inviscid solution yields only a correction of $O(Re^{-1/2})$. For bodies with no-slip boundaries, on the other hand, $\nabla \mathbf{u} : \mathbf{T}$ is still $O(1)$ outside the boundary layer, but inside the boundary layer $\nabla \mathbf{u} : \mathbf{T} = O(Re)$. When integrated over the boundary layer, which is $O(Re^{-1/2})$ in radial thickness, this produces an $O(\sqrt{Re})$ contribution to the total dissipation,

$$\text{boundary-layer contribution viscous dissipation} \sim \sqrt{Re} \cdot \text{const},$$

and the corresponding estimate for C_D is

$$\boxed{C_D \sim \frac{\text{const}}{\sqrt{Re}}} \tag{10--262}$$

(assuming that the flow does not separate). Thus, in the more common no-slip problem, the dominant contribution to the dissipation comes from the boundary layer. Although the estimate of dissipation that is due to the inviscid or potential flow does not change, it is of little use, and an estimation of drag by means of the rate of viscous dissipation still requires that we determine the complete boundary-layer solution.

One other aspect of the present problem to reconsider is the bubble shape. So far in this section we have assumed that the bubble is exactly spherical. However, in general we would expect a bubble to deform in a flow, and we have shown earlier that the shape can be determined from the normal-stress condition,

$$\mathbf{n} \cdot \mathbf{n} \cdot \mathbf{T} - \mathbf{n} \cdot \mathbf{n} \cdot \widehat{\mathbf{T}} = \sigma(\nabla \cdot \mathbf{n}), \tag{10--262}$$

Laminar Boundary-Layer Theory

which represents a balance between the normal-stress components that are due to the fluid motion and capillary forces that are due to interfacial or surface tension. For a bubble with negligible density and viscosity relative to the surrounding fluid, the normal stress inside is just the pressure, that is,

$$\mathbf{n} \cdot \mathbf{n} \cdot \widehat{\mathbf{T}} = -p_{bubble}. \tag{10-263}$$

In *general*, if we regard the bubble shape as unknown, we must solve for the velocity, pressure, and bubble shape *simultaneously*, using the equations of motion, the usual boundary conditions

$$\left. \begin{array}{c} \mathbf{u} \cdot \mathbf{n} = 0 \\ \mathbf{n} \cdot \mathbf{t} \cdot \mathbf{T} = 0 \end{array} \right\} \text{ at } r = r_s(\theta), \tag{10-264}$$

$$u_r \to -\cos\theta, \; u_\theta \to \sin\theta \text{ as } r \to \infty, \tag{10-265}$$

plus the normal-stress condition (10–262). Unless the bubble is only weakly deformed from spherical, this requires a numerical approach because the problem is highly nonlinear, not only in the equations of motion for $Re \neq 0$ but also in all of the boundary conditions, as these involve the unit normal and unit tangent vectors at an interface of unknown shape. If we assume that the shape is approximately spherical, however, we can use the domain perturbation method (Chap. 4) to analytically calculate the first departure from sphericity by using the solution for the velocity and pressure fields for a spherical bubble. At the same time, we can obtain a criterion in terms of dimensionless parameters for the bubble to remain nearly spherical.

To do this, let us suppose that the bubble is nearly spherical, with a shape described in dimensionless terms by

$$\boxed{r_s = 1 + \varepsilon f(\theta) + \cdots +,} \tag{10-266}$$

where ε is a small parameter, to be determined, and $f(\theta)$ is an unknown function that describes the departure from sphericity at $O(\varepsilon)$, assuming that the shape will be axisymmetric. We showed in Chap. 7 that the unit normal and tangent vectors, calculated in terms of $f(\theta)$, are

$$\mathbf{n} = [1 + O(\varepsilon^2)]\mathbf{i}_r - \left[\varepsilon\frac{\partial f}{\partial \theta} + O(\varepsilon^3)\right]\mathbf{i}_\theta, \tag{10-267}$$

$$\mathbf{t} = \left[\varepsilon\frac{\partial f}{\partial \theta} + O(\varepsilon^3)\right]\mathbf{i}_r + [1 + O(\varepsilon^2)]\mathbf{i}_\theta, \tag{10-268}$$

and the curvature function $\nabla \cdot \mathbf{n}$ appearing in the normal-stress balance is

$$\nabla \cdot \mathbf{n} = 2 - \varepsilon\left[2f + \frac{1}{\sin\theta}\frac{\partial}{\partial \theta}\left(\sin\theta\frac{\partial f}{\partial \theta}\right)\right] + O(\varepsilon^2). \tag{10-269}$$

Thus the boundary conditions (10–262)–(10–264) become

$$\boxed{\begin{array}{c} T_{rr} + p_{bubble} + O(\varepsilon) \\ = \dfrac{\sigma}{\rho U_\infty^2 a}\left\{2 - \varepsilon\left[2f + \dfrac{1}{\sin\theta}\dfrac{\partial}{\partial \theta}\left(\sin\theta\dfrac{\partial f}{\partial \theta}\right) + O(\varepsilon^2)\right]\right\}, \end{array}} \tag{10-270}$$

$$\left. \begin{array}{c} u_r + O(\varepsilon) = 0 \\ \tau_{r\theta} + O(\varepsilon) = 0 \end{array} \right\} \text{ at } r = 1 + O(\varepsilon). \tag{10-271}$$

G. The Boundary Layer on a Spherical Bubble

It can be seen from (10–271) that the boundary conditions (10–264) revert, at the leading order of approximation, to the boundary conditions (10–196) and (10–197). It follows that the velocity and pressure fields at this same level of approximation are simply the solutions for a spherical bubble that were obtained earlier. The normal-stress condition has not been required at all to obtain these results, but now, with the shape assumed to be unknown, the normal-stress condition provides a means to determine the $O(\varepsilon)$ departure from sphericity. The dimensionless parameter in (10–270) is known as the Weber number,

$$We \equiv \frac{\rho U_\infty^2 a}{\sigma}. \tag{10–272}$$

With the initial estimates of the velocity and pressure fields for a spherical bubble, we can calculate the left-hand side of (10–270). The normal-stress component is

$$T_{rr} = -p + \frac{1}{Re}\left(2\frac{\partial u_r}{\partial r}\right). \tag{10–273}$$

To leading order for $Re \gg 1$, we have seen already that the velocity field is given by the potential flow solution [see (10–258)], and this is also true of the velocity gradients, except in the boundary-layer region where the $O(Re^{-1/2})$ correction to u_r and u_θ produces a $O(1)$ modification in the gradient of the tangential velocity normal to the wall. In this region, however,

$$\left(\frac{\partial u_r}{\partial r}\right)_{r=1} = \left.\frac{\partial V}{\partial Y}\right|_{Y=0} = -3\cos\theta + O(Re^{-1/2}). \tag{10–274}$$

The pressure in the boundary layer is

$$p = p_0(\theta) + O(Re^{-1/2}) \tag{10–275}$$

because $\partial p_0/\partial Y = 0$ [see (10–234)], and thus $p_0(\theta)$ is determined from the pressure distribution in the outer, inviscid flow in the limit $(r-1) \ll 1$. In general, this pressure distribution consists of both a dynamic and a static part, as was also true in the corresponding problem at low Reynolds number (see Chap. 7). To calculate the dynamic part, we utilize the leading-order, inviscid flow equations for the outer region:

$$\frac{\partial p}{\partial r} = -u_r\frac{\partial u_r}{\partial r} - \frac{u_\theta}{r}\frac{\partial u_r}{\partial \theta} + \frac{u_\theta^2}{r}, \tag{10–276}$$

$$\frac{1}{r}\frac{\partial p}{\partial \theta} = -u_r\frac{\partial u_\theta}{\partial r} - \frac{u_\theta}{r}\frac{\partial u_\theta}{\partial \theta} - \frac{u_r u_\theta}{r}, \tag{10–277}$$

with u_r, and u_θ given by the inviscid flow solution (10–155) and (10–156). If we denote the far-field pressure on the horizontal plane through the center of the bubble as p_∞, we obtain, on integration,

$$p_{fluid} = p_\infty + \frac{1}{2} - \frac{9}{8}\sin^2\theta + p_{static}. \tag{10–278}$$

The hydrostatic pressure variation at the bubble surface, in dimensional terms, is

$$p'_{static} = -(\rho g a)\cos\theta. \tag{10–279}$$

Hence,

$$p_{fluid} = p_\infty + \frac{1}{2} - \frac{9}{8}\sin^2\theta - \frac{\rho g a}{\rho U_\infty^2}\cos\theta. \tag{10–280}$$

Laminar Boundary-Layer Theory

For the spherical bubble, however, we have calculated the hydrodynamic drag (10–260). Hence, at steady state, the balance between buoyancy and drag is

$$\frac{4}{3}\pi a^3 \rho g = 12\pi \mu a U_\infty, \tag{10-281}$$

and it follows that the coefficient of the hydrostatic term in (10–280) is

$$\frac{a\rho g}{\rho U_\infty^2} = \frac{9}{Re}. \tag{10-282}$$

The implication of (10–282) is that the hydrostatic variation in pressure at the bubble surface is negligible for $Re \gg 1$ relative to the dynamic pressure variation in (10–278). Finally, collecting the results (10–273)–(10–282), we can rewrite the normal-stress condition (10–270) in the form

$$\left[p_B - p_\infty - \frac{1}{2} + \frac{9}{8}\sin^2\theta + O\left(Re^{-1}\right)\right]$$
$$= \frac{1}{We}\left\{2 - \varepsilon\left[2f + \frac{1}{\sin\theta}\frac{\partial}{\partial\theta}\left(\frac{\partial f}{\partial\theta}\sin\theta\right)\right]\right\}. \tag{10-283}$$

It is now evident that for f to be $O(1)$, as we assumed in writing (10–266), the small parameter must be

$$\varepsilon \equiv We. \tag{10-284}$$

In some ways, this is the most important result of the analysis. To ensure that the shape is near spherical, we require $We \ll 1$.

To determine the bubble shape at $O(We)$, we rewrite (10–283) to include (10–284) and change independent variables from θ to $\mu = \cos\theta$ to obtain

$$\frac{d}{d\mu}\left[(1-\mu^2)\frac{df}{d\mu}\right] + 2f = \left(-p_B + p_\infty + \frac{1}{2} + \frac{2}{We} - \frac{9}{8}\sin^2\theta\right). \tag{10-285}$$

Rewriting the right-hand side in terms of Legendre polynomials,

$$\frac{d}{d\mu}\left[(1-\mu^2)\frac{df}{d\mu}\right] + 2f = \frac{3}{4}P_2(\mu) + \left(\frac{2}{We} - p_B + p_\infty - \frac{1}{4}\right)P_0(\mu). \tag{10-286}$$

We require a solution of (10–286), which is bounded at $\mu = \pm 1$ and satisfies the two constraints

$$\int_{-1}^{1} f(\mu)d\mu = 0, \tag{10-287}$$

$$\int_{-1}^{1} f(\mu)\mu d\mu = 0. \tag{10-288}$$

The first of these derives from the condition of constant bubble volume, whereas the second requires that the center of volume of the deformed bubble remain fixed at the origin

The Boundary Layer on a Spherical Bubble

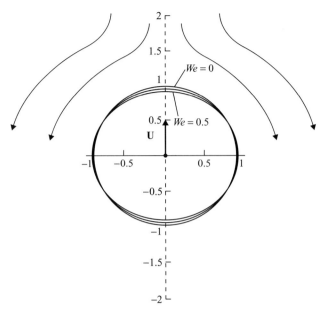

Figure 10–12. The shape of a rising gas bubble as a function of the Weber number, as predicted by small-deformation, boundary-layer theory. $We = 0, 0.25, 0.5$.

of the coordinate system. Solving (10–286), subject to (10–287) and (10–288), we find that

$$f = -\frac{3}{16} P_2(\mu), \qquad (10\text{–}289)$$

$$p_B - p_\infty - \frac{1}{4} = \frac{2}{We}. \qquad (10\text{–}290)$$

The latter determines the pressure rise across the bubble surface expressed in terms of p_∞. Combining (10–289) with (10–266), we see that the bubble shape is

$$r_s = 1 + We\left(-\frac{3}{16} P_2(\mu)\right) + O\left(We^2\right). \qquad (10\text{–}291)$$

Closer examination of (10–291) shows that the bubble is deformed into an oblate ellipsoid of revolution. A sketch of the bubble shape for several small values of We is given in Figure 10–12.

Moore[32] generalized the boundary-layer analysis for a spherical bubble to include the deformation to an oblate ellipsoidal shape. His analysis is not reproduced here, but it is worth recording the final result for the drag coefficient, which takes the form

$$C_D = \frac{24}{Re} G(\chi)\left[1 + \frac{H(\chi)}{Re^{-1/2}}\right]. \qquad (10\text{–}292)$$

Here χ is the ratio of major to minor axes of the ellipsoid and $G(\chi)$ and $H(\chi)$ are functions of χ. Numerical values for G and H were tabulated by Harper[33] for $1.0 \leq \chi \leq 4.0$ and are reproduced here in Table 10–1.

Laminar Boundary-Layer Theory

Table 10–1. The functions $G(\chi)$ and $H(\chi)$*

χ	$G(\chi)$	$H(\chi)$	χ	$G(\chi)$	$H(\chi)$
1.0	1.000	−2.211	2.6	4.278	1.499
1.1	1.137	−2.129	2.7	4.565	1.884
1.2	1.283	−2.025	2.8	4.862	2.286
1.3	1.437	−1.899	2.9	5.169	2.684
1.4	1.600	−1.751	3.0	5.487	3.112
1.5	1.772	−1.583	3.1	5.816	3.555
1.6	1.952	−1.394	3.2	6.155	4.013
1.7	2.142	−1.186	3.3	6.505	4.484
1.8	2.341	−0.959	3.4	6.866	4.971
1.9	2.549	−0.714	3.5	7.237	5.472
2.0	2.767	−0.450	3.6	7.620	5.987
2.1	2.994	−0.168	3.7	8.013	6.517
2.2	3.231	+0.131	3.8	8.418	7.061
2.3	3.478	+0.448	3.9	8.834	7.618
2.4	3.735	+0.781	4.0	9.261	8.189
2.5	4.001	+1.131			

* Reproduced from Ref. 33.

NOTES

1. S. Kaplan and P. A. Lagerstrom, Asymptotic expansions of Navier–Stokes solutions for small Reynolds numbers, *J. Math. Mech.* **6**, 585–593 (1957); S. Kaplan, Low Reynolds number flow past a circular cylinder, *J. Math. Mech.* **6**, 595–603 (1957); I. Proudman and J. R. A. Pearson, Expansions at small Reynolds numbers for the flow past a sphere and a circular cylinder, *J. Fluid Mech.* **2**, 237–262 (1957).
2. L. Prandtl, Uber Fliissigkeitsbewegung bei sehr kleiner Reibung, *in Proceedings of the Third International Mathematics Congress*, Heidelberg. Reprinted in Ludwig Prandtl: Gesammelte Abhandlungen zur angewandten Mechanik, Hydro- und Aerodynamik (Collected works), ed. by W. Tollmien, H. Schlichting and H. Goortler, vol. II, Springer, Berlin, 575–584 (1904).
3. H. Schlichting, *Boundary-Layer Theory*, (6th ed.) (McGraw-Hill, New York, 1968), see Chap. 7, pp. 118–121; H. Schlichting and K. Gersten, *Boundary Layer Theory*, (8th ed. Revised) (Springer-Verlag, New York, 2000).
4. See Ref. 3, Chap. 1, Section f.
5. S. Goldstein, Fluid mechanics in the first half of this century, *Annu. Rev. Fluid Mech.* **1**, 1–28. (1969).
6. An important assumption inherent in the derivation of (10–5) is that $\rho = $ constant. If $\rho \neq$ constant, vorticity can be generated internally when $g \wedge \nabla \rho \neq 0$. This is one reason why motions of density stratified fluids are often qualitatively different from the motion of a constant density fluid in the same domain.

    The subject of *geophysical fluid dynamics* encompasses two effects not usually considered in other areas of fluid mechanics: rotation of the fluid and the existence of density stratifications in the oceans and atmosphere. Together, these lead to very important and fascinating qualitative changes in the fluid motions. A useful account of some of the phenomena encountered in rotating and stratified fluids can be found in Chaps. 15 and 16 of D.J Tritton, *Physical Fluid Dynamics* 2nd ed. (Oxford University Press, Oxford, 1988).

    Other recent books on geophysical fluid dynamics and buoyancy effects in flow include J. Pedlosky, *Geophysical Fluid Dynamics*, (Springer-Verlag, New York, 1987) and J. S. Turner, *Buoyancy Effects in Fluids*, (Cambridge University Press, Cambridge, 1979).
7. A complete compilation of results from the subject of theoretical hydrodynamics – the motion of an inviscid fluid satisfying Euler's equations [namely, Eqs. (10–1) and (10–2), but with the

Notes

viscous term deleted] – can be found in the classic book; L. M. Milne-Thompson, *Theoretical Hydrodynamics*, (5th ed.) (Macmillan, London, 1967).

8. The problem of start-up flow for a circular cylinder has received a great deal of attention over the years because of its role in understanding the inception and development of boundary-layer separation. An insightful paper with a comprehensive reference list of both analytical and numerical studies is S. I. Cowley, Computer extension and analytic continuation of Blasius' expansion for impulsive flow past a circular cylinder, *J. Fluid Mech.* **135**, 389–405 (1983).
9. One proposal, which also contains references to earlier work on this problem, is F. T. Smith, A structure for laminar flow past a bluff body at high Reynolds number, *J. Fluid Mech.* **155**, 175–91 (1985).
10. B. Fornberg, Steady viscous flow past a circular cylinder up to Reynolds number 600, *J. Comput. Phys.* **61**, 297–320 (1985).
11. S. B. Pope, *Direct numerical simulation, in Turbulent Flows*, (Cambridge University Press, Cambridge, 2000).
12. The original papers describing the behavior of solutions of the boundary-layer equations at a separation point are S. Goldstein, On laminar boundary-layer flow near a position of separation, *Q. J. Mech. Appl. Math.* **1**, 43–69 (1948); K. Stewartson, On Goldstein's theory of laminar separation, *Q. J. Mech. Appl. Math.* **11**, 399–410 (1958).

    A more modem view of boundary-layer separation including the structure of boundary-layer solutions near a separation point can be found in C. Williams, Incompressible boundary-layer separation, *Annu. Rev. Fluid Mech.* **9**, 113–44 (1977); F. T. Smith, Steady and unsteady boundary-layer separation, *Annu. Rev. Fluid Mech.* **18**, 197–220 (1986).
13. A very readable account of higher order approximations in boundary-layer theory may be found in the textbook, M. Van Dyke, *Perturbation Methods in Fluid Mechanics* (annotated edition) (Parabolic Press, Stanford, CA, 1975) or in a review paper by the same author M. Van Dyke, Higher-order boundary-layer theory, *Annu. Rev. Fluid. Mech.* **1**, 265–92 (1969).
14. H. Blasius, Grenzschichten in fliissigkeiten mit kleiner reibung, *Zeit. Math. Phys.* **56**, 1–37 (1908) (English translation in NACA TM 1256).
15. G. F. Carrier and C. C. Lin, On the nature of the boundary-layer near the leading edge of a flat plate, *Q. Appl. Math.* **6**, 63–68 (1948).
16. K. Stewartson, On the impulsive motion of a flat plate in a viscous fluid, *Q. J. Mech. App. Math.* **4**, 182–198 (1951).
17. see Ref. 7.
18. D. R. Hartree, On an equation occurring in Falkner and Skan's approximate treatment of the equations of the boundary-layer, *Proc. Cambridge Phils. Soc.* **33**, 223–39 (1937); K. Stewartson, Further solutions of the Falkner–Skan equation, *Proc. Cambridge Phils. Soc.* **50**, 454–465 (1954).
19. An interesting extension of these ideas was discovered by Stewartson (Ref. 17) who showed that there is another branch of solutions for $0 > p > -0.1998$ that is physically acceptable in the sense that $f' \to 1$ from below but that contains a region for η near zero where $f'(\eta) < 0$. It was originally suggested by Stewartson that these solutions may be relevant to the postseparation region of the flow, assuming that the separation process (which occurs at $p = -0.1998$) produces a decrease in the adverse pressure gradient so that $p > -0.1998$ is applicable in this region. Later analysis, by Stewartson and others, has analyzed the details of interaction between the boundary layer and the exterior flow in the vicinity of a separation point. Much of this work is summarized in a fascinating review paper: K. Stewartson, D'Alembert's paradox, *SIAM Rev.* **23**, 308–43 (1981).
20. An example of the pressure distribution predicted by potential-flow theory and the experimentally measured distribution for a case in which the boundary-layer separates was shown in Fig. 10–3.
21. Note that the scaling factor used in Fig. 10–3 is that usually adopted by experimentalists, (1/2), and thus differs by a factor of 2 from (10–128).
22. L. Howarth, On the calculation of steady flow in the boundary-layer near the surface of a cylinder in a stream, *Aeronautical Research Council (UK) RM, 1632*, (1935).
23. H. Schlichting, *Boundary-Layer Theory* (6th ed.) (McGraw-Hill, New York, 1968).
24. A series of experiments was carried out by Acrivos and co-workers that showed that the transition from a steady symmetric recirculating wake flow downstream of a cylindrical body to an unsteady

wake with periodic vortex shedding could be delayed from an expected Reynolds number of about 40 to a value of $Re = 200$ by insertion of a thin plate (known as a splitter plate) on the downstream axis of symmetry at the end of the circulating region: A. Acrivos, L. G. Leal, D. D. Snowden, and F. Pan, Further experiments on steady separated flows past bluff objects, *J. Fluid Mech.* **34**, 25–48 (1968).
25. These ideas are developed in more detail in the following paper: L. G. Leal, Vorticity transport and wake structure for bluff bodies at finite Reynolds number, *Phys. Fluids A: Fluid Dynamics* **1**: 124–131 (1989). Additional insight into the evolution of a separated wake for a solid, circular cylinder may be obtained at large, but finite, Reynolds number from numerical solutions such as those in W. M. Collins and S. C. R. Dennis, Flow past an impulsively started circular cylinder, *J. Fluid Mech.* **60**, 105–127 (1973).
26. The original reference to this work (in German) is given by Schlichting (Ref. 3), p. 148. The presentation here is similar to that given by Schlichting.
27. F. W. Scholkemeier, Die laminare reibungsschicht an rotationssymmetrischen korpen, *Arch. Math.* **1**, 270–7 (1949).
28. Experimental observations of bubble shape and other aspects of bubble motions are given in many research papers. One quite comprehensive reference is D. Bhaga and M. E. Weber, Bubbles in viscous liquids: Shapes, wakes and velocities, *J. Fluid Mech.* **105**, 61–85 (1981). This and some of the older work on this problem are summarized in the book R. Clift, J. R. Grace, and M. E. Weber, *Bubbles, Drops and Particles* (Academic, New York, 1978).
29. M. Abramowitz and I. A. Stegun, *Handbook of Mathematical Functions* (Dover, New York, 1965).
30. V. G. Levich, Motion of gas bubbles with high Reynolds numbers, *Zh. Eksp. Teor. Fiz.* **19**, J8–24 (1949). An excellent summary of theoretical and experimental work (up to 1972) on the motion of bubbles and drops is, J. F. Harper, The motion of bubbles and drops through liquids, *Adv. Appl. Mech.* **12**, 59–129 (1972).
31. D. W. Moore, The boundary layer on a spherical gas bubble, *J. Fluid Mech.* **16**, 161–76 (1963).
32. D. W. Moore, The velocity of rise of distorted gas bubbles in a liquid of small viscosity, *J. Fluid Mech.* **23**, 749–66 (1965).
33. J. F. Harper, The motion of bubbles and drops through liquids, *Adv. Applied Mech.* **12**, 59–129 (1972).

PROBLEMS

Problem 10–1. Qualitative Questions

(a) The vorticity generated at a no-slip boundary is $O(Re^{1/2})$ in magnitude for $Re \gg 1$, and the vorticity generated at a zero-shear-stress surface is $O(\kappa u_s) = O(1)$. We claim that the boundary layer thickness is $O(Re^{-1/2})$ in both cases. We have further suggested that the boundary-layer thickness can be ascertained by considering the balance between convection parallel to the boundary and diffusion normal to the boundary. Is the magnitude of the vorticity an issue with respect to this argument or not? If not or if so, please explain.

(b) The calculation of drag by means of Eq. (10–247) can be done by use of the potential-flow solution to calculate the viscous dissipation Φ for a bubble, but not for a solid sphere. Explain in detail what goes wrong for the solid sphere case,

$$\Phi \equiv \nabla \mathbf{u} : \mathbf{T}.$$

(c) The *amplitude* of deformation of a bubble is $O(We)$ at finite Reynolds numbers, but is $O(Ca) = O(\mu u_c/\sigma)$ for a low-Reynolds-number flow. The proof of this statement for finite Reynolds number is in Chap. 10. Show how this proof changes to lead to $O(Ca)$ for $Re \ll 1$.

(d) Discuss the qualitative nature of the boundary layer at a liquid–liquid interface that has finite curvature as the ratio of the viscosities, μ_A/μ_B, of the two fluids is varied from

Problems

0 to $O(1)$. What is the magnitude of the vorticity generated? What is the scaling of the boundary-layer thickness with Reynolds number? How does the problem change as the curvature of the interface is increased? How large can the curvature be before the boundary-layer analysis breaks down?

Problem 10–2. Inviscid, Potential Flow Past a Half Cylinder. Consider inviscid, potential flow past the half cylinder depicted in the figure. Calculate the force (life and drag) on the object, assuming that the flow does not separate. If the flow does separate at 90°, what happens to the lift and drag?

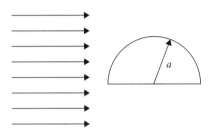

Problem 10–3. Spherical-Cap Bubble. The shape of a gas bubble rising through a liquid at large Reynolds and Weber numbers ($We \equiv \rho U^2 a/\sigma \gg 1$) is very nearly a perfect spherical cap, with an upper surface that is spherical and a lower surface this is flat.

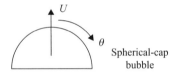

Suppose that σ is sufficiently small, i.e., We is sufficiently large, that surface tension plays no role in determining the bubble shape, except possibly locally in the vicinity of the rim where the spherical upper surface and the flat lower surface meet. Further, suppose that the Reynolds number is sufficiently large that the motion of the liquid can be approximated to a first approximation, by means of the potential-flow theory. Denote the radius of curvature at the nose of the bubble as R(at $\theta = 0$). Show that a self-consistency condition for existence of a spherical shape with radius R in the vicinity of the stagnation point, $\theta = 0$, is that the velocity of rise of the bubble is

$$U = \frac{2}{3}\sqrt{gR}.$$

(*Hint*: Do not forget that p at the bubble surface is *total* pressure.)

Problem 10–4. Boundary Layer on a Flat Plate in an Accelerating Flow, $u_\infty = \lambda x$. Consider flow past a flat plate in the throat of a 2D channel as depicted in the figure. If the free-stream velocity is given by $u_\infty = \lambda x$, where x is the distance from the leading edge and λ is a constant, show that the flow in the boundary layer on the plate is governed by an ODE. Solve for the streamfunction numerically. How does the boundary-layer thickness grow with x? How does the shear stress vary with x?

Laminar Boundary-Layer Theory

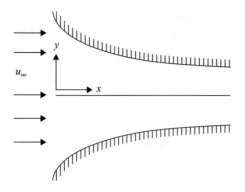

Problem 10–5. Boundary-Layer on a Flat Plate in an Accelerating Flow, $u_\infty = \lambda x^{1/2}$. Consider flow past the flat plate in the throat of a 2D channel as depicted in Problem 10–4. If the free-stream velocity is given by $u_\infty = \lambda x^{1/2}$ where x is the distance from the leading edge and λ is some constant, develop an *ordinary* differential equation that governs the flow in the boundary layer. How does the boundary layer thickness grow with x? Solve for the streamfunction numerically. What is the shear stress on the plate? How would your approach to this problem change if the free-stream velocity were given as $u_\infty = \lambda(x^{1/2} - x^{3/2})$? One approach would have you use the self-similar solution to the first problem as the leading approximation for this more general problem. What would be the problem for the second-order term in such a scheme?

Problem 10–6. Boundary Layer in a Decelerating Flow. Consider the boundary-layer problem for a flat plate when the potential flow yields a velocity

$$u_e(x) = U_0(1 - \alpha x^n).$$

A self-similar solution is not possible. Why? However, a solution can be generated as a power series in x with coefficients that are functions that depend on a similarity variable.

(a) Obtain the governing equations and boundary conditions for these coefficient functions. (The first should be the Blasius equation.)
(b) Use integral methods to estimate the point of boundary-layer separation for $x > 0$.

Problem 10–7. Boundary Layer in a Decelerating Flow. Consider flow past a flat plate where the x velocity outside the boundary layer is given by

$$\mathbf{u}_\infty = \lambda \mathbf{x}^{-1/2}.$$

(a) Render the Navier–Stokes equations dimensionless for this problem and determine the conditions under which a boundary-layer description is valid.
(b) Show that the boundary-layer equations admit a self-similar solution, and determine the rate of boundary-layer growth with x. Obtain the similarity transformation and resulting ODE with boundary conditions. (Derive these results yourself, rather than simply trying to adapt the solutions that are carried out in the body of the chapter.)

Problem 10–8. Boundary Layer on a Moving Web. A flat plate is pulled through a wall with a constant velocity U_0 into a Newtonian fluid of kinematic viscosity ν. Assume that $U_0/\nu \gg 1$.

Derive the governing DEs and boundary or matching conditions for the boundary-layer motion that occurs in the immediate vicinity of the plate surface. Show that this problem can be reduced to solving an ODE by similarity transformation.

Problems

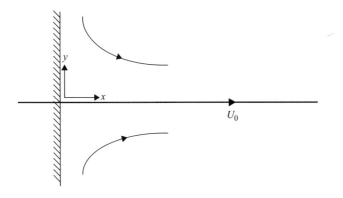

Problem 10–9. Translating Flat Plate. Consider the high-Reynolds-number laminar boundary-layer flow over a semi-infinite flat plate that is moving parallel to its surface at a constant speed U in an otherwise quiescent fluid. Obtain the boundary-layer equations and the similarity transformation for $f'(\eta)$. Is the solution the same as for uniform flow past a semi-infinite stationary plate? Why or why not? Obtain the solution for f' (this must be done numerically). If the plate were truly semi-infinite, would there be a steady solution at any finite time? (*Hint*: If you go "far" downstream from the leading edge of the flat plate, the problem looks like the Rayleigh problem from Chap. 3). For an arbitrarily chosen time T, what is the regime of validity of the boundary-layer solution?

Problem 10–10. Mixing Layer. Two semi-infinite streams of the same fluid, but with different uniform, parallel velocities, U and αU, are brought into contact. Owing to the action of viscosity, a boundary layer will develop over which the initial discontinuity in velocity will be smoothed out as illustrated in the figure. Derive the boundary-layer equations for

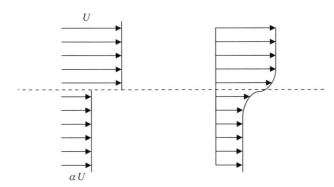

this shear layer assuming that the streamline dividing the upper and lower fluid streams remain flat and parallel to the flow direction upstream of the contact point. A similarity transformation is possible, so derive the equation for the similarity function. Be sure to state carefully the boundary conditions satisfied by this function. In general, would you expect the streamline dividing the upper and lower fluids to remain flat and parallel to the upstream flow or not? Explain what you think might happen. (*Hint*: think "pressure.")

Problem 10–11. Jeffrey–Hamel Flow. Consider converging flow (that is, flow inward toward the vertex) between two infinite plane walls with an included angle 2α, as shown in the figure. This is known as Jeffery–Hamel flow.

Laminar Boundary-Layer Theory

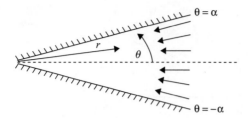

The flow is best described in cylindrical coordinates (r, θ, z) with z being normal to the plane of flow. We assume $u_z = 0$ and there is no z dependence, $\partial/\partial z(\ ) = 0$. Also, we assume that the flow is entirely radial in the sense that fluid moves along rays $\theta = $ const toward the vertex, that is, $u_\theta = 0$. (Obviously a solution of this form will apply for only large r, away from the opening of the vertex.)

(a) Show that $u_r = f(\theta)/r$.
(b) Obtain the governing DE for $f(\theta)$:

$$\frac{2}{\nu} f \frac{df}{d\theta} + \frac{d^3 f}{d\theta^3} + 4\frac{df}{d\theta} = 0.$$

(c) Assume no-slip conditions apply at the walls and obtain a sufficient number of boundary conditions or other constraints on $f(\theta)$ to have a well-posed problem for the determination of $f(\theta)$.
(d) Introduce a normalized angle ϕ and a normalized flow variable $F(\phi)$ such that

$$\phi = \theta/\alpha \text{ and } F(\phi) = \alpha f/q,$$

where q is the volume flow rate per unit width in the z direction, to obtain the dimensionless equation

$$Re F \frac{dF}{d\phi} + \frac{d^3 F}{d\phi^3} + 4\alpha^2 \frac{dF}{d\phi} = 0$$

plus boundary conditions and auxiliary conditions on $F(\phi)$; $(Re \equiv 2q\alpha/\nu)$.

(e) Obtain an approximate solution for the creeping-flow limit $Re \to 0$.
(f) Obtain an approximate solution for $\alpha \to 0$ (the lubrication limit).
(g) Consider the solution for $Re \to \infty$. Show that the solution reduces in this case to

$$F(\phi) = -\frac{1}{2}.$$

The solution obviously does not satisfy the no-slip condition on the solid boundaries, $\phi = \pm 1$. Explain why. Outline how a solution could be obtained for $Re \to \infty$ by use the method of matched asymptotic expansions, which satisfies the boundary conditions on the walls. Be as detailed and explicit as possible, including actually setting up the equations and boundary conditions for the region near the wall. An exact solution is possible. The following integral may be useful:

$$\int \frac{d\xi}{\left(\xi - \frac{1}{2}\right)\sqrt{\xi + 1}} = \sqrt{\frac{2}{3}} \log \frac{\sqrt{\xi + 1} - \sqrt{3/2}}{\sqrt{\xi + 1} + \sqrt{3/2}}.$$

Calculate the drag on the plates.

One may also obtain an approximate solution by means of similarity transform by use of local Cartesian variables x and y, parallel and perpendicular to either of the walls.

Problems

For this purpose, the solution $F(\phi) = -1/2$ provides an expression for $u_e(x)$. Use the following formulas if necessary:

$$\int \frac{d\xi}{\sqrt{(\xi-1)^2(\xi+2)}} = \frac{2}{\sqrt{3}} \tanh^{-1}\left(\frac{\sqrt{2+\xi}}{\sqrt{3}}\right) = C, \quad \tanh^{-1}\left(\sqrt{\frac{2}{3}}\right) = 1.146.$$

Problem 10–12. Higher-Order Approximations for the Blasius Problem. The classical boundary-layer theory represents only the first term in an asymptotic approximation for $Re \gg 1$. However, in cases involving separation, we do not seek additional corrections because the existence of a separation point signals the breakdown of the whole theory. When the flow does not separate, we can calculate higher-order corrections, and these provide useful insight and results. In this problem, we reconsider the familiar Blasius problem of streaming flow past a semi-infinite flat plate that is oriented parallel to a uniform flow.

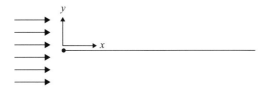

(a) The Blasius function $f(\eta)$ for large η has the asymptotic form

$$f(\eta) \sim \eta - \beta_1 + \text{terms that are exponentially small}.$$

Show that the second term in the outer region is $O(Re^{-1/2})$ in magnitude and that the second-order outer problem is equivalent to potential flow past a semi-infinite, parabolically shaped body.

(b) In general, the second-order *inner* solution satisfies a boundary-layer-like equation in which account is taken of the fact that the pressure distribution in the boundary layer is modified by the second-order correction to the *outer* flow. Derive governing equations and boundary conditions for the second-order *inner* solution for the Blasius problem. Show that the solution for the second-order term in the outer region, corresponding to a semi-infinite flat plate, yields the matching condition

$$\frac{\partial \Psi_2}{\partial Y}(x, \infty) = 0,$$

where Ψ_2 is the streamfunction for the second-order inner problem, and thus demonstrate that the second-order *inner* solution is

$$\Psi_2(x, Y) \equiv 0.$$

(Note that $u \equiv \partial \Psi / \partial Y$ and $V \equiv -\partial \Psi / \partial x$, where Y and V are the usual stretched boundary-layer variables. The preceding solution of Ψ_2 implies that the $O(Re^{-1/2})$ change in the pressure distribution for a semi-infinite flat plate produces *no* change in the boundary-layer region.)

(c) For a *finite* flat plate, the second-order problem is modified because the condition

$$\psi_2(x, 0) = -\beta_1 \sqrt{2x}$$

(here ψ_2 is the streamfunction for the second-order *outer* problem) applies only for $0 < x \le 1$ ($x = 1$ is the end of the plate if the plate length is used as a characteristic length scale). For $x > 1$, an approximation is that the *displacement thickness* associated with

Laminar Boundary-Layer Theory

the wake is a *constant* rather than a parabolic function of x as in the region $0 < x \le 1$. Thus a solution can be obtained for ψ_2 in this case, which leads to the matching condition

$$\lim_{Y \gg 1} \Psi_2(x, Y) = \frac{\beta_1}{\sqrt{2\pi}} \frac{1}{\sqrt{x}} \log \frac{1+\sqrt{x}}{1+\sqrt{x}}.$$

Show that an approximate solution can be obtained for $\Psi_2(x, Y)$ in the form of a series expansion in x with coefficients that are functions of the Blasius similarity variable η. Determine the governing DEs and boundary conditions for the first tow terms in this expansion. It can be shown that

$$C_D \equiv \frac{\text{drag}}{\frac{1}{2}\rho U_\infty^2 L} \sim \frac{1.328}{\sqrt{Re}} + \frac{C_1}{Re} + \cdots +,$$

and if enough terms are retained in the approximate solution for Ψ_2, the constant C_1 can be approximated as $C_1 = 4.12$. You need not calculate C_1 in detail, but you should explain the preceding result for C_D and explain in principle how to calculate C_1. What value does C_1 have for a semi-infinite flat plate? Does this surprise you? Explain.

Problem 10–13. Wake Far Downstream of a 2D Body in a Uniform Flow. The boundary-layer approximation, in which derivatives in the primary flow direction are assumed to be small compared with derivatives across the flow, finds application in a number of problems far from a source of mass or momentum in an unbounded fluid. One example is in the analysis of the wake far downstream of a body in a uniform streaming flow. As an example, let us consider a two-dimensional cylindrical body that has a cross section that is symmetric about the $y = 0$ plane, but is otherwise arbitrary, placed in a Newtonian fluid that is moving with a constant velocity $\mathbf{U} = U\mathbf{i}$, as depicted in the figure.

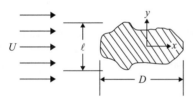

Very near to the body, the velocity field is very complicated, in part because of the complex geometry of the body, and differs considerably from the free-stream form $\mathbf{U}$, which would occur everywhere if the body were not present. However, far downstream ($x \gg 1$) the velocity field gradually returns to its undisturbed state; that is, the deviations from the undisturbed velocity field $\mathbf{U}$ become small. Under these circumstances, the actual velocity field, which we denote as $\mathbf{u}$, can be expressed as

$$\mathbf{u} = \mathbf{U} - \mathbf{u}' \text{ with } |\mathbf{u}'| \ll |\mathbf{U}| \quad (x \text{ large}). \tag{1}$$

The field $\mathbf{u}'$ is called the disturbance velocity field because it represents the difference between the undisturbed field $\mathbf{U}$, which would exist if the body were not present, and the actual velocity in the presence of the body. It is observed that the velocity field $\mathbf{u}$ is completely symmetric about $y = 0$ (all x). Furthermore, when x is large, the major deviations from the free-stream velocity $\mathbf{U}$ occur in the vicinity of the x axis ($y = 0$); that is, $\mathbf{u}'$ is nonzero only near $y = 0$ for large x, and this region is called the *wake-flow region*.

(a) One interesting feature of the wake-flow region is that one can directly determine the force acting on the body by knowing (measuring) the disturbance velocity component $u'_x \equiv \mathbf{i}_x \cdot \mathbf{u}'$ as a function of y. Employ the approximate macroscopic balance equation(s) to derive an expression for the streamwise (x) component of the force (F_x) acting on

Problems

the body in terms of $U = \mathbf{i}_x \cdot \mathbf{U}$ and u'_x at a distance downstream station. Assume that $\mathbf{i}_y \cdot \mathbf{u} \equiv 0$ for large values of x. Show all of your work, including your choice of a control volume surface, and carefully state all assumptions that you may have had to employ. You should find that

$$F_x = c_1 \int_{y=0}^{\infty} u'_x(2U - u'_x) dy, \tag{2}$$

in which c_1 is a constant. Note that the y component of the net force acting on the body is zero because of the symmetry imposed about the x axis ($y = 0$). Show that $F_y \equiv 0$, using the macrobalance approach. Because the drag (F_x) is a fixed number for a given body and a given value of U, it follows from (2) that

$$\int_{y=0}^{\infty} u'_x(2U - u'_x) dy = \text{const, independent of } x \text{ for large values of } x.$$

Hence, neglecting the small term that is quadratic in $|u'_x|$, we obtain

$$\int_{y=0}^{\infty} 2u'_x U dy = \text{const (independent of } x\text{)}; \quad x \gg 1,$$

or

$$\int_{y=0}^{\infty} 2u'_x dy = \text{const} = \frac{F_x}{c_1 U}. \tag{3}$$

(b) Observations lead us to believe, in addition to the previously described features of the flow, that

$$\left| \frac{\partial^2 u'_x}{\partial y^2} \right| \gg \left| \frac{\partial^2 u'_x}{\partial x^2} \right| \tag{4}$$

in the wake-flow region. Use this fact, as well as (1), to show that an approximate solution for u'_x can be obtained from

$$U \frac{\partial^2 u'_x}{\partial x} = v \frac{\partial^2 u'_x}{\partial y^2} \quad (x \gg 1), \tag{5}$$

subject to boundary conditions

$$\begin{aligned} u'_x &\to 0, \quad |y| \to \infty \quad (x \gg 1), \\ \frac{\partial u'_x}{\partial y} &= 0, \quad y = 0 \quad (x \gg 1), \end{aligned} \tag{6}$$

and condition (3).

(c) Suppose that you decide to nondimensionalize equations (5) and (6). What length and velocity scales would you use? Do both of these choices continue to make sense as you move far away from the body? (Note that the body would appear as a thin line as we move to very large values of x.) Assume that

$$u'_x = c_2 x^m f(\eta),$$

where

$$\eta = \frac{y}{x^n} c_3$$

and c_2 and c_3 are constants. Determine the coefficients m and n, and derive the governing equation for f. With appropriate choices for the constants c_2 and c_3, you should find that

$$f'' + \frac{1}{2}\eta f' + \frac{1}{2}f = 0. \tag{7}$$

(*Hint*: The coefficient m cannot be zero. Why not?) What boundary conditions must f satisfy? Why should you have expected a similarity solution in this problem?

(d) Solve for the similarity function f.

(e) Show that the constant that appears in the similarity function $f(\eta)$ can be related simply to the force magnitude F_x.

Problem 10–14. Two-Dimensional Jet. Consider the 2D flow created when fluid is force with high velocity through a very narrow slit in a wall as pictured in the figure (this flow is referred to as a 2D symmetric jet).

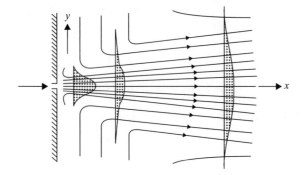

As the fluid leaving the slit comes into contact with the surrounding fluid, the action of viscosity slows it down while causing some of the surrounding fluid to be accelerated. Hence, with increasing x, the maximum velocity decreases while the width of the jet increases.

(a) Show that this process must occur so that the total momentum in the x direction is conserved; that is, show that

$$J = \rho \int_{-\infty}^{\infty} [u(x,y)]^2 dy = \text{const (independent of } x\text{)}.$$

(b) When the average velocity of the fluid leaving the slit divided by the kinematic velocity is large, the jet is observed to be long and narrow. Under these conditions the equations governing the flow in the jet can be reduced to the boundary-layer-like equations. What is the pressure gradient in the jet? What are the appropriate boundary conditions? (Note that it is more convenient to use conditions at the center of the jet and at one edge rather than conditions at both edges of the jet.)

(c) Rewrite your equation in terms of the nondimensionalized streamfunction Ψ, defined as

$$\frac{\partial \Psi}{\partial Y} = u, \quad \frac{\partial \Psi}{\partial x} = -V,$$

where variables are considered to be nondimensionalized as in the boundary-layer theory in the limit of $Re \to \infty$.

(d) Show that, by introduction of the similarity transformation,

$$\Psi = x^m f(\eta), \quad \eta = \frac{Y}{x^n},$$

the governing equation can be reduced to

$$f''' + 2ff'' + 2f'^2 = 0.$$

Find m and n as part of your solution. What are the boundary conditions and momentum constraints in terms of $f(\eta)$? What is the rate of decrease of the centerline velocity with x, and how fast does the width of the jet increase?

Problems

(e) This equation can be integrated twice to find $f(\eta)$; after applying the boundary conditions, we obtain

$$f(\eta) = \alpha \tan h(\alpha \eta).$$

What are the corresponding velocity components u and V? Relate α to J, the total momentum of the fluid leaving the slit.

(f) Find an expression for the half-width of the jet defined as the distance from the centerline to the point where u equals 1 percent of its centerline value. Find an expression for the local *Reynolds number as a function of x*.

Problem 10–15. Flow of a Power-Law Fluid Past a Flat Plate. Consider the uniform flow of a power-law fluid past a thin flat plate at high Reynolds number. We wish to determine the flow field and drag on the plate. For this problem, it is sufficient to model the flow with the equations

$$\frac{\partial u}{\partial x} + \frac{\partial v}{\partial y} = 0, \tag{1}$$

$$\rho\left(u\frac{\partial u}{\partial x} + v\frac{\partial u}{\partial y}\right) = -\frac{\partial p}{\partial x} + \frac{\partial}{\partial x}\left(\eta\frac{\partial u}{\partial x}\right) + \frac{\partial}{\partial x}\left(\eta\frac{\partial u}{\partial y}\right), \tag{2}$$

$$\rho\left(u\frac{\partial v}{\partial x} + v\frac{\partial v}{\partial y}\right) = -\frac{\partial p}{\partial y} + \frac{\partial}{\partial x}\left(\eta\frac{\partial v}{\partial x}\right) + \frac{\partial}{\partial y}\left(\eta\frac{\partial v}{\partial y}\right), \tag{3}$$

where p is pressure, (u, v) and (x, y) are the velocities and coordinates parallel and perpendicular, respectively, to the plate, and ρ is the density of the fluid. The "viscosity" of the fluid, η, is approximately given by

$$\eta = k\left|\frac{\partial u}{\partial y}\right|^{n-1}, \tag{4}$$

where k and n are constants.

(a) Nondimensionalize the equations. What is the Reynolds number Re for this problem?
(b) For $Re \gg 1$, what is the outer velocity field for the singular asymptotic expansion?
(c) What are the boundary-layer equations for a power-law fluid?
(d) Given the rescalings used to derive the boundary-layer equation, how does the thickness of the boundary layer vary along the plate? Estimate the drag on the plate.
(e) Briefly describe how you would determine the velocity profile in the boundary layer?

Problem 10–16. 2D Flow Near a Stagnation Point. Consider a flow impinging on a horizontal flat plate, as illustrated in the figure. Assume that far away from the plate the flow is given by

$$u(x, y) = \alpha x$$
$$v(x, y) = -\alpha y,$$

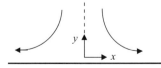

and of course on the plate the velocities vanish because of the no-slip condition. This flow is an extremely good model for the fluid motion near a silicon wafer in a chemical vapor deposition reactor, and we will use the following analysis in Chap. 11 to estimate heat and mass transfer coefficients in a chemical vapor deposition reactor.

Laminar Boundary-Layer Theory

(a) Show that an exact solution of the steady Navier–Stokes equation is
$$u(x, y) = \alpha x f'(\eta), \quad v(x, y) = -(v\alpha)^{1/2} f(\eta),$$
where $\eta = (\alpha/v)^{1/2} y$ and
$$f''' + ff'' + 1 - f'^2 = 0$$
with the appropriate boundary conditions. What are the boundary conditions?

(b) Numerically determine and plot f and f'.

(c) Does a boundary layer exist, and, if so, how thick is it? How could you have estimated this quickly from dimensional analysis?

11

Heat and Mass Transfer at Large Reynolds Number

We saw in Chapter 10 that the boundary-layer structure, which arises naturally in flows past bodies at large Reynolds numbers, provides a basis for approximate analysis of the flow. In this chapter, we consider heat transfer (or mass transfer for a single solute in a solvent) in the same high-Reynolds-number limit for problems in which the velocity field takes the boundary-layer form. We saw previously that the thermal energy equation in the absence of significant dissipation, and at steady state, takes the dimensionless form

$$\mathbf{u} \cdot \nabla \theta = \frac{1}{Pe} \nabla^2 \theta, \tag{11-1}$$

where

$$\theta \equiv \frac{T - T_\infty}{T_0 - T_\infty},$$

$$Pe \equiv Re\, Pr.$$

In Chap. 9, we considered the solution of this equation in the limit $Re \ll 1$, where the velocity distribution could be approximated by means of solutions of the creeping-flow equations. When $Pe \gg 1$, we found that the fluid was heated (or cooled) significantly in only a very *thin thermal boundary layer* of $O(Pe^{-1/3})$ in thickness, immediately adjacent to the surface of a no-slip body, or $O(Pe^{-1/2})$ in thickness if the surface were a slip surface with finite interfacial velocities. We may recall that the governing convection–diffusion equation for mass transfer of a single solute in a solvent takes the same form as (11–1) except that θ now stands for a dimensionless solute concentration, and the Peclet number is now the product of Reynolds number and Schmidt number,

$$Sc \equiv \nu/D,$$

where D is the molecular diffusion coefficient for the solute. In this chapter, we discuss the analysis in the first several sections in terms of the heat transfer problem, but we should remember that the analysis applies equally well to the single-component mass transfer problem as well, provided only that the boundary conditions are applicable and that there is no appreciable velocity normal to the surface as the solute moves either toward or away from the surface. These latter effects were mentioned at the beginning of Chap. 9, and they will be discussed here in more detail in Section G.

In this chapter, we consider heat transfer from heated (or cooled) bodies in the dual limit $Re \gg 1$ and $Pe \gg 1$.[1] For simplicity, we restrict most of the discussion to solid bodies with no-slip surfaces and also assume that the body geometry is 2D. These conditions can both be removed following the lead of the high-Peclet-number results in Chap. 9.

Heat and Mass Transfer at Large Reynolds Number

Because $Re \gg 1$, we assume that the velocity field in the vicinity of the body is characterized by a momentum boundary layer. In addition, because $Pe \gg 1$, we also expect the temperature distribution to be characterized by a thin thermal boundary layer. Although both boundary layers will be thin, we should not necessarily expect the scaling of Chaps. 9 and 10 to be preserved. Note that the condition $Pe \gg 1$ does not limit the range of allowable values for the Prandtl number, Pr. Indeed, if Pr is small, the Peclet number can still be large provided that the Reynolds number is large enough. However, the magnitude of Pr does determine the relative size of Re and Pe. If $Pr \gg 1$, then $Pe \gg Re \gg 1$. On the other hand, when $Pr \ll 1$, then $Re \gg Pe \gg 1$. In fact, we noted in Sub-section A.3 of Chap. 9 that fluids with large Pr are not very common, whereas the Schmidt number for most liquids is large. On the other hand, the Prandtl number can be quite small in practice for gases especially, whereas the Schmidt number is rarely less than $O(1)$. With the exception of Section G, which is specifically concerned with mass transfer, the remaining sections of this chapter are presented in the context of heat transfer. However, the "high-Pr" theory is really primarily of direct application in the mass transfer context. Of course, in a more general context, we have already noted that the asymptotic theories provide a basis for interpolating to intermediate values of Pr (or Sc) (illustrated here in Section E).

Although the dependence of the thermal boundary-layer thickness on the independent parameters Re and Pr (or Pe) remains to be determined, we may anticipate that the magnitudes of Re and Pe will determine the *relative* dimensions of the two boundary layers. If $Pe \gg Re \gg 1$, both the momentum and thermal layers will be thin, but it seems likely that the thermal layer will be much the thinner of the two. Likewise, if $Re \gg Pe \gg 1$, we can guess that the momentum boundary layer will be thinner than the thermal layer. In the analysis that follows in later sections of this chapter, we consider both of the asymptotic limits $Pr \to \infty (Pe \gg Re \gg 1)$ and $Pr \to 0 (Re \gg Pe \gg 1)$. We shall see that the relative dimensions of the thermal and momentum layers, previously anticipated on purely heuristic grounds, will play an important and natural role in the theory.

The primary focus of our analysis, as in Chap. 9, is a prediction of the correlation between the Nusselt number, Nu, representing the dimensionless total heat flux, and the independent dimensionless parameters Re and Pr (or Pe). The engineering literature abounds with such correlations, determined empirically from experimental data.[2] The general form of these correlations is

$$Nu = c Re^a Pr^b. \tag{11-2}$$

For laminar flow conditions and $Re \gg 1$, the coefficient a is found to be

$$a = 0.5,$$

whereas the coefficient b varies from approximately 0.3 to 0.5, depending on the range of values considered for Pr and the boundary conditions at the surface of the body. Again, we limit our presentation to bodies with no-slip surfaces. In this case, the coefficient c is generally $O(1)$ and *dependent* on only the geometry of the body. We shall prove, from the asymptotic *formulation* of the problem (that is, without solving any detailed equations), that

$$b = \frac{1}{2} \text{ in the limit } Pr \to 0,$$

$$b = \frac{1}{3} \text{ in the limit } Pr \to \infty.$$

Let us begin by considering the governing equation and boundary conditions for $Re \gg 1$ and $Pe \gg 1$.

A. GOVERNING EQUATIONS ($Re \gg 1$, $Pe \gg 1$, WITH ARBITRARY Pr OR Sc NUMBERS)

We start with the nondimensionalized thermal energy equation, in the form (11–1), where spatial variables are scaled by use of the dimension of the body as a characteristic length scale. The geometry of the body is assumed to be arbitrary, but 2D for simplicity. However, if the flow separates, the boundary-layer analysis that we eventually obtain applies only to the portion of the body surface where the momentum boundary layer remains attached. It is simplest to restrict attention to cases in which the boundary layer does not separate at all. In addition, if we wish to apply our results to the corresponding mass transfer problem, we assume that the blowing parameter is small enough that that phenomena can be neglected. At the end of the chapter (Section G) we return to this question and consider the effects of blowing or suction on the problem.

The Prandtl number may be either large or small, but we assume that $Pe \gg 1$, as previously stated. In this limit, (11–1) reduces to the familiar form

$$\mathbf{u} \cdot \nabla \theta = 0, \tag{11–3}$$

in which conduction terms are neglected altogether and the solution in regions where the streamlines are open is

$$\theta = \theta(\psi) = 0. \tag{11–4}$$

It is evident from our previous analyses of heat transfer at large Pe (Chap. 9) that this solution and the governing equation (11–3) is valid as a first approximation in some *outer* region away from the body surface.

This outer solution $\theta \equiv 0$ must be supplemented near the body surface by a solution that includes the effects of conduction and thus satisfies the condition $\theta = 1$ at the boundary. For large Peclet numbers, $Pe \gg 1$, such a solution will be valid in a thin thermal boundary layer. Following the procedures of earlier chapters, the characteristic length scale for this region is determined by *rescaling*. The primary change from the preceding analyses is that the thermal boundary-layer thickness depends *independently* on the two parameters Re and Pr rather than depending on the combined parameter $Pe = RePr$ only. This occurs because the velocity distribution in the region near the body surface is characterized by a length scale, $\ell = a\, Re^{-1/2}$, that depends explicitly on the Reynolds number. Thus the Reynolds number and the Prandtl number "appear" *independently* in the thermal energy equation rather than only in the form of the combined parameter Pe. In contrast, for the low-Reynolds-number limit considered earlier, the velocity field in the region relevant to the thermal energy equation is characterized by the length scale of the body, and it is only Pe that appears in the thermal problem.

We begin by considering the inner thermal boundary-layer limit for arbitrary $Pr \sim O(1)$, but with $Pe \gg 1$. The velocity distribution in the vicinity of the body surface is characterized by the length scale

$$\ell \sim a(Re)^{-1/2},$$

independent of the form of the temperature distribution. Thus, as a first step, we rescale the thermal energy equation in a manner consistent with the momentum boundary-layer scaling; that is, we introduce

$$Y = y\sqrt{Re}, \quad V = v\sqrt{Re} \tag{11–5}$$

into the thermal energy equation (11–1), now expressed in terms of local Cartesian, boundary-layer coordinates. The result is

$$u(x,Y)\frac{\partial \theta}{\partial x} + V(x,Y)\frac{\partial \theta}{\partial Y} = \frac{1}{Pr}\frac{\partial^2 \theta}{\partial Y^2} + O(Re^{-1}Pr^{-1}). \qquad (11\text{–}6)$$

We wish to solve this equation subject to the boundary conditions

$$\theta = 1 \quad \text{at} \quad Y = 0, \ x > x^*, \qquad (11\text{–}7a)$$

$$\theta \to 0 \quad \text{for} \quad Y \gg 1, \qquad (11\text{–}7b)$$

$$\theta = 0 \quad \text{at} \quad x = x^*, \ Y > 0. \qquad (11\text{–}7c)$$

The conditions (11–7a) and (11–7c) correspond to the assumption that the surface of the body is heated (or cooled) to a constant temperature T_0 beyond a certain position denoted as x^*. Upstream of x^*, the body is not heated. Hence, at steady state in the present boundary layer limit, both the body surface and the fluid remain at the ambient temperature T_∞ for $x < x^*$. If the body is heated over its whole surface, then $x^* = 0$. In this case, the "leading edges" of the thermal and momentum boundary layers are coincident.

Before going any further, there is one aspect of the rescaling (11–5) that requires discussion. It was stated that this rescaling is relevant to $Pr \sim O(1)$, and this is clearly consistent with our earlier discussion of the role of the Prandtl number in determining the relative dimensions of the thermal and momentum boundary layers. In particular, for $Pr \sim O(1)$, the two parameters Re and Pe are of comparable magnitude, and we should expect the width of the momentum and thermal layers to be similar. It is not surprising in this case that the momentum boundary-layer scaling should apply to the thermal energy equation. However, our goal is to solve the thermal problem for a wide range of Prandtl numbers, and it is not so obvious that the rescaling (11–5) and governing equation (11–6) should apply to the limiting cases $Pr \to 0$ or $Pr \to \infty$, where our previous arguments suggested that the scales of the two boundary layers might be quite different.

The fact is that the thermal boundary-layer equation (11–6) is relevant for the whole range of possible values for Pr, provided only that $Re, Pe \gg 1$, as stated earlier. In the case of $Pr \to \infty$, our earlier discussion suggested that the thermal layer (the region in which θ decreases from its surface value, $\theta = 1$, to its free-stream value, $\theta = 0$) would be thin compared with the dimensions of the momentum layer [the region in which u increases from its surface value, $u = 0$, for a no-slip body to its potential-flow limit, $u_e(x)$]. But if this is true, the whole of the thermal layer will lie "inside" the momentum layer, and it should not be surprising if the scaling (11–5) is relevant. The less obvious case is the limit $Pr \ll 1$. In this case, it was suggested earlier that the extent of the thermal layer would be large relative to the momentum layer. Nevertheless, the boundary-layer velocity distributions are still relevant in the following sense. As Pr decreases in magnitude, the thermal layer lies increasingly in the part of the momentum boundary layer that corresponds to large values of Y. To put it another way, the thermal layer is increasingly controlled by the potential-flow velocity distribution. However, regardless of the magnitude of Pr, we assume that $Pe \gg 1$! Hence, no matter how wide the thermal region is compared with the momentum boundary layer, it always remains asymptotically thin compared with the dimension of the body. Thus, no matter how small the Prandtl number becomes, it is only *the limit of the potential-flow velocity profile* $u_e(x)$ that is relevant to the thermal energy equation.

B. Exact (Similarity) Solutions for Pr (or Sc) ~ O(1)

In particular, at leading order in Re^{-1}, the matching condition for the momentum boundary-layer solution,

$$\lim_{Y \gg 1} u(x, Y) \Leftrightarrow \lim_{y \to 0} u(x, y) = u_e(x) \text{ for } Re \to \infty, \quad (11\text{–}8)$$

ensures that this limit of the potential-flow solution is *identical* to the outer limit ($Y \gg 1$) of the boundary-layer solution. Thus, for all Pr, no matter how small, the use of the momentum boundary-layer scaling remains valid in (11–6), and there is a smooth transition to the limiting case $Pr \to 0$, provided only that $Pe \gg 1$.

We deduce from these arguments that the Reynolds number dependence of the Nusselt number can be determined directly from (11–5). We recall that the definition of the Nusselt number, in terms of the dimensionless temperature gradient, is

$$Nu \equiv \int_{x=x^*}^{x=x_L} -\left(\frac{\partial \theta}{\partial y}\right)_{y=0} dx, \quad (11\text{–}9)$$

where x_L is the end of the heated surface (usually the "trailing edge" of the body) and $y = 0$ corresponds to the body surface. Hence, as shown earlier, it follows from dimensional analysis of the thermal energy equation (11–1) that

$$Nu = Nu(x^*, x_L; Re, Pr). \quad (11\text{–}10)$$

For all $Re, Pr \sim O(1)$, the value of the number on the right-hand side of (11–10) is also $O(1)$. Here, however, we consider the limit $Re \gg 1$. Hence, applying the momentum boundary-layer scaling (11–5) to the definition (11–9), we see that

$$Nu = \sqrt{Re} \int_{x^*}^{x_L} -\left(\frac{\partial \theta}{\partial Y}\right)_{Y=0} dx. \quad (11\text{–}11)$$

The temperature gradient $(\partial \theta / \partial Y)_{Y=0}$ is determined by the governing equation and boundary conditions (11–6) and (11–7). Thus,

$$\frac{\partial \theta}{\partial Y} = \frac{\partial \theta}{\partial Y}(x, Y; Pr),$$

and it follows that

$$Nu = \sqrt{Re} F(x^*, x_L; Pr), \quad (11\text{–}12)$$

where

$$F(x^*, x_L; Pr) \equiv -\int_{x^*}^{x_L} \left(\frac{\partial \theta}{\partial Y}\right)_{Y=0} dx. \quad (11\text{–}13)$$

B. EXACT (SIMILARITY) SOLUTIONS FOR Pr (OR Sc) ~ O(1)

To proceed further, it is necessary to solve the thermal boundary-layer problem defined by (11–6) and (11–7). However, in spite of the fact that we have a *linear* DE for θ, it is generally difficult to solve because of the complex and variable coefficients $u(x,Y)$ and $V(x, Y)$. We

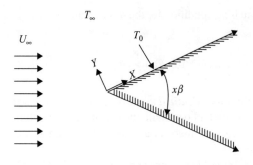

Figure 11–1. A schematic representation for heat transfer from a semi-infinite wedge with surface temperature T_0 (dimensionless $\theta \equiv 1$) and ambient temperature $T_\infty (\theta = 0)$.

shall see shortly that we can obtain approximate closed-form solutions irrespective of the geometry of the body (provided, of course, that the basic assumptions of an "attached" boundary layer remain valid) for the two asymptotic limits $Pr \to \infty$ and $Pr \to 0$. For arbitrary $Pr \sim O(1)$, it is possible to solve (11–6) in analytic form for only a few very specific bodies.

In particular, let us consider the class of body geometries for which the *momentum* boundary-layer equations allow similarity solutions to be obtained. Thus we consider streaming flow past a semi-infinite wedge with included angle $\pi \beta$, as sketched in Fig. 11–1. Then, provided that $x^* = 0$, so that the leading edge of the thermal layer is coincident with the leading edge of the momentum layer and the surface temperature distribution is of the form

$$\theta_s(x) \approx x^m, \tag{11–14}$$

it can be proven that similarity solutions also can be obtained for θ in the form

$$\boxed{\theta = \theta_s(x) H(\eta),} \tag{11–15}$$

where η is the similarity variable from the fluid mechanics problem $[\eta \equiv Y/g(x)]$.

Here we consider, in detail, only the simplest case with $m = 0$ and $\beta = 0$, namely, a semi-infinite flat plate with $\theta_s = 1$ for all x. In this case, the velocity field is given by the Blasius solution

$$u = f'(\eta), \quad V = g'(\eta f' - f), \tag{11–16}$$

with

$$\eta \equiv \frac{Y}{g(x)} \quad \text{and} \quad g(x) = \sqrt{2x}. \tag{11–17}$$

Now, let us assume that a similarity solution also exists for θ in the form

$$\boxed{\theta = H(\eta).} \tag{11–18}$$

The similarity variable η, (11–17), is the same as for the Blasius solution. Substituting (11–18) into (11–6) leads to the similarity equation

$$\boxed{\frac{d^2 H}{d\eta^2} + Pr\, f(\eta) \frac{dH}{d\eta} = 0,} \tag{11–19}$$

where we have used the fact that $gg' = 1$. As expected, the similarity transformation (11–18) leads to a second-order ODE for H. Hence, in order for the similarity solution to be workable, it is evident that the boundary condition (11–7b) and the initial condition (11–7c) must

C. The Asymptotic Limit, Pr (or Sc) ≫ 1

Figure 11–2. The temperature profile for heat transfer from a horizontal flat plate at large Reynolds number (large Peclet number) for several values of the Prandtl number, $0.01 \leq Pr \leq 100$.

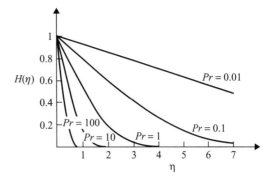

collapse into a single condition for H. It is for this reason that we require the leading edge of the thermal and momentum boundary layers to coincide (that is, $x^* = 0$). In this case, the boundary conditions on $H(\eta)$ become

$$H(0) = 1,$$
$$H(\eta) = 0 \quad \text{as} \quad \eta \to \infty. \tag{11-20}$$

A formal solution satisfying (11–19) and (11–20) is obtained easily, namely,

$$H(\eta) = 1 - \frac{\int_0^\eta e^{-Pr \int_0^t f(s)ds} dt}{\int_0^\infty e^{-Pr \int_0^t f(s)ds} dt}. \tag{11-21}$$

However, the Blasius function $f(\eta)$ is available only as the numerical solution of the Blasius equation, and it is thus inconvenient to evaluate this formula for $H(\eta)$. A simpler alternative is to numerically integrate the Blasius equation and the thermal energy equation (11–19) simultaneously. The function $H(\eta)$, obtained in this manner, is plotted in Fig. 11–2 for several different values of the Prandtl number, $0.01 \leq Pr \leq 100$. As suggested earlier, it can be seen that the thermal boundary-layer thickness depends strongly on Pr. For $Pr \gg 1$, the thermal layer is increasingly thin relative to the Blasius layer (recall that $f \to .99$ for $\eta \sim 4$). The opposite is true for $Pr \ll 1$.

Apart from the special cases discussed in this section, the thermal boundary-layer equation (11–6) can only be solved analytically for the two asymptotic limits, $Pr \ll 1$ and $Pr \gg 1$. This is the subject of the next two sections.

C. THE ASYMPTOTIC LIMIT, Pr (OR Sc) ≫ 1

Let us begin with the limit $Pr \gg 1$. As stated at the beginning of this chapter, this limit is actually of more direct application to the corresponding single-component mass transfer problem in which large values of the Schmidt number are common. The problem is now to obtain an approximate asymptotic solution of (11–6) and (11–7). Although the problem defined by (11–6) and (11–7) is derived in Section A as the *inner* boundary-layer approximation of the overall problem for $Pe \gg 1$, $Re \gg 1$, its solution for arbitrary geometry and Pr either large or small is itself a singular-perturbation problem. In other words, for these limiting cases, the domain consists of the *global* outer region where the governing

Heat and Mass Transfer at Large Reynolds Number

equation and solution at the leading order of approximation are (11–3) and (11–4), plus the boundary-layer region governed by (11–6) and (11–7), which itself splits into two parts.

In the high-Pr case, we obtain the leading-order term in an asymptotic expansion, for the part of the domain where the momentum boundary-layer scaling is applicable, by taking the limit $Pr \to \infty$ in (11–6). The result is

$$\mathbf{u} \cdot \nabla \theta = 0, \tag{11-22}$$

now expressed in terms of momentum boundary-layer variables, but with the same solution as before,

$$\theta = 0, \tag{11-23}$$

for all parts of the momentum boundary-layer domain that contain open streamlines. Clearly, this approximate solution still does not satisfy the boundary condition $\theta = 1$ at the body surface. In the limit $Pr \to \infty$, the conduction term in (11–6) is neglected, and the resulting solution pertains to only the *outer portion* of the momentum boundary layer, away from the body surface. Thus, as suggested earlier, we see that the solution of the thermal boundary-layer equation (11–6) for large Pr is itself a singular perturbation (matched asymptotic expansion) problem.

From a physical point of view, the fact that conduction is neglected in (11–6) for $Pr \to \infty$ is indicative of the fact that the momentum boundary-layer length scale is *not* relevant in the vicinity of the body surface for this limit. This is, in fact, demonstrated nicely by the exact similarity solution of the preceding section. Turning again to Fig. 11–2, we see that the thermal region of heated fluid becomes thinner and thinner in the momentum boundary-layer variables as Pr is increased. Clearly, the momentum boundary-layer dimension is not an appropriate measure of the distance over which the temperature changes from its surface value to the free-stream value $\theta = 0$. Viewed in another way, it is evident from Fig. 11–2 that the temperature gradient for large Pr is not independent of Pr when calculated with momentum boundary-layer variables. To obtain the form of (11–6) that applies to the region in the immediate vicinity of the body surface, it is necessary to rescale so that the dimension of the thermal region in terms of the new independent variables is $O(1)$.

We proceed formally. Thus we introduce the rescaled variable

$$\tilde{Y} = Y Pr^\alpha \tag{11-24}$$

(where $\alpha > 0$) into (11–6). As before, the change of variables from Y to $\tilde{Y}$ corresponds to the introduction of a new characteristic length scale. Equation (11–6) becomes

$$Pr^{2\alpha-1} \frac{\partial^2 \theta}{\partial \tilde{Y}^2} = u\left(x, \frac{\tilde{Y}}{Pr^\alpha}\right) \frac{\partial \theta}{\partial x} + Pr^\alpha V\left(x, \frac{\tilde{Y}}{Pr^\alpha}\right) \frac{\partial \theta}{\partial \tilde{Y}}. \tag{11-25}$$

The objective is to choose α such that the conduction term is retained in this rescaled equation, even in the limit $Pr \to \infty [\tilde{Y} \sim O(1)]$. Examining (11–25), we see formally that it is the limit of u and V for small values of the second argument that is relevant to the limit $Pr \gg 1$. Thus, to determine the appropriate forms for u and V, we consider a Taylor series expansion for small values of the momentum boundary-layer variable Y, namely,

$$u\left(x, \frac{\tilde{Y}}{Pr^\alpha}\right) \sim \left(\frac{\partial u}{\partial Y}\right)_{Y=0} \frac{\tilde{Y}}{Pr^\alpha} + \frac{1}{2}\left(\frac{\partial^2 u}{\partial Y^2}\right)_{Y=0} \frac{\tilde{Y}^2}{Pr^{2\alpha}} + \cdots + . \tag{11-26}$$

C. The Asymptotic Limit, Pr (or Sc) ≫ 1

Here, we have set $u|_{Y=0} = 0$ because of the no-slip condition at the body surface. The coefficient of the second term, $(\partial u/\partial Y)_{Y=0}$, is the dimensionless shear stress, which is a function of position on the body surface. We denote this function as $\lambda(x)$, that is,

$$\left.\frac{\partial u}{\partial Y}\right|_{Y=0} = \lambda(x). \tag{11–27}$$

The coefficient of the second term is

$$\left.\frac{\partial^2 u}{\partial Y^2}\right|_{Y=0} = -u_e(x)\frac{du_e}{dx}. \tag{11–28}$$

Thus, for $Pr \gg 1$, we can approximate u in the form

$$u\left(x, \frac{\tilde{Y}}{Pr^\alpha}\right) \sim \lambda(x)\frac{\tilde{Y}}{Pr^\alpha} - \frac{1}{2}u_e(x)\frac{du_e}{dx}\frac{\tilde{Y}^2}{Pr^{2\alpha}} + \cdots +. \tag{11–29}$$

We can then determine the corresponding form for V from (11–29) by using the continuity equation:

$$V\left(x, \frac{\tilde{Y}}{Pr^\alpha}\right) \sim -\frac{1}{2}\lambda'(x)\frac{\tilde{Y}^2}{Pr^{2\alpha}} + \frac{1}{6}\frac{d}{dx}\left(u_e\frac{du_e}{dx}\right)\frac{\tilde{Y}^3}{Pr^{3\alpha}} + \cdots +. \tag{11–30}$$

The geometry of the body has not been specified, except for the restriction to two dimensions and the assumption that the velocity field in the vicinity of the body is given in terms of a boundary-layer solution (that is, no separation). Indeed, the only way that the geometry is reflected in (11–25), (11–29), or (11–30) is by means of the functions $\lambda(x)$ and $u_e(x)$. We shall see that the leading-order approximation to the solution of (11–25) can be obtained for *arbitrary* $\lambda(x)$ – that is, for *arbitrary* geometry.

To obtain α, we substitute (11–29) and (11–30) into (11–25). The result is

$$Pr^{3\alpha-1}\frac{\partial^2\theta}{\partial Y^2} = \left(\lambda\tilde{Y} - \frac{1}{2}u_e\frac{du_e}{dx}\frac{\tilde{Y}^2}{Pr^\alpha} + \cdots +\right)\frac{\partial\theta}{\partial x}$$
$$- \left[\frac{1}{2}\lambda'\tilde{Y}^2 - \frac{1}{6}\frac{d}{dx}\left(u_e\frac{du_e}{dx}\right)\frac{\tilde{Y}^3}{Pr^\alpha} + \cdots +\right]\frac{\partial\theta}{\partial\tilde{Y}}. \tag{11–31}$$

We pick α so that the conduction term on the left balances the largest convection terms on the right, namely,

$$\alpha = \frac{1}{3}.$$

Thus,

$$\boxed{\tilde{Y} = YPr^{1/3},} \tag{11–32}$$

and the resulting form for the leading-order approximation to (11–31) is

$$\boxed{\frac{\partial^2\theta}{\partial\tilde{Y}^2} = \lambda(x)\tilde{Y}\frac{\partial\theta}{\partial x} - \frac{\lambda'(x)}{2}\tilde{Y}^2\frac{\partial\theta}{\partial\tilde{Y}} + O\left(Pr^{-1/3}\right).} \tag{11–33}$$

Heat and Mass Transfer at Large Reynolds Number

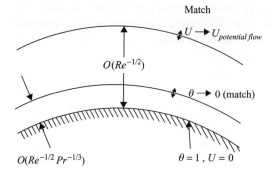

Figure 11–3. A schematic showing the two-layer structure of the thermal-momentum boundary layer for large Pe, with $Re \gg 1$ and $Pr \gg 1$. In this case, the distance over which θ changes to its free-stream value is vanishingly small compared with the dimension of the momentum boundary layer.

This is the governing equation for the leading-order approximation in an asymptotic expansion for θ in the innermost region near the body surface. The boundary condition (11–7a) applies to this leading approximation for θ,

$$\boxed{\theta = 1 \quad \text{at} \quad \tilde{Y} = 0.} \tag{11–34a}$$

The solution is also required to match (11–23), which is the leading-order approximation in the outer part of the momentum boundary layer:

$$\boxed{\theta = 0 \quad \text{for} \quad \tilde{Y} \gg 1 \quad (Pr \to \infty).} \tag{11–34b}$$

A schematic showing the relative scales of the thermal and momentum layers in this limit of $Pr \gg 1$ is shown in Fig. 11–3.

The form of the correlation among Nu, Re, and Pr can now be obtained for the present case $Re \gg 1$ and $Pr \gg 1$. The starting point is (11–11). The rescaling (11–32) now provides an asymptotic estimate for $\partial \theta / \partial Y |_{Y=0}$, namely,

$$\frac{Nu}{Re^{1/2}} = -Pr^{1/3} \int_{x^*}^{x_L} \left(\frac{\partial \theta}{\partial \tilde{Y}} \right)_{\tilde{Y}=0} dx. \tag{11–35}$$

The dimensionless gradient $\left(\partial \theta / \partial \tilde{Y} \right)_{\tilde{Y}=0}$ is a function of x that is independent of Re or Pr. Hence,

$$\boxed{\frac{Nu}{Re^{1/2} Pr^{1/3}} = F(x^*, x_L), \quad Re \gg 1, \quad Pr \gg 1.} \tag{11–36}$$

This is the primary result of the present section. The constant on the right-hand side depends on only the geometry of the body, that is, $\lambda(x)$ at the leading order of approximation. We shall see shortly that the equation and boundary conditions (11–33) and (11–34) can be solved analytically for arbitrary $\lambda(x)$.

First, however, it is important to note that the form of the correlation (11–36), though *independent* of the geometry of the body, is *dependent* on the boundary conditions at its surface. In particular, let us suppose that the no-slip boundary condition is replaced with a condition that allows $u(x, 0) \neq 0$ at the body surface. In this case, instead of (11–29), the form for u is

$$u\left(x, \frac{\tilde{Y}}{Pr^\alpha}\right) = u(x, 0) + \lambda(x) \frac{\tilde{Y}}{Pr^\alpha} + \cdots +. \tag{11–37}$$

C. The Asymptotic Limit, Pr (or Sc) ≫ 1

Hence, when this form (and the corresponding expression for V) is substituted into (11–25), the value of α required for balancing conduction with the largest convection term now becomes

$$\alpha = \frac{1}{2}.$$

Thus, for any problem with $u(x, 0) \neq 0$, the form of the correlation for Nu changes to

$$\boxed{\frac{Nu}{Re^{1/2} Pr^{1/2}} = G(x^*, x_L)} \tag{11-38}$$

for $Re \gg 1$, $Pr \gg 1$. We consider an example of this type in the Problems section at the end of this chapter.

Let us now return to (11–33), which is valid for any arbitrary (no-slip) 2D body in the limit $Re \gg 1$, $Pr \gg 1$. As noted earlier, the geometry of the body enters at this leading order of approximation only through the function $\lambda(x)$. The somewhat surprising fact is that we can solve (11–33) for *arbitrary* $\lambda(x)$. Hence we can obtain a *universal* first approximation for θ and for the Nusselt number correlation (11–36).

To solve (11–33) for arbitrary $\lambda(x)$ we introduce a similarity transformation of the form

$$\boxed{\theta = \theta(\eta), \quad \text{where} \quad \eta = \frac{\tilde{Y}}{g(x)}.} \tag{11-39}$$

The function $g(x)$ is to be determined as part of the solution and must depend explicitly on $\lambda(x)$. To satisfy the boundary conditions (11–34), as well as the initial condition (11–7c), it is necessary that

$$g(x^*) = 0. \tag{11-40}$$

When the similarity transformation (11–39) is introduced into (11–33), the result is

$$\frac{d^2\theta}{d\eta^2} + \eta^2 \frac{d\theta}{d\eta}\left(\lambda g^2 g' + g^3 \frac{\lambda'}{2}\right) = 0. \tag{11-41}$$

Hence, if a similarity solution of the form (11–39) exists, it must be possible to determine $g(x)$ such that the coefficient of the second term in (11–41) is a constant. As usual, the value of this constant can be chosen for convenience (without any effect on the solution).

Let us suppose that the constant is equal to 3. Then the constraint on g for existence of a similarity solution can be written in the form of a first-order ODE for g^3, namely,

$$\boxed{\frac{\lambda}{3}\frac{dg^3}{dx} + \frac{1}{2}\frac{d\lambda}{dx}g^3 = 3.} \tag{11-42}$$

To see whether this equation has a solution satisfying the condition (11–40), we can rewrite it in the form

$$\frac{dg^3}{dx} + g^3\left(\frac{d\ln\lambda^{3/2}}{dx}\right) = \frac{9}{\lambda}. \tag{11-43}$$

The homogeneous equation has a solution

$$\frac{c}{\lambda^{3/2}}.$$

777

Hence, the general solution of (11–42) is

$$g^3 = \frac{c}{\lambda^{3/2}} + \frac{9}{\lambda^{3/2}} \int_{x^*}^{x} \sqrt{\lambda(t)}\, dt. \tag{11–44}$$

To satisfy the condition (11–40),

$$c = 0,$$

so that

$$g(x) = \frac{9^{1/3}}{\sqrt{\lambda(x)}} \left[\int_{x^*}^{x} \sqrt{\lambda(t)}\, dt \right]^{1/3}. \tag{11–45}$$

The governing equation for the similarity function $\theta(\eta)$ now takes the form

$$\frac{d^2\theta}{d\eta^2} + 3\eta^2 \frac{d\theta}{d\eta} = 0, \tag{11–46}$$

with boundary (matching) conditions

$$\theta(0) = 1, \tag{11–47a}$$
$$\theta(\infty) = 0. \tag{11–47b}$$

Equation (11–46) can be integrated twice, with respect to η, to obtain

$$\theta = c_1 \int_{\eta}^{\infty} e^{-t^3}\, dt + c_2.$$

Hence, applying the boundary conditions (11–47), we find that

$$\theta = \frac{\int_{\eta}^{\infty} e^{-t^3}\, dt}{\int_{0}^{\infty} e^{-t^3}\, dt} = \frac{1}{\Gamma(4/3)} \int_{\eta}^{\infty} e^{-t^3}\, dt. \tag{11–48}$$

This completes the similarity solution for arbitrary $\lambda(x)$. The temperature profile given by (11–48) is plotted in Fig. 11–4.

The only remaining task is to use the solution (11–48) with $g(x)$ given by (11–45) to evaluate the geometry-dependent coefficient on the right-hand side of (11–36). For this purpose, we first evaluate $(\partial \theta / \partial \tilde{Y})_{\tilde{Y}=0}$:

$$-\left(\frac{\partial \theta}{\partial \tilde{Y}}\right)\bigg|_{\tilde{Y}=0} = -\frac{d\theta}{d\eta}\bigg|_{\eta=0} \left(\frac{\partial \eta}{\partial \tilde{Y}}\right) = -\frac{1}{g(x)} \frac{d\theta}{d\eta}\bigg|_{\eta=0}. \tag{11–49}$$

Now,

$$\left(\frac{d\theta}{d\eta}\right)_{\eta=0} = -\frac{1}{\Gamma(4/3)}. \tag{11–50}$$

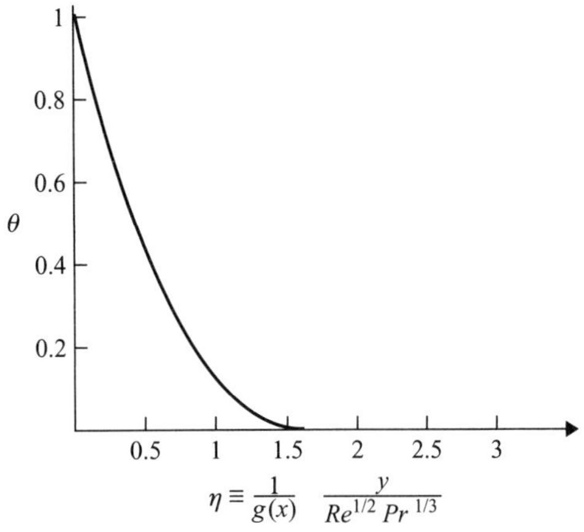

C. The Asymptotic Limit, Pr (or Sc) ≫ 1

Figure 11–4. The self-similar temperature profile for the thermal boundary layer in the limit $Re \gg 1$, $Pr \gg 1$.

$$\eta \equiv \frac{1}{g(x)} \frac{y}{Re^{1/2} Pr^{1/3}}$$

Hence,

$$-\left(\frac{\partial \theta}{\partial \tilde{Y}}\right)_{\tilde{Y}=0} = \frac{\sqrt{\lambda(x)}}{9^{1/3}\Gamma(4/3)} \left[\int_{x^*}^{x} \sqrt{\lambda(t)} dt\right]^{-1/3}. \tag{11-51}$$

To obtain $F(x^*, x_L)$, this must be integrated over the body surface, that is,

$$F(x^*, x_L) \equiv -\int_{x^*}^{x_L} \left(\frac{\partial \theta}{\partial \tilde{Y}}\right)_{\tilde{Y}=0} dx = \frac{3^{1/3}}{2\Gamma(4/3)} \left[\int_{x^*}^{x_L} \sqrt{\lambda(t)} dt\right]^{2/3}. \tag{11-52}$$

The results in (11–51) and (11–52) are, of course, completely general for arbitrary $\lambda(x)$. Here, for illustration purposes, we evaluate them only for the special case of a flat plate (parallel to the free stream) where, according to the Blasius solution from Chap. 10,

$$\lambda(x) = \frac{0.332}{\sqrt{x}}.$$

In this case,

$$-\left(\frac{\partial \theta}{\partial \tilde{Y}}\right)_{\tilde{Y}=0} = \frac{(0.332)^{1/3}}{9^{1/3}\Gamma(4/3)} \frac{1}{x^{1/4}} \frac{1}{(4/3)^{1/3}(x^{3/4} - x^{*3/4})^{1/3}}$$

$$= \frac{0.339}{x^{1/2}} \left[\frac{1}{1 - (x^*/x)^{3/4}}\right]^{1/3}. \tag{11-53}$$

Hence,

$$\frac{Nu}{Re^{1/2} Pr^{1/3}} = F(x^*, x_L) = 0.677 \left(x_L^{3/4} - x^{*3/4}\right)^{2/3}. \tag{11-54}$$

Here, x_L is the dimensionless position corresponding to the end of the heated portion of the plate surface and x^* is the position of the beginning of the heated zone. If the plate length is finite, so that the plate length is used as a characteristic length scale, and the whole of the plate surface is heated, then

$$x^* = 0 \quad \text{and} \quad x_L = 1$$

Heat and Mass Transfer at Large Reynolds Number

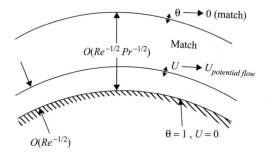

Figure 11–5. A schematic representation showing the two-layer structure of the thermal-momentum boundary layer for large Pe, with $Re \gg 1$ and $Pr \ll 1$. In this limit, the distance required for θ to approach its free-stream value is much larger than the distance required for the velocity to approach the potential-flow form. Thus convection within the thermal boundary layer is dominated by the outer limit of the momentum boundary-layer velocity distribution (or equivalently, the inner limit as we approach the body surface of the potential-flow distribution).

so that

$$Nu = 0.677 Re^{1/2}[Pr^{1/3} + O(1)]. \qquad (11\text{–}55)$$

The geometry-dependent coefficient in this case is 0.677. The reader will probably no longer be surprised to find that this coefficient is $O(1)$ in magnitude.

D. THE ASYMPTOTIC LIMIT, Pr (OR Sc) $\ll 1$

Now let us turn to the solution of the thermal boundary-layer equation (11–6) for the limit $Pr \to 0$. Although the thermal and momentum boundary layers both remain very thin compared with the dimension of the body, this is the limit, as seen in Fig. 11–5, where the thermal layer (the region where θ decreases from the surface value $\theta = 1$ to the free-stream value $\theta = 0$) is of much greater extent than the momentum layer (the region where u depends explicitly on Y). From a physical point of view, the limit $Pr \ll 1$ can be thought of as corresponding to increasing the thermal conductivity for fixed values of the other parameters. Hence, as $Pr \to 0$, the relative efficiency of heat transfer by conduction goes up, and we require larger velocities before the transport by convection becomes as important as conduction. Hence the thermal region increases in width relative to the momentum boundary-layer scaling as the Prandtl number is decreased.

If we proceed formally, then we obtain the governing equation for the first term in an asymptotic expansion for small Prandtl number for the region in which the momentum boundary-layer scaling is appropriate by taking the limit $Pr \to 0$ in (11–6). The result is

$$\frac{\partial^2 \oplus}{\partial Y^2} = 0. \qquad (11\text{–}56)$$

For convenience, we denote the dimensionless temperature as $\oplus$ in this part of the domain to differentiate it from the temperature distribution elsewhere. It can be seen from (11–56), that the heat transfer process in this region is dominated at leading order by pure conduction. If we integrate (11–56), the temperature profile is thus predicted to take the linear form

$$\oplus = a(x)Y + b(x).$$

D. The Asymptotic Limit, Pr (or Sc) ≪ 1

Hence, applying the boundary condition (11–7a), we obtain

$$b \equiv 1,$$

$$\boxed{\oplus = 1 + a(x)Y.} \tag{11-57}$$

It is evident, on further examination, that the solution (11–57) *cannot* be uniformly valid within the thermal boundary layer because it cannot be made to satisfy the matching condition (11–7b) for any choice of the coefficient $a(x)$. Hence (11–56) and the solution (11–57) are only valid in the part of the thermal layer that lies closest to the body surface, and, to obtain $a(x)$, we must seek a second solution, valid in the outer part of the thermal boundary layer, with $a(x)$ determined from matching.

To obtain an asymptotic approximation in the outer part of the thermal region, we seek a rescaling of (11–6) – that is, a new nondimensionalization – in which both conduction and convection terms are retained in the governing equation for the leading-order term in the asymptotic expansion for $Pr \ll 1$. We have seen, by examination of Fig. 11–2, that the extent of the thermal layer for decreasing values of Pr depends strongly on Pr. Thus, so too does $\nabla \theta$, $\nabla^2 \theta$, and so on. From this perspective, the objective of rescaling is to find a new characteristic length scale such that the dimensionless distance over which θ changes, in terms of the newly scaled variables, is $O(1)$, that is, independent of Pr. To rescale, we introduce the formal change of variables,

$$\boxed{\overline{Y} = Pr^\alpha Y \quad (\alpha > 0).} \tag{11-58}$$

Hence, for $Pr \ll 1$, changes in $\overline{Y}$ of $O(1)$ can be seen to correspond to very large changes in Y of $O(Pr^{-\alpha})$!

In the preceding section for $Pr \gg 1$, we rescaled both u and V also in recognition of the fact that these quantities have a characteristic magnitude proportional to $Pr^{-\beta}$ $(\beta > 0)$ when we restrict ourselves to a region within $O(Pr^{-1/3})$ of the body surface. This rescaling, however, was not applied directly to u and V. Instead, a Taylor series approximation was introduced for u and V at small values of Y, and the rescaling of u and V was accomplished implicitly by rescaling Y. It should be noted, however, that the result was equivalent to a *direct* rescaling of u and V of the form

$$u = \tilde{u} \, Pr^{-1/3},$$

$$V = \tilde{V} \, Pr^{-2/3}.$$

Here, for illustrative purposes, we proceed in the other way, by directly rescaling u and V in accord with (11–58), that is,

$$\overline{u} = Pr^\beta u\left(x, \frac{\overline{Y}}{Pr^\alpha}\right), \tag{11-59a}$$

and, hence, by continuity,

$$\overline{V} = Pr^{\alpha+\beta} V\left(x, \frac{\overline{Y}}{Pr^\alpha}\right). \tag{11-59b}$$

Now, in the limit $Pr \to 0$,

$$\overline{u} = Pr^\beta u(x, \infty) = Pr^\beta u_e(x). \tag{11-60}$$

Heat and Mass Transfer at Large Reynolds Number

But because $u_e(x)$ is already a function of $O(1)$, it appears that the appropriate scale for u must be

$$\beta \equiv 0. \tag{11-61}$$

Let us suppose that this result is correct. In this case, the rescaling of (11–6) requires only the transformations (11–58) and (11–59b), with $\alpha \neq 0$. Expressing (11–6) in terms of $\overline{Y}$ and $\overline{V}$, we find that

$$u\left(x, \frac{\overline{Y}}{Pr^\alpha}\right) \frac{\partial \theta}{\partial x} + \overline{V}\left(x, \frac{\overline{Y}}{Pr^\alpha}\right) \frac{\partial \theta}{\partial \overline{Y}} = Pr^{2\alpha-1} \frac{\partial^2 \theta}{\partial \overline{Y}^2}. \tag{11-62}$$

Now, in the limit $Pr \to 0$,

$$u\left(x, \frac{\overline{Y}}{Pr^\alpha}\right) \to u_e(x) \tag{11-63}$$

and, thus, by means of continuity,

$$\overline{V}\left(x, \frac{\overline{Y}}{Pr^\alpha}\right) \to -\frac{du_e}{dx}\overline{Y}. \tag{11-64}$$

The value of the rescaling parameter α is determined by the condition that the limiting form of (11–62) should retain both conduction and convection in the limit $Pr \to 0$. Hence, in view of (11–63) and (11–64), it can be seen from (11–62) that

$$\boxed{\alpha = \frac{1}{2}.} \tag{11-65}$$

Thus, the governing equation for the leading-order term in an asymptotic expansion for this outer part of the thermal boundary layer is

$$\boxed{u_e(x)\frac{d\theta}{dx} - u_e'\overline{Y}\frac{\partial \theta}{\partial \overline{Y}} = \frac{\partial^2 \theta}{\partial \overline{Y}^2}.} \tag{11-66}$$

It may be noted that the scaling parameter α and the form of the governing equation (11–66) in this case are independent of whether the boundary condition at the body surface involves slip or no-slip.

The boundary conditions for (11–66) are (11–7c), plus the condition

$$\boxed{\theta \to 0 \quad \text{for} \quad \overline{Y} \gg 1} \tag{11-67a}$$

as a matching condition with the global outer solution (11–4), and the matching condition

$$\lim_{Y \gg 1} \oplus (x, Y) \Leftrightarrow \lim_{\overline{Y} \ll 1} \theta\left(x, \overline{Y}\right) \quad \text{for} \quad Pr \to 0. \tag{11-67b}$$

The left-hand side of this condition contains, at leading order, just the solution (11–57). It is convenient to express this solution in terms of the rescaled variables for the *outer* part of the thermal boundary layer, namely,

$$\oplus(x, Y) = a(x)\frac{\overline{Y}}{Pr^{1/2}} + 1. \tag{11-68}$$

D. The Asymptotic Limit, Pr (or Sc) ≪ 1

However, in this form, it is clear that $a(x) \equiv 0$. If $a(x) \neq 0$, then it would be necessary that the solution for $\theta(x, \overline{Y})$ contain a term of $O(Pr^{-1/2})$. But $\theta(x, \overline{Y}) = O(1)$ at the leading order of approximation, as one can see from the boundary value (11–7a). Hence, the *first approximation* in the asymptotic expansion for $\oplus$ in the innermost region of the thermal boundary layer reduces to just

$$\oplus = 1 \qquad (11\text{–}69)$$

and the matching condition (11–67b) can be treated as a boundary condition for the leading-order approximation to θ, namely,

$$\boxed{\theta(x, 0) = 1.} \qquad (11\text{–}67\text{c})$$

Physically, in the limit $Pr \to 0$, the width of the thermal layer is so much greater than the momentum layer that θ does not vary at all across the momentum layer at the leading order of approximation. A sketch showing the multiscale structure of the thermal boundary region in this limit of $Pr \ll 1$ is shown in Fig. 11–5.

The problem for the first term in an asymptotic solution for the temperature distribution θ in the outer part of the thermal layer is thus to solve (11–66) subject to the conditions (11–67a), (11–67c), and (11–7c). Again, we see that the geometry of the body enters implicitly through the function $u_e(x)$ only. As in the high-Pr limit, a general solution of (11–66) is possible even for an arbitrary functional form for $u_e(x)$. Before we move forward to obtain this solution, however, a few comments are probably useful about the solution (11–69) for the innermost part of the boundary layer immediately adjacent to the body surface.

The most important point to note is that the leading-order solution $\oplus = 1$ does not imply that the heat flux at the body surface is zero, even though this might at first appear to be the case. In general, the condition of matching between the inner and outer approximations within the thermal boundary layer requires not only (11–67c) but also that the spatial derivatives of θ and $\oplus$ should match at each level of approximation for $Pr \to 0$. Thus, in particular,

$$\lim_{\overline{Y} \ll 1} Pr^{1/2} \frac{\partial \theta}{\partial \overline{Y}} \Leftrightarrow \lim_{Y \gg 1} \left(\frac{\partial \oplus}{\partial Y} \right) \quad \text{for} \quad Pr \to 0. \qquad (11\text{–}70)$$

Now, $\partial \theta / \partial \overline{Y} = O(1)$ at the leading order of approximation. Hence, it follows that the dimensionless temperature gradient within the innermost region should be $O(Pr^{1/2})$. But the solution (11–69) represents only the first term in an asymptotic expansion. The condition (11–70) suggests strongly that

$$\oplus_{inner} = 1 + Pr^{1/2} \oplus_1 (x, Y) + \cdots + . \qquad (11\text{–}71)$$

However, if this is true, then it can be seen from (11–6) that the governing equation for $\oplus_1$ is still

$$\frac{d^2 \oplus_1}{\partial Y^2} = 0, \qquad (11\text{–}72)$$

with the general solution

$$\oplus_1 = c(x) Y + d(x). \qquad (11\text{–}73)$$

The boundary condition (11–7a) at $Y = 0$ now becomes

$$\oplus_1 = 0 \quad \text{at} \quad Y = 0, \qquad (11\text{–}74)$$

so that $d(x) \equiv 0$. The matching condition (11–70) then requires that

$$\left.\frac{\partial \theta}{\partial \overline{Y}}\right|_{\overline{Y} \ll 1} \Leftrightarrow \left(\frac{\partial \Theta_1}{\partial Y}\right)_{Y \gg 1} \quad \text{for} \quad Pr \to 0. \tag{11–75}$$

Thus $c(x) \equiv (\partial \theta / \partial \overline{Y})_{\overline{Y} \ll 1}$, and it follows from (11–71) that

$$\boxed{\Theta_{inner} = 1 + Pr^{1/2} \left(\frac{\partial \theta}{\partial \overline{Y}}\right)_{\overline{Y} \ll 1} Y + \cdots + .} \tag{11–76}$$

Hence, to evaluate the heat flux at the body surface, it is sufficient to determine the solution $(\partial \theta / \partial \overline{Y})|_{\overline{Y} \ll 1}$ in the *outer* part of the thermal layer. It also follows from this discussion that

$$\frac{Nu}{Re^{1/2}} \equiv -\int_{x^*}^{x_L} \left(\frac{\partial \Theta}{\partial Y}\right)_{Y=0} dx = -Pr^{1/2} \int_{x^*}^{x_L} \left(\frac{\partial \theta}{\partial \overline{Y}}\right)_{\overline{Y} \ll 1} dx.$$

Hence,

$$\boxed{\frac{Nu}{Re^{1/2} Pr^{1/2}} = -\int_{x^*}^{x_L} \left(\frac{\partial \theta}{\partial \overline{Y}}\right)_{\overline{Y} \ll 1} dx = K(x^*, x_L)} \tag{11–77}$$

for $Re \gg 1$, $Pe \ll 1$, but with $Pe = RePr \gg 1$. Thus, to obtain a leading-order approximation for the Nusselt number, we must solve the differential equation (11–66) subject to the boundary conditions (11–67).

One important point, before finishing this problem by solving the equation (11–66), is to note that, unlike the limit $Pr \gg 1$, both the scaling (11–65) of the boundary-layer thickness and the form of the boundary-layer equation are independent of the boundary conditions on the velocity at the body surface. That is, the same scaling holds whether the boundary is a no-slip or slip surface, and, in fact, the complete correlation (11–77) depends on only the shape of the body.

The dependence of the coefficient $K(x^*, x_L)$ on the geometry of the body is specified by means of the function $u_e(x)$. However, as in the high-Pr limit, it is possible to obtain a general solution of (11–66) and (11–67) for arbitrary geometry [that is, arbitrary $u_e(x)$]. The method of analysis is once again to use a similarity transformation:

$$\theta = \theta(\eta) \quad \text{with} \quad \eta = \frac{\overline{Y}}{g(x)}. \tag{11–78}$$

Again, the function $g(x)$ must be determined subject to the condition $g(x^*) = 0$.

If the similarity transformation (11–78) is introduced into (11–66), the result is

$$\frac{d^2\theta}{d\eta^2} + \eta \frac{d\theta}{d\eta} (u_e g g' + u'_e g^2) = 0, \tag{11–79}$$

and a necessary condition for existence of a similarity solution is that it be possible to determine $g(x)$ in terms of $u_e(x)$ such that the coefficient in parentheses is constant. For

D. The Asymptotic Limit, Pr (or Sc) ≪ 1

convenience, we choose this constant equal to 2. Examination of (11–79) indicates that the condition on g can be written in the form of a first-order ODE for g^2,

$$\frac{dg^2}{dx} + 2g^2 \frac{d \ln u_e}{dx} = \frac{4}{u_e}. \tag{11–80}$$

Hence a necessary condition for existence of a similarity solution is that g^2 satisfy (11–80), i.e.,

$$g^2 = \frac{c}{u_e^2} + \frac{4}{u_e^2} \int_{x^*}^{x} u_e(t)\, dt. \tag{11–81}$$

The constant c is determined from the boundary condition that $g^2(x^*) = 0$. Thus,

$$c = 0,$$

and

$$g(x) = \frac{2}{u_e} \left[\int_{x^*}^{x} u_e(t)\, dt \right]^{1/2}. \tag{11–82}$$

With this form for $g(x)$, the boundary condition (11–67a) and the initial condition (11–7c) are both expressed in terms of the single condition

$$\theta(\infty) = 0. \tag{11–83}$$

The governing, similarity form of the equation for $\theta(\eta)$ is now

$$\frac{d^2\theta}{d\eta^2} + 2\eta \frac{d\theta}{d\eta} = 0. \tag{11–84}$$

Hence, integrating twice with respect to η and applying the condition (11–83), plus the condition (11–67b) in the form

$$\theta(0) = 1, \tag{11–85}$$

we obtain the general solution

$$\theta = \frac{\int_\eta^\infty e^{-t^2}\, dt}{\int_0^\infty e^{-t^2}\, dt}. \tag{11–86}$$

The integral coefficient in the denominator is just $\Gamma(3/2)$, which has a numerical value $\sqrt{\pi}/2$. Hence,

$$\theta = \frac{2}{\sqrt{\pi}} \int_\eta^\infty e^{-t^2}\, dt \equiv \operatorname{erfc}(\eta), \tag{11–87}$$

where erfc(η) is the *complimentary error function*:

$$\operatorname{erfc}(\eta) = 1 - \operatorname{erf}(\eta),$$

Heat and Mass Transfer at Large Reynolds Number

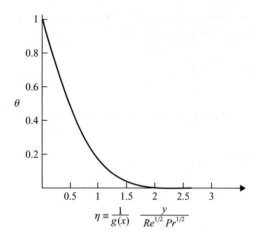

Figure 11–6. The self-similar temperature profile for the thermal boundary layer with $Re \gg 1$, $Pr \ll 1$.

where

$$\text{erf}(\eta) \equiv \frac{2}{\sqrt{\pi}} \int_0^{\eta} e^{-t^2} dt.$$

This temperature profile is plotted in Fig. 11–6.

The only remaining task is to use the solutions (11–87) and (11–82) to evaluate the coefficient $K(x^*, x_L)$ in the correlation (11–77) for the Nusselt number as a function of Re and Pr. The integrand is

$$\left(\frac{\partial \theta}{\partial \overline{Y}} \right)_{\overline{Y} \ll 1} = \left. \frac{d\theta}{d\eta} \right|_{\eta=0} \frac{\partial \eta}{\partial \overline{Y}} = \frac{1}{g(x)} \left. \frac{d\theta}{d\eta} \right|_{\eta=0}.$$

But

$$\left. \frac{d\theta}{d\eta} \right|_{\eta=0} = -\frac{2}{\sqrt{\pi}},$$

and thus

$$\left. \left| \frac{\partial \theta}{\partial \overline{Y}} \right|_{\overline{Y}=0} \right. = -\frac{u_e}{2} \left[\int_{x^*}^{x} u_e(t) dt \right]^{-1/2} \frac{2}{\sqrt{\pi}}. \tag{11–88}$$

Hence,

$$K(x^*, x_L) \equiv \int_{x^*}^{x_L} \left(\frac{\partial \theta}{\partial \overline{Y}} \right)_{\overline{Y}=0} dx = \frac{2}{\sqrt{\pi}} \left[\int_{x^*}^{x_L} u_e(t) dt \right]^{1/2} \tag{11–89}$$

The results in (11–88) and (11–89) are for general geometries. Here, we again consider explicitly only the specific case of a horizontal flat plate for which

$$u_e(x) \equiv 1. \tag{11–90}$$

E. Use of the Asymptotic Results at Intermediate Pe (or Sc)

In this case, the local heat flux is

$$\left|\frac{\partial \theta}{\partial \overline{Y}}\right|_{\overline{Y}=0} = -\frac{0.564}{\sqrt{x-x^*}} = -\frac{0.564}{\sqrt{x}} \quad \text{if } x^* \equiv 0. \tag{11–91}$$

Then

$$K(x^*, x_L) = \frac{2}{\sqrt{\pi}}(x_L - x^*)^{1/2}, \tag{11–92}$$

or, in the case in which

$$x^* \equiv 0 \text{ and } x_L \equiv 1,$$

we find that

$$K(x^*, x_L) \equiv \frac{2}{\sqrt{\pi}} = 1.128. \tag{11–93}$$

E. USE OF THE ASYMPTOTIC RESULTS AT INTERMEDIATE Pe (OR Sc)

Before proceeding to other topics, it is worthwhile to reflect briefly on the results of Sections 11.C and 11.D. In particular, let us focus our attention on the two results in (11–53) and (11–91) for the local temperature gradient in the case of a horizontal flat plate, with $x^* \equiv 0$. These correspond to correlations for the *local* dimensionless heat flux

$$\frac{Nu_x}{Re^{1/2}Pr^{1/2}} = \frac{0.564}{\sqrt{x}} \quad \text{for } Pr \ll 1, \tag{11–94}$$

$$\frac{Nu_x}{Re^{1/2}Pr^{1/3}} = \frac{0.339}{\sqrt{x}} \quad \text{for } Pr \gg 1. \tag{11–95}$$

It may be noted that the functional dependence of the local heat flux on x is the same as the shear stress. This is a consequence of the fact that the thickness of the thermal boundary layer varies with x in the same way as the momentum boundary-layer thickness. Furthermore, the form of the correlations for large and small Prandtl numbers are also quite similar. However, this latter observation may be somewhat misleading. In the case $Pr \gg 1$, the heat flux increases to very large values, proportional to $Pr^{1/3}$ as Pr increases. On the other hand, the heat flux for $Pr \ll 1$ decreases as $Pr^{1/2}$ for $Pr \to 0$. Clearly the heat transfer process is much less efficient for small Pr than for large Pr.

From a practical point of view, it frequently will be the case that the Prandtl number is neither very small nor very large. For example, for most gases, $Pr \sim O(1)$, whereas the Prandtl number for relatively simple, low-molecular-weight liquids like water is $Pr \sim O(5)$. Thus an obvious question is whether the asymptotic results for $Pr \to 0$ and $Pr \to \infty$ can provide any useful guidance for such cases. The answer is that a useful estimate of the rate of heat transfer can be obtained for the whole range of values of Pr by simple interpolation between the two asymptotic results. For example, let us again consider the case of a horizontal flat plate, for which the local Nusselt number is given by (11–94) and (11–95) for the two limiting cases $Pr \to 0$ and $Pr \to \infty$, respectively. In this case, we plot $Nu_x\sqrt{x}/Re^{1/2}$, corresponding to (11–94) and (11–95), versus Pr on a log–log graph. The

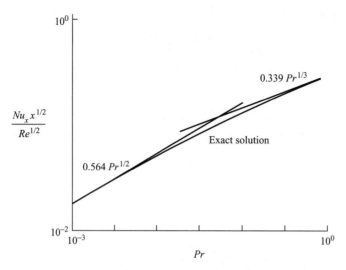

Figure 11–7. A comparison of the asymptotic forms for the local Nusselt number correlation for $Pr \ll 1$ and $Pr \gg 1$, with the result from the exact similarity solution for arbitrary Prandtl number. (*Note:* All results are for the special case of a horizontal flat plate.)

result shown in Fig. 11–7 is two straight lines, one with slope 1/2 and the other with slope 1/3, that intersect at $Pr \sim O(1)$. Also shown on the same plot is the *exact* result for this problem, obtained from the similarity solution in Section B,

$$-\frac{\partial \theta}{\partial Y}\sqrt{x} = -\left|\frac{dH}{d\eta}\right|_{\eta=0} = \frac{1}{\sqrt{2}\int_0^\infty e^{-Pr\int_0^t f(s)ds}dt}. \qquad (11\text{–}96)$$

We see that the maximum difference between this exact result and the nearest of the two asymptotes is about 20% at the point where the two asymptotes intersect.

We derive two basic conclusions from the comparison between the exact and asymptotic results in Fig. 11–7. First, the transition region between the low- and the high-Pr-number limits is fairly narrow, and the asymptotes thus provides a good approximation of the exact results for a surprisingly wide range of values for Pr. Second, the comparison suggests strongly the likelihood of obtaining a reasonable estimate for Nu at intermediate values of Pr for more complex problems by a smooth interpolation between the two asymptotes.

F. APPROXIMATE RESULTS FOR SURFACE TEMPERATURE WITH SPECIFIED HEAT FLUX OR MIXED BOUNDARY CONDITIONS

All of the results obtained in the preceding sections of this chapter have been based on the assumption that the temperature at the body surface is uniform. However, there are many important cases for which this condition does not hold. For example, in many practical applications it is the heat flux at the surface that is held constant, rather than the surface temperature. One example of such a problem is a wire (resistance) heater. For this type of problem, the objective of theory is no longer the heat flux (or total rate of heat transfer), as this is already known, but the temperature distribution in the fluid and especially at the surface of the body (where one might wish to determine the maximum temperature, for example).

Other problems exist to which neither the temperature nor the heat flux is known, but some combination of these. One example is the flow of a gas containing a condensable

F. Approximate Results for Surface Temperature

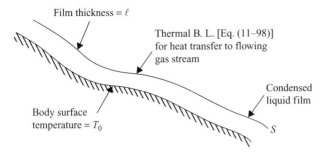

Figure 11–8. A schematic representation of the local surface conditions for heat transfer from a solid body with surface temperature T_0 to a gas stream when there is a condensed liquid film on the body surface. As explained in the text, this leads to an approximate boundary condition on the surface S of mixed type (11–98).

component past a cooled solid surface, when a liquid film is deposited on the body. In this case, an exact solution for the temperature distribution in the gas (and thus the rate of heat transfer between the body and the gas), would require solving a coupled problem in which the velocity and temperature fields are determined in both the liquid film and the gas. However, an approximate solution is possible, provided the liquid film is thin enough so that the transfer of heat across the film is dominated by conduction. In this case, at steady state, the rate of heat transfer across the film is balanced by the transport of heat from the outer surface of the film to the gas, and this leads to a *mixed-type boundary condition* for the gas-phase heat transfer process. To see this, let us suppose that the temperature at the surface of the cooled body is fixed at a value T_0 and that the width of the liquid film is ℓ. Then the steady-state heat flux balance is

$$k_L \frac{T_o - T_s}{\ell} = -k_G \left.\frac{\partial T}{\partial \eta}\right|_s, \qquad (11\text{--}97)$$

where k_L and k_G are the liquid- and gas-phase thermal conductivities, T_s denotes the surface temperature (unknown) at the outer surface of the liquid film, and $(\partial T/\partial \eta)_s$ is the temperature gradient normal to this surface in the gas. Rearranging, we obtain

$$\boxed{-\left.\frac{\partial T}{\partial \eta}\right|_s + \frac{k_L}{k_G \ell} T_s = \frac{k_L}{k_G \ell} T_0.} \qquad (11\text{--}98)$$

The right-hand side is known from the specified temperature at the solid. Hence (11–98) is a boundary condition of the so-called *mixed type* for the temperature distribution in the gas. A schematic representation of the situation is shown in Fig. 11–8. Both T_s and $(\partial T/\partial \eta)_s$ are unknown, and we thus seek both the surface temperature and the local heat flux from the solution. Note, however, that it is enough to determine one of these two quantities directly in our solution. The other can then be obtained from (11–98).

There are at least two approaches that we can take to solve problems in which either the heat flux or the mixed-type condition is specified as a boundary condition. If it is desired to determine the temperature distribution throughout the fluid, then we must return to the governing thermal boundary-layer equation (11–6) – assuming that $Re, Pe \gg 1$ – and develop new asymptotic solutions for large and small Pr, with either $\partial T/\partial \eta|_{y=0}$ or a condition of the mixed type specified at the body surface. The problem for a constant, specified heat flux is relatively straightforward, and such a case is posed as one of the exercises at the end of this chapter. On the other hand, in many circumstances, we might be concerned with determining only the temperature distribution on the body surface [and thus $\partial T/\partial \eta|_s$ from (11–98) for the mixed-type problem], and for this there is an even simpler approach that can be derived directly from the results of Sections C and D. This involves the derivation of an integral equation for the surface temperature.

To see how this can be done, let us pose a generalization of (11–6) and (11–7) as follows:

$$u\frac{\partial \theta}{\partial x} + V\frac{\partial \theta}{\partial Y} = \frac{1}{Pr}\frac{\partial^2 \theta}{\partial Y^2}, \quad (11\text{–}99)$$

with

$$\begin{aligned}\theta &= 0 \quad \text{at} \quad Y = \infty, \\ \theta &= 0 \quad \text{for} \quad Y > 0, \quad x = x^*, \\ \theta &= \theta_s(x) \quad \text{for} \quad Y = 0, \quad x > x^*.\end{aligned} \quad (11\text{–}100)$$

Of course, if our objective were to calculate the complete temperature distribution in the fluid, we would never pose the problem in this form because $\theta_s(x)$ is unknown and thus not satisfactory as a boundary condition. However, our goal here is to determine $\theta_s(x)$ for cases in which $\partial \theta/\partial Y|_{Y=0}$ or $[(\partial \theta/\partial Y) + (k_L/k_Gl)\theta]_{Y=0}$ is specified, and the objective in writing the problem in the form (11–99)–(11–100) is to show that we can evaluate $\theta_s(x)$ directly from the asymptotic solutions obtained earlier for a *constant* surface temperature, $\theta_s = 1$. The key to transforming from our original solutions to a formula for $\theta_s(x)$ satisfying (11–99)–(11–100) is that the governing equations and boundary conditions for θ are all linear.

It is convenient to proceed formally in a series of steps.

(1) First, we consider the problem in which $\theta_s(x) = 0$ for $0 \le x \le x^*$, and $\theta_s(x) = 1$ for $x > x^*$, For arbitrary 2D bodies, we have already solved this problem for the two limits $Pr \to 0$ and $Pr \to \infty$, In particular, let us define $K(x; x^*, Pr)$ as

$$-\frac{\partial \theta}{\partial Y}\bigg|_{Y=0} \equiv K(x; x^*, Pr). \quad (11\text{–}101)$$

Then, it can be seen from (11–51) and (11–88) that

$$K(x, x^*, Pr) = \begin{cases} \dfrac{Pr^{1/3}\sqrt{\lambda(x)}}{9^{1/3}\Gamma(4/3)}\left[\displaystyle\int_{x^*}^{x}\sqrt{\lambda(t)}\,dt\right]^{-1/3} & \text{for } Pr \to \infty \\[2ex] \dfrac{Pr^{1/2}u_e(x)}{\sqrt{\pi}}\left[\displaystyle\int_{x^*}^{x}u_e(t)\,dt\right]^{-1/2} & \text{for } Pr \to 0 \end{cases}. \quad (11\text{–}102)$$

(2) Second, if we consider the trivial generalization in which the surface temperature distribution is

$$\theta_s(x) = 0, \quad 0 \le x \le x^*,$$
$$\theta_s(x) = \theta_s(x^*) \quad \text{for } x \ge x^*,$$

where $\theta_s(x^*)$ is a constant value not equal to unity, the solution for $(\partial \theta/\partial Y)_{Y=0}$ is now

$$-\left(\frac{\partial \theta}{\partial Y}\right)_{Y=0} = K(x, x^*, Pr)\theta_s(x^*). \quad (11\text{–}103)$$

F. Approximate Results for Surface Temperature

(3) Third, we consider the case of a surface temperature distribution of the form

$$\theta_s(x) = 0, \quad 0 \le x < x^*,$$
$$\theta_s(x) = \theta_s(x^*), \quad x^* \le x \le x^* + \Delta x^*,$$
$$\theta_s(x) = 0, \quad x > x^* + \Delta x^*.$$

The solution, in terms of $(\partial \theta / \partial Y)_{Y=0}$ for this problem, is just a linear combination of the solutions to two problems of the type considered in the second step, namely,

$$\text{Problem 1:} \begin{cases} \theta_s(x) = 0, \; 0 \le x < x^* \\ \theta_s(x) = \theta_s(x^*), \; x \ge x^* \end{cases},$$

$$\text{Problem 2:} \begin{cases} \theta_s(x) = 0, \; 0 \le x < x^* + \Delta x^* \\ \theta_s(x) = -\theta_s(x^*), \; x \ge x^* + \Delta x^* \end{cases}.$$

Hence,

$$-\left(\frac{\partial \theta}{\partial Y}\right)_{Y=0} = K(x, x^*, Pr)\theta_s(x^*) - K(x, x^* + \Delta x^*, Pr)\theta_s(x^*),$$

$$-\left(\frac{\partial \theta}{\partial Y}\right)_{Y=0} = \left[K(x, x^*, Pr) - K(x, x^* + \Delta x^*, Pr)\right]\theta_s(x^*). \qquad (11\text{–}104)$$

In the limit as the increment $\Delta x^* \to 0$, this can be written in the equivalent form

$$-\left(\frac{\partial \theta}{\partial Y}\right)_{Y=0} = -\frac{\partial K}{\partial x^*}\theta_s(x^*)dx^*, \quad (\Delta x^* \to 0). \qquad (11\text{–}105)$$

(4) Finally, let us suppose that $\theta_s(x)$ is an arbitrary but smooth function of x and that we can approximate this function by breaking the surface region into a large number of discrete intervals in which $\theta_s(x)$ can be approximated as a constant, corresponding, for example, to the average value over the interval. Then the solution, in terms of $-(\partial \theta / \partial Y)_{Y=0}$, for each individual interval is of the type shown in Step (3). Thus the *discrete* approximation to the whole solution is a sum of solutions of the type from Step (3) for each subinterval, namely,

$$-\frac{\partial \theta}{\partial Y}\bigg|_{Y=0} = \sum_k \{K(x, k\Delta x^*, Pr) - K[x, (k+1)\Delta x^*, Pr]\}\bar{\theta}_s(k\Delta x^*), \qquad (11\text{–}106)$$

where the sum on k is such that $(k+1)\Delta x^* = x_L$ for the largest value of k.

Now, in the limit $\Delta x^* \to 0$, the discrete approximation to the surface temperature distribution passes smoothly to the continuous function $\theta_s(x)$, and the summation in (11–106) becomes an integral over x. Thus,

$$-\left(\frac{\partial \theta}{\partial Y}\right)_{Y=0} = -\int_0^x \left(\frac{\partial K}{\partial t^*}\right) \theta_s(t^*) dt^*. \qquad (11\text{--}107)$$

Now let us suppose that we are given the heat flux on the body surface as a function of x,

$$-\left(\frac{\partial \theta}{\partial Y}\right)_{Y=0} = g(x), \qquad (11\text{--}108)$$

Substituting this into (11–107), we obtain

$$\int_0^x \frac{\partial K}{\partial t^*}(x, t^*, Pr)\theta_s(t^*) dt^* = g(x). \qquad (11\text{--}109)$$

This is a *linear integral equation* (of the first kind) that now can be solved directly, in principle, to determine the unknown surface temperature distribution $\theta_s(x)$. The main advantage of this formulation, relative to solving the whole problem (11–6) with (11–108) as a boundary condition, is that (11–109) can be solved to determine $\theta_s(x)$ directly, without any need to determine the temperature distribution elsewhere in the domain. This latter problem is only 1D, in spite of the fact that the original problem was fully 2D.

If instead of (11–108) we have boundary conditions of the mixed type, a similar result is obtained. In particular, let us suppose that

$$-\frac{\partial \theta}{\partial Y}\bigg|_{Y=0} + c_1 \theta_s = c_2 f(x), \qquad (11\text{--}110)$$

where $f(x)$ is known. Then, substituting into (11–107), we obtain

$$\int_0^x \frac{\partial K}{\partial t^*}(x, t^*, Pr)\theta_s(t^*) dt^* + c_1 \theta_s(x) = c_2 f(x). \qquad (11\text{--}111)$$

In this case, we again have a linear integral equation (of the second kind) for the unknown surface temperature distribution.

The solutions to (11–109) or (11–111) can be carried out either numerically or, because the equations are linear, by an assortment of approximate analytical techniques. An attempt is not made to present these solution methods here. Instead, to preserve space, the interested reader is referred to any of the standard textbooks on linear integral equations. A classic is the book *Linear Integral Equations* by W. V. Lovitt, which is available from Dover in paperback.[3]

Let us conclude by briefly summarizing the material of this section in the following manner. If we wish to determine θ everywhere in the domain, we would have to solve the differential equation (11–6) – either exactly or by means of an asymptotic approximation– subject to either of the boundary conditions (11–108) or (11–110) at the body surface. If, on the other hand, we wish to determine only $\theta_s(x)$, then it is advantageous to solve the *boundary integral* (11–109) or (11–111). This converts the original 2D problem into a 1D problem and allows $\theta_s(x)$ to be calculated directly without the necessity of determining θ everywhere in the domain.

G. LAMINAR BOUNDARY-LAYER MASS TRANSFER FOR FINITE INTERFACIAL VELOCITIES

It was previously indicated that the analysis of the previous sections of this chapter could apply equally well to heat transfer problems or to the single-solute mass transfer problem with θ interpreted as the dimensionless concentration, $\theta \equiv (c - c_\infty)/(c_0 - c_\infty)$, with c representing the mass fraction of solute and the Schmidt number substituted for the Prandtl number. The key assumption in this assertion is that the interfacial velocity generated that is due to the transfer of solute to or from the body surface is small enough to play a negligible role in both the fluid motion and as a convective contribution to the mass transfer rate. In this section, we consider how the problem changes if this assumption is not satisfied.[4]

If we assume that the concentration of solute is constant at the surface of the body, then the dimensionless version of the mass transfer problem is formally identical to that previously, solved, namely the governing equation (11–1) with the boundary conditions $\theta = 1$ at the body surface and $\theta \to 0$ at infinity, plus the initial condition $\theta = 1$ at the leading edge of the boundary layer $x = 0$. The primary difference is that the normal velocity at the surface of the body is now nonzero with a magnitude given in dimensionless form by

$$v = -\frac{B}{Pe}\frac{\partial \theta}{\partial y} \quad \text{at } y = 0 \tag{11–112}$$

from Chap. 9, Subsection A.3. Following the usual custom in boundary-layer analyses, we use y to designate a local "Cartesian" coordinate normal to the body surface. In view of the fact that the Reynolds number is large, we expect the usual momentum boundary-layer scaling (11–5) to apply so that the governing equation takes the form (11–6)

$$u(x,Y)\frac{\partial \theta}{\partial x} + V(x,Y)\frac{\partial \theta}{\partial Y} = \frac{1}{Sc}\frac{\partial^2 \theta}{\partial Y^2} + O\left(Re^{-1}Sc^{-1}\right), \tag{11–113}$$

with the boundary condition

$$\theta = 1 \quad \text{at} \quad Y = 0, \tag{11–114a}$$

the matching condition

$$\theta \to 0 \quad \text{for} \quad Y \gg 1 \text{ in the limit } Re \to \infty. \tag{11–114b}$$

and the initial condition

$$\theta = 0 \quad \text{at} \quad x = 0. \tag{11–114c}$$

Applying the same scaling to the expression (11–112), we find

$$\boxed{V(x,0) = -\frac{B}{Sc}\left(\frac{\partial \theta}{\partial Y}\right)_{Y=0}.} \tag{11–115}$$

A necessary condition for self-consistency in assuming that the usual boundary-layer scaling is applicable is that $V(x, 0)$ remain bounded.

The sign of the "blowing" parameter,

$$B \equiv \frac{c_0 - c_\infty}{1 - c_0}, \tag{11–116}$$

determines whether there is blowing or suction. "Blowing" corresponds to positive values of B and the range of possible values is $0 < B < \infty$. On the other hand, "suction" corresponds to negative values and the range of possible values is $-1 < B < 0$. For nonzero values of B, there are three asymptotic limits that have been considered, namely $B \to \infty$ or $B \to -1$

Heat and Mass Transfer at Large Reynolds Number

with $Sc = O(1)$, and $Sc \to \infty$ with $B = O(1)$. The statement that Sc or $B = O(1)$ does not mean that they may not be numerically large, only that these are fixed parameters as far as the asymptotic limit is concerned. We consider only the case $Sc \to \infty$ with $B = O(1)$ here.

The limit $Sc \to \infty$ implies that the normal velocity $V(x, 0)$ is asymptotically small regardless of the size of B. Thus the change in $V(x, 0)$ is too small to affect the leading-order boundary-layer velocity distributions. Nevertheless, we shall see that the mass transfer rate is still changed. The other two limits $B \to \infty$ or $B \to -1$ both correspond to $V(x, 0) \to \infty$, which means that the velocity profiles will change and there is an intimate coupling between the momentum and mass transfer equations.

The analysis for $Sc \to \infty$ closely follows the large-Prandtl-number analysis of Section C. The solution of (11–113) is a singular perturbation with a leading-order outer solution $\theta = 0$ and an inner solution that satisfies a boundary-layer equation that is obtained by rescaling according to

$$\tilde{Y} = Y Sc^\alpha. \tag{11-117}$$

The rescaled equation is

$$Sc^{2\alpha-1} \frac{\partial^2 \theta}{\partial \tilde{Y}^2} = u\left(x, \frac{\tilde{Y}}{Sc^\alpha}\right) \frac{\partial \theta}{\partial x} + Sc^\alpha V\left(x, \frac{\tilde{Y}}{Sc^\alpha}\right) \frac{\partial \theta}{\partial \tilde{Y}}. \tag{11-118}$$

We can still use a Taylor series representation of the velocity components for $Y \ll 1$. Retaining the same notation as used in Section C, we obtain

$$u\left(x, \frac{\tilde{Y}}{Sc^\alpha}\right) \tilde{\lambda}(x) \frac{\tilde{Y}}{Sc^\alpha} - \frac{1}{2} u_e(x) \frac{du_e}{dx} \frac{\tilde{Y}^2}{Sc^{2\alpha}} + \cdots +, \tag{11-119}$$

where

$$\left.\frac{\partial u}{\partial Y}\right|_{Y=0} = \lambda(x).$$

We can then determine the corresponding form for V by using the continuity equation and the boundary condition (11–116):

$$V\left(x, \frac{\tilde{Y}}{Sc^\alpha}\right) \sim -\frac{B}{Sc}\left(Sc^\alpha \frac{\partial \theta}{\partial \tilde{Y}}\right)_{\tilde{Y}=0} - \frac{1}{2}\lambda'(x) \frac{\tilde{Y}^2}{Sc^{2\alpha}} + \frac{1}{6}\frac{d}{dx}\left(u_e \frac{du_e}{dx}\right) \frac{\tilde{Y}^3}{Sc^{3\alpha}} + \cdots + . \tag{11-120}$$

Hence, with the largest terms only reatained, the governing equation for the leading-order approximation for the concentration distribution in the inner region is

$$Sc^{3\alpha-1} \frac{\partial^2 \theta}{\partial Y^2} = [\lambda \tilde{Y} + O(Sc^{-\alpha})] \frac{\partial \theta}{\partial x} - \left[B\left(\frac{\partial \theta}{\partial \tilde{Y}}\right)_{\tilde{Y}=0} + \frac{1}{2}\lambda' \tilde{Y}^2 + O(Sc^{-\alpha})\right] \frac{\partial \theta}{\partial \tilde{Y}}. \tag{11-121}$$

Thus, as before, $\alpha = 1/3$, and the leading-order inner problem reduces to

$$\frac{\partial^2 \theta}{\partial Y^2} = \lambda \tilde{Y} \frac{\partial \theta}{\partial x} - \left[B\left(\frac{\partial \theta}{\partial \tilde{Y}}\right)_{\tilde{Y}=0} + \frac{1}{2}\lambda' \tilde{Y}^2\right] \frac{\partial \theta}{\partial \tilde{Y}}. \tag{11-122}$$

with

$$\theta = 1 \quad \text{at} \quad \tilde{Y} = 0, \tag{11-123a}$$

$$\theta \to 0 \quad \text{as} \quad \tilde{Y} \to \infty \quad \text{(matching)}, \tag{11-123b}$$

$$\theta = 0 \quad \text{at} \quad x = 0. \tag{11-123c}$$

G. Laminar Boundary-Layer Mass Transfer for Finite Interfacial Velocities

We have already solved this equation for the analgous heat transfer problem in which $B = 0$ by similarity transformation. Perhaps surprisingly, the exact same similarity transformation works here. Hence, to solve (11–122) for arbitrary $\lambda(x)$ and B, we introduce the similarity transformation

$$\theta = \theta(\eta), \quad \text{where } \eta = \frac{\tilde{Y}}{g(x)}, \tag{11–124}$$

with

$$g(x) = \frac{9^{1/3}}{\sqrt{\lambda(x)}} \left[\int_0^x \sqrt{\lambda(t)} dt \right]^{1/3}. \tag{11–125}$$

The resulting DE for $\theta(\eta)$ is

$$\frac{\partial^2 \theta}{\partial \eta^2} + \left(3\eta^2 + B \left. \frac{\partial \theta}{\partial \eta} \right|_{\eta=0} \right) \frac{\partial \theta}{\partial \eta} = 0, \tag{11–126}$$

and the boundary, matching, and initial conditions become

$$\theta(0) = 1,$$
$$\theta(\infty) = 0. \tag{11–127}$$

The derivation of a formal solution of (11–126) and (11–127) is straightforward. Because $\theta'(0)$ is just a number, which we can denote as

$$b \equiv -\theta'(0), \tag{11–128}$$

we can simply integrate (11–126) twice with respect to η and apply the boundary conditions (11–127). The result is

$$\theta = 1 - \frac{\int_0^\eta \exp(Bbs - s^3) ds}{\int_0^\infty \exp(Bbs - s^3) ds}. \tag{11–129}$$

Of course, the solution is not complete because the constant b remains as an unknown. This is critical, because b corresponds to the mass flux at the body surface. To evaluate b, we differentiate (11–129) and evaluate the derivative at $\eta = 0$. The result is

$$\frac{1}{b} = \int_0^\infty \exp(Bbs - s^3) ds. \tag{11–130}$$

This solution of this equation has several asymptotic forms:

$$b = 6^{1/3}(1+B)^{-1/3} \quad \text{as} \quad B \to -1,$$
$$b = (\Gamma(4/3))^{-1} = 1.2 \quad \text{as} \quad B \to 0, \tag{11–131}$$
$$b = 0.7245(Bb)^{1/4} \exp(-0.3849(Bb)^{3/2}) \quad \text{as} \quad B \to \infty.$$

If we put the developments of this section together, it is clear that the correlation between the Sherwood number for $Re \gg 1$, $Sc \gg 1$ is

$$Sh \sim Re^{1/2} Sc^{1/3} Bb.$$

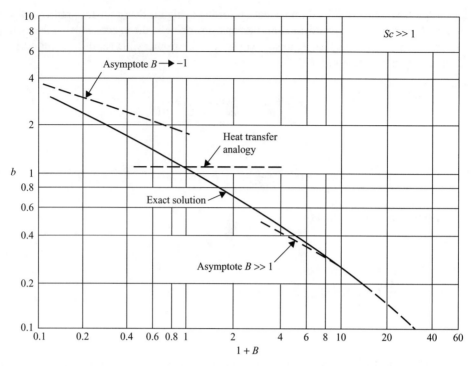

Figure 11–9. Forced convection for large Sc; b is defined by Eq. 11–128 (source: Ref. 4).

A plot of b versus $(1 + B)$ is shown in Fig. 11–9. The result at $B = 0$ is the result calculated in Section C. For $B < 0$, we see that b (and thus the rate of mass transfer) is increased. The asymptotic behavior is already achieved at $(1 + B) \approx 0.2$. On the other hand, for $B > 0$, the rate of mass transfer is decreased, and the asymptotic behavior is achieved by $(1 + B) \approx 4$. It can be seen from Fig. 11–9 that the rate of mass transfer is quite sensitive to B. In particular, the rate of mass transfer from a body can be much different from the rate of heat transfer, even when the Reynolds number and the Schmidt number for the mass transfer problem are identical to the values of the Reynolds and Prandtl numbers for the heat transfer case. This is a consequence of the fact that the transport process right at the body surface, in the absence of any mass flux across the surface (as in the heat transfer case), is dominated by a diffusion process. Hence even a rather tiny convective transport contribution can be extremely important.

Although the effects of blowing/suction that are due to mass transfer at the body surface thus play an important role in determining the rate of mass transfer, it is worthwhile to emphasize once again that the transport rate in this limit $Sc \to \infty$ was still assumed to be small enough that the flow is not modified at the leading order of approximation. We have seen that the normal velocity relevant to the mass transfer boundary layer is

$$V \approx -\frac{B}{Sc}\left(\frac{\partial \theta}{\partial Y}\right)_{Y=0} \approx -\frac{B}{Sc^{2/3}}\left(\frac{\partial \theta}{\partial \tilde{Y}}\right)_{\tilde{Y}=0} = BbSc^{-2/3}.$$

Because $V = Re^{1/2}v = O(1)$ in the usual boundary-layer analysis (see Chap. 10), we can neglect the effect of the "blowing" velocities on the velocity profiles provided that

$$BbSc^{-2/3} \ll 1.$$

One additional comment is that this analysis is easily generalizable to the situation in which the boundary is a slip surface (e.g., an interface of the same shape). The Taylor series

Problems

approximations for the velocity components, (11–119) and (11–120), must be changed to reflect the modified boundary conditions following the similar analysis from Chap. 9.

NOTES

1. The analysis in this chapter largely follows the original developments of J. D. Goddard and A. Acrivos, Asymptotic expansions for laminar forced-convection heat and mass transfer, Part 2, Boundary-layer flows, *J. Fluid Mech.* **24**, 339–66 (1966).
2. R. H. Perry and D. W. Green, *Perry's Chemical Engineers Handbook* (7th ed.) (McGraw-Hill, New York, 1997).
3. W. Y. Lovitt, *Linear Integral Equations* (Dover, New York, 1950; originally published in 1924 by McGraw-Hill).
4. A. Acrivos, The asymptotic form of the laminar boundary-layer mass-transfer rate for large interfacial velocities, *J. Fluid Mech.*, **12**, 337–57 (1962).

PROBLEMS

Problem 11–1. Similarity Solutions. If $u_e = x^m$, find the most general functional form for the surface temperature, $\theta_s = \theta_s(x)$, that allows a similarity solution of the thermal boundary-layer equation in two dimensions for large $Re \gg 1$ and $Pr = 0(1)$.

Problem 11–2. Specified Heat Flux. We have considered the development of thermal boundary-layer theory for a 2D body with a constant surface temperature θ_0. We have also discussed a method to determine the *surface* temperature when the heat flux is specified. In this problem, we wish to solve for the leading-order approximation to the temperature distribution in the fluid for an arbitrary 2D body when the heat flux is specified as a *constant*.

(a) Solve this problem *directly* using matched asymptotic expansions for $Pr \gg 1$ and $Pr \ll 1$.
(b) Suppose the heat flux is $Q = Q(x)$. Is there any combination of $u_e(x)$ and $Q(x)$ for which similarity solutions can be obtained? Prove your answer.

Problem 11–3. High-Pr Thermal Boundary Layer for a 2D Body with a Slip Surface. We have considered the thermal boundary-layer problem in this chapter for an arbitrary 2D body with no-slip boundary conditions for $Re \gg 1$ and Pr (or Sc) either arbitrarily large or small. If we assume that we have a body of the exact same shape, but the surface of which is a "slip surface" (e.g., it is an interface, so that the surface tangential velocity is not zero), the form of the correlation between Nusselt number and Pr will change for $Pr \gg 1$. Solve this problem, i.e., derive the governing boundary-layer equation, and show that it has a similarity solution. What is the resulting form of the heat transfer correlation among Nu, Re, and Pr?

Problem 11–4. Consider Problem 8 from Chap. 10. If you have not solved this problem, you will need to do so before tackling the following additional questions.

(a) Assume that the plate surface temperature is T_0, which is larger than the ambient temperature of the fluid, T_∞. Assume that $U/(k/\rho C_p) \gg 1$. Use the method of matched asymptotic expansions to obtain the governing DEs and boundary conditions in the limit $Pr \gg 1$. Solve the resulting equation for the temperature distribution and determine the relationship between Nu and Pr.

(b) Consider the same problem, but for $Pr \ll 1$. If necessary, you may assume that the *tangential* velocity u in the outer potential-flow region is $O\left(Re^{-1/2}\right)$ for $Re \gg 1$. The normal velocity at the outer edge of the boundary layer can be obtained from the solution of Problem 8 from Chap. 10.

Problem 11–5. Mass Transfer from a Spherical Bubble at High Re. We consider a single bubble that we approximate as being spherical in shape that is rising because of buoyancy through a quiescent fluid. We assume that the bubble is large enough, or the fluid is inviscid enough that the Reynolds number, based on the rise velocity U, is large. The bubble contains a solute A that is being transported through its surface into the bulk fluid. The concentration of A far from the bubble is zero. The concentration of A inside the bubble is assumed to be independent of time (although the bubble is losing A, the total amount of A inside the bubble is large compared with the amount that is lost because of mass transfer in the time available for mass transfer). We also assume that the bubble radius is fixed independent of time and that any interfacial velocity effects are negligible. Consider the asymptotic problems of mass transfer at both large and small values of the Schmidt number. Derive the concentration boundary-layer equation for these two cases. To do this, you will need to determine a uniformly valid first approximation to the velocity field (see Section G, Chap. 10). You may be able to solve the resultant boundary-layer equations, but all that you are required to do here is to derive them, and hence determine the form of the correlation between Nu and Sc. Comment on the differences between the two limiting cases.

Problem 11–6. Mass and Heat Transfer for Chemical Vapor Deposition. Consider the following model for chemical vapor deposition (CVD) on a surface. A reactive species is transported toward the surface by a 2D flow near a stagnation point, as illustrated in the figure. Far away from the surface the flow is given by

$$u(x, y) = \alpha x,$$

$$v(x, y) = -\alpha y.$$

(You may have solved for the flow field in Problem 16, Chap. 10. If not, you will need to solve 10–16 before attempting this problem). Also, far from the surface the concentration of the reactive species is C_∞ and the temperature of the fluid is T_∞. At the surface, the species is consumed by a first-order surface reaction at the rate $r = kC(x, 0)$.

(a) What are the dimensionless groups that completely characterize the consumption of the species on the surface, assuming isothermal operation?
(b) The uniformity of deposition is of course determined by the rate of reaction on the surface, and therefore we need to compute the variation in concentration on the surface. What is $C(x,0)$ assuming $Sc \gg 1$ in the limit of fast and slow reactions? (You will need to quantify the meaning of fast and slow reactions.)
(c) Describe the experimental measurements and the subsequent data analysis of the preceding CVD system that might be used to determine the reaction rate constant k.
(d) The reaction is exothermic and temperature variations on the surface might significantly affect the value of the rate constant on the surface. Neglecting the temperature dependence of the rate constant for the moment and assuming the heat of reaction is ΔH, compute the temperature field on the surface for fast and slow reactions and $Pr \gg 1$. For what conditions are the temperature variations negligibly small?
(e) Suppose the temperature did significantly affect the rate constant, but not the physical properties of the fluid. Describe clearly how you would determine $C(x,0)$ and $T(x,0)$. (Do not actually carry out the computations.)

Problems

Problem 11–7. Mass Transfer with Finite Interfacial Velocities. In Section G, we considered the problem of mass transfer at large Reynolds and Schmidt numbers from an arbitrary 2D body with a no-slip boundary condition imposed at the particle surface. We noted that the form of the solution would be different if the tangential velocity at the body surface were nonzero, i.e., $u_s(x) \neq 0$. Determine the form of the mass transfer boundary-layer equation for this case, and solve it by using a similarity transformation. What conditions, if any, are required of $u_s(x)$ for a similarity solution to exist?

12

Hydrodynamic Stability

The subject of *hydrodynamic stability* theory is concerned with the response of a fluid system to random disturbances. The word "hydrodynamic" is used in two ways here. First, we may be concerned with a *stationary* system in which flow is the result of an instability. An example is a stationary layer of fluid that is heated from below. When the rate of heating reaches a critical point, there is a spontaneous transition in which the layer begins to undergo a steady convection motion. The role of hydrodynamic stability theory for this type of problem is to predict the conditions when this transition occurs. The second class of problems is concerned with the possible transition of one flow to a second, more complicated flow, caused by perturbations to the initial flow field. In the case of pressure-driven flow between two plane boundaries (Chap. 3), experimental observation shows that there is a critical flow rate beyond which the steady laminar flow that we studied in Chap. 3 undergoes a transition that ultimately leads to a turbulent velocity field. Hydrodynamic stability theory is then concerned with determining the critical conditions for this transition.

For both types of problem, we can view the mathematical problem as one of determining the consequence of adding an initial perturbation in the velocity, pressure, temperature, or solute concentration fields to a basic unperturbed state. If the perturbation grows in time, the original unperturbed state is said to be unstable. If, on the other hand, the perturbation decays in time, the system will eventually revert to its original unperturbed state and it is said to be stable. This may occur for some range of the dimensionless parameters of the problem, but not for others. The value of the parameter separating stable and unstable regimes is called the "critical" value. The question of stability or instability may depend on the magnitude and type of the perturbation that is introduced. The most important transitions that occur are usually those that result from *infinitesimally small perturbations*. A system that is unstable to an arbitrarily small perturbation of any type will undergo a spontaneous and unavoidable instability, because it is impossible to shield a real physical system from arbitrarily small perturbations. The analysis of the response to infinitesimal perturbations is known as *linear* stability theory. A stationary state or flow that is *linearly unstable* is not physically realizable in the laboratory as it is impossible to avoid arbitrarily small disturbances.

The designation "linear" to describe the theoretical study of the fate of small initial disturbances is due to the fact that the dynamics of such small disturbances can be described by means of a linear approximation of the Navier–Stokes, continuity, and other transport equations. Because the governing equations are linear, analytic theory is often possible but this requires that the unperturbed state or flow, whose stability we wish to study, be known analytically. Furthermore, this base solution must be quite simple for even the linear approximation of the equations to be analytically tractable. In practice this reduces significantly the number of problems in which complete analytic results are possible and also explains why hydrodynamic stability theory has been particularly successful in analyzing problems

A. Capillary Instability of a Liquid Thread

in which the unperturbed state is one of zero motion. Although the same basic theoretical framework of linear stability theory can be applied to other more complex problems, the resulting problem must often be solved numerically.

An important limitation of the so-called linear stability theory is that the approximate linearized versions of the Navier–Stokes and transport equations are based on the assumption that the perturbations to the velocity, pressure, or temperature fields are small enough that nonlinear terms, i.e., those whose magnitude is proportional to products such as $|\mathbf{u}^2|$ or $|\mathbf{u}\theta|$, can be neglected compared with linear terms, i.e., those with magnitude proportional to $|\mathbf{u}|, |\theta|, |p|$, etc. Because the initial perturbations are arbitrarily small, this condition is clearly satisfied for these perturbations. However, if the system is unstable, the amplitude of these initial disturbances will grow and eventually the linearized approximation of the full Navier–Stokes and transport equations will no longer remain valid. Hence, though linear stability theory can tell us if the system is stable or unstable to infinitesimal disturbances, it *cannot* predict the ultimate flow that results from this instability. As a consequence of this, and also because some systems may be stable to linear disturbances but unstable to disturbances of larger amplitude, a considerable amount of research has also focused on so-called *nonlinear* stability theory, i.e., the response of the system when the perturbations are of finite amplitude. The latter work is closely related to other branches of nonlinear mathematics, especially *bifurcation* theory.

In the present chapter, we focus our attention on problems that are amenable to linear stability analysis.[1] The reader who is interested in pursuing nonlinear stability theory and/or the related topics in bifurcation theory will find many texts on these topics. Among current books we may note several listed at the end of this chapter,[2] but this is a quickly evolving area and the number of texts is also growing rapidly. In spite of our limitation to linear stability theories, we will see that there are many examples of particular importance to chemical engineers. Our goal is not an exhaustive list of all such possible problems, or even an exhaustive study of every aspect of the problems that we do consider. Rather, we hope to give the reader an introduction to this topic that will allow further study as necessary, as well as alerting you to some important examples that are common in technological applications. We begin with several problems for isothermal systems. Then we turn to problems of thermal convection, for which we must first discuss the Boussinesq approximation of the full equations for buoyancy-driven flows.

A. CAPILLARY INSTABILITY OF A LIQUID THREAD

We begin with a classic problem that is related to many important technological applications. This concerns the stability of a stationary cylinder of fluid of constant radius a. We have previously discussed this problem from a qualitative physical point of view in Chap. 2, as an example of a so-called capillary flow, in which fluid motion is driven by spatial gradients of the curvature of an interface between two immiscible liquids or between a liquid and a gas. That discussion is not repeated here, but the reader is encouraged to review what was already discussed in Chap. 2 before proceeding with the remainder of this section. The problem is related to the primary mechanism of dispersion of one immiscible fluid into another, in which the mixing flow is generally designed to produce highly stretched drops that break up by means of a slightly more complicated version of the capillary instability that we consider here. There are, in fact, many important examples of capillary-driven breakup processes for liquid threads in engineering applications involving two immiscible fluids or a fluid and a gas.

We assume for present purposes that we have a cylinder of Newtonian liquid of density ρ and undeformed radius a that is infinitely long, and is surrounded by air. We denote the interfacial tension between the liquid and air as γ. In the present analysis, no account will be taken of gravitational effects in analyzing the stability of the cylindrical shape. The contribution of hydrostatic pressure variations that is due to the perturbation of the cylinder shape will be small relative to those caused by capillary effects provided that $ga^2\rho/\gamma \ll 1$. The contributions of variations in the dynamic pressure that are due to any relative translational motion between the cylinder and the air will be small compared again with variations that are due to capillary effects, provided that $\rho U^2 a/\gamma \ll 1$, where U is the magnitude of the relative velocity. Finally, we assume that the system is completely isothermal and that the interface is free of any surface-active contaminants.

The infinitely long cylinder with no motion of the interface or of the fluid within the cylinder is, of course, a possible equilibrium configuration, in the sense that it is a surface of constant curvature so that the stationary constant-radius fluid satisfies all of the conditions of the problem, including the Navier–Stokes and continuity equations (trivially), as well as all of the interface boundary conditions including especially the normal-stress balance, which simply requires that the pressure inside the cylinder exceed that outside by the factor γ/a. The question for linear stability theory is whether this stationary configuration is stable to infinitesimal perturbations of the velocity, the pressure, or the shape of the cylinder.

The governing equations for any flow, including a perturbation to the unperturbed state are the Navier–Stokes and continuity equations,

$$\rho\left(\frac{\partial \mathbf{u}'}{\partial t'} + \mathbf{u}' \cdot \nabla' \mathbf{u}'\right) = -\nabla' p' + \mu \nabla'^2 \mathbf{u}', \tag{12-1}$$

$$\nabla' \cdot \mathbf{u}' = \mathbf{0}. \tag{12-2}$$

Primes are used to remind the reader that these variables are all dimensional. In general, the interface including the perturbation in shape can be represented as the points (defined with respect to a cylindrical coordinate system), where the function

$$F \equiv r' - a[1 + f(z', \phi, t')] = 0.$$

In the following analysis, we consider only *axisymmetric perturbations*. It can be shown that the nonaxisymmetric perturbations are less unstable than axisymmetric perturbations (see Problems section). Hence, we describe the interface shape as the points where the function

$$F \equiv r' - a[1 + f(z', t')] = 0. \tag{12-3}$$

Clearly this occurs at

$$r' = a[1 + f(z', t')],$$

which is an axisymmetric perturbation of the original cylinder of radius a. The function f can be quite general, except, of course, that the perturbation of shape must conserve volume. The boundary conditions at the interface are the kinematic condition,

$$\mathbf{u}' \cdot \mathbf{n} = -\frac{1}{|\nabla' F|}\frac{\partial F}{\partial t'} \quad \text{at } r' = a(1+f), \tag{12-4}$$

the condition of zero shear stress (assuming the viscosity of the gas is negligible compared with that of the liquid),

$$\mathbf{t} \cdot \mathbf{n} \cdot \mathbf{T}' = 0 \quad \text{at } r' = a(1+f), \tag{12-5}$$

and the normal-stress balance

$$p' - p'_{air} - (\boldsymbol{\tau}' \cdot \mathbf{n}) \cdot \mathbf{n} = \gamma(\nabla' \cdot \mathbf{n}) \quad \text{at } r' = a(1+f). \tag{12-6}$$

A. Capillary Instability of a Liquid Thread

We will assume that all perturbations are infinitesimal in magnitude, including the perturbation of the interface shape. Hence,

$$f = \varepsilon \hat{f} \ll 1. \tag{12-7}$$

Hence, because

$$\mathbf{n} = \nabla' F / |\nabla' F|,$$

we can easily show that

$$\mathbf{n} = \mathbf{i}_r - \varepsilon a (\partial \hat{f} / \partial z') \mathbf{i}_z + O(\varepsilon^2), \tag{12-8a}$$

$$\mathbf{t} = \varepsilon a (\partial \hat{f} / \partial z') \mathbf{i}_r + \mathbf{i}_z + O(\varepsilon^2). \tag{12-8b}$$

We can now write the boundary conditions (12–4)–(12–6) in component form,

$$u'_r - \varepsilon a \frac{\partial \hat{f}}{\partial z'} u'_z = \varepsilon a \frac{\partial \hat{f}}{\partial t'} + O(\varepsilon^2) \quad \text{at } r' = a(1 + \varepsilon \hat{f}), \tag{12-9}$$

$$\tau'_{rz} + \varepsilon a \frac{\partial \hat{f}}{\partial z'} (\tau'_{rr} - \tau'_{zz}) + O(\varepsilon^2) = 0 \quad \text{at } r' = a(1 + \varepsilon \hat{f}), \tag{12-10}$$

$$p' - p'_{air} - \tau'_{rr} + \varepsilon a \frac{\partial \hat{f}}{\partial z'} (\tau'_{rz} + \tau'_{zr}) + O(\varepsilon^2)$$

$$= \frac{\gamma}{a} \left[1 - \varepsilon \hat{f} - \varepsilon a^2 \frac{\partial^2 \hat{f}}{\partial z'^2} + O(\varepsilon^2) \right] \quad \text{at } r = a(1 + \varepsilon \hat{f}). \tag{12-11}$$

Here, we have assumed that the perturbation in the shape of the cylinder is small, but the equations and boundary conditions are otherwise still in a general form. In particular, we have made no assumptions about the magnitude of any of the disturbance quantities other than the magnitude of the change in cylinder shape.

The next step is to nondimensionalize the governing equations and boundary conditions. We recognize the cylinder radius as an appropriate choice for a characteristic length scale, but the choice for a characteristic velocity scale is not so obvious. Thus we nondimensionalize by using

$$\ell_c = a, \quad \text{and} \quad t_c = \ell_c / u_c$$

but leave the characteristic velocity unspecified for the moment, in hopes that an appropriate choice will become evident from the nondimensionalized equations. There are two possible choices for the characteristic pressure, either $p_c = \rho u_c^2$ or $p_c = \mu u_c / a$, depending on whether the appropriate Reynolds number, $Re \equiv u_c a / \nu$, is large or small. We begin by considering the stability of the liquid thread for $Re \gg 1$. Hence, starting with the normal-stress condition,

$$\rho u_c^2 (p - p_{air}) - \frac{\mu u_c}{a} \left[2 e_{rr} - \varepsilon \frac{\partial \hat{f}}{\partial z} (2 e_{rz} + 2 e_{zr}) + O(\varepsilon^2) \right]$$

$$= \frac{\gamma}{a} \left[1 - \varepsilon \hat{f} - \varepsilon \frac{\partial^2 \hat{f}}{\partial z^2} + O(\varepsilon^2) \right] \quad \text{at } r = 1 + \varepsilon \hat{f}, \tag{12-12}$$

where $e_{ij} \equiv (\partial u_i / \partial x_j + \partial u_j / \partial x_i)/2$ is the dimensionless rate-of-strain tensor.

Now, we have seen already in Chap. 2 that the mechanism for capillary-driven motion can be viewed as being a consequence of the pressure gradients that are associated with

Hydrodynamic Stability

spatial variations of the interface curvature. Thus the dominant balance in (12–12) is between the pressure difference on the left and the capillary term on the right. This tells us that

$$u_c \equiv (\gamma/\rho a)^{1/2}, \tag{12–13}$$

and the full nondimensionalized equations for $Re \equiv a(\gamma/\rho a)^{1/2}/\nu \gg 1$ are

$$\frac{\partial \mathbf{u}}{\partial t} + \mathbf{u}\cdot\nabla\mathbf{u} = -\nabla p + \frac{1}{Re}\nabla^2 \mathbf{u}, \tag{12–14}$$

$$\nabla\cdot\mathbf{u} = 0, \tag{12–15}$$

with

$$u_r - \varepsilon\frac{\partial \hat{f}}{\partial z}u_z = \varepsilon\frac{\partial \hat{f}}{\partial t} + O(\varepsilon^2) \quad \text{at } r = 1+\varepsilon\hat{f}, \tag{12–16}$$

$$\tau_{rz} + \varepsilon\frac{\partial \hat{f}}{\partial z}(\tau_{rr}-\tau_{zz}) + O(\varepsilon^2) = 0 \quad \text{at } r = 1+\varepsilon\hat{f}, \tag{12–17}$$

$$(p - p_{air}) - \frac{1}{Re}\left[2e_{rr} - \varepsilon\frac{\partial \hat{f}}{\partial z}(2e_{rz} + 2e_{zr}) + O(\varepsilon^2)\right]$$

$$= \left[1 - \varepsilon\hat{f} - \varepsilon\frac{\partial^2 \hat{f}}{\partial z^2} + O(\varepsilon^2)\right] \quad \text{at } r = 1+\varepsilon\hat{f}. \tag{12–18}$$

1. The Inviscid Limit

As stated at the outset, our goal is to investigate the fate of an infinitesimal perturbation to the base state, which in this case is a stationary circular cylinder of constant radius in which the pressure inside differs from the air pressure by σ/a (i.e., the dimensionless pressure difference is one). Hence we consider an initial perturbation in all of the independent variables (including the shape function f, which we have already introduced as $\varepsilon\hat{f}$):

$$u_r = \varepsilon\hat{u}_r, \quad u_z = \varepsilon\hat{u}_z, \quad (p - p_{air}) = 1 + \varepsilon\hat{p}, \quad f = \varepsilon\hat{f}. \tag{12–19}$$

The evolution of this initial perturbation is governed by the equations and boundary conditions (12–14)–(12–18). However, because the magnitude of this perturbation is small, we can obtain approximate equations by substituting (12–19) into (12–14)–(12–18), and retaining only the terms that are $O(\varepsilon)$. As stated above, we also consider only the limiting case $Re \gg 1$, and thus we also neglect all terms of $O(Re^{-1})$. The final result for the governing equations is

$$\frac{\partial \hat{u}_r}{\partial t} = -\frac{\partial \hat{p}}{\partial r}, \tag{12–20}$$

$$\frac{\partial \hat{u}_z}{\partial t} = -\frac{\partial \hat{p}}{\partial z}, \tag{12–21}$$

$$\frac{1}{r}\frac{\partial}{\partial r}(r\hat{u}_r) + \frac{\partial \hat{u}_z}{\partial z} = 0. \tag{12–22}$$

The limit in this case can be seen to reduce the equations to the linearized stability equations for an inviscid fluid. As a consequence, not all of the interface boundary conditions can be satisfied. Our experience from Chap. 10 shows that we should not expect the solution to satisfy the zero-shear-stress condition, which will come into play only if we were to

A. Capillary Instability of a Liquid Thread

consider a boundary layer at the interface at the next level of approximation in Re^{-1}. Hence we require that our solution satisfy the kinematic and normal-stress conditions only, namely (12–16) and (12–18). We can use the method of domain perturbations to convert these conditions from $r = 1 + \varepsilon \hat{f}$ to $r = 1$. Because

$$\hat{u}_r = \hat{u}_r|_{r=1} + \left(\frac{\partial \hat{u}_r}{\partial r}\right)_{r=1} \varepsilon \hat{f} + O(\varepsilon^2), \quad \text{and} \quad \hat{u}_z = \hat{u}_z|_{r=1} + \left(\frac{\partial \hat{u}_z}{\partial r}\right)_{r=1} \varepsilon \hat{f} + O(\varepsilon^2),$$

we can see by substituting into (12–16) and (12–18) that the $O(\epsilon)$ approximation to these conditions, consistent with the linear analysis of the $O(\epsilon)$ perturbations, is

with

$$\hat{u}_r = \frac{\partial \hat{f}}{\partial t} \quad \text{at } r = 1, \tag{12-23}$$

$$\hat{p} = -\hat{f} - \frac{\partial^2 \hat{f}}{\partial z^2} \quad \text{at } r = 1. \tag{12-24}$$

The equations (12–20)–(12–24) are the so-called linear stability equations for this problem in the inviscid fluid limit. We wish to use these equations to investigate whether an arbitrary, infinitesimal perturbation will grow or decay in time. Although the perturbation has an arbitrary form, we expect that it must satisfy the linear stability equations. Thus, once we specify an initial form for one of the variables like the pressure $\hat{p}$, we assume that the other variables take a form that is consistent with $\hat{p}$ by means of Eqs. (12–20)–(12–24). Now the obvious question is this: How do we represent a disturbance function of *arbitrary* form? For this, we take advantage of the fact that the governing equations and boundary conditions are now linear, so that we can represent any smooth disturbance function by means of a Fourier series representation. Instead of literally studying a disturbance function of arbitrary form, we study the dynamics of all of the possible Fourier modes. If *any mode* is found to grow with time, the system is unstable because, with a disturbance of infinitesimal amplitude, every possible mode will always be present.

Now the most convenient starting point is to assume that the cylinder suffers a infinitesimal perturbation in its shape of the form[3]

$$\hat{f} = Ce^{\sigma t} \sin kz. \tag{12-25}$$

Although we are interested in the response of the cylinder to a perturbation of arbitrary shape, such a perturbation can be constructed as a Fourier sine series, and it is sufficient to look at the stability of a single mode for all possible values of k. If the shape-perturbation mode grows for any k, the system is unstable. Now, if $\hat{f}$ has the form (12–25), we see from the normal-stress condition (12–24) that $\hat{p}$ must be equal to

$$\hat{p} = C(k^2 - 1)e^{\sigma t} \sin kz \quad \text{at } r = 1. \tag{12-26}$$

We can calculate the form for $\hat{p}$ throughout the flow domain by noting that, because

$$\frac{\partial \hat{\mathbf{u}}}{\partial t} = -\nabla \hat{p},$$

and because the fluid is incompressible, (12–22), the pressure must be a harmonic function

$$\nabla^2 \hat{p} = 0. \tag{12-27}$$

The most general solution of (12–27) that is bounded at $r = 0$ and periodic in z with wavelength $2\pi/k$ is

$$\hat{p} = (A_k(t) \sin kz + B_k(t) \cos kz) I_0(kr). \tag{12-28}$$

Hydrodynamic Stability

Here, I_0 is the modified Bessel function of the first kind.[4] Because the boundary value at $r = 1$ is given by (12–26), we see that

$$B_k = 0,$$

$$A_k(t) = C\left[\frac{k^2 - 1}{I_0(k)}e^{\sigma t}\right], \qquad (12\text{–}29)$$

so that

$$\hat{p} = C(k^2 - 1)e^{\sigma t}\sin kz\,\frac{I_0(kr)}{I_0(k)}. \qquad (12\text{–}30)$$

Then, according to (12–20) and (12–21),

$$\hat{u}_z = -C(k^2 - 1)\frac{e^{\sigma t}}{\sigma}k\,\cos kz\,\frac{I_0(kr)}{I_0(k)}, \qquad (12\text{–}31)$$

$$\hat{u}_r = -k(k^2 - 1)C\frac{e^{\sigma t}}{\sigma}\sin kz\,\frac{I_1(kr)}{I_0(k)}. \qquad (12\text{–}32)$$

These forms for **u** satisfy the continuity condition (12–22), and thus the only remaining condition that must be satisfied is the kinematic condition (12–23). If we substitute for $\hat{u}_r$ and $\hat{f}$, this condition can be written in the form

$$C\frac{e^{\sigma t}}{\sigma}\sin kz\left(-k(k^2 - 1)\frac{I_1(k)}{I_0(k)} - \sigma^2\right) = 0. \qquad (12\text{–}33)$$

The condition for a nontrivial solution (i.e., a solution other than $C = 0$) is that the growth-rate coefficient must take the value

$$\sigma^2 = k(1 - k^2)\frac{I_1(k)}{I_0(k)}. \qquad (12\text{–}34)$$

So, for a given value of k, there is only a single value of σ^2 (denoted as the *eigenvalue*) for which the solution is not just $\hat{f}, \hat{p}, \hat{u}_r$, and $\hat{u}_z = 0$.

From (12–34), we can see that $\sigma^2 > 0$ for $k^2 < 1$ (i.e., σ is real with one positive and one negative root). Hence all "modes" with $k^2 < 1$ are unstable, whereas those with $k^2 > 1$ correspond to oscillatory modes (in time) and are stable. The actual values of the positive root for σ for various values of k are shown in Table 12–1. If we convert to dimensional variables $\sigma' = (u_c/\ell_c)\sigma$ and $k' = k/a$, we obtain

$$(\sigma')^2 = \frac{\gamma}{\rho a^3}(k'a)(1 - k'^2 a^2)\frac{I_1(k'a)}{I_0(k'a)}. \qquad (12\text{–}35)$$

We see that there is a fastest-growing linear mode near $k = 0.7$. If all initial perturbations with $k^2 < 1$ started with the same initial amplitude, the mode with wavelength $\lambda \approx 2\pi/0.7$ would have the largest amplitude, and we might expect to see it as the dominant wavelength

A. Capillary Instability of a Liquid Thread

Table 12–1. Growth rates for capillary instability of an inviscid fluid

k	σ
0	0
0.2	0.1382
0.4	0.2567
0.6	0.3321
0.7	0.3433
0.8	0.3269
0.9	0.2647
0.95	0.1992
1	0

in experiments. Indeed, to the extent that this mode remains the largest into the finite-amplitude regime, it may be expected to determine the size of drops that will be formed when the varicose waves represented by (12–25) lead to pinch-off and breakup of the liquid cylinder. However, we see that there is a whole range of wave numbers k that exhibit very similar growth rates, and thus any initial differences in amplitude will be important over a significant period of time. This suggests that we might be able to exert some control on the size of the drops at pinch-off by introducing a dominant initial amplitude perturbation in the wavelength range of comparable growth rates.

Capillary breakup of a thread of liquid is the source of the drops in one type of ink-jet printer. In this case, the thread emanates from a nozzle, and the capillary waves increase in amplitude along the thread from the nozzle to the point of break-up. If we think a little about the analysis that we have just performed of the growth of capillary waves on the surface of a stationary cylinder of fluid, it should be clear that it also applies to a cylinder in which the liquid is translating uniformly as is true here, with the exception of a short region at the exit of the nozzle where the velocity distribution within the cylinder adjusts from its form inside the nozzle to a uniform, plug flow. The preceding linear stability analysis suggests that we could exert considerable influence on the size of the drops by means of the ratio of the interfacial tension to the density, γ/ρ, and also by introducing a largest-amplitude disturbance at the nozzle. In practice this is done by introducing a time-dependent oscillation of the pressure in the upstream reservoir. It should be remembered, however, that we cannot expect the linear stability analysis to provide quantitative information about the dominant wavelength (and thus the drop size) at pinch-off. By the time a visible perturbation mode has appeared, the assumption of an infinitesimal perturbation will not be strictly valid any longer, and we should expect nonlinear effects to be important long before actual break-up.

A final important comment is that the analysis shows that disturbances with a wavelength longer that $2\pi a$ will grow, but perturbations that are shorter than this will not grow. The explanation for the existence of this critical wavelength for instability was given already in Chap. 2. The result, $2\pi a$, obtained originally by Rayleigh, was based on strictly geometric arguments and is known as the *Rayleigh criteria for stability*. For shorter wavelengths, the tendency for instability that is due to the variations in *radius* with z is dominated by the competing curvature variations in the axial direction. We may expect, in this problem, that perturbations with wavelengths greater than $2\pi a$ will be unstable for viscous as well as inviscid threads, but that the growth rates will be slower as the viscosity of the cylindrical thread is increased.

2. Viscous Effects on Capillary Instability

We now turn to the case of a cylinder in which viscous effects play a dominant role. In this case, the initial analysis is the same as that of the preceding section, up to the point of nondimensionalizing. Now the appropriate choice for a characteristic pressure is $p_c = \mu u_c/a$, and hence, instead of (12–12), we have

$$\frac{\mu u_c}{a}(p - p_{air}) - \frac{\mu u_c}{a}\left[2e_{rr} - \varepsilon\frac{\partial \hat{f}}{\partial z}(2e_{rz} + 2e_{zr}) + O(\varepsilon^2)\right]$$

$$= \frac{\gamma}{a}\left[1 - \varepsilon\hat{f} - \varepsilon\frac{\partial^2 \hat{f}}{\partial z^2} + O(\varepsilon^2)\right] \quad \text{at } r = 1 + \varepsilon\hat{f}. \tag{12–36}$$

It follows that the appropriate choice for u_c is

$$u_c = \frac{\gamma}{\mu}, \tag{12–37}$$

and the nondimensionalized and linearized $[O(\varepsilon)]$ stability equations and boundary conditions are

$$Re\frac{\partial \hat{\mathbf{u}}}{\partial t} = -\nabla\hat{p} + \nabla^2\hat{\mathbf{u}}, \tag{12–38}$$

$$\nabla \cdot \hat{\mathbf{u}} = 0, \tag{12–39}$$

with

$$\hat{u}_r = \frac{\partial \hat{f}}{\partial t} \quad \text{at } r = 1, \tag{12–40}$$

$$\hat{\tau}_{rz} = 0 \quad \text{at } r = 1, \tag{12–41}$$

$$\hat{p} - 2\frac{\partial \hat{u}_r}{\partial r} = -\hat{f} - \frac{\partial^2 \hat{f}}{\partial z^2} \quad \text{at } r = 1. \tag{12–42}$$

The Reynolds number now takes the form

$$Re \equiv \frac{u_c a}{\nu} = \frac{\rho\gamma a}{\mu^2}.$$

In this case, it is convenient to express $\hat{f}$ in the more general form

$$\hat{f} = C\exp(ikz + \sigma t). \tag{12–43}$$

Then, as before, the pressure perturbation is a harmonic function and thus, corresponding to (12–43), must be expressible in the form

$$\hat{p} = \Pi_0 \exp(ikz + \sigma t)I_0(kr), \tag{12–44}$$

where Π_0 is a constant, related to C, the value of which remains to be determined. Given (12–44),

$$\nabla\hat{p} = \Pi_0 k\exp(ikz + \sigma t)[I_1(kr)\mathbf{i_r} + iI_0(kr)\mathbf{i_z}], \tag{12–45}$$

A. Capillary Instability of a Liquid Thread

and it can be shown that the velocity field can be expressed in the form

$$\boxed{\hat{\mathbf{u}} = \left\{-ikrU(r)\mathbf{i_r} + \frac{1}{r}\frac{\partial}{\partial r}[r^2 U(r)]\mathbf{i_z}\right\} \exp(ikz + \sigma t),} \qquad (12\text{--}46)$$

where the function $U(r)$ remains to be determined.

The first thing to note is that the proposed solution (12–46) satisfies the continuity equation (12–39) for any choice of $U(r)$. Of course, $U(r)$ must also satisfy the equation of motion (12–38). To determine the appropriate form for $U(r)$, let us first consider the r component of this equation:

$$Re\frac{\partial \hat{u}_r}{\partial t} = -\frac{\partial \hat{p}}{\partial r} + \frac{\partial}{\partial r}\left(\frac{1}{r}\frac{\partial}{\partial r}(r\hat{u}_r)\right) + \frac{\partial^2 \hat{u}_r}{\partial z^2}. \qquad (12\text{--}47)$$

If we substitute for $\hat{u}_r$ and $\partial \hat{p}/\partial r$ from (12–45) and (12–46) and factor out the common exponential factor, we obtain

$$\exp(ikz + \sigma t)\left\{ReU\sigma + i\Pi_0 \frac{I_1(kr)}{r} - \frac{d^2 U}{dr^2} - \frac{3}{r}\frac{dU}{dr} + k^2 U\right\} = 0. \qquad (12\text{--}48)$$

In fact, we also obtain precisely the same result from the z component of (12–38). So we see that (12–46) will satisfy the linearized equation of motion provided we choose $U(r)$ to satisfy the nonhomogeneous ODE

$$\frac{d^2 U}{dr^2} + \frac{3}{r}\frac{dU}{dr} - (k^2 + Re\sigma)U = i\Pi_0 \frac{I_1(kr)}{r}.$$

It can be shown that a solution of this equation that is bounded at $r = 0$ is

$$U = i\left\{A\frac{I_1[(k^2 + Re\sigma)^{1/2}r]}{r} - \frac{\Pi_0}{\sigma Re}\frac{I_1(kr)}{r}\right\}, \qquad (12\text{--}49)$$

where A is a constant of integration. It remains now to satisfy the boundary conditions (12–40)–(12–42).

The condition (12–40) requires that

$$\sigma = k\left[AI_1(\beta) - \frac{\Pi_0}{\sigma Re}I_1(k)\right], \qquad (12\text{--}50)$$

where for convenience we use $\beta^2 \equiv k^2 + \sigma Re$. The tangential-stress condition (12–41) requires that

$$\left(\frac{d^2 U}{dr^2} + \frac{3}{r}\frac{dU}{dr} + k^2 U\right)\exp(ikz + \sigma t)|_{r=1} = 0. \qquad (12\text{--}51)$$

Hence, substituting for $U(r)$, we obtain an algebraic relationship between the two constants A and Π_0,

$$A = \frac{\Pi_0}{\sigma Re}\frac{2k^2}{k^2 + \beta^2}\frac{I_1(k)}{I_1(\beta)}. \qquad (12\text{--}52)$$

Hydrodynamic Stability

With this result for A, Eq. (12–50) for the growth-rate coefficient σ becomes

$$\boxed{\sigma^2 = \frac{\Pi_0}{Re}\left(\frac{k^2 - \beta^2}{k^2 + \beta^2}\right)[kI_1(k)].} \qquad (12\text{–}53)$$

Finally, to obtain a prediction for stability, we must apply the normal-stress condition (12–42) to determine Π_0. After some algebra, (12–42) leads to the condition

$$\boxed{\Pi_0 I_0(k)\left\{1 + \frac{2k^2}{\sigma Re}\frac{I_1'(k)}{I_0(k)}\left[1 - \frac{2k\beta}{k^2+\beta^2}\frac{I_1(k)}{I_1(\beta)}\frac{I_1'(\beta)}{I_1'(k)}\right]\right\} = k^2 - 1,} \qquad (12\text{–}54)$$

where I_1' stands for the derivative of I_1. An expression for σ solely in terms of k and Re can be obtained by eliminating Π_0 between Eqs. (12–53) and (12–54). The result is

$$\sigma^2 Re\left(\frac{k^2+\beta^2}{k^2-\beta^2}\right)\frac{I_0(k)}{kI_1(k)}\left\{1 + \frac{2k^2}{\sigma Re}\frac{I_1'(k)}{I_0(k)}\left[1 - \frac{2k\beta}{k^2+\beta^2}\frac{I_1(k)}{I_1(\beta)}\frac{I_1'(\beta)}{I_1'(k)}\right]\right\} = k^2 - 1.$$

This equation can be used to evaluate $\sigma = \sigma(k, Re)$ and thus ascertain the stability as a function of k and Re.

The preceding calculation makes no assumption about the value of the Reynolds number, Re, except that it is finite. A much simpler limiting result can be obtained[5] in the limit $Re \ll 1$. To obtain this result, we first note that $\sigma^2/(k^2 - \beta^2) = (k^2 - \beta^2)/Re^2$. Then the preceding equation can be rewritten as

$$(k^4 - \beta^4) - 2k^2(k^2+\beta^2)\frac{I_1'(k)}{I_0(k)}\left[1 - \frac{2k\beta}{k^2+\beta^2}\frac{I_1(k)}{I_1(\beta)}\frac{I_1'(\beta)}{I_1'(k)}\right] = Re(k^2-1)\frac{kI_1(k)}{I_0(k)}. \qquad (12\text{–}55)$$

Now, the limiting form for this equation for $Re \ll 1$ has been derived in the book on hydrodynamic stability by Chandrasekhar,[5] and we follow his analysis. First, we note that, for $Re \ll 1$, we can approximate β as

$$\beta \sim k + \alpha \quad \text{where } \alpha \equiv \sigma Re/2k \ll 1.$$

Using this result, we expand all of the terms on the left-hand side of (12–55) to first order in α. The result is

$$4k^3\alpha - 4k^4\alpha \frac{I_1'(k)}{I_0(k)}\left[\frac{I_1''(k)}{I_1'(k)} - \frac{I_1'(k)}{I_1(k)}\right] + O(\alpha^2).$$

After some further manipulation, using known properties of the modified Bessel functions and putting in the definition of α, we can write this as

$$2k^3\sigma\left\{\frac{I_0(k)}{I_1(k)} - \left(1 + \frac{1}{k^2}\right)\frac{I_1(k)}{I_0(k)}\right\} Re + O(Re^2).$$

A. Capillary Instability of a Liquid Thread

Table 12–2. Growth rates versus wave number for capillary instability of a highly viscous thread

k	σ	$(-k^2)/6$
0	0.1667	0.1667
0.1	0.1650	0.1650
0.5	0.1249	0.1250
0.8	0.0598	0.0600
1.0	0	0

Hence, substituting this into the left-hand side of (12–55) and solving for the growth-rate coefficient σ, we obtain

$$\sigma = \frac{1-k^2}{2\left\{k^2\dfrac{I_0(k)^2}{I_1(k)^2} - (1+k^2)\right\}} + O(Re). \tag{12-56}$$

In dimensional terms, this is

$$\sigma' = \left(\frac{\gamma}{\mu a}\right)\left(\frac{1-k'^2 a^2}{2\{k'^2 a^2[I_0(k'a)/I_1(k'a)]^2 - (1+k'^2 a^2)\}} + O(Re)\right),$$

which may be compared with the result (12–34) for an inviscid fluid. The sign of σ depends on k. Not surprisingly, we find that σ is positive (i.e., the system is unstable) only for $k < 1$. The addition of even a strong viscous effect does not change the range of wavelengths that are unstable, only the rate of growth of the unstable modes and their dependence on k, as we can see by comparing the preceding result with (12–34).

One peculiar feature of the result (12–56) is that the denominator is approximately equal to 6 over the whole range of unstable wavelengths, $0 \le k < 1$, varying monotonically from 6.000 at $k = 0$ to 6.037 for $k = 1$. Thus a reasonable (though *ad hoc*) approximation to (12–56) is

$$\sigma \simeq \frac{1}{6}(1-k^2).$$

This shows that the fastest-growing linear mode (i.e., the maximum value of σ) now occurs at $k = 0$. Indeed, exact values of σ as functions of k are given in Table 12–2. Clearly for this case of dominant viscous effects, there is no finite wavelength mode of maximum instability (i.e., no intermediate value of k with a maximum in the growth rate). The maximum growth rate occurs for waves of infinite wavelength. Thus the capillary thread will not break into drops at distances comparable with the diameter of the cylinder, as was the case for an inviscid cylinder, but rather will have a tendency to break at distant places. This behavior is similar to what is seen for very viscous polymeric threads, though it should be cautioned that there are additional reasons why this might occur if the cylinder were viscoelastic in addition to being very viscous.

3. Final Remarks

The instability of an isolated cylindrical thread of liquid previously analyzed is but one of a large number of related instabilities. The most direct extension is to consider the thread to be immersed in a second immiscible liquid, which was analyzed many years ago by Tomotika.[6] Not surprisingly, the same range of wavelengths is unstable, $0 \le k' < 2\pi a$, but the growth rates and the fastest-growing linear mode both depend on the viscosity ratio. More directly related to the technological application of wire or fiber coating is the stability of an annular layer of liquid on the outer surface of a solid cylinder. This was first analyzed by means of linear stability theory by Goren.[7] Subsequent investigators[8] have extended this analysis to consider the effect of flow in the coating layer (say due to gravitational drainage), with the remarkable result that the film may actually be stabilized against breakup into drops by a (nonlinear) coupling between the growth of the instability and the flow. Other extensions[9] include the effect of surfactant at the liquid interface (which slows the growth process by means of Marangoni effects associated with the concentration gradients that result from "wave" growth at the interface) and the effects of non-Newtonian rheology. Another whole class of problems is the stability of a viscous fluid layer on the inside surface of a cylindrical tube.[10] The same basic instability occurs, and may be relevant to a number of technological problems, such as the film associated with the motion of an elongated drop in a cylindrical channel or the stability of the mucous membrane in the lung.[11] The basic instability mechanism for all of these problems is the same, as already described qualitatively in Chap. 2. The changes in geometry, the addition of surfactant, or the presence of either another liquid or a solid wall cannot inhibit the linear instability, but only control the growth rates of the disturbance and its dependence on wavelength. Several of these problems are presented as homework examples at the end of this chapter.

B. RAYLEIGH–TAYLOR INSTABILITY (THE STABILITY OF A PAIR OF IMMISCIBLE FLUIDS THAT ARE SEPARATED BY A HORIZONTAL INTERFACE)

Another mechanism for possible instability of a fluid is the coupling of gravity with variations of the fluid density parallel to $\mathbf{g}$. Specifically, if we have a stationary fluid layer in which the density increases with height (i.e., with increase of z in a reference frame such that $\mathbf{g} = -g\mathbf{i}_z$), our common experience tells us that we may expect a spontaneous motion that tends to over-turn the density variation. The "mechanism" can be understood in (overly) simplistic terms. In particular, if a fluid element is displaced vertically up by some perturbation, it finds itself surrounded by a more dense fluid, and this provides an additional buoyancy force that pushes it further in the upward direction. Conversely, if the perturbation leads initially to a downward displacement, the fluid finds itself surrounded by less-dense fluid and there is a net downward body force that pushes it further in the downward direction. In both circumstances the system is seen to be unstable. A small displacement yields a force that tends to propagate the initial perturbation. On the other hand, if the density distribution were inverted, a small upward displacement would place the fluid element at a position where it is surrounded by less-dense fluid and the net buoyancy force on it is downward so that the fluid element experiences a "restoring" force that diminishes the perturbation, and the system is stable. We recognize, of course, that these arguments are like "cartoons" that give the essence of the correct idea but are much too simplified to extract any detailed information.

If we consider the simplest case of a pair of superposed fluids, each of constant but different density, we expect that when the more-dense fluid is above the less-dense fluid, there is the potential for instability. The basic question is whether this configuration is always unstable, or whether there is any possibility that the system might be stabilized by the presence of either a finite interfacial tension or finite viscosity in the bulk fluids.

The simplest fluid system that is capable of exhibiting this type of gravitational instability is a pair of "unbounded," incompressible, isothermal, immiscible fluids that are separated by a horizontal interface. We may denote the density and viscosity of the upper fluid as ρ_2, μ_2, while that of the lower fluid is denoted as ρ_1, μ_1, and the interfacial tension is γ.

B. Rayleigh–Taylor Instability

We are interested in the case in which $\rho_2 > \rho_1$, where the system is potentially unstable. Although we have established the principle of nondimensionalizing as a way of identifying dimensionless parameters, in the present case it is not clear what choice to make for any of the characteristic variables, and so we proceed in this case with the dimensional equations and boundary conditions.

We analyze the stability of this configuration[12] by using a Cartesian coordinate system with z being perpendicular to the interface and positive in the upward direction. Hence, $\mathbf{g} = -g\mathbf{i}_z$ and in each of the two fluids, there is a hydrostatic pressure distribution, i.e.,

$$p_1' = -\rho_1 g z' + \varepsilon \hat{p}_1'; \quad p_2' = -\rho_2 g z' + \varepsilon \hat{p}_2', \qquad (12\text{--}57\text{a})$$

to which we can add an arbitrary constant because the fluids are assumed to be incompressible. The primed symbol signifies that the variable is dimensional, and the "caret" represents a perturbation variable. It is convenient to use the symbol ε to represent the magnitude of the perturbation quantities. Because the fluid is assumed to be stationary in the base state, the velocity components consist of only the perturbation velocity,

$$u_i' = \varepsilon \hat{u}_i', \quad v_i' = \varepsilon \hat{v}_i', \quad \text{and} \quad w_i' = \varepsilon \hat{w}_i', \qquad (12\text{--}57\text{b})$$

where i can be either 1 or 2, representing the two-fluids. In this section, we assume that the viscosities and the densities are fixed. The only other quantity that must be perturbed if we are to study the instability of the two-fluid system is the interface shape. We assume that the interface is initially flat and horizontal. Hence we can represent its perturbed shape as being those points where a function F is zero,

$$F(x', y', z', t') \equiv z' - \varepsilon \hat{h}'(x', y', t'). \qquad (12\text{--}58)$$

This has the usual benefit that the unit normal vector to the interface is then

$$\mathbf{n} \equiv \pm \frac{\nabla F}{|\nabla F|},$$

where the $+$ or $-$ is a matter of the sign convention (care must be taken when using $\mathbf{n}$ in the normal-stress boundary condition at the interface).

The analysis now follows that in the previous section, at least qualitatively. First we derive a set of linearized equations of motion and boundary conditions by neglecting all terms that are $O(\varepsilon^2)$ or higher order. As far as the equations of motion are concerned, it is enough to derive them in one of the two fluids. We choose fluid 1. The linearized $[O(\varepsilon)]$ Navier–Stokes equations are thus

$$\rho_1 \frac{\partial \hat{u}_1'}{\partial t'} = -\frac{\partial \hat{p}_1'}{\partial x'} + \mu_1 \nabla'^2 \hat{u}_1', \qquad (12\text{--}59\text{a})$$

$$\rho_1 \frac{\partial \hat{v}_1'}{\partial t'} = -\frac{\partial \hat{p}_1'}{\partial y'} + \mu_1 \nabla'^2 \hat{v}_1', \qquad (12\text{--}59\text{b})$$

$$\rho_1 \frac{\partial \hat{w}_1'}{\partial t'} = -\frac{\partial \hat{p}_1'}{\partial z'} + \mu_1 \nabla'^2 \hat{w}_1'. \qquad (12\text{--}59\text{c})$$

From continuity,

$$\frac{\partial \hat{u}_1'}{\partial x'} + \frac{\partial \hat{v}_1'}{\partial y'} + \frac{\partial \hat{w}_1'}{\partial z'} = 0. \qquad (12\text{--}60)$$

Now, we want to consider the stability of an infinitesimal disturbance of random form, which satisfies these equations. We envision the two fluids as being of infinite extent in the

Hydrodynamic Stability

x' and y' directions, as well as extending to $+z'$ and $-z'$, respectively. There is no constraint on the form of the disturbance in x' and y'. However, the velocity and pressure disturbances must satisfy boundary conditions at the interface $z' = \varepsilon \hat{h}'(x', y', t')$. This suggests that we consider disturbances in the form

$$\hat{u}'_1 = u'_1(z') \exp(i\alpha'_x x' + i\alpha'_y y') \exp(\sigma' t'),$$
$$\hat{v}'_1 = v'_1(z') \exp(i\alpha'_x x' + i\alpha'_y y') \exp(\sigma' t'),$$
$$\hat{w}'_1 = w'_1(z') \exp(i\alpha'_x x' + i\alpha'_y y') \exp(\sigma' t'), \quad (12\text{–}61)$$
$$\hat{p}'_1 = p'_1(z') \exp(i\alpha'_x x' + i\alpha'_y y') \exp(\sigma' t').$$

An *arbitrary* disturbance form in the x' and y' directions could be expressed as a sum of the Fourier modes of wave number α'_x and α'_y, but because the governing equations are linear with coefficients that are independent of x', y', it is enough to consider the stability of these disturbance quantities one mode at a time, for arbitrary values of α'_x and α'_y. The functions of z' must be chosen to satisfy boundary conditions on the fluid interface. The stability is determined by the sign of the real part of σ'. The reader is reminded that the primes on all of the symbols mean that they are dimensional.

To obtain governing equations for $u'_1(z)$, $v'_1(z)$, etc., we substitute (12–61) into (12–59) and (12–60). It is convenient to use the shorthand symbol D' to stand for the derivative with respect to z'. After factoring out the common factor $\exp(i\alpha'_x x' + i\alpha'_y y') \exp(\sigma' t')$ that appears in every term, we find that the result is

$$\rho \sigma' u'_1 = -i\alpha'_x p'_1 + \mu_1 \left[D'^2 - (\alpha'^2_x + \alpha'^2_y) \right] u'_1, \quad (12\text{–}62\text{a})$$

$$\rho_1 \sigma' v'_1 = -i\alpha'_y p'_1 + \mu_1 \left[D'^2 - (\alpha'^2_x + \alpha'^2_y) \right] v'_1, \quad (12\text{–}62\text{b})$$

$$\rho_1 \sigma' w'_1 = -D' p'_1 + \mu_1 \left[D'^2 - (\alpha'^2_x + \alpha'^2_y) \right] w'_1, \quad (12\text{–}62\text{c})$$

$$i\alpha'_x u'_1 + i\alpha'_y v'_1 + D' w'_1 = 0. \quad (12\text{–}63)$$

Now, it is convenient to combine these four coupled equations to obtain a single higher-order equation for one of functions u'_1, v'_1, w'_1, and p'_1. To do this, we multiply (12–62a) by $i\alpha'_x$ and (12–62b) by $i\alpha'_y$, and then add them together:

$$\alpha'^2 p'_1 = \rho_1 \sigma'(i\alpha'_x u'_1 + i\alpha'_y v'_1) - \mu_1(D'^2 - \alpha'^2)(i\alpha'_x u'_1 + i\alpha'_y v'_1),$$

where we use $\alpha'^2 \equiv \alpha'^2_x + \alpha'^2_y$. However, then, in view of (12–63), this can be written as

$$\alpha'^2 p'_1 = -\rho_1 \sigma' D' w'_1 + \mu_1(D'^2 - \alpha'^2) D' w'_1. \quad (12\text{–}64)$$

Finally, we can combine (12–62c) and (12–64) to eliminate the pressure function $p'_1(z)$. The result is

$$\boxed{[-\rho_1 \sigma' + \mu_1(D'^2 - \alpha'^2)](D'^2 - \alpha'^2) w'_1 = 0.} \quad (12\text{–}65\text{a})$$

A similar equation can also be derived for fluid 2 (with disturbance functions defined just as in (12–61),

$$\boxed{[-\rho_2 \sigma' + \mu_2(D'^2 - \alpha'^2)](D'^2 - \alpha'^2) w'_2 = 0.} \quad (12\text{–}65\text{b})$$

It is not necessary to repeat all of the details of the derivation.

B. Rayleigh–Taylor Instability

Finally, to complete the statement of the linear stability problem, we require boundary conditions. Far from the interface, as $z' \to \pm\infty$, we expect the disturbance velocities to decay to zero:

$$\hat{u}'_1, \hat{v}'_1, \hat{w}'_1, \hat{p}'_1 \to 0 \quad \text{as } z' \to -\infty, \quad \text{and} \quad \hat{u}'_2, \hat{v}'_2, \hat{w}'_2, \hat{p}'_2 \to 0 \quad \text{as } z' \to \infty. \tag{12–66}$$

In addition there are the usual boundary conditions at the fluid interface. The first is continuity of the tangential velocity components

$$\varepsilon \hat{u}'_1 = \varepsilon \hat{u}'_2 \quad \text{and} \quad \varepsilon \hat{v}'_1 = \varepsilon \hat{v}'_2 \quad \text{at } \hat{z} = \varepsilon \hat{h}'(x', y', t') \tag{12–67}$$

Because the shape function $\hat{h}'$ is unknown, we transform the boundary conditions to the undisturbed interface position at $z' = 0$ by using the domain perturbation technique, which was introduced in previous chapters. Hence we can express $\varepsilon \hat{u}'_1$ at $z' = \varepsilon \hat{h}'$ in terms of its value at $z' = 0$ by using a Taylor series approximation:

$$\varepsilon \hat{u}'_1|_{z=h'} = \varepsilon \hat{u}'_1|_{z=0} + \varepsilon^2 (\partial \hat{u}'_1/\partial z')|_{z'=0}\,\hat{h}' + O(\varepsilon^3). \tag{12–68}$$

However, we see that the only term on the right-hand side of (12–68) that is linear in ε is the value of the function at $z' = 0$. Hence, for the linearized stability problem, we can simply express all of the boundary conditions directly at $z' = 0$. Hence, the boundary conditions (12–67) become

$$\boxed{u'_1(0) = u'_2(0) \quad \text{and} \quad v'_1(0) = v'_2(0).} \tag{12–69}$$

The kinematic boundary condition is

$$\hat{w}'_1 = \hat{w}'_2 = \frac{\partial \hat{h}'}{\partial t'} \quad \text{at } z' = \varepsilon \hat{h}'(x', y', t').$$

In view of the form of $\hat{w}'_1$ (and $\hat{w}'_2$) [see (12–61c)], we see that the disturbance shape function must also be expressible in the form

$$\hat{h}' = H \exp(i\alpha'_x x' + i\alpha'_y y') \exp(\sigma' t'), \tag{12–70}$$

where H is a constant to be determined. Hence the kinematic condition can now be expressed in the form

$$\boxed{w'_1(0) = w'_2(0) = \sigma' H.} \tag{12–71}$$

Last, there are the normal- and tangential-stress conditions. The general form of the normal-stress condition [keeping in mind (12–57)] is

$$\hat{p}'_2 - \hat{p}'_1 + (\rho_1 - \rho_2) g \hat{h}' + 2\left(\mu_1 D' \hat{w}'_1 - \mu_2 D' \hat{w}'_2\right) - \gamma \nabla_2'^2 \hat{h}' = 0 \quad \text{at } z' = \varepsilon \hat{h}'. \tag{12–72}$$

The sign convention for the capillary term in (12–72) comes from (12–6) and (12–58), plus the fact that we use the $+$ sign in the definition of $\mathbf{n}$ in terms of the ∇F. The linearized form [i.e., neglecting $O(\varepsilon^2)$ and higher-order terms in the domain perturbation approximation] of (12–72) is

$$\boxed{p'_2(0) - p'_1(0) + (\rho_1 - \rho_2) g H + 2(\mu_1 D' w'_1 - \mu_2 D' w'_2)|_{z=0} + \gamma \alpha'^2 H = 0.} \tag{12–73}$$

The tangential-stress condition for a clean, isothermal interface is

$$(\hat{\tau}'_1)_{zx} = (\hat{\tau}'_2)_{zx} \quad \text{and} \quad (\hat{\tau}'_1)_{zy} = (\hat{\tau}'_2)_{zy} \quad \text{at } z' = \varepsilon \hat{h}'$$

815

Hydrodynamic Stability

These conditions can be expressed in the linearized form

$$\mu_1(i\alpha'_x w'_1 + D'u'_1)|_{z'=0} = \mu_2(i\alpha'_x w'_2 + D'u'_2)|_{z'=0},$$
$$\mu_1(i\alpha'_y w'_1 + D'v'_1)|_{z'=0} = \mu_2(i\alpha'_y w'_2 + Dv'_2)|_{z'=0}. \quad (12\text{--}74)$$

The stability problem is thus to solve (12–65a) and (12–65b), subject to the boundary conditions (12–69), (12–71), (12–73), and (12–74), plus the condition that the disturbance quantities vanish at infinity. This is a classic *eigenvalue* problem. One possible solution is the null solution – all disturbance quantities equal to zero. For given values of the dimensional parameters, there is a single value of the growth-rate coefficient for which the solution is nontrivial. This is the *eigenvalue*, and the corresponding solution is the *eigensolution*. The stability depends on the sign of the real part of the eigenvalue.

1. The Inviscid Fluid Limit

We shall see that the fact that the fluids are viscous does not play a critical role in determining whether a pair of fluids with different density is unstable or not. We begin in this section by solving the linear stability problem in the limit where both fluids are assumed to be inviscid. Then we will return in Subsection 2 to consider how the problem is changed when the fluid viscosity is not neglected.

In the inviscid limit, the general linear stability problem takes the following simpler form. First, the governing equations, (12–65a) and (12–65b), are reduced to a pair of second-order DEs:

$$(D'^2 - \alpha'^2)w'_1 = 0,$$
$$(D'^2 - \alpha'^2)w'_2 = 0. \quad (12\text{--}75)$$

Because the fluids are approximated as inviscid, neither the "no-slip" conditions (12–69) nor the continuity of shear-stress conditions (12–74) can be imposed. Hence the solutions of (12–75) satisfy the kinematic condition in the form (12–71) and the normal-stress condition (12–73), with the viscous-stress contribution neglected:

$$p'_2(0) - p'_1(0) + (\rho_1 - \rho_2)gH + \gamma \alpha^2 H = 0. \quad (12\text{--}76)$$

The problem (12–75), with homogeneous boundary conditions (12–71) and (12–76), is a classic *eigenvalue* problem.

Now the solutions of (12–75) are a pair of exponentials in each fluid, one growing and one decaying with distance from the interface. The exponentially growing solutions are inconsistent with the boundary conditions (12–66) as $z' \to \pm\infty$. Hence the most general acceptable solutions of (12–75) are

$$w'_1 = C'_1 \exp(\alpha' z'), \quad (z' < 0),$$
$$w'_2 = C'_2 \exp(-\alpha' z'), \quad (z' > 0). \quad (12\text{--}77)$$

Now, according to the kinematic boundary condition,

$$C'_1 = C'_2, \quad (12\text{--}78a)$$

$$C'_1 = \sigma' H. \quad (12\text{--}78b)$$

B. Rayleigh–Taylor Instability

To apply the normal-stress condition, we must calculate the pressure. According to (12–64),

$$p'_1 = -\frac{\rho_1 \sigma'}{\alpha'^2} D' w'_1 \quad \text{and} \quad p'_2 = -\frac{\rho_2 \sigma'}{\alpha'^2} D' w'_2. \tag{12-79}$$

Substituting (12–77)–(12–79) into the normal-stress condition, we find

$$\frac{\sigma'}{\alpha'^2}(\rho_2 + \rho_1)\alpha' \sigma' H + [(\rho_1 - \rho_2)g + \gamma \alpha'^2]H = 0.$$

However, this equation can be solved for the growth-rate coefficient

$$\boxed{\sigma'^2 = \frac{((\rho_2 - \rho_1)g - \gamma \alpha'^2)\alpha'}{\rho_2 + \rho_1}.} \tag{12-80}$$

Clearly, if the interfacial tension is zero, the system is unstable for all wavelength disturbances. This is not surprising. If we have a higher-density fluid overlaying a lower-density fluid, and no viscosity or interfacial tension to inhibit motion, the system will overturn given any infinitesimal disturbance. This was suggested by the qualitative, "cartoonlike" description of the instability mechanism at the beginning of this section. We note also that the growth-rate coefficient σ' grows monotonically with increase of the wave number α'. Larger wave numbers correspond to shorter-wavelength perturbations of the interface shape. As the wavelength decreases, the "hydrostatic" pressure gradient that drives the growth of the perturbation increases and hence σ' increases. On the other hand, when we add a finite interfacial tension, the two-fluid system is unstable only if

$$\alpha'^2 < \frac{g(\rho_2 - \rho_1)}{\gamma}. \tag{12-81}$$

We see from (12–80) and (12–81) that short-wavelength disturbances are stabilized by the interfacial tension as these are associated with large interface curvature. We can recall from the earlier discussions of capillary flows (for example in Chap. 2) that interfacial tension tends to reduce any displacements of a flat interface at a rate that is proportional to the magnitude of the interface curvature. When the rate of decay of a perturbation exceeds the rate of growth that is due to gravitational forces, the system is stable to displacements of that particular wavelength, but when the decay rate is slower it is unstable. This is the balance of terms that is inherent in (12–80) and is the reason that short-wavelength disturbances to the interface shape can be stabilized by interfacial tension. There is a fastest-growing wavelength between $\alpha' = 0$ and α' given by (12–81). For convenience in illustrating these points, we can nondimensionalize (12–80) by introducing a characteristic time and length scale according to

$$\sigma' = \frac{\sigma}{t_c} \quad \text{and} \quad \alpha' = \frac{\alpha}{\ell_c}; \quad \text{where } t_c \equiv \left[\frac{\gamma}{g^3(\rho_2+\rho_1)(\delta_2-\delta_1)^3}\right]^{1/4} \quad \text{and}$$

$$\ell_c \equiv \left[\frac{\gamma}{g(\rho_2+\rho_1)(\delta_2-\delta_1)}\right]^{1/2}, \quad \text{with } \delta_2 \equiv \frac{\rho_2}{\rho_2+\rho_1}, \quad \delta_1 \equiv \frac{\rho_1}{\rho_2+\rho_1}$$

The dimensionless form for (12–80) is then

$$\boxed{\sigma^2 = \alpha - \alpha^3.}$$

Hydrodynamic Stability

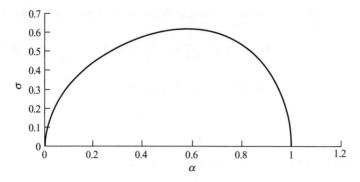

Figure 12–1. The dimensionless growth-rate coefficient σ as a function of the dimensionless wave number α for an inviscid fluid; $\sigma' = [g^3(\rho_2 + \rho_1)(\delta_2 - \delta_1)^3/\gamma]^{1/4}\sigma$ and $\alpha' = [g(\rho_2 + \rho_1)(\delta_2 - \delta_1)/\gamma]^{1/2}\alpha$.

A plot of σ versus α is given in Fig. 12–1. In this dimensionless framework, $\alpha = 1$ is the critical wave number corresponding to (12–81), and the maximal growth-rate coefficient is $\sigma = 2/3\sqrt{3}$ at $\alpha = 1/\sqrt{3}$.

For a two-fluid system of *infinite* extent in the x and/or y direction, the wavelengths of the disturbance can be as large as desired, and there will always be unstable modes in an inviscid fluid system for which $\rho_2 > \rho_1$. However if the two-fluid system is laterally confined (for example, it is in a container), the available range of values of α will be limited, and it is possible that the system will be stable if the density difference is not too large or the interfacial tension is large. Of course, we have not incorporated any boundary conditions on the sidewalls to explicitly account for container walls and so we realize that the analysis is not strictly applicable to this case. Nevertheless, though the predicted growth rates are not strictly correct for a finite fluid domain, the inference of stability in a region of finite extent is correct.

2. The Effects of Viscosity on the Stability of a Pair of Superposed Fluids

The most obvious question is whether the results of the preceding section are representative of the behavior of real fluids with finite viscosity. To address this question, we can return to the governing equations (12–65a) and (12–65b), plus the boundary conditions, including all of the viscous terms.

In each of the two fluids, the governing equation can be written in the form

$$\boxed{\left[1 - \frac{\nu_i}{\sigma'}(D'^2 - \alpha'^2)\right](D'^2 - \alpha'^2)w'_i = 0,} \qquad (12\text{–}82)$$

where $i = 1$ or 2 depending on which fluid we are considering. This equation is fourth order, and thus there are four independent solutions that we can determine by assuming that $w' = e^{pz'}$ and solving the characteristic equation for the four independent roots for p. This general solution is

$$\boxed{\begin{aligned} w'_i = {} & A'_i \exp(\alpha' z') + B'_i \exp(-\alpha' z') + C'_i \exp\left[\left(\alpha'^2 + \frac{\sigma'}{\nu_i}\right)^{1/2} z'\right] \\ & + D'_i \exp\left[-\left(\alpha'^2 + \frac{\sigma'}{\nu_i}\right)^{1/2} z'\right]. \end{aligned}} \qquad (12\text{–}83)$$

B. Rayleigh–Taylor Instability

Again we require that the disturbance decays far from the interface; hence

$$w'_1 \to 0 \quad \text{as } z' \to -\infty, \quad \text{and} \quad w'_2 \to 0 \quad \text{as } z' \to \infty. \tag{12–84}$$

At the interface, $z' = 0$, these solutions must satisfy the conditions:
(a) the kinematic condition (12–71),

$$w'_1(0) = w'_2(0) = \sigma' H; \tag{12–85a}$$

(b) continuity of tangential velocity, (12–69), which can be seen from (12–63) to imply that

$$D'w'_1|_{z'=0} = D'w'_2|_{z'=0}; \tag{12–85b}$$

(c) continuity of tangential stress (12–74), which can be written in the form

$$\mu_1(D'^2 + \alpha'^2)w'_1 = \mu_2(D'^2 + \alpha'^2)w'_2; \tag{12–85c}$$

and, finally, the normal-stress balance, (12–73). If we substitute for the pressure terms by using (12–64) (and the corresponding equation for p'_2), we can write the normal-stress condition in the form

$$\frac{\sigma'}{\alpha'^2}(\rho_1 D'w'_1 - \rho_2 D'w'_2) - \left[\frac{1}{\alpha'^2}(D'^2 - \alpha'^2) - 2\right](\mu_1 D'w'_1 - \mu_2 D'w'_2)$$
$$+ [\gamma \alpha'^2 + (\rho_1 - \rho_2)g]H = 0. \tag{12–85d}$$

We have carried out all of the analysis of this section with the equations and boundary conditions in dimensional form. However, in view of the rather large number of dimensional parameters that appear in (12–82)–(12–85), there is a definite advantage in nondimensionalizing. The problem is to choose characteristic scales because these must be intrinsic, and formed from the available dimensional parameters, rather than explicitly evident. The available parameters are $\rho_1, \rho_2, \mu_1, \mu_2, g,$ and γ. Given that the mechanism for motion is the body force that is due to gravity, it would be surprising if the characteristic velocity did not involve g. In fact, a combination of dimensional parameters with units of velocity is $(g\nu)^{1/3}$. The decision to use ν_1 or ν_2 in this definition is completely arbitrary. Hence we take

$$u_c = (g\nu_1)^{1/3}, \quad \ell_c = (\nu_1^2/g)^{1/3}, \quad \text{and} \quad t_c = \ell_c/u_c = (\nu_1/g^2)^{1/3}. \tag{12–86a}$$

and define a dimensionless wave number and growth-rate factor

$$\alpha \equiv \alpha' \ell_c \quad \text{and} \quad \sigma \equiv \sigma' t_c. \tag{12–86b}$$

Note that we denote dimensionless variables without the prime. Finally, it will be convenient to denote the ratio of kinematic viscosities as

$$\lambda \equiv \nu_2/\nu_1$$

and the ratio of densities as

$$\delta_1 \equiv \frac{\rho_1}{\rho_1 + \rho_2} \quad \text{and} \quad \delta_2 \equiv \frac{\rho_2}{\rho_1 + \rho_2}.$$

Hydrodynamic Stability

The solution and boundary conditions (12–83) and (12–85) can now be written in *dimensionless* form:

$$w_1 = A_1 \exp(\alpha z) + B_1 \exp(-\alpha z) + C_1 \exp[(\alpha^2 + \sigma)^{1/2} z]$$
$$+ D_1 \exp[-(\alpha^2 + \sigma)^{1/2} z], \qquad (12\text{–}87a)$$

$$w_2 = A_2 \exp(\alpha z) + B_2 \exp(-\alpha z) + C_2 \exp[(\alpha^2 + \sigma\lambda)^{1/2} z]$$
$$+ D_2 \exp[-(\alpha^2 + \sigma\lambda)^{1/2} z], \qquad (12\text{–}87a)$$

with

$$w_1(0) = w_2(0) = \sigma h \quad \text{(where } h \equiv H/\ell_c\text{)}, \qquad (12\text{–}88a)$$

$$Dw_1|_{z=0} = Dw_2|_{z=0} \quad \text{(where } D \equiv \partial/\partial z \text{ and } z \equiv z'/\ell_c\text{)}, \qquad (12\text{–}88b)$$

$$\delta_1(D^2 + \alpha^2)w_1 = \delta_2\lambda(D^2 + \alpha^2)w_2, \qquad (12\text{–}88c)$$

$$(\delta_1 Dw_1 - \delta_2 Dw_2) - \frac{\alpha^2}{\sigma}\left[\frac{1}{\alpha^2}(D^2 - \alpha^2) - 2\right](\delta_1 Dw_1 - \delta_2\lambda Dw_2)$$
$$+ \frac{\alpha^2}{\sigma}\left[\frac{1}{Ca}\alpha^2 + (\delta_1 - \delta_2)\right] h = 0. \qquad (12\text{–}88d)$$

Here the capillary number that appears in (12–88d) is

$$Ca \equiv \frac{(g\nu_1)^{1/3}\nu_1(\rho_1 + \rho_2)}{\gamma}.$$

Now, in view of the conditions, (12–84), we see that $B_1 = D_1 = A_2 = C_2 = 0$, and thus

$$w_1 = A_1 \exp(\alpha z) + C_1 \exp(q_1 z), \quad \text{where } q_1 \equiv (\alpha^2 + \sigma)^{1/2}, \qquad (12\text{–}89a)$$

$$w_2 = B_2 \exp(-\alpha z) + D_2 \exp(-q_2 z), \quad \text{where } q_2 \equiv (\alpha^2 + \sigma\lambda^{-1})^{1/2}. \qquad (12\text{–}89b)$$

Then, applying the boundary conditions of stress and velocity continuity at the interface (12–88a–d) to the general solutions (12–89), we obtain a set of four algebraic equations for $A_1, B_2, C_1,$ and D_2:

$$A_1 + C_1 = B_2 + D_2, \qquad (12\text{–}90a)$$

$$A_1\alpha + C_1 q_1 = -B_2\alpha - D_2 q_2, \qquad (12\text{–}90b)$$

$$\delta_1\left[2\alpha^2 A_1 + (\alpha^2 + q_1^2) C_1\right] = \delta_2\lambda\left[2\alpha^2 B_2 + (\alpha^2 + q_2^2) D_2\right], \qquad (12\text{–}90c)$$

$$\delta_1(\alpha A_1 + q_1 C_1) + \delta_2(\alpha B_2 + q_2 D_2) - (\alpha^2/\sigma)\left[\delta_1\left\{-2A_1\alpha + q_1 C_1\left[-3 + (q_1^2/\alpha^2)\right]\right\}\right.$$
$$\left. + \delta_2\lambda^{-1}\left\{-2B_2\alpha + q_2 D_2\left[-3 + (q_2^2/\alpha^2)\right]\right\}\right] + (\alpha^2/2\sigma^2)[(\alpha^2/Ca) \qquad (12\text{–}90d)$$
$$+ (\delta_1 - \delta_2)](A_1 + C_1 + B_2 + D_2) = 0.$$

B. Rayleigh–Taylor Instability

Therefore, we see that the boundary conditions at the interface lead to a set of four linear algebraic equations for the constants A_1, C_1, B_2, D_2. The condition for existence of a nontrivial solution of this set of algebraic equations is that the determinant of the coefficient matrix must equal to zero. This condition leads to a complicated algebraic equation relating the dimensionless growth-rate coefficient σ to the dimensionless wave number α for specified values of the fluid viscosities, the fluid densities, and the interfacial tension. As usual, stability is determined by the sign of the real part of σ.

An explicit analytic relation for σ was obtained many years ago for the general case by application of this determinant condition.[13] Today, we can use one of the commercial computer algebra packages to evaluate the roots for σ with the material parameters and the wave number specified. Here we consider the somewhat simpler problem in which the kinematic viscosities of the two fluids are equal. Although this assumption simplifies the algebra, it is sufficient to expose the key fact that the influence of viscous effects is confined to quantitative changes in the growth or decay rates rather than any qualitative change in the basic instability.

When $\nu_1 = \nu_2$ (i.e., $q_1 = q_2 = q$), the set of algebraic equations, (12–90a)–(12–90d), can be written in the form

$$\begin{vmatrix} 1 & 1 & -1 & -1 \\ \alpha & q & \alpha & q \\ 2\delta_1\alpha^2 & \delta_1(2\alpha^2+\sigma) & -2\delta_2\alpha^2 & -\delta_2(2\alpha^2+\sigma) \\ \delta_1 M + N & \delta_1 L + N & \delta_2 M + N & \delta_2 L + N \end{vmatrix} \begin{Vmatrix} A_1 \\ C_1 \\ B_2 \\ D_2 \end{Vmatrix} = 0, \quad (12\text{–}91)$$

where

$$L = q\left[1 - \frac{\alpha^2}{\sigma}\left(-3 + \frac{q^2}{\alpha^2}\right)\right], \quad M \equiv \left(\alpha + \frac{2\alpha^3}{\sigma}\right),$$

$$q = (\alpha^2 + \sigma)^{1/2}, \quad \text{and} \quad N \equiv \frac{\alpha^2}{2\sigma^2}\left\{\frac{\alpha^2}{Ca} - (\delta_2 - \delta_1)\right\}.$$

The condition that the determinate of coefficients equals zero leads to a moderately complicated-looking algebraic equation:

$$2\alpha^2(\delta_1 - \delta_2)^2(L - M)(\alpha - q) - \sigma\{4\delta_1\delta_2 L + M[\alpha(\delta_1 - \delta_2)^2 - q(\delta_1 + \delta_2)^2] \\ + 2N(\delta_1 + \delta_2)(\alpha - q)\} = 0. \quad (12\text{–}92)$$

To utilize this equation, we specify δ_1, δ_2, and Ca, and then calculate the roots σ that satisfy the equation for each value of the wave number α. If any of the roots has a positive real part, the system is unstable. If the real parts of all roots are negative, on the other hand, the system is stable. The wave number with the largest real part for σ is the fastest-growing infinitesimal disturbance.

We consider two cases for fixed $\delta_2 - \delta_1 = 0.5$ (because $\delta_1 + \delta_2 = 1$, this means that $\delta_2 = 3/4$ and $\delta_1 = 1/4$), the first with $Ca = \infty$ (i.e., interfacial tension $\gamma \equiv 0$), and the other with $Ca = 1$. The results for the largest real value of the dimensional growth rate σ' as a function of dimensional wave number α' are plotted in Fig. 12–2. A plot is also made of the result

$$\sigma' = [(\delta_2 - \delta_1)g\alpha']^{1/2}$$

for an inviscid fluid from (12–80) with $\gamma = 0$. It is necessary to use dimensional variables for this plot because the scaling (12–86) clearly does not make sense when the viscosities are equal to zero.

In the limit $\gamma = 0$ and/or $Ca = \infty$, we see that all wave numbers are unstable in both cases. *The only change from including the viscosity of the fluids is that the growth rates are reduced relative to those for inviscid fluids.* The addition of viscous effects cannot

Hydrodynamic Stability

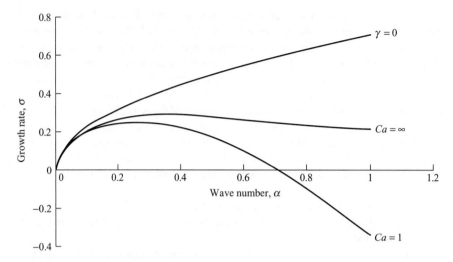

Figure 12–2. The largest real value of the dimensional growth rate σ' as a function of dimensional wave number α' for a pair of fluids with $\lambda = 1$, $\delta_2 = 0.75$, $\delta_1 = 0.25$; for $Ca = \infty$ and $Ca = 1$. Also shown is the growth rate for an inviscid fluid with $\gamma = 0$ calculated from (12–80).

influence the basic mechanism for instability, but only provide some viscous resistance to (and slowing of) the ensuing flow.

The result for $Ca = 1$ is again qualitatively unchanged by the addition of viscous effects. There is still a range of wave numbers in which the instability is stabilized by capillary effects and a range of large-wavelength disturbances that is unstable. The growth and decay rates are modified by the addition of viscous effects but the range of stable and unstable wave numbers remains the same. The wavelength of the fastest-growing unstable mode is modified by the viscosity, though not the existence of such a mode.

3. Discussion

A key idea from the preceding analysis, as well as the analysis of capillary instability in section A, is that viscous effects often cannot change the conditions for instability of a rest state, but only moderate the rate of growth or decay of disturbances. In such cases, analysis of the stability (or instability) of the inviscid limit can be extremely useful in identifying the conditions for instability. In view of the fact that the inviscid analysis is very much simpler, this is an important observation.

The problem as outlined in the preceding sections, had been basically done in the last half of the nineteenth century,[14] before the input of Sir Geoffrey Taylor,[15] for whom the instability is now commonly named along with Rayleigh. It is clear that the stability of a pair of fluids of different density that is accelerated perpendicular to the interface can be treated exactly as previously outlined. We have only to replace the gravitational acceleration **g** with the actual acceleration that the fluid pair experiences. Taylor's contribution was to consider this latter problem for a pair of fluids of different density. In this case, we can analyze the problem by writing the equations of motion with respect to a moving reference frame. The essential point is that, with respect to the accelerating frame, the body force per unit mass, **g**, is reduced (or increased) by an amount **a**. The dimensional growth-rate coefficient is

$$\sigma'^2 = \hat{g}\alpha \left[\frac{\rho_2 - \rho_1}{\rho_2 + \rho_1} - \frac{\gamma \alpha'^2}{g(\rho_2 + \rho_1)} \right], \tag{12–93}$$

C. Saffman–Taylor Instability at a Liquid Interface

where $\hat{\mathbf{g}} \equiv \mathbf{g} - \mathbf{a}$. If $(\mathbf{g} - \mathbf{a}) > 0$, the condition for stability is precisely as before with the same neutral stability condition, but with altered growth or decay rates. If $(\mathbf{g} - \mathbf{a}) < 0$, the stability is reversed with respect to ρ_2 and ρ_1. Thus, when two superposed fluids of different density are accelerated perpendicular to their interface, this surface is unstable or stable depending on whether the acceleration is directed from the more-dense fluid to the less-dense fluid or vice versa.

A final remark concerns the stability of an accelerated interface that is spherically symmetric rather than flat as in the preceding analysis. This problem was actually considered for the case of a spherical bubble at the end of Chap. 4. In that context, it is relevant to the fate of a bubble that may be either expanding or contracting because of a change in the external pressure. The results obtained are consistent with those in the present section in the sense that the interface is unstable when $d^2 R/dt^2 > 0$, i.e., when it is accelerated from the less-dense fluid toward the more-dense fluid. During bubble expansion, $d^2 R/dt^2 < 0$, but during the collapse phase there is a period of time when bubble collapse is being arrested by the increase of pressure within the bubble where $d^2 R/dt^2 > 0$ and instability is expected (this is observed experimentally by the resultant breakup into a cloud of smaller bubbles). Of course, the growth rates will be changed relative to that for a plane interface because of the decrease of interfacial area and the resultant change in wavelength and amplitude of any disturbance that this causes. A more remarkable fact, which is not "explained" by the analogy with the basic Rayleigh-Taylor problem, is that instability is also possible when $d^2 R/dt^2 < 0$, provided $(\dot{R}/R)^2$ is large enough. These results are discussed in detail in Chap. 4.

C. SAFFMAN–TAYLOR INSTABILITY AT A LIQUID INTERFACE

A problem that is somewhat analogous to the instability of an accelerating interface occurs when two superposed viscous fluids are forced by gravity and an imposed pressure gradient through a porous medium. This problem was analyzed in a classic paper by Saffman and Taylor.[16] If the steady state is one of uniform motion with velocity V vertically upwards and the interface is horizontal, then it can be shown that the interface is stable to infinitesimal perturbations if

$$\boxed{\left(\frac{\mu_2}{k_2} - \frac{\mu_1}{k_1}\right) V + (\rho_2 - \rho_1)g > 0} \qquad (12\text{–}94\text{a})$$

and unstable if

$$\boxed{\left(\frac{\mu_2}{k_2} - \frac{\mu_1}{k_1}\right) V + (\rho_2 - \rho_1)g < 0,} \qquad (12\text{–}94\text{b})$$

where 1 is the upper fluid and 2 is the lower fluid. Here, k_i stands for the permeability of the porous media for the fluid i. To prove this statement, we first need to introduce Darcy's law for flow in a porous medium in order to define the permeability, and then carry out a classic linear stability analysis.

1. Darcy's Law

We have not previously discussed flows in porous media. The subject is itself a very large one, driven in part by numerous applications, and there have been a number of well-known textbooks written just on this one topic.[17] We do not pretend here to say anything comprehensive about the subject, but only to briefly discuss the basis for Darcy's law, which is the simplest model for flow of a single-phase Newtonian fluid through a porous material.

Now, of course, there are a great variety of materials that are classified as porous media, and each is characterized by a different microstructure. However, qualitatively, we can characterize a porous media as a solid material having a huge number of small-scale random, interconnected pathways for flow. We assume that we have a single-phase Newtonian fluid moving with local velocities that are small enough through pores with a characteristic dimension that is small enough that the flow at the level of a single pore is in the creeping-flow regime. Hence, in principle, if we were given a detailed description of the geometry of the porous material, we could obtain a solution of the creeping-flow equations for the detailed velocity fields in all of the pore spaces (and, indeed, for idealistic models of a porous media such as spherical grains arranged on a periodic lattice this has been done using boundary integral techniques[18]). However, for many purposes, this is not only extremely complex, but the result includes far more information than we need. Frequently, what we seek is an averaged description of the flow, with a scale of resolution that is far larger then the individual grain or pore level description. Darcy's law provides such an averaged description of the flow.

Quite detailed derivations have been developed for Darcy's law.[19] These derivations provide an expression for the volume-averaged velocity over a length scale that is large compared with the individual pore length scale, but still small compared with the overall scale of gradients of the averaged velocity (or pressure) within the porous medium. There is, in fact, a strong analogy between the definitions of the volume-averaged velocity and pressure in a porous medium, and the volume-averaged derivation of the equations of motion for a fluid by means of the continuum approximation. However, one difference in this case is that we assume that the microscale equations governing flow are the linear creeping-flow equations. Hence, when we consider the averaged velocity in a porous medium, it is not surprising that it is *linearly* related to the gradient of the averaged total pressure, i.e.,

$$\mathbf{u}' = -\frac{k}{\mu} \operatorname{grad}(p' + \rho g z') = \operatorname{grad} \phi'. \qquad (12\text{--}95)$$

As usual in this chapter, primed variables are dimensional. Here z' is the "vertical" axis with **g** acting in the negative z' direction. Equation (12–95) is, in fact, *Darcy's law*. The fluid density and viscosity are denoted as ρ and μ. The constant k is known as the permeability and represents the role of the geometry of the porous medium, as well as the nature of the solid–fluid interaction. Qualitatively, k increases as the density of the solid phase is decreased and decreases as it is increased. The porous medium that is described by (12–95) is "isotropic" in the sense that the same volumetrically averaged velocity is obtained for a given magnitude of the total pressure gradient independent of direction. We also assume that the porous medium is homogeneous in the sense that the permeability is not a function of position (as it is in many actual porous materials). We do not present any more detailed discussion of the derivation of Darcy's law. In particular, without a detailed micromechanical derivation, we cannot say anything more about the dependence of the permeability on the geometric features of the porous medium, nor anything about the dependence on specific fluid properties. Instead, we simply regard the permeability as an empirically determined characteristic of the porous material, which is specified to us.

In many applications, we may wish to describe the motion of two fluids, which may be miscible (i.e., the interfacial tension is zero, though the fluids may have different viscosities or other properties) or immiscible (oil and water, for example) in a porous medium. At the microscale, within the pores, there is a well-defined interface separating two immiscible

C. Saffman–Taylor Instability at a Liquid Interface

fluids. Two miscible fluids will remain separated for only a finite time and a spontaneous "mixing" process will occur at the pore scale so that the "interface" between them within any pore will be diffuse (i.e., characterized by a length scale that is finite relative to the characteristic scale of the pores, for example). However, our interest is in a description of the motion of the two fluids with a scale of resolution that is relevant for Darcy's law; namely, much larger than the individual pore scale. Thus the "interface" between the fluids must be recognized as an averaged description of a surface separating the two fluids. The actual transition between a bulk, homogeneous fluid A and a second bulk, homogeneous fluid B will generally occur over a length scale that encompasses a number of pores both because of micromixing for miscible fluids and because the heterogeneity of the pore structure produces "dispersion" or spreading of the position of the actual microscale interface (for both miscible and immiscible fluids) because of differences in the rate of motion within different pores. However, for the Darcy scale description of motion, this can still be treated as a surface of discontinuity between A and B, provided the "thickness" of this transition region remains small compared with the scale of resolution for Darcy's law (one can think of this as the length scale for volume averaging if Darcy's law is viewed as having been derived from the microscale description by means of a volume-averaging procedure). The Saffman–Taylor instability describes the stability of this surface to perturbations in the shape. Obviously the analysis breaks down as the length scale of these perturbations becomes comparable with the pore scale of the porous medium.

It may be noted that (12–95) essentially replaces the macroscale equations of motion with an algebraic relation between the pressure gradient (or the gradient of the potential function ϕ'). It is not surprising therefore that the Darcy velocity field cannot satisfy boundary conditions involving either the tangential velocity components or viscous stresses at either solid (i.e., nonpermeable) boundaries or at the macroscale interface between two fluids in a porous medium. The only boundary conditions that can be imposed are the kinematic condition at an impermeable solid boundary or the kinematic condition plus the normal-stress condition for the pressure and capillary terms, but excluding the viscous-stress contributions, at an interface. To see that the latter conditions are possible, we note that the relationship between $\mathbf{u}'$ and the gradient of a potential is equivalent mathematically to potential-flow theory (already discussed in Chap. 10). In particular, because we have assumed that the permeability is a constant, and we know that the pressure is a volume average of a harmonic function, it follows from Darcy's law that

$$\nabla' \cdot \mathbf{u}' = 0, \qquad (12\text{–}96)$$

and hence,

$$\nabla'^2 \phi' = 0. \qquad (12\text{–}97)$$

The potential function ϕ' can satisfy a Neumann-type boundary condition involving $\nabla \phi' \cdot \mathbf{n}$. At an interface, a pair of potential functions can be obtained for the two fluids that satisfy *continuity* of the normal velocity. Furthermore, because

$$p' = -\frac{\mu}{k}\phi' - \rho g z', \qquad (12\text{–}98)$$

we also satisfy *continuity* of pressure (including the capillary pressure jump if the interfacial tension is nonzero).

For problems that involve boundaries or interfaces where we need to satisfy more general boundary conditions, involving either the tangential velocity component or viscous stresses, a more general theory is needed instead of (12–95). The most commonly used generalization is due to Brinkman, though this is valid only for highly porous materials

Hydrodynamic Stability

with a low volume fraction of solid.[20] However, for our present purposes, it is sufficient to limit our considerations to flows governed by Darcy's law.

2. The Taylor–Saffman Instability Criteria

Therefore, as previously stated, we assume that we have a pair of superposed Newtonian fluids. We denote the vertical upward coordinate as z, with x and y being parallel to the plane of the horizontal interface that separates the two fluids. We fix the fluid interface as being *initially* at $z = 0$. There is no advantage for this particular problem in carrying out the analysis in terms of dimensionless variables. Hence we simply retain dimensional equations throughout. For simplicity, we therefore drop the use of primes to denote dimensional variables. *All variables in what follows are dimensional.*

We should note that this description of the system as being two fluids separated by a flat interface already has inherent in it the spatial averaging to a scale of resolution that is much larger than the individual pore level of description. The volume-averaged velocity in each fluid is determined by Darcy's law. As noted earlier, we assume that the fluids (and the interface between them) move with a uniform velocity V in the positive z direction. It is therefore convenient to consider the problem with respect to a moving reference frame that is fixed at the unperturbed fluid interface, i.e., we introduce $\bar{z}$, which is related to the original "laboratory" frame of reference as

$$\bar{z} \equiv z - Vt. \qquad (12\text{--}99)$$

From this perspective of the moving coordinate $\bar{z}$, the two fluids at infinity appear to be stationary.

We assume that the flat interface between the two fluids, now designated as $\bar{z} = 0$, is perturbed with an arbitrary infinitesimal perturbation of shape. As usual, for a linear stability analysis, we consider only a single Fourier mode in each of the x and y directions, with the wave number (or wavelength) as a parameter in the stability analysis. Hence we consider a perturbation of the form

$$F(x, y, \bar{z}, t) = \bar{z} - \varepsilon h(x, y, t) \quad \text{with} \quad h(x, y, t) = \bar{h} \exp[i(\alpha_x x + \alpha_y y)] \exp(\sigma t). \qquad (12\text{--}100)$$

We shall see at the end of our analysis that it would be sufficient to consider a 2D perturbation of the shape, but for now, we consider the more general form (12–100). Again, it is important to keep in mind that the interface described by (12–100) is an averaged surface separating the two fluids, not the actual interface that exists at the individual pore level description. The matter of *effective* interface properties (such as an effective interfacial tension) is then a somewhat complicated issue. In the present analysis, we neglect all effects of interfacial tension between the two fluids. In cases of practical significance, such as the flow induced in an underground oil reservoir during oil recovery operations, surfactants are often added to drive the interfacial tension at the pore level to the smallest values that can be achieved. If the fluids are miscible (i.e., $\gamma = 0$, though with different viscosities and densities), the macroscale (i.e., volume-averaged) motion of the two fluids and interface in the porous media is known as a "miscible" displacement. The present analysis is directly applicable to this case.

In the present analysis, we seek to determine the fate of a perturbation to the interface shape when there is a mean, uniform velocity V in the upward direction normal to the interface. Hence, in the laboratory reference frame

$$\mathbf{u}_i = V \mathbf{i_z} + \varepsilon \hat{\mathbf{u}}_i, \; p_i = P + \varepsilon \hat{p}, \; \text{where } i = 1 \text{ or } 2. \qquad (12\text{--}101)$$

C. Saffman–Taylor Instability at a Liquid Interface

However, in the moving frame, in which the undisturbed interface corresponds to $\bar{z} = 0$, we see only the perturbation flow, $\varepsilon \hat{\mathbf{u}}_i$. We assume that the latter also satisfies Darcy's law and can thus be related to a disturbance potential function

$$\hat{\mathbf{u}}_i = \nabla \hat{\phi}_i, \tag{12-102}$$

where

$$\nabla^2 \hat{\phi}_i = 0. \tag{12-103}$$

We require that the disturbance velocity decay far from the interface (i.e., as $z \to \pm\infty$). Furthermore, at the fluid interface the normal component of the disturbance velocity satisfies the kinematic condition,

$$\frac{\partial \hat{\phi}_1}{\partial \bar{z}} = \frac{\partial \hat{\phi}_2}{\partial \bar{z}} = \bar{h}\sigma \exp[i(\alpha_x x + \alpha_y y)] \exp(\sigma t), \tag{12-104}$$

and the normal-stress condition reduces to the requirement that the pressure be continuous at the interface.

Now, given the expression (12–100) for the disturbance shape function at the interface, an appropriate form for the solutions of (12–103) for the disturbance potential functions is

$$\begin{aligned}\hat{\phi}_1(x, y, \bar{z}, t) &= F_1(\bar{z}, t) \exp[i(\alpha_x x + \alpha_y y)], \\ \hat{\phi}_2(x, y, \bar{z}, t) &= F_2(\bar{z}, t) \exp[i(\alpha_x x + \alpha_y y)],\end{aligned} \tag{12-105}$$

with

$$\frac{\partial^2 F_i}{\partial \bar{z}^2} - (\alpha_x^2 + \alpha_y^2) F_i = 0. \tag{12-106}$$

Hence, incorporating the solution of (12–106) into (12–105), together with the far-field condition that the disturbance functions should decay for large $|\bar{z}|$, we have

$$\begin{aligned}\hat{\phi}_1(x, y, \bar{z}, t) &= G_1(t) \exp[i(\alpha_x x + \alpha_y y)] \exp\left[-\left(\alpha_x^2 + \alpha_y^2\right)^{1/2} \bar{z}\right], & \bar{z} > 0, \\ \hat{\phi}_2(x, y, \bar{z}, t) &= G_2(t) \exp[i(\alpha_x x + \alpha_y y)] \exp\left[\left(\alpha_x^2 + \alpha_y^2\right)^{1/2} \bar{z}\right], & \bar{z} < 0.\end{aligned} \tag{12-107}$$

Finally, we have to apply the boundary conditions at the interface. These conditions are strictly applied at the deformed interface $\bar{z} = \varepsilon h$. However, the domain perturbation argument from the Rayleigh–Taylor section shows that the boundary conditions for the linearized disturbance problem can equally well be applied at the unperturbed surface, $\bar{z} = 0$.

Thus the kinematic boundary condition (12–104) yields

$$G_1(t) = -G_2(t) = -\frac{\bar{h}\sigma}{\left(\alpha_x^2 + \alpha_y^2\right)^{1/2}} \exp(\sigma t). \tag{12-108}$$

Hence

$$\hat{\phi}_1 = -\frac{\bar{h}\sigma}{\left(\alpha_x^2 + \alpha_y^2\right)^{1/2}} \exp(\sigma t) \exp[i(\alpha_x x + \alpha_y y)] \exp\left[-\left(\alpha_x^2 + \alpha_y^2\right)^{1/2} \bar{z}\right], \tag{12-109a}$$

$$\hat{\phi}_2 = \frac{\bar{h}\sigma}{\left(\alpha_x^2 + \alpha_y^2\right)^{1/2}} \exp(\sigma t) \exp[i(\alpha_x x + \alpha_y y)] \exp\left[\left(\alpha_x^2 + \alpha_y^2\right)^{1/2} \bar{z}\right]. \tag{12-109b}$$

Hydrodynamic Stability

The final condition that must be satisfied is pressure continuity at the interface. We can see from (12–98) that the pressure can be calculated in terms of the complete potential function as

$$p_1 = -\frac{\mu_1}{k_1}\phi_1 - \rho_1 g\bar{z},$$
$$p_2 = -\frac{\mu_2}{k_2}\phi_2 - \rho_2 g\bar{z}. \quad (12\text{–}110)$$

Here, $\phi_1 = V\bar{z} + \varepsilon\hat{\phi}_1$ and $\phi_2 = V\bar{z} + \varepsilon\hat{\phi}_2$. The relevance of the hydrostatic pressure contribution is its *difference* across the interface. Hence it is sufficient to calculate it at $\bar{z} = \varepsilon h$, rather than calculating its absolute value by taking account of the fact that the contributions to both p_1 and p_2 change as the interface moves vertically. Hence, at the interface, the condition of pressure continuity gives

$$\frac{\mu_1}{k_1}\left\{V\bar{z} - \frac{\varepsilon\bar{h}\sigma}{(\alpha_x^2+\alpha_y^2)^{1/2}}\exp(\sigma t)\exp[i(\alpha_x x + \alpha_y y)]\exp[-(\alpha_x^2+\alpha_y^2)^{1/2}\bar{z}]\right\} + \rho_1 g\bar{z}$$
$$= \frac{\mu_2}{k_2}\left\{V\bar{z} + \frac{\varepsilon\bar{h}\sigma}{(\alpha_x^2+\alpha_y^2)^{1/2}}\exp(\sigma t)\exp[i(\alpha_x x + \alpha_y y)]\exp[-(\alpha_x^2+\alpha_y^2)^{1/2}\bar{z}]\right\} + \rho_2 g\bar{z},$$
$$(12\text{–}111)$$

where

$$\bar{z} = \varepsilon\bar{h}\exp[i(\alpha_x x + \alpha_y y)]\exp(\sigma t) \ll 1.$$

This gives

$$\frac{\mu_1}{k_1}\left[V - \frac{\sigma}{(\alpha_x^2+\alpha_y^2)^{1/2}}\right] + \rho_1 g = \frac{\mu_2}{k_2}\left[V + \frac{\sigma}{(\alpha_x^2+\alpha_y^2)^{1/2}}\right] + \rho_2 g. \quad (12\text{–}112)$$

Solving this equation for the growth-rate factor σ, we obtain

$$\boxed{\left(\frac{\mu_1}{k_1} + \frac{\mu_2}{k_2}\right)\frac{\sigma}{(\alpha_x^2+\alpha_y^2)^{1/2}} = (\rho_1 - \rho_2)g + \left(\frac{\mu_1}{k_1} - \frac{\mu_2}{k_2}\right)V.} \quad (12\text{–}113)$$

We see that $\sigma > 0$ if $(\rho_1 - \rho_2)g + (\frac{\mu_1}{k_1} - \frac{\mu_2}{k_2})V > 0$ and the system is *unstable*. Otherwise the system is stable. According to (12–113), when the system is unstable, the growth rate increases monotonically with the wave number. However, we might expect that short-wavelength disturbances, corresponding to large values of the wave number, are stabilized by interfacial tension effects that are not considered in the present analysis.

We note that the system will be unstable even if the densities are equal, when the ratio of the viscosity to the permeability is larger for the upper fluid than for the lower fluid. Recall that the direction of motion is in the positive z direction. Hence if a fluid with a larger ratio of μ/k is being displaced by a fluid with a lower ratio, the interface between the two fluids will be unstable. Experimentally, what is seen is that the small linear perturbations of the interface grow into large fingers of the lower fluid that protrude across the moving interface into the upper fluid. This fingering motion, which is initiated by the Saffman–Taylor instability, can have very important consequences. For example, in the displacement of oil with water in an underground reservoir, the Saffman–Taylor instability leads to fingers of the water that protrude out in front of the mean oil–water interface, and hence arrive at a producing well "prematurely," causing water to be "produced" and oil to be left behind in the reservoir.

D. Taylor–Couette Instability

The mechanism for the Saffman–Taylor instability is most easily discussed in the absence of any density difference between the two fluids. We see from Darcy's law that a fluid with a large value of k/μ will move faster under the action of a given pressure gradient than a fluid with smaller k/μ. In the present situation, we have two fluids that move with the same velocity V. The lower fluid has a larger value of k/μ, and hence the pressure gradient required is smaller than the pressure gradient in the upper fluid where the value of k/μ is smaller. Now let us suppose that the interface is initially perturbed into a sinusoidal oscillation of the interface height above and below the position of the mean surface. Where the interface is deformed upward, we can imagine that the lower fluid finds itself momentarily in a region where the pressure gradient in the direction of motion is larger than in the lower fluid region. This tends to make this region move faster than the mean velocity V, and this accentuates the perturbation. On the other hand, where the interface dips below $\bar{z} = 0$, the upper fluid finds itself in a region in which the pressure gradient is smaller. Hence the fluid moves at a velocity that is lower than V, and again the perturbation gets bigger. Of course, as pointed out earlier, this "cartoon like" description of the instability mechanism is highly oversimplified and is only intended to provide a qualitative idea of why the system is unstable, as the analysis shows that it is.

D. TAYLOR–COUETTE INSTABILITY

In this section, we study the stability of Couette flow between a pair of concentric rotating cylinders. The steady flow was considered earlier in Section C of Chap. 3.

The general problem of the stability of various types of shear flow has occupied a great deal of research for nearly a century. There are two basic classes of problem. First, is the instability of a parallel shear flow such as simple shear flow or either 2D or axisymmetric Poiseuille flow, in which the problem of stability is a very subtle balance of viscous and inertia effects. Although the basic flow is very simple in these cases, the analysis of instability is difficult and involved, so much so that complete books have been written on this subject alone.[21] At the end of this chapter, we will return to a very brief discussion of this class of problems.

The second class of shear-flow stability problems is those in which the base flow is not unidirectional. Couette flow is an example of this class of problems. Another is the flow in a curved channel, or the boundary-layer flow along a curved surface. In these problems, there is a potential for instability that is due to the *centrifugal force* that arises because of the centripetal acceleration as the fluid moves along the curved streamlines of the undisturbed flow. These problems are more nearly analogous to Rayleigh–Taylor instability, in which there is a gradient in the body force that can produce instability. However, we shall see that a fundamental distinction from all of the problems studied to this point is that the Taylor–Couette flow can be stabilized by viscous effects. In all of the preceding examples, viscous effects decrease the rate of growth of unstable disturbances, but they play no role in determining whether a particular disturbance is stable or unstable.

In the case of Couette flow, the base flow whose stability we wish to study, was shown in Chap. 3, Section C, to be

$$u'_r = u'_z = 0, \, u'_\theta = C'_1 r' + \frac{C'_2}{r'}, \qquad (12\text{--}114)$$

where

$$C'_1 \equiv \frac{\Omega_2 R_2^2 - \Omega_1 R_1^2}{R_2^2 - R_1^2}; \; C'_2 \equiv \frac{(\Omega_1 - \Omega_2) R_1^2 R_2^2}{R_2^2 - R_1^2}.$$

Hydrodynamic Stability

The variables in (12–114) including the coefficients C'_1 and C'_2 are all dimensional, as indicated by the fact that they are primed. We see that the fluid moves in circular paths and thus the fluid particles undergo a centripetal acceleration. However, the centrifugal force associated with this centripetal acceleration does not produce a secondary flow, because it is exactly balanced by a radial pressure gradient,

$$\frac{\partial p'}{\partial r'} = \rho \frac{u_\theta'^2}{r'} = \rho \left(C_1'^2 r' + \frac{2C'_1 C'_2}{r'} + \frac{C_2'^2}{r'^3} \right). \tag{12–115}$$

The corresponding pressure is

$$p' = \rho \left(C_1'^2 \frac{r'^2}{2} + 2C'_1 C'_2 \ln r' - \frac{C_2'^2}{2r'^2} \right). \tag{12–116}$$

As noted already in Chap. 3, the balance between centrifugal force and radial pressure gradient produces a potentially unstable arrangement that can be thought of as resulting from a gradient of angular momentum.

A qualitative description of the mechanism for instability can be understood as follows. A fluid element that is displaced radially outward, carrying its angular momentum with it, will find itself in a position in which the radial pressure gradient is either larger or smaller than it was at its initial position (where the centrifugal force is in balance with the pressure gradient). Depending on the distribution of angular momentum (and thus on the new pressure gradient), this will tend to either allow a further displacement and instability or a net restoring force that pushes the fluid back to its initial position. *It should be emphasized that the role of viscous effects in this problem is fundamentally different from either the capillary instability or the Rayleigh–Taylor problem.* In the latter two problems, viscous effects could not change the basic mechanism for instability. Hence, the addition of viscosity to the problem could only change the rate of growth or decay of a perturbation, but never change an unstable mode to a stable mode. On the other hand, in the case of Couette flow, an unstable configuration for an inviscid fluid may be stabilized by viscous effects that diffuse angular momentum, and thus tend to reequilibrate the displaced element to the angular momentum and radial pressure gradient at its new position before the imbalance between the local pressure gradient and the angular momentum can lead to instability.

To analyze the linear stability of a Couette flow, we begin with the Navier–Stokes and continuity equations in a cylindrical coordinate system. The full equations in dimensional form can be found in Appendix A. We wish to consider the fate of an arbitrary infinitesimal disturbance to the base flow and pressure distributions (12–114) and (12–116). Hence we consider a linear perturbation of the form

$$\begin{aligned} u'_r &= \varepsilon \hat{u}'_r, \\ u'_\theta &= u'_\theta + \varepsilon \hat{u}'_\theta, \\ u'_z &= \varepsilon \hat{u}'_z, \\ p' &= p' + \varepsilon \hat{p}'. \end{aligned} \tag{12–117}$$

We assume that the velocity and pressure fields are axisymmetric even in the disturbance fields. Although comparison with experiment shows this assumption to be valid for the linear perturbations that are unstable, it eventually breaks down far enough beyond the initial critical point in terms of the cylinder rotation rates. The corresponding linearized [i.e., $O(\varepsilon)$] equations of motion and continuity in dimensional form are

$$\rho \left[\frac{\partial \hat{u}'_r}{\partial t'} - 2 \left(C'_1 + \frac{C'_2}{r'^2} \right) \hat{u}'_\theta \right] = -\frac{\partial \hat{p}'}{\partial r'} + \mu \left(\nabla'^2 \hat{u}'_r - \frac{\hat{u}'_r}{r'^2} \right), \tag{12–118a}$$

D. Taylor–Couette Instability

$$\rho\left(\frac{\partial \hat{u}'_\theta}{\partial t'} + 2C'_1 \hat{u}'_r\right) = \mu\left(\nabla'^2 \hat{u}'_\theta - \frac{\hat{u}'_\theta}{r'^2}\right), \tag{12–118b}$$

$$\rho\frac{\partial \hat{u}'_z}{\partial t'} = -\frac{\partial \hat{p}'}{\partial z'} + \mu \nabla'^2 \hat{u}'_z, \tag{12–118c}$$

$$\frac{\partial (r' \hat{u}'_r)}{\partial r'} + \frac{\partial (r' \hat{u}'_z)}{\partial z'} = 0, \tag{12–118d}$$

where

$$\nabla'^2 \equiv \frac{1}{r'}\frac{\partial}{\partial r'}\left(r'\frac{\partial}{\partial r'}\right) + \frac{\partial^2}{\partial z'^2}.$$

These linearized disturbance equations can be expressed in dimensionless form by scaling using the characteristic quantities

$$\ell_c = R_1, \; t_c = \Omega_1, \; u_c = R_1 \Omega_1 \tag{12–119}$$

Following the precedent of previous sections, dimensionless velocities are now indicated without primes. The caret continues to indicate that the quantity in question is a "disturbance" flow variable. Finally, we denote the dimensionless $(\hat{u}_r, \hat{u}_\theta, \hat{u}_z)$ as $(\hat{u}, \hat{v}, \hat{w})$. The result is

$$\frac{\partial \hat{u}}{\partial t} - 2\left(C_1 + \frac{C_2}{r^2}\right)\hat{v} = -\frac{\partial \hat{p}}{\partial r} + \frac{1}{Re}\left(\nabla^2 \hat{u} - \frac{\hat{u}}{r^2}\right), \tag{12–120a}$$

$$\frac{\partial \hat{v}}{\partial t} + 2C_1 \hat{v} = \frac{1}{Re}\left(\nabla^2 \hat{v} - \frac{\hat{v}}{r^2}\right), \tag{12–120b}$$

$$\frac{\partial \hat{w}}{\partial t} = -\frac{\partial \hat{p}}{\partial z} + \frac{1}{Re}\nabla^2 \hat{w}, \tag{12–120c}$$

$$\frac{\partial (r\hat{u})}{\partial r} + \frac{\partial (r\hat{w})}{\partial z} = 0, \tag{12–120d}$$

where $C_1 \equiv C'_1/\Omega_1$, $C_2 \equiv C'_2/R_1^2 \Omega_1$, and $Re \equiv R_1^2 \Omega_1/\nu$.

The linear stability analysis largely follows that of the preceding sections. We assume that

$$\begin{aligned}\hat{u} &= u(r)\cos\alpha z \; \exp(\sigma t), \\ \hat{v} &= v(r)\cos\alpha z \; \exp(\sigma t), \\ \hat{w} &= w(r)\sin\alpha z \; \exp(\sigma t). \end{aligned} \tag{12–121}$$

It will be noted that we express the Fourier mode in the z direction in terms of sin and cos rather than as an exponential. The fact that $\hat{u}$ appears as $\cos \alpha z$ and $\hat{w}$ appears as $\sin \alpha z$ is a consequence of the continuity equation. For convenience, let us denote $D \equiv \partial/\partial r$ and $L \equiv D^2 + (D/r) - (1/r^2)$. Then (12–120a) and (12–120c) become

$$\sigma u(\cos\alpha z)e^{\sigma t} - 2\left(C_1 + \frac{C_2}{r^2}\right)v(\cos\alpha z)e^{\sigma t} = -\frac{\partial \hat{p}}{\partial r} + \frac{1}{Re}(Lu - \alpha^2 u)(\cos\alpha z)e^{\sigma t}, \tag{12–122a}$$

$$\sigma w(\sin\alpha z)e^{\sigma t} = -\frac{\partial \hat{p}}{\partial z} + \frac{1}{Re}\left(Lw + \frac{w}{r^2} - \alpha^2 w\right)(\sin\alpha z)e^{\sigma t}. \tag{12–122b}$$

Hydrodynamic Stability

It we cross-differentiate (12–122a) and (12–122b) and subtract the resulting two equations, we can eliminate the pressure and obtain a single higher-order equation:

$$-\sigma u\alpha + 2\left(C_1 + \frac{C_2}{r^2}\right)\alpha v - \sigma Dw = \frac{1}{Re}[-\alpha Lu + \alpha^3 u - L(Dw) + \alpha^2 Dw]. \tag{12–123}$$

On the other hand, the continuity equation after (12–121) is substituted is

$$rDu + u + \alpha rw = 0. \tag{12–124}$$

and this equation can be rewritten in the form

$$\alpha Dw = -D^2 u - \frac{1}{r}Du + \frac{u}{r^2} = -Lu.$$

Combining this equation with (12–123), we can eliminate w. After a little rearranging, the resulting equation can be expressed as

$$\boxed{(L - \alpha^2 - \sigma Re)(L - \alpha^2)u = 2\alpha^2 Re\left(C_1 + \frac{C_2}{r^2}\right)v.} \tag{12–125}$$

Equation (12–120b), which we have not used yet, gives

$$\boxed{(L - \alpha^2 - \sigma Re)v = 2ReC_1 u.} \tag{12–126}$$

Although we could easily eliminate u to obtain a single sixth-order equation for v, we will see that there is some advantage in maintaining (12–125) and (12–126) as separate equations.

To solve these disturbance flow equations, we must apply boundary conditions. These are

$$\boxed{\begin{aligned} u &= 0 \quad \text{at } r = 1 \quad \text{and } r = \kappa (\equiv R_2/R_1), \\ v &= 0 \quad \text{at } r = 1 \quad \text{and } r = \kappa, \\ Du &= 0 \quad \text{at } r = 1 \quad \text{and } r = \kappa. \end{aligned}} \tag{12–127}$$

The last of these conditions is the same as $w = 0$, as we can see from the continuity equation (12–124). The condition on v comes from the kinematic condition at the walls of the Couette device, and the conditions on u and w are the no-slip conditions.

1. A Sufficient Condition for Stability of an Inviscid Fluid

A very important result can be obtained very simply in the case of an inviscid fluid. In the present problem, this corresponds to considering the disturbance equations in the limit $Re \to \infty$:

$$-\sigma(L - \alpha^2)u = 2\alpha^2\left(C_1 + \frac{C_2}{r^2}\right)v,$$
$$-\sigma v = 2C_1 u,$$

or, combining,

$$(L - \alpha^2)v = \frac{4\alpha^2 C_1}{\sigma^2}\left(C_1 + \frac{C_2}{r^2}\right)v.$$

D. Taylor–Couette Instability

In fact, we can go one step further and write this equation in the form

$$D\left[\frac{1}{r}D(rv)\right] - \alpha^2\left(\frac{1}{\sigma^2}\frac{F}{r} + \frac{1}{r}\right)rv = 0, \qquad (12\text{–}128)$$

with the boundary condition

$$rv = 0 \quad \text{at } r = 1 \text{ and } r = \kappa. \qquad (12\text{–}129)$$

The coefficient F that appears in (12–128) is

$$F \equiv \frac{1}{r^3}\frac{d}{dr}\Gamma^2, \quad \text{where } \Gamma \equiv \omega r^2 \text{ and } \omega \equiv \overline{C}_1 + \frac{\overline{C}_2}{r^2}. \qquad (12\text{–}130)$$

We shall discuss the physical significance of Γ shortly.

First, let us see what we can say about *stability* for the inviscid fluid. The key is to note that (12–128) and (12–129) are problems of the so-called *Sturm–Liouville* type. This means that we can characterize the sign of the growth-rate factor σ^2 based on the sign of F. Before drawing any conclusions, it may be useful to briefly review the general Sturm–Liouville theory. The latter relates to the properties of the general second order ODE,

$$\frac{d}{dx}\left(p(x)\frac{du}{dx}\right) - q(x)u + \lambda\rho(x)u = 0, \qquad (12\text{–}131)$$

with homogeneous boundary conditions

$$u(0) = u(1) = 0. \qquad (12\text{–}132)$$

The coefficients ρ and q are assumed to be continuous, and p is continuously differentiable. Furthermore, p and ρ are assumed to be positive, and q is nonnegative in the interval $0 \le x \le 1$. The values of λ for which the problem (12–131) and (12–132) has a nontrivial solution (i.e., a solution other than $u = 0$) are called the eigenvalues. It is shown in the general Sturm–Liouville theory that all eigenvalues are positive and real, $\lambda > 0$. Now if we compare (12–128) and (12–131),

$$p = r^{-1}, \; q = \alpha^2 r^{-1}, \; \rho = \alpha^2 r^{-1}, \text{ and } \lambda = -F\sigma^{-2}.$$

Thus,

(i) If $F > 0$, σ^2 must be real and negative for nontrivial solutions to exist for (12–128). Thus $\sigma = i\sigma_I$, and the flow is oscillatory but *neutrally stable*, in the sense that the amplitude of an infinitesimal perturbation does not grow with time.

(ii) If $F < 0$, σ^2 must be real and positive for nontrivial solutions to exist for (12–128). Thus $\sigma^2 = \sigma_R^2 > 0$, and one root is $\sigma_R > 0$. This means that the flow is *unstable*.

(iii) If $F > 0$ some places but $F < 0$ at other places, then, according to the general Sturm–Liouville theory, σ^2 has both positive and negative real parts and the flow is *unstable*.

Hydrodynamic Stability

Hence, in an inviscid fluid, the flow is *unstable* if, in any region,

$$\boxed{\frac{d}{dr}(\Gamma^2) < 0.} \qquad (12\text{–}133a)$$

but the flow is *stable* if

$$\boxed{\frac{d}{dr}(\Gamma^2) > 0} \qquad (12\text{–}133b)$$

in the entire fluid region. When we consider a viscous fluid, we may anticipate that the latter condition remains sufficient for *stability*. On the other hand, it is possible that the flow may *not* be unstable even if $d\Gamma^2/dr < 0$. We shall take up this latter question shortly. First, however, it may be useful to revisit the physical mechanism for instability now that we have the criteria (12–133a) and (12–133b).

A physically based explanation of the stability criterion (12–133b) was originally proposed by Rayleigh (1916) without benefit of the theory that we have developed. In effect, Rayleigh's proposal is the qualitative explanation that was given at the beginning of this section. We can formalize this explanation somewhat by introducing a theorem, known as the Kelvin circulation theorem. According to this theorem, if a fluid is inviscid, the circulation $2\pi r' u'_\theta$ along any fluid ring is constant. We have defined a dimensionless circulation, $\Gamma = ru_\theta$. We recall that variables without a prime are dimensionless and those with a prime are dimensional. According to the Kelvin theorem, $\Gamma = $ const for a given ring of fluid.

Now, at equilibrium,

$$\text{centripetal acceleration per unit mass} = \frac{u'^2_\theta}{r'} = \frac{\Gamma^2}{r^3}\left(R_1\Omega_1^2\right). \qquad (12\text{–}134)$$

This means, as already stated, that an inward force $[(\Gamma^2/r^3)(R_1\Omega_1^2)]$ (per unit mass) is necessary to keep the fluid element from undergoing radial motion. The steady-state pressure gradient

$$\frac{\partial p'}{\partial r'} = \frac{\rho \Gamma^2}{r^3}\left(R_1\Omega_1^2\right) \qquad (12\text{–}135)$$

is just sufficient for this purpose. Now suppose we have a ring at $r' = r'_1$ with $\Gamma_1 = r_1(\overline{u}_\theta)_{r=r_1}$ that is displaced outward to a new radial position $r' = r'_2 > r'_1$. The centripetal acceleration at position 2 is

$$\rho\frac{\Gamma_1^2}{r_2^3}\left(R_1\Omega_1^2\right), \qquad (12\text{–}136)$$

because Kelvin's circulation theorem tells us that $\Gamma = ru_\theta = $ const for a given ring of fluid. On the other hand, the pressure gradient at r_2 is

$$\rho\frac{\Gamma_2^2}{r_2^3}\left(R_1\Omega_1^2\right). \qquad (12\text{–}137)$$

Therefore, the system is stable if the pressure gradient exceeds the centripetal acceleration,

$$\Gamma_2^2 > \Gamma_1^2 \qquad (12\text{–}138a)$$

or

$$d\Gamma^2/dr > 0. \qquad (12\text{–}138b)$$

D. Taylor–Couette Instability

The criteria (12–138) for stability of an inviscid fluid had actually been obtained by Rayleigh using qualitative arguments long before any detailed analysis had been done. Rayleigh stated the condition for stability as

$$\Omega_2 R_2^2 \geq \Omega_1 R_1^2. \tag{12-139}$$

We can easily demonstrate that (12–138) and (12–139) are equivalent. If we begin with the definition of Γ in (12–130), then (12–138b) can be re-expressed as

$$C_1 \left(C_1 + \frac{C_2}{r^2} \right) > 0. \tag{12-140}$$

Now the Rayleigh condition ensures that $C_1 > 0$. Although C_2 may be negative, it is always less negative than C_1 is positive. Furthermore,

$$1 \leq r \leq R_2/R_1.$$

Hence the Rayleigh condition (12–139) ensures that the condition (12–140) is satisfied. From the Rayleigh form of the stability criteria we see that

$$\begin{aligned}&\text{(a)} \quad \Omega_1 = 0 \quad \text{(stable for all } \Omega_2\text{),}\\ &\text{(b)} \quad \Omega_2 = 0 \quad \text{(unstable for all } \Omega_1\text{),}\\ &\text{(c)} \quad \text{if } \Omega_1 \text{ and } \Omega_2 \text{ have opposite signs, then unstable.}\end{aligned} \tag{12-141}$$

More generally, *stability* requires that the cylinders be rotated in the same direction with the angular velocity of the outer cylinder exceeding that of the inner cylinder by the ratio $(R_2/R_1)^2$

2. Viscous Effects

We have seen in the preceding section that stability of an inviscid fluid in Couette flow requires that the Rayleigh condition be satisfied. Clearly, if the Rayleigh criterion is satisfied, the flow will also be stable for a fluid with finite viscosity. The diffusion of angular momentum in a viscous fluid will tend to *stabilize* the flow. In the framework of the qualitative explanation that was given in the introduction to this section, viscous diffusion provides a mechanism for the mismatch in angular momentum caused by the displacement of a fluid ring to be dissipated before it can lead to an instability. Hence an already stable configuration for an inviscid fluid will not be destabilized by the addition of viscosity. On the other hand, it is possible that unstable configurations for an inviscid fluid could be stabilized, thus postponing the onset of instability. Figure 12–3 shows, the instability bounds according to the Rayleigh criteria, as well as the actual stability boundary for a viscous fluid, calculated from the thin-gap theory that is subsequently presented. It is convenient to think of the criterion for instability as specifying the extent to which one may exceed $\Omega_1 = 0$ before instability sets in.

The full stability problem is given by the equations and boundary conditions (12–125)–(12–127). Although it can be solved, we consider here the somewhat simpler limit of a narrow gap that was originally analyzed by Taylor (1923) many years ago.[22] Taylor showed that, when the gap $(R_2 - R_1)$ is small compared with the mean cylinder radius, $(R_2 + R_1)/2$, we can find an explicit analytical solution by assuming (based on experimental evidence) that the growth-rate factor σ at the point of transition between stable and unstable states is equal to zero. In the preceding analyses, explicit expressions were obtained for σ, which turned out to be real. However, in general, σ may be real (thus giving $\sigma = 0$ at the "neutral" state), or it may be complex so that the neutral state (i.e., the state separating growing

Hydrodynamic Stability

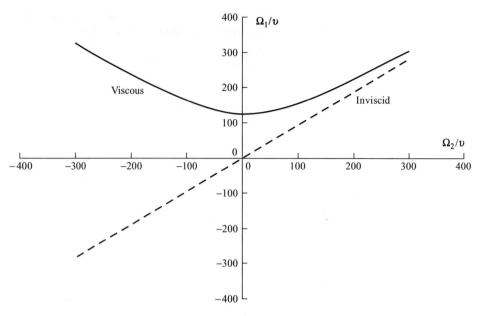

Figure 12–3. The stability bound according to the Rayleigh criteria (12–139), and the stability bound for a viscous fluid calculated from the thin-gap approximation (12–158). In this figure, we have assumed that $R_2 = 8$ cm and $R_1 = 7$ cm.

and decaying solutions) is characterized by the *real* part of σ going to zero and $\sigma \to i\sigma_I$. Complex values of σ indicate that the linear disturbance evolves in an oscillatory fashion in time, whereas real σ corresponds to dynamics that is monotonic in time. That is, instabilities set in as a stationary, secondary flow. A system that has real σ is said to exhibit the principle of "exchange of stabilities." Because the knowledge that $\sigma = 0$ at the neutral state generally simplifies the analysis, a key feature of many stability theories is first to establish whether the principle of the exchange of stabilities is satisfied or not. Often this can be done analytically. However, in the present case, we simply follow the precedent of Taylor without proof.

The assumption of a narrow gap means that in the dimensionless framework of equations (12–125)–(12–127), the range of r within the gap is

$$r = 1 + \delta, \text{ where } \delta \ll 1.$$

Hence the operator L in (12–125) and (12–126) can be approximated as

$$L \equiv D^2 + (D/r) - (1/r^2) \sim D^2[1 + O(\delta)] \sim D^2 \text{ in the limit } \delta \to 0. \quad (12\text{--}142)$$

It is also convenient to express (12–125) and (12–126) in *dimensional* form because the characteristic scales that were used in nondimensionalizing these equations are not the most appropriate choice for the thin-film problem (for example we used $\ell_c = R_1$, whereas the gap width is a much more appropriate choice for a characteristic length scale in the narrow gap problem). Hence, reversing the nondimensionalization (12–119) and introducing the approximation (12–142), we have

$$\nu \left(D'^2 - \alpha'^2 - \frac{\sigma'}{\nu} \right) (D'^2 - \alpha'^2) u' = 2\alpha'^2 \left(C_1' + \frac{C_2'}{r'^2} \right) v', \quad (12\text{--}143a)$$

$$\nu \left(D'^2 - \alpha'^2 - \frac{\sigma'}{\nu} \right) v' = 2C_1' u'. \quad (12\text{--}143b)$$

D. Taylor–Couette Instability

A final approximation for the narrow-gap limit can be made on the right-hand side of (12–143a). If we introduce $s \equiv (r' - R_1)/d$, where $d \equiv R_2 - R_1$, then

$$C_1' + \frac{C_2'}{r'^2} \sim \Omega_1 \left[1 - \left(1 - \frac{\Omega_2}{\Omega_1}\right) s \right]. \tag{12–144}$$

Now we nondimensionalize by using characteristic scales relevant to the narrow-gap limit,

$$\ell_c^* = d, \quad \hat{\alpha} = \alpha' d, \quad \hat{\sigma} = \sigma'(d^2/\nu), \quad \hat{D} = D' d^2. \tag{12–145}$$

Then, in dimensionless form, (12–143a) and (12–143b) become

$$(\hat{D}^2 - \hat{\alpha}^2 - \hat{\sigma})(\hat{D}^2 - \hat{\alpha}^2) u = \frac{2d^2 \Omega_1}{\nu} \hat{\alpha}^2 \{1 - [1 - (\Omega_2/\Omega_1)]s\} v,$$

$$(\hat{D}^2 - \hat{\alpha}^2 - \hat{\sigma}) v = \frac{2d^2}{\nu} C_1' u.$$

We may note that we have dropped the primes off of the velocity components u and v without having stated how we would scale these quantities. However, because the equations and boundary conditions are linear and homogeneous in these quantities, their nondimensionalization is completely arbitrary without any impact on the form of the equations.

Finally, because the equations are linear, there is no point in having combinations of dimensional parameters appearing in both equations. Instead, we define

$$\hat{u} \equiv \left(\frac{\nu}{2d^2 \Omega_1 \hat{\alpha}^2}\right) u. \tag{12–146}$$

Then

$$(\hat{D}^2 - \hat{\alpha}^2 - \hat{\sigma})(\hat{D}^2 - \hat{\alpha}^2) \hat{u} = \left[1 - \left(1 - \frac{\Omega_2}{\Omega_1}\right) s\right] v, \tag{12–147a}$$

$$(\hat{D}^2 - \hat{\alpha}^2 - \hat{\sigma}) v = -T \hat{\alpha}^2 \hat{u}, \tag{12–147b}$$

where the dimensionless parameter T is known as the Taylor number and defined as

$$T \equiv -\frac{4 C_1' \Omega_1}{\nu^2} d^4 \sim \frac{2 \Omega_1 (\Omega_1 - \Omega_2) R_1 d^3}{\nu^2}.$$

The differential operator is

$$\hat{D} \equiv \frac{d}{ds}$$

and the boundary conditions (12–127) are

$$\hat{u} = v = \hat{D}\hat{u} = 0 \quad \text{at } s = 0, 1. \tag{12–148}$$

Now, following the assumption of Taylor that the principle of exchange of stabilities is satisfied, so that $\sigma \equiv 0$ at the neutral stability point, the objective of the stability analysis is to obtain nontrivial solutions of (12–147) and (12–148) with $\sigma \equiv 0$ for various values of $\hat{\alpha}$ and then determine the minimum value of T as a function of $\hat{\alpha}$. This minimum of T is the critical value of the Taylor number for transition to instability. The corresponding value of $\hat{\alpha}$ is known as the critical wave number and represents the (dimensionless) wave

Hydrodynamic Stability

number of the most unstable linear mode, in the sense that as we slowly increase T the first infinitesimal mode that becomes unstable is the one with wave number $\hat{\alpha}$.

The system (12–147) and (12–148) is not *self-adjoint* and so the best we can do is an approximate solution. The analysis that we present is due to Chandrasekhar.[1] Because $v = 0$ at $s = 0$ and 1, we represent v in a sin series of the form

$$v = \sum_{m=1}^{\infty} c_m \sin m\pi s. \qquad (12\text{–}149)$$

Having chosen this approximate form for v, the equation (12–147a) becomes

$$(\hat{D}^2 - \hat{\alpha}^2)^2 \hat{u} = (1 + \beta s) \sum_{m=1}^{\infty} c_m \sin m\pi s, \qquad (12\text{–}150)$$

and we can solve this equation for $\hat{u}$ subject to the four boundary conditions on $\hat{u}$ and $\hat{D}\hat{u}$. The coefficient $\beta \equiv [(\Omega_2/\Omega_1) - 1]$. With $\hat{u}$ determined in this way, and v given by (12–149), Eq. (12–147b) then leads to a secular equation for T.

The general solution of (12–150) for $\hat{u}$ is

$$\hat{u} = \sum_{m=1}^{\infty} \frac{c_m}{(m^2\pi^2 + \hat{\alpha}^2)} \left[A_1^{(m)} \cosh \hat{\alpha} s + B_1^{(m)} \sinh \hat{\alpha} s + A_2^{(m)} s \cosh \hat{\alpha} s \right.$$
$$\left. + B_2^{(m)} s \sinh s + (1 + \beta s) \sin m\pi s + \frac{4\beta m\pi}{m^2\pi^2 + \hat{\alpha}^2} \cos m\pi s \right]. \qquad (12\text{–}151)$$

Application of the four boundary conditions on $\hat{u}$ then gives

$$A_1^{(m)} = -\frac{4\beta m\pi}{m^2\pi^2 + \hat{\alpha}^2},$$

$$B_1^{(m)} = \frac{m\pi}{\Delta}[\hat{\alpha} + \delta_m(\sinh\hat{\alpha} + \hat{\alpha}\cosh\hat{\alpha}) - \gamma_m \sinh\hat{\alpha}],$$

$$A_2^{(m)} = -\frac{m\pi}{\Delta}[\sinh^2\hat{\alpha} + \delta_m\hat{\alpha}(\sinh\hat{\alpha} + \hat{\alpha}\cosh\hat{\alpha}) - \gamma_m\hat{\alpha}\sinh\hat{\alpha}],$$

$$B_2^{(m)} = \frac{m\pi}{\Delta}[(\sinh\hat{\alpha}\cosh\hat{\alpha} - \hat{\alpha}) + \delta_m\hat{\alpha}^2\sinh\hat{\alpha} - \gamma_m(\hat{\alpha}\cosh\hat{\alpha} - \sinh\hat{\alpha})],$$

where

$$\Delta \equiv \sinh^2\hat{\alpha} - \hat{\alpha}^2,$$

$$\delta_m \equiv \frac{4\beta}{m^2\pi^2 + \hat{\alpha}^2}[(-1)^{m+1} + \cosh\hat{\alpha}],$$

$$\gamma_m \equiv (-1)^{m+1}(1 + \beta) + \frac{4\beta}{m^2\pi^2 + \hat{\alpha}^2}\hat{\alpha}\sinh\hat{\alpha}.$$

Finally, substituting (12–149) and (12–151) into (12–147b) yields

$$\sum_{n=1}^{\infty} c_n(n^2\pi^2 + \hat{\alpha}^2)\sin n\pi s = \hat{\alpha}^2 T \sum_{m=1}^{\infty} \frac{c_m}{(m^2\pi^2 + \hat{\alpha}^2)} \left[A_1^{(m)} \cosh\hat{\alpha} s + B_1^{(m)} \sinh\hat{\alpha} s \right.$$
$$+ A_2^{(m)} s \cosh\hat{\alpha} s + B_2^{(m)} s \sinh s + (1 + \beta s)\sin m\pi s$$
$$\left. + \frac{4\beta m\pi}{m^2\pi^2 + \hat{\alpha}^2} \cos m\pi s \right].$$

$$(12\text{–}152)$$

What remains is to determine the coefficients c_m. If exact results could be obtained for these coefficients satisfying Eq. (12–152), we would have an exact solution for the original

D. Taylor–Couette Instability

problem (12–147) and (12–148) with $\hat{\sigma} = 0$. Of course, this is not possible. Instead, what we try to do is obtain an approximate solution in an integral sense for the coefficients c_m. We do this by multiplying (12–152) by $\sin n\pi s$ and integrating over s from 0 to 1. The result is a set of linear homogeneous equations for the coefficients. The condition that the approximate solution not be trivial is that not all of the coefficients $c_m = 0$. This requires that the determinate of the coefficient matrix be equal to zero. Symbolically,

$$\sum_{n=1}^{\infty} X_{mn} c_n = 0 \quad \text{for } m = 1, 2, 3, \ldots, \infty,$$

and the condition that the determinate of coefficients equals zero provides an infinite-order characteristic equation of the form

$$\det X_{mn} = \left\| E_{mn}(\hat{\alpha}) - \frac{1}{2}(n^2\pi^2 + \hat{\alpha}^2)^3 \frac{\delta_{mn}}{\hat{\alpha}^2 T} \right\| = 0. \tag{12–153}$$

The objective from (12–153) is to determine T as a function of $\hat{\alpha}$. Because we cannot deal with it as an infinite-order equation, we might try to treat it approximately by setting the determinate equal to zero of the matrix formed from the first $n \times n$ rows and columns. We would then hope that we could begin with a small value of n and that the result for the lowest positive root would converge rather quickly for increasing n. In fact, Chandrasekhar has shown that a reasonable first approximation to the solution of (12–153) can be obtained by simply setting the (1,1) element of X_{mn} equal to zero. Although this would seem to be an extremely crude approximation, it turns out to give quite a reasonable approximation to the critical Taylor number T_c and the critical wave number $\hat{\alpha}_c$ in the range $0 \le \Omega_2/\Omega_1 \le 1.5$.

The result of setting $X_{11} = 0$ is a simple algebraic equation for T as a function of $\hat{\alpha}$, namely,

$$T = \frac{2}{2+\beta} \left(\frac{(\pi^2 + \hat{\alpha}^2)^3}{\hat{\alpha}^2 \{1 - 16\hat{\alpha}\pi^2 \cosh^2(\hat{\alpha}/2)/[(\pi^2 + \hat{\alpha}^2)^2 (\sinh \hat{\alpha} + \hat{\alpha})]\}} \right). \tag{12–154}$$

The critical Taylor number according to this approximate result is the minimum of T over the range of possible values for $\hat{\alpha}$, i.e., the value of $\hat{\alpha}$, where

$$\frac{dT}{d\hat{\alpha}} = 0 \quad \text{and} \quad \frac{d^2 T}{d\hat{\alpha}^2} > 0.$$

The result is

$$T_c = \frac{2}{2+\beta}(1715) = \frac{3430}{1 + (\Omega_2/\Omega_1)} \quad \text{at} \quad \hat{\alpha}_c = 3.12. \tag{12–155}$$

According to this approximate condition, Couette flow will be stable for

$$T \le T_c = \frac{3430}{1 + (\Omega_2/\Omega_1)}.$$

Hydrodynamic Stability

If we substitute in the definition for the Taylor number for this thin-gap approximation, this can be written in the form

$$\frac{4\Omega_1}{\nu^2}\left[\frac{(\Omega_2 R_2^2 - \Omega_1 R_1^2)}{(R_2^2 - R_1^2)}\right](R_2 - R_1)^4 \geq \frac{3430\Omega_1}{\Omega_1 + \Omega_2}. \qquad (12\text{--}156)$$

The Rayleigh criterion for the stability of an inviscid fluid was

$$\Omega_2 R_2^2 - \Omega_1 R_1^2 \geq 0. \qquad (12\text{--}157)$$

The condition (12–156) for a viscous fluid, on the other hand, can be expressed in the form

$$\Omega_2 R_2^2 - \Omega_1 R_1^2 \geq \frac{3430}{\Omega_1 + \Omega_2}\frac{(R_2^2 - R_1^2)}{(R_2 - R_1)^4}\frac{\nu^2}{4}. \qquad (12\text{--}158)$$

If $\Omega_2 = 0$, for example, the Rayleigh criterion indicates that Couette flow is unstable for any Ω_1. On the other hand, the approximate stability criterion (12–158) for a viscous fluid says that the system will remain stable for

$$\Omega_1^2 \leq 3430\frac{(R_2^2 - R_1^2)}{R_1^2(R_2 - R_1)^4}\frac{\nu^2}{4}.$$

Increased kinematic viscosity ν increases the stability of the flow. This role of viscous effects is fundamentally distinct from the preceding examples, in which the inclusion of viscosity was found to modify the growth rates of unstable disturbances, but not to modify the conditions for linear instability.

We have noted in the introduction to this section that there are a number of problems in which there is an analgous destabilization of a base flow that is due to centrifugal forces. Among these is the so-called Dean instability, which destabilizes the pressure gradient driven viscous fluid flow between a pair of parallel, but curved, walls (i.e., with $\partial p/\partial \phi \neq 0$). Closely related to this problem is the Taylor–Dean problem in which there is both an azimuthal pressure gradient, as in the Dean problem, and also relative rotation of the boundaries in the ϕ direction as in the Taylor–Couette case. One experimental approximation of this problem is realized when a vertical barrier to azimuthal flow is inserted into a Couette device at a fixed azimuthal angle.[23] Finally, a very well-known example of a centrifugally driven instability is the Gortler instability, which produces roll cells aligned in the mean flow direction within a boundary layer on a concave wall. We will consider simplified versions of these problems in the problems section. We will see that the stability theory is closely related to that for the Taylor–Couette problem.

E. NONISOTHERMAL AND COMPOSITIONALLY NONUNIFORM SYSTEMS

In all of the sections of this chapter so far, we have considered problems that exhibit instabilities for single-component, isothermal fluids. However, in many applications, the fluids may be multicomponent materials and may also be nonisothermal. This generally means that both the bulk and interfacial material properties will be nonuniform. Although we might be tempted to immediately go back and try to extend the results of the preceding sections to account for this fact, the derivation of such extended results would be quite

E. Nonisothermal and Compositionally Nonuniform Systems

complex in most instances, and the net result for the problems considered to this point would tend to be quantitative rather than qualitative changes in the results. This is because the basic mechanisms for instability and for stabilization of instability in these particular problems are not qualitatively changed by the addition of spatial gradients in material properties.

The more important problems for us to consider here are situations in which the presence of temperature or solute concentration gradients produce fluid motions, when otherwise no motion would exist. In this context, the presence of temperature or solute concentration gradients adds a potential new mechanism for instability. Examples include the addition of buoyancy forces due to nonuniform densities and the addition of Marangoni stresses at fluid interfaces, both of which are capable of driving fluid motions in systems that could otherwise exist in a stationary (i.e., motionless) state. We have previously discussed the Marangoni mechanism for driving fluid motions because of gradients in interfacial tension. However, in all of the previous chapters and sections of this book, we have assumed that the density was constant (the "forced" convection approximation). But, there is a whole additional class of problems in which flows are driven by buoyancy forces that are created because the density is nonuniform. The basic idea is that hydrostatic pressure gradients cannot balance spatially nonuniform body forces unless the gradients in density are precisely parallel to **g**, and in all other cases the fluid must be in motion. This general class of flows is known as "natural convection." A classic example is the well-known Rayleigh–Benard problem that we will consider in the next section. In that problem, we consider a stationary layer of fluid that is heated from below. Assuming that the density decreases with increase of the temperature, the stationary system thus consists of a layer with light fluid at the bottom and heavier fluid toward the top. If there is a small perturbation that disrupts the balance between hydrostatic pressure and the body forces, there is a clear possibility that these density gradients may drive a so-called "natural" convective motion in the fluid layer, with the lighter fluid rising up and the heavier fluid moving down.

In the next three sections, we consider three classical, but idealized problems that are intended to illustrate new sources for instability associated with nonuniform material properties. Once we recognize the potential for instability and we understand the mechanisms that can potentially stabilize these new mechanisms, we can qualitatively apply these ideas (and the corresponding analysis) to new situations, including an evaluation of the significance of the isothermal, uniform property assumption for cases like those evaluated in the preceding sections.

If we are going to consider problems involving fluid motions with nonuniform fluid properties that are due to the existence of nonuniform temperature or solute concentration profiles, the exact problem would involve the Navier–Stokes, continuity, and transport equations (either for heat or solute) in nearly their most general forms. Based on the discussion of the preceding chapters, the only approximations that seem completely obvious are to assume that the motions are slow enough that the fluids can be treated as incompressible and to assume that the Brinkman number is small enough to neglect viscous dissipation. This still leaves a very difficult, highly nonlinear problem. If we focus on nonisothermal problems, the governing equations for a single-component incompressible fluid are still in the form (2–108)–(2–110). Fortunately, under most circumstances involving moderate changes of temperature or solute concentration, there is a much simpler approximation of these equations that is known as the *Boussinesq approximation.* To explain this approximation, we focus on the case of a single-component nonisothermal fluid. The generalization to the case of compositionally nonuniform systems in which fluid properties are influenced by solute concentrations is a straightforward extension from the thermal problem.

Hydrodynamic Stability

The Boussinesq Equations

We begin with the dimensional form of the continuity, momentum, and thermal energy equations for an incompressible Newtonian fluid, simplified only to the extent that we neglect viscous dissipation:

$$\rho\left(\frac{\partial \mathbf{u}'}{\partial t'} + \mathbf{u}' \cdot \nabla'\mathbf{u}'\right) = \rho \mathbf{g} - \nabla' p' + \mu \nabla'^2 \mathbf{u}' + \nabla'\mu \cdot \left(\nabla'\mathbf{u}' + \nabla'\mathbf{u}'^T\right), \quad (12\text{–}159)$$

$$\frac{1}{\rho}\left(\frac{\partial \rho}{\partial t'} + \mathbf{u}' \cdot \nabla'\rho\right) + \nabla' \cdot \mathbf{u}' = 0, \quad (12\text{–}160)$$

$$\rho C_p\left(\frac{\partial T}{\partial t'} + \mathbf{u}' \cdot \nabla' T\right) = \nabla' \cdot (k \nabla' T). \quad (12\text{–}161)$$

The primes on $\mathbf{u}'$, p', and ∇' simply indicate that the variables are dimensional. Because the temperature T is assumed to be nonuniform, all of the material properties, ρ, μ, k, and C_p, also depend on spatial position and thus $\nabla'\mu$ and $\nabla' k \neq 0$. For convenience, we assume that the fluid has an ambient or reference temperature T_0, and we denote the material properties evaluated at this temperature as ρ_0, μ_0, k_0, and C_{p0}. The complexity of the problem represented by (12–159)–(12–161) is formidable. In the absence of an external imposed flow, the fluid motion is completely dependent on the density (that is, temperature) distribution in the fluid, which in turn depends on the velocity field. The equation of motion and the thermal energy equation are intimately coupled. Furthermore, the equations are very strongly nonlinear; even the viscous and conduction terms in (12–159) and (12–161) are now nonlinear in view of the dependence of ρ, μ, and k on T and the coupling between $\mathbf{u}'$ and T.

To make progress, we need to consider carefully the role of the body-force term in (12–159). It is clear from the preceding discussion that natural convection is due to the body-force term $\rho\, \mathbf{g}$. However, we have already seen that nonzero $\rho\, \mathbf{g}$ does not necessarily produce motion. In particular, *when ρ is constant*, the constant body force can be balanced completely by a hydrostatic variation of pressure with no motion. It is the *variation* in ρ *with spatial position* that leads to fluid motion. To account for these physical facts, it is convenient to express (12–1) in an alternative form by subtracting the hydrostatic terms, namely,

$$0 = -\nabla' p_h' + \rho_0\, \mathbf{g}. \quad (12\text{–}162)$$

Then, expressing p' in (12–159) as the sum of a hydrostatic and dynamic contribution,

$$p' = p_h' + p_d', \quad (12\text{–}163)$$

and subtracting (12–162) from (12–159), we obtain

$$\rho\left(\frac{\partial \mathbf{u}'}{\partial t'} + \mathbf{u}' \cdot \nabla'\mathbf{u}'\right) = (\rho - \rho_0)\mathbf{g} - \nabla' p_d' + \mu \nabla'^2 \mathbf{u}' + \nabla'\mu \cdot \left(\nabla'\mathbf{u}' + \nabla'\mathbf{u}'^T\right). \quad (12\text{–}164)$$

Now, if $\rho - \rho_0 \neq 0$, then $\mathbf{u}'$ must be *nonzero* to satisfy (12–164). It is evident in this form that the magnitude of the driving force for fluid motion is proportional to the magnitude of gradients of the density difference $\rho - \rho_0$.

Of course, the introduction of (12–162) and (12–163) into (12–159) to obtain (12–164) is analogous to the introduction of dynamic pressure to derive Eq. (2–91) for systems of *constant* density. Indeed, the commonly used description of p_d' as the dynamic pressure contribution derives from this previous analysis. In the present case, however, this designation is somewhat misleading. The contribution p_d' is actually the difference between the total pressure at a point and the hydrostatic pressure that would occur at the same point in

E. Nonisothermal and Compositionally Nonuniform Systems

the fluid if the density were ρ_0 everywhere. Thus, it includes both a *dynamic* contribution and a *static* contribution because of the fact that the density actually differs from ρ_0.

Of course, (12–164) is still exact, and the system of equations (12–164), (12–160), and (12–161) is no easier to solve than the original system of equations. To produce a tractable problem for analytic solution, it is necessary to introduce the so-called *Boussinesq approximation*, which has been used for many of the existing analyses of natural and mixed convection problems. The essence of this approximation is the assumption that the temperature variations in the fluid are small enough that the material properties ρ, μ, k, and C_p can be approximated by their values at the ambient temperature T_0, except in the body-force term in (12–164), where the approximation $\rho = \rho_0$ would mean that the fluid remains motionless.

The Boussinesq approximation can be formalized in the following way.[24] Let us assume that the maximum value of $T - T_0$ is small in the sense that the various material properties can be approximated in the linear forms

$$\frac{\rho_0}{\rho} = 1 + \beta(T - T_0) + \cdots +, \tag{12–165a}$$

$$\frac{\mu_0}{\mu} = 1 + \alpha(T - T_0) + \cdots +, \tag{12–165b}$$

$$\frac{k_0}{k} = 1 + \gamma(T - T_0) + \cdots +, \tag{12–165c}$$

$$\frac{C_{p0}}{C_p} = 1 + \delta(T - T_0) + \cdots +. \tag{12–165d}$$

If we substitute these expressions into (12–160), (12–161), and (12–164), we obtain

$$[1 + \beta \Delta T + O(\Delta T)^2]\left[\frac{\partial(\beta \Delta T)}{\partial t'} + \mathbf{u}' \cdot \nabla'(\beta \Delta T) + O(\Delta T)^2\right] + \nabla' \cdot \mathbf{u}' = 0, \tag{12–166a}$$

$$\rho_0 C_{p0}[1 - \beta \Delta T + O(\Delta T)^2][1 - \delta \Delta T + O(\Delta T)^2]\left(\frac{\partial T}{\partial t'} + \mathbf{u}' \cdot \nabla' T\right)$$
$$= k_0 \nabla' \cdot \{[1 - \gamma \Delta T + O(\Delta T)^2]\nabla' T\}, \tag{12–166b}$$

$$\rho_0[1 - \beta \Delta T + O(\Delta T)^2]\left(\frac{\partial \mathbf{u}'}{\partial t'} + \mathbf{u}' \cdot \nabla' \mathbf{u}'\right)$$
$$= -\rho_0[\beta \Delta T + O(\Delta T)^2]\mathbf{g} - \nabla' p'_d + \mu_0[1 - \alpha \Delta T + O(\Delta T)^2]\nabla'^2 \mathbf{u}'$$
$$- \mu_0 \nabla'[\alpha \Delta T + O(\Delta T)^2] \cdot (\nabla' \mathbf{u}' + \nabla' \mathbf{u}'^T), \tag{12–166c}$$

where $\Delta T \equiv T - T_0$. Now, let us suppose that

$$\beta \Delta T, \, \gamma \Delta T, \, \alpha \Delta T, \, \delta \Delta T \ll 1. \tag{12–167}$$

The coefficients β, γ, α, and δ are generally quite small ($\beta \sim 10^{-3}/°C$ for liquids), and thus these conditions do not necessarily imply very small values for ΔT. In any case, if the inequalities (12–167) are satisfied, it is clear that first approximations of (12–166a) and (12–166b) are

$$\nabla' \cdot \mathbf{u}' = 0, \tag{12–168}$$

$$\rho_0 C_{p0}\left(\frac{\partial T}{\partial t'} + \mathbf{u}' \cdot \nabla' T\right) = k_0 \nabla'^2 T. \tag{12–169}$$

843

We see that the errors in approximating (12–166a) and (12–166b) by (12–168) and (12–169) are $O(\Delta T)$ times the appropriate material coefficient β, γ, or δ, relative to the terms retained. The corresponding approximation to (12–166c) requires a little more care. Superficially, the prescription leading to (12–168) or (12–169) would seem to suggest *neglecting all* terms of $O(\Delta T)$, including the buoyancy term, thus leaving only the inertia, viscous, and dynamic pressure terms evaluated with the material properties at the ambient temperature.

However, a little thought will indicate that the inertia, viscous, and dynamic pressure terms in (12–166c) cannot be *larger* than the buoyancy term because the flows that are responsible for the *existence* of these terms are *driven* by the buoyancy forces. On the other hand, they cannot all be smaller either and still satisfy (12–166c). Thus, to approximate (12–166c) at the same level that is inherent in (12–168) and (12–169), we keep the first approximation to the inertia and viscous terms (that is, the terms independent of ΔT), plus the buoyancy term at $O(\Delta T)$, which *must* be the same magnitude as the *largest* of the inertia and/or viscous terms. The result is

$$\rho_0 \left(\frac{\partial \mathbf{u}'}{\partial t'} + \mathbf{u}' \cdot \nabla' \mathbf{u}' \right) = -\rho_0 \beta \Delta T \mathbf{g} - \nabla' p_d' + \mu_0 \nabla'^2 \mathbf{u}'. \quad (12\text{–}170)$$

The approximation leading to (12–170) may at first seem quite arbitrary. In particular, it appears that we have kept $O(1)$ approximations to the viscous, inertia, and pressure gradient terms but an $O(\Delta T)$ approximation to the buoyancy term. However, this is misleading. Because the motion in natural convection is driven by the buoyancy term, the characteristic magnitude of the *velocities* is actually proportional to $\beta \Delta T g$ raised to some power. Hence it will turn out that the terms *retained* in (12–170) are all of the same order of magnitude in general, and the terms neglected are all $O(\Delta T)^m$ smaller, where m is some positive exponent.[25]

Equations (12–168)–(12–170) are known as the *Boussinesq equations of motion* and will form the basis for the natural convection stability analyses in this chapter. In fact, the Boussinesq approximation has been used in much of the published theoretical work on natural convection flows. Although one should expect quantitative deviations from the Boussinesq predictions for systems in which the temperature differences are large (greater than $\sim 10°C$–$20°C$), it is likely that the Boussinesq equations remain qualitatively useful over a considerably larger range of temperature differences. In any case, although the Boussinesq equations represent a very substantial simplification of the exact equations, the essential property of *coupling* between the thermal and velocity fields is preserved, and, even in the Boussinesq approximation, the solution of natural convection problems is more complicated than the forced convection heat transfer problems that we encountered earlier.

We have already noted that the general class of flows driven by buoyancy forces that are created because the density is nonuniform is known as "natural convection." If we examine the Boussinesq approximation of the Navier–Stokes equations, (12–170), we can see that there are actually two types of natural convection problems. In the first, we assume that a fluid of ambient temperature T_0 is heated at a bounding surface to a higher temperature T_1. This will produce a nonuniform temperature distribution in the contiguous fluid, and thus a nonuniform density distribution too. Let us suppose that the heated surface is everywhere horizontal. Then there is a steady-state solution of (12–170) with $\mathbf{u}' = 0$, and the body-force terms balanced by a modification to the hydrostatic pressure distribution, such that

$$\nabla' p_d' = -\rho_0 \beta \Delta T \mathbf{g}. \quad (12\text{–}171)$$

However, if the heated surface makes an angle $\alpha(\mathbf{x}_s) \neq 0$ for some or all positions on the surface, then the component of the body force normal to the surface can at least locally be balanced by a hydrostatic pressure gradient normal to the wall, but no such balance is

F. Natural Convection in a Horizontal Fluid Layer Heated

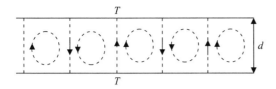

Figure 12–4. The Rayleigh–Benard configuration for buoyancy-driven convection in a horizontal fluid layer that is heated from below.

possible parallel to the wall, and the existence of a nonzero body-force term in (12–170) means that there must be motion of the fluid parallel to the wall for any *arbitrarily* small temperature nonuniformtiy. Of course, the existence of the stationary solution in the case of a horizontal surface does not mean that this solution will be realized in practice. If the lower plate is heated, then the fluid configuration consists of heavy fluid overlaying a lighter fluid and it is clear that this is potentially unstable. However, we shall see that, unlike the case of a heated nonhorizontal surface in which a temperature difference of any magnitude will produce motion in the fluid, the horizontal configuration is unstable only if the temperature difference exceeds a certain critical value.

F. NATURAL CONVECTION IN A HORIZONTAL FLUID LAYER HEATED FROM BELOW – THE RAYLEIGH–BENARD PROBLEM

We have argued in the proceding section that, if the temperature gradient at every point in the fluid is in the same direction as gravity, it is possible to have a pure conduction mode of heat transfer with *no* induced motion in the fluid. The best-known prototype occurs in an "unbounded" layer of fluid between two horizontal plane surfaces with heating from below. This is the famous Rayleigh–Benard problem. Convection occurs in this case only because the basic pure conduction state is unstable above some critical value of the temperature gradient. For small temperature differences between the two horizontal boundaries, the fluid does not move, and heat transfer is by conduction. However, as the temperature difference is increased, a critical condition is eventually reached where the motionless base state becomes unstable, and there is an abrupt and spontaneous transition to a convective mode of heat transfer. The analysis here largely follows the developments that are due to Chandrasekhar.[1]

1. The Disturbance Equations and Boundary Conditions

The Rayleigh–Benard configuration is sketched in Fig. 12–4. We assume that the fluid is between two infinite plane surfaces that are separated by a distance d. The lower surface is at a constant temperature T_1 and the upper one is at a lower temperature T_0. Our starting point is the time-dependent Boussinesq equations, (12–160), (12–169), and (12–170), but now in *dimensionless* form,

$$\nabla \cdot \mathbf{u} = 0, \qquad (12\text{–}172)$$

$$\frac{\partial \theta}{\partial t} + \mathbf{u} \cdot \nabla \theta = \frac{1}{Pr}\left(\frac{v_0}{u_c d}\right)\nabla^2 \theta, \qquad (12\text{–}173)$$

$$\frac{\partial \mathbf{u}}{\partial t} + \mathbf{u} \cdot \nabla \mathbf{u} = -\nabla p + \left(\frac{v_0}{u_c d}\right)\nabla^2 \mathbf{u} + \frac{\beta(T_1 - T_0)gd}{u_c^2}(1-\theta)\mathbf{i}_z. \qquad (12\text{–}174)$$

Here d is the characteristic length scale, the characteristic velocity scale is denoted as u_c but is left unspecified for the moment, the characteristic time scale is d/u_c, and

$$\theta \equiv \frac{T_1 - T}{T_1 - T_0}. \qquad (12\text{–}175)$$

This definition is convenient because $\theta = 0$ at $z = 0$, and $\theta = 1$ at $z = 1$. Pr is the Prandtl number, ν_0/κ_0. Note that $\mathbf{g} = -g\mathbf{i}_z$.

If we examine (12–174) more closely, we see that there are several "possible" choices for the characteristic velocity. An apparently convenient choice is to designate $u_c = \nu_0/d$ as this appears to reduce the number of parameters to the greatest extent. This choice of characteristic velocity is really saying that the characteristic time scale is the diffusion time d^2/ν_0. The governing equations, (12–173) and (12–174), then become

$$\frac{\partial \theta}{\partial t} + \mathbf{u} \cdot \nabla \theta = \frac{1}{Pr}\nabla^2 \theta, \tag{12-176}$$

$$\frac{\partial \mathbf{u}}{\partial t} + \mathbf{u} \cdot \nabla \mathbf{u} = -\nabla p + \nabla^2 \mathbf{u} + Gr(1-\theta)\mathbf{i}_z. \tag{12-177}$$

The new dimensionless parameter that appears in these equations is known as the Grashof number,

$$Gr \equiv \frac{\beta(T_1 - T_0)gd^3}{\nu_0^2}.$$

The first thing to notice is that there is a steady-state solution for *arbitrary* values of the dimensionless parameters, namely,

$$\boxed{\begin{aligned} \mathbf{u} &= 0, \\ \theta &= \overline{\theta} = z, \\ \nabla p &= \nabla \overline{p} = +Gr(1-z)\mathbf{i}_z. \end{aligned}} \tag{12-178}$$

In this mode, heat transfer is strictly by *conduction*. Because the temperature gradient is vertical, the body force can be completely balanced by a hydrostatic pressure gradient, and the solution (12–178) is valid over the whole range of possible values for $T_1 - T_0$.

In fact, *experimental* studies show that the static layer is actually maintained up to some finite, critical value of $T_1 - T_0$, but beyond this point there is an *abrupt* and *spontaneous* transition to convective motion. The most clearcut evidence of this transition, in addition to flow visualization, is an abrupt increase in the slope of the heat transfer rate versus the temperature difference $T_1 - T_0$. It is notable that the abrupt transition from conduction to convection occurs, for a given fluid and a given fluid layer depth, at a value of $T_1 - T_0$ that is independent of any precautions that may be taken to isolate the system from external disturbances. This suggests that the static fluid layer is *unstable* to disturbances of arbitrarily small magnitude.

The most important feature to try to predict theoretically is the critical condition for instability, as this marks the boundary between relatively inefficient heat transfer by conduction and the onset of convection. To do this, we consider the fate of an arbitrary, fully 3-D initial perturbation of the base state (12–178), namely,

$$\boxed{\begin{aligned} \mathbf{u} &= (\varepsilon \hat{u}, \ \varepsilon \hat{v}, \ \varepsilon \hat{w}), \\ \theta &= \overline{\theta} + \varepsilon \hat{\theta}, \\ p &= \overline{p} + \varepsilon \hat{p}, \end{aligned}} \tag{12-179}$$

where the amplitude of the perturbation is denoted by ε.

The governing equations for the disturbance variables are just the Boussinesq equations, (12–172), (12–176), and (12–177). However, in view of the fact that $\varepsilon \ll 1$, these equations

F. Natural Convection in a Horizontal Fluid Layer Heated

can be *linearized* by neglect of all quantities that are quadratic in ε. Thus, substituting (12–179) into the Boussinesq equations and linearizing, we obtain (in Cartesian component form)

$$\frac{\partial \hat{u}}{\partial x} + \frac{\partial \hat{v}}{\partial y} + \frac{\partial \hat{w}}{\partial z} = 0, \quad (12\text{–}180)$$

$$\frac{\partial \hat{\theta}}{\partial t} + \hat{w} = \frac{1}{Pr} \nabla^2 \hat{\theta}, \quad (12\text{–}181)$$

$$\frac{\partial \hat{u}}{\partial t} = -\frac{\partial \hat{p}}{\partial x} + \nabla^2 \hat{u}, \quad (12\text{–}182)$$

$$\frac{\partial \hat{v}}{\partial t} = -\frac{\partial \hat{p}}{\partial y} + \nabla^2 \hat{v}, \quad (12\text{–}183)$$

$$\frac{\partial \hat{w}}{\partial t} = -\frac{\partial \hat{p}}{\partial z} + \nabla^2 \hat{w} - Gr\hat{\theta}. \quad (12\text{–}184)$$

The boundary conditions depend, of course, on the nature of the flat surfaces at $z = 0$ and 1. We shall return to conditions for the velocity components shortly. It may be noted that the Rayleigh–Benard problem has historically been stated in a slightly different form, which is equivalent to rescaling $\hat{\theta}$ to $\widetilde{\theta} \equiv \hat{\theta}/Pr$. The difference is that (12–181) and (12–184) then become

$$Pr\frac{\partial \widetilde{\theta}}{\partial t} + \hat{w} = \nabla^2 \widetilde{\theta}, \quad (12\text{–}185)$$

$$\frac{\partial \hat{w}}{\partial t} = -\frac{\partial \hat{p}}{\partial z} + \nabla^2 \hat{w} - Ra\widetilde{\theta}, \quad (12\text{–}186)$$

where

$$Ra \equiv Gr\,Pr$$

is known as the Rayleigh number. Obviously, the formulations in terms of Gr and Ra are entirely equivalent, and the choice is just a matter of preference.

To analyze (12–180)–(12–184), it is convenient to combine them in order to reduce the number of dependent variables. This can be done in a straightforward way. First, we can combined (12–182) and (12–183) by differentiating with respect to x and y, respectively, and adding. The result is

$$\frac{\partial}{\partial t}\left(\frac{\partial \hat{u}}{\partial x} + \frac{\partial \hat{v}}{\partial y}\right) = -\nabla_1^2 \hat{p} + \nabla^2\left(\frac{\partial \hat{u}}{\partial x} + \frac{\partial \hat{v}}{\partial y}\right), \quad (12\text{–}187)$$

where

$$\nabla_1^2 \equiv \frac{\partial^2}{\partial x^2} + \frac{\partial^2}{\partial y^2}.$$

However, using continuity, we can write this as

$$\left(\frac{\partial}{\partial t} - \nabla^2\right)\frac{\partial \hat{w}}{\partial z} = \nabla_1^2 \hat{p}. \quad (12\text{–}188)$$

We have thus reduced the original set of five equations and five unknowns to three equations, (12–181), (12–184), and (12–188), for three variables, $\hat{w}$, $\hat{p}$, and $\hat{\theta}$. To eliminate

$\hat{p}$ we can combine (12–184) and (12–188). First, we differentiate (12–188) with respect to z,

$$\left(\frac{\partial}{\partial t} - \nabla^2\right)\frac{\partial^2 \hat{w}}{\partial z^2} = \nabla_1^2\left(\frac{\partial \hat{p}}{\partial z}\right), \tag{12–189}$$

and then we operate on (12–184) with ∇_1^2,

$$\left(\frac{\partial}{\partial t} - \nabla^2\right)\nabla_1^2 \hat{w} = -\nabla_1^2\left(\frac{\partial \hat{p}}{\partial z}\right) - Gr\nabla_1^2\hat{\theta}. \tag{12–190}$$

Then, adding (12–189) and (12–190), we obtain

$$\left(\frac{\partial}{\partial t} - \nabla^2\right)\nabla^2 \hat{w} = -Gr\nabla_1^2\hat{\theta}. \tag{12–191}$$

We now have two equations, (12–181) and (12–191), for $\hat{\theta}$ and $\hat{w}$. Finally, these two equations can be combined to eliminate $\hat{\theta}$. To do this, we operate on (12–191) with $[\partial/\partial t - (1/\Pr)\nabla^2]$,

$$\left(\frac{\partial}{\partial t} - \frac{1}{\Pr}\nabla^2\right)\left(\frac{\partial}{\partial t} - \nabla^2\right)\nabla^2 \hat{w} = -Gr\nabla_1^2\left(\frac{\partial}{\partial t} - \frac{1}{\Pr}\nabla^2\right)\hat{\theta}.$$

Hence, substituting on the right-hand side by using (12–181), we obtain

$$\boxed{\left(\frac{\partial}{\partial t} - \frac{1}{\Pr}\nabla^2\right)\left(\frac{\partial}{\partial t} - \nabla^2\right)\nabla^2 \hat{w} = Gr\nabla_1^2 \hat{w}.} \tag{12–192}$$

Thus the problem of describing the time-dependent evolution of an arbitrary initial disturbance reduces to the solution of a single sixth-order PDE for the disturbance velocity component $\hat{w}$. Once $\hat{w}$ is known, the other disturbance quantities can be calculated. For example, $\hat{\theta}$ can be obtained from (12–181), and $\hat{p}$, $\hat{u}$, and $\hat{v}$ can then be determined by means of (12–180), (12–182), and (12–183). The objective here is to use (12–192) to examine the conditions for growth or decay of an arbitrary initial disturbance in $\hat{w}$. Because the governing equation is linear and separable, it will clearly have a solution of the form

$$\boxed{\hat{w} = W(x, y, z)e^{\sigma t},} \tag{12–193}$$

with

$$\boxed{\left(\sigma - \frac{1}{\Pr}\nabla^2\right)(\sigma - \nabla^2)\nabla^2 W = Gr\nabla_1^2 W.} \tag{12–194}$$

As in the previous cases discussed in this chapter, the stability is then equivalent to determining the *sign* of the real part of the growth rate σ. It is clear, on examination of (12–194), that σ generally depends on Gr, $\Pr$, and the spatial form of the disturbance function $W(x,y,z)$. The question of representing an arbitrary initial disturbance is best considered in

F. Natural Convection in a Horizontal Fluid Layer Heated

the context of an overall solution scheme for (12–194). Because this equation is linear and separable, the solution W can be expressed in the general form

$$W(x, y, z) = f(z)F(x, y). \tag{12-195}$$

Further, it is clear on examination of (12–194) that any solution for F must satisfy the constraint

$$\nabla_1^2 F = \text{const } F.$$

In fact, such a solution is just

$$F = e^{i(\alpha_x x + \alpha_y y)}, \tag{12-196}$$

in which case

where
$$\nabla_1^2 F = -\alpha^2 F \tag{12-197}$$

$$\alpha^2 = \alpha_x^2 + \alpha_y^2. \tag{12-198}$$

But (12–196) is nothing more than a normal Fourier component with wave numbers α_x and α_y. As in the previous sections of this chapter, it is sufficient simply to consider the dynamics of a single Fourier mode with the wave number α as a parameter. Although the growth rate σ will clearly depend on α, we can obtain a sufficient condition for instability by finding the maximum value of the Grashof number for which $Re(\sigma) \equiv 0$ over all possible values of α, $0 \le \alpha \le \infty$.

We should note that the linear theory cannot provide any information on the spatial pattern of any unstable mode, but only information on its length scale by means of the wave number α. Any combination of α_x and α_y that yields the critical value of the combined parameter α is consistent with the linear theory. To make predictions about the spatial pattern that might be observed in an experiment (for example, whether the unstable mode is "roll cells," "hexagonal cells," or some other form) requires a theory that includes nonlinear terms corresponding to finite-amplitude disturbances.[26] Although it is sometimes true that the experimentally observed wavelength is similar to the wavelength of the most unstable mode from linear theory, this is somewhat of an accident. By the time a disturbance flow pattern is large enough to be observed experimentally, it is definitely beyond the allowable amplitude for the linear theory, which allows only for "infinitesimal" disturbances. Nonlinear effects may lead to a significant change in the wavelength of the observable mode by the time it grows to an amplitude that is strong enough to allow for experimental observation.

The functional form for $f(z)$ must satisfy the ODE we obtain by substituting (12–195) and (12–197) into (12–194), that is,

$$\left\{ \left[\sigma - \frac{1}{\Pr}(D^2 - \alpha^2) \right] [\sigma - (D^2 - \alpha^2)](D^2 - \alpha^2) + Gr\alpha^2 \right\} f = 0, \tag{12-199}$$

as well as boundary conditions for f that are to be specified at $z = 0$ and 1. Note that D^2 in (12–199) is shorthand for the second derivative, d^2/dz^2. The boundary conditions on f depend on the physical assumptions that are made about the bounding surfaces at $z = 0$

Hydrodynamic Stability

and 1. Because these surfaces are assumed to be nondeforming, the kinematic condition is simply

$$\boxed{f = 0 \quad \text{at } z = 0, \ 1.} \tag{12–200}$$

However, the DE for f is sixth order. Thus we require at least four additional boundary conditions. One physically plausible assumption is that the boundaries at $z = 0, 1$ are isothermal. In this case,

$$\hat{\theta} = 0 \quad \text{at } z = 0, \ 1,$$

and we can see from (12–191) that this implies

$$\boxed{[\sigma - (D^2 - \alpha^2)](D^2 - \alpha^2)f = 0 \quad \text{at } z = 0, 1.} \tag{12–201}$$

The third pair of boundary conditions on f depends on whether the boundaries at $z = 0$ and 1 are solid or free surfaces. At a solid surface, $\hat{u} = \hat{v} = 0$, and it therefore follows from continuity that

$$\frac{\partial \hat{w}}{\partial z} = 0, \quad z = 0, \ 1.$$

From (12–193) and (12–195), we see that this implies

$$\boxed{Df = 0 \quad \text{at a solid boundary.}} \tag{12–202}$$

On the other hand, at a free surface, where the fluid layer is bounded above (or below) by an inviscid fluid, the boundary condition of shear-stress continuity is approximated as

$$\hat{\tau}_{xz} = \hat{\tau}_{yz} = 0.$$

Using the definition for τ for a Newtonian fluid, we find that these conditions become

$$\frac{\partial \hat{u}}{\partial z} = \frac{\partial \hat{v}}{\partial z} = 0$$

or

$$\frac{\partial}{\partial z}\left(\frac{\partial \hat{u}}{\partial x} + \frac{\partial \hat{v}}{\partial y}\right) = 0.$$

Hence, by continuity,

$$\frac{\partial^2 \hat{w}}{\partial z^2} = 0,$$

and this implies that

$$\boxed{D^2 f = 0 \quad \text{at a free surface.}} \tag{12–203}$$

The most common configuration for a fluid layer that is heated from below is probably to have a solid boundary at $z = 0$ and a free surface (gas–liquid interface) at $z = 1$. However, there is a major simplification of the analysis if both surfaces are free, and the results are qualitatively similar regardless of which of the boundary conditions, (12–202) or (12–203), is applied at which boundary.

F. Natural Convection in a Horizontal Fluid Layer Heated

The simplification for two free surfaces is that f and *all* even derivations of f are zero at $z = 0, 1$. To see this, we note that the conditions (12–200) and (12–203), when substituted into the condition (12–201), require that

$$D^4 f = 0 \quad \text{at } z = 0, 1. \tag{12–204}$$

Hence, from the governing equation, (12–199), it can be seen that $D^6 f = 0$ at $z = 0, 1$, and so on, by differentiating (12–199) to higher-order derivatives of f with respect to z.

Thus, for two free surfaces, the *eigenvalue* problem for σ is reduced to finding a nontrival solution of Eq. (12–199), subject to the six boundary conditions, (12–200), (12–203), and (12–204). In particular, let us suppose that we specify Pr, Gr, and α^2 (the wave number of the normal model of perturbation). There is then a single eigenvalue for σ such that $f \neq 0$. If $\text{Real}(\sigma) < 0$ for all α, the system is stable to infinitesimal disturbances. On the other hand, if $\text{Real}(\sigma) > 0$ for *any* α, it is unconditionally unstable. Stated in another way, the preceding statements imply that for any Pr there will be a certain value of Gr such that all disturbances of any α decay. The largest such value of Gr is called the *critical* value for linear stability.

2. Stability for Two Free Surfaces

The stability problem with two free surfaces was first solved in 1916 by Lord Rayleigh. Because f and all even derivatives of f are zero at $z = 0$ and $z = 1$, a fundamental solution of the problem is

$$f_n = \sin(n\pi z), \tag{12–205}$$

where n is any integer, $n \geq 1$. This satisfies all of the boundary conditions, (12–200), (12–203), and (12–204). The governing equation, (12–199), becomes

$$\left[\sigma + \frac{1}{Pr}(n^2\pi^2 + \alpha^2)\right][\sigma + (n^2\pi^2 + \alpha^2)](n^2\pi^2 + \alpha^2) - Gr\alpha^2 = 0.$$

Multiplying by Pr, we can rewrite this equation in the form first derived by Rayleigh:

$$(\sigma Pr + \gamma)(\sigma + \gamma)\gamma - \alpha^2 Ra = 0, \tag{12–206}$$

where the Rayleigh number was defined previously in conjunction with (12–186) and

$$\gamma \equiv (n^2\pi^2 + \alpha^2). \tag{12–207}$$

Clearly, for Pr, γ, α, and Ra fixed, (12–206) is a *characteristic* equation for the eigenvalues, i.e., the growth-rate coeffieicent, σ. For $Ra \neq 0$, the solution is

$$\sigma = \frac{1}{2Pr}\left[-\gamma(Pr+1) \pm \sqrt{\gamma^2(Pr-1)^2 + \frac{4\alpha^2 Ra Pr}{\gamma}}\right]. \tag{12–208}$$

In the limit $Ra \equiv 0$, there is no buoyancy to sustain an initial disturbance, and we see that there are two real, negative values for σ:

$$\sigma_1 = -\frac{\gamma}{Pr} \quad \text{and} \quad \sigma_2 = -\gamma \quad \text{for} \quad Ra \equiv 0. \tag{12–209}$$

As expected, in this case the system is stable.

Hydrodynamic Stability

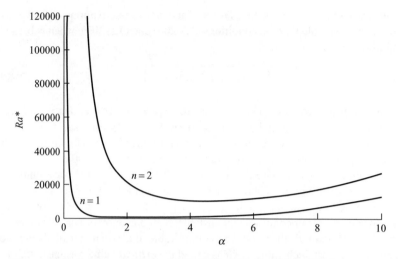

Figure 12–5. Stability criteria for the Rayleigh–Benard problem. The two curves shown are the neutral stability curves for the modes $n = 1$ and $n = 2$. The region above the curve for $n = 1$ is unstable, whereas that below is stable. The critical Rayleigh number is 657.511 at a critical wave number of 2.221

More generally, because $Pr > 0$ and $\gamma > 0$, it can be seen from (12–208) that the two eigenvalues are always *real*, and this means that an initial disturbance will either grow or decay monotonically with time. Thus the neutral state, separating stable and unstable conditions, corresponds to

$$\boxed{\sigma = 0.} \qquad (12\text{–}210)$$

It is interesting to note that the growth rates predicted from (12–208) generally depend on Pr, but the condition of neutral stability is seen from (12–206) to be independent of Pr in the present case. Going back to (12–206), it follows from (12–210) that the value of Rayleigh number that gives neutral stability is

$$\boxed{Ra^* = \frac{(n^2\pi^2 + \alpha^2)^3}{\alpha^2} \quad \text{for} \quad n = 1, 2, 3, \ldots,} \qquad (12\text{–}211)$$

For a given wave number, the *least* stable mode corresponds to $n = 1$ (this yields the smallest value of Ra^*). If we plot Ra^* versus α for $n = 1$, as in Fig. 12–5, we obtain the so-called neutral stability curve. For a given α, any value of Ra that exceeds $Ra^*(\alpha)$ corresponds to an *unstable* system, whereas any smaller value is stable. The *critical* Rayleigh number, Ra_c, for linear instability is the minimum value of Ra^* for all possible values of α, and the corresponding value of $\alpha = \alpha_{\min}$ is known as the critical wave number.

We can obtain an analytic expression for Ra_c from (12–211) by finding the value of α such that $dRa^*/d\alpha^2 = 0$. On differentiating (12–211), we find that

$$\boxed{\begin{aligned} \alpha_c &= \sqrt{\frac{\pi^2}{2}} = 2.2214, \\ Ra_c &= \frac{27}{4}\pi^4 = 657.511. \end{aligned}} \qquad (12\text{–}212)$$

F. Natural Convection in a Horizontal Fluid Layer Heated

Thus, for $Ra < 657.511$, the system is stable to arbitrary small disturbances. However, any real system that satisfies the assumptions of the analysis will be unstable for $Ra > 657.511$, as it is impossible to avoid an infinitesimal disturbance that contains the α_c component.

We have noted earlier that the case of a fluid layer with two free surfaces is qualitatively representative of the general nature of the solution for more realistic boundary conditions. Of course, the change from a slip to no-slip boundary condition does tend to stabilize the flow somewhat so that the critical Rayleigh number increases in the order

$$\underset{\substack{\text{Two free}\\\text{boundaries}}}{Ra_c} \quad < \quad \underset{\substack{\text{One free}\\\text{boundary}}}{Ra_c} \quad < \quad \underset{\substack{\text{Two no-slip}\\\text{boundaries}}}{Ra_c}.$$

A detailed analysis in the latter case shows, in fact, that

$$\underset{\substack{\text{Two no-slip}\\\text{boundaries}}}{Ra_c} = 1707.762 \text{ with } \alpha_c = 3.117. \tag{12–213}$$

One may also question the assumption of isothermal boundaries. In reality, rather than strictly isothermal conditions, with fixed temperatures T_0 and T_1, we might expect to have a finite heat flux to the surroundings. Suppose that the "surroundings" consist of a pair of reservoirs, the bottom one at a constant temperature T_1 and the upper one at a constant temperature T_0. Then, if we assume local equilibrium at the two boundaries (which could be either solid or a fluid interface),

$$-k\frac{\partial T}{\partial z'} = -h_0 (T - T_1) \text{ at } z' = 0 \quad \text{and} \quad -k\frac{\partial T}{\partial z'} = h_1(T - T_0) \text{ at } z' = d,$$

where h_0 and h_1 are heat transfer coefficients. The right-hand sides represent heat transferred to (or from) the surrounding reservoir, which must be balanced at local equilibrium by the heat flux from (or to) the fluid layer to the boundary by means of conduction. Thus, in dimensionless terms,

$$\frac{k}{d}\frac{\partial \theta}{\partial z} = h_0\theta \text{ at } z = 0, \quad \text{and} \quad -\frac{k}{d}\frac{\partial \theta}{\partial z} = h_1(\theta - 1) \text{ at } z = 1,$$

and the temperature perturbation satisfies a mixed-type boundary condition

$$\frac{\partial \hat{\theta}}{\partial z} = \left(\frac{h_0 d}{k}\right)\hat{\theta} = Bi_0\,\hat{\theta} \text{ at } z = 0 \quad \text{and} \quad -\frac{\partial \hat{\theta}}{\partial z} = \left(\frac{h_1 d}{k}\right)\hat{\theta} = Bi_1\,\hat{\theta} \text{ at } z = 1. \tag{12–214}$$

Here, $Bi \equiv hd/k$ is known as the Biot number. In this case, the isothermal limit corresponds to $Bi \to \infty$, and it turns out that the system is less stable (lower Ra_c) for finite Bi. It may also be noted that if the boundaries were heated to temperatures T_0 and T_1 by a heater that maintains a constant heat flux rather than maintaining a fixed temperature, the boundary conditions for the temperature perturbation would be (12–214) with $Bi_i = 0$.

3. The Principle of Exchange of Stabilities

The case of two free surfaces allows a straightforward analytic solution, including a closed-form expression for the growth-rate factor σ. The other possible sets of boundary conditions are more difficult. Frequently, however, instead of trying to achieve a complete solution, we will be satisfied with trying to determine the conditions corresponding to the neutral stability point. We follow that objective here. It may not be obvious to the reader that this objective

Hydrodynamic Stability

is any simpler than solving the full problem. There is one class of stability problems for which it is a major simplification, and that is problems that satisfy the so-called "principle of exchange of stabilities." In this case, the eigenvalue is known to be real. Then the neutral stability condition is that $\sigma \equiv 0$. In general, we only know that the real part of σ is zero at the neutral point. If we can derive a closed-form analytic formula for σ, as we have been able to do for almost every problem to date, we can see directly whether σ is real or complex, but there is no real point because we can evaluate the formula for σ directly to establish the relationships between the independent parameters at the neutral point. However, for other problems, there is a major simplification if we can establish ahead of time whether σ is real or not.

It is worthwhile illustrating the proof of the principle of exchange of stabilities for the Rayleigh–Benard problem. Not only will this allow us to discuss the derivation of instability criteria for the case of no-slip boundaries, but the approach to proving this principle can also be applied to other problems.

We begin at the point of Eq. (12–199), but now written in the alternative form

$$(D^2 - \alpha^2 - \sigma)(D^2 - \alpha^2)f = \alpha^2 Ra\theta,$$
$$(D^2 - \alpha^2 - \sigma Pr)\theta = -f. \tag{12-215}$$

We use the boundary conditions in the forms

$$z = 0 \begin{cases} f = 0, & \text{either } Df \quad \text{or} \quad D^2 f = 0, \\ \theta = 0 \end{cases} \tag{12-216a}$$

$$z = 1 \begin{cases} f = 0, & \text{either } Df \quad \text{or} \quad D^2 f = 0. \\ \theta = 0 \end{cases} \tag{12-216b}$$

Now let us suppose that σ is complex with eigensolutions θ and f and σ^* is the complex conjugate of σ with corresponding eigensolutions θ^*, f^* (these are the complex conjugates of θ and f). We can then combine (12–215) in the following form:

$$f(D^2 - \alpha^2 - \sigma^*)(D^2 - \alpha^2)f^* + \alpha^2 Ra\theta^*(D^2 - \alpha^2 - \sigma Pr)\theta = 0.$$

Thus,

$$f[D^4 - (2\alpha^2 + \sigma^*)D^2 + (\alpha^4 + \alpha^2\sigma^*)]f^* + \alpha^2 Ra\theta^*(D^2 - \alpha^2 - \sigma Pr)\theta = 0. \tag{12-217}$$

Now we integrate this equation from $z = 0$ to $z = 1$, evaluating the various terms by integration by parts. First,

$$\int_0^1 f D^4 f^* dz = f D^3 f^* \Big|_0^1 - \int_0^1 Df (D^3 f^*) dz. \tag{12-218a}$$

The first term is zero because $f = 0$ at $z = 0, 1$. Integrating once again, we find

$$\int_0^1 f D^4 f^* dz = -Df(D^2 f^*) \Big|_0^1 + \int_0^1 (D^2 f)(D^2 f^*) dz. \tag{12-218b}$$

Again the first term is zero, either because $Df = 0$ (no − slip) or $D^2 f^* = 0$ (zero shear). But now we note that $hh^* = |h|^2$ for any complex number or function. Furthermore, because f^* is the complex conjugate of f, it follows that $D^2 f^*$ is the complex conjugate of $D^2 f$. Hence

$$\int_0^1 f D^4 f^* dz = \int_0^1 |D^2 f|^2 dz \equiv J_2. \tag{12-218c}$$

F. Natural Convection in a Horizontal Fluid Layer Heated

By means of similar arguments, the following additional results can be obtained

$$\int_0^1 f\, D^2 f^* dz = -\int_0^1 |Df|^2 dz \equiv -J_1, \qquad (12\text{–}219\text{a})$$

$$\int_0^1 ff^* dz = -\int_0^1 |f|^2 dz \equiv J_0, \qquad (12\text{–}219\text{b})$$

$$\int_0^1 \theta^* D^2 \theta\, dz = -\{\int_0^1 |D\theta|^2 dz\} \equiv -J_4, \qquad (12\text{–}219\text{c})$$

$$\int_0^1 \theta \theta^* dz = \int_0^1 |\theta|^2 dz \equiv J_3. \qquad (12\text{–}219\text{d})$$

Thus (12–217) becomes

$$J_2 + (2\alpha^2 + \sigma^*)J_1 + (\alpha^4 + \alpha^2 \sigma^*)J_0 = \alpha^2 Ra[J_4 + (\alpha^2 + \sigma\, Pr)J_3]. \qquad (12\text{–}220)$$

We equally well could have combined (12–215) in the opposite sense so that

$$f^*[D^4 - (2\alpha^2 + \sigma)D^2 + (\alpha^4 + \alpha^2 \sigma)]f + \alpha^2 Ra\theta(D^2 - \alpha^2 - \sigma^* Pr)\theta^* = 0.$$

The integral equalities (12–218) and (12–219) then give

$$J_2 + (2\alpha^2 + \sigma)J_1 + (\alpha^4 + \alpha^2 \sigma)J_0 = \alpha^2 Ra[J_4 + (\alpha^2 + \sigma^* Pr)J_3]. \qquad (12\text{–}221)$$

Now adding (12–220) and (12–221), we get (after dividing by 2)

$$J_2 + 2\alpha^2 J_1 + \alpha^4 J_0 - \alpha^2 Ra(J_4 + \alpha^2 J_3) = -\sigma_{Real}(J_1 + \alpha^2 J_0 - \alpha^2 Pr\, Ra\, J_3). \qquad (12\text{–}222)$$

Now, the J_i are all positive definite. Hence, from (12–222), we can reach the conclusion that

$$\boxed{\sigma_{Real} < 0 \quad \text{if} \quad Ra \le 0.} \qquad (12\text{–}223)$$

Of course, this conclusion is not very interesting as it only says that the system is stable if we cool it from below! However, if we subtract (12–221) from (12–222) and then divide by $2i$ we obtain the more important result

$$\sigma_{Imag}[J_1 + \alpha^2 J_0 + \alpha^2 Ra\, Pr\, J_3] = 0.$$

Thus,

$$\boxed{\sigma_{Imag} = 0 \quad \text{if} \quad Ra \ge 0.} \qquad (12\text{–}224)$$

Thus for a fluid layer heated from below, σ is real, and we can simply set $\sigma = 0$ at the neutral state, and this greatly simplifies the problem of finding Ra_c corresponding to the marginal stability state, when a simple analytic solution is not possible.

4. Stability for Two No-Slip, Rigid Boundaries

The linear stability problem when one or both of the boundaries above and below a fluid layer are rigid, no-slip surfaces is difficult to solve. Thus the principle of exchange of stabilities, which we have just proven, is a major advantage because we can set $\sigma = 0$ at the neutral stability point (i.e., at the transition between stable and unstable conditions).

Hydrodynamic Stability

In this case, the governing equation (12–199) becomes

$$\boxed{(D^2 - \alpha^2)^3 f = -\alpha^2 Ra\, f.} \qquad (12\text{–}225)$$

In view of the symmetry of the problem it is advantageous to assume that the boundaries are at $z = \pm 1/2$. Then the boundary conditions are

$$\boxed{\begin{aligned} f &= 0 \quad \text{at} \quad z = \pm 1/2, \\ Df &= 0 \quad \text{at} \quad z = \pm 1/2, \\ (D^2 - \alpha^2)^2 f &= 0 \quad \text{at} \quad z = \pm 1/2. \end{aligned}} \qquad \begin{aligned}(12\text{–}226)\\(12\text{–}227)\\(12\text{–}228)\end{aligned}$$

The latter assumes that the boundaries are isothermal.

Now, the problem (12–225)–(12–228) is an eigenvalue problem. In this case, given α^2, a nonzero solution will exist only a for certain value of Ra. Because we have already set $\sigma = 0$, the critical value Ra_{crit} is then the minimum of Ra^* for all possible α.

Now, because the governing equation (12–225) has constant coefficients we can seek a general solution as a superposition of solutions of the form

$$f = \exp(\pm qz). \qquad (12\text{–}229)$$

If we substitute (12–229) into (12–225), we obtain a characteristic equation for q:

$$\boxed{(q^2 - \alpha^2)^3 = -\alpha^2 Ra.} \qquad (12\text{–}230)$$

To write a solution of this equation, it is convenient to introduce a new variable τ such that

$$\alpha^2 Ra^* \equiv \alpha^6 \tau^3.$$

In other words,

$$\tau \equiv (Ra^*/\alpha^4)^{1/3}. \qquad (12\text{–}231)$$

Then the roots of (12–230) are

$$\begin{aligned} q^2 &= -\alpha^2(\tau - 1), \\ q^2 &= \alpha^2 \left[1 + \frac{\tau}{2}(1 \pm i\sqrt{3})\right], \end{aligned} \qquad (12\text{–}232)$$

or, taking square roots, we have six roots

$$\begin{aligned} \pm i q_0, &\quad \text{where} \quad q_0 = \alpha(\tau - 1)^{1/2}, \\ \pm q, &\quad \text{where} \quad q = q_r + i q_i, \\ \pm q^*, &\quad \text{where} \quad q^* = q_r - i q_i, \end{aligned} \qquad (12\text{–}233)$$

with

$$q_r = \alpha \left[\frac{1}{2}\sqrt{1 + \tau + \tau^2} + \frac{1}{2}\left(1 + \frac{1}{2}\tau\right)\right]^{1/2},$$

$$q_i = \alpha \left[\frac{1}{2}\sqrt{1 + \tau + \tau^2} - \frac{1}{2}\left(1 + \frac{1}{2}\tau\right)\right]^{1/2}.$$

We can consider either even or odd solutions based on the solution form (12–229). The *even solution* takes the form

$$f = A_0 \cos q_0 z + A \cosh q z + A^* \cosh q^* z. \qquad (12\text{–}234)$$

F. Natural Convection in a Horizontal Fluid Layer Heated

Table 12–3. Rayleigh number at the neutral stability point ($\sigma = 0$) as a function of the dimensionless wave number

α	Ra
0	∞
1	5854.5
2	2177.4
3	1711.28
3.117	1707.76
4	1879.3
5	5459.3

Then
$$Df = -A_0 q_0 \sin q_0 z + Aq \sinh qz + A^* q^* \sinh q^* z, \qquad (12\text{–}235)$$

$$(D^2 - \alpha^2)^2 f = A_0(q_0^2 + \alpha^2)^2 \cos q_0 z + A(q^2 - \alpha^2)^2 \cosh qz + A^*(q^{*2} - \alpha^2)^2 \cosh q^* z. \qquad (12\text{–}236)$$

Now
$$(q_0^2 + \alpha^2)^2 = \alpha^4 \tau^2,$$
$$(q^2 - \alpha^2)^2 = \frac{1}{2}\alpha^4 \tau^2 (-1 \pm i\sqrt{3}),$$

so (12–236) can be rewritten in the alternative form
$$(D^2 - \alpha^2)^2 f = \frac{1}{2}\alpha^4 \tau^2 [2A_0 \cos q_0 z + (i\sqrt{3} - 1)A \cosh qz - (i\sqrt{3} + 1)A^* \cosh q^* z]. \qquad (12\text{–}237)$$

Now the boundary conditions (12–225)–(12–228) give a 3×3 matrix equation for the coefficients A_0, A, A^*. Namely

$$\begin{vmatrix} \cos \frac{1}{2}q_0 & \cosh \frac{1}{2}q & \cosh \frac{1}{2}q^* \\ -q_0 \sin \frac{1}{2}q_0 & q \sinh \frac{1}{2}q & q^* \sinh \frac{1}{2}q^* \\ \cos \frac{1}{2}q_0 & \frac{1}{2}(i\sqrt{3} - 1) \cosh \frac{1}{2}q & -\frac{1}{2}(i\sqrt{3} + 1) \cosh \frac{1}{2}q^* \end{vmatrix} \begin{vmatrix} A_0 \\ A \\ A^* \end{vmatrix} = 0. \qquad (12\text{–}238)$$

The condition for nontrivial solutions is that the determinate of the coefficient matrix must be equal to zero. After some algebraic manipulation (which can be done with one of the computer algebra programs), this leads to

$$-q_0 \tan \frac{1}{2}q_0 = \frac{(q_r + q_i\sqrt{3}) \sinh q_r + (q_r\sqrt{3} - q_i) \sinh q_i}{\cosh q_r + \cos q_i}. \qquad (12\text{–}239)$$

This is a very complicated transcendental equation relating α and $\tau \equiv (Ra^*/\alpha^4)^{1/3}$, which must be solved numerically. We specify α and then use this equation to determine τ, and then from τ we calculate Ra^*. Because $\sigma = 0$, the relationship of Ra^* versus α defines the neutral stability curve for the even solutions. A set of numerically calculated values is given in Table 12–3. The critical Rayleigh number is the minimum value of Ra^* as a function of α. It can be seen from Table 12–3 that this is 1707 and occurs at a critical wave number of 3.117. This is the result that was stated earlier as (12–113).

In addition to the even solution (12–234), there is also an odd solution of the form

$$f = B_0 \sin q_0 z + B \sinh qz + B^* \sinh q^* z$$

that can also be made to satisfy all of the boundary conditions by following the same method previously outlined. However, this solution is much more stable. In this case the minimum Rayleigh number is 17,610 at a wavenumber of 5.365. Clearly, the stability is controlled by the even solution, and 1707 is the global critical Rayleigh number.

G. DOUBLE-DIFFUSIVE CONVECTION

A seemingly simple extension of the Rayleigh–Benard problem is to consider the instability of a fluid layer when the density gradient is established as a consequence of gradients of two independent entities, such as temperature and a solute, or two solutes. The most highly studied example comes from the field of physical oceanography, in which vertical density gradients are established in the oceans or seas by gradients of temperature and salinity. In this case, the problem is also known as thermohaline convection. In this case the contributions to the density gradient may be opposing; e.g., evaporation renders the water more salty near the upper surface, but the fluid is also hotter at the top because of radiant heating. Another common system for laboratory investigations is a system containing two solutes, most frequently sugar and salt.

In any case, if we have a stationary fluid layer with a vertical density profile, it might appear that this problem is highly redundant and that one should be able to anticipate the behavior by applying the Rayleigh–Benard results. From this we would conclude that the system is stable if the density decreases in moving vertically upward. Furthermore, in the case in which the density is larger at the top, the system may be stabilized by diffusion processes. However, it turns out that the Rayleigh–Benard picture can be far from the truth in many instances. A striking indication of this is that a fluid layer can be unstable with convection setting in spontaneously, even if the overall density decreases upward, so that the system would be stable according to the Rayleigh–Benard analysis for a system with a single entity (either temperature or the concentration of a solute) influencing the density.

The key to understanding this statement and the overall behavior of a system with two entities contributing to the density gradient is to note that instabilities that cannot be accounted for by the Rayleigh–Benard theory require (1) that the density profile for one of the two entities be unstable even though the overall density profile is apparently stable, and (2) that the diffusivity for transport of the two entities must be significantly different. This is the source of the name "double-diffusive convection." Borrowing from the oceanographic literature, we can identify two distinct cases.

The first is known as the "finger regime". In this case we have warm, salty water overlaying cooler, fresher water. Let us suppose that the overall density profile has the density decreasing as we go from the bottom to the top of the fluid layer. The contribution to the density profile produced by the salinity gradient is unstable in the sense that the salty water at the top has a higher density from the fresher water below it. However, the contribution that is due to the temperature gradient is stable and, because the overall density decreases with height, the temperature gradient contribution to the density profile is stronger. In this case, an instability leading to the spontaneous onset of convection may occur even though the overall density profile appears to be stable. Qualitatively, the explanation for this lies in the difference in the thermal and mass diffusivities. A fluid element that is displaced upward

G. Double-Diffusive Convection

by some perturbation process will quickly equilibrate thermally with its new surroundings, but will still have less salt. This means that it is lighter than its surroundings and there is a net buoyancy force that causes it to continue to move upwards. When this happens, the system will spontaneously undergo convection and it is unstable. Of course, this physical description is just a "cartoon" of a complex situation, and it must be confirmed and refined by analysis. We note that the temperature profile is stabilizing in the sense that initially warmer water is on top, and it is the fact that the diffusion of heat occurs on a time scale that is much faster than the diffusion of salt that is responsible for *the essential destabilizing process*. Furthermore, this simple physical picture suggests that the resultant motion will be a cellular convection (not oscillatory in time) and the principle of exchange of stabilities should be operational for this case. Of course, this speculation must again be confirmed by detailed analysis. Finally, we should note that the key generic feature identifying this "finger regime" is that the unstable contribution to the density gradient is contributed by the entity (salt) that diffuses more slowly (i.e., $D_{salt} < \kappa$).

The second case is known as the "diffusive regime". In this case, for the oceanographic application, we have cold, fresh water overlaying warmer, saltier water below. The contribution of the salinity gradient to the overall density gradient is stable, whereas the contribution that is due to the temperature gradient is unstable. In this case, the unstable contribution to the overall density profile is associated with the entity that has the larger diffusivity. Again, to gain qualitative understanding, we can consider the fate of a fluid element that is displaced upward by some perturbation. In this case, the "displaced" fluid element will lose heat to the surroundings but retains its original salinity (more or less), and this produces a net buoyancy force that drives the fluid element back toward its initial position. However, because the transport of heat occurs at a finite rate, the fluid element will always be a little cooler that its surroundings, and this produces an excess in the restoring force relative to what we would have in the salinity gradient alone, and this may cause the fluid element to overshoot its initial position of neutral equilibrium and then undergo an oscillatory motion. If the magnitude of these oscillations grows, the system is unstable. Of course, in this case, the motion is oscillatory, not monotonic, in time, and the principle of exchange of stabilities will not be valid. Again, the driving mechanism for instability is the difference in the diffusive transport rate for the two entities that control the density distribution.

Much of this behavior can be predicted by means of a classic linear stability analysis that is quite similar to the analysis of the Rayleigh–Benard problem of the preceding section. We consider again a fluid layer of width d. We assume that the density distribution is determined by a temperature profile and a solute concentration profile (though the transition to two solutes is completely straightforward, as we will see). Because many of the steps of deriving governing equations for the linear stability theory are similar to the analysis of Rayleigh–Benard instability in the preceding section, we will try to avoid repetition as much as possible without being unclear.

The basic starting point is again the governing equations (12–160), (12–161), and (12–164). To these equations, we must add a species transport equation for our solute

$$\frac{\partial c}{\partial t'} + \mathbf{u}' \cdot \nabla' c = \nabla' \cdot (D \nabla' c). \qquad (12\text{–}240)$$

Here, D is the species diffusivity. In deriving the Bousinesq approximation of the Navier-Stokes equations, we must remember that the material properties now depend both on the temperature and the solute concentration, e.g.,

$$\frac{\rho_0}{\rho} = 1 + \beta(T - T_0) - \beta'(c - c_0) + O((\Delta T)^2, (\Delta c)^2). \qquad (12\text{–}241)$$

Hydrodynamic Stability

Hence, the (dimensional) Boussinesq equations become

$$\rho_0 \left(\frac{\partial \mathbf{u}'}{\partial t'} + \mathbf{u}' \cdot \nabla' \mathbf{u}' \right) = -\rho_0(\beta \Delta T - \beta' \Delta c)\mathbf{g} - \nabla' p'_d + \mu_0 \nabla'^2 \mathbf{u}', \qquad (12\text{--}242)$$

$$\nabla' \cdot \mathbf{u}' = 0, \qquad (12\text{--}243)$$

$$\rho_0 C_{p0} \left(\frac{\partial T}{\partial t'} + \mathbf{u}' \cdot \nabla' T \right) = k_0 \nabla'^2 T, \qquad (12\text{--}244)$$

$$\frac{\partial c}{\partial t'} + \mathbf{u}' \cdot \nabla' c = D_0 \nabla'^2 c, \qquad (12\text{--}245)$$

where the material properties are evaluated at the ambient (or reference) temperature and concentration, T_0 and c_0. These equations can be nondimensionalized following the analysis of the Rayleigh–Benard problem. If we again choose the characteristic velocity

$$u_c = \nu_0/d,$$

the result is

$$\nabla \cdot \mathbf{u} = 0, \qquad (12\text{--}246)$$

$$\frac{\partial \theta}{\partial t} + \mathbf{u} \cdot \nabla \theta = \frac{1}{Pr} \nabla^2 \theta, \qquad (12\text{--}247)$$

$$\frac{\partial \chi}{\partial t} + \mathbf{u} \cdot \nabla \chi = \frac{1}{Sc} \nabla^2 \chi, \qquad (12\text{--}248)$$

$$\frac{\partial \mathbf{u}}{\partial t} + \mathbf{u} \cdot \nabla \mathbf{u} = -\nabla p + \nabla^2 \mathbf{u} + \left(Gr(1 - \theta) - \hat{Gr}(1 - \chi) \right) \mathbf{i}_z. \qquad (12\text{--}249)$$

Here, the dimensionless concentration and temperature are

$$\chi \equiv \frac{c_1 - c}{c_1 - c_0} \quad \text{and} \quad \theta \equiv \frac{T_1 - T}{T_1 - T_0}. \qquad (12\text{--}250)$$

The dimensionless parameters that appear are the Prandtl and Schmidt numbers,

$$Pr \equiv \nu_0/\kappa_0, \text{ and } Sc \equiv \nu_0/D_0 \qquad (12\text{--}251\text{a})$$

and a pair of Grashof numbers

$$Gr \equiv \frac{\beta(T_1 - T_0)gd^3}{\nu_0^2} \quad \text{and} \quad \hat{Gr} \equiv \frac{\hat{\beta}(c_1 - c_0)gd^3}{\nu_0^2}. \qquad (12\text{--}251\text{b})$$

As in the case of the Rayleigh–Benard problem, there is a steady-state solution of these equations:

$$\boxed{\begin{aligned} \mathbf{u} &= 0, \\ \theta &= \overline{\theta} = z, \\ \chi &= \overline{\chi} = z, \\ \nabla p &= \nabla \overline{p} = +(Gr - \hat{Gr})((1-z)\mathbf{i}_z). \end{aligned}} \qquad (12\text{--}252)$$

G. Double-Diffusive Convection

To determine the stability of this stationary state, we consider the fate of an arbitrary, fully 3D initial perturbation of the base state (12–252), namely,

$$\boxed{\begin{aligned}\mathbf{u} &= (\varepsilon\hat{u}, \varepsilon\hat{v}, \varepsilon\hat{w}), \\ \theta &= \bar{\theta} + \varepsilon\hat{\theta}, \\ \chi &= \bar{\chi} + \varepsilon\hat{\chi}, \\ p &= \bar{p} + \varepsilon\hat{p},\end{aligned}} \tag{12-253}$$

where the amplitude of the perturbation is denoted by ε.

The governing linear stability equations for the disturbance variables are (in Cartesian component form)

$$\frac{\partial \hat{u}}{\partial x} + \frac{\partial \hat{v}}{\partial y} + \frac{\partial \hat{w}}{\partial z} = 0, \tag{12-254}$$

$$\frac{\partial \hat{\theta}}{\partial t} + \hat{w} = \frac{1}{Pr}\nabla^2\hat{\theta}, \tag{12-255}$$

$$\frac{\partial \hat{\chi}}{\partial t} + \hat{w} = \frac{1}{Sc}\nabla^2\hat{\chi}, \tag{12-256}$$

$$\frac{\partial \hat{u}}{\partial t} = -\frac{\partial \hat{p}}{\partial x} + \nabla^2\hat{u}, \tag{12-257}$$

$$\frac{\partial \hat{v}}{\partial t} = -\frac{\partial \hat{p}}{\partial y} + \nabla^2\hat{v}, \tag{12-258}$$

$$\frac{\partial \hat{w}}{\partial t} = -\frac{\partial \hat{p}}{\partial z} + \nabla^2\hat{w} - Gr\hat{\theta} + \hat{Gr}\hat{\chi}. \tag{12-259}$$

The boundary conditions depend, of course, on the nature of the flat surfaces at $z = 0$ and 1. We shall return to conditions for the velocity components shortly.

Alternatively, the double-diffusion problem has traditionally been analyzed in terms of the Rayleigh numbers

$$Ra = Gr\,Pr \quad \text{and} \quad \hat{Ra} = \hat{Gr}\,Sc.$$

As noted earlier in conjunction with the Rayleigh–Benard analysis, this is equivalent to redefining a dimensionless temperature and concentration

$$\hat{\theta} \text{ to } \widetilde{\theta} \equiv \hat{\theta}/Pr \quad \text{and} \quad \widetilde{\chi} \equiv \hat{\chi}/Sc.$$

The governing equations are then (12–254), (12–257), (12–258), plus

$$Pr\frac{\partial \widetilde{\theta}}{\partial t} + \hat{w} = \nabla^2\widetilde{\theta}, \tag{12-260}$$

$$Sc\frac{\partial \widetilde{\chi}}{\partial t} + \hat{w} = \nabla^2\widetilde{\chi}, \tag{12-261}$$

$$\frac{\partial \hat{w}}{\partial t} = -\frac{\partial \hat{p}}{\partial z} + \nabla^2\hat{w} - Ra\widetilde{\theta} + \hat{Ra}\widetilde{\chi}. \tag{12-262}$$

Hydrodynamic Stability

The so-called finger case of hot, salty water over cold, fresh water corresponds to $Ra < 0$ and $\hat{R}a < 0$, and in this case the salt gradient is "unstable" (in the sense that it corresponds to more-dense fluid overlaying less-dense fluid) whereas the temperature gradient is stable. On the other hand, the diffusive case of cold, fresh water over hot, salty water has $Ra > 0$ and $\hat{R}a > 0$, and it is the temperature gradient that tends to produce instability whereas the salinity gradient is stable.

To proceed, we now combine Eqs. (12–254), (12–257), and (12–258) to eliminate $\hat{u}$, $\hat{v}$, and $\hat{p}$, thus producing

$$\left(\frac{\partial}{\partial t} - \nabla^2\right)\nabla^2 \hat{w} = -Ra\nabla_1^2 \hat{\theta} + \hat{R}a \nabla_1^2 \hat{\chi}, \tag{12–263}$$

$$\left(Pr\frac{\partial}{\partial t} - \nabla^2\right)\hat{\theta} = \hat{w}, \tag{12–264}$$

$$\left(Sc\frac{\partial}{\partial t} - \nabla^2\right)\hat{\chi} = \hat{w}. \tag{12–265}$$

If we then introduce the disturbance functions

$$\boxed{\begin{aligned}\hat{\theta} &= e^{\sigma t} F(x, y) h(z),\\ \hat{\chi} &= e^{\sigma t} F(x, y) k(z),\\ \hat{w} &= e^{\sigma t} F(x, y) f(z),\end{aligned}} \tag{12–266}$$

we can express these in the form

$$\boxed{\begin{aligned}(D^2 - \alpha^2 - \sigma)(D^2 - \alpha^2)f &= -\alpha^2 Ra\, h + \alpha^2 \hat{R}a\, k, &(12\text{–}267)\\ (D^2 - \alpha^2 - \sigma Pr)h &= -f, &(12\text{–}268)\\ (D^2 - \alpha^2 - \sigma Sc)k &= -f. &(12\text{–}269)\end{aligned}}$$

For arbitrary values of Ra and $\hat{R}a$, the principle of "exchange of stabilities" does not hold. We therefore limit our analysis to the case of two free surfaces, for which we can obtain closed-form analytic results. Although idealistic, this case is adequate to demonstrate the most important qualitative features of the double-diffusive convection problem.

As in the Rayleigh–Benard problem, the boundary conditions for two free surfaces are

$$\boxed{f = D^2 f = 0 \quad \text{at } z = 0, 1,} \tag{12–270}$$

and if we assume that the surfaces are isothermal and also isosalinity boundaries so that

$$h = k = 0 \quad \text{at } z = 0, 1, \tag{12–271}$$

then

$$\boxed{D^4 f = 0 \quad \text{at } z = 0, 1.} \tag{12–272}$$

Thus we assume that the eigensolutions are sums of terms of the form

$$f = \sum_n A_n f_n \text{ where } f_n = \sin n\pi z, \tag{12–273}$$

G. Double-Diffusive Convection

and it is sufficient to consider just a single term for all possible values of n. Clearly in view of (12–268) and (12–269), it follows from (12–273) that it is also sufficient to consider

$$h_n = k_n = \sin n\pi z. \qquad (12\text{–}274)$$

If we substitute into (12–267)–(12–269) we obtain

$$(k^2 + \sigma)k^2 f_n = \alpha^2 (Ra\, h_n - \hat{Ra}\, k_n),$$
$$(k^2 + \sigma Pr)h_n = f_n, \qquad (12\text{–}275)$$
$$(k^2 + \sigma Sc)k_n = f_n,$$

where

$$k^2 \equiv n^2\pi^2 + \alpha^2.$$

We can eliminate h_n and k_n to obtain a characteristic equation that is cubic in σ:

$$(k^2 + \sigma)k^2(k^2 + \sigma Pr)(k^2 + \sigma Sc) = \alpha^2[Ra\,(k^2 + \sigma Sc) - \hat{Ra}\,(k^2 + \sigma Pr)].$$

Rearranging, we find that this becomes

$$\boxed{\begin{aligned}& Sc\, Pr\, \sigma^3 + (Pr + Sc + Sc\, Pr)k^2 \sigma^2 \\ & \quad + \left[k^4(Pr + Sc + 1) + \frac{\alpha^2}{k^2}(Pr\,\hat{Ra} - Sc\, Ra)\right]\sigma \\ & \quad + [k^6 + \alpha^2(\hat{Ra} - Ra)] = 0.\end{aligned}} \qquad (12\text{–}276)$$

It will be convenient for what follows to rewrite (12–276) in the shorthand form

$$A\sigma^3 + B\sigma^2 + C\sigma + D = 0.$$

Instead of seeking a general solution to this cubic equation, we can see what conclusions are possible by using the fact that we are interested in characterizing the conditions for transition between stable and unstable states by means of the condition of neutral stability, where the real part of σ is equal to zero. First we find the real and imaginary parts of (12–276) with

$$\sigma = \sigma_{Real} + i\sigma_{Imag} \quad \text{and} \quad \sigma_{Real} = 0.$$

The real part is

$$-B\sigma_{Imag}^2 + D = 0, \qquad (12\text{–}277)$$

and the imaginary part is

$$-A\sigma_{Imag}^3 + C\sigma_{Imag} = 0. \qquad (12\text{–}278)$$

Now there are two possible cases, either $\sigma_{Imag} = 0$ or $\sigma_{Imag} \ne 0$. If $\sigma_{Imag} = 0$, the unique conclusion from (12–277) and (12–278) is that D must be equal to zero, namely

$$(Ra - \hat{Ra})_{\sigma=0} = \frac{(n^2\pi^2 + \alpha^2)^3}{\alpha^2}.$$

Hydrodynamic Stability

The minimum occurs for $n = 1$ and $\alpha^2 = \pi^2/2$, and is

$$(Ra - \hat{Ra})_{\sigma=0} = \frac{27\pi^4}{4}. \tag{12-279}$$

This is essentially just the neutral stability condition for direct-mode buoyancy-driven convection. In the Ra, $\hat{Ra}$ plane, this is a straight line that defines a possible stability boundary. Note, however, that we do not get conditions for validity of the assumption $\sigma_{Imag} = 0$. For this, we must examine the other case, $\sigma_{Imag} \neq 0$.

Let us therefore assume $\sigma_I \neq 0$. In this case, we see from (12–277) and (12–278) that

$$\sigma_{Imag}^2 = \frac{D}{B} \quad \text{and} \quad \sigma_{Imag}^2 = \frac{C}{A}, \tag{12-280}$$

respectively. We can draw two conclusions from (12–280). First, if $\sigma_{Imag} \neq 0$, it follows that σ_{Imag}^2 is positive definite and real, and thus a necessary condition for existence of $\sigma_{Imag} \neq 0$ and $\sigma_{Real} = 0$ is that

$$C > 0, D > 0. \tag{12-281}$$

Second, where $\sigma_{Imag} \neq 0$, the ratio of coefficients D/B and C/A must be equal. After some algebra, this condition can be written in the form

$$Ra = \left(\frac{1+Sc}{1+Pr}\right)\left(\frac{Pr}{Sc}\right)^2 \hat{Ra} + \left(\frac{Pr}{Sc} + 1\right)\left(\frac{1}{Sc} + 1\right)\frac{(n^2\pi^2 + \alpha^2)^3}{\alpha^2}.$$

This is the relationship between Ra and $\hat{Ra}$ on the neutral curve when $\sigma_I \neq 0$. The minimum value of Ra (or $\hat{Ra}$) at the neutral point occurs again for $n = 1$ and $\alpha^2 = \pi^2/2$, and the relationship

$$Ra_c = \left(\frac{1+Sc}{1+Pr}\right)\left(\frac{Pr}{Sc}\right)^2 \hat{Ra} + \left(\frac{Pr}{Sc} + 1\right)\left(\frac{1}{Sc} + 1\right)\frac{27\pi^4}{4} \tag{12-282}$$

defines another straight line in the Ra, $\hat{Ra}$ plane that will be a stability boundary whenever $\sigma_{Imag} \neq 0$.

Now, to determine when either (12–279) or (12–282) defines the stability boundary, we must establish conditions when $\sigma_{Imag} \neq 0$. To do this, we simply examine the conditions for C and $D > 0$. These are

$$Ra - \hat{Ra} < \frac{k^6}{\alpha^2} = \frac{27\pi^4}{4}; \ Ra - \frac{Pr}{Sc}\hat{Ra} < \frac{27\pi^4}{4}\left(\frac{1 + Sc + Pr}{Sc}\right). \tag{12-283}$$

If these conditions hold, then instability first appears as an oscillation when the Rayleigh numbers satisfy (12–282). It should be remembered that the conditions (12–283) pertain only to the question of whether σ_{Imag} is zero or nonzero at the point of neutral stability. It says nothing about whether the growth of disturbances will be oscillatory or not in the unstable regions where $\sigma_{Real} > 0$.

Before the stability diagram defined by (12–279) and (12–282) is considered, a few comments are in order. First, we can always satisfy the conditions (12–281) when $Pr > 0$ and fixed and $Sc \to \infty$. In this case,

$$Ra_c|_{Sc \to \infty} = \frac{27\pi^4}{4} = Ra_c|_{Rayleigh-Benard},$$

G. Double-Diffusive Convection

as expected. Second, the condition for $\partial \rho/\partial z = 0$ is

$$Ra = \left(\frac{Pr}{Sc}\right) \hat{Ra}. \qquad (12\text{--}284)$$

To see this, we note from (12–241) that an approximate expression for the overall density gradient can be written in the form

$$\frac{\partial \rho}{\partial z} = \rho_0 \left(-\beta \frac{\partial T}{\partial z} + \hat{\beta}\frac{\partial c}{\partial z}\right) = \rho_0 \hat{\beta}\frac{\partial c}{\partial z}\left(1 - \frac{Ra}{\hat{Ra}}\frac{Sc}{Pr}\right). \qquad (12\text{--}285)$$

However, in case the conditions (12–282) hold,

$$\frac{Ra_c}{\hat{Ra}}\frac{Sc}{Pr} = \left(\frac{1+Sc}{1+Pr}\right)\left(\frac{Pr}{Sc}\right) + \left(1 + \frac{Sc}{Pr}\right)\left(\frac{1}{Sc}+1\right)\frac{27\pi^4}{4}\frac{1}{\hat{Ra}} \qquad (12\text{--}286)$$

and $\partial\rho/\partial z \neq 0$. For example, if we consider the limit $\hat{Ra} \to \infty$, it follows that

$$\frac{Ra_c}{\hat{Ra}}\frac{Sc}{Pr} = \left(\frac{1+Sc}{1+Pr}\right)\left(\frac{Pr}{Sc}\right) \le 1,$$

because $Pr/Sc < 1$. However, if $\hat{Ra}, Ra > 0$, this corresponds to a case in which $\partial\rho/\partial z \le 0$. This demonstrates that the system can exhibit an oscillatory instability even when it appears on the basis of the overall density gradient to be gravitationally stable (this is the diffusive regime).

On the other hand, suppose that (12–283) does not hold. Then we see from (12–279) that

$$\frac{Ra_c}{\hat{Ra}}\frac{Sc}{Pr} = \frac{Sc}{Pr} + \frac{27\pi^4}{4}\frac{1}{\hat{Ra}}\frac{Sc}{Pr}. \qquad (12\text{--}287)$$

Now suppose that $Sc/Pr > 1$ (true, for example, for "heat" and "salt"), but that $Ra, \hat{Ra}$ are negative (this means that both T and c are increasing with increasing z). Then

$$\frac{Ra_c}{\hat{Ra}}\frac{Sc}{Pr} = \frac{Sc}{Pr} - \frac{27\pi^4}{4}\frac{1}{|\hat{Ra}|}\frac{Sc}{Pr},$$

and for $|\hat{Ra}|$ sufficiently large,

$$\frac{Ra_c}{\hat{Ra}}\frac{Sc}{Pr} \ge 1.$$

Again, $\partial\rho/\partial z < 0$, but the system can still exhibit instability (in this case for the "finger" regime).

A general stability diagram can be drawn based on the preceding results. This is most conveniently represented in the Ra versus $\hat{Ra}(Pr/Sc)$ plane. The result is shown as Fig. 12–6. The straight line with slope 1 through the origin [i.e., $Ra = \hat{Ra}(Pr/Sc)$] denotes the conditions at which the overall density gradient is equal to zero. In addition, the two curves (12–279) and (12–282) are shown. The former is a straight line (XZ) that crosses the Ra axis (for $\hat{Ra} = 0$) at $27\pi^4/4$, and the $\hat{Ra}(Pr/Sc)$ axis at $-(27\pi^4/4)(Pr/Sc)$. The latter is also a straight line (XW) that intercepts the Ra axis at $(27\pi^4/4)[(1 + Pr/Sc)(1 + 1/Sc)]$. To the left of the intersection point of these two curves, the principle of exchange of stabilities is valid and σ is real, whereas to the right σ is complex and instability occurs as unstable oscillations.

From this stability diagram, let us now consider the various regions. The system is stable below the boundary ZXW and unstable above it. In the lower-right-hand quadrant, both contributions to the density gradient are stable and so the system is obviously stable.

Hydrodynamic Stability

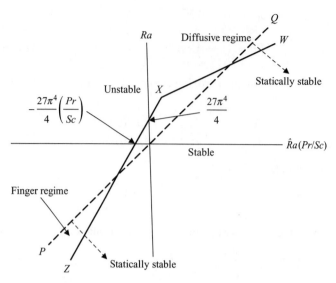

Figure 12–6. A sketch of the stability diagram for double-diffusive convection. The stability boundary is the solid line ZXW. The dashed line PQ denotes the boundary for static stability where the net density gradient is equal to zero. It is assumed in this sketch that $Pr/Sc < 1$.

In the upper-left-hand quadrant, both contributions to the density gradient correspond to heavier fluid over lighter fluid, and the system is unstable when the sum $Ra - \hat{Ra} > 27\pi^4/4$, just as in the case when the instability is driven by density gradients of a single component. Note that σ is real in this case for all combinations of contributions that are due to the two components. The interesting quadrants are the upper right, in which temperature (or whatever species is associated with Ra if it is not heat) is "destabilizing" and salinity (or whatever species is associated with $\hat{Ra}$) is "stabilizing" (this is the "diffusive" quadrant), and the lower left in which the roles of heat and salinity are reversed (this is the "finger" quadrant).

Let us consider the upper-right-hand ("diffusive") quadrant first. To the left of the crossing point X, the problem is just a standard buoyancy-driven convection in which the destabilizing effect of the temperature gradient is increased enough to compensate for the stabilizing influence of the salinity gradient. To the right of the point X, instability occurs by means of unstable oscillations, and we see that these can occur with the overall density gradient both positive, $\partial \rho / \partial z > 0$, and negative, $\partial \rho / \partial z < 0$. In fact, the whole region below the dashed line extension of ZX would be stable in a single-component system because the density gradient is not large enough to drive a transition by means of instability to a direct buoyancy-driven "convection" flow. The mechanism for instability in this case (the "diffusive" regime) was already discussed at the beginning of this section. The present analysis gives us the specific criteria for instability. The lower-left-hand quadrant is similar. In this case, for most of the quadrant, the contribution of the salinity gradient to the overall density gradient is large enough to drive a buoyancy-driven convective instability even based on the Rayleigh–Benard criteria for a single-component system. However, there is a region of instability where the overall density gradient is either negative or smaller than required for onset of buoyancy-driven convection, but the system is still unstable. This is the so-called finger regime. The key to understanding the double-diffusive convection problem is to realize that it is the difference in diffusion rates that creates the possibility for an initially stable system to be transformed to an unstable one by the mechanisms discussed at the beginning of this section.

H. MARANGONI INSTABILITY

Another type of convective instability for a horizontal fluid layer is known as "Marangoni instability." This type of instability can be observed whenever there is at least one of the surfaces of the fluid layer that is an interface. The mechanism was described qualitatively in Chap. 2. This mechanism involves motions driven by gradients of interfacial or surface tension created by gradients of temperature. The most common observation is for a fluid layer that is heated from below, and thus it was at first mistaken for Rayleigh–Benard instability. In fact, it is almost certain that the original experimental observations of convective instabilities by Benard, which were originally analyzed by Rayleigh as being due to buoyancy-driven motion, were actually observations of Marangoni instability in spite of the fact that they led to Benard's name as associated with buoyancy-driven convection.

The only uncertainty in this statement is that the observations of Benard were qualitative rather than quantitative. However, Benard worked with a thin fluid layer with a free upper surface. He observed that the ultimate pattern of the convection motion was hexagonal cells with motion up in the middle and down at the hexagonal edges, whereas later "nonlinear" theory of the expected pattern for buoyancy-driven convection was that it should be parallel roll cells, with motion up and down at alternative cell boundaries. Equally significantly, though Benard did not leave behind quantitative data that could be used to calculate a critical Rayleigh number, it is clear that the depths of his fluid layers were very small ($\sim \frac{1}{2}$ mm) and that the Rayleigh number for his experiments could not have been larger than $O(1)$, far below the theoretical predictions of a critical Rayleigh number for buoyancy-driven convection. Later, Pearson[27] pointed out that paint films often display a steady cellular circulatory flow of the same general type as that observed in the case of fluid layers with a free surface that are heated from below. However, this motion cannot be due to buoyancy-driven convection because circulation is observed whether the free surface is on the top or on the underside of the fluid layer, i.e., even when **g** is reversed. Other related observations were also reported by Sternling and Scriven,[28] which could not be explained as due to buoyancy-driven motions.

In this section, we consider the classic problem of a fluid layer of depth d, with an upper surface that is an interface with air that is maintained at an ambient temperature T_0. The fluid layer is heated from below, and we shall assume that the lower fluid boundary is isothermal with temperature T_1 ($> T_0$). This problem sounds exactly like the Rayleigh–Benard problem with a free upper surface. However, we consider the fluid layer to be very thin (i.e., d small) so that the Rayleigh number, which depends on d^3, is less than the critical value for this configuration. Nevertheless, as previously suggested, the fluid layer may still undergo a convective motion that is due to Marangoni instability.

The mechanism for Marangoni instability, as described already in Chap. 2, is that a perturbation flow brings slightly warmer fluid from the bottom of a fluid layer, which is initially warmer at the bottom and cooler at the free surface, and thus creates a local perturbation of the interface temperature, leading to *surface-tension gradients*. Because the interface is cooler away from the perturbed region, the surface tension is higher there, and this has the tendency to sustain the perturbation by driving fluid motion along the surface away from the perturbed spot, thus pulling more heated fluid up from the bottom of the fluid layer. As fluid moves across the interface it is cooled to the temperature of the surroundings and, in any case, must eventually be drawn back toward the bottom of the fluid layer to satisfy mass conservation requirements. A "cartoon" sketch of a cellular convection pattern that could be driven by this mechanism is shown in Fig. 12–7. Whether the whole motion keeps going or dies out with time then depends on whether the surface-tension-induced force at the interface is great enough to keep the cellular flow going against the resisting forces caused by viscous stresses and thermal diffusion, which tends to homogenize the temperature field. We expect therefore that the critical parameter for instability will reflect this competition

Hydrodynamic Stability

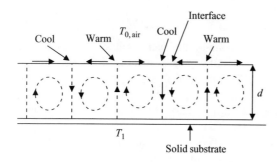

Figure 12–7. A schematic for Marangoni instability. Warm liquid is convected up to the surface, which causes the temperature at the upper surface warmer than the ambient air temperature T_0. The resultant surface-tension gradient drives motion in the same direction, thus producing instability.

between viscous and thermal damping and the surface-tension-induced motion. It is, of course, not obvious yet that the fluid layer could be stable to buoyancy-driven motions but still unstable to Marangoni instability. To see that this is possible, we must carry out a detailed analysis to determine the critical condition for Marangoni instability.

The governing equations for the linear stability theory are the same as for the Rayleigh–Benard problem, namely (12–215), except that it is customary to drop the buoyancy terms because these are of secondary importance for very thin fluid layers where Marangoni instabilities are present but $Ra \ll 1$. Furthermore, we state without proof (see Problems section) that the exchange of stabilities is valid. Thus $\sigma \equiv 0$ at the neutral state. Assuming that

$$\hat{w} = f(z)e^{i(\alpha_x x + \alpha_y y)},$$
$$\hat{\theta} = h(z)e^{i(\alpha_x x + \alpha_y y)}, \tag{12-288}$$

as before, we then find that the governing equations (with $Ra \equiv 0$, $\sigma \equiv 0$) are

$$(D^2 - \alpha^2)^2 f = 0,$$
$$(D^2 - \alpha^2)h = -f. \tag{12-289}$$

The boundary conditions at the lower solid boundary are the same as for the Rayleigh–Benard problem,

$$f = 0, \ Df = 0 \text{ (no-slip)}, \ h = 0 \text{ (isothermal)} \quad \text{at } z = 0. \tag{12-290}$$

Apart from the assumption that the normal velocity at the upper interface is zero,

$$f = 0 \quad \text{at } z = 1, \tag{12-291}$$

the boundary conditions at the upper interface are different from the Rayleigh–Benard problem and must be treated carefully because it is the coupling between the heat transfer and the presence of interfacial-tension gradients at this surface that is responsible for the Marangoni instability.

As already noted we assume that the gas above the fluid layer is held at a constant temperature T_0. Assuming local thermal equilibrium, we have seen [Eqs. (12–214)] that the thermal boundary condition for the disturbance temperature at the upper surface can be expressed in the nondimensional form

$$-\frac{\partial \hat{\theta}}{\partial z} = Bi \, \hat{\theta} \quad \text{at } z = 1.$$

H. Marangoni Instability

Now, if

$$Bi \to \infty, \hat{\theta} \to 0,$$

i.e., the interface approaches closely to an isothermal surface with constant temperature T_0. Because the disturbance temperature *perturbation* approaches zero, there is no source for surface-tension variations along the interface and the system will be stable. If, on the other hand,

$$Bi \to 0, \frac{\partial \hat{\theta}}{\partial z} \to 0.$$

In this case, the interface approaches an adiabatic condition of no heat transfer to the surroundings. Thus the temperature is not forced to equal the reservoir value T_0, but can vary by an amount determined by the heat transfer rates within the fluid layer (i.e., the maximum ΔT on the interface will be $T_1 - T_0$, the same as the imposed ΔT across the fluid layer. This is the case of greatest instability. Substituting the expression for $\hat{\theta}$ into the local equilibrium condition, the boundary condition becomes

$$\boxed{-Dh = (Bi)h \quad \text{at } z = 1.} \tag{12-292}$$

Finally, the shear-stress boundary condition (2–141) for an interface incorporates the Marangoni contribution

$$O = [(\tau - \hat{\tau}) \cdot \mathbf{n}] \cdot \mathbf{t}_i + (\text{grad}_s \gamma) \cdot \mathbf{t}_i.$$

If we express the interfacial tension in the approximate form

$$\gamma = \gamma_0 [1 - \gamma_T (T - T_0)],$$

with γ_T being a material constant. Then

$$\text{grad}_s \gamma = -\gamma_0 \gamma_T [\nabla - \mathbf{n}(\mathbf{n} \cdot \nabla)] T.$$

Hence, for the present problem,

$$\tau_{xz} = \mu \frac{\partial u}{\partial z} = -\gamma_0 \gamma_T \frac{\partial T}{\partial x}; \quad \tau_{yz} = \mu \frac{\partial v}{\partial z} = -\gamma_0 \gamma_T \frac{\partial T}{\partial y} \quad \text{at } z' = d. \tag{12-293}$$

If we differentiate the first of these conditions with respect to x and the second with respect to y and add them together,

$$\mu \frac{\partial}{\partial z} \left(\frac{\partial u}{\partial x} + \frac{\partial v}{\partial y} \right) = -\gamma_0 \gamma_T \nabla_1^2 T \quad \text{at } z' = d.$$

By means of the continuity equation, this can also be written as

$$\mu \frac{\partial^2 w}{\partial z^2} = \gamma_0 \gamma_T \nabla_1^2 T \quad \text{at } z' = d. \tag{12-294}$$

Hydrodynamic Stability

Finally, if we substitute (12–288) into (12–294), and express the resulting equation in dimensionless form, we obtain the third boundary condition at $z = 1$:

$$\boxed{D^2 f|_{z=1} = \alpha^2 Ma\, h(1).} \tag{12–295}$$

The dimensionless parameter that appears in this boundary condition is known as the Marangoni number and is defined as

$$\boxed{Ma \equiv \frac{d\gamma_0 \gamma_T (T_1 - T_0)}{\mu \kappa}.} \tag{12–296}$$

The eigenvalue problem that must be solved in this case to determine the conditions for instability of the fluid layer is (12–289)–(12–292) and (12–295). Because we have already put $\sigma = 0$, we pick α and Bi and find the corresponding value of Ma such that nontrivial solutions exist for $f(z)$ and $h(z)$. The critical condition for instability is thus to determine the minmum Ma as a function of α (with Bi being fixed for any particular configuration).

We can begin with a general solution of (12–289a). This can be written either as exponentials or in terms of the hyperbolic functions sinh and cosh. The latter turns out to be more convenient. Hence,

$$f = a \sinh \alpha z + b \cosh \alpha z + cz \sinh \alpha z + dz \cosh \alpha z, \tag{12–297}$$

where the four undetermined constants are denoted as a, b, c, and d. To this solution, we can immediately apply the three boundary conditions $f = Df = 0$ at $z = 0$ from (12–290), and $f = 0$ at $z = 1$ from (12–291). The result is

$$f = a \left(\sinh \alpha z + \frac{\alpha \cosh \alpha - \sinh \alpha}{\sinh \alpha} z \sinh \alpha z - \alpha z \cosh \alpha z \right). \tag{12–298}$$

We can now substitute (12–298) for f in (12–289b), and solve for $h(z)$ subject to the boundary conditions $h = 0$ at $z = 0$ from (12–290) and $Dh + (Bi)h = 0$ at $z = 1$ from (12–292). The result, after a considerable amount of algebra, is

$$\begin{aligned} h = a \Bigg\{ & \frac{4}{3\alpha} z \cosh \alpha z + \left(\frac{\alpha \cosh \alpha - \sinh \alpha}{4\alpha \sinh \alpha} \right) z^2 \cosh \alpha z - \frac{1}{4} z^2 \sinh \alpha z \\ & - \left(\frac{\alpha \cosh \alpha - \sinh \alpha}{4\alpha^2 \sinh \alpha} \right) z \sinh \alpha z - \Bigg[\frac{\alpha^2 \cosh^2 \alpha + \alpha \sinh \alpha \cosh \alpha + \sinh^2 \alpha}{4\alpha^2 \sinh \alpha (\alpha \cosh \alpha + Bi \sinh \alpha)} \\ & + \frac{Bi(\alpha^2 + \alpha \sinh \alpha \cosh \alpha + \sinh^2 \alpha)}{4\alpha^2 \sinh \alpha (\alpha \cosh \alpha + Bi \sinh \alpha)} \Bigg] \sinh \alpha z \Bigg\}. \end{aligned} \tag{12–299}$$

Now there is one constant left and one boundary condition, (12–295). Of course, one way to satisfy (12–295) is by means of the trivial solution that can be thought of as setting $a = 0$ in both (12–298) and (12–299). Instead, we seek a condition that leads to a nontrivial

H. Marangoni Instability

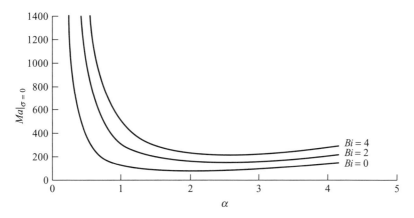

Figure 12–8. Neutral stability curves from Eq. (12–300) for three different values of Bi. The critical Marangoni number for each Bi is the minimum value over the range of possible values for α.

solution. This determines the eigenvalues $Ma_{\sigma=0}$ for each α and Bi. Again, a considerable amount of algebra leads to an analytic expression for this eigenvalue:

$$Ma|_{\sigma=0} = \frac{8\alpha(\alpha \cosh\alpha + Bi\ \sinh\alpha)(\alpha - \sinh\alpha \cosh\alpha)}{\alpha^3 \cosh\alpha - \sinh^3\alpha}. \qquad (12\text{–}300)$$

Now, for each value of Bi, we can plot the neutral stability curve, as shown in Fig. 12–8 for $Bi = 0, 2,$ and 4. The critical Marangoni numbers for these three cases are approximately 80, 160, and 220. As noted earlier, the system is stabilized by increase of Bi because this leads toward an isothermal interface, and thus cuts the available Marangoni stress to drive convection. The critical wave numbers for these three cases are, respectively, 2.0, 2.3, and 2.5.

It has been estimated that the Rayleigh number in Benard's original experiments was only about $O(1)$, but that the Marangoni number was in the range $O(50\text{–}100)$. Thus, as stated earlier, it is highly likely that Benard actually observed convection that was due to Marangoni instability in his experiments, rather then buoyancy driven convection as was assumed at the time. Of course, in most real systems of a fluid layer with a free upper surface that is heated from below, both buoyancy-driven and surface-tension-driven convection are possible (the exception being a fluid layer in a microgravity environment). We thus get one or the other depending on whether Ma or Ra gets to its critical value first as we raise ΔT across the fluid layer. In practice, the factor that normally determines what occurs is the depth of the fluid layer. It will be noted that $Ra \sim d^3$ whereas $Ma \sim d$. Thus, in a "thin" fluid layer such as in Benard's experiments ($d <\approx 1$cm), we would generally expect to see Marangoni instability, but as the depth of the layer increases buoyancy-driven convection is favored.

Finally, we would note that a more difficult problem is the analysis of Marangoni instability at the interface between two liquids. An important example is interfacial instabilities caused by gradients of surface active solute. This problem was studied originally by Sternling and Scriven, who were trying to explain the observation that a container with two liquids, one of low density overlaying another of higher density, will remain stationary if we have two pure fluids, but will exhibit vigorous mixing and emulsification if one of the fluids contains a solute. For example, if we have a layer of pure water on the bottom with a layer of toluene containing 10% methanol on the top, the water will remain clear, but the organic phase becomes turbid as water is spontaneously emulsified into the organic phase.

Hydrodynamic Stability

The mechanism for this clearly cannot be buoyancy driven in the absence of temperature gradients, because the water is heavier than the toluene/methanol mixture. The mathematical problem of Marangoni instability at the interface between two liquids is daunting. We would need to consider transport processes in both liquid phases; hence we would have a sixth-order system of equations like (12–291) in each liquid. Furthermore, it turns out that the exchange of stabilities does not hold, thus σ is complex, and because we must then consider equations for both the real and imaginary parts, we would end up with a 24th-order algebraic system. Although Sternling and Scriven did not solve this problem, they were able to give some rules for when Marangoni driven convection should be expected. This is important as the rates of mass transfer are greatly increased between two fluids when there is interfacial mixing (Marangoni convection) present.

I. INSTABILITY OF TWO-DIMENSIONAL UNIDIRECTIONAL SHEAR FLOWS

An important problem is to analyze the stability of fluid flows. With the exception of the Taylor–Couette and Saffman–Taylor problems, this chapter has focused on stability questions when the base state of the system was one with no motion (or rigid-body motion), so that instability addresses the conditions for spontaneouis onset of flow. An equally valid question is whether a particular flow, such as Poiseuille flow in a pipe (or any of the other flows that we have analyzed in previous chapters of this book), is stable, especially to infinitesimal perturbations as linear instability determines whether the particular flow is actually realizable in experiments. This question was first mentioned back in Chapter 3 when we analyzed simple unidirectional flow problems and noted that solutions such as Poiseuille's solution for flow through a tube was a valid solution of the Navier–Stokes equations for all Reynolds numbers, even though common experience tells us that beyond some critical Reynolds number there is a transition to turbulent flow in the tube.

Not surprisingly, a great deal of work has been done to try to theoretically analyze the stability of nontrivial fluid flows. In fact, some books on hydrodynamic stability are devoted strictly (or largely) to this topic.[21] In part, this is because the problems can be extremely difficult to analyze, the mechanisms for instability are very much more subtle than the problems analyzed in earlier sections of this chapter, and in some cases (most notably, laminar Poiseuille flow in a pipe) the problem remains at least partially unresolved in the sense that observation and analytic, linear stability results do not coincide. The basic requirement for analytic theory is that a closed-form analytic solution must be known if we are to study its stability. Furthermore, as the dimensionality of this solution increases, the complexity of the linear stability problem increases. Hence, a great deal of the existing literature has focused on the stability of steady, unidirectional flows, or, more recently, general linear (homogeneous) flows, or mild departures from these flows (including for example, weak evolution of the flow in the streamwise direction). One way to circumvent the absence of closed-form analytic solutions for more complicated flows, at least in principle, is to carry out the analysis of the linear stability problem numerically, assuming that an exact solution is known for the base flow. There has also been work on analyzing the stability of numerically generated flows in a linear stability framework, but this can be difficult to interpret because it is difficult to control numerical errors in the base-flow solution well enough to establish response to "infinitesimal" disturbances.

The conclusion from this discussion is that the linear stability of flows is a complicated topic, even if we restrict ourselves to a simple class of problems such as steady, 2D, unidirectional flows, and it is not possible to provide a comprehensive summary in the space available. Instead, we consider only an introduction to the stability for this limited class of problems, with the intention of giving a qualitative sense of the analysis and some very basic results. In particular, we summarize the theory for an inviscid fluid, which has been

I. Instability of Two-Dimensional Unidirectional Shear Flows

known since the time of Rayleigh, and derive the famous Orr–Sommerfeld equation, which governs the linear stability for a viscous fluid. However, the analysis of this equation is too involved to present here. The reader who is interested in knowing more about this topic is urged to consult one of the several texts that treat this topic in detail.[21]

1. Inviscid Fluids

a. The Rayleigh stability equation. We begin by considering the stability of an inviscid fluid that is undergoing a steady, 2D unidirectional flow. We assume that the velocity profile for this basic flow can be expressed in the form

$$\mathbf{U}' = U'(z')\mathbf{i} \quad \text{for} \quad 0 \leq z' \leq d.$$

The two boundaries at $z' = 0$ and d can be either rigid walls or a free surface. In fact, we do not need to be so restrictive in our description of the problem. One or both of the boundaries may also be at infinity. The function $U(z')$ can be considered as an arbitrary function of z'. The equations governing $\mathbf{U}'$ and the linear disturbance flow are the dimensional Navier–Stokes equations, (12–1) and (12–2), but, in this case, with the viscous terms neglected.

It is convenient to nondimensionalize. We must identify a characteristic length scale ℓ_c and velocity scale u_c. For the former we choose d, though we recognize that this would need to be modified if the fluid is unbounded. For the characteristic velocity scale, it is traditional in this field to choose the maximum value of $\mathbf{U}'$. In addition, we assume that the characteristic time is $t_c = \ell_c/u_c$, and the characteristic pressure is $p_c = \rho u_c^2$. Then the governing equations, which are the continuity equation and the inviscid Navier–Stokes equations (usually called the Euler equations), can be written in the form

$$\frac{\partial \mathbf{u}}{\partial t} + \mathbf{u} \cdot \nabla \mathbf{u} = -\nabla p; \quad \nabla \cdot \mathbf{u} = 0. \tag{12–301}$$

We denote the basic flow in dimensionless form as

$$\mathbf{U} = U(z)\mathbf{i} \quad \text{for} \quad 0 \leq z \leq 1, \tag{12–302}$$

which we assume satisfies (12–301) and all boundary conditions exactly, but is otherwise arbitrary.

To consider the stability of the basic flow (12–302), we introduce an infinitesimal perturbation to both the velocity and the pressure fields,

$$\mathbf{u} = U(z)\mathbf{i} + \varepsilon \hat{\mathbf{u}}, \quad p = P + \varepsilon \hat{p}, \tag{12–303}$$

and, following the precedent of the previous stability calculations, we can study the linear stability problem by considering a normal mode decomposition of $\hat{\mathbf{u}}$ and $\hat{p}$. Hence, we write

$$\hat{\mathbf{u}}(\mathbf{x}, t) = \hat{\mathbf{u}}(z) \exp[i(\alpha_x x + \alpha_y y)] \exp(\sigma t),$$
$$\hat{p}(\mathbf{x}, t) = \hat{p}(z) \exp[i(\alpha_x x + \alpha_y y)] \exp(\sigma t). \tag{12–304}$$

In general, σ is expected to be complex, $\sigma = \sigma_R + i\sigma_I$. If $\sigma_R > 0$ for some choice α_x and α_y, the system is unstable for this mode, whereas it is stable if $\sigma_R < 0$.

For convenience, we denote the components of $\hat{\mathbf{u}}$ as $\hat{\mathbf{u}} = (\hat{u}, \hat{v}, \hat{w})$. Thus, if we substitute (12–303) and (12–304) into (12–301), retaining only the terms that are linear in ε and remembering that the basic flow is an exact solution, we obain

$$\begin{aligned}
(\sigma + i\alpha_x U)\hat{u} + U'\hat{w} &= -i\alpha_x \hat{p}, \\
(\sigma + i\alpha_x U)\hat{v} &= -i\alpha_y \hat{p}, \\
(\sigma + i\alpha_x U)\hat{w} &= -D\hat{p}, \\
i\alpha_x \hat{u} + i\alpha_y \hat{v} + D\hat{w} &= 0.
\end{aligned} \tag{12–305}$$

Hydrodynamic Stability

If the boundaries are rigid, we can assume that

$$\hat{w} = 0 \quad \text{at } z = 0, 1. \tag{12-306}$$

As usual, in order that the solution be well behaved at infinity, α_x and α_y are real.

Finally, without loss of generality, we can express $\sigma = -i\alpha_x c$, where c may now be complex, and then we can express (12–305) in the form that is most common in the literature for this class of stability problem:

$$\begin{aligned} i\alpha_x(U-c)\hat{u} + U'\hat{w} &= -i\alpha_x \hat{p}, \\ i\alpha_x(U-c)\hat{v} &= -i\alpha_y \hat{p}, \\ i\alpha_x(U-c)\hat{w} &= -D\hat{p}, \\ i\alpha_x \hat{u} + i\alpha_y \hat{v} + D\hat{w} &= 0. \end{aligned} \tag{12-307}$$

The advantage of this latter notation is that it emphasizes the propagating wavelike structure of the disturbance (12–304). For example, we can rewrite the expression (12–304) for $\hat{u}$ in the form

$$\hat{u}(\mathbf{x}, t) = \hat{u}(z) \exp[i(\alpha_x x + \alpha_y y - \alpha_x c_R t)] \exp(\alpha_x c_I t), \tag{12-308}$$

so that it now appears as a propagating wave in the direction $(\alpha_x, \alpha_y, 0)$ with phase velocity $\alpha_x c_R / (\alpha_x^2 + \alpha_y^2)^{1/2}$. We refer to this as a 3D disturbance in the sense that it propagates obliquely to the undisturbed flow direction. There is some important subtlety associated with the singularity in the disturbance flow equations where $U - c_R = 0$, but we ignore this point for the moment. Stability is clearly now associated with the sign of the imaginary part of c.

Before considering the question of stability, it is useful to note that the general 3D disturbance flow problem, defined by (12–306)–(12–308), can be transformed to a form that is mathematically identical to the special case of a 2D disturbance with $\alpha_y = \hat{v} = 0$. To motivate this transformation, we first combine Eqs. (12–307a) and (12–307b) by multiplying (12–307b) by α_y/α_x and adding. The result is

$$i(U-c)(\alpha_x \hat{u} + \alpha_y \hat{v}) + U'\hat{w} = -i\left(\frac{\alpha_x^2 + \alpha_y^2}{\alpha_x}\right)\hat{p}. \tag{12-309}$$

Now, let us define

$$\alpha \equiv (\alpha_x^2 + \alpha_y^2)^{1/2} \quad \text{and} \quad \alpha\tilde{u} \equiv \alpha_x \hat{u} + \alpha_y \hat{v} \tag{12-310a}$$

and rescale the pressure according to

$$\hat{p} \equiv \frac{\alpha_x}{\alpha} \tilde{p}. \tag{12-310b}$$

Then (12–309) can be written in the form

$$i\alpha(U-c)\tilde{u} + U'\hat{w} = -i\alpha \tilde{p}, \tag{12-311a}$$

while (12–307c) and (12–307d) become

$$i\alpha(U-c)\hat{w} = -D\tilde{p}, \tag{12-311b}$$

$$i\alpha \tilde{u} + D\hat{w} = 0. \tag{12-311c}$$

The change of variables (12–310) is a version of Squire's transformation, in recognition of Squire who first discovered it.

The transformed problem (12–311) is mathematically equivalent to the special case of a 2D disturbance with $\alpha_y = \hat{v} = 0$. Inherent in Squire's transformation is the fact that, for

I. Instability of Two-Dimensional Unidirectional Shear Flows

each unstable 3D mode, there is a 2D mode (i.e., propagating in the mean-flow direction) that is more unstable. To see this, let us suppose that the eigenvalue from (12–311) is $c = f(\alpha) = f[(\alpha_x^2 + \alpha_y^2)^{1/2}]$. Now, the growth rate of a 3D disturbance with total wave number α is $\alpha_x c_I$. However, for each such disturbance there is a corresponding 2D disturbance with the same wave number α that has a growth rate αc_I that is larger because $\alpha > \alpha_x$ if $\alpha_y \neq 0$. The important conclusion is that we need to consider only 2D disturbances in establishing a sufficient condition for stability.

The governing equations (12–307) for a 2D disturbance ($\alpha_y = \hat{v} = 0$) can be combined to obtain a single higher-order equation that can be used to study the stability conditions. First, we note that the continuity equation (12–307d) can be satisfied by introducing a function ϕ, such that

$$\hat{u} = D\phi, \quad \hat{w} = -i\alpha_x \phi. \tag{12–312}$$

The introduction of ϕ is equivalent to introducing a *stream function* for this 2D flow problem. If we introduce these expressions for $\tilde{u}$ and $\hat{w}$ into (12–307a), we see that

$$\hat{p} = U'\phi - (U-c)D\phi. \tag{12–313}$$

Then, finally, by substituting (12–312) and (12–313), we can express eq. (12–307c) entirely in terms of ϕ:

$$\boxed{(U-c)(\phi'' - \alpha_x^2 \phi) - U''\phi = 0.} \tag{12–314}$$

This equation is known as the *Rayleigh stability equation*. Together with the boundary conditions

$$\boxed{\alpha_x \phi = 0 \quad \text{at } z = 0, 1,} \tag{12–315}$$

it defines a classic eigenvalue problem for c as a function of α_x with $U(z)$ given.

b. The Inflection-point theorem. A key result was obtained from (12–314) more than 100 years ago. It is known as the *Rayleigh inflection-point theorem*. According to this theorem, a *necessary* condition for instability of an inviscid fluid is that the basic velocity profile has an inflection point. We assume that $c_I \neq 0$ so that (12–314) is not singular. This also means that we are assuming that the system is unstable. The classic proof of the inflection-point theorem is to multiply (12–314) by the complex conjugate function ϕ^* and integrate across the fluid from $z = 0$ to $z = 1$. If we integrate the term involving $\phi^* D^2 \phi$ by parts, this leads to

$$\int_0^1 (|D\phi|^2 + \alpha_x^2 |\phi|^2) dz + \int_0^1 \frac{U''}{U-c} |\phi|^2 dz = 0. \tag{12–316}$$

Now, the imaginary part of (12–316) is

$$c_I \int_0^1 \frac{U''}{|U-c|^2} |\phi|^2 dz = 0. \tag{12–317}$$

It follows that U'' must change sign at least once across the fluid – i.e., there has to be an inflection point in the undisturbed velocity profile.

The existence of an inflection point is therefore a *necessary* condition for instability, but it is not a sufficient condition. Indeed, there are some quite simple velocity profiles with one or more inflection points that are actually stable. Nevertheless, the inflection-point theorem is a useful starting point in assessing the potential of an inviscid flow for instability. If there is no inflection point, an inviscid fluid will be stable. Subsequent researchers have

Hydrodynamic Stability

also established *sufficient* conditions for instability, and this work has been reviewed in a number of places.[21] Space limitations preclude additional discussion here.

2. Viscous Fluids

There are two possibilities for the role of viscous effects. One is that the system is unstable in the absence of viscous effects, but the latter stabilizes the system whenever the Reynolds number is below some critical value (which depends, of course, on the problem). The second is that the inviscid fluid is stable, but viscous effects act in such a way as to produce instability.

There is no question that the stability of a viscous fluid undergoing some flow is an exceedingly important problem. However, compared with the case of an inviscid fluid, progress in understanding the stability of a viscous fluid has been very slow. The basic equation for linear stability of a *unidirectional flow* was derived approximately 100 years ago by Orr[29] and Sommerfeld,[30] but the optimal tools for analytic progress involve the use of asymptotic approximation methods that were not developed for nearly 50 years. Much of the literature over the past century has concentrated on the development of approximate methods of analysis, but the story is still not complete, especially in cases in which viscosity is destabilizing. In this subsection, we derive the Orr–Sommerfeld equation. However, the complexities of the solutions of this equation are such that we cannot go much further. In the present chapter, we use it to discuss sufficient conditions for *stability*. To learn more, the interested reader is encouraged to study one of the specialized textbooks on hydrodynamic stability theory.

a. The Orr–Sommerfeld equation. In fact, the derivation of the celebrated Orr–Sommerfeld equation is "alarmingly" straightforward and follows closely the derivation of Rayleigh's stability equation in the. Again, the analysis applies to a steady, 2D unidirectional shear flow. The difference is only that we retain viscous terms in the governing equations. Hence, instead of (12–301), we start with

$$\frac{\partial \mathbf{u}}{\partial t} + \mathbf{u} \cdot \nabla \mathbf{u} = -\nabla p + \frac{1}{Re}\nabla^2 \mathbf{u}; \quad \nabla \cdot \mathbf{u} = 0, \qquad (12\text{–}318)$$

where $Re \equiv u_c \ell_c / \nu$ is the Reynolds number. The velocity and pressure fields again take the form

$$\mathbf{u} = U(z)\mathbf{i} + \varepsilon \hat{\mathbf{u}}, \quad p = P + \varepsilon \hat{p}. \qquad (12\text{–}319)$$

However, in this case $U(z)$ can no longer be considered as an arbitrary function, but must satisfy the unidirectional flow equations with $\nabla p = \partial p/\partial x = $ const.

To derive the Orr–Sommerfeld equation, we substitute (12–319) into (12–318) and linearize by retaining terms that are linear in $\varepsilon \ll 1$. If we consider a typical disturbance mode in the form

$$\hat{\mathbf{u}}(\mathbf{x}, t) = \hat{\mathbf{u}}(z)\exp[i(\alpha_x x + \alpha_y y - c_R t)]\exp(c_I t), \qquad (12\text{–}320a)$$

$$\hat{p}(\mathbf{x}, t) = \hat{p}(z)\exp[i(\alpha_x x + \alpha_y y - c_R t)]\exp(c_I t), \qquad (12\text{–}320b)$$

we obtain the linearized disturbance equations:

$$i\alpha_x(U-c)\hat{u} + U'\hat{w} = -i\alpha_x \hat{p} - Re^{-1}\left[D^2 - (\alpha_x^2 + \alpha_x^2)\right]\hat{u},$$
$$i\alpha_x(U-c)\hat{v} = -i\alpha_y \hat{p} - Re^{-1}\left[D^2 - (\alpha_x^2 + \alpha_x^2)\right]\hat{v},$$
$$i\alpha_x(U-c)\hat{w} = -D\hat{p} - Re^{-1}\left[D^2 - (\alpha_x^2 + \alpha_x^2)\right]\hat{w}, \qquad (12\text{–}321)$$
$$i\alpha_x \hat{u} + i\alpha_y \hat{v} + D\hat{w} = 0.$$

For rigid no-slip boundaries, the boundary conditions are

$$\hat{u} = \hat{v} = \hat{w} = 0 \quad \text{at } z = 0, 1. \qquad (12\text{–}322)$$

I. Instability of Two-Dimensional Unidirectional Shear Flows

The Equations (12–321) are an obvious generalization of the inviscid, linearized disturbance equations (12–307), and it is therefore not surprising that Squire's theorem turns out to be applicable. To see this, we apply Squire's transformation (12–310), plus the one additional condition

$$Re = (\alpha/\alpha_x)\tilde{Re}, \tag{12-323}$$

to the 3D disturbance flow equations (12–321). Again, this reduces the equations to an equivalent 2D problem:

$$[D^2 - \alpha^2 - i\alpha\tilde{Re}(U-c)]\tilde{u} = \tilde{Re}U'\hat{w} + i\alpha\tilde{Re}\tilde{p}, \tag{12-324a}$$

$$[D^2 - \alpha^2 - i\alpha\tilde{Re}(U-c)]\hat{w} = \tilde{Re}D\tilde{p}, \tag{12-324b}$$

$$i\alpha\tilde{u} + D\hat{w} = 0, \tag{12-324c}$$

with

$$\tilde{u} = \hat{w} = 0 \quad \text{at } z = 0, 1. \tag{12-325}$$

These equations are the same form as (12–321) with $\alpha_y = \hat{v} = 0$ and thus define a mathematically equivalent 2D problem. According to Squire's theorem, to determine the minimum critcal Reynolds number for stability, it is sufficient to consider only 2D disturbances. This follows immediately from the fact $\alpha \geq \alpha_x$ so that $\tilde{Re} \leq Re$.

Because the stability analysis can now be focused exclusively on 2D disturbances, we can again use the "stream-function" transformation (12–312) to express the linear stability problem in terms of a single equation for the disturbance function ϕ. In this case,

$$\hat{p} = U'\phi - (U-c)D\phi + (i\alpha_x Re)^{-1}(D^2 - \alpha_x^2)D\phi,$$

and thus

$$\boxed{(i\alpha_x Re)^{-1}\left(D^2 - \alpha_x^2\right)^2 \phi - (U-c)\left(D^2 - \alpha_x^2\right)\phi + U''\phi = 0,} \tag{12-326}$$

which is the so-called Orr–Sommerfeld equation. The boundary conditions are now

$$\boxed{\phi = D\phi = 0 \quad \text{at } z = 0, 1.} \tag{12-327}$$

b. A sufficient condition for stability. A *sufficient condition* for stability of steady, 2D unidirectional flows can be obtained from the Orr–Sommerfeld equation following an analysis that is due to Synge.[31] We multiply (12–236) by the complex conjugate ϕ^* and integrate the resulting equation across the fluid layer. Integrating by parts where necessary, and using the boundary conditions (12–237), we obtain

$$I_2^2 + 2\alpha_x^2 I_1^2 + \alpha_x^4 I_0^2 = -i\alpha_x QRe + i\alpha_x Rec\left(I_1^2 + \alpha_x^2 I_0^2\right),$$

where

$$I_n^2 \equiv \int_0^1 |D^n\phi|^2 dz,$$

$$Q \equiv \int_0^1 \left[U|D\phi|^2 + (\alpha_x^2 U + U'')|\phi|^2\right]dz + \int_0^1 U'(D\phi)\phi^* dz.$$

Now, because Q and c are complex, this equation can be split into a real and an imaginary part,

$$\left(I_2^2 + 2\alpha_x^2 I_1^2 + \alpha_x^4 I_0^2\right) - \alpha_x Q_I Re + \alpha_x Rec_I \left(I_1^2 + \alpha_x^2 I_0^2\right)$$
$$+ i\left[\alpha_x Q_R Re - \alpha_x Rec_R \left(I_1^2 + \alpha_x^2 I_0^2\right)\right] = 0.$$

Hence,

$$c_I = \frac{-\left(I_2^2 + 2\alpha_x^2 I_1^2 + \alpha_x^4 I_0^2\right) + \alpha_x Q_I Re}{\alpha_x Re \left(I_1^2 + \alpha_x^2 I_0^2\right)},$$

where

$$Q_I = \frac{1}{2}i \int_0^1 U'[\phi(D\phi^*) - (D\phi)\phi^*]dz.$$

Now

$$|Q_I| \le \int_0^1 |U'||\phi||D\phi|dz \le k\,(I_0 I_1), \quad \text{where } k = \max_{0 \le z \le 1} |U'|$$

and the second inequality represents an application of Schwarz's inequality theorem. But now, this gives an upper bound for c_I,

$$c_I \le \frac{k I_0 I_1 - (\alpha_x Re)^{-1}\left(I_2^2 + 2\alpha_x^2 I_1^2 + \alpha_x^4 I_0^2\right)}{\left(I_1^2 + \alpha_x^2 I_0^2\right)}.$$

Because instability occurs when $c_I > 0$, it follows that a sufficient condition for *stability* is

$$k I_0 I_1 - (\alpha_x Re)^{-1}\left(I_2^2 + 2\alpha_x^2 I_1^2 + \alpha_x^4 I_0^2\right) < 0$$

or

$$\alpha_x Re < \frac{\left(I_2^2 + 2\alpha_x^2 I_1^2 + \alpha_x^4 I_0^2\right)}{k I_0 I_1}.$$

This condition would, in principle, allow a specific estimate for the critical Reynolds number for stability as a function of α_x, assuming that the eigenfunctions of the Orr–Sommerfeld equation had not been calculated. In the absence of explicit solutions for ϕ, the right-hand side could be estimated with trial functions to provide an upper bound on the critical value of Re. For present purposes, we simply note that the argument of Synge proves that there is a critical Reynolds number for stability of a steady, 2D unidirectional flow.

We recognize that the discussion here of the stability of simple flows is too truncated to allow the reader more than a glimpse of the subject. We encourage any reader who wants to learn more of this subject to consult one of the excellent texts that have been written on this topic.

NOTES

1. P. G. Drazin and W. H. Reid, *Hydrodynamic Stability* (Cambridge University Press, Cambridge, 1982); S. Chandrasekhar, *Hydrodynamic and Hydromagnetic Stability* (Dover, New York, 1992).
2. D. D. Joseph, *Stability of Fluid Motions 1 and 2*, Tracts in Natural Philosophy Series (Springer-Verlag, New York, 1976); G. E. Swaters, *Introduction to Hamiltonian Fluid Dynamics and Stability Theory*, Vol. 102 of Chapman and Hall/CRC Monographs and Surveys Series (CRC, New York, 1999); P. G. Drazin, *Introduction to Hydrodynamic Stability* (Cambridge University Press, Cambridge, 2002); B. Straughan, *Energy Method, Stability, and Nonlinear Convection* (2nd Ed.) (Springer-Verlag, New York, 2003).

Notes

3. Alternatively, we could choose to express the shape function in the more general form $\hat{f} = C \exp(imz + \partial t)$. Because the cylinder is considered to be infinitely long, the use of $\exp(imz)$ instead of $\sin mz$ (or $\cos mz$) is completely arbitrary and will not change the results of the analysis.
4. M. Abramowitz and I. A. Stegun, *Handbook of Mathematical Functions* (9th printing) (Dover, New York, 1972).
5. S. Chandrasekhar, *Hydrodynamic and Hydromagnetic Stability* (Dover, New York, 1992).
6. S. Tomotika, On the instability of a cylindrical thread of a viscous liquid surrounded by another viscous liquid, *Proc. R. Soc. London Ser. A* **150**, 322 (1935).
7. S. L. Goren, On the instability of an annular thread of fluid, *J. Fluid Mech.* **12**, 309–19 (1962).
8. D. Quere, Fluid coating on a fiber, *Annu. Rev. Fluid Mech.* **31**, 347–84 (1999).
9. B. J. Carroll and J. Lucassen, Effect of surface dynamics on the process of droplet formation from supported and free liquid cylinders, *J. Chem. Soc. Faraday Trans.* **70**, 1228–39 (1974).
10. See Ref. 7. Also see P. Hammond, Nonlinear adjustment of a thin annular film of viscous fluid surrounding a thread of another within a circular cylindrical pipe, *J. Fluid Mech.* **137**, 363–84 (1983).
11. J. B. Grotberg, Pulmonary flow and transport phenomena, *Annu. Rev. Fluid Mech.* **26**, 529–71 (1994).
12. The analysis in this section is similar to that in Ref. 5.
13. W. J. Harrison, The influence of the viscosity on the oscillations of superposed fluids, *Proc. London Math Soc.*, **6**, 396–405 (1908).
14. Lord Rayleigh, Investigation of the character of the equilibrium of an incompressible heavy fluid of variable density, *Proc. London Math. Soc.* **14**, 170–7 (1883). (also in *Scientific Papers* (1900), Vol. 2, 200–7).
15. G. I. Taylor, The instability of liquid surfaces when accelerated in a direction perpendicular to their planes, *Proc. R. Soc. London Ser. A* **202**, 192–6 (1950).
16. P. G. Saffman and G. I. Taylor, The penetration of a fluid into a porous medium or Hele–Shaw cell containing a more viscous fluid, *Proc. R. Soc. London Ser. A* **245**, 312–29 (1958).
17. J. Bear, *Dynamics of Fluids in Porous Media* (American Elsevier, New York (1972; also available from Dover, New York, 1988; H. Brenner, *Transport Processes in Porous Media* (McGraw-Hill, New York, 1987); R. A. A. Greenkorn, *Fluid Phenomena in Porous Media: Fundamentals and Applications in Petroleum, Water and Food Production* (Marcel Dekker, New York, 1983); A. Bejan, I. Dineer, S. Lorente, A. F. Miguel, and A.H. Reis, *Porous and Complex Flow Structures in Modern Technologies* (Springer-Verlag, New York, 2004).
18. A. A. Zick and G. M. Homsy, Stokes flow through periodic arrays of spheres, *J. Fluid Mech.* **115**, 13–26 (1982); A. S. Sangani and A. Acrivos, Slow flow through a periodic array of spheres, *Int. J. Multiphase Flow* **8**, 343–60 (1982); G. Liu and K. E. Thompson, A domain decomposition method for modelling Stokes flow in porous materials, *Int. J. Numer. Meth. Fluids* **38**, 1009–25 (2002).
19. See the first two references in Ref. 17.
20. L. Durlofsky and J. F. Brady, "Analysis of the Brinkman equation as a model for flow in porous media," *Phys. Fluids* **30**, 3329–41 (1987).
21. The majority of Drazin and Reid, Ref. 1. See also R. Betchov and W. O. Criminale, Jr., *Stability of Parallel Flows* (Academic, New York, 1967). There are also good descriptions at a more readable level in the book by Drazin, Ref. 2, and in chapters in C.-S. Yih, *Fluid Mechanics, A Concise Introduction to the Theory* (McGraw-Hill, New York, 1969) and Ronald L. Panton, *Incompressible Flow* (2nd ed.), (Wiley, New York, 1995).
22. G. I. Taylor, Stability of a viscous liquid contained between two rotating cylinders, *Philos. Trans. R. Soc. Ser. A* **223**, 289–343 (1923).
23. Y. L. Joo and E. S. G. Shaqfeh, Observations of purely elastic instabilities in the Taylor–Dean flow of a Boger fluid, *J. Fluid Mech.* **262**, 27–74 (1994).
24. The reader who is interested in a more detailed account of the development of the Boussinesq approximation may refer to E. A. Spiegel and G. Veronis, "On the Boussinesq approximation for a compressible fluid, *Astrophysical J.* **131**, 442–7 (1960); J. M. Mihaljan, A rigorous exposition of the Boussinesq approximations applicable to a thin layer of fluid, *Astrophys. J.* **136**, 1126–33 (1962).

25. In reality, the order of the error is $(\Delta T)^m$ multiplied by the appropriate material coefficient.
26. A. Schlüter, D. Lortz and F. Busse, "On the stability of steady finite-amplitude convection," *J. Fluid Mech.* **23**, 129–4 (1965); F. H. Busse, "Non-linear proportions of thermal convection," *Rep. Prog. Phys.* **41**, 1929–1937 (1987).
27. J. R. A. Pearson, On convection cells induced by surface tension, *J. Fluid Mech.*, **4**, 489–500 (1958).
28. C. V. Sternling and L. E. Scriven, Interfacial turbulence: hydrodynamic instability and the Marangoni effect, *AIChE J.* **5**, 514 (1959); L. E. Scriven and C. V. Sternling, On cellular convection driven by surface tension gradients: effects of mean surface tension and surface viscosity, *J. Fluid Mech.* **19**, 321 (1964).
29. W. Orr, The stability or instability of the steady motions of a perfect liquid and of a viscous liquid, *Proc. R. Irish Acad. Ser. A* **27**, 9–68, 69–138 (1907).
30. A. Sommerfeld, Ein Beitrag zur Hydrodynamischen Erklaerung der Turbulenten Fluessigkeitsbewegungen, in *Proceedings of the Fourth International Congress of Mathematicians*, Vol. III, 116–124 (1908).
31. J. L. Synge, Hydrodynamical instability, *Semicentenn. Pub. Am. Math. Soc.* **2**, 227–69 (1938).

PROBLEMS

Problem 12–1. Capillary Instability – Nonaxisymmetric Modes. Prove that the axisymmetric mode is the "most unstable" mode for capillary instability of a liquid cylinder in air.

Problem 12–2. Capillary Instability of a Varicose Sheet. We have seen that a cylinder of fluid surrounded by air is unstable because of growing capillary waves. An equivalent problem is the stability of capillary waves on a round laminar jet when the velocity profile across the jet is uniform. We now wish to consider whether the same type of instability is relevant to a fluid sheet of finite width d and infinite lateral extent (the sheet has an interface above and below). Equivalently, we could also ask whether a planar (2D) laminar jet is unstable to capillary wave growth. Assume that the sheet is subject to an infinitesimal 1D wavelike disturbance that is symmetric about the center plane and corresponds to an initial sheet thickness $h = h_0(1 + \varepsilon \sin kx)$. Prove whether the sheet is linearly stable or unstable to this type of disturbance.

Problem 12–3. Capillary Instability for a Thread in a Second Immiscible Fluid. In this problem, we consider the effect on capillary instability if, instead of being surrounded by air, the thread of liquid is surrounded by a second, viscous immiscible fluid that is assumed to be unbounded in the radial direction. Derive a condition from which you could, in principle, calculate the growth-rate parameter for an axisymmetric disturbance, $\sigma = \sigma(k, Re, \lambda)$ where k is the axial wave number and λ is the ratio of the external fluid viscosity to the viscosity of the liquid thread. This condition can be simplified if either $Re \ll 1$ or the thread is inviscid (though viscous effects still remain in the outer fluid). Evaluate σ for several k values in each of these two cases. What is the qualitative effect of the second viscous fluid? For example, is the range of unstable k values changed? Is the fastest-growing linear mode changed relative to the case of a thread in air?

Problem 12–4. The Capillary Instability of an Annular Fluid Film. Two problems that are closely related to the classic problem of capillary instability of a iquid thread in air are the stability of an annular fluid film that may either coat the surface of a wire or coat the inside surface of a circular tube. We assume in both cases that the unperturbed thickness of the fluid film is a constant, independent of position. There are many applications for both of these problems. For example, the former is related to the stability of a fluid film in a wire

Problems

coating application, whereas the latter is qualitatively relevant to the stability of the mucous coating in the airways of an animal or person.

(a) First consider the stability of the wire coating configuration to axisymmetric perturbations of the shape of the film when the fluid is assumed to be *inviscid*. Assume that the unperturbed outer radius of the liquid film is a and that the radius of the cylindrical wire is sa. Specifically, obtain an analytical result for the dimensionless growth-rate coefficient $\sigma[\equiv \sigma'/(\gamma/\rho a^3)]$ as a function of the axial wave number k and the ratio of the radius of the wire to the outer radius of the film s.

(b) Now consider the case of a viscous film. Derive an equation from which you could evaluate the growth-rate coefficient as a function of k, s, and Re. Show that this equation can be simplified considerably in the limit $Re \ll 1$. Evaluate σ in this limiting case for several values of k and at least two values of s (Interesting choices should be $s = 0.1$ and $s = 0.9$. If you are energetic, you might also want to calculate for some intermediate values of s).

(c) Finally, consider the case in which the fluid film is inside the circular cylinder. You should seek to show that the analysis of parts (a) and (b) can be applied by means of a simple transformation. To show this, you may wish to go back and review all of the equations and boundary conditions for the preceding problem. What changes? (*Hint*: not much.) Discuss the stability, assuming for simplicity that the film is inviscid as in part (a).

Problem 12–5. The Effect of an Insoluble Surfactant on Capillary Instability of a Liquid Thread. In this problem, we consider the effect of an insoluble surfactant on the capillary stability of a cylindrical thread of liquid in air. It is well known that the surface tension of the air–liquid interface is decreased by the presence of surfactant to a degree that depends on the surface concentration of the surfactant. If the surfactant were uniformly dispersed on the surface, the surface tension would be decreased relative to its value for a clean interface, and this fact would lead to a decrease in the growth rates. If this were the only effect of surfactant on the stability of the thread, it would not be particularly interesting. An experimentalist would most likely measure the equilibrium value of the surface tension, and, being unaware of whether there were surfactant present or not, would simply use this value to calculate growth rates. There is, however, an additional effect of surfactant that can be very important, and this is due to the effect of Marangoni stresses caused when the flow within the liquid thread produces a nonuniform concentration distribution of surfactant at the interface. It is this effect that we consider in this problem. We assume that changes in the surfactant concentration on the deforming gas–liquid interface are governed by the convection–diffusion equation, (2–162),

$$\frac{\partial \Gamma}{\partial t} + \nabla_S \cdot (\Gamma \mathbf{u}_s) + \Gamma(\nabla_s \cdot \mathbf{n})(\mathbf{u} \cdot \mathbf{n}) = D_s \nabla_s^2 \Gamma.$$

The initial surfactant concentration is denoted as Γ_{eq} and the corresponding surface tension is γ_0. We further assume that the surfactant concentration is low enough that we can approximate the relationship between Γ and γ as linear, hence yielding the interface-stress balance in the form of Eq. (2–171):

$$(\mathbf{T} - \hat{\mathbf{T}}) \cdot \mathbf{n} = \gamma_{eq} \mathbf{n}(\nabla_s \cdot \mathbf{n}) - \beta \gamma_s \{\Gamma'[\mathbf{n}(\nabla_s \cdot \mathbf{n})] - \nabla_s \Gamma'\}.$$

(a) Derive the governing equations and boundary conditions by using cylindrical coordinates. The only changes relative to the discussion in the chapter will be in the boundary conditions and in the associated surface transport equation for the surfactant. You may assume that the problem is axisymmetric. Now obtain the linear stability equations. Note that Marangoni stresses can play no role if the fluid is inviscid, so we must retain the viscous terms.

(b) Assuming that there is an infinitesimal, axisymmetric perturbation of the shape of the thread of the form (12–43), use these equations and boundary conditions to generate the characteristic equation relating σ to k, Re, Pe_s, and $\alpha \equiv \beta \gamma_s \Gamma_{eq}/\gamma_{eq}$, corresponding to Eq. (12–54).

(c) Now, following the analysis from the text, consider the limiting case of a very viscous thread. Does the presence of surfactant stabilize or destabilize the thread?

Problem 12–6. The Linear Stability of a Spherically Symmetric Fluid Interface to Radial Accelerations. The classical Rayleigh–Taylor analysis that is described in Section B examines the stability of a plane interface between two fluids of different density to accelerations normal to the interface and shows that the interface is unstable or stable, depending on whether the acceleration is directed from the heavier fluid to the lighter fluid, or vice versa. In this problem, we consider the related problem of a spherically symmetric interface that is subjected to radial accelerations. This is a generalization of the problem of an expanding or contracting gas bubble that was considered in Chap. 4.

We suppose that we have a fluid of density ρ_1 contained within a spherical interface of radius R. This spherical body of fluid is surrounded by an unbounded body of fluid of density ρ_2. The interfacial tension is denoted as γ. To facilitate the analysis, we consider the fluids to be incompressible and also inviscid. You may assume that the change in radius $R(t)$ is specified (and thus too $\dot{R}$ and $\ddot{R}$). We wish to analyze the linear stability of the spherical interface to perturbations of shape of the form (4–298),

$$f_1 = a_{kl}(t) P_k^l(\cos\theta) e^{il\phi}.$$

In particular, derive an equation for the amplitude function $a_{kl}(t)$ and from that an equation of the form (4–312). Specifically consider the cases when $\rho_2 > \rho_1$, $\gamma = 0$ and $\rho_2 < \rho_1$, $\gamma = 0$.

Problem 12–7. The Dean Problem. A problem related to the Taylor–Couette problem is the instability of pressure-gradient-driven flow in a curved channel. An example of this is the flow that is due to a pressure gradient in the azimuthal direction between curved walls that are sections of a pair of nonrotating concentric cylinders of radius R_1 and R_2, respectively.

(a) Determine the steady velocity profile, assuming that there is a constant pressure gradient, $\partial p/\partial \theta = -G$, and that $u_r = u_z = 0$, $u_\theta = V(r)$.
(b) Consider the stability of this basic flow to an axisymmetric disturbance of the form (12–121). Derive a pair of equations for the dimensionless radial and azimuthal velocity functions, analogous to (12–125) and (12–126). What are the boundary conditions? (State clearly how you nondimensionalize the problem, and specifically what you choose as characteristic scales.)
- *Note*: Let the dimensionless perturbation velocities be denoted as $(u_r, u_\theta) = (u, v)e^{\sigma t + i\alpha z}$.
(c) Apply the narrow-gap approximation. It will be convenient to express the results in terms of

$$y \equiv (r - R_0)/(R_2 - R_1),$$

Problems

where R_0 is the mean radius of the cylinders $R_0 = (R_2 + R_1)/2$. Thus, $-1/2 \le y \le 1/2$. You should find that $V(r)$ reduces to plane Poiseuille flow,

$$V(r) \sim \frac{3}{2} V_m (1 - 4y^2),$$

where V_m is the mean value of $V(r)$, and you should obtain a pair of governing stability equations that are similar to (12–147). In particular, it should be possible to write these equations in the form

$$(D^2 - \alpha^2 - \sigma)(D^2 - \alpha^2)u = (1 - 4y^2)v,$$
$$(D^2 - \alpha^2 - \sigma)v = -\alpha^2 \beta y u, \tag{1}$$

where β is a dimensionless constant proportional to the square of the Reynolds number,

$$\beta \sim \left(\frac{V_m d}{\nu}\right)^2, \quad \text{where } d \equiv R_2 - R_1.$$

The boundary conditions are

$$u = Du = v = 0 \quad \text{at } y = \pm 1/2. \tag{2}$$

(d) It has not been proven, but is generally assumed that the principle of exchange of stabilities is valid for this problem (i.e. $\sigma = 0$ at the neutral stability point). The eigenvalue problem obtained by putting $\sigma = 0$ in Eq. (1) then provides a basis for determining the relationship between the dimensionless parameter β and the wave number α at the neutrally stability point

An approximate solution method follows the procedure outlined in Subsection D2. Carry out this procedure to determine a condition equivalent to (12–153). What result do you get from the one-term approximation corresponding to (12–154)?

Problem 12–8. Rayleigh–Taylor Stability of Superposed Fluids Confined by a Vertical Cylinder. Consider a pair of inviscid, incompressible fluids of density ρ_1 and ρ_2, respectively, with the lighter fluid (ρ_1) below the heavier fluid (ρ_2) in a long vertical rigid cylinder of radius R. The interfacial tension of the interface is denoted as γ and in its rest configuration is horizontal and flat. We showed in the analysis of Rayleigh–Taylor instability of an infinite plane interface that short-wavelength disturbances are stabilized by the interfacial tension of the interface, but that a system consisting of a heavy fluid resting above a light fluid is always unstable to disturbances of large-enough wavelength. However, when we confine fluid in a cylinder of finite radius, this restricts the maximum possible wavelength and it is possible that this may stabilize the system.

We wish to consider the linear stability of the two-fluid system inside the cylinder to an infinitesimal perturbation of shape. To analyze this problem, you should use cylindrical coordinates (r, θ, z) with the positive z axis being vertically upward. The disturbance to the interface shape can be expressed in the form

$$F = z - [1 + f(r, \theta)] = 0 \quad (r \le R)$$

and $f(r, \theta)$ analyzed as a superposition of normal modes

$$f(r, \theta) = \sum_{n=0}^{\infty} f_n(r, \theta),$$

where

$$f_n(r, \theta) \sim J_n(kr) \cos n\theta \, e^{\sigma t}.$$

Hydrodynamic Stability

Determine the characteristic equation for σ. Deduce that there is stability if

$$R^2 g(\rho_2 - \rho_1) < \gamma (S_{1,1})^2,$$

where $S_{n,m}$ is the *mth* positive zero of J_n^1.

Problem 12–9. Rayleigh–Taylor Instability for a Pair of Superposed Fluids that are Bounded Above and Below by an Impermeable Wall. In this problem, we wish to determine the effect of a pair of horizontal bounding walls that exist at a finite distance above and below a horizontal fluid interface on Rayleigh–Taylor instability. We suppose that the interface is located at $z' = 0$ and that the two fluids are inviscid, with the density of the fluid below the interface ρ_1 being less than the density of the fluid above the interface ρ_2. The boundary above the interface is located at $z' = h_2$, whereas that below the interface is at $z' = -h_1$. Show that

$$\sigma'^2 = \frac{\sinh\alpha' h_1 \sinh\alpha' h_2 [(\rho_2 - \rho_1)g - \gamma\alpha'^2]\alpha'}{\rho_1 \cosh\alpha' h_1 \sinh\alpha' h_2 + \rho_2 \cosh\alpha' h_2 \sinh\alpha' h_1}.$$

What conclusions can you draw from this result?

Problem 12–10. Capillary Instability of a Gas Cylinder in a Liquid. Consider an infinitely long cylindrical interface of radius R that is filled with a gas and surrounded by an exterior liquid ($r > R$). The surface tension is denoted as γ and the density of the liquid is ρ. We consider an initial infinitesimal perturbation of the shape

$$F = r' - R(1 + \hat{f}),$$

where $\hat{f}$ can be expressed as a superposition of normal modes

$$\hat{f} = \sum \hat{f}_n \quad \text{with } \hat{f}_n \equiv C e^{\sigma t} e^{i(kz + n\theta)}.$$

Show that

$$\sigma^2 = -\frac{\gamma}{R^3 \rho} \frac{\alpha K_n'(\alpha)}{K_n(\alpha)} (1 - \alpha^2 - n^2).$$

What conclusions do you draw about the stability?

Problem 12–11. Marangoni Instability (The Principle of Exchange of Stabilities). Following the procedure that was outlined in Section F for the Rayleigh–Benard problem, prove that the "principle of exchange of stabilities" is valid for the Marangoni instability problem (Section H).

Problem 12–12. A Variational Principle for the Rayleigh–Benard Problem. There are a number of ways to obtain exact or approximate solutions for the eigenvalue problems that arise in linear stability theory. Here we illustrate an approximate method of estimating the critical Rayleigh number for the Rayleigh–Benard problem, based on the derivation of a variational (or extremum) principle. We illustrate the use of the principle by considering the case of two no-slip, rigid boundaries, where the governing equations and boundary conditions are given by Eqs. (12–225)–(12–228). We begin by noting that (12–225) can be written in the alternative form

$$F \equiv (D^2 - \alpha^2)^2 f, \tag{1}$$

$$(D^2 - \alpha^2) F = -\alpha^2 Ra\, f, \tag{2}$$

Problems

with boundary conditions

$$f = 0 \quad \text{at } z = \pm 1/2, \tag{3}$$
$$Df = 0 \quad \text{at } z \pm 1/2, \tag{4}$$
$$F = 0 \quad \text{at } z = \pm 1/2. \tag{5}$$

(a) Multiply Eq. (2) by F and derive an exact integral formula for $\alpha^2 Ra$ by integrating this equation across the gap from $z = -1/2$ to $z = 1/2$.

(b) The formula obtained in (a) can be written in an alternative form that is obtained by integrating by parts. The result is

$$\alpha^2 Ra = \frac{\int_{-1/2}^{1/2} [(DF)^2 + \alpha^2 F^2] dz}{\int_{-1/2}^{1/2} [(D^2 - \alpha^2) f]^2 dz}. \tag{6}$$

The formula is exact. If we have the exact solutions for f and F, we could use them to calculate Ra as a function of α.

(c) We want to use the result (6) to estimate $Ra = Ra(\alpha)$ by means of approximate trial functions for F [and thus f by solution of (1) for this particular F] so that the trial functions satisfy the boundary conditions exactly but do not solve the full pair of coupled equations (1) and (2). Denote formula (6) in the symbolic form

$$\alpha^2 Ra = \frac{I_1}{I_2}.$$

Now consider a perturbation of F that we can denote as δF. This perturbation of F will produce a perturbation of I_1 and I_2 and thus of the predicted value of Ra. We denote the changes in I_1, I_2 and Ra as δI_1, δI_2 and δRa. Hence

$$\alpha^2 \delta Ra = \frac{\delta I_1}{I_2} - \frac{I_1}{I_2^2} \delta I_2 = \frac{1}{I_2}(\delta I_1 - \alpha^2 Ra \, \delta I_2).$$

Show, by integration by parts, that

$$\delta I_1 = -2 \int_{-1/2}^{1/2} \delta F[(D^2 - \alpha^2) F] dz,$$

$$\delta I_2 = 2 \int_{-1/2}^{1/2} (\delta F) f \, dz.$$

Hence

$$\alpha^2 \delta Ra = -\frac{2}{I_2} \int_{-1/2}^{1/2} \delta F[(D^2 - \alpha^2) F + \alpha^2 Ra F] dz.$$

Conclusion: This proves that the change in Ra that is due to a variation in F will be zero if F (and f) are solutions of the original problem, because the integrand is then equal to zero. It can be shown (though we will not do it here) that the stationary point where δRa is zero is actually a minimum with respect to the class of functions F (and f) that satisfy the boundary condition of the original problem, but are not solutions of the sixth-order differential equation (12–225).

(d) Use the result of (c) to estimate values for Ra as a function of α by choosing a trial function for F of the form

$$F = 1 + A \cos \pi z + \cos 2\pi z.$$

Hydrodynamic Stability

The coefficient A is arbitrary. The corresponding trial function f should be obtained by solving Eq. (1) subject to the boundary conditions (3) and (4). The best estimate for Ra will be obtained by minimizing the integral in (6) with respect to the value of A. Compare the approximate values for $\alpha = 2, 3$, and 4 with the exact values from Table 12–3.

Problem 12–13. Raleigh–Benard Convection – Nonisothermal Boundaries. The assumption of isothermal boundaries in the buoyancy-driven convection instability problem is generally an oversimplification. A more realistic picture is that the upper and lower surfaces are in contact with reservoirs that are maintained at a constant temperature so that the thermal boundary conditions are better approximated as

$$-k\frac{\partial T}{\partial z'} = -h_1(T - T_1) \quad \text{at } z' = 0,$$

$$-k\frac{\partial T}{\partial z'} = h_2(T - T_0) \quad \text{at } z' = d,$$

where h_1 and h_2 are a pair of heat transfer coefficients that may be different at the top and bottom surfaces. We assume that the heat flux in the absence of any perturbations is set at a level that produces the unperturbed temperatures T_1 and T_0 at the lower and upper boundaries. If we introduce nondimensionalization and consider a perturbation to the temperature field following the analysis in the text, then

$$\theta = \frac{T_1 - T}{T_1 - T_0} \quad \text{and} \quad \theta = \bar{\theta} + \varepsilon \hat{\theta}, \quad \text{where } \bar{\theta} = z.$$

The boundary conditions for the perturbation temperature field can be written in the form

$$\frac{\partial \hat{\theta}}{\partial z} = Bi_1 \hat{\theta} \quad \text{at } z = 0,$$

$$\frac{\partial \hat{\theta}}{\partial z} = -Bi_2 \hat{\theta} \quad \text{at } z = 1.$$

The dimensionless parameters that appear in these equations are Biot numbers $Bi \equiv hd/k$. It will be noted that the isothermal boundary condition, i.e., $\hat{\theta} = 0$, corresponds to the limit $Bi \to \infty$. The condition of a fixed heat flux, $\partial \hat{\theta}/\partial z = 0$, can be associated with the limit $Bi \to 0$ (though it must be remembered that the unperturbed heat flux is assumed to be maintained at the level required to produce the unperturbed temperatures T_1 and T_0).

(a) Demonstrate that the "principle of exchange of stabilities" is valid for this problem.
(b) Prove that there is there a variational principle for this problem. (See Problem 12–12). You should assume that $f = 0$ at $z = \pm 1/2$ and that either $Df = 0$ or $D^2 f = 0$ at $z \pm 1/2$. What is the formula, corresponding to Eq. (6) in Problem 12–12, from which estimates of $Ra = Ra(\alpha)$ may be made?
(c) We considered the isothermal limit, $Bi \to \infty$, both in the text and in Problem 12–12. Use the variational principle to estimate Ra_c in the case $Bi_1 = Bi_2 = 0$, with $f = Df = 0$ at the upper and lower boundaries. Compare the result to that obtained in Problem 12–12. Does the difference in these two results make qualitative sense?

Problem 12–14. Rayleigh–Benard Convection – One Free and One Rigid Boundary. We have noted in the text that there is an odd (or antisymmetric) solution to the problem and two rigid isothermal boundaries in addition to the even solution that we worked out in detail. The minimum Rayleigh number for the odd solution has been shown to be 17,610 at a wave

Problems

number $\alpha = 5.365$. It is possible from the result for this case to deduce that the critical Rayleigh number when one boundary is free and one is rigid is

$$Ra_c = 1101 \text{ with } \alpha_c = 2.682.$$

Justify this result.

Problem 12–15. Stability of a Fluid Layer in the Presence of Both Marangoni and Buoyancy Effects. A fluid layer is heated from below. It has a rigid, isothermal boundary at the bottom, but its upper surface is a nondeforming fluid interface. There are now two potential mechanisms for instability when the fluid is heated from below; buoyancy-driven and surface-tension-gradient-driven convection. Determine the eigenvalue problem (i.e., the ODE or equations and boundary conditions) that you would need to solve to predict the linear instability conditions. Is the principle of exchange of stabilities valid? Discuss how you would approach the solution of this eigenvalue problem.

Problem 12–16. The Effect of Shear Flow on Rayleigh–Benard Instability. We wish to consider the effect of shear on the Rayleigh-Benard buoyancy-driven instability. The problem that we will analyze is identical to that outlined in Section H (for a pair of rigid boundaries) except that there is a simple shear flow in a direction that we can designate as x, driven by motion of the upper boundary, i.e.,

$$\mathbf{u} = \frac{U_0 z}{d}\mathbf{i}_x.$$

(a) Derive the governing, linearized sixth-order DE and boundary conditions for the z component of the velocity.
(b) Obtain from this equation and boundary conditions the governing problem for $f(z)$, where

$$w = f(z) e^{i(\alpha_x x + \alpha_y y)} e^{-i\alpha_x c t}.$$

The fact that we call the exponent of the time-dependent term $-i\alpha_x ct$ instead of just σt is just for convenience and does not change the actual problem at all from the forms analyzed in the chapter. You should find

$$\{[i\alpha_x Pr Re(c-z) + (D^2 - \alpha^2)][i\alpha_x Re(c-z) + (D^2 - \alpha^2)](D^2 - \alpha^2) + Ra\,\alpha^2\}f = 0,$$

where $Re \equiv U_0 d/\nu$, plus boundary conditions.

(c) Note that, if $\alpha_x = 0$, the equation obtained in (b) reduces exactly to the problem solved to determine the critical Rayleigh number in the absence of any shear. If the eigenvalue problem is solved for $\alpha_x \neq 0$, the critical Rayleigh number is increased relative to the case $\alpha_x = 0$. What physical interpretation can you make from these facts?

Problem 12–17. Buoyancy-Driven Instability of a Fluid Layer in a Porous Medium Based on Darcy's Law. We consider the classical Rayleigh-Benard problem of a fluid layer that is heated from below, except in this case, the fluid is within a porous medium so that the equations of motion are replaced with the Darcy equations, which were discussed in Subsection C1 of this chapter. Hence the averaged velocity within the porous medium is given by Darcy's law

$$\mathbf{u}' = -\frac{k}{\mu} \text{grad}\,(p' + \rho g z') = \text{grad}\,\phi',$$

Hydrodynamic Stability

where k is known as the permeability of the porous medium. If the permeability is a constant, as we shall assume here, it follows from Darcy's law that

$$\nabla \cdot \mathbf{u}' = \mathbf{0}$$

or

$$\nabla^2 \phi' = 0.$$

We assume that the fluid still satisfies the Bousinesq approximation in that all properties are fixed except for the density, which decreases linearly with increase of the temperature.

Now, let us assume that we have a fluid layer with a depth d and that the temperature is fixed at the upper and lower boundaries:

$$T = T_1 \quad \text{at } z' = 0, \quad \text{and} \quad T = T_0 \quad \text{at } z' = d.$$

We also assume that these boundaries are impermeable, so that the Darcy velocity satisfies the kinematic condition $\mathbf{u}' \cdot \mathbf{n} = 0$. No other boundary conditions can be satisfied by the velocity, as explained in Subsection C.1. Finally, the thermal energy equation is unchanged from its usual form,

$$\frac{\partial T}{\partial t'} + \mathbf{u}' \cdot \nabla' T = \kappa \nabla'^2 T,$$

where κ is the thermal diffusivity.

Analyze the linear stability of the fluid layer. The analysis is similar to the classical Rayleigh–Benard problem, but simpler because the Navier–Stokes equations are replaced with Darcy's law. It can be solved analytically. In place of the Rayleigh number, you should find that the stability depends on a modified Rayleigh number,

$$Ra_D = \frac{\beta \Delta T g d k}{\kappa \nu}.$$

which is sometimes called the Darcy–Rayleigh number. Your analysis should lead you to the result that the value of Ra_D that gives neutral stability is

$$Ra_D^* = \frac{(n^2 \pi^2 + \alpha^2)^2}{\alpha^2}.$$

Hence the critical value of $(Ra_D)_c$ is $\pi^2/4$.

Problem 12–18. Buoyancy-Driven Instability of a Fluid Layer in a Porous Medium Based on the Darcy–Brinkman Equations. A more complete model for the motion of a fluid in a porous medium is provided by the so-called Darcy–Brinkman equations. In the following, we reexamine the conditions for buoyancy-driven instability when the fluid layer is heated from below. We assume that inertia effects can be neglected (this has no effect on the stability analysis as one can see by reexamining the analysis in Section H) and that the Boussinesq approximation is valid so that fluid and solid properties are assumed to be constant except for the density of the fluid. The Darcy–Brinkman equations can be written in the form

$$0 = -\nabla' p' - \frac{\mu}{k} \mathbf{u} + \mu \nabla^2 \mathbf{u} + \rho \mathbf{g}$$
$$\nabla' \mathbf{u} = 0,$$

together with the thermal energy equation in its standard form,

$$\frac{\partial T}{\partial t'} + \mathbf{u}' \cdot \nabla' T = \kappa \nabla'^2 T.$$

Problems

In the present case, we again assume that the upper and lower boundaries are impermeable and isothermal with temperatures

$$T = T_1 \quad \text{at } z' = 0, \quad \text{and} \quad T = T_0 \quad \text{at } z' = d.$$

In this case, the solution can also satisfy boundary conditions on the horizontal velocity components. It is convenient, for illustrative purposes because an analytic solution can be obtained, to assume that these boundaries are free (i.e., zero shear stress).

Analyze the stability for this problem. You should begin by nondimensionalizing. You should find that there are three dimensionless parameters,

$$Ra = \frac{\beta \Delta T g d^3}{\kappa \nu}, \quad \frac{k}{d^2}, \quad Pr = \frac{\nu}{\kappa}.$$

When the dimensionless permeability k/d^2 is equal to zero, the problem reduces to the Darcy approximation that was analyzed in the previous problem. On the other hand, for $k/d^2 \to \infty$, the problem is the classical Rayleigh–Benard case for a single viscous fluid. With the notation of Section H, the governing equation at the neutral stability point becomes

$$\left[(D^2 - \alpha^2) - \frac{d^2}{k}\right](D^2 - \alpha^2)f(z) = \alpha^2 Ra\theta(z),$$

$$(D^2 - \alpha^2)\theta(z) + f(z) = 0,$$

$$f = D^2 f = \theta = 0 \quad \text{at } z = 0, 1.$$

Plot your results for the critical Rayleigh number, Ra_c, as a function of k/d^2. How small does k/d^2 have to be before we reach the Darcy limit, and how large before the result reduces to the classical Rayleigh–Benard case?

APPENDIX A

Governing Equations and Vector Operations in Cartesian, Cylindrical, and Spherical Coordinate Systems

RECTANGULAR CARTESIAN COORDINATES

$$d\mathbf{x} = dx\mathbf{e}_x + dy\mathbf{e}_y + dz\mathbf{e}_z$$

Standard Vector Operations

$$\nabla \psi = \frac{\partial \psi}{\partial x}\mathbf{e}_x + \frac{\partial \psi}{\partial y}\mathbf{e}_y + \frac{\partial \psi}{\partial z}\mathbf{e}_z,$$

$$\nabla \cdot \mathbf{A} = \frac{\partial A_x}{\partial x} + \frac{\partial A_y}{\partial y} + \frac{\partial A_z}{\partial z},$$

$$\nabla^2 \psi = \nabla \cdot (\nabla \psi) = \frac{\partial^2 \psi}{\partial x^2} + \frac{\partial^2 \psi}{\partial y^2} + \frac{\partial^2 \psi}{\partial z^2},$$

$$\nabla \wedge \mathbf{A} = \left(\frac{\partial A_z}{\partial y} - \frac{\partial A_y}{\partial z}\right)\mathbf{e}_x + \left(\frac{\partial A_x}{\partial z} - \frac{\partial A_z}{\partial x}\right)\mathbf{e}_y + \left(\frac{\partial A_y}{\partial x} - \frac{\partial A_x}{\partial y}\right)\mathbf{e}_z,$$

$$\nabla \mathbf{A} = \frac{\partial A_x}{\partial x}\mathbf{e}_x\mathbf{e}_x + \frac{\partial A_y}{\partial x}\mathbf{e}_x\mathbf{e}_y + \frac{\partial A_z}{\partial x}\mathbf{e}_x\mathbf{e}_z + \frac{\partial A_x}{\partial y}\mathbf{e}_y\mathbf{e}_x + \frac{\partial A_y}{\partial y}\mathbf{e}_y\mathbf{e}_y + \frac{\partial A_z}{\partial y}\mathbf{e}_y\mathbf{e}_z$$

$$+ \frac{\partial A_x}{\partial z}\mathbf{e}_z\mathbf{e}_x + \frac{\partial A_y}{\partial z}\mathbf{e}_z\mathbf{e}_y + \frac{\partial A_z}{\partial z}\mathbf{e}_z\mathbf{e}_z,$$

$$\nabla \cdot \mathbf{B} = \left(\frac{\partial B_{xx}}{\partial x} + \frac{\partial B_{yx}}{\partial y} + \frac{\partial B_{zx}}{\partial z}\right)\mathbf{e}_x + \left(\frac{\partial B_{xy}}{\partial x} + \frac{\partial B_{yy}}{\partial y} + \frac{\partial B_{zy}}{\partial z}\right)\mathbf{e}_y$$

$$+ \left(\frac{\partial B_{xz}}{\partial x} + \frac{\partial B_{yz}}{\partial y} + \frac{\partial B_{zz}}{\partial z}\right)\mathbf{e}_z.$$

Equations of Motion

Symbols of velocity components:

$$\mathbf{u} = u(x, y, z, t)\mathbf{e}_x + v(x, y, z, t)\mathbf{e}_y + w(x, y, z, t)\mathbf{e}_z.$$

Continuity equation:

$$\nabla \cdot \mathbf{u} = \frac{\partial u}{\partial x} + \frac{\partial v}{\partial y} + \frac{\partial w}{\partial z} = 0.$$

Appendix A: Governing Equations and Vector Operations

x momentum:
$$\rho\left(\frac{\partial u}{\partial t} + u\frac{\partial u}{\partial x} + v\frac{\partial u}{\partial y} + w\frac{\partial u}{\partial z}\right) = \rho g_x - \frac{\partial p}{\partial x} + \mu\left(\frac{\partial^2 u}{\partial x^2} + \frac{\partial^2 u}{\partial y^2} + \frac{\partial^2 u}{\partial z^2}\right).$$

y momentum:
$$\rho\left(\frac{\partial v}{\partial t} + u\frac{\partial v}{\partial x} + v\frac{\partial v}{\partial y} + w\frac{\partial v}{\partial z}\right) = \rho g_y - \frac{\partial p}{\partial y} + \mu\left(\frac{\partial^2 v}{\partial x^2} + \frac{\partial^2 v}{\partial y^2} + \frac{\partial^2 v}{\partial z^2}\right).$$

z momentum:
$$\rho\left(\frac{\partial w}{\partial t} + u\frac{\partial w}{\partial x} + v\frac{\partial w}{\partial y} + w\frac{\partial w}{\partial z}\right) = \rho g_z - \frac{\partial p}{\partial z} + \mu\left(\frac{\partial^2 w}{\partial x^2} + \frac{\partial^2 w}{\partial y^2} + \frac{\partial^2 w}{\partial z^2}\right).$$

Viscous stresses:
$$\tau_{xx} = 2\mu\frac{\partial u}{\partial x}, \quad \tau_{yy} = 2\mu\frac{\partial v}{\partial y}, \quad \tau_{zz} = 2\mu\frac{\partial w}{\partial z},$$
$$\tau_{xy} = \tau_{yx} = \mu\left(\frac{\partial u}{\partial y} + \frac{\partial v}{\partial x}\right), \quad \tau_{yz} = \tau_{zy} = \mu\left(\frac{\partial v}{\partial z} + \frac{\partial w}{\partial y}\right),$$
$$\tau_{zx} = \tau_{xz} = \mu\left(\frac{\partial w}{\partial x} + \frac{\partial u}{\partial z}\right).$$

CYLINDRICAL COORDINATES

$$x = r\cos\theta, \quad y = r\sin\theta, \quad z = z,$$
$$r = \sqrt{x^2 + y^2}, \quad \theta = \tan^{-1}\frac{y}{x}, \quad z = z,$$
$$\mathbf{e}_r = \mathbf{e}_x \cos\theta + \mathbf{e}_y \sin\theta,$$
$$\mathbf{e}_\theta = -\mathbf{e}_x \sin\theta + \mathbf{e}_y \cos\theta,$$
$$\mathbf{e}_z = \mathbf{e}_z,$$
$$d\mathbf{x} = \frac{dr}{h_1}\mathbf{e}_r + \frac{d\theta}{h_2}\mathbf{e}_\theta + \frac{dz}{h_3}\mathbf{e}_z,$$
$$h_1 = 1, \quad h_2 = \frac{1}{r}, \quad h_3 = 1.$$

Standard Vector Operations

$$\nabla\psi = \frac{\partial\psi}{\partial r}\mathbf{e}_r + \frac{1}{r}\frac{\partial\psi}{\partial\theta}\mathbf{e}_\theta + \frac{\partial\psi}{\partial z}\mathbf{e}_z,$$

$$\nabla\cdot\mathbf{A} = \frac{1}{r}\frac{\partial}{\partial r}(rA_r) + \frac{1}{r}\frac{\partial A_\theta}{\partial\theta} + \frac{\partial A_z}{\partial z},$$

$$\nabla^2\psi = \nabla\cdot(\nabla\psi) = \frac{1}{r}\frac{\partial}{\partial r}\left(r\frac{\partial\psi}{\partial r}\right) + \frac{1}{r^2}\frac{\partial^2\psi}{\partial\theta^2} + \frac{\partial^2\psi}{\partial z^2},$$

$$\nabla\wedge\mathbf{A} = \left(\frac{1}{r}\frac{\partial A_z}{\partial\theta} - \frac{\partial A_\theta}{\partial z}\right)\mathbf{e}_r + \left(\frac{\partial A_r}{\partial z} - \frac{\partial A_z}{\partial r}\right)\mathbf{e}_\theta$$
$$+ \left[\frac{1}{r}\frac{\partial}{\partial r}(rA_\theta) - \frac{1}{r}\frac{\partial A_r}{\partial\theta}\right]\mathbf{e}_z,$$

Cylindrical Coordinates

$$\nabla \mathbf{A} = \frac{\partial A_r}{\partial r} \mathbf{e}_r \mathbf{e}_r + \frac{\partial A_\theta}{\partial r} \mathbf{e}_r \mathbf{e}_\theta + \frac{\partial A_z}{\partial r} \mathbf{e}_r \mathbf{e}_z$$

$$+ \frac{1}{r}\left(\frac{\partial A_r}{\partial \theta} - A_\theta\right) \mathbf{e}_\theta \mathbf{e}_r + \frac{1}{r}\left(\frac{\partial A_\theta}{\partial \theta} + A_r\right) \mathbf{e}_\theta \mathbf{e}_\theta + \frac{1}{r}\frac{\partial A_z}{\partial \theta} \mathbf{e}_\theta \mathbf{e}_z$$

$$+ \frac{\partial A_r}{\partial z} \mathbf{e}_z \mathbf{e}_r + \frac{\partial A_\theta}{\partial z} \mathbf{e}_z \mathbf{e}_\theta + \frac{\partial A_z}{\partial z} \mathbf{e}_z \mathbf{e}_z,$$

$$\nabla \cdot \mathbf{B} + \left[\frac{1}{r}\frac{\partial}{\partial r}(rB_{rr}) + \frac{1}{r}\frac{\partial B_{\theta r}}{\partial \theta} + \frac{\partial B_{zr}}{\partial z} - \frac{B_{\theta\theta}}{r}\right] \mathbf{e}_r$$

$$+ \left[\frac{1}{r}\frac{\partial}{\partial r}(rB_{r\theta}) + \frac{1}{r}\frac{\partial B_{\theta\theta}}{\partial \theta} + \frac{\partial B_{z\theta}}{\partial z} + \frac{B_{\theta r}}{r}\right] \mathbf{e}_\theta$$

$$+ \left[\frac{1}{r}\frac{\partial}{\partial r}(rB_{rz}) + \frac{1}{r}\frac{\partial B_{\theta z}}{\partial \theta} + \frac{\partial B_{zz}}{\partial z}\right] \mathbf{e}_z.$$

Equations of Motion
Symbols for velocity components:

$$\mathbf{u} = u_r(r, \theta, z, t)\, \mathbf{e}_r + u_\theta(r, \theta, z, t)\, \mathbf{e}_\theta + u_z(r, \theta, z, t)\, \mathbf{e}_z.$$

Continuity equation

$$\nabla \cdot \mathbf{u} = \frac{1}{r}\left[\frac{\partial}{\partial r}(ru_r) + \frac{\partial u_\theta}{\partial \theta}\right] + \frac{\partial u_z}{\partial z} = 0.$$

r momentum:

$$\rho\left(\frac{\partial u_r}{\partial t} + u_r\frac{\partial u_r}{\partial r} + \frac{u_\theta}{r}\frac{\partial u_r}{\partial \theta} + u_z\frac{\partial u_r}{\partial z} - \frac{u_\theta^2}{r}\right)$$

$$= \rho g_r - \frac{\partial p}{\partial r} + \mu\left\{\frac{\partial}{\partial r}\left[\frac{1}{r}\frac{\partial}{\partial r}(ru_r)\right] + \frac{1}{r^2}\frac{\partial^2 u_r}{\partial \theta^2} - \frac{2}{r^2}\frac{\partial u_\theta}{\partial \theta} + \frac{\partial^2 u_r}{\partial z^2}\right\}$$

θ-momentum:

$$\rho\left(\frac{\partial u_\theta}{\partial t} + u_r\frac{\partial u_\theta}{\partial r} + \frac{u_\theta}{r}\frac{\partial u_\theta}{\partial \theta} + u_z\frac{\partial u_\theta}{\partial z} + \frac{u_r u_\theta}{r}\right)$$

$$= \rho g_\theta - \frac{1}{r}\frac{\partial p}{\partial \theta} + \mu\left\{\frac{\partial}{\partial r}\left[\frac{1}{r}\frac{\partial}{\partial r}(ru_\theta)\right] + \frac{1}{r^2}\frac{\partial^2 u_\theta}{\partial \theta^2} + \frac{2}{r^2}\frac{\partial u_r}{\partial \theta} + \frac{\partial^2 u_\theta}{\partial z^2}\right\}.$$

z momentum:

$$\rho\left(\frac{\partial u_z}{\partial t} + u_r\frac{\partial u_z}{\partial r} + \frac{u_\theta}{r}\frac{\partial u_z}{\partial \theta} + u_z\frac{\partial u_z}{\partial z}\right)$$

$$= \rho g_z - \frac{\partial p}{\partial z} + \mu\left[\frac{1}{r}\frac{\partial}{\partial r}\left(r\frac{\partial u_z}{\partial r}\right) + \frac{1}{r^2}\frac{\partial^2 u_z}{\partial \theta^2} + \frac{\partial^2 u_z}{\partial z^2}\right].$$

Viscous stresses:

$$\tau_{rr} = 2\mu\frac{\partial u_r}{\partial r}, \quad \tau_{\theta\theta} = 2\mu\frac{1}{r}\left(\frac{\partial u_\theta}{\partial \theta} + u_r\right), \quad \tau_{zz} = 2\mu\frac{\partial u_z}{\partial z},$$

Appendix A: Governing Equations and Vector Operations

$$\tau_{r\theta} = \tau_{\theta r} = \mu \left[r \frac{\partial}{\partial r}\left(\frac{u_\theta}{r}\right) + \frac{1}{r}\frac{\partial u_r}{\partial \theta} \right], \quad \tau_{\theta z} = \tau_{z\theta} = \mu \left(\frac{1}{r}\frac{\partial u_z}{\partial \theta} + \frac{\partial u_\theta}{\partial z} \right),$$

$$\tau_{zr} = \tau_{rz} = \mu \left(\frac{\partial u_z}{\partial r} + \frac{\partial u_r}{\partial z} \right).$$

SPHERICAL COORDINATES

$$x = r \sin\theta \cos\phi, \quad y = r\sin\theta \sin\phi, \quad z = r\cos\theta,$$

$$r = \sqrt{x^2+y^2+z^2}, \quad \theta = \tan^{-1}\frac{\sqrt{x^2+y^2}}{z}, \quad \phi = \tan^{-1}\frac{y}{x},$$

$$\mathbf{e}_r = \sin\theta\cos\phi\,\mathbf{e}_x + \sin\theta\sin\phi\,\mathbf{e}_y + \cos\theta\,\mathbf{e}_z,$$

$$\mathbf{e}_\theta = \cos\theta\cos\phi\,\mathbf{e}_x + \cos\theta\sin\phi\,\mathbf{e}_y - \sin\theta\,\mathbf{e}_z,$$

$$\mathbf{e}_\phi = -\sin\phi\,\mathbf{e}_x + \cos\phi\,\mathbf{e}_y$$

$$d\mathbf{x} = \frac{dr}{h_1}\mathbf{e}_r + \frac{d\theta}{h_2}\mathbf{e}_\theta + \frac{d\phi}{h_3}\mathbf{e}_\phi,$$

$$h_1 = 1, \quad h_2 = \frac{1}{r}, \quad h_3 = \frac{1}{r\sin\theta}.$$

Standard Vector Operations

$$\nabla\psi = \frac{\partial\psi}{\partial r}\mathbf{e}_r + \frac{1}{r}\frac{\partial\psi}{\partial\theta}\mathbf{e}_\theta + \frac{1}{r\sin\theta}\frac{\partial\psi}{\partial\phi}\mathbf{e}_\phi,$$

$$\nabla\cdot\mathbf{A} = \frac{1}{r^2}\frac{\partial}{\partial r}(r^2 A_r) + \frac{1}{r\sin\theta}\frac{\partial}{\partial\theta}(\sin\theta A_\theta) + \frac{1}{r\sin\theta}\frac{\partial A_\phi}{\partial\phi},$$

$$\nabla^2\psi = \nabla\cdot(\nabla\psi) = \frac{1}{r^2}\left[\frac{\partial}{\partial r}\left(r^2\frac{\partial\psi}{\partial r}\right) + \frac{1}{\sin\theta}\frac{\partial}{\partial\theta}\left(\sin\theta\frac{\partial\psi}{\partial\theta}\right) + \frac{1}{\sin^2\theta}\frac{\partial^2\psi}{\partial\phi^2}\right],$$

$$\nabla\wedge\mathbf{A} = \frac{1}{r\sin\theta}\left[\frac{\partial}{\partial\theta}(\sin\theta A_\phi) - \frac{\partial A_\theta}{\partial\phi}\right]\mathbf{e}_r + \frac{1}{r}\left[\frac{1}{\sin\theta}\frac{\partial A_r}{\partial\phi} - \frac{\partial}{\partial r}(rA_\phi)\right]\mathbf{e}_\theta$$

$$+ \frac{1}{r}\left[\frac{\partial}{\partial r}(rA_\theta) - \frac{\partial A_r}{\partial\theta}\right]\mathbf{e}_\phi,$$

$$\nabla\mathbf{A} = \frac{\partial A_r}{\partial r}\mathbf{e}_r\mathbf{e}_r + \frac{\partial A_\theta}{\partial r}\mathbf{e}_r\mathbf{e}_\theta + \frac{\partial A_\phi}{\partial r}\mathbf{e}_r\mathbf{e}_\phi$$

$$+ \frac{1}{r}\left(\frac{\partial A_r}{\partial\theta} - A_\theta\right)\mathbf{e}_\theta\mathbf{e}_r + \frac{1}{r}\left(\frac{\partial A_\theta}{\partial\theta} + A_r\right)\mathbf{e}_\theta\mathbf{e}_\theta + \frac{1}{r}\frac{\partial A_\phi}{\partial\theta}\mathbf{e}_\theta\mathbf{e}_\phi$$

$$+ \frac{1}{r}\left(\frac{1}{\sin\theta}\frac{\partial A_r}{\partial\phi} - A_\phi\right)\mathbf{e}_\phi\mathbf{e}_r + \frac{1}{r}\left(\frac{1}{\sin\theta}\frac{\partial A_\theta}{\partial\phi} - \cot\theta A_\phi\right)\mathbf{e}_\phi\mathbf{e}_\theta$$

$$+ \frac{1}{r}\left(\frac{1}{\sin\theta}\frac{\partial A_\phi}{\partial\phi} + A_r + \cot\theta A_\theta\right)\mathbf{e}_\phi\mathbf{e}_\phi,$$

Spherical Coordinates

$$\nabla \cdot \mathbf{B} = \left[\frac{1}{r^2} \frac{\partial}{\partial r}(r^2 B_{rr}) + \frac{1}{r \sin\theta} \frac{\partial}{\partial \theta}(\sin\theta\, B_{\theta r}) + \frac{1}{r \sin\theta} \frac{\partial B_{\phi r}}{\partial \phi} - \frac{B_{\theta\theta}}{r} - \frac{B_{\phi\phi}}{r} \right] \mathbf{e}_r$$

$$+ \left[\frac{1}{r^2} \frac{\partial}{\partial r}(r^2 B_{r\theta}) + \frac{1}{r \sin\theta} \frac{\partial}{\partial \theta}(\sin\theta\, B_{\theta\theta}) + \frac{1}{r \sin\theta} \frac{\partial B_{\phi\theta}}{\partial \phi} + \frac{B_{\theta r}}{r} - \frac{\cot\theta}{r} B_{\phi\phi} \right] \mathbf{e}_\theta$$

$$+ \left[\frac{1}{r^2} \frac{\partial}{\partial r}(r^2 B_{r\phi}) + \frac{1}{r \sin\theta} \frac{\partial}{\partial \theta}(\sin\theta\, B_{\theta\phi}) + \frac{1}{r \sin\theta} \frac{\partial B_{\phi\phi}}{\partial \phi} + \frac{B_{\phi r}}{r} + \frac{\cot\theta}{r} B_{\theta\phi} \right] \mathbf{e}_\phi.$$

Equations of Motion

Continuity equation:

$$\nabla \cdot \mathbf{u} = \frac{1}{r^2} \frac{\partial}{\partial r}(r^2 u_r) + \frac{1}{r \sin\theta} \frac{\partial}{\partial \theta}(\sin\theta\, u_\theta) + \frac{1}{r \sin\theta} \frac{\partial}{\partial \phi}(u_\phi) = 0.$$

r momentum:

$$\rho \left(\frac{\partial u_r}{\partial t} + u_r \frac{\partial u_r}{\partial r} + \frac{u_\theta}{r} \frac{\partial u_r}{\partial \theta} - \frac{u_\theta^2 + u_\phi^2}{r} + \frac{u_\phi}{r \sin\theta} \frac{\partial u_r}{\partial \phi} \right)$$

$$= \rho g_r - \frac{\partial p}{\partial r} + \mu \left[\frac{1}{r^2} \frac{\partial}{\partial r}\left(r^2 \frac{\partial u_r}{\partial r} \right) + \frac{1}{r^2 \sin\theta} \frac{\partial}{\partial \theta}\left(\sin\theta \frac{\partial u_r}{\partial \theta} \right) \right.$$

$$\left. + \frac{1}{r^2 \sin^2\theta} \frac{\partial^2 u_r}{\partial \phi^2} - \frac{2 u_r}{r^2} - \frac{2}{r^2} \frac{\partial u_\theta}{\partial \theta} - \frac{2 u_\theta \cot\theta}{r^2} - \frac{2}{r^2 \sin\theta} \frac{\partial u_\phi}{\partial \phi} \right].$$

θ momentum:

$$\rho \left(\frac{\partial u_\theta}{\partial t} + u_r \frac{\partial u_\theta}{\partial r} + \frac{u_\theta}{r} \frac{\partial u_\theta}{\partial \theta} + \frac{u_r u_\theta}{r} - \frac{\cot\theta\, u_\phi^2}{r} + \frac{u_\phi}{r \sin\theta} \frac{\partial u_\theta}{\partial \phi} \right)$$

$$= \rho g_\theta - \frac{1}{r} \frac{\partial p}{\partial \theta} + \mu \left[\frac{1}{r^2} \frac{\partial}{\partial r}\left(r^2 \frac{\partial u_\theta}{\partial r} \right) + \frac{1}{r^2 \sin\theta} \frac{\partial}{\partial \theta}\left(\sin\theta \frac{\partial u_\theta}{\partial \theta} \right) \right.$$

$$\left. + \frac{1}{r^2 \sin^2\theta} \frac{\partial^2 u_\theta}{\partial \phi^2} + \frac{2}{r^2} \frac{\partial u_r}{\partial \theta} - \frac{u_\theta}{r^2 \sin^2\theta} - \frac{2 \cos\theta}{r^2 \sin^2\theta} \frac{\partial u_\phi}{\partial \phi} \right].$$

ϕ momentum:

$$\rho \left(\frac{\partial u_\phi}{\partial t} + u_r \frac{\partial u_\phi}{\partial r} + \frac{u_\theta}{r} \frac{\partial u_\phi}{\partial \theta} + \frac{u_r u_\phi}{r} + \frac{\cot\theta\, u_\theta u_\phi}{r} + \frac{u_\phi}{r \sin\theta} \frac{\partial u_\phi}{\partial \phi} \right)$$

$$= \rho g_\phi - \frac{1}{r \sin\theta} \frac{\partial p}{\partial \phi} + \mu \left[\frac{1}{r^2} \frac{\partial}{\partial r}\left(r^2 \frac{\partial u_\phi}{\partial r} \right) + \frac{1}{r^2 \sin\theta} \frac{\partial}{\partial \theta}\left(\sin\theta \frac{\partial u_\phi}{\partial \theta} \right) \right.$$

$$\left. + \frac{1}{r^2 \sin^2\theta} \frac{\partial^2 u_\phi}{\partial \phi^2} - \frac{u_\phi}{r^2 \sin^2\theta} + \frac{2}{r^2 \sin\theta} \frac{\partial u_r}{\partial \phi} + \frac{2 \cos\theta}{r^2 \sin^2\theta} \frac{\partial u_\theta}{\partial \phi} \right].$$

Stresses:

$$\tau_{rr} = 2\mu \frac{\partial u_r}{\partial r},$$

$$\tau_{\theta\theta} = 2\mu \left(\frac{1}{r} \frac{\partial u_\theta}{\partial \theta} + \frac{u_r}{r} \right),$$

$$\tau_{\phi\phi} = 2\mu \left(\frac{1}{r \sin\theta} \frac{\partial u_\phi}{\partial \phi} + \frac{u_r}{r} + \frac{u_\theta \cot\theta}{r} \right),$$

Appendix A: Governing Equations and Vector Operations

$$\tau_{r\theta} = \tau_{\theta r} = \mu \left[r \frac{\partial}{\partial r} \left(\frac{u_\theta}{r} \right) + \frac{1}{r} \frac{\partial u_r}{\partial \theta} \right],$$

$$\tau_{\theta\phi} = \tau_{\phi\theta} = \mu \left[\frac{\sin\theta}{r} \frac{\partial}{\partial \theta} \left(\frac{u_\phi}{\sin\theta} \right) + \frac{1}{r \sin\theta} \frac{\partial u_\theta}{\partial \phi} \right],$$

$$\tau_{\phi r} = \tau_{r\phi} = \mu \left[\frac{1}{r \sin\theta} \frac{\partial u_r}{\partial \phi} + r \frac{\partial}{\partial r} \left(\frac{u_\phi}{r} \right) \right].$$

APPENDIX B

Cartesian Component Notation

Cartesian component notation offers an extremely convenient shorthand representation for vectors, tensors, and vector calculus operations. In this formalism, we represent vectors or tensors in terms of their typical components. For example, we can represent a vector **A** in terms of its typical component A_i, where the index i has possible values 1, 2, or 3. Hence we represent the position vector **x** as x_i, and the (vector) gradient operator ∇ as $\frac{\partial}{\partial x_i}$. Note that there is nothing special about the letter that is chosen to represent the index. We could equally well write x_j, x_k, or x_m, as long as we remember that, whatever letter we choose, its possible values are 1, 2, or 3. The second-order identity tensor **I** is represented by its components δ_{ij}, defined to be equal to 1 when $i = j$ and to be equal to 0 if $i \neq j$. The third-order alternating tensor ε is represented by its components ε_{ijk}, defined as

$$\varepsilon_{ijk} = \begin{cases} 0 & \text{if any of } i, j \text{ or } k \text{ is equal} \\ +1 & \text{if } ijk = 123, 231, 312 \\ -1 & \text{if } ijk = 321, 213, 132 \end{cases}.$$

Using this notation, we can represent the standard products between two vectors **A** and **B** by using the index summation convention as

$$\text{Scalar product} \quad \mathbf{A} \cdot \mathbf{B} \rightarrow A_i B_i,$$

$$\text{Vector product} \quad \mathbf{A} \wedge \mathbf{B} \rightarrow \varepsilon_{ijk} A_j B_k,$$

$$\text{Dyadic product} \quad \mathbf{A}\mathbf{B} \rightarrow A_i B_j.$$

According to the summation convention, we must sum over any repeated index over all possible values of that index. So the scalar product produces a scalar that is equal to $A_1 B_1 + A_2 B_2 + A_3 B_3$, whereas the vector product is a vector, the ith component of which is $\varepsilon_{ijk} A_j B_k$ (so, for example, the component in the 1 direction is $A_2 B_3 - A_3 B_2$), and the dyadic product is a second-order tensor with a typical component $A_i B_j$ (if we consider all possible combinations of i and j, there are clearly nine independent components).

The standard gradient operations can be represented in the same way:

$$\nabla \psi \rightarrow \partial \psi / \partial x_i,$$

$$\nabla \cdot \mathbf{A} \rightarrow \partial A_i / \partial x_i,$$

$$\nabla^2 \psi = \nabla \cdot (\nabla \psi) \rightarrow \frac{\partial}{\partial x_i}\left(\frac{\partial \psi}{\partial x_i}\right) = \frac{\partial^2 \psi}{\partial x_i^2},$$

Appendix B: Cartesian Component Notation

$$\nabla \wedge \mathbf{A} \to \varepsilon_{ijk} \frac{\partial A_j}{\partial x_k},$$

$$\nabla \mathbf{A} \to \frac{\partial A_j}{\partial x_i} \quad \text{(the } ij \text{ component corresponding to the ordered pair of unit vectors } \mathbf{e}_i \mathbf{e}_j \text{)},$$

$$\nabla \cdot \mathbf{B} \to \frac{\partial}{\partial x_i} B_{ij} \quad \text{(here B is a second-order tensor)}.$$

One particularly useful feature of the Cartesian index notation is that it provides a very convenient framework for working out vector and tensor identities. Two simple examples follow:

(1) Show that $\nabla \cdot (\psi \mathbf{A}) = \nabla \psi \cdot \mathbf{A} + \psi \nabla \cdot \mathbf{A}$.

We can write the left-hand side in Cartesian index notation as

$$\nabla \cdot (\psi \mathbf{A}) \to \frac{\partial (\psi A_i)}{\partial x_i}$$

Now the indicated differentiation in the index notation can be carried out by use of the chain rule,

$$\frac{\partial (\psi A_i)}{\partial x_i} = \frac{\partial \psi}{\partial x_i} A_i + \psi \frac{\partial A_i}{\partial x_i}.$$

Now we can identify the two terms on the right-hand side as

$$\nabla \psi \cdot \mathbf{A} \quad \text{and} \quad \psi \nabla \cdot \mathbf{A}.$$

It is important to note that vector relations such as the identity that we have proven in this example are invariant to the coordinate system. So, even though we have proven this result by using Cartesian index notation, the result is valid in all coordinate systems (e.g., Cartesian, cylindrical, spherical, etc.)

(2) Show that $\nabla \cdot (\phi \mathbf{AB}) = \nabla \cdot (\phi \mathbf{A}) \mathbf{B} + (\phi \mathbf{A}) \cdot \nabla \mathbf{B}$ (here, both **A** and **B** are vectors). This time,

$$\nabla \cdot (\phi \mathbf{AB}) \to \frac{\partial (\phi A_i B_j)}{\partial x_i} = \frac{\partial (\phi A_i)}{\partial x_i} B_j + \phi A_i \frac{\partial B_j}{\partial x_i}.$$

We can easily identify the right-hand side as $\nabla \cdot (\phi \mathbf{A}) \mathbf{B} + (\phi \mathbf{A}) \cdot \nabla \mathbf{B}$.

Index

Air hockey table, (*See* Thin-films), 325
Alternating tensor, 30, 526
Analytic approximations, 205
    asymptotic solutions, 205
    macroscopic balances, 205
Angular momentum, 29
Archimedes principle, 39
Asymptotic approximations, 204
    linearization via asymptotic methods, 222
Asymptotic convergence, 207, 218
    contrast to power series convergence, 217
Asymptotic expansions, 216
    domain perturbation, 232, 272, 396, 539
    gauge function, 217, 221, 246
    general considerations, 216
    matched, 169, 213
    multiple time-scale methods, 250, 260, 264
    regular, 207, 218, 227
        equivalence to successive approximations, 209
    singular, 169, 213, 218
        gauge functions, 217, 221, 246
        reduction in the order of the differential equation, 219, 244, 646
    two-time scale expansions, 264
    uniqueness, 217
Asymptotic methods, 131
    approximating the governing equations, 205, 222
    condition of matching, 215, 219, 246
        matching versus patching, 219, 246
        mismatch, 215
    domain perturbation method, 232
    rescaling, 212, 245
        stretching, 212
    role of nondimensionalization and rescaling, 209, 245
Axisymmetric pure straining flow, 533
    (*See also* Extensional flow, axisymmetric,) 555

Benard's original experiments, 871
Bernoulli's equation, 707
Bessel equation, 138, 180
Bessel functions, 138, 180
    J_0 and Y_0, 138
        expansion for small argument, 181

    orthogonality, 139
    modified, I_n, K_n, 329, 608, 806
Biaxial extensional flow, 471, 641
Biharmonic equation, 448, 620
Billiard ball gas model, 15
Biot number, 495, 853
Bi-spherical coordinates, 576
Blasius equation, 715
Blasius series, 728
    axisymmetric bodies, 733, 737
    cylindrical bodies, 728
Blasius solution, 714
Blood flow, 175
Body forces, 25
Body torque/couple, 30
Bond number, 368
Boundary conditions, 17, 65
    at a fluid interface
        conservation of thermal energy, 68
        gas-liquid interface, 358, 485
        interface deformation, 74
        simplifications in the thin-film approximation, 359
        surface force balances, 76, 77
        with surfactant, 89
    at infinity in an unbounded fluid, 66
    dynamic, 69
        Navier slip condition, 70
        no-slip condition, 69
    heat transfer
        constant surface temperature, 790
        mixed-type boundary conditions, 595, 615, 789
        specified heat flux, 788
        uniform surfaces temperature, 594
    kinematic, 67, 75
        at an interface, 75
        with phase transformation, 67, 153
    Navier slip condition, 70
    normal stress balance at a fluid interface, 79
        Young-Laplace equation, 79
    slip
        at a solid surface (Navier slip), 70
        at an interface, 72, 73
        for polymer melts, 73

Index

Boundary conditions (*cont.*)
   stress balance, 78
      with surfactant, 95
   stress conditions, 76
   tangential (shear) stress balance at a fluid interface, 84
   thermal, 68
      latent heat, 69
Boundary effects (*See also* Wall effects)
   sphere toward a wall, lubrication, 320
   surface roughness, 324
Boundary-integral methods, 538, 545, 550, 564
   problems in a bounded domain, 568
      Green's functions, 569
      point force near a solid wall, 569
   problems involving a fluid interface, 565
      with insoluble surfactant, 568
   rigid body in an unbounded domain, 565
Boundary-layer theory (laminar flow)
   axisymmetric bodies, 733
   boundary and matching conditions, 708
   bubble, 739
   concentration boundary layer, 245
   curvature terms, 247, 249, 706
   cylindrical bodies – Blasius series, 725
   favorable and adverse pressure gradient, 727
   flat plate-Blasius solution, 713
   governing equations, 706
      for an axisymmetric body, 736
      independence of cross-sectional shape in 2D, 706
      the role of body geometry, 709
   higher-order theory, 711
   Howarth transformation, 730
   Prandtl's approach, 698
   pressure distribution, 707
   pulsatile flow, high frequency, 209
   scaling, 705
   semi-infinite wedge, Falkner–Skan problem, 719
   separation, 709, 731
   spherical bubble, 739
   stagnation point flow, 723, 729
Boundary-layers (*See also* Thermal boundary layer; Boundary layer theory (laminar flow)), 213
   curvature terms, 247, 249
   in pulsatile tube flow, 213
   in the air hockey problem, 336
   mass transfer in a catalyst pellet, 242, 245
   thermal, 161
Boussinesq approximation, 110, 404, 841, 859
   Boussinesq equations, 844, 846
Brinkman model for porous media, 825
Brinkman number, 121, 219, 596, 615
Brownian motion, 53
   rotational diffusivity, 53
Bubbles (*See also* Cavitation bubbles)
   boundary conditions, 485
   boundary-layer theory, 739
   drag on a spherical bubble, $Re \ll 1$, 485
   drag, $Re \gg 1$,
      ellipsoidal bubble, 753
      spherical bubble, 749
   dynamics, 250
   growth, 255
   Marangoni effects, $Re \ll 1$, 486, 491
   oscillation of volume, 260
   Rayleigh–Plesset equation, 250, 253
   shape deformation, $Re \gg 1$
      $We \ll 1$, 749, 753
   shape instability, 269
   spherical bubble, translating $Re \ll 1$, 669
   surfactant effects on buoyancy driven motion, 95, 491
   thermocapillary-driven motion, 86
   thermocapillary motions, $Re \ll 1$, 86, 486
      drag in the presence of a temperature gradient, 489
      terminal velocity in the presence of a temperature gradient, 489
Buoyancy driven convection (Rayleigh–Benard instability), 845
Buoyancy force, 39, 80
   neutral buoyancy, 39, 470
   on a sphere, 470

Capillary flows, 39, 80, 371
   capillary instability of a liquid thread (*See* "Stability"), 82, 801
   drop breakup, 82
   drop shape relaxation, 80
   film (drop) spreading at low Bond number, 371
   interface flattening, 81
Capillary instability of a liquid thread (*See* Stability), 82, 801
Capillary number, 478, 494, 820
   role in drop deformation, 478, 537
Capillary pressure, 80
   jump, spherical drop, 480
Capillary spreading, 371
Cartesian component notation, 22
Catalyst pellet, diffusion inside with fast reaction, 242
   effectiveness factor, 249
Cauchy equations of motion, 29, 33
Cavitation bubbles, 250
   equilibrium radius, 255
   periodic pressure oscillation, 260
      at resonant frequency, 261
   Rayleigh–Plesset equation, 250, 253
   sonoluminescence, 250
   stability of equilibrium radius, 250, 255
   stability to nonspherical disturbances, 269
      near a wall, 269
      Rayleigh–Taylor instability, 277
   step change of pressure, 254
Centripetal acceleration, 127, 358
   centrifugal force, 127, 359
Characteristic scales, 115, 209
   heat transfer, 769
      large Peclet number, 647, 663, 666, 671
         closed streamline, 676
         length scales, 647, 648, 668
      small Peclet number, 603
         length scales, 607, 621, 623, 636

Index

length scale, 116
pressure
  at large Reynolds number, 433, 698
  at small Reynolds number, 431
temperature, 159, 596
  when the heat flux is specified, 615
time, 136
  for diffusion-based processes, 137, 150, 247
velocity, 116
Circular cylinder
  boundary layer theory, 725
  pressure distribution, $Re \gg 1$, 710
  separation, boundary-layer, 731
Circular tube, 166
  Poiseuille flow, 121, 229
  Poiseuille flow, startup, 135
  pulsatile flow, 175, 205
    asymptotic approximation, large frequency, 209
      boundary-layer, 213
      near-wall region, 211
    asymptotic approximation, small frequency, 182, 206
    effects of inertia, 176, 182
    phase lags due to inertia, 179, 182
  slightly curved, 224
Closed streamline flow, heat transfer, 662, 671
Closure approximations, 64
Complementary complex velocity fields, 179
Complex fluid, 52, 56
  emulsion, or immiscible blend, 57
  origin of non-Newtonian behavior, 52, 55
  polymer liquids, 57
  solution of rod-like molecules, 62
  suspension of rigid spheres, 58
Complimentary error function, 746
Computational fluid dynamics (CFD), 204
Concentration boundary layer, 245
Concentric rotating cylinder problem (*See* Couette flow), 125, 295
Conduction, heat transfer, 113, 137, 152, 600
  circular cylinder, 604
  far-field temperature distribution, 631
  sphere, 600
Conservation of
  angular momentum, 29
  energy, 31
  linear momentum, 26
  mass, 18
Constitutive equations, 36
  continuum mechanical approach, 46, 59
  coordinate invariance, 43
  heat flux vector, 42
  material objectivity, 43
  molecular modeling approach, 61
  Newtonian fluid, 37, 45, 48
  non-Newtonian fluids, 52, 59
  simple fluid, 61
  solution of rod-like molecules, 52, 62
  Stokesian fluid, 60
  stress, in a moving fluid, 46
  stress, in a stationary fluid, 38

suspension of rod-like particles, 62
Contact angle, 71, 372
Contact line (moving), 70, 71, 370
Continuity equation, 18, 19
  general form, curvilinear coordinates, 658
Continuum approximation, 13
  boundary conditions, 17
  continuum hypothesis, 13, 14
  continuum point variables, 14
  convective vs. molecular transport mechanisms, 15
    conduction of heat, 42
    convective flux, 16, 42
    diffusive flux, 16
    momentum diffusion; stress, 17
Control volume, 18, 19
Convected time derivative, 20, 21
Convection
  buoyancy driven (Rayleigh Benard instability), 845
  double diffusive, 858
  Marangoni driven, 867
  Thermohaline, 858
Coordinate invariance, 43
Coordinate systems, 112, 114, 661
  bi-spherical, 576
  curvilinear, 114, 128, 446–462
  cylindrical, 114, 115
  polar cylindrical, 225
  scale factors/metrics, 114, 446
  spherical, 472
Coordinate transformations
  Cartesian to spherical, 633
Coriolis force, 359
Corner flows
  hinged plates, 454
  immersion of a solid plate into a liquid, 453
  "stirring" flow, 454
    Moffatt eddies, 457
  "wiper-blade" on a solid surface (Taylor), 452
Correlations, Heat and Mass Transfer
  across a fluid interface, $Pe \gg 1$, 668
  arbitrary two-dimensional body
    heat transfer, general linear flow, $Pe \ll 1$, 642
    heat transfer uniform flow, $Pe \ll 1$, 629, 633
    heat transfer, uniform flow, $Pe \gg 1$, 644, 656, 659
    heat transfer, uniform flow, $Re \gg 1$;
      $Pr = O(1)$. 771
      $Pr \gg 1$;
        no-slip boundary conditions, 776
        slip boundary conditions, 776
      $Pr \ll 1$, 784, 786
  bubble, spherical
    mass transfer, uniform flow, $Re \ll 1$, $Pe \gg 1$, 670
  closed streamlines, heat transfer, $Pe \gg 1$, 671
  cylinder, rotating in shear flow, $Pe \gg 1$, 678
  dependence on surface boundary conditions, $Pe \gg 1$, 668
  flat plate heat transfer
    $Re \gg 1$, $Pr \gg 1$, 779
    $Re \gg 1$, $Pr \ll 1$, 786

Index

Correlations, Heat and Mass Transfer (*cont.*)
  forced convection heat transfer
    general, 597, 768
    large Reynolds number, 771
      $Pr \gg 1$, 776
      $Pr \ll 1$, 784
    small Reynolds number, 597
  interpolation between asymptotes, 643
    heat transfer at intermediate Pr, 787
  mass transfer with finite blowing, 793
  sphere
    conduction, 601
    general linear flow, $Pe \ll 1$, 641
    rotating sphere in shear, $Pe \gg 1$, 679
    simple shear flow, $Re \ll 1$, $Pe \ll 13$, 640
    uniform flow, $Pe \ll 1$, $Re \ll 1$
      for Nu, with specified surface temperature, 613, 614
      for surface temperature, heat flux specified, 616
    uniform flow, $Pe \gg 1$ $Re \ll 1$, 644, 652, 656
Corrugated channel, 233
Couette flow, 125, 295
  curvature effects, 133
  instability, 134, 829
  rheological application, 132
  Taylor number, 134
  thin gap approximation, 131
Couple/Torque
  body, 30
  surface, 30
Creeping flow, 429
  axisymmetric body
    drag formula for an arbitrary axisymmetric body in an axisymmetric flow, 446–462
  boundary-integral methods, 538, 545, 550, 564
    problems in a bounded domain, 568
      Green's functions, 569
      point force near a solid wall, 569
    problems involving a fluid interface, 565
      with insoluble surfactant, 568
    rigid body in an unbounded domain, 565
  bubble, translating, 669
  corner flows (*See* Corner flows), 451
  cylinder in simple shear flow, 673
  cylinder in uniform flow, Stokes paradox, 619
  drop in a general linear flow, 537
  drop in a uniform flow/sedimentation, 477
  Faxen's law, 571
  force on a body in an arbitrary flow, 572
  fundamental solutions, 545
    point force near a solid wall, 569
    Stokeslet solution for a point force singularity, 545, 547
    Stokes' (force) dipole, 551, 552
      Stresslet and Rotlet solutions, 553
    Stokes' (force) qudrapole, 551, 552
      potential dipole solution, 553
  general solutions
    eigenfunction expansions, 524
    axisymmetric flow in spherical coordinates, 462
    two-dimensional flow in Cartesian coordinates, 449
    two-dimensional flow in cylindrical coordinates, 450
    in terms of vector harmonic functions, 525
  integral representation of Ladyzhenskaya, 547
    double-layer potential, 550
    single-layer potential, 550
  Lamb's solution, 524
  sphere in linear shear and axisymmetric extensional flow, 534, 555
  Stokes' problem, 529, 554
  surface distribution of point forces, 545, 550
  via distributions of singularities, 550
    axisymmetric body, 551
    slender body theory, 545, 560
governing equations, 430
  in terms of streamfunction, 448
image singularities, 569
in uniform flow, 464
  far-field velocity distribution (the Stokeslet velocity field), 464, 465
  formula for the streamfunction, 465
linearity, consequences of, 434
  drag on mirror image bodies (flow reversal), 434
  lateral migration, proof that it is impossible in a creeping flow, 438
  lift on a translating, rotating sphere in simple shear flow, 436
  resistance matrices, existence in creeping flow, 439
method of reflections, 576
prolate spheroid in uniform flow, 557
quasi-steady approximation, 432
  Reynolds number, as a ratio of two time scales, 433
  Strouhal number, 432
reciprocal theorem, 571
representation theorem; the solution in terms of harmonic functions, 526
resistance matrices, 439
rotating sphere, 528
slender body theory, 545, 560
  as a line distribution of point force singularities, 551
sphere in axisymmetric extensional flow, 470, 555
sphere in a general linear flow, 530
  in a simple shear flow, 534
    angular velocity, 536
    torque on the sphere, 536
sphere in a uniform flow, 466, 529, 554
  fore-aft symmetry, 469
  terminal velocity in sedimentation, 470
  vorticity distribution, 468
Stokes flow, 466, 529, 554, 625
Stokes' law, 468
  form drag, 470
  friction drag, 470
Stokeslet, 466

Index

superposition principle, 437, 440, 471, 536
translation of a drop, 477
   with surfactant, 490
two-dimensional flows, 449
two-spheres, 577
vorticity in creeping flow, 468
Crystal growth, 386
Curvature, (principle radii) of, 79
   (*See* Interface)
Curved tube, flow through, 224
Curvilinear coordinate systems, 114, 128, 657
   curvature terms, 128, 247
   cylindrical coordinates, 114, 115
   metrics, 114, 677
   scale factors, 114, 657
Cylindrical coordinate system, 114
   circular cylindrical coordinates, 115

D'Alembert's paradox, 700–726
Damköhler number, 243
Dean instability, 840
Dean problem, 224
Deborah number, 54, 576
Deviatoric stress, 13, 46
Diffusion, 16
   characteristic time scales, 137, 247
   from a d dimensional pulse-like source, 363
      solution for large times (self-similar form), 363
   interface diffusion, 495
   molecular transport mechanism & the continuum hypothesis, 16
   momentum, 137
   nonlinear diffusion equations, 362
   species diffusion equation, 243, 362
   sphere, in a sphere with fast reaction, 242
   surface diffusivity of surfactants, 93
Dilation rate, 24
Dimensionless parameters, 116, 117
Disjoining pressure, 376
   Hamaker constant, 377
Dissipation, 33, 34, 120
   Brinkman number, 121, 219
   drag on a spherical bubble, 749
   effect on rheometry, 223
   gas bubbles, high Reynolds number, 273
   relationship to drag, 747, 749
Distinguished limit, 403, 414, 504
Disturbance flow, 465, 529
Divergence theorem, 19
Domain perturbations, 232, 272
   bubble deformation, 271
   drop deformation, 477, 483, 539
   fundamental ideas, 397
      calculating interface shapes, 481
   shallow cavity, 396, 413
   wavy-wall channel, 232
Double diffusive convection (*See* Stability), 858
Double-layer potential, 550
Drag coefficient
   bubble, $Re \gg 1$,
      deformed, 753
      spherical, 749
   cylinders, $Re \gg 1$, 725
   flat plate, $Re \gg 1$, 717
Drag force
   arbitrary axisymmetric body in an axisymmetric flow, $Re \ll 1$, 446–462
   bubble, $Re \gg 1$, 749
      spherical, 753
      with deformation, 753
      with thermocapillary effects, $Re \ll 1$, 489
   drag via reciprocal theorem, $Re \ll 1$, 571
   flat plate, $Re \gg 1$, 717
   form drag, 16, 470, 726
   friction drag, 726
   inertial corrections, 573, 626
   mirror image bodies, $Re \ll 1$, 434
   momentum defect, relation to drag, 626
   potential flow, 726
   prolate spheroid, $Re \ll 1$, 559
   slender body, $Re \ll 1$, 560, 563
   solid sphere in uniform flow, $Re \ll 1$
      inertia correction at small Re, 626
      Stokes drag, 468, 555
   spherical bubble, $Re \ll 1$, 485
   spherical drop, $Re \ll 1$, 483
   Stokes law, 468, 555
   two-spheres, $Re \ll 1$, 577
   via the mechanical energy balance, 747, 749
Drops
   breakup, 82
      end-pinching, 83
      in a general linear flow, 537
      steady-state drop shape, 543
      transient response, 542
   inertia for small Re, effect on drop shape, 484
   Hadamard-Rybczynski solution, 482
   Marangoni effects, 486, 491
   shape in linear shear flow, $Re \ll 1$, 543
   shape in uniaxial extension, $Re \ll 1$, 543
   shape relaxation, 80
      characteristic time scale, 542
   spreading due to capillary forces, 374
   spreading due to gravity, 40, 369
   stability of shape, $Re \ll 1$, 485
   surfactant effect on buoyancy driven motion, 95, 490
      drag for a rising drop, 501
      fast adsorption kinetics, Biot $\ll$ 1, 503
         drag on a drop, 509
      governing equations and boundary conditions, 493
      incompressible surfactant limit, 497
         no-slip, immobilized interface, 497
      insoluble surfactant limit, 497
      spherical cap configuration, 497, 500
      terminal velocity, 501
   thermocapillary-driven motion, 86
   translation through a quiescent fluid, $Re \ll 1$, 477
      drag, 483
      shape/deformation, 483, 485
      terminal velocity, 470, 484

Index

Dyadic product, 28
Dynamic similarity, principle of, 431

Eccentric rotating cylinder, 295
  narrow gap/lubrication, 297
    lubrication forces, 303
    pressure distribution, 303
    slight eccentricity (definition), 297
Effective diffusivity, porous media, 242
Effective viscosity of a suspension, 475
Effectiveness factor in a catalyst pellet, 249
Eigenvalue problems, 816, 856, 870
  eigenfunctions, 138, 461
  eigensolutions, 137, 149
  eigenvalues, 138, 149, 461
Einstein viscosity law for a dilute suspension of spheres, 473
Elasticity, entropy as a source, 59
Elasticity parameter, interface, 495
Energy
  conservation of, 31
    First Law of Thermodynamics, 32
  internal energy, 31
  kinetic energy, 31
  mechanical energy balance, 33
  thermal energy balance, 33
  transport, molecular, 31
Enthalpy, specific, 33
Entropy inequality, 31, 35
Equations of motion
  Cauchy, 29, 33
  Navier–Stokes, 49
Error function, 145
  complementary error function, 785–793
Euler equations for an inviscid fluid, 859–873
Euler's equation, 459
Eulerian frame of reference
  acceleration, 253
  time derivative, 22
Extensional flow, axisymmetric, 470, 471
  biaxial, 471
  uniaxial, 471

Falkner–Skan solution, 719
  governing equation, 722
  Hartree solutions, 724
  physical interpretation, 722
  reduction to stagnation point flow, 723
  relevance to separation, 725
Ferrofluid, 30
Faxen's law ($Re \ll 1$), 571
Fick's law for surfactant diffusion, 93
Film leveling, 379
  with both gravity and capillary forces present, 380
Film rupture, 377, 381
Film stabililty, 377
Flat plate
  boundary-layer (Blasius) solution, 713
  drag and drag coefficient, 717
  leading edge singularity, 717
Floquet theory, 279

Fokker–Planck equation, 62
Force dipole solution, $Re \ll 1$, 551, 552
Force quadrapole solution, $Re \ll 1$, 551, 552
Forced convection approximation, 593, 594
Form drag, 470
Fourier–Bessel series, 139
Fourier's law of heat conduction, 42, 43, 45, 174
Free-boundary problems, 477
Frumkin adsorption isotherm, 91
Fundamental solutions
  creeping flow equations, 545, 547
  point source of heat in a uniform flow, 606, 610
  point source of heat in simple shear flow, 637

Gamma function, 367, 655
Gauge functions, 217
Gegenbauer polynomials, 461
General linear flow, 533, 640
Gibbs adsorption isotherm, 90
Gibb's elasticity, 95
Goldstein transformation, 608
Gortler instability, 840
Graetz problem, 164
Grashof number, 846, 860
Gravitational spreading, 367
Gravity currents, 41
Gravity number, 377
Green's functions, flow near a solid wall, $Re \ll 1$, 568

Hadamard–Rybczynski solution, 482, 485
Hagen–Poiseuille Law, 123
Hamaker constant, 377
Hamiltonian function, 267
Harmonic functions, 527
Harmonic oscillator, 260
  damped, 276
  forced, resonance, 261
  secular terms, 263
Heat equation, 137
Heat flux vector, 32, 42
Heat transfer (*See also* Correlations, heat and mass transfer)
  across a fluid interface, 666
  across regions of closed streamlines, $Pe \gg 1$, 662
  analogy with mass transfer, 598, 614, 643, 767
  arbitrary three-dimensional body in uniform flow
    $Pe \ll 1$, arbitrary Re, 627
    $Pe \gg 1$, $Re \ll 1$, 656, 659
      axisymmetric bodies, 661
      bodies with closed streamlines, 656, 662
      two-dimensional bodies, 659
  boundary conditions
    constant surface temperature, 594, 790
    mixed boundary conditions, 789
    specified heat flux, 595, 615, 788
  bubble or drop, $Re \ll 1$, $Pe \gg 1$, 668
  closed streamline flows, $Pe \gg 1$, 662
    general principles, 662, 671, 679
    rotating cylinder in simple shear flow, 672
  conduction, 113, 137, 152, 600, 604
    sphere, 600

Index

cylinder, $Pe \ll 1$
   rotating, in shear flow, $Pe \gg 1$, 672
exact solutions, $Re \gg 1$, 771
forced convection approximation, 593, 594
open versus closed streamlines, $Pe \gg 1$, 645
phase change with heat transfer, 152
point source of heat in simple shear, 637
point source of heat in uniform flow, 606, 610
scaling and dimensionless parameters, 596, 597, 606, 666
   Nusselt number, 597, 771
   Peclet number, 596, 767
   Prandtl number, 596, 768
sphere
   conduction, 600
   in general linear flow, $Pe \ll 1$, 640
   in generalized shear flow, $Re \ll 1$, $Pe \gg 1$, 663
   in simple shear flow, $Pe \ll 1$, $Re \ll 1$, 633, 679
   in uniform flow
      $Re \ll 1$, $Pe \ll 1$, 602
         specified heat flux, 615
         specified surface temperature, 602
      $Re \ll 1$, $Pe \gg 1$, 645
strong convection limit, $Pe \gg 1$ (*See* Thermal boundary layer)
small Reynolds number, 643, 644
thermal boundary layer (*See* Thermal boundary layer)
   effect of dynamic boundary conditions, 666, 776
   large Reynolds number
      asymptote, $Pr \ll 1$, 773
      asymptote, $Pr \gg 1$, 780
   small Reynolds number, 643, 647
unidirectional flows, 157
Henry's Law, 669
Homogeneous fluid, 47
Howarth transformation, 730
Hydrodynamic interactions between bodies, $Re \ll 1$, 576
Hydrodynamic stability (*See* "Stability"), 800
Hydrostatic pressure, 39
   hydrostatic pressure gradients, flows driven by, 40
Hyperbolic flow, 641

Image singularities in creeping flow, 569
Incompressible fluid, 24, 50
Inertial effects in uniform flow past a sphere, 616
Inertial effects on the force on a body, 573, 575
Inertial migration, 439
Instability (*See* Stability)
Integral equations, 564
   for surface temperature in thermal boundary layers, 790
   of the first kind, 565
Interface, 73
   boundary conditions, 359
      in the presence of mass transfer, 666
      simplification for a thin film, 359
   curvature (in terms of interface shape), 271, 360, 539
   diffusion of surfactant, 93, 495
   elasticity parameter, 495
   Gibbs elasticity, 95
   Marangoni number, 495
   porous media, in a, 825
   rheology, 89
   shape, 74
   surface excess quantities, 67
   surface force balance, 76, 77
   surfactant, 89
   viscosity, 89
   unit normal vector (in terms of interface shape), 271, 359, 539
Interfacial tension, 72, 76
   origin of, 77
Internal energy, 31
Irreducible tensors, 528, 531
   second order, 531
   third order, 532
Irrotational flow, 700–726
Isotropy, 44, 47

Journal bearing (*See* Eccentric rotating cylinders), 297

Kelvin circulation theorem, 834
Kinematic boundary condition, 67
   at an interface, 75
   with phase transformation, 67, 153
Kinematic viscosity, 51
Kinetic energy, 31

Ladyzhenskaya solution for $Re \ll 1$, 547
Lagrangian acceleration, 253
Lagrangian frame of reference
   acceleration, 253
   time derivative, 22, 170
Langmuir adsorption isotherm, 91, 494
Latent heat, 69, 154
Lateral migration, 438
   deformable particle or drop, 439
   inertial effect, 439
   non-Newtonian effect, 439
Legendre functions, 460
Legendre polynomials, 460, 609
   orthogonality property, 461
   relationship to Gegenbauer polynomials, 461
Legendre's equation, 460
Leibnitz rule, 361
Limit point, 255
Linear algebraic equations, 821
   condition for existence of non-trivial solutions, 821, 857
Linear flow, general 2D, 641
Linear integral equations, 792
Linear momentum, 26
Linear oscillator, 276
Linear (simple) shear flow, 118
Linearity, consequences for creeping flow, 434
Linear stability theory, 258, 800
   (*See* Stability)
   equilibrium size of cavitation bubbles, 256
   thin film subject to van der Waals forces, 378

Index

Local solutions, corners, 452
Lubrication
  approximation, 307
    as the first approximation in a singular perturbation theory, 308
      matching, 313
  boundary curvature, 307
  characteristic scales, 297
    pressure, 300
  configurations, 307, 308
  journal bearing, (*See* Eccentric rotating cylinder), 297
  lubrication/thin film equations, 300, 306, 312
    general derivation, 308
  Reynolds lubrication equation, 294, 301, 313, 317
    derivation, eccentric cylinder problem, 300
    derivation, general case, 311
  slider block, 315
  sphere near a wall, 320
    vanishing gap width, 324
      the role of surface roughness, 324
      van der Waals forces, 325

Mach number, 24
Magnetohydrodynamics, 25
Marangoni effects/flows, 85
  inclusion in boundary-integral calculations, 568
  instability (*See* Stability), 87, 812, 867
  thermocapillary-driven motion of bubble/drops, 86, 486
    due to temperature gradients, thermocapillary flows, 486
    due to surfactant concentration gradients, 491
  shallow cavity flows, 385, 405
Marangoni number, 27, 870
Mass (conservation of), 18, 251
Mass transfer (*See also* correlations), 593, 598
  across a fluid interface, $Pe \gg 1$, 598, 666
    general principles, 666
  analogy with heat transfer, 598, 614, 643, 767
    boundary conditions, 598, 599
    Schmidt number vs. Prandtl number, 598
    Sherwood number vs. Nusselt number, 599
  diffusion
    solute, 601
  finite interfacial velocity, "blowing" and "suction", 599
    blowing number, 599, 793
    boundary layer theory, $Re \gg 1$, 785–793
  Peclet number, 598
  surfactant, 503
  translating bubble/drop for $Pe \gg 1, Re \ll 1$, 668
Matching, 215, 219, 246
  mismatch, 215
Material control volume, 19
Material objectivity, 43
Material point, 19
Material time derivative, 20, 21
Mathieu's equation, 278
Mechanical energy balance, 33
  use to calculate drag, 747

Memory, fluids with, 57
Micro-gravity, 486
Moffat eddies 457
Molecular modeling, 61
  closure approximations, 64
  Fokker–Planck equation, 62
  solution or suspension of rods, 62
Momentum, conservation of, 26, 29
Momentum defect in a wake, 626
Momentum transport, 17, 140
  diffusion, 140
  molecular, 15
Moving contact line, 70, 71, 370
Multiple time-scale perturbations, 250, 260, 264
Multi-pole expansion, 551

Natural convection, 111, 594, 844
Navier slip condition, 70
  slip coefficient, 70
  slip length, 70
Navier–Stokes equations, 49
  in terms of streamfunction, 449
    spherical coordinates, 472
  in terms of vorticity, 448
  multiplicity of solutions for $Re \gg 1$, 697
  solution procedures, 111
Newtonian fluids, 37, 45, 48
Newton's Law of Mechanics, 25
  angular momentum, 29
  second law, 25
  third law, 28
Nonautonomous equation, 278
Nondimensionalization, 115, 136, 140, 209
  characteristic scales, 115, 140
  dimensionless parameters, 116, 117
  leading to the creeping flow equations, 430
    characteristic pressure, viscous flow, 431
    quasi-steady approximation (Strouhal number), 432
  rescaling, 245
Non-isothermal systems
  Boussinesq approximation, 841, 859
Nonlinear Diffusion Equations, 365
  solution by similarity transformation, 362
Non-Newtonian fluids, 52
  Deborah number, 54, 576
  force on a body ($De \ll 1$), 575
  memory effects, 57
  origin of non-Newtonian behavior, 52, 55
  shear-thinning, 55
  simple fluid, 61
  solution of rod-like molecules, 52, 62
  Stokesian fluid, 60
    viscoelastic liquids, 57
  Weissenberg number, 54, 58
Non-Newtonian corrections to force on a body via reciprocal theorem, 573, 575
Normal stress balance at an interface, 79
No-slip condition, 69
Numerical methods, 111, 204
Nusselt number, 597, 640, 771

Index

One-dimensional flow, 110, 112, 253
Order symbols, 47, 131, 136
   big "O", 131, 207, 217
   little O, 217, 237
Orr–Somerfeld equation (See also "Stability"), 862–873
Oseen's analysis, Whitehead paradox, 619

Patching, 219, 246
Pathlines, streamfunction, 447
Peclet number, 160, 407, 494, 596, 598, 634, 675, 767
   interfacial Peclet number, 495
Phase change, in solidification, 152
   kinematic condition, 153
   latent heat, 154
Phase lags due to inertia in time-dependent flow, 179, 217
Point force in creeping flow, 545, 547
   distributions of, 545, 551
   near a solid wall, 569
Poiseuille flow
   in a circular tube, 121, 229
   start-up, 135
   two-dimensional, 119
Porous media
   Brinkman model, 825
   catalyst pellet, 242
   Darcy's Law, 823, 824
      interface, 825
      permeability, 823, 824
   dispersion, 825
   effective diffusivity, 242
   Saffman–Taylor instability, 823
Potential dipole solution, $Re \ll 1$, 553
Potential flow theory
   boundary conditions, 700
   comparison with experiments, 701
   d'Alemberts paradox, 701
   for a horizontal flat plate, 713
   for a semi-infinite wedge, 722
   for uniform flow past a circular cylinder, 701, 726
   for uniform flow past a sphere, 734
   governing equations, 700
   role in boundary layer theory, 700, 704
Power series expansion, 217
   convergence, 217
   domain of convergence, 218
Prandtl, contribution to boundary layer theory, 698
Prandtl number, 596, 768, 860
Pressure
   characteristic pressures at large and small Reynolds number, 433, 698
   dynamic, 51
   hydrostatic, 39
   mechanical definition, 50
   thermodynamic, 38
Pressure distribution for flow past a cylinder at $Re \gg 1$, 710
   potential flow, circular cylinder, 727
Principle of exchange of stabilities, 836, 853
Prolate spheroid, uniform flow, $Re \gg 1$, 557
   drag force, 559
Pseudo-vector (or tensor), 43, 441, 525
   common examples, 526
Pulsatile flow, tube, 175, 205
   asymptotic approximation, large frequency, 209
      near-wall region, 211
      boundary-layer, 213
   asymptotic approximation, small frequency, 182, 206
   effects of inertia, 176, 182
   phase lags due to inertia, 179, 182

Quasi-steady approximation, $Re \ll 1$, 432

Rate of strain tensor, 33, 132
Rayleigh criteria for capillary stability, 807
Rayleigh inflection point theorem, 875
Rayleigh number, 847, 861
Rayleigh problem, 142
Rayleigh–Benard instability (See Stability), 845
Rayleigh–Plesset equation, 250, 253
Rayleigh–Taylor instability (See Stability), 277, 812
Reciprocal theorem of creeping flow theory, 571
Regular asymptotic expansion, 207, 218, 227
Reflections, method of, 576
Rescaling, 212, 245, 606, 622, 636, 648
   stretching, 212, 648
Resistance matrices/tensors in creeping flows, 439
   bodies near a plane wall, 443
   bodies with ellipsoidal symmetry, 443
   bodies with spherical symmetry, 442
   symmetry properties, 442
   trajectory equations, in terms of resistance tensors, 444
Reynolds (lubrication) equation, 294, 301, 313, 317
Reynolds number, 124, 130, 239, 336, 429, 431, 484, 634
   as a ratio of two time scales, 433
Reynolds transport theorem, 22, 26
Rheology
   experiments, 126, 132
   rheometer geometries, 126
Rheometry, 53, 54
   effect of viscous dissipation, 219, 223
   rheometer flow, 60
Rotating reference frame, 359
Rotlet, 553

Saffman–Taylor instability, (See Stability), 823
Scale factors, 114
Schmidt number, 598, 767, 860
Screw-like body, coupling of translation and rotation, $Re \ll 1$, 443
Secondary flow, 231
Secular solutions/terms, 263, 280
Sedimenting sphere
   terminal velocity, 470
Semi-infinite wedge, 719
   boundary-layer (Falkner–Skan) solution, 723
   heat transfer, 772
   potential flow solution, 722

Index

Separation
　mechanisms for separation, 732
　prediction via boundary-layer theory, 710
　predictions for a circular cylinder, 731
　predictions for a sphere, 738
　relationship to drag for $Re \gg 1$, 726
　role in validity of boundary-layer theory, 709
　role of pressure distribution, 725
　separation point, defined in 2D, 709, 726
　via Falkner–Skan equation, 725
Separation of variables, 135, 149
Shallow cavity flows, 385
　enclosed cavity, 386
　　impermeable end walls, volume flux integral constraint, 387
　　Reynolds equation, 388
　　the end regions, 386, 402
　　singular perturbation, 387
　thermocapillary flow, 385, 404
　　analogy with crystal growth, 385
　　analysis via the domain perturbation procedure, 413
　　analysis via thin-film solution procedure, 410
　　characteristic velocity, 405, 407
　　heat transfer coefficient, 409
　　Marangoni stress, 385, 405
　with a deformable interface, 391
　　analysis via classical thin-film analysis, 392
　　　small deformation, 393
　　analysis via domain perturbations, 396
　　the end regions, 395, 402
Shear flow, simple
　startup of, 148
Shear-rate, 45, 641
Shear-thinning fluids, 55
Sherwood number, 599
Shooting method, 716
Similarity solutions, 142, 369
　conditions for existence, 152
　　in a 2D boundary-layer problem, 720, 724
　　necessary, 147
　　sufficient, 147
　diffusion from a pulse-like initial source, 363
　　linear diffusion equation, 363
　　nonlinear diffusion equation, 362, 365
　flat plate – Blasius solution, 714
　for film rupture, 381
　mass transfer boundary layer with finite interfacial velocity, 795
　momentum boundary layer on a bubble, 744
　nonlinear diffusion equations, 362, 365
　Rayleigh problem, 142, 151
　semi-infinite wedge – Falkner–Skan solution, 719
　　heat transfer, 772
　solidification at a plane interface, 152
　spreading film/drop (capillary), 374
　spreading film/drop (gravitational), 369
　start-up of simple shear at short times, 148, 150
　thermal boundary-layer, $Re \gg 1$
　　exact solutions, $Pe \sim O(1)$, 771
　　$Pr \gg 1$, arbitrary two-dimensional body
　　　no-slip boundary conditions, 777
　　$Pr \ll 1$, arbitrary two-dimensional body, 784
　thermal boundary-layer, $Re \ll 1$, 669
　　arbitrary solid, uniform flow $Pe \gg 1$
　　　axisymmetric body shapes, 661
　　　2D body shapes, 660
　　sphere, $Pe \gg 1$, 652
　　spherical bubble, $Pe \gg 1$, 670
Similarity transformation, 142, 146
Simple fluid, 61
Simple (linear) shear flow, 118, 219
　dissipation, 219
　start-up, 148
Single-layer potential, 550
Singular asymptotic expansion, 218
　reduction in order of the differential equation, 219
　singular limit, 219
Slender-body theory, $Re \ll 1$, 545, 560
　force ratio, translation parallel and perpendicular to the axis, 560, 563
Slider block, lubrication, 315
Slip
　"apparent" slip, 73
　at a fluid-fluid interface, 72
　at a moving contact line, 70
　Navier slip condition, 70
　relationship to adhesion, 72
　with polymeric liquids, 73
Solenoidal vector field, 24
Sonoluminescence, 250
Solidification, 152
Species transport equation, 243
Sphere
　(See also Heat transfer)
　axisymmetric extensional flows, $Re \ll 1$, 470
　diffusion with fast reaction, 242
　drag, $Re \ll 1$, 466
　force in an arbitrary flow, $Re \ll 1$, 572
　general linear flows, $Re \ll 1$, 530, 555, 640
　　in a simple shear flow, $Re \ll 1$, 534, 555, 640
　in a tube, 307
　inertia effects at small Re, 616
　　(See also Uniform flow, sphere)
　lift, rotating, $Re \ll 1$, 436
　lubrication forces near a plane wall, 324
　rotating, $Re \ll 1$, 528
　steady translation, $Re \ll 1$, 466, 529, 554
　Stokes flow, 16, 466, 529, 554, 625
　terminal velocity, $Re \ll 1$, 470
　translation toward a plane wall (Lubrication), 320
　two translating spheres, 577
　uniform flow, $Re \ll 1$, 466
Spherical harmonics, 274
Spreading drop
　driven by capillary forces, 374
　driven by gravitational forces, 40, 369
Spreading film, (See also Thin films with a free surface), 367
　capillary spreading (small Bond number limit), 371
　Tanner's law, 373

Index

gravitational spreading (large Bond number limit), 367
gravity currents, 367
spontaneous spreading, 373
Squire's transformation, 874, 877
Stability, 124
   bubble shape, 269
   capillary instability, 82, 256, 801
      condition for neglect of gravitational effects, 802
      inviscid limit, 804, 816
         fastest growing linear mode, 806
         governing equations, 804
         growth rate coefficient, 806
         stability criteria, 806
      Rayleigh criteria for stability, 807
      related problems, 811
      viscous effects, 808
         governing equations, 808
         growth rate coefficient, general, 810
         growth rate coefficient, $Re \ll 1$, 811
         stability criteria, 811
   cavitation bubble radius, 256
   Couette flow, (*See* Taylor Couette instability)
   Dean instability, 840
   double diffusive convection instability, 858
      diffusive regime, 859
      finger regime, 858
      governing equations, 859
      two free surfaces, 862
         general stability diagram, 865
   exchange of stabilities, 259
   Gortler instability, 840
   linear stability analysis, 258, 378, 800
      disturbance of arbitrary form, superposition of Fourier modes, 805, 849
      equations for an inviscid fluid, 804, 816
   Marangoni instability, 87, 867
      at the interface between two liquids, 867–871
      Bernard's original experiments, 867, 871
      critical Marangoni number, 871
      mechanism, 867
   non-autonomous equations, 278
   non-linear stability, 801
   Rayleigh–Benard instability, 845
      governing equations, 845
      principle of exchange of stabilities, 853
      two free surfaces, 851
         critical Rayleigh number and wave number, 852
         growth rate coefficient, 851
         non-isothermal boundaries, 853
      two no-slip, rigid boundaries, 855
   Rayleigh–Taylor instability, 812
      acceleration effect (Taylor), 822
         governing equations, 813
         growth rate coefficient, 822
      inviscid fluid, 816
         bounded domain, 818
         growth rate coefficient, 817
         stability criteria, 817
      mechanism, 812

      viscous effects, 818, 821
   Saffman–Taylor instability, 823
      growth rate coefficient, 828
      mechanism, 829
      stability criteria, 823
   Shear flows, 2D unidirectional, 865–872
      Orr–Sommerfeld equation, 862–873
      inviscid fluids, 873
         Rayleigh inflection-point theorem, 875
         Rayleigh stability equation, 873, 875
      viscous fluids,
         Orr–Sommerfeld equation, 862–873, 876
         sufficient condition for stability, 877
   Taylor–Couette instability, 134, 829
      inviscid fluid, a sufficient condition for stability, 832
      Rayleigh criteria, 835
      mechanism, 830, 834
      viscosity, the role in instability, 830, 835
         principle of exchange of stabilities, 836
         Taylor number, 837
            critical value, 839
   Taylor–Dean instability, 840
   thermohaline convection (*See* Double diffusive convection instability), 858
   thin film with van der Waals forces, 377
      on the bottom of a solid substrate, 379, 457
      on the upper side of a solid substrate, 378
         instability of a bounded film, 360
         stability criteria, 320, 379, 381
Stagnation point flow, 723, 729
Startup flow
   flow in a circular tube, 135
   simple shear flow, 148
      at short times, 148, 150
Statics, 37
Stefan number, 157
Stokesian fluid, 60
Stokes law, 468, 555
   correction due to inertia, 626
Stoke's paradox, 621
   for flow past a cylinder at $Re \ll 1$, 621
   for heat transfer from a cylinder, 605
Stokeslet, 466
   fundamental solution for a point force, $Re \ll 1$, 545, 547
Streamfunction, 444
   creeping flow equations, in terms of streamfunction, 448
   Navier–Stokes equations, in terms of the streamfunction, 448
   velocity components, spherical coordinates, 459
   volume flux, relationship, 447
Streamlines, open versus closed, 645, 656, 662
Stress
   deviatoric stress, 46
   state of, 29
   static fluid, 38
   tensor, 28, 38
      symmetry, 31
   vector, 17, 26

909

Index

Stress equilibrium, principle of, 27
Stress singularities, 717
Stresslet, 553
Strouhal number, 177, 431, 432
Sturm–Louiville Theory, 137, 833
Successive approximation, 209
Superposition principle, 204
   for creeping flows, 437, 440, 471
Surface diffusivity, 93
Surface energy flux vector (*See* heat flux vector), 32, 37
Surface excess quantities, 13, 67
Surface forces (*See also* stress vector or stress tensor), 25, 37
Surface tension, 72, 76, 358
   origin of, 77
Surfactant (Surface Active Agents), 89
   chemical nature of a surfactant for oil/water systems, 491
   effect on the buoyancy-driven motion of bubbles or drops, 95, 490
      fast adsorption kinetics, Biot $\gg$ 1, 497, 503
      governing equations and boundary conditions, 493
      incompressible surfactant limit, 497
         no-slip; immobilized interface, 497
      insoluble surfactant limit, 497
         spherical cap configuration, 497
            cap angle, 500
            drag (force), 501
            terminal velocity, 501
   Fick's law for surfactant diffusion, 93
   Frumkin adsorption isotherm, 91
   Gibbs adsorption isotherm, 90
   Gibb's elasticity, 95
   insoluble surfactant, in boundary integral calculations, 568
   Langmuir adsorption isotherm, 91, 494
      linearized, low concentrations, 496
   maximum surfactant concentration, 496
   surfactant mass conservation, 93
   transport equation, 92
   transport processes, 492, 495, 503
Suspensions, 52, 64
   bulk property definitions, 474
   bulk stress, 474
   viscosity, Einstein's law, 473, 475
Symmetry, stress tensor, 31

Tangential Stress balance, 84
Tanner's law, 373
Taylor–Couette instability, (*See* Stability), 829
Taylor–Dean instability, 840
Taylor dispersion, 157, 164, 166
   convection velocity, 174
   dispersion coefficient, 166
   effective diffusivity, 166, 167, 174
   effective thermal conductivity, 174
Taylor number, 134
Tensor
   dyadic product, 28
   irreducible, 528, 531
      arbitrary second order tensor in terms of irreducible parts, 531
      arbitrary third order tensor in terms of irreducible parts, 532
   rate of strain, 33, 132
   stress, 28, 38
   thermal conductivity tensor, 43
   velocity gradient tensor, 33
   vorticity, 33
Terminal velocity, sedimentation, 470
Thermal boundary layer, 647, 664
   across a fluid interface, 666
      general principles, 666
      spherical bubble (mass transfer), 668
   arbitrary body shape, uniform flow $Re \ll 1$, $Pe \gg 1$, 656
      axisymmetric bodies, 661
      bodies with closed streamlines, 656, 662
      general form, thermal boundary layer equations, 659
      two-dimensional bodies, 659
   arbitrary two-dimensional body, $Re \gg 1$, $Pe \gg 1$
      $Pr \gg 1$, 773
      $Pr \ll 1$, 780
   bubble or drop, $Re \ll 1$, $Pe \gg 1$, 668
   exact similarity solutions, $Re \gg 1$, 771
   governing equations, $Re \ll 1$
      no-slip surfaces, 651, 659
      interface or slip surfaces, 667, 669
   governing equations, $Re \gg 1$, $Pe \gg 1$
      general principles, 769
      Pr (or Sc) of $O(1)$, 770
      Pr (or Sc) $\gg 1$
         no-slip boundary conditions, 776
         slip boundary conditions, 777
      $Pr \ll 1$, 782
   heat flux, specified
      equations for surface temperature, 595, 597
      $Pe \gg 1$, $Re \gg 1$, 788
   mass transfer with finite blowing, 793
   mixed boundary conditions, $Pe \gg 1$, $Re \gg 1$, 789
      use of constant temperature solutions to derive integral equations for surface temperature, 790
   scaling, 647, 648
      boundary layer thickness, $Re \gg 1$, $Pe \gg 1$, 768, 769
      $Pr \gg 1$, 774
      $Pr \ll 1$, 781
      dependence on boundary conditions, 663, 666, 776
      finite interfacial velocity-blowing, 794
   sphere
      general two-dimensional shear slows, 664
      uniform flow, $Re \ll 1$, $Pe \gg 1$, 645
   spherical bubble, $Re \ll 1$, $Pe \gg 1$, 668
   Squire's transformation, 874, 877
Thermal conductivity, 44
   thermal conductivity tensor, 43

Index

Thermal diffusivity, 159
Thermal energy balance, 33, 68
Thermal energy equation, 153, 594
    general form, curvilinear coordinates, 658
Thermal wake, 611, 654
Thermocapillary effects, 84, 486
    in shallow cavity flows, 385, 404
    in the motion of bubbles and drops, 86, 486
Thermodynamics
    First law, 32
    Second law, 35, 49
Thermohaline convection, (*See* Double diffusive convection instability), 858
Thin films with a free surface (interface), 294, 355
    acceleration effects, 381
    applications, 355
    boundary conditions at the fluid interface, 357, 359
    characteristic velocities, 360, 368, 374, 405
    curvature effects, 358, 360
    derivation of governing equations
        basic equations and interface boundary conditions, 355
        body force terms, and angle of film inclination, 357
        characteristic velocity scales, 357, 360
        dynamical equation for the interface shape function, 360
        simplification of interface boundary conditions for a thin-film, 359
    dynamical equation for interface shape, 360
        analogy with nonlinear diffusion equations, 362
    rupture, 377
        similarity solution to breakthrough, 384
    shallow cavity flows, 385
    spreading films
        capillary (small Bond number), 371
            characteristic velocity, 374
            role of the equilibrium contact angle, 372
            spontaneous spreading, 373
            Tanner's law, 373
        gravitational (large Bond number), 367
            characteristic velocity, 368
            conditions at the front of the drop/film, 371
            similarity solution, 369
    with van der Waals forces, 376
        film rupture, 377, 381
        characteristic timescale, 384
        topological singularity, 382
        linear stability, 377
            criteria for stability, 320, 379, 381
            on the underside of a solid boundary, 379, 457
Thin gap approximation, 294, 297, 355
    air hockey problem (lift force on the air hockey puck), 325
    in the lubrication limit, $Re \ll 1$, 328
    lift on the disk, 345
    uniform blowing approximation, 332
        $\tilde{Re} \ll 1$, lubrication, 334
        $\tilde{Re} \gg 1$, boundary layers, 336
    approximation, 297

characteristic pressure, 300
characteristic scales, 297, 309
    governing equations, 300, 312
        pressure distribution, 312
        solution procedure, 311
    journal bearing problem, 297
    lubrication (*See* Lubrication), 294
Topological singularity, 382
Torque
    arbitrary body for $Re \ll 1$, 441
    circular cylinder in simple shear flow, 673
Transport theorem, Reynolds, 22, 26
Turbulent flow, 111
Two-dimensional linear flow, 641, 664
Two-time scale expansions, 264

Uniaxial extensional flow, 471, 641
Unidirectional flow, 45, 110
    definition, 112, 113
    governing equation, 114
    unidirectional versus one-dimensional flow, 253
    wavy wall channel, parallel to grooves, 233
Uniform (plug) flow, 210
Uniform flow, inertial effects
    past a cylinder, 619
        Stokes' paradox, 621
    past a sphere, 616
        bounded domains, 621
        drag, inertia effect, 626
        inertia effects, 616
        Whitehead paradox, 617, 619
Unit normal vector, 15, 359
    at the surface of a thin film, 360
    for a deformed bubble, 271
    in terms of interface shape, 74, 271, 480, 539

Van der Waals forces, 376
    approaches for inclusion with a thin-film, 376
        disjoining pressure, 376
        molecular mechanism, 376
Vector fields
    general representation theorem, 445
    solenoidal, 24, 445
Vector harmonic functions, 525, 527
    decaying harmonics, 527
    growing harmonics, 527
    vector equality, 525
Vector potential function for vorticity, 445
Vector/tensor parity, 525
Velocity gradient tensor, 33
Viscoelastic fluids, 57
Viscosity
    bulk viscosity, 49
    kinematic, 51
    shear viscosity, 48
Viscous dissipation (*See* Dissipation), 120, 219
    Brinkman number, 219
    effect on rheometry, 223
Vortex line stretching, 699
Vortex street, 702

Index

Vorticity
    analogy with heat transport in two dimensions, 699, 702
    diffusion, 699
    on a free surface, 739
    production
        at a fluid interface, 739
        at a no-slip surface, 739
        vortex stretching, 699
    role in separation, 733
    Stoke's flow, 468
    tensor, 33
    transport equation, 698
    vector, 47
    wake, 626

Wakes
    closed streamline wakes, $Re \ll 1$, 656
    momentum defect, 626
    thermal, 654
    vorticity at $Re \ll 1$, 626
Wall effects, 320, 443, 568
Wavy wall channel, 232
    parallel to the grooves, 233
    perpendicular to the grooves, 237
Weber number, 751
Wedge (*See* Semi-infinite wedge), 719
Weissenberg number, 54, 58
Whitehead's paradox, 602, 603
    for flow at low Re, 617, 619
    for heat transfer at low Peclet number, 602, 603, 612
    Oseen's explanation, 619

Young equation, 72
Young–Laplace equation, 79